ASME PRESS

ROBOTICS AND MANUFACTURING

RECENT TRENDS IN RESEARCH AND APPLICATIONS

VOLUME 6

Proceedings of the Sixth International Symposium on Robotics and Manufacturing (ISRAM '96)
May 28–30, 1996
Montpellier, France

EDITORS

MOHAMMAD JAMSHIDI
NASA ACE Center, University of New Mexico

FRANÇOIS PIN
Oak Ridge National Laboratory

PIERRE DAUCHEZ
LIRMM, France

NEW YORK ASME PRESS 1996

ISBN 0-7918-0047-4
ISSN 1052-4150
Library of Congress Cataloging-in-Publication Number 88-8138

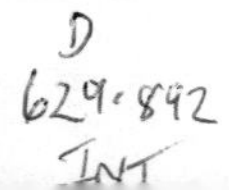

This volume is dedicated to **Professor Mohamed Mansour** for his pioneering works on continuous and discrete-time control systems and his dedicated services to the international control community.

Mohamed Mansour was born in Damietta, Egypt, on August 30, 1928. He received the B.Sc. and M.Sc. degree in electrical engineering from the University of Alexandria, Egypt, in 1951 and 1953, respectively, and the Dr. Sc. Techn. degree in electrical engineering from ETH Zurich, Switzerland, in 1965. He was Assistant Professor in Electrical Engineering at Queen's University, Canada, from 1967-1968. He has been Professor and Head of the Department of Automatic Control at ETH Zürich since 1968, he was Dean of Electrical Engineering from 1976-1978 and Head of the Institute of Automatic Control and Industrial Electronics, ETH Zürich, during 1976-1978, 1980-1982, 1984-1986, and 1989-1990. He was Visiting Professor at IBM Research Lab., San Jose, California, from September to December 1974; at the University of Florida, Gainesville, from January to March 1975; at the University of Illinois, Urbana, from August to December 1981; at the University of California, Berkeley, from January to March 1983; at the Australian National University, from October to November 1989 and from February to March 1992; at the Tokyo Institute of Technology, from October to November 1992; at the University of Miami, from January to April 1994; at the Australian National University, from January to March 1995; at the Tokyo Institute of Technology, Nippon Steel Chair of Intelligent Control, from July to September 1995. He was President of the Swiss Society of Automatic Control from 1979-1985; Member of the Council and Treasurer of IFAC (International Federation of Automatic Control) from 1981-1993; President of the 4th IFAC/IFIP Conference on "Digital Computer Applications for Process Control" in 1974; Chairman of the International Program Committee of the IFAC Symposium on "Computer Aided Design of Control Systems" in 1979; Chairman of the International Program Committee of the 4th IFAC/IFORS Symposium on Large-Scale Systems "Theory and Applications" in 1986; Co-Chairman of the IOTA/IFAC Symposium on "Dynamics of Controlled Mechanical Systems" in 1988; Vice-Chairman of International Program Committee of the IFAC/IFORS/IMACS Symposium on Large Scale Systems "Theory and Applications" in 1989; Co-Chairman of the International Program Committee of the first IFAC Symposium on "Design Methods of Control Systems" in 1991; Chairman of the "International Workshop on Robust Control", Ascona, in 1992; Chairman of the International Program Committee of "Fundamentals of Sampled-Data Systems" in 1992; Chairman of the Meeting of Professors of Automatic Control in Western Europe "WEPAC '92" in 1992; Vice-Chairman of the IFAC Education Committee from 1978-1981; Member of the Senate of the Swiss Academy of Natural Sciences from 1979-1985; Member of the Scientific Committee of INTERKAMA from 1980-1989; Member of the Evaluation Committee of the Research Program "Control of Complex Systems" of the German Research Council from 1986-1989; Member of the Scientific Committee on Fuzzy Control of the Government of North Rhein Westfalia from 1991-present; Delegate of IFAC to the United Nations in Geneva from 1982-1990; Chairman of the Committee of "Awards and Nominations" of the IEEE Control Systems Society from 1989-1990; Chairman of the Committee for Eng. Sciences of Third World Academy of Sciences (TWAS) from 1991-present; Delegate of TWAS to the United Nations from 1991-present. His awards include Silver Medal of ETH Zurich, Switzerland 1965, Fellow IEEE 1985, Honorary Professor of the Gansu University of Technology, PR China 1986, Honorary Professor of the Guangxi University, PR China 1987, Associate Fellowship of the Third World Academy of Science 1989, IFAC Outstanding Service Award 1990, Gullemin-Cauer Award IEEE CAS Society 1992 IFAC Advisor Award 1993, Honorary Membership of the Swiss Society of Automatic Control 1994. His fields of interest are: control systems—especially stability theory and digital control—stability of power systems, and digital filters, Main Research Results are Stability Criteria, Mansour-Matrix for Discrete Systems (Applications: Stability, Identification, Model Reduction, Lattice Filters for one & multidimensional Systems), and Strong Theorem for Robust Stability. His hobbies are Languages, Philosophy and Religions, Ping Pong, and Tennis.

TABLE OF CONTENTS

PREFACE

This volume constitute a report on the research papers presented for the 6th International Symposium on Robotics and Manufacturing held in Montpellier, France from May 28-30, 1996. The fields of robotics and manufacturing continue to be critical areas for productivity and economic competitiveness of nations around the world. The applications of these fields are practically endless. Robots are now being effectively used in medical surgery and medicine in general.

The sixth symposium brought together scientists and technologists from numerous countries around the world. Government, industry, and academia were fully represented here.

The editors would like to take this opportunity and thank all the authors for their contributions to this volume and ISRAM '96. We would like to thank Professor Mohamed Mansour, Professor Bernard Roth, Professor Michael Athans, Dr. Philippe Coiffet, Professor A. H. Mamdani, Professor A. Titli, Professor H. Unbehauen, Professor Madan Singh, Dr. Larry Fogel, Dr. Darwish Al-Gobaisi, Dr. Raimondo Paletto, Professor Lotfi Zadeh, and Professor Hans Zimmermann for their valuable speeches and contributions to the technical program of this Symposium. The diligent efforts of Dr. François Pierrot, Co-chair of the organizing committee are greatly appreciated. We wish to thank the sincere help received from Ms. Annie MacFarlane (NASA ACE Center, University of New Mexico, USA), Ms. Karen Harber (Oak Ridge National Laboratory, USA), Ms. Corine Zicler (LIRMM, France) for their invaluable assistant in making this meeting a success.

Mohammad Jamshidi
Albuquerque, NM USA

François Pin
Oak Ridge, TN, USA

Pierre Dauchez
Montpellier, France

ISRAM '96

Chair: **François G. Pin**
Building 7601, MS 6305 ,
Oak Ridge National Laboratory
P. O. Box 2008 Oak Ridge, TN 3783-6305 USA
Tel: 423-574-6130 Fax: 423-574-4624 email: pin@ornl.gov

Co-Chairs: **Rachid Alami** LAAS-CNRS, France, email: rachid@laas.fr
Shinichi Yuta Tsukuba University, Japan, email: yuta@is.tsukuba.ac.jp
René Zapata LIRMM, France, email: zapata@lirmm.fr

Members

Organizing Committee

Organizing Chair: **Pierre Dauchez**
LIRMM
161 rue Ada 34392
Montpellier Cedex 5, France
Tel: +33-67-41-85-61 Fax: +33-67-41-85-00
email: dauchez@lirmm.fr

Co-Chair: **François Pierrot,** France, email: pierrot@lirmm.fr

WAC European Coordinators: **Philippe Fraisse**, France, email: fraisse@lirmm.fr
Corine Zicler, France, email: zicler@lirmm.fr

INTELLIGENT DECENTRALIZED PLANNING AND COMPLEX STRATEGIES FOR NEGOTIATION IN FLEXIBLE MANUFACTURING ENVIRONMENTS

DIRK ANSORGE
TU München, Institut für Werkzeugmaschinen und Betriebswissenschaften (iwb), D-85609 Aschheim,
email: as@iwb.mw.tu-muenchen.de

ANDREAS KOLLER
TU München, Institut für Informatik, Lehrstuhl für Robotik und Echtzeitsysteme, D-81667 München,
email: koller@informatik.tu-muenchen.de

ABSTRACT
The use of autonomous systems with local planning intelligence permits the optimization of task-processing in flexible production environments. Integration of local planning intelligence into the general task-planning process is necessary if the overall planning process is to be handled productively. For this purpose the coordinating instance distributes the tasks to the individual systems by means of its own, purpose-developed negotiation mechanism. Adequate approximate planning of the coordinating instance and negotiation capabilities provide the autonomous systems with maximum scope for performing their planning and decision-making tasks.

KEYWORDS: decentralized planning, coordination, negotiation agents, distributed knowledge base, negotiation protocols,

1 Introduction

The tasks of the work required in the Special Research Project (SFB) 331[1] include the development of autonomous systems, and above all boosting the efficiency of these systems. In addition to simply executing tasks, efforts are now being made to create local intelligence in autonomous systems with a view to enable local planning tasks to be carried out. As a result, it will become possible to use the autonomous systems' local data in particular, for the rectification of malfunctions.

2 Integration of autonomous systems into a general task-planning process

In order to optimize the overall process in a production plant, it is necessary to integrate the various autonomous systems into an overall, coordinated task-planning process. Its main task is to plan orders in approximate terms, to initiate the execution of tasks and to monitor the progress of work in the production process. It is crucially important that the autonomous systems' capabilities for independent task-planning and for dealing with

[1] This work is supported by *Deutsche Forschungsgemeinschaft* within the *Sonderforschungsbereich 331, „Informationsverarbeitung in autonomen, mobilen Handhabungssystemen“*, projects Q1 and Q6

malfunctions are used as effectively as possible. However, this renders dedicated advance planning of a task at a global level, in terms of its schedule and procedure, inadvisable. Instead, a general task-planning process must allow sufficient planning scope for the autonomous systems, so that their planning and decision-making capability are not impaired and that global objectives, such as meeting deadlines, can nevertheless be achieved.

3 Concept for a global coordinating instance

Following the evaluation of the various central and decentral approaches to control and instrumentation technology, it emerged that autonomous systems can best be integrated by hybrid means. This both renders it possible to pursue global objectives and provides scope for exploiting the autonomous systems' planning and operative capabilities.

3.1 Approximate concept

In order to allow the autonomous systems maximum planning scope, tasks should be distributed on the basis of a negotiation mechanism and not according to rigid directives. The negotiation protocol used for this purpose is based on the contract net protocol [1] which was extended by *Levi & Hahndel (1992)* [2] and *Reinhart & Pischeltsrieder (1995)* [3] specifically for the distributed planning of autonomous systems as part of SFB 331.

A coordinating instance for performing the tasks and satisfying the requirements as effectively as possible is being developed at the iwb. In its initial form, it consists of two parts: global planning and negotiation management. Only the global planning aspect is presented here.

3.2 Integration of decentral planning intelligence into the coordinating instance's global task-planning process

For local planning intelligence to be used to optimum effect, the global planning process has to accommodate flexible schedules and exhibit scope to rearrange of the sequence of work processes performed by the autonomous systems. In addition, it provides freedom to manoeuvre if the dedicated allocation of work processes to machines only takes place in the negotiation process, and not during planning at coordination level. However, in order to approximately estimate capacity during the planning stage, there is the restriction that the work processes are loosely assigned to capacity groups *(Fig. 3.2.1)*. A capacity group consists of the same type of autonomous systems.

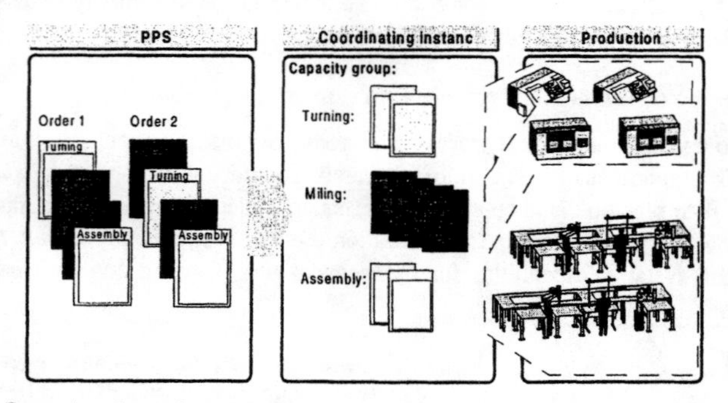

Fig. 3.2.1 Formation of capacity groups

In the first planning stage, the order data provided for example by a Production Planning System are scheduled as part of global planning, working backwards from the order's final deadline. An order-specific extrapolation factor by which the duration of the work processes is multiplied is now determined, taking account of various criteria such as the scheduled start of task and end of task dates, the average processing time of the work processes and the current production capacity situation.

The resulting time spans form a planning period within which a work process (WP) can be freely repositioned. The overall scheduling consistency is upheld if any overlaps between the planning periods for successive work processes are initially prevented *(Fig. 3.2.2)*.

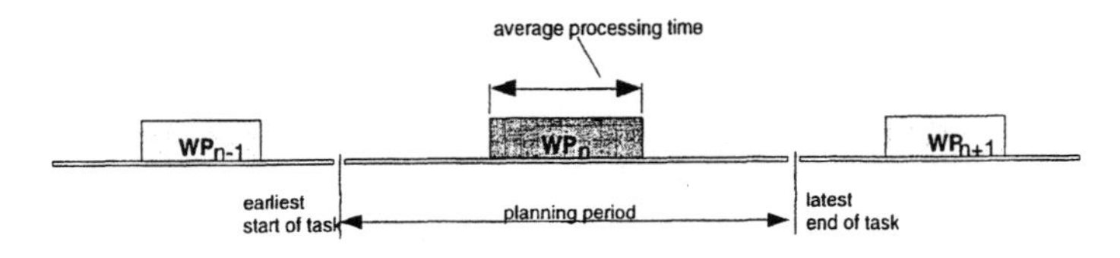

Fig. 3.2.2 Technical and scheduling consistency of work processes (WP)

Since the exact position of a work process is not known at this point in time, for purposes of establishing how much planning capacity is required it is assumed that the tasks will be evenly distributed throughout the entire planning period. If the average processing duration of a work process is deemed to be capacity 1, the capacity requirement F is obtained from the product of the average processing duration and the capacity with the value 1. For the duration of the planning period, the required capacity k is then obtained from the quotient of the capacity requirement F and the planning period (*Fig. 3.2.3*).

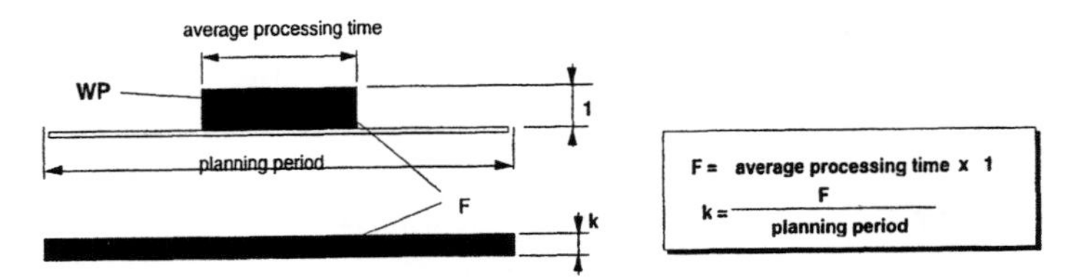

Fig. 3.2.3 Distribution of capacity throughout planning period

Since scheduling bottlenecks may nevertheless occur in spite of extrapolation of the work processes, they are explicitly localized and resolved by extrapolating the planning periods of the work processes affected. This makes it possible to calculate the capacity requirements for all capacity groups at any time t by totaling the capacity values (k) of the corresponding work processes.

4 The Communication Model of the SFB 331

Trying to increase the scope of planning for the autonomous systems leads to higher requirements in the field of the communication abilities of the autonomous systems. If the systems are allowed to optimize their local planning, it becomes necessary that the systems take care by themselves for the coordination of their own planning with the planning of the other autonomous systems or with the coordinating instance. To realize this, it is necessary to integrate negotiation mechanisms into the autonomous systems. Our current model of the communication connections between the different autonomous systems of the SFB 331 has the following structure:

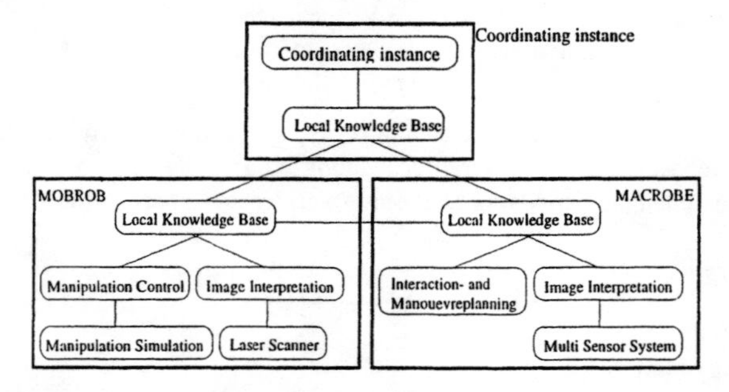

Fig. 4.1 Structure of the manufacturing environment of the SFB 331

In the center of our communication structure you find three instances of local knowledge bases, which together form our distributed knowledge base [4] [5] [6]. This knowledge base has been developed at the chair of robotics and real-time systems of the Institute of Computer Science at the TU München. The three main autonomous systems which have been developed in the SFB 331 are the coordinating instance, which has been described above, the autonomous vehicle MACROBE, and the mobile robot MOBROB. Each of these autonomous systems has its own local knowledge base, in which - among other information - their specific model of the environment is stored.

As the distributed knowledge base represents the central communication unit, it is the obvious medium to integrate intelligent protocols for negotiation into it.

5 Negotiation Concepts

The concepts for negotiation in our current work concern the following three fields:

- examination of a pool of tasks
- support of special cooperation tasks between mobile systems
- support of path planning for the mobile systems of our factory environment

In this article only the first aspect will be discussed.

5.1 The Pool of Tasks

Up to now the distribution of tasks was relatively fixed. That means that tasks were given for example by the coordinating instance to a specific autonomous system. This happened by placing the full task data into the knowledge base and specifying the receiving autonomous system. The system was informed by the active components of our knowledge base [7]. The concepts of the knowledge base guaranteed complete and correct transportation of the data and its transformation into the format of the autonomous system. The next step in our development will make this concept more flexible:

- tasks are no longer given to a specific autonomous system. The autonomous systems themselves decide whether they are interested in getting a certain task.
- negotiation protocols are used to decide which autonomous system takes the task.

- the autonomous systems work at more than one task at a time. This enables them to use local planning abilities to optimize their work.

To realize negotiations in our factory environment, in a first step we implemented a Contract Net Protocol. Systems which are interested in offering a certain task create an instance of the class *invitation to bid.* The data is sent to all local knowledge bases using the concepts of our distributed knowledge base [4]. Every system which is interested in working on this task generates an instance of the type *offer.* These offers are sent to the inviting system. The inviting system judges the offers and chooses one system to receive a particular task. Depending on the problems which may arise while rating the offers and choosing a system, the steps of invitation and offering may be repeated. In its first realization an instance of the class *invitation to bid* has the following structure:

instance of the class invitation to bid
identifier
invitating system
status of invitation
earliest time to start the task
latest time to end the task
average duration of task
type of task
Evaluation of the task: *(bonus/malus points)*
parameters according to type of task
duration of validity of invitation

Depending on the type of task the attributes contain a lot of information which is needed to solve the tasks. These parameters were developed during the practical utilization of the systems and are not relevant to the principles of negotiations discussed here. The inviting system puts its demands for the two time stamps into the attributes and fills in its evaluation. The *bonus points* are comparable to a priority system. The name was chosen to make clear that in the context of autonomous systems there is no obligation to respect such priorities. Thus this figure is only a proposal which shows the level of interest of the inviting system. Besides, these points show a kind of reward, depending for example on how fast the task is solved. *Malus points*, on the other hand, are comparable to a penalty. This field may also contain a sort of a function, depending on the type of fault of the autonomous system. It may depend on whether the task is only finished late or is not finished at all. Thus *malus points* show the risk which the offering systems takes, if it accepts the task. Beside these aims, the system of points fulfils the following tasks:

- the inviting system is able to keep lists of the offering systems and their reliability. This will be taken into consideration or may be used to learn about the other systems.

- the offering systems are able to learn about their own planning capabilities, using the *malus point* system as a kind of a corridor, which should be reached. The number should stay within a certain limit. Scoring too many points means that the system is planning to riskily, scoring never any points means the system is saving its ressources.

Similar point systems are integrated into the class *offers*. The point system in this class has the additional significance of enabling the autonomous systems to tell their estimated costs to the inviting system.

5.2 Agents for Negotiation

The main difficulty in introducing intelligent negotiation protocols in the SFB 331 arises from the fact that the autonomous systems taking part in these negotiations are very heterogeneous. For example the mobile system MACROBE has been designed to find its path in an environment which is only roughly known to it. However, to choose a task or a sequence of tasks MACROBE has to rely on a simple priority-oriented system. The mobile robot MOBROB on the other hand is very autonomous on the planning level. Its sensorial capabilities on the other hand are limited, as the vehicle itself is track-bounded and uses landmarks. In order to increase the scope of planning for the autonomous systems we will improve the strict assignment of the tasks to single autonomous systems. Therefore we developed a theoretical concept based on our knowledge base which adds so called *agents for negotiation* to each autonomous system. These *negotiation agents* build a homogenous layer and enable our autonomous systems to lead complex negotiations. In this article only the first realisation for the practical use within the SFB 331 is discussed. The first system for which we developed a simple negotiation agent was the autonomous vehicle MACROBE. For this purpose an application program has been developed that is able to handle the data structures for *invitations to bid* and which is also able to create offers for MACROBE. The negotiation agent on the other hand supplies MACROBE with task data using its normal interface to its local knowledge base. Using the negotiation agent MACROBE gets as powerful in negotiation as its negotiation agent.

6 Conclusion

We are currently working on integrating negotiation mechanisms into our factory environment and have integrated the first simple Contract Net based protocols into our autonomous sytems. We proved their applicability by building a first negotiation agent for our system. The next task is to include the negotiation management of the coordinating instance and to connect the autonomus systems with the coordinating instance. We will then test the efficiency of the whole system in the different conditions of the production enviroment.

7 References

[1] Smith R.G.: The Contract Net Protocol: High Level Communication and Control in a Distributed Problem Solver. In: IEEE Transaction on computers. Vol. C-29 (1980) 12.

[2] Levi P., Hahndel S.: Restriktionsbasiertes Verhandlungskonzept für eine dezentrale, kooperative Aktionsplanung. In: Rembold, U. et. al., Autonome Mobile Systeme, 8. Fachgespräch, Karlsruhe, 1992, S. 93-105.

[3] Reinhart G., Pischeltsrieder K.: Flexible Electrically-Powered Transport Vehicles in Future Production Structures. In: Rembold, U. (Hrsg.); Dillmann, R. (Hrsg.); Hertzberger, L.O. (Hrsg.); Kanade, T. (Hrsg): Proceedings of the Int. Conf. on Intelligent Autonomous Systems (IAS4), Karlsruhe. Amsterdam, IOS Press, 1995. S. 15-25.

[4] Schweiger J., Ghandri K. and Koller A.: "Concepts for A Distributed Real-Time Knowledge Base for Teams of Autonomous Systems", In: Proceedings of IEEE/RSJ/GI International on Intelligent Robots and Systems. Federal Armed Forces University Munich, Germany, September 1994 S.1508-1515.

[5] Seidl U.: "Teammanagement für eine verteilte Wissensbasis", Diplomarneit, TU München, 1993, November 1993.

[6] Wirth M.: "Konzepte für eine verteilte Wissensbasis für Fertigungsumgebungen". In: Beiträge 8. Fachgespräch über Mobile Systeme AMS 91, 1992, TU Karlsruhe, Rembold, U.

[7] Schweiger et al., Handbuch zur Q6-Wissensbasis Shell, TU München, Juni 1995

A PRODUCTION PLANNING MODEL TO SELECT MACHINES AND TOOLS, AND ALLOCATE OPERATIONS IN FMS

A. ATMANI
Tregaskiss Ltd.
Windsor, Ontario N0R 1L0

R.S. LASHKARI
Department of Industrial & Manufacturing Systems Engg.
University of Windsor
Windsor, Ontario N9B 3P4

ABSTRACT

This paper presents a production planning model to determine the optimal machine selection, tool assignment, and operation allocation in a flexible manufacturing system. It is assumed that there is a set of CNC machines with known processing capabilities, that a set of part types with known process plans is selected for manufacture, and that each operation of a part type may be assigned to various machines using specified tools. The model determines the optimal plan by minimizing the total costs of operations, material handling and set-ups.

KEYWORDS: Operation allocation, tool assignment, FMS planning, integer programming

INTRODUCTION

In recent years, the need for flexibility, efficiency, and quality has imposed major changes on manufacturing industries, giving prominence to such issues as reducing manufacturing lead time to satisfy customers, flexibility to adapt to market changes, and improvement in productivity and reduction of production costs in order to maintain market shares. In response, flexible manufacturing systems (FMS) have emerged as a viable answer to the problems of flexibility and efficiency. However, proper pre-planning is a key condition to FMS success in dealing with problems of poor utilization of equipment and tools, and costly and frequent setups. Thus, the difficult task facing the production planner is the optimal selection of machines and the assignment of part operations to the selected machines.

In this paper we propose a production planning model for determining the optimal selection of machines and tools, and operations allocation in a flexible manufacturing system.

There have been many contributions to the literature on operation allocation in FMS. A brief review of some of the relevant works in this area is given here. Stecke [1] formulated the loading problem in a flexible manufacturing system as a nonlinear mixed integer program with the objective of balancing the workloads. A branch and bound algorithm to solve the problem was then developed by Berrada and Stecke [2].

Sarin and Chen [3] presented a mixed integer programming formulation of the machine loading-tool allocation problem in FMS, minimizing the total machining costs which are dependent on machine-tool combinations. Lashkari et al. [4] extended the formulation of the operation allocation problem to include the aspects of refixturing and limited tool availability.

Damodaran et al. [5] developed mathematical models for allocating operations to machines in single-cell and multiple-cell environments. Kim and Yano [6] presented a model of the loading problem in FMS with unequal workload targets, and developed a new branch and bound procedure to efficiently solve the model. Stecke and Raman [7] presented a queueing network production planning model to determine the optimal machine workload assignments in an FMS. Liang [8] proposed a two-stage approach to the joint problem of part selection, machine loading, and machine speed selection problem in FMS. In the first stage, the mathematical model solves the part selection and machine loading problem, whereas in the second stage it determines the optimal cutting speed for all job-tool-machine combinations. Modi and Shanker [9] presented a generalized loading problem with the objective of part movement minimization in an FMS with machine, tool, and process plan flexibilities.

Atmani et al.[10] introduced a 0-1 integer programming model for the simultaneous solution of the cell formation and operation allocation problems in cellular manufacturing, considering multiple process plans for each part type, and multi-option operation-to-machine assignment. Atmani [11] presented a production planning model for determining the optimal machine selection and operation allocation in FMS. Kuhn [12] presented a loading problem in FMS involving the assignment to the machine tools of all operations and associated cutting tools required for part types selected for production. A heuristic algorithm was also developed to solve the model.

MATHEMATICAL MODEL

It is assumed that there is a set of CNC machines labelled $j=1, \ldots, m$, and a set of part types labelled $i=1, \ldots, n$, where the part type i is to be produced in a batch size of b_i over the planning period. A part type i is processed under a given process plan (or route). The operations associated with the part type i are represented by the indices $s=1, \ldots, S(i)$. Each operation of the part type i requires a specific tool t, and may be performed on more than one machine. The set of machines that are capable of performing operation s of part type i is given by J_{is}. We use the abbreviation (tj) to indicate that tool t is assigned to machine j.

The machine-tool-operation allocation problem involves the selection of CNC machines, selection of tools to be assigned to the machines, and the assignment of operations of each part type to the selected tool-machine combinations, such that the total costs of operations, material handling, and setup are minimized, and that the constraints imposed on the system are satisfied.

To begin the development of the model, we introduce the zero-one decision variables $X_{si}(tj)$, where $X_{si}(tj)$ assumes the value of one if operation s of part type i is assigned to the combination (tj), and zero otherwise. The elements of the objective function are as follows.

The operation cost is given by $OC(X_{si}(tj))$:

$$OC(X_{si}(tj))=\sum_{i=1}^{n} b_i \sum_{s=1}^{S(i)} \sum_{j \in J_{is}} CO_{sij} X_{si}(tj) \tag{1}$$

where b_i is the known batch size of part type i and CO_{sij} is the cost of operation s of a unit of part type i on machine j using tool t. The material handling cost is given by $TC(X_{si}(tj))$:

$$\bar{TC}(X_{si}(tj)) = \sum_{i=1}^{n} b_i \sum_{s=1}^{S(i)-1} \sum_{j \in J_{is}} \sum_{\hat{j} \in J_{i(s+1)}} TC_{ij\hat{j}} \, X_{si}(tj) X_{(s+1)i}(\hat{t}\hat{j}) \tag{2}$$

where $TC_{ij\hat{j}}$ is the cost of moving a unit of part type i from machine j to machine $\hat{j}$. It is noted that operation s of the part type i is performed on (tj), while operation $s+1$ is performed on $(\hat{t}\hat{j})$. As $TC(X_{si}(tj))$ is a nonlinear function, we use the following linearization technique [13] which prescribes replacing Eq. (2) with:

$$TC(L_{si}(tj\hat{t}\hat{j})) = \sum_{i=1}^{n} b_i \sum_{s=1}^{S(i)-1} \sum_{j \in J_{is}} \sum_{\hat{j} \in J_{i(s+1)}} TC_{ij\hat{j}} L_{si}(tj\hat{t}\hat{j}) \tag{3}$$

where $L_{si}(tj\hat{t}\hat{j})$ is a new 0-1 variable satisfying the following two sets of constraints:

$$X_{si}(tj) + X_{(s+1)i}(\hat{t}\hat{j}) - 2L_{si}(tj\hat{t}\hat{j}) \geq 0 \qquad \forall i, s \in \{1,2,..,S(i)-1\}, \; t \in \tau, j \in J_{is}, \hat{j} \in J_{i(s+1)} \tag{4}$$

$$X_{si}(tj) + X_{(s+1)i}(\hat{t}\hat{j}) - L_{si}(tj\hat{t}\hat{j}) \leq 1 \qquad \forall i, s \in \{1,2,..,S(i)-1\}, \; t \in \tau, j \in J_{is}, \hat{j} \in J_{i(s+1)} \tag{5}$$

Finally, the setup cost is given by $SC(M_j)$:

$$SC(M_j) = \sum_{j=1}^{m} SC_j M_j \tag{6}$$

where SC_j is the setup cost on machine j (which may include the costs related to change and calibration of tools, change of control program, etc.), and M_j is an auxiliary 0-1 variable which assumes a value of 1 when machine j is selected, and 0 otherwise. Combining all the terms, the objective function of the machine-tool-operation allocation problem is given below:

$$Minimize\quad Z=\sum_{i=1}^{n} b_i \sum_{s=1}^{S(i)} \sum_{j\in J_{is}} CO_{sij} X_{si}(tj) + \sum_{i=1}^{n} b_i \sum_{s=1}^{S(i)-1} \sum_{j\in J_{is}} \sum_{\hat{j}\in J_{i(s+1)}} TC_{ij\hat{j}} L_{si}(tj\hat{t}\hat{j}) + \sum_{j=1}^{m} SC_j M_j \qquad (7)$$

The constraints imposed on the system are as follows. The first two sets of constraints, Eqs.(8) and (9), ensure that each tool is assigned to only one machine, once a machine is selected; however, a machine may have more than one tool assigned to it:

$$\sum_{j=1}^{m} Y(tj) = 1 \qquad \forall t \qquad (8)$$

$$\sum_{t\in T} Y(tj) \geq M_j \qquad \forall j \qquad (9)$$

where $Y(tj)$ is an auxiliary 0-1 variable, which assumes a value of 1 if tool t is assigned to machine j, and 0 otherwise.

The next set of constraints, Eq.(10), allows for operations to be assigned to a machine-tool combination, once the combination is properly selected:

$$\sum_{i=1}^{n} \sum_{s=1}^{S(i)} X_{si}(tj) \geq Y(tj) \qquad \forall tj \qquad (10)$$

The next set of constraints, Eq.(11), will ensure that the magazine capacity of a machine is not overloaded:

$$\sum_{t\in T} S_t\, Y(tj) \leq U_j M_j \qquad \forall j \qquad (11)$$

where S_t is the number of tool slots tool t occupies in the magazine, and U_j is the number of tool slots available in machine j's magazine.

The next set of constraints is designed to ensure that the operations assigned to machine j, once it is selected, will not exceed it's capacity:

$$\sum_{t\in T} \sum_{i=1}^{n} b_i \sum_{s=1}^{S(i)} \xi_{si}(tj) X_{si}(tj) \leq A_j\, M_j \qquad \forall j \qquad (12)$$

where $\xi_{si}(tj)$ is the time to perform operation s of part type i on machine j using tool t.

The next set of constraints ensures that the tool life of tool t, i.e., T_t, is not exceeded:

$$\sum_{i=1}^{n} b_i \sum_{s=1}^{S(i)} \xi_{si}(tj)\, X_{si}(tj) \le T_t Y(tj) \qquad \forall tj \tag{13}$$

The linearization constraints are reproduced below as Eqs. (14) and (15), and the last set of constraints, Eq.(16), ensures integrality of the variables.

$$X_{si}(tj) + X_{(s+1)i}(\hat{t}\hat{j}) - 2L_{si}(tj\hat{t}\hat{j}) \ge 0 \qquad \forall i, s \in \{1,2,..,S(i)-1\},\ t \in \tau, j \in J_{is}, \hat{j} \in J_{i(s+1)} \tag{14}$$

$$X_{si}(tj) + X_{(s+1)i}(\hat{t}\hat{j}) - L_{si}(tj\hat{t}\hat{j}) \le 1 \qquad \forall i, s \in \{1,2,..,S(i)-1\},\ t \in \tau, j \in J_{is}, \hat{j} \in J_{i(s+1)} \tag{15}$$

$$L_{si}(tj\hat{t}\hat{j}),\ X_{si}(tj),\ Y(tj),\ M_j \in \{0,1\} \qquad \forall (i,s), t \in \tau,\ j \in J_{is}, \hat{j} \in J_{i(s+1)} \tag{16}$$

The problem may now be summarized as:

Minimize the objective function, Eq. (7)
Subject to constraints (8) to (16).

AN ILLUSTRATIVE EXAMPLE

To demonstrate the applicability of the model, a numerical example is prepared; however, due to space limitation it is not presented here. Interested readers may contact the authors for a complete copy of the example and the solution.

CONCLUSIONS

The paper presents a production planning model for the problem of machine-tool selection and operation allocation in FMS. The model determines the optimal machine-tool combinations, and assigns the operations of the part types to the machines, while minimizing the processing, transportation and setup costs. It considers the constraints imposed on the system by such factors as the tool magazine capacity, tool life, and machine capacity.

ACKNOWLEDGEMENT

The research reported here is supported by a grant from the Natural Sciences and Engineering Research Council (NSERC) of Canada.

REFERENCES

1. Stecke, K. E. "Formulation and solution of nonlinear integer production planning problems for flexible manufacturing systems." *Management Science*. 29(1983), 273-288.

2. Berrada , M. and Stecke, K. E. "A branch and bound approach for machine load balancing in flexible manufacturing system." *Management Science.* 32(1986), 1316-1335.
3. Sarin, S.C. and Chen, C.S. " The machine loading and tool allocation problem in a flexible manufacturing systems." *Intl. J. of Production Research.* 25(1987), 1081-1094.
4. Lashkari, R.S., Dutta, S.P., and Padhye, A.M. " A new formulation of operation allocation problem in flexible manufacturing systems: mathematical modelling and computational experience." *Intl. J. of Production Research.* 25(1987), 1267-1283.
5. Damodaran, V., Lashkari, R. S., and Singh, N. "A production planning model for cellular manufacturing systems with refixturing considerations." *Intl. J. of Production Research*, 30(1992), 1603-1615.
6. Kim, Y-D. and Yano, C.A. "A new branch and bound algorithm for loading problems in flexible manufacturing systems." *Intl. J. of Flexible Manufacturing Systems*, 6(1994), 361-382.
7. Stecke, K. E. and Raman, N. "Production planning decision in flexible Manufacturing systems with random material flow." *IIE Transactions,* 26(1994), 2-17.
8. Liang, M. "Integrating machining speed part selection and machine loading decisions in flexible manufacturing systems." *Computers and Industrial Engineering*, 26(1994), 599-608.
9. Modi, B.K. and Shanker, K. " Models and solution approaches for part movement minimization and load balancing in FMS with machine, tool and process plan flexibilities." *Intl. J. of Production Research*, 33(1994), 1791-1816.
10. Atmani, A., Lashkari, R.S. and Caron, R. J. "A mathematical programming approach to joint cell formation and operation allocation in cellular manufacturing." *Intl. J. of Production Research*, 1(1995), 1-15.
11. Atmani, A. "A production planning model for flexible manufacturing systems with setup cost consideration." *Proc. of 17th Intl. Conf. on Computers and Industrial Engineering*, 29(1995), 723-727.
12. Kuhn, H. "A heuristic algorithm for the loading problem in flexible manufacturing systems." *Intl. J. of Flexible Manufacturing Systems*, 7(1995), 229-254.
13. Taha, A.H. *Operations Research*, 5th Ed. Macmillan Publishing Co., New York (1992).

Notation

$\mathbf{i} \in \{1, 2, ..., n\}$	part types
$\mathbf{s} \in \{1, 2, ..., S(i)\}$	operations of part type **i**
$\mathbf{j} \in \{1, 2, ..., m\}$	machines
$\mathbf{J_{is}}$	set of machines on which operation s of part type **i** can be performed.
τ	set of tools required for operations.
$\mathbf{M_j}$	auxiliary variable for machine selection; equals **1** if machine **j** is selected for production, **0** otherwise.
$\mathbf{Y(tj)}$	auxiliary variable for tool/machine allocation; equals **1** if tool **t** is assigned to machine **j**, **0** otherwise.
$\mathbf{L_{si}(tj\hat{t}\hat{j})}$	linearization variable; equals **1** if operations **s** and **s+1** of part type **i** are assigned to **(tj)** and $(\hat{t}\hat{j})$, respectively, **0** otherwise.
$\mathbf{X_{si}(tj)}$	operation allocation variable; equals **1** if operation **s** of part type **i** is performed on machine **j** using tool **t**, **0** otherwise.
$\boldsymbol{\xi_{si}(tj)}$	time to perform operation **s** of part type **i** on machine **j** using tool **t**
$\mathbf{b_i}$	batch size of part type **i**
$\mathbf{CO_{si}(tj)}$	cost of operation **s** of a unit of part type **i** on machine **j** using tool **t**.
$\mathbf{TC_{ij\hat{j}}}$	cost of moving a unit of part type **i** from machine **j** to machine $\hat{j}$.
$\mathbf{SC_j}$	setup cost on machine **j**
$\mathbf{A_j}$	capacity of machine **j** during the planning period (hours).
$\mathbf{U_j}$	capacity of machine **j**'s tool magazine (number of slots).
$\mathbf{T_t}$	tool life of tool **t** (hours).

Robotic Control via Stimulus Response Learning

Andreas Birk
Universität des Saarlandes, c/o Lehrstuhl Prof. W.J. Paul
Postfach 151150, 66041 Saarbrücken, Germany
cyrano@cs.uni-sb.de, http://www-wjp.CS.Uni-SB.DE/~cyrano/

Abstract

Stimulus response learning is a new approach to enable robots to learn an adaptive and robust control-mechanism. Learning in this paradigm is an evolutionary process, leading to a behavioral model of any given environment. The model is build up in a special data-structure as collection of regularities between sensor-data and activation of effectors. This process is guided by two simple and universal i.e., independent of the robot and environment, heuristics for rating the usefulness of basic elements of the model. We present a system learning from scratch to control a robot-gripper in a blocks-world via a camera. The system learns movements in the plane and to manipulate building-blocks. The most challenging experiments were done in noisy real world set-ups, using unprocessed high-resolution images.

Keywords: Evolutionary Algorithm, Robotics, Adaptive Behavior, Machine Learning

1 Introduction

The ability to learn is one of the most important features of intelligence. It enables cognitive systems to adapt to un-foreseen circumstances and to improve with practise. Though these tasks appear to be very easy to humans, automatic adaption and improvement of robots is still a very open research field.

Therefore, we developed *stimulus response learning* as a new machine learning approach suited for robotic control. We believe, that cognition can only emerge in systems interacting with a sufficiently complex environment. Or — abusing an ancient Roman slogan — "mens sana in corpere sano".

Though our interests focus on understanding cognition and building a humanoid artificial intelligence in long term, *stimulus response learning* is an interesting approach in regard to industrial applications. Because a learning control-mechanism allows noisy and changing environments, the execution of more various tasks and low-priced imprecise hardware.

2 Learning of Control

We wish robots to be able to learn to do any given task in any given environment. So, we must supply our robot with a mechanism for investigating his environment, a suitable storage for his discoveries and a mechanism for applying his knowledge learned.

2.1 How to store Discoveries

Our robot makes a discovery when he finds useful regularities between sensor-data and activation of effectors. The meaning of "useful" depends on the task(s) the robot is expected to do and will be discussed later. Discoveries are the basic drive for the evolution of the robot's model of the world, which tells him what to do in a certain situation to achieve a certain result.

The world-model is built up in a special data-structure, a dynamic directed graph. The nodes are so-called **stimulus response rules (SRR)**, simple behavioral rules made of tests on sensor-data and actions i.e., activation of effectors. In its most general form a SRR is a **TOTE** (Fig.1), an abbreviation for $test_1$, operate, $test_2$, exit. If $test_1$ — the so-called **condition** — holds, the TOTE can be executed. This means, the operate or **action** is repeatedly activated until $test_2$ — the so-called **feedback** — holds. After this, the exit or **result** should be fulfilled. This test represents the state of the environment after execution of the TOTE and is used for planning.

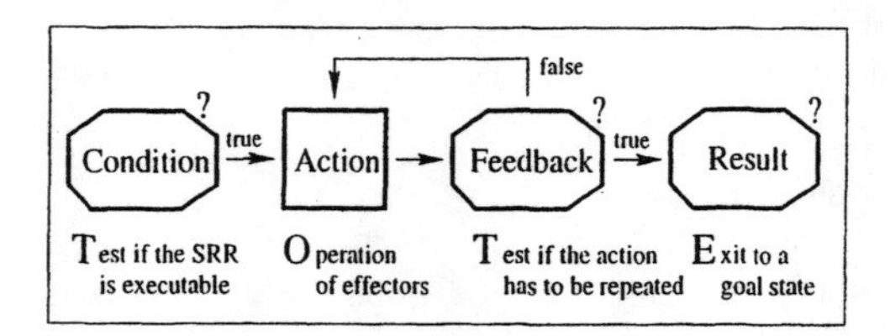

Figure 1: The TOTE as the most general form of a SRR

In the graph representing the world-model, a directed edge (s_1, s_2) stands for a possible consecutive execution of s_1 and s_2. This means, the system witnessed frequently that the condition of s_2 holds after execution of s_1. An important point is, that this kind of implication is purely based on statistics.

2.2 The Evolution of the World-Model

The robot evolves the directed graph representing his world-model in a step-wise manner. In a time-step t the robot chooses randomly between **creation** of a new SRR or edge and **training** of the current world-model $G_t = (V_t, E_t)$ with V_t is the set of SRRs at time-step t and $E_t \subset V_t \times V_t$ is the set of edges.

In doing so, a SRR-pointer, the so-called **standpoint** is used to mark the SRR that was executed last. It represents the system's present position in the world.

Learning as operations on the graph takes mainly place in direct graph-theoretic neighborhood of the standpoint i.e., on $M_{sp} = \{s \in V_t | (\text{standpoint}, s) \in E_t\}$, the set of SRRs with an edge leading to from the standpoint. This set is so to say the population of our evolutionary algorithm. Therefore, it is not necessary to process the whole world-model learned so far, but a very small subset to breed new knowledge.

A new SRR s is created by using parts of SRRs from M_{SP}. In doing so, the systems focuses on SRRs from M_{SP} with good records according to the — soon defined — quality-measures.

Creation of a new edge is done by searching a SRR s with holding condition in the current world-model G_t. If the search is successful, an edge from the standpoint to s is included in G_{t+1} and s is executed.

A training-step is done as follows. A SRR s with holding condition is randomly chosen among M_{sp} and executed. Training is necessary to get the statistical records for the quality-measures.

After every system-step, no matter if training or creation, all bad — with respect to the quality measures — SRRs and edges are deleted.

2.3 The Quality-Measures

The quality-measures are designed to rate the usefulness of the robot's discoveries. They are a kind of universal fitness-function in the evolution of the world-model. They use two simple concepts — **reliability** and **applicability** — and are based on the fact that a SRR or an edge represents an assumption:

The execution of a SRR is **successful** if afterwards **1.)** the condition is not holding i.e., the action has changed the state of the environment, and **2.)** the feedback became true before the action was repeated more than constant rep_{max} times, and **3.)** the result is holding. An edge $e = (s_1, s_2)$ is **successful** if the execution of its target-SRR s_2 is successful.

The **reliability** of a SRR s, respectively edge e is defined as

number of successful executions of s (or e) per number of executions of s (or e)

The **applicability** of a SRR s, respectively edge e is defined as

number of executions of s (or e) per lifetime of s (or e)

where lifetime is the number of time-steps of the system since creation of s (or e).

The **quality** of a SRR s, respectively edge e is reliability times applicability:

number of successful executions of s (or e) per lifetime of s (or e)

2.4 Using the World-Model

Using a graph as world-model planning becomes very easy. As mentioned before, a SRR-pointer *standpoint* models the system's current position in the world. It is always set on the last executed SRR. Given a desirable state g of the environment the system can search its world-model for a SRR s where the result of s holds on g. To achieve the goal g the system can execute the SRRs along the shortest path from the standpoint to s.

3 Learning Eye-Hand-Coordination

3.1 The Task

According to the Swiss psychologist Piaget [3] child development proceeds in stages. The first one is the sensorimotor stage, characterized by eye-hand-coordination. This can be compared to the task of learning to control a robot-arm via a camera (Fig.2):

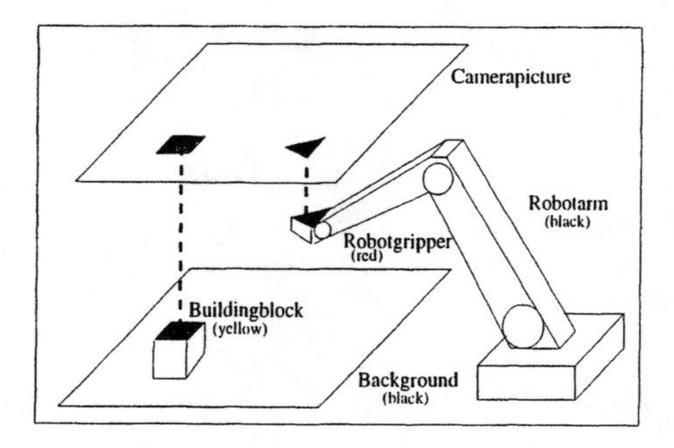

Figure 2: The set-up

A camera looks perpendicularly on a black background with some colored building-blocks on it. A red robot-gripper, the so-called hand, can be moved by one of the actions `north`, `south`, `west`, or `east` in one of the four directions of the plane. Furthermore, the hand can `grip` the building block if its right under the hand and carry it around. It can release the block with the action `ungrip`.

3.2 The World-Model

The world-model for moving the hand is a kind of grid. It consists of SRRs of the following form: "If I see the hand at position P, I can take (hand moving) action A and the hand will be visible at position P'." Four edges are leading from every SRR — except those on the edges of the grid — to its neighbors in the four directions of the plane. If the system is told to get the hand to a position P in the current video image, the system searches a SRR s with a result representing P and executes the actions of the SRRs of the shortest path from the *standpoint* to s.

The complete world-model for hand-movements and manipulating building-blocks consists of two such grids. The first one is the above described "eye-hand-grid". The second one describes how building-blocks can be carried around. The two grid are interconnected by SRRs for — proper — grasping and un-grasping building-blocks i.e., `grip` is only used if and only if the block is under the hand, `ungrip` is only used if and only if the block is held.

The hardness of the task to learn this world-model is twofold. First, the system has to learn appropriate representations of its hand and blocks dependent on the kind of tests used. We will present shortly results from experiments with two different kind of tests. Second, the system has to find out what is important to look at. For example the system has to discover that hand-movements are independent of block-positions, but grasping is not.

3.3 Teststrings and Gary L. Drescher

In one class of experiments the camera picture is sectioned as a grid. In each field of the grid the most frequent color in it is determined and written to a vector of so-called **input-nerves**.

As tests we use strings as follows. Every position in the string corresponds to a position in the vector of input-nerves. A character in the string is either a color or a so-called **joker**. A teststring is fulfilled if and only if every character is either equal to the current value of the corresponding input-nerve or a joker. Test-strings are created with three universal operators:

Adaption produces a test-string S as "snapshot" of the current environment i.e., the entries of S are set equal to the entries in the input-nerves.

Mutation produces a new test-string S from an existing one S' by replacing an entry in S' by a joker at a random position.

Crossover produces a new test-string S from two existing ones S' and S'' by copying the head of S' and the tail of S'' with respect to a random position j.

The operators work on a population of tests from SRRs from M_{sp}, the set of SRRs with an edge leading to from the standpoint. In doing so, they focus on SRRs with high fitness according to the quality-measures.

Gary L. Drescher presents in [2] an own approach to learn eye-hand-coordination. His best run on a Thinking Machines CM2 (16K processors, 512 Mbyte main memory) ended after one day with memory overflow. His system learned approximately 70% of the desired world-model. A corresponding world-model is found with *Stimulus Response Learning* in 25 seconds on a SUN Sparc 10 completely. The total amount of memory used is less than 250 Kbyte.

Furthermore, Drescher's experiments are done in simulations only. All experiments with *Stimulus Response Learning* were done in real world set-ups as well. In doing so, the system was very successful in dealing with noise and errors.

3.4 Unprocessed Real-World Images

In the most challenging class of experiments a real-world set-up is used with video images with resolutions up to 300×300 pixel under changing illumination conditions (weather). Furthermore, no vision processing is used. So, pictures of the hand and the building-blocks are disturbed by reflexes, shadows, changing brightness etc.

In this class of experiments, tests are represented as programs in a simple turtle-graphics language with commands to move the turtle, to draw lines, and to do a for-loop. A test is fulfilled if and only if the output of the program is contained in the current video image. Instead of adaption, mutation, and crossover, following operators are used to create turtle-tests:

Conc : Takes two programs and concatenates them to one.

Split : Splits a program into two new ones at a randomly chosen place.

Hill-climbing on constants : Performs a hill-climbing-step on a randomly chosen constant in a program to minimize a **picture-distance-function**.

The picture-distance-function is a heuristic to measure the similarity of two images. Details can be found in [1].

In simulations the system learned very fast hand-movements and grasping. In doing so, it represented the hand as red triangle and building-blocks as rectangles. Due to this promising results we thought adaption to a real world set-up to be easy. We expected usage of vision filters and some fuzziness of tests to be sufficient. But a series of experiments showed, that in the real world the camera input contains no triangles or rectangles. The hand and the building-blocks have no simple geometric representation. Noise due to reflexes and shadows is always present.

An unexpected solution of this problem was found by the system itself. We started a run using unprocessed real world images as input to see what would happen. After few hours the system started to move and grip successfully. It invented a kind of edge-detection. The system represented the hand and building-blocks via programs testing parts of the contour. Following analysis of the camera input revealed that these lines are hardly disturbed in our set-up. Therefore, they can be used by the system to determine the position of the hand and building-blocks. This is realized by the system with the help of the *quality-measures.*

This result was achieved not only once. In every run done so far hand movements and grasping were learned using this edge-detection. The complete world-model for these tasks is learned in approximately 50 hours on average. In doing so, the run-time is dominated by the speed of the robot arm.

4 Conclusion

We presented *stimulus response learning* as a new approach to enable robots to learn an adaptive and robust control-mechanism. This control-mechanism is based on a behavioral model of the world, that is learned with an evolutionary algorithm. The basic algorithm is neither dependent on the robot nor its environment. We presented experiments to learn eye-hand-coordination i.e., to control a robot-arm via a camera. Our results are not from simulations only, but as well from real-world set-ups.

References

[1] Andreas Birk. Learning geometric concepts with an evolutionary algorithm. In *Proc. of The Fifth Annual Conference on Evolutionary Programming.* The MIT Press, Cambridge, 1996.

[2] Gary L. Drescher. *Made-up minds, A constructivist approach to artificial intelligence.* The MIT Press, Cambridge, 1991.

[3] Jean Piaget. *Gesammelte Werke.* Klett-Cotta, Stuttgart, 1991.

On the Use of Nonlinear Observability Conditions in Determining Test Situations for a Mobile Robot Localizer

Ph. Bonnifait and G. Garcia
Laboratoire d'Automatique de Nantes, URA CNRS 823
Ecole Centrale de Nantes/Université de Nantes, 1 rue de la Noë
44072 Nantes, France. Philippe.Bonnifait@lan.ec-nantes.fr

ABSTRACT

The convergence analysis of a robot localizer based, for instance, on odometry and goniometric measurements of known landmarks, requires simulations. Studying nonlinear observability highlights situations where any localization system based on the same continuous model may undergo difficulties, whatever the technological and algorithmical solutions. We check the convergence of an Extended Kalman Filter through simulations in these situations. Moreover, this study gives informations on the behaviour of the filter when one or more beacons are hidden.

KEYWORDS : mobile robot localization, nonlinear observability, localizer convergence, Kalman filtering.

INTRODUCTION

The process of designing a localizer for a mobile vehicle, especially when using both relative and absolute sensors, can be seen as equivalent to designing a (generally nonlinear) observer of the state of the vehicle. As exemplified by the literature, this general aspect is most often not taken into account. One writes and implements a localization software tailored for the system at hand, taking directly into account technological constraints (discrete measurements rather than continuous, asynchronous rather than synchronous, etc.) In our case, the system is based on odometry plus the measurements of asynchronous azimuth angles of known landmarks. Another system may continuously track target landmarks and provide continuous measurements (or at least high frequency discrete measurements). Nevertheless, the continuous system is the same in both cases.

In the paper, we show that analyzing the observability of the continuous system helps determine situations in which any observer may (or sometimes will) undergo problems. In practice, these situations are not necessarily easy to determine intuitively. Moreover, extensive simulation tests, defined more or less randomly, may not detect these situations.

CONTINUOUS SYSTEM STATE-SPACE MODELLING

The location of a vehicle can be represented by $X=(x,y,\theta)$ where (x,y) is the position of the middle of the wheel-base and θ the heading angle (fig. 1). The observations are the azimuth angles λ_i of "m" landmarks B_i, the coordinates of which are (x_i,y_i). Let $U=[u_1,u_2]^t$ be the control vector where u_1 and u_2 are the translational and rotational speeds :

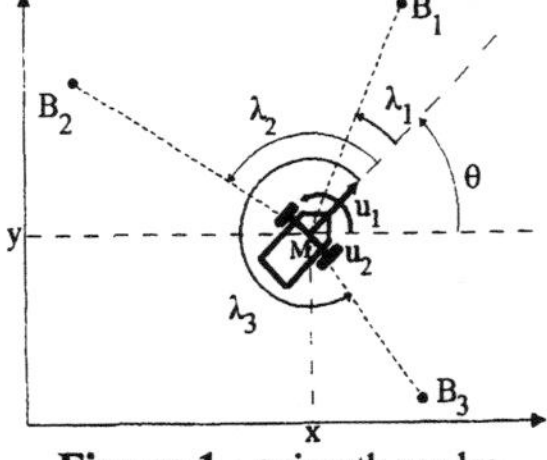

Figure 1 : azimuth angles

$$\begin{cases}\dot{x}=u_1.\cos\theta\\ \dot{y}=u_1.\sin\theta \Leftrightarrow \dot{X}=f(X,U)\\ \dot{\theta}=u_2\end{cases} \quad (1)$$

$$\lambda_i=\operatorname{atan2}(y_i-y,x_i-x)-\theta=g_i(X) \; (1\le i\le m) \quad (2)$$

Before the robot moves, a static localization is computed in the reference frame, using at least three beacons. So, we suppose $m \geq 3$ in the sequel.

If we suppose that the azimuth angles are continuously available, we concatenate them in an observation vector λ ($\dim(\lambda)=m$). Nevertheless, while the vehicle is moving, one or even two landmarks can be hidden or too far from the sensor to be detected. In such a case, the dimension of the output vector λ decreases. If it does not change in the time interval $[t_0,T]$, the state-space description of the system is given by :

$$\text{for all t in } [t_0, T] \begin{cases} \dot{X}(t) = f(X(t), U(t)) & x \in \Re^3 \\ \lambda(t) = g(X(t)) & \lambda \in \Re^p \text{ with } p \leq m \end{cases} \tag{3}$$

SYSTEM OBSERVABILITY

Observability concepts and rank condition

For nonlinear systems (see equation (3)), Hermann and Krener [4] related observability to the concept of indistinguishability of states with respect to the inputs (nonlinear observability depends on the inputs, contrary to the linear case).

An initial state X_0 is ***observable***, if one can find an input which distinguishes X_0 from all other possible states. If the knowledge of the inputs and observations over time allows to distinguish X_0 not among all possible states, but just among states in a given neighborhood of X_0, X_0 is said to be ***weakly observable***. Additionally, if distinguishing X_0 from other states requires the knowledge of the inputs and outputs only on a finite interval of time $[t_0,T]$, then X_0 is said to be ***locally observable***.

Another concept of nonlinear observability is exposed in [5]. A system is observable if, and only if, the state of the system subject to input U can be expressed as a function of the observation, the input and their derivatives with respect to time :

$$X=\varphi\left(\lambda,\dot{\lambda},...,\lambda^{(k_1)},U,\dot{U},...,U^{(k_2)}\right) \tag{4}$$

Roughly, it corresponds to the Local Weak Observability (LWO) of Hermann and Krener.

A sufficient condition allows to conclude to the LWO of a system, by computing the rank of an observability matrix **O** [4] [5] :

$$\mathbf{O}=\left[dg_1 \;\; dL_f g_1 \;\; dL_f^2 g_1 \;\; ... \;\; dg_p \;\; ... \;\; dL_f^2 g_p\right] \in \mathbf{M_{3x(3.p)}} \text{ with } L_f g=\frac{\partial g}{\partial x}f=\dot{y} \tag{5}$$

where dg is the gradient vector of g. The Lie derivative of g with respect to f shows that the model is taken into account in **O**.

Finally, the ***sufficient*** condition is : *If rank(**O**)=3 then the system is LWO.*

Application to the system (Robot+Sensor+Beacons)

This section will highlight some situations where the sufficient rank condition does not allow to conclude to the LWO of the system.

Three beacons. In this part, we suppose p=3 (see equation (3)), which is the normal situation for our localizer. In such a case, the determination of the state X, given the three outputs λ_i, involves inverting g, using the inverse function theorem.

A geometrical solution, for determining the location of the mobile given the outputs only, is to use the fact that two measures, λ_i and λ_j, constrain the robot to lie on a given circle defined by the two beacons and the angle $(\lambda_i-\lambda_j)$. The three circles available intersect at the location of the robot. The heading is determined by the value of one azimuth angle.

Part 1 : application of the inverse function theorem. Let us consider the Jacobian :

$$\mathbf{O_1} = [dg_1 \;\; dg_2 \;\; dg_3] \tag{6}$$

One can notice that $\mathbf{O_1}$ (a sub-matrix of **O,** equation (5)) does not depend on the input.
Inverse function theorem : *X can be expressed as a function of the outputs* λ_i, *if the determinant of* $\mathbf{O}_1$ *is different from zero.*

The curve det($\mathbf{O}_1$)=0 is the circle (called (C) on fig. 2) defined by the three beacons.

The difficulty for this case, can be interpreted geometrically. Indeed, the three circles defined before merge in (C) and so X can't be determined.

Since we can invert g anywhere except on (C), we can conclude that the system is **LWO uniformly** (since U does not appear) anywhere except, *perhaps*, on (C).

Part 2 : study of the singularity (C). In the part before, the observability theory was not used. We will here apply the rank condition to the matrix **O** of equation (5).

If we consider the Lie derivatives, we notice that $L_f g_i(X)$ can be written as :

$$L_f g_i(X) = u_1 \cdot \psi(x, y, \theta, x_i, y_i) - u_2 \tag{7}$$

So, if u_1=0 (no translational speed) for all states X and all inputs u_2 :

$$dL_f g_i = dL_f^2 g_i = \vec{0} \Rightarrow \mathbf{O} = \left[dg_1 \ dg_2 \ dg_3 \ \vec{0} \ ... \ \vec{0} \right] \tag{8}$$

Thus, in this case, rank(**O**)=rank($\mathbf{O}_1$) and therefore, on (C), rank(**O**)<3. We can't assert that the system is not observable since the rank condition is only sufficient. We will see later that, in this case, our observer is not able to drive an initial estimation error to zero.

If we now suppose that the robot moves on (C) ($u1 \neq 0$), we could calculate $dL_f g_i$ and $dL_f^2 g_i$ and extract other sub-matrices, but the computations become quickly untractable. Since we do not know whether the system is observable on (C) when u_1 is different from zero, what we will do is test our observer for a mobile running on this possibly difficult trajectory.

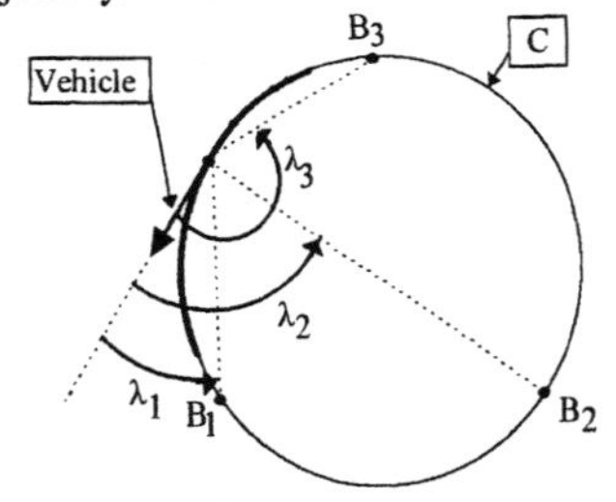

Figure 2 : problematic path with 3 beacons

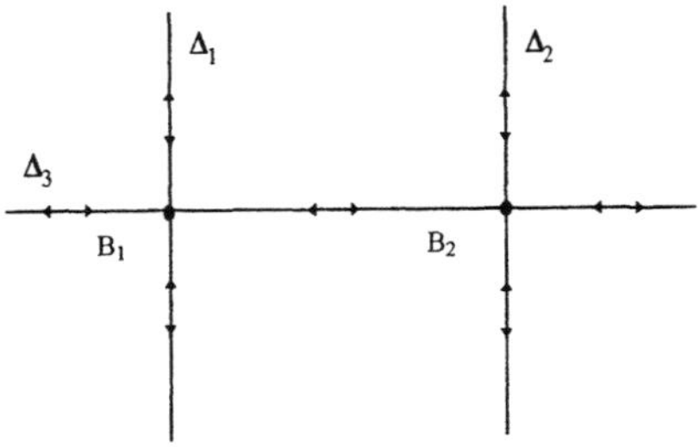

Figure 3 : problematic paths with 2 beacons

Two beacons. In this part, we suppose the sensor only detects two landmarks. We have p<3. Therefore, state observation can't be obtained using only the outputs. The evolution model and the input are here of crucial importance.

$$\mathbf{O} = \left[dg_1 \quad dL_f g_1 \quad dL_f^2 g_1 \quad dg_2 \quad dL_f g_2 \quad dL_f^2 g_2 \right] \in \mathbf{M_{3x6}} \tag{9}$$

If u1=0, using expression (7), the observation matrix reduces to :

$$\mathbf{O} = \left[dg_1 \quad dg_2 \quad \vec{0} \ \vec{0} \ \vec{0} \ \vec{0} \right] \tag{10}$$

Therefore, rank(**O**)<3 and we find a new problematic situation. With two beacons, it is clear that, when the vehicle is motionless, the filter cannot estimate three parameters.

In the sequel, we suppose $u1 \neq 0$. Consider the determinants (denoted s_i) of the 20 sub-matrices of **O**. We are trying to find (X,U)=(x, y, θ, u_1, u_2) such that (S) {s_1=0,..., s_{20}=0}.
(S) is a nonlinear differential system and the computations are untractable. So, we consider two new variables :

$$\frac{\partial u_1}{\partial t} = v_1 \quad \frac{\partial u_2}{\partial t} = v_2 \tag{11}$$

We solve (S) with the unknowns $(x,y,\theta,u_1,u_2,v_1,v_2)$. Using the symbolic computation software Maple, we obtain (some) solutions to which we apply the constraints (11). Finally, the problematic situations are when the robot is tracking one of the lines of figure 3, whatever its translational speed.

It should be noticed that we have been able to check that the states defined by other straight lines trough B_1 or B_2 satisfy the rank condition for non-zero translational speeds.

One beacon. In this case, the observation is a scalar. The observation matrix is now square. After computations, we find that det(**O**) is **always** equal to zero whatever the state and the input. This means that **O** has some columns or some rows which are linear combinations of the others. In this case, the system is generically non observable.

REAL SYSTEM STATE ESTIMATION

Algorithm principle

In fact, the real system differs from the continuous one, because the observer will have a discrete-time form, so that estimation will occur at discrete points in time.

Moreover, we have supposed that the three azimuth angles were provided simultaneously and continuously (here continuously would be at the sampling rate). In fact, the sensor is a CCD detector which rotates with a constant speed. The landmarks (light sources) are detected one at a time and asynchronously, since the angular interval between two landmarks depends on the position and movement of the mobile.

We use incremental encoders mounted on the wheels to measure the control vector U.

The principle of the algorithm presented below relies on such a prediction and uses absolute azimuth angles to update the location of the robot [1] [2]. This multisensor approach takes advantage of sensor redundancy and complementarity. A lot of variants of this mixed solution (using range-finders, inertial sensors...) can be found in the literature [6] [7] [8].

Discrete evolution model

A discrete evolution model for the continuous system can be obtained by the odometric technique. Odometric equations express the elementary displacements of the robot as a function of the elementary rotations of the wheels (Δq_1 and Δq_2) provided by the optical encoders attached to the left and right wheel of the vehicle.

Let r denote the radius of the wheels and e, the wheel base. When the robot moves, the encoders measure Δq_1 and Δq_2 between time instant t_{k-1} and t_k. Then, we can compute ΔD_k, the distance traveled by point M, and $\Delta\theta_k$ the elementary rotation :

$$\Delta D_k = r\cdot(\Delta q_1 + \Delta q_2)/2 \text{ and } \Delta\theta_k = r\cdot(\Delta q_1 - \Delta q_2)/e \tag{12}$$

Suppose we know X_k, we can compute X_{k+1} by [3] :

$$\begin{cases} x_{k+1} \approx x_k + \Delta D_k.\cos(\theta_k + \Delta\theta_k/2) \\ y_{k+1} \approx y_k + \Delta D_k.\sin(\theta_k + \Delta\theta_k/2) \\ \theta_{k+1} = \theta_k + \Delta\theta_k \end{cases} \Leftrightarrow X_{k+1} = \mathbf{F}(X_k, U_k) \tag{13}$$

Discrete Kalman filtering formulation

Let us consider now the nonlinear stochastic system with state-space description:

$$\begin{cases} X_{k+1} = \mathbf{F}(X_k, U_k) + \alpha_k \\ \lambda_k = g_i(X_k) + \beta_k \quad \text{with } i = 1, 2, \text{ or } 3 \end{cases} \tag{14}$$

One can note that the observation equation is now scalar, but **non-stationary**. α_k and β_k are, respectively, system and observation noises. α_k represents the effects of slippage or dragging on the ground, plus the effects of errors on robot parameters such as r and e.

Between two estimation phases, the Extended Kalman Filter has a "high" frequency (as compared to the frequency of the absolute readings) state and error prediction phase.

SIMULATIONS

The principle of our simulations is to have the robot follow a given reference trajectory. While the robot moves, we compute the corresponding encoder and sensor measures. The filter is input these values corrupted by noise.

Study of the filter behaviour on the circle defined by the three beacons

The LWO of the system has not been proved on the circle (C). In the following simulation, we compare the estimations of a filter with no initial error, and the same one, running on the same noisy measures, but with an erroneous initial estimate (with position and heading errors, in bold on figure 4. The test trajectory is in bold on figure 2).

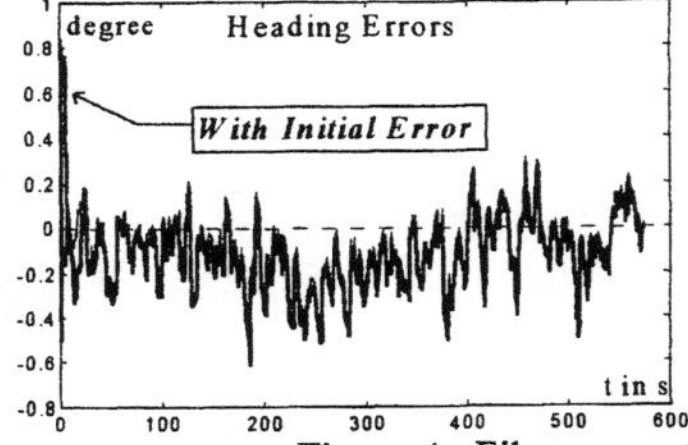

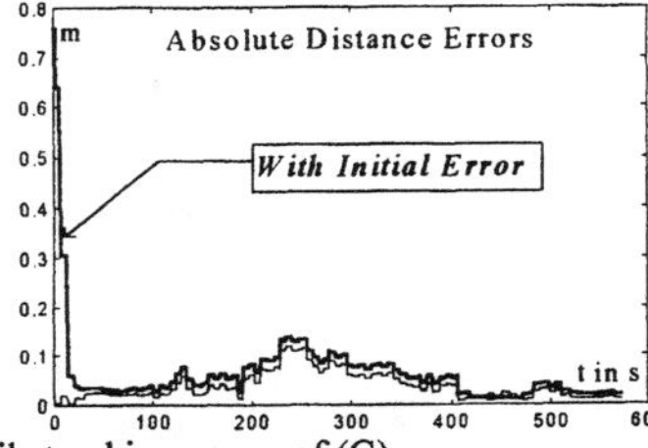

Figure 4 : Filter errors for a mobile tracking an arc of (C).

The simulations show that the filter with initial error converges quickly and its outputs become identical to the outputs of the filter with no initial error. This kind of behaviour indicates the convergence of the filter. Furthermore, after few iterations, the heading error has zero mean (which shows that the filter has no bias) and remains bounded.

We have also simulated a motionless vehicle on the circle (remind that this is a problematic situation). The observer does not converge towards the real position and heading, when an initial error occurs. Nevertheless, the filter outputs never diverge.

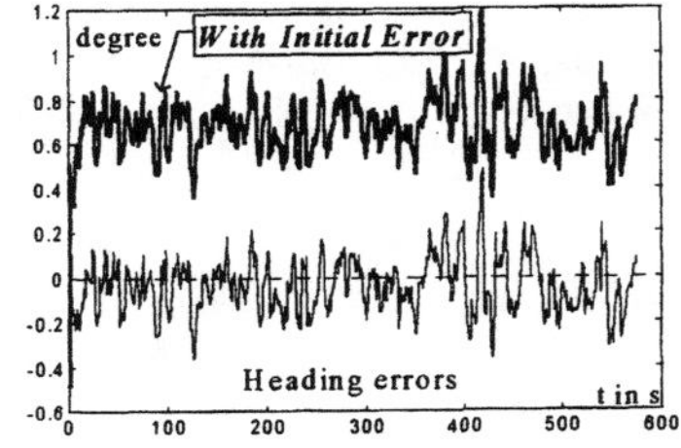

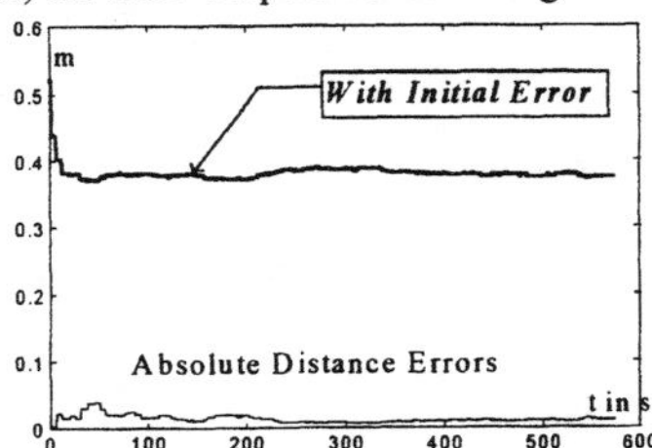

Figure 5 : Filter errors for a motionless mobile on (C).

We conclude that if the vehicle is motionless on (C), the observer we propose does not converge but does not diverge : if the robot stops on (C) with an estimation error, this error will remain bounded. In practice, such a trajectory is unlikely to occur. In a typical application, the robot moves in the triangle formed by the three beacons. When out of this triangle, it typically uses a new set of closer beacons.

Study of the filter behaviour with two beacons

We report below the results of two simulations, in which the robot moves in the direction of a beacon. This situation is especially difficult for the filter, because the azimuth angle of a beacon is a constant, whatever the position of the robot on the line.

When the straight line is not perpendicular to (B_1,B_2), the filter converges (see figure 6). Remind that we have proved the LWO of the system on these trajectories.

Additionally, we can see that the filter does not converge on the perpendicular line. We have noticed that the estimate converges towards a line trough B_1, but different from the real one. This line depends on the initial error and on the initial state covariance matrix.

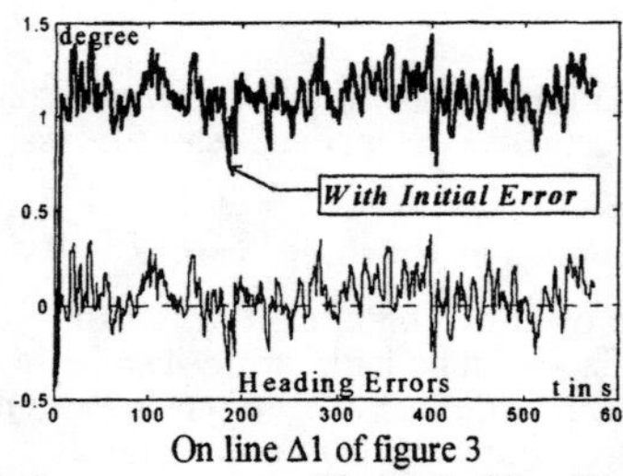

On line $\Delta 1$ of figure 3

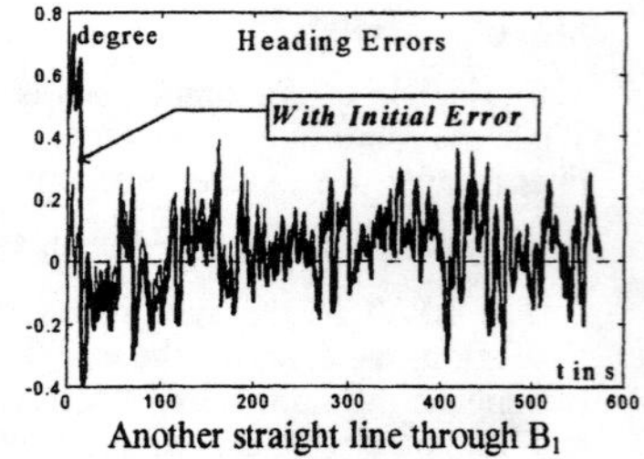

Another straight line through B_1

Figure 6 : Filter heading errors with two beacons.

Study of the filter behaviour with one beacon

The non-observability with one beacon is clear : heading and distance errors increase unboundedly. Yet, we analyze the outputs of two processes of the same filter, with no initial error in both cases. The two simulations use the same odometric measures of a robot tracking a straight line at a constant speed. We can conclude that, even if the observer diverges with one beacon, the divergence rate is larger without the exteroceptive measures.

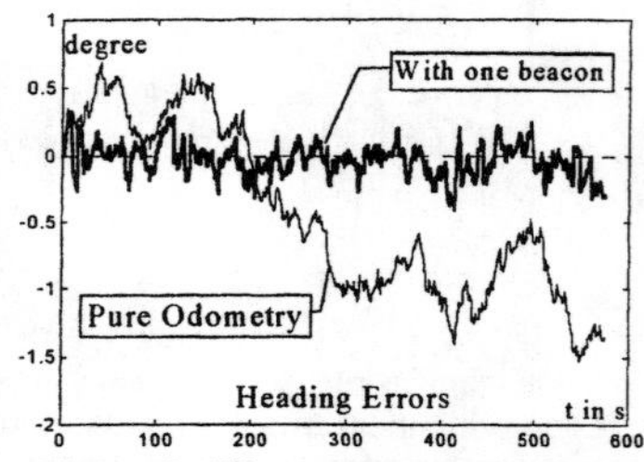

Figure 7 : Filter errors on a straight line.

CONCLUSION

The practical interest of studying the nonlinear observability is to determine situations where any localization system based on the same continuous system will undergo difficulties, whatever the technological and algorithmical solutions. Nevertheless, this step cannot guarantee that there exists an observer which computes the state, when the state is claimed to be observable. So, filter convergence should be checked anyway through simulations

The approach we have developed can be applied to different systems such as range finders, or inertial sensors. Yet, it does not give any information on the accuracy of the observer. As a matter of fact, precision depends on the number of landmarks and on their configuration.

REFERENCES

1. G. Garcia, Ph. Bonnifait and J.-F Le Corre, "A multisensor fusion localization algorithm with self-calibration of error-corrupted mobile robot parameters", ICAR, Barcelona, pp 391, September 1995.
2. Ph. Bonnifait and G. Garcia. " A Multisensor Localization Algorithm for Mobile Robots and its Real-Time Experimental Validation ", IEEE ICRA, Minneapolis, April 1996.
3. C. Ming Wang, "Location estimation and uncertainty analysis for mobile robots", IEEE ICRA, Philadelphia, April 1988.
4. R. Hermann and A.J. Krener, "Nonlinear controllability and observability", IEEE Transactions On Automatic Control, vol. AC-22, n° 5, October 1977.
5. S. Diop and M. Fliess, "Nonlinear observability, identifiability, and persistent trajectories", 30th IEEE CDC, pp 714-719, Brighton, England, December 1991.
6. C. Durieu, J. Opderbecke and G. Allègre, "A data fusion application for location of a mobile robot using an odometer and a panoramic laser telemeter", Intel. Auto. Systems 3, Pittsburgh, Feb 1993.
7. T. Nishizawa, A. Ohya and S. Yuta, "An implementation of on-board position estimation for mobile robot", IEEE ICRA, pp 395-400, Nagoya, May 1995.
8. J. Horn and G. Schmidt, "Continuous localization of a mobile robot based on 3D-laser-range-data, predicted sensor images, and dead-reckoning", Robotics and Autonomous Systems n°2-3, May 1995.

IMPROVING REACTIVE NAVIGATION USING PERCEPTION TRACKING

REINHARD BRAUNSTINGL

Department of Mechanics, Technical University of Graz
Kopernikusgasse 24, A-8010 Graz / Austria
E-Mail: reinhard@mech.tu-graz.ac.at

ABSTRACT

Local navigation of mobile robots relies on the presently available sensor information. In case a sensor loses contact with a previously detected but still near object, this information does not represent any longer the situation the robot is in. A fuzzy controller using this information thus may generate wrong commands. Tracking the perceptual information of a sensor while the robot moves on results in a more robust controller input enabling the robot to move on with incomplete or without sensor information for a short time. The paper at hand presents a method to track the perception of a sensor in order to generate a so-called virtual perception, which then is used to guide the robot. Two simulated experiments demonstrate that the method can be applied to different types of mobile robots being equipped with different sensors.

KEYWORDS: Local Navigation, Mobile Robot, Ultrasonic Sensors

1. INTRODUCTION

When navigating a mobile robot equipped with ultrasonic sensors, the inherent uncertainties of this type of sensor have to be dealt with. It is well known that these sensors provide accurate distance information but rather imprecise information on the exact position of a detected obstacle. There are techniques to overcome this disadvantage which use repeated measurements from different positions while taking into account the characteristics of the sensor. This procedure yields a grid-based or feature-based representation, i.e. a map of the robot's environment with quite good resolution [1], [2]. The map can be used subsequently to generate appropriate control commands [3]. Another, more direct approach has been proposed by [4], where the representation of the environment is replaced by a representation of the robot's perception, which serves as a fuzzy description of the environment and subsequently is used as input for a fuzzy system. Thus, quick reaction of the robot, as it is necessary for reactive navigation, is made possible with minimum effort. The method has been proved to be efficient in practical tests with a real robot. However, in an omnidirectional robot, it requires 360-degrees perception at all times, since only the presently available perceptual information of the ultrasonic sensors is used.

This requirement of constantly available perception of all obstacles is the weak point of the fuzzy controller mentioned above. If this precondition is not fulfilled, the controller may fail, which is illustrated basically in fig. 1. In fig. 1a a small obstacle enters a blind sector between two sensors. A controller relying only on the presently available perceptual information will not take into account the presence of the obstacle any longer and certainly not generate appropriate commands. Car-like mobile robots that don't even have to cover every sector with a sensor due to their kinematic constraints can encounter similar problems. If such a robot loses perception altogether - for example while rounding a corner (fig. 1b) - it has no more information to generate correct control commands.

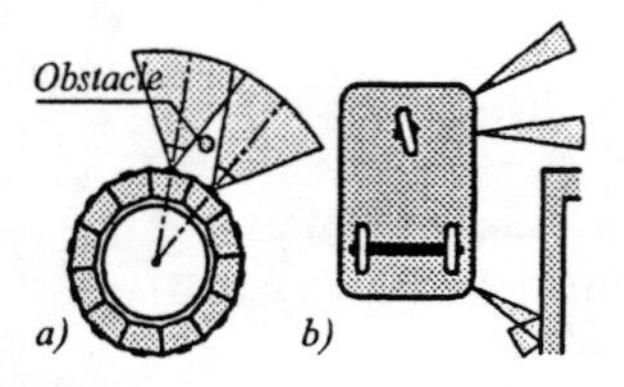

Figure 1. Disadvantages of purely reactive navigation. a) blind spots between two sensors. b) large blind areas of a car-type mobile robot. Both cases cause loss of perception of a previously detected object.

The work at hand presents the idea of a perception memory to overcome the disadvantages mentioned above. To achieve this, the real perception of every sensor is replaced by a virtual perception which is updated with the data of the sensor. The generation of control commands is based on this virtual perception instead of the real perception. In case an ultrasonic sensor loses contact with an obstacle the virtual perception can be tracked, i.e. its changing with respect to time is calculated, while the robot moves on. Thus, the robot may travel a short distance without any information from its sensors, just like a human being may close his eyes for a short time while walking.

The following pages will briefly explain firstly how fuzzy logic can be used to perform perception based local navigation, and secondly how the perception can be tracked in case a sensor loses contact with an object. Finally, examples of simulated experiments with two different robots will be presented.

2. PERCEPTION BASED NAVIGATION

The raw sensor data provided by ultrasonic sensors yield a somehow fuzzy notion of the environment. A fuzzy system is perfectly suited to directly process such imprecise data and can therefore be used as a controller for quick reactive navigation, since the need of complex and time-consuming data processing is eliminated. The input for such a fuzzy controller can be provided by a so-called perception vector [4] which is illustrated in fig 2.

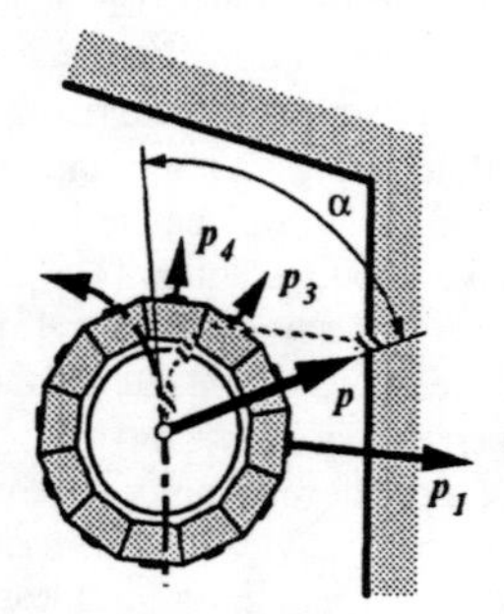

Figure 2. The vector of general perception $\boldsymbol{p}$ describes a situation in a fuzzy way.

The perception of each ultrasonic sensor is represented by a vector $\boldsymbol{p}_i$ whose length is a function of the distance measured by the sensor and whose direction coincides with the sensor axis. Several of these vectors can be combined to form a general perception vector $\boldsymbol{p}$ representing the situation the robot is in in a fuzzy way. In fig. 2 this vector points to the right and a little bit in front of the robot, which corresponds to the location of the two walls: right and front. A fuzzy controller can easily use this description of the robot's perception. To achieve wall following behavior, for example, it is sufficient to keep the length of the perception vector constant and its direction α perpendicular to the direction of motion. This can be achieved by setting up rules like e.g.

```
IF perception MEDIUM and RIGHT_BACK turn RIGHT
IF perception HIGH and RIGHT_FRONT turn HARD_LEFT
IF perception_change HIGH BRAKE
```

However, if a sensor fails to detect an obstacle even for a short time, the fuzzy system generates wrong responses due to incorrect input data. The method of perception tracking described below overcomes this difficulty by providing a virtual perception in case a sensor loses contact with an object.

3. PERCEPTION TRACKING

To be as general as possible let us look at a robot of arbitrary shape equipped with ultrasonic sensors. Any such sensor may be located at a position $\boldsymbol{u}$ with its axis pointing to the direction $\boldsymbol{s}$ (fig. 3). Furthermore we assume a virtual perception coordinate system located at E, pointing to a *direction of attention* $\boldsymbol{a}_1$. A frame r represents the robot's position and orientation, $\boldsymbol{x}$ and φ. Its state of motion is given by the velocity $\boldsymbol{v}$ of the reference point and the angular velocity $\omega_{r/w} = \dot{\varphi}$ of the robot with respect to a fixed frame w.

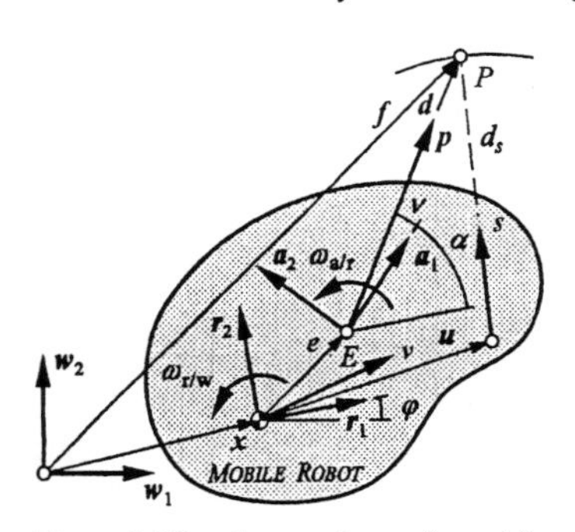

Figure 3. Virtual perception p of a mobile robot. An ultrasonic sensor passes its perception data to the virtual perception system. In case the real sensor loses perception the virtual perception can be tracked as the robot moves on.

An ultrasonic sensor receives an echo from a point located on an arc at a distance d_s from the sensor. If we assume this point P to be on the sensor axis, a sensor located at E will detect P at a distance d and at an angle ν. Defining a so-called perception function

$$p = f(d, \nu) \tag{1}$$

the distance measurements of an ultrasonic sensor can be transformed into a virtual perception $\boldsymbol{p}$. This function assigns a perception value to a distance d taking into account the angle between $\boldsymbol{a}_1$ and $\boldsymbol{p}$, thus making it possible to assign different weight to points detected in different directions. Furthermore, data from different types of ultrasonic sensors can thus easily be combined easily into one perception.

Now let us track the virtual perception taking into account the robot's movement. Starting from $\boldsymbol{f} = \boldsymbol{x} + \boldsymbol{e} + \boldsymbol{d} = \text{const.}$, assuming that P does not move we can express $\dot{\boldsymbol{d}}$ as

$$\dot{\boldsymbol{d}} = -\dot{\boldsymbol{x}} - \omega_{r/w} \times \boldsymbol{e} = -\boldsymbol{v}_E \,. \tag{2}$$

Using the inverse function of eq. (1) $d = g(p, \nu)$ we get

$$\boldsymbol{d} = g\left(\boldsymbol{a}_1 \cos\nu + \boldsymbol{a}_2 \sin\nu\right). \tag{3}$$

Differentiating equation (3), using $\dot{\boldsymbol{a}}_1 = \boldsymbol{a}_2(\dot{\varphi} + \dot{\alpha})$ and $\dot{\boldsymbol{a}}_2 = -\boldsymbol{a}_1(\dot{\varphi} + \dot{\alpha})$, and eq. (2) together with $\boldsymbol{v}_E = (\boldsymbol{v}_E \boldsymbol{a}_1)\boldsymbol{a}_1 + (\boldsymbol{v}_E \boldsymbol{a}_2)\boldsymbol{a}_2$ we get two differential equations

$$
\begin{aligned}
&\dot{p}\frac{\partial g}{\partial p}\cos\nu+\dot{\nu}\left(\frac{\partial g}{\partial \nu}\cos\nu-g\sin\nu\right)=-\boldsymbol{v}_E\boldsymbol{a}_1+g(\dot{\varphi}+\dot{\alpha})\sin\nu\\
&\dot{p}\frac{\partial g}{\partial p}\sin\nu+\dot{\nu}\left(\frac{\partial g}{\partial \nu}\sin\nu+g\cos\nu\right)=-\boldsymbol{v}_E\boldsymbol{a}_2-g(\dot{\varphi}+\dot{\alpha})\cos\nu\,.
\end{aligned}
\tag{4}
$$

After carrying out some calculations taking into account $\boldsymbol{v}_E\boldsymbol{a}_1=\boldsymbol{v}_E(\boldsymbol{r}_1\cos\alpha+\boldsymbol{r}_2\sin\alpha)$ and $\boldsymbol{v}_E\boldsymbol{a}_2=\boldsymbol{v}_E(-\boldsymbol{r}_1\sin\alpha+\boldsymbol{r}_2\cos\alpha)$ and setting $\boldsymbol{v}_E=\dot{\boldsymbol{x}}+\boldsymbol{\omega}_{r/w}\times\boldsymbol{e}$ according to eq. (2) these equations yield the derivatives of angle and length of the perception vector:

$$
\begin{aligned}
&\dot{\nu}=\frac{1}{g}\left[(\dot{\boldsymbol{x}}+\boldsymbol{\omega}_{r/w}\times\boldsymbol{e})(\boldsymbol{r}_1\sin(\alpha+\nu)-\boldsymbol{r}_2\cos(\alpha+\nu))\right]-\omega_{r/w}-\omega_{a/r}\\
&\dot{p}=-\frac{1}{\partial g/\partial p}\left[(\dot{\boldsymbol{x}}+\boldsymbol{\omega}_{r/w}\times\boldsymbol{e})(\boldsymbol{r}_1\cos(\alpha+\nu)+\boldsymbol{r}_2\sin(\alpha+\nu))+\frac{\partial g}{\partial \nu}\dot{\nu}\right],
\end{aligned}
\tag{5}
$$

where $\omega_{a/r}=\dot{\alpha}$ is the angular velocity of the virtual perception coordinate system relative to the robot. The virtual perception is obtained by integration of eq. (5).

4. REACTIVE NAVIGATION AND RESULTS

Now the method of perception tracking shall be demonstrated using first a simulation of an omnidirectional and then of a car-type mobile robot. Since this simulation has already been successfully used in the past to develop mobile robot controllers, major differences are not to be expected when perception tracking will be tested in the real robots. Although the robots are quite different, perception based navigation makes it possible to use the same fuzzy controller as presented in [4] in both cases. The local navigation task is also the same as defined there, namely to follow a wall.

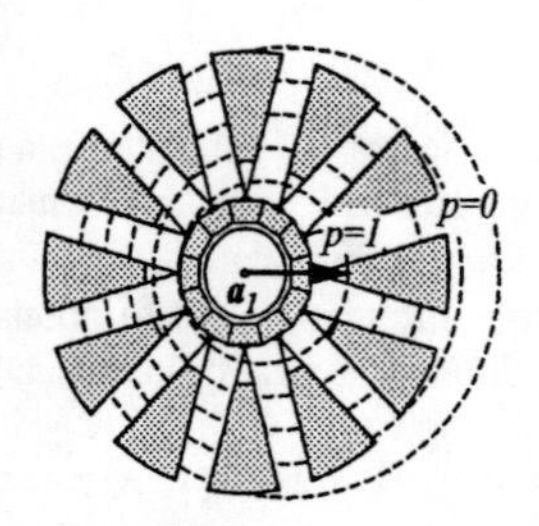

Figure 4. The omnidirectional robot with its 12 sensors and the area covered by the perception function. The dotted ellipses are lines of constant perception.

4.1 Omnidirectional Robot

First the mobile robot simulator was set up with the data of the omnidirectional robot VEA-1 of the Spanish research center IKERLAN (fig. 4). The sonic cones of the real robot overlap slightly, but we narrowed the cones in the simulation, thus provoking loss of perception more frequently to illustrate the problem more clearly. The virtual perception system was placed at the center of the robot with the direction of attention $\boldsymbol{a}_1$ coinciding with the path tangent. The inverse perception function used was

$$
g(p,\nu)=d_{\min}\frac{1-\varepsilon}{1-\varepsilon\cos\nu}\left[n-p(n-1)\right],
\tag{6}
$$

resulting in the curves of constant perception being ellipses of eccentricity ε. Looking in the direction of attention $\boldsymbol{a}_1$, $p=1$ corresponds to an obstacle at a distance $d_{\min}$, $p=0$ is mapped to a distance $n\,d_{\min}$.

Fig. 5a demonstrates what may happen during a wall following experiment without perception tracking. When approaching the corner, the robot starts turning right correctly. However, at the same time the orientation of the front sensor changes, which results in loss of perception of the wall in front of the robot. The robot turns back - correctly, given the purely reactive controller, since it has only a notion of the wall to its left now. This

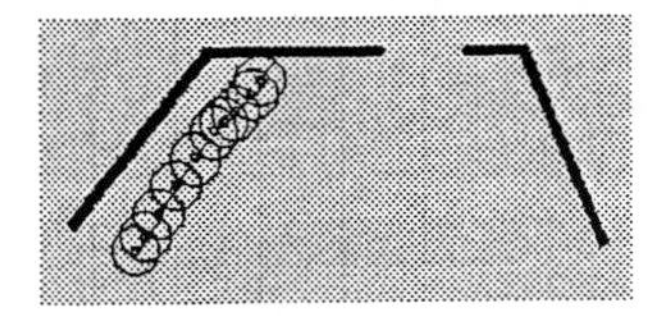

Figure 5a. Local navigation without perception tracking. Loss of perception while turning right causes the robot to turn left again, thus hitting the wall.

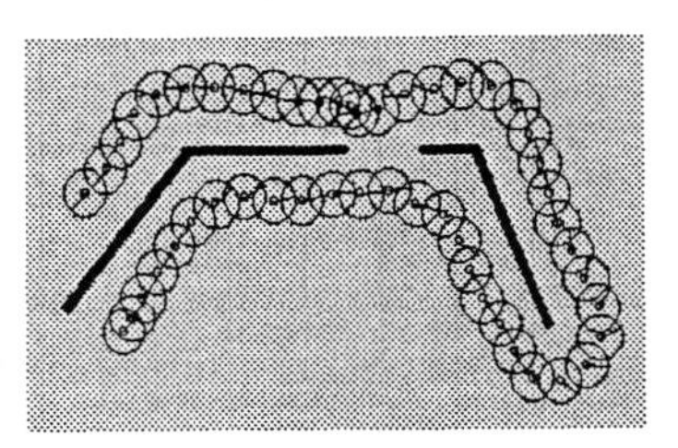

Figure 5b. Perception tracking enables the robot to round a corner even if a sensor only once gets a short glimpse of perception. The robot remembers the perception of the wall and acts accordingly.

results in detecting the front wall again and the robot finds itself in a deadlock situation which may last until the robot hits the wall. However, using a perception vector - composed of the virtual perceptions of each sensor - as input of the fuzzy controller, the robot is able to remember the perception of the corner. In fig. 5b a virtual perception is obtained by integrating eq. (5), thus taking into account the robot's movement, in case a sensor loses perception. Now the robot is able to round the corner without difficulty, although, for a short time, its sensors do not receive any echo.

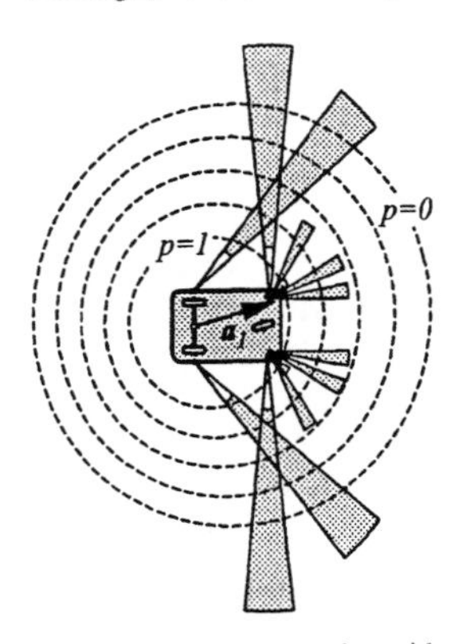

Figure 6. The car-type robot with its ultrasonic sensors and the area covered by the perception function. The dotted ellipses are lines of constant perception.

4.2 Car-like Robot

This experiment was repeated using the data of a very different robot, ROMEO-3R of the University of Sevilla, which has a car-like kinematic. It is equipped with ultrasonic sensors of different types and ranges and has relatively large blind sectors in between them (fig. 6). The same controller was used as in the example above, but the membership function of the adjective „left" of perception was changed so that its maximum was at 70° instead of 90°. Furthermore the size of the elliptical perception function was increased due to the bigger range of the sensors. The virtual perception system was placed at the center of the robot's rear axle and the direction of attention a_1 was kept parallel to the front wheel. This example shows well, how different ultrasonic sensors can be combined to contribute to a general perception of the robot's surroundings.

Without perception tracking (fig. 7a) the robot is not able to proceed along the wall. It encounters the same problem as described in the example above. Although it rounds one corner successfully and starts perceiving the second one with one of its digital front sensors, the robot loses perception this time immediately when turning right. It turns back left a little bit, gets an echo from the wall just to lose it again when turning right. These improper control actions are repeated until the robot hits the wall which does not happen if the perception of the wall is tracked after a sensor has lost contact. In fig. 7b the robot also loses perception of the wall when starting to turn right. However, this time it tracks the perception it had before and the controller acts as if the robot continuously perceived the wall. Similar situations occur sometimes during the wall following operation, but if only one sensor gets an echo from a wall for a short time, the robot is able to navigate with such limited information.

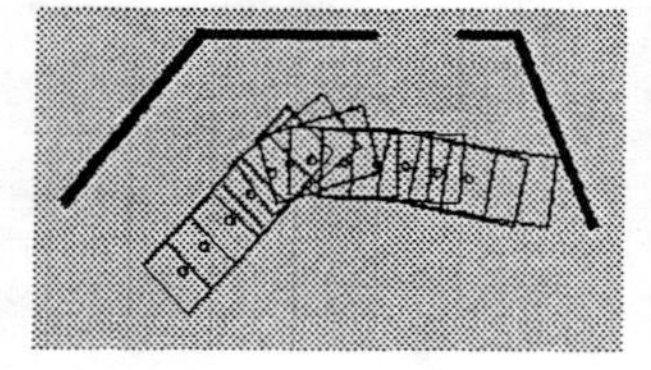

Figure 7a. Loss of perception while turning right causes the robot to turn left again, thus hitting the wall.

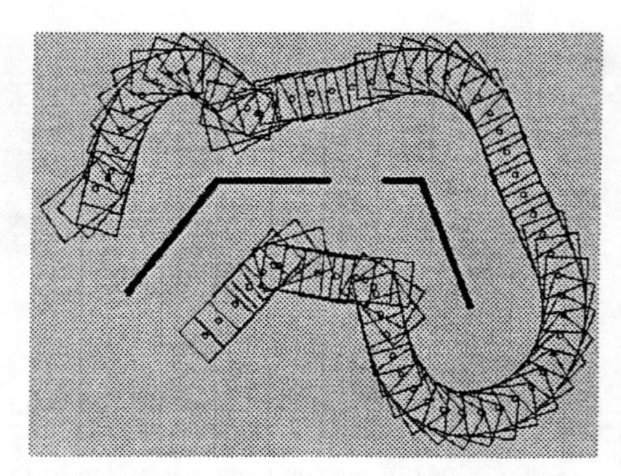

Figure 7b. Perception tracking enables the robot to round a corner even if a sensor only once gets a short glimpse of perception.

5. CONCLUSION

Perception based local navigation relies on the fact that perception is a correct representation of the situation the robot is in. In case an object cannot continuously be detected by a sensor, this precondition is not fulfilled and the robot's reactions may not be correct. In this paper a simple method is presented as to how a virtual perception of an object may be obtained, tracked and thus remembered if a sensor loses contact with that object. This perception can be used as input for a local navigation controller just like „real" perception. The method is demonstrated using simulated experiments with two different robots. The next step will be to carry out experiments with the real robots.

REFERENCES

1. Oriolo G., G. Ulivi and M. Vendittelli, Motion Planning With Uncertainty: Navigation on Fuzzy Maps, in: *4th IFAC Symposium on Robot Control*, pp.71-78, Capri, Italy, Sept., 1994.
2. Leonard J., H.F. Durrant-Whyte, *Directed Sonar Sensing for Mobile Robot Navigation*, Boston, Kluwer Academic Publishers, 1992.
3. Borenstein J. and Y. Koren, The Vector Field Histogram - Fast Obstacle Avoidance for Mobile Robots, in: *IEEE Transactions on Robotics and Automation, Vol. 7, No. 3*, pp. 278-288, June, 1991.
4. Braunstingl R., P. Sanz and J.M. Ezkerra, Fuzzy Logic Wall Following of a Mobile Robot Based on the Concept of General Perception, in: *Proc. of the Seventh International Conference on Advanced Robotics, Vol 1*, pp.367-376, Sant Feliu de Guixols, Spain, Sept., 1995.

A SYSTEM FOR PERFORMING SITE CHARACTERIZATION FOR TEST RANGES CONTAINING UNEXPLODED ORDNANCE

Edward Brown
Wright Laboratory WL/FIVC
Tyndall Air Force Base, Florida 32403

Carl D. Crane III
Center for Intelligent Machines and Robotics
University of Florida, Florida 32611

ABSTRACT

Wright Laboratory has been tasked by the Naval Explosive Ordnance Technical Division, (NAVEODTECHDIV) to develop robotic platforms to perform characterization of areas set aside for ordnance testing. These areas require the identification and removal of the unexploded ordnance before they can be utilized for safe, productive use.

The characterization task is performed by autonomously sweeping a designated area with the Autonomous Tow Vehicle (ATV). The ATV makes use of several advanced technologies. A hybrid navigation and guidance system using an external Kalman filter delivers vehicle position based on information from a Differential Global Positioning System (DGPS) and an Inertial Navigation System (INS). Sophisticated path planning algorithms, and an intelligent software architecture provide a measure of autonomy.

Keywords: autonomous vehicles, unexploded ordnance, magnetometers, ground penetrating radar, path planning, Kalman Filter, GPS, INS, robotics, robotic vehicle

1. THE UXO PROBLEM

The need for a characterization system is driven by the recent Base Realignment and Closure movement in the Department of Defense. Millions of acres of former test ranges, impact zones, and hazardous waste sites must be rendered safe before they are turned over for public use. Systems, such as the Subsurface Ordnance Characterization System (SOCS) are needed to quickly, safely, and accurately identify and locate subsurface ordnance and hazardous waste.

2. OBJECTIVES

The purpose of the SOCS development was to prove the concept of an autonomous, self propelled, data gathering system. The SOCS performs autonomous surveys of areas of interest, providing time and position stamps for sensor data samples. SOCS then stores this sensor data for later analysis. The product of this development effort will be a technology transfer package that will be used by private contractors to perform hazardous cleanup of public lands. The SOCS mission is to operate as a test platform and will be used to investigate the use of new and promising subsurface sensing devices. Coupling this platform with a standard test range provides a means of comparing the performance capabilities of each tested sensor directly.

3. SYSTEM DESCRIPTION

The characterization system is composed of the Autonomous Tow Vehicle (ATV), the Multiple Sensor Platform (MSP), and the Mobile Command Station (MCS). The ATV performs autonomous surveys of designated areas and provides the data collection system with time and position information. The MSP acts as a non-magnetic instrument carrier for testing sensor performance, and a platform for data collection. The MCS acts as the base station for control of the vehicle by the operator. It contains the operator interface, GPS base station, and the data analysis and display computers.

The ATV consists of several integrated subsystems; the vehicle itself, the vehicle electronics subsystem, which provides for computer control of the vehicle, the navigation system, which provides time and position information, generates the path plan, and controls the vehicle during path execution, the communication system, which provides telemetry information for the GPS system, a video channel from the vehicle to the mobile command station, and a two way data link that transmits and receives status and command information from the operator, and finally, the data collection subsystem, which controls the sensors aboard the sensor platform, collects the sensor data during survey operations, and stores the data for later analysis.

VEHICLE DESCRIPTION

The ATV chassis was based on the John Deere Gator. The vehicle has been modified to accommodate computer control and to reduce the overall magnetic signature. The Gator was chosen for a variety of reasons. One of the important factors in selecting the Gator was the ease with which it could be converted to computer control. Since the gearing was simple, it uses a centrifugal pulley system, no actuation was required to accommodate gear changes. Another factor was the rear cargo bed, which provided a convenient and simple platform for the integration of the necessary electronics, navigation, and communication equipment. Figure 1 shows its current configuration with the MSP.

Figure 1 - SOCS

In order to allow computer control, the vehicle's throttle, braking, and steering mechanisms have been slightly modified to accept actuating motors and absolute encoders. They were modified so that they could remain functional during manual operations. The actuation and information from the encoders from these mechanisms are then controlled by proportional, integrated, differential, PID, servo controllers which in turn takes direction from the guidance system.

NAVIGATION SYSTEM

The navigation system provides the means for the autonomous survey of a given area. The operator must provide information regarding the boundary of the area of interest and any obstacles contained within that area. Given this information the system will perform the following: (1) autonomously navigate the vehicle from its current position to the edge of the field to be surveyed; (2) plan an efficient path which targets 100% coverage of the field with a user specified overlap for each swath; (3) autonomously executes the planned path, collecting sensor data while avoiding collisions with expected or unexpected obstacles. These tasks are performed by three subsystems: the Path Planner, the Positioning System, and the Path Executioner.

PATH PLANNER

The area to be surveyed is assumed to contain regions where the vehicle is prohibited from operating. Buildings, trees, telephone poles, lakes, and other obstacles are represented as polygonal shapes and stored in an area map. A path planner is used to generate an efficient path from the starting position to a position at the beginning of the path for the area to be swept, from an area that has just been swept to another survey area, or back to the starting point[1].

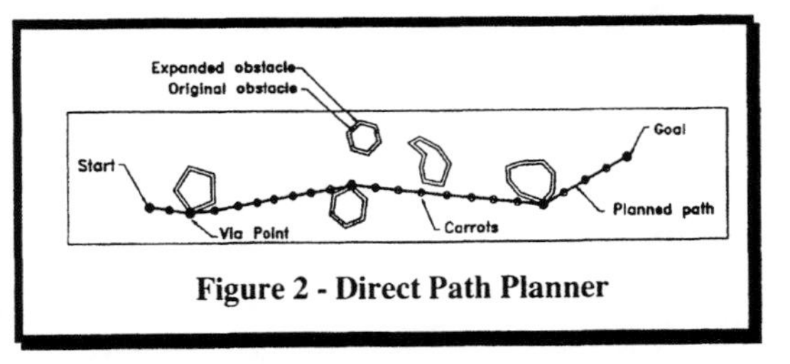

Figure 2 - Direct Path Planner

Direct Path Planning creates a direct path from the vehicle start position to a goal position. The first step in planning an efficient path is to expand all known obstacles in the local map so that the vehicle can be treated as a point. Figure 2 shows the polygon obstacles and their expansion. The amount of expansion is equal to the radius of a circle which circumscribes the vehicle. The A* search algorithm is then applied to determine the shortest path to the goal as follows: using the obstacle vertices which are visible from the current location, a cost is calculated using the sum of the distance to each vertices and the distance from there to the goal via a straight line. The lowest cost choice is selected as the first via point and the process is then repeated using all visible vertices from there. The process is complete when a straight line connects the current via point to the goal.

Once the vehicle arrives at a goal position near the field to be swept, a second method of path planning is used to generate a field-sweep path as shown in Figure 3.

A field is modeled by an N-sided polygon. Two adjacent vertices, A and B, are chosen and used to generate parallel rows across the field. The rows are separated by a user defined swath width, L, which represents the width of the detection system, plus a desired overlap. Point A is the start position for field sweeping. Point B is required to be the point next to A such that motion from A to B is clockwise motion around the boundary. The line segment AB corresponds to row #1.

The endpoints of the rows are used to define the path to be followed, where K is the number of rows to be swept for the field. Each row is checked for intersection with the obstacles loaded into the database. Wherever an intersection is encountered, an alternate route around each side of the obstacle is examined. The shortest detour is incorporated into the total field sweep path.

The current sweep pattern used for the survey vehicle is the "Half Field Method," see Figure 3. Row #1 is swept followed by the middle row M, where M=K/2. Next, row #2 is swept followed by the middle row (M+1). This pattern is continued until the entire field has been swept. This method does not require a row be swept more than once except when K is odd. In such cases, row M is swept twice.

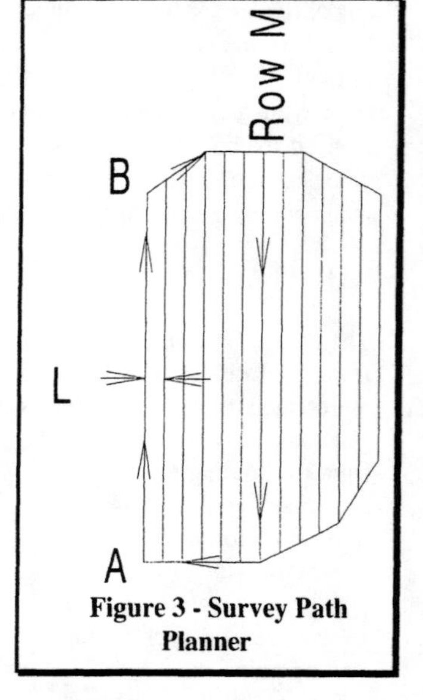

Figure 3 - Survey Path Planner

POSITIONING SYSTEM

To successfully navigate along a pre-planned path, the ATV must have some means of accurately and consistently determining its position and orientation. This problem is being addressed by the application of an inertial navigation system, INS, integrated with a differential global positioning system, GPS.

The ATV uses the Modular Azimuth Position System, MAPS for inertial navigation. The MAPS is a completely self contained, strapped down, laser gyro system. Given an initial position, the MAPS makes use of its three ring laser gyros and three accelerometers to determine relative position, angular orientation, and velocities. Position and orientation data from the MAPS are made available at a rate of 12.0 Hz. The MAPS makes use of velocity updates to damp velocity errors that cause drift in the position accuracy over time.

The GPS system implemented uses P and C code RF signals, which are transmitted from orbiting satellites, to determine position and velocity data at a rate of 0.5 Hz. A single GPS unit is subject to a number of errors outside the control of the receiving unit. The military introduces two forms of random errors into the signal known as selective availability and anti-spoofing. Under these conditions, the GPS only delivers absolute position to within 80.0 meters. To increase the system's accuracy, a method known as differential GPS is being applied: A GPS receiver is placed at a known pre-surveyed location, the base station. A second remote GPS receiver is placed on the moving vehicle. Using prior knowledge of its position, the base station receiver can determine the systematic or bias errors from the incoming signal. The corrections are then transmitted to the remote vehicle, where they are then used to reduce errors. Position data have been found to be accurate in the range of 2-10 centimeters 85% of the time using this method.

The integration of the GPS with the MAPS has greatly increased the overall system performance. The two systems complement each other well in that the MAPS provides continuous data at high rates while the GPS system is not subject to drift. The external software Kalman filter uses models of the navigation instruments used, and an error history of these same instruments to predict current vehicle position. This system provides a robust means for acquiring position during intermittent dropouts or spurious instrument errors.

PATH EXECUTIONER

The path executioner continuously generates a recommended steering angle and vehicle velocity that keeps the survey vehicle on the pre-planned path until the goal is reached. The path is used to obtain sub-goals at 3.0 meter intervals. Around corners, sub-goals are placed on the interior of the path to ensure a smooth transition between segments. PID control is used to steer the vehicle directly towards the current sub-goal. The desired vehicle velocity is a linear function of the distance to the next turn and the sharpness of that turn.

DATA COLLECTION SYSTEM

The data collection system uses several file systems to store configuration, real time, and target information. The Interplatform Data Set, IDS, contains all of the configuration and environmental data and describes: sensor descriptions and geometric locations, weather, terrain, time, date, and location of the test area. The Semi-Raw Data Set, SRD, contains a header describing a sensor. The data field contains a time record and an associated sensor sample. A separate SRD is then generated for each sensor type and is keyed to the IDS by the sensor description and the time and date. An additional SRD is generated by the navigation system that contains records for time and position. This file is then used during post processing to determine position for the sensor data for any given time. The Standard Data Set, STD, is generated following post processing and an analysis of the SRD file. The STD is then composed of target information: position, orientation, and an estimate of the ordnance type and class. These file formats are defined and described in The Simultaneous Data Collection and Processing System Interplatform Data Set[2].

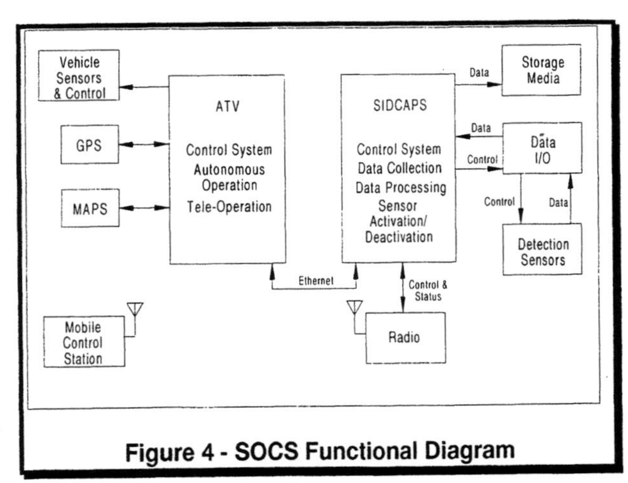

Figure 4 - SOCS Functional Diagram

Figure 4 shows a functional diagram of the Data Collection System and its interface with the ATV Navigation System. The figure shows that the Data Collection System both controls the functioning of the sensors as well as directing the data stream from the sensors. In addition to directing and manipulating the data stream, the Data Collection System acquires time and position from the ATV Navigation System. The details of the interaction between the different subsystems may be found in the SOCS Interface Design Document [3].

MULTIPLE SENSOR PLATFORM

The Multiple Sensor Platform was developed to provide a structure for an array of cesium magnetometers and a hanger for a Ground Penetrating Radar, GPR. The platform is shown in Figure 1. The GPR is suspended below the platform frame via a pinned hanger. That hanger structure contains an encoder which measures the relative angular displacement from the platform frame which is also written to an SRD file for GPR position resolution. The magnetometers are hung from the rear of the platform via an articulated beam. The beam can be secured in various pitch attitudes with respect to the platform. In addition the magnetometer sensor head clinching mechanism secure the sensor with variations in roll.

4. SUMMARY

The SOCS was developed for autonomous operations because of the difficulty of accurately controlling the position and orientation of a remote vehicle, and because of the inherently hazardous conditions in which it must operate. The ATV provides a time and position stamp for all sensor data collected aboard the Multi-Sensor Platform. The integration of vehicle control and sensor data collection makes the SOCS system unique. Although the SOCS system cannot be used in all terrain, it does provide a unique capability for the vast majority of test range and impact zone sites..

5. REFERENCES

1. Rankin, A.L., "Path Planning and Path Execution Software for an Autonomous Nonholonomic Robot Vehicle, " Master's Thesis, University of Florida, 1993.

2. WINTEC Inc., "The Simultaneous Data Collection and Processing System Interplatform Data Set," USAF Contract #F08637-94-C6042, June 1995

3. WINTEC Inc., "SOCS Interface Design Document," USAF Contract #F08637-94-C6042, June 1995.

SYSTEM DESCRIPTION OF THE UNMANNED GROUND VEHICLE TECHNOLOGY TEST BED

Arthur Bradley
Dynetics, Inc.

Jeff Todd
Dynetics, Inc.

ABSTRACT

The Technology Test Bed is an unmanned ground vehicle system capable of teleoperated driving utilizing either radio frequency or fiber optic control links. The system has been designed for remote reconnaissance and is equipped with a high-resolution targeting camera, forward-looking infrared, global positioning system, eyesafe laser range finder, inertial navigation system, and acoustic sensor array. The Technology Test Bed system consists of a Mobile Base Unit, Operator Control Unit, and an optional portable repeater station.

KEYWORDS: technology test bed, mobile base unit, operator control unit, remote terminal, distributed control, unmanned ground vehicle

INTRODUCTION

As the technological battlefield continues to develop, it is becoming possible to keep soldiers our of harm's way by allowing robotic devices to perform what were previously dangerous manned missions. By using sophisticated sensors, the vehicles are able to provide the operator with a remote presence allowing for an improved awareness of the battlefield. Unmanned vehicles offer unique advantages including the capability to carry heavy equipment, operate for an extended duration without sleep or food, and perform their duties without fear or hesitation.

A new addition to the community of unmanned ground vehicles (UGVs) is the Technology Test Bed (TTB). The TTB is a state-of-the-art test bed on which advanced sensors can be easily tested. It consists of a Mobile Base Unit (MBU), Operator Control Unit (OCU), and an optional radio frequency (RF) repeater station. Figure 1 shows the entire TTB system. The MBU is teleoperated out to the remote reconnaissance site by the operator inside the OCU. The TTB is capable of remote operation using either RF or fiber-optic links with a maximum range of 10 km. The MBU is currently equipped with a wide array of sensors including a high-resolution targeting camera, forward-looking infrared (FLIR), global positioning system (GPS), eyesafe laser range finder (LRF), inertial navigation system (INS), and an acoustic detection system (ADS).

TR-96-0782-NC

Figure 1. TTB System

MOBILE BASE UNIT

The MBU is a vehicle capable of both day and night teleoperated driving, as well as remote reconnaissance. It is equipped with separate driving and reconnaissance, search, target acquisition (RSTA) masts, a dual redundant command distribution bus controlled by a 68040-based system processor, a communication subsystem, and 10 microcontroller remote terminals (RTs) that perform the low-level control on the vehicle's actuators and sensors. Figure 2 is a photo of the MBU performing reconnaissance with the RSTA mast partially elevated.

TR-96-0783-NC

Figure 2. MBU Vehicle

MBU Platform

The MBU is built around a standard 1037 carrier-type high-mobility multi-wheeled vehicle (HMMWV). The selection of the HMMWV as the MBU platform offers several important advantages. First, the HMMWV is a well-established vehicle, removing the burden from the UGV engineer of having to design a reliable platform. Second, the U.S. military has a large inventory of HMMWVs, making spare parts and replacement vehicles inexpensive and readily available. It should be noted that the MBU and OCU are so modular that they can be assembled around any appropriate HMMWV without drilling a single hole in the body of the vehicle. Also, the HMMWV is a rugged platform capable of traversing a wide variety of terrain (sand, mud, snow, etc.). Finally, the HMMWV is large enough to operate as either a manned or an unmanned vehicle, allowing an operator to manually drive the vehicle when desired.

Sensor Suite

The MBU is currently equipped with sensors mounted on two mast assemblies. The first mast (driving mast) has two color video cameras and two light-enhancing black and white cameras. The color cameras are used for driving during daylight hours, while the enhancing cameras are for night driving. Having two cameras (either day or night) allows for operation in a stereo vision mode if desired. The stereo vision provides the operator with what appears to be a three-dimensional image. When operating in monovision, the second camera offers a level of redundancy in case of a single camera failure. The driving mast has the ability to pan ± 135° from its center position.

The second mast is for RSTA and is equipped with a high-resolution color targeting camera, eyesafe LRF, FLIR, INS, GPS, and an ADS. The RSTA sensors are used for reconnaissance/target detection over an area of interest. The mast provides smooth, adjustable pan and tilt movement, as well as fixed speed elevation control. The RSTA mast is capable of panning ±270°, tilting ±45°, and elevating up to 3 m above the stowed position. Figure 3 shows the MBU's driving and RSTA mast assemblies.

TR-96-0784-NC

Figure 3. MBU's Driving and RSTA Mast Assemblies

Distributed Control via 1553 Bus

Control of the MBU is achieved using a distributed control approach. Each control task on the vehicle is accomplished using dedicated closed-loop microcontrollers (RTs). The RTs are used to perform closed-loop control of the vehicle's actuators and sensors and continually monitor the actuators/sensors status and report the appropriate information back to the system processor. The MBU control electronics consist of six self-contained RT modules, four modified RTs built on Versa Modular Eurocard (VME) boards, and two VME-based amplifier boards. The RTs are designed to be nonspecific, programmable resources easily reconfigured for a variety of applications. They offer a set of common resources that include eight digital outputs, eight digital inputs, eight analog outputs, eight analog inputs, three pulsewidth modulated outputs, three high-speed inputs, and a serial test port.

The six RT modules are the steering/transmission RT, brake/throttle RT, drive safety RT, engine vehicle interface RT, power distribution RT, and communications RT. The first two

RTs provide for closed-loop control of the steering, transmission, brake, and throttle electrohydraulic actuators. The drive safety RT monitors the state of the vehicle and has the ability to stop the vehicle if a problem is detected. Control of basic HMMWV functions (lights, horn, speedometer, etc.) is achieved with the engine vehicle interface RT. The power distribution RT controls the switching of power to the vehicle's subsystems, and the communications RT controls the communications subplate assembly. Four other RTs require additional signal conditioning or control logic circuitry; therefore, they are built on conduction-cooled VME boards. The major functions of the four modified RTs are to control the driving and RSTA masts, multiplex the MBU video, and provide programmable RS-232 and RS-422 serial interfaces. Figure 4 shows both the RT module and VME form factor RT.

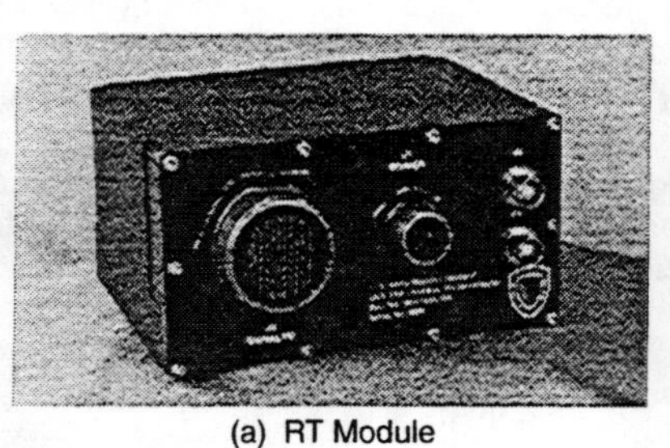

(a) RT Module

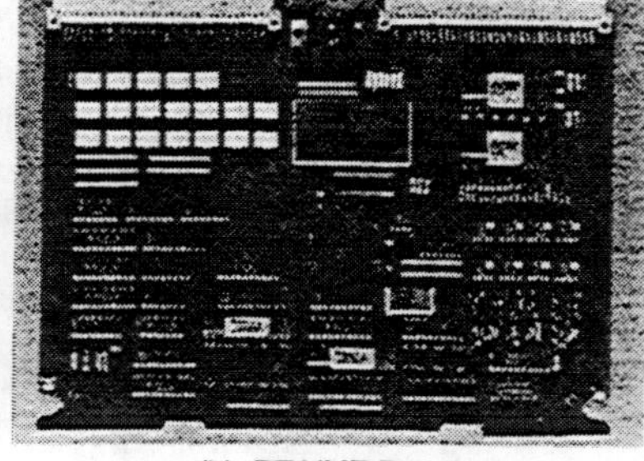

(b) RT VME Board

TR-96-0785-NC

Figure 4. Module and VME Form Factor RTs

A military standard MIL1553B bus serves as the link by which the RTs and MBU system processor communicate with each other. The 1553 bus has a 1-Mbit/s data rate, Manchester encoding noise immunity, and is expandable up to 32 RTs. The bus is dual redundant, meaning that both primary and secondary buses route to every RT around the entire vehicle. This redundancy allows the loss or failure of a single bus without affecting system operation.

System Processor/Real-Time Software

The MBU contains a 68040-based system processor operating under VxWorks. The real-time multitasking operating system is used to run several different processes concurrently. The system processor has five major functions to perform that are divided into three separate multitasking processes. The first function is to decipher the information in the OCU serial packets and convert the data into appropriate vehicle commands. These vehicle commands are not bus-specific commands, but rather generic vehicle commands, e.g., "turn on the lights" or "switch to FLIR video." The second function is to filter the incoming vehicle commands based on the current state of the vehicle. The vehicle state is determined by the status information returned from each RT on the 1553 bus. The status information is updated at a rate of 25 Hz. The five possible vehicle states are: parked, stealth, secured, stopped, driving, or backing. Erroneous vehicle commands issued by the operator are filtered by the system processor to prevent system damage. For example, the processor prevents the vehicle from shifting directly from forward into reverse. The third function the processor performs is to send out the appropriate commands over the 1553 bus to the RTs. Commands sent over the bus are RT specific and cover a wide range of possible vehicle commands, e.g., RT 1 honk the horn, RT 7 pan the driving mast, etc. The fourth processor function is to use the RT status information to send MBU status packets back to the operator. The final function is to serve as a safety monitor. The processor monitors the MBU subsystems, and in the case of a critical error, brings the vehicle to a safe state (stopped and requiring operator reset).

OPERATOR CONTROL UNIT

The OCU is the second vehicle in the TTB system and serves as the operator control station. The vehicle consists of a communication subsystem, OCU computer, and an array of operator interfaces. It is built around a missile carrier-type HMMWV, offering the same advantages as discussed with the MBU, specifically: availability, mobility, reliability, and adequate size.

OCU Communication Subsystem

As discussed earlier, the TTB system is capable of operating using either an RF or fiber-optic link between the OCU and MBU. The RF communication system consists of a dual-channel telemetry transmitter and two video/telemetry receivers at the OCU. The MBU has complementary hardware. The two video receivers each have two subcarriers associated with them. One of the receivers handles video on the primary carrier, downlink data on one subcarrier, and left downlink audio on the other. The other video receiver also has a video signal on the primary carrier, redundant downlink data on one subcarrier, and the right downlink audio on the second subcarrier. The dual-channel telemetry transmitter is used to send uplink audio on the primary carrier and uplink data on the subcarrier. By having two video/telemetry receivers, the TTB communication system has a level of redundancy that allows for the loss of a single receiver/transmitter pair.

OCU Computer

The OCU computer is a 80286-based machine used to encode OCU commands and decode/display MBU vehicle status information. The computer is enclosed in a rugged industrial case and has an LCD display and numerous pushbuttons on the front panel. The main functions of the computer are to monitor the operator interfaces, update the LCD display, form OCU command packets, and receive MBU status packets.

Operator Interfaces

The primary purpose of the OCU is to provide the operator with a mobile but functional work area in which he/she can remotely control the MBU. A variety of operator interfaces are used to control the MBU including a joystick, T-bar, computer panel, flat panel display, and headphones with accompanying microphone. The joystick is used to control the pan/tilt motion of the driving and RSTA mast assemblies. The T-bar provides the operator with controls for steering, throttle, brake, transmission, and a few miscellaneous driving functions. A wide selection of sensor controls are available to the operator via the computer pushbutton panel. Audio and video feedback are provided to the operator by the color, flat-panel display and headphones. A photo illustrating the operator work area is given as Figure 5.

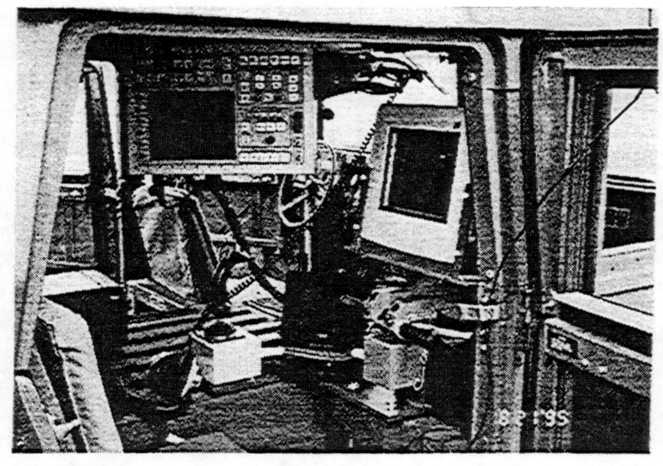

TR-96-0786-NC

Figure 5. Operator Work Area in the OCU

CONCLUSION

The TTB is a unique addition to the existing array of research and development UGVs. It offers both teleoperated driving and remote reconnaissance through the use of technologically advanced sensors and distributed vehicle control. The primary mission of the early TTBs is to serve as rugged, reliable, and easily expandable test beds on which advanced sensor suites can be tested. By serving as a test bed, the TTB allows the military to fully investigate the requirements for a useful and practical UGV. Obvious missions for which future TTBs might be used include enemy scouting, serving as a tireless sentry, or operating as a leader in a completely unmanned leader-follower convoy. The TTB was not created to replace man on the battlefield, but rather to assist in removing the soldier from unnecessary danger.

A LEARNING STEREO-HEAD CONTROL SYSTEM

L. BERTHOUZE, S. ROUGEAUX and Y. KUNIYOSHI
Autonomous Systems Section,
Electrotechnical Laboratory
Umezono 1-1-4, Tsukuba, Ibaraki 305, Japan
email: {berthouz, rougeaux, kuniyosh}@etl.go.jp

F. CHAVAND
Equipe Systemes Complexes,
Centre d'Etudes Mecaniques d'Ile-de-France
40 rue du Pelvoux, Courcouronnes CE1455, Evry 91020, France
email: chavand@iup.univ-evry.fr

ABSTRACT

Active vision tasks (such as tracking or saccadic moves) require gaze control which, in turn, depends on both the optical system and the mechanical structure. The knowledge of the image-joint jacobian is not straightforward and results traditionally from calibration procedures. Physiological studies have shown a real-time adaptation in humans' visual system, that can be modelled by a feedback-error-learning (FEL) scheme. In this paper, we propose a FEL-based modular control system that makes a gaze-platform learn its control *while* tracking moving objects. The framework used to build the learning system is described along with experiments on ESCHeR - Etl Stereo Compact Head for Robot vision -. As far as the authors are aware, this is the first implementation of a real-time learning controller on an active vision system.

KEYWORDS: active vision, adaptive control, vestibulo-ocular reflex.

INTRODUCTION

For purposive robots in a complex and dynamic world, the most basic visual functions required are target detection, tracking and fixation. In order to address these issues, an active vision system based on a 4DOF gaze-platform equipped with foveated wide angle lenses has been recently implemented [1]. These so-called *human-like* lenses exhibit a wide field of view along with a space-varying resolution so that they facilitate both detection and close observation. However, the strong optical distortions (resulting from spatial varying resolution [2]) combined with the redundancies of the mechanical gaze-platform make an efficient control law difficult to be explicited. It is hence desirable to make the plant learn by itself the commands that produce satisfactory results.

LEARNING THE INVERSE MODEL

Finding the input that produces the desired output involves the knowledge of the inverse-model. In the following sections, we briefly describe two opposite approaches (direct inverse modeling and feedback error learning) and we justify our choice.

Direct Inverse Modeling

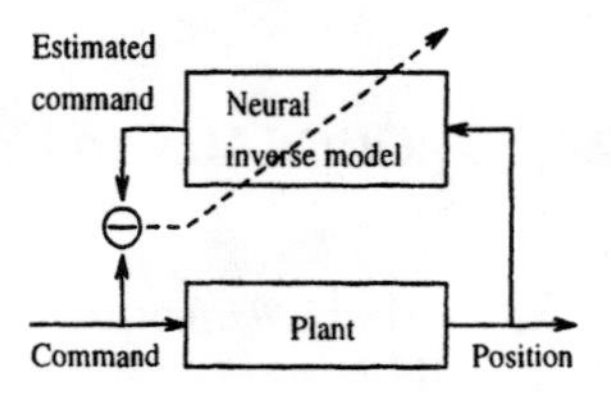

Figure 1. Direct inverse modeling.

Before being used as a feedforward controller, the inverse model is learned using the architecture shown in Figure 1. A set of motor commands is sent to the plant. The neural inverse model (NIM) receives the resulting position as its input. The difference between actual and NIM's estimated commands is computed and used as a error signal for modifying NIM's adaptive weights using a Widrow-Hoff rule (approximation of a gradient descent).

Feedback Error Learning [3]

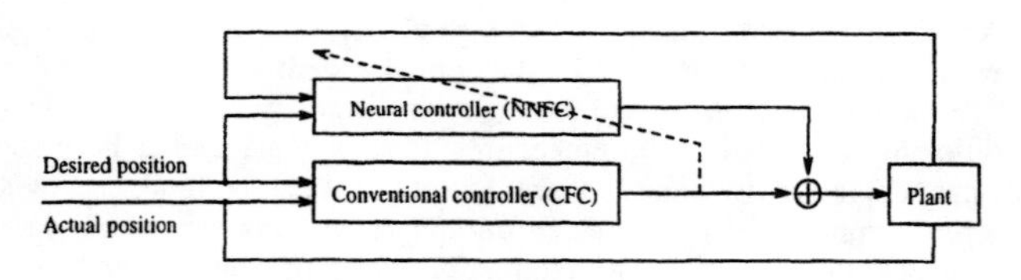

Figure 2. Feedback error learning.

The output of a conventional feedback controller (CFC) that models a linear approximation of the inverse-model, is propagated (dotted arrow in Figure 2) into a neural network feedback controller (NNFC) as the error signal. NNFC estimated motor command and CFC's output are summed so that NNFC does not mimic CFC but rather acquires the nonlinearities of the plant.

Summary

The direct inverse modeling approach is very simple but exhibits critical drawbacks : (1) as far as we know, there is no mathematical guarantee that the model will follow the desired behaviour; (2) this is a two-stages approach (the inverse-model cannot be used during the training phase); (3) the choice of pertinent training sets is not obvious, especially when the transfer function is not unique (kinematic redundancies)[4].

Unlike direct-inverse-modeling's approach, feedback error learning is achieved *online*. The convergence is proven (using German's theorem and Lyapunov's second method [5]) assuming (1) a guaranteed asymptotic convergence of the learning phase (role of the gain of the linear controller) and (2) a very small and positive-definite learning rate.

For the sake of flexibility and speed, we have chosen the feedback-error-learning approach to build a modular architecture. In the following section, we describe the overall control architecture after a brief presentation of the vision system.

EXPERIMENTAL FRAMEWORK

ESCHeR, the gaze-platform

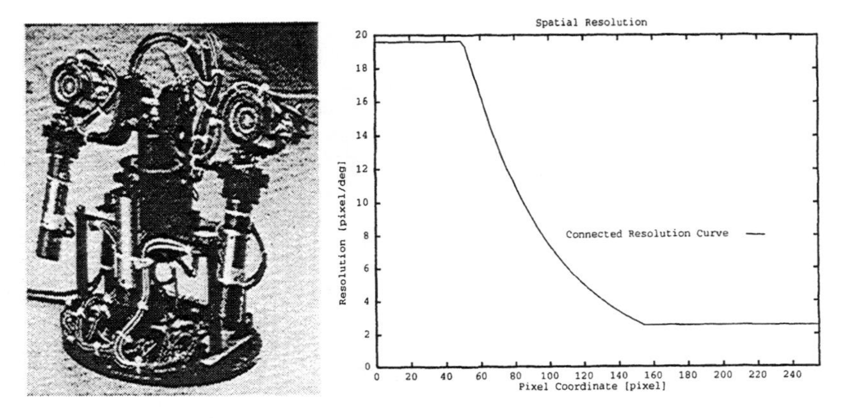

Figure 3. ESCHeR: the vision head and its spatial resolution curve.

Designed for the purpose of studying active vision, ESCHeR (Figure 3) is a four DOF stereo-head consisting of:

- Four DC motors mounted with encoders allowing to drive pan, tilt and vergence for both cameras. Each joint is driven by the *servo-controller*, a transputer-based architecture which receives the asynchronous command issued from the FEL controller and converts it into a smooth, 500 Hz interpolated command for the motor.
- Two CCD cameras equipped with the human-like wide angle foveated lenses described in [2].

The redundancies between pan and vergence joint along with the high distortions of the lenses contribute to the high non-linearity of the image-to-joint transfer function.

Overall control architecture

As shown in Figure 4, our control architecture consists of a closed-loop in which the control box learns the image-to-joint transfer function. In order to reduce the complexity of the learning phase, each of ESCHeR's four joints is independantly controlled by feedback-error-learning. Pan (respectively tilt) controller couples horizontal (respectively vertical) coordinates of both images. However, unlike animal vestibulo-ocular reflex (VOR) which stabilizes retinal images during head motion by generating compensatory (equal and opposite) eyes's moves, pan and vergence controls are not explicitely bound. In our implementation, pan is controlled so that it minimizes the difference between left and right image horizontal coordinates. Optionnal inputs are externally computed image coordinates that correspond to a switch in the focus of attention. They are null in tracking mode.

Each CFC is a very crude linear controller with low gain (about 10^{-4} rads/pixel). This gain results from a tradeoff between (1) the convergence of each controller (the linear controller guides the learning of the adaptive feedback controller (NNFC)) and (2) the stability of the whole architecture (when gains are too high, delays may induce significant disturbances).

The neural controller consists of a three-layer network. Both input and output layers

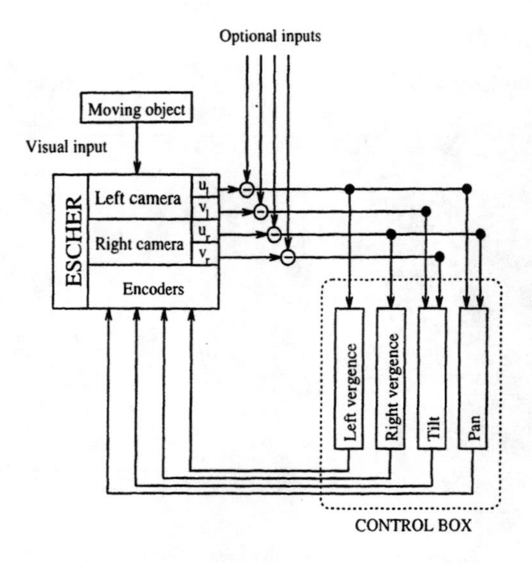

Figure 4. Overall control architecture: each box is ruled by a feedback-error-learning scheme.

are linear for normalizing. The hidden layer is nonlinear (arctangents). The modification of NNFC's weights is performed by a Newton-like method.

Experimental results

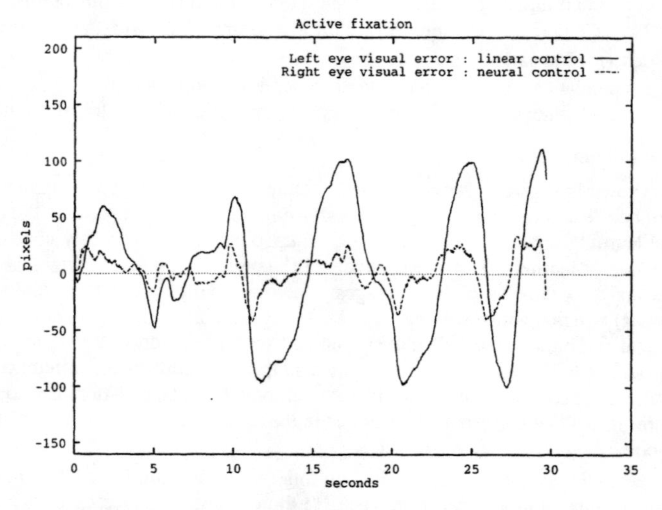

Figure 5. Tracking experiment after neural component has become dominant.

In our experiments, a bright object is swinging on a pendulum in front of ESCHeR. The target position is computed by a simple thresholding method and fed to the control box every 33ms. The platform then start tracking following the crude linear assumption. After about 20s of *learning by tracking*, the neural part of each controller becomes dominant. As shown by Figure 5, the variance of the visual error (quadratic sum of each coordinate) is significantly reduced in the right eye which is under adaptive control.

In Table 1, we show the standard deviation of the visual error with different setups. It clearly appears that the redundancy induced by the control of both the pan and the vergences is learned by the system as a nonlinearity. The residual visual error is mainly due to the 50 ms delay introduced by image acquisition and processing. Although the residual error may appear significant when evaluated in pixels, it never exceeds 2 degrees of angular error (conversion made using ESCHeR's spatial resolution curve shown in Figure 3).

Pan control	Vergence control	Std dev
Not controlled	linear	80
	neural	22
Controlled	linear	50
	neural	13

Table 1. Standard deviation of the visual error during tracking experiments with various setups.

CONCLUSION

Whereas humans continuously learn, robots are traditionally rigidly preprogrammed. However, in case of nonlinearities and redundancies, the analytical determination of an efficient control law is not necessarily straightforward. Instead, a system that learns is potentially more capable. In this paper, we have described a modular control system based on feedback-error-learning that evolves in performance while operating: starting with a crude knowledge of its image-to-joint jacobian, ESCHeR, a gaze-platform equipped with highly distorted lenses, adapts its control while tracking a moving object. Prior random explorations are not necessary since the initial approximation guides the learning and ensures the asymptotic convergence of the control. The effectiveness of our system is shown through experimental results and without any prediction, a precision of 2 degrees is achieved.

This scheme is currently extended to address more complex active vision tasks. In particular, we intend to make both the stereo-head platform and a robot manipulator learn how to cooperate for dectecting, tracking and grasping moving objects.

REFERENCES

1. Kuniyoshi, Y., Kita, N., Rougeaux, S., and Suehiro, T., "Active Stero Vision System with Foveated Wide Angle Lenses", *2nd Asian Conference on Computer Vision, Singapore* (1995) 359-363.
2. Kuniyoshi, Y., Kita, N., Sugimoto, K., Nakamura, S., and Suehiro, T., "A Foveated Wide Angle Lens for Active Vision", *IEEE International Conference on Robotics and Automation, Nagoya, Japan* (1995) 2982-2985.
3. M. Kawato, M., Furukawa, K., Suzuki, R., "A Hierarchical Neural-Network Model for Control and Learning of Voluntary Movement", Biological Cybernetics 57 (1987) 169-185.
4. Jordan, M.I., "Motor Learning and the Degree of Freedom Problem", *Attention and Performance,* Jeannerod M (1990), 796-836.
5. Gomi, H., Kawato, M., "Adaptive Feedback Control Models of the Vestibulocerebellum and Spinocerebellum", *Biological Cybernetics* 68 (1992) 105-114.

VISION-BASED HANDLING WITH A MOBILE ROBOT

STEFAN BLESSING

TU München, Institut für Werkzeugmaschinen und Betriebswissenschaften (iwb), 80290 München, Germany,
e-mail: **bl@iwb.mw.tu-muenchen.de**

STEFAN LANSER, CHRISTOPH ZIERL

TU München, Institut für Informatik, Forschungsgruppe Bildverstehen (FG BV), 80290 München, Germany,
e-mail: **{lanser,zierl}@informatik.tu-muenchen.de**

ABSTRACT

Mobile systems become more and more important in the area of modern manufacturing. In order to handle an object with a manipulator mounted on an autonomous mobile system (AMS) within a changing environment, the object has to be identified and its 3D pose relative to the manipulator has to be determined with sufficient accuracy, because in general its exact position is not known à priori. The object recognition unit of the presented system accomplishes this 3D pose estimation task using a single CCD camera mounted in the gripper exchange system of a mobile robot. The reliability of the results is checked by an independent fault-detection unit. A recovery unit handles most of the possible faults autonomously increasing the availability of the system.

KEYWORDS: vision-based handling, autonomous mobile system, 3D object recognition, fault-detection and recovery.

INTRODUCTION

The automation of manipulation tasks in manufacturing environments is often based on industrial robots. To make this automation more profitable, a *mobile* robot can be used to perform manipulations at different places, wherever it is needed. Within a joint research project[1] towards the development of autonomous mobile systems located at the *TU München* the mobile robot MOBROB has been developed to fulfil manipulation tasks autonomously, even in a changing environment.

The required autonomy increases the demands on the sensors of such systems, because both the position of the autonomous mobile system (AMS) and the pose of the object to be grasped are affected by uncertainty. In order to handle an object with a manipulator mounted on an AMS, the object has to be identified and its 3D pose relative to the manipulator has to be determined with sufficient accuracy. The presented vision-based object recognition system uses images taken from the camera mounted in the gripper exchange system of the mobile robot (see Fig. 1).

The system architecture is shown in Fig. 2. The vision-based *object recognition unit* consists of a recognition and a localization module described in the following section.

[1]This work was supported by *Deutsche Forschungsgemeinschaft* within the *Sonderforschungsbereich 331, "Informationsverarbeitung in autonomen, mobilen Handhabungssystemen"*, projects L9 and M2.

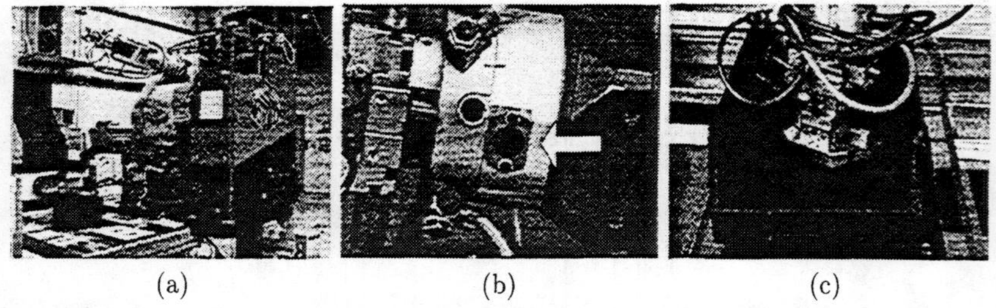

(a) (b) (c)

Figure 1. (a) MobRob (*Mobile Robot*) at the *Institut für Werkzeugmaschinen und Betriebswissenschaften (iwb)* with (b) a CCD camera mounted in the gripper exchange system of the robot. (c) Vision-based grasping of a workpiece.

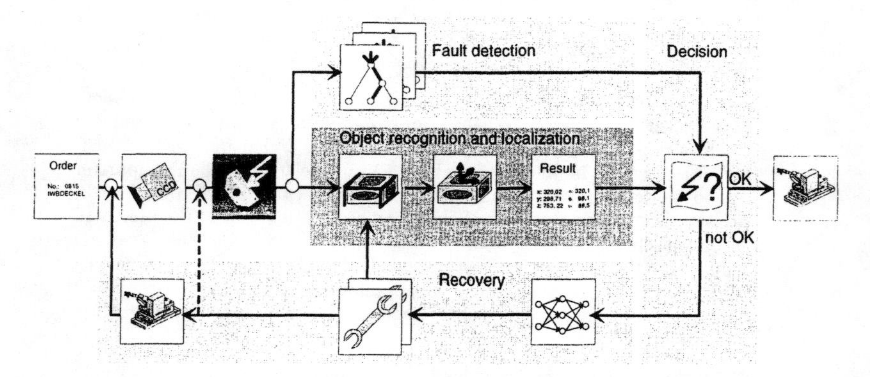

Figure 2. Closed-loop architecture of the presented grasping system.

The *fault-detection module* of the *error-handling unit* presented in the subsequent section analyzes the grabbed image and the result of the object localization. If there are any faults obstructing the correct handling of the object the *recovery module* tries to clear the fault autonomously increasing the availabilty of the system.

VISION-BASED OBJECT RECOGNITION

In this section the *object recognition unit* of the proposed system is briefly described. Based on a single image from a calibrated CCD camera it identifies objects and computes their 3D pose relative to the robot manipulator. For a general introduction into this field of research see e.g. [1] or [9].

Calibration

In order to obtain the 3D object pose from the grabbed video image the internal camera parameters (mapping the 3D world into pixels) as well as the external camera parameters (pose of the CCD camera relative to the manipulator) have to be determined with sufficient accuracy.

Camera Model. The camera model describes the projection of a 3D point P_W in the scene into the 2D pixel $[c, l]^T$ of the video image of the CCD camera. The proposed

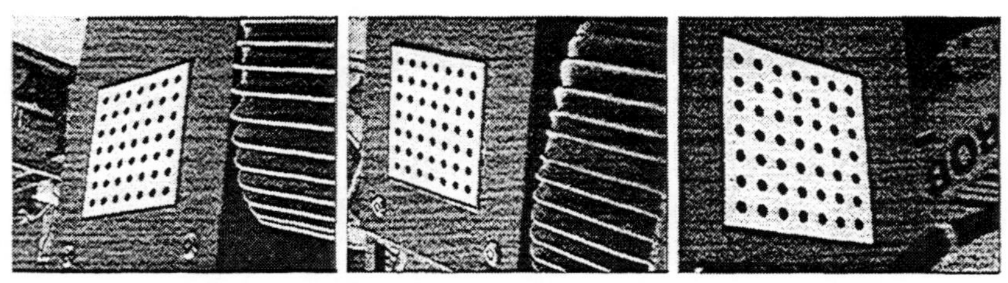

Figure 3. The calibration table mounted on the mobile robot seen from different viewpoints with known relative movements of the manipulator.

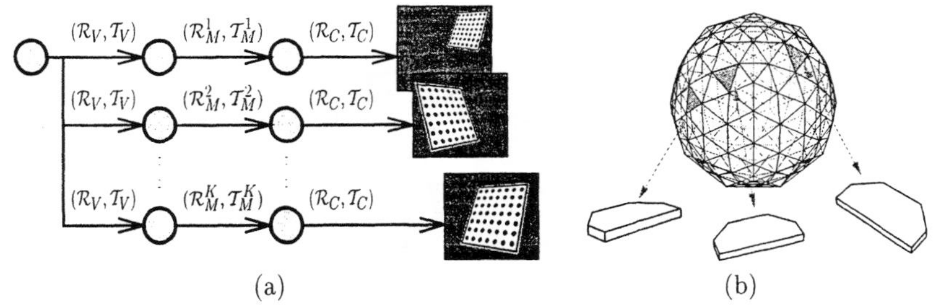

Figure 4. (a) Estimation of the camera pose $(\mathcal{R}_C, \mathcal{T}_C)$ relative to the *tool center point* (*hand-eye calibration*) based on known relative movements $(\mathcal{R}_M^k, \mathcal{T}_M^k)$ of the manipulator. (b) Each triangle of the tesselated Gaussian sphere defines a 2D view of an object.

approach uses the model of pinhole camera with radial distortions [11]: It includes a rotational matrix $\mathcal{R}$ describing the orientation, a vector $\mathcal{T}$ describing the position of the camera in the world (external parameters), and the internal parameters b (*effective focal length*), κ (*distortion coefficient*), S_x and S_y (scaling factors), and $[C_x, C_y]^T$ (*image center*).

Internal Camera Parameters. In the first stage of the calibration process the internal camera parameters b, κ, S_x, S_y, and $[C_x, C_y]^T$ are computed by simultanously evaluating images of a 2D calibration table with N circular marks taken from K different viewpoints, see Fig. 3. This *multiview calibration* [5] minimizes the distances between the projected 3D midpoints of the marks and the corresponding 2D points in the video images. The 3D pose of the camera $\mathcal{R}, \mathcal{T}$ is estimated during the minimization process. Thus, only the model of the calibration table itself has to be known à priori.

Hand-Eye Calibration. Once the internal camera parameters have been determined the 3D pose of the camera relative to the *tool center point* is estimated in the second stage of the calibration process (*hand-eye calibration*).

In the case of a camera mounted on the manipulator of a mobile robot the 3D pose of the camera $(\mathcal{R}, \mathcal{T})$ is the composition of the pose of the robot $(\mathcal{R}_V, \mathcal{T}_V)$, the relative pose of the manipulator $(\mathcal{R}_M, \mathcal{T}_M)$, and the relative pose of the camera $(\mathcal{R}_C, \mathcal{T}_C)$, see Fig. 4(a). Performing controlled movements $(\mathcal{R}_M^k, \mathcal{T}_M^k)$ of the manipulator similar to

[12] $(\mathcal{R}_C, \mathcal{T}_C)$ can be determined by minimizing

$$e(\vec{x}) = \sum_{k=1}^{K} \sum_{i=1}^{N} \|\tilde{s}_i^k \times c_i(M_i, \vec{x}, \mathcal{R}_M^k, \mathcal{T}_M^k)\|^2 \longrightarrow \min$$

with $\tilde{s}_i^k$ the normalized vector of the line of sight through the 2D point $\tilde{m}_i^k$ in the kth video image and $c_i(M_i, \vec{x}^k, \ldots)$ the 3D midpoint of a mark on the calibration table transformed in the camera coordinate system, for details see [5].

Since the used 2D calibration table is mounted on the mobile robot itself, the manipulator can move to the different viewpoints for the *multiview calibration* automatically. Thus, the calibration can be accomplished in only a few minutes.

3D Pose Estimation

Our object recognition system uses à priori known models of 3D objects, which are generated in a offline process, and a single intensity image of the scene. The pose estimation is performed in two steps: First, hypotheses of visible objects and their rough pose are generated by a recognition module. In a second step, these hypotheses are verified and refined by a localization module. For details see [4].

In case of multiple instances of the same object appearing in a scene, this process can be iterated. After each iteration the image features already mapped to previously detected objects are eliminated.

Model Generation. Using a tesselated Gaussian sphere each object is represented by a set of up to 320 normalized perspective 2D views, see Fig. 4(b). This model generation process is based on a geometric model of the environment described in [10]. The underlying boundary representation (B-Rep) of a polyhedral 3D object can be derived from a CAD modeler. The comparability between the highly detailed CAD model and the extracted image features (which are limited by the resolution of the CCD camera) is ensured by simulating the image preprocessing on the model features.

Object Recognition. The aim of the object recognition module is to identify objects and to determine their rough 3D pose by searching for the appropriate 2D model view matching the image. This is done by establishing correspondences between image lines extracted from the CCD image and model lines from a 2D view of an object.

First, a set of *associations* is built. An association is defined as a quadruple (I_j, M_i, v, c_a) where I_j is an image feature, M_i is a model feature, v is one of the 320 2D model views of an object, and c_a is a confidence value of the correspondence between I_j and M_i. This value can be obtained by traversing aspect-trees [7] or by a simple geometrical comparison of the features incorporating topological constraints.

In order to select the "correct" view the associations are used to build *hypotheses* $\{(object, \mathcal{A}_i, v_i, c_i)\}$. For each 2D view v_i all corresponding associations with sufficient confidence are considered. From this set of associations the subset of associations $\mathcal{A}_i$ with the highest rating forming a *consistent labeling* of image features is selected. The confidence value c_i depends on the confidence values of the included associations and the percentage of mapped model features. The result of the described recognition process is a ranked list of possible hypotheses (see Fig. 5(a)) which are verified and refined by the localization module.

Object Localization. In the case of a successful verification the localization module computes a modified hypothesis where some correspondences may be changed based on the *viewpoint consistency constraint.* This refinement of correspondences can be

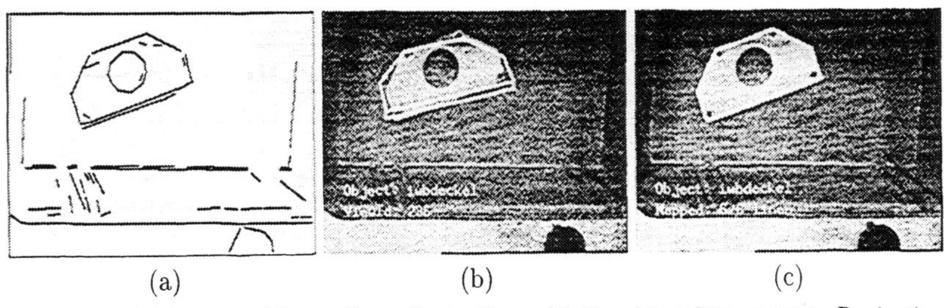

Figure 5. (a) Extracted image lines of a toolbox with the object IWBDECKEL. Projection of object IWBDECKEL into the original video image according to the (b) initial and (c) refined pose estimation.

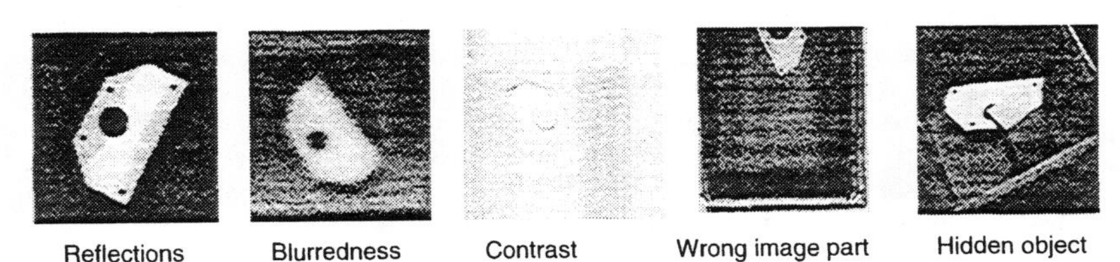

Figure 6. Typical problems encountered by vision-based systems in a manufacturing environment.

accelerated by computing specific search spaces in the video image [3]. By aligning model and image lines the final object pose (R, t) with full 6 DOF is computed using a weighted least squares technique similar to [6].

If only coplanar features are visible which are seen from a large distance compared to the size of the object (Fig. 5), the 6 DOF estimation is quite unstable because some of the pose parameters are highly correlated. In this case, à priori knowledge of the orientation of the manipulator with respect to the ground plane of the object might be used to determine two angular degrees of freedom. Naturally, this approach decreases the flexibilty of the system. Tilted objects cannot be handled any longer. A more flexible solution is the use of a second image of the scene taken from a different viewpoint with known relative movement of the manipulator (*motion stereo*). By simultanously aligning the model to both images the flat minimum of the 6 DOF estimation can be avoided. Note, that for well structured objects with some not coplanar model features a 6 DOF estimation based on a single video image yield good results as well.

ERROR-HANDLING

In a manufacturing environment there are some factors aggravating vision-based pose estimation. The fault-detection module detects these faults also using additional information not available to the object recognition unit. In case of a detected failure the recovery module is activated in order to overcome the problem autonomously.

Typical Problems in a Manufacturing Environment

Applying vision-based object recognition methods in a manufacturing environment is affected by a wide range of disturbing factors, e.g. reflections and shadows due to

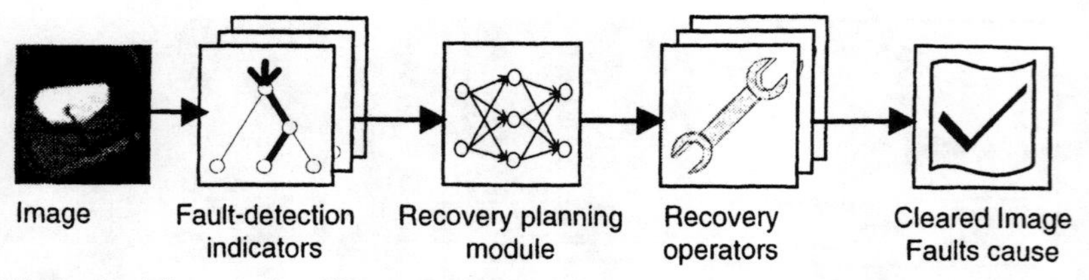

Figure 7. The error-handling unit consist of fault-indicators and recovery operators connected by the recovery planning module.

specific illumination conditions and surface characteristics, or objects, from which only a fraction is visible, see Fig. 6. In general, these faults result in additional or missing edges in the image obstructing the interpretation. This may lead to different pose hypotheses all compatible with the extracted image features or to a complete failure of the object recognition unit. Most of these spurious hypotheses can be detected by exploiting external information like the expected distance to the object. On the other side, it is very difficult to determine the reason for a failure as listed in Fig. 6. However, in some specific environments, e.g. a toolbox with a homogenous surface, low-level indicators analyzing selected image characteristics can be used to detect these problems.

Recovery Strategies

Corresponding to the various faults listed in Fig. 6 the system can choose between different strategies to clear a detected fault:

- Considering the next pose hypothesis (*object recognition unit*)
- Adapting parameters for the image preprocessing (*object recognition unit*)
- Adapting parameters for the image interpretation (*object recognition unit*)
- Adapting the aperture or focus of the CCD camera (*robot guiding system*)
- Moving the manipulator to a more suitable position (*robot guiding system*)
- Reporting to the external error-handling (*manufacturing control system*)

A controlled movement of the manipulator can be used to increase the accuracy of a successful pose estimation as well, see the previous section.

Structure of the error-handling unit

Most of the faults spoiling the object recognition are a superposition of several faults what makes detection of these faults even more difficult. The definition of some basic faults leads to specialized error-handling modules. These modules consist of an indicator to detect a specific fault and a recovery operator to clear it, see Fig. 7.

Based on the results of all indicators analyzing the current image as well as the result and confidence of the localization, a decision is made, whether the results are reliable enough to grasp the object. Otherwise the detected fault has to be cleared by the system. In this case, based on the results obtained, the *recovery planning module* generates a plan to clear the fault. The plan is executed by the recovery operators, which are together with an indicator part of a fault-specific error-handling module.

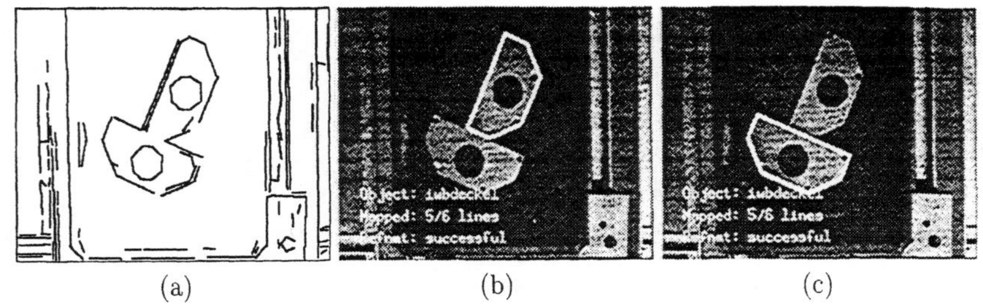

Figure 8. Successful localization of two workpieces of the type IWBDECKEL in a toolbox: (a) the detected image lines, (b) the first IWBDECKEL found in the image, and (c) the other IWBDECKEL found.

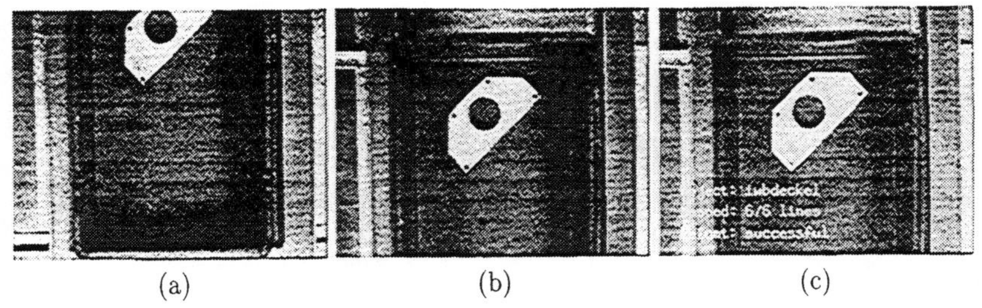

Figure 9. Example for a successful error-handling: (a) no reliable pose estimation of the workpiece due to invisible image features, (b) video image after a controlled movement of the camera mounted on the manipulator, and (c) final pose estimation of the workpiece.

ROBUST POSE ESTIMATION WITH ERROR-HANDLING

In Fig. 2 the structure of the whole system is shown. The sequence is started by the *robot guiding system* [8], instructing the *object recognition unit* to detect and localize an object. After grabbing a video image the object recognition unit generates hypotheses about the 3D pose of the object as described in the second section. At the same time, the fault-detection indicators analyze the image and forward their results to the decision module. Considering the results of the object recognition and the fault-detection indicators, a decision is made, whether the pose estimation is reliable and can be returned to the robot guiding system, see the previous section. Fig. 8 shows an example for a successful 3D pose estimation.

If the result of the pose estimation is considered to be unreliable a recovery plan is generated automatically. Depending on the chosen recovery plan, the next hypotheses of the object recognition unit are tested, the object recognition is re-parameterized or a request to move the manipulator or to adapt the camera parameters is sent to the robot guiding system. This sequence is iterated (closed-loop) until a reliable pose estimation for grasping the object is found or no internal error-handling is possible, see Fig. 9.

In case of a successful pose estimation the 3D simulation system USIS [2] is activated to perform a collision-free grasp planning for the manipulation process. Finally, this online generated robot program is downloaded and executed, see Fig. 1(c).

SUMMARY AND FUTURE WORK

An architecture for a grasping system on an autonomous mobile robot was presented consisting of a *vision-based object recognition unit* and an explicit *error-handling unit*. Based on à priori known models the object recognition unit identifies objects in single video images and determines their 3D pose (all 6 DOF). The aim of the error-handling unit is to detect and to clear possible failures increasing the availability of the whole system.

Future research will be focused on improving the fault indicators and the corresponding recovery strategies. For intelligent error-handling a database should be integrated in the error-handling system, storing information from all recognitions and error-handling procedures. With this knowledge-based error detection the faults then will be detected and their reasons concluded more reliable.

REFERENCES

1. T. Y. Young (Ed.). *Handbook of Pattern Recognition and Image Processing: Computer Vision*, volume 2. Academic Press, Inc., 1994.
2. D. Kugelmann. Autonomous Robotic Handling Applying Sensor Systems and 3D Simulation. In *IEEE International Conference on Robotics and Automation*, volume 1, pages 196–201. IEEE Computer Society Press, 1994.
3. S. Lanser and T. Lengauer. On the Selection of Candidates for Point and Line Correspondences. In *International Symposium on Computer Vision*, pages 157–162. IEEE Computer Society Press, 1995.
4. S. Lanser, O. Munkelt, and C. Zierl. Robust Video-based Object Recognition using CAD Models. In U. Rembold, R. Dillmann, L.O. Hertzberger, and T. Kanade, editors, *Intelligent Autonomous Systems IAS-4*, pages 529–536. IOS Press, 1995.
5. S. Lanser and Ch. Zierl. Robuste Kalibrierung von CCD-Sensoren für autonome, mobile Systeme. In R. Dillmann, U. Rembold, and T. Lüth, editors, *Autonome Mobile Systeme*, Informatik aktuell, pages 172–181. Springer-Verlag, 1995.
6. D. G. Lowe. Fitting Parameterized Three-Dimensional Models to Images. *IEEE Trans. on Pattern Analysis and Machine Intelligence*, 13(5):441–450, 1991.
7. O. Munkelt. Aspect-Trees: Generation and Interpretation. *CVGIP: Image Understanding*, 61(3):365–386, May 1995.
8. K. Pischeltsrieder. *Steuerung autonomer mobiler Roboter in der Produktion*. iwb Forschungsberichte. Springer-Verlag, 1996. To appear.
9. A. R. Pope. Model-Based Object Recognition. Technical report 94-04, University of British Columbia, January 1994.
10. N. O. Stöffler and T. Troll. Model Update by Radar- and Video-based Perceptions of Environmental Variations. In *International Symposium on Robotics and Manufacturing*. ASME Press, New York, 1996. To appear.
11. R. Y. Tsai. An Efficient and Accurate Camera Calibration Technique for 3D Machine Vision. In *Computer Vision and Pattern Recognition*, pages 364–374. IEEE Computer Society Press, 1986.
12. C. C. Wang. Extrinsic Calibration of a Vision Sensor Mounted on a Robot. *Transactions on Robotics and Automation*, 8(2):161–175, April 1992.

A FUZZY COLLISION AVOIDANCE FOR ROBOTS IN 3D-SPACE

ULRICH BORGOLTE
FernUniversität Hagen
Department of Electrical Engineering
Chair of Process Control
Frauenstuhlweg 31
D 58644 Iserlohn, Germany
email: ulrich.borgolte@fernuni-hagen.de

ABSTRACT

In this paper, a fuzzy control method doing online collision detection as well as collision avoidance in multi robot systems is presented. It detects the danger of collisions and generates collision free trajectories while the robots are moving. Their dynamic behaviour and constraints are considered, too. The states of the robots are expressed with linguistic variables, the reactions of the robots due to a danger of collision are determined by rules based on membership functions. In a last step, the linguistic reactions are defuzzified to get crisp values for motor control. The advantage of this method is the straight-forward implementation of expert knowledge into a multi robot control system. This includes the adaption to different types of robot kinematic.

KEYWORDS: collision avoidance, multi robot systems, fuzzy control

INTRODUCTION

Within the last decade, robotic applications have been introduced to many assembly processes. But there is still a lot of work to realize the *factory of the future.* From the robotic point of view, special attention have to be directed to high level programming, multi sensor integration, and coordinated motion. In fact it is still difficult to manage two or more industrial robots working together under time-variant conditions.

Even if global path planning has been carried out carefully for all the kinematics present in the workcell [1][2][3], an online supervision of the interacting robots is inevitable to manage situations which can not be preplanned (e.g. error stop condition, unforeseen objects ...).

In this paper, the algorithms are given for a pair of robots in a multi-robot system, where the computational time needed, permits a real time application. The performance of this approach is demonstrated by a simulation example. The algorithm presented is based on the Stanford-Arm. The hand including gripper and load is considered by safety factors. It has been shown that the algorithm is applicable to robots with six rotational axes, if the collision avoidance is restricted to the x-y-plane. Its practical application to robot systems is currently in action in our laboratories.

To detect the danger of collision between the robot j and some other robot k, a so-called Virtual Hindrance Robot (VHR) is constructed from the relationship between robot j and robot k, where robot j has to avoid collisions with robot k. The VHR has the same geometry as the robots j and k. The difference between this virtual robot and the robot j, together with the relative velocity between this systems, gives a valuation for the time until a collision may occur. The states of the VHR (position, velocity, acceleration) are expressed with fuzzy terms. A rule base associates reactions to the states, regarding the dynamic behaviour of the robots. The reactions for every axis are defuzzified and given to the feedback control of robot j.

PROBLEM DESCRIPTION

In this paper, robots with the main axes *rotation-translation-translation* are considered. The axes are denoted as φ, r, and z, respectively (fig. 1). The method proposed enables an easy way to adapt different kinematic structures, thus it is not restricted to a special type of robot.

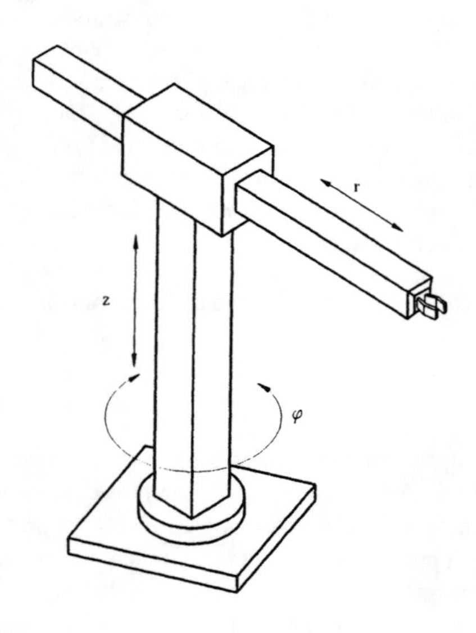

Figure 1: Robot configuration

Within a system of r robots, we are considering the robots j and k. Robot k has the right-of-way, i.e. robot j has to avoid collisions. With this, a fuzzy inference engine is assigned to robot j, getting inputs from the axis controllers of both robots. If a predefined danger of collision is reached, the output of the inference engine will change the desired output vector for the axis control of robot j.

Considering the given kinematic structure, collision avoidance can be done by changing the height of the third link (z), decreasing the length of the third axis (r), or by rotating away from the arm of robot k (φ). As the third link is of fixed length, it just slides forward and backward with respect to the second link. Thus a reduction of r (i.e. a reduction of the distance between z-axis and TCP) results in an increasing of its backside part. This has to be considered, and a decision is made whether the TCP or the backside part (BP) of the third link is taken into account. A thorough overview on the geometric considerations, including the computation of the distance, is given in [4]. The architecture of the complete system is shown in fig. 2. First of all, from the data coming from the robot controllers (i.e. state space vectors), geometric information is extracted. This is given to the fuzzy control. The output from the fuzzy control, a vector of correcting values, is given back to the controller of robot j. This vector is added to the desired position vector.

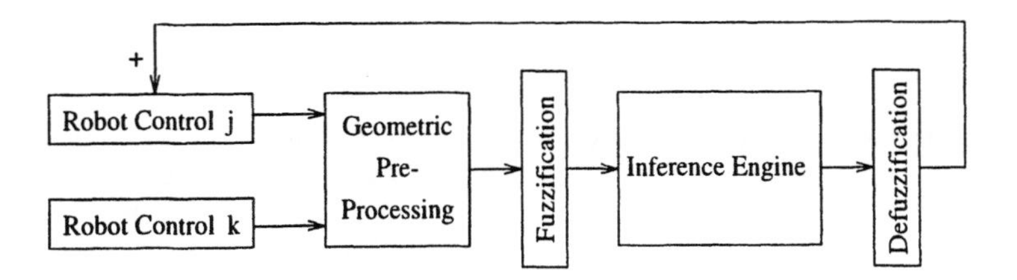

Figure 2: System architecture with data flow

ARCHITECTURE

The following process parameters are inputs to the fuzzy control:

Parameter	Description
τ	Indicator: TCP or BP (backpoint of 3rd link) to be considered
φ	Angle of relevant part of 3rd link (TCP or BP)
z	z-axis of robot j
r	r-axis of robot j
$\Delta\varphi$	Indicator: direction of free space for φ-axis of robot j
Δz	z-distance between robots j and k
d	Distance of robot arms
d'	Distance velocity

Output of the fuzzy control are three modification values for the main axes of robot j:

Parameter	Description
$\delta\varphi$	Modification of φ-axis of robot j
δz	Modification of z-axis of robot j
δr	Modification of r-axis of robot j

To make the module independent of a special geometric configuration, the input variables are normalized. Corresponding, the output is adapted to the special configuration.

The rule base consists of 144 rules. Eighteen are coping with the z-axis, 12 with r-axis, 8 with φ-axis. This corresponds to the priority of collision avoiding movements with this type of robots. The first choice is the z-, the second one the r-, and the last one the φ-axis. The remaining 106 rules are handling special and borderline cases. To give an impression of the rules, just some typical examples:

IF d is VeryClose AND Δz is CloseAbove AND d' is Worse THEN δz is VeryHigh
IF r is InFront AND φ is Greater THEN δr is Retract
IF r is InFront AND d' is Worse AND φ is Greater THEN $\delta\varphi$ is Left

From the different types of inference and defuzzification methods in common use, the Algebraic Product and Center of Gravity are chosen. These have been proved to serve well with PID control [5].

SIMULATION EXAMPLE

Figure 3 shows a typical situation for collision avoidance. Robot 1 (on the left) turns around (-0.8,0.0) and starts at (-0.8,1.2), robot 2 turns around (1.0,0.0) and starts at (0.8,-1.0). The left window shows the top view of the final situation without collision avoidance (i.e. after execution of the movements). The distance of the arms is shown in the upper right window, the z-axes in the lower right one. Both robots are at $z = 1.0$, therefore a collision occurs (where $d = 0$).

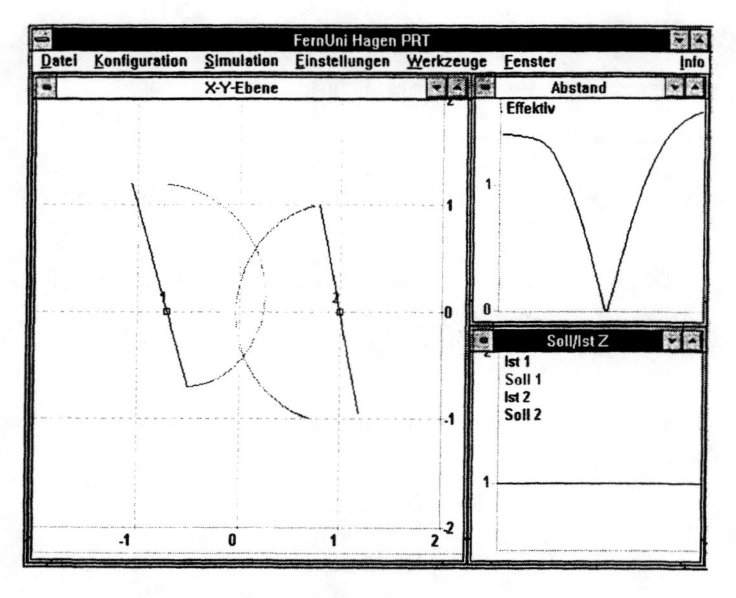

Figure 3: Example of collision avoidance

Figure 4 shows the z-axis of robot 1, if this is the only one doing collision avoidance. The upper line is the new trajectory, the lower one the planned. Figure 5 shows the z-axes of the two robots, if both are avoiding collisions. The upper line is the axis of robot 1, the lower that of robot 2. The distances of the arms are shown in fig. 6, where the lower line represents no collision avoidance, the intermediate one collision avoidance just by robot 1, and the upper one by both robots.

The simulation is done with path interpolation as well as dynamic modeling and nonlinear decoupling and control. The computation of the fuzzy control during the application phase is neglectible with regard to the interpolation cycle of the robot control. The fuzzy control was developed using the toolbox *FuzzySoft* under Windows. The simulation has been written in Microsoft Visual C/C++.

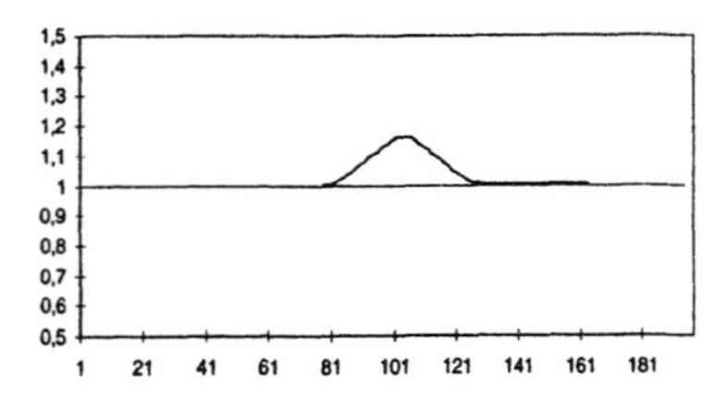

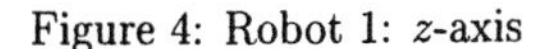

Figure 4: Robot 1: z-axis

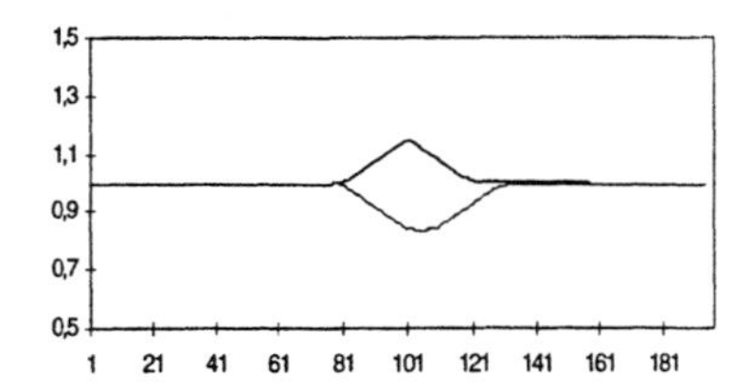

Figure 5: Robots 1/2: z-axes

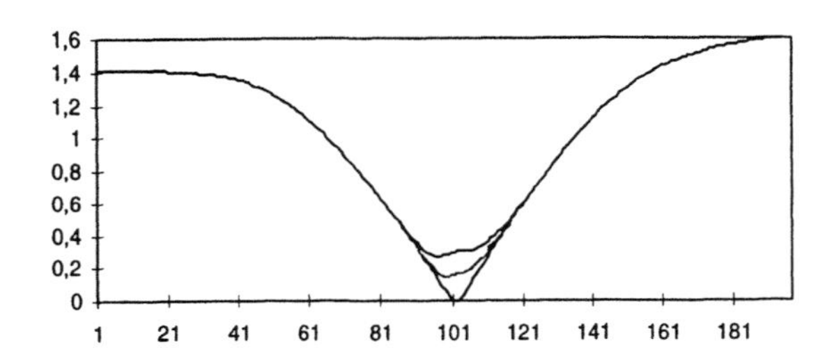

Figure 6: Distances with different methods

CONCLUSION

A new approach for the solution of the online collision avoidance problem has been presented. Based on the states of the devices involved, a Virtual Hindrance Robot is constructed. The relationship with this gives a valuation for the danger of collision. The relation is expressed with fuzzy terms. A rule base associates reactions to the states, regarding the dynamic behaviour of the robots. The reactions for every axis are given back to the control of the real robots. The method is applicable to in principle any kind of kinematic structure.

REFERENCES

1. Khatib, O. "Real-time obstacle avoidance for manipulators and mobile robots", *International Journal of Robotics Research*, Vol. 5, No. 1, (1986), 90–98.
2. Lumelsky, V.J. "Effect of kinematics on motion planning for planar robot arms moving amidst unknown obstacles", *IEEE Journal of Robotics and Automation*, Vol. RA-3, (1987), 207–223.
3. Adolphs, P., Tolle, H. "Collision-free real-time path-planning in time varying environment", Proc. Int. Conf. on Intelligent Robots and Systems (IROS'92), (1992), 445–452.
4. Borgolte, U. "An online algorithm for three degrees-of-freedom collision avoidance". Proc. 2nd Singapore Int. Conf. on Intelligent Systems (SPICIS'94), (1994), B454–B459.
5. Jager, R. *Fuzzy Logic in Control.* Thesis Technische Universiteit Delft (1995).

Situation Parallelism / Calculus Parallelism for Distributed Control / Command in Robotics.

S. Belot, C. Moreno
Université d'Evry Val d'Essonne - CERMA, LaMI.

A. Abadia
Universidad del Valle.

ABSTRACT

This paper is a contribution to the conception of distributed control systems. It uses notions like *Real Time* and *Distributed Artificial Intelligence*, it introduces the notion of *Multi-Protocols Gateway* and *Distributed Concept*. We describe the relative nature of the distribution through two modelisation methods. The first one is implemented on the command of a type SCARA robot. The second one is a simulation of a serial robot implemented on a Real Time Object kernel.

KEYWORDS: Real Time, Distributed Artificial Intelligence, Multi-Protocols Gateway, Distributed Concept.

INTRODUCTION

The achievement of a distributed control depends on the desired granularity of the distribution. The more the granularity is low, the more we tend to a *calculus parallelism* and the more the granularity is high, the more we tend to a *parallelism of situation*.

These two aspects are complementary when they are distant enough. The parallelism of situation actually defines macroscopic entities which cooperate to solve a high level problem whereas the calculus parallelism allows to achieve efficiently low level operations induced by the upper levels. However, the problematic is not to achieve efficiently operations but to define high level entities.

We introduce a study on a real system in which the axes of a serial robot are considered to be autonomous and can communicate (fig. 1). The presence of heterogeneous systems and networks justifies the presentation of a multi-protocols gateway. Real time allows a distribution of the control inside of a single system by specifying tasks which cooperate. That supposes a previous analysis bringing out the different subsystems which intervene as well as their degree of cooperation. This preliminary phase substantially constrains the representation paradigm. In particular, the control of a manufacturing robot corresponds to a precise way of representing it, by a centralized or a distributed model.

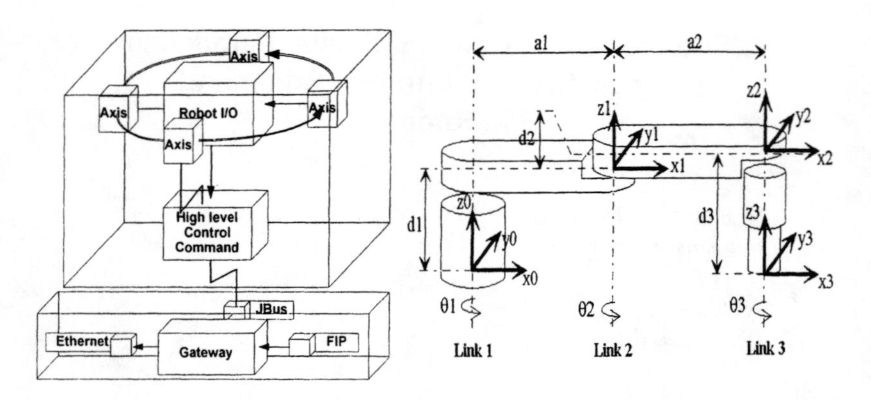

figure 1. type SCARA serial robot (4 axes), heterogeneous systems, heterogeneous protocols

DISTRIBUTED CONTROL

The modelisation of a mechanical system is classically realized by following an analytic process in which the formalism is adapted to the mathematical analysis of the problem. However, this method is something of not adapted to exploitation. In particular, it is often necessary to have enough capacity for calculation and the real time control/command notion becomes then quite questionable. The solutions often derive from computer science and it is difficult to modify the vision we have about the problem in order to limit its arithmetic aspects.

We expose two implementations which show how relative the distribution notion is. The first study uses a classical centralised model and looks for a way to distribute it ; we will see that this method in fact corresponds to a calculus parallelism and that it proves nothing about the system's ability to be distributed. A second study resumes the modelisation by insisting on the *Concept* and its distribution. Then the computer science, and Distributed Artificial Intelligence in particular [6], contributes to its expression and not only to the implementation efficiency.

Distribution of equations

The geometric, cinematic and dynamic inverse models are calculated for each axis. The result equations are distributed at a processus level so that each axis is separately controled. The expressions of the mathematical and physical constraints are also locally achieved for each axis.

The study is independent of the kind of model we use, so we can reduce it to the study of the geometric model. Here are the three equations for the three axes (fig. 1) :

$$d_3 = d_1 + d_2 - z \; ; \qquad (1)$$

$$\theta_2 = \arccos\left((x^2 + y^2 - a_1^2 - a_2^2) / (2.a_1.a_2) \right) ; \qquad (2)$$

$$\theta_1 = \text{arctg}\,(y / x) - \text{arctg}\,(a_2.\sin(\theta_2) / (a_1 + a_2.\cos(\theta_2))) . \qquad (3)$$

Mathematic precedences. In an analytic point of view it is quite possible to study each equation apart from the others. The equation (3) is calculated from the angle of the axis 2, but a precedence in the calculus is nowhere explicitly mentioned. So that if we implemente this system of equations, we must take care to calculate well the equation (2) before the equation (3), which constitutes a synchronisation point when we distribute the calculus.

Mathematic limits. We define the mathematic limits for each axis according to mathematic impossibilities. For example we easily see that the equation (2) is only calculable if the expression in the calculus of *arccos* is included in the domain [-1,1]. The mathematic limits can then be expressed by making an analysis of the equations which have a specific meaning at physical level. We can content ourselves with the redundancy of physical observations to confirm the mathematic conclusions but we will see that it is also interesting to think only about the physical observations.

Physical limits. These limits act on the reachable domain. They correspond to the presence of physical stops on the robot or other mecanism which restrict its movements. These limits can be related to each axis for the distribution.

Concept distribution

In the precedent study, the physical properties of the robot were emerging from equations. The model is centralized and its distribution is just possible at calculus level. Here we attach a great importance to the physical properties as properties emerging from behaviors of communicating entities [3]. Then the calculus gets its legitimacy within a limited analytic context : « the calculus is a substitute for a missing vision of the real world ». In particular, if we can not see the goal, we can search for a way to simulate the presence of this goal *-virtual perception-* by defining a reference and computing the absolute position of the robot.

Moving analysis. This analysis is fundamental, we show how, through a totally natural process, it is possible to achieve a distributed model. It relies on the moving technique used by all living being [2]. The moving entity first defines a goal by locating it visually. This localisation defines the distance to the goal and the entity attempts to minimize it. The goal is an attractor that pull the entity to it.

Imagine that we want to take an object on a table (our goal). The distance we will try to minimize is the one between the object and our hand, and our sight allows us to check it to adapt our approach. The object becomes an attractor for the hand, and all the elements bound to the hand are also attracted (to the point of stress).

About the calculus precedences. In a previous chapter, we have seen the necessity to take into account the presence of the axis 2 in the formula of the axis 1 for the implementation of the model. Now if we take a look at its physical interpretation, we see that the precedence notion is closely linked to the « attraction zone » notion. If we suppose that we pull the prehensile (axis 3) into the direction of the object to take, it becomes the release mechanism of the ensemble movement. The other axes follow the movement by solidarity to the axis 3. This is the physical justification of precedence in calculus.

Study of the links. We can define a mechanical link in differents ways. For example the mechanic will use a symbolic close to the physical representation, the electronics engineer will differentiate the passive elements from the active ones and the computer scientist will represent a stream of messages subjected to some processings. Those three representations are complementary. The connections of the links generate a complex running stemming from the different reactions to the streams of informations [7]. The whole can be represented by a graph of dependencies in which the nodes are active elements (they use the information) and the bonds are passive elements (they propagate the information).

Links interactions. The attraction zone and graph of dependencies notions allow a good representation of the robot. The physical relations between the links define a static and non oriented graph.

The non oriented graph defines the structural level of the articular system. The connections between the links define the whole streams of possible informations which achieve the system behaviors. We can deduct from a specified attraction zone a graph of dependencies composed of several roots (fig 2). These roots correspond to the set of elements located in the attraction zone.

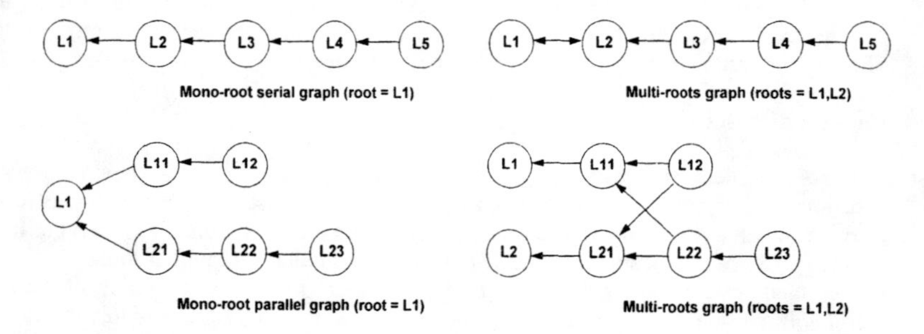

figure 2. graph of dependencies. A graph may be serial, parallel or mixed. The presence of several roots may generate some cycles and crossings in the graph.

The graph of dependencies is not static, the attraction zone may develop and generate new roots (flexible systems). The streams of informations spread from the roots to the terminal links (ex : the floor). The evaluation of the distance to the goal is the release mechanism of the information stream. The analysis of the interactions between the links make three graph types appear : a Static and Non Oriented Graph (SNOG) which defines the structural level, a Dynamic Oriented Graph (DOG) which defines the dependencies and a dual graph of the graph of dependencies which defines the information stream.

Real time implementation. The link autonomy such as its capacities of communication make it an entity close to the Agent concept [1]. The system dynamic sets high time constraints. The link behaviors are released by a message reception. Each link is an asynchronous process activated by the arrival of a message (fig. 3). The control processes are deterministic and the processing times are very short. The system response time depends on the length of the information stream (i.e. the depth in the DOG of streams) for a specified attraction zone. We can define a maximum bound which corresponds to the greater stream length. The input/output stream connections of each control process are achieved from the SNOG. The DOG of dependencies is defined from the roots in the attraction zone and must contain no circuit (fig. 4), then the control system is deterministic.

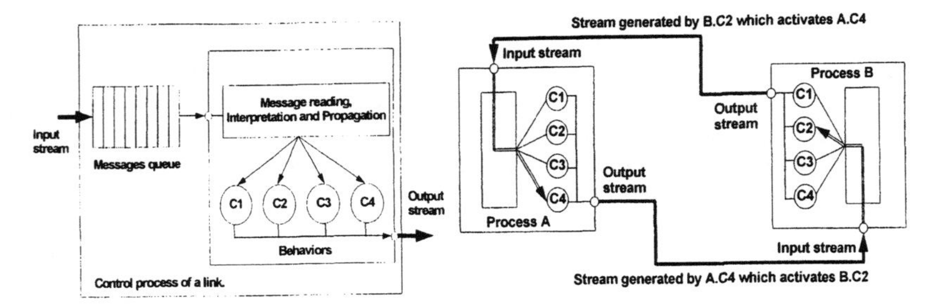

Figure 3. representation of a link control process.

Figure 4. example of a graph with a circuit.

Communications and heterogeneous systems

Problematic. The implementation of a control process is subjected to different kinds of constraints [4] : physical, logical, temporal and geographic. The control architecture of our SCARA robot is based on heterogeneous systems interconnection from low level networks (FIP, MODBUS) and supervision networks like TCP/IP. The whole platform inputs/outputs are distributed and it is just possible to reach them throughout an opened connectivity. Our approach for the control of the site is developped from the notion of distributed virtual automation constituted of different portions of heterogeneous equipments and its running is homogenized by a gateway [5]. The virtual automation may be remote configurated, controlled and managed throughout one of these connections.

Results. The gateway allows to connect different kinds of networks. Each kind of network is tackled in a different way according to its nature and function. The only considered constraint is the type of the final installation. Indeed the use of a Master-Slave structure for the virtual automation allows a greater transparency of the gateway towards the interconnected equipments. The temporal needs and the gateway's complexity required the use of a real time kernel.

Example of an implementation with a classical model (fig. 5)

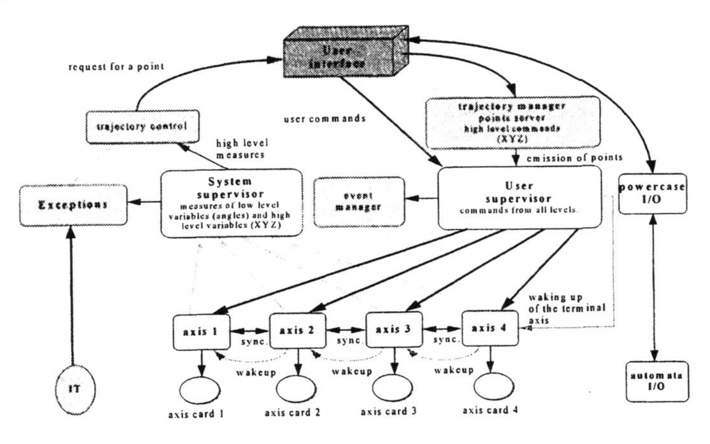

Figure 5. Real time architecture. Processes organisation.

We use a real time architecture orthogonal to DOS in which we have synchronous and asynchronous processes, synchronous mailing (based on the ADA Rendez-vous principle) and asynchronous mailing as well as a real time graphic interface. The axes control is implemented from parallelized geometric and cinematic models (fig. 5). Two measures and commands management processes, the *system supervisor* and the *user supervisor*, allow to access to articular and work space data. The whole axes control processes, the exception management process and the two measures and commands control processes form the basic kernel of this architecture. We developped around this kernel a system which follows through the trajectory by constant speed as well as an interface which opens a control access from a JBUS network. The multi-protocols gateway allowed to widen this access to the more used networks (FIP, Ethernet TCP/IP, CAN, ...).

SIMULATION OF THE DISTRIBUTED MODEL

While we were achieving the implementation on the serial robot, we made a second implementation with the distributed model on a simulation platform. This work allows to validate the notion of distributed Concept. The simulation platform is a Real Time Object kernel. This tool contributes to a good representation of links from the structural graph of the robot. To put the information streams DOG into service is moreover made easier by the presence of multiple communication chanels (i.e. several chanels for a single object).

There are two types of links : rotation link and translation link. There is a basic object for each of the three possible views (mechanical, electrical and informatical view). These three objects are assembled into an active object which composes the link reactivity. For the simulation, the object in charge of the interface with the real world (motor and sensor) is associated with an object which transmits the rotation or translation order to a graphic interface representing the robot. The measures are a direct loop of the order.

CONCLUSION AND PROSPECTS

A distributed model is to be elaborated for a parallel robot in order to validate completely the distributed Concept notion. The choice of a distributed solution is part of a process which tends to represent the different elements of a system as naturally as possible. This choice requires to think about the dependencies which exist between the different elements of the system. These interactions make information streams appear which characterize the system. This solution allows to generalize the problem of a model implementation in a real time environment for which the processes and data are not necessary centralized on a single computer. The control system will adapt itself to an environment in evolution.

REFERENCES

1. Agha, « Research Directions in Concurrent Object-Oriented Programming », *The MIT Press*, 1993.
2. Braitenberg, V., « Vehicles, Experiments in Synthetic Psychologiy », *The MIT Press*, Psychology, 1986.
3. Coiffet P., « L'automatisation intégrale du travail est-elle possible ? », *Colloque International Productique Robotique du Sud Europe Atlantique*, Juin 1995.
4. ISO TC184 SC5 WG2 TCCA, « Requirements for Systems Supporting Time Critical Communications », *ISO TC184 SC5 WG2*, N292, 26-3-1992.
5. Jacquelin B., Moreno C., « L'interconnexion de réseaux locaux industriels hétérogènes », *Automation Europe 96*, [à paraître].
6. Labidi S., Lejouad W., « De l'Intelligence Artificielle Distribuée aux Systèmes Multi-Agents », *Rapport de recherche INRIA*, n° 2004, Août 1993.
7. Mataric M. J., « Interaction and Intelligent Behavior », *PhD Thesis MIT*, 1994.

A GENERALIZED METHOD FOR OPTIMAL MODELING IN ROBOTIC APPLICATIONS*

E. J. Bernabeu, J. Tornero & M. Mellado
Departamento de Ingeniería de Sistemas,
Computadores y Automática (DISCA),
Universidad Politécnica de Valencia, P.O.B. 22012
E-46071 Valencia, SPAIN

ABSTRACT

This paper introduces a generalized technique for the geometric modeling of objects in industrial robotic systems. The presented method finds the envelope of minimum volume needed to model an object. The envelopes are generated through the application of an iterative process based on the Hough Transform. For improving the computational cost, a procedure, based on clustering, is also introduced.

KEYWORDS: Hough transform, 3D geometric modeling, minimum volume, generalized modeling, clustering, planning applications.

INTRODUCTION

Nowadays, for running, under real-time restrictions, some robotic applications, like collision detection and avoidance, trajectory or path planning, is needed to dispose a geometric model of the robotic system. In this way, we introduced in this paper a method, able to generate automatically a minimum volume for enveloping every element presented in an industrial robotic environment.

The volumes generated by this method are a special type of spherical volumes called spherically-extended polytopes (s-tope). These volumes are frequently used in nowadays researches [1], [2], [3], [4], because of their robustness when representing different elements in a robotic system.

Depending on the level of the accuracy required in the representation, this method is able to generate from the simplest sphere to the most complex s-tope.

This geometric modeler is divided into two hierarchical levels. The top level is based on the Downhill Simplex method [5], and the low level consists of the application of the Hough Transform [7,8].

In order to reduce the computational cost of this method, a technique based on point clustering is introduced too. This technique is used to provide the Downhill Simplex level more informed initial conditions of searching.

---

* This work was partially supported by the CICYT project TAP 95-1086-C02-01

OPTIMIZATION FOR THE MINIMUM VOLUME APPROXIMATION

The volume approximation problem implies that a set of N points from the surface of the actual object is known. For example, the set of points can be detected by means of sensors, artificial vision (clustering), taken from CAD tools, processed with some method of sensor data fusion. A model of the object is obtained with the constraint that all these points have to be included in its volume.

The presented optimal geometric modeler generates s-tope models. A s-tope $S_{0\text{-}n}$ is defined in function of a set of spheres $s_0,\dots,s_n$, each one represented by a 4-tupla (center coordinates $c_i=(x_i,y_i,z_i)$ and radius r_i), by the linear combination of them:

$$S_{0-n} = s_0 + \sum_{i=1}^{n} \lambda_i (s_i - s_0), \qquad \lambda_i \geq 0, \ \sum_{i=1}^{n} \lambda_i = 1 \tag{1}$$

Properties like excellent computational cost in distance computation, collision-detection, etc., and details of s-topes can be obtained in different papers [2,3]. Examples of s-topes are sphere, bi-sphere, defined by $S_{01}=s_0+\lambda_1(s_1-s_0)$, with $0\leq\lambda_1\leq1$, tri-sphere, defined by $S_{012}=s_0+\lambda_1(s_1-s_0)+\lambda_2(s_2-s_0)$, with $0\leq\lambda_1$, $0\leq\lambda_2$, $\lambda_1+\lambda_2=1$.

The complexity of the optimal generation depends on the complexity of the volume used for the approximation. For example, if a sphere is used, 4 parameters are required (3 for center and one for radius), for a bi-sphere, 8 parameters are required. Obviously, the simplest case implies low accuracy while the most complex volume makes possible to get a high accurate model of the real object. A good trade-off between complexity and accuracy can be reached with a family of sphere-based objects.

The optimization problem must minimize a volume function that depends on m parameters, having N constraints (one for each point to envelop). The presented method for obtaining the minimum volume consists of two hierarchical levels: the top level optimizes a q (=m-k) parameters function (sphere centers) without constraints. This level can be solved with the Downhill Simplex optimization method [5]. The low level minimizes the volume function for k parameters (sphere radii) considering the N constraints. This level consists of the application of the Hough Transform.

As the geometric modeler is based on a volume minimization, we need to compute the s-tope defined by the above-mentioned m parameters. The volume of a s-tope is described in terms of distinguishing, first of all, the upper and lower part of the s-tope. The upper part consists of curved and planar patches bounded by circular arc and straight lines segments (edges). A vertical ray is extending downwards from each point of an edge, in this way, every edge has associated a vertical "wall". These walls decompose the space below the upper envelope into curved prisms. The volume of such a prism is computed in O(1) time, and the number of prisms is linearly related to the number of faces of the s-tope. Adding these elementary volumes, the volume of the space below the upper envelope is obtained. Then, doing the same, the volume of the space below the lower envelope is obtained too. The wanted volume will be obtained by subtracting the computed volume of both spaces. The volume of a bi-sphere is computed more easily. Considering that it is a revolution envelope, its volume can be described in terms of

$$Vs_{01} = \frac{2}{3}\pi\left(r_0^3 + r_1^3\right) + \frac{4}{3}\gamma_{01}\left(r_0^3 - r_1^3\right) + \frac{\pi}{4}\left\|c_1 - c_0\right\| \cos\gamma_{01}\left(r_0 + r_1\right) \tag{2}$$

with $\sin\gamma_{01} = \dfrac{r_0 - r_1}{\left\| c_1 - c_0 \right\|}$ and $\|.\|$ representing the Euclidean norm.

At the top level, the Downhill Simplex method works in a q-dimensional space with a variable $C=(c_0,\ldots,c_n)$, the set of centers of the spheres. It starts from a simplex consisting of q+1 points, C_0 and $C_i=C_0+\delta e_i$, i=1,...,q, where C_0 is the initial condition, e_i are q unit linearly independent vectors and δ is a parameter chosen according to the problem's characteristic length scale. For the first step, low level procedure is invoiced to compute volume function at each one of the q+1 points. In every step, downhill simplex method makes a reflection of the point of the simplex where function is highest through the opposite face of the simplex to a lower point. Hence, only for this new point the volume function should be evaluated. If no better point is reached by this way, expanding or contracting the point in its moving direction can be checked. If no improvement has been got, the whole simplex can contracts itself in all directions, pulling itself in around the point with lowest volume function value.

On the other hand, the low level, using the information provided by the top level, i.e., the centers of the spheres, obtains the s-tope of minimum volume needed to envelop the input points. Obtaining the minimum s-tope means to find the suitable radii of every provided center. Later, the volume will be computed and returned to the top level. At this level, based on the Hough Transform, all input p_i must be projected to the (n+1)-dimensional structure formed by $c_0,\ldots,c_n$, i.e., a normal point must be computed:

$$p_i^{\perp} = c_0 + \sum_{j=1}^{n} \lambda_{ij}(c_j - c_0) \qquad (3)$$

This point is computed solving the following linear equation system:

$$\left(p_i^{\perp} - p_i\right)\cdot\left(c_j - c_0\right) = 0, \quad j = 1,\ldots,n \qquad (4)$$

where $\cdot$ means inner product of vectors. Therefore, a set of parameters $(\lambda_{i1},\ldots,\lambda_{in})$ is obtained. One parameter more is required to express the point p_i related to this (n+1)-dimensional structure:

$$d_i = \left\| p_i^{\perp} - p_i \right\| \qquad (5)$$

With this process, every point p_i is transformed to the vector $(\lambda_{i1},\ldots,\lambda_{in},d_i)$. The $p_i=(\lambda_{i1},\ldots,\lambda_{in},d_i)$ verifying $\lambda_{ij} > 0, j = 1,\ldots,n$, $\sum_{j=1}^{n} \lambda_{ij} < 1$ are called Half-Way Points (HWP). The rest of p_i, called Outer-Way Points (OWP), determine d_i^{min}, i=0,...,n

- For $p_i=(\lambda_{i1},\ldots,\lambda_{in},d_i)$ such that $\lambda_{ij} \le 0, j = 1,\ldots,n$ find $d_0^{min} = \max\left(\left\| p_i - c_0 \right\|\right)$
- For $p_i=(\lambda_{i1},\ldots,\lambda_{in},d_i)$ such that $\lambda_{ij} \ge 1$, $\lambda_{ij} \ge 1 + \sum_{k=1}^{n} \lambda_{ik}$ $k \ne j$, get $d_j^{min} = \max\left(\left\| p_i - c_j \right\|\right)$

where $d_j^{min}, j = 0,\ldots,n$ represent the minimal radii for centers $c_0,\ldots,c_n$ to generate a s-tope that envelops all the OWP. Note that $d_j^{min} \ge 0, j = 0,\ldots,n$ must be always verified.

The HWP $p_i=(\lambda_{i1},\ldots,\lambda_{in},d_i)$ are converted into hyper-planes by means of the equation:

$$\frac{r_j - r_0}{1} = \frac{d_i - r_0}{\lambda_{ij}} \Rightarrow r_j = \left(1 - \frac{1}{\lambda_{ij}}\right) r_0 + \frac{d_i}{\lambda_{ij}}, j = 1,\ldots,n \qquad (6)$$

where $r_0,\ldots,r_n$ are the unknown radii of the centers $c_0,\ldots,c_n$. These radii define hyper-planes that are called Hough Hyper-Planes (HHP). HHP are characterized by:

- The slopes of the straight lines in the plane defined by $r_0,\dots,r_n$ are always negative. All HHP generated from a HWP $p_i=(\lambda_{i1},\dots,\lambda_{in},d_i)$,

$$\lambda_{ij} > 0, j=1,\dots,n, \sum_{j=1}^{n}\lambda_{ij} < 1 \Rightarrow \frac{1}{\lambda_{ij}} > 1 \rightarrow 1-\frac{1}{\lambda_{ij}} < 0, j=1,\dots,n \quad (7)$$

- The HHP determine a half hyper-space in $r_0,\dots,r_n$:

$$r_j \geq \left(1-\frac{1}{\lambda_{ij}}\right)r_0 + \frac{d_i}{\lambda_{ij}}, j=1,\dots,n \quad (8)$$

where $(r_0,\dots,r_n)$, at the half hyper-space, generate a s-tope that envelops all HWP.

For the OWP, new hyper-planes are defined, with $r_j = d_j^{min}, j=0,\dots,n$ (9), leading to new half hyper-spaces $r_j \geq d_j^{min}, j=1,\dots,n$ (10).

The half hyper-space defined by Eq. (8) and (10) represent all the valid values for the radii of the spheres of the s-tope with centers $c_0,\dots,c_n$ to envelop all the input points. Obviously, it can be shown that for a set of values $(r_0,\dots,r_n)$, which makes Eq. (8) or (10) to fail, there exists at least one point p_i that is not covered by the s-tope defined.

The set of values $r_0,\dots,r_n$ that verifies Eq. (8) and (10) and at least one of equations expressed in Eq. (6) and (9) is called the Minimum Volume S-Tope Locus (MVSTL), that is the locus where the s-tope with minimum volume is going to lay.

The minimum volume s-tope for enveloping all input points is generated by some point located at the MVSTL. In this way, a search, sweeping the MVSTL, must be done to find the $(r_0,\dots,r_n)$ that originates the minimum volume s-tope.

For example, if a bi-sphere with minimum volume is intended to be obtained (n=1):

- Every p_i to envelop is represented by means of (λ_i,d_i)
- For $p_i=(\lambda_i,d_i)$ that $\lambda_i \leq 0$ and $\lambda_i \geq 1$ find $d_0^{min} = \max(\|p_i - c_0\|)$, $d_1^{min} = \max(\|p_i - c_1\|)$
- The HHP will be straight lines defined by equation: $r_1 = \left(1-\frac{1}{\lambda_i}\right)r_0 + \frac{d_i}{\lambda_i}$
- A MVSTL (Fig. 1) is represented by thick stroke line.

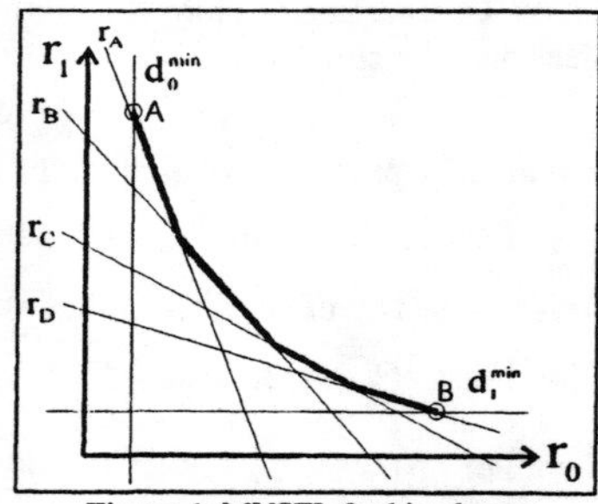

Figure 1: MVSTL for bi-spheres

Computational Cost Optimization

The Downhill Simplex method of optimization has the problem that if there is many internal iterations, the computational cost is considerably increased. Translating this problem into our minimization volume application. If the initial conditions of optimization C_0, are not appropriated, the computational cost required is too high, because of the important number of iterations. On the other hand, if the initial conditions are not suitable, the probability of reaching a local minimum is also high. Providing to Downhill Simplex method an intelligent C_0, the computational cost will be reduced.

The informed C_0 are obtained by the application of a non-hierarchical clustering technique [9]. This method, giving a set of point to envelop S, and the s-tope required, acts

1. Select n points from S as initial centers of clustering.
2. The rest of the points in S, are assigned to the closer cluster (closer center cluster).
3. Compute the new centers clusters, as the center mass of the cluster.
4. Repeat 2,3 until the cluster does not change.

The initial centers are selecting according to the following criteria: for generating a bi-sphere, select the farthest points in S, for a tri-sphere, select the three points that originates the triangle of maximum area, etc. Figure 2 shows the computation cost of the clustering procedure, and the comparison of the computational costs, giving intelligent initial conditions or not, in a bi-sphere generation process. From the figure, it is easy to conclude the important reduction in the computational cost.

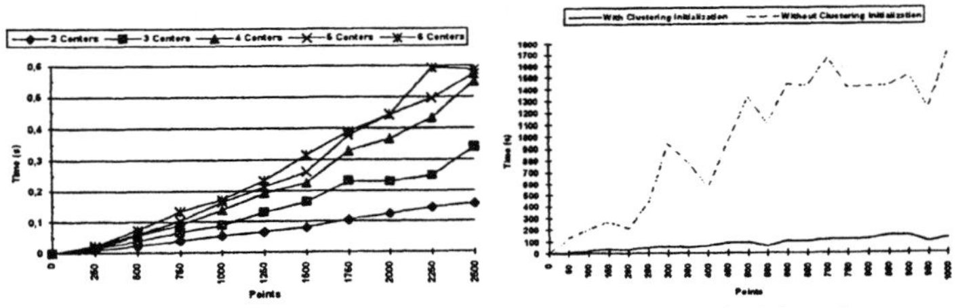

Figure 2: Cost optimized by the application of non hierarchical clustering

APPLICATION TO ROBOTIC SYSTEMS

Given the nature of robotic systems, a multi-level hierarchical model can be adopted. The hierarchical structure is expanded by considering different models according to the degree of accuracy required. In this way, a robot can be modeled using a global model, that is, the model envelops the whole robot, (see fig. 3) through envelopes of different accuracy, and/or using a link model. Obviously, this approach is applied to any object in the robotic system environment.

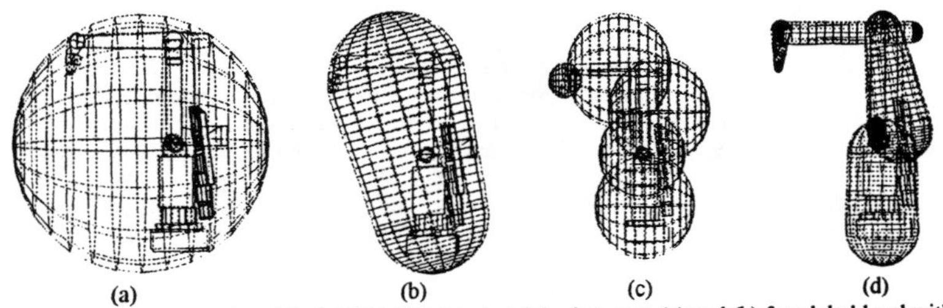

(a) (b) (c) (d)

Figure 3: Hierarchical model of ABB IRB L6 industrial robot-arm: (a) and (b) for global level with sphere and bi-sphere; (c) and (d) for link level with spheres and bi-spheres.

The use of this multi-level hierarchical modeling technique has been used successfully in many robotics application, such as collision-detection or obstacle-avoidance. For example, in the collision-detection application, cost can be reduced in a 90 % [3].

When a multi-level hierarchical model of the robotic system is obtained, the configuration space can be computed too. For trajectory or path planning, any of the two most popular techniques for planning, i.e., graph searching and artificial fields, can be used.

In the configuration space modeled with s-topes, the technique of artificial potential fields for path planning takes the best, because distance computation is very fast. Figure 4 shows two examples of path planning with artificial potential fields.

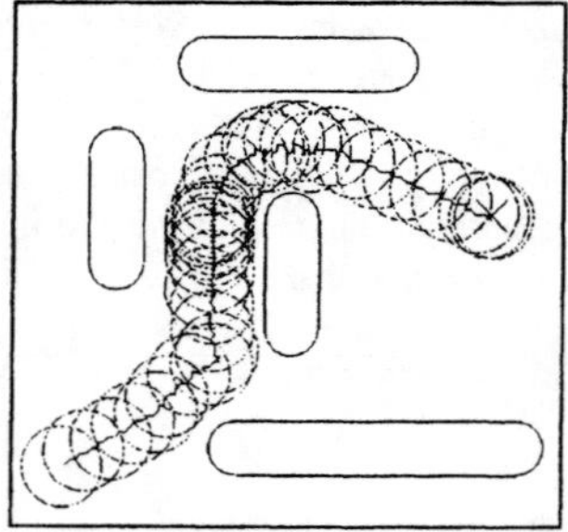

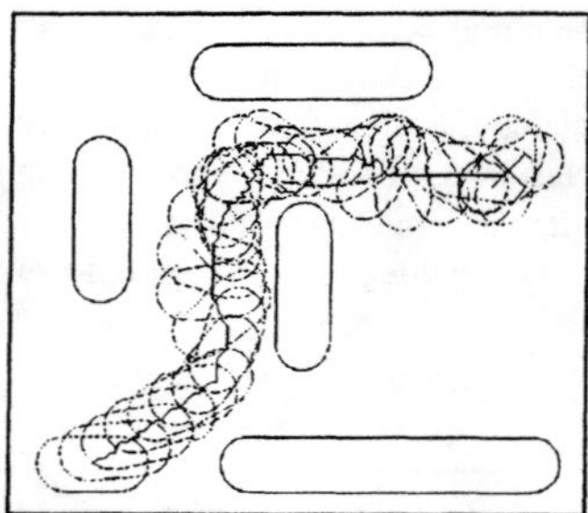

Figure 4: Path planning for a robot mobil modeled with a sphere (on the left) and a bi-sphere with rotation (on the right).

CONCLUSIONS

An automatic method for modeling objects in a multi-level hierarchical way has been presented. This method only needs a set of characteristic points taken from the object to envelop and the interval of accuracy required. For improving its computational cost a clustering procedure has been also introduced.

The general geometric modeler tool allows to use any type of technique for the planning of trajectory or paths in robotic systems.

REFERENCES

1. Tornero J., Hamlin G.J., Kelley R.B. "Spherical-Object Representation and Fast Distance Computation for Robotic Applications", IEEE Int. Conf. on Robotics & Automation, (2), 1602-1608, 1991.
2. Hamlin G.J., Kelley R.B., Tornero J. "Efficient Distance Calculation using Spherically-Extended Polytope (S-Tope) Model". IEEE Int. Conf. on Robotics & Automation, Nice, (3) 2502-2507, 1992.
3. Tornero J., Bernabeu E.J. "An Intelligent Procedure for Collision Detection in Robotic Systems". IFAC Symposium on Intelligent Component and Instrument for Control Applications, 97-102, 1992.
4. Del Pobil A.P., Serna M.A., Llovet J. "A New Representation for Collision Avoidance and Detection". IEEE Int. Conf. on Robotics & Automation, (1), 246-251, 1992.
5. Gilbert E.G., Johnson D.W., Keerthi S.S. "A Fast Procedure for Computing the Distance Between Complex Objects in 3-D Space". IEEE Journal of Robotics & Automation (4), 193-203, 1988.
6. Nelder & Mead. Computer Journal. (7), 1965.
7. Bernabeu E.J., Tornero J. "An Optimal and Automatic Geometric Modeler of Real Objects in Robotic Systems Using the Hough Transform Method". 2° Int. Symp. on Methods and Models in Automation and Robotics, 621-626,1995.
8. Bernabeu E.J., Tornero J. "An Automatic Method for the Geometric Modeling of Robotic Systems Based on the Hough Transform". The 11th ISPE/IEE/IFAC Int. Conf. on CAD/CAM, Robotics and Factories of the Future, 758-763, 1995.
9. Jain A.K., Dubes R.C. *Algorithms for Clustering Data*. Michigan State University. Prentice Hall 1988.

AN ADAPTIVE SLIDING MODE CONTROL ALGORITHM WITH UPDATED SWITCHING SURFACE PARAMETERS FOR ROBOTIC MANIPULATORS

NURDAN BEKİROĞLU, YORGO İSTEFANOPULOS

Boğaziçi University, Electrical and Electronic Engineering Dept., Turkey

ABSTRACT

In this work, the tracking performance and the robustness property of an adaptive sliding mode control (ASMC) algorithm for robotic manipulators, are both improved by using updated switching surface (USS) parameters without causing chattering and parameter drift problems. The performances of ASMC and ASMC-USS algorithms are compared and evaluated via computer simulations for a two joint robotic manipulator.

KEYWORDS: adaptive sliding mode control, updated switching surface parameters, robotic manipulators

INTRODUCTION

During the sliding motion of SMC, predesigned sliding surface determines the system performance and provides insensitivity to perturbations and disturbances. However most of the sliding surfaces proposed so far have been designed without considering the initial tracking errors. For this reason, an important problem in SMC has been reducing the distance between the initial tracking errors and the sliding surface appropriately, during the reaching period. Using the conventional design methods, one can only make sure that the system reaches the sliding surface in finite time but the reaching mode behavior is otherwise unspecified. This implies that the reaching mode in general does not possess the disturbance invariance property which is valid for the sliding mode. Therefore the distance between the initial tracking errors and the sliding surface should be eliminated or at least reduced to improve the robustness of SMC. On the other hand, since the chattering problem is caused by the non-ideal reaching, the reaching mode behavior is important from the view-point of chattering as well as tracking performance and robustness.

The USS approach [1,2] proposed by the authors, forces tracking errors of the system to follow the behaviors of certain reference error models and the initial tracking errors are taken into account while choosing the initial values of the USS parameters. Since the distance between the initial tracking errors and the sliding surface is reduced effectively (even eliminated for some cases), considerable improvements are observed in the tracking performance and the robustness property of the SMC algorithm used, without causing the chattering problem. In general, the USS approach can be combined with various SMC or ASMC algorithms and applied to a large class of high-order multi-input imprecise nonlinear systems.

The problem of controlling uncertain robotic manipulators subjected to a large class of state dependent input disturbances has been solved in [3] without using *a priori* knowledge about the manipulator. In the present paper, the ASMC algorithm of [3] is augmented with the USS approach and the resulting algorithm is evaluated with respect to the tracking performance, the robustness property, the chattering and the parameter drift problems. The tracking problem and its solution by the ASMC algorithm are discussed in the second section. Augmenting this algorithm with the USS approach is explained in the third section. Based on the computer simulation results obtained for a two joint robotic manipulator, comparison of the ASMC and the ASMC-USS algorithms, is presented in the fourth section. Finally in the last section, the results are summarized.

TRACKING PROBLEM AND ITS SOLUTION BY ADAPTIVE SMC

Consider the n-joint robotic manipulator subjected to state dependent input disturbances

$$\mathbf{M}(\theta)\,\ddot{\theta} + \mathbf{C}(\theta,\dot{\theta})\,\dot{\theta} + \mathbf{g}(\theta) + \mathbf{f}(\dot{\theta}) = \tau(t) + \mathbf{d}(\theta,\dot{\theta}) \tag{1}$$

where $\theta, \dot{\theta}, \ddot{\theta} \in R^n$ are the angular position, velocity and acceleration vectors. Moreover $\mathbf{M} \in R^{nxn}$, $\mathbf{C}\,\dot{\theta} \in R^n$, $\mathbf{g} \in R^n$, $\mathbf{f} \in R^n$, $\tau \in R^n$ and $\mathbf{d} \in R^n$ represent the symmetric positive definite inertia matrix, the effects of centripetal and Coriolis torques, gravitation and friction, the control torques and the disturbances respectively. By suitably defining $\mathbf{C} \in R^{nxn}$, the skew-symmetry property can be satisfied $\mathbf{s}^T(\dot{\mathbf{M}} - 2\mathbf{C})\,\mathbf{s} = 0$ for an arbitrary vector $\mathbf{s} \in R^n$. Furthermore there exist unknown constants $\rho_j \in R^+$ such that

$$\|\mathbf{M}(\theta)\| \le \rho_1 \quad ; \quad \|\mathbf{C}(\theta,\dot{\theta})\| \le \rho_2 \|\dot{\theta}\| \quad ; \quad \|\mathbf{g}(\theta)\| \le \rho_3 \tag{2}$$

$$\|\mathbf{f}(\dot{\theta})\| \le \rho_4 + \rho_5 \|\dot{\theta}\| \quad ; \quad \|\mathbf{d}(\theta,\dot{\theta})\| \le \rho_6 + \rho_7 \|\theta\| + \rho_8 \|\dot{\theta}\| \tag{3}$$

where $\|.\|$ shows induced norm for matrices and Euclidian norm for vectors. Bounded desired trajectories for the robotic manipulator are given by $\theta_d, \dot{\theta}_d, \ddot{\theta}_d \in R^n$. The control problem is to design a suitable torque vector τ for the system (1) which ensures that both $\tilde{\theta} = \theta - \theta_d$ and $\dot{\tilde{\theta}} = \dot{\theta} - \dot{\theta}_d$ converge to zero asymptotically as $t \to \infty$.

Theorem 1: For the manipulator given in (1), if $\mathbf{s} \in R^n$ is defined as $\mathbf{s} = \dot{\theta} - \dot{\theta}_r$ where $\dot{\theta}_r \in R^n$ is chosen in such a manner that bounded $\mathbf{s}$ implies the boundedness of $\theta, \dot{\theta}, \dot{\theta}_r, \ddot{\theta}_r$ and if the control law is defined as:

$$\tau = -\mathbf{K}_d\,\mathbf{s} - (\hat{\rho}_1 \|\ddot{\theta}_r\| + \hat{\rho}_2 \|\dot{\theta}\|\,\|\dot{\theta}_r\| + \hat{\rho}_3 + \hat{\rho}_4 \|\dot{\theta}\| + \hat{\rho}_5 \|\theta\|)\ \mathrm{sgn}(\mathbf{s}) \tag{4}$$

where $\mathbf{K}_d$ is a positive definite matrix; $\mathrm{sgn}(\mathbf{s}) = [\,\mathrm{sgn}(s_1).....\mathrm{sgn}(s_n)\,]^T$ and the control parameters $\hat{\rho}_k(t) \in R$ $(k = 1,...,5)$ are evaluated by integrating

$$\dot{\hat{\rho}}_1(t) = \eta_1 \|\ddot{\theta}_r\|\,\|\mathbf{s}\|_1 \quad ; \quad \dot{\hat{\rho}}_2(t) = \eta_2 \|\dot{\theta}\|\,\|\dot{\theta}_r\|\,\|\mathbf{s}\|_1 \tag{5}$$

$$\dot{\hat{\rho}}_3(t) = \eta_3 \|\mathbf{s}\|_1 \quad ; \quad \dot{\hat{\rho}}_4(t) = \eta_4 \|\dot{\theta}\|\,\|\mathbf{s}\|_1 \quad ; \quad \dot{\hat{\rho}}_5(t) = \eta_5 \|\theta\|\,\|\mathbf{s}\|_1 \tag{6}$$

where $\eta_k \in R^+$ $(k = 1,...,5)$ and $\|\mathbf{s}\|_1 = |s_1| + ... + |s_n|$, then $\mathbf{s}$ and $\hat{\rho}_k(t)$ are bounded. Furthermore $\mathbf{s} \to 0$ as $t \to \infty$.

Proof: Consider the Lyapunov function given below [3]:

$$V = \frac{1}{2}\mathbf{s}^T\,\mathbf{M}\,\mathbf{s} + \frac{1}{2}\sum_{k=1}^{5} \frac{(\bar{\rho}_k - \hat{\rho}_k(t))^2}{\eta_k} \tag{7}$$

where $\hat{\rho}_k(t)$ are the estimates of $\bar{\rho}_k$ defined as $\bar{\rho}_1 = \rho_1$, $\bar{\rho}_2 = \rho_2$, $\bar{\rho}_3 = \rho_3 + \rho_4 + \rho_6$, $\bar{\rho}_4 = \rho_5 + \rho_8$, $\bar{\rho}_5 = \rho_7$. Using the definition of $\mathbf{s}$ and (1), one obtains:

$$\mathbf{M}\,\dot{\mathbf{s}} = \mathbf{M}\,(\ddot{\theta} - \ddot{\theta}_r) = \tau + \mathbf{d} - \mathbf{C}\,(\mathbf{s} + \dot{\theta}_r) - \mathbf{g} - \mathbf{f} - \mathbf{M}\,\ddot{\theta}_r \tag{8}$$

If (7) is differentiated with respect to time and (8) is replaced in $\dot{V}$ and if the skew-symmetry property and the relations for $\|\mathbf{s}\|_1$, (2), (3) and $\bar{\rho}_k$ are used, then

$$\dot{V} \le \mathbf{s}^T\tau + \|\mathbf{s}\|_1\,(\bar{\rho}_1 \|\ddot{\theta}_r\| + \bar{\rho}_2 \|\dot{\theta}\|\,\|\dot{\theta}_r\| + \bar{\rho}_3 + \bar{\rho}_4 \|\dot{\theta}\| + \bar{\rho}_5 \|\theta\|) + \sum_{k=1}^{5} \frac{(\bar{\rho}_k - \hat{\rho}_k)(-\dot{\hat{\rho}}_k)}{\eta_k} \tag{9}$$

is obtained. Replacing (4), (5) and (6) in (9), one gets $\dot{V} \le -\mathbf{s}^T \mathbf{K}_d \mathbf{s} \le 0$. This implies that $\mathbf{s}$ and $\hat{\rho}_k(t)$ are bounded in (7). Since θ, $\dot{\theta}$, $\dot{\theta}_r$ and $\ddot{\theta}_r$ are bounded, $\dot{\mathbf{s}}$ in (8) is also bounded. Thus $\ddot{V}$ is bounded and $\dot{V}$ is uniformly continuous [3]. Hence it follows from Barbalat's Lemma that $\dot{V} \to 0$, which implies $\mathbf{s} \to 0$ as $t \to \infty$. □

During the sliding mode $\mathbf{s} = 0$, the dynamic behavior of the manipulator is completely determined by the sliding surface structure. If $\dot{\theta}_r = \dot{\theta}_d - \Lambda \tilde{\theta}$ where $\Lambda = \mathrm{diag}[\lambda_1 \ldots\ldots \lambda_n]$, $\lambda_i \in R^+$ $(i = 1,\ldots,n)$, then both $\tilde{\theta}$ and $\dot{\tilde{\theta}}$ converge to zero as $t \to \infty$. In Theorem 1, the existence of ρ_j $(j = 1,\ldots,8)$ is necessary to guarantee stability of the closed-loop system. However these constants are not required to evaluate the control torques (4). In fact the adaptation algorithm given in (5) and (6), estimates these constants indirectly and the estimated parameters $\hat{\rho}_k(t)$ are fed to the control torques. Since $\mathrm{sgn}(\mathbf{s})$ in (4) may lead to the chattering problem, it can be replaced by $\mathrm{sat}(\mathbf{s})$ where $\mathrm{sat}(\mathbf{s}) = [\ \mathrm{sat}(s_1)\ldots\ldots\mathrm{sat}(s_n)\ ]^T$,

$$\mathrm{sat}(s_i) = \left\{ 1 \ \text{ for } \ \varepsilon_i < s_i \ ; \ \frac{s_i}{\varepsilon_i} \ \text{ for } \ -\varepsilon_i \le s_i \le \varepsilon_i \ ; \ -1 \ \text{ for } \ s_i < -\varepsilon_i \right\} \tag{10}$$

and $\varepsilon_i \in R^+$ are small constants. Because of this smoothing process, however, $\mathbf{s}$ can not become zero exactly and the parameter drift problem is observed in the adaptation algorithm. In order to overcome this problem, one has to use a stopping criterion in the adaptation rules (5) and (6). The criterion proposed in this study, is to replace $\dot{\hat{\rho}}_k(t)$ values with zero when their magnitudes are very small ($|\dot{\hat{\rho}}_k| \le \delta_k$, $\delta_k \in R^+$ are small constants).

ASMC WITH UPDATED SWITCHING SURFACE PARAMETERS

Although the ASMC algorithm [3] reviewed above achieves satisfactory performance without using *a priori* information about the robotic manipulator, the initial tracking errors are not considered in the design of the sliding surface. Thus it is possible to improve the performance of this algorithm further by using our USS approach [1,2]. The main contribution of the USS approach is that one obtains improved tracking precision and robustness without causing chattering and parameter drift problems.

Theorem 2: Consider the manipulator (1) and the switching functions $\mathbf{s} = \dot{\theta} - \dot{\theta}_r$ where $\dot{\theta}_r = \dot{\theta}_d - \Lambda(t)\,\tilde{\theta}$, $\Lambda(t) = \mathrm{diag}[\lambda_1(t)\ldots\ldots\lambda_n(t)]$, $\lambda_i(t) = \lambda_{mi} - \hat{\lambda}_i(t)$, $\lambda_{mi} \in R^+$ and $\hat{\lambda}_i(t) \in R$ $(i = 1,\ldots,n)$ are obtained by integrating

$$\dot{\hat{\lambda}}_i(t) = -\gamma_i\, \tilde{\theta}_i\, (\ E_i + 2\, \tilde{\theta}_{mi}\ ) \tag{11}$$

where $\gamma_i \in R^+$ are the gains chosen by the designer, $E_i = \tilde{\theta}_i - \tilde{\theta}_{mi}$ are the errors and

$$\dot{\tilde{\theta}}_{mi} + \lambda_{mi}\, \tilde{\theta}_{mi} = \dot{\tilde{\theta}}_i \tag{12}$$

are the reference error models. If the control law (4) and the adaptation rules (5), (6) are used, then $\mathbf{s} \to 0$, $\tilde{\theta} \to 0$ and $\dot{\tilde{\theta}} \to 0$ as $t \to \infty$. Furthermore $\lambda_i(t)$ and $\hat{\rho}_k(t)$ are bounded.

Proof: Considering the switching functions in $\mathbf{s}$ separately

$$\dot{\tilde{\theta}}_i + \lambda_i(t)\, \tilde{\theta}_i = s_i \tag{13}$$

as linear time-varying systems and (12) as reference models, the following error equations can be obtained by subtracting (12) from (13):

$$\dot{E}_i + \lambda_{mi} E_i = s_i - \dot{\tilde{\theta}}_i + (\lambda_{mi} - \lambda_i(t)) \tilde{\theta}_i \quad (14)$$

Since the left hand sides have stable structures, the ideal inputs s_i $(i = 1,...,n)$ are:

$$s_i = \dot{\tilde{\theta}}_i + (\lambda_i(t) - \lambda_{mi}) \tilde{\theta}_i \quad (15)$$

When these expressions are compared to (13), one observes that the only difference comes from the fact that some constant parameters are subtracted from $\lambda_i(t)$. These subtracted constant parameters do not affect the rules given in (11) but change arbitrarily chosen initial conditions of the updated parameters $\lambda_i(t)$. Therefore without depending on the initial conditions of $\lambda_i(t)$, the switching functions (13) have appropriate structures to force $\tilde{\theta}_i$ and $\dot{\tilde{\theta}}_i$ to follow $\tilde{\theta}_{mi}$ and $\dot{\tilde{\theta}}_{mi}$ respectively. In other words, it is guaranteed that E_i and $\dot{E}_i$ are bounded and converge to zero as $t \rightarrow \infty$. Based on this fact, it is clear form (13) and (14) that $\tilde{\theta}_i$ and $\tilde{\theta}_{mi} = \tilde{\theta}_i - E_i$ are bounded and converge to zero as $t \rightarrow \infty$. Therefore θ_i are bounded and it is deduced from (11) that $\dot{\hat{\lambda}}_i(t)$ are bounded and converge to zero as $t \rightarrow \infty$, which imply that $\lambda_i(t)$ and $\dot{\theta}_r$ are bounded. Then choosing (7) as the Lyapunov function and following the same steps given in the proof of Theorem 1, one may show that $\dot{V} \leq 0$. Hence **s** and $\hat{\rho}_k(t)$ are bounded in (7); $\dot{\tilde{\theta}}_i$ and $\dot{\theta}_i$ are bounded in (13), which imply that $\ddot{\theta}_r$ is also bounded. Since **s**, $\hat{\rho}_k(t)$, θ, $\dot{\theta}$, $\dot{\theta}_r$ and $\ddot{\theta}_r$ are all bounded, $\dot{\mathbf{s}}$ in (8) and thus $\ddot{V}$ are also bounded. This leads to uniformly continuous $\dot{V}$. Based on Barbalat's Lemma $\dot{V}$, **s** and $\dot{\tilde{\theta}}$ in (13) converge to zero. Thus, both $\tilde{\theta} \rightarrow 0$ and $\dot{\tilde{\theta}} \rightarrow 0$ as $t \rightarrow \infty$. □

COMPARATIVE ANALYSIS VIA COMPUTER SIMULATIONS

Consider the two joint robotic manipulator [4] whose structure is given in (1):

$$M_{11} = (m_1 + m_2)\, l_1^2 + m_2 l_2^2 + 2 m_2 l_1 l_2 \cos(\theta_2) + 5 \quad (16)$$

$$M_{12} = M_{21} = m_2 l_2^2 + m_2 l_1 l_2 \cos(\theta_2) \quad ; \quad M_{22} = m_2 l_2^2 + 5 \quad (17)$$

$$[\mathbf{C}\dot{\theta}]_1 = -(\dot{\theta}_2^2 + 2\dot{\theta}_1\dot{\theta}_2)\, m_2 l_1 l_2 \sin(\theta_2) \quad ; \quad [\mathbf{C}\dot{\theta}]_2 = m_2 l_1 l_2 \sin(\theta_2)\dot{\theta}_1^2 \quad (18)$$

$$g_1 = ((m_1 + m_2)\, l_1 \cos(\theta_2) + m_2 l_2 \cos(\theta_1 + \theta_2))\, g \quad ; \quad g_2 = m_2 l_2 \cos(\theta_1 + \theta_2)\, g \quad (19)$$

$$f_i = \dot{\theta}_i + \mathrm{sgn}(\dot{\theta}_i)\,(0.5 + (15 - 0.5)\exp(-|\dot{\theta}_i| / 0.5)) \quad ; \quad d_i = \theta_i \quad i = 1,2 \quad (20)$$

where $m_1 = 0.5$ kg, $m_2 = 6.25$ kg, $l_1 = 1$ m, $l_2 = 0.8$ m, $g = 9.81$ m/sec^2. The desired angular positions are given as $\theta_{di}(t) = 1.5 - 0.5\cos(1.5t)$ rad; initial conditions of the system are $\theta_i(0) = 0.5$ rad, $\dot{\theta}_i(0) = -0.1$ rad/sec and the integration time is $\Delta t = 0.003$ sec. Moreover the gains and the initial conditions are chosen as $\eta_k = 0.5$ and $\hat{\rho}_k(0) = 1$; the boundary layer parameters are taken as $\varepsilon_i = 0.02$ and the parameters used in the stopping criterion are selected to be $\delta_1 = \delta_5 = 0.2$ and $\delta_2 = \delta_3 = \delta_4 = 0.1$. In the ASMC algorithm, $\mathbf{K}_d = \mathrm{diag}[\,150 \;\; 100\,]$ and $\Lambda = \mathrm{diag}[\,4 \;\; 4\,]$, while in the ASMC-USS algorithm, $\mathbf{K}_d = \mathrm{diag}[\,1000 \;\; 1000\,]$, $\lambda_{mi} = 20$, $\tilde{\theta}_{mi}(0) = 0.01$, $\gamma_i = 50$ and $\hat{\lambda}_i(0) = 19.999$. Moreover $\hat{\lambda}_i(t)$ values are replaced by zero, when their magnitudes are very small (≤ 0.05).

In Fig.1 and Fig.2, position and velocity errors of the ASMC and the ASMC-USS algorithms are compared. It is clear that the tracking precision is much better in the ASMC-USS algorithm compared to the other one. In Fig.3, since $| s_i |$ values are very small in the ASMC-USS algorithm, one may choose large values for the elements of $\mathbf{K}_d$ without causing very large initial torques and initial velocity errors. Furthermore torques applied to the joints are compared for both of the algorithms in Fig.4. Although large element values for $\mathbf{K}_d$ are chosen in the ASMC-USS algorithm, the initial torques are not larger than the initial torques observed in the ASMC algorithm. In fact the torque behaviors are quite similar. Moreover small $| s_i |$ values of ASMC-USS algorithm result in suitable control parameters $\hat{\rho}_k(t)$ shown in Fig.5. Thus the parameter drift and the chattering problems are eliminated easily. The switching surface parameters λ_i of the ASMC algorithm are chosen to be constant at 4, while $\lambda_i(t)$ of the ASMC-USS algorithm start from small values and increase until they reach to the appropriate values as shown in Fig.6.

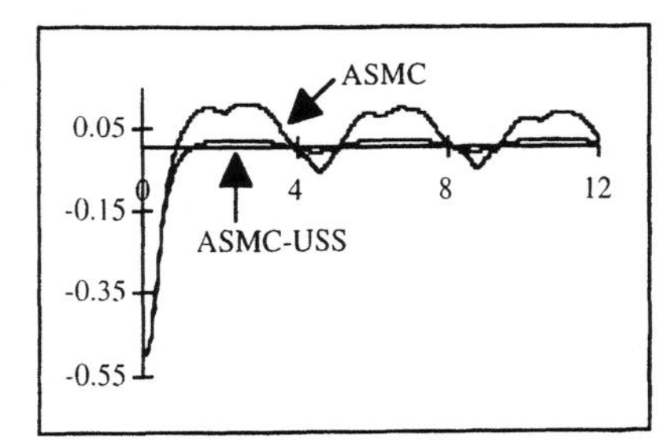

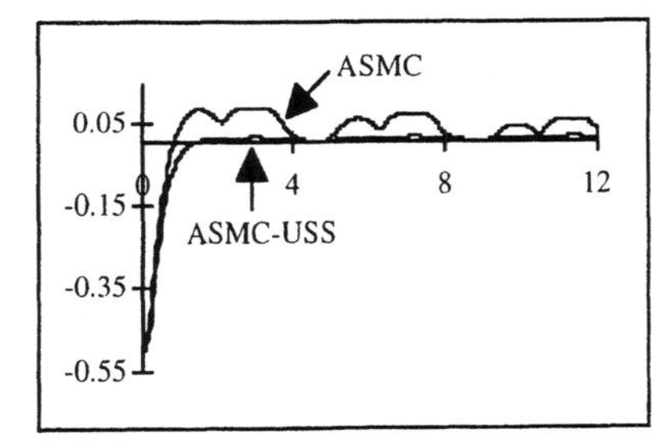

Figure 1. In ASMC and ASMC-USS algorithms, $\tilde{\theta}_1$ and $\tilde{\theta}_2$ (right) behaviors (rad) versus time (sec).

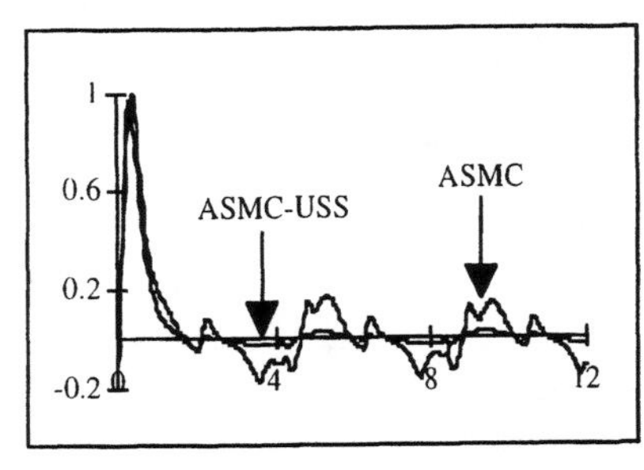

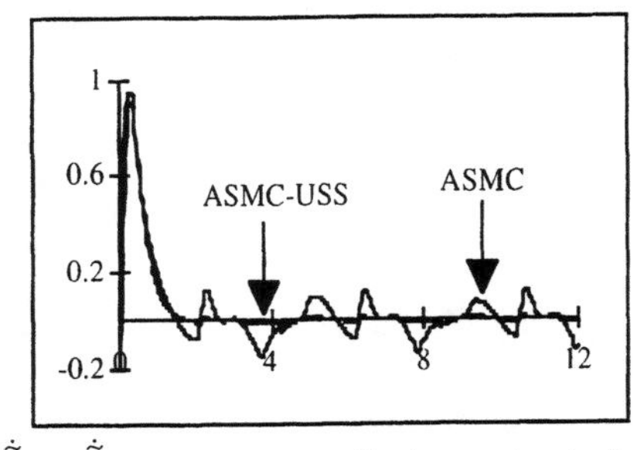

Figure 2. In ASMC and ASMC-USS algorithms, $\dot{\tilde{\theta}}_1$ and $\dot{\tilde{\theta}}_2$ (right) behaviors (rad/sec) versus time (sec).

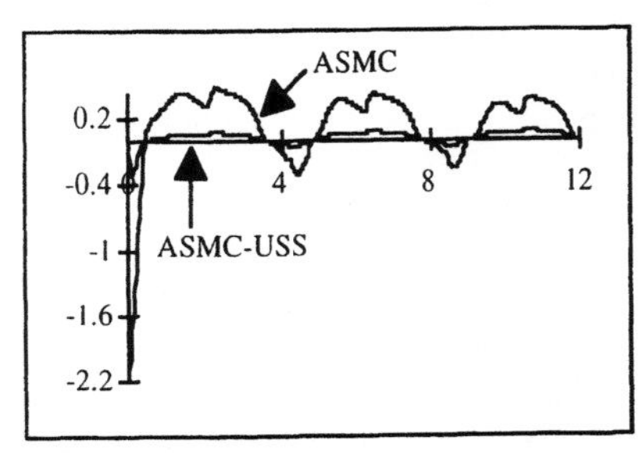

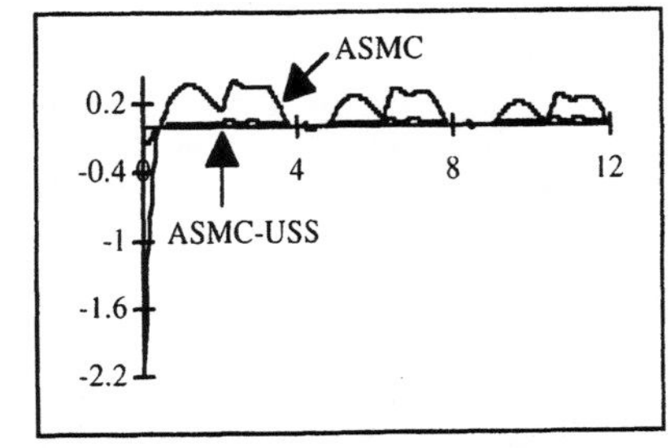

Figure 3. In ASMC and ASMC-USS algorithms, s_1 and s_2 (right) behaviors versus time (sec).

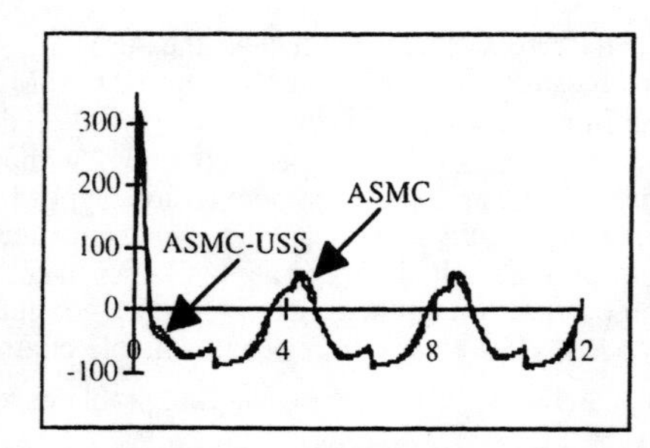

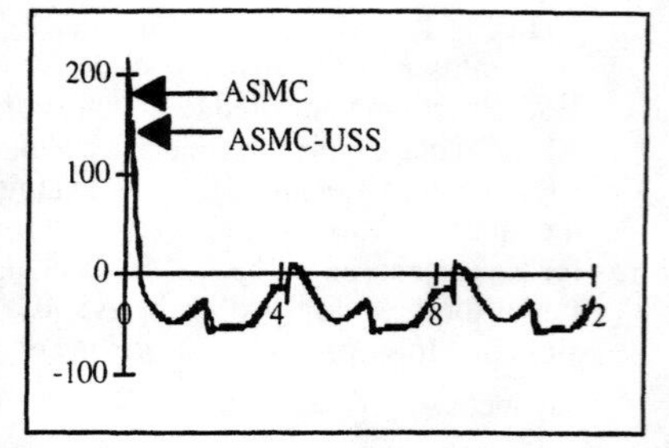

Figure 4. In ASMC and ASMC-USS algorithms, τ_1 and τ_2 (right) behaviors (N.m) versus time (sec).

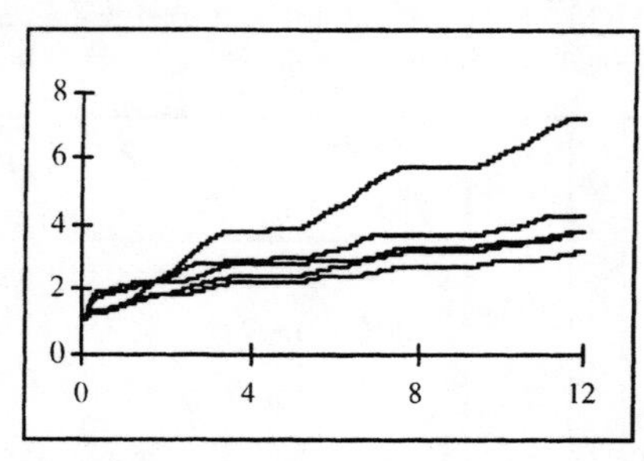

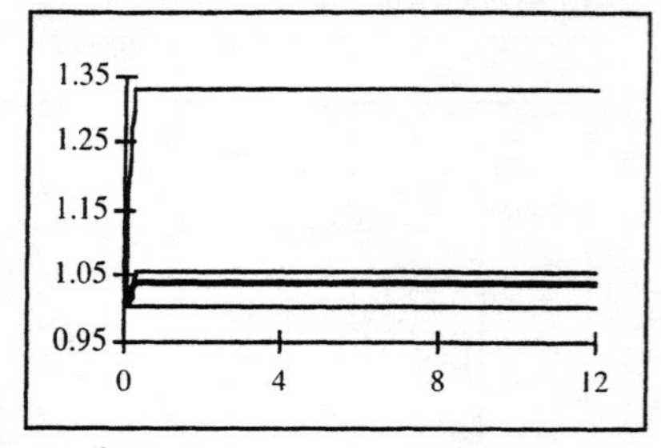

Figure 5. In ASMC and ASMC-USS (right) algorithms, $\hat{\rho}_k$ (k = 1,...,5) behaviors versus time (sec).

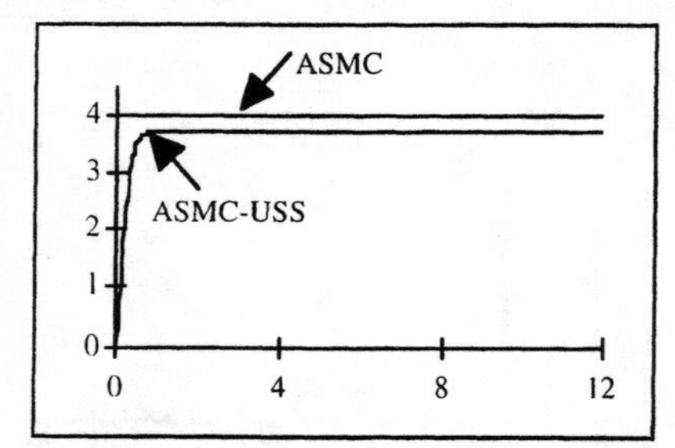

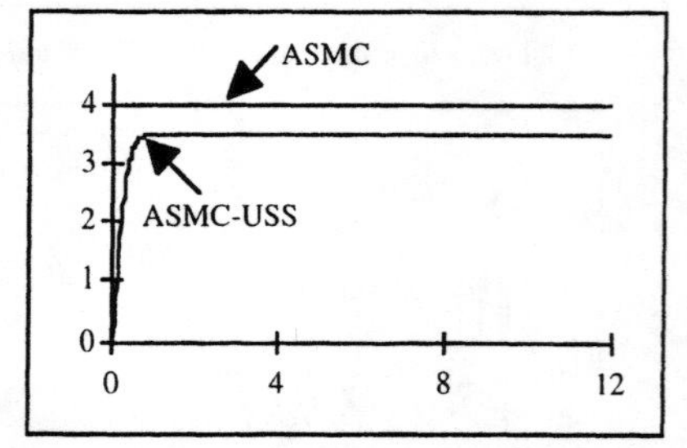

Figure 6. In ASMC and ASMC-USS algorithms, λ_1 and λ_2 (right) behaviors versus time (sec).

CONCLUSION

In this paper, the tracking performance and the robustness property of an ASMC algorithm [3] for robotic manipulators, are both improved by using updated switching surface parameters without causing chattering and parameter drift problems.

Acknowledgement The authors would like to acknowledge the facilities provided by the Mechatronics Research and Application Center of Boğaziçi University.

REFERENCES

1. Bekiroğlu, N., H.I. Bozma, Y. İstefanopulos "Model Reference Adaptive Approach to Sliding Mode Control." *American Control Conference'95*, Seattle (June, 1995), 1028-1032.
2. Bekiroğlu, N., Y. İstefanopulos "Comparison of Two Sliding Mode Control Algorithms and Their Model Reference Adaptive Versions." *Automatics and Informatics'95*, Sofia (Nov., 1995), 290-297.
3. Su, C-Y., Y. Stepanenko "Adaptive Sliding Mode Control of Robot Manipulators: General Sliding Manifold Case." *Automatica* 30 (9) (1994), 1497-1500.
4. Leung, T-P., Q-J. Zhou, C-Y. Su "An Adaptive Variable Structure Model Following Control Design for Robot Manipulators." *IEEE Transactions on Automatic Control* 36 (3) (March, 1991), 347-353.

FAST MOTION PLANNING IN DYNAMIC ENVIRONMENTS WITH THE PARALLELIZED Z^3–METHOD

BORIS BAGINSKI

Technische Universität München, Institut für Informatik,
Orleansstr. 34, 81667 Munich, Germany,
e-mail: baginski@informatik.tu-muenchen.de

ABSTRACT

We present a method to plan collision free paths for robots with any number of degrees of freedom in dynamic environments. The method proved to be very efficient as it ommits a complete representation of the high dimensional search space. Its complexity is linear in the number of degrees of freedom. A preprocessing of the geometry data of the robot or the environment is not required. With the time as an additional dimension of the search space it is possible to use the method in known dynamic environments or for multiple robots sharing a common work space. The method allows efficient parallel planning of independent sub-tasks to increase its performance.

KEYWORDS: path planning, dynamic environments, randomized algorithms, parallelized atgorithms

INTRODUCTION

The method presented in this paper is part of our research on intelligent and autonomous robot systems. An important component of such a system is a path planner that creates collision free and physically possible motions between positions that were planned on a higher, more abstract level. A path planner for practical use must be able to work fast in any kind of environment and with any load of the robot. In addition, there are often dynamic changes in the environment that make it necessary to take time into consideration for the motion planning.

The pose of a robot can be described by the vector of its joint positions. One point in the space of possible joint values (*configuration space* or *c-space*) is the precise description of a robot's position. The solution of the path planning task for an n-joint robot is to find a curve between start and goal in the n-dimensional *c-space* [8]. In a dynamic environment the *c-space* is extended by the dimension time. The solution has to be found in the $n+1$-dimensional *c-space-time*. This extension is not homogenous, as the dimension time is bound to strictly monotonous increase, while the other dimensions allow any motions, only constrained by the robot's dynamics.

In this paper we present the Z^3-method for path planning as a sequential algorithm for static environments. It is then extended to dynamic environments and the way we parallelized it is introduced. The efficiency is demonstrated in two scenarios.

THE Z^3–METHOD

The only requirement for the Z^3–method (ZZZ is a german abbreviation of *goal–directed and randomized planning in temporally changing environments*) is a geometric and kinematic model of the robot and the environment. The method is a further development of the ZZ-method by B. Glavina [6]. It consists of two hierarchically coupled components. The lower level is an efficient goal–directed planner that only uses local information to try to pass obstacles. The upper level is a randomized planner that uses the local planner and combines the results.

Local Goal–Directed Planning

To avoid exponential complexity with respect to the number of dimensions no complete representation of the search space is constructed. In contrast, only one dimensional subspaces are explored. The goal directed search moves linear from start to goal. The motion is calculated in discrete steps, collision tests are performed in short distances. The stepwidth is calculated from the tolerance that was added to the robot's geometry model and assures collision free continuous motion between two test points [6].

If a collision with an obstacle occures, an avoiding movement (*slide step*) is tried. First a point that is very close to the obstacle's surface is calculated with a depth-limited bisection. Then directions that are orthogonal to the desired direction and orthogonal to each other are computed. These directions and the respective reverse directions are the possible avoiding directions. In the n-dimensional case this results in $2(n-1)$ possible directions. Figure 1 (a) illustrates this calculation.

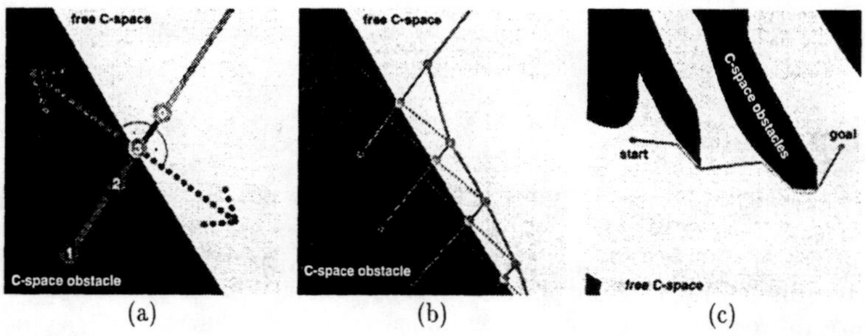

Figure 1. (a) *The last collision free position of the linear movement is* **0**, *the next step would be* **1** *but collides. With a bisection (points* **2** *and* **3***) a point very close to the surface is calculated and orthogonal avoiding directions are computed.* (b) *A sequence of successful slide steps along a c-space obstacle.* (c) *An example for a path planned with the local planner.*

The stepwidth for the slide step is now calculated to not exceed the discretizing stepwidth. Then the avoiding points that lead closer to the goal are checked for collision. If none of the allowed avoiding points is collision free the local planner terminates without success (dead end). The restriction on steps that lead closer to the goal is necessary to avoid 'fluttering' that results in an infinite loop.

After a successful slide step the linear motion in the direction of the goal is tried again, and again it may be necessary to calculate a slide step. Figure 1 (b) shows a sequence of slide steps along an obstacle's surface. In many cases the proposed combination of goal directed steps and slide steps can find a solution for the path planning task by considering local information only, see Figure 1 (c). The complexity of the local planner is linear with respect to the number of degrees of freedom. This number determines the number of avoiding directions that have to be tested. In static environments time is irrelevant and the local planner can retry a failed task in the reverse direction from goal to start.

Global Randomized Planning

First of all, it is tried to solve the task with the local planner. If this fails, random collision free subgoals are used to create partial tasks to solve the whole task through combination. There are two parameters to be chosen for the global planner, first the maximum number M of subgoals that are used to find solutions, and second the maximum number m of subgoals that are allowed on a path from start to goal. Tests show that most of all path planning tasks can be solved with one or very few subgoals along the path. These tests proved that it is more efficient to restart the planner with M new random subgoals instead of allowing any number ($< M$) of subgoals on the path. It is better to check the possible paths with one subgoal first, then the paths with two subgoals and so on. As far as possible unnecessary local planning shall be avoided. The following algorithm follows this conception:

```
check the direct path from start to goal, if its possible then SUCCESS
create M random subgoals as the initial members of set U (set of unconnected points)
initialize the set S (points that can be reached from the start) with the starting point.
Loop from 1 to m
    initialize the set S_new as empty set
    Loop for all points s_i out of S
        Loop for all points u_j out of U
            try to connect s_i with u_j with the local planner
            if this is successful:
                try to connect u_j with the goal with the local planner
                if this is possible:  SUCCESS
                if this is not possible:
                    remove u_j from U
                    add u_j to S_new
        end Loop
        remove s_i from S
    end Loop
    make S equal with S_new
    if S is empty, then NO SOLUTION - RETRY WITH NEW SUBGOALS
end Loop
NO SOLUTION - RETRY WITH NEW SUBGOALS
```

This algorithm creates a tree, growing from the start into the set of subgoals. For every height level of the tree possible connections to the goal are tested before the growth continues. In static environments the global planning can start from the goal as well. This results in a faster reduction of the number of unconnected points and thus in faster planning. The presented algorithm evaluates all paths with up to m subgoals. This testing is complete, if a path exists for the given random subgoal distribution it is found.

Consideration of Time

Time can be included as an additional dimension in the local planning process if the starting time is known and the status of the system can be calculated at any later time, i.e. the dynamics of the environment and the robot are known. Starting the local planner from a known point in *c-space-time* allows to plan a path and an arrival time. For the application in reality the transition from linear steps to slide steps (sudden change of direction) must be considered with care. This can be achieved by correcting the time of some earlier steps to slow down the motion appropriately.

The time can be included in the global planning as well. The subgoals are not fixed in time when they are created. The time is fixed when the subgoal is reached by the local planner. Not being fixed in time, the subgoals are not guaranteed to be collision free, but the random generation can estimate the time heuristically and thus reduce the danger of colliding subgoals. The time is propagated forward in the growing tree, every subgoal is reached only once with the local planner yielding a subgoal arrival time. In the end, a goal arrival time can be returned. It is not possible to plan reverse from goal to start in dynamic environments, neither local nor global.

Parallelized Planning

There are of course several possibilities to parallelize the Z^3-method. The most obvious way is to run several instances of the local planner in parallel. The tasks of the local planer are almost independent of each other, only connected at start and end positions. The inner loop of the algorithm presented above is the key to efficient parallelization of the local planning. All leaves of the tree that was constructed from the start can be tested with all unconnected subgoals simultaneously.

The number of computing nodes is in general neither dynamic nor unlimited, but a fixed configuration parameter, e.g. the number of workstations available in a local cluster. Thus we propose a *scheduling* algorithm to control the available computing nodes. The idea is to assure the highest possible load and to check the most promising connections in the subgoal graph first. The scheduling algorithm fills up the computing nodes respectivly. If a computing node gets available (a previous run of the local planner has terminated), the connection between the pair of positions in the subgoal graph that leads closest to a global solution is tested next. This best pair is calculated based on all the knowledge of the subgoal graph that is available at that time. If, for example, a point that was unconnected before, gets connected to the tree – and thus to the start – the connection to the goal is tested immediately.

PRACTICAL RESULTS

The experiments shown in this chapter were performed on Hewlett Packard Unix Workstations. The geometry simulation is based on an automatically created hierarchy of hull bodies that allows very efficient collision testing. Up to now, our system is only capable to plan paths for robots in static environments. More results of sequential planning are shown in [2].

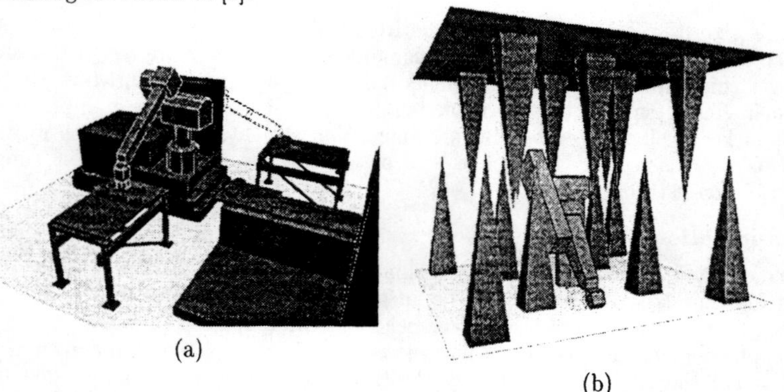

Figure 2. *MOBROB scenario (a) and PYRAMID scenario (b). See text for details.*

Sequential Planning in the Scenario MOBROB

This scenario (Figure 2 a) shows a manipulator on a mobile platform in an industrial environment. For the experiments only the six joints of the manipulator are used (the platform is static). The six dimensional *c-space* contains 57.6% free space. The simulation system needs an average time of 0.4 ms to check a position for collision. The experiments are performed with $M = 25$ total subgoals and a maximum of $m = 4$ subgoals on one path. 5,000 tasks are created by combining random positions that are very close to obstacle surfaces, so these tasks can be described as random pick-and-place-tasks. The following results are measured:

- 100% of the tasks are solved
- the average run time is 0.064 sec, max. is 2.56 sec
- there is an average of 0.042 subgoals per planned path. This shows the efficiency of the local planner, that can solve almost all tasks
- the local planner is used 1.16 times on an average.

The results in this very realistically modelled scenario prove that the Z^3-method is a method of choice for path planning in real applications with hard constraints on time. Even in the rare cases where the global planner becomes necessary the planning times are usually below one second.

Parallel Planning in the Scenario PYRAMIDS

The way of parallelizing the Z^3-method by scheduling the local planning requires little communication, compared with the high computational efforts for the kinematic and geometric simulation. For this reason we chose a cluster of workstations as the development environment. In the shared file system, all computing nodes have access to the same data. The processes are controlled and interconnected through *PVM* (Parallel Virtual Machine [9]).

The pyramid scenario is created as an example where local planning with slide steps fails very often, due to the large number of obstacles very close to the robot. The task shown in Figure 2 (b) (start position: solid drown robot, goal: wire frame robot) requires a complete turn by 350 degrees in the workspace. The six dimensional *c-space* contains 37.5% free space. The simulation system needs an average time of 0.44 ms to check a position for collision. The sequential implementation solves this task in an average time of 31.2 sec (100 runs).

The parallel planner is started 100 times with $M = 50$ total subgoals and $m = 5$ maximum number of subgoals on one path. We use 30 workstations running the local planner in parallel, controlled by one master workstation, running the scheduling algorithm. The average planning time including all communication is 7.5 sec. Compared to the sequential algorithm, this is a speedup by the factor 4.

The results achieved with the parallel planner are preliminary. The scheduling strategy implemented now suffers one major drawback: the direct connection between start and goal is tested first in one computing node, and only if this fails, multiple instances of the local planner start working. It will increase the performance to start searching connections via subgoals immediately when the planning begins, as there are idle computing nodes available.

DISCUSSION AND CONCLUSION

We presented a method that is able to plan paths in dynamic environments. Its complexity is linear in the number of degrees of freedom, time being an additional degree of freedom. Some problems and objections, as well as ongoing and future work, are discussed in this chapter.

The path planned with the Z^3-method is not optimal. The use of subgoals results in sharp corners and detours in the path. For static environments we use a local polygon optimizer that improves the quality of the pathes very efficiently [4].

One problem of the local planner is that the slide steps are very close to the obstacles, yielding 'dangerous paths' in uncertain environments. We try to integrate dynamic *protection shields* to guarantee a safety distance whenever this is possible. This can be done without increasing the planning complexity.

Further investigations are done in developing a new local planer that is especially efficient for hyperredundant manipulators. By *shrinking* and *expanding* the model of the robot along its trajectory its possible to find solutions in much more cases than with the slidesteps, with the same linear computational complexity. First results are very promising [1], and an integration into parallel planning is straightforward.

In static environments the Z^3-method does not use the 'experiences' of prior planning. The 'subgoal–tree' is removed when a solution is found. Other path planning algorithms that represent the free part of the *c-space* with a graph, e.g. [5, 7], show better results for repeated planning in the same environment. In our opinion, the dominating criterium is the constant efficiency in unknown and dynamic environments that is a property of the Z^3-method. The only planner we know with comparable good results is the *R*andommized *P*ath *P*lanner [3], but the 'random walks' that escape local minima will not use the whole free space in a way the random subgoals do.

To summarize, the Z^3-method is a reliable and versatile concept. The global planner can be used in other areas as well and is not bound with robotics. All planning tasks that can not be solved in one step and where the decomposition is not obvious can be adressed with this randomized and parallizable strategy.

REFERENCES

1. Boris Baginski. Local motion planning for manipulators based on shrinking and growing geometry models. In *Proceedings of IEEE Conference on Robotics and Automation*, Minneapolis, April 1996.
2. Boris Baginski. The Z^3–method for fast path planning in dynamic environments. In *Proceedings of IASTED Conference Aplications of Control and Robotics*, pages 47–52, Orlando, Florida, January 1996.
3. J. Barraquand and J.-C Latombe. A monte-carlo algorithm for path planning with many degrees of freedom. In *Proceedings of IEEE Conference on Robotics and Automation*, pages 1712–1717, Cincinnati, Ohio, May 1990.
4. Stefan Berchtold and Bernhard Glavina. A scalable optimizer for automatically generated manipulator motions. In *Proccedings of IEEE/RSJ/GI International Conference on Intelligent Robots and Systems IROS'94*, pages 1796–1802, Munich, September 1994.
5. Martin Eldracher. Neural subgoal generation with subgoal graph: An approach. In *Proceedings of World Conference on Neural Networks WCNN '94*, pages II–142 – II–146, 1994.
6. Bernhard Glavina. *Planung kollisionsfreier Bewegungen für Manipulatoren durch Kombination von zielgerichteter Suche und zufallsgesteuerter Zwischenzielerzeugung.* PhD thesis, Technische Universität München, February 1991.
7. Lydia Kavraki and Jean-Claude Latombe. Randomized preprocessing of configuration space for fast motion planning. In *Proceedings of IEEE Conference on Robotics and Automation*, pages 2138–2145, San Diego, California, May 1994.
8. Jean-Claude Latombe. *Robot Motion Planning.* Kluver Academic Publishers, 1991.
9. V. Sunderam. PVM: A framework for parallel distributed computing. *Concurrency: Practice and Expirience*, 2(4), December 1990.

Acoustic World Modeling

M.Caccia, G.Veruggio
Consiglio Nazionale delle Ricerche - Istituto Automazione Navale
Via De Marini, 6 16149 Genova - ITALY
e-mail: max@ian.ge.cnr.it

R.Cristi
Department of Electrical and Computer Engineering
Naval Postgraduate School
Monterey, CA 93943 - USA
e-mail: cristi@ece.nps.navy.mil

Abstract

This paper presents an EKF-based method to estimate a simplified World Model (i.e. the internal representation of its position and speed and of its 2-D environment model) of an Unmanned Underwater Vehicle (UUV) on the basis of range measurements taken by an high frequency pencil beam profiling sonar. For the purposes of this study, the environment is represented as a set of reflecting surfaces.

Introduction

For autonomous execution of commands given by the human operator, an intelligent underwater vehicle must be able to manage uncertainty in sensor data and environment modeling. To do this, it needs to understand its surroundings and must therefore have an internal representation of its relations with the real world (World Model).

In this paper, the authors consider a World Model that only comprises the robot's kinematic status (linear and angular position and speed) and a geometric representation of the operating environment (i.e. entities and the description of their distribution in space); this representation does not include information about physico-chemical parameters, nor decisions and actions the robot has made.

In particular, this work focuses on fine localization and reconstruction of basic environment entities during UUV missions carried out in partially structured environments. A typical example of such missions could be a survey at constant depth of a harbour area or the depths around an offshore platform, where the environment may be delimited by constructions such as piers, ships at berth, platforms.

Underwater robotics is characterized by a high degree of uncertainty both in "a priori" knowledge of the operating environment (currents, seabed profile, physical-chemical water parameters, etc.) and in sensor data. The on-board inertial and acoustic positioning systems of a small UUV cannot determine vehicle position and speed with high precision [1]. The environment may be sensed by visual or sonar devices, or a combination of these. In certain cases, a vision-based approach might be unattractive due to murky water, poor visibility and the need for lighting. In these conditions, acoustic monitoring methods perform better, but their sampling frequency is lower.

In this research the environment is sensed by a high frequency pencil beam profiling sonar Tritech ST-1000 (see [2] for a brief description of performance). Experimental results indicate that sonar data describe the external environment primarily in terms of reflecting surfaces and their intersections (edges) [3]. This suggests that the environment is described as a set of linear approximations of reflecting surfaces (segments in the case of a 2-D representation of the environment) which can be represented and reconstructed in the way shown in Section 1.

Section 2 describes the World Model state equations, which consist of the linear kinematic model for vehicle position and orientation and the parameters of reflecting surfaces. Furthermore, this section shows how the measurement equation affects vehicle position and the parameters of detected reflecting surface.

The problems of acoustic localization and environment model estimation have been tackled separately by the authors in [4][5] and a potential field-based approach has been presented in [6] whereby the same sonar measurements are used to estimate the vehicle's position and speed and the environment model. Section 3 shows that it is possible to estimate the World Model by means of an EKF-based algorithm. A key role in this process is played by the Referee module, which determines the reflecting surface each sonar echo comes from.

To conclude, Section 4 shows simulation results based on tested models of the Tritech ST-1000 and of IAN's prototype UUV Roby.

Reflecting surfaces representation and reconstruction

The 2-D operating environment can be described as a collection of linear reflecting surfaces (segments). The geometric parameters of a segment are position, orientation, length (coordinates of the extremes) and their uncertainties.

The segment is the best line fitting a set of sonar pings belonging to the same reflecting surface.

Given the sonar range measurement ρ and the estimated position $\hat{\underline{p}}$ with its covariance matrix P, the corresponding sonar ping is $\hat{\underline{x}} = \hat{\underline{p}} + T\,\rho$ where T is a rotation matrix function of vehicle heading and sonar bearing.

The estimated sonar ping $\hat{\underline{x}}$ has a covariance matrix S=P+R where R is the sonar error covariance matrix. Let us associate to each sonar ping a unitary mass distributed according to the gaussian distribution described by S.

Where m sonar pings belong to the same reflecting surface, this can be reconstructed as being the rigid body made up of the mass distributions associated to the pings. This rigid body is characterized by a mass, a center of mass, which specifies its position, and a matrix of inertia I.

Be λ_m and λ_M the eigenvalues of I ($\lambda_m < \lambda_M$) and $\underline{v}_m$ and $\underline{v}_M$ the corresponding eigenvectors (of unitary length), $\underline{v}_m$ indicates the Principal Axis of Inertia and the orientation of the estimated linear reflecting surface is $\alpha = \text{atan2}(\underline{v}_m(2),\underline{v}_m(1))$ (see [7] for the case where the mass distribution covariance matrix S is 0: $\hat{\underline{x}}$ is a rigid point). The resulting segment is the one that best fits the sonar measurements in the sense of the LMS and the minimum eigenvalue λ_m measures the mean square position error in the direction orthogonal to the segment. In particular $\lambda_m = m\,\sigma_m^2$, where m is the segment mass and σ_m^2 is the covariance of the estimated position along the direction orthogonal to the segment. The corresponding diagonal matrix is: $\begin{bmatrix} \sigma_m^2 & 0 \\ 0 & \sigma_M^2 \end{bmatrix}$ and represents the uncertainty on the estimated reflecting surface position.

Let us consider a frame positioned in the segment center of mass with axis x coinciding with the Principal Axis of Inertia and the density function $f_{xy}(x,y) = \frac{1}{2\pi\sigma_m\sigma_M} e^{-\frac{1}{2}\left(\frac{x^2}{\sigma_M^2}+\frac{y^2}{\sigma_m^2}\right)}$

The density function of segment orientation $w = \mathrm{arctg}(y/x)$, $-\frac{\pi}{2} < w < \frac{\pi}{2}$ is [8]:

$$f_w(w) = \frac{\frac{\sigma_m}{\pi\sigma_M}}{\frac{\sigma_m^2}{\sigma_M^2}\cos^2(w) + \sin^2(w)}$$

This shows that the segment inertia matrix also contains information about segment orientation uncertainty. In this way, a linear reflecting surface is basically described by 3 parameters: its center of mass coordinates xg and yg and its orientation α. Mass m is a heuristic measurement of the reconstructed surface reliability [5].

When a candidate reflecting surface is sufficiently reliable, it is added to the World Model; where its reliability decreases under a fixed threshold it is removed from the World Model [9].

UUV World Model and measurement equation

In this paper UUV World Model is intended as the internal representation (estimation) of a vehicle's position and speed and of its operating environment, which is represented as a collection of linear reflecting surfaces as seen in the previous section.

The environment is sensed by a pencil beam profiling sonar mounted on the vehicle: the sonar bearing β and the vehicle orientation ψ in an earth-fixed reference frame are supposed to be measured with high precision.

The World Model state considers:

vehicle position : $\underline{p} = [x, y]^T$ where x and y are vehicle coordinates in an earth-fixed reference frame;

vehicle speed: $\underline{v} = [u, v]^T$ where u and v are vehicle's velocities in a vehicle-fixed reference frame (surge and sway);

n reflecting surfaces: $\underline{s}_i = [\alpha_i, xg_i, yg_i]^T$ where α_i is the surface orientation, and xg_i and yg_i are the coordinates of its center of mass in an earth-fixed reference frame

So the state of the model is: $\underline{x} = \left[\underline{p}^T, \underline{v}^T, \underline{s}_1^T, \ldots, \underline{s}_n^T\right]^T$ with $\underline{x} \in \Re^{4+3n,1}$ where n is the number of reflecting surfaces.

This model is a linear system:

$$\underline{x}(k+1) = A_k\,\underline{x}(k) + G\left(\underline{a}(k) + \underline{w}(k)\right)$$

where $\underline{a}$ is the measured UUV acceleration along the surge and sway axes and $\underline{w}$ is a gaussian noise of covariance matrix $Q_k = \text{diag}\{\text{cov}(w1)\ \text{cov}(w2)\}$,

Matrix $G=\begin{bmatrix} 0 & 0 & 1 & 0 & \underline{0}_3^T & \dots & \underline{0}_3^T \\ 0 & 0 & 0 & 1 & \underline{0}_3^T & \dots & \underline{0}_3^T \end{bmatrix}^T$ and matrix $A_k = \begin{bmatrix} F_k & 0 \\ 0 & I_{3n} \end{bmatrix}$ with

$F_k = \begin{bmatrix} I_2 & \begin{matrix} \cos(\psi_k)dt & -\sin(\psi_k)dt \\ \sin(\psi_k)dt & \cos(\psi_k)dt \end{matrix} \\ 0 & I_2 \end{bmatrix}$ where ψ_k is the measured vehicle orientation at time k

Where the i-th reflecting surface has been pinged by the sonar, the **measurement** equation is:

$$\rho(k) = h_k^i(\underline{x}(k)) + v(k)$$

where ν is the measurement error and is modeled as a gaussian noise with covariance R_k=cov(ν).

The measurement function is: $h_k^i(\underline{x}(k)) = \dfrac{[xg_i(k) - x(k)]\sin(\alpha_i(k)) - [yg_i(k) - y(k)]\cos(\alpha_i(k))}{\sin(\alpha_i(k) - \beta_k)}$

Extended Kalman Filter - based World Model Estimation

Having determined that the sonar echo at time k comes from the i-th reflecting surface, it is possible to define the matrix $H_k^i \equiv \left.\dfrac{\partial h_k^i(\underline{x})}{\partial \underline{x}}\right|_{\underline{x}=\hat{\underline{x}}(k/k-1)}$ and to compute the estimated World Model state $\hat{\underline{x}}(k/k)$ based on an EKF technique. So only the estimated vehicle position and speed and the parameters of the hit surface are corrected on the basis of the innovation, while the estimated parameters of the other surfaces do not change (they are equal to the predicted ones).

The position and speed estimates are only corrected in the direction orthogonal to the detected reflecting surface and only the distance between this surface and the vehicle (and not their respective positions) is observable [2].

In order to apply this method it is necessary to decide which reflecting surface has been detected by the sonar at time k. This is the task of the Referee module.

At time k the measured sonar range $\rho(k)$ and the estimated state $\hat{\underline{x}}(k/k-1)$ are known. So the Referee computes the error δ_l in the case the l-th reflecting surface in the map has been hit: $\delta_l(k) = \rho(k) - h_k^l(\hat{\underline{x}}(k/k-1))$

Its expected square value is $E\{\delta_l^2(k)\} = H_k^l P(k/k-1) H_k^{l\,T} + R_k$ where P is the state covariance matrix.

We define $\mathcal{S}$ as the set of reflecting surfaces (both reliable and candidate) which could have been hit at time k:

$$\mathcal{S} = \{\underline{s}_j : \delta_j^2(k) \le E\{\delta_j^2(k)\}\}$$

The estimated hit reflecting surface is:

$\underline{s}_h$ with $h = \arg\min_{j \in \mathcal{S}} \{\delta_j^2(k)\}$

In the case of $\mathcal{S} = \varnothing$, the sonar measurement cannot be associated to any reflecting surface in the map: $\hat{\underline{x}}(k/k)$ is equal to the prediction $\hat{\underline{x}}(k/k-1)$ and the observer tries to build a new segment.

The resulting algorithm is:

if < $\underline{s}_h$ = NULL > then
    $\hat{\underline{x}}(k/k) = \hat{\underline{x}}(k/k-1)$
    < try to build a new candidate segment >
else if < $\underline{s}_h$ is a candidate reflecting surface > then
    $\hat{\underline{x}}(k/k) = \hat{\underline{x}}(k/k-1)$
    < add the sonar plot at time k to $\underline{s}_h$ >
else // < $\underline{s}_h$ is a reliable reflecting surface >
    < apply EKF to compute $\hat{\underline{x}}(k/k)$ >

Simulative Results

To validate the algorithm we simulated a vehicle moving in a 20 x 20 meters tank at constant depth. The UUV hydrodynamic model and equations of motion draw on the model presented by Fossen in [10]; the hydrodynamic parameters are those computed for CNR-IAN's prototype Roby. Figure 1 shows the real vehicle trajectory and map.

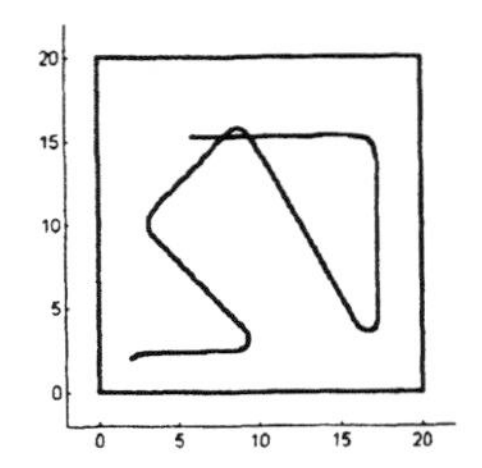

Figure 1

Figure 2 shows a partially known environment model (the information about surface orientations is poor). The EKF is initialized with the vehicle supposed to be still and an initial position error of 1 m in both directions x and y.

In the case the system has no acceleration measurements, Figure 3 shows the estimated trajectory and environment model. It is clear that the estimated position is only corrected in the direction orthogonal to the reflecting surface detected by the sonar. Figure 4 shows how the estimate of the trajectory and environment model improves when the measurements of linear acceleration with an error lower than 1 cm/s^2 are considered.

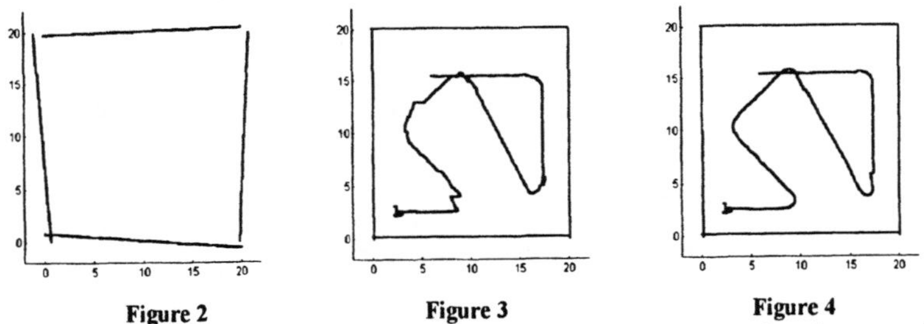

Figure 2 **Figure 3** **Figure 4**

In the case of partially known environment (see Figure 5, where a surface is not a-priori known), the estimated trajectory is less precise, but the algorithm is able to reconstruct the unknown surface and to correct the estimate of the others. The reconstructed surface is not used to update the vehicle position. Figure 6 and 7 show the results in the cases respectively of no acceleration measurements and of measured linear acceleration.

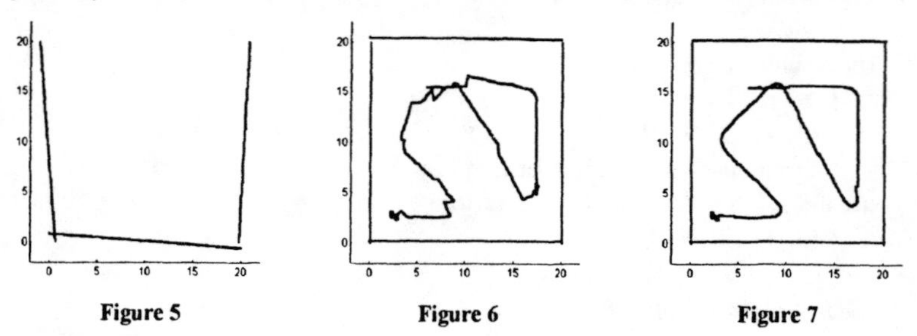

Figure 5 Figure 6 Figure 7

Conclusions

The problems of localizing a vehicle within known, partially known or unknown environments have been addressed. They are all formulated on the basis of the same state space model, which includes the orientation of the reflecting surfaces. Several problems remain to be solved, mainly the integration of a linear approximation of the environment with a more general occupancy-cells model. Also, we currently extending our study to 3-D contexts.

Acknowledgments

Prof. Roberto Cristi gratefully acknowledges the interest of Dr. Teresa McMullen of the Office of Naval Research for providing sponsorship for his research work.

References

[1] M.D.Ageev, B.A.Kasatkin, A.Scherbatyuk (1995). Positioning of an Autonomous Underwater Vehicle. Proceedings of the International Program Development in Undersea Robotics & Intelligent Control, pp.15-17. Lisboa, Portugal.

[2] R.Cristi, M.Caccia, G.Veruggio (1996). Motion Estimation and Modeling of the Environment For Underwater Vehicles. 6th IARP Workshop on Underwater Robotics. Toulon, France

[3] B.A.Moran, J.J.Leonard, C.Chryssostomidis (1993). Geometric Shape from Sonar Ranging. Proceedings of the 8th International Symposium on Unmanned Untethered Submersible Technology, pp. 370-383. Durham, NH.

[4] R.Cristi (1994). Navigation and Localization in a Partially Known Environment. Proceedings of the 1994 Symposium on Autonomous Underwater Vehicle Technology, pp.263-267. Cambridge, MA

[5] R.Bono, M.Caccia, E.Spirandelli, G.Veruggio (1994). 2-D Map and Object Detection from Multiple Sonar Scans. Proceedings of the Autonomous Vehicles in Mine Countermeasures Symposium, pp. 6.77-6.81. Monterey, CA.

[6] R.Cristi, M.Caccia, G. Veruggio, A.J.Healey (1995). A Sonar Based Approach to AUV Localization. Proceedings of the 3rd IFAC Workshop on Control Applications in Marine Systems. Trondheim, Norway.

[7] D.H.Ballard, C.M.Brown (1982). Computer Vision. Prentice Hall, U.S.A.

[8] A.Papoulis (1965). Probability, Random Variables, and Stochastic Processes. McGraw-Hill. New York, U.S.A.

[9] M.Caccia, R.Bono, G.Veruggio (1995). 2-D Acoustic World reconstruction. Proceedings of the 9th International Symposium on Unmanned Untethered Submersible Technology, pp. 299-305. Durham, NH.

[10] T.I.Fossen (1994). Guidance and Control of Oceans Vehicles. John Wiley & Sons, UK

IMPEDANCE MATCHING CONTROL OF A COMPLIANT MECHANICAL ONE DEGREE OF FREEDOM IN A CONSTRAINED ENVIRONMENT TASK

O. CHAMPOUSSIN, J.P. SIMON, G.A. OMBEDE, A. JUTARD

Laboratoire d'Automatique Industrielle bât 303
Institut National des Sciences Appliquées de Lyon
20 Avenue Albert Einstein / F-69621 Villeurbanne Cedex / France
Tel (33) 72 43 81 98 / Fax (33) 72 43 85 35 / eMail champou@lai1.insa-lyon.fr

ABSTRACT

An impedance model of a mechanical one degree of freedoom in dynamic interaction with the environment is proposed.This model is based on the scattering wave approach, in which the power is considered in terms of exchange between the robot and the environment. A matched impedance control approach is presented through power waves which are linear combination of force(torque) and motion generally choosen as independant variables. The power reflection coefficient explained in the frequency domain, and a matching condition for the controller is deduced, and implemented on a one DOF link mechanically in interaction with the environment. This simple control algorithm does not use a force sensor, and assume the coupled link environment stability relatively to the incertaincy of the model parameters.

KEYWORDS: impedance, matching, power, compliance, robot.

INTRODUCTION

The manufacturing tasks and assembly operation require mechanical interactions with the environment or with the object being manipuled. These contact tasks are caracterized by a dynamic interaction between the robot and the environment which often cannot be predicted accurately. The magnitude of the mechanical work exchanged, during the contact, may vary drastically and cause a significant alteration of performance of the robotic control system.

There exist any tasks, common in a variety of industrial manufacturing process such as robot cutting, surface grinding, polishing and deburring, co-operation between assembly or carrying robot, during which the interaction forces must be accommoded rather than resisted.

A variety of different approaches of manipulator comportment, during compliant motion, have been proposed : Raiberg and Graig [1], Hogan [2], Anderson and Spong [3], Kazerouni [4], Liu and Goldenberg [5].

According to Hogan, the robot behaves as a mass-spring-dashpot system whose parameters (Inertia-Stiffness-Damping) can be specified arbitrarily. Hogan also suggests to use mass-spring-dashpot as the target impedance model, knowing that typically the contact force and the resulting motion are generally related by a second order differential equation.

When the contact tasks are modelled as a mechanical impedance, e.g. inertia, damping and stiffness terms, compliant motion is given to this form of task, derived from the fact that the work environment deformes during contact. In these cases, the cartesian position error and applied forces of the end effector are considered, but the model of this manipulator interaction with the environment does not take account of the notion of power in terms of exchange between the robot and the environment via the robot mechanism during the task [6], [7], [8].

The objective of matched impedance control is to optimize the exchange power between the robot and environment in spite of interaction force which varies due to incertainty on the location of the point of contact and environment's properties.

Basically two problems occur in compliant motion. The first problem is how to make qualitative or quantitative specifications on contact task's requirement. The second problem is to find an appropriate control strategy for robot relatively to these tasks.

When studying a physical system, the following question is indispensable: what is the appropriate formalism which well describes the system?. In fact, significant properties can be masked by a formalism not well adapted. The approach consists in choosing physical independant variables which are generally measurable to characterize the system. In robotic control strategies design, the force(torque) and/or position(motion) are generally choosen as the best independant variables to use for the analysis. However, one can equally choose any linear transformation of these variables as long as the transformation is not singular. For this reason, as our main interest is the power relations in a system, we define mechanical powers waves in (1) noted "a" and "b" which are the result of just one of an infinite number of linear transformations.

$$a = \frac{C_m + Z_c\Omega_e}{2\sqrt{Re(Z_c)}} \quad b = \frac{C_m - Z_c^*\Omega_e}{2\sqrt{Re(Z_c)}} \tag{1}$$

Cm is the actuator torque, Zc the controller actuator impedance, Ω_e the load motion, Re(Zc) the real part of Zc, * is the conjugate in the Laplace transform.

A similar power waves where first introduced in electrical circuit by Pentfield [9] for the discussion of noise performance in amplifiers. Kurokawa [10] late used these waves to measure the noise in linear amplifiers. Our objectives in this paper is first to modelize any mechanical control structure as a Norton(Thevenin) circuit. Second, we discussed the physical meaning of the mechanical power waves defined in (1) as well as the scattering matrix based on this concept for a one-DOF mechanical structure. Third, we implement in this one DOF mechanical structure a simple compliant control algorithm based on the matching between impedance mechanical structure and environment impedance.

MODELIZATION OF A 1-DOF LINK CONTROL IN A NORTON EQUIVALENT CIRCUIT

We are interested, given the closed loop system control configuration proposed by Goldenberg [11] (fig 3), to implement a matched target impedance of a robot coupled with the environment. The methodology is based on replacing the robot-environment model by an analogous equivalent electrical circuit to analyse its comportment in a frequency domain, in the way of the wave scattering approach [10], [12].

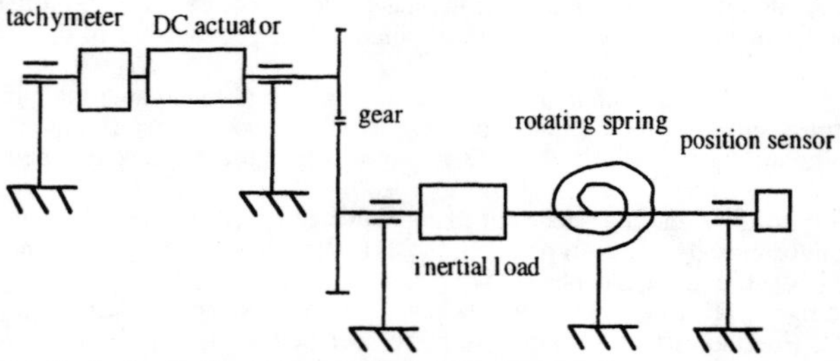

fig 1 : basic diagram of the link drived by a DC motor with gear.

Our link consist of a DC actuator, a gear between the actuator and the environment constitued by an inertial load and a rotational spring, and two sensors for the position and velocity of the link-environment coupling. The DC actuator is driven by a voltage amplifier, and the self effect of the rotor is neglected in the low frequency field of the control signal.

The block diagram of the servo system studied is (fig 2). In this diagram, K_θ is the position transducer, C(s) the controller, A the voltage power amplifier gain, K_c K_e are the torque and electric constants of the actuator, R the resistance, n the gear coefficient, J_{md} is the actuator and tachymeter inertia, J_{en} and K_{en} are the inertia and spring environment.The desired and real environment position are θ_c and θ_e, Ω_m the actuator velocity, C_m, C_r and C_e are actuator, perturbation, and environment torques.

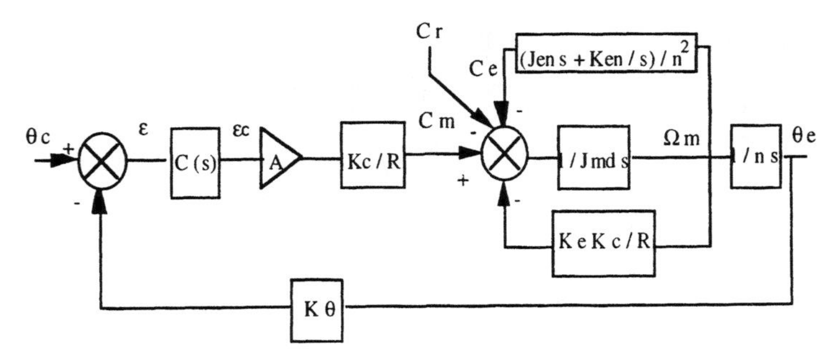

fig.2: block diagram control of the link

In order to study the power repartition in this system, let's establish an equivalent block diagram where the input is the desired motion $\Omega_{in} = \frac{1}{K_\theta} \frac{d\theta_c}{dt}$. We have (fig 3) :

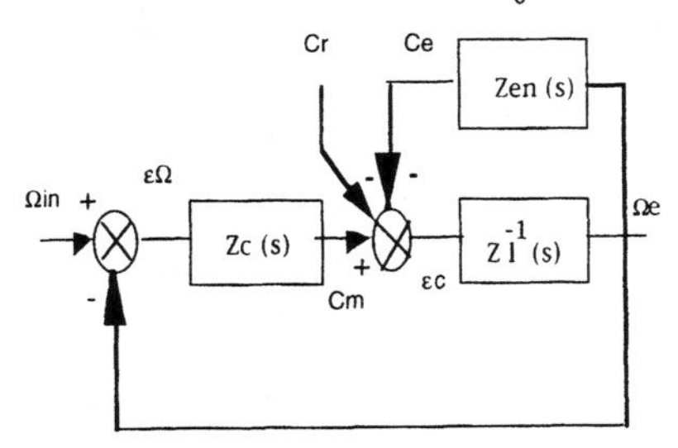

fig.3: control block diagram of the link with desired motion input.

with : $Z_c(s) = \frac{K_\theta K_c A}{n R} \frac{C(s)}{s}$ $\quad Z_l(s) = J_{md} s + \frac{K_e K_c}{R}$

$$Z_{en}(s) = \frac{1}{n^2}\left(J_{en} s + \frac{K_{en}}{s}\right)$$

Using the analogy between electrical circuit, the basic equivalent diagram control is equivalent to the following Norton circuit(fig.4).

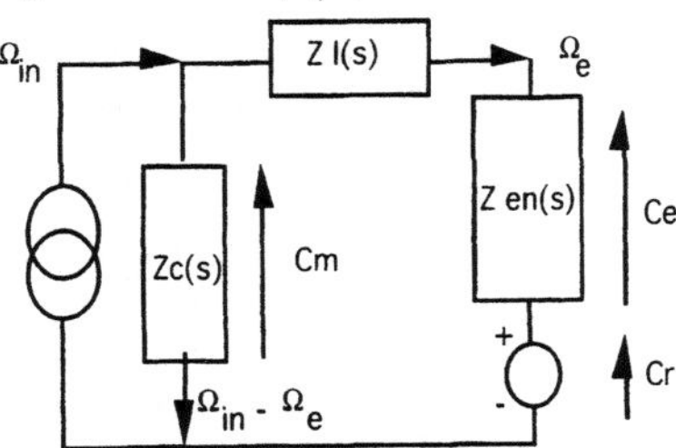

Fig.4 : Norton equivalent circuit of the diagram control.

With a Norton equivalent diagram, we deduce the following Thevenin equivalent circuit (fig 5).The approach of this section shows that any control mechanical system can be modelized as a Norton(Thevenin) equivalent circuit with a flow source(effort source) and an internal impedance Zc. These models are interesting when the circuit is analyzed in terms of relation between effort and flow, or equivalently in terms of power.

PHYSICAL MEANING OF THE POWER REPARTITION

Now we can discuss about the mechanical power waves defined in (1). For this purpose, let's consider the following Thevenin equivalent circuit of linear generator in which the link impedance is $Z_{lc} = Z_l + Z_c$.

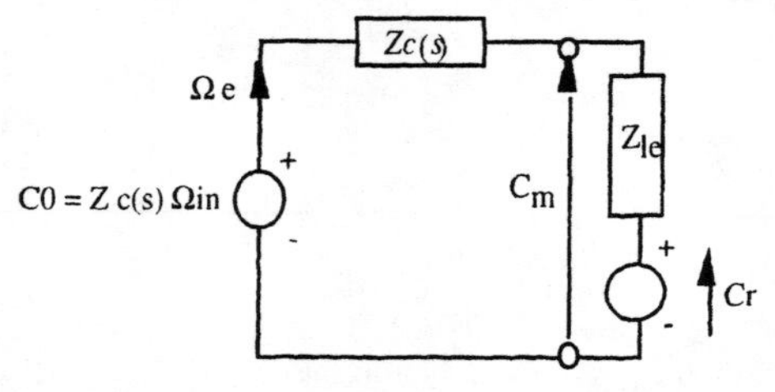

Fig.5 : Thevenin equivalent circuit of the diagram control.

The controller-actuator impedance Z_c is considered as the internal impedance of the source. Then for no external perturbation torque, C_o is the actuator torque for an environment load impedance infinite, i.e the maximum torque that the actuator can supply with $\Omega_e = 0$.

When the link is in contact with the environment, we write the following equation :

$$C_0 = Z_c\Omega_e + C_m$$

Suppose the environment impedance Z_{en} infinite, then the motion Ω_e becomes nul, and considering the magnitude of the input torque, we obtain:

$$|C_0| = |Z_c||\Omega_{in}| = C_{m\,max}$$

This is the maximal torque that the actuator can supply without moving. In fact the controller actuator structure is composed by an voltage amplifier, and then physically there is a limit current i_{sat} which imposes a limit torque $C_{m\,sat} = K_c\, i_{sat}$ generally less than $C_{m\,max}$:

$$|C_0| = |Z_c||\Omega_{in}| \le C_{m\ sat} \le C_{m\ max}$$

The mechanical power into the load Z_{lc} is given by:

$$P_m = Re(Z_{le})\ |\Omega_e|^2$$

where Ω_e is the motion of the load.With : $|\Omega_e| = \left|\dfrac{C_0}{(Z_{le} + Z_c)}\right|$

we obtain :

$$P_m = Re(Z_c)\left|\frac{C_o}{(Z_{le} + Z_c)}\right|^2$$

Let's now consider the mechanical power waves defined in (1). The incident mechanical power wave is "a":

$$a = \frac{C_m + Z_c\Omega_e}{2\sqrt{Re(Z_c)}} = \frac{C_0}{2\sqrt{Re(Z_c)}}$$

The square of the magnitude is: $$a * a = \frac{|C_o|^2}{4\,Re(Z_c)} \qquad (2)$$

The reflected mechanical power wave "b" is:

$$b = \frac{C_m - Z_c^*\Omega_e}{2\sqrt{Re(Z_c)}} = \frac{(Z_{le}\ -\ Z_c^*)\ \Omega_e}{2\sqrt{Re(Z_c)}}$$

For a matched mechanical system we say that all the incident power a^*a is consumed in the actuator during the link-environment motion. The reflected power b^*b is then nul, and we deduce the matching condition : $$Z_c = Z_{le}^* \qquad (3)$$

At the matching the corresponding maximum mechanical power into the load is:

$$P_{m\ max} = \frac{|C_0|^2}{4\,Re(Z_c)}$$

As C_o is the maximum torque of the actuator, $P_{m\ max}$ represents the available power of the actuator. The exchangeable power between the actuator and the load is the maximum power that the actuator power can supply : $P_{ex}=P_{m\ max}$. We note that: $P_{ex}= a^*a$.

Now, let's consider the difference between incident and reflected power. With the "a" and "b" power wave, it is easy to see that : (4)

$$a^*a - b^*b = \frac{(C_m + Z_c\Omega_e)^*(C_m + Z_c\Omega_e)}{4\ Re(Z_c)} - \frac{(C_m - Z_c{}^*\Omega_e)^*(C_m - Z_c{}^*\Omega_e)}{4\ Re(Z_c)} = Re(C_m\Omega_e{}^*)$$

The right hand side of (4) is the mechanical power which is actually transfered from the actuator to the load Z_{le}. This actual mechanical power is the power consumed to this load. We interpret equations (2) and (4) as follow.

The actuator is sending the incident mechanical power a*a towards a load regardless of the load impedance. If the load impedance is matched, i.e (3) is satisfied, the net power consumed to the load for the motion is the exchangeable mechanical power a*a. When the load impedance is not matched, the net power consumed into the load is equal to a*a - b*b. A part of the incident mechanical power a*a is reflected back to the source. This reflected mechanical power is b*b. Associated with these incident mechanical power and reflected mechanical power, there are incident power waves "a" and reflected power waves " b " respectively.

The reason why we do not consider the incident and reflected mechanical power directly for the analyse lies on the fact that there is not linear relation between powers but there is a linear relation between powers waves "a" and "b".

POWER REFLECTION COEFFICIENTS

In a mechanical system, when we consider two independant variables such as force(torque), and motion, the ratio is an impedance. Similarly, since we have two quantities "a" and "b", let's define the ratio $\rho = b/a$ and call it the power wave reflection coefficient. Using the relation $C_m = Z_{le}\Omega_e$, we have:

$$\rho = \frac{Z_{le} - Z_c^*}{Z_{le} + Z_c}$$

The power reflection coefficient is:

$$|\rho|^2 = \left|\frac{Z_{le} - Z_c^*}{Z_{le} + Z_c}\right|^2$$

As P_{ex} represents the incident mechanical power and P_{abs} the net power consumed into the load, we have the following relation:

$$P_{abs} = \left(1 - |\rho|^2\right)P_{ex}$$

Then the reflected mechanical power P_r is:

$$P_r = |\rho|^2\,P_{ex}$$

When the matching conditions are satisfied the power reflection coeficient become zero.

CONTROLLER PARAMETERS

The objective is to obtain the average net power consumed into the load (coupled link-environment) equal to the average incident power of the source.

$$P_{ex} = |a|^2 = P_{abs}$$

We have explained in (3) the matching conditions which optimize the power transfert from the source to the environment :

$$Z_c = (Z_l + Z_e)^* = Z_{le}^*$$

with $Z_{le}(s) = \left(\frac{J_{en}}{n^2} + J_{md}\right)s + \frac{K_e K_c}{R} + \frac{K_{en}}{n^2 s} = J_{le}s + B_{le} + \frac{K_{le}}{s}$

In truth the operateur $Z_c(s)$ must have the same Laplace structure as $Z_{le}(s)$:

$$Z_c(s) = \frac{K_\theta K_c A}{nR}\ \frac{C(s)}{s} = A_c\left(M_c \cdot s + B_c + \frac{K_c}{s}\right)$$

which gives :

$$C(s) = \left(M_c \cdot s^2 + B_c \cdot s + K_c\right)$$

with

$$A_c = \frac{K_\theta K_c A}{nR}$$

Practically to conjuge $Z_{le}(s)$ we use the fact that in sinusoïdal control signal, the Laplace conjugate operator may be : p* = -p, then :

$$A_c B_c = B_{le} \quad \text{and} \quad (A_c \cdot M_c + J_{le}) \cdot \omega = \frac{K_{le} + A_c K_c}{\omega}$$

The second relation gives whatever ω :

$$A_c M_c = -J_{le} \quad \text{and} \quad A_c K_c = -K_{le}$$

This means that the controler may work with a periodic but not necessary sinusoïdal signals to achieve the impedance matching control of the mechanical one DOF.

APPLICATION

This compliant control is numerically implemented on the one DOF servo link described on fig(1), in which : $K_e = K_c = 3.1\ 10^{-2}$ Nm/A, n = 20, R = 16 Ω, $K_\theta = 3.23$ rd/s, A = 1. The environment spring $K_{en} = 0.17$ Nm/rd is measured, and the link environment inertia is estimated at $J_{le} = 21\ 10^{-7}$ kgm^2. The viscous friction of the global mechanical structure is estimated less than $(K_e K_c)/R$, and the dry friction is not considered. Practically for the adjustement, a gain $\alpha = 0.5$ is added on the controller in all the manipulations.The torque actuator is obtained from the current actuator.In every manipulations the maximal actuator current is small comparatively to the saturation current i_{sat}= 1 A.

The graph 1-a-b-c show for a sinusoïdal input θ_c at 0.5 Hz, the response of the link θ_e, the actuator torque, and the average actuator power window between t and t+T :

$$P_w(t) = \frac{1}{T} \int_t^{t+T} P_m \cdot dt$$

This average actuator power grows up to its maximal value quite in the first period T of the input signal. It means that all the available power of the source is directly consumed for the motion in the actuator. This available power sets up in a minimum time, the first one period.

In the same experimental condition, rigid obstacle on the link trajectory prevents to follow the specified trajectory (graph 2 a-b-c).In the constrained part of the trajectory the actuator torque is 50% greater than in the non constrained part, but without oscillation on the obstacle. This means that the coupled link-environment consumes only the necessary power to accomplish the task even when the task is perturbed, and then the dynamic reaction of the environment does not affect the servo link which remains stable in all the trajectory.

Graph 3 a-b, illustrate another experiment with an input step of 142° (static gain is about 0.35), and a rigid obstacle on the trajectory. The constrained part of the actuator torque is 30% greater than the no constrained part, but in the torque non constrained transition zone the torque is more unstable than in the constrained in which it is stabilized by the contact with the obstacle.The stability of the coupled link-environment is assumed by the compliant control.

For a 0° constant input position, graph 4 a-b show the link reaction to an external perturbation torque. The stability is alway assumed with a large constrained environment position ($\theta_e$max = 25°) and a small variation of actuator torque.

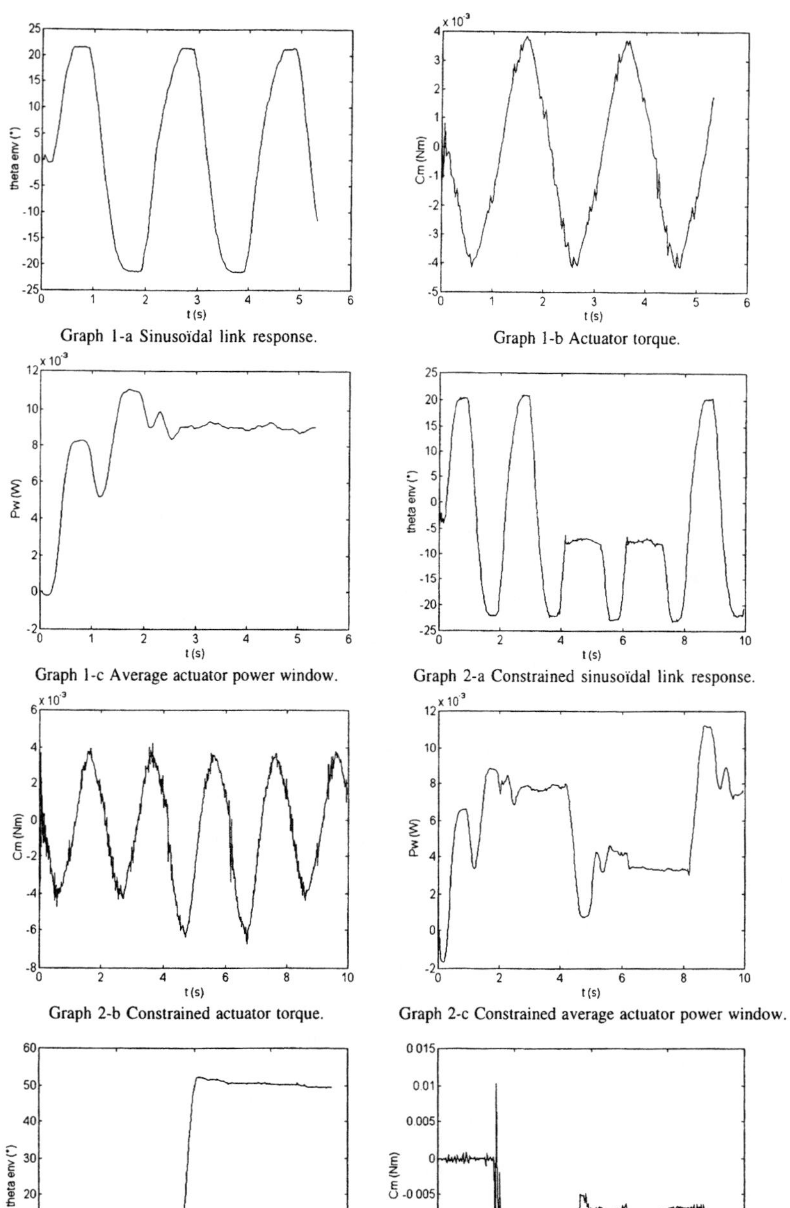

Graph 1-a Sinusoïdal link response.

Graph 1-b Actuator torque.

Graph 1-c Average actuator power window.

Graph 2-a Constrained sinusoïdal link response.

Graph 2-b Constrained actuator torque.

Graph 2-c Constrained average actuator power window.

Graph 3-a Constrained step link response.

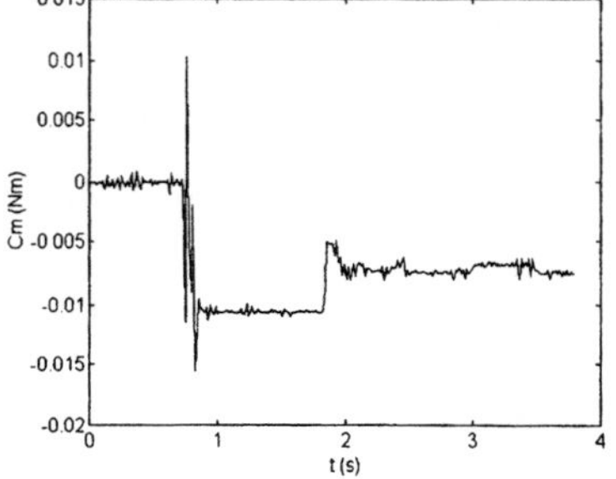

Graph 3-b Constrained actuator torque.

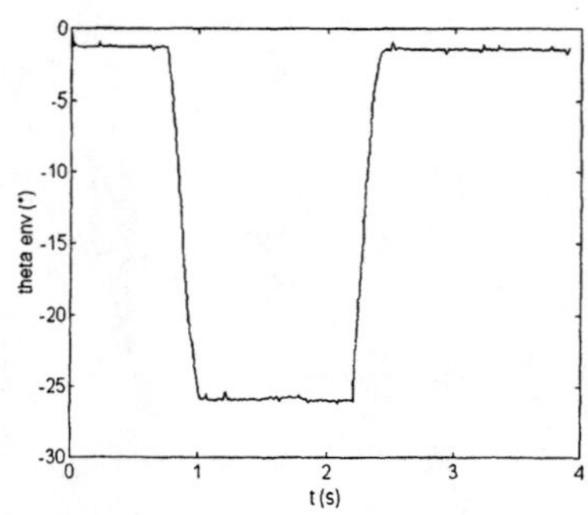

Graph 4-a Link response under external perturbation.

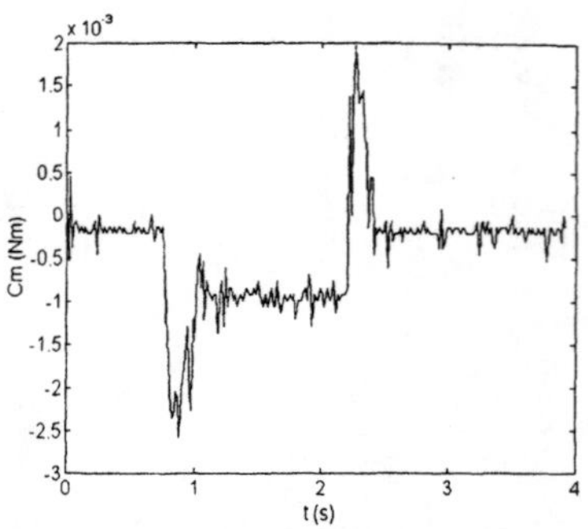

Graph 4-b Actuator torque under external perturbation.

CONCLUSION

The concept of power wave and the physical meaning have been discussed and developed for a one DOF link coupled to an inertial and stiff environment. It results when the servo link and the environment are mechanically matched by the control law that the coupling requires and consumes only the necessary power for the execution of the constrained or not mechanical task.Then the stability of the mechanical dynamic coupling is sassumed whatever the constrains applied on the link. The implementation of the control law request the knowledge of the link-environment parameters, but the incertainty does not affect excessively the compliancy of the control.However these parameters must be estimated in real time for the efficiency of the control, particulary in case of variable environment parameters. This will be the topic of some future works.

REFERENCES

[1] Raibert M.H., and Craig J.J. "Hybrid Position / Force control of manipulators", ASME Journal of Dynamic Systems, Measurement, and Control, Vol 102, June 1981, pp. 418-432.

[2] Hogan N. "Impedance Control : An Approach to Manipulation", ASME Journal of Dynamic System, Measurement, and Control, VOL 107, 1985, pp 1-24.

[3] Anderson R.J., Spong M.W. "Hybrid Impedance Control of Robotic Manipulators", IEEE Conference on Robotics and Automation, 1987, pp 1073-1080.

[4] Kazerooni H., Transaction of the ASME, Journal of Dynamic Systems Measurement and Control, september 1989, Vol 111, PP 416-425.

[5] Liu G.J., Goldenberg A.A., "Robust Hybrid Impedance Control of Robot Manipulator", IEEE Int. Conf. on Robotics and Automation, 1991, pp 287-292.

[6] Simon J.P., Betemps M., Jutard A., "Application of the Wave Scatter Theory to the Impedance Model of a Robot", INRIA-SIAM Int, Conf, on Mathematical and Numerical Aspect of Propagation Phenomena, Strasbourg April 23-26, 1991.

[7] Simon J.P., Betemps M., Jutard A., "Matching impedance model of a constrained robot-environment task using scattering S matrix ", IFACS-IMACS-IEEE International Workshop on Motion Control for Intelligent Automation, Vol 2, pp 31-37, Perugia Italy, 27-29 October 1992.

[8] Ombede G.A, Simon J.P, Betemps M, Jutard A, "Optimal Efficiency of a Robot Environment Interaction Task in a Matching Approach", Ro Man Sy' 94, Tenth CISM-IFToMM Symposium on Theory and Practice of Robots and Maniplators, Gdansk, Poland, 12-15 September 1994.

[9] Penfield P. "Noise in resistance amplifiers" IEEE Trans. on microwaves theory and techniques. vol. CT-7, june 1960, pp 166-170.

[10] Kurokawa K., " Power Waves and Scattering Matrix", IEEE on Transaction microwaves Theory and Techniques. March 1965, pp 194-203.

[11] Goldenberg A.A., "Implementation of Force and Impedance Control in Robot Manipulators", IEEE Int, Conf, on Robotic and Automation, Philadelphia, 1988.

[12] Rivier E., Sardos R. "La matrice S", Edition Masson 1982.

LASER-BASED SEGMENTATION AND LOCALIZATION FOR A MOBILE ROBOT *

José A. Castellanos Juan D. Tardós

Dpto. de Informática e Ingeniería de Sistemas
Centro Politécnico Superior, Universidad de Zaragoza
María de Luna 3, 50015 Zaragoza, SPAIN
Email: joseac@prometeo.cps.unizar.es

ABSTRACT

In this paper we present a technique to segment the data obtained by a laser range finder mounted on a mobile robot navigating in an structured indoor environment. Using this segmentation and an apriori map of the environment we find the localization of the mobile robot by application of a constraint-based matching scheme. A probabilistic method is used to represent the uncertainty and partiallity of the geometric information involved.

KEYWORDS: Laser-Based Segmentation, Probabilistic Model, Constraint-Based Localization, Feature-Based Method.

INTRODUCTION

An important problem to be solved in mobile robotics is that of developping an efficient method to localize the robot without modifing its environment. A related problem is that of mapmaking, providing the robot with knowledge about its surroundings. These problems have given rise to a great variety of solutions using different types of external sensors mounted on a mobile robot. We are interested in feature-based methods, in which a set of features (e.g. line segments) are extracted from the sensed data and then matched against the corresponding features in the model. Leonard et al. [8] give a wide overview on the use of sonar range finders to build the map of the environment of the mobile robot and to localize it inside such an environment. Drumheller, [5] proposes a matching method by which range data from a sonar rangefinder could be used to determine the location of a mobile robot inside a

*This work was partially supported by CICYT, project TAP94-0390 and by a research grant FPI PN94-29099535.

room. Crowley [4] also proposes a feature-based method for a 2D environment, using range from a rotating sonar. Finally, Horn et al. [7] propose a localization system for a free navigating mobile robot based on a matching between vertical planar surfaces extracted from a laser rangefinder readings with predicted surfaces from a model of the environment.

In this work, we propose a segment-based method combined with a probabilistic method to build the local map of the environment of a mobile. This information is subsequently used to localize the robot.

UNCERTAIN GEOMETRIC INFORMATION

One of the fundamental aspects to suceed in reasoning with uncertain geometric features is choosing an appropiate representation for the location of geometric entities of diverse nature.

Representation of Uncertain Geometric Features

In this work, we use the 2D version of the Symmetries and Perturbations Model (SP-model) [11], which combines the use of probability theory to represent the imprecision in the location of a geometric element, and the theory of symmetries to represent the partiallity due to characteristics of each type of geometric element. A reference E is associated to every geometric element $\mathcal{E}$. Its location is given by a *location vector* $\mathbf{x}_{WE} = (x,\ y,\ \phi)^T$, respect to a base reference, W, composed of two Cartesian coordinates and an angle. The estimation of the location of an element is denoted by $\hat{\mathbf{x}}_{WE}$, and the estimation error is represented locally by a *differential location vector* $\mathbf{d}_E$ relative to the reference attached to the element. Thus, the true location of the element is:

$$\mathbf{x}_{WE} = \hat{\mathbf{x}}_{WE} \oplus \mathbf{d}_E$$

where $\oplus$ represents the composition of location vectors (the inversion is represented with $\ominus$).

To account for the symmetries of the geometric element, we assign in $\mathbf{d}_E$ a null value to the degrees of freedom corresponding to them, because they do not represent an effective location error. We call *perturbation vector* the vector $\mathbf{p}_E$ formed by the non null elements of $\mathbf{d}_E$. Both vectors can be related by a row selection matrix B_E that we call *self-binding matrix* of the geometric element:

$$\mathbf{d}_E = B_E^T \mathbf{p}_E \quad ; \quad \mathbf{p}_E = B_E \mathbf{d}_E$$

Based on these ideas, the SPmodel represents the information about the location of a geometric element $\mathcal{E}$ by a quadruple $\mathbf{L}_{WE} = (\hat{\mathbf{x}}_{WE}, \hat{\mathbf{p}}_E, C_E, B_E)$, where:

$$\mathbf{x}_{WE} = \hat{\mathbf{x}}_{WE} \oplus B_E^T \mathbf{p}_E \ ; \ \hat{\mathbf{p}}_E = E[\mathbf{p}_E] \ ; \ C_E = Cov(\mathbf{p}_E)$$

Transformation $\hat{\mathbf{x}}_{WE}$ is an estimation taken as base for perturbations, $\hat{\mathbf{p}}_E$ is the estimated value of the perturbation vector, and C_E its covariance. When $\hat{\mathbf{p}}_E = 0$, we say that the estimation is *centered*.

Uncertain Location of a Laser Point. A laser point is characterized by a pair (ρ_k, ϕ_k) of polar coordinates with respect to the sensor, L. An uncertain location $\mathbf{L}_{LP_k} = (\hat{\mathbf{x}}_{LP_k}, \hat{\mathbf{p}}_{P_k}, C_{P_k}, B_{P_k})$, is assigned to each of these points, where:

$$\hat{\mathbf{x}}_{LP_k} = (\rho_k \, cos\phi_k, \rho_k \, sin\phi_k, \phi_k)^T$$

$$\hat{\mathbf{p}}_{P_k} = (\hat{d}_x, \, \hat{d}_y) \; ; \; C_{P_k} = \begin{pmatrix} \sigma^2_{xP_k} & 0 \\ 0 & \sigma^2_{yP_k} \end{pmatrix} \; ; \; B_{P_k} = \begin{bmatrix} 1 & 0 & 0 \\ 0 & 1 & 0 \end{bmatrix}$$

with $\sigma^2_{xP_k} = \sigma^2_\rho$ and $\sigma^2_{yP_k} = \rho^2_k \sigma^2_\phi$, where σ_ρ represents the sensor range error and σ_ϕ represents the positioning error of the laser beam.

Uncertain Location of a Laser Segment. A laser segment is represented by an uncertain location $\mathbf{L}_{LE} = (\hat{\mathbf{x}}_{LE}, \hat{\mathbf{p}}_E, C_E, B_E)$, with respect to the sensor, L, and its observed length: $\mathbf{l}_E = \{\hat{l}_E, \sigma^2_{l_E}\}$ where:

$$\hat{\mathbf{p}}_E = (\hat{d}_y, \, \hat{d}_\phi) \; ; \; C_E = \begin{pmatrix} \sigma^2_{yE} & \sigma_{y\phi E} \\ \sigma_{y\phi E} & \sigma^2_{\phi E} \end{pmatrix} \; ; \; B_E = \begin{bmatrix} 0 & 1 & 0 \\ 0 & 0 & 1 \end{bmatrix}$$

Integration of Uncertain Geometric Information

The estimation of the location of an object or feature from a set of geometric observations is a nonlinear problem, due to the existence of orientation terms, that can be solved using the extended Kalman filter or the extended information filter [2]. We use the *information filter* formulation because it simplifies the analysis of the influence of each observation on the estimation of the object location. Let $\mathbf{x}$ be the state vector whose value is to be estimated, and let there be n observations $\mathbf{y}_k$ of $\mathbf{x}$, where $k \in \{1, \ldots, n\}$, affected by white Gaussian noise:

$$\hat{\mathbf{y}}_k = \mathbf{y}_k + \mathbf{u}_k \; ; \; \mathbf{u}_k \sim N(0, S_k)$$

Let each observation $\mathbf{y}_k$ be related to $\mathbf{x}$ by an implicit non-linear function of the form $\mathbf{f}_k(\mathbf{x}, \mathbf{y}_k) = 0$. Since $\mathbf{f}_k$ is nonlinear, we use a first order approximation:

$$\mathbf{f}_k(\mathbf{x}, \mathbf{y}_k) \simeq \mathbf{h}_k + H_k(\mathbf{x} - \hat{\mathbf{x}}) + G_k(\mathbf{y}_k - \hat{\mathbf{y}}_k)$$

where:

$$\mathbf{h}_k = \mathbf{f}_k(\hat{\mathbf{x}}, \hat{\mathbf{y}}_k) \; ; \; H_k = \left.\frac{\partial \mathbf{f}_k}{\partial \mathbf{x}}\right|_{(\hat{\mathbf{x}}, \hat{\mathbf{y}}_k)} \; ; \; G_k = \left.\frac{\partial \mathbf{f}_k}{\partial \mathbf{y}}\right|_{(\hat{\mathbf{x}}, \hat{\mathbf{y}}_k)}$$

The estimation $\hat{\mathbf{x}}_n$ of the state vector and its covariance P_n after integrating the n measurements are:

$$\hat{\mathbf{x}}_n = P_n M_n \; ; \; P_n^{-1} = \sum_{k=1}^{n} F_k \; ; \; M_n = -\sum_{k=1}^{n} N_k$$

where:

$$F_k = H_k^T (G_k S_k G_k^T)^{-1} H_k \; ; \; N_k = H_k^T (G_k S_k G_k^T)^{-1} h_k$$

Matrix P_n^{-1}, the inverse of the covariance matrix is denominated *information matrix of the estimation*, while matrix F_k is denominated *information matrix of the observation.*

LOCAL MAP BUILDING

Considering that a mobile robot is placed in some location $\mathbf{x}_{WR} = (x, y, \phi)^T$ we call *local map* the static segment-based environment of the robot, that can be build using an external sensor, such as a laser rangefinder.

Segmentation of Laser Data

Segmentation of laser data is carried out in two steps: first we find regions formed by a unique polygonal line and then we extract each of the segments of such a line.

Finding Homogeneous Regions. The segmentation of laser data may be simplified if we first find those regions which are build up by a unique polygonal line. Each of these regions will be called an *homogeneous region*. In order to get the homogeneous regions we focus on the parameter ρ_k of each laser point and on a distance criterium. Due to the geometry of the problem (figure 1 left) a method which considers a variable threshold is required.

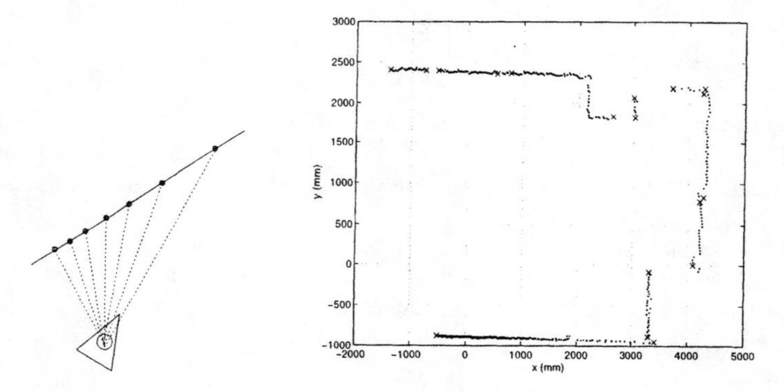

Figure 1: Distance between laser points (left). Homogeneous regions endpoints (X-shaped marks) for a sample laser scan (right).

We state the problem of finding homogeneous regions as a *tracking* problem [2] where the target is the endpoint of the laser beam. We propose a Kalman filter approximation combined with a statistical test based on the Mahalanobis distance [2] to decide whether point r_{k+1} belongs to the same or different homonegeous region that point r_k. We state the Kalman filter in the following way:

$$\mathbf{x}_{k+1} = \begin{bmatrix} 1 & 1 \\ 0 & 1 \end{bmatrix} \mathbf{x}_k + \mathbf{w}_k \quad ; \quad \mathbf{z}_k = \begin{bmatrix} 1 & 0 \end{bmatrix} \mathbf{x}_k + \mathbf{v}_k$$

where $\mathbf{x}_k = (\rho_k,\ \dot{\rho}_k)^T$. We assume white, gaussian and independent noise, with zero mean and known covariance. The measurement noise, $\mathbf{v}_k$, is related to the sensor range noise, and the state noise, $\mathbf{w}_k$, is related to the maximum accuracy obtainable in differenciating gaps in the laser data. In figure 1 (right) we represent the endpoints of the obtained homogeneous regions for a given sample scan.

Estimating Endpoints of Segments. A given homogeneous region consists of a set of connected segments. We apply an iterative line fitting algorithm, widely used in computer vision [1], and other laser data segmentation approaches [6]. The procedure might be sketched as follows:

1. Estimate the infinite edge, $\mathbf{L}_{LE}$, defined by the endpoints, $\mathbf{L}_{LP_1}$ and $\mathbf{L}_{LP_2}$, of the initial region.

2. For all the intermediate points, $\mathbf{L}_{LP_k}$, calculate the Mahalanobis distance between the point, P_k, and the edge, E, by the equation:

$$D_k^2 = \frac{\hat{y}_{EP_k}}{\sigma^2_{y_{EP_k}}} = \frac{\hat{y}_{EP_k}}{\sigma^2_{y_E} + 2\hat{x}_{EP_k}\sigma_{y\phi_E} + \hat{x}^2_{EP_k}\sigma^2_{\phi_E} + \sigma^2_{x_{P_k}} sin^2\hat{\phi}_{EP_k} + \sigma^2_{y_{P_k}} cos^2\hat{\phi}_{EP_k}}$$

   where $B_{EP_k} = [0 \ 1 \ 0]$ is the binding matrix of the pairing, $\hat{\mathbf{x}}_{EP_k} = (\hat{x}_{EP_k}, \ \hat{y}_{EP_k}, \ \hat{\phi}_{EP_k})^T$ is the relative location vector between the point and the edge, and $\sigma^2_{y_{EP_k}}$ is the covariance [1] of $\hat{y}_{EP_k}$.

3. If $Max(D_k^2) \leq D^2_{m,\alpha}$, where $D^2_{m,\alpha}$ is a threshold value, obtained from the χ^2_m distribution, such that the probability of rejecting a good matching is α and $m = rank(B_{EP_k})$, then the point may be considered compatible with the edge and the process is stopped; otherwise we have a breakpoint. Then we recurse the process considering the two new regions.

Integrating Uncertain Points into Edges

Fitting an edge (i.e. a straight line) to each of the set of points comprised between each pair of endpoints is achieved by application of the information filter. Let $\mathbf{L}_{LE}$ and $\mathbf{L}_{LP_k}$ represent the estimated location of the edge, E, and of a point, P_k, respectively. Consider the perturbation vector of the edge, $\mathbf{p}_E$, as the state to be estimated, and the perturbation vector of each observed point, $\mathbf{p}_{P_k}$, as the measurements. The implicit measurement function is given by the fact that the location of the observed point and the location of the edge must coincide, up to the symmetries of the pairing. Then, we have:

$$\begin{aligned}
\mathbf{f}_k(\mathbf{x}, \mathbf{y}_k) &= B_{EP_k}\mathbf{x}_{EP_k} \\
&= B_{EP_k}(\ominus B_E^T \mathbf{p}_E \oplus \hat{\mathbf{x}}_{EP_k} \oplus B_{P_k}^T \mathbf{p}_{P_k}) \\
&= B_{EP_k}(\ominus B_E^T \mathbf{x} \oplus \hat{\mathbf{x}}_{EP_k} \oplus B_{P_k}^T \mathbf{y}_k) \\
&= 0
\end{aligned}$$

Solving, with $\hat{\mathbf{x}}_{EP_k} = (\hat{x}_{EP_k}, \ \hat{y}_{EP_k}, \ \hat{\phi}_{EP_k})^T$, we have:

$$\begin{aligned}
\mathbf{h}_k &= B_{EP_k}\hat{\mathbf{x}}_{EP_k} &&= \hat{y}_{EP_k} \\
H_k &= -B_{EP_k} J_{1\oplus}\{0, \hat{\mathbf{x}}_{EP_k}\} B_E^T &&= -[\ 1 \quad \hat{x}_{EP_k}\ ] \\
G_k &= B_{EP_k} J_{2\oplus}\{\hat{\mathbf{x}}_{EP_k}, 0\} B_{P_k}^T &&= [\ sin\hat{\phi}_{EP_k} \quad cos\hat{\phi}_{EP_k}\ ]
\end{aligned}$$

[1] The covariance of a nonlinear function $y = f(x)$ is calculated as $Cov(y) = JCov(x)J^T$, where J is the Jacobian of the function.

Fusionating Similar Oriented Edges

Consider two consecutive edges, $\mathbf{L}_{LE_1}$ and $\mathbf{L}_{LE_2}$. The Mahalanobis distance between them is given by:

$$D^2 = (B_E \hat{\mathbf{x}}_{E_1E_2})^T (B_E Cov(\mathbf{x}_{E_1E_2}) B_E^T)^{-1} (B_E \hat{\mathbf{x}}_{E_1E_2})$$

where, B_E is the binding matrix of the pairing, which, in this case, has the same value as the self-binding matrix of an edges, $\hat{\mathbf{x}}_{E_1E_2}$ is the estimated relative location vector between the edges, and $Cov(\mathbf{x}_{E_1E_2})$ is the covariance matrix of this location vector. Then, if $D^2 \leq D^2_{m,\alpha}$, with $m = rank(B_E)$, we consider that the edges are similar enough to be fusionated. The new edge is estimated by application of the information filter, being the perturbation vector of the new edge, $\mathbf{p}_E$, the state to be estimated, and the perturbation vector of each fusionated edge, $\mathbf{p}_{E_k}$, the measurements. Then, we have:

$$\begin{aligned} \mathbf{f}_k(\mathbf{x}, \mathbf{y}_k) &= B_E \mathbf{x}_{EE_k} \\ &= B_E(\ominus B_E^T \mathbf{p}_E \oplus \hat{\mathbf{x}}_{EE_k} \oplus B_E^T \mathbf{p}_{E_k}) \\ &= B_E(\ominus B_E^T \mathbf{x} \oplus \hat{\mathbf{x}}_{EE_k} \oplus B_E^T \mathbf{y}_k) \\ &= 0 \end{aligned}$$

Solving this expression, with $\hat{\mathbf{x}}_{EE_k} = (\hat{x}_{EE_k}, \hat{y}_{EE_k}, \hat{\phi}_{EE_k})^T$, we have:

$$\begin{aligned} \mathbf{h}_k &= B_E \hat{\mathbf{x}}_{EE_k} &&= \begin{bmatrix} \hat{y}_{EE_k} \\ \hat{\phi}_{EE_k} \end{bmatrix} \\ H_k &= -B_E J_{1\oplus}\{0, \hat{\mathbf{x}}_{EE_k}\} B_E^T &&= -\begin{bmatrix} 1 & \hat{x}_{EE_k} \\ 0 & 1 \end{bmatrix} \\ G_k &= B_E J_{2\oplus}\{\hat{\mathbf{x}}_{EE_k}, 0\} B_E^T &&= \begin{bmatrix} cos\hat{\phi}_{EE_k} & 0 \\ 0 & 1 \end{bmatrix} \end{aligned}$$

Estimating Lengths of Segments

The length of a segment is a function of the relative location of its endpoints.

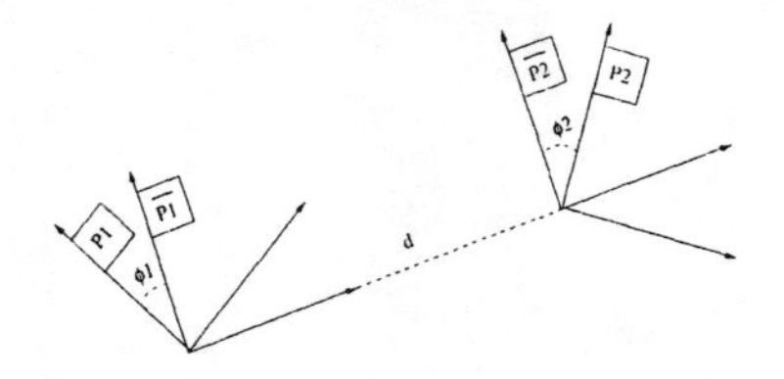

Figure 2: Aligned references of two 2D points.

Considering two 2D points whose location is represented by references P_1 and P_2 respectively (figure 2), we observe that their relative location is defined by one parameter: their distance d which is a non-linear function of $\hat{\mathbf{x}}_{P_1P_2} = (\hat{x}_{P_1P_2}, \hat{y}_{P_1P_2}, \hat{\phi}_{P_1P_2})^T$.

In [10] Neira proposes a method consisting in finding two *aligned references* $\bar{P}_1$ and $\bar{P}_2$ which equivalently describe the location of the points, such that d is a linear function of $\mathbf{x}_{\bar{P}_1\bar{P}_2}$. The aligning transformations, $\mathbf{x}_{P_1\bar{P}_1} = (0,\ 0,\ \phi_1)^T$ and $\mathbf{x}_{P_2\bar{P}_2} = (0,\ 0,\ \phi_2)^T$, belong to the symmetries of a point. The estimated relative location between $\bar{P}_1$ and $\bar{P}_2$ is given by:

$$\hat{\mathbf{x}}_{\bar{P}_1\bar{P}_2} = \ominus(\hat{\mathbf{x}}_{LP_1} \oplus \mathbf{x}_{P_1\bar{P}_1}) \oplus (\hat{\mathbf{x}}_{LP_2} \oplus \mathbf{x}_{P_2\bar{P}_2})$$

Thus the estimated length of the segment and its observed covariance are given by:

$$\begin{aligned}
\hat{l}_E &= d &&= \sqrt{\hat{x}^2_{\bar{P}_1\bar{P}_2} + \hat{y}^2_{\bar{P}_1\bar{P}_2}} \\
\sigma^2_{l_E} &= \sigma^2_d &&= \sigma^2_{xP_1} cos^2\hat{\phi}_{P_1\bar{P}_1} + \sigma^2_{yP_1} sin^2\hat{\phi}_{P_1\bar{P}_1} \\
& &&= \sigma^2_{xP_2} cos^2\hat{\phi}_{P_2\bar{P}_2} + \sigma^2_{yP_2} sin^2\hat{\phi}_{P_2\bar{P}_2}
\end{aligned}$$

In figure 3 we show the set of segments obtained by the previous algorithm for a sample case.

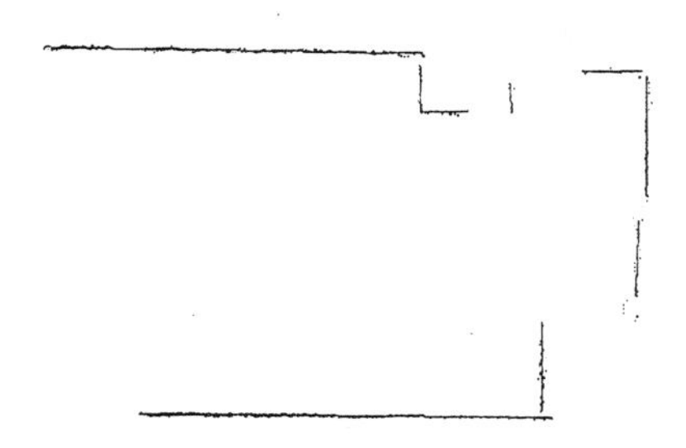

Figure 3: Result of the segmentation algorithm considering a laser scan with a 240 degrees field of view.

Mobile Robot Self-Localization

Determining the location of a robot is an important problem in navigating an autonomous vehicle in a structured indoor environment. In a two dimensional space, the location of a mobile robot can be represented by a location vector $\mathbf{x}_{WR} = (x, y, \phi)^T$. We calculate the mobile robot localization by the method used in [3] where localization is stated as a matching between observed segments, obtained by a laser rangefinder, and model segments, stored in a database, representing the structure of the environment. Reduction of the matching problem complexity is achieved by application of geometric constraints. Estimation of the robot localization is carried out using an estimation method, the information filter, that finds a transformation such that the error between each transformed model feature and its corresponding observed feature is minimal in some sense [9]. Figure 4 summarizes the resultant localization of the mobile robot. With the obtained localization we also present the standard deviations of each parameter, which indicate the precision of the obtained location vector.

CONCLUSIONS

We have proposed a method to process the information obtained by a laser sensor, in order to build a local map of the environment of the mobile robot, obtaining a

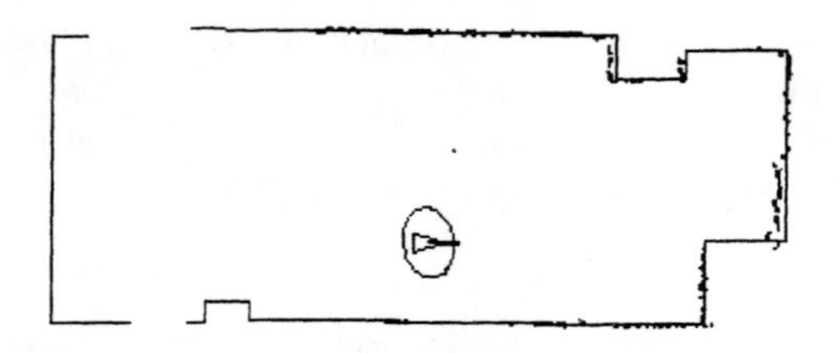

Figure 4: Laser points superimposed to the map of the room in which the robot is located. Obtained location is $\mathbf{x}_{WR} = (x, y, \phi)^T = (2353.025\ mm, 4582.061\ mm, 1.570554\ rad)^T$, with $\sigma_x = 4.093\ mm$, $\sigma_y = 4.955\ mm$ and $\sigma_\phi = 2.262\ mrad$.

set of segments, characterized by an uncertain location and an estimated length. This information have been used to localize the robot in a structured environment. We have solved the localization problem as a matching problem by application of a constraint-based matching scheme. Experimental results show the accuracy obtained by application of the previous scheme to the localization problem. Future work will consider higher level features of the map, such as corners, to obtain a more robust location of the robot. Integration with a vision subsystem is also one of our future goals.

References

[1] D.H. Ballard and C.M. Brown. *Computer Vision*. Prentice Hall, Englewood Cliffs, N.J., 1982.

[2] T. Bar-Shalom and T.E. Fortmann. *Tracking and Data Association*. Academic Press In., 1988.

[3] J.A. Castellanos, J.D. Tardós, and J. Neira. Constraint-based mobile robot localization. In *Proc. of the 1996 IEEE Int. Workshop on Advanced Robotics and Intelligent Machines*, Salford, Great Britain, 1996. To Appear.

[4] J.L. Crowley. Navigation for an intelligent mobile robot. *IEEE Journal of Robotics and Automation*, 1(1), 1985.

[5] M. Drumheller. Mobile robot localization using sonar. *IEEE Trans. on Pattern Analysis and Machine Intelligence*, 9(2):325–332, 1987.

[6] J. Gonzalez, A. Stentz, and A. Ollero. An iconic position estimator for a 2d laser rangefinder. In *Proc. 1992 Int. Conf. Robotics and Automation*, pages 2646–2651, Nice, France, 1992.

[7] J. Horn and G. Schmidt. Continuous localization of a mobile robot based on 3d-laser-range-data, predicted sensor images and dead-reckoning. *Journal of Robotics and Autonomous Systems, Special Issue "Research on Autonomous Mobile Robots"*, 14:99–118, 1995.

[8] J.J. Leonard and H.F. Durrant-Whyte. *Directed Sonar Sensing for Mobile Robot Navigation*. Kluwer Academic Publishers, London, 1992.

[9] J. Neira, J. Horn, J.D. Tardós, and G. Schmidt. Multisensor mobile robot localization. In *Proc. 1996 IEEE Int. Conf. on Robotics and Automation*, Minneapolis, USA, 1996. To Appear.

[10] J. Neira, L. Montano, and J.D. Tardós. Constraint-based object recognition in multisensor systems. In *Proc. 1993 IEEE Int. Conf. on Robotics and Automation*, pages 135–142, Atlanta, Georgia, 1993.

[11] J.D. Tardós. Representing partial and uncertain sensorial information using the theory of symmetries. In *Proc. 1992 IEEE Int. Conf. on Robotics and Automation*, pages 1799–1804, Nice, France, 1992.

ROBOT SHAPING: THE HAMSTER EXPERIMENT

Marco Colombetti, Giuseppe Borghi*
PM-AI&R Project
Dipartimento di Elettronica e Informazione Politecnico di Milano, Italy
{colombet, borghi}@elet.polimi.it
http://www.elet.polimi.it/people/{colombet, borghi}

Marco Dorigo
IRIDIA
Université Libre de Bruxelles, Belgium
mdorigo@ulb.ac.be
http://iridia.ulb.ac.be/dorigo/dorigo.html

ABSTRACT

In this paper we present an example of the application of a technique, which we call *robot shaping*, to designing and building learning autonomous robots. Our autonomous robot (called HAMSTER[1]) is a multi-sensor mobile robot that performs the task of collecting "food" and bringing it to its "nest". Its control architecture is based on the behavioral paradigm. The behavioral modules are implemented as classifier systems and are learned by a reinforcement learning technique exploiting the Bucket Brigade and an extended version of the Genetic Algorithm. The chief features of HAMSTER are that it combines innate (i.e., prewired) and learned behaviors, and that training was carried out in a simulated environment and then transferred to the real robot.

KEYWORDS: Learning; classifier systems; autonomous robots

1. INTRODUCTION

Our work is concerned with designing and building learning autonomous robots. Programming an autonomous robot so that it reliably acts in a dynamic environment is a difficult thing to do. This is due to such problems as the lack of necessary information at design time, the unpredictability of environmental events, and the inherent noise of the robot's sensors and actuators. For example, a robot may have to move in a cluttered environment with an *a priori* unknown topology; moreover, people may be moving around, and the robot may be equipped with noisy sensors, like sonars and dead-reckoning, to sense its surroundings. It is clear that a *learning* autonomous robot, that is, an autonomous robot that can acquire knowledge by interaction with the environment and subsequently adapt and/or change its behavior in the course of its life, could greatly simplify the work of its designer. A learning robot need not be given all the details of the environment in which it is going to act: it will acquire them by direct interaction and exploration. Also, its sensors and actuators need not be finely tuned: they will adapt to the specific task requirements and environmental conditions.

---

* Currently Ph.D. student at the Università di Padova, Italy.

1 HAMSTER : Highly Autonomous System TrainEd by Reinforcement.

Our work finds its roots in both behavior-based robotics and in reinforcement learning [1]. The control system of our robots comprises, according to behavior-based robotics, functional modules which implement behaviors (like avoiding obstacles, searching for objects, and the like). Such behavioral modules are learned by a reinforcement learning technique which we call *robot shaping* [2]. By this term we denote the use of reinforcement learning as a means to translate suggestions coming from an external *trainer* [3] into an effective control strategy that allows a robot to achieve a goal. The important point, which differentiates our approach from most of the current research on learning autonomous agents, is that the trainer plays a fundamental role in the robot learning process.

In a robot shaping setup (see Fig. 1) the main roles are played by: (i) the autonomous agent (robot); (ii) the trainer; and (iii) the environment. We shall exemplify our approach by means of the HAMSTER example.

The HAMSTER robot (Fig. 2) is composed of a *shell* (i.e., its "anatomy"), a set of *sensors* and *actuators*, and a learning *control system*. The learning control system communicates with the sensors and actuators through a sensorimotor interface using binary messages. At each control cycle, sensor messages are sent by the sensors (both real and virtual – see below) to the learning control system, which in turn sends action messages to the actuators. Action messages are translated into real actions performed in the environment, which are externally viewed as the robot's behavior. The trainer compares the behavior of the robot with a predefined *target behavior*, and then sends a *reinforcement* message to the learning system. Using such reinforcement, the strength of the associations between sensor messages and action messages are modified, thus changing HAMSTER's behavior in accordance with the target behavior. This feedback cycle is executed every time HAMSTER performs an action.

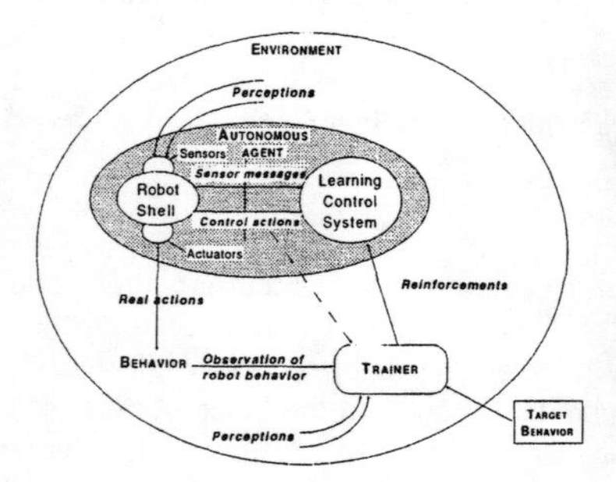

Figure 1. *Robot shaping setup.*

Figure 2. *A picture of HAMSTER.*

HAMSTER's target behavior is to bring "food" to its "nest". The chief features of HAMSTER are that it combines "innate" (i.e., prewired) and learned behaviors, and that training was carried out in a simulated environment and then transferred to the real robot. The innate behavior is a pre-programmed obstacle avoidance function, the output of which is composed with the output of the learned behaviors as explained in Section 2. To keep the trainer as simple as possible and to reduce the learning time, we chose to train the agent in a simplified simulated environment with no obstacles. The learned behavioral modules were then transferred to the real HAMSTER for experimentation.

The remainder of the paper is organized as follows. Section 2 presents HAMSTER's environment, virtual sensors and behaviors. Section 3 describes some experiments. Finally, in Section 4 some concluding remarks are presented.

2. HAMSTER AND ITS ENVIRONMENT

HAMSTER moves in an office-like environment lightened by both artificial and natural lights. The terrain is smooth, and people can interfere with the robot's movements. HAMSTER's environment, a room of size 14 × 13.3 m, includes various obstacles: glass and brick walls, doors, and columns (see Fig. 3). Moreover, the environment includes food pieces (i.e., violet cylinders with a diameter of 30 cm and a height of 70 cm), which slide on the floor when pushed by HAMSTER, and a nest (the area close to a red cylinder with a diameter of 30 cm and a height of 130 cm).

HAMSTER's shell is Robuter, a commercial platform produced by RoboSoft (see Fig. 4). It is 102 cm long, 68 cm wide and 44 cm high. The configuration we used has a belt of 24 Polaroid sonars, surrounding the whole platform, and a color vision camera on top of the robot shell, which covers a horizontal visual field of 64 degrees. HAMSTER is also endowed with a dead-reckoning system (i.e., a sensor that estimates the robot's position and heading), which approximately identifies the position of the robot with respect to a fixed reference frame. Motion is produced by two motors acting on two independent wheels. To allow HAMSTER to retain food while moving, we mounted two rigid metal bars, acting as a pair of arms, on its front. Data coming from the physical sensors are used to compute the output of a set of virtual sensors, as presented below.

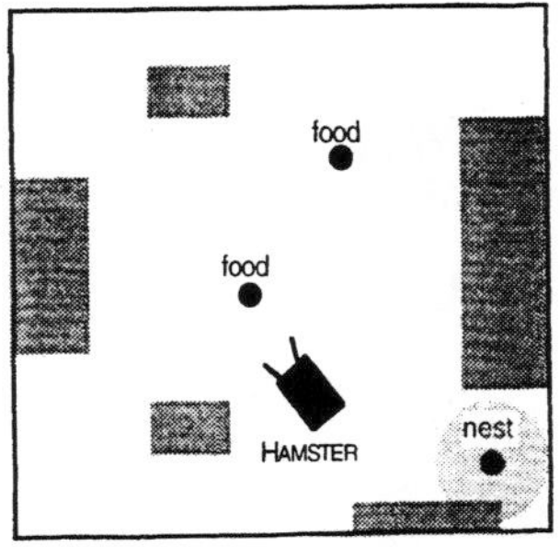

Figure 3. *Map of HAMSTER's environment.*

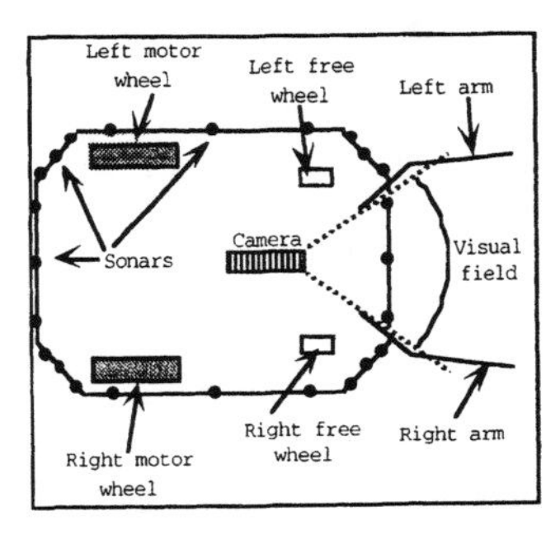

Figure 4. *A sketch of HAMSTER.*

2.1 Virtual sensors

The virtual Food-Sensor detects the presence of a food piece in front of HAMSTER and supplies information about its angular position relative to the robot heading. A horizontal strip in the lower part of the color camera frame is vertically divided into five sectors, and in each of them the percentage of violet pixels is computed. A sector is said to contain a food piece if such percentage is greater than a fixed threshold. A five-bit message is then produced associating a bit to each sector (the bit is on if the corresponding sector contains a piece of food). Notice that if there is a close-by food piece in front of HAMSTER it is possible that more than one bit be on. If at least one of the Food-Sensor bits is on, then another virtual sensor, called Presence-Of-Food-Sensor, is set to on.

The virtual Nest-Sensor supplies information about the angular position of the nest relative to the robot heading. It generates a four-bit message encoding the nest's angular position (the interval [0, 360] degrees is divided into sixteen 22.5 degree sectors). If the nest is visible by the color camera then the angular position is computed in a similar way as described above for the Food-Sensor. Otherwise the Nest-Sensor computes the four-bit message using dead-reckoning data. The same data are used to set the In-Nest-Sensor when HAMSTER reaches the nest. The (X,Y) values of the dead-reckoning system is set to zero every time HAMSTER is in the nest. The angular value ϑ is also set arbitrarily to zero because its initial value is irrelevant to the aim of the Nest-Sensor.

The sonar sensors are used to measure the distance among HAMSTER and obstacles or food pieces. The Bump-Sensor is a two-bit virtual sensor. Its output depends on the minimum of the values returned by the 3 frontal sonar sensors. Bump-Sensor bits are 00 if this value is greater than 1 m, 11 if the minimum value is less than 0.3 m, 01 and 10 for the intermediary cases. The Proximity-Sensor is a one-bit sensor which is on if at least one Bump-Sensor bit is on, that is, if an object or a food piece are closer than 1 m.

2.2 Behavior architecture

HAMSTER's goal is to learn the HoardFood behavior: to collect food (violet cylinder) and to bring it into the nest (near the red cylinder). The architecture of HAMSTER's behaviors is depicted in Figure 5. The HoardFood behavior is obtained by coordinating the execution of four basic behaviors: GetFood, ReachNest, LeaveNest and AvoidObstacles. Each basic behavior proposes a motion vector for HAMSTER. The proposals of GetFood, ReachNest and LeaveNest are mutually exclusive, and one of them is chosen by the coordinator module on the basis of the current situation. For example, ReachNest's proposal is chosen if (Presence-Of-Food AND Bump AND NOT In-Nest) is true. AvoidObstacles adds the chosen motion vector to its own. The direction of the motion vector is used as target for the low level controller, while the robot's speed is directly chosen by AvoidObstacles on the basis of the minimum value returned by the sonars. The basic behaviors are shortly described in the following.

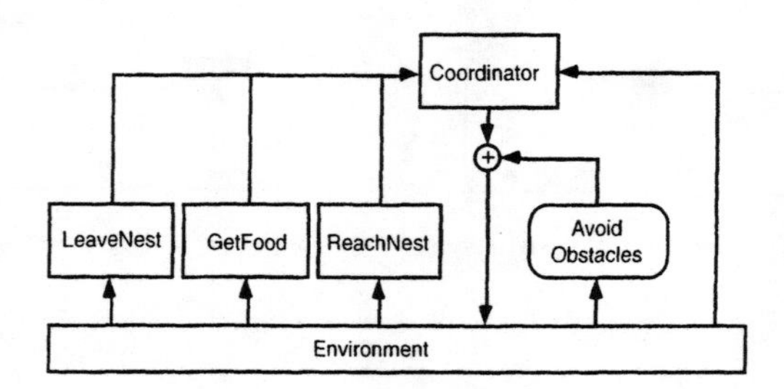

Figure 5. *HAMSTER's controller architecture .*

AvoidObstacles. It is an innate behavior, that is, a programmed behavior implementing an obstacle avoidance algorithm based on the artificial potential approach (see for example [4]). This behavior normally uses all the 24 sonar sensors to compute a feasible moving direction for HAMSTER. If Presence-Of-Food-Sensor and Proximity-Sensor are on (i.e., there is food in front of HAMSTER) then the three frontal sensors are inhibited, allowing HAMSTER to approach the food. Such inhibition process is learned and controlled by the GetFood and ReachNest behaviors.

LeaveNest. Active when the robot is in the nest: the robot leaves the nest, in such a way that any previously captured food piece is left in the nest.

GetFood. Active when the robot is out of the nest and no food is captured: search and get a piece of food.

ReachNest. Active when the robot is out of the nest and a piece of food is captured: reach the nest, pushing the piece of food.

HAMSTER learns by means of ALECSYS ([5]; [6]; [7]), a learning classifier system (LCS) which exploits two learning algorithms: the *Bucket Brigade* [8] and the *Genetic Algorithm* [9]. An LCS is a parallel production system in which rules are randomly initialized and then modified by the two learning algorithms. These learning algorithms exploit reinforcement information which is provided, on a step-by-step basis, by the trainer. Basically, the trainer compares the observed behavior of the learning robot with a predefined target behavior and provides a positive or a negative reinforcement accordingly.

3. EXPERIMENTS

Using ALECSYS we developed a controller in a simulated obstacle-free environment, and then transferred the controller onto the real robot to run some experiments in the real environment. The reason for doing so was that we wanted to keep the reinforcement function as simple as possible. For example, the reinforcement for approaching food was simply proportional to the decrease in the distance between the robot and the food piece; in an environment with obstacles, a reasonable reinforcement function should have taken into account also the effect of obstacle avoidance on the robot's behavior. As we did not have a suitable obstacle-free real environment, training had to be performed in a simulated environment. The coordinator and all the behaviors are learned, except *AvoidObstacles* which is programmed. Each behavior is learned separately keeping the other ones fixed.

Performance, computed as the number of moves necessary to accomplish the task, was averaged over a set of sample situations. More precisely, we chose three different initial situations (see Fig. 6), all including two food pieces, and ran 10 trials for each of them, recording the total number of cycles necessary to complete the hoarding of both food pieces (each cycle lasted about 1 second).

Finally, we computed the mean and the standard deviation of such data (Table I). In fact, in situation c, HAMSTER was unable to accomplish the task five times out of ten, due to the difficulty of getting the piece of food in front of the initial robot position and then avoiding the close obstacle; for this situation, the data reported are relative to the five successful trials only.

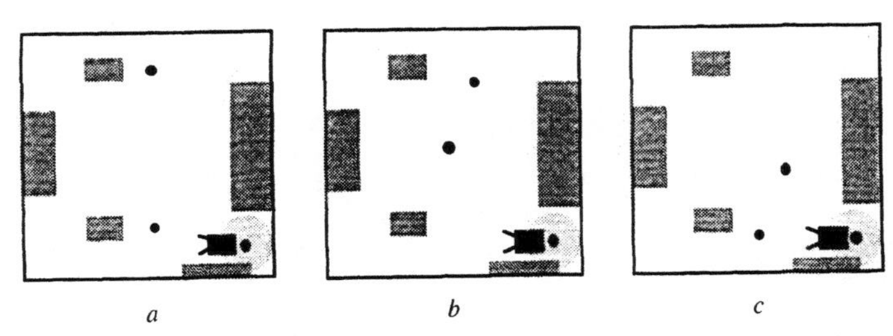

Figure 6. *The three experimental situations. Rectangular shaded areas are obstacles, black circles are food pieces. HAMSTER is shown at its initial location, that is in the nest (shaded circle).*

It appears that the behaviors learned in the simulated environment transfer well to the real robot. This is due to the fact that the interaction of HAMSTER with its environment is rather simple, and does not critically depend on environmental features that are difficult to simulate.

TABLE I. ***Global performance index of HAMSTER in the real environment.***

	Situation a	Situation b	Situation c
Mean	365.3	450	313.5
Standard deviation	27.77	127.30	34.33

4. CONCLUSIONS

The HAMSTER project shows that it is feasible to implement a robot's controller starting from both innate and learned behavioral modules. However, to keep the reinforcement program as simple as possible it was necessary to carry out training in environments where the innate behaviors are not active. In our case, this required carrying out training in a simulated environment, and then transferring the learned controller to the physical robot for the final assessment.

Whether the global performance of Hamster is to be considered acceptable is difficult to say. In a real application, some minimum performance level would have been established in advance; in our case, we can only say that an informal observation of the behavior gave us the impression that Hamster was performing reasonably well.

REFERENCES

1. Dorigo M. & U. Schnepf (1993). Genetics-based Machine Learning and Behavior Based Robotics: A New Synthesis. *IEEE Transactions on Systems, Man, and Cybernetics*, **23**, 1, 141–154.
2. Dorigo M. & M. Colombetti (1994a). Robot Shaping: Developing Autonomous Agents through Learning. *Artificial Intelligence*, **71**, 2, 321–370.
3. Dorigo M. & M. Colombetti (1994b). The role of the trainer in reinforcement learning. *Proceedings of MLC-COLT '94 Workshop on Robot Learning*, S. Mahadevan *et al.* (Eds.), July 10th 1994, New Brunswick, NJ, 37–45.
4. Latombe J.C. (1991). *Robot Motion Planning*, Kluwer, Dordrecht, The Netherlands.
5. Dorigo M. & E. Sirtori (1991). ALECSYS: A Parallel Laboratory for Learning Classifier Systems. *Proceedings of the Fourth International Conference on Genetic Algorithms*, San Diego, California, R.K. Belew & L.B. Booker (Eds.), Morgan Kaufmann, 296–302.
6. Dorigo M. (1993). Genetic and Non-Genetic Operators in Alecsys. *Evolutionary Computation*, **1**, 2, 151–164, MIT Press.
7. Dorigo M. (1995). ALECSYS and the AutonoMouse: Learning to Control a Real Robot by Distributed Classifier Systems. *Machine Learning*, **19**, 3, 209–240.
8. Holland J.H. (1980). Adaptive algorithms for discovering and using general patterns in growing knowledge bases. *International Journal of Policy Analysis and Information Systems*, **4**, 2, 217–240.
9. Holland J. H. (1975). *Adaptation in natural and artificial systems*, The University of Michigan Press, Ann Arbor, Michigan.

MODELING AND IDENTIFICATION OF NON-GEOMETRIC PARAMETERS IN SEMI-FLEXIBLE PARALLEL ROBOTS

Louise Cléroux and Clément M. Gosselin
Département de Génie Mécanique, Université Laval
Québec, Québec, Canada, G1K 7P4

ABSTRACT

The modeling and identification of parallel manipulators with flexible joints and links is addressed in this paper. A general model is developed to take into account the flexibilities, and an algorithm is proposed for the identification of mechanisms from measured static poses. Because of their inherent stiffness, parallel manipulators are often suggested for tasks where heavy loads are to be handled. However, deflections can occur which affect the pose of the moving platform. The objective of this work is thus to improve the static accuracy of parallel manipulators.

KEYWORDS: robots, flexible, parallel, modeling, identification

INTRODUCTION

Parallel manipulators are being considered for applications where heavy loads must be handled. Although parallel mechanisms are inherently very stiff, elastic deformations may occur under large loads. Therefore, flexibilities must be studied in order to improve the static precision of such mechanisms. Several authors have proposed methods for the calibration of the rigid body parameters (see for instance: [4, 7, 8, 13]). Other researchers have incorporated the structural flexibilities in the geometric model of the manipulator [10, 15]. Still others have focused on dynamic semi-flexible models in a control oriented perspective [2, 9, 12]. Experimental results have been reported to justify the efforts committed to the development of representative semi-flexible models. The gain in absolute precision obtained from rigid body geometric calibrations has been said to be in the range of 75% to 80% [1, 10]. An additional 16% increase in absolute precision was obtained in [1] when the calibration of current industrial robots incorporated flexibilities. Incremental flexible models have also been proposed [1, 3, 8, 11, 14]. In these models, gross motion is described with the Denavit-Hartenberg convention while, for example, additional homogeneous transformations express local corrections. All of these models have dealt with serial manipulators exclusively. Only a few authors [5] have addressed the flexibilities in parallel robotic mechanisms.

In this paper, the model presented in [3] is revisited and several of its limitations are eliminated. The gradient is now used in the identification algorithm. Also, the model is now capable of handling any series of joints and links, and any type of robotic structure. This allows the model to be applied to parallel robotic mechanisms. The semi-flexible static model combines rigid body geometric effects, a beam model for each link of the manipulator, and a spring model for each of the joints. All the beam modes are considered in addition to the joint springs. The Denavit-Hartenberg convention is used after the parallel mechanism has been decomposed into a set of serial structures. This paper is organized as follows: the first section introduces the numerical simulator of the geometric model. Then, the semi-flexible geometric model is presented. Finally, an algorithm is proposed for the identification of the non-geometric parameters, observability issues are discussed and identification results are presented for a planar parallel three-degree-of-freedom manipulator.

NUMERICAL SIMULATOR

A numerical simulator has been developed to generate data on parallel flexible robotic mechanisms. The simulator is based on the decomposition of the mechanism into a set of serial structures connecting the base to the moving platform. In fact, this principle can also be applied to other robotic systems such as cooperating robots, as illustrated in Fig. 1.

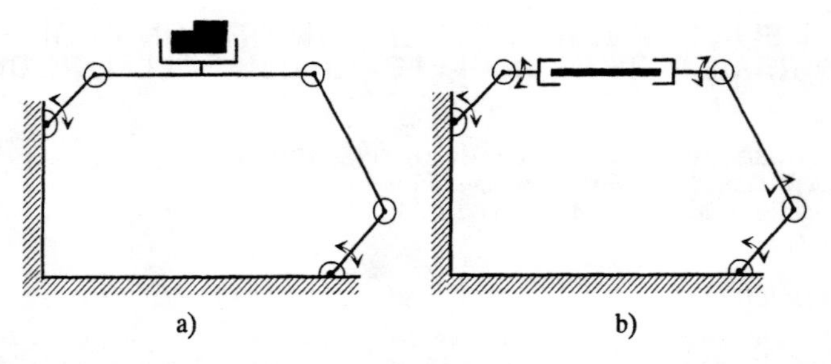

Figure 1: Two Identical Situations from a Geometrical Point of View: a) A Parallel Manipulator with Two Legs; b) Two Robots Working in Cooperation.

Assuming that the platform (Fig. 1a) or the manipulated object (Fig. 1b) is rigid, the system can be described as a set of serial structures (legs). The Denavit-Hartenberg convention is used for the geometric description of each of these serial structures. A single fixed frame ($\{0\}$ in Fig. 2) is arbitrarily identified and serves as the first frame in the description of each of the serial structures. To ensure the closure of the kinematic loops, a single last frame ($\{R\}$ in Fig. 2) is also used. It is chosen to represent the pose of the manipulated object or platform. With this decomposition, the manipulator is modeled as a set of serial structures subject to constraints.

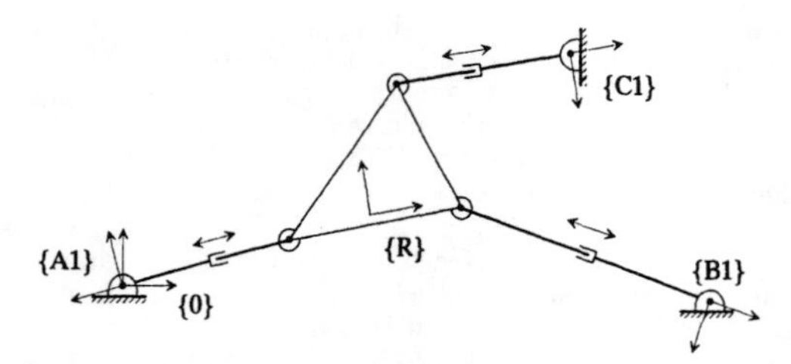

Figure 2: Decomposition Scheme for a Planar Parallel 3 DOF Robot.

Furthermore, since it is desired to take into account the effect of gravity and of the external forces and torques on the serial structures, the distribution — among the legs — of the forces and torques applied to the platform must be computed in order to satisfy the equilibrium equations of the complete system. The algorithm described in [6] is used in order to calculate, from the weight of the platform and the external wrenches applied to it, the forces and torques transferred to each leg of the parallel manipulator.

Simulator's Flow Chart

The simulator's flow chart is presented in Fig. 3. First, the pose (position and orientation) of the moving reference frame $\{R\}$ must be specified with respect to the fixed frame $\{0\}$ (step 1). Then, the full description of each serial structure starting at frame $\{0\}$ and ending at frame $\{R\}$ must be given (step 2). The joint coordinates can then be computed, using a rigid model (step 3). There is no real distinction between the actuated and free joints. With this first approximation of the configuration of the mechanism, it is possible to distribute the wrenches applied to the platform among the serial chains connecting the base to the platform (step 4). The Newton-Gauss algorithm is then used — with a numerical gradient — to adjust the joint coordinates in order to minimize the error between the estimated wrenches at the platform and the desired ones (step 5). The joint coordinates thereby obtained are then used to recompute the distribution of the loads in the mechanism and this iterative process is repeated until the platform has been stabilized under the loads.

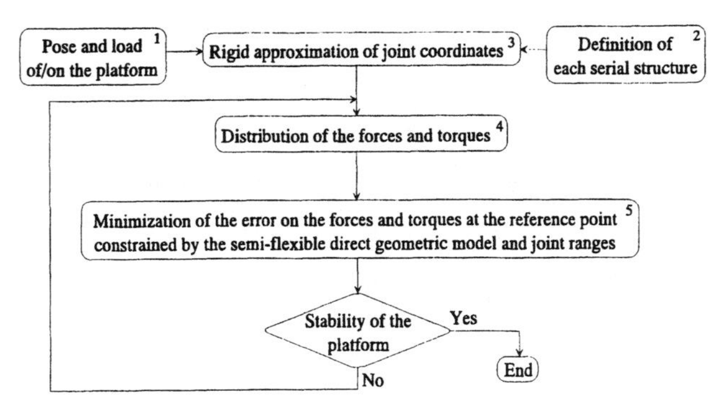

Figure 3: Simulator's Flow Chart.

SEMI-FLEXIBLE GEOMETRIC MODEL

The flexibilities included in the model described here are of two types: i) the flexibilities at the joints and ii) the flexibilities of the links. Hence, the semi-flexible model comprises three main components:

- the DENAVIT-HARTENBERG MODEL which defines the nominal geometry of each of the serial structures of the manipulator;
- a LINEAR SPRING MODEL at each joint to account for the joint flexibilities;
- an EQUIVALENT BEAM MODEL at each link to account for the deformations of the link.

Linear Spring Model

The joints of the robotic mechanisms considered here can be either of the revolute or prismatic type. The joint deformation due to the flexibility is then given by

$$ {}^{i}_{i'}\boldsymbol{T} = \begin{cases} \mathbf{Rot}(\delta\theta_i, z_i) & \text{with } \delta\theta_i = {}^{i}f_6\ \kappa_i \quad \text{if joint } i \text{ is a revolute} \\ \mathbf{Trans}(\delta d_i, z_i) & \text{with } \delta d_i = {}^{i}f_3\ \kappa_i \quad \text{if joint } i \text{ is a prismatic} \end{cases} \tag{1} $$

where ${}^{i}f_3$ is the force along the joint i (third component of the local wrench) in N; ${}^{i}f_6$ is the torque at joint i (sixth component of the local wrench) in Nm; κ_i is the flexibility of joint i in rad/Nm (revolute joint) or in Rad/N (prismatic joint); **Rot**(,) and **Trans**(,) are homogeneous transformations (4×4 matrices) corresponding to a pure rotation or a pure translation respectively.

Equivalent Beam Model

Within the limitations of Castigliano's theorem, it is possible to establish an equivalent beam model accounting for all possible deformations (tension/compression, horizontal and vertical bending, and torsion) caused by external forces and torques, and the mechanism's own weight. Furthermore, it is possible to represent the beam deformations with a homogeneous transformation while imposing a first order approximation on the orientation deformations as follows:

$$ {}^{i''}_{i'''}\boldsymbol{T} = \begin{bmatrix} 1 & -\varphi z_i & \varphi y_i & \delta x_i \\ \varphi z_i & 1 & -\varphi x_i & \delta y_i \\ -\varphi y_i & \varphi x_i & 1 & \delta z_i \\ 0 & 0 & 0 & 1 \end{bmatrix} \tag{2} $$

$$\delta x_i = \text{fct}(S_i) \quad (3) \qquad \varphi x_i = \text{fct}(J_i) \quad (6)$$
$$\delta y_i = \text{fct}(Iz_i) \quad (4) \qquad \varphi y_i = \text{fct}(Iy_i) \quad (7)$$
$$\delta z_i = \text{fct}(Iy_i) \quad (5) \qquad \varphi z_i = \text{fct}(Iz_i) \quad (8)$$

where S_i is the area of the section, Iy_i and Iz_i are the moments of inertia and J_i is the torsion inertia of the equivalent beam of link i.

Complete Model

The complete model is the result of the integration of the Denavit-Hartenberg model, the linear spring model and the equivalent beam model. The total transformation can be written as

$$ {}^{i}_{i+1}\mathbf{\Gamma} = {}^{i}_{i'}T \cdot \mathbf{Trans}(a_i, x_i) \cdot {}^{i''}_{i'''}T \cdot \mathbf{Rot}(\alpha_i, x_i) \cdot \mathbf{Trans}(d_{i+1}, z_{i+1}) \cdot \mathbf{Rot}(\theta_{i+1}, z_{i+1}) \quad (9)$$

IDENTIFICATION OF THE NON-GEOMETRIC PARAMETERS

The objective is to use a flexible model to improve the accuracy of parallel robotic mechanisms. However, for a given manipulator, the structural parameters of the links as well as the flexibility of the joints must be identified in order to be able to use the model.

The function used for the identification is the pose error: the difference between the theoretical and measured (or simulated) poses of the last frame. The variables to be identified are the non-geometric parameters. The pose error is given by

$$\Phi = \left({}^{0}_{R}\mathbf{\Gamma} \cdot {}^{0}_{R}\hat{\mathbf{\Gamma}}^{-1} \right) - \mathbf{I}_4 \quad (10)$$

where ${}^{0}_{R}\mathbf{\Gamma}$ is the homogeneous transformation of the measured (or simulated) pose of the last frame $\{R\}$ with respect to the fixed frame $\{0\}$; ${}^{0}_{R}\hat{\mathbf{\Gamma}}$ is the homogeneous transformation of the estimated pose of the last frame $\{R\}$ with respect to the fixed frame $\{0\}$, based on the estimate of the non-geometric parameters; $\mathbf{I}_4$ is the 4×4 identity matrix.

The identification is accomplished using the Newton-Gauss technique where each of the relevant elements of the above matrix is used to generate one equation in the system of nonlinear equations. Since each serial structure is used separatly to obtain a pose of the last frame $\{R\}$ through its own flexibilities, the closure of the kinematic loops must be imposed. Hence, a function based on the pose error as defined above has been written. However, in the latter function, the poses being compared are the desired pose and the estimated pose through each of the serial structures. Thus, the number of equations is multiplied by s, where s is the number of chains connecting the base to the platform. If the manipulator comprises three such chains, one has

$$\mathbf{b} = [\, \varphi_{111} \cdots \varphi_{114}\, \varphi_{121} \cdots \varphi_{124}\, \varphi_{131} \cdots \varphi_{134}\, \varphi_{211} \cdots \varphi_{234}\, \varphi_{311} \cdots]^T = 0 \quad (11)$$

where φ_{ijk} is the element corresponding to line j and column k of the matrix Φ (equ. (10)) obtained through the ith serial structure.

The latter system of equations is written assuming that only one pose of the last frame is considered. However, in a practical application, several situations (poses and loads) may be used, which will increase the number of equations in the system and improve the stability. Following the Newton-Gauss procedure with an analytical gradient, the correction of the estimates of the parameters is computed, at each iteration, as

$$\Delta \boldsymbol{p} = -\mathbf{J}^I \cdot \mathbf{b} \qquad \text{with} \qquad \mathbf{J} = \frac{\partial\, \mathbf{b}}{\partial\, \boldsymbol{p}} \quad (12)$$

where **b** is the vector$(12fs \times 1)$ of pose error equations; $\boldsymbol{p}$ is the vector $((\sum_{i=1}^{s} 5n_i) \times 1)$ of parameters to be estimated; **J** is the Jacobian matrix of the system of equations; $\mathbf{J}^I$ is the generalized inverse of matrix **J** defined as $(\mathbf{J}^T\mathbf{J})^{-1}\mathbf{J}^T$; n_i is the number of frames between $\{0\}$ and $\{R\}$ of the ith serial structure; s is the number of serial structures and f is the number of different situations used.

Obervability Issues

Throughout the tests, an *observability* problem was revealed. Indeed, a parameter representing the area of the section of a beam (S_i) is used to compute the deformations associated with the tension/compression of the links. However, it is well known that such a deformation is always much smaller than the deformations caused by bending and torsion. Hence, all the non-geometric parameters can be identified even when the parameters representing the area of the section are far away from their real values. Eliminating this particular parameter from the outset is therefore the best solution since it has very little effect on the pose error. Moreover, since all four beam parameters are in fact related through the geometry of the section of the equivalent beam, it is possible to identify the parameters associated with the area of the sections *a posteriori* from the other three beam parameters or, alternatively, simply disregard the parameters representing the area of the sections.

Furthermore, using the equations as described above directly for the identification of the non-geometric parameters of a planar three-degree-of-freedom parallel manipulator poses another observability problem. Indeed, since this manipulator is constrained to move on a plane, out-of-the-plane deformations cannot be modeled and hence, some of the parameters of the general model cannot be identified. Additionally, not all equations contained in (11) can be used since there is no out-of-the-plane pose error possible. The presence of free joints also has an effect on the number of parameters to be identified. Indeed, since these joints are free to rotate, the spring model is not applicable.

Results

Six situations (pose and load) have been arbitrarily chosen in order to test the identification algorithm for the planar three-degree-of-freedom manipulator. For each situation, the identification has been repeated fifty times and each time the first estimate of the non-geometric parameters was randomly selected withing a ±25% range around the real values. The results are summarized in Table I.

Table I: Statistics on the Identification of the Non-Geometric Parameters.

Residual Error	Max. on Posi. (meters)	Min. on Posi. (meters)	Max. on Orient. (degrees)	Min. on Orient. (degrees)
Situation #1	$1,4 \times 10^{-3}$	$8,6 \times 10^{-5}$	$6,8 \times 10^{-2}$	$5,4 \times 10^{-2}$
Situation #2	$2,0 \times 10^{-3}$	$6,0 \times 10^{-4}$	$1,2 \times 10^{-2}$	$1,6 \times 10^{-3}$
Situation #3	$2,6 \times 10^{-3}$	$0,0$	$1,2 \times 10^{-2}$	$5,6 \times 10^{-3}$
Situation #4	$1,1 \times 10^{-3}$	$3,6 \times 10^{-4}$	$5,2 \times 10^{-2}$	$4,4 \times 10^{-2}$
Situation #5	$2,8 \times 10^{-3}$	$2,1 \times 10^{-3}$	$3,7 \times 10^{-1}$	$2,8 \times 10^{-1}$
Situation #6	$5,3 \times 10^{-3}$	$8,8 \times 10^{-4}$	$1,7 \times 10^{-1}$	$1,6 \times 10^{-1}$

From the results of Table I, several issues can be discussed. The residual error on the orientation of the platform is generally small since aproximating the orientation to the 10th of a degree is usually sufficient. On the other hand, the residual error on the position can be of the order of a few millimeters. In many cases, this is clearly unacceptable. Detailed interpretation of the results shows that in about 60% of the tests, the residual error on the position is smaller than one millimeter. Additionally some situations seem to produce better results than others which suggests that a constructive method allowing to choose the situations in order to improve the accuracy should be developed.

CONCLUSION

The modeling and identification of parallel manipulators with flexible joints and links has been addressed in this paper. A semi-flexible geometric model has been developed based on the nominal geometry of the manipulator and the flexibilities of the joints and the links, represented respectively as springs and elastic beams. The objective of this work was to improve the static precision of parallel manipulators, which are often suggested for tasks where heavy loads are to be handled. Results obtained for a planar three-degree-of-freedom parallel manipulator have demonstrated the capability to numerically simulate the direct semi-flexible geometric model of parallel manipulators and the success of the identification algorithm for the non-geometric parameters. Further work is currently in progress in order to improve the precision of the identification algorithm. The influence of loads, configurations and number of situations used are being studied.

ACKNOWLEDGEMENTS

The authors would like to thank the Natural Sciences and Engineering Research Council of Canada (NSERC), the Fonds pour la Formation des Chercheurs et l'Aide à la Recherche du Québec (FCAR), the Fondation de l'Ecole Polytechnique de Montréal and Université Laval for their financial support.

REFERENCES

[1] Caenen, J.L. and Angue, J.C., 1990, 'Identification of Geometric and Non Geometric Parameters of Robots', *Proceedings of the 1990 IEEE International Conference on Robotics and Automation*, pp. 1032–1037.

[2] Chang, L.-W. and Hamilton, J.F., 1991, 'Dynamics of Robotic Manipulators with Flexible Links', *ASME Journal of Dynamic Systems, Measurement, and Control*, Vol. 113, March, pp. 54–59.

[3] Cléroux, L., Gourdeau R. and Cloutier, G.M., 1995, 'A Semi-Flexible Geometric Model for Serial Manipulators', *Robotica*, Vol. 13, pp. 385–395.

[4] Everett, L.J. and Hsu, T.W., 1988, 'The Theory of Kinematic Parameter Identification for Industrial Robots', *ASME Journal of Dynamic Systems, Measurement, and Control*, March, pp. 96–100.

[5] Fattah, A., Angeles, J. and Misra, A.K., 1994, 'Direct Kinematics of a 3-DOF Spatial Parallel Manipulator with Flexible Legs', *Proceedings of the ASME Mechanisms Conference*, DE-Vol. 72, pp. 285–291.

[6] Gosselin, C. M., 1993, 'Parallel Computation Algorithms for the Kinematics and Dynamics of Parallel Manipulators', *Proceedings of the 1993 IEEE International Conference on Robotics and Automation*, Atlanta, pp. 883–888.

[7] Hayati, S.A., Tso, K. and Roston, G., 1988, 'Robot Geometry Calibration', *Proceedings of the 1988 IEEE International Conference on Robotics and Automation*, pp. 947–951.

[8] Hsu, T.W. and Everett, L.J., 1985, 'Identification of the Kinematic Parameters of a Robot Manipulator for Positional Accuracy Improvement', *ASME Conference on Computers in Engineering*, pp. 263–267.

[9] Jonker, B., 1990, 'A Finite Element Dynamic Analysis of Flexible Manipulators', *The International Journal of Robotics Research*, Vol. 9, No. 4, pp. 59–74.

[10] Judd, R.P. and Knasinski, A.B., 1990, 'A Technique to Calibrate Industrial Robots with Experimental Verification', *IEEE Transactions on Robotics and Automation*, Vol. 6, No. 1, pp. 20–30.

[11] Meghdari, A., 1991, 'A Variational Approach for Modelling Flexibility Effects in Manipulator Arms', *Robotica*, Vol. 9, pp. 213–217.

[12] Piedboeuf, J.-C., Hurteau, R. and Ziarati, K., 1992, 'Logiciel de simulation et de commande pour les robots flexibles', *Conférence Canadienne et Exposition: Automatisation Industrielle*, pp. 13.13–13.17.

[13] Stone, H.W., 1987, *Kinematic Modelling, Identification, and Control of Robotic Manipulators*, Kluwer Academic Publishers.

[14] Tang, S.C. and Wang, C.C., 1987, 'Computation of the Effects of Link Deflections and Joint Compliance on Robot Positioning', *Proceedings of the 1987 IEEE International Conference on Robotics and Automation*, pp. 910–915.

[15] Whitney, D.E., Lozinski, C.A. and Rourke, J.M., 1986, 'Industrial Robot Forward Calibration Method and Results', *ASME Journal of Dynamic Systems, Measurement and Control*, March, pp. 1–8.

EXPERIMENTAL RESULTS OF OPERATIONAL SPACE CONTROL ON A DUAL-ARM ROBOT SYSTEM

FABRIZIO CACCAVALE
Dipartimento di Informatica e Sistemistica
Università degli Studi di Napoli Federico II
Via Claudio 21, 80125 Napoli, Italy

JÉRÔME SZEWCZYK
Laboratoire de Robotique de Paris
Université Paris 6
10/12 Avenue de l'Europe, 78140 Velizy, France

ABSTRACT

This paper is aimed at reporting experimental work performed on a system of two cooperative industrial robots grasping a common object. An operational space formulation is proposed for describing both absolute and relative variables, and proper force and velocity mappings are derived. Based on such a formulation a hybrid force/position control scheme is designed. Numerical results of a case study developed on the laboratory set-up are discussed.

KEYWORDS: cooperative robots, operational space, force/position control

INTRODUCTION

In the latest years there has been a great deal of attention in the robotics community toward cooperative manipulator systems. The challenge is to exploit the superior potential of multiple-armed over single-armed robot systems in terms of manipulability of a commonly held object. In view of the mechanical constraints imposed by the object, one should account for the kineto-static relationship between end-effector contact forces and velocities and object forces and velocities [1].

One important point in coordinated motion of two robot manipulators is the definition of meaningful operational space variables in order to allow a clear description of a given task. As pointed out in [2], such variables do not in general coincide with the variables originating from a kineto-static description. Establishing an operational space formulation is also useful for approaching the control problem in an effective fashion.

Several works have been devoted to control of multiarm systems, such as [3–10] to mention only a few. Their common denominator is to achieve control of both the absolute motion of the object and the relative forces exerted between each pair of manipulator end effectors.

The present work reports the results of an experimental investigation on a laboratory set-up constituted by two cooperative industrial robots, whose key feature is the availability of an open control architecture [11] allowing implementation of control algorithms through a PC. A kinematics description is derived which allows the control problem to be formulated in the operational space [12]. A hybrid force/position control scheme is designed for the case of a deformable object. A case study is developed to demonstrate the performance of the above control scheme on a typical coordinated motion for the dual-arm system.

OPERATIONAL SPACE FORMULATION

Consider a system of two planar arms, whose end effectors tightly grasp a common object. Let

$$\boldsymbol{x}_i = (\boldsymbol{p}_i^{\mathrm{T}} \quad \vartheta_i)^{\mathrm{T}} \qquad i = 1, 2 \tag{1}$$

denote the (3×1) vector of each end-effector position and orientation; $\boldsymbol{x}_i$ and all subsequent quantities are expressed with reference to a common base frame.

In order to derive an operational space formulation, it is natural to consider three variables to describe the absolute position and orientation of the object with respect to the base frame, and another three variables to describe the relative position and orientation between the two end-effectors. As for the absolute variables, a simple choice is [13]

$$\boldsymbol{x}_a = (\boldsymbol{p}_a^{\mathrm{T}} \quad \vartheta_a)^{\mathrm{T}} \tag{2}$$

where $\boldsymbol{x}_a = \frac{1}{2}(\boldsymbol{x}_1 + \boldsymbol{x}_2)$. As for the relative variables, it is worth choosing

$$\boldsymbol{x}_r = (\rho \quad \alpha - \vartheta_a \quad \vartheta_r)^{\mathrm{T}} \tag{3}$$

where ρ is the length of the vector difference $\boldsymbol{p}_r = \boldsymbol{p}_2 - \boldsymbol{p}_1$, α is the angle expressing the orientation of $\boldsymbol{p}_r$ with respect to the base frame, and $\vartheta_r = \vartheta_2 - \vartheta_1$. Notice that $\boldsymbol{x}_r$ describes the mechanically constrained directions of the operational space; in fact, for a tightly grasped rigid object $\boldsymbol{x}_r$ is constant, whereas for a nonrigid object, $\boldsymbol{x}_r$ is variable. This issue confirms the effectiveness of the above description of relative variables.

Eqs. (2) and (3) reveal that both absolute and relative variables can be expressed as a function of the position and orientation of the two end effectors defined in (1). This can be conveniently achieved in terms of differential kinematics relationships as follows. Let $\boldsymbol{v} = (\dot{\boldsymbol{x}}_1^{\mathrm{T}} \quad \dot{\boldsymbol{x}}_2^{\mathrm{T}})^{\mathrm{T}}$ denote the (6×1) stacked vector of end-effector velocities. It can be shown that the (6×1) stacked vector of operational space velocities (absolute and relative) $\dot{\boldsymbol{x}} = (\dot{\boldsymbol{x}}_a^{\mathrm{T}} \quad \dot{\boldsymbol{x}}_r^{\mathrm{T}})^{\mathrm{T}}$ can be expressed as a function of $\boldsymbol{v}$ by the relationship

$$\dot{\boldsymbol{x}} = \boldsymbol{W}\boldsymbol{v}, \tag{4}$$

with

$$\boldsymbol{W} = \begin{pmatrix} \frac{1}{2}\boldsymbol{I}_3 & \frac{1}{2}\boldsymbol{I}_3 \\ \frac{1}{2}\boldsymbol{L} - \boldsymbol{A} & \frac{1}{2}\boldsymbol{L} + \boldsymbol{A} \end{pmatrix} \quad \boldsymbol{A} = \begin{pmatrix} \frac{p_{rx}}{\rho} & \frac{p_{ry}}{\rho} & 0 \\ -\frac{p_{ry}}{\rho^2} & \frac{p_{rx}}{\rho^2} & 0 \\ 0 & 0 & 1 \end{pmatrix} \quad \boldsymbol{L} = \begin{pmatrix} 0 & 0 & 0 \\ 0 & 0 & -1 \\ 0 & 0 & 0 \end{pmatrix} \tag{5}$$

where $\boldsymbol{I}_3$ is the (3×3) identity matrix, and the matrix $\boldsymbol{W}$ has full rank as long as $\rho \neq 0$.

Let $\boldsymbol{f} = (\boldsymbol{f}_a^{\mathrm{T}} \quad \boldsymbol{f}_r^{\mathrm{T}})^{\mathrm{T}}$ denote the vector of generalized forces associated with the operational space velocity $\dot{\boldsymbol{x}}$ in (4), and $\boldsymbol{h} = (\boldsymbol{h}_1^{\mathrm{T}} \quad \boldsymbol{h}_2^{\mathrm{T}})^{\mathrm{T}}$ denote the vector of generalized forces at the contact with the two end effectors. In view of the duality between forces and velocities coming from the principle of virtual work in mechanics, from (4) it follows that

$$\boldsymbol{h} = \boldsymbol{W}^{\mathrm{T}}\boldsymbol{f}. \tag{6}$$

It is worth noticing that the forces $\boldsymbol{f}_a$ and $\boldsymbol{f}_r$ depend on the particular choice of operational space variables as in (2) and (3), and thus they do not coincide with the commonly termed external and internal forces as in, e.g., [1]. In particular the mapping between the contact forces $\boldsymbol{h}$ and the external forces $\boldsymbol{h}_e$ is established by

$$\boldsymbol{h}_e = \boldsymbol{\Phi h} \tag{7}$$

where $\boldsymbol{\Phi}$ is the (3×6) grasp matrix given by

$$\boldsymbol{\Phi} = \begin{pmatrix} 1 & 0 & 0 & 1 & 0 & 0 \\ 0 & 1 & 0 & 0 & 1 & 0 \\ \frac{1}{2}p_{ry} & -\frac{1}{2}p_{rx} & 1 & -\frac{1}{2}p_{ry} & \frac{1}{2}p_{rx} & 1 \end{pmatrix}, \tag{8}$$

under the assumption that object's center of mass is located in the middle of the line segment connecting the two contact points.

Without loss of generality, assume that each arm has three degrees of freedom, i.e., $\boldsymbol{q}_i$ is the (3×1) vector of joint variables. Thus, the differential kinematics for the two arms can be written in compact form as

$$\boldsymbol{v} = \boldsymbol{J}(\boldsymbol{q})\dot{\boldsymbol{q}} \tag{9}$$

where $\boldsymbol{q} = (\,\boldsymbol{q}_1^{\mathrm{T}} \quad \boldsymbol{q}_2^{\mathrm{T}}\,)^{\mathrm{T}}$ and $\boldsymbol{J} = \mathrm{diag}\{\boldsymbol{J}_1, \boldsymbol{J}_2\}$, being $\boldsymbol{J}_1$ and $\boldsymbol{J}_2$ the Jacobians of the two arms, and $\boldsymbol{O}_3$ the (3×3) null matrix. Similarly to above, the dual mapping between the vector of contact forces $\boldsymbol{h}$ and the vector of joint torques $\boldsymbol{\tau} = (\,\boldsymbol{\tau}_1^{\mathrm{T}} \quad \boldsymbol{\tau}_2^{\mathrm{T}}\,)^{\mathrm{T}}$ is given by

$$\boldsymbol{\tau} = \boldsymbol{J}^{\mathrm{T}}(\boldsymbol{q})\boldsymbol{h}. \tag{10}$$

In order to obtain the overall velocity and force mappings between the joint space and operational space, it is sufficient to combine Eqs. (4) and (9), and Eqs. (6) and (10), respectively, leading to:

$$\dot{\boldsymbol{x}} = \boldsymbol{W}\boldsymbol{J}\dot{\boldsymbol{q}} \tag{11}$$

$$\boldsymbol{\tau} = \boldsymbol{J}^{\mathrm{T}}\boldsymbol{W}^{\mathrm{T}}\boldsymbol{f}. \tag{12}$$

HYBRID FORCE/POSITION CONTROL

In order to design a control scheme for the above dual-arm robot system, it is convenient to consider the equations of motion in the operational space [12]. These can be written as

$$\boldsymbol{\Lambda}\ddot{\boldsymbol{x}} = \boldsymbol{f}_c - \boldsymbol{f}_o \tag{13}$$

provided that Coriolis, centrifugal, friction and gravity terms have been properly compensated for both the manipulators and the object. In (13) $\boldsymbol{f}_c$ denotes the driving forces and $\boldsymbol{f}_o$ are the generalized reaction forces due to interaction between the dual-arm system dynamics and the object.

In view of (13), the operational space inertia matrix is given by

$$\boldsymbol{\Lambda} = \boldsymbol{W}^{-T}\boldsymbol{J}^{-T}\boldsymbol{M}\boldsymbol{J}^{-1}\boldsymbol{W}^{-1} + \boldsymbol{\Lambda}_o, \tag{14}$$

where $\boldsymbol{\Lambda}_o = \mathrm{diag}\{\boldsymbol{\Lambda}_o^a, \boldsymbol{O}_3\}$, being $\boldsymbol{\Lambda}_o^a$ is the object inertia matrix with respect to the absolute operational variables, and $\boldsymbol{M} = \mathrm{diag}\{\boldsymbol{M}_1, \boldsymbol{M}_2\}$, being $\boldsymbol{M}_1$ and $\boldsymbol{M}_2$ the joint space inertia matrices of the two arms.

In order to design a hybrid force/position control scheme [14], it is convenient to regard the driving forces as given by two different contributions, based on a positional and a force control action respectively

$$f_c = f_{cp} + f_{cf}. \tag{15}$$

According to the inverse dynamics concept for controlling nonlinear and coupled systems, the positional control action can be chosen as

$$f_{cp} = \Lambda \Sigma (\ddot{x}_d + K_v \dot{e} + K_p e) \tag{16}$$

where $\Sigma = \text{diag}\{I_3, O_3\}$ is a matrix accomplishing a proper selection of unconstrained operational space directions to be motion controlled, K_p and K_v are two positive definite diagonal matrices, and $e = x_d - x$ is the operational space error between the desired x_d and actual x variables. Notice that in (16) Λ is assumed to be invertible (J is nonsingular).

Assuming that force sensors are available at both end effectors giving measurements of contact forces h_m, the resulting operational space forces can be computed via (8) as $f_m = W^{-T} h_m$. The force control action forces can be chosen as

$$f_{cf} = \bar{\Sigma} \left(f_d + K_f e_f + K_i \int_0^t e_f d\sigma \right) \tag{17}$$

where $\bar{\Sigma} = I_3 - \Sigma$, K_f and K_i are positive definite diagonal matrices, and $e_f = f_d - f_m$ is the error between the desired f_d and actual f_m forces.

Notice that the matrix $\bar{\Sigma}$ in (17) selects the last three components of force control action which are associated with the constrained operational space directions. Moreover ,it is worth observing that f_{cf} is mapped into contact forces belonging to the null space of the grasp matrix Φ in (7), i.e.,

$$\Phi W^{\mathrm{T}} f_{cf} = 0, \tag{18}$$

as can be verified via (5) and (8). This implies that the commanded force control action has no effect on the object motion.

EXPERIMENTAL RESULTS

Experiments have been performed on a set-up of two industrial robots SMART-3 S manufactured by Comau (Fig. 1); the two robot manipulators have six joints, but one of them is mounted on a sliding track providing an extra degree of freedom. Only joints 2, 3 and 5 of each robot have been used in order to realize a planar two three-degree-of-freedom arm system; all the other joints have been kept mechanically braked. Each joint is actuated by brushless motors via gear trains; shaft absolute resolvers provide motor position measurements.

The original controller of the SMART-3 S is the C3G 9000, a VME-based system comprised of two processing boards; namely, the Robot CPU (RBC) and the Servo CPU (SCC). The robots are controlled by an "open" version of the C3G 9000; a bus-to-bus communication link is established with a standard personal computer [14]. The control algorithms have been implemented as C modules running on a 486DX2/66 PC connected to the controller. At each sampling interval, the PC reads the actual joint positions and computes the motor current set-points based on the user algorithm; the current set-points are then passed to the motors through the communication link.

The object carried by the two robots consists of a double transversal aluminum bar grasping two small brass bars, each of them attached to the force/torque sensor mounted on the corresponding robot via a tool adapter. Such a system simplifies object mounting/dismounting operations, but introduces a nonnegligible object deformability.

Two force/torque F/T 130/10 sensors manufactured by ATI have been used to measure the contact forces.

The hybrid force/position control scheme based on (15)–(17) has been implemented on the set-up described above at a sampling rate of 500 Hz. The gains in (16) have been chosen as $\boldsymbol{K}_p = (2\pi f_0)^2 \boldsymbol{I}_6$ and $\boldsymbol{K}_v = 4\pi f_0 \zeta \boldsymbol{I}_6$, with $f_0 = 7.5$ Hz and $\zeta = 0.8$. The gains in (17) have been chosen as $\boldsymbol{K}_f = \mathrm{diag}\{0, 0, 0, 4, 2, 1.5\}$ and $\boldsymbol{K}_i = \mathrm{diag}\{0, 0, 0, 0.04, 0.001, 0.001\}$.

Coriolis, centrifugal, friction and gravity terms have been compensated for both the arms and the object in order to achieve (13).

An operational space motion is assigned from the initial configuration of the dual-arm system specified by $\boldsymbol{x}_a = (\,0.9562 \quad 0.7155 \quad 1.5657\,)^T$ and $\boldsymbol{x}_r = (\,0.0660 \quad -1.6415 \quad 3.1454\,)^T$ where lengths are expressed in [m] and angles in [rad]. The commanded path is divided into three portions; first, a 0.5657 straight-line displacement for $\boldsymbol{p}_a$ with a rotation of $\pi/3$ for ϑ_a is commanded for a 3 s duration; then the manipulator is kept at rest for 2 s in its final location; and finally, the first portion is executed backward. For the three portions, initial and final configurations are interpolated using 5th-order polynomial time laws such that null initial and final velocities and accelerations are obtained. The operational space relative variables $\boldsymbol{x}_r$ have to be kept constant and equal to the above initial values. The desired force $\boldsymbol{f}_d$ has been set equal to zero.

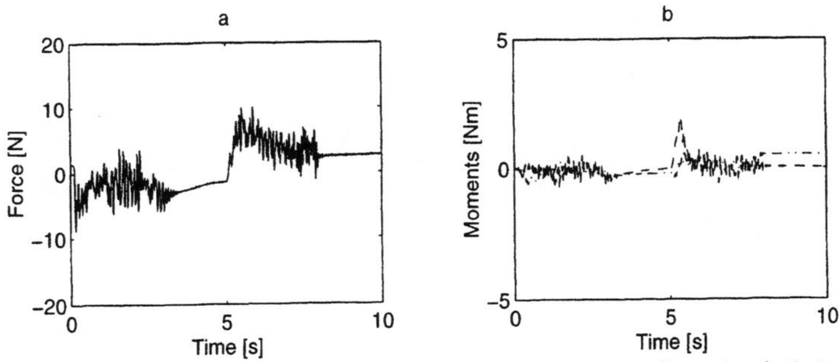

Figure 1. Object reaction force (a) and moments (b): *solid*—1st component (linear force), *dashed*—2nd component component (momentum), *dashdot*—3rd component (momentum).

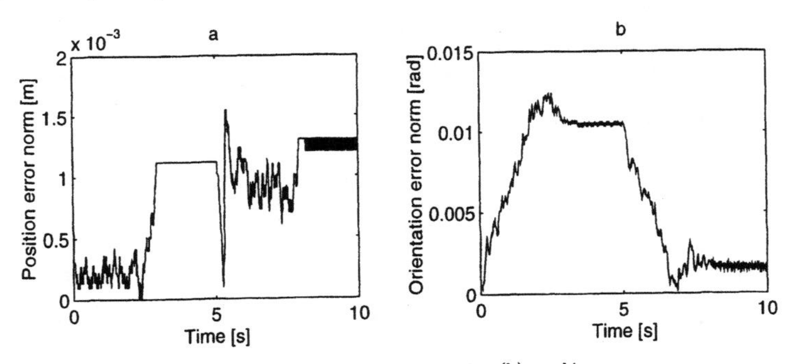

Figure 2. Time history of norm of position (a) and orientation (b) tracking errors.

Fig. 1 illustrates the time history of the last three components of $\boldsymbol{f}_m$ obtained with the proposed control law. The results show that the first two components of reaction forces are kept limited. Notice that at the beginning of the last portion of the commanded path a sudden change of the reaction force is experienced; this can be attributed to backlash, causing an impact on the gear teeth at the joints when motion is inverted.

In order to fully evaluate the performance of the system, it is worth reporting also tracking errors on absolute and relative operational space variables. Fig. 2 illustrates the time history of the norm of tracking errors, respectively on absolute position and absolute orientation. It can be seen that such errors are small.

CONCLUSION

This paper has presented experimental results of controlling a dual-arm robot system when a suitable operational space formulation is adopted. The two arms move in a plane and grasp a common deformable object; it has been confirmed in practice that a hybrid force/position control strategy gives good performance, in that it succeeds in keeping the object reaction forces close to the desired values. Future work will be devoted to testing other control strategies, as well as to extending the investigation to spatial motion of the system including the use of kinematic redundancy.

Acknowledgements

This work was supported partly by the *EC Programme on Human Capital and Mobility* within the ERNET funding framework and partly by *Ministero dell'Università e della Ricerca Scientifica e Tecnologica* within the 40% and 60% funding framework.

REFERENCES

1. M. Uchiyama and P. Dauchez, "Symmetric kinematic formulation and non-master/slave coordinated control of two-arm robots," *Advanced Robotics*, vol. 7, pp. 361–383, 1993.
2. P. Chiacchio, S. Chiaverini, and B. Siciliano, "Cooperative control schemes for multiple robot manipulator systems," *Proc. 1992 IEEE Int. Conf. on Robotics and Automation*, Nice, F, pp. 2218–2223, 1992.
3. J.Y.S. Luh and Y.F. Zheng, "Constrained relations between two coordinated industrial robots for motion control," *Int. J. of Robotics Research*, vol. 6, n. 3, pp. 60–70, 1987.
4. T.E. Alberts and D.I. Soloway, "Force control of a multi-arm robot system," *Proc. 1988 IEEE Int. Conf. on Robotics and Automation*, Philadelphia, PA, pp. 1490–1495, 1988.
5. S.A. Hayati, "Position and force control of coordinated multiple arms," *IEEE Trans. on Aerospace and Electronic Systems*, vol. 24, pp. 584–590, 1988.
6. T.J. Tarn, A.K. Bejczy, and X. Yun, "Design of dynamic control of two cooperating robot arms: Closed chain formulation," *Proc. 1987 IEEE Int. Conf. on Robotics and Automation*, Raleigh, NC, pp. 7–13, 1987.
7. T. Yoshikawa and X. Zheng, "Coordinated dynamic hybrid position/force control for multiple robot manipulators handling one constrained object," *Proc. 1990 IEEE Int. Conf. on Robotics and Automation*, Cincinnati, OH, pp. 1178–1183, 1990.
8. P. Chiacchio, S. Chiaverini, L. Sciavicco, and B. Siciliano, "Dynamic force/motion control of cooperative robot systems," *Proc. 1990 ASME Winter Annual Meet.*, Dallas, TX, DSC-vol. 26, pp. 121–126, 1990.
9. A.J. Koivo and M.A. Unseren, "Reduced order model and decoupled control architecture for two manipulators holding a rigid object," *ASME J. of Dynamic Systems, Measurement, and Control*, vol. 113, pp. 646–654, 1991.
10. J.T. Wen and K. Kreutz-Delgado, "Motion and force control of multiple robotic manipulators," *Automatica*, vol. 28, pp. 729–743, 1992.
11. F. Dogliani, G. Magnani, and L. Sciavicco, "An open architecture industrial controller," *Newsl. of IEEE Robotics and Automation Society*, vol. 7, no. 3, pp. 19–21, 1993.
12. O. Khatib, "Inertial properties in robotic manipulation: An object level framework," *Int. J. of Robotics Research*, vol. 13, pp. 19–36, 1995.
13. P. Chiacchio, S. Chiaverini, and B. Siciliano, "Dexterous reconfiguration of a two-arm robot system," *Proc. 3rd IEE Int. Conf. on Control*, Edinburgh, GB, pp. 347–351, 1991.
14. M.H. Raibert and J.J. Craig, "Hybrid position/force control of manipulators," *ASME J. of Dynamic Systems, Measurement, and Control*, vol. 103, pp. 126–133, 1981.

Radiation Hardened Multiplexing Systems, a Necessary Aspect of the Radiation Hardening of Robots

Simon Coenen, Marc Decréton
SCK·CEN Nuclear Research Centre
Mol, Belgium

ABSTRACT

The radiation hardening of advanced remote manipulators requires a suitable rad hard replacement for each sensitive component. Especially in the case of intelligent systems with many sensors, multiplexing is one of the crucial issues in the hardening process. The specifications of such a multiplexer are discussed for different applications. Several multiplexing techniques are illustrated with results of tests performed in representative conditions, ranging from low radiation environments when inspecting waste storage sites up to extreme conditions appearing in fusion reactor maintenance. A state of the art on emerging technologies in microelectronics is also given with their relevance to radiation hardened multiplexing systems.

KEYWORDS: radiation hardening, multiplexing, robotics

INTRODUCTION

Remote manipulators are widely used in nuclear maintenance, refurbishing and dismantling tasks, in cases where high radiation or contamination risks impede human intervention. To increase operational safety and efficiency, and to reduce operator stress, a certain level of intelligence and computer assistance is considered essential. A semi-autonomous control approach, sometimes called telerobotics, is used where low level control aspects such as repetitive operations and collision avoidance are handled by the computer, while the operator can concentrate on the higher level decisions.

Increasing the number of sensors and control structures however increases also the number of lines between the manipulator and its controller. When a high reliability and ease-of-use for the operator are required, a higher degree of redundancy of data is needed thus increasing further the number of lines. Umbilical management becomes very difficult and on its turn decreases the reliability of the system. The obvious solution here is multiplexing. A trade-off between the number of lines, reliability, and technical limitations has to be found to provide a satisfying solution.

For use in a nuclear environment, the manipulator has to be hardened against radiation. Usually, sensitive parts like sensors are replaced with suitable rad hard transducers and signal processing is kept remote. Depending on the application such replacements are often available. The multiplexing circuit has to be hardened as well and is one of the electronic components where radiation hardening is unavoidable.

This paper will first outline some general specifications for multiplexing systems for use in various radiation environments. Several multiplexing techniques are discussed and illustrated with results from representative tests. Before concluding, a state of the art on technologies in microelectronics is given with their possible benefits for the radiation hardening of multiplexing systems.

SPECIFICATIONS

The main goal of a multiplexing system is obviously to reduce the number of cables between the signal source and its measuring or controlling system and hence ease the umbilical management. To gain in the reduction of cables, a 8 to 1 multiplexer has to be considered as a minimum. Signal conditioning has to be incorporated as well. Most of the signals that are generated by sensors are very low level analog voltage signals, which risk to be severely influenced by spike voltage peaks from the robot itself or from the environment, or to disappear into the noise generated by the long cables. Either pre-amplification or digitalisation can overcome this problem. Digitising the signals is to be preferred, since digital signals are more easily transferred over long cables and thus a large improvement in signal integrity is obtained.

For use in radiation environment, a hardening of the multiplexing circuit is essential. The ambient conditions in typical nuclear teleoperation depend however strongly on the type of task. Apart from conditions typically found also in industrial environments, such as high temperatures or under water operation, ionising radiation presents here a particular challenge. The radiation tolerance however is not always critical. Inspection in waste storages for instance does not require high radiation hardening and standard components can be used. Spent fuel handling, in-core manipulation and some decommissioning tasks on the other hand, can sometimes involve very high gamma doses. Four areas of radiation conditions can be roughly defined which give an indicative radiation scale for typical nuclear teleoperated applications. These areas are described in table 1.

TABLE I. Indicative radiation scale for typical nuclear teleoperated applications.

Area	Dose rate (Gy/h)	Total dose (Gy)	Type of application
A	< 0.01	< 10	Maintenance work in low radiation environment; light decontamination; teleoperation is alternative to hands-on work; no radiation hardening required
B	< 10	< 1000	Most decontamination work; intervention in primary loop; some cell work; teleoperation is required; special material needed; rad hard electronics available
C	< 1000	< 1E6	Reactor vessel intervention; dismantling work; hot cell tasks; material and electronics choice heavily limited
D	> 1000	> 1E6	In-core fusion reactor maintenance; fuel manipulations; critical material choice; no electronics available

III. MULTIPLEXING TECHNIQUES

Various possibilities to build rad hard multiplexing circuits are open, each with their advantages and disadvantages. Each solution will however depend greatly on the type of application and will often be a trade-off between reliability, desired functionality, and cost.

A. Commercially available components

When the required radiation tolerance is low, in applications as described above in area A, no real radiation hardening is needed, and hence commercially available components can be used. Most commercially available components provide threshold damage levels of around 10 Gy [1], [2] (100 Gy for some specific applications).

B. Space graded components

For applications in area B, space graded or rad hard replacements can be defined for most electronic components. Most of the commercially available rad hard components are specified up to 3 or 10 kGy. These figures however may not be considered absolute, because in many cases the degradation of these components is very dependent on dose rate and environmental conditions such as temperature. The exact environmental conditions of the tests used to specify the radiation tolerance are not always known, and they will often be different from the conditions encountered in the practical application.

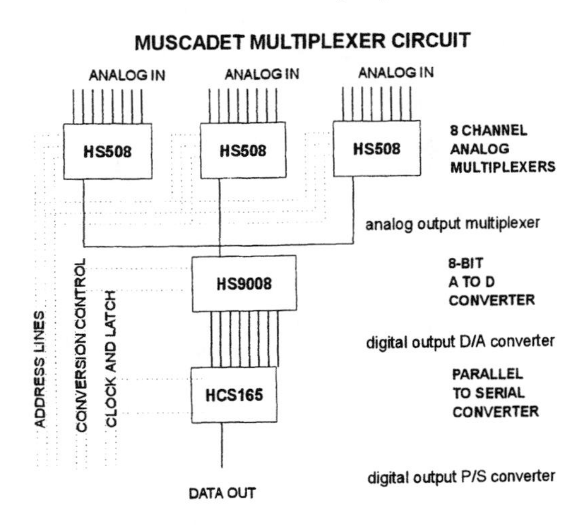

Figure 1. Schematical diagram of prototype 24 to 1 multiplexer

SCK·CEN has performed some irradiation tests on radiation tolerant circuits, provided by Harris Semiconductor for instance [3]. A 24 to 1 multiplexing circuit was build using 3 analog multiplexers (type HS508ARH), 1 analog to digital converter (type HS9008RH) and one parallel to serial shift register (type HCS165MS). The block diagram of the circuit is shown in figure 1. The HS508ARH multiplexer was build with a dielectrically isolated (DI) CMOS technology, which eliminates the possibility of latchup. Hardened gate and field oxides provide excellent total dose hardness. The HS9008RH A/D converter was build using Harris' AVLSIRA process, while the HCS165MS was fabricated with the 1.2 μm silicon on sapphire (SOS) process, which provides a high level of total dose immunity and extreme single event upset immunity.

The experiment, called MUSCADET [4], was performed from November 1994 till March 1995 in the GEUSE spent fuel irradiation facility [4], situated at SCK·CEN at a dose rate of 110 Gy/h in a dry air atmosphere and an ambient temperature of 22 °C. The experiment consisted of several campaigns, during which each separate component was tested. During each campaign, different parameters of the components, such as supply current, output levels and functionality were monitored on-line. During the last irradiation campaign, the 24 to 1 prototype multiplexer was irradiated. A resistor-tree network was used to present input voltages in a linear range from 0 to 5 V to the 24 input lines of the multiplexer. Each input line can be selected separately and is presented to the input of the analog to digital converter. After conversion, the digital output is latched into the parallel to serial converter, which then presents an 8 bit sequence to the output line. During irradiation, the following parameters were measured: the analog output signals from the 3 multiplexers for all 24 input lines, the digital output from the analog to digital converter and the digital output from the parallel to serial converter. Figures 2 to 4 show the results of part of these measurements: the measurements shown are the analog output of the multiplexer channel 15 (set at 3 V), and the digital output of the A/D converter and the parallel to serial converter. For convenience, the digital output on figures 3 and 4 is represented by the decimal value of the 8 bit output code, 0 representing 0 V and 255 representing 5 V.

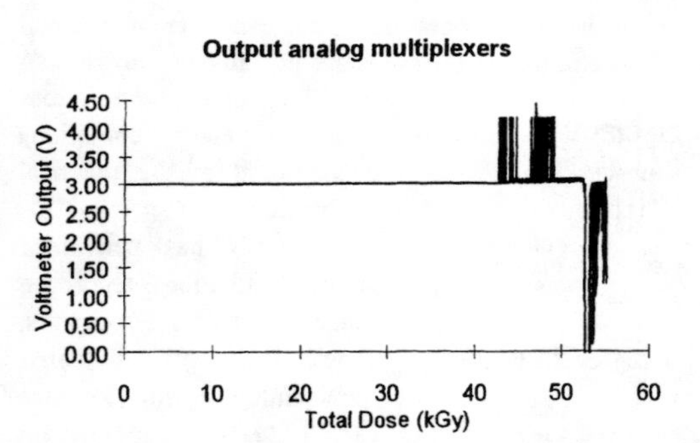

Figure 2. Total dose hardness of analog multiplexers HS508ARH

Figure 2 shows the output of the analog multiplexers versus total dose for an input value of 3 V (channel 15). The input remained stable at the output up to a total dose of 42 kGy. At that point, the multiplexer started to fail: channel selection did not function properly anymore. At a total dose of 52 kGy, the same multiplexer completely broke down and showed a short circuit, leading to failure and breakdown of the other multiplexers as well.

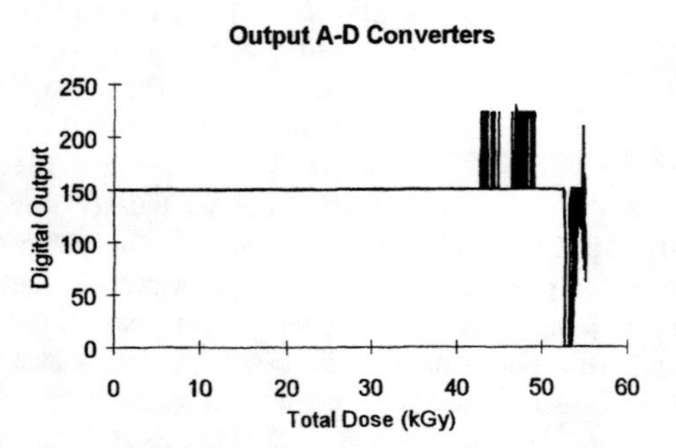

Figure 3. Total dose hardness of A/D converter HS9008RH

The analog to digital converter, shown in figure 3 kept functional during the whole experiment: the digitised value of 3 V, which was 150 remains stable at the output of the converter. Even the fluctuations in output due to the breakdown of one of

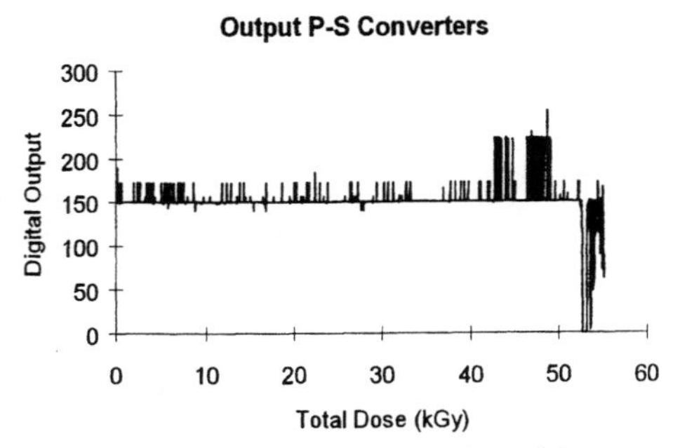

Figure 4. Total dose hardness of parallel to serial converter HCS165MS

the multiplexers are clearly present.

The parallel to serial converter kept functioning properly as well during the complete test: the digital value of 150 is clearly presented as output. As shown in figure 4, this circuit showed to be sensitive to environmental noise which explains the sporadic faulty measurements (less than 1 % faulty measurements). Environmental noise spikes sometimes cause the circuit to latch 1 or 2 bit which are then evidently lost.

C. Experimental technologies

Applications in area C requires the total dose hardness of specific electronic components to be at least 1 MGy. Here no suitable solution is available yet. New emerging technologies such as silicon on insulator (SOI), gate all around (GAA) and Gallium Arsenide (GaAs) could eventually bring a solution. An overview and state of the art on these technologies will be outlined later in the paper.

D. Discrete components

When extreme conditions are encountered (area D), a different approach is required. The use of discrete components such as discrete bipolar transistors and robust switching elements such as relais, as well as a design of the circuit which takes into account the radiation induced degradation of each separate component has to be considered. Using such an approach, two simple multiplexer circuits were build and irradiated up to a total dose of 8 MGy [5]. Irradiation was performed in the CMF spent fuel facility [4] at SCK·CEN, with a dose rate of 20 kGy/h in an inert atmosphere at 65 °C. A low switching rate multiplexer (switching rate 10 Hz) based on an RS flip-flop using a Schmitt trigger and a few relays (see figure 5) was first designed and modelled using Spice simulations [6]. Results from previous irradiation experiments were included in the Spice model to maintain the functionality of the circuit, even at extremely high total doses. The circuit was monitored during irradiation and started to fail at a total dose of 1 MGy. Post irradiation examination (total dose 7 MGy) showed that

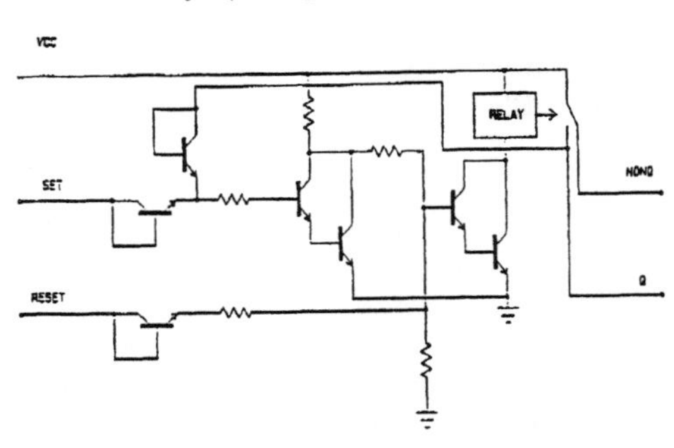

Figure 5. RS Flip-flop based on Schmitt Trigger

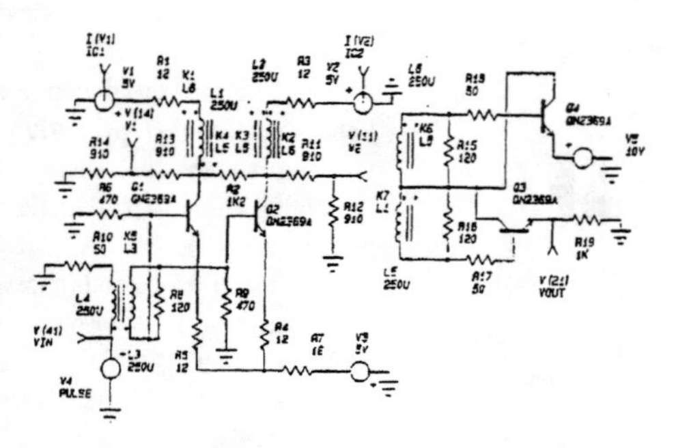

Figure 6. High switching rate multiplexer channel

corrosion effects had lead to a short circuit which was causing the failure. After cleaning, the circuit appeared to be functioning as expected by the simulations. Since this circuit was rather limited in switching time, an alternative fast switching circuit was built using a balanced transistor chopper (figure 6).

This circuit was also modelled and irradiated at the same time as the low speed circuit. Failure occurred at a total dose of 2 MGy, but here again after cleaning the corroded parts of the circuit, functionality was maintained at a total dose level of 8 MGy.

This approach clearly shows that a good knowledge of the degradation of each component, incorporated in a radiation hardened design of the circuit is quite suitable to build rad hard applications even to very high total dose values. The same technique can be used as well with application specific integrated circuits (ASIC): the combination of a radiation hardened processing technique and a radiation hardened design could possibly increase the total dose tolerance of such a circuit with one or two orders of magnitude.

IV. OVERVIEW OF RAD HARD TECHNOLOGIES

Throughout the literature, extensive data on the radiation tolerance of electronic components can be found. In general, three aspects are considered: single event upset, dose rate events and total dose effects. Single event upsets are usually of great importance for spacecraft electronics, dose rate effects are of more interest to military applications. For the nuclear applications, total dose effects are the main concern: since maximum dose rates, except for some very specific applications, are less than 1 kGy/h, dose rate effects can be neglected.

Standard non-hardened CMOS components offer usually a radiation hardness up to 10 Gy [1], [2]. However the manufacturers of these components do not offer any hardness assurance of their products. Commercially available rad-hard CMOS products offer hardness assurance up to 10 kGy. Silicon on Sapphire (SOS) techniques are presently used by Harris to provide circuits hardened to 10 kGy or more. SOS techniques offer not only a higher radiation resistance than ordinary MOS-devices, but also allow for high speed applications [11].

Experimental hardened CMOS circuits have been reported in the literature at levels of 100 kGy and more. The Harris AVLSIRA process, a bulk CMOS process has been reported to show only small changes in behaviour up to a total dose of 100 kGy [7]. Silicon on Insulator (SOI) techniques have been reported with total dose hardness of 1 MGy [8]. Gate all around (GAA) techniques are an improvement of the SOI technique and show promising results [9], [10]. Gallium Arsenide (GaAs) technologies show to be very hard against total dose effects: minor radiation damage is measured at 1 MGy in GaAs FET's [14]. Bipolar integrated circuits (TTL an I2L) are relatively hard to total dose effects. Most circuits can be operated up to total dose levels of 10 kGy. Modern bipolar technologies like emitter coupled logic (ECL) and collector diffused isolation (CDI) remain hard to above 100 kGy [15]. Figure 7 gives an overview of the different technologies and their total dose hardness.

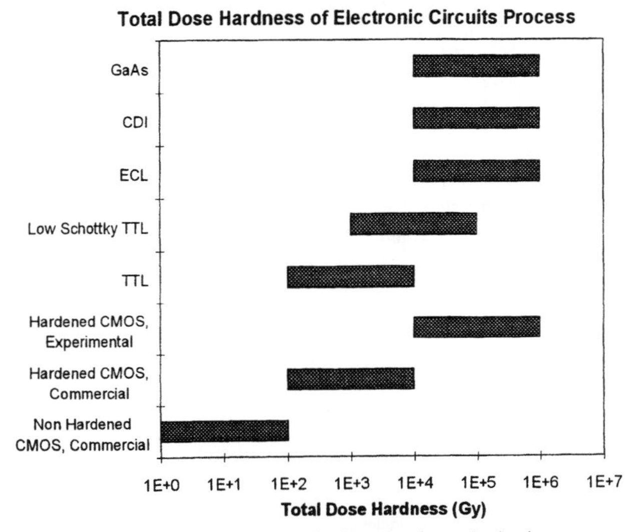

Figure 7. Overview of total dose hardness for electronic circuits process

Other approaches for radiation hardening have been studied as well. It is well known that specific biasing and polarisation conditions for most electronic devices can increase or decrease their total dose hardness. Using a redundancy of components to alternate the supplied/unsupplied phases the total dose level of 68HC05 microcontrollers has been increased by a factor of 20 [12]. The combination of a radiation hardened process and a radiation hardened design in the form of application specific integrated circuits (ASIC) could also allow to increase the total dose hardness of dedicated circuits with one or two orders of magnitude. A U.H.F. complementary bipolar process using dielectric isolation, used as a possible basis for an ASIC, has been tested up to 30 kGy, without major degradation [13].

V. CONCLUSION

Within the radiation hardening process of robotics systems, the radiation hardening of electronic components and, in particular multiplexing components, has to be considered as well. An increasing number of sensors will render umbilical management difficult and therefore decreases the overall reliability of the system. Various possibilities for multiplexing are open, but each solution will depend on the radiation tolerance requirements of the application. Inspection and waste surveillance do not require radiation tolerant electronics. Decontamination work and some work in hot cell require a low radiation tolerance which can be provided by most space graded components. Typical teleoperation work requires a total dose assurance up to 10 kGy, which is commercially available with major producers of electronic components. Extreme cases of radiation levels require new technologies, which are now at an experimental state, to be developed up to a commercial level. Another solution can be found in combining a rad hard technology and a rad hard design in the form of an ASIC. Further research on new technologies and on design techniques is necessary to allow for reliable radiation tolerant components, at very high total doses.

VI. ACKNOWLEDGEMENT

The work on radiation tolerant multiplexing techniques has been performed under partial support of the European Commission, particularly in the framework of the TELEMAN/ENTOREL Project (Contract F12T-CT-0011) and the ITER - T252 task of the Programme for Fusion Technology (Contract ERB-5000-CT95-0070).

VII. REFERENCES

[1] A.E. Benemann, "A Guide to Total Dose Effects on Electronic Components", TELEMAN/ENTOREL, TM41, Siemens-03-31-93-02, October 1993

[2] A. Holmes-Siedle, L. Adams, *Handbook of Radiation Effects*, Oxford Science Publications, Oxford University Press, 1993

[3] S. Coenen, D. Van Beckhoven, "MUSCADET: Gamma Irradiation Experiment on Multiplexing Components, Irradiation Report", SCK·CEN -Note G04004 - 27/95 - 100 SC/vv

[4] S. Coenen, "Les Installations d'Irradiation Gamma au SCKCEN et les Essais de Tenue aux Radiation pour la Fusion et le Démantèlement", *L'Onde Electrique*, Vol. 75 N° 3, pp. 53 - 57, May 1995.

[5] M. Decréton, S. Coenen, F. Du Mortier, F. Vos, "GASTAFIORE II, Gamma Irradiation of Sensors Transducers and Fibre Optics for Remote Handling", SCKCEN-Note TEL/F6010/91-126/MD, July 1991.

[6] P. Leszkow, M. Decréton, "Radiation Tolerant Techniques of Multiplexing Analogue Signals in a Nuclear Environment", *Proceedings of the RADECS-91 Conference*, pp. 254 - 258, Montpellier (France), Sept. 9-12, 1991.

[7] S.L. Thomas et al., "Measurements of a radiation hardened process: Harris AVLSIRA", *Nuclear Instruments and Methods in Physics Research* A-342, pp 164-168, 1994.

[8] J.L. Leray et al., "CMOS/SOI Hardening at 100 Mrad", *IEEE Transactions on Nuclear Science* Vol. 37, N° 6, December 1990.

[9] E. Simoen et al., "D.C and Low Frequency Noise Characteristics of Gamma-irradiated Gate-All-Around Silicon-On-Insulator MOS Transistors", *Solid State Electronics*,Vol. 38, N° 1 pp. 1-8, 1995.

[10] J.P. Colinge, A. Terao, "Effects of Total-Dose Irradiation on Gate All-Around (GAA) Devices", *IEEE Transactions on Nuclear Science*, Vol. 40, N° 2, pp. 78-82, April 1993.

[11] G.C. Messenger, M.S. Ash, *The Effects of Radiation on Electronic Systems*, Van Nostrand Reinhold Company, New York, 1986

[12] A. Giraud, M. Robiolle, "Fault Tolerant embedded Computers and Power Electronics for Nuclear Robotics", *Proceedings of the ANS 6th Topical Meeting on Robotics and Remote Systems*, pp. 143 - 150, Monterey, California, February 1995.

[13] M. Mellote et al., "Total Dose Response of a U.H.F. Ccomplementary-Bipolar Process using Dielectric Isolation (DI/SOI)", *RADECS 3 Conference*, September 1995, Arcachon, France.

[14] R. Zuleeg, K. Lehovec, "Radiation Effects in Gallium Arsenide Junction Field Effect Transistors", *IEEE Transactions on Nuclear Science*, NS-27 (5), pp. 1343-1354, October 1980.

[15] A. Benemann et al., "The Effects of Radiation on Electronic Devices and Circuits", *Kerntechnik* 55, N° 5, pp. 261 - 267, 1990

ROBOT CONTROL STRATEGY FOR CAMERA GUIDANCE IN LAPAROSCOPIC SURGERY

A. Casals, J. Amat, R. Sarrate
Dep. of Automatic Control and Computer Engineering.
Universitat Politècnica de Catalunya (UPC).
Pau Gargallo nº 5, 08028-Barcelona (Spain).
Fax: 34-3-4017045, email: casals@esaii.upc.es

ABSTRACT

A robotic system to automatically guide the camera in laparoscopic surgery is presented. The goal of this work is to generate the adequate camera control strategies to track the working scene during a surgical procedure. The autonomous behaviour to the camera guidance releases the surgeon from the continuous attention of its manual control or avoids the need of an assistant. The system is based on the computer vision analysis of the laparoscopic image that allows to identify either a scene's relevant point, or the surgical instruments. The adequate robot guidance requires to define a smart strategy to foresee the surgeon needs and to move the camera smoothly at the convenient speed in every different situation.

KEYWORDS: laparoscopic surgery, medical robotics, vision based control

INTRODUCTION

In the last years, significant research efforts have been done in the development of technology for minimally invasive surgery [1]. An analysis of robotics in minimally invasive surgery is done in [2], while in [3] a study of the problems in laparoscopic surgery is presented as well as their possible solutions. The main research efforts done up to now, in this kind of surgery, to aid the surgeon and to decrease intervention risks have been oriented mainly in two directions. First, the increase of the tele-perception capabilities, either with the introduction of 3D imagery or the perception of forces and touch. And second, the development of robotic systems to hold and move the laparoscope.

Some robotic manipulators under human control have been developed and tested. In [4] a six degrees of freedom manipulator controlled by the surgeon by means of a foot pedal is described. Other manipulators have already been designed and tested using different interfaces to receive the surgeon orders, either through oral commands or by means of head movements [5] or teleoperated [6]. A feasibility study in the use of image processing and other human-machine telerobotic command interfaces have been published [7].

This paper analyzes the control problems of a robotic system designed to automatically guide, by means of computer vision [8], the laparoscope during a surgical procedure. The first clinical trials in real conditions were different colecystectomies carried out in february 1996 in the Adrià Surgical Centre in Barcelona by the surgeon Dr. Enric Laporte. In a first phase an industrial robot, Scara type (fig. 1), has been used, but in the next future a specific robot will be developed, so as to reduce the system cost as well as to minimize the occupied space within the operating theater.

THE ROLE OF THE ASSISTANT ROBOT

Laparoscopic surgery requires the introduction into the patient's abdomen of several instruments, the surgical tools, as well as the TV camera microoptics, the laparoscope, to provide the surgeon with the visualization of the intraabdominal working space.

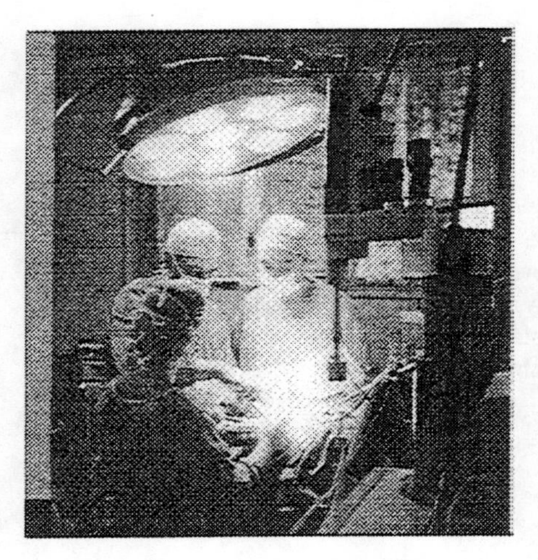
Fig. 1 The experimental robotic system scenario

The whole robotic system developed consists of the robot holding the laparoscope, and the vision system that processes the visualized image in order to generate the robot control orders, thus avoiding the need of manual assistance. The robotic system constitutes an auxiliary aid to the surgeon providing a smart continuous assistance in camera guidance. This smart behaviour consists on keeping the image centered, as stable as possible, within the surgeon working area. This guidance avoids the need of explicit orders from the surgeon, and so allows to reduce the total intervention time. The robot guidance is based on the error analysis of the selected image elements position with respect to the image central point.

The laparoscope movements are restricted to four degrees of freedom due to the constrains imposed by the operation through the trocar. This behaviour corresponds to that of a passive cylindrical joint located in the abdominal wall entering point. Due to the elasticity of the patient's abdominal walls, the above-mentioned cylindrical joint has two additional passive degrees of freedom, thus providing the required four d.o.f. The resulting four degrees of freedom are three rotations φ, ψ and θ and the prismatic movement ρ, as shown in figure 2.

Although the laparoscope rotation θ can be useful for a better centering of the image when oblique optics are used, it produces a rotation of the visualized image on the screen with respect to the surgeon reference working frame, when the laparoscope utilized has a frontal vision. This rotation demands a mental effort from the surgeon to keep his or her operating ability. For this reason many surgeons, among them those of the research team working in this project, impose the avoidance of this rotation. It is not easy to manually maintain θ fixed , but on the other way, the removal of this d.o.f. simplifies mechanically the link between the robot and the laparoscope in the automatic mode. Thus, the system requires only three degrees of freedom.

The whole system structure is shown in figure 3.

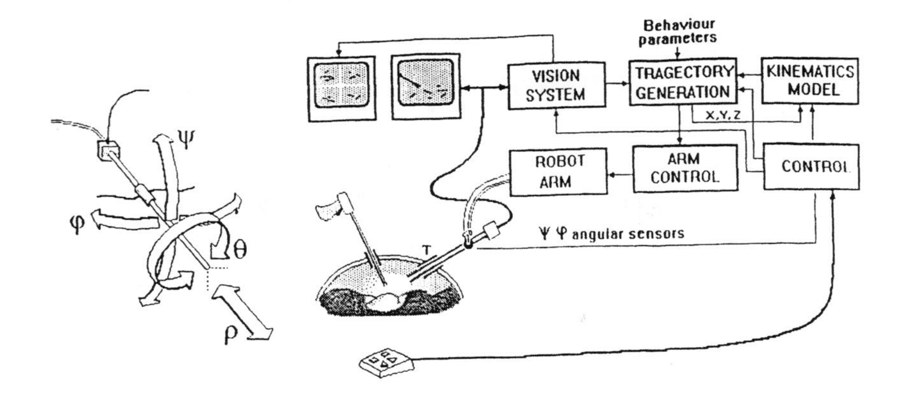

Fig. 2 Laparoscope degrees of freedom

Fig. 3 System structure

THE VISION SYSTEM

The vision system works from two main algorithms oriented to the detection of the elements to be tracked by the camera. The first detects and tracks the surgical instruments, while the second locates a relevant enough element on the scene, selected by the surgeon.

The first algorithm is based on the extraction of some features obtained from the surgical instruments. To assure the absolute reliability of the system the surgical instruments have been marked with strips that facilitate their detection and improve their location. The third dimension needed to control zooming is inferred from the 2D image by measuring the size of the instruments marks.

On the other hand, there are some situations during the intervention in which it is convenient to fix the laparoscope position so as to be able to freely move the instruments or to change them without affecting the laparoscope. In this case, the possible patient movements can change the selected point of view. The system must compensate these movements. In such situations the system changes to the manual mode and the surgeon orders the robot to memorize the laparoscope position to maintain the selected field of view.

Under these circumstances the second vision algorithm applies. The algorithm takes as reference features for the camera positioning the most relevant lines of the scene contained in a window placed in the center of the image.

The vision system developed allows to locate the selected elements once each 120 ms. This sampling time is enough to guarantee the generation of the required data for the robot control in real time.

CONTROL STRATEGY

The problem to solve, to adequately guide the robot during an intervention, is to establish the control strategy that dynamically determines the laparoscope position which is really efficient to the surgeon at every instant.

Operating modes

The usual operating mode is automatic, that means that the robot is directly controlled from the vision system orders. From the study of the movements that the assistant uses to perform under the surgeon requirements and for the different situations appearing in a diverse set of surgical interventions, we have established a serie of rules.
When the guidance criteria is to keep the camera field of view over a given area of interest the system operates in manual mode.
The surgeon need to change the zoom at his will in some special situations, requires a third mode.

Automatic mode. In this operating mode there are three different situations. In all of them the centering of the scene that results more efficient to the surgeon has been shown to obey the same common criteria. These criteria have been used to define the system behaviour.

- tracking of a scene with two tools: In this case the scene center point is considered to be the central point of the segment defined by the two tools ends points.
- tracking of a scene containing a working tool and an auxiliary steady one: In this case, the image center zone is also a point located in the segment defined by the two tools end point, but this point is closer to the tool that moves faster.
- tracking of a scene containing only one working tool: In this case the point selected as image center is not exactly the tool tip, but it is shifted a step forward.

The control criteria is based on the calculation of the position error function. The error considered is the distance between the evaluated interest center point and the image center.

Manual mode. The manual control strategy is not conceived as a manual robot guidance supplying the robot coordinates at the surgeon will. This kind of control would take a relatively important amount of time during the intervention. In spite of this, the manual operating mode is an aid to maintain a given point of view or to move towards some points predefined by the surgeon during a previous exploration. In both cases, the fixed observation points also require an image servostabilization to compensate the camera movements that can be produced due to the movements of the trocar, consequence of the patient's movements.
The surgeon control of the observation points to be stored or their posterior retrieval is carried out by means of a foot pedal. Further to the image monitor, a second screen allows to visualize the images taken from the memorized point of views, thus facilitating its way back to them.

Zoom control. An additional control possibility is to provide the surgeon the capability to change the zoom according to his needs. This control is done by means of two more push-buttons that enable the surgeon to increase or to decrease the field of vision, either in manual mode, over the image fixed point, or in automatic mode, tracking the area of interest.

Kinematic chain analysis

The kinematic chain that starts at the robot end effector and finishes at the laparoscope tip consists of two articulated links.

The first link is an extension of the robot end-effector that is used to move the robot away from the operating table, in order to let some free space to the surgical team. The only effect of this link is a translation of the robot end-effector axis.

The second link is the laparoscope itself. The laparoscope is coupled to the previous link through an universal joint having two degrees of freedom. This link is guided, on one extreme by the universal joint and in an intermediate point by the trocar, that allows its displacement and the three rotations φ, ψ and θ shown in figure 2.

A given point in the working space that has the coordinates ${}^{W}X$, ${}^{W}Y$ and ${}^{W}Z$ in the reference frame corresponding to the laparoscope tip, corresponds to a point ${}^{R'}X = X - X_0$, ${}^{R'}Y = Y - Y_0$ and ${}^{R'}Z = Z - Z_0$ (figure 4), in the reference frame defined in the trocar, being X_0, Y_0 and Z_0 the displacement produced by the extensor link utilized.

At every different instant these coordinates verify, assuming the trocar position is considered fixed, the following relations:

$$\frac{{}^{R'}X}{{}^{W}X} = \frac{{}^{R'}Y}{{}^{W}Y} = \frac{{}^{R'}Z}{{}^{W}Z} = \frac{\sqrt{{}^{R'}X^2 + {}^{R'}Y^2 + {}^{R'}Z^2}}{\sqrt{{}^{W}X^2 + {}^{W}Y^2 + {}^{W}Z^2}} = \frac{L-l}{l}$$

where L is the laparoscope length and l is the length of the laparoscope penetration.

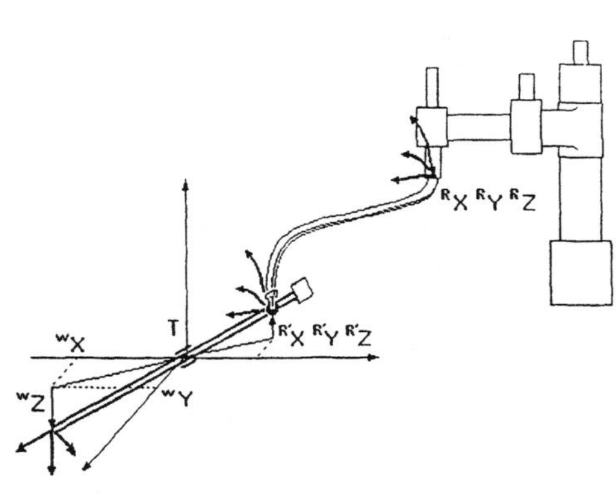

Fig. 4 The robotics kinematic chain

This expression allows to calculate the mechanical gain produced by the trocar depending on the laparoscope penetration length, data required to stabilize the guidance system.

The assumption that the trocar is fixed although it can really move is not a major problem since its movements produce variations in l relatively low. These low variations do not cause appreciable changes on the mechanical gain producing only displacements in the axis $\Delta^{W}X$ and $\Delta^{W}Y$ that the closed loop control has to compensate.

Closed loop behaviour

The error function obtained by the vision system can not be used directly as a control signal to move the robot due to the continuous image oscillations that this guidance would produce. For that reason a filtering of the error signal is required.

According to the surgeon kind of movements two different strategies are considered. Frequently the surgeon works over a local area with small amplitude movements. In these situations the robot arm should track the global movement trend (i.e., the lower frequency components) filtering out higher frequency components.

However, the surgeon must be able to drastically change the operation area. In this case, the robot should react accordingly to move quickly the camera.

These requirements have been satisfied using a Butterworth filter. The class of Butterworth IRR filters is suficient were requirements are loosely defined. The filter verifies:

- The filter can be customized by choosing the appropriate order n, and cutoff frequency f_c. High values of n increase the smoothness of the response but increases the delay. Similarly, lower values of f_c increase the signal smoothness. A second order filter proves enough for the application. The cutoff frequency can be adjusted to set the dynamics of the camera motion.
- The instantaneous error between the actual interest area position and the filter output is continuously measured. Whenever the absolute value of this error exceeds an adjustable threshold, the filter is reinitialized to the new instruments or interest point position to avoid moving out of the field of view. The threshold sets the maximum allowable movement amplitude that can be achieved working in a local area.

The interval of operational cut off frequencies ranges from 0,025Hz to 0,25Hz.

RESULTS

The development of this project has followed two successive phases. In a first phase we have developed and set up the vision system which identifies and tracks with high reliability the surgeon tools. In this phase a specific image processing hardware has been developed.

After the development of the visual tracking system and closing the loop with the robot, the next step has been to establish the control criteria according to the surgeons needs and preferences.

After some experimentation first with the surgeons training mock-up, and later on with chickens, until the system behaviour has been considered as very satisfactory by the surgeons of our working team, we have carried out different real clinical trials.

Acknowledges

We would like to acknowledge the valuable contribution of Prof. Josep Grané and also to the surgical team. We wish also to greatly thanks the task of Xavier Valls in the implementations and set up of the system.

This work is funded by the CIRIT, the Catalan Research Agency.

REFERENCES

1. H. H. Rinisland, "Basics of Robotics and Manipulators in Endoscopic Surgery" *Endoscopic Surgery and Allied Technologies*, Vol. 1, n. 3 June 1993 pp. 154-159
2. J. Troccaz, Robots in Surgery" *The 7th International Symposium on Robotics Research*, Munich, Germany October 21-24 1995 (in press)
3. A. Casals, J. Amat, E. Laporte, "Robotics Aids for Laparoscopic Surgery Problems" *The 7th International Symposium on Robotics Research*, Munich, Germany October 21-24 1995 (in press)
4. N. I. Narwell, D. R Weker and Y. Wang "A Force Controllable Macro-Micro Manipulator and its Application to medical robotics". JPL Computer Motion Inc. report 1994
5. P. A. Finlay, "Small is Beautiful". 1st. European Conference on Medical Robotics, Robomed'94, Barcelona, SPAIN June 20-22, 1994 pp,. 13-15.
6. D. Hurteau et al. "Laparoscopic Surgery Assisted by a Robotic Cameraman: Concepts and experimental results, IEEE International Conference on Robotics and Automation, 1994. pp. 2286-2289.
7. "An overview of Computer-Assisted surgery at IBM. G. *Robotics Research: IFRR*, 1994. pp. 533-544
8. A. Casals, J. Amat and E. Laporte "Automatic Guidance of an Assistant Robot in Laparoscopic Surgery" 1996 IEEE Int. Conference on Robotics and Automation, Minneapolis USA, 24-26 April, 1996 (in press)

DESIGN AND CONTROL OF AN ANTHROPOMORPHIC SERVOPNEUMATIC FINGER JOINT

DIPL.-ING. A. CZINKI
Department of Fluid Power Transmission and Control(IFAS), RWTH Aachen

DR.-ING. Y.-S. HONG
Korean Institute of Science and Technology(KIST), Seoul

PROF. DR-ING. H. MURRENHOFF
Department of Fluid Power Transmission and Control(IFAS), RWTH Aachen

ABSTRACT

The paper surveys the development of a servopneumatic finger joint, that is going to be the basis for a dextrous mechanical hand. The latter is intended to permit flexible manipulation of arbitrary objects in a hostile and disordered, three-dimensional environment. As a first step towards this servopneumatic hand, a finger joint prototype was built up. Main focus of its design has been the integration of the drive unit, the required angular and pressure sensors and the necessary valves into the finger digit. The report presents a brief description of the servopneumatic finger joint, along with the results of the prototype functional evaluation. It outlines the finger joint performance and provides a preview on the aspired servopneumatic hand.

KEYWORDS: servopneumatic finger joint, semi-rotary drive,
puls-modulated switching valves

INTRODUCTION

In general, industrial robot grippers are either designed with regard to the shape of a certain workpiece, or they are single-degree of freedom systems operated in an open control loop. In contrast to this, mechanical hands are seeking a generalised dexterity that might once allow them to replace human labour in sophisticated assembling, inspection and maintenance applications. This prospect is especially important as work in hazardous or remote environment is considered.

However, the substitution of human labour by mechanical hands is only possible if they offer a flexibility that is comparable to the human hand. Due to the necessity for a dextrous human hand like manipulator, intensive research was conducted in the field of grasping theory/L1, P1/, the design of mechanical hands/B1, M1/ and the control of multi-degree of freedom systems/C1, L2, N1/. Nevertheless, the dexterity of the human hand is - and will be for decades to come - unattainable. Besides insufficient sensory information and a lack of reasonable control strategies, mechanical hands are limited in their dexterity by inadequate drive systems. Electric drives, frequently applied as finger joint drives, are either very weak or unpleasantly heavy. In this context the application of servopneumatic drives might offer an interesting alternative. Pneumatic drives provide favourable qualities such as inherent overload security, good controllability and a good power-weight ratio. The latter makes them well suited for an integration into the finger digits and the use as direct drives, so that the installation space required for energy transmission can be reduced and the systems rigidy can be, due to the omission of tendons and cables, improved. The advantageous characteristics of controlled pneumatic drives, listed above, have been the motivation for the development of the direct driven, servopneumatic finger joint, that is presented in this paper.

DRIVE CONCEPT

Since a mechanical hand will consist of 9 to 20 degrees of freedom, reliability and easy maintenance of the drive units are primary objectives for the finger joint design. Additionally, the drive units must allow fine manipulation, which demands for low masses and low friction along with small clearances in the kinematic chain.

Pneumatic drives are well known to fulfil these demands in the context with linear motion. However, by the use of a suitable transmission, pneumatic drives can also provide high precision swivelling and rotary motions. In order to prove the suitability of pneumatic drives for the generation of precise swivelling motion, a variation of the rodless belt cylinder called 'belt motor' was developed at IFAS. It was successfully applied as a drive for a SCARA robot /S1/. With regard to the particularly high requirements of the finger joint application concerning positioning accuracy and minimisation, the belt motor concept was reconsidered and especially adapted to the requirements of a mechanical hand. Figure 1 shows the drive concept resulting from this work.

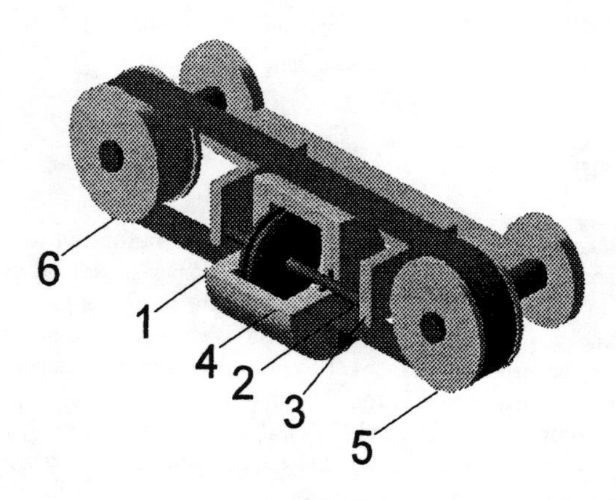

Figure 1: Drive concept

Actuation forces are generated by a standard pneumatic cylinder (1). Its piston rod(2) is fixed stationary on the systems housing(3). Thus an admission of pressure to one of the cylinder chambers causes the cylinder cage (4) to move. This linear motion is transformed into a rotation by a belt/pulley system, where one of the pulleys is used as a drive pulley (5) and the other (6) is utilised for the deflection of the belt.

FINGER JOINT DESIGN

As a first step towards a servopneumatic hand, the feasibility of the drive concept was investigated on a single finger joint(figure 2). Main objectives of the prototype have been the integration of the drive unit, the sensors and the valves into a test bench. Since the finger joint design is modular, it offers the option to build up and investigate various different finger joint configurations by simply adding additional finger joints. Furthermore, the modularity will minimise the manufacturing costs, simplify the control architecture and will reduce the expenditure necessary for maintenance.

Figure 2: Servopneumatic finger joint

FINGER JOINT CONTROL

Control of the joint is achieved by the application of four pulse-width modulated on-off-valves. They provide high flow rates in combination with low space requirements and thus offer an interesting alternative to the application of servovalves. However, the applications in a controlled pneumatic system require the ability of fine-regulation of the mass flow, which leads to high demands concerning the valve dynamics.

For the finger joint prototype micro valves with a nominal flow of about 8 l/min and a switching time of about 10 ms are applied. As their dynamics are insufficient, they are driven under over-excited conditions with more than triple voltage. In order to avoid electrical overload, a current control was incorporated. By this means, the switching time could be reduced to 1 ms.

For the control, a three loop state controller with an additional parallel branch is applied(figure 3). Basically, it consists of a non-linear control element in the main branch and a velocity and pressure feedback in parallel branches. The value of speed is derivated by differentiation of the angular sensor signal. Neglecting frictional and gravitational influences, the pressure difference between the cylinder chambers is proportional to the acceleration of the drive. This allows the substitution of the acceleration signal by the pressure signal. By this means, the problems arising from a repeated numerical differentiation can be avoided. An integrating element in a parallel branch improves the control accuracy. In order to avoid over-swings, this branch is only active in a certain range of deviation.

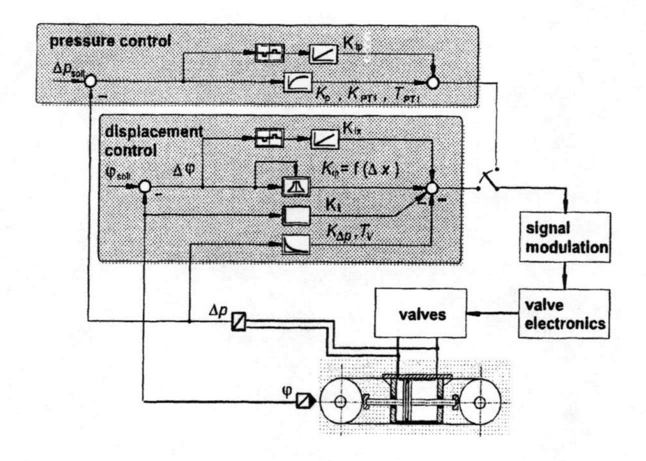

Figure 3: Controller layout

Figure 4 shows the finger joint response on a 10 to 100° step function. Basically, the plot documents the good dynamics and accuracy of the finger joint drive. The maximum finger joint velocity during a 90° finger joint rotation is higher than 300°/s. A 90° finger joint rotation takes less than 0.4 seconds. The cut-out of the displacement, shown in figure 4, proves the high accuracy of the drive. After a short hunting, the displacement deviation is down to one increment of the sensor system resolution, which corresponds to 0.03°. During the investigations in pressure and displacement control mode, the finger joint was able to prove its good dynamics and its high positioning accuracy.

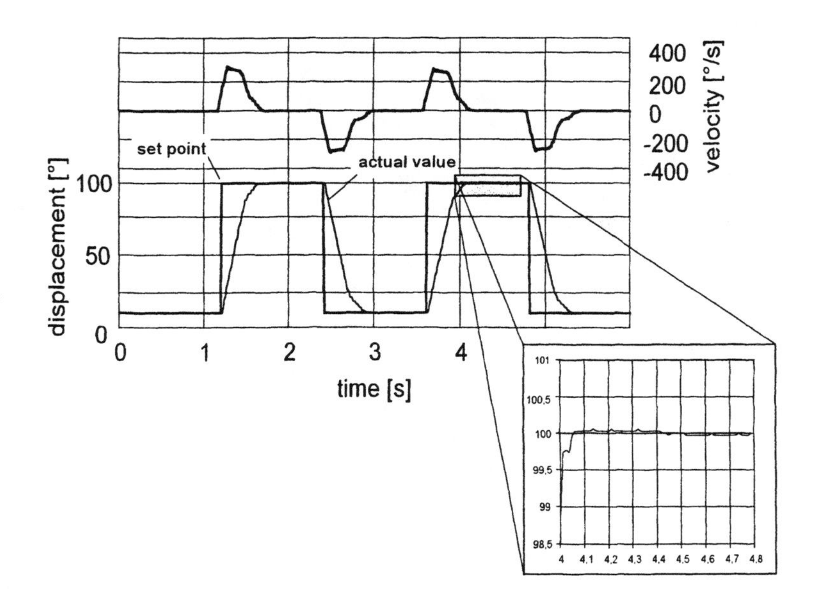

Figure 4: Step function response

OUTLOOK

Although the results of the finger joint investigations are very promising, the development of the finger joint prototype has to be regarded only as a first step towards a successful dextrous servopneumatic hand design. Future work will have to focus on:

- an enhancement of the drive performance
- an advanced degree of miniaturisation
- association of several finger joints to a servopneumatic finger and a servopneumatic hand, respectively
- investigations concerning the behaviour of this multi-degree of freedom system in compliance motion

and others.

SUMMARY

The paper surveys the development of a single-degree of freedom servopneumatic finger joint. With respect to the demands required for the use in a dextrous mechanical hand, a servopneumatic drive was developed. By the use of a linear pneumatic cylinder in combination with a belt/pulley system, a specifically space saving and powerful rotatory drive design was achieved. In contrast to many other anthropomorphic manipulators, the drive unit, the complete set of sensors and the valves are all integrated into the finger digit. With this configuration, an essentially autonomous finger module became possible. By the integration of the drive into the finger digit, the space required for the transmission of the actuation forces from the drive to the finger joint are minimised, and accuracy and controllability are improved. As a result of this work a substantially new drive concept, particularly adapted to the requirements of mechanical hands can be provided.

REFERENCES

B1	Bologni, L. Caselli, S. Melchiorri, C.	Design issues for the U. B. robotic hand, NATO ARW on Robots with dedoundancy: Design, Sensing and Control, Salo, Italy, 1988
C1	Cutkosky, M. R. Kao, I.	Computing and Controlling the Compliance of a Robotic Hand, IEEE Robotics and Automation Conference, San Francisco, USA, 1986
J1	Jacobsen, S. C Iversen, I. K. Knutti, D. F. Biggers, K.B.	Design of the UTAH/MIT dextrous hand, Proceedings of IEEE International Conference on Robotics and Automation, Philadelphia, USA, 1988
L1	Li, Z. Hsu, P. Sastry, S.	Grasping and coordinated Manipulation by a multifingered Robot Hand, The international Journal of Robotics Research, Vol. 8, No. 4, August 1989
L2	Loucks, C. S. Johnson, V. J.	Modelling and Control of the Stanford/JPL hand, Proceedings of IEEE International Conference on Robotics and Automation, Raleigh, USA, 1990
M1	Mason, M. T. Salisbury, J. K.	Robot Hands and the Mechanics of Manipulation, MIT Press, Cambridge, USA, 1985
N1	Narasimhan, S. Siegel, D. M. Hollerbach, J.	CONDOR: A revised Architecture for Controlling the Utah/MIT hand, IEEE Conference on Robotics and Automation, Philadelphia, USA, 1988
S1	Schillings, K. Tappe, P.	Adaptive and nonlinear Controller Concepts for a servopneumatic direct driven SCARA-robot, Third International Symposium on Measurement and Control in Robotics, Torino, Italy, 1993
P1	Park, Y. C. Starr, G. P.	Optimal Grasping using a multifingered Robot Hand, Proceedings of IEEE International Conference on Robotics and Automation, 1990

PRELIMINARY DEVELOPMENTS OF A CONTROL SCHEME FOR A 11-DOF ROBUST UNDERACTUATED ARTICULATED HAND

***Eric Dégoulange, **Clément M. Gosselin**

**LIRMM - Université Montpellier II*
161, rue Ada - 34392 Montpellier Cedex 5
France

***Département de Génie Mécanique*
Université Laval
Québec, Québec
G1K 7P4, Canada

ABSTRACT

This paper discusses the control of a three-finger underactuated articulated hand which is being developed at Laval University and which is designed for industrial applications. Since the hand is underactuated, classical grasping analyses and control schemes cannot be used. Several control srategies have been developed in order to handle the underactuation, which resulted in the definition of a particular combined position/force control scheme. Associated with the mechanical behavior of the underactuated fingers, this control scheme allows the grasping of all kinds of objects with a control of the forces applied to them. The prototype of the hand being under construction, the control scheme was tested on a finger simulator which includes a complete model of the underactuated finger and a complete model of the actuators. Following the presentation of the project, the control simulator is described and the principle of the combined position/force control is addressed. Some preliminary simulation results are then given in this paper.

KEYWORDS: articulated hand, underactuation, position/force control

1 Introduction

Recent breakthroughs in the field of manufacturing robotics, combined with the improved performance of the computers have made possible the execution of complex coordinated motions in real time. It then clearly appeared that, in order to extend the application domain of conventional serial manipulators, the ability of the robots to manipulate objects had to be improved. Ever since, the mechanics of grasping has become an important research subject, with two major aspects: the study of the problems related to the grasping of objects by the robot manipulators and the study of mechanical architectures and control strategies adapted to a robot hand.

Research has led to the design of several prototypes of articulated hands, ranging from simple one-degree-of-freedom grippers which are simple to control but have a low grasping potential, to the "human-like" model [1] allowing most of the possible grasping configurations, but which is relatively complex to control because of the large number of degrees of freedom to actuate.

The work presented here addresses the design and control of a 11-dof underactuated articulated hand, allowing most of the grasping configurations with a reduced complexity of the control scheme (only 5 actuators for the 11-dof).

The project and the specifications are presented in the first section of the paper. The various components of the fingers are then described. The third section presents the control simulator developed for testing the particular combined position/force control scheme which has been defined. The principle of this combined position/force control is then addressed, together with a brief review of control schemes of other existing hands. Finally, the last part of the paper presents simulation results.

2 Scope of the project

Researchers at Laval University are currently working on a project aiming at the design and control of a powerful and robust articulated hand intended to execute complex manipulation tasks in hostile industrial environments where human safety is threatened. One of the foreseen applications pertains to the maintenance of high voltage electric distribution lines, where the replacement of insulators still requires a human intervention. In this application, the hand will be mounted at the tip of a robot manipulator remotely controlled by an operator. The hand under development is therefore designed to lift and/or grip a 600 N load.

Preliminary studies have led to the design of a three-finger articulated hand. Each finger has 3 phalanges and the hand has a total of 11 degrees of freedom. This design is original in that the system is underactuated since the 11 dof are controlled by only 5 actuators (1 actuator per finger and 2 for the coupled self rotation of the fingers). The hand uses electric motors and ball screw assemblies ensure the reduction. Power is transmitted to the fingers by means of linkages. The fingers are 16.5 cm long and 4.0 cm wide. The ratio between the length of the successive phalanges is approximately 0.7 which leads to the optimum kinematic dexterity [2]. On the base, fingers are located at the vertices of a 10 cm equilateral triangle. The proposed solution allows most of the grasping configurations [3] with a minimum weight, low cost, and low size given the reduced number of actuators.

During a grasping task, the fingers move as a single rigid body in free space until one of the phalanges is in contact with the object. The phalanx in contact stops its motion but the following phalanges keep moving until the next phalanx establishes contact and so on. The particular mechanical structure allows a firm grasp of the object once in contact, since the links adapt their position to the shape of the object.

The concept and mechanical design of the underactuated finger is to be patented and hence, no detailed figure can be included in this paper.

3 Description of the components used for the hand

The fingers are moved by 3 electric motors (one by finger, via the ball screw assemblies). They are DC Brushless motors from Kollmorgen Inland Motor. The continuous stall torque delivered by the motors is about 30 N.cm (at zero speed), and will allow the fingers to exert forces up to the required 600 N. In order to know the position of the motor's rotor, each Inland motor is equipped with an optical encoder (2048 counts/rotation). The motors use PWM current amplifiers which allow to directly control the motor torque.

Because of the underactuation, the knowledge of the finger's configuration, as well as the position of the contact points along the phalanges, are important data which are used in the computation of the forces applied to the object. Therefore, rotating potentiometers have been added inside the phalanges and tactile sensors have been mounted on the internal side of each finger. The tactile sensors, made by Interlink, are only used to detect the contact points of the object along the phalanges and not to determine the magnitude of the contact forces.

4 Description of the simulator

The hand being under construction, a simulator has been developed for one finger in order to define and test the control algorithm to be used with the prototype. This simulator has been developed in C language and contains three main parts as shown in **Figure 1**.

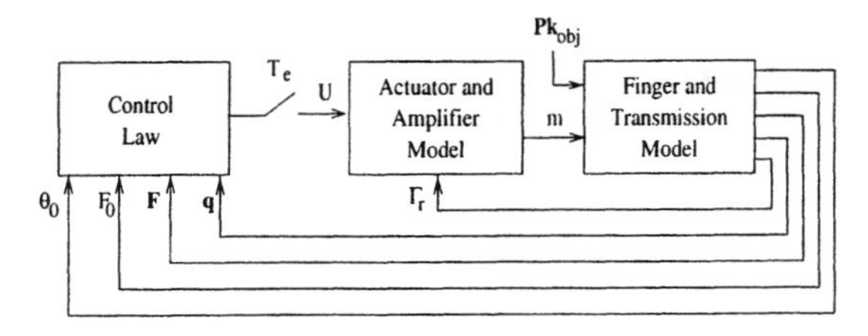

Figure 1: Principle of the control simulator.

The signal U, issued by the control law block, is sampled at T_e (sampling period of the system) and is sent to the actuator model block. This block includes a complete model of the actuator (electrical and mechanical model of the motor with dynamic equations, and electrical model of the PWM amplifier). Sampling and saturation phenomena are also taken into account. Variable m is the output of this block and represents the angular position of the motor. This position is sent to the last block of the simulator which includes a model of the transmission and a model of the underactuated finger. The contact between the link and the object is modeled as a high stiffness spring. The object is assumed to be a circle, and the vector $\mathbf{Pk}_{obj}= [P_{obj}\ k_{obj}]^T$ represents the position of the center and the radius (P_{obj}), and the stiffness (k_{obj}). Based on this data, the model determines the force vector $\mathbf{F}=[F_1\ F_2\ F_3]^T$ and the position vector $\mathbf{P}=[P_1\ P_2\ P_3]^T$ at the contact point on each phalanx, as well as the configuration $\mathbf{q}=[q_1\ q_2\ q_3]^T$ of the finger. Variables θ_0 and F_0 represent respectively the angular value of the linkage located at the base of the finger and the force exerted at the bar which transmits the power to the finger. Finally, Γ_r is the resistant torque applied to the motor. This external torque is equivalent to the force exerted by the finger on the object, including the friction forces.

Because of the complexity of the mechanical structure, the model of the finger is not general but can handle most of the loading cases. A similar model will have to be implemented on the real controller in order to determine the force vector **F**. But in this case, **q**, **P**, θ_0 and F_0 will be given by the sensors.

5 Description of the control scheme

At this point, most of the research has led to task description and definition of control schemes for fully actuated hands such as the UTAH/MIT Hand, the Standford/JPL Hand, the Hyuma Hand, the UB Hand II, the Anthrobot-2 Hand, and several others.

One of the most important problems is the stability of the grasping. The control vectors, defined in a frame attached to the object, usually describe the position and the orientation of the object. Since the system is redundant with respect to the task, an additional vector can be defined, corresponding to the internal force in the object. The control of this force ensures the stability of the grasp. The forces exerted by each finger can be related to the resultant force at the object by a grasping matrix noted **G** [4]. Another interesting point is that stability is improved if the system exhibits some compliance. It is more interesting to introduce this compliance at the control level. Salisbury proposed such a control scheme called "Active Stiffness Control" [5][6]. Other complex control schemes based on the notion of combined position/force control have been developed [7][8][9] for the types of articulated hand mentioned above.

Similarly, some underactuated hands such as the USC/Belgrade Hand and the UPenn Hand have been developed. For underactuated hands, the apparent redundancy with respect to the task cannot be used. The control of such hands then relies on the definition of strategies using a basic position control scheme with force detection and/or brake action [10]. Therefore complex manipulations cannot be performed.

In the case of the hand developed at Laval University, several control strategies have been developed in order to handle the underactuation. They resulted in the definition of a particular combined position/force control scheme, as shown in **Figure 2**.

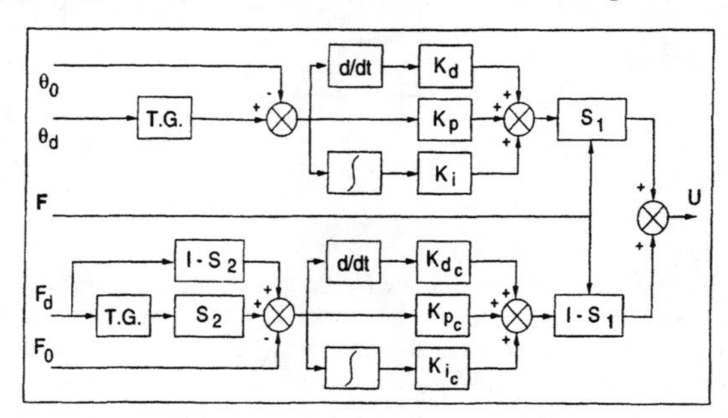

Figure 2: Block diagram of the combined position/force control.

Position control is performed during the approach phase of the fingers until contact with the object is realized. Due to the underactuation, it is impossible to constrain the configuration of the fingers in free space. On the other hand, their mechanical structure allows a firm grasp of the object once in contact, since the links adapt their position to the object's shape. Force control is then used to apply the desired grasping forces. The force applied to the object is determined by the model of the finger, the motor torque and the tactile data.

Both position and force control use a PID type control law. S_1 allows to select one of the two servo loops. If S_1=1, position control is used while if S_1=0, force control is used. The value of S_1 is a function of the vector **F**. Similarly, S_2 allows to either apply the desired force in one step or use a force generation to reach the desired force. Moreover, the controlled variables of the force control loop can be the force F_0, the equivalent torque Γ_0 or, interestingly, the force on one of the phalanges (one component of **F**).

6 Simulation and results

Several simulations have been performed and only a few typical results obtained using the combined position/force control are reported here. The simulation consists in fixing a desired joint position θ_d so that the object is in contact before $\theta_0 = \theta_d$. The desired grasping effort is defined by the torque Γ_d at the base of the finger, set to 5 N.m. **Figure 3** presents the initial and final configurations for a given loading case as well as the simulation parameters.

The results are shown in **Figure 4**. Figure 4-a compares the evolution of the angle θ_0 initially computed by the trajectory generation (dotted line) with the actual evolution of the angle (solid line) using the combined position/force control. It can be observed that the trajectory described by θ_0 starts to deviate from the reference trajectory when the force control loop is activated (at time t=0.493s).

Similarly, figure 4-b shows the evolution of the torque Γ_0 for the combined position/force control (solid line) and for a simple position control (dotted line). In the latter case, it can be observed that the torque is larger than the desired value. Figures 4-c and 4-d respectively present the evolution of the angles and the evolution of the forces at the phalanges. The dashed line corresponds to the phalanx near the base, the dotted line to the following phalanx and the solid line to the distal phalanx. One can also see the instant at which each phalanx establishes contact with the object (t1=0.184s, t2=0.354s, t3=0.493s). It is worth mentioning that the three forces remain low during the closing phase, until the three phalanges are in contact with the object. This natural mechanical behavior of the finger is essential to ensure a complete grasp of the object without pushing it away.

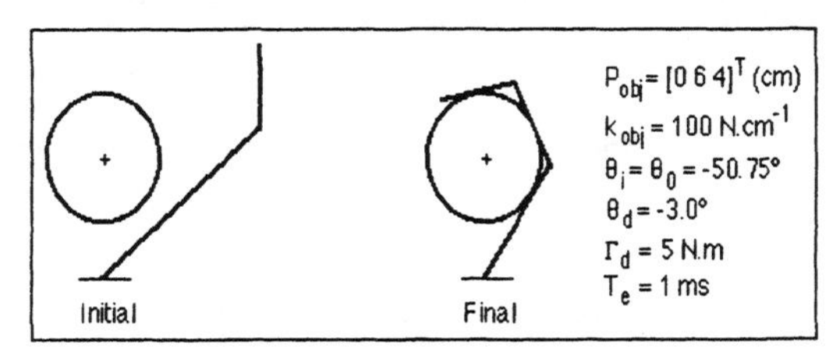

Figure 3: Initial and final configurations of the finger.

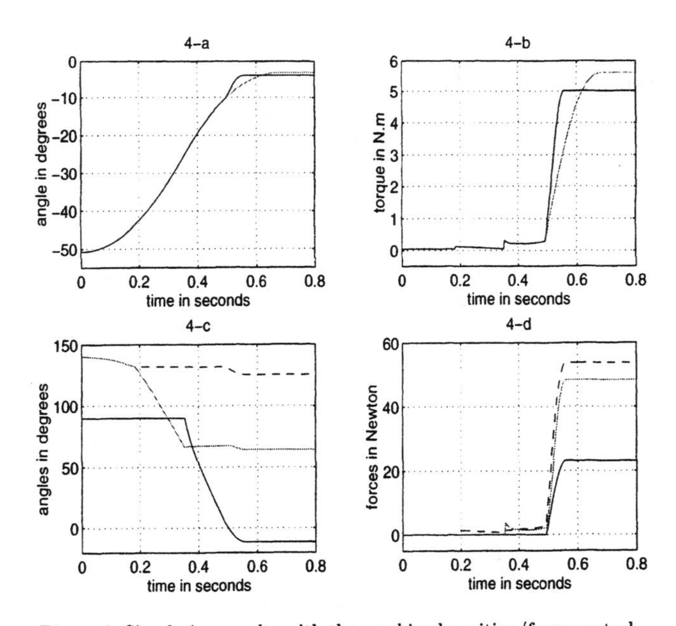

Figure 4: Simulation results with the combined position/force control.

Once the desired torque is reached, and the system is stabilized, perturbations consisting in moving the object towards the finger with constant speed (0.5 cm.s^{-1}) have been introduced, between time t=0.65s and t=0.75s. When a simple position control with force detection is used, the forces increase and remain high (figure 5-a) with the risk of breaking the object. On the other hand, the combined position/force control allows to compensate for the variations of the forces applied to the object, as shown in figure 5-b. This is another advantage of the proposed control scheme.

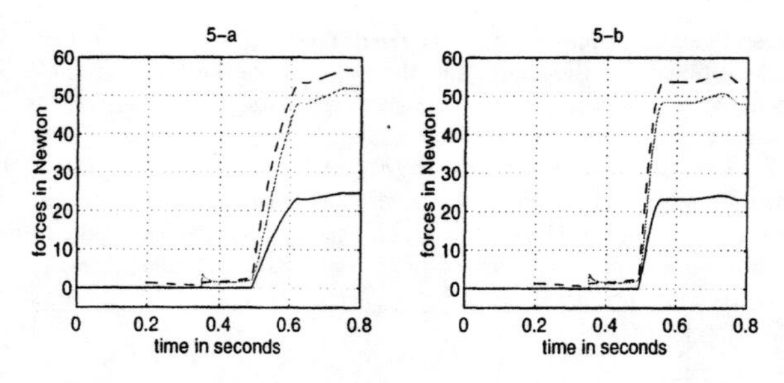

Figure 5: Comparison of the responses with a moving object.

7 Conclusion

In this paper, the first simulation results of a control scheme for a new underactuated hand developed at Laval University have been presented. The combined position/force control scheme proposed here combines the advantages of the position control (in free space) and the force control (in constrained space). The next step of this work consists in implementing this control scheme on the real controller and evaluate the performance on the experimental setup.

Acknowledgements: This work has been completed under research grants from l'Institut de Recherche en Santé et Sécurité du Travail du Québec (IRSST). The authors would also like to thank Thierry Laliberté for his work on the mechanical design of the Laval University Hand.

References

1. S.C. Jacobsen et al., 'The UTAH/MIT Dextrous Hand: Work in Progress', *The International Journal of Robotics Research*, Vol. 3, No. 4, pp. 21-50, 1984.
2. C.M. Gosselin and J. Angeles, 'A Global Performance Index for the Kinematic Optimization of Robotic Manipulators', *ASME Journal of Mechanical Design*, Vol. 113, No. 3, pp. 220-226, 1991.
3. S.C. Jacobsen et al., 'High Performance, High Dexterity, Force Reflective Teleoperator', *ANS Winter meeting*, Washington, 1990.
4. M.T. Mason and J.K. Salisbury, 'Robot Hands and the Mechanics of Manipulation', *The MIT Press*, Cambridge, 1985.
5. J.K. Salisbury, 'Active Stiffness Control of a Manipulator in Cartesian Coordinates', *Proceedings of the 19th IEEE Conference on Decision and Control*, Albuquerque, pp. 95-100, 1980.
6. J.K. Salisbury and J.J. Craig, 'Articulated Hands: Force Control and Kinematic Issues', *The International Journal of Robotics Research*, Vol. 1, No. 1, pp. 4-17, 1982.
7. J.P. Merlet, 'Commande par retour d'efforts', *Technique de la robotique*, Tome 2, Hermes, 1988.
8. S. Narasimhan et al., 'CONDOR: A Computational Architecture for Robots', in *Dextrous Robot Hands*, (S.T. Venkataraman and T. Iberall eds.), Springer-Verlag, New-York, pp. 117-135, 1990.
9. T. Okada, 'Computer Control of Multijointed Finger System for Precise Object-Handling', *IEEE Journal of Systems, Man and Cybernetics*, Vol. 12, No. 3, pp. 289-299, 1982.
10. M. Saliba and C.W. de Silva, 'An Innovative Robotic Gripper for Grasping and Handling Research', *Proceedings of the International Conference on Industrial Electronics, Control and Instrumentation*, Kobe, pp. 975-979, 1991.

Gravity Counter Balancing of a Parallel Robot for Antenna Aiming

G R Dunlop and T P Jones
Department of Mechanical Engineering, University of Canterbury,
PB 4800 Christchurch, New Zealand.
EMail: R.Dunlop@Canterbury.ac.NZ

Abstract

A three degree of freedom spherical coordinate robot used to aim an antenna has been modified so as to reduce the parallel mechanism to the two angular degrees of freedom. This allows the antenna to be aimed anywhere within the visible hemisphere without encountering a singularity. Only two servo mechanisms are needed to aim the antenna. When the antenna is scanned along the horizon, the servo torque requirements vary rapidly and are maximised. The use of counterweights to reduce the torque requirements is examined, and results are presented to show the practicality of such an antenna aiming system based on this parallel robot.

INTRODUCTION

The parallel robot approach to antenna aiming[1,2,3,4] was developed as a way of overcoming the "keyhole" problem[3,5,6] which is associated with the first axis of standard 2 dof (degrees of freedom i.e. elevation and bearing in this case) antenna aiming systems. The first axis is the vertical axis of the common Alt-Az (elevation or altitude over azimuth) antenna mounting system, the horizontal axis of the less common x-y mounting system, or the north-south axis of an astronomical mounting system[3]. The singularities in these 2 dof serial mechanisms may arise either from mechanical interference or from dynamic positioning limitations[6].

An early attempt[2,3] to use parallel mechanisms to solve the keyhole problem used a 6 dof Stewart platform in which the extra 4 dof were used to maximise the antenna pointing stiffness i.e. to stay away from the singularities where an uncontrolled dof is gained. Later work[4,7] resulted in a triple servomotor 3 dof parallel mechanism for spherical coordinate positioning. The kinematics of this 3 dof parallel mechanism (shown in fig. 1) are straightforward[7]. Since only the 2 angular coordinates were required to aim the antenna, the radial variation was not required so the radius was held constant by a central strut as shown in fig. 2. The result is a 2 dof parallel mechanism suitable for aiming an antenna anywhere within the visible hemisphere i.e. anywhere above the horizon. The mechanism is designed to be free of singularities within this workspace, and only 2 servomotors mounted on the base are required to aim the antenna. It is worth noting that all conventional 2 dof serial aiming mechanisms must have the first servo drive axis strong enough to carry not only the antenna and the counter weights, but also the second axis servomotor and gearbox. The parallel mechanism can have the servomotors and gearboxes mounted directly on the base where the mass does not contribute to the load.

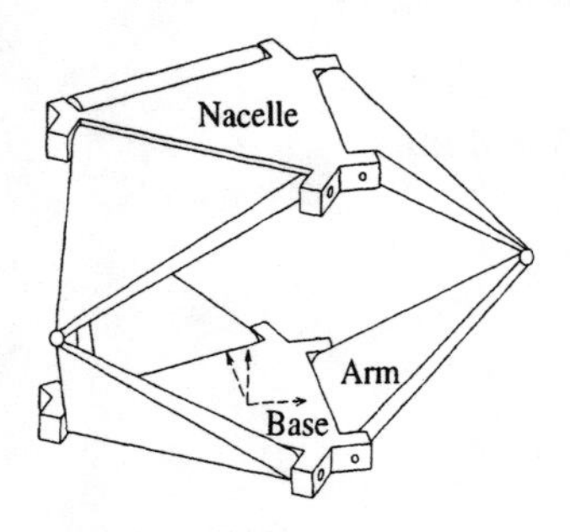

Figure 1. Three axis parallel link mechanism requires 3 servomotors. $\{M,n,g,\Sigma f\}=\{3,8,9,15\}$

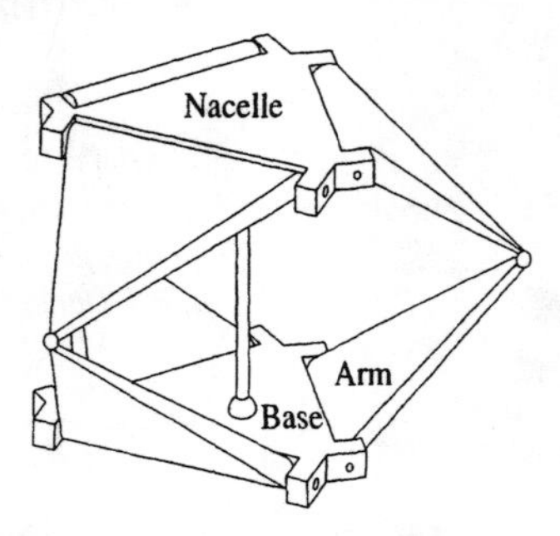

Figure 2. Two axis parallel link mechanism requires 2 servomotors. $\{M,n,g,\Sigma f\}=\{3,9,11,21\}$

THE PARALLEL MECHANISM

The 3 dof parallel mechanism shown in fig. 1 has a mobility of 3 as calculated from Kutzbach's criteria: $M = 6(n-g-1) + \Sigma f$ where n is the number of links, g the number of joints and Σf the sum of all the joint freedoms. The 2 dof mechanism shown in fig. 2 also has a mobility of 3 but one of these corresponds to rotation of the central strut about its longitudinal axis (i.e. unobservable) which has no effect on the other 2 mobilities which correspond to the antenna aiming angles. From a theoretical point of view, the ball joint at one end of the strut can be replaced by a Hooke or Cardin joint. The rotational freedom of the central strut is removed so that the parameters are $\{M,n,g,\Sigma f\}=\{2,10,12,20\}$ and the 2 remaining mobilities are those required. In practice, it is just as easy to use the ball joint.

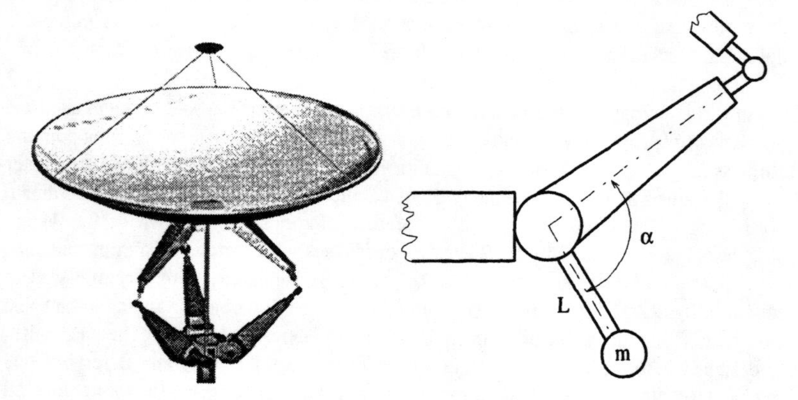

Figure 3. The parallel mechanism antenna mount.

Figure 4. An arm showing the position of the counter weight (1 of 3 arms).

It was found that gravitational loading on the 2 dof parallel antenna aiming system shown in fig. 3 produced the largest torque demand on the servo mechanisms when the antenna was aimed at the horizon. The worst case loading on the servos occurs when the antenna sweeps 360° in azimuth at 0° elevation. The torque requirements for a realistic 2m diameter antenna are plotted in fig. 5 which shows the torque requirements lie in the range of +360Nm to -200Nm. This is to be compared[8] to a similar conventional Alt-Az system which required ±1500Nm to aim exactly at a weather satellite passing near the zenith at a maximum elevation of 89.9°. The conventional antenna system incorporated counterweights so the torque requirement was dynamic in nature and increased as the singularity through the zenith was approached. The closer the tracking path to the zenith, the larger the torque requirement. It should be noted that since the approach to the singularity can be set as close as desired, the peak torques for the Alt-Az system can be arbitrarily set to any value for exact tracking. In practice, the tracker must keep the satellite within the beam so the aiming criteria can be relaxed from the exact tracking condition[6]. The figures used here are for one of the new generation of 44GHz satellite down link and a 2m antenna dish.

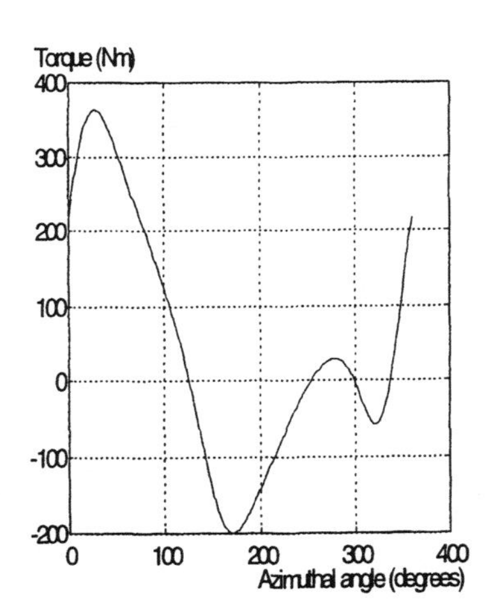

Figure 5. Motor torque versus azimuthal angle, no counter balancing, 0° elevation.

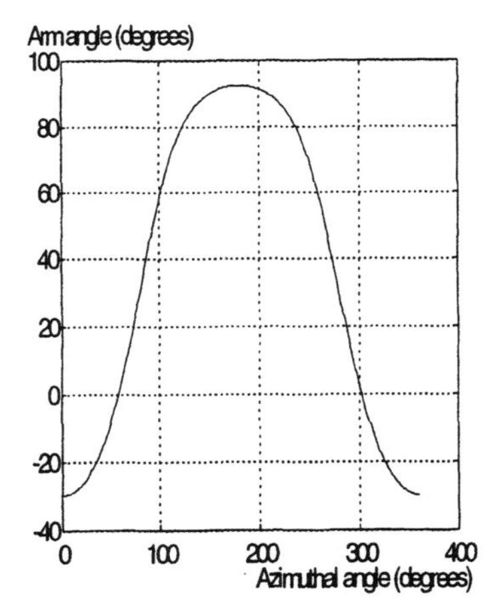

Figure 6. Arm angle versus azimuthal angle, with 0° elevation.

COUNTER BALANCING

It was experimentally observed that the servomotors tended to overload as the antenna arm moved close to the horizon (the step motors dropped out of sync.). It was also observed that the load could be reduced by attaching a fixed pendulum counter weight to each arm as shown in fig. 4. The counter weights were identical for each driven arm but different for the passive arm. Experimental optimisation proved difficult so numerical optimisation was used once the mathematical model had been experimentally confirmed at several measurement positions. The plot of torque versus azimuth is shown in fig. 5 for the worst case of 0° elevation. This data was combined with that of the plot shown in fig. 6 of the actuator arm angle (measured up from the horizon) versus azimuth so as to produce a plot of the torque requirement versus angle which is shown in fig. 7. This particular plot was a sensitive way to show the torque changes which occurred as the counterweight masses and angles (c.f. fig. 4) were varied.

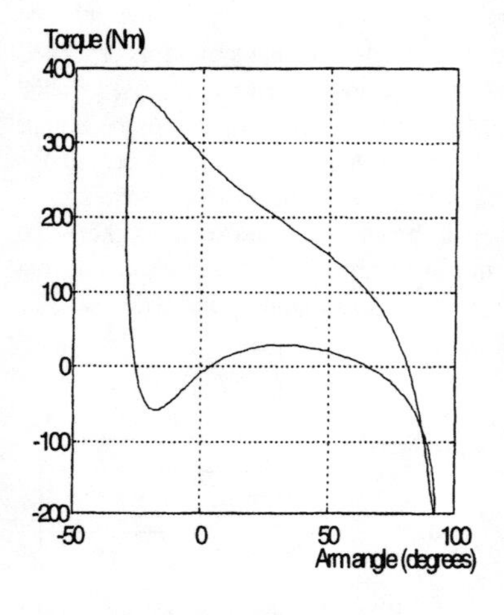

Figure 7. Motor torque versus arm angle with no counter balancing, 0° elevation.

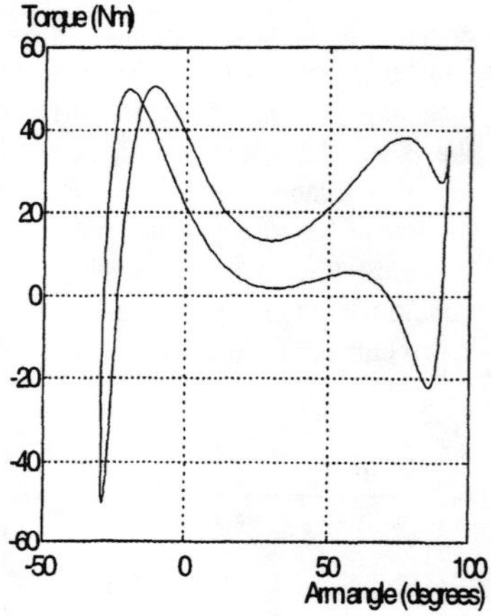

Figure 8. As for fig. 7 but with counter balancing, 0° elevation.

RESULTS

The many intermediate results obtained while manually altering the parameters are not presented here but the final optimisation result is presented in fig. 8. The final mass and angle values obtained for the counter weights are:

for the 2 driven arms: 40.5kg at 0.5m and 170°,
and for the single passive arm: 53.0kg at 0.5m and 159°.

The plot of the torque versus arm angle shows that the torque requirement is now in the range of ±50Nm. This is a quite significant reduction from the original +360Nm to -200Nm required without counterweights, and it is a huge reduction from the ±1500Nm required for the conventional Alt-Az system (however note the earlier comments about approaching within one beam width of the zenith).

During the manual optimisation process, it was observed that altering the angle between the counter weights and the arms changed the position of the minimum located at around 90° of arm angle while altering the mass values had the greatest effect on the minimum and maximum around -30° of arm angle. The two effects were not decoupled so the process was necessarily interactive with the result obtained being quite sensitive to the values chosen for the counterweights.

For completeness, the torque requirements at 30° and 60° elevation have also been computed. These results are plotted in figs. 9 and 10 respectively and clearly show that the 0° elevation plot still represents the worst case. The values for the second servo motor are identical for the torque versus arm angle plots, and are moved by 120° of azimuth.

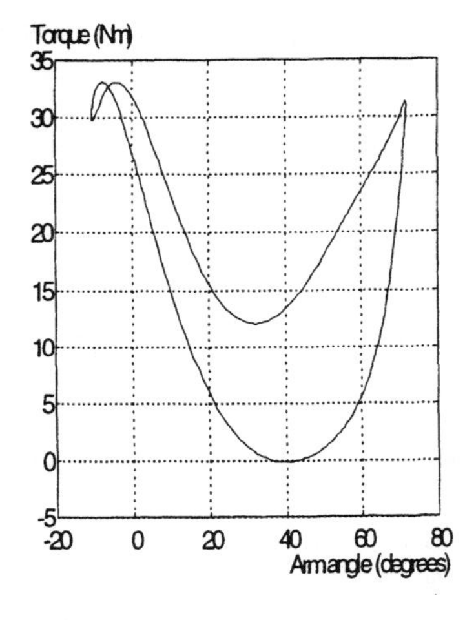

Figure 9. Motor torque versus arm angle with counter balancing, 30° elevation.

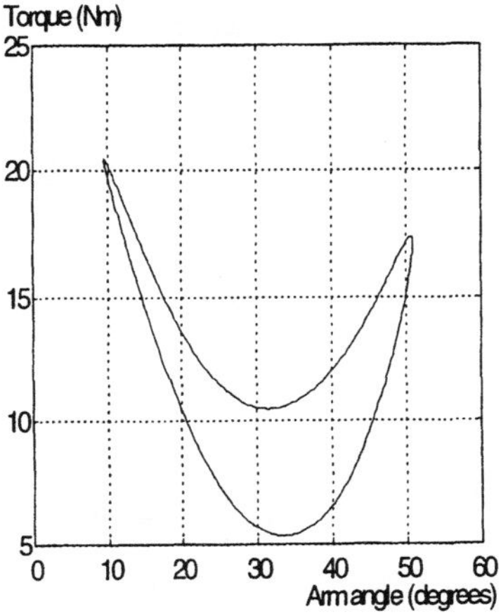

Figure 10. As for figs. 7 and 8 but with an elevation of 60°.

It is also worth noting in passing that the torque curves can be offset so that only positive or negative torques are required to move the mechanism. The appropriate choice of counter weights then allows a unidirectional servo motor to be used which requires a less expensive power amplifier. The unidirectional loading also minimises gearbox backlash.

CONCLUSION

A 2 dof parallel system for aiming antennas has been developed and a system of counter weights designed to reduce the servo motor torque requirements. For the example examined, the torque reduction was approximately 7 times. When compared to a conventional serial Alt-Az antenna aiming system for the particular case chosen, the torque requirements were reduced by a factor of 30.

The exact torque reduction is less important than the removal of the singularity in the workspace of the aiming system since the torque theoretically tends to infinite as the arm moves past the singularity. The addition of counterweights in the parallel mechanism has made a worthwhile reduction to the torque demands since these were largely due to static gravitational effects. The inertia and hence dynamic loading will be increased by the counterweights but the torque requirements are quite small for satellite tracking with this singularity free parallel mechanism.

REFERENCES

[1] Eliss P J 1987 Tracking Antenna Mount, European Patent Application No. 87306732.6

[2] Dunlop G R and Afzulpurkar N V 1988 Six degree of freedom parallel link robotic mechanism: Geometrical design considerations, Proc. IMC Conf , Chch, NZ, pp29/1-8.

[3] Dunlop G R, Ellis P J and Afzulpurkar N V 1993 The satellite tracking keyhole problem: a parallel mechanism mount solution, IPENZ Trans, Vol 20, No 1/EMCh, pp1-7.

[4] Dunlop G R, Lintott A B and Jones T P 1992 Linear and Spherical Parallel Robots with Three DOF, Proc. First Australian Workshop on the Theory of Machines & Mechanisms, Melbourne University Press, ISBN 0-7325-0594-1, pp175-201.

[5] CCIR 1978 Mobile Services: Recommendations and reports of the CCIR XIVth plenary assembly, Kyoto, Volume VIII, Rep. 594-1, pp381-399.

[6] Crawford P S and Brush R J H 1995 Trajectory optimisation to minimise pointing error, IEE Computing & Control Engineering Journal, Vol 6, No 2, pp61-67.

[7] Dunlop G R, Lintott A B and Jones T P 1994 Three DOF parallel robots for linear and spherical movements, ISRAM'94, Intelligent Automation and Soft Computing, Eds. Jamshidi M, and Nguyen CC, TSI Press, ISBN 0-962-7451-4-6, Vol 1, pp655-660.

[8] Dunlop G R and Jones T P 1995 A novel alternative mechanism for driving a satellite antenna dish, submitted for publication in the IMechE Systems & Control Engineering Journal.

THE NEXT GENERATION MUNITIONS HANDLER ADVANCED TELEROBOTICS TECHNOLOGY DEMONSTRATOR PROGRAM

T. E. Deeter , G. J. Koury, M. B. Leahy Jr

AFMC Robotics and Automation Center of Excellence
Technology and Industrial Support Directorate
Robotics and Automation Branch
San Antonio Air Logistics Center, Kelly AFB, TX 78241
ti-race@sadis05.kelly.af.mil

T. P. Turner

AFMC Air Force Seek Eagle Office
Munitions Materiel Handling Equipment Focal Point
Eglin AFB, Florida 32542
turnert@skmail.eglin.af.mil

ABSTRACT

Munitions are currently loaded on fighter aircraft using 1950s technology. These labor intensive methods, while adequate for a second wave forward deployed military, are not optimal for supporting global reach global power projection into the 21st century. Given the dynamic nature of the flightline environment and the dexterity requirements for arming and securing munitions, a fully automated solution is not technically feasible. More importantly, the munitions handling community is not looking to replace human operators, but rather enhance their efficiency while improving working conditions and reducing workload. The solution lies in the incorporation of emerging telerobotics technology into the munitions handling process. The Next Generation Munitions Handler (NGMH) Advanced Telerobotics Technology Demonstrator (ATTD) will enable the first realistic evaluation of emerging telerobotics technologies for flightline applications. The project history, system design and programmatics of this on-going United States Air Force and Navy joint program are discussed.

KEYWORDS: telerobotics, robotics, omni-directional, man amplification

INTRODUCTION

In recent years, the Air Force Munitions Materiel Handling Equipment Focal Point (MMHE), has commissioned several studies to examine the current and future roles of robotics and automation technologies in flightline operations. Few of these studies progressed beyond the point of paper analysis and even fewer resulted in flightline evaluation of prototype hardware. There was a basic lack of knowledge as to which robotic and automation technologies were applicable to the tasks. The net result was a set of very narrow focus technology investigations with minimal contribution to the flightline knowledge base.

Why the difficulty? There were many contributing factors, not the least of which was the general focus on the development of fully automated systems to perform the tasks required during flightline operations. The flightline is a very dynamic and unstructured environment that does not lend itself well to 'hard' automation. Development of machine intelligence to account for the environment was, and still is, a very complex challenge requiring significant basic and applied research. The gap between manual and fully autonomous systems can not be crossed in a single leap. A more evolutionary process is mandated. Telerobotics is now recognized as the key enabling technology for incorporating the advantages of robotics and automation into flightline operations.

The central tenant of telerobotics is the effective blending of the skills of human operators with the tireless precise positioning ability of a machine. In our context, telerobotics is not a point solution, but a family of human system interface options ranging from teleoperation to supervisory control. The particular brand of telerobotics is determined by the appropriate mixture of human/machine abilities necessary to satisfy task requirements. Unfortunately, telerobotics is more of a custom concept than an off-the-shelf solution. And in today's acquisition climate, custom one-off solutions are not affordable. To correct this deficiency, the United States Air Force Materiel Command's Robotics and Automation Center of Excellence (RACE) embarked on an initiative to reduce the life cycle costs of telerobotic systems. The Unified Telerobotic Architecture Project (UTAP) provides a standard framework of devices and interfaces which define a system capable of affordably addressing a wide range of applications [2,3,4,5].

UTAP is a four phase project designed to transition a vision of interoperable, affordable, commercially available telerobotic control systems into reality. Results from the first two phases provide the philosophical design underpinnings for revisiting the flightline environment. The operational command's need to enhance the affordability of munitions handling tasks presents the user sponsored application. Technological evolution and user requirements converged to create the first implementation of the unified telerobotics architecture, the Next Generation Munitions Handler (NGMH) Advanced Telerobotics Technology Demonstrator (ATTD).

The intent of the NGMH ATTD is three-fold. First, provide a platform for operators to evaluate the generic flightline applicability and robustness of robotics and automation technologies. Second, provide Headquarters Air Combat Command (HQ ACC) with valuable information on the utility of robotics as specifically applied to munitions handling operations. Third, validate the UTAP as a viable basis for developing affordable telerobotic solutions to complex AF applications.

The objective of this paper is to present an overview of the NGMH ATTD program. Section one highlights the program history and provides background on the munitions loading process. Section two describes the system design and the necessary technology advances. The ATTD incorporates a large reach, heavy payload with high precision, redundant manipulator on an omni-wheeled platform. It will be able to respond to human directional inputs without noticeable delay and position 2500 lbs within 1 mm of commanded location. The NGMH is pushing the envelope in the fields of hydraulic actuators and controls, force sensing, and human machine interfaces. Recent advances in those areas will be presented along with references for more detailed investigation. Future program milestones are the subject of section three which is followed by a short conclusion. Graphic simulations and a video support the briefing.

PROGRAM HISTORY

The majority of munitions handling operations involve the use of the MJ-1A/B Aerial Stores Lift Truck, known as the "jammer". The jammer is standard equipment for loading munitions, fuel tanks, pylons, and special weapons weighing up to 3,000 lbs.[1] The jammer is a diesel powered, self-propelled vehicle incorporating hydraulics to perform the heavy lifting required for munitions handling. Guidance of the vehicle is performed by the driver who is seated at the rear of the vehicle, far removed from the point of action.

A three member load crew is employed to perform the munition operations. The driver and two additional members are responsible for performing the detailed loading instructions in a quick, efficient, yet safe manner. The procedures for loading weapons on aircraft vary with the type of munition and also with the type of aircraft. The general procedures can be broken down into four basic steps; build up, transportation, installation and final hook up. A quick review is provided. Additional details are in [6].

The first step, build-up, occurs in a dedicated munitions build up area. During this step, the weapon receives suspension lugs, fuse assemblies, fin assemblies, and any other wiring necessary for installation/delivery. Once complete, the weapon is transported to the loading area.

Step 2, transportation, is accomplished by securing the weapon to a munitions trailer which is then towed to the loading area. For most applications the loading area is configured to support an Integrated Combat Turnaround (ICT). An ICT is the rapid retrieval and relaunch of combat aircraft and is practiced in a confined area to simulate the hard aircraft shelters used in high threat combat situations. Steps 3 and 4 are performed here. Munitions loading starts with the jammer transporting the weapon from the trailer to the vicinity of the aircraft attachment point. During all jammer operations, communications between the driver and the other load crew members is constant. Due to the noise levels on the flightline, verbal communication between the driver and the other crew members is nearly impossible, so all communication is accomplished through hand signals. As the munition nears the loading station, fidelity of each movement is crucial. The crew member closest to the attach point utilizes separate fine motion controls for the final insertion of the weapon attach lugs into the bomb rack or missile rail. The munition is locked into place and the two loaders continue with final attachments (step 4) while the jammer is safely driven away from the area.

The Air Force has utilized this effective loading procedure for decades. However, as with most systems, there is room for improvement. Barring serious aircraft mechanical or electrical failure, the installation (loading) and final hook up of the weapons is the most time consuming part of the relaunch process. Therefore, efforts to improve ICT times must concentrate on these steps. Step 4 is beyond the capability of current robotics and automation technology, therefore the NGMH ATTD attacks the limitations of the installation process.

Several problems beleaguer the installation process. What at first appears to be a trivial task of mating two parts together in a desired configuration, becomes complex when tight part tolerances and limited visibility are factored in. The jammer system offers little in terms of user feedback and provides no ability to coordinate joint motions. All motion is accomplished through separate actuation of each individual joint. The load crew is unable to sense the forces being exerted on the weapon and must adjust the weapon based on vision alone. In many cases, this is insufficient. For instance, the loading of missiles onto missile rails involves aligning three missile lugs with rail attach points, inserting the lugs, then sliding the missile along the rail to lock into place. The jammer provides a vertical heavy lift capability, but the sliding motion takes place in the near horizontal plane. The tolerances between the missile and missile rail can be very tight, leaving very small allowances in misalignment. This typically results in the missile binding in the rail without any visual evidence.

Now add to those restrictions the constraints imposed when loading an aircraft such as the Air Force F-22. The primary weapon bay is in the belly of an aircraft whose very limited ground clearance prevents the jammer from driving underneath the aircraft and reaching the weapons bay, significantly adding to loading difficulty. The driver becomes a safety spotter for the members performing the mating operation, but his vantage point is restricted, and his usefulness is questionable. The driver represents little more than a platform input device - that which moves the vehicle in response to an external command.

SYSTEM DESIGN

A common first reaction to these sets of problems is to redesign the entire system. However, modifying all the munitions and launch racks in the inventory is not an affordable option. Process breakdown studies reveal it is theoretically possible to eliminate the driver if the ability of the other crew members to interact with the jammer is enhanced. Thus, the driver is free to perform other tasks that require his skills, thereby enhancing our personnel utilization and decreasing aircraft turnaround time. Eliminating the need for a dedicated driver is the primary driver in the NGMH system design.

The NGMH ATTD system design utilized the talents of an integrated product team composed of operators and technologists from the military and the national laboratories. The RACE is providing technical direction. Oak Ridge National Laboratory (ORNL) was selected as the lead laboratory. The initial phase of the ATTD was the detailed conceptual design which leveraged the initial NGMH feasibility study conducted by the University of Utah [7]. The ORNL system design exploits several leading edge technologies including a fully omnidirectional platform, precision hydraulics under very high loads, high resolution Force/Torque sensing for the insertion tasks, joint coordination with real time redundancy resolution, hand-on-the-system telerobotics and man amplification in the range of 100 to 1. Variations of these technologies have been demonstrated in one form or another, however, the objectives of this system are truly unique and push the edge of the envelope.

While the main objective is to provide the load crew with a better tool, the ATTD is also constricted by the requirement not to exceed current mobility specifications, i.e. footprint and weight. The kinematically redundant arm consists of 8 or 9 degrees of freedom configurations to provide the optimal reach, obstacle avoidance and joint limit avoidance necessary to load the most difficult weapon stations on all fighter aircraft regardless of the configuration. The system will be outfitted with a sensor suite allowing safe operation in and around multi-million dollar aircraft.

The system is designed to operate with a 2500 lb. munition and achieve 1 mm, 0.2^{0} incremental accuracy. This accuracy has never been demonstrated for a redundant heavy payload hydraulic robot. In addition to the precision specifications, the controller must also compensate for the munitions inertia, minimize impact forces and ensure a stable transition between free space and the launcher surface. The resultant hydraulic control system is one of the ATTD's major technology challenges.

The human system interface is another major challenge. The man amplification aspect of the system allows the operator to manipulate the 2500 lb. munition as if it were a 25 lb. barbell. This amplification coupled with the systems kinematics allows the operator to manipulate the munition with small inputs to a device similar to a flightstick that interprets these inputs and executes the desired movement of the munitions. This reduces the numerous repositioning routines currently hampering the loading process. A successful user interface will allow the user to manipulate the munition from a variety of locations. The user will be able to transport and load from a seated position or from a position in close proximity to the munition.

Research and development efforts are addressing the major challenges. Initial tests have been very positive and we are confident the technologies will be mature enough to support the ambitious program schedule. Further technology refinements necessary to support a full scale acquisition program are discussed in [6].

PROGRAMMATICS

The NGMH ATTD is fully funded by a joint consortium of military services and agencies. ORNL is currently manufacturing the manipulator and designing the configuration of the omni-directional platform. The manipulator will be completed in late 1996 at which time validation testing begins. Validation testing will progress from gross load manipulation throughout the workspace and preliminary man-machine interface analysis to a full scale loading procedure with real launch racks and inert munitions. The mobility platform will be fabricated by the second quarter of 1997 at which time a four month integration will begin. Upon completion of the integration and laboratory validation, extensive field testing will be completed by the Navy and the Air Force. The field testing will include operation of the system by current munitions loading experts. The ATTD is designed to allow rapid reconfiguration of control and operator interface modalities so the evaluators are exposed to the full range of technically feasible solutions. This is the only means to capture the operator feedback critical to the decision to proceed with a potential prototype and production unit. Upon completion of a successful test program, a full scale acquisition program will be initiated to produce the NGMH and bring the munitions loading process to the state-of-the-art. Technologies developed for the NGMH ATTD also have the potential to revolutionize a wider range of military and civilian applications [6].

CONCLUSION

The Next Generation Munitions Handler (NGMH) Advanced Telerobotics Technology Demonstrator (ATTD) will enable the first realistic evaluation of emerging telerobotics technologies for flightline applications. This innovative joint program has entered the fabrication phase and is scheduled for field evaluations in 1997. Successful testing will pave the way for a full acquisition program which enhances munitions loading affordability and effectiveness and supports global reach and global power projection into the 21st century.

REFERENCES

1. Ground Support Equipment for Aerial Stores Handling and Aircraft Maintenance, *Engineering Specifications, Issue No. XVI*, Standard Manufacturing Company, Dallas, TX, 1990.
2. *Generic Telerobotic Architecture for C-5 Industrial Processes*, JPL Final Report, August 1993.
3. Leahy, M. B. Jr, and B.K. Cassiday, "RACE pulls for Shared Control", *Proceedings of SPIE Conference on Cooperative Intelligent Robotics in Space III*, Boston 1992.
4. Leahy, M. B. Jr, and S.B. Petroski, "Telerobotics for Depot Applications", *Proceedings of NAECON 93*, Dayton, May 1993.
5. Leahy, M. B. Jr, and S.B. Petroski, "Telerobotics for Depot Modernization", *Proceedings of the AIAA Conference on Intelligent Robots in Field, Factory, Service and Space (CIRFFS)*, Houston, TX, March 1994.
6. Leahy, M.B. Jr, and Neil Hammel "The Next Generation Munitions Handler Prototype Acquisition Campaign: Targets and Courses of Action", Air Command and Staff College research, May 1995.
7. Morgenthaler, D.G. "Robotics Application to Munitions Operations," Center for Engineering Design, University of Utah, February 1994.

FEED-IN-TIME CONTROL BY INCURSION

Daniel M. DUBOIS
Institute of Mathematics, UNIVERSITY OF LIEGE
avenue des Tilleuls 15, B-4000 LIEGE (Belgium)
*Fax */32/41/669489 - e-mail dubois@lema.ulg.ac.be*

ABSTRACT

This paper deals with an innovative mathematical tool for modelling, simulating and controlling systems in automation engineering. Classically, feedback processes are based on recursive loops where the future state of a system is computed from the present and past states. With the new concept of incursion, an inclusive recursion, the future state of a system is taken into account for computing this future state in a self-referential way. The future state is computed from the mathematical model of the system. With incursion, numerical instabilities in the simulation of finite difference equations can be stabilised. The incursive control of systems can also stabilise feedback loops by anticipating the effect of the control what I call a feed-in-time control. In this short paper, the particular case of the modelling, simulation and control of a robot arm in a working space is studied. The highly non-linear model is based on recursive finite difference equations which give rise to instabilities, bifurcations and fractal chaos. The control by incursion, called feed-in-time control, of such a robot arm stabilises its trajectory. Numerical simulations show that the robot arm reaches set points in a few steps in any point of the working space.

KEYWORDS: feed-in-time, incursion, control, robotics, chaos, fractal, stability, instability, finite difference equations, non-linear systems.

INTRODUCTION

The paper will discuss the potential innovative power of the incursion in automation engineering with the presentation of a typical example in robotics. The incursion is a new concept developed by the author [3,4,5,6], which deals with an extension of the recursion. A recursive process is always depending on the present or past states of a system. With incursion, the process can depend on past and present states of the system but also to its potential future states.

Classical control [1] of any system by feedback is based on recursive processes in which the current control function u(t) at the present time step t is a function of the deviation (the error) between the set point x_r and the state x(t) resulting from the preceding control u(t-Δt) at the preceding time step t-Δt, the effect of which giving the new state x(t+Δt) at the next time step t+Δt. So a time delay Δt is present in this feedback loop which can give rise to instabilities or critical oscillations around the set point. Let us remark that numerical instabilities can also appear in the simulation of discretised differential equations when the time step Δt of the discretisation is too large. It was demonstrated that such instabilities can disappear even for large time step Δt with incursion in using backward and forward derivatives, for example, for linear and non-linear discrete oscillating systems [4,5,6,9]. This would permit to simulate discretised equation systems in real time for automation engineering. When multiple iterates are generated at each step, the incursion is a hyperincursion [7,5,9], an extension of hyper recursion.

This paper deals with a feed-in-time control in which the control function u(t) at time step t takes into account of its effect on the state of the system $x(t+\Delta t)$ at the future time step $t+\Delta t$. This future state is estimated from a mathematical model of the dynamics of the system. In other words, the control function is computed in anticipating its effect on the system at the next time step. The formal model of this feed-in-time control is given by the following simplified example. Let us consider the incursive control of a system represented by discrete equations [4]:

$$u(t) = u_r + c(x).(x(t+\Delta t) - x(t))/\Delta t \quad (1a)$$
$$x(t+dt) = x(t) + \Delta t.[f(x(t),p) + b.u(t)] \quad (1b)$$

where u(t) is the incursive control function, u_r is a function of the explicit set-point or reference signal r(t) of the variable, x(t) the state variable of the system, f the recursive function of the model of the system depending on the variable x(t) and on the parameter p, b is a known parameter. The function c(x) can be explicitly defined as a function of the following derivative of the function f(x,p):

$$c(x) = A + B.df(x,p)/dx \quad (1c)$$

where A and B are constants. Let us remark that with non-linear functions f(x,p) of at least of degree two in x, the control u(t) is non-linear. Such an incursive control was applied to stabilise the chaos in the Pearl-Verhulst map $f(x,p) = p.x.(1 - x)$ [3,4]. Equation 1c gives $c(x)=A-B.(p-2.p.x)$. Without set point ($u_r = 0$), the incursive control can stabilise a system in an unstable regime. Indeed, with unstable or chaotic systems, the incursive control changes the unstable regime to a stable one which becomes the implicit set point [4,8,9]. The incursive method is general, the variable x and the function f can be given by a vector and a matrix, as demonstrated for non-linear Lotka-Volterra equations and for the inverse cinematic problem for a robot arm [9]. This paper will present a new application which will show this stabilisation of the chaos in the case of another model of a robot arm.

The incursive control u(t) is a forward derivative which depends on the measured value of the variable at the present time t and its unknown value at the next future time step $t+\Delta t$. In adaptive control, such a future state is estimated [2]. The originality of the incursive control deals with the way to compute the future state without estimation. The variable $x(t+\Delta t)$ in eq. 1a can be explicitly replaced by eq. 1b by self-reference as follows:

$$u(t) = u_r + c(x).(f(x(t),p)) + b.u(t)) \quad (2a)$$

and an incursive equation is still obtained because the control u(t) is self-referential. It is important to notice that the incursive control takes into account its action at the future time step and can be transformed to the following recursive control:

$$u(t) = [u_r + c(x).f(x(t),p)]/[1- c(x).b] \quad (2b)$$

This paper deals with an application in robotics. The arm of a robot is considered in a two-dimensions working space. A mathematical model given by finite difference equations takes into account the angles of the two members of the arm and the set point is defined in polar co-ordinates. The incursive control of such a system stabilises the chaos of the recursive model by a feed-in-time.

CONTROLLING THE CHAOS OF A ROBOT ARM

This paper deals with a special problem in robotics, that is the inverse cinematic problem of computing the iterates n=1,2,3,... of two angles $\alpha(n)$ and $\beta(n)$ of a robot arm with two members of length l from initial conditions $\alpha(0)$ and $\beta(0)$ to a set point given by polar co-ordinates (ρ_r, γ_r) in the working space (x,y) as shown in Figure 1.

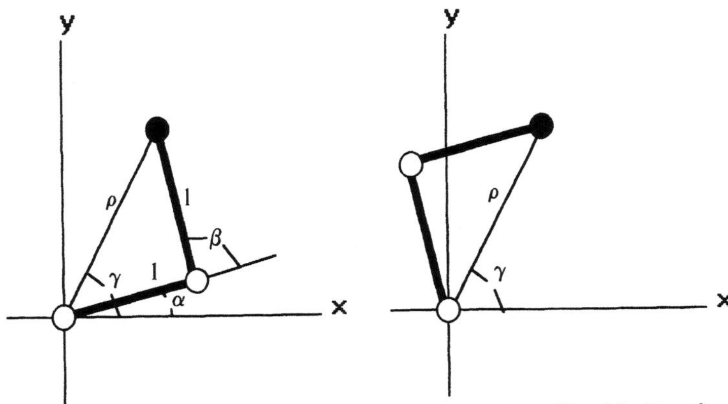

Figures 1a-b: (a) The arm of the robot is constituted of two members of length l with angles α and β. The extremity of the arm is defined by the polar co-ordinates (ρ,γ). (b) For the same polar co-ordinates, there are two symmetrical solutions.

From the following discrete recursive equation system

$$\alpha(t+\Delta t)=\alpha(t)+\Delta t.[\ \rho_r/2-l.\cos(\gamma_r-\alpha(t)]/c \tag{3a}$$
$$\beta(t+\Delta t)=\beta(t)+\Delta t.[l.\sqrt{}(2.(1+\cos(\beta(t))-\rho_r]/c \tag{3b}$$

the iterates $\alpha(t+\Delta t)$ and $\beta(t+\Delta t)$ are computed from their preceding values at time step t, starting with the initial state of the arm $\alpha(0)$ and $\beta(0)$ in function of the final state defined by the polar co-ordinates (ρ_r,γ_r). The value of the constant c depends on the length and time units. The angles are given in radians. The advantage of this model is the fact that the two equations are independent of each other: the two control motors of each member of the arm can move independently. The second equation is only a function of the radial set point which depends only on the angle $\beta(t)$. Let us remark that the man moves a hand in estimating with the eyes the radial distance of an object to be taken and the rotation angle. When this system reaches at a steady state, we have

$$\alpha(t+\Delta t)=\alpha(t) \text{ and } \beta(t+\Delta t)=\beta(t) \tag{4a}$$

so that $\rho_r=2.l.\cos(\gamma_r-\alpha(t))$ and $\rho_r=l.\sqrt{(2.(1+\cos(\beta(t)))}$ (4b)

The criteria of stability of the steady state are given by

$$|d\alpha(t+\Delta t)/d\alpha(t)| = |1-\Delta t.l.\sin(\gamma_r-\alpha(t))/c| < 1 \quad (5a)$$
$$|d\beta(t+\Delta t)/d\beta(t)| = |1-\Delta t.l.\sin(\beta(t))/[c.\sqrt{(2.(1+\cos(\beta(t))}]| < 1 \quad (5b)$$

For $\Delta t<c/l$, the system is only stable for some values of the angles. Figure 2a shows that the arm reaches a set point but with transient bifurcations for the angle $\alpha(t)$. The set point ($\rho_r=0,\gamma_r=0$) is particularly unstable and gives rise to chaos for $\beta(t)$ as shown in Figure 2b.

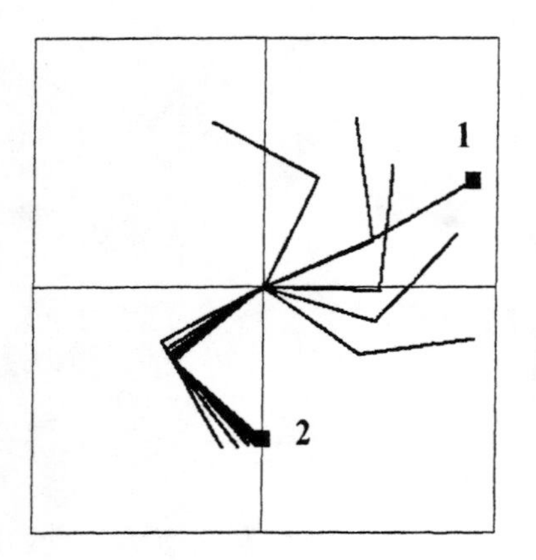

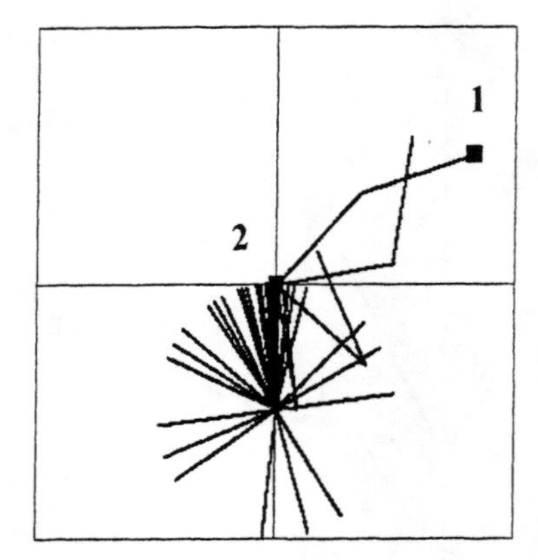

Figures 2a-b: (a). The robot arm reaches the set point 2, starting from the point 1, after transient bifurcations for $\alpha(t)$. (b).The robot arm gives rise to chaos for $\beta(t)$ at the set point ($\rho_r=0,\gamma_r=0$).

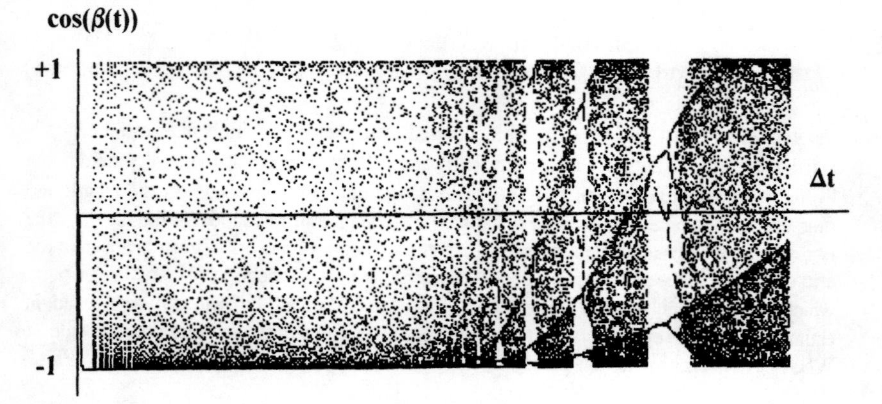

Figure 2c: Chaos diagram of $\cos(\beta(t))$ as a function of $0<\Delta t<2.c/l$ for ($\rho_r=0,\gamma_r=0$).

Figure 2c gives the chaos diagram of $\cos(\beta(t))$ for $0<\Delta t<2.c/1$. Typical fractal chaos is present for $\Delta t>c/l$. The feed-in-time control by incursion stabilises such a chaotic behaviour. Indeed, let us introduce the following incursive control for eqs. 3a-b from eqs. 1a-c:

$$u(t)=[-\sin(\gamma_r-\alpha(t))].[\alpha(t+\Delta t)-\alpha(t)] \quad (6a)$$
$$v(t)=[-\sin(\beta(t)/\sqrt{(1+\cos(\beta(t))}].[\beta(t+\Delta t)-\beta(t)] \quad (6b)$$
$$\alpha(t+\Delta t)=\alpha(t)+\Delta t.[\ \rho_r/2\text{-}l.\cos(\gamma_r-\alpha(t)]/c+u(t) \quad (6c)$$
$$\beta(t+\Delta t)=\beta(t)+\Delta t.[l.\sqrt{(2.(1+\cos(\beta(t))}-\rho_r]/c+v(t) \quad (6d)$$

where the control functions u and v depend on the present and future values of the angles at time t and $t+\Delta t$. These control functions are discrete forward time derivatives and play the role of an anticipation for the next future step, what is called a feed-in-time. Classically a feedback control takes into account the discrete backward time derivatives which are given by the outputs of the system at times t and $t-\Delta t$. It can be remarked that the control functions are equal to zero when the system reaches its steady state. The role of the anticipative incursion is to increase the stability of the system for larger time steps Δt and to enhance the transient path to the set point which can be quicker. Figures 3a-b give the numerical simulations of this feed-in-time system for the same cases as in Figures 2a-b. The chaotic behaviour disappears. It is important to remark that the time step for figures 2a-b and 3a-b was rather large, $\Delta t=0.9.c/l$. In taking $\Delta t=2.c/l$, the uncontrolled system given by eq. 3a-b is completely chaotic. In the same conditions, the incursive system 6a-b-c-d is stable and moreover the robot arm reaches any set point in a few steps.

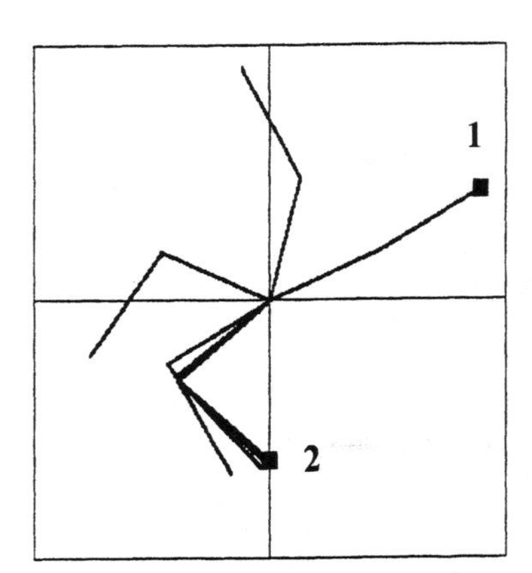

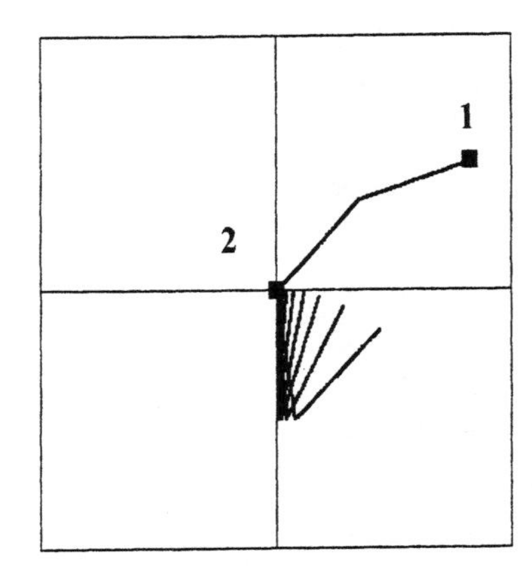

Figures 3a-b: (a). Feed-in-time control by incursion for the arm starting at the same point 1 with the same set point 2 as in the figure 2a. With the incursive control, the set point is reached after a few time steps. (b). Feed-in-time control at the set point (0,0) corresponding to figure 2b. The chaos has disappeared and the robot arm reaches the set point again in a few time steps..

The figures 4a-b show two simulations with time successive set points defining trajectories in the working space with $\Delta t=2.c/l$. Each set point is obtained by only one iteration.

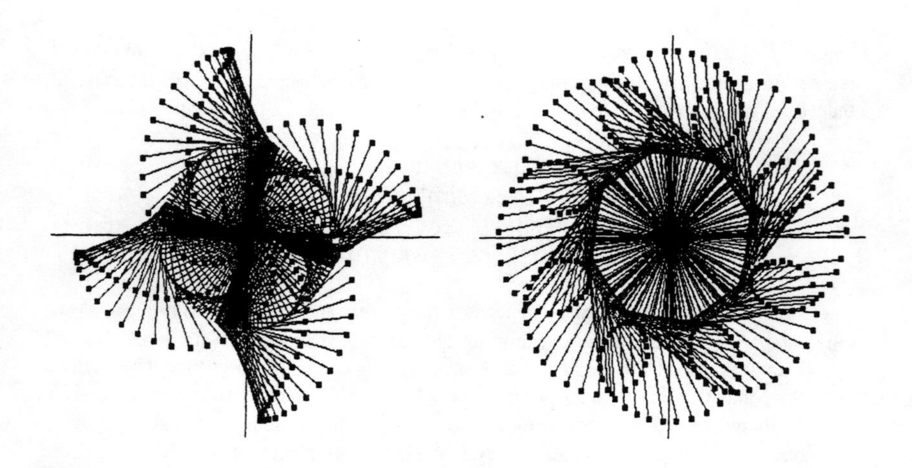

Figures 4a-b: Feed-in-time control of a robot arm with set points defined by a trajectory in the working space. Only one time step by set point is computed.

CONCLUSION

In conclusion, the feed-in-time concept could be a new innovative tool for modelling, simulating and controlling of systems in automation engineering with the introduction of incursion as an extension of the recursion in taking into account the future states in defining self-referential systems.

REFERENCES

1. Craig J. J. *Introduction to Robotics, Mechanics and Control,* Addison-Wesley Publ. Company, (1989).
2. Portier B., Oppenheim G. "Adaptive Control of Nonlinear Dynamic Systems: Study of a Nonparametric Estimator." *J. Syst. Eng.* (1993), 1,40-50.
3. Dubois D. M. *The Fractal Machine.* Presses Universitaires de Liège (1992), 375 p.
4. Dubois D. M. "Total Incursive Control of Linear, Non-linear and Chaotic Systems." *In* G. Lasker (ed.): Advances in Computer Cybernetics. Int. Inst. for Advanced Studies in Syst. Res. and Cybernetics, vol. II, (1995), 167-171.
5. Dubois D. M. "Introduction of the Aristotle's Final Causation in CAST: Concept and Method of Incursion and Hyperincursion". In F. Pichler, R. Moreno Diaz, R. Albrecht (Eds.): Computer Aided Systems Theory - EUROCAST'95. Lecture Notes in Computer Science, 1030, Springer-Verlag Berlin Heidelberg, (1996), 477-493.
6. Dubois D. M. "A Semantic Logic for CAST related to Zuse, Deutsch and McCulloch and Pitts Computing Principles". In F. Pichler, R. Moreno Diaz, R. Albrecht (Eds.): Computer Aided Systems Theory - EUROCAST'95. Lecture Notes in Computer Science, 1030, Springer-Verlag Berlin Heidelberg, (1996) 494-510.
7. Dubois D. M., Resconi G. *HYPERINCURSIVITY: a new mathematical theory.* Presses Universitaires de Liège (1992), 260 p.
8. Dubois D. M., Resconi G. "Holistic Control by Incursion of Feedback Systems, Fractal Chaos and Numerical Instabilities." *CYBERNETICS AND SYSTEMS'94*, edited by R. Trappl, World Scientific (1994), 71-78.
9. Dubois D. M., Resconi G. *Advanced Research in Incursion Theory applied to Ecology, Physics and Engineering.* COMETT European Lecture Notes in Incursion. Edited by A.I.Lg., Association des Ingénieurs de l'Université de Liège, D/1995/3603/01, (1995), 105 p.

END MILLING FORCE ANALYSIS UNDER FUZZY LOGIC CONTROL

M.T. Dos Santos

Escola Politécnica da Universidade de São Paulo -EPUSP, Departamento de Engenharia Mecanica, Av.Prof. Mello Moraes,2231 05508-900, São Paulo - SP, Brasil (e-mail: mteixeira@iai.es).

C.R. Peres

Universidade Federal de Santa Catarina. GRUCON, Florianópolis, Santa Catarina, Brasil (e-mail: clodeinir@iai.es),

S.R. Torrecillas, J.R. Alique, A. Alique, C. González.

Instituto de Automática Industrial, N-III, km22,800, La Poveda, 28500, Arganda Del Rey, Madrid, España(e-mail: ros@iai.es).

ABSTRACT

This work shows a Supervisory Fuzzy Control System (SYFLOCS) developed for end-milling. This system was developed on the Cutting Theory. The researchers sought for a solution to the problems in end milling by measuring the spindle motor current and controlling the cutting parameters. Considered in this system are the spindle power, tool life, rigidity, endurance limit, and superficial finish to calculate an optimum Metal-Removal Rate (MRR). The theory of tool wear is reviewed and a new method to monitor this variable is proposed. The experimental application is shown and the theory is validated.

KEYWORDS: Fuzzy Logic Control, Metal-Removal Rate, Tool Wear, End-Milling.

INTRODUCTION

The determination of the optimum machining parameters, like feedrate and spindle speed in metal cutting, is an important aspect in an economic manufacturing process [1] [2]. Many authors have been researching for an optimum point for the cutting parameters in end milling [3][4][5] and their research showed very good results with model simulation and some practical implementation. The proposal of our research group is based in the Metal-Removal Rate (MRR) theory [3] and in a Supervisory Fuzzy Logic Control System (SYFLOCS) [6][7][8]. The MRR theory was reviewed and developed by Arggawal [3] and it was applied in the high speed milling of aluminium.

The theoretical model for the cutting forces was developed and based in the Tlusty's theory [4]. The end milling force model was applied to the MRR equation, thus this equation was used to compute the new feedrate and the new spindle speed for each new cutting depth during a rough machining. In this manner the MRR was maintained at an optimum. The research gives the analytical expression for obtain the most efficient cutting parameters for the variations that occurs during the end milling of aluminium.

Significant limitations for the machining process are the power of the machine tool and the endurance of the cutting, so these important variables are considered in an analytical model that this paper describes. This model is used to further improve the productivity of the SYFLOCS.

The high complexity of the process is considered, thus the strict nonlinearities, the limited knowledge of its physical essence, variations in its characteristics and also the inadequacies of the sensorial system were observed and it was concluded that the techniques of classical control are not the most suitable for it.

The utilisation of the SYFLOCS for improve performance of chip-cutting machining gives another option, and some very promising results have been reached and these are described in our latest publications [9][10]. It has guaranteed the satisfactory velocity parameters for the mechanical variations that can arise in the process, as well as for variations in the block material and tool wear. The supervisor system measures the force (F) by means a current sensor in the spindle axis motor. The error (ΔF) and the first derivation (Δ^2F) are the inputs to the algorithm of control with concern for the established reference force; and the variations in the feed speed (Δf), and in the spindle speed (Δs) are the outputs. The control system changes the cut parameters (s and f) for different conditions that arise during end-milling to maintain the optimum cutting force for the optimum MRR.

Next section describes the relationship between force and MRR theory. The Supervisory Fuzzy Control System is described and the Experimental Applications are shown.

CUTTING FORCE AND METAL-REMOVAL RATE

For a better knowledge of the end milling machine and to provide the mathematical and physical bases for this process we used the Tlusty Machine Dynamic Theory [3] [4] to model the resultant force in the end-milling process. The model suggested in this paper was studied for small tool diameters between 19 and 50 mm. Figure 1 shows end-milling with two teeth. Thus, we developed the Tlusty Theory for this case and the Eq.(1) was obtained.

$$Fr = 2.088 \cdot K_s f_t d_A \cos\beta \tag{1}$$

where **Fr** is the resultant radial force on the end mill (N), $\mathbf{K_s}$ is the specific force (N/mm$^{2)}$, $\mathbf{f_t}$ is the feed per tooth (mm), $\mathbf{d_A}$ is the axial depth (mm), and β is the helical angle.

This expression is near enough to computing the resultant cut force for a full end milling. We considered a constant thickness chip, so we made the maximum thickness chip equal to the feed per tooth. This approach produces a sensible error when we computed the specific force at the cut.

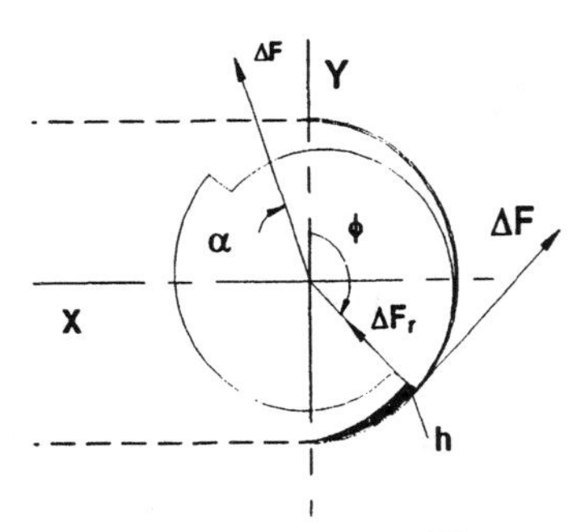

Figure 1. Two flute end -milling.

The MRR was examined by Arggawal. He described the relation between MRR and the spindle power and then defined the endurance limit chip load to various types of cut in aluminium alloys. The MRR allows the calculation of instantaneous productivity for end-milling. Therefore we would be able to determine and to control the best instantaneous parameters for this process. The MRR is described mathematically by the expression:

$$MRR = d_A \cdot d_R \cdot i \cdot n \cdot N \tag{2}$$

where **MRR** is the Metal-Removal Rate (mm³/min), $\mathbf{d_A}$ is the axial depth of cut (mm), $\mathbf{d_R}$ is the radial depth of cut (mm), **i** is the chip load or feed per tooth (mm/tooth/rev), **n** is the number of teeth or flutes, and **N** is the spindle speed (rpm).

The average spindle power cut per cubic millimetres of metal removal per minute (unit kilowatts) was conservatively assumed to be equal to 1.48×10^{-5}. Therefore the kilowatts consumed at the cut can be written as follows:

$$P_c = 1.48 \times 10^{-5} MRR \tag{3}$$

$$P_c = 1.48 \times 10^{-5} d_A \cdot d_R \cdot i \cdot n \cdot N \tag{4}$$

where **Pc** is the spindle power (kW).

The constant that defines the penetration of the cut is as follows

$$K = \frac{d_A \cdot d_R}{d_c^2} \tag{5}$$

where **K** is constant, $\mathbf{d_c}$ is the cutter diameter (mm).

By substituting Eq.(5) into Eq.(4), we obtain

$$P_c = 1.48 \times 10^{-5} \cdot K \cdot d_c^2 \cdot n \cdot N \tag{6}$$

Arggawal describes a method for defining spindle and feedrate requirements. It was considered the following limiting factors: tool life (60 min or more), endurance limit, surface finish and rigidity. "The tool mustn't fail with fatigue forces on the cutter and the workpiece should be within the material limits to provide desired surface finish; and, finally, the tool must be rigid enough not to create excessive vibrations and/or chatter marks on the workpiece".

For High Speed Steel (HSS) the corrected endurance limit was found to be equal to $S_e = 88530\,\text{kPa}$. By using end milling theory we obtain for this tool the following expressions for endurance limit chip load (maximum feed per tooth), Eq.(7) and its correspondent endurance limit spindle power at the cut Eq.(8).

$$i_e = \frac{S_e \cdot d_a^4}{2921490 \cdot K \cdot d_c^2 \cdot n \cdot (l - \frac{d_A}{2})} \quad (7)$$

$$P_e = \frac{S_e \cdot N \cdot d_a^4}{1.9738 \times 10^{11} \cdot (l - \frac{d_A}{2})} \quad (8)$$

where i_e is the endurance limit chip load, $\mathbf{P_e}$ is the endurance limit spindle power (kW), **Se** is the corrected endurance limit (kPa), **N** is the spindle speed (rpm), $\mathbf{d_{cs}}$ is the toolshank diameter (mm), $\mathbf{d_c}$ is the end mill diameter (mm), $\mathbf{d_A}$ is the axial depth cut (mm), **l** is the flute length (mm).

The relationship between the resultant force and MRR is shown in Eq.(9). This expression shows that this relationship is direct. Therefore, when the resultant force is constantly maintained the MRR will be maintained at a constant too. We can see in Eq.(2) the MRR changes with the changes in the depths, chip load, number of teeth and spindle speed. In our practical experience we use a workpiece with a variable axial depth of cut to study the behaviour of SYFLOCS. Thus, there are just two variables that we should control to maintain a constant MRR these are the feedrate and spindle speed, and the force is the measured variable. This is shown in the Experimental Application.

$$MRR = 3.4865 \times 10^{-6} \cdot d_{cs} \cdot N \cdot F_r \quad (9)$$

What we propose in practical experience is to show that the SYFLOCS developed is capable of maintaining a constant MRR through the control of the force. Conversely, we expect to show that in ideal machining conditions these values keep to constant proportional factor, and when some deficiencies appear in the process, like tool wear, this relationship changes.

SUPERVISORY FUZZY LOGIC CONTROL SYSTEM (SYFLOCS)

The purpose of the SYFLOCS is to maintain the cutting force as near as possible to a given reference force. This reference force must take into consideration the type of milling cut and MRR. The supervisory control system has to control some unreliable conditions as well as the impossibility of measuring some variables in the milling process.

Fuzzy control algorithms

Our experiments with control schemes are based on the rather universal structure of the fuzzy logic controller [9]. Its knowledge base is obtained, in essence, from the experience of skilled operators [7]. The output vector is:

$$\underline{u}^T = [\Delta f \;\; \Delta s] \quad (10)$$

where Δf and Δs represents the feed and spindle speed variation.

The algorithm includes the control engineering sound practice, dealing with the process dynamics and compensating the process delay by means of the introduction of the error derivative (difference), its input vector is:

$$\underline{y}^T = [\Delta F \quad \Delta^2 F] \tag{11}$$

where ΔF represents the cutting force error (deviation from set point value), and the $\Delta^2 F$ is the first difference of the error.

The fuzzy partition of the variables universes of discourse are shown in Figure 2. The units of Δs and Δf are expressed in percent of the initial values programmed to the CNC, s_r and f_r, while ΔF and $\Delta^2 F$ are given, for the sake of simplicity, in amperes, because the cutting force is measured by means of a current sensor. The rule base for the algorithm, represented in an array form is shown in Table I.

Table I. Rule Base.

$\Downarrow \Delta^2 F$ / $\Rightarrow \Delta F$	N		C		P	
	Δf	Δs	Δf	Δs	Δf	Δs
N	NG	PG	NP	PP	CE	CE
C	NP	PP	CE	CE	PP	PP
P	CE	CE	PP	NP	PG	NG

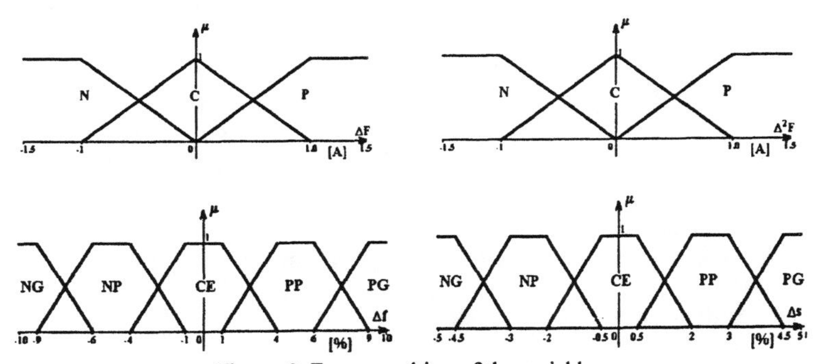

Figure 2. Fuzzy partition of the variables.

Our fuzzy controller follows rather classic standards. Its actual kernel is a look-up table pre-calculated (off line) by means of a general purpose shell, FURAL [11]. For calculating the Cartesian product we employed the compositional operator "sup-min" (Zadeh) and the defuzzification strategy was the Center of Gravity.

Scheme of the supervisory control system

The reference values (set points) for the spindle speed (s^*) and feed rate (f^*) are generated on line by the controller, represented by the personal computer (options for operators settings are, of course, allowed). The PC is fed with actual values of these speeds, s^o and f^o, as well as those of the cutting force set point F^* and measured value F. Spatial position $\underline{x}^*$ is fed by the operator and, for our experiments, was fixed: constant vertical position of the tool (cutting depth depending on the shape of the workpiece) and unidirectional roughing.

At each sampling instant k, the cutting force error and error first difference are calculated as

$$\Delta F(k) = F^* - F(k) \tag{12}$$

$$\Delta^2F(k) = \Delta F(k) - F(k\text{-}1) \quad (13)$$

With these values, as well as those of s and f the corresponding values of $\Delta f(k)$ and $\Delta s(k)$ are obtained from the look-up tables, a 2x19x19 array. The set point values generated by our controllers were

$$f^*(k) = f(k\text{-}1) + K_f f(k) \quad (14)$$

$$s^*(k) = s_r + K_s \Delta s(k) \quad (15)$$

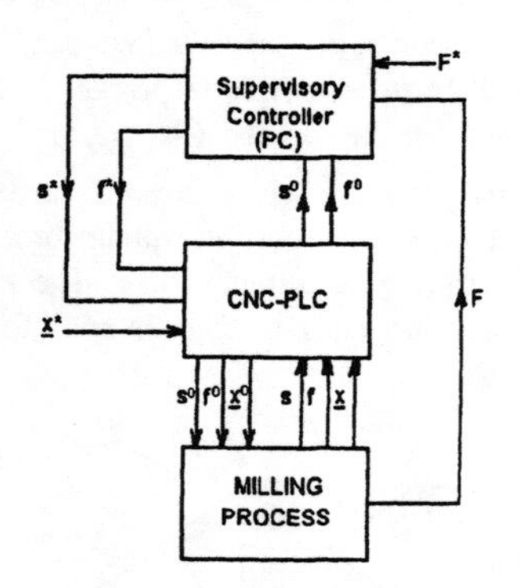

Figure 3. Scheme of the control system.

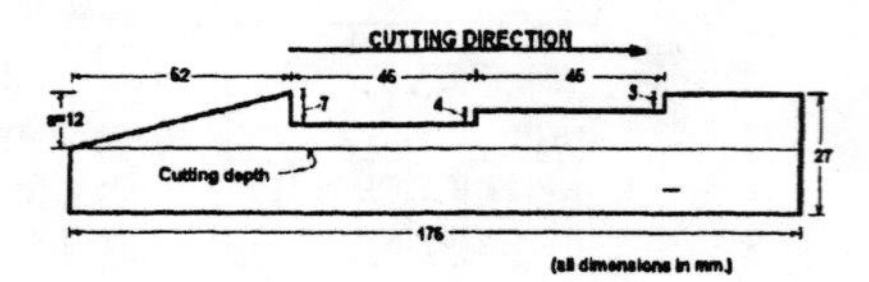

Figure 4. Workpiece.

The gains K_f and K_s were introduced for tuning purposes. Also security bounds for s and f (s_{lim} and f_{lim}) for chatter and damage prevention were implemented. A DURAL (aluminum alloy) workpiece was employed for the tests. Its quite irregular profile was chosen, intentionally: one ramp and three step shaped disturbances in the cutting depth, figure 4.

EXPERIMENTAL APPLICATION

In the Experimental Application we give an example that shows an increase in MRR when the SYFLOCS is used in the end-milling operation. Figure 5-a observes the end-milling behaviour without SYFLOCS, and in figure 5-b with SYFLOCS. It's noted in figure 5-a that the cutting force and MRR reach their optimum only at the last step of the workpiece for a constant feedrate and spindle speed. Figure 5-b observes the SYFLOCS behaviour employing greater effort to keep the reference force and MRR. The changes in the feedrate were limited to 300 mm/min to avoid bad machining conditions.

We have observed that there are differences in the SYFLOCS behaviour when machining the same workpiece and under the same technological parameters, when the state of the tool state changes. Comparing figures 6-a and 6-b we observe these variations for different tool states. Great amplitude differences are observed in the vibrations frequency of the measured force and the differences of the machine time (2.5 times) in each case. It is noted that in general, great amplitude of frequency vibrations of the measured force correspond to tool wear which produces superficial imperfection in the workpiece. The example shows the system behaviour for two types of tool states.

It is important to note that there is not any type of vibration in the new tool, however great vibration amplitudes are evident in the worn tool, especially in the last step of the workpiece where the greatest cutting depth is reached.

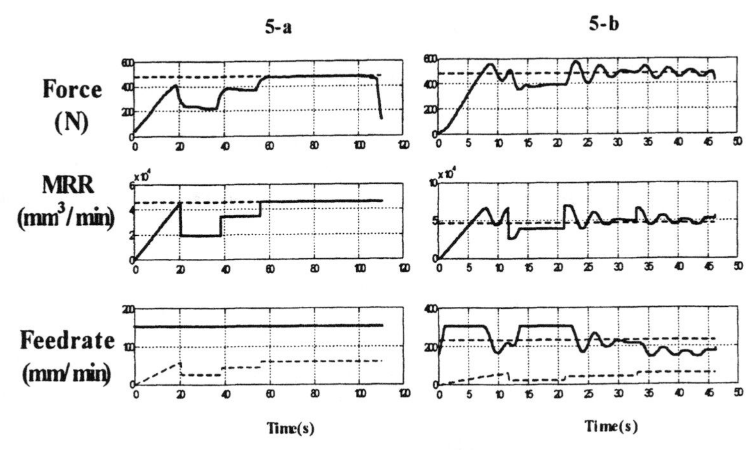

Figure 5. (a) End-milling without SYFLOCS; **(b)**End-milling with SYFLOCS.
Cutting Parameters: $s^0 = 827$ rpm, $f^0 = 152$ mm/min, tool diameter =25 mm, F^0=500 N, $MRR^0 = 45.736$ mm^3/min.

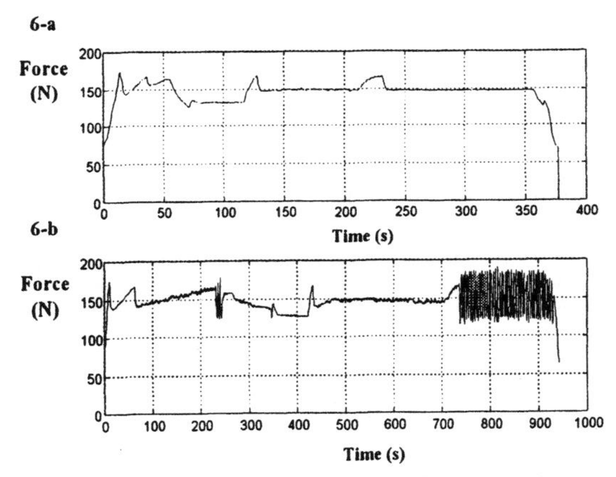

Figure 6. (a) Milling using new tool;(**b)** Milling using worn tool.
Cutting parameters: $s^0 = 1200$ rpm, $f^0 = 100$ mm/min, tool diameter = 25mm.

CONCLUSION

The knowledge acquired in our Experimental Application should be used in failure detection, which will infer about the tool state and inform of its wear grade after each machining. To focus the problem in this way is very interesting because it always takes into in consideration the dynamic of the process, as the supervisory control is continually acting upon it.

Much research continues improving the models where new algorithms of approximation are used to control and monitor the cutting process

[12][13][14][15][16][17], but a completely trustworthy model is not a reality yet. The SYFLOCS is our proposal to control and to detect failures in cutting process. The system provided the correct feedrate and spindle speed to give an optimum MRR. It's noted in figure 5-b that there is a reduction in the machining time when it is compared with figure 5-a. The oscillations in the force control for machining using an optimum MRR is the problem that we are solving. New algorithms have been studied to correct delays in the system (Center Machining). The Adaptive Fuzzy Control Algorithm could be a suitable solution for the random variations in the end-milling process, because of the great differences of conditions that are presented by the cutting process. The SYFLOCS is applied and the end-milling time variation and oscillation in the force control, figure 6-b, are detected and related to the tool state. Detecting the wear by means of the machining time in a process controlled by a supervisor is a particularly interesting idea because does not limit the system only to control algorithm behaviour.

This control has the objective of minimising vibrations and optimising the cutting time. This implicates the softness answers to smooth variation, 'tool wear', and strongest answers to abrupt variations, 'broken'. Therefore we have been getting goods results to achieving a more economical cutting process, where the tool wear and superficial finish are used by the SYFLOCS.

ACKNOWLEDGEMENT

This work is done within the project of investigation "System of hierarchical control of a machine based on technical tool of artificial intelligence (intelligent machining)",which is aided by CICYT, Spain.

REFERENCES

1. Levi, R., Rosseto, S. "Machining Economics and Tool Life Variation, Part I and Part II", Journal of Engineering for Industry, pp. 393-402 (1978).
2. Iwata, K., Morotsu, Y.I., Iwatsubo, T., Fujii, S. "A Probabilistic Approach to the Determination of the Optimum Cutting Conditions", Journal of Engineering for Industry, pp. 1099-1107 (1972).
3. King, Robert I., "Handbook of High Speed Machining Technology, pp. 197-240, New York, 1985.
4. Tlusty, J. and P.MacNeil, "Dynamics of Cutting Forces in the End Milling", Annals of CIRP, Vol.24/1/1975. pp. 21-25
5. Altintas. "Direct Adaptive Control of End Milling Process". Int. J. Mach. Tools Manufact. Vol. 34, No. 4,461-72,1994.
6. Kim, M.W. Cho, K. Kim. "Application of the fuzzy control strategy to adaptive force control a non-minimum phase end milling operations". Int. J.Mach Tools Manufact., Vol. 34, No. 5, pp. 677-696,1994.
7. Lee, C. "Fuzzy Logic in Control Systems: Fuzzy Logic Controller" -Part I, IEEE Trans. on SMC, Vol. 20, No. 2, pp. 404-418,1990.
8. Dubois. "An application of fuzzy arithmetic to the optimisation of industrial machining processes", .Math Modelling, Vol. 9, No. 6, 461-475,1987.
9. Haber, R. H., Peres, C. R, Alique, J. R., Salvador T. R, Alique, A., "Two Approaches for a Fuzzy Supervisory Control System of a Vertical Milling Machine" VI IFSA - World Congress, Vol. 1, 397-400 São Paulo, Brazil (1995).
10. Dos Santos, M.T., Peres, C.R., Ros, S.T., Alique, J.R, Alique, A.. "End-Milling Process Stress Monitoring by Current Sensor", CSME'96, Canada (1996).
11. Pérez, R., Ortiz, A. "Regulador Fuzzy" (in Spanish), Internal Report, Instituto Superior Politécnico "J.A.Mella", Santiago de Cuba (1987)
12. Nolzen, H. "Fault Diagnostic and Wear Detection in the Control Loop of Milling Machine", SAFEPROCESS, Helsinki (1994).
13. Konrad, H., Isermann, R., Oette, H.U. "Supervision of Toll Wear and Surface Quality During and Milling Operations", IFAC Workshop "Intelligent Manufacturing System", Austria (1994).
14. Elbestawi, M.A:, Papazafiriu, T.A., Du, R.X. "In Process Monitoring of Tool Wear in Milling Using Cutting Force Signature", Int. J. Mach. Tools Manufact", Vol.31, No.1, pp.55-75, Elsenvier Science Ltd., Great Britain (1991).
15. Usui, E., Shirakashi, T., Kitagawa, T. "Analytical Prediction of Cutting Tool Wear", Wear, 100, pp. 129-151 (1984).
16. Teitenberg, T.M., Bayoumi, A.E., Yucesan, G. "Tool Wear Modelling Through and Analytic Mechanistic Model of Milling Process", Wear, 154,pp.287-304 (1992).
17. Lin, S.C., Yang, R.J. "Force-Based Model for Tool Wear Monitoring in Face Milling", Int. J. Mach. Tools Manufact", Vol.35, No.9, pp. 1201-1211, Elsenvier Science Ltd., Great Britain (1995).

Influence of Dynamics in Statically Stable Walking Machines

P. Gonzalez de Santos. M.A. Jimenez and M.A. Armada
Instituto de Automatica Industrial-CSIC
La Poveda, 28500 Arganda del Rey, Madrid, Spain

ABSTRACT

This paper addresses how dynamic effects modify the measurement of the static stability of a discontinuous gait. For this study, a dynamic planar model of a four-legged walking machine was derived. Then, both the longitudinal and dynamic stability margins were computed and compared. Final results show that the static stability margin is an adequate measurement for studying stability, even at high velocities and accelerations.

KEYWORDS: Walking machines, legged locomotion, static stability, dynamic stability.

INTRODUCTION

In the past few decades, theoretical investigation on walking machines has produced many results in machine configurations, force distribution and control, dynamic modeling, and, of course, gait generation.

At the very beginning of gait research, researchers studied discontinuous gaits, going on immediately to study continuous gaits. Continuous gaits seemed more interesting because they are observed in nature. The most widely used gait is the wave gait employed by almost all mammals and insects at low speed. These gaits were mathematically stated and simulated, and displayed very good walking features. Nevertheless, discontinuous gaits present some features that have been ignored for years. For instance, discontinuous gaits present better characteristics than wave gaits in terms of stability, speed and power consumption. These gait features have been investigated by the authors of this paper during recent years, but such work has been confined to static stability, or rather, has avoided dynamic aspects [1]. For instance, discontinuous gaits can achieve a better longitudinal stability margin (LSM) than wave gaits and also greater speed for small duty factors (β), but this is for rectangular speed profiles in joints, which means a very large (infinite) joint acceleration. The main question at this point is what does remain true in gait features if dynamic effects are incorporated.

APPROACH

To study dynamic effects in a walking machine when performing discontinuous gaits, a planar dynamic model was derived. The reasons for creating a two-dimensional dynamic model are twofold. First, the components of force and motion of an animal perpendicular to the plane of motion during walking are usually relatively small [5], [7]. Second, this paper compares the longitudinal stability margin (LSM) with the dynamic stability margin (DSM), and both are stability measurements along the longitudinal axis of the machine. Therefore, a planar model is sufficient for this paper's purposes.

The LSM is defined as the minimum distance of the vertical projection of the body's center of gravity to the front and rear boundaries of the support polygon (defined by the feet in support). This measurement was introduced early in the study of static stability. Later, an extension of this margin was stated to consider dynamic aspects. The idea is to consider the DSM as the distance from the pressure center to the front and rear boundaries of the support polygon. This is the point projected along the direction of the resultant force (gravitational, inertial and external forces) acting on the mass center. This is the usual definition of DSM, although some authors use a slight variation [4].

Discontinuous gaits perform motion sequentially. At any given time, just one element (leg or body) is moving, and body motion is realized with all feet in support. That means that dynamic effects occur only during the body propelling phase.

For a static stable gait, the LSM in the direction of motion is given by (See Fig. 2):

$$LSM = Min\left[\left|\frac{x_{b0_1} + x_{b0_2}}{2}\right|, \left|\frac{x_{b0_3} + x_{b0_4}}{2}\right|\right] \tag{1}$$

where (x_{b0_i}, z_{b0_i}) are the compoments of leg i in the levelled body reference frame, $|m|$ is the absolute value of m and Min is the Minimum function.

The DSM can be computed as:

$$DSM = Min\left[\left|\frac{x_{b0_1} + x_{b0_2}}{2} - z_{b0_1} \cdot \frac{f_x}{f_z}\right|, \left|\frac{x_{b0_3} + x_{b0_4}}{2} - z_{b0_1} \cdot \frac{f_x}{f_z}\right|\right] \tag{2}$$

Note that all feet are in the same plane, thus z_{b0_i} is the same for every foot.

To derive the dynamic model, the RIMHO walking machine was considered. This machine is shown in Fig. 1, and its features may be found in [2].

Figure 1. The RIMHO Walking Machine

DYNAMIC PLANAR MODEL

The planar model of the quadruped is shown in Fig. 3. The main body is modelled as a rigid beam. A pair of legs is attached to the front end of the body, and the other two legs are attached to the rear end of the body. For the sake of simplicity, the legs are modelled as massless. Because the legs are of the pantographic type, foot forces and motions can be expressed as linear combinations of forces and motion along the axes of the body reference frame.

To derive the model, three reference frames are defined: the earth reference frame (x_e, z_e), the body reference frame (x_b, z_b) and an additional levelled reference frame located at the center of the body (x_{b0}, z_{b0}). The center of the body is assumed to be coincident with the center of gravity (cg), and the location of the cg is given by (x, z) in the earth reference frame.

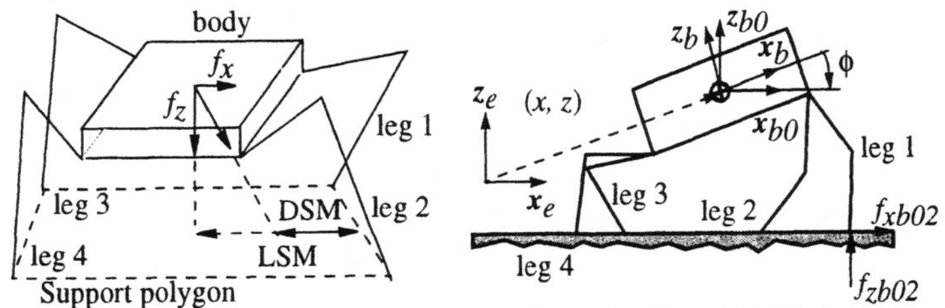

Figure 2. Definition of LSM and DSM

Figure 3. Planar Model of the Quadruped

To derive the equations of motion, the Newton-Euler method is first applied. For a two-dimensional rigid solid, these equations are reduced to:

$$\vec{F} = m\vec{a} \tag{3}$$

$$\vec{N} = J\dot{\vec{\omega}} \tag{4}$$

Where m is the mass of the body, F is the force acting at the cg, a is the resultant body acceleration, N is the total moment acting on the body caused by the foot forces, J is the moment of inertia and ω is the angular velocity of the body.

Taking into account the fact that the forces acting on the body are those exerted by the feet, then equation (3) can be split for each component into:

$$m\ddot{x}(t) = \sum_{i=1}^{4} f_{xb0_i}(t) \tag{5}$$

$$m\ddot{z}(t) = \sum_{i=1}^{4} f_{zb0_i}(t) - mg \tag{6}$$

where (f_{xb0}, f_{zb0}) is the foot force in the levelled body reference frame and i denotes the foot number.

Equation (4) may be written as:

$$J\ddot{\phi}(t) = \sum_{i=1}^{4} f_{xb0_i}(t) \cdot z_{b0_i}(t) - \sum_{i=1}^{4} f_{zb0_i}(t) \cdot x_{b0_i}(t) \tag{7}$$

where (x_{b0_i}, z_{b0_i}) is the position of foot i in the levelled body reference frame.

Equations (5) to (7) give the behavior of the body for given foot forces, but for control purposes it is necessary to know the foot forces needed to accomplish a desired body motion. To accomplish that, the inverse dynamic model must be derived.

The desired body motion is specified in terms of force, and previous equations may be rewritten as:

$$m\ddot{x}(t) = F_x(t) \tag{8}$$

$$m\ddot{z}(t) = F_z(t) \tag{9}$$

$$J\ddot{\phi}(t) = \Phi(t) \tag{10}$$

where $F_x(t)$, $F_z(t)$ and $\Phi(t)$ are the desired forces and moments. For instance, if a vertical body motion is required without any body rotation, the right-hand terms of equations (8) to (10) should be: $F_x(t) = 0$ and $\Phi(t) = 0$ for any required $F_z(t)$.

Substituting conditions (8)-(10) into equations (5)-(7) yields:

$$\sum_{i=1}^{4} f_{xb0_i}(t) = F_x(t) \tag{11}$$

$$\sum_{i=1}^{4} f_{zb0_i}(t) = mg + F_z(t) \tag{12}$$

$$\sum_{i=1}^{4} f_{xb0_i}(t) \cdot z_{b0_i}(t) = \sum_{i=1}^{4} f_{zb0_i}(t) \cdot x_{b0_i}(t) + \Phi(t) \tag{13}$$

Equations (11)-(13) constitute a linear equation system of three equations in eight unknowns. The indeterminacy of this system can be reduced by assuming that forces along the x-axis are the same for all four feet. This requirement guarantees that the interaction force between any two feet is zero. The interaction force is defined as the component of the difference between the two-foot forces directed along the line joining the two feet. The zero interaction force constraint was introduced by Waldron, and it prevents foot slippage because it prevents feet from fighting one another [6]. For a two-dimensional model, the zero interaction force constraint is equivalent to considering all horizontal foot forces equal. Thus, this last condition along with equation (11) gives:

$$f_{xb0_i}(t) = F_x(t)/4 \tag{14}$$

Therefore, the (11)-(13) equation system may be reduced to a system of two equations (12)-(13) in four unknowns f_{zb0_i}, which can be written in matrix form as:

$$A \cdot f_z = w \tag{15}$$

where

$$A = \begin{bmatrix} 1 & 1 & 1 & 1 \\ x_{b0_1}(t) & x_{b0_2}(t) & x_{b0_3}(t) & x_{b0_4}(t) \end{bmatrix} \tag{16}$$

$$f_z = \left[f_{zb0_1} \; f_{zb0_2} \; f_{zb0_3} \; f_{zb0_4} \right]^T \tag{17}$$

$$w = \begin{bmatrix} m \cdot g + F_z(t) & \sum_{i=1}^{4} f_{xb0_i}(t) \cdot z_{b0_i}(t) - \Phi(t) \end{bmatrix}^T \quad (18)$$

An infinite number of solution sets exists for this system, and thus it is possible to apply optimality concepts to find a unique solution. Optimization methods have been proposed by several authors under different conditions. Interested readers may find a short description of this topic in [4]. For this paper's purposes, the method described by Klein and Chung in [3] will be used.

This method uses the general solution of any underdeterminate linear system that is given by:

$$f_z = A^+ \cdot w + (I - A^+ \cdot A) \cdot p \quad (19)$$

where A^+ is the pseudoinverse of A and p is an arbitrary vector with the same dimension as f_z. The total solution of (19) is the one nearest to a given vector p. In [3] the authors define several possible values for p. In this paper the initial value $p = [0\ 0\ 0\ 0]^T$ will be used. This choice means that the solution is given just by the pseudoinverse matrix, but using the full solution results in some simulation shortcomings. Keeping the simple solution makes simulations easier and faster. The drawback is that the solution can yield negative foot forces, which are inadmissible because the walking machine does not grasp the ground but simply stands on it. Nevertheless, this phenomenon has never been detected in simulation.

After all these premises, the simulation can be performed in the following steps:

1. Define the foot positions in the earth reference frame (x_{ei}, z_{ei}).
2. Define the initial body position in the earth reference frame (x, z).
3. Compute the foot positions in the levelled body reference frame (x_{b0i}, z_{b0i}) and the body reference frame (x_{bi}, z_{bi}). These positions are given by:

$$\begin{bmatrix} x_{b0_i} \\ z_{b0_i} \end{bmatrix} = \begin{bmatrix} x_{e_i} - x \\ z_{e_i} - z \end{bmatrix} \quad (20)$$

and

$$\begin{bmatrix} x_{b_i} \\ z_{b_i} \end{bmatrix} = R_z(t) \begin{bmatrix} x_{e_i} - x \\ z_{e_i} - z \end{bmatrix} \quad (21)$$

where $R_z(t)$ is the rotation matrix.

4. Define the desired body trajectory in terms of forces and/or accelerations. For instance, a straight line along the x-axis is defined as $F_z(t) = 0$ and $\Phi(t) = 0$ for any desired $F_x(t)$. At this point, it is also necessary to define initial conditions for velocity and acceleration.
5. Apply the inverse dynamic model to find the foot forces required to move the body in the desired trajectory under the desired conditions. This model gives the foot forces in the levelled body reference frame.
6. Compute the forces in the body reference frame. These forces are the actuator forces and they are given by:

$$\begin{bmatrix} f_{xb_i} \\ f_{zb_i} \end{bmatrix} = R_z(t) \begin{bmatrix} f_{xb0_i} \\ f_{xb0_i} \end{bmatrix} \tag{22}$$

7. Compute the direct dynamic model to find the position of the body. With this position, it is possible to repeat the procedure to find the body position in the next sample period.

The results should be those specified in equations (8)-(10) as the desired motion of the body, but because of the optimization introduced in (19), they may differ from expected results.

SIMULATION RESULTS

The main aim of this paper is to show how the stability margin is affected by dynamic effects when performing discontinuous gaits. In a discontinuous gait, the body is only propelled forward with all four feet in contact with the ground. This motion is performed following a straight line along the x axis. This trajectory may be defined for simulation purposes as:

$$F_x(t) = m \cdot a_x(t); \quad F_z(t) = 0; \quad \Phi(t) = 0 \tag{23}$$

where

$$a_x(t) = \begin{cases} 0; & t \le t_1 \\ a; & t_1 < t \le t_2 \\ 0; & t_2 < t \le t_3 \\ -a; & t_3 < t \le t_4 \\ 0; & t \ge t_4 \end{cases} \tag{24}$$

is the acceleration required to accelerate the body, move it at a constant velocity, and finally decelerate until motion has fully stopped.

Data utilized in the simulation are as follows:

Vehicle mass: m = 50 kg

Moment of inertia: J = 0.5 $kg.m^2$

Initial body position: $(x_0, z_0) = (2, 1)$ m

Body velocities: $(\dot{x}_0, \dot{z}_0) = (0, 0)$ $m.s^{-1}$

Maximum body acceleration: $a = 0.5$ $m.s^{-2}$

Initial foot positions: $(x_{e_1}, z_{e_1}) = (3.25, 0)$ m

$(x_{e_2}, z_{e_2}) = (4, 0)$ m

$(x_{e_3}, z_{e_3}) = (0.75, 0)$ m

$(x_{e_4}, z_{e_4}) = (1.5, 0)$ m

The results obtained after applying the algorithm described above are shown in Fig. 4 and Fig. 5. These results are the same as those defined in equations (8)-(10), which means that the optimization derived in (19) works well.

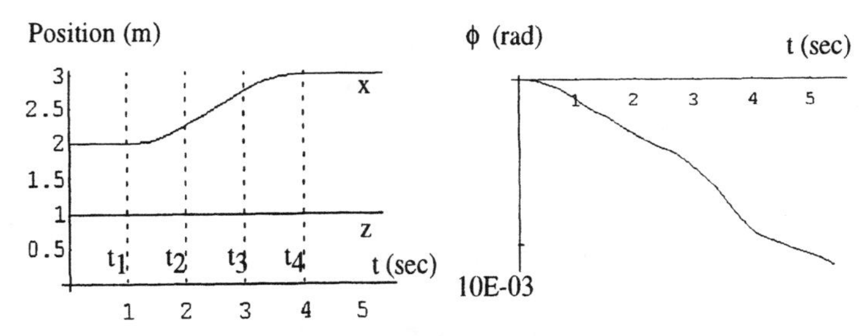

Figure 4. Body position when following a straight line

Figure 5. Body attitude when following a straight line

Fig. 6 shows the foot forces required to accomplish the desired motion. All of them are positive; therefore, the solution is acceptable.

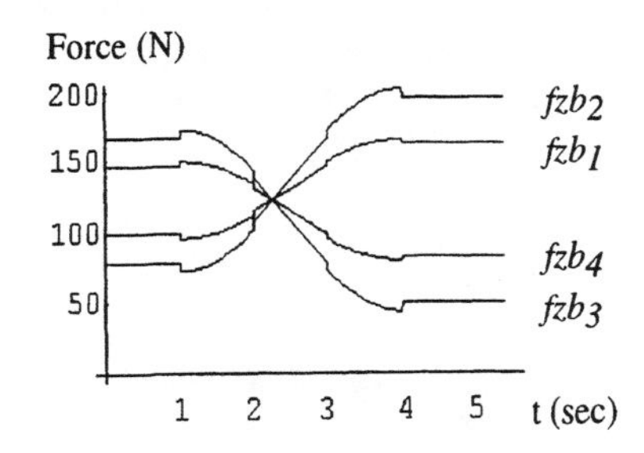

Figure 6. Foot forces to accomplish the trajectory

Finally, Fig. 7 shows the LSM in a solid line and the DSM in a dashed line along the specified trajectory. The two stability margins coincide when no acceleration/deceleration occurs. During acceleration/deceleration periods, the DSM differs from the LSM, but it is always greater than or equal to the LSM.

Thus, if the stability margin along the entire duty cycle is considered, the LSM is the smaller margin. Therefore, if a gait with a given LSM is specified, the dynamic effects are not going to decrease it, and the LSM, considered as the minimum along the locomotion cycle, remains the same.

distance (m)

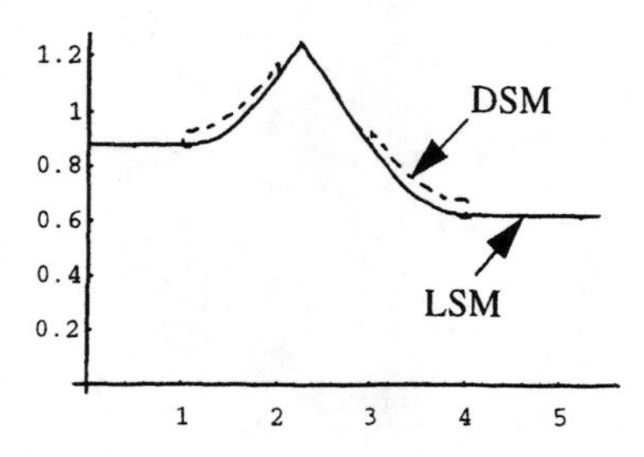

Figure 7. LSM and DSM

CONCLUSIONS

To study the effects of dynamics when performing discontinuous gaits under static stability, a planar dynamic model was derived for simulation purposes. The behavior of this model was checked, the foot forces required to follow a given trajectory were computed, and the static and dynamic stability margins were investigated.

The main result is that dynamic aspects do not change the properties of static stability margins, and the criterium of considering the LSM as a stability measurement may be valid at high velocity and acceleration, i.e., when dynamic effects appear.

ACKNOWLEDGMENT

We gratefully acknowledge the support of CICYT (Spain) under grant TAP94-0783.

REFERENCES

1. Gonzalez de Santos, P. and Jimenez, M.A. "Generation of Discontinuous Gaits for Quadruped Walking Machines," Journal of Robotic Systems, 12(9), pp. 599-611, September 1995.
2. Jimenez, M.A., Gonzalez de Santos, P. and Armada, M.A. "A Four-Legged Walking Test Bed," 1st IFAC International Workshop on Intelligent Autonomous Vehicles, Hampshire, U.K., pp. 8-13, April 1993.
3. Klein, C.A. and Chung, T.S. "Force Interaction and Allocation for the Legs of a Walking Vehicle,"IEEE Journal of Robotics and Automation, Vol. RA-3, No. 6, December 1987.
4. Lin, B.S., and Song, S.M. "Dynamic Modeling, Stability and Energy Efficiency of a Quadrupedal Walking Machine," IEEE International Conference on Robotics and Automation, Atlanta, Georgia, May 2-6, pp. 367-373, 1993.
5. Pandy, M.G., Kumar, V., Berme, N. and Waldron, K.L. "The Dynamics of Quadrupedal Locomotion," ASME Journal of Biomechanical Engineering, vol. 110, pp. 230-237, August 1988.
6. Waldron, K.J. "Force and Motion Management in Legged Locomotion," IEEE Journal of Robotics and Automation, Vol. RA-2, No. 4, December 1986
7. Wong, H.C. and Orin, D.E., "Dynamic Control of a Quadruped Standing Jump," IEEE International Conference on Robotics and Automation, Atlanta, Georgia, May 2-6, pp. 346-351, 1993.

ESTIMATING CONTACT UNCERTAINTIES USING KALMAN FILTERS IN FORCE CONTROLLED ASSEMBLY

STEFAN DUTRE, SABINE DEMEY and JORIS DE SCHUTTER
K.U. Leuven, Dept. of Mech. Eng., Div. PMA, Celestijnenlaan 300B, 3001 Heverlee, Belgium

JAYANTHA KATUPITIYA
School of Mech. and Manf. Eng., Univ. of New South Wales, Sydney NSW 2052, Australia

JAN DE GEETER
SCK•CEN Belgian Research Centre for Nuclear Energy, Boeretang 200, B-2400 Mol, Belgium

ABSTRACT

This paper presents a general approach to identify geometrical uncertainties and changes in contact situations during force controlled assembly operations. The problem of identification, which refers to the estimation of exact relative positions and orientations of contacting objects, is formulated and solved using contact models and Kalman Filters. The Kalman Filter uses the measured contact forces and velocities as its measurement vector. The problem of identifying the changes in contact situations, which is called monitoring, is solved by carrying out a statistical test on the sum of normalized and squared innovations of the Kalman Filter, within a moving window. Experimental results of a peg into hole assembly operation are presented.

KEYWORDS: force controlled assembly, kinematic contact models, uncertainty identification, model validation, Kalman Filter.

INTRODUCTION

Uncertainties in a force controlled assembly operation result from the lack of precise knowledge as to how the manipulated objects are contacting each other. This involves both *geometrical* and *topological* uncertainties [2]. The geometrical uncertainties are the position and orientation of each contact normal, whereas the topological uncertainties are the number of contacts and the type of each contact (vertex-line, surface-curve, curve-curve, etc.). The number of uncertainties depends on the geometry of the objects and the current contact topology.

Determining the contact uncertainties allows us to correct the contact model to match the real contact situation, and hence to carry out accurate assembly motions. The approach presented in this paper is based on the use of virtual contact manipulators [2, 3] to model expected contact situations and their uncertainties, and an

Extended Kalman Filter [7, 8] to estimate the geometrical uncertainties, also called *identification*. Using a Kalman Filter (KF) has a number of advantages over the energy based identification method described in [6]: (i) a KF contains stochastical information of the measurements and the initial states. As a result, information about the accuracy of estimated states is available at each time step; (ii) measurements from different sensors can easily be combined; (iii) a KF combines measurements of different time steps. In some cases, this may result in more identifiable uncertainties.

Validation of the correctness of the expected contact topology, i.e. determination of the topological uncertainties, is called *monitoring*. It is based on the statistical tests carried out on the KF estimation process [1]. The failure of the statistical test indicates that the measurement model – which contains the contact model – used in the KF is no longer valid.

Next section introduces the measurement model of the Kalman Filter and describes how this measurement model is used for estimation of the geometrical uncertainties. Changes in the contact situation are monitored by statistically verifying the quality of the KF estimates. The following section describes the insertion strategy used in a real peg into hole insertion, and models the different contact situations that arise during the insertion task. The next section presents experimental results of the off-line identification and monitoring of these contact situations and the last section draws some conclusions.

IDENTIFICATION AND MONITORING

This section presents a general approach for the *identification* of first order geometric uncertainties, and the *monitoring* of contact situation changes in force controlled assembly operations. The identification problem is model based, i.e. it relies explicitly on kinematic contact models, represented by the twist and wrench Jacobians of virtual contact manipulators. A comprehensive coverage of modelling contact kinematics by virtual contact manipulators is given in [2]. The identification problem is formulated in terms of a Kalman filter (KF) [7, 8].

Identification. This subsection explains how first order geometric uncertainties are identified using a Kalman filter.

The first order geometric uncertainties $\boldsymbol{v}$ are the states of the KF. As explained in [2], these uncertainties correspond to uncertainties on the initial estimates of the joint values of the contact manipulator. They result from the unknown initial relative position and orientation of the object and the environment. The nominal evolution of the joints of the contact manipulators can be computed from the motion carried out by the robot as explained in [2]. Hence the KF only estimates the uncertainties, not the nominal evolution, and the uncertainties are supposed to be constant. Thus a static Kalman filter is sufficient, i.e. the prediction stage is trivial (prediction of uncertainty at time instant i equals the uncertainty estimated at instant $i-1$).

Let the twist vector $\mathsf{t} = [\boldsymbol{v}^T\ \boldsymbol{\omega}^T]^T$ contain translational and rotational velocities $\boldsymbol{v}$ and $\boldsymbol{\omega}$ of the manipulated object. Similarly, contact forces and torques $\boldsymbol{f}$ and $\boldsymbol{n}$ are grouped into a wrench vector $\mathsf{w} = [\boldsymbol{f}^T\ \boldsymbol{n}^T]^T$. Define the measurement vector $\boldsymbol{z} = [\mathsf{t}^T\ \mathsf{w}^T]^T$.

Let $\boldsymbol{J}$ and $\boldsymbol{G}$ be the so-called twist and wrench Jacobians of the contact model containing a set of, respectively, twist and wrench basis screws which span the unconstrained and constrained vector space [2]. These Jacobians $\boldsymbol{J}$ and $\boldsymbol{G}$ are both functions of the geometrical uncertainties, $\boldsymbol{v}$ to be estimated by the KF. The measurement equation of the KF, i.e. the relation between the state vector $\boldsymbol{v}$ and the measurements z, is derived from the reciprocity condition, i.e. the actual twist should be reciprocal to any wrench which belongs to the modelled wrench space (spanned by wrench Jacobian $\boldsymbol{G}$), and, similarly, the actual wrench should be reciprocal to any twist which belongs to the modelled twist space (spanned by twist Jacobian $\boldsymbol{J}$). Mathematically this is expressed as:

$$\boldsymbol{G}(\boldsymbol{v})^T \mathsf{t} = 0 + \rho_f, \qquad \boldsymbol{J}(\boldsymbol{v})^T \mathsf{w} = 0 + \rho_v, \tag{1}$$

or:

$$\mathcal{H}(\boldsymbol{v})\boldsymbol{z} = 0 + \rho, \tag{2}$$

with:

$$\mathcal{H}(\boldsymbol{v}) = \begin{bmatrix} \boldsymbol{G}^T(\boldsymbol{v}) & \boldsymbol{O}_{n_f \times 6} \\ \boldsymbol{O}_{n_x \times 6} & \boldsymbol{J}^T(\boldsymbol{v}) \end{bmatrix}. \tag{3}$$

n_x and n_f are respectively the number of degrees of freedom and the number of constraints of the manipulated object. $\rho = [\rho_f^T \ \rho_v^T]^T$ is assumed to be a $((n_f + n_x) \times 1)$ Gaussian noise vector with covariance matrix $\boldsymbol{R}$. The measurements $\boldsymbol{z}$ are supposed to contain Gaussian measurement noise with constant covariance matrix $\boldsymbol{R}_z$. This propagates to the covariance of the noise on the relation (2) using:

$$\boldsymbol{R} = \mathcal{H}(\boldsymbol{v}) \ \boldsymbol{R}_z \ \mathcal{H}^T(\boldsymbol{v}). \tag{4}$$

It is possible to consider only *twist based* identification, corresponding to the first part of (1), or only *wrench based* identification, corresponding to the second part of (1). The composition and dimensions of $\mathcal{H}$ change accordingly. Measurement equation (2) is a nonlinear function of the uncertainties and the measurements.

This measurement model is linearized around the currently available estimate of $\boldsymbol{v}$, i.e. $\hat{\boldsymbol{v}}$, using:

$$\boldsymbol{H} = \left.\frac{\partial \mathcal{H}}{\partial \boldsymbol{v}}\right|_{\boldsymbol{v}=\hat{\boldsymbol{v}}} \boldsymbol{z}. \tag{5}$$

This involves the derivative of the Jacobians $\boldsymbol{J}$ and $\boldsymbol{G}$ with respect to $\boldsymbol{v}$ [5].

At time instant i, the innovations $\boldsymbol{\eta}_i$, are defined as:

$$\boldsymbol{\eta}_i = \mathcal{H}_i \ \boldsymbol{z}_i. \tag{6}$$

$\mathcal{H}_i$ is obtained by substituting $\hat{\boldsymbol{v}}_{i-1}$ for $\boldsymbol{v}$ in (3). The innovations indicate to what extent the motion and force measurements deviate from perfect reciprocity w.r.t. the model, given the current estimate $\hat{\boldsymbol{v}}_{i-1}$.

Let $\boldsymbol{P}_{i-1}$ denote the error covariance matrix of $\hat{\boldsymbol{v}}_{i-1}$. The covariance matrix of $\boldsymbol{\eta}_i$, $\boldsymbol{S}_i$, and the Kalman gain matrix $\boldsymbol{K}_i$—which guarantees minimum variance—are given by:

$$\boldsymbol{S}_i = \boldsymbol{R}_i + \boldsymbol{H}_i \boldsymbol{P}_{i-1} \boldsymbol{H}_i^T, \tag{7}$$

$$\boldsymbol{K}_i = \boldsymbol{P}_{i-1} \boldsymbol{H}_i^T \boldsymbol{S}_i^{-1}. \tag{8}$$

$\boldsymbol{H}_i$ and $\boldsymbol{R}_i$ are obtained by substituting $\hat{\boldsymbol{v}}_{i-1}$ for $\boldsymbol{v}$ in (5) and (4). $\boldsymbol{K}_i$ is used to estimate the contact uncertainties $\hat{\boldsymbol{v}}_i$ from $\hat{\boldsymbol{v}}_{i-1}$, and the corresponding error covariance matrix $\boldsymbol{P}_i$ from $\boldsymbol{P}_{i-1}$:

$$\hat{\boldsymbol{v}}_i = \hat{\boldsymbol{v}}_{i-1} + \boldsymbol{K}_i \boldsymbol{\eta}_i, \tag{9}$$

$$\boldsymbol{P}_i = (\boldsymbol{I} - \boldsymbol{K}_i \boldsymbol{H}_i) \boldsymbol{P}_{i-1}. \tag{10}$$

An initial guess of the states and their covariance, $\hat{\boldsymbol{v}}_0$ and $\boldsymbol{P}_0$, is determined by other sensors (vision, etc.) or specified by the user.

The estimation of the uncertainties need not necessarily be done on-line. For example, the alignment operation, described in a later section, is halted for a short period of time until the measurement data have been processed by the KF and the uncertainties are determined. The alignment operation can then resume, now with the full knowledge of the uncertainties. However, on-line uncertainty identification is the final goal.

Monitoring. The term monitoring refers to the detection of a change in contact situation. The KF fails to give statistically good estimates when the measurement model and hence the current contact situation is no longer valid.

The statistical test used is the χ^2-test on the sum of the k latest normalized innovations squared (SNIS) [1]. At instant i the SNIS is:

$$\epsilon_i = \sum_{j=i-k+1}^{i} \boldsymbol{\eta}_j^T \boldsymbol{S}_j^{-1} \boldsymbol{\eta}_j. \tag{11}$$

k is chosen by the user as a trade-off between the speed of detection and the false alarm rate.

The SNIS (11) is χ^2 distributed with the number of degrees of freedom equal to $k(n_f + n_x)$, i.e. k times the number of measurement equations, i.e. $6k$ if both twist and wrench identification is used. This number reduces to k times n_f in case of twist based identification, and k times n_x in case of wrench based identification. A confidence interval of e.g. 95% can be used for the detection of a contact situation change.

PEG INTO HOLE INSERTION

The experiment involves insertion of a cylindrical peg in a hole in a poorly structured environment, including finding the hole and aligning the axes of peg and hole. The insertion strategy consists of a sequence of different contact situations which are described in [6]. The task starts in free space. Then, a curve-face contact is established. Subsequently a three point contact configuration and finally a cylindrical contact are reached. This section describes the models of the first three contact situations. The possible transient contacts between the curve-face contact and the three point contact are not modelled.

Free space. This situation is modelled by the twist and wrench Jacobians $\boldsymbol{J}_{fs}$ and $\boldsymbol{G}_{fs}$ [2]:

$$\boldsymbol{J}_{fs} = \boldsymbol{I}, \qquad \boldsymbol{G}_{fs} = \boldsymbol{O}. \tag{12}$$

This contact situation has no geometrical uncertainties.

Curve-face contact. The bottom rim of the peg contacts the plane of the hole. This is modelled by means of a five degrees-of-freedom virtual manipulator (Fig. 1). This contact situation has two identifiable geometrical uncertainties: (i) the position of the contact point on the bottom rim of the peg, represented by the rotation angle θ of the peg about its own axis. If the contact point is in the $Y-Z$ plane, θ equals zero. (ii) the angle α between the peg axis and the contact normal.

If r is the peg radius, and if s_α and c_α are respectively the sine and the cosine of α, the twist and wrench Jacobians $\boldsymbol{J}_{cf}$ and $\boldsymbol{G}_{cf}$ [2] are (for θ equals zero):

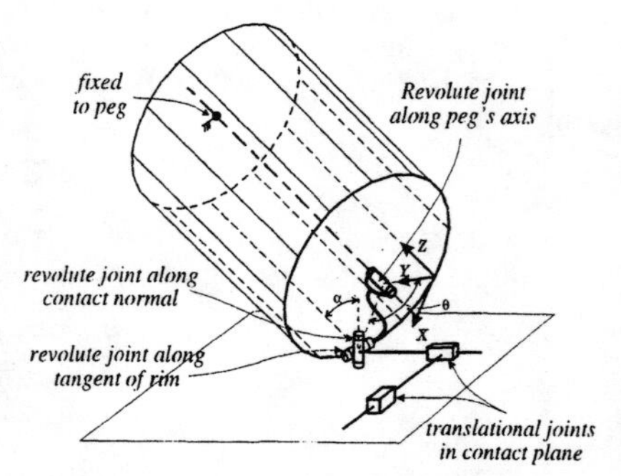

Figure 1: *Curve-Face* with a five joint kinematic model.

$$
\boldsymbol{J}_{cf} = \begin{bmatrix} -r & 0 & 0 & 1 & 0 \\ 0 & 0 & 0 & 0 & -c_\alpha \\ 0 & 0 & 0 & 0 & s_\alpha \\ 0 & 1 & 0 & 0 & 0 \\ 0 & 0 & s_\alpha & 0 & 0 \\ 1 & 0 & c_\alpha & 0 & 0 \end{bmatrix}, \quad \boldsymbol{G}_{cf} = \begin{bmatrix} 0 \\ s_\alpha \\ c_\alpha \\ 0 \\ 0 \\ 0 \end{bmatrix}. \tag{13}
$$

Three point contact. There are three distinct contacts:

1. *surf*: the contact between the surface of the peg and the rim of the hole.
2. *rim1* and *rim2*: the two contacts between the bottom rim of the peg and the rim of the hole. These two contacts are positioned symmetrically with respect to the plane through the peg's axis and through *surf*.

Each of the three point contacts is modelled by means of a five degrees-of-freedom virtual manipulator [2] (Fig. 2). These manipulators form three parallel connections between the hole and the peg, and constrain the peg's motion freedom in the same way as the contacts. For a detailed description of the alignment motion and the twist and wrench Jacobians see [3]. The three point contact has two identifiable geometrical uncertainties: the error on the alignment angle β (i.e. the angle between the axes of peg and hole), and the rotation angle ϕ about the peg axis.

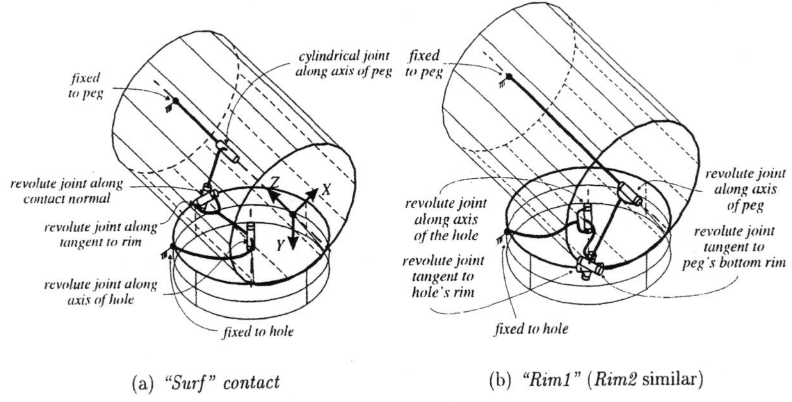

(a) *"Surf" contact* (b) *"Rim1"* (*Rim2* similar)

Figure 2: Kinematic contact models of three point contact

EXPERIMENTAL RESULTS

Test Setup. A KUKA IR 361/8 robot is equipped with a six component flexible force/torque sensor of which the deformations are measured. Its end effector holds a peg with radius $r = 100mm$. The experiment is executed with a task frame based compliant motion robot controller [4]. The measured twists and wrenches are stored during execution and are used by the KF for off-line identification and monitoring.

Identification. Fig. 3 (a) shows the result of the estimation of the contact uncertainties during the curve-surface contact. Because the wrench measurements were heavily influenced by friction, only twist measurements were used. θ is not identifiable from twist measurements only if the peg moves over the plane in a direction perpendicular to the X axis (Fig. 1) and if $\theta \approx 0$, as was the case during the experiment. Hence only results for α are presented. The initial guess of α is 0.873 *rad*. The KF converges to about 1 *rad*, which is consistent with the real situation. The contact situation changes around 22 seconds and as a result the estimations are no longer valid.

Fig. 3 (b) shows the uncertainties identified during the peg alignment process while maintaining a three-point contact. The KF estimates the error on the alignment angle β, whereas the nominal evolution of the alignment angle is derived from the motion specification. The initial guess for the error is 0.174 *rad*. The initial guess of the angle ϕ about the axis of the peg is also 0.174 *rad*. The measurement vector consists again of twists only. The dashed line in the top graph shows the actual alignment angle β. The solid line shows the alignment angle β estimated by the KF. These angle estimates are accurate to within 0.5 of a degree. The bottom graph shows the rotation angle about the axis of the peg. This angle is around zero and is consistent with the experimental situation.

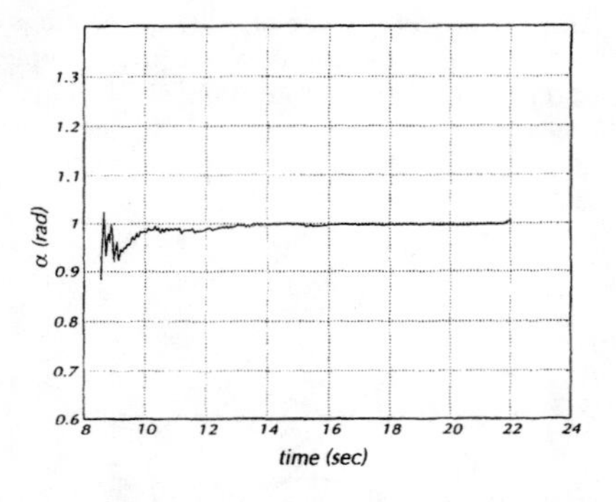

(a) Angle between axis of the peg and the contact normal (α).

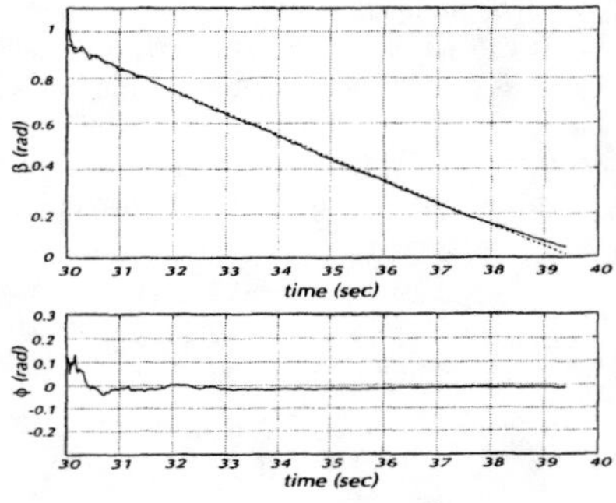

(b) Top: alignment angle β. Bottom: rotation angle about the peg's axis.

Figure 3: Identified geometrical uncertainties of the curve-face contact and the three point contact.

Monitoring. Fig. 4 shows the distribution of the χ^2 variable formed by equation (11) for the successive contact situations. The two dotted lines show the 5% and 95% confidence values for ϵ_i. The window width k is 10 and the number of measurement equations is 6, 1 ($n_f = 1$) and 3 ($n_f = 3$), for respectively the free space, the curve-face and the three point contact situation, since only twist measurements are used for the latter two. The value of ϵ_i should lie below the dotted line of 95% confidence for the estimates of the KF to be acceptable. When ϵ_i goes above the 95% line, the contact situation is statistically no longer valid. After 8 seconds ϵ_i jumps above the 95%

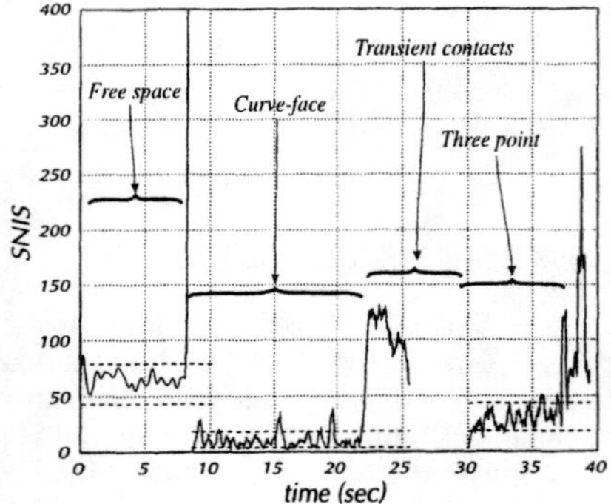

Figure 4: *The sum of normalized innovations squared (SNIS) within a moving window of 10 samples for the successive contact situations. The dotted lines show the 5% and 95 % confidence boundaries.*

confidence line which means that the transition to the curve-face contact takes place. From now on the model of the curve-face contact is valid. ϵ_i is low until $t = 22$ *sec*. Then ϵ_i goes above the 95% confidence line, which means that the bottom edge of the slanted peg falls into the hole. The controller is asked to generate a three point contact which is established at $t = 30$ *sec*. The contacts during this transition are not modelled. The alignment starts, and ends when the identified alignment error is small, i.e. at $t = 39$ *sec*. In a completely aligned configuration the three point contact is not stable. In fact, due to clearance the three point contact is already no longer valid shortly before complete alignment; at $t = 38$ *sec* ϵ_i exceeds the 95% confidence line.

CONCLUSION

This paper presents a new and general approach that can be used in a real assembly situation to determine contact uncertainties and contact situation changes. The method is based on recursive Kalman filtering techniques. It contains stochastical information about the measurements and the initial states and gives information about the accuracy of the estimated states at each time step. The changes in contact situations are monitored by statistically verifying the quality of the KF estimates. Experimental results have verified the theoretical basis.

ACKNOWLEDGEMENTS

This work was sponsored by the Belgian Program on Interuniversity Attraction poles initiated by the Belgian State —Prime Minister's Office— Science Policy Programming (IUAP-50). The scientific responsibility is assumed by its authors.

REFERENCES

[1] Yaakov Bar-Shalom and Xiao-Rong Li. *Estimation and Tracking, Principles, Techniques, and Software*. Artech House, 1993.

[2] H. Bruyninckx, S. Demey, S. Dutré, and J. De Schutter. Kinematic models for model based compliant motion in the presence of uncertainty. *Int. J. Rob. Research*, 14(5):465–482, 1995.

[3] H. Bruyninckx, S. Dutré, and J. De Schutter. Peg-on-hole: A model based solution to peg and hole alignment. In *Proc. IEEE Int. Conf. Rob. Automation*, pages 1919–1924, Nagoya, Japan, 1995.

[4] J. De Schutter and H. Van Brussel. Compliant robot motion I-II. A formalism for specifying compliant motion tasks. *Int. J. Rob. Research*, 7(4):3–33, 1988.

[5] S. Dutré. Calculating the derivative of a Jacobian for serial and parallel manipulators. *Internal report 95R49*, 1995.

[6] S. Dutré, H. Bruyninckx, and J. De Schutter. Contact identification and monitoring based on energy. In *Proc. IEEE Int. Conf. Rob. Automation*, Minneapolis, 1996.

[7] Arthur Gelb. *Applied Optimal Estimation*. The Analytic Sciences Corporation, 3rd edition, 1978.

[8] R. E. Kalman. A new approach to linear filtering and prediction problems. *Trans. ASME, J. Basic Engineering*, 82:34–45, 1960.

A FRAMEWORK FOR ROBOT MODELLING, SIMULATION AND TEST

P. Déplanques, L. Yriarte, R. Zapata, J. Sallantin

LIRMM - UMR 9928 CNRS - Universite Montpellier II
161 rue Ada 34392 Montpellier Cedex 5 FRANCE
email: {deplanqu, yriarte, zapata, js}@lirmm.fr
tel: (33) 67 41 85 85 - fax: (33) 67 41 85 00

ABSTRACT

Complex systems modelling and testing is an important issue for robotics, as these systems become more and more sophisticated, integrating embedded modules with their own physical resources and control software. There is a need for a generic methodology, integrating representations and tools for each step of robot design, from specification to test.

Our purpose is to develop a framework validating these concepts on mobile robots. It provides services to specify a robot's structure, control and missions within a normative representation language. Then, it allows to evaluate the behaviour of the robot, through its simulation as well as real scale testing, provides tools for failure explanation, and verify if it is fitted for a given mission.

KEYWORDS: test, simulation, modelling, behaviour, agent

INTRODUCTION

In this paper, we present the state of our work on the METEAUR[1] project, dedicated to testing the autonomy of mobile robots. This work comes as a general methodology, with a set of tools for knwoledge representation, modelling, and temporal reasoning.

Considerations on modelling and test

Modelling a system becomes necessary when we need to predict its behaviour in a context that never occurred before. A simple observation of the system's behaviour is not sufficient neither to explain why it acted this way so far, nor to predict what it is going to do next. Even worse, there is no way to ensure that the parameters used for the observation were relevant, nor to raise important parameters that could have been omitted.

For that matter, a modelling framework should :

- Provide a generic description of the related domain. Particular systems are modelled according to the specification of their domain, allowing different models to be compared on the same basis, by referring to this generic level.

[1] acronym for ***ME**thodology for **TE**sting the **AU**tonomy of **R**obots*, DRET contract #93414/DRET.

- Describe and classify the contexts in which the system evolves, so that the system can be partially validated, for a given context type.
- Express the behaviour of the system in a description language that must be explicitly specified, regarding to the system's own model, the context type, and the generic domain knowledge.

Modelling and testing a robot

When it comes to robotics, specific problems occur.

- **In the modelling process :** Robotics involves many different fields, including physics, electronics, and computing. Even if notions on each of these topics are well defined, such as physical components, sensors and actuators, or software architectures, their integration in a whole complex system is uneasy to model.

  Modelling the interaction between a robot and its environment is a major issue for robotics. Providing an extensive model of all the entities likely to influence any robot's behaviour would be unrealistic, as the environment modelling process is dependant on what the robot was designed for. On the other hand, having specific and unrelated models for every robot would not allow reuse and relative testing.

- **In the testing process :** A robot must be considered not only as a complex physical object, but also as an agent with a behaviour that can be nondeterministic. It can also be difficult to decide wether an unpredictable behaviour is due to its control software or to physical failures.

  For that matter, testing should rely on robot's structure as well as on it's expected behaviour. Structural testing should ensure that a robot has the relevant configuration for fulfilling its mission, in a given type of environment. Behavioural testing compares mission specification, simulation, and effective behaviour, in order to define contexts in which the robot can be validated.

Despite that, there is a need for robot testing in terms of functional autonomy. The notion of autonomy, applied to robotics, is the ability of a given robot to perform missions for which it is designed, regarding the constraints inherent to the mission environment. When this environment is dynamic and partially unknown, autonomy is related to intelligence, as the ability of an agent to deal with an unknown environment in order to fulfill its goals.

ARCHITECTURE OF A TEST PLATFORM

The test platform developed for this project as shown in figure1 relies on a layered architecture, featuring the static, dynamic and temporal representations required for adressing the problems raised above.

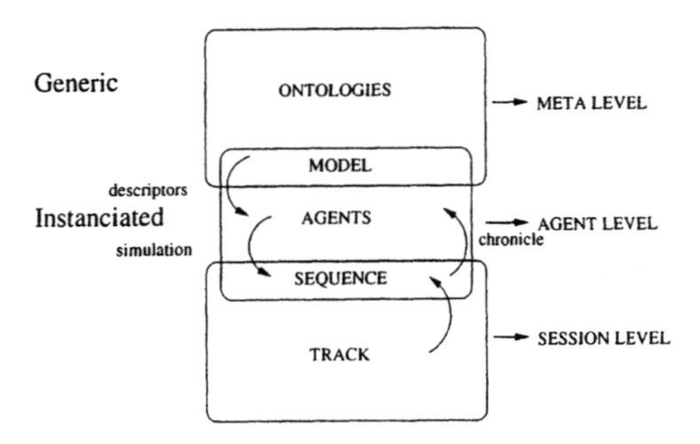

Figure 1: Global Architecture

Each level holds a software representation of the robot, that is queried from the upper level, and used for specifying the lower level.

Model types and related tests

The approach underlying this layered architecture relies on the abstraction level of the different models. Each model level can be queried for testing, and updated according to the test results.

Meta level. This level holds a static (structural) representation of a robot. It contains all available knowledge about robotics, as well as general considerations about physics and geometry, represented with an ontology[2]. This ontology is used to build a model allowing reasoning about the structure of the robot, without taking its behaviour in account. The type of environment in which the robot evolves is also modelled in order to choose which parameters will be taken in account for simulation.

Tests at the meta level adress queries like:

- What components does this robot use for motion ? How is energy transmitted to these components ?
- What is the structure used to describe this robot's environment ? Which properties have used to describe the object called "wall-1" in this structure ?
- Which components and software tasks are involved in the system realizing the "avoiding" functionnality ?

Agent level. This level holds a dynamic model of a robot. It features a multi-agent simulator, where each agent stands for a system referenced in the static model, and simulates the functionnality performed by this system. The interaction of the agents provides a simulation of the global robot behaviour. The state of the whole system is represented with a set of parameters, consulted and modified by the agents. These parameters are choosen when the dynamic model is specified, among properties from the static model.

Tests at the agent level consists in defining an initial configuration for the robot within its environment, and watch the global evolution of the system, for simulation and real-scale test. Different configurations of the robot can be tested in simulation by changing the agent configuration.

Session level. This level holds a temporal representation of the robot's behaviour. A session (simulated or real-scale) is referenced with its initial configuration (nature and properties of the objects in the environment) and the temporal evolution of the system's state. This evolution is archived as the sequence of values taken by the state variables.

Testing a session consists in extracting elementary behaviours from the sequences. These observed behaviours are used to validate functionnalities at the agent level.

SAMPLE APPLICATION

We used the METEAUR platform to test a small mobile robot, RAT, dedicated to the experimentation of fast local navigation algorithms. This robot is a car-like vehicule, powered by an electric motor activating the rear wheels. Its perception devices are three ultra-sonic sensors placed on the front, and an odometer on the front left wheel. Its mission is fast motion without collision, implanted with a control task for reactive avoiding based on the Deformable Virtual Zones (DVZ) method.

Our work in this case consisted in defining an ontology for robotics, use this ontology in a structural model of robot RAT, make a distributed simulator and tracking tools according to the specifications we provided.

Ontology of robotics

The ontology of robotics in figure2 sets five points of view on the domain: Agent assigns a behaviour and a nominal environment to a system, Subsystem sets relations of aggregation and recursive decomposition on systems designed to fullfill functionnalities, Component includes all physical objects, Control holds notions of tasks and software architecture, and Flow relates to transmission of data, motion, energy between tasks and components.

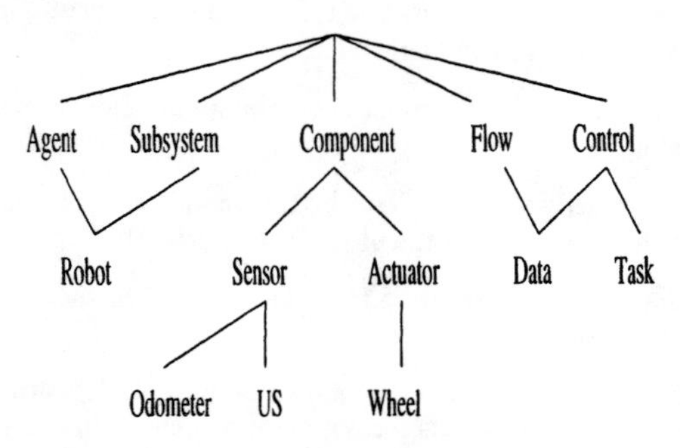

Figure 2: Points of view in ontology

The ontology of environment features physical notions like geometry, space and time, as defined in[2]. For instance, we specify the class "Robot" inheriting from "Agent" and "SubSystem", with a property "Wheelbase" under the dimension "length" defined in the ontology of environment.

```
(Define-Class Robot (?X)
"A robot is an autonomous agent and a decomposable subsystem"
```

```
  :Def (And (Agent ?X) (SubSystem ?X))
  :Axioms (
   (=>  (Robot ?X)
        (Has-Property ?X Wheelbase)
   )
))
```

Decomposition of robot RAT's physical subsystems in figure3 is part of RAT's static model.

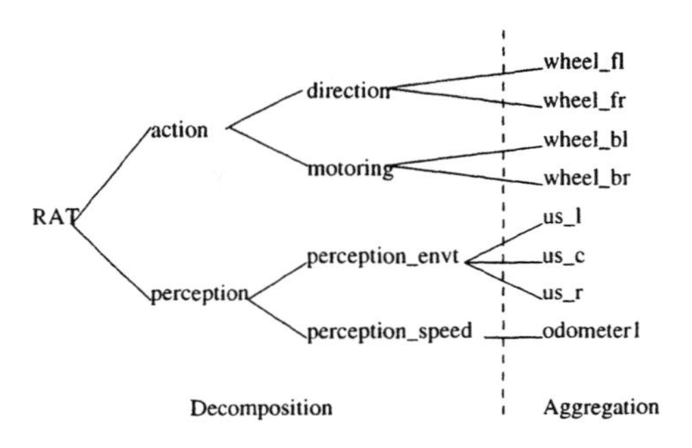

Figure 3: System decomposition

RAT's model also includes task model and data transmission, as well as an environmental model suited for small mobile robots.

Simulation

Figure4 shows the architecture of RAT's simulator. Each agent stands either for a part of the structural model, or for a salient agent or property in the environment (such as collision). Descriptors are parameters related to properties of the model evolving with time, such as the robot's velocity.

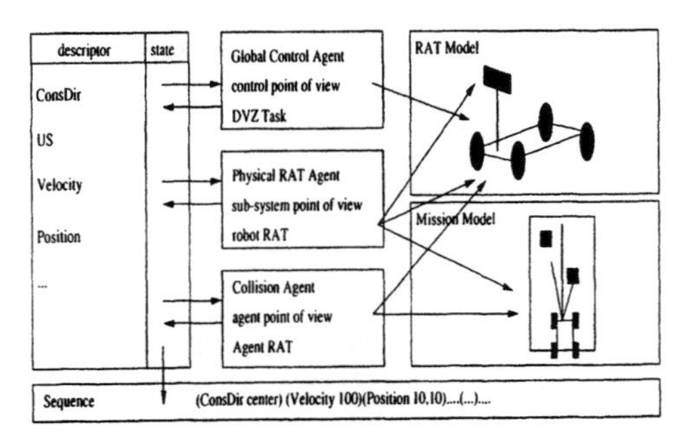

Figure 4: Multi-Agent simulation

RAT is modelled here with 3 agents : "Global control" stands for the control software, "Physical RAT" models physical evolution and interaction of the robot's components, and "Collision" verifies that the robot mission is fulfilled.

Any action performed by an agent generates a state, which is a time-stamped value for a descriptor, like in : `(1500 ConsMotor slow) (1600 Velocity 120)`.

Behaviour analysis

Elementary behaviors are expressed with Chronicles[1]. A chronicle is a pattern on the descriptors values, related to a typical situation we try to recognize in a sequence. A chronicle features a set of temporal variables, a set of descriptors and a set of assertions. Assertions are constraints on the temporal variables and the descriptors. Assertions can be either Hold, Event or Constraint. Hold assertions are verified when a descriptor keeps a given value between two time points. Event assertions are verified when a descriptor changes from one value to another at a given time point. Constraints assertion define precedence between time points. A chronicle is matched on a sequence when an assignation of its temporal variables satisfying all its assertions can be found.

This sample chronicle expresses that agent Global Control notices an unefficient stop (more than one second braking before stop).

```
Chronicle Late_Stop {
        t1 - t0 > 1000
        t1 - t2 >= 0
        Event(ConsMotor:(=slow,=brake),t0)
        Hold(ConsMotor=brake,(t0,t1))
        Event(Velocity:(>0,<=0),t2)}
```

CONCLUSION

A platform have been realized, and experimented on a mobile robot, RAT. The platform features at the ontological level an ontolingua specification and a prolog KB for the structural model. The agent level contains a blackboard-type multi-agent simulator for RAT, able to generate sequences of a simulation, and a set of chronicles describing some aspects of RAT's behaviour. Tools at the temporal level features a chronicle matcher, and a classifier.

Using this framework on an existing robot like RAT have driven its conceptors to refine its specifications and define conditions of use. Modelling of other robots is planned, as well as integration into the framework of other tools like inductive learning and agent model generation.

References

[1] M. Ghallab. Past and future chronicles for supervision and planning. In J.P. Haton, editor, *Proceedings of the 14th Int. Avignon Conference*, pages 23–34. EC2 and AFIA, Juin 1994.

[2] T.R. Gruber and G.R Olsen. An ontology for engineering mathematics. In *Fourth International Conference on Principles of Knowledge Representation and Reasoning*, Bonn, Germany, 1994. Gustav Stresemann Institut.

OPTIMAL PLANNING OF COLLISION-FREE MOVEMENTS OF ROBOT ARMS USING EXTERIOR AND EXACT PENALTY METHODS

F. Danes and G. Bessonnet
Laboratoire de mécanique des solides, URA CNRS
Université de Poitiers-UFR Sciences - S.P.2M.I.
Boulevard 3, Téléport 2, BP 179, 86960 Futuroscope Cedex, France

ABSTRACT

The study deals with optimal motion planning of robotic manipulators along unspecified paths. Unilateral constraints are imposed on phase variables in order to avoid: (i) excessive bending of the arm, (ii) overly high values of the joint velocities, (iii) collision with obstacles in the workcell. This problem is dealt with by applying the Pontryagin Maximum Principle and by using an exact penalty method which accounts for the phase constraints. This method, usually developed in the field of mathematical programming, is described and compared to the exterior penalty method already used in robotics. To illustrate their differences, simulation examples implementing each of these techniques are presented.

KEYWORDS: obstacle avoidance, dynamic optimization, penalty methods.

1 - INTRODUCTION

For robot manipulators used in the manufacturing industry, performing repetitive transfers is one of the main task for which they are designed. For many years, studies have been conducted on optimal dynamic path planning so as to improve the quality of their movements. The earliest works consisted in minimizing the motion travel time [1,2]. Since this performance criterion leads to bang-bang actuating inputs, and possibly to damaging jerks of the mechanical system, a mixed time-energy criterion, ensuring actuating input continuity and therefore smoother movements, has been introduced [3,4,5,6]. This minimization problem must be completed by a set of unilateral state constraints which respect technological limitations of the articulated system (intrinsic constraints) and define the free workspace in which the optimized trajectory must be situated (extrinsic constraints) in order to avoid collisions with the environment. Generally, these constraints are dealt with by using techniques similar to interior [1] or exterior [5,6,7] penalty methods, commonly practiced in mathematical programming, which allows one to solve the constrained problem by minimizing a sequence of unconstrained problems. To be implemented, the interior penalty method needs a no colliding initial path which can be hazardous to compute. Conversely, the exterior penalty method is active only if the constraints are infringed, thus, trajectories still remain slightly colliding. In this study, we compare avoiding obstacle by applying the exterior penalty method and the exact penalty method. The latter is a technique developed in the field of mathematical programming. Its application, which is to our knowledge new in robotics, does not present the inconveniences related to the two previous techniques since it allows to obtain, from a colliding initial path, a movement that entirely respects the state constraints. By using the optimality conditions stated by the Pontryagin Maximum Principle, the dynamic optimization problem is converted into a two point boundary value problem solved by means of a finite difference method implemented by the routine D02RAF of the Fortran Nag Library. Simulation examples are presented. They concern a four-link vertical robot described by a full dynamic model.

2 -PERFORMANCE CRITERION AND CONSTRAINTS

Here, we will consider a multibody system articulated as a n degree of freedom open kinematic chain. Dynamic model, with Hamiltonian formulation, can be written under the form of the generic 2n-vector state equation [4,6]:

$$t \in [0,T], \qquad \dot{x}(t) = f(x(t), u(t)), \tag{1}$$

where x represents the 2n-vector phase variables and u is the n-vector of joint actuating inputs.
This state equation goes together with the boundary conditions:

$$x(0) = x^0 \ , \ x(T) = x^T \tag{2}$$

which specify the initial and final state of the mechanical system.
In order to respect actuator technological limitations, joint actuating inputs are submitted to bounds such as:

$$|u_i(t)| \leq u_i^{max} \quad , \quad \forall t \in [0,T], \ \forall \ i \in \{1,n\} \tag{3}$$

As in previous works [4,5,6], the performance criterion that is minimized is an integral functional of the type:

$$J(u,T) = \int_0^T L(x(t),u(t))dt \tag{4}$$

where the Lagrangian L is defined as:

$$L(x(t),u(t)) = (1-\mu) + \frac{\mu}{2}\sum_{i=1}^{n}\left[u_i(t) / u_i^{max}\right]^2 \quad , \mu \in]0,1[\tag{5}$$

The criterion (4,5) allows one to optimize the travel time T and the integral of reduced quadratic robot controls. The factor μ assigns a relative weight to both previous quantities. It is worth noting that the quadratic dependence of L with respect to the quadratic inputs ensures the continuity of the last ones. This property results in smoothing the motion.
The mechanical system is submitted to phase constraints limiting both joint motions and velocities. These intrinsic constraints are defined by the inequalities:

$$k = 1,\dots,n \quad \begin{cases} g_k^i(x) \equiv x_k - x_k^{max} \leq 0 \ , \ g_{n+k}^i(x) \equiv x_k^{min} - x_k \leq 0 \\ g_{2n+k}^i(x) \equiv \dot{x}_k - \dot{x}_k^{max} \leq 0 \ , \ g_{3n+k}^i(x) \equiv \dot{x}_k^{min} - \dot{x}_k \leq 0 \end{cases} \tag{6}$$

Collision-free conditions defining the free workspace are introduced as extrinsic constraints under the general form:

$$k = 1,\ldots,m \quad , \quad g_k^e(x) \le 0 \tag{7}$$

The content of the functions g_k^e will be specified in section 5.
We express this double set of unilateral constraints under the form of the 4n+m scalar state inequalities:

$$g_k(x(t)) \le 0 \quad , \quad k \in \{1,\ldots,4n+m\} \tag{8}$$

By using penalty methods, the constrained optimization problem (1), (2), (3), (4), (8) can be recast into an unconstrained problem. This transformation is carried out in the next section by implementing two different penalty methods.

3 - EXTERIOR AND EXACT PENALTY METHODS

Penalty methods are commonly used in mathematical programming [8,9]. They consist in penalizing the cost function in such a way that its minimization implies the respect of inequality constraints. These techniques can be applied to optimal control problems [6,10] similar to the dynamic optimization problem stated in the previous section.
Consider the set of phase constraints (8) and define the penalized criterion:

$$J^*(u,T) = \int_0^T \{L(x(t),u(t)) + r\ G(x(t))\}dt \tag{9}$$

where r is a penalty multiplier and G stands for a penalty function depending on the state x through the phase constraints g_k. Explicit expression of G depends on the penalty method used. We define it for the so-called exterior and exact penalty methods described in the next subsections.

3.1 - Exterior Penalty Method

The principle of this technique consists in penalizing the criterion only when the constraints are infringed. Thus, the penalty function G in (9) must satisfy:

$$G(x) = 0 \quad \text{if } g_k(x) \le 0, \forall k \,; \quad G(x) > 0 \quad \text{otherwise.} \tag{10}$$

Following [5,8,9,11], we define G under the form:

$$G(x) = \sum_k [\max_{t \in [0,T]} (0, g_k(x(t)))]^2 \tag{11}$$

So, the criterion J^* minimizes the positive values of the constraints when they are infringed.
The penalty multiplier r defines a relative weight of G with respect to the Lagrangian L. Therefore, since J^* minimizes the integral of the product r.G in the sum L+r.G , the minimal contribution of G to J^* will be all the weaker as r will have higher values. Thus, as in mathematical programming it has been proved [8] that the function G vanishes as r tends towards infinity. Thus, the solution in the limit satisfies the state

constraints g_k . But, in practice, the multiplier r is limited to a finite value, therefore the constraints are not exactly satisfied. The optimal path will continue to collide slightly. This drawback is avoided by introducing corrective parameters θ_k. The technique which results is the so-called exact penalty method.

3.2 - Exact Penalty Method

Exact penalty method allows one to penalize the performance criterion even if the constraints are respected according to the values of positive parameters θ_k:

$$G(x) = 0 \quad \text{if } g_k(x) \leq -\theta_k, \forall k \; ; \quad G(x) > 0 \quad \text{otherwise.} \tag{12}$$

As advocated in [9,11], we consider the following penalty function:

$$G(x) = \sum_k [\max_{t \in [0,T]} (0, g_k(x(t)) + \theta_k)]^2 \tag{13}$$

At first, r and θ_k are equal to zero. Contrarily to the previous method, r is increased at the appropriate rate towards a given finite value. Then, there exists, under some conditions [9], a value of each θ_k for which each constraint is strictly respected all along the path. We obtain the searched value of θ_k by an iterative scheme [9], such as:

$$\theta_k{}^{(i+1)} = \theta_k{}^{(i)} + \max(\max_{t \in [0,T]} g_k(x(t)^{(i)}) \, , -\theta_k{}^{(i)}) \tag{14}$$

Thus, when the constraint is initially respected, the constraint function remains negative, so the criterion is not penalized. In the opposite case, the criterion is more or less penalized until $\max_{t \in [0,T]} g_k(x(t))$ vanishes.

The parameters θ_k symbolize the enlargement of the constraints. Indeed, by using the exterior penalty method with a high value of r the constraints g_k are almost respected. The exact penalty method is similar to the exterior penalty method applied to the virtual constraints $g_k + \theta_k$ which correspond to the constraints g_k enlarged. Then, these virtual constraints can be almost respected. So the real constraints g_k are exactly respected when the suitable enlargements θ_k, automatically generated by the recursive formula (14), are obtained.

4 - REFORMULATING THE DYNAMIC OPTIMIZATION PROBLEM

Application of the Pontryagin Maximum Principle allows one to convert the optimal transfer problem defined by (1), (2), (3), (9) into an ordinary differential problem.
Let U be the set of admissible controls defined by inequalities (3).
Introduce the Hamiltonian:

$$w \in \Re^{2n}, \quad H(x,u,w) = w^T f(x,u) - L(x,u) - r\, G(x) \tag{15}$$

Pontryagin's maximum principle states that if (x,u,T) is an optimal control process, then [12]:

(i) There exists a 2n-vector function w(t) which is a solution of the adjoint system:

$$\forall t \in [0,T], \quad \dot{w}(t) = -(\partial H(x(t),u(t),w(t)) / \partial t)^T \tag{16}$$

(ii) The optimal control u(t) fulfills the Hamiltonian maximality condition:

$$\forall t \in [0,T], \quad H(x(t),u(t),w(t)) = \max_{v \in U} H(x(t),v,w(t)) \tag{17}$$

(iii) At final instant T, the Hamiltonian is equal to zero:

$$H(x(T),u(T),w(T)) = 0 \tag{18}$$

As in [4, 6], an explicit expression of the control u is derived from (17):

$$u_i(t) = \mathrm{Sat}((u_i^{max})^2 w_{n+i}(t) / \mu), \quad i = 1,\dots,n \tag{19}$$

where Sat is the saturation function defined as:

If $|x| < x^{max}$, $\mathrm{Sat}(x) = x$ else $\mathrm{Sat}(x) = x^{max}\mathrm{sign}(x)$.

Substituting the expression (19) of u in the state equation (1) and the adjoint system (16), the unknown vector functions x and w appear as a solution of a 4n-dimensional differential system:

$$t \in [0,T], \quad \begin{cases} \dot{x} = \mathcal{F}_1(x,w) \\ \dot{w} = \mathcal{F}_2(x,w) \end{cases} \tag{20}$$

which must satisfy the boundary conditions (2), the unknown final time T being determined by the pointwise condition (18).

As in [6], after having rescaled the running time as $\tau = t / T$ and introduced the (4n+1)th differential equation $\dot{T}(\tau) = 0$, the problem is transformed into a (4n+1)-dimensional differential system submitted to 4n+1 boundary conditions including the condition (18). This ordinary differential two-point boundary value-problem is numerically solved by using the routine D02RAF of the Fortran Nag Library.

5 - SIMULATION EXAMPLE

The following results concern the no colliding optimal path planning of a 4R vertical robot (fig.1). The optimal trajectory is submitted to sixteen intrinsic constraints of type (6). So as to avoid an excessive bending of the arm, a seventeenth phase constraint is formulated. This constraint prevents the arm from colliding with himself by compelling its last segment to be, if the worst comes to the worst, vertical:

$$g_{17}^i(x) \equiv x_2 + x_3 + x_4 - \frac{\pi}{2} \leq 0$$

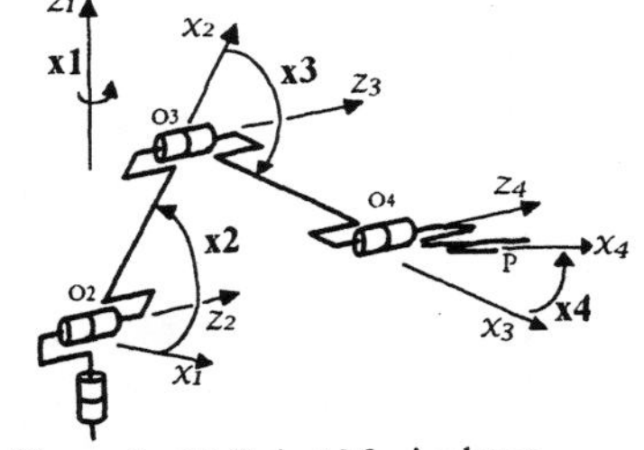

Figure 1. 4R Robot Manipulator

To these intrinsic constraints is added an obstacle avoidance extrinsic constraint. The obstacle we will consider is a cylinder whose axis is vertical, parallel to (O,z) and located in x=0 (m) and y=0.3 (m). Its faces, whose radius is 0.1(m), are positioned in z=0; 0.7 (m).

When a collision occurs between the arm and the obstacle, the constraint function g^e is the area separating the colliding part of the arm from the upper face of the cylinder (fig.2). This constraint function allows the arm to turn around the outside of the cylinder.

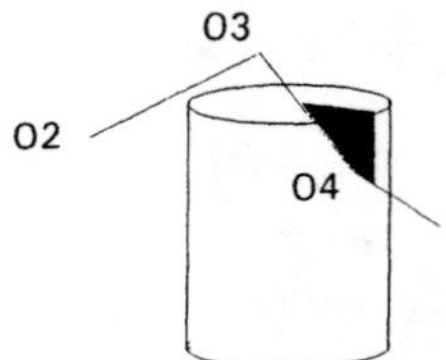

Figure 2. Upper Cross Section

If the robot does not collide, the value of g^e is equal to zero when the exterior penalty method is used and it is negative. On the other hand, g^e is equal to the opposite of the shortest distance from the arm to the obstacle when the exact penalty method is at stake.

The three following paths are obtained for a unique value of the penalization factor μ, the only difference between them is due to the penalization. Their initial and final states are:

$$x(0) = (0°,0°,10°,10°,0\ ,0\ ,0\ ,0) \qquad x(T) = (150°,0°,10°,10°,0\ ,0\ ,0\ ,0)$$

These trajectories correspond to:

- an initial path (#1) only submitted to intrinsic constraints with a weak penalization by exterior penalty method (fig.3).
- an optimal path (#2) with constraints dealt with by the exterior penalty method (fig.4).
- an optimal path (#3) with constraints dealt with by the exact penalty method (fig.5).

White segments symbolize the colliding parts of the articular chain.

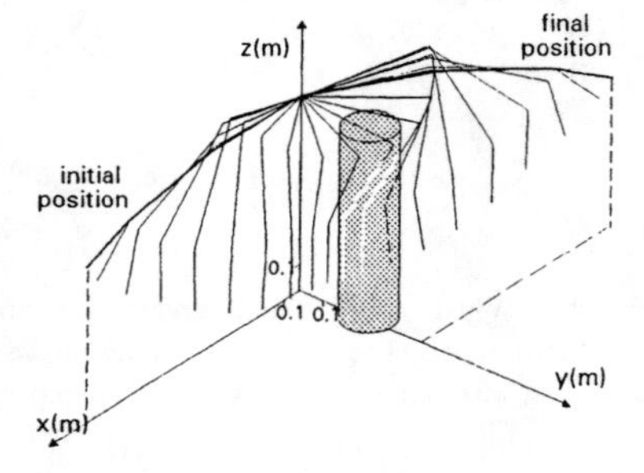

Figure 3. Partly Constrained Guess Optimal Trajectory

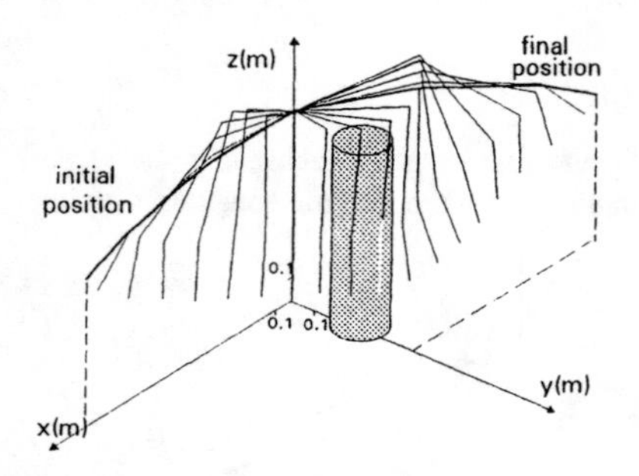

Figure 4. Exterior Penalty Method: optimal constrained trajectory.

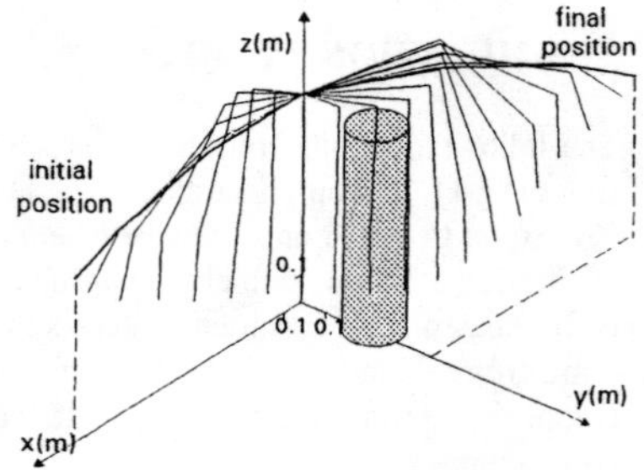

Figure 5. Exact Penalty Method: optimal collision-free trajectory.

We can observe that the two methods are efficient, however path #2 collides slightly, as foreseen by the theory, while path #3 really avoids the obstacle.

The difference of penalization produced by each of the two penalization methods entails a transfer time modification and so the actuator inputs (fig.6a,b) and the joint velocities (fig.7a,b) are more or less saturated.

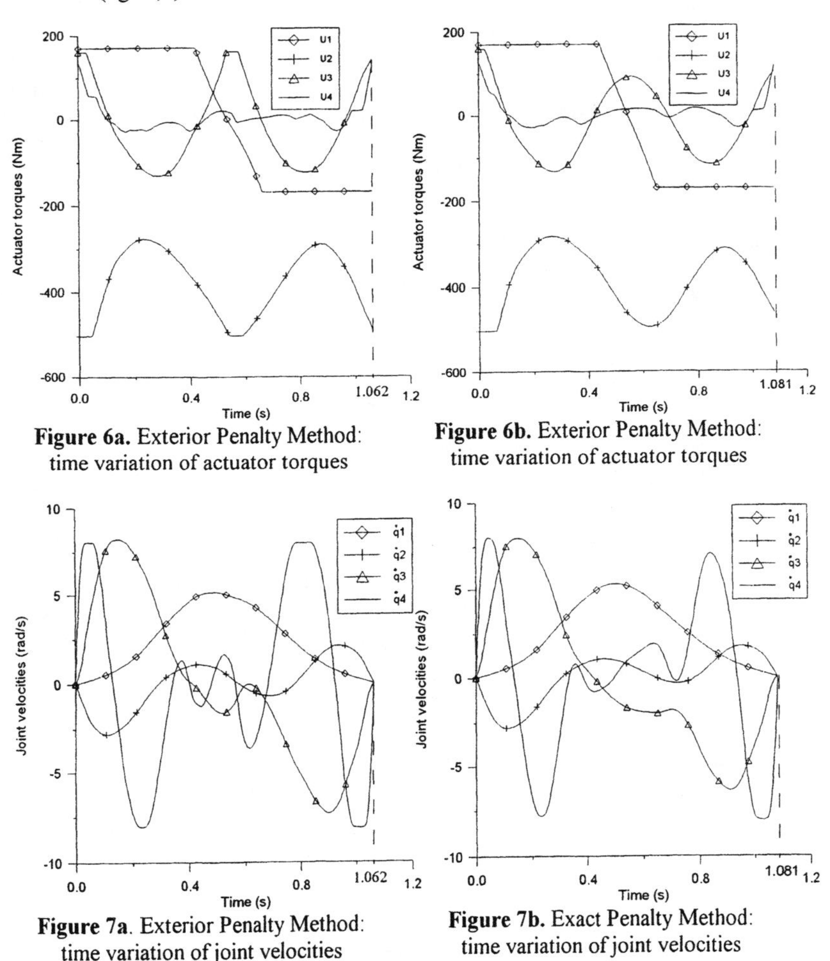

Figure 6a. Exterior Penalty Method: time variation of actuator torques

Figure 6b. Exterior Penalty Method: time variation of actuator torques

Figure 7a. Exterior Penalty Method: time variation of joint velocities

Figure 7b. Exact Penalty Method: time variation of joint velocities

By varying the coefficient μ, the same travel time can be obtained by the exterior and exact penalty methods. Subsequently, the control saturation is similar in both cases but the obstacle remains perfectly avoided with the exact penalty method while the path calculated by the exterior penalty method still collides slightly.

6 - CONCLUSION

The study presented is aimed at accounting for phase constraints in optimal motion planning of robot arms. The constraints define the following: obstacle avoidance and bounds which phase variables are submitted to. They are dealt with by means of two different penalty methods. The first one, which theoretically leads to a collision-free path, only reduces the collision in practice. So, the constraints must be enlarged to be

strictly respected. Here is the interest of the second one. With the exact penalty method, the enlargement of the constraints is automatically generated according to the importance of the collision. Thus, with given real obstacles and phase variables bounds, an optimal collision-free path can be directly obtained.
The computation technique proposed in this paper is designed for off-line obstacle-free trajectory planning. For a four active joint vertical robot, of which the full dynamic model is accounted for, it takes about 20 minutes on a Vax 4000-500 Dec computer to compute a no colliding trajectory from a fully colliding guess trajectory. Future work will be devoted to the improvement of the implementation of the exact penalty method for obstacle avoidance in a cluttered environment. Computation will be carried out on a fast Alpha-type workstation.

REFERENCES

1. Shiller Z., Dubowsky S. (1989) "*Robot Path Planning with Obstacles, Actuator, Gripper, and Payload Constraints*" International Journal of Robotics Research, Vol. 8, N°6, pp. 3-18.
2. Bobrow J. E., Dubowsky S., Gibson J. S. (1985) "*Time-Optimal Control of Robotics Manipulators Along Specified Paths*" International Journal of Robotic Research, Vol. 4, N°3, pp. 3-17.
3. Khouki A., Haman Y. (1990) "*Optimal Time Energy Trajectory Planning for Robots with Hard Constraints*" Control Theory and Advanced Technology, Vol. 6, N°3, pp. 417-439.
4. Bessonnet G. (1992) "*Optimisation Dynamique des Mouvements Point à Point de Robots Manipulateurs*" Thesis, Poitiers University.
5. Danes F., Bessonnet G. (1995) "*Planification Dynamique de Trajectoires Non Collisionnelles en Robotique*" Actes du 12e Congrès Français de Mécanique, Strasbourg, 4-8 Sept. 1995,pp. 353-356.
6. Bessonnet G., Jutard-Malinge A.D. (1995) "*Optimal Free Path Planning of Robot Arms Submitted to Phase Constraints*" IASTED Interational Conference on Robotic and Manufacturing, pp. 51-56, Cancun.
7. Wang D., Haman Y. (1992) "*Optimal Trajectory Planning of Manipulators with Collision Detection and Avoidance*" International Journal of Robotics Research, Vol. 11, N°5, pp. 460-468.
8. Fiacco A. V.,McCormick G. P. (1967) "*The Slacked Unconstrained Minimization Technique for Convex Programming*" Journal of Applied Mathematics, Vol. 15, N°3, pp. 505-515.
9. Arora J. S., Chahande A. I., Paeng J. K. (1991) "*Multiplier Methods for Engineering Optimization*" International Journal for Numerical Methods in Engineering, Vol. 32, pp. 1485-1525.
10. Lele M. M., Jacobson D. H. (1969) "A Proof of the Convergence of the Kelley-Bryson Penalty Function Technique for State-Constrained Control Problems" Journal of Mathematical Analysis and Applications, Vol. 26, pp. 163-169.
11. Fletcher R. (1975) "*An Ideal Penalty Function for Constrained Optimization*" Journal of the Institute of Mathematical Optimization, Vol. 15, pp. 319-342.
12. Pontryagin L., Boltiansky V., Gamkrelidze A., Mishchenko E. (1962) "*The Mathematical Theory of Optimal Processes*" Wiley Interscience, New-York.

DATA

Center of mass positions:	$O_2G_2=0.16$	$O_3G_3=0.16$	$O_4G_4=0.14$	(m)
Link lengths:	$O_2O_3=0.45$	$O_3O_4=0.44$	$O_4P=0.22$	(m)
Link masses:	$m_2=0.77$	$m_3=0.36$	$m_4=0.16$	(kg)
Moments of inertia:				
$I_{O2Z0}=1.5$	$I_{O2X2}=0.75$	$I_{O3X3}=0.15$	$I_{O4X4}=0.1$	($kg.m^2$)
	$I_{O2Y2}=6$	$I_{O3Y3}=3$	$I_{O4Y4}=0.45$	
	$I_{O2Z2}=7$	$I_{O3Z3}=3$	$I_{O4Z4}=0.45$	
Bounds for actuating inputs:				
$U_1^{max}=170$	$U_2^{max}=500$	$U_3^{max}=160$	$U_4^{max}=150$	(N.m)

Bounds for phase variables: ($q_i^{min} = - q_i^{max}$, $q_i^{min} = - q_i^{max}$)

$q_1^{max} = 360°$ $q_2^{max} = 70°$ $q_3^{max} = 140°$ $q_4^{max} = 120°$ $\dot{q}_i^{max} = 8$ rad/s

WORKSPACE ANALYSIS OF A 4R REGIONAL ARM BASED ON SINGULARITY CLASSIFICATION

YANN DELMAS (†,‡), CATHERINE BIDARD (†)

†*Service de Téléopération et Robotique, CEA/DTA/DPSA, CEN-FAR, FRANCE*

‡*Ecole Nationale Supérieure des Arts et Métiers, LMS URA CNRS 1776, Paris, FRANCE*

ABSTRACT

In this paper, the workspace of a redundant 4R regional structure with joint limits is obtained using a method based on the search for singularities and their classifications. We pursue the analysis of the workspace in order to display the connectivity among the different regions of the workspace.

KEYWORDS: redundant manipulators, self-motion, workspace

Introduction

Numerous remote manipulation for nuclear maintenance require enhanced motion capabilities which may be achieved using redundant degrees of freedom. The Robotics and Teleoperation Laboratory of C.E.A is developping a redundant arm. However while the workspace is extended by redundant degrees of freedom, the connectivity structure of the workspace becomes generally more complex. Dealing with connectivity rupture could be managed using convenient trajectory planning, but this cannot be done when the arm is controlled by a master arm. Thus, we developped a method for determining the workspace of serial redundant manipulators which allows the study of its connectivity [1]. In the present paper we apply this method to obtain the workspace of a redundant 4R regional structure with joint limits and information on its connectivity. The Reachable workspace envelopp of a regional structure with four revolute joints has been studied using algebraic technics [2] for a general architecture without joints limits. Numerical techniques have been used to study the dexterity space [3], but don't deal explicitly with the redundancy aspect. Moreover, connectivity structure of the workspace is not handled by these methods. The 4R regional arm, which has a special configuration of joints axes, is presented in the first part. The method for generating the limits of the workspace reachable by the wrist center is briefly presented in the second part. In the third part we illustrate its application to the 4R regional structure and give the final result. In the fourth part we study the connectivity among the regions of the workspace delimited by the image of the singular configurations.

Figure 1: 4R regional architecture

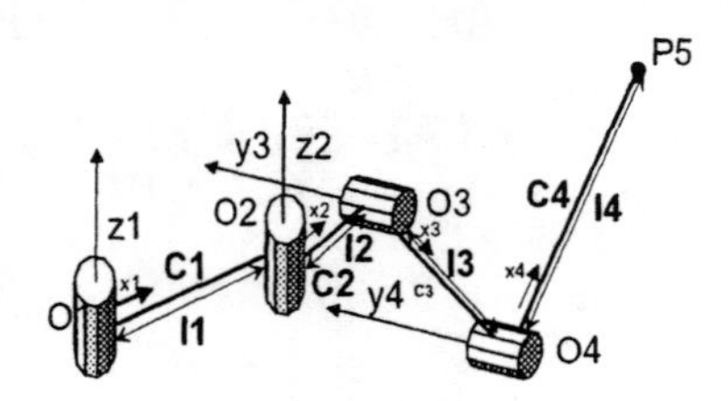

Kinematic structure of the 4R regional structure

The arm is a serial arm composed of two planar RR subchains in two orthogonal plans (figure(1)). Its task is to place the center P5 of a spherical wrist.

The length of the linsk are : l1=0.31m, l2=0.162m, l3=0.29m, l4=0.468m and the joint limits are defined as $\theta_2 \in [\pi/3, \pi], \theta_3 \in [\pi/3, \pi], \theta_4 \in [-\pi, -\pi/3]$.

A 1-dof joint motion of the manipulator exist that keeps the end effector point P5 fixed. This motion is called the *self-motion* of the manipulator.It can be parametrized by the angle θ_1 which fixes the plane $\Pi = (O2, x2, z2)$.

The *augmented inverse kinematic* gives the configuration of the manipulator for a end-effector position $\underline{X}_{P5} = (x, y, z)^t$ and a value of the redundant parameter θ_1.

For given $\underline{X}_{P5}$ and θ_1, there are two solutions for the second joint coordinate θ_2 :

$$\theta_{2(1)} = \arctan(y - l1\sin(\theta_1), x - l1\cos(\theta_1)) - \theta_1 \;\Big|\; \theta_{2(2)} = \theta_2^1 + \pi$$

For each θ_2, there are two solutions called elbow up configurations ($\theta_4 > 0$) and elbow down configurations ($\theta_4 < 0$) :

$$\left\{\begin{array}{lcl} \theta_{3(i,1)} & = & \alpha_{(i)} - \delta_{(i)} \\ \theta_{4(i,1)} & = & \pi - \beta_{(i)} > 0 \end{array}\right. \text{ or } \left\{\begin{array}{lcl} \theta_{3(i,2)} & = & \alpha_{(i)} + \delta_{(i)} \\ \theta_{4(i,2)} & = & -\pi + \beta_{(i)} < 0 \end{array}\right.$$

with :

$$\begin{array}{lcl|lcl} \alpha_{(i)} & = & \arctan(-z, R2_{(i)}) & R2_{(i)} & = & x\cos(\theta_1 + \theta_{2(i)}) - y\sin(\theta_1 + \theta_{2(i)}) \\ \beta_{(i)} & = & \arccos(\frac{l3^2 + l4^2 - R1^2_{(i)}}{2l3l4}) & & \ldots & -l1\cos(\theta_{2(i)}) - l2 \\ \delta_{(i)} & = & \arccos(\frac{R1^2_{(i)} + l3^2 - l4^2}{2l3l4}) & R1^2_{(i)} & = & R2^2_{(i)} + z^2, \end{array}$$

where i stands for the index of the θ_2 solution. Thus there are up to four solutions.

Determation of the limits of the workspace

Principle of the method - Our method to obtain the workspace of the manipulator is based on an enumeration and a classification of all the configurations where the motion of the end effector point P5 is locally limited. This is a two step method :

- **1** - The first step is to select the *singular configurations* with restricted operational motion of the point P5 from an exhaustive enumeration of candidate configurations.
- **2** - The second step is to extract the *limit singular configurations* which are unescapable by a reconfiguration of the manipulator on a regular configuration on its self-motion passing through the singular configuration.

Among the limit configurations, we may distinguish will find the global limits of the workspace and the inner limits. The inner limits are local limits where the manipulator must move back to recover mobility.

Determination of singular configurations - The manipulator is in a *kinematic singular configuration* when the regional jacobian of the manipulator, with respect to the joint variables, is rank defficient.

To consider the influence of the joint limits, we search for *semi-singular configurations* [4] which are defined by two conditions:
- some joints are at a limit and the sub-chain formed by the free joints is in a singular configuration,
- the contribution of the joints at a limit can't compensate the missing directions.
This test is called *semi-singular test* [1].

Classification of singular configurations - We use a second order local model of the self-motion around a configuration, with respect to the self-motion parameters [1], to determine the existence of a feasible self-motion around the two types of singular configurations. For the kinematic singularities, we obtained a necessary condition on the coeffecients of the second order local model of the self-motion [5, 1] which classified the singular configurations as branch, isolated or special singularities. For the semi-singularities, depending on the number of joint at a limit we use a first order or a second order analysis.

All these tests are automated. The algorithms take as input a base of the kernel of the jacobian matrix, the vector of hessian of the direct geometric map with respect to joint parameters and the definition of the tangent to the singular manifolds. We have shown that through a proper decomposition of the singular configuration domains in distinct sub-domains delimited by their intersections, it is possible to test only one configuration in each sub-domain for fixed joint limits. The nature of a singularity or semi-singularity domain can change only after meeting another singular configuration domain [6]. This approach greatly reduces the cost of evaluation.

Application to the 4R regional structure

Kinematic singular configurations - Six factors, $f_i, (i = 1, 6)$ can be extracted from the four sub-determinants of the jacobian.

$$\begin{aligned} f1 &= l2 + l3\cos(\theta_3) + l4\cos(\theta_3+\theta_4) \\ f2 &= l3\cos(\theta_3) + l4\cos(\theta_3+\theta_4) \\ f3 &= \sin(\theta_2) \end{aligned}$$

$$\begin{aligned} f4 &= \cos(\theta_3+\theta_4) \\ f5 &= l1\cos(\theta_2) + l2 + l3\cos(\theta_3)\ldots \\ &\cdots +l4\cos(\theta_3+\theta_4) \\ f6 &= \sin(\theta_4) \end{aligned}$$

Figure 2: Decomposition of the singular domain for $\underline{\theta} = [0, 0, \theta_3, 0]$

Singularity	Decomposition on θ_3	Type
$\underline{\theta} = [0, 0, \theta_3, 0]$	$[-\pi, -s_0]$	crossable
	$[-s_0, -b_1]$	non crossable
	$[-b, -\pi/2]$	crossable
	$[-\pi/2, \pi/2]$	non crossable
	$[\pi/2, b_1]$	crossable
	$[b_1, s_0]$	non crossable
	$[b_1, \pi]$	crossable

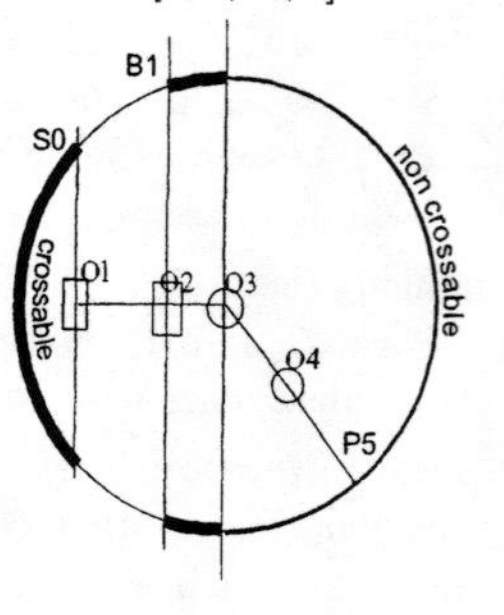

The kinematic singular configurations are defined implicitely by the combinaison :

$$\{f1 = 0, f5 = 0\}, \{f1 = 0, f6 = 0\}, \{f3 = 0, f6 = 0\}, \{f2 = 0, f4 = 0, f6 = 0\}.$$

We have enumerated all the combinations which cause the jacobian to be rank deficient and extract the explicit descriptions (or parametrized descriptions) of the singular configuration domains. After a decomposition in non-intersecting sub-domains we have tested one configuration in each sub-domain to determine which are limits from those which are escapable singularities. Due to space limitation, this will not be exhaustively presented in this paper.

To illustrate the method, let consider the kinematic singular configuration surface

Figure 3: Limit positions of the workspace

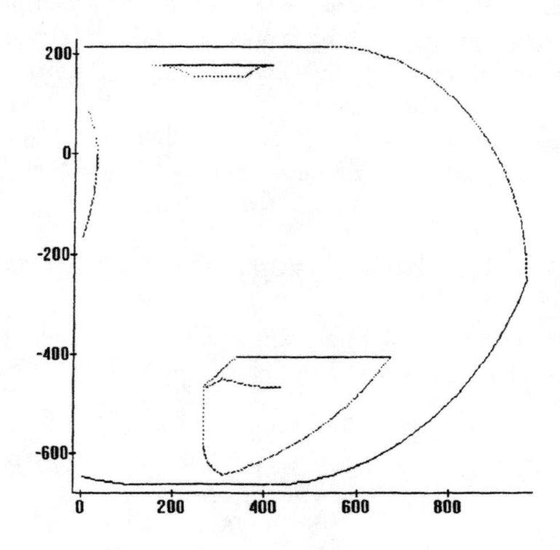

defined by $\underline{\theta} = [\theta_1, 0, \theta_3, 0]$ where the factors f3 and f6 vanish. The surface is decomposed into 7 sub-domains. We classify them as crossable if a self-motion is possible. The Table figure (2) gives the decomposition, where $s_0 = \arccos(-\frac{l1+l2}{l3+l4})$ and $b_1 = \arccos(-\frac{l2}{l3+l4})$. This decomposition is determined by the values of θ_3 where one of the other factors vanishs, such as f5=0 for $\theta_3 = s_0$. For a fixed θ_1, the operational point P5 sweeps a circle in the plane (x1,z1) center on O3 with radius (l3+l4) (figure (2)).

Limits of the workspace - Using this method we obtain the limit positions of the point P5 displacement in the workspace. The workspace can be plotted in the (x,z) plane as it is revolute around the z1 axis. The figure 3 shows the limit configurations of the workspace. From the enumeration, it is possible to determine the singular condition associated with each of them. This is of great interest for the analysis of the influence of parameters such as joint limits.

Connectivity Structure of the workspace

The connectivity of the workspace is related to the path planning problem and the inverse kinematic problem. The number and the topology of the self motion associated with a position varies on the workspace. For the 4R manipulator without joint limits, there can be up to height self-motion manifolds for a position. The joints limits decompose them into distinct manifolds. Thus, distinct families of solution to the inverse kinematic problem exist. It is possible to decompose the workspace in distinct domains(W-sheets) inside which the number and the topology of self-motions is constant (propriety of homotopy)[7]. The domains are delimited by the image of kinematic singular configurations without joint limits, and also by the image of semi-singular configurations with joint limits. Thus it is possible to study the the connectivity structure amongst the W-sheets and their pre-images, called C-bundles, which are associated with a type of inverse kinematic solution. Due to the homotopy propriety, it is possible to reduce the analysis to discret points on each side of the separating surfaces in the operational space. This reduces drasticly the evaluation of the connectivity structure. For the 4R manipulator, we used the augmented inverse kinematic map to scan the redundant parameter θ_1. We identified the values associated with the value $\theta_4 = 0$ or $\theta_4 = \Pi$ to get the number of distincts sub domains.

Comparing the modification in the self-motion sub-domain on each side of separating surface we determine the connectivity graph of the manipulator. The figure (4-b) presents the graph with respect to the decomposition of the workspace in the plane (x,z(figure (4-a)) as there is no joint limits for the first joint.

Conclusion

We developed a method to classify the kinematic singular configurations and semi-singular configurations into crossable or non-crossable. We applied this method to obtain the external and inner limits of the workspace of a 4R regional manipulator. The inner limits are representative of dead-end branch in some direction of motion.

Figure 4: (a): Workspace decomposition, (b): Connectivity graph

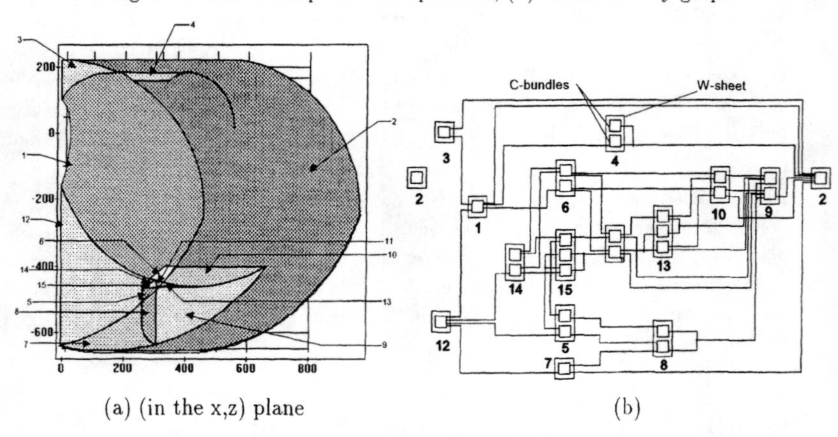

(a) (in the x,z) plane (b)

All the singular configurations evaluated during the classification process can be used to decompose the operational workspace into C-sheets where the self-motion topology is the same for all the points inside. This organizes the analysis of the self-motion structure above the workspace. From this analysis, we obtained a connectivity graph of the workspace which is significant of the complexity of the workspace. The enumeration process is still tedious but the analysis is now feasable owing, to the automation of classification tests. Our method gives information on the influence of design parameters and has been used to improve the arm design by removing inner limits.

References

[1] Y. Delmas and C. Bidard, *Determination of the workspace of redundant manipulators with joint limits*, Ninth World Cong. on the Theory of Machines and Mechanisms, 1995.

[2] M.Ceccarelli and A. Vinciguerra, *On the workspace of general 4r manipulators*, The Int. Jour. of Robotics Research, 14(14), April 1995.

[3] J. Rastegar, *Workspace analysis of 4r manipulators with various degrees of dexterity*, Jour. of Mech. Trans. and Auto. in design, 110, March 1988.

[4] C.L. Lück and S. Lee. *Self-motion topology for redundant manipulators with joint limits*, Proc. of IEEE int. Conf. on Robotics and Automation, 1993.

[5] J.C. Kieffer, *Selected Topics in Mechanisms and Robotics: Singularities, inverse Kinematics, and Collision detection*, PhD thesis, University of Illinois at Chicago, 1990.

[6] P. Borrel, *Contribution à la modélisation géométrique des robots manipulateurs, application à la conception assitée par ordinateur*, Thèse d'État,USTL, Montpellier, 1986.

[7] J.W.IV Burdick, *Kinematic Analysis and Design of Redundant Robot Manipulators*, PhD thesis, Stanford University, 1988.

Dense Reconstruction Using Fixation and Stereo Cues

Reyes Enciso, Zhengyou Zhang and Thierry Viéville
projet RobotVis INRIA
06902 Sophia-Antipolis (FRANCE)

ABSTRACT

In this paper, we investigate the issue of accurate estimation of the three-dimensional (3D) coordinates of a static scene from real images, combining fixation and stereo cues. We may need to compute 3D data in many applications: vehicle positioning and maneuver, object observation and recognition, moving or fixed obstacle avoidance, 3D mapping for surveillance, etc. More specifically, we discuss the idea of using fixation to recover the 3D coordinates of some points in the robotic frame to help an *uncalibrated* camera to reconstruct a static scene.

KEYWORDS: Monocular system, Fixation, Correlation, Projective structure, Dense reconstruction.

INTRODUCTION

In this paper we address the problem of computing a dense 3D reconstruction by cooperation of stereo and fixation cues. Fixation allow us to compute the coordinates of the fixation point in a world frame of reference. In our example this frame is attached to the center of the linear degree of freedom. Stereo cues allow us to compute a disparity map of the scene. We study how to transform this disparity map to a projective reconstruction of the scene. Fixating at least five points we can relate the projective reconstruction to the Euclidean one.

Abbott et Ahuja [1] have already address the problem of reconstructing a surface by dynamic integration of focus, camera vergence and stereo. The limitation of the approach is that the surface must be continuous. The second limitation is the choice of the weight for the different criteria in the minimization process. In [3] they avoid the first limitation by selecting several fixation points. The technique used by Krotkov described in [7] based on the cooperation of focus, stereo and vergence modules however provides poor final results.

THE HARDWARE

The Arges Monocular System has a linear degree of freedom with a resolution of 0.1 mm, using a slow screw driven control. A Computer controlled color CCD Camera is used (an Acom1 PAL with a f=5.9 to 47.2 mm zoom-lens, a motor iris F1.4 to F22 and a numerical autofocus on 10bits). The camera is mounted in a Pan-tilt turret, with a resolution of 3.086 minutes of arcs, a 4 lbs capacity and a speed up to 300 deg/sec, using constant current bipolar motor drivers, via a rs232C interface.

The kinematic of our system

In the picture **Figure** 1, we have represented the world reference frame and the pan and tilt angles used to define the kinematic of the robot.

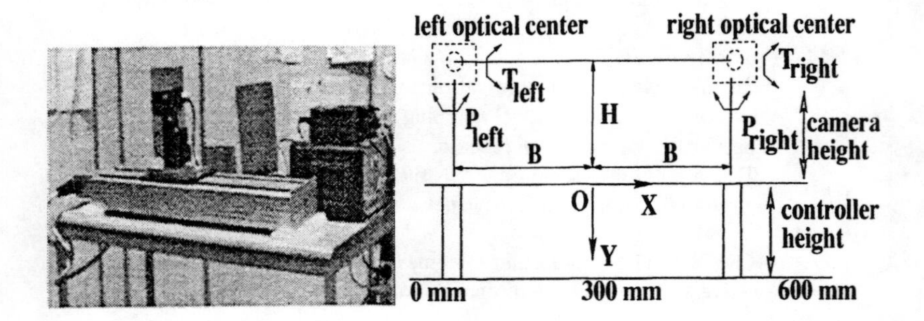

Figure 1: The monocular system Arges. The world reference frame is attached to the center **O** of the linear degree of freedom at the controller height .

The **pan** left and right angles $\mathcal{P}_l, \mathcal{P}_r$ correspond to a rotation about **Y** axis; the **tilt** angles $\mathcal{T}_l, \mathcal{T}_r$ to a rotation about the **X** axis. The relation between the two rotation axes is known : pan is on tilt. Let us assume tilt is fixed and to note $\mathcal{P} = \frac{\mathcal{P}_l+\mathcal{P}_r}{2}$ and $\mathcal{V} = \frac{\mathcal{P}_l-\mathcal{P}_r}{2}$ to simplify the equations. Then, $\mathcal{P}_l = \mathcal{P} + \mathcal{V}$ and $\mathcal{P}_r = \mathcal{P} - \mathcal{V}$. The optical center of the right camera world coordinates are $(B, -H, z_r)$ and the left one are $(-B, -H, z_l)$ with $z_l, [z_r]$ is the distance of the left [right] optical center to the left [right] retinal plane.

Using this notations, we can write the displacement matrices as :

$$\begin{aligned} \mathbf{D}_{right} &= \left(\begin{array}{c|c} R(\mathbf{Y}, \mathcal{P} - \mathcal{V}) \cdot R(\mathbf{X}, \mathcal{T}) & \begin{array}{c} B \\ -H \\ z_r \end{array} \\ \mathbf{0}_{1x3} & 1 \end{array} \right) \\ \mathbf{D}_{left} &= \left(\begin{array}{c|c} R(\mathbf{Y}, \mathcal{P} + \mathcal{V}) \cdot R(\mathbf{X}, \mathcal{T}) & \begin{array}{c} -B \\ -H \\ z_l \end{array} \\ \mathbf{0}_{1x3} & 1 \end{array} \right) \end{aligned} \quad (1)$$

with the hypothesis that the zoom is a displacement through the **Z** axis.

Direct kinematic While fixating a point with world coordinates (X, Y, Z), if the optical center in the left retinal frame has coordinates (0, 0, l) and in the right retinal frame has (0,0, r), we obtain :

$$\left\{ \begin{array}{rcl} X & = & -\frac{B\cos(\mathcal{P})}{\sin(\mathcal{V})} \\ Y & = & \frac{H}{\cos(\mathcal{T})} + \frac{B\tan(\mathcal{T})}{\tan(\mathcal{V})} \\ Z & = & \frac{B\cos(\mathcal{P})}{\sin(\mathcal{V})} \end{array} \right. \quad (2)$$

THE ALGORITHM

Our dense reconstruction technique from two images consists of the following steps:

Step 1 : Determine the epipolar geometry

We use the robust algorithm described in [9]. As a consequence, we have a set of matched points at our disposal. The corner detector used is described in [4]. We create a set of candidate matches by correlation. To disambiguate the candidate matches we use a *relaxation technique*. We define a measure called the **strength** of the match that defines the *goodness* of the match according to the number of matches in the neighborhood, the correlation score and the average distance between the pairs of matches in the set. Using a Lest-Median-of-Squares technique we can detect the outliers in the correspondences and compute a robust estimation of the fundamental matrix (minimizing the distance for each correspondence to the epipolar lines).

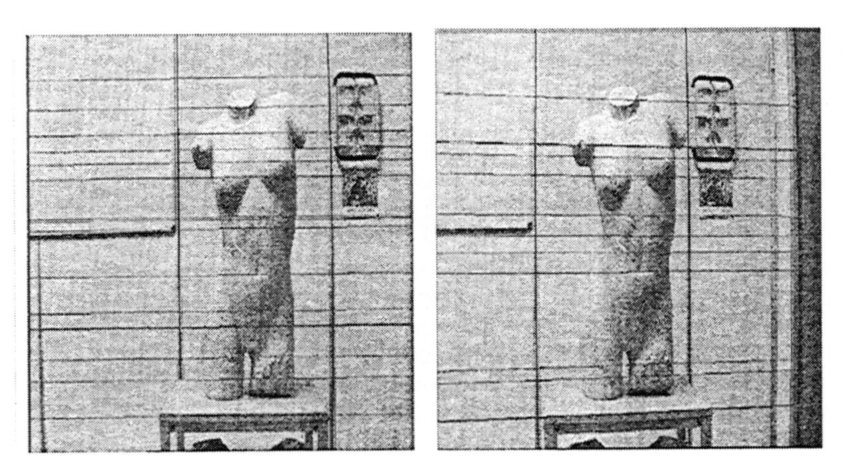

Figure 2: Input images (768x576) and some epipolar lines.

Step 2 : Fixating a set of points

Choose at least five points and *fixate* them with the active vision system to obtain their 3D Euclidean coordinates. These points are called the Euclidean reference points.

The goal is to center the image point that we want to fixate in both, left and right image. In the experiment the points were chosen interactively and centered by correlation in both images independently. We then compute, the Euclidean reference point coordinates (X_i, Y_i, Z_i) using Equation (2). In the **Figure** 3 we show the ten 2D points used in the experiment and the two original images.

Figure 3: Input images of size (768x576) and the points of reference.

Step 3 : Computing the disparity by stereo

First, we rectify the images as described in [8]. That means, once the epipolar geometry computed, we can make the rows of the image coincide with the epipolar lines to make easier the disparity computation.

The disparity map is computed using a Zero Mean Normalized Cross-Correlation method (see [2] for a comparative study) with a 9×9 pixels correlation window. Once the disparity d computed and the images rectified, the 2D coordinates of each correspondence are (u_r, v_r) in the left image and $(u_r + d_i, v_r)$ in the right image. A good estimator of the 3D projective coordinates of the match is $(u_r, v_r, d_i, 1)$. In the **Figure** 4 we show the rectified images and the disparity map.

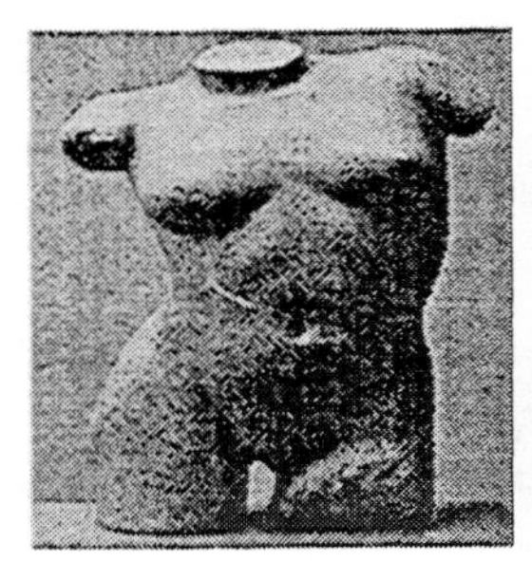 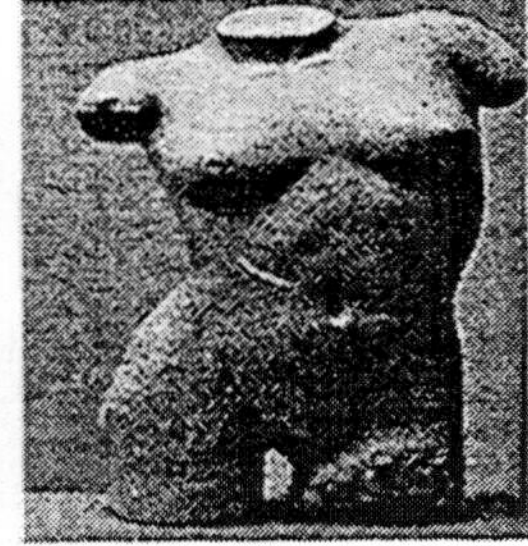 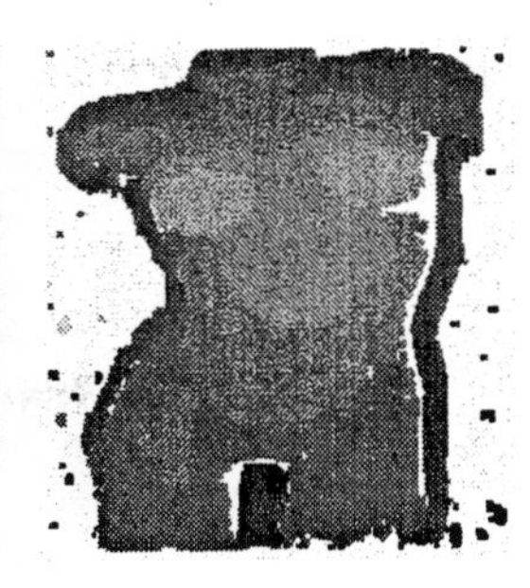

Figure 4: Rectified images and the corresponding disparity map. All of them (256x256) pixels.

Step 4 : From Projective to Euclidean Reconstruction

Transform the reconstructed projective structure into a Euclidean one using the obtained Euclidean reference points. Our output is, a dense Euclidean reconstruction of the scene.

Computing the Projective Reconstruction We call $\tilde{\mathbf{P}}$ and $\tilde{\mathbf{P}}'$ the two Camera Projection Matrices relating the 3D projective coordinates of a point to the 2D homogeneous coordinates of the point matches. As we know the disparity map, we can define :

$$[u_r, v_r, 1]^t = \underbrace{\begin{bmatrix} 1\,0\,0\,0 \\ 0\,1\,0\,0 \\ 0\,0\,0\,1 \end{bmatrix}}_{\tilde{\mathbf{P}}} [u_r, v_r, d_i, 1]^t, [u_r + d_i, v_r, 1]^t = \underbrace{\begin{bmatrix} 1\,0\,1\,0 \\ 0\,1\,0\,0 \\ 0\,0\,0\,1 \end{bmatrix}}_{\tilde{\mathbf{P}}'} [u_r, v_r, d_i, 1]^t \quad (3)$$

using $(u_r, v_r, d_i, 1)^t$ as projective coordinates of the points [5].

From the Projective Reconstruction to the Euclidean one. We can relate the Euclidean reference points $(X_i, Y_i, Z_i, 1)$ to the projective points by a collineation (a 4×4 matrix) called the *projective distortion matrix* $\tilde{\mathbf{D}}$:

$$[u_r, v_r, d_i, 1]^t \;=\; \tilde{\mathbf{D}}[X_i, Y_i, Z_i, 1]^t.$$

Let the Euclidean camera projection matrices be $\tilde{\mathbf{M}}$ and $\tilde{\mathbf{M}}'$. Following the pinhole model, we have:

$$\begin{aligned} s\,[u_i,\ v_i,\ 1]^t &= \tilde{\mathbf{M}}\,[X_i,\ Y_i,\ Z_i,\ 1]^t &= \tilde{\mathbf{P}}\tilde{\mathbf{D}}\,[X_i,\ Y_i,\ Z_i,\ 1]^t, \\ s'\left[u_i',\ v_i',\ 1\right]^t &= \tilde{\mathbf{M}}'\,[X_i,\ Y_i,\ Z_i,\ 1]^t &= \tilde{\mathbf{P}}'\tilde{\mathbf{D}}\,[X_i,\ Y_i,\ Z_i,\ 1]^t. \end{aligned}$$

Solving this system we can reconstruct any match. Another possibility is to compute the Euclidean Reconstruction for each match (any point where the disparity is defined) using Equation (4). We have used the second possibility.

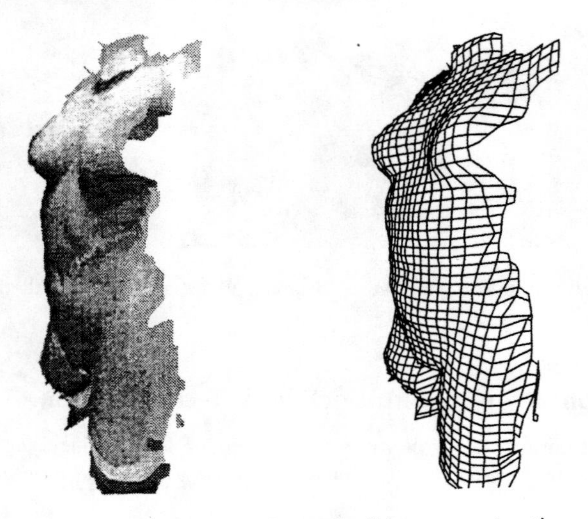

Figure 5: Dense Euclidean reconstruction.

References

[1] A. L. Abbott and N. Ahuja. Surface reconstruction by dynamic integration of focus, camera vergence, and stereo. In *Proc. Second Int'l Conf. Comput. Vision*, pages 532–543, Tampa, FL, Dec. 1988.

[2] P. Aschwanden and W. Guggenbühl. Experimental results from a comparative study on correlation-type registration algorithms. In *ISPRS Workshop*, Bonn, Germany, 1992.

[3] S. Das and N. Ahuja. Multiresolution image acquisition and surface reconstruction. In *Proceedings of the 3rd Proc. International Conference on Computer Vision*, pages 485–492, Osaka, Japan, Dec. 1990. IEEE Computer Society Press.

[4] R. Deriche and T. Blaszka. Recovering and characterizing image features using an efficient model based approach. In *Proceedings of the International Conference on Computer Vision and Pattern Recognition*, pages 530–535, New-York, June 1993. IEEE Computer Society, IEEE.

[5] F. Devernay and O. Faugeras. From projective to euclidean reconstruction. RR 2725, INRIA, Nov. 1995.

[6] O. Faugeras. What can be seen in three dimensions with an uncalibrated stereo rig. In G. Sandini, editor, *Proceedings of the 2nd European Conference on Computer Vision*, volume 588 of *Lecture Notes in Computer Science*, pages 563–578, Santa Margherita Ligure, Italy, May 1992. Springer-Verlag.

[7] E. Krotkov and R. Bacjcsy. Active vision for reliable ranging: Cooperating focus, stereo and vergence. *The International Journal of Computer Vision*, 11(2):187–203, 1993.

[8] L. Robert. *Perception stéréoscopique de courbes et de surfaces tridimensionnelles. Applications à la robotique mobile*. PhD thesis, École Polytechnique, Paris, France, Mar. 1993.

[9] Z. Zhang, R. Deriche, O. Faugeras, and Q.-T. Luong. A robust technique for matching two uncalibrated images through the recovery of the unknown epipolar geometry. *Artificial Intelligence Journal*, 78(1-2):87–119, 1994. Appeared in October 1995, also INRIA Research Report No.2273, May 1994.

[10] Z. Zhang, O. Faugeras, and R. Deriche. Calibrating a binocular stereo through projective reconstruction using both a calibration object and the environment. In R. Mohr and C. Wu, editors, *Proc. Europe-China Workshop on Geometrical modelling and Invariants for Computer Vision*, pages 253–260, Xi'an, China, Apr. 1995.

ROBINSPEC : A MOBILE WALKING ROBOT FOR THE SEMI-AUTONOMOUS INSPECTION OF INDUSTRIAL PLANTS

L. FORTUNA, A. GALLO, G. GIUDICE, G. MUSCATO

Dipartimento Elettrico Elettronico e Sistemistico
Università degli Studi di Catania
viale A. Doria 6 - 95125 CATANIA ITALY
TEL: +39-95-339535 FAX: +39-95-330793
e-mail: giomus@dees.unict.it

ABSTRACT

In the paper a new service robot designed for the inspection of industrial plants, is described. The robot is moved by means of three legs each connected to the surface with two electro-magnets that allows the robot to operate also on vertical surfaces or upside-down. A brief description of the mechanic, electronic and control systems is given.

KEYWORDS: Mobile robots, Walking-robots, Fuzzy control, Autonomous systems , Service Robots, Mechatronics.

INTRODUCTION*

ROBINSPEC is a new mobile service robot that has been designed and built at the DEES System and Control Laboratory of the University of Catania and is actually at the test phase.
Recently the term *Service robot* has been defined in order to classify all that kind of robots that are designed for the execution of useful work for humans and equipment different from the classical applications of industrial robots [1],[2],[5]. The production of Service robots is at the beginning and most of the robots are actually built as research prototypes. Nowadays the low cost of high-power computing systems and of complex sensors, like vision based systems, allow robotic technologies to be applied also in applications different from the traditional ones. Among them several service robots have been realized for solving inspection problems [3],[4].
The robot ROBINSPEC has been designed in order to solve inspection problems in dangerous or hard-to-reach environments such has storage tanks, long or high pipe-lines, factory chimneys, distillation columns etc. Actually such inspection problems are usually solved by using human operators, but very often they have to operate in hazardous environment with high health risks. Moreover in most cases such

* This work has been supported by "Contributo finanziario Regione Siciliana P.O.P. Sicilia 90/93"

inspections have to be done in environment that are not easy to reach by an human operator. The availability of an autonomous system which could be used for inspection in all these situation is then needed. Another possible application of ROBINSPEC consists in the remote positioning of sensors for non-destructive materials testing .
In the following sections a brief description of the mechanical structure, the electronic system and the control strategies will be discussed.

MECHANICAL STRUCTURE

A picture of the robot structure is reported in Fig. 1. The robot is moved by means of three legs each connected to the surface with two electro-magnets that allows the robot to operate also on vertical surfaces or upside-down.

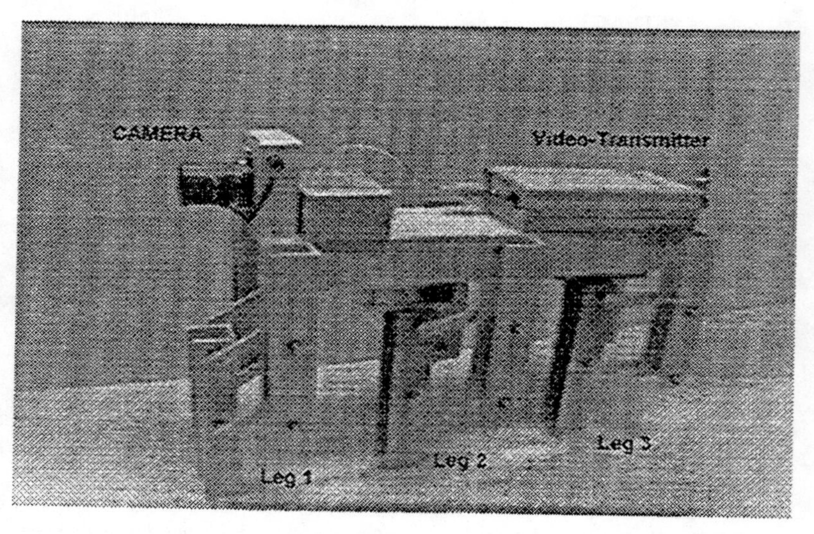

Fig. 1 : The ROBINSPEC system (without electro-magnets)

The main actuators for the three legs are three independent DC motors connected to a gear speed reducer and to a potentiometer for position measurement. The skeleton of the robot is made of aluminium. Figure 2 reports a particular of one leg with the position of the electro-magnets.

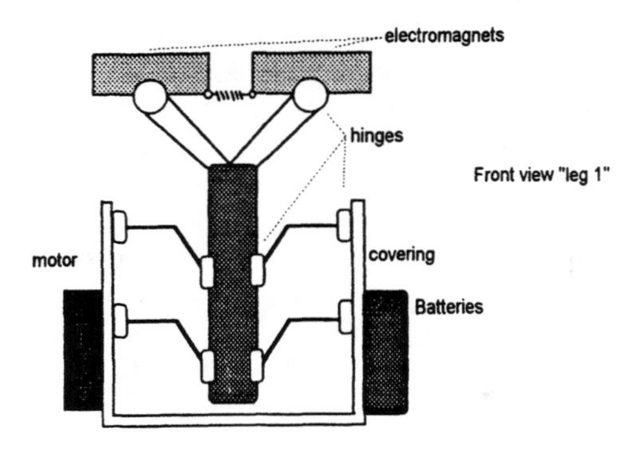

Fig. 2 Particular of one leg (front view)

The motion of ROBINSPEC consists of seven different steps as reported in Fig. 3. The adoption of legs allows also to avoid small obstacles like bolts, flanges etc.
The central legs will be equipped with an auxiliar motor that permits the rotation of the system. A video camera is positioned on the robot by using a stepper motor which allows to change the view angle.

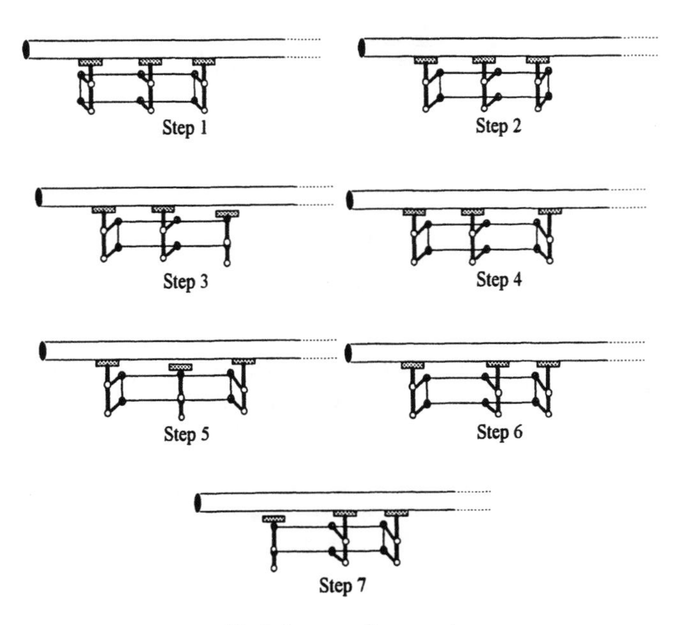

Fig. 3. Sequence of movement.

CONTROL

ROBINSPEC has been designed to work in a semi-autonomous way. This means that even if all the operations and movements of the robot can be tele-controlled from a base station , for several simple tasks ROBINSPEC can operate autonomously. For example, during the inspection of the wall of a storage tank the robot can inspect automatically the whole surface of the tank requiring the help of the operator only for not scheduled operations. The measurements obtained from the sensors and the images captured from the camera are radio-transmitted to the base station in order to be elaborated or monitored from the operator. In Fig. 4 a block diagram of the whole system is reported.

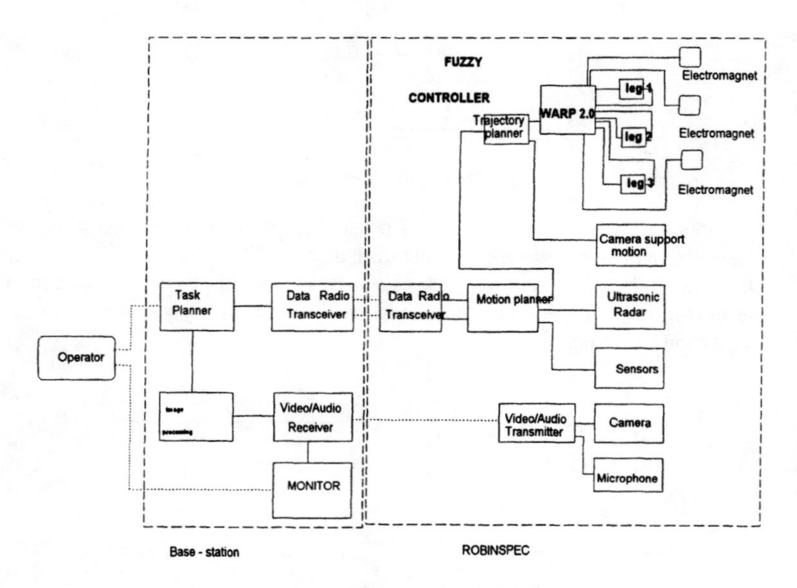

Fig. 4 Block diagram of the ROBINSPEC system

Base-station

The base station is based on a 486 Personal computer with an image acquisition board. The station is connected via a serial interface to a radio transceiver in order to establish a data link with the robot. The image acquisition board is linked with a video radio receiver which allows the remote images from ROBINSPEC to be acquired and processed. At present the image processing algorithm allows to detect the presence of rust in the surface. An extra audio channel is available to give to the operator an *audio feedback* of the remote operations by using a microphone on the robot.

Robinspec

The heart of ROBINSPEC is the motion planner module which is linked to most of the function of the system. This module, implemented using an ST9 microcontroller, on the basis of the commands received from the base station and on the measurements of the ultrasonic radar, communicates to the trajectory planner the references for the three legs. The module is also responsible for the orientation of the

support of the camera. The ultrasonic radar is implemented by using three POLAROID sensors placed in the front part of the system and allows to detect unplanned obstacles that could be present along the trajectory.
The three legs are controlled by using an ST6 microcontroller interfaced with a WARP 2.0 SGS/Thompson Fuzzy processor. The microcontroller is used to acquire the position of the legs by using three A/D channels connected to the potentiometer. Once the signals have been acquired, the error and the rate of the error with respect to a reference trajectory are computed and given as inputs to the Fuzzy rule processor. The WARP 2.0 (Weight Associative Rule Processor) is a digital Fuzzy processor with 8bit I/O able to elaborate up to 256 fuzzy rules, 8 inputs, 4 outputs with a clock frequency of 40 MHz. A simplified block diagram of WARP is reported in Fig. 5.

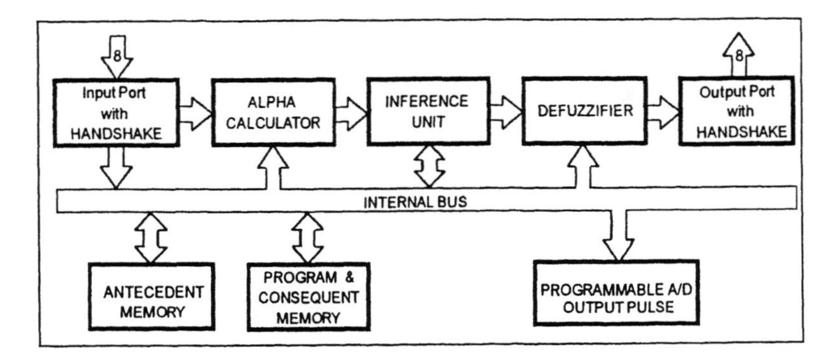

Figure 5. Block diagram of the Fuzzy processor WARP 2.0

The fuzzy processor compute the fuzzy rules and gives the outputs to three PWM modulators connected to the three DC motors. A fourth input channel given to WARP allows to change the reference of the three legs to modify the inclination of the system. The ST6 microcontroller gives also the outputs to the three blocks of electromagnets.

CONCLUSIONS

The design of ROBINSPEC was an engineering problem that involved several different disciplines. Among these particular attention has been devoted to the mechanics of the system (optimization of weights, kinematic) , the electronics , the control (Fuzzy control of the legs and of the electro-magnets), telecommunication (radio-link with the base station), image processing and software engineering.

The robot ROBINSPEC is actually at the end of the first phase of the design which includes the following main steps :

- optimization of the walking sequence with tuning of the parameters of the leg Fuzzy controller;

- optimization of the current of the electro-magnets in order to minimize energy consumption while guaranteeing surface adhesion;
- optimization of the interface on the base station;

At the present the trajectory of the robot is telecontrolled from the base-station. The second phase of the design will consist in making ROBINSPEC able to automatically generate motion planning, minimizing the intervention of the operator.

REFERENCES

1. R.D. Schraft, "Mechatronics and Robotics for Service Applications", IEEE Robotics and Automation Magazine, Vol. 1, N. 4, Dec. 1994.

2. S. Asami, "Robots in Japan: Present and Future", IEEE Robotics and Automation Magazine, Vol. 1, N. 2, June, 1994.

3. H. Schempf, B. Chemel, N. Everett, "Neptune: Above-Ground Storage Tank Inspection Robot System", IEEE Robotics and Automation Magazine, Vol. 2, N.2, June 1995.

4. E. Byler, W. Chun, W. Hoff, D. Layne, "Autonomous Hazardous Waste Drum Inspection Vehicle", IEEE Robotics and Automation Magazine, Vol. 2, N. 1, March 1995.

5. L. Fortuna, G. Muscato, G. Nunnari, A. Pandolfo, A. Plebe, "Application of Neural Control in Agriculture: An Orange Picking Robot", *Acta Horticulturae*, 1995.

ADMISSIBLE PATHS AND TIME-OPTIMAL MOTIONS FOR A WHEELED MOBILE ROBOT

JEAN-YVES FOURQUET and MARC RENAUD
LAAS-CNRS, 7 Avenue du Colonel Roche,
31077 Toulouse Cedex

ABSTRACT

This paper deals with the problem of finding the time-optimal motion along a predefined path that satisfies the non-holonomic *rolling without slipping* constraint for a class of wheeled mobile robots described by their dynamical model. Admissible paths are first described. Then, an algorithm is proposed that gives the optimal motion.

KEYWORDS : nonholonomic mobile robot, dynamics, time-optimality, imposed paths.

1. INTRODUCTION

In this paper, we are concerned with time-optimal open-loop strategies for wheeled mobile robots. We use the following conventions. The robot's *configuration* is defined by the nx1 matrix $\mathbf{q}$ of generalized coordinates which belongs to the generalized space $\mathbb{R}^n$. The *path*, which corresponds to changes as a function of a *geometric parameter* s (curvilinear abscissa for example), is distinguished from the *motion*, which corresponds to changes as a function of *time* t. Thus, a (generalized) *path* is a curve $\mathbf{q}(s)$ in the configuration space and among these paths an *admissible path* is a path which verifies the nonholonomic constraint. On the other hand, a *motion* - or *trajectory* - is the *time-history of the configuration* $\mathbf{q}(t)$. To a given trajectory corresponds a unique path whereas, in general, to a given path correspond an infinity of trajectories for controlled mechanical systems.

Recently, open-loop strategies for wheeled mobile robots in unconstrained environment have given rise to an abundant literature.

The first kind of contributions deals with the following path planning problem: *given any initial and final configurations* $\mathbf{q}^0$ *and* $\mathbf{q}^f$, *find one curve in the configuration space that links the former to the latter and which verifies the nonholonomic constraints* [1, 2, 3, 4].

The second kind of works deals with the following *time-optimal motion planning* problem for the dynamical model of wheeled mobile robots: *given any initial and final states* $\mathbf{x}^0$ *and* $\mathbf{x}^f$, *find the time-history of the torques (accelerations as a first approximation) given by the driven wheels so as to reach the latter from the former time-optimally by taking into account both kinematic and dynamic constraints such as the bounds on the torques* [5, 6, 7].

In this paper, we consider a third kind of problem closely related to these works. In fact, when an admissible path is obtained, it remains to define the evolution along this path as a function of time i.e. *the motion along the path.* Here, we consider that a good motion is a time-optimal one. Then, the problem at hand writes as follows : *Let be given an admissible path, the constraints that its extreme points have to be reached at given velocities (without any loss of generality, these velocities will be fixed to zero), the dynamical constraints acting on the robot, what is the time-optimal motion along this path ?*

The approach we present applies to a great number of mechanical systems, and particularly to nonholonomic mechanical systems. Here, we focus on the HILARE mobile robot [6]. This robot has two driven wheels at the rear and a castor at the front. The configuration is defined by $\mathbf{q} = (\mathbf{q}_1\ \mathbf{q}_2\ \mathbf{q}_3)^t = (x\ y\ \theta)^t$ in $\mathbb{R}^3$ (x and y denote the coordinates of the midpoint on the rear axle and θ the orientation of the robot in the plane). It is submitted to the nonholonomic *rolling without slipping* constraint:

$$d\mathbf{q}_1 \sin \mathbf{q}_3 - d\mathbf{q}_2 \cos \mathbf{q}_3 = 0 \tag{1}$$

that gives the well-known *kinematic* control system :

$$\begin{aligned} \frac{d\mathbf{q}_1}{ds} &= \mathbf{v}_1(s) \cos \mathbf{q}_3(s) \\ \frac{d\mathbf{q}_2}{ds} &= \mathbf{v}_1(s) \sin \mathbf{q}_3(s) \\ \frac{d\mathbf{q}_3}{ds} &= \mathbf{v}_2(s) \end{aligned}$$

that is

$$\frac{d\mathbf{q}}{ds} = A(\mathbf{q}(s))\ \mathbf{v}(s) \tag{2}$$

Concerning dynamics, we consider that the torques are bounded in the same way for both driven wheels and that angular acceleration of each of them is directly obtained from the torque by multiplying by a scaling factor. With the preceding assumptions, the HILARE's dynamic model writes:

$$\begin{aligned} \dot{\mathbf{x}}_1 &= \tfrac{1}{2}(\mathbf{x}_4 + \mathbf{x}_5) \cos \mathbf{x}_3 \\ \dot{\mathbf{x}}_2 &= \tfrac{1}{2}(\mathbf{x}_4 + \mathbf{x}_5) \sin \mathbf{x}_3 \\ \dot{\mathbf{x}}_3 &= \tfrac{r}{d}(\mathbf{x}_4 - \mathbf{x}_5) \\ \dot{\mathbf{x}}_4 &= \mathbf{u}_1 \\ \dot{\mathbf{x}}_5 &= \mathbf{u}_2 \end{aligned}$$
$$\mathbf{u} \in \mathcal{U} \ \text{ with } \mathcal{U} = \{\ u \mid |\mathbf{u}_i| \leq A\ ;\ i = 1,2\}$$

with $\mathbf{x} = (\mathbf{x}_1\ \mathbf{x}_2\ \mathbf{x}_3\ \mathbf{x}_4\ \mathbf{x}_5)^t = (\frac{x}{r}\ \frac{y}{r}\ \theta\ \omega_r\ \omega_l)^t$; $\mathbf{u} = (\mathbf{u}_1\ \ \mathbf{u}_2)^t = (a_r\ \ a_l)^t$;

where ω_r and a_r (ω_l and a_l) are respectively the angular velocity and angular acceleration of the right (left) driven wheels. Then, by denoting $\mathbf{p} = (\mathbf{x}_4\ \ \mathbf{x}_5)^t$, the dynamic model writes :

$$\begin{aligned} \dot{\mathbf{q}} &= B(\mathbf{q})\mathbf{p} \qquad && \mathbf{p} \in \mathbb{R}^2 \\ \dot{\mathbf{p}} &= \mathbf{u} && \mathbf{u} \in \mathcal{U} \subset \mathbb{R}^2 \end{aligned} \tag{3}$$

2. ADMISSIBLE PATHS

During the last five years, several techniques have been rised so as to solve the nonholonomic path planning problem. Most of them are based on adequate changes of input and state variables that convert the kinematic control systems into canonical forms (chained, power, flat forms,...) [2, 3]. In fact, the *relatively simple* kinematics of HILARE allows to solve the path planning problem without requiring the machinery of canonical forms. For illustrating purpose, we briefly list some possible choices[1]. By taking $\mathbf{v}_2(s) = 1$ i.e. $\mathbf{q}_3(s) = s$, we have :

a) $\mathbf{v}_1(s) = K$ (K is a constant) : circle of radius K with center at the origin; $\mathbf{v}_1(s) = Ks$: involute of circle. By variation of the constants K and s^f, these choices allow to reach all the points of a dimension 2 manifold in the configuration space $\mathbb{R}^3$.

b) $\mathbf{v}_1(s) = \alpha s + \beta$ where α and β are constants allow to reach every triple $(\mathbf{q}_1^f, \mathbf{q}_2^f, \mathbf{q}_3^f)$ (with $\mathbf{q}_3^f \neq 0$) from the origin by variation of α, β and s^f.

3. MINIMUM TIME CONTROL FOR AN IMPOSED PATH

The problem can be formulated as follows: determine control $\mathbf{u} : t \in \left[t^0, t^f\right] \rightarrow \mathbf{u}(t) \in \mathcal{U}$ allowing the robot configuration $\mathbf{q}(t)$ to evolve in a minimum time, that is, by minimizing the cost function:

$$\mathcal{J} = \int_{t^0}^{t^f} dt \; = \; t^f - t^0 \tag{4}$$

from an initial imposed configuration $\mathbf{q}(t^0) = \mathbf{q}^0$ to a final imposed configuration $\mathbf{q}(t^f) = \mathbf{q}^f$ along the *imposed path*, the starting and destination configurations being reached at a zero velocity $\dot{\mathbf{q}}(t^0) = \dot{\mathbf{q}}(t^f) = 0$.

3.1. Problem reduction

Here, the computations are given for HILARE in the case where the input $\mathbf{v}(s)$ is given[2] as it appears natural from the path planning step described in part 2.

The derivation rules of composite functions allow computation of:

$$\dot{\mathbf{v}} = \frac{d\mathbf{v}}{ds}\dot{s} \; ; \; \dot{\mathbf{q}} = \frac{d\mathbf{q}}{ds}\dot{s} \tag{5}$$

and from (2) and (3):

---

[1]Without any loss of generality, we suppose that the initial configuration $\mathbf{q}^0$ is the origin and $s \in [0, s^f]$

[2]Of course, the computations can be done in a similar way if the data is the configuration as a function of s i.e. $\mathbf{q} = \mathbf{q}(s)$. In this case, derivatives up to second order have to be computed and additional singularities may appear in the computations.

$$A(\mathbf{q})\mathbf{v}\dot{s} = B(\mathbf{q})\mathbf{p} \tag{6}$$

gives :

$$\mathbf{v}\dot{s} = C\mathbf{p} \;\; ; \;\; \text{with} \;\; C = r\begin{pmatrix} \frac{1}{2} & \frac{1}{2} \\ \frac{1}{d} & -\frac{1}{d} \end{pmatrix} \tag{7}$$

The preceding equations allow to consider $\dot{\mathbf{q}}$ as a function of s and $\dot{s}$ and $\dot{\mathbf{p}}$ as a function of s, $\dot{s}$ and $\ddot{s}$. As $\dot{s}$ is positive, the formulation of the dynamic model leads to the following equation in s, $\dot{s}$ and $\ddot{s}$:

$$\mathbf{d}(s)\ddot{s} + \delta(s)\dot{s}^2 = \mathbf{u} \tag{8}$$

where :

$$\mathbf{d}(s) = \begin{pmatrix} \mathbf{d}_1(s) \\ \mathbf{d}_2(s) \end{pmatrix} = \frac{1}{r}\begin{pmatrix} \mathbf{v}_1(s) + \frac{d}{2}\mathbf{v}_2(s) \\ \mathbf{v}_1(s) - \frac{d}{2}\mathbf{v}_2(s) \end{pmatrix} \tag{9}$$

$$\delta(s) = \begin{pmatrix} \delta_1(s) \\ \delta_2(s) \end{pmatrix} = \frac{1}{r}\begin{pmatrix} \frac{d}{ds}(\mathbf{v}_1(s) + \frac{d}{2}\mathbf{v}_2(s)) \\ \frac{d}{ds}(\mathbf{v}_1(s) - \frac{d}{2}\mathbf{v}_2(s)) \end{pmatrix} \tag{10}$$

Now, we have to solve the following optimal control problem :

$$\text{Min} \int_{t^0}^{t^f} dt$$

$$s(t^0) = s^0 \; ; \; s(t^f) = s^f \; ; \; \dot{s}(t^0) = \dot{s}(t^f) = 0$$
$$\mathbf{d}(s)\ddot{s} + \delta(s)\dot{s}^2 = \mathbf{u} \; ; \; \mathbf{u} \in \mathcal{U}$$
$$\mathcal{U} = \{ \, u \mid |\mathbf{u}_i| \leq A \; ; \; i = 1,2\}$$

Now we are able to use the so-called *phase-plane techniques* so as to deal with our problem in a way similar to the one used for robotic manipulators [8], [9] [10] [11]. Here, we only recall the mains facts concerning these works.

First, the equation (8) is rewritten by taking $\mathbf{y} = (\mathbf{y}_1 \;\; \mathbf{y}_2)^t = (s \;\; \dot{s})^t$ and a new control variable $w = \ddot{s}$ in order to "put" the nonlinear dynamics in the new control domain $\mathcal{W}(\mathbf{y})$ which is state-dependent.
Then determining the *vector control* $\mathbf{u}$ reduces to the simpler determination of the *scalar control* $w = \ddot{s} : t \in \left[t^0, t^f\right] \rightarrow w(t) = \ddot{s}(t) \in \mathcal{W}(s, \dot{s}) \subset \mathbb{R}$ where the new control w belongs to a new admissible control set $\mathcal{W}(s, \dot{s})$ defined by the following 4 inequalities[3]:

$$U_{i_{min}} \leq \mathbf{d}_i(s)w + \delta_i(s)\dot{s}^2 \leq U_{i_{max}}; \quad i = 1,2 \tag{11}$$

Depending on the couple "path-robot" at hand i.e., depending on the values taken by the $\mathbf{d}_i(s), \delta_i(s)$ over the interval $[s^0, s^f]$, we always arrive at a more or less peculiar version of the following optimal control problem:

[3]Here, $U_{i_{min}} = -A$ and $U_{i_{max}} = A$ but the case where boundaries $U_{i_{min}}$ and $U_{i_{max}}$ depend on $\mathbf{q}$ and $\dot{\mathbf{q}}$ offers no additional difficulties.

$$\text{Min} \int_{t^0}^{t^f} dt$$

$$\begin{aligned} \dot{\mathbf{y}}_1 &= \mathbf{y}_2 \\ \dot{\mathbf{y}}_2 &= w \end{aligned} \tag{12}$$

$$\mathbf{y}_1(t^0) = s^0 \; ; \; \mathbf{y}_1(t^f) = s^f$$
$$\mathbf{y}_2(t^0) = 0 \; ; \; \mathbf{y}_2(t^f) = 0$$

$$\begin{aligned} \mathbf{y} &\in \mathcal{R} \subset \mathbb{R}^2 \\ w &\in \mathcal{W}(\mathbf{y}) \end{aligned}$$

Thus, we are in the following theoretical framework:

Time-optimal control problem for a "double integral" plant such that the (scalar) control is bounded in a state-dependent domain $\mathcal{W}(\mathbf{y})$ *and the state belongs to a subset* $\mathcal{R}$ *of the state space* $\mathbb{R}^2$.

3.2. Facts from optimal control theory

We will say that a subinterval $(\tau_1, \tau_2) \subset [t^0, t^f]$ is an *interior interval* if the state constraints are verified in the sense of the strict inequality for all $t \in (\tau_1, \tau_2)$. On the contrary, a subinterval $[\tau_1, \tau_2]$ is called a *boundary interval* if a constraint is verified in the sense of equality (or *saturated*) for all $t \in [\tau_1, \tau_2]$. Then, the following properties can be obtained :

Property 1 *The optimal control* $w^\star$ *take its value on one or the other bound of its domain* $\mathcal{W}(\mathbf{y})$ *for every time in an interior interval; such arcs of motion are traditionally called* bang *arcs.*

It is a *local* information about the structure of optimal solutions. It takes a *global* nature when there is no state constraints:

Property 2 *If* $\mathcal{R} = \mathbb{R}^2$ *then the optimal control is bang-bang with only one switch.*

For the problems with state constraints, it remains to study the boundary intervals and the switching laws from an interval to another. The Green-Stokes theorem allows to decide the way the optimal arcs are connected one to the other. It gives the following result :

Property 3 *The phase plane motion that gives the minimum time between two given states is the one which is the* highest *in the admissible region. In particular, if a piece of the boundary* $\partial\mathcal{R}$ *of the region is such that a phase motion can traverse it (it may be the case when* $\mathbf{d}_i \equiv 0$ *[11]) then the highest phase motion may go up it.*

3.3. General structure of the algorithms

The algorithms considered for robotic manipulators, and in particular the one by Slotine & Yang [10], make use of the local structure of the optimal solutions from:

- the bang-bang structure of optimal interior intervals,
- the characterization of those points of the Maximum Velocity Curve (MVC) [8] (the MVC is the locus of points where $\mathcal{W}$ is reduced to a single value ; this curve constitutes for a part the boundary $\partial\mathcal{R}$ of $\mathcal{R}$) which may belong to an optimal solution.

Slotine & Yang build bang arcs (*limit curves*) from this particular points and connect them to the bang arcs starting from $\mathbf{y}^0 = (s^0\ 0)^t$ and $\mathbf{y}^f = (s^f\ 0)^t$ and obtained respectively by forward and backward integration.

Moreover, it has been shown that this strategy has to be changed in order to take into account the existence of boundary intervals resulting from the verifications of the cases where : $\exists i$ s.t. $\mathbf{d}_i(s) \equiv 0\ \forall s \in (s_a, s_b)$ [11].

In the following part, we illustrate this approach on an example simple enough to allow symbolic computation for the most part.

4. EXAMPLE

We "pick" a path in the "canonical class" defined by $\mathbf{v}_1(s) = \alpha s + \beta$ and $\mathbf{v}_2(s) = 1$ (with $\alpha = -1.6309$; $\beta = 2.1309$; $s^f = 1.70$ i.e. $\mathbf{q}^f = (1.205\ 0.431\ 1.700)^t$; $r = 1$, $d = \frac{2}{\pi}$ and $A = 1$). The initial configuration is the origin in $\mathbb{R}^3$ (see fig. 1, 2).

Here, it is easy to show that the MVC is (for s such that both $\mathbf{d}_1(s)$ and $\mathbf{d}_2(s)$ are different from zero ; that is outside the *zero-inertia points* [10]), given by :

$$\dot{s}(s) = \left(\frac{|\mathbf{d}_1(s)| + |\mathbf{d}_2(s)|}{|\Delta(s)|} \right)^{\frac{1}{2}} \tag{13}$$

with : $\delta_1 = \delta_2 = \alpha \neq 0$ and $\Delta = \alpha(\mathbf{d}_1 - \mathbf{d}_2)$.

In this case, the initial bang motion from $\mathbf{y}^0$ reaches the MVC, follows it (during this time interval both actuators are saturated), then reach $\mathbf{y}^f$. The corresponding optimal phase plane motion is shown at figure 3.

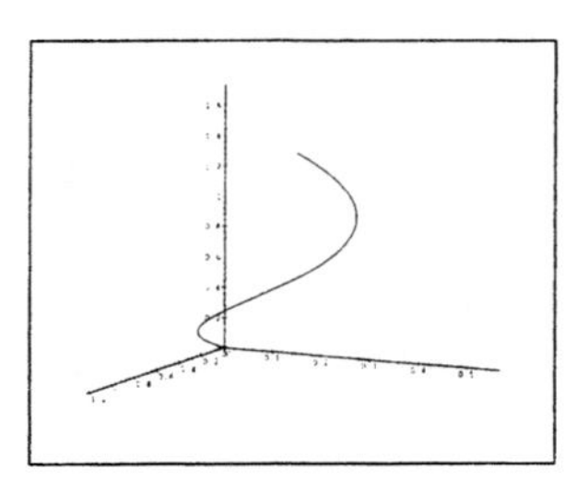

Figure 1: path in (x,y,θ)

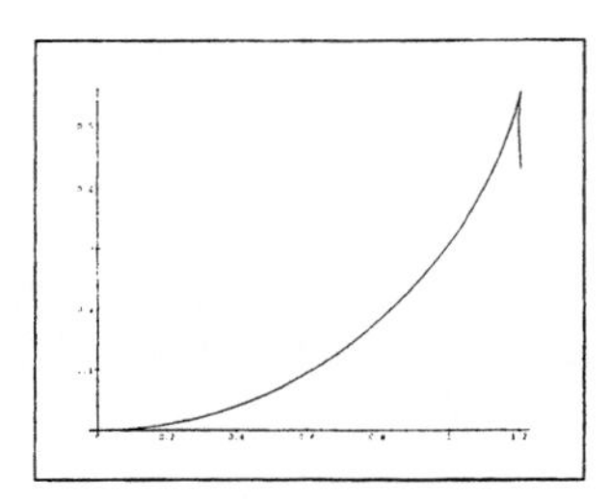

Figure 2: projection of the path in the plane (x,y)

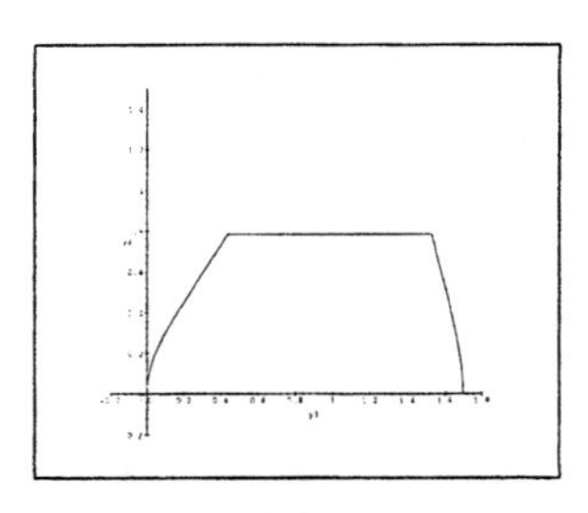

Figure 3: optimal phase plane motion

References

[1] G. Jacob. *Motion Planning by piecewise constant or polynomial inputs* IFAC Nonlinear Control Systems Design Symposium, pages 628-633, June 1992.

[2] D. Tilbury, R. Murray, and S. Sastry. *Trajectory generation for the N-trailer problem using Goursat normal form.* In proceedings of the IEEE Conference on Decision and Control, pages 971-977, San Antonio, Texas, 1993.

[3] P. Rouchon, M. Fliess, J. Levine and P. Martin. *Flatness and motion planning : the car with n trailers* in proceedings of Second European Control Conference, Groningen, pages 1518-1522, 1993.

[4] P. Soueres, J-Y. Fourquet and J-P. Laumond. *Set of reachable positions for a car* IEEE Transactions on Automatic Control, 39(8), pages 1626-1631, August 1994.

[5] P. Jacobs, A. Rege and J-P. Laumond. *Non-holonomic Motion Planning for Hilare-like Mobile Robots* International Symposium on Intelligent Robotics, Bangalore, 1991.

[6] S. Fleury, P. Soueres, J-P. Laumond and R. Chatila. *Primitives for smoothing mobile robot trajectories* IEEE International Conference on Robotics and Automation, Atlanta, pages 832-839, 1993.

[7] D. B. Reister and F. G. Pin *Time-Optimal Trajectories for Mobile Robots with two Independently Driven Wheels* International Journal of Robotic Research, vol. 13, n. 1, pp 38-54, February 1994.

[8] J.E. Bobrow, S. Dubowsky, and J.S. Gibson. *Time-optimal control of robotic manipulators along prespecified paths.* International Journal of Robotics Research, 4(3):3–17, Fall 1985.

[9] K.G. Shin and N.D. Mckay. *Minimum-time control of robotic manipulators with geometric path constraints.* IEEE Transactions on Automatic Control, AC-30(6):531–541, June 1985.

[10] J.J.E. Slotine and H.S. Yang. *Improving the efficiency of time-optimal path-following algorithms.* IEEE Transactions on Robotics and Automation, 5(1):118–124, February 1989.

[11] M. Renaud and J-Y. Fourquet. *Time-Optimal Motions of Robot Manipulators Including Dynamics* The Robotics Review 2. O. Khatib, J.J. Craig and Tomás Lozano-Pérez editors, MIT Press, pages 225-259, 1992.

HIERARCHICAL PETRI NETS FOR MANUFACTURING SYSTEMS

P. FINOTTO, D.CRESTANI

Laboratoire d'Informatique de Robotique et de Microélectronique de Montpellier

161, rue Ada F - 34392 Montpellier Cedex 5

Tel. : (33) 67 41 85 81 Fax : (33) 67 41 85 00

E-mail : <name> @ lirmm . fr

ABSTRACT

Since their creation, Petri Nets have often been applied to Manufacturing System Design. Due to the important size of systems to model, numerous modular methods have been developed in order to make design and validation easier. We present in this paper a bottom-up methodology which provides a natural and powerful method to the designer. This method allows to built progressively the whole system model, with possible re-use of validated components.

KEYWORDS : Petri Nets, Manufacturing System, Modular Design, Discrete Events Systems

INTRODUCTION

Considering the continuous increase in the complexity of Manufacturing Systems (M.S.) design, the use of models and methods is necessary in order to make design and validation easier. Several models are commonly used in this field, of which the most used are queuing networks, Max. Algebra, Petri nets, Function Charts For Control Systems (F.C.C.S.), and State Machines.

Some of them, like queuing network [1] or F.C.C.S. [2], are more dedicated to specific fields. They are powerful to model particular systems. But their validation capacity are limited, because simulation seems to be the only one solution to treat large scale systems. On the contrary, models like Max. Algebra [3] or State Machines offer large capacity of validation, based on well formed theories. But they are limited in their descriptive power. Petri nets [2] present a high modelling capacity, including sequences, parallelism, synchronisation and conflicts, and at the same time a lot of analysis means [4].

But the practical use of Petri nets, in an industrial environment, seems difficult because of the important size of information to be handle.

Therefore we propose in this paper a methodology whose aim is to take into account the modular and strong connecting aspect of M.S., while allowing a progressive analysis.

We first present the different methods commonly used for modular design. Then, in Part 2, we define our method showing the interest of its use. In the third section, we present an example dealt with this model, before a conclusion.

1. OVERVIEW ON MODULAR DESIGN WITH PETRI NETS

The goals of hierarchical methods in design are multiple. First, a system to be modelled is splitted into sub-systems whose sizes are smaller than the initial system one's. Such methods take advantage of the naturally modular aspect of Manufacturing Systems. A second interest in hierarchical methods is the ability to validate step by step the design. The whole model is built progressively, validating each stage. Moreover, respecting rules, some methods guaranties to obtain an accurate model which need no more validation.

1.1. Substitution method

First, the studied system is modelled into a high abstraction Petri net, in which places represent complex actions, or transitions represent complex events. Then a refinement process is performed where one place or one transition is replaced by a more detailed Petri net. In other words, a net is substituting for a place, or a transition. This step can be repeated again and again, until the system to be modelled was detailed enough. Examples are given on figure 1, where the two kinds of substitution are illustrated.

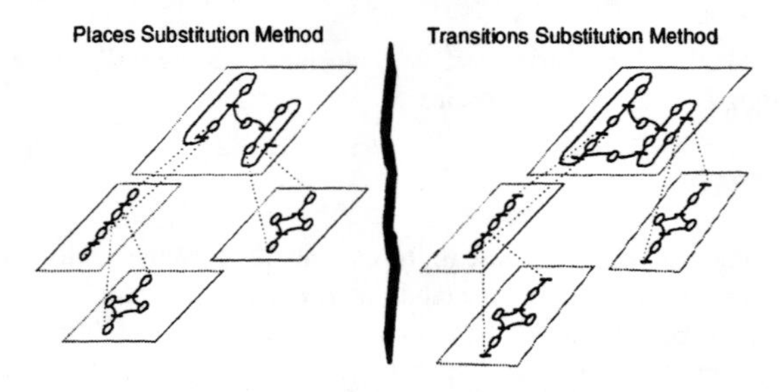

Figure 1 : Examples of substitution methods concepts

Respecting some rules, usual properties, safeness and boundedness are kept like in [5] for place substitution, or in [6] for transition substitution. In [7], starting from a safe, alive and reversible net, places are replaced by basic modules which allow to describe sequences, parallelism, conflicts or mutual exclusion.

Such approaches allow to obtain correct Petri net without any validation. They are powerful to describe slightly connected sub-systems, but links between sub-net cannot be represented easily, so this approach does not suit to strongly connected systems as those often find in M.S..

1.2. Sharing methods

The system is first modelled as a set of independent sub-systems. The control of each sub-system is represented in one Petri net, apart from others. Relations between subsystems are represented by common places or transitions, which are shared by several sub-nets, as shown in figure 2.

From a validation point of view, global structural properties can be deduced from local one's. In [8], synthesis rules allow to calculate global structural invariants from invariants of each sub-net.

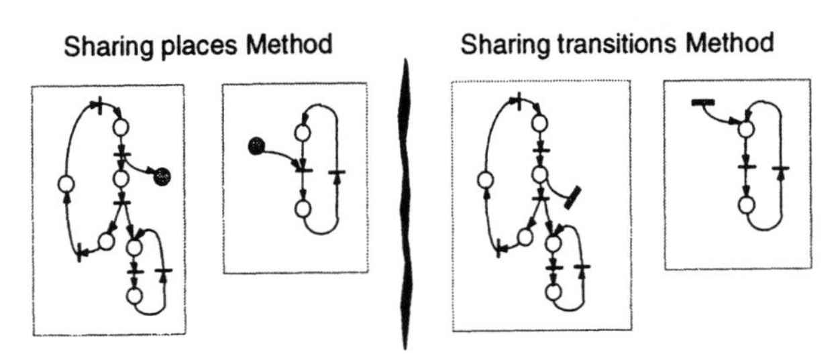

Figure 2 : Examples of sharing methods concepts

The most important interest of sharing methods is to naturally represent connections between sub-nets. However, validation results are fewer than for substitution approaches.

2. THE PROPOSED METHOD

The complexity of M.S. is due, less to the complexity of sub-systems which compose the M.S., than to the complexity of relations between those sub-systems. So, substitution methods are not very efficient for strongly connected systems. The proposed method is close to sharing one's. But a problem in the use of sharing methods is that there is no methodology to guide the designer in his work. So, we propose a hierarchical methodology making up the global system step by step.

2.1. The global process

Examining the usual structure of a M.S., it's modular and hierarchical aspect appears. A machine can be viewed as a set of actuators and sensors which have their own behaviour. So, from a logical point of view, a machine is the coordination of actuators and sensors. Similarly, a production cell can be considered as the coordination of a set of machines and product flows. Our method is inspired by such an organisation.

We will first define the vocabulary used to present the method :

- An *operative component* is a physical or logical entity of the system to be modelled. For example an actuator (physical), or an actuators coordination (logical).
- An *operative system* is a set of operative components. For example, a set of actuators and sensors.
- A *control unit* is a Synchronised Petri net, which describes an operative component behaviour.

- A *control system* is the Petri net yielded by aggregating a set of control units.
- A *basic operative system* is either a system with a very simple control (for example a jack), or a more complex system previously modelled. In this case, a validated control system is available.
- A *primitive control system* is a control system which models one basic operative system. So, it holds only one control unit.

Note : Any control system holds at least one control unit.

The hierarchical formalisation is based upon the following level concepts. Each control system or control unit holds an integer representing its *level*. So levels define a partial order upon control units and upon control systems. Into a level, control systems or control units are sorted using an unique integer index. So, each control unit *U* or control system S of a model can be located in the design hierarchy with an unique 2-uple < level , position_in_level >, as $U_{i,j}$, or $S_{i,j}$.

In order to illustrate the design process, a simple example is presented on figure 3. This example treat of a 3 positions shunting, obtained by using two 2 positions shunting.

For a start, the behaviour of each basic physical constituent is described in *one control unit*. At this level, conditions associated to transitions represent the external environment control over each net. Here three *control units* are associated respectively to shunting1, shunting2, and presence sensor.

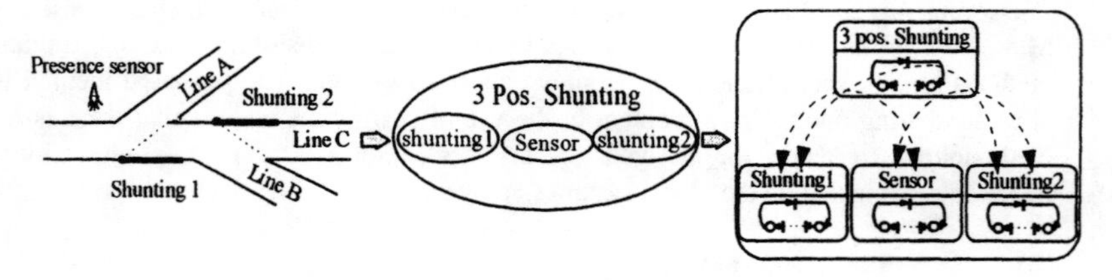

Figure 3 : Example of a 3 positions shunting

Then these basic constituents are gathered into a new *control system* corresponding to a system entity. For the example, the "3 positions shunting" *control system* is the set of shunting1, shunting2, and presence sensor. To this new set, is associated one *control unit*, which models the coordination of the different sub-systems. This coordination must represent the synchronisation between shuntings and sensor. So this coordination Petri net must contain links between the three sub-nets (figure 3).

This gathering step should be repeated until the whole system has been described.

2.2. Links between sub-nets

The aim of the method is to further a natural process for the designer. When a new level is added, some conditions are replaced by arcs coming from upper level control units. Similarly, a coordination control unit has to know some states or events from the control units it coordinates. So, arcs link low level control units to up level control units.

We propose to represent links between levels with arcs linking transitions of low level nets to places of up level nets, as shown in figure 4.

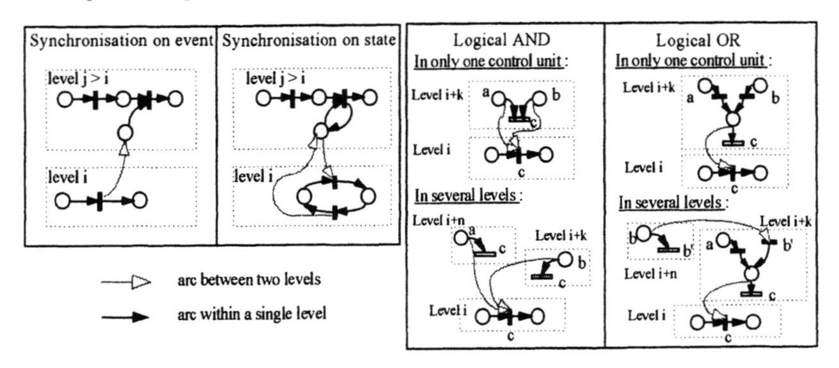

Figure 4 : Synchronisation arcs

With such a solution, no token get in or get out places from low level control units. It means that a validate control unit could be use for another application without doing any changes. Moreover, this method allows to check both events and states synchronisation structures, and to represent all the basic logical structures (and, or) (figure 4).

It is equivalent to say that there are arcs between level i and j, and to say that some transitions of level i are duplicated in level j, or shared by level i and level j.

3. A MODELLING EXAMPLE

Considering the previous example, figure 5 shows the Petri nets associated to the two shuntings and to the presence sensor. Each of them is the lonely control unit forming the primitive control system Shunting1, or Shunting2, or Presence sensor.

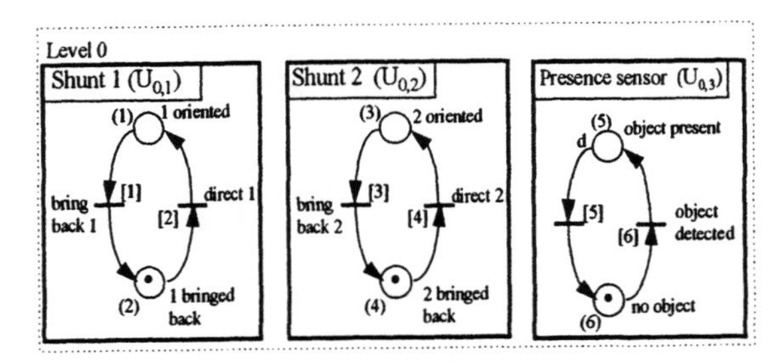

Figure 5 : Primitive control systems

The "3 Positions Shunting" control system is made up of the 3 previous control systems, and the control unit "3 Positions Shunting" (figure 6). When a pallet comes, the two shuntings must be oriented so as to direct the pallet towards the line A, B or C. The line is determined with the conditions associated to transitions 7,8 and 9. So this system could be included into a greater system, in which these conditions would be replaced by arcs coming from upper levels.

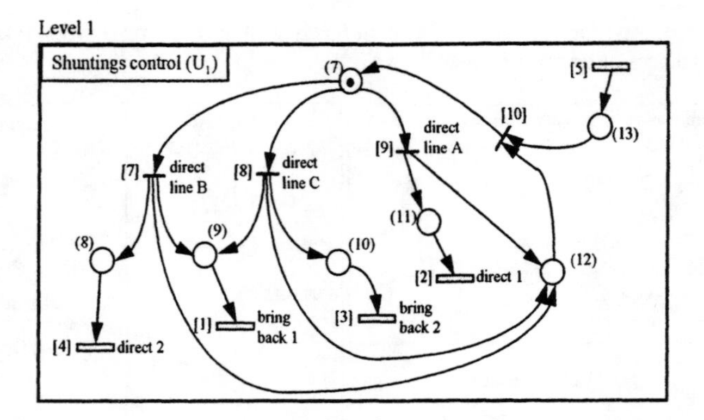

Figure 6 : Coordination Control Unit

4. CONCLUSION

We have presented, in this paper, a full method for Manufacturing Systems control design. This method allows the designer to model his system in a modular manner, and a natural way. Starting from the modular structure of the operative part, the whole control system is built progressively. At each stage, a new Petri Net is created, and represents the coordination of the nets previously defined. The modelling power of this method has been shown by representing all the basic structures. Finally, a validated control system can be re-used in another application without doing any changes.

REFERENCES

1. Bel G., Dubois D. "Modélisation et Simulation de Systèmes Automatisés de Production." APII *0399-0516-1985/01.*
2. David R., Alla H. *Petri Nets and Grafcet: Tools for Modelling Discrete-Events Systems.* Prentice-Hall, London (1992).
3. Baccelli F., Cohen G., Olsder G.J., Quadrat J.P *Syncronization and Linearity - An Algebra for Discrete Event Systems.*, Wiley, New York (1992).
4. Hollinger D. "Utilisation pratique des réseaux de Petri dans la conception des systèmes de production" *T.S.I. 0752-4072/85/06* AFCET (1985).
5. Valette R. "Analysis of Petri nets by stepwise refinements" *Journal of Computer and System Science, Vol. 18,* (1979), 35-46.
6. Suzuki I., Murata T. "A method for stepwise refinements and abstraction of Petri nets" *Journal of Computer and System Science, Vol. 27, n° 1,* (1983), 51-76.
7. M.C Zhou, F. DiCesare, A.A Desrochers "A Top-Down Modular Approach to Systematic Synthesis of Petri nets Models for Manufacturing Systems" *Proc. of IEEE Robotics and Automation Conference* (1989) 534-539.
8. Agerwala T., Choed-Amphai Y., "A Synthesis rule for concurrent systems" *Proc. Of Design Automation Conference* (1978) 305 - 311.

MODELING OF THE 2-DELTA 6-DOF DECOUPLED PARALLEL ROBOT

A. GOUDALI, J.P. LALLEMAND and S. ZEGHLOUL
Laboratoire de Mécanique des Solides. U.R.A. C.N.R.S 861
SP2MI - B.P. 179
86960 FUTUROSCOPE Cédex

ABSTRACT

In this paper, an effective method is proposed to establish an explicit relationship between the end effector operational co-ordinates and the active and passive joint variables of a 6 d.o.f parallel robot (2-Delta). Thus, two complete geometrical models are introduced. A simulation of the 2-Delta robot is also presented on a C.A.D. Robotics system. This simulation has allowed us to validate the cohesion of our calculations, and to show the the behavior of the internal and external structures. Finally, an approach is proposed to study the influence of small clearances of the passive joint on the precision of the position and rotation of the effector. This approach is based on a concept similar to that of Yoshikawa's manipulability.

KEYWORDS
Parallel robot, uncoupled, modelization, workspace, elliposid of clearance.

INTRODUCTION

Parallel robots have been the subject of several studies due to the considerable interest that they have demonstrated by their lightness, rigidity and rapidity. All these particular properties open the door to all spatial applications where mass problems are particulary crucial. They are also used as robot end effector. The first parallel robot structure dates from 1939 when Willard [1] proposed a parallel structure to paint cars. In 1949, Gough proposed an articulated machine to test tires. Next Stewart [2] suggested the utilisation of this structure as a movement generator for flight simulators. It was also used by Rebouled [3], and by Merlet [4] as a compliant wrist of a robot. Amongst spatial operators of 3 d.o.f., the Delta structure designed by R. Clavel [5] constitutes technological innovation. Other 3 d.o.f. parallel structures have been developed. We can mention the parallel robotY-Star developed by Hervé [6], the robot Speed R-Man developed by Reboulet [7].

We will put forward the 2-Delta robot which in it's principle, constitutes a new and original structure. This robot is entirely parallel and possesses 6 d.o.f. Its features lie in mechanical uncoupling of translation and orientation motions. This property is what differentiates it in comparison to the other traditional 6 d.o.f parallel robots or the one issued from the Delta structure [8]. This article partially treats the different studies that have been carried out on the 2-Delta robot and which have been presented in [9].

PRINCIPLE OF THE 2-DELTA STRUCTURE

The 2-Delta structure (Figure. 1) is composed of two overlapped Delta structures. The first, which controls the displacements parallel to the fixed base (0), is the structure of the initial Delta robot proposed by Clavel [5]. The second structure controls orientations of the platform (3). The two structures are independent and allow the mechanical uncoupling of the translation and the rotation of the end effector.

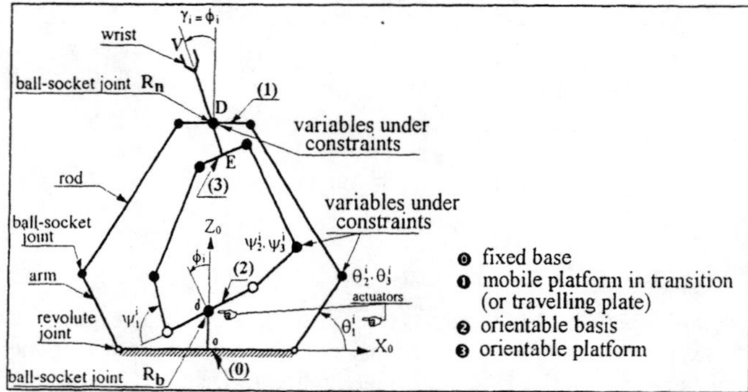

Figure. 1. Principle of the 2-Delta structure

GEOMETRIC MODELIZATION OF THE 2-DELTA STRUCTURE

The modelization of parallel robots is quite specific. There is no existing systematic and simple model, therefore it is necessary to find the well adapted methods of each structure. In the case of the 2-Delta robot, we propose an analytic modelization.

The 2-Delta robot possesses a significant number of passive joint variables (26 variables if parallelograms are not segments). Therefore, it is necessary to study the influence of the joint limits constraint on the end effector workspace.

The complete **D**irect **G**eometrical **M**odel allows us to calculate the position and the orientation of the gripper as well as passive joint variables of the external and the internal structure (figure. 1), when articular variables control are known. To determine this model, firstly, we must calculate the complete geometric model of the external structure represented by relationship (1), then we calculate the vector O'E which is used in the complete geometric inverse model of the internal structure represented by relationship (2). The inverse operation gives the complete **G**eometrical **I**nverse **M**odel. The exploitation of these two models and the study of the joint limit of different variables have allowed us to effectively examine the position and orientation workspace (with or without constraints on the passive joint variables) of the 2-Delta robot.

Complete **D.G.M**. of the external structure : Complete **G.I.M**. of the internal structure :

$$\begin{cases}(X_{OV}, \gamma_i) = f(\theta_1^i, \phi_i) \\ (\theta_3^i, \theta_2^i) = g(X_{OV}(\theta_1^i, \phi_i), \gamma_i(\theta_1^i, \phi_i))\end{cases} \quad (1) \qquad \begin{cases}(\psi_3^i, \psi_2^i) = g(X_{O'E}) \\ (\psi_1^i) = f(\psi_2^i(X_{O'E}), \psi_3^i(X_{O'E}))\end{cases} \quad (2)$$

Where :

X_{OV} : vector whose components represent coordinates of the point V of the gripper referred to the platform (O) frame.

O'E: vector whose components represent coordinates of the point E of the platform (3) referred to the platform (2) frame.

θ_1^i : active joint variables of the platform (1).

θ_1^i, θ_3^i : passive joint variables of the external structure.

$\phi_i(\gamma_i)$: orientation joint variables of the of the platform (2)

ψ_2^i, ψ_3^i : passive joint variables of the internal structure.

Figure 2 shows simulation obtained on a SMAR simulator [10].This simulation allows us to validate the models (complete **D.G.M** and complete **G.I.M**.)

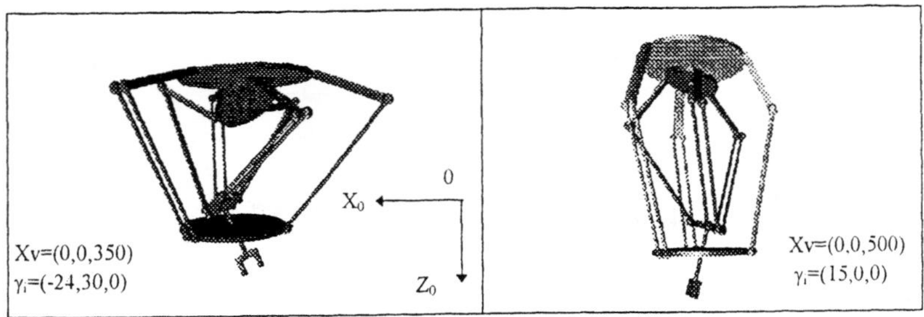

Figure.2 Morphology of the 2-Delta robot in different postures

GENERATION OF THE POSITION AND ORIENTATION WORKSPACES

To determine the position or the orientation workspace of the 2-Delta robot, we will use the technique of discretization which consists, in this case, of the definition of the position workspace, of fixing the orientation of the end effector and finally seeking all positions that can be reached by the end effector, regardless of the cross section chosen. In the case of the generation of workspace orientation, we will consider the end effector position to be fixed. Then, we will seek all possible orientations. We have separated this space into two distinct subspaces :

workspace without joint limits or theoretic space.

workspace with joint limits or space with constraints.

Relationship between space of theoretic position and space with constraints

The introduction of limits on the passive articular variables reduces the workspace of the 2-Delta structure in comparison to the theoretic space. We have attempted to investigate the proportion that represents space with constraints in comparison to theoretic space. Figure 3 illustrates these two spaces.

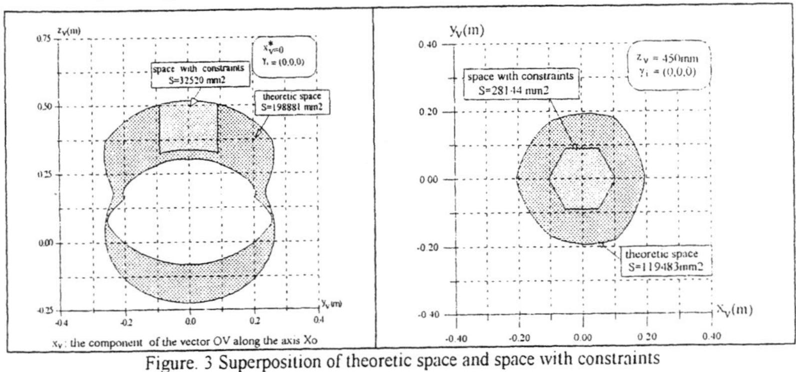

Figure. 3 Superposition of theoretic space and space with constraints

Cross section of theoretic orientation space :

The theoretic orientation space (figure. 4) defines all possible orientation of the end effector without considering any limitations on passive joints. In order to obtain a represention example in the plain, we will fix one of the three orientation angles of the gripper (γ_3 = 0), then will seek all possible orientations for a given position of the gripper.

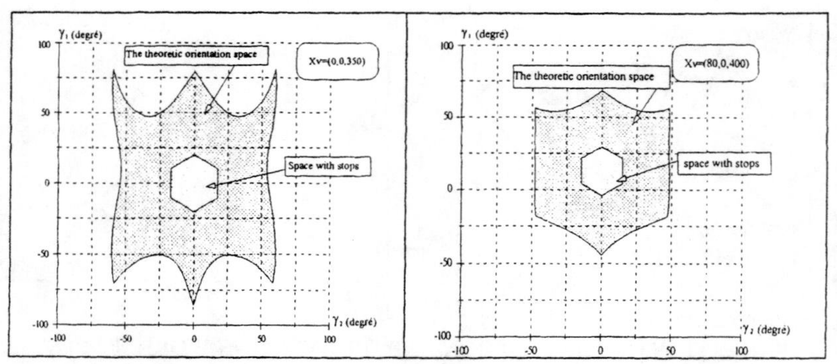

Figure. 4 Superposition of the orientation and with stops in theoretic space

MODELIZATION OF THE CLEARANCES IN POSITION AND IN ORIENTATION

In this paragraph, we consider the influence of the small variations of the rod lengths of the 2-Delta around their face values on translation and rotation vectors of the end effector. We will consider that these variations express the existence of localized clearance in the passive joints of the rods. We will define afterwards the ellipsoids of clearance based on the concept of the manipulability ellipsoids [10].

Problem statement and simplifying hypotheses

In order to reduce the problem complexity, it seems useful to propose some simplifying hypotheses concerning the geometry of the robot, as well as its physical parameters. We suppose that :

- each parallelogram is made of one rod.
- all simulations are carried out taking into consideration that active joints are without clearances in each chosen configuration, and that the variation of the position and the orientation of the effector are due solely to variations of rod lengths which act as linear actuators.

We will separate the modelization of clearance into two parts :

- position clearance model which gives position variation of the effector solely for clearance at the passive joint of each rod of the external structure.
- orientation clearance model which gives orientation variation of the effector solely for clearance at the passive joint of the internal structure.

Position clearance model

We define the position clearance model by :

$$\delta X_V = J \delta L_2$$

Where :

$\delta Xv = (\delta x_V, \delta y_V, \delta z_V)$: represents variations of the position of the effector.

$\delta L_2 = (\delta L_2^1, \delta L_2^2, \delta L_2^3)$: represents elementary length variations of the rods found in the external structure (the index j represents the chain number).

J: is a jacobian matrix of the partial derivatives of X_V with respect to L_2.

J is given by : $J_{ij} = \dfrac{\partial f_i (L_2)}{\partial L_2^j}$ $i = 1, \ldots 3$ and $J = 1, \ldots 3$

Most of these expressions are complex. Thus we must use the symbolic calculation software (Maplle V) to obtain these expressions

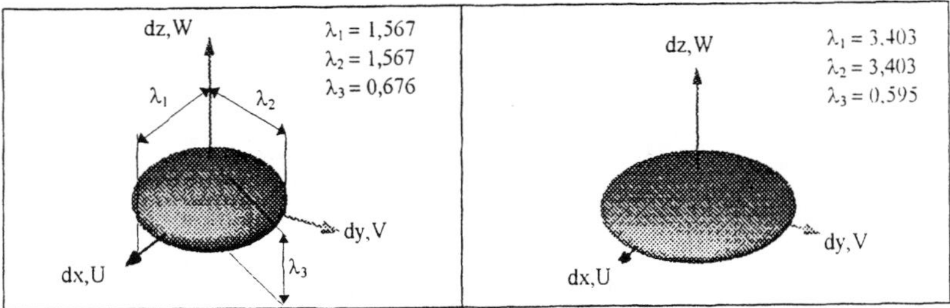

Figure.7 Dimensions and morphology of the ellipsoides of position clearance on the axis z_0

Figure 7 represents the ellipsoid of clearance defined by $\delta X_V^T (JJ^T)^{-1} \delta X \leq 1$ When the effector moves along the axis z_0, we notice that this axis constitutes an axis of revolution for these ellipsoids (the first two eigen values of each matrix $(JJ^T)^{-1}$ always remain equal ($\lambda_1 = \lambda_2$) whatever the position. This leads to an equitable distribution of the clearance in the plain $[x_0, y_0]$. Similarly, we observe that the volume of the ellipsoids of clearance increases considerably as the position of the effector moves along the axis z_0. This is due to the vicinity of the singular configurations. On the other hand, for postures of the wrist following the axis x_0 or the axis y_0 (that is not shown here), the ellipsoids change in dimension and in form in relation to the robot configuration. In fact, the clearance in these postures does not have the same effect in all directions.

The exploitation and the application of these results are very large. They allow, for example, a designer to know in advance the value of error in a position of the end effector according to the structure configuration. They also allow us to select the best configurations to execute a task inside the workspace.

Orientation clearance model

In this paragraph, we will deal with the clearance at the passive joint of the internal structure on the gripper orientation. To obtain this model, we will define the direct geometric model which gives coordinates of the gripper in realtion with the parameters of the internal structure. Therefore we will show it through the relationship :

$$\text{D.G.M.} \Rightarrow G(X_v, \phi_i, \psi_1^i, \gamma_i, L_3, L_4) = 0$$

Which allows us to write $\delta\gamma_i = JL_i$ where J is the jacobien matrix (2*3).

We have studied ellipses of clearance in different positions. We present here two ellipses obtained for two extreme positions on the axis z_0 (figure. 8)

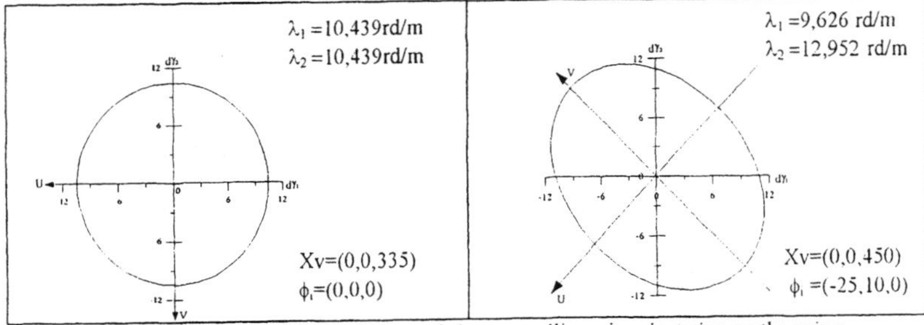

Figure.8 Dimensions and morphology of clearance ellipses in orientation on the axis z_0

We notice in general that eigenvalues which represent dimensions of clearance ellipse are roughly equal whatever the configuration. This shows that the uncertainty in orientation is almost the same in all postures (except postures close to single configuration). If we consider the obtained ellipses along the axis z_0 for orientation to be zero, we will observe that the surface of orientation uncertainty is represented by a circle since the two eigenvalues are equal, which shows that no direction is penalized. On the other hand, for configurations away from the axis z_0, we notice a dissymmetry represented by a change of ellipse morphology, which explains the uncertainty difference due to the clearance that is not the same in all directions.

This study shows that the existence of passive joints, which in general characterize parallel robots, can be a source of error and perturbation and susceptible to producing position and orientation imprecisions of the effector. In the case of the 2-Delta robot, we have shown, for distanced postures of singular configurations, that the clearance has more influence on the position of the effector than on its orientation. In this section we presents a methodology that allows one to identify and select configurations so that the effect of the clearance is minimized. Similarly the introduction of the ellipsoids of clearance (or ellipses of clearance) allows for a good understanding of the general volume of the uncertainty in fuction of the configuration.

CONCLUSION

The main objective of this study consisted in examining a new paralell robot family, characterized by the uncoupled mechanics between the translation and rotation of the wrist, and proposing approaches to solve problems in design, modelization, joint limits and influence of the clearance.

REFERENCES

1. Willard, Pollard and Evanston *"Position-Controlling Apparatus"* Application April 22, 1938. Serial N° 203,634 Renewed June 14, 1940.
2. D. Stewart *"A Platform with six degrees of freedom"* IME, Proc., Vol. 80, Part I, 15, p. 371 386, 1965.
3. C. Reboulet *"Modélisation des robots parallèles",* Techniques de la Robotique, Tome I, p. 257-284, Hermès, Paris, 1988.
4. J.P. Merlet "Commande par retour d'effort", Thèse Université de Paris VI, 1986. 4.
5. R. Clavel *"Une nouvelle structure de manipulateur parallèle pour la robotique légère"* , RAIRO APII, Vol. 23, n° 6, p. 501-519, 1989.
6. J.M. Hervé, F. Sparacino, *"Structural synthesis of parallel robot generating spatial translation".* ICAR 91, Pise, juin 1991.
7. C. Reboulet, C. Lambert, N. Nombrail, N. Delpech. *"Etude et mise en oeuvre du robot SPEED-R-MAN".* Rapport intermédiaire MRT/DERA, décembre 1991.
8. F. Pierrot, A. Fournier, P. Dauchez, *"Towards a fully parallel 6 d.o.f. robot for high speed, application",* Proc. of the 1991 IEEE ICRA, Saramento, californie, April 1991.
9. A. Goudali "Contribution à l'Etude d'un Nouveau Robot Parallèle 2-Delta à Six Degrès de Liberté avec découplage". Thèse de l'Université de Poitiers 1995.
10. S. Zeghloul, *"Développement d'un système de C.A.O.-Robotique intégrant la planification de tâches et la syntèse de sites robotisés".* Thèse de Doctorat d'Etat, Poitiers 1991.
11. T. Yoshikawa, *"Manipulability of Robotic Mecanisms".* Proceeding of the 2nd Int. Symposium of Research, 439-446 (1984).

DEVELOPMENT AND IMPLEMENTATION OF REAL-TIME CONTROL MODULES FOR REDUNDANT MANIPULATORS

K. Glass **R. Colbaugh**

New Mexico State University, Las Cruces, NM 88003, USA
kglass@nmsu.edu, colbaugh@nmsu.edu

ABSTRACT

This paper considers the problem of controlling the motion of redundant manipulators to achieve a desired set of (possibly simultaneous) tasks. The paper develops two real-time control modules for implementation with the Sequential Modular Architecture for Robotics and Teleoperation (SMART) which has been developed at Sandia National Laboratories. SMART provides a "generic" framework within which individual subsytems can be combined, simulated, and implemented for real-time control of complex robotic systems. Control modules developed in the paper include an adaptive gravity compensation controller and a real-time sensor-based collision avoidance strategy. The adaptive gravity compensation controller is a position regulation strategy that requires no information regarding the manipulator dynamic model or payload and can be proven to be passive. The whole-arm collision avoidance algorithm is implemented within a damped-least-squares inverse kinematic formulation. The example modules are demonstrated through computer simulation on a 15 degree-of-freedom manipulator.

1. INTRODUCTION

The desire for increased performance, versatility, autonomy, and reliability has motivated the study of *kinematically redundant manipulators*, which possess more kinematic degrees-of-freedom (DOF) than are (nominally) necessary to complete a given task. For example, the U.S. Department of Energy (DOE) has proposed utilizing kinematically redundant manipulators in many of its environmental restoration and waste management operations, such as waste storage tank remediation and waste handling and inspection within repositories. NASA is evaluating the use of highly redundant systems for deploying sensors in remote inspection tasks. In these space and waste management applications robots will be required to perform complex tasks in harsh, remote, and unstructured environments. The interest in redundant robots is motivated by the expectation that the extra DOF can provide the additional dexterity and maneuverability necessary to operate in unstructured settings. Note, however, that achieving this anticipated improvement in performance requires the development of control systems that can be adapted to different manipulators performing a variety of operations in complex and uncertain environments.

Completing tasks within these environments may require multiple robots, a variety of sensing devices, and input from different sources. The flexibility of the robot workcell will then be defined by how rapidly workcell devices can be reconfigured to complete a given task. For example, in a pipe cutting operation, two manipulators may cooperate to hold a common payload while another mechanism (perhaps teleoperated) performs the cutting. Once the cut is complete, the task of disposing of the pipe may require one of the original manipulators to interact with a mobile manipulator which will transport the waste to a disposal site. Many researchers have studied the control architecture problem for robots, and a number of different approaches have been introduced. High-level interfaces to existing controllers have been suggested primarily to provide for efficient task planning, programmability, and peripheral device interfacing [e.g. 1]. Generic robot controllers that can be used in place of industrial robot controllers have been proposed in [e.g., 2,3]. Although these architectures have been successfully implemented to achieve a variety of individual tasks, the flexibility of these architectures for use in different scenarios is often limited by how rapidly new systems can be synthesized. One approach to solving this "reconfigurable control" problem is to develop software modules that define particular robot behaviors/tasks and then facilitate the combining of these behaviors/tasks by using a generic real-time control architecture. One such architecture is SMART (Sequential Modular Architecture for Robotics and Teleoperation) which has been developed at Sandia National Laboratories [4]. SMART provides a framework within which individual subsytems can be combined, simulated, and implemented for real-time control of complex robotic systems.

In this paper, recent results are presented on the development of an intelligent control system for kinematically redundant manipulators. The proposed control scheme permits good performance to be achieved despite uncertainties in the task and environment. In particular, this paper presents two real-time control modules for implementation with SMART. These modules include an adaptive gravity compensation controller and a sensor-based real-time collision avoidance strategy. Simulation results for controlling the behavior of a 15 DOF manipulator performing a waste remediation task within the SMART environment are presented.

2. BACKGROUND

The underlying theory used to develop SMART relies on representing robotic systems using vector-network theory. To design new control loops, each hardware and software component that comprises the loop is represented by a *passive* subnetwork. Each module in SMART represents a distinct computable element that is updated independently of the other elements in the network. Before information is passed between modules a change of basis on the output of the module is performed. This change of basis and the modules' passivity facilitates the arbitrary combining of modules to ensure that the feedback interconnection of all of the modules remains passive. Another consequence of this design is that individual subnet modules can be distributed to different CPUs for execution, and the overall system is guaranteed to be stable. Maintaining stability results in safe, reliable, and predictable system operation. Additionally, the distribution of control software across multiple CPUs allows complex hardware systems and software algorithms to be configured, controlled, and tested with little concern for computational overhead.

To date, SMART has been used for adding behaviors and configuring workcells with mechanisms that have existing industrial position controllers. Typically, these controllers solve the position regulation or point-to-point motion control problem. The position regulation problem for manipulators involves designing a control strategy to cause the manipulator to evolve from its initial state to some desired final state, where typically the final velocity is to be zero. Note that the capability to accurately and robustly solve the position regulation problem is fundamental for any robot control system. It is well-known that simple proportional-derivative feedback controllers are capable of globally asymptotically stable regulation of rigid robots, provided that the effects of gravity on the manipulator are compensated. Ordinarily the requisite gravity compensation is achieved by including a gravity model in the control scheme; however this requires precise *a priori* knowledge of both the structure and the parameter values for this model, including the effects of any payload. Particularly restrictive is the need for information concerning the payload, because in typical tasks many different payloads are encountered and it is unrealistic to assume that the properties of all payloads are accurately known. Additionally, this requirement limits the modularity and portability of the controllers, so that implementation of these schemes within SMART would mean that individual modules would have to be developed for each potential robotic system. In order for a generic position regulation controller to be implemented as a SMART module, a passive gravity compensation scheme (based on the controller proposed in [5]) has been developed that requires no manipulator gravity model or payload information.

Once the manipulator position controller is in place, the inverse kinematic problem for redundant manipulators can be formulated so that additional tasks can be performed that take advantage of the dexterity of the robot. One fundamental problem that can be solved using redundant manipulators is the whole-arm collision avoidance problem [6]. This problem can be solved within a Damped-Least-Squares (DLS) framework for any manipulator whose joint-space is of higher dimension than its task-space. Additionally, the DLS formulation of inverse kinematics can be proven passive and is thus readily implementable as a SMART module.

3. REAL-TIME CONTROL MODULES

3.1 Adaptive Gravity Compensation

Consider the manipulator dynamic model written in terms of generalized coordinates $\mathbf{x} \in \Re^n$:

$$\mathbf{F} = H(\mathbf{x})\ddot{\mathbf{x}} + V_{cc}(\mathbf{x}, \dot{\mathbf{x}})\dot{\mathbf{x}} + \mathbf{G}(\mathbf{x}) \tag{1}$$

where $\mathbf{F} \in \Re^n$ is the generalized force associated with $\mathbf{x}$, $H \in \Re^{n\times n}$ is the manipulator inertia matrix, $V_{cc} \in \Re^{n\times n}$ quantifies Coriolis and centripetal acceleration effects, and $\mathbf{G} \in \Re^n$ is the vector of gravity forces. It is well-known that the dynamics (1) possesses considerable structure. For example, for any set of generalized coordinates $\mathbf{x}$, the matrix H is bounded, symmetric and positive-definite, the matrices $\dot{H}$ and V_{cc} are bounded in $\mathbf{x}$, depend linearly on $\dot{\mathbf{x}}$, and are related according to $\dot{H} = V_{cc} + V_{cc}^T$. The vector of gravity forces $\mathbf{G}$ is the gradient of a potential energy function and is bounded with bounded first partial derivatives [7].

Consider now the following adaptive position regulation scheme

$$\mathbf{F} = \mathbf{f}(t) + k_1\gamma^2\dot{\mathbf{e}} + k_2\gamma^2\mathbf{e} \quad , \quad \dot{\mathbf{f}} = \beta(\dot{\mathbf{e}} + \frac{k_2}{k_1\gamma}\mathbf{e}/(1+\parallel \mathbf{e} \parallel)) \tag{2}$$

where $\mathbf{f}(t) \in \Re^n$ is the adaptive element and k_1, k_2, γ, β are positive scalar constants. The adaptive element $\mathbf{f}$ is intended to compensate for gravity torques $\mathbf{G}$, and β is the adaptation gain for $\mathbf{f}$. Application of this controller to the manipulator dynamics yields the following closed-loop dynamics:

$$H\ddot{\mathbf{e}} + V_{cc}\dot{\mathbf{e}} + k_1\gamma^2\dot{\mathbf{e}} + k_2\gamma^2\mathbf{e} + \mathbf{G}(\mathbf{x}_d) - \mathbf{G}(\mathbf{x}) = -\phi \tag{3a}$$

$$\dot{\phi} = \beta(\dot{\mathbf{e}} + \frac{k_2}{k_1\gamma}\mathbf{e}/(1+\parallel \mathbf{e} \parallel)) \tag{3b}$$

where $\phi = \mathbf{f} - \mathbf{G}(\mathbf{x}_d)$. The convergence properties of this controller are established in the following lemma.

Lemma: If γ is chosen large enough then the adaptive scheme (2) ensures that (1) evolves with all signals uniformly bounded and so that $\mathbf{e}, \dot{\mathbf{e}}$ converge asymptotically to zero.

Proof: Let $U(\mathbf{x})$ denote the gravitational potential energy of the manipulator and define the following Lyapunov function candidate:

$$V = \frac{1}{2}\dot{\mathbf{e}}^T H \dot{\mathbf{e}} + \frac{1}{2}k_2\gamma^2\mathbf{e}^T\mathbf{e} + \frac{k_2}{k_1\gamma}\mathbf{e}^T H\dot{\mathbf{e}}/(1+\parallel \mathbf{e} \parallel) + \frac{1}{2\beta}\phi^T\phi$$
$$+ U(\mathbf{x}) - U(\mathbf{x}_d) + \mathbf{G}^T(\mathbf{x}_d)\mathbf{e} \tag{4}$$

Differentiating (4) along (3) and simplifying yields

$$\dot{V} \leq -(k_1\gamma^2 - \frac{k_2}{k_1\gamma}\lambda_{max}(H)\frac{k_2 k_{cc}}{k_1\gamma}) \parallel \dot{\mathbf{e}} \parallel^2 -(\frac{k_2^2\gamma}{k_1} - \frac{k_2 M}{k_1\gamma})\frac{\parallel \mathbf{e} \parallel^2}{1+\parallel \mathbf{e} \parallel} + k_2\gamma\frac{\parallel \mathbf{e} \parallel \parallel \dot{\mathbf{e}} \parallel}{(1+\parallel \mathbf{e} \parallel)^{1/2}}$$

where M is a positive scalar constant satisfying $M \parallel \mathbf{x}_d - \mathbf{x} \parallel \geq \parallel \mathbf{G}(\mathbf{x}_d) - \mathbf{G}(\mathbf{x}) \parallel$ $\forall \mathbf{x}_d, \mathbf{x}$ (the boundedness of the partial derivatives of $\mathbf{G}$ ensures that such an M exists). Routine manipulation reveals that, if γ is chosen large enough, then $\dot{V} \leq 0$; the claims of the lemma then follow using standard arguments [5]. ■

We are now in a position to establish the passivity of the proposed controller as an easy consequence of the preceding stability analysis.

Theorem: The proposed adaptive regulation scheme (2) is passive.

Proof: Let $\mathbf{q} = \dot{\mathbf{e}} + \frac{k_2}{k_1\gamma}\mathbf{e}/(1+\parallel \mathbf{e} \parallel)$ and consider the manipulator error dynamics (3a),(3b). From elementary passivity theory it is known that if each subsystem in the system (3a),(3b) is passive then the feedback interconnection is also passive [8]. Consider first the adaptive block (3b), viewed as a map from $\mathbf{q} \to \phi$. Direct calculation reveals that

$$\frac{d}{dt} \parallel \phi \parallel^2 = 2\phi^T\dot{\phi} = 2\beta\phi^T\mathbf{q}$$

so that

$$\int_0^\infty \phi^T\mathbf{q} = \frac{1}{2\beta}\int_0^\infty \frac{d}{dt} \parallel \phi \parallel^2 dt \geq -\frac{1}{2\beta} \parallel \phi(0) \parallel^2 \tag{5}$$

and passivity can be concluded.

Next, consider the closed-loop dynamics block (3a). Define $V^* = V - \frac{1}{2\beta} \parallel \phi \parallel^2$ and note that, using the calculations in the Lemma, we can conclude

$$\dot{V}^* = \dot{V} - \frac{1}{\beta}\phi^T\dot{\phi} = \dot{V} - \phi^T\mathbf{q} \tag{6}$$

Now $V^* \geq 0$ and $\dot{V} \leq 0$, so that elementary passivity theory permits the conclusion that the map $-\phi \to \mathbf{q}$ is passive [8] (indeed, it is easy to show that this map is strictly passive). Thus the adaptive regulation scheme is a feedback interconnection of two passive blocks and is therefore passive. ■

3.2 DLS Collision Avoidance

Consider the problem of maneuvering a (kinematically redundant) manipulator in a congested workspace in such a way that the desired end-effector trajectory is tracked and, at the same time, collisions between the manipulator links and workspace obstacles are avoided. Given the adaptive controller developed in the previous section, which is capable of ensuring that the robot joint-space position $\theta \in \Re^n$ closely tracks any desired trajectory $\theta_d(t)$, this problem becomes one of *kinematic control* and involves specifying $\theta_d(t)$ so that the end-effector trajectory tracking/collision avoidance objectives are accomplished simultaneously.

Let $\mathbf{x} \in \Re^m$ denote the position/orientation of the manipulator end-effector, and recall that $m < n$ because the robot is kinematically redundant. We can augment the end-effector task with additional user-specified tasks $\mathbf{y}$ so that the forward kinematics and differential kinematics for the manipulator can be written as

$$\mathbf{z} = \begin{bmatrix} \mathbf{x} \\ \mathbf{y} \end{bmatrix} = \begin{bmatrix} \mathbf{f}(\theta) \\ \mathbf{g}(\theta) \end{bmatrix} = \mathbf{h}(\theta) \quad , \quad \dot{\mathbf{z}} = \begin{bmatrix} J_e(\theta) \\ J_c(\theta) \end{bmatrix} \dot{\theta} = J(\theta)\dot{\theta} \qquad (7a),(7b)$$

where $\mathbf{z} \in \Re^{(r+m)}$ is the configuration vector, $\mathbf{f} : \Re^n \rightarrow \Re^m$, $\mathbf{g} : \Re^n \rightarrow \Re^r$ with r the number of kinematic functions to be controlled as the additional task, $J_e \in \Re^{m\times n}$ is the end-effector Jacobian matrix, and $J_c \in \Re^{r\times n}$ is the Jacobian matrix associated with the additional task.

The desired behavior of the manipulator can now be specified by defining the desired configuration vector $\mathbf{z}_d(t) = [\mathbf{x}_d^T(t) \mid \mathbf{y}_d^T(t)]^T$. The DLS approach to computing the desired joint-space trajectory $\theta_d(t)$ corresponding to $\mathbf{z}_d$ is to calculate $\dot{\theta}_d$ from

$$\dot{\theta}_d = [J^T J + W_v]^{-1} J^T[\dot{\mathbf{z}}_d + K\mathbf{e}] \qquad (8)$$

and then integrate it to obtain θ_d. In (8), $\mathbf{e} = \mathbf{z}_d - \mathbf{z}$ is the tracking error, $K \in \Re^{(r+m)\times(r+m)}$ is a symmetric positive-definite position feedback gain that corrects for linearization error and other unmodeled effects, and the importance of low joint velocities is reflected in the choice of the symmetric positive-definite weighting matrix $W_v \in \Re^{n\times n}$. The DLS formulation of the control scheme gives optimal solutions to (7b) that are robust to singularities. This is achieved by optimally reducing the joint velocities which may induce minimal errors in the task performance by modifying the task trajectories.

In order to incorporate the DLS inverse kinematics scheme as a SMART module, the passivity of the map (8) needs to be assured. The jacobian matrix is a transformation that is power conserving, and as such no power can be gained or dissipated through this transformation. The DLS inverse kinematics algorithm proposed here is implemented as in [9]; this formulation maintains this power conserving property and is therefore passive.

In light of the fact that the DLS inverse kinematics routine is passive, providing SMART with a sensor-based collision avoidance behavior is accomplished through two steps. The first step utilizes an efficient recursive algorithm to formulate the requirement that no robot link should contact any workspace obstacle. This requirement can be expressed as a set of task-space inequality constraints of the form

$$\mathbf{g}(\theta) \geq 0 \qquad (9)$$

where $\mathbf{g} : \Re^n \rightarrow \Re^p$ concisely quantifies the collision avoidance requirement. One method of constructing these inequalities for a general manipulator moving in a workspace cluttered with convex obstacles is described in [6].

The second step in the collision avoidance scheme is to ensure satisfaction of (9) while tracking the desired end-effector trajectory. This objective is achieved by slightly modifying the DLS configuration control strategy to include inequality constraints. The inequality constraint (9) can be incorporated into the DLS control framework by introducing the constraint error e_{s_i} defined as follows: $e_{s_i} = 0$ if $g_i \geq 0$, and $e_{s_i} = -g_i$ if $g_i < 0$. Then the DLS formulation (8) can be written as

$$[J^T J + W_v + \sum_{i=1}^{p} w_i J_i^T J_i]\dot{\theta}_d = J^T(\dot{\mathbf{z}}_d + K\mathbf{e}) + \sum_{i=1}^{p} w_i J_i^T e_{s_i} \tag{10}$$

where $J_i = \partial g_i / \partial \theta \in \Re^{1 \times n}$ and the w_i are nonnegative scalar weights whose magnitudes indicate the relative importance of inequality $g_i \geq 0$. The determination of g_i, and thus e_{s_i}, is accomplished directly using the output of a sensor model.

3.3 Implementation Results

The robot position regulation controller (2) and the collision avoidance strategy (10) are now applied to a 15-DOF manipulator in a computer simulation study. The simulation environment consists of a Personal Iris Graphics Workstation, the *Hydra* graphics rendering package, and SMART. The objective of the simulation example is to illustrate the capability of SMART to be used for point-to-point motion control of a large manipulator running the sensor-based collision avoidance algorithm for the task of avoiding the four pipes depicted in Figure 1. The manipulator used in this simulation is a large 9 DOF (macro) positioning manipulator with a 6 DOF (micro) Schilling manipulator at the distal end. The DH parameters for the manipulator are given in Table 1; throughout this example, the unit of length is inches and the unit of angle is degrees. The manipulator is populated with sensors that are modelled to produce an output that approximates the results given in [10].

i	1	2	3	4	5	6	7	8	9	10	11	12	13	14	15
α_{i-1}	0.0	0.0	90.0	90.0	90.0	0.0	0.0	90.0	90.0	90.0	90.0	0.0	0.0	-90.0	-90.0
a_{i-1}	0.0	0.0	7.5	7.45	7.45	114.0	114.0	7.45	7.45	7.45	6.65	33.0	19.0	5.25	0.0
d_i	$d_1(t)$	25.0	162.0	0.0	0.0	0.0	0.0	113.0	0.0	18.0	0.0	0.0	0.0	0.0	11.0

i	1	2	3	4	5
θ_i	0.0	$\theta_2(t)$	$\theta_3(t)$ + 180.0	$\theta_4(t)$ + 180.0	$\theta_5(t)$ + 90.0
i	6	7	8	9	10
θ_i	$\theta_6(t)$	$\theta_7(t)$ + 90.0	$\theta_8(t)$ + 180.0	$\theta_9(t)$ + 180.0	$\theta_{10}(t)$ + 180.0
i	11	12	13	14	15
θ_i	$\theta_{11}(t)$ + 90.0	$\theta_{12}(t)$	$\theta_{13}(t)$	$\theta_{14}(t)$ - 90.0	$\theta_{15}(t)$

Table 1: DH parameters for manipulator

In this simulation, beginning from the initial end-effector position of $\mathbf{z}(0) = [60 \quad 252 \quad 80]^T$ the manipulator is commanded to visit each of the four setpoints $\mathbf{z}(10) = [-70 \quad 0 \quad 0]^T$, $\mathbf{z}(20) = [300 \quad 0 \quad 100]^T$, $\mathbf{z}(30) = [0 \quad -250 \quad 30]^T$, $\mathbf{z}(40) = [-300 \;\; 0 \;\; 80]^T$ (defined relative to the initial position), while simultaneously avoiding the four pipes. The position regulation scheme (2) is implemented as a tracking controller [5] to ensure that θ closely tracks θ_d, and the DLS control approach to obstacle avoidance (10) is implemented with e_{s_i} generated using the output of the sensor model. The adaptive gain $\mathbf{f}$ is set to zero initially, while the remaining controller parameters are set as follows: $k_1 = 10, k_2 = 20, \gamma = 5$, and $\beta = 100$. The weights for the DLS control strategy (10) are chosen as $K = 100$ and $W_v = 1$. Figure 2a shows the evolution of the trajectory in the x, y, and z directions, respectively, and Figure 2b shows (the negative of) e_i for the collision avoidance subtask. From these Figures, it is seen that the end-effector task and the collision avoidance requirement are satisfied simultaneously.

4. ACKNOWLEDGEMENTS

The research described in this paper was supported in part through contracts with Sandia National Laboratories (SNL) and the Department of Energy (WERC). The authors would also like to acknowledge the many helpful conversations with Bob Anderson of SNL on many aspects of this work.

5. REFERENCES

1. Miller, D.J. and R.C. Lennox, "An Object-Oriented Environment for Robot System Architectures", *Proc. IEEE International Conference on Robotics and Automation*, Sacramento, CA, April 1990
2. Sollach, E.M. and A.A. Goldenberg, "Real-Time Control of Robots: Strategies for Hardware and Software Development", *Robotics and Computer-Integrated Manufacturing*, Vol. 6, 1989
3. Stewart, D.B., D.E. Schmitz, and P.K. Khosla, "Implementing Real-Time Robotic Systems Using CHIMERA II", *Proc. IEEE International Conference on Robotics and Automation*, Sacramento, CA, April 1990
4. Anderson, R.J., "How to Build a Modular Robot Control System using Passivity and Scattering Theory", *Proc. IEEE International Conference on Decision and Control*, New Orleans, LA, December 1995
5. Colbaugh, R., K. Glass, and E. Barany, "Adaptive Regulation of Manipulators Using Only Position Measurements", to appear *International Journal of Robotics Research*, 1996
6. Glass, K., R. Colbaugh, D. Lim, and H. Seraji, "Real-Time Collision Avoidance For Redundant Manipulators", *IEEE Transactions on Robotics and Automation*, Vol. 11, No. 3, 1995
7. Spong, M. and M. Vidyasagar, *Robot Dynamics and Control*, Wiley, New York, 1989
8. Slotine, J-J. and W. Li, *Applied Nonlinear Control*, Prentice Hall, Englewood Cliffs, New Jersey, 1991
9. Anderson, R.J., *SMART Class Notes*, Intelligent Systems and Robotics Center, Sandia National Laboratories, Albuquerque, NM 1995
10. Novak, J.L. and J.T. Feddema, "A Capacitance-Based Proximity Sensor for Whole Arm Obstacle Avoidance", *Proc. IEEE Conference on Robotics and Automation*, Nice, France, May 1992

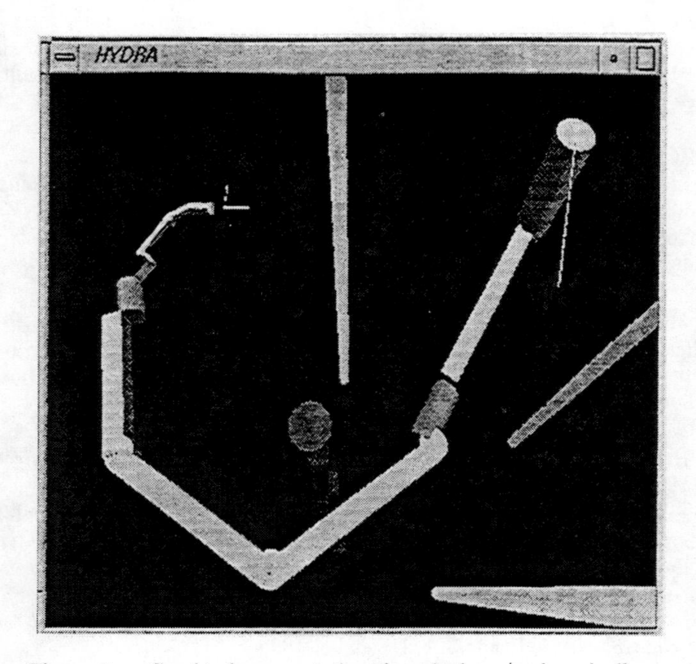

Figure 1: Graphical representation of manipulator/tank workcell

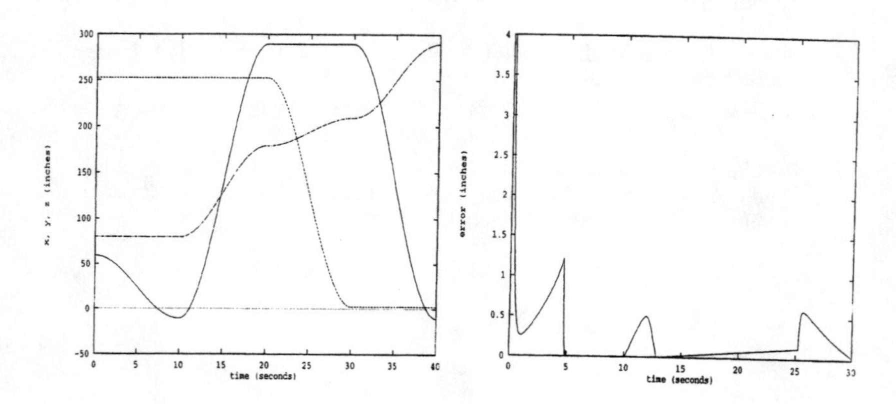

Figure 2a: Response of x, y, and z coordinates in SMART simulation

Figure 2b: Response of collision avoidance error coordinate (e_i) in SMART simulation

ONBOARD ELECTRONIC SYSTEMS FOR NUCLEAR ENVIRONMENTS

A. GIRAUD
LETI (CEA - Advanced Technologies)
CEA/Saclay F91191 Gif-sur-Yvette Cedex
tel : (33-1)69086430 fax : (33-1) 69086859
e_mail : agiraud@heron.saclay.cea.fr

ABSTRACT

To answer needs of nuclear civil industries, embedded rad-tolerant electronic systems are often used to put in place of complex wiring while offering added safety on data processing and transferring.
Studies have been done in CEA/LETI to reach this goal for a total dose level of up to 1 kGy even 3 kGy. Soon, use of only rad-hardened components was not considered for price and supplying reasons.
An original methodology based on the knowledge of behavior of industrial components under γ radiation have been defined to select them.
After, some design rules have been established to take into account variations of the most significant parameters of the components under radiation.
In order to validate these principles, a new concept of rad and partially fault tolerant embedded system, MICADO, has been realized and certified for at least 3 kGy.
A small industrial production fit out remote handling engines of French nuclear operators.
Increase of total dose level to 10 kGy or more is going to be effective with always use of standard components.

KEYWORDS: hardening, fault-tolerant, robotics, embedded system, remote handling, recovery, CMOS components

INTRODUCTION

Nuclear industries for civil fields like reprocessing and decommissioning, but also operations on hazardous areas are mainly at the beginning of research on semi-autonomous nuclear robotics.

As the result of such developments, embedded electronic systems could not be ignored.

It seems necessary to define as preamble to this communication what exactly is an embedded electronic system and the main constraints in which it is put through.

Embedded (onboard) electronic system

Depending of the required autonomy, two main categories can be envisaged.

The first includes small systems mainly design to reduce wiring and own enough autonomy to control data processing and transferring. Graceful degradation should be necessary to assume at least a coming back to repair and to avoid dead loss of robots in hazardous areas. Typical applications are for remote handling robots.

The second requests higher autonomy level with more computing power and is mainly useful for wheeled or legged robots.

But the boundary between them is not so obvious. When it is possible, embedded software in the first category could assume some servoing actions to reduce bauds rates with control rooms (for instance, to drive an arm when it is associated with a traveling crane). In any case, up to now and probably also for many years, most of the projects involve embedded systems of the first class.

General concept: an embedded system can be seen as a set of functional blocks like

•Power supply to convert and stabilize voltage from electrical net or batteries. Some local safety controls are included.

•Power electronics to drive actuators. Vulcain, an hardened amplifier for DC motors will be presented during this conference by M. Marceau (CEA/LETI)

•Digital and anolog data input/output, data encoders

•Communication with transducers for material or immaterial links

•Central unit with processors and associated memories, timers, ...

•External memories

•Digital and anolog data input and output

Main constraints: the well-known constraints available for embedded systems like low consumption, low thermal dissipation and reduced dimensions exist with more criticity for nuclear systems.

Influence of radiation, mostly total dose effect up to 10 kGy (dose rate up to 50 Gy is neglected), impose new criteria in the choice of basic components and in the way to design systems. Timelessness, availability and dependability but also waterproof decontaminable crates to confine electronics systems are now requested to assume average life-time of most of the applications with minimum time to repair.

METHODOLOGY FOR THE CHOICE OF COMPONENTS

While CMOS technology is not the best choice in comparison with bipolar for rad tolerance point of view, it permits to fit with most of above conditions.

Some efforts had been done by space and military projects by hardening this technology [1]. Furthermore, the newest CMOS technologies, by use of thicker lithography, increase their intrinsic total dose tolerance.

What is important to know is that for a level of up to 10 kGy, standard CMOS technology fulfill for a large part of I/O modules [2].

However, for high integrated circuits, like processors, these results are not fully true. That is the same for components of mixed technology.

Because radiation tests have shown that it was often possible to find a native rad tolerant component, a original methodology for the choice of the components has been developed to reduce prices of systems and limit use of rad-hardened lines

Principles of the methodology

To make profitable selection of components, it is necessary to gather information on the varied components able to assume specified functions. To do that, the proposed steps are explained below:

•Provisioning of the largest number of existing market references with common electrical and mechanical compatibility to permit easy change. Data bases are a significant help, but also, sometimes, radiation tests results given by the manufactors themselves.

•Purchasing of at least 5 of each one with a preference when it is possible for CMOS technology

•Designing or using industrial test-beds to measure functionality for each component

•Testing under radiation two non supplied components of each reference. Components are kept out at regular total dose levels to verify them on the test-beds. Scheduling of the operations is important to give coherent and comparable results. Dose rate is around 60 Gy/h with a ^{60}Co source.

•Rejecting components unable to support up to 3 times the total dose expected for them on real conditions.

•Designing specific disposal to perform tests in normal use conditions

•Testing two components of each remaining reference. Dose rate is around 10 Gy/h with a ^{60}Co source. Irradiation is stopped when functionality is lost.

•Measuring roughly the time necessary to retrieve functionality at room conditions. This time gives an order of idea on the possibility to recover partial or whole functionality

When these tests are finished, it is possible to select components able to be used in the design of a prototype of the embedded system [3].

Provisioning of components

In any way, prototyping is the first step to validate concepts, and, if necessary or requested, to test under radiation the whole embedded system.

To industrialize the prototype, some important data have to be known: life-time and total dose for final applications, number of systems, maintenance and repairing policies, timelessness of electronic and particularly integrated components, ..

The next two main situations could be seen:

•Use of rad-hardened component. Total dose resistance is warranted by manufacturors. End of production is generally scheduled because these components are mainly used in spatial and military projects. As often there is no other manufacturors, stock is necessary to prevent a future lack.

•Use of industrial components. No lifetime or manufacturing design is warranted. Stock is strongly recommended for components with narrow safety margin on total dose tolerance or when at least two manufacturers have not been found.

In any case, scheduled tests permit to follow the evolution of the total dose effect on the newest components.

For its own needs, DEIN/LETI have a stock of selected industrial components. Each batch is accepted or rejected according to some laws not mentioned here.

METHODOLOGY FOR THE DESIGN

The design of an hardened embedded system with industrial components is based on the association of components which had been evaluated for tolerance to the total dose effect.

To increase lifetime under radiation, some design rules have to be added to minimize some damages and to assist recovery phenomenon.

Drifts adjustments

The important number of parameters able to drift under radiation oblige to limit adjustments to the most significant of them

First one is the slope down of the current gain of bipolar transistors. To compensate, it is necessary to select, when it is possible the transistors with the highest gain, or to introduce at the design time some additional amplifier stages.

Second one is the slope down of the threshold voltage for MOS transistors. To compensate, it is necessary to increase control voltage band and, when it is possible, use transistors in switching mode.

Damage reducing

Possibilities to reduce damages under radiation can be exploited by considering some particular states of the components.

So, saturation state, during time where transistors are not working, reduce gain slope of bipolar transistors. For MOS transistors during the same time, non supplied state is preferable. More, alternative running (supplied and non supplied switching) largely increases total dose tolerance

Recovery phenomenon

One of the well-known recovery methods is heating. It can be used for components after irradiation with a temperature above 150°C.

Recovery has been also observed if damage reducing conditions are respected. In many cases, recovery is faster when components stay under radiation. Nevertheless, recovery does not restitute native total dose resistance.

Next step would be to mix these methods.

Redundancy

Studies have been shown that even in a same batch, characteristics of the components are very different.

So, when total dose margin is too narrow, probability of failures is so high that it is necessary to specify some spares modules for the most critical of them.

An important result is fault-tolerance for modules which have spares. Graceful degradation could limit or reduce repairing time and increase availability of the embedded system [4].

Shielding

When weight and dimensions are not a real problem, use of shields reduce effects of radiation and then allow increase total dose tolerance of the embedded system.

MICADO, AN EMBEDDED SYSTEM FOR NUCLEAR APPLICATIONS

The design of the system shown in Figure 1 takes into account the previous principles. After achieving of prototypes, an industrial set, shown in Figure 2, have been produced for French nuclear operators. Final tests and validation on traveling crane inside irradiated cells are taking place from now until end of June.

To optimize volume, standardization and easy change, choice had been made to use 3U size and 96 way connector for the back plane bus. CPU drives modules with a common soft control bus. Exchange rate is limited, but secured because no failures can be exported from the CPU. For all others exchanges, back plane bus is used to suppress all direct links

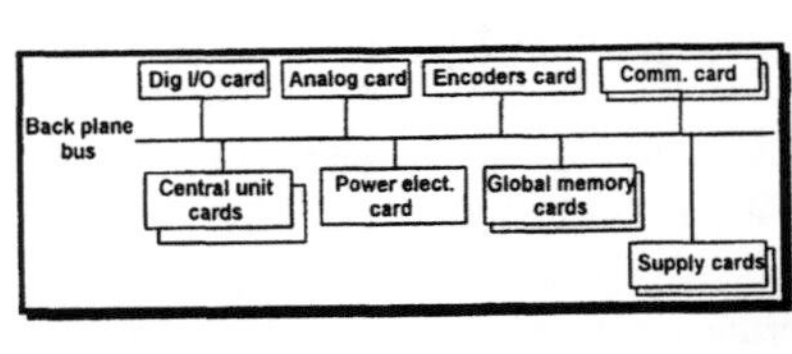

Figure 1 : architecture of Micado

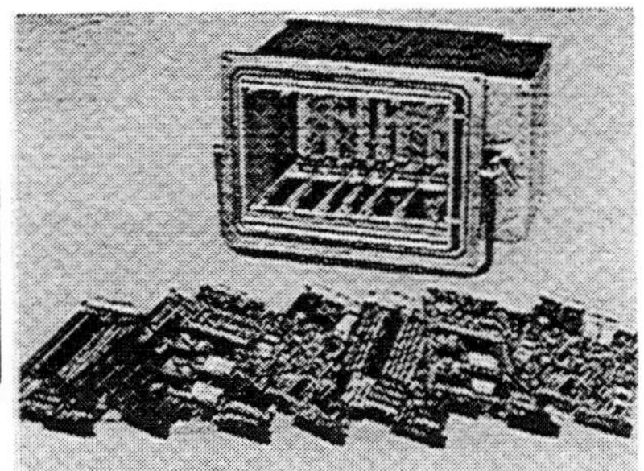

Figure 2 : Micado, industrial product

All the modules except central unit use intrinsic hardened components associated with drift compensations. Total dose tests of each module had overrun the level of 3 kGy. Watchdogs take care of each module. When they are effective, they put the module in a safe position or swap on a spare if it had been designed for.

The architecture of CPU, shown in Figure 3, use recovery phenomenon with alternative running on microcontrollers of family 68HC05. A factor 20 in total dose tolerance had been stepped over [5].

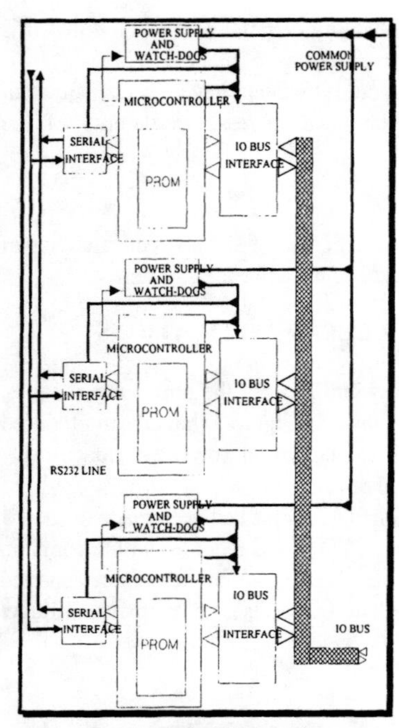

Figure 3 : CPU module

The central unit, expanding to 4 basic units, is a good answer to graceful matching with environment (radiation), computing power and redundancy.

In normal mode, only one of the microcontrollers is supplied and executes tasks. After a one minute run, it questions the bus until it finds a newly supplied microcontroller able to assume continuation of work. If there is one, context and activity are commuted. The power supply of the previous microcontroller is cut off for one minute.

If there is a failure on the active microcontroller, fault confinement is achieved through isolation from the bus and cut off of its power supply. The safety position of the I/O cards is turned on.

The application is recovered a few seconds later by doing a cold restart on one of the others microcontrollers [6].

To maintain a minimum performance level, when there is only one microcontroller a watch-dog sets a sleeping time after ten minutes work. Degraded operation is preferable to no operation at all.

Built to resist to total dose effects, this architecture is fault tolerant with dynamic reconfiguration and graceful degradation[4].

External memory is added for data diagnosis, rollback points and journaling.

Tests under radiation had been done in 1994. Total dose were up to 3 kGy with a rate of 10 Gy/h].

To optimize recovery and availability for critical applications, global memory could be used to update activity table of the microcontrollers.

MICADO, PERSPECTIVES FOR THE NEAR FUTURE

As it had been shown, embedded electronic systems in nuclear applications can use a great number of standard components and tolerate a total dose up to 3 kGy. Moreover, these components are in the way to be generalized for space and military nuclear activities.

These newest technologies will be especially accepted if in one hand, reliability and availability increase trust of final users and in the other hand, an soft evolution of existing systems allow higher total dose level like 10 kGy or even 20 kGy and a little more computing power.

Developments are in progress on MICADO to answer to these new needs.

Upgrade to 10 kGy

For all modules except microcontrollers and some RAM, this level is not a problem.

Newest batches of microcontrollers often own an intrinsic tolerance to the total dose of up to 1.4 kGy in fully supplied state.

Two simple actions could take place:

•By a process of irradiation and annealing at room conditions, selection of the best of microcontrollers. In this case, around 20 % of total dose tolerance is lost.

•Use of alternative running to extend total dose tolerance over 10 kGy and use of a rad-hardened memory for common memory

Upgrade to 20 kGy and above

The actual design of MICADO does not support such a level of total dose.

Use of alternative running is extended to all modules even I/O modules. To keep in place modularity and limited size while doubling number of components, use of CMS technology is required. Fault tolerance will also increase availability.

For the moment, use of actual microcontrollers is not enough. But, some processors own such an intrinsic total dose tolerance that, with alternative running, expected level is reachable.

The process of thermal recovery on parts of components during non supplied state could also increase radiation tolerance. The design of this process is not yet finished.

Rad-hardened technology

To go above, 100 kGy or 1 Mgy, standard technology even with new design will not reach these levels.

Actual rad-hardened technologies are often obsolete and in any case too expensive.

Today, a new one called DMILL technology designed by CEA/LETI and now industrialized could be expected. Some developments and tests are scheduled for this year.

REFERENCES

1., J.L Leray, O. Mousseau, ;"Les problèmes de durcissement de l'électronique" Revue scientifique et technique de la DAM n 10 (1994)(CEA French Atomic Agency).

2. F. Joffre, J. Buisson, "Instrumentation pour l'étude des effets de la dose cumulée" RADECS Proceeding 1991 pp 569-573

3. F. Joffre "Méthodologie de conception de systèmes industriels durcis" L'Onde électrique nov-déc 1993 vol 73

4. D. P. SIEWIOREK "Reliability and availability techniques" Digital Press Bedford MA (1982) pp 63-169

5. F. Joffre "Procédé pour la prolongation de la durée de fonctionnement ..." Patent no 90-07287 of June 1990 in Europe nd USA.

6 F. Joffre "Calculateur embarqué à tolérance de pannes" Revue de la société des ingénierus de l'ESE no 145 (Avril/Mai 1992) p 38

GENETIC LEARNING FOR ADAPTATION IN AUTONOMOUS ROBOTS

John J. Grefenstette

Naval Research Laboratory, Washington, DC 20375-5337, U.S.A.
gref@aic.nrl.navy.mil

ABSTRACT

This paper deals with problems arising in robots that are expected to perform autonomously for extended periods. An important problem for such systems is how to adapt the robot's rules of behavior in response to changes in its own capabilities. If the robot finds that some sensors or some basic actions are no longer available, perhaps due to a problem with its hardware or due to some undetected environmental cause, then the robot must learn new rules for performing its mission that use whatever remaining capabilities are still available. This paper presents an approach called *Anytime Learning* that enables a robot to gracefully adapt to its current capabilities.

KEYWORDS: Genetic Algorithms, Evolutionary Computation, Robot Learning

INTRODUCTION

An important problem arising in robots that are expected to perform autonomously for extended periods is how to adapt the robot's rules of behavior in response to unexpected changes in its own systems. For example, suppose the robot periodically checks its sensors and its ability to perform basic actions. If it finds that some sensors or actions are no longer available, perhaps due to a problem with the robot's hardware or due to some undetected environmental cause, then it must learn new rules for performing its mission that use whatever remaining capabilities are still available.

We have developed an approach to this problem that we call *Anytime Learning* [2][7]. In this approach, the robot interacts both with the external environment and with an internal simulation. The robot's execution module controls the robot's interaction with the environment, and includes a monitor that dynamically modifies the robot's internal simulation model based on the monitor's observations of the actual robot and the sensed environment. The robot's learning module continuously tests new strategies for the robot against the simulation model, using a genetic algorithm [4] to evolve improved strategies, and updates the knowledge base used by the execution module with the best available results. Whenever the simulation model is modified due to some observed change in the robot or the environment, the genetic algorithm is restarted on the modified model. The learning system operates indefinitely, and the execution system uses the results of learning as they become available.

This paper presents a case study of this approach on a surveillance task in which a robot is assigned to stay within observation distance of another mobile agent. The robot uses Anytime Learning to adapt to changes in its own basic action set over time. A simulation study shows that the approach yields effective adaptation to a variety of partial system failures. Experiments are now under way with Nomadic 200 mobile robots.

THE ANYTIME LEARNING MODEL

The Anytime Learning model addresses the problem of adapting a robot's behavior in response to changes in its operating environment and its capabilities. The outline of the approach is shown in Figure 1. There are two main modules in the Anytime Learning model. The *execution module* controls the robot's interaction with its environment. The *learning module* continuously tests new strategies for the robot against a simulation model of the environment. When the learning module discovers a new strategy that, based on simulation runs, appears to be likely to improve the robot's performance, it updates the rules used by the execution module. The execution module includes a *monitor* that measures aspects of the operational environment and the robot's own capabilities, and dynamically modifies the robot's internal simulation model based on these observations. When the monitor modifies the simulation because of an environmental change, it notifies the learning system to restart its learning process on the new simulation.

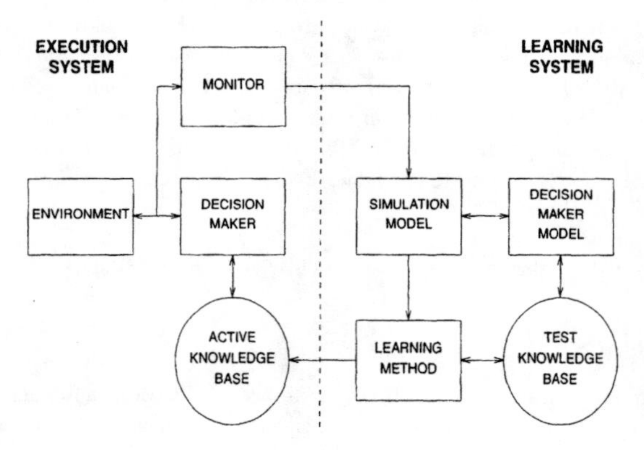

Figure 1: The Anytime Learning Model

This general architecture may be implemented using a wide variety of execution modules, learning methods, and monitors. The key characteristics of the approach are:

- Learning continues indefinitely. This is unlike most machine learning methods, which employ a training phase, followed by a performance phase in which learning is disabled.
- The learning system experiments on a simulation model. For most robotic applications, experimenting with the physical robot may be time-consuming or dangerous. Using a simulation models permits the safe use of learning methods that consider strategies that may occasionally fail.
- The simulation model is updated to reflect changes in the real robot or environment.

This final point reflects our assumption that the robot designer generally has at least partial knowledge of the robot and the environment. Knowledge that is relatively certain can be embodied in the fixed part of the simulation. Such knowledge might include certain fixed characteristics of the physical environment (e.g., gravity), as well as some aspects of the robot's design and performance. On the other hand, the robot designer should also identify those aspects of the environment and the robot's capabilities that are uncertain, and include these as changeable parts of the simulation module.

ADAPTATION CASE STUDY

As an example of the Anytime Learning model, we consider a surveillance task in which a robot is assigned to keep close to a specified tracking distance of another mobile agent. We focus on how well the tracking robot is able to perform its mission even when it suffers partial system failures.

The Performance Task: Tracking

In this study, an autonomous mobile robot is assigned to stay within observation distance of another mobile agent, as shown in Figure 2. The task is divided into *episodes* which begin with the target placed at a random distance and bearing from the tracker robot. The target performs a random walk combined with obstacle avoidance. The tracker's performance is measured by its mean squared error from the ideal tracking distance. It is considered a mission failure if the tracker collides with the target agent or with the walls surrounding the tracking area.

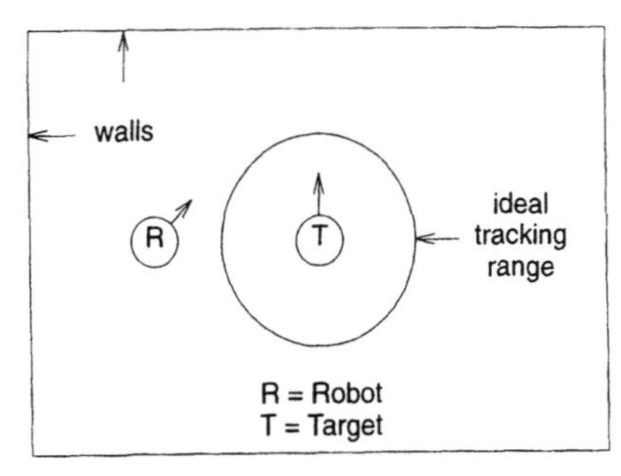

Figure 2: The Tracking Task

Execution Module

The execution module for the tracker robot includes a rule-based system that operates on reactive rules. A typical rule might be:

IF range = [35, 45] AND bearing = [340, 35] THEN SET turn = -24 (Strength 0.8)

For this task, five abstract sensors are defined for the robot: the range and bearing from it to the nearest object, the target's heading with respect to the target's bearing, and the range and bearing from the tracker to the target. These abstract sensors are derived from information available through the actual sensors on the Nomadic 200 mobile robots.

The abstract sensor values are all discretized. The bearing values are partitioned into intervals of five degrees. The range is partitioned into intervals of five inches from 0 to 150 inches. The heading is partitioned into 45 degree segments.

The learned actions are velocity mode commands for controlling the translational rate and the steering rate. The translation is given as -4 to 16 inches/sec in 4 inch/sec intervals. The steering command is given in intervals of 4 degrees/sec from -24 to 24 degrees/sec.

The rule *strength* is set by the learning system to estimate the quality of the rule. The execution module uses rule strengths to resolve conflicts among multiple rules that match the current sensors readings, but suggest different actions. In such cases, rules with higher strength are favored. See [1] for details.

The Monitor

In the current study, the monitor periodically measures a *capability list* for the robot at time t:

$$capability(t) = \langle\ maxright(t)\ ,\ maxleft(t)\ ,\ maxforward(t)\ ,\ maxback(t)\ \rangle$$

that indicates (1) the robot's maximum turning rate to the right; (2) the robot's maximum turning rate to the left; (3) the robot's maximum forward translation rate; and (4) the robot's maximum backward translation rate. We are not concerned at this point with the exact method used to compute the capability list; the important point is that the monitor is required only to identify symptoms of problems, not the causes.

When the monitor notices a change in the capability list, it modifies the simulation used by the learning system by setting the parameters of the simulation. In this case, the simulation has parameters that limit the robot's motion to the values in the current capability list.

Learning Module

The learning module uses SAMUEL[1], a learning program that uses genetic algorithms and other competition-based heuristics to improve its decision-making rules. Each individual in SAMUEL's genetic algorithm is an entire rule set, or strategy, for the robot. We have previously reported on using SAMUEL to learn simple robot behaviors such as navigation and collision avoidance [8][9]. In this study, a parallel version of SAMUEL used 12 workstations to perform parallel fitness evaluations [3].

When the monitor notices a change in the capability list, the learning module is restarted on the modified simulation model. At this time, the learning system re-initializes the population of strategies in the genetic algorithm by finding nearest neighbors from the *case base* consisting of previously learned strategies. Strategies in the case base are indexed by the capability list in place at the time the strategy was learned. Using previously learned strategies to initialize the population allows the system to very quickly adapt to situations that are similar to those seen before. See [7] for more details.

Experimental Design

The experimental design for this study was as follows: The tracking robot began with a set of default rules for tracking (with a success rate of about 30%). The learning system was initialized with an empty case base, and a simulation model that included the full capabilities of the robot. The experiment was run for a 12-hour period. At 15 minute intervals, the capabilities of the tracking robot were changed, simulating an intermittent partial system failure. At this point, the learning system initialized half of the population using the nearest neighbors in the case base, measured by similarity to the current capability list. The other half of the population was initialized with the initial default strategy.

During this experiment the robot encountered 12 distinct cases, occurring in random order over the 12 hour period. This allowed us to observe how quickly the system could adapt to both new and previously encountered situations.

Experimental Results

Table 1 shows some of the results of the experiment. The row labeled *N* indicates the number of times the case occurred during the experiment. The row labeled *Default* shows how well the robot performs on that case when using the default rule set. This shows the expected performance without any learning. The *Full* row shows how well the robot performs when using the strategy learned under "ideal" conditions (e.g., with full robot capability) in the default simulation model. The row labeled *First* column shows how well the tracking robot performed by the end of the first 15 minute period for that case. The *Final* row shows how well the robot performed on the indicated case by the end of the experiment, after seeing the case *N* times.

In all cases, the system was able to learn strategies that significantly outperformed the default strategy. The higher values in the *First* row than in *Full* row show the benefits of altering the simulation model on the basis of the observed capability lists. The higher values in *Final* row than in *First* row show the benefits of using previously learned cases to initialize the genetic algorithm, since the system was able to do better when it observed the same situation multiple times.

Table 1

Case	1	2	3	4	5	6	7	8	9	10	11	12
N	7	7	5	5	5	5	4	2	2	2	2	2
Default	30.5	27.2	29.1	27.4	30.5	30.6	32.2	29.5	29.8	29.8	30.0	29.1
Full	86.5	20.5	45.9	84.6	34.0	21.4	18.9	37.8	16.6	33.4	32.3	32.8
First	85.2	72.3	73.1	85.4	77.5	73.4	45.5	71.8	56.8	55.2	62.3	67.1
Final	86.5	78.8	73.2	85.8	82.3	77.9	59.0	75.5	57.3	62.3	62.3	68.0

Qualitative Observations

In general, the robot was able to adapt to changes in its capabilities. In some cases, this required the discovery of entirely different approaches to achieving the goals of the task. For example, when faced with the loss of its abilities to turn to the right, the robot discovered a set of rules that performed right turns by executing a tight, partial rotation to the left, exiting the spin at the correct moment to effectively produce a right turn. A similar but symmetric behavior was learned when faced with the loss of ability to turn to the left. When both right and left turns were partially restricted, the robot learned to stay a little further away from the target, allowing it to remain close to the nominal tracking distance using only slight turns. Overall, the robot was able to continue to perform its mission, at a somewhat reduced level of efficiency, even in the face of partial reductions in system capabilities.

RELATED WORK

Others have also explored combinations of case-based methods and genetic algorithms. Kelly and Davis [5] and Skalak [10] use a genetic algorithm to select instances or weight attributes in nearest-neighbor retrieval systems. Zhou [11] describes classifier systems for simulated robots that recall similar cases (stored as classifier rules) when faced with a new environment. If no relevant past cases exist, then the standard classifier system algorithm learns a new solution. In non-genetic approaches to robotic learning, Ram and Santamaria [6] use *continuous case-based reasoning* to perform tasks such as autonomous robotic navigation. They learn cases which provide information for the navigation system to deal with specific environments encountered.

CONCLUSION AND FUTURE WORK

The simulation results reported here illustrate the potential utility of the Anytime Learning approach to address the problem of adaptation in autonomous robots. Along with previous studies [2][7], these results show that genetic algorithms are particularly well-suited for Anytime

Learning, since previously learned strategies can be used to initialize the genetic population, leading to rapid adaptation to recurring situations. Experiments are currently under way that will measure the effectiveness of this approach on the Nomadic 200 robots. Future work will also focus on simultaneously adapting to both environmental changes and to changing robot capabilities. We expect this approach will eventually lead to robust and adaptive robots capable of long-range autonomous operation in dynamic real-world environments.

References

[1] Grefenstette, J. J., Ramsey, Connie L., and Schultz, Alan C., (1990). "Learning sequential decision rules using simulation models and competition," *Machine Learning*, **5(4)**, 355-381

[2] Grefenstette, J. J. and C. L. Ramsey (1992). "An approach to anytime learning," *Proceedings of the Ninth International Conference on Machine Learning,* 189-195, San Mateo, CA: Morgan Kaufmann.

[3] Grefenstette, J. J. (1995). "Robot learning with parallel genetic algorithms on networked computers," *Proceedings of the 1995 Summer Computer Simulation Conference (SCSC '95),* 352-357, Ottawa, Ontario, Canada.

[4] Holland, J. H. (1975). *Adaptation in Natural and Artificial Systems.* Univ. Michigan Press, Ann Arbor, 1975.

[5] Kelly, J. D. and L. Davis (1991). "A hybrid genetic algorithm for classification," *Proceedings of the 12th International Joint Conference on Artificial Intelligence* 645-650.

[6] Ram, A. and J. C. Santamaria (1993). "Continuous case-based reasoning," *Case-Based Reasoning: Papers from the 1993 Workshop,* Tech. Report WS-93-01, 86-93. AAAI Press, Washington, D.C.

[7] Ramsey, C. L. and J. J. Grefenstette (1993). "Case-based initialization of genetic algorithms," *Proceedings of the Fifth International Conference on Genetic Algorithms* 84-91.

[8] Schultz, Alan C. and Grefenstette, John J. (1992). "Using a genetic algorithm to learn behaviors for autonomous vehicles," *Proceedings of the of the AIAA Guidance, Navigation and Control Conference*, Hilton Head, SC, August 10-12, 1992.

[9] Schultz, Alan C. (1994). "Learning robot behaviors using genetic algorithms," *Intelligent Automation and Soft Computing: Trends in Research, Development, and Applications, v1*, Mohammad Jamshidi and Charles Nguyen, editors, Proceedings of the First World Automation Congress (WAC '94), 607-612, TSI Press: Albuquerque.

[10] Skalak, D. B. (1993). "Using a genetic algorithm to learn prototypes for case retrieval and classification," *Case-Based Reasoning: Papers from the 1993 Workshop,* Tech. Report WS-93-01 211-215. AAAI Press.

[11] Zhou, H. H. (1990). "CSM: A computational model of cumulative learning," *Machine Learning 5(4),* 383-406.

FAULT HANDLING IN FLEXIBLE MACHINING CELLS

PER GULLANDER
Department of Production Engineering,
Chalmers University of Technology, S-412 96 Göteborg, Sweden

ANDERS ADLEMO
Department of Computer Engineering,
Chalmers University of Technology, S-412 96 Göteborg, Sweden

SVEN-ARNE ANDRÉASSON
Department of Computing Science,
Chalmers University of Technology, S-412 96 Göteborg, Sweden

ABSTRACT

Designing and implementing control systems for manufacturing systems are complicated tasks, especially since the systems must meet high demands concerning cost, flexibility, quality, safety and availability. To simplify the development process, a reference architecture is preferably used as a basis. The aim of the research presented in this paper is to define an architecture that would yield control systems with high flexibility and availability. An architecture for cell control systems is first presented. The main objective of this architecture is to increase the flexibility with respect to new product types and new manufacturing equipment. Typical errors that occur in machining cells are then described on the basis of a hierarchical decomposition of the system into error-prone components. Finally, an extended version of the architecture is outlined, the aim of which is to yield control systems for machining cells that are not only highly flexible but highly available as well.

KEYWORDS: Flexible Manufacturing Systems, Cell Control System, Fault Handling, Error Recovery, Fault Tolerance, Availability.

INTRODUCTION

Manufacturing systems of the future must meet demands for high availability, low development and operation cost, high productivity, high product quality and high operational safety. A key attribute in manufacturing systems to meet these demands is *flexibility*. A system can be flexible towards changes that occur on different time scales. On a long time scale, flexibility permits incorporation of new equipment. On a shorter time scale, it allows for new products. On an even shorter time scale, flexibility allows for mixed production of several product types. Flexibility on a very short time scale yields a fault-tolerant system with improved availability. Different aspects of fault tolerance in manufacturing systems can be found in e.g. [1], [6], [12].

To support the design and implementation of control systems, a *reference architecture* should be used that describes how the control system should be designed and implemented. Such an architecture has been developed in a research project at Chalmers University of Technology[3]. The main objective of this architecture is to reduce costs and increase the control system's flexibility with respect to changes, such as new products to be manufactured, new machine tools or new processes. However, this architecture considers only fault-free production and does not explicitly include functions for error handling, which are very important for obtaining systems with high availability. This paper presents preliminary research results to extend this architecture to encompass *error handling capabilities* as well.

The next section explains various concepts regarding system dependability. The two sections that follow outline the reference architecture and describe typical errors in machining cells. Finally, error handling within the framework of the architecture, as well as necessary extensions to the architecture, are discussed and conclusions are given.

DEPENDABILITY CONCEPTS

Dependability, which is a measure of the trustworthiness of the delivery of services from a system, can be divided into three different aspects known as impairments, means and attributes [14]. *Impairments* are undesired, but not unexpected, circumstances. *Means* are the methods, tools and solutions that enable (i) the ability to deliver a service on which reliance can be placed and (ii) the possibility to reach confidence in this ability. *Attributes,* e.g. reliability and availability, enable (i) the quantification of service quality resulting from the impairments and the means and (ii) the properties that are expected that the system will express.

The impairments are further divided into faults, errors and failures. A *fault* is the original cause of any impairment, such as a defective component or a programmer's mistake. A fault causes a latent error. An *error* is that part of the system which is liable to lead to a failure, i.e. an impairment that has taken effect, such as using a defective component. A *failure* occurs when the delivered service deviates from the specified service, i.e. an error produces erroneous data that affects the delivered service. A failure on one level (level *n*) in a system can be viewed as a fault on the next higher level (level *n-1*). This implies that there is a chain, as depicted in Figure 1.

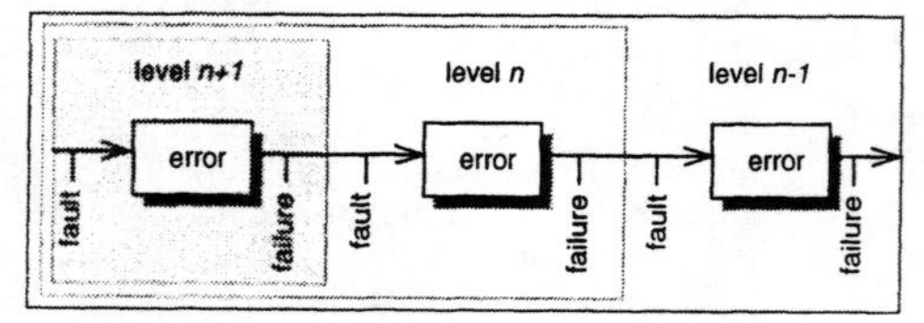

Figure 1. The relationship between faults, errors, and failures.

Designing reliable systems involves the selection of a failure response that combines several reliability techniques. First, the fault must be detected and diagnosed:

- *Fault detection* can be divided into two major classes. *On-line detection* provides real-time detection capabilities, while *off-line detection* means that the device cannot be used during the test.
- *Diagnosis* can be required if the fault detection technique does not provide information about the failure location and/or its properties.

The next stage is to respond to the error, either by masking it or by using some recovery procedure. *Fault masking* is a technique for hiding the effects of a failure. In *hierarchical fault masking,* higher levels mask the further propagation of a failure from lower levels. In *group fault masking,* a group of devices together mask the further propagation of a failure. *Recovery* is done after the detection of a failure to eliminate the effects of the failure. There are several techniques for achieving this:

- *Retry,* i.e. a second attempt of an operation; may be successful particularly if the cause of a first-try failure is transient and causes no damage.
- *Restart* of a system can be necessary if too much damage has been caused by a failure or if a system is not designed to handle recovery.

- *Reconfiguration* can be used to replace a failed component with a spare or to isolate the failed component.
- *Repair* of a failed and replaced device can be either *on-line repair* or *off-line repair*. After the repair, *reintegration* of a failed device can be carried out.

Fault confinement is achieved by limiting the spread of fault effects to only one area of the system, thereby preventing the contamination of other areas.

AN ARCHITECTURE FOR FLEXIBLE CONTROL SYSTEMS

The architecture that is briefly presented in this section describes the design of a generic cell control system; a more detailed description can be found in [3]. In the work towards developing this architecture, several inter-related problem areas were identified: model of the manufactured products, model of the manufacturing resources, the control algorithm, the operator interface and system development methodology. The main features of the architecture are: (i) physical separation of generic functions from functions specific to the products and the resources currently in use; (ii) separation of product models from resource models; (iii) storage of product and resource models in a database; and (iv) a modular structure of the control system software.

Products (also called parts further on) and resources are described using *data models* that are implemented in a database accessible from all the control system modules. The resource data model describes capabilities and constraints of the manufacturing resources. The product data model formally describes (i) all the possible sequential combinations of operations to manufacture the product, (ii) information on how to handle and move the part, (iii) product type information and (iv) error handling procedures (exceptions) [5]. These exceptions are triggered by flags set by the controller module that performs the error diagnosis, or are manually set by an operator.

In the general case, the control system comprises several co-operating modules, each responsible for specific functions; see Figure 2. For each physical device in the manufacturing system, there exists a corresponding module in the cell controller called the *internal resource*, which executes resource-specific tasks, keeps track of the current state of the manufacturing resource and manages proprietary communication. Depending on the type of

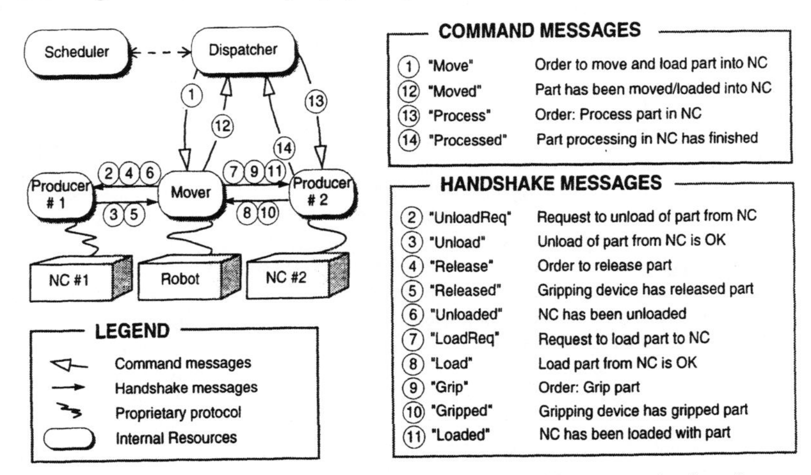

Figure 2. A generic message-passing structure [10]. The scenario includes unloading of a part from producer # 1, transportation to producer # 2, loading, and finally processing of the part.

equipment that the internal resources control, they are sorted into *producers*, *movers* and *locations*. Functions for scheduling and coordination, also including deadlock handling and optimization, are performed by the *scheduler* and *dispatcher* modules, respectively. By matching the product models with the resource models, the scheduler dedicates a resource to each operation in the operation list and develops a schedule (or supervisor) that specifies all possible ways in which the set of products can be manufactured in the cell [9]; see Figure 3.

The controller modules have a client/server relationship and synchronize their activities using *messages*. A generic model describing system behavior and the message-passing between the modules during normal, error-free operation is depicted in Figure 2 [10]. The dispatcher is provided with a homogeneous message-based interface to all physical devices.

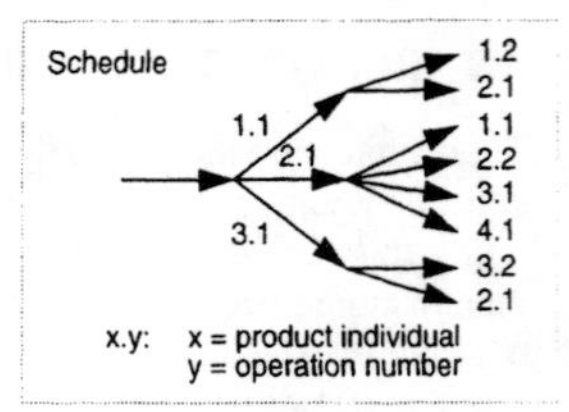

Figure 3. The schedule includes all possible ways to manufacture the products in a specific cell.

ERRORS IN MACHINING CELLS

To be able to refine the architecture presented to include error handling capabilities as well, errors that occur in manufacturing cells were studied and are described below. As suggested by [7], errors are context-dependent and should be classified in close relation to a generic classification of manufacturing operations. The error types presented in this paper are typically present in *machining cells,* i.e. those having no assembly operations. Nonetheless, many of the errors are likely to exist in other types of manufacturing systems as well.

Errors can be classified in many ways. One way is to distinguish between unrecoverable and recoverable errors. While recovery can be achieved from some errors either easily or after investing substantial amounts of money and time, no recovery at all can be made from other errors. In the "computing community", errors often affect only information entities, while errors in the "manufacturing community" often mean physical damage and recovery is therefore more difficult. Another classification scheme is based on the main cause of errors (i.e. the faults): (i) design errors, (ii) component failures, (iii) human errors and (iv) external factors [11]. It is interesting to note that, in a study of 45 FMS installations, 94% of the disturbances were classified as being either design errors or component failures [11].

However, in many situations, it is not possible to identify the exact cause of an error, and it might sometimes not even be necessary to identify the exact cause of the error in order to recover from it [8]. Hence, in this paper, a classification scheme was used that is more relevant when designing control systems. This classification does not trace the causes of an error back to the original error, e.g. in the design of fixtures or control programs. Instead, the errors are classified on the basis of the system in which they appear. Errors and their originating faults can be located in any of the *subsystems* of a cell: (i) the resources, i.e. producers (machine tools and measuring stations), movers (robots) and locations (buffer storages); (ii) operators; (iii) products/pallets; (iv) control system; (v) other subsystems, e.g. communication network or safety system. Each such subsystem is in itself composed of several components that might cause failures, e.g. part feeders, clamping devices, sensors and doors. According to case study results ([11], [13]), the most common errors were caused in (i) machining centers and other NC machine tools (unsuccessful loading of programs, coolant disturbances and tool changing malfunctioning), (ii) storage equipment, (iii) gantry or robot (positioning errors, mechanical failures, and robot stoppages) and (iv) the control computer (computer crash or program errors). The remainder of this section presents the system components and their typical errors.

The *resources* of the cell, e.g. machine tools, lathes and measuring stations, consist of a multitude of subsystems, e.g. robots, doors, sensors, fixtures, motors, tools and tool changer, coolant, control system with NC-programs, registers (storing tool and reference point data) and communication hardware and software. All these components may contain errors that can generate failures to the resource in question, e.g. a door or a clamping device that does not open/close properly, a tool that is worn out or broken or a fault in an NC program.

Even in highly automated manufacturing systems, certain functions must be performed by *human operators* [4]. Regardless of the operators level of training, experience and skill, their performance will vary and lead to errors in the system [16]. An error could be the omission of a step in a task or an entire task. An error could also be that the task is performed incorrectly with regard to accuracy, sequence, time or quality. The operator(s) can cause errors to the system either directly (e.g. manual operations) or via any of the subsystems, e.g. by damaging the product or positioning the product wrongly, inspection errors, faulty handling of equipment or input of faulty data into the control system.

The *cell control system* itself is an integral part of the manufacturing system. It is obvious that adding control automation to a system will not by default decrease the number of disturbances, but the goal for such a control system should be to have the capacity to recover from more errors than it might cause. Failures in the cell control system can originate from any of its subsystems, e.g. computer hardware, operating system, databases or controlling applications.

A distinction can be made between serious errors, such as a disk crash or a database shutdown, and less serious errors, which can be defined as discrepancies between the actual state in the controlled system and the state of the internal model/representation of the system [15], e.g. if a robot holds a product that differs from the one specified in the database or if the control system "thinks" that a machine tool is empty when it is actually loaded with a part.

Errors can also be caused by changes (outside the tolerances) of the *product* dimensions and by changes in product positioning and handling. There may exist a multitude of reasons for such product errors. Operations previously performed on the product — manual as well as automatic — may have failed for some reason: worn-out tools were perhaps used or the operator might have dropped or damaged the product. However, errors in products are usually not detected until they cause errors in the resource that handles the erroneous product, e.g. a robot.

FAULT HANDLING IN MACHINING CELLS

The ultimate goal of any manufacturing system is to maintain the specified services by recovering automatically from the errors that might occur in the system. Errors should therefore be anticipated to make it possible to prepare procedures for error recovery. However, the system designer cannot foresee all possible scenarios of failures. Furthermore, error recovery is often better performed by the operator because of the difficulty of implementing automatic recovery [4]. Also, it is not always possible, nor necessary, to maintain the same level of service after the occurrence of an error. Usually, a degraded functionality is better than no functionality at all [2]; see Figure 4. In this section, different aspects of fault handling are discussed in relation to the architecture presented earlier.

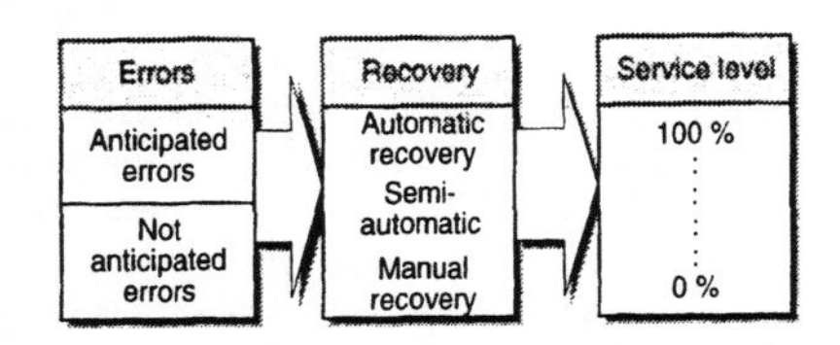

Figure 4. The goal is to maintain the same service level through automatic error recovery.

Error detection, diagnosis, recovery and confinement

There exist feasible techniques for the *detection* of errors, but these techniques all require high observability of the system [15]. There are generally two ways in which an error in a subordinated system can be detected: (i) the superior system is actively informed about the error, e.g. by receiving a message, or (ii) the superior system detects the error as a discrepancy between expected and actual state, e.g. sensor data are not as expected, there is lack of response within a specified time span (i.e. a time-out), there is incorrect response or an operator hears or visually detects a discrepancy.

The main difficulty of *diagnosis* is that the system designer cannot foresee all possible scenarios of failures. Another difficulty is the fact that one error is usually followed by many more errors caused by the original error — the problem is to determine which of all the detected errors is the original error. The error diagnosis is performed using information stored in a database or in an expert system. The diagnosis could be performed manually by an operator, automatically by an error diagnosis module or through a combination in which the operator is provided with some kind of support [2].

Many errors cannot be handled automatically, but must be handled by operators or maintenance personnel. The operator can perform error detection, error diagnosis and error recovery as a support for the internal resources and the dispatcher. Knowledge-based expert systems can be used to support the operator who performs these operations [17], e.g. to locate and make a recovery from an error or to restart the system in a correct way [18]. Minor disturbances, however, might be handled by the control system in order to provide at least a degraded service level even when (maintenance) personnel are not present, e.g. during production at night.

To reduce the number of errors that must be handled on one level of control, the propagation of errors from one system should be restricted, i.e. some kind of *error confinement* must be present. By implementing error handling in this way, (i) the complexity of error handling at a superior control level will be reduced and (ii) error handling procedures on a superior level could be made general to all specific errors that exist in a manufacturing equipment, thus increasing the flexibility of the control system. In the architecture presented earlier, errors should be confined to the internal resources' level of control if they cannot be confined to the lowest level. Otherwise, the error must be handled by the dispatcher. To be able to confine errors in this way, there must, in a general case, exist functions for error detection, diagnosis and recovery, both in the internal resources and in the dispatcher module.

The architecture must consider both control system errors and errors appearing in the resources (possibly caused by a product or an operator error). Recovery from product errors that are detected in a measuring station as part of a product's normal operation sequence can either be implemented as a separate exception stored in the database together with the normal operation list or, which is more likely, as an alternative branch in the operation list. Production stops caused by such safety systems have the effect of suddenly reducing the functionality of some parts of the system. Since this situation is very similar to an error situation, such stops can be handled in much the same way. *Control system errors* that are serious (e.g. control computer crashes) cannot be avoided, but the effects can (and must) be reduced, e.g. by making it possible to maintain some degraded level of production without the control system. The number of less serious errors can be reduced, e.g. by using sensors to verify that the contents of the robot gripper is the correct product. *Resource errors* can be restricted to one single resource, but can also involve more than one resource. If, for example, an error occurs when two resources are working together (e.g. when a part is loaded into a machine tool), both the internal resources detect, diagnose, recover from the error, and perhaps also send an "Error" message to the dispatcher. If the dispatcher receives error messages from both a mover and a producer, it must determine the location of the error on the basis of error type information stored in a database.

Example of error handling

An example of a machine tool error is used to illustrate the principles for error handling in our architecture. Errors that occur in a mover or in a measuring station are treated in much the same way as errors in machine tools. That is, if the error cannot be confined to the local control system, it will be detected by the corresponding internal resource, which will diagnose the error, update the database and, if possible, recover from the error. If complete recovery cannot be made from the error, it will send an "Error" message to the dispatcher.

A broken tool in a machine could be detected by the machine's local controller, but the error cannot usually be fully recovered from at this level. The control system's internal resource for the specific machine (see Figure 2) will detect the error, either by receiving a specific error message (if the machine could detect and/or diagnose the error itself) or by a time-out. If the error has not been diagnosed by the machine tool, it should be diagnosed by the internal resource. When the error has been diagnosed, the error type will be stored in the database – in the product and/or in the resource model.

Errors can be recovered from by the internal resources using recovery procedures (called exceptions) stored in the product and resource models in the database. Which exception to use is determined by the error type stored in the database. The error can be completely confined to this level of control only if the error recovery procedure affects neither the resource models (e.g. the machine capability) nor any product model. If the error could not be diagnosed, or if a diagnosed error could not be recovered from by the internal resource itself, the internal resource will actively inform (i.e. send a message to) the dispatcher that an error has occurred. Since the broken tool affects the service provided by the machine, the internal resource updates the machine's tool capability (stored in the database) to make it correspond to the tools currently available to the machine and to send an "Error" message to the dispatcher. This actual capability will then be used by the scheduler to develop a new schedule. In this way, production can proceed despite the broken tool. Some products will be automatically re-routed to other machines, while other products may be temporarily routed out of the cell since they cannot be manufactured with the reduced functionality.

The dispatcher detects errors either by receiving an error message or through a time-out event, e.g. in the case of failure of an internal resource. If the error has not been diagnosed by the subordinated systems, the dispatcher will initiate an error diagnosis, either automatically or by an operator that is supported with information from the database and/or an expert system.

When the error has been diagnosed by the dispatcher, an error recovery procedure is performed. This may involve sending a rescheduling request to the scheduler. If, for example, a machine has changed its capabilities, or if a machine tool is temporarily unavailable, the new schedule is based on the new machine's tool capability. If a product is damaged, the product's error recovery procedure is used instead of its normal operation list when the new schedule is developed.

CONCLUSIONS

This paper briefly described an architecture that was primarily designed with the objective of increasing the flexibility of a control system in a manufacturing cell. However, in order for such an architecture to be realistic, it is not sufficient for it to only be flexible with respect to new product types or new manufacturing equipment. The architecture must also incorporate functions for error handling, i.e. detection, diagnosis and recovery. The aim of the research presented in this paper is to extend the original architecture to handle error situations as well. On the basis of a hierarchical decomposition of the system into fault-prone components, typical errors that might occur in a manufacturing cell were described. The original architecture implicitly includes features that support the implementation of fault-handling functionality, e.g. (i) the product's operation lists can include alternative routes, (ii) the operation lists do not include any specification of the resources that should be used in order to make this decision as late as possible in time and (iii) it is possible to issue a reschedule request whenever needed in order to obtain a schedule that is based on the system's actual status. However, to be able to handle errors, functions for error detection, diagnosis and recovery must be added to the architecture, the message-passing scheme must be extended with error messages and the database must store error recovery procedures and information about the error types diagnosed.

ACKNOWLDEGEMENTS

This work was partially supported by the Swedish National Board for Industrial and Technical Development (NUTEK) under grant number 9304792-2.

REFERENCES

1. Adlemo, A., Andréasson, S.-A. "Fault Tolerance in Flexible Manufacturing Systems." *Modern Manufacturing, Information Control and Technology*. Edited by M. B. Zaremba and B. Prasad. Springer Verlag, London, U.K. (1994), 287-326.
2. Adlemo, A., Jonsson, E., Andréasson, S.-A. "Dependability Improvement in Computerized Manufacturing Systems." *Intelligent Automation and Soft Computing: Trends in Research, Development and Applications*. Edited by M. Jamshidi *et al.* TSI Press, Albuquerque, U.S.A. Vol. 1 (1994), 1-6.
3. Adlemo, A., *et al.* "Towards a Truly Flexible Manufacturing System." *Journal of Control Engineering Practice,* 3(4) (April, 1995), 545-554.
4. Adlemo, A., *et al.* "Operator Control Activities in Flexible Manufacturing Systems." Accepted for publication in the *International Journal of Computer Integrated Manufacturing*, (1996/97).
5. Andréasson, S.-A., *et al.* "Database Design for Machining Cell Level Product Specification." *The 1995 IEEE 21st International Conference on Industrial Electronics Control and Instrumentation, IECON'95.* Orlando, U.S.A., Vol. 1 (1995), 121 - 126.
6. Chintamaneni, P. R., *et al.* "On Fault Tolerance in Manufacturing Systems." *IEEE Network*, 2(3), (May, 1988), 32-39.
7. O'Connor, R. F., *et al.* "Identification, Classification and Management of Errors in Automated Component Assembly Tasks." *International Journal of Production Research*, 31(8) (1993), 1853-1863.
8. DiCesare, F., *et al.* "Extending Error Recovery Capability in Manufacturing by Machine Reasoning." *IEEE Trans. Systems, Man, and Cybernetics,* 23(1) (January/February, 1993), 221-228.
9. Fabian, M., *et al.* "Dynamic Products in Control of an FMS Cell", *Symposium on Information Control Problems in Manufacturing Technology, INCOM'95*. Beijing, China, (October, 1995). 139-144.
10. Gullander, P., *et al.* "Generic Resource Models and a Message-Passing Structure in an FMS Controller." *The 1995 IEEE International Conference on Robotics and Automation, ICRA'95.* Nagoya, Japan, (1995), 1447-1454.
11. Järvinen, J., *et al.* "Causes and Safety Effects of Production Disturbances in FMS Installations: A Comparison of Field Survey Studies in the U.S.A. and Finland." *International Journal of Human Factors in Manufacturing*, 6(1) (Winter,1996), 57-72.
12. Kopetz, H., *et al.* "Distributed Fault Tolerant Real-Time Systems: the Mars Approach." *IEEE Micro*, 9(1) (February, 1989), 25-40.
13. Kuivanen, R. "Disturbance Control in Flexible Manufacturing." *International Journal of Human Factors in Manufacturing*, 6(1) (Winter, 1996), 41-56.
14. Laprie, J.-C. (Ed.), *Dependability: Basic Concepts and Terminology*, Springer Verlag, London, U.K. (1992).
15. Loborg, P. "Error Recovery in Automation – An Overview." *Proceedings of Robotics Workshop*. Linköping, Sweden. (1993).
16. Park, K. S. "Human Reliability." *Handbook of Industrial Engineering*. 2nd ed., Edited by G. Salvendy, John Wiley & Sons Inc., New York, U.S.A. (1992), 991-1004.
17. Smith, R., Gini, M. "Error Management for Robot Programming." *Journal of Intelligent Manufacturing,* Vol. 3 (1992), 59-73.
18. Stahre, J., Johansson, A. "Operator Decision Support in Manufacturing Systems — Error Detection and Restart using Expert Systems." *Proceedings of Robotics Workshop.* Linköping, Sweden, (1993).

A NEW FLEXIBLE EXCEPTION HANDLING APPROACH FOR AUTONOMOUS MOBILE SERVICE ROBOTS

D. GLÜER and G. SCHMIDT *
Department of Automatic Control Engineering (LSR)
Technische Universität München; D-80290 München, Germany

ABSTRACT

While in many robot applications knowledge for exception handling is formulated through rules, this paper outlines a different approach based on *relations*. Its main advantage is the clear representation of dependent and/or highly redundant pieces of knowledge. A therapy for recovery is generated by queries to the knowledge base. Results of a query are enforced by cooperating modules in the supervisory level of a hierarchical robot control system. Experimental validation of this exception handling approach has been performed with the autonomous mobile robot MACROBE.

KEYWORDS: exception handling, recovery, relation, relational algebra, mobile robot, autonomy, Petri Net, agent.

INTRODUCTION

Mobile robots can perform a variety of complex service tasks in indoor or outdoor scenarios. With respect to maximum system autonomy and successful completion of a given service task, a mobile robot needs to cope with typical unexpected events or exceptions. Therefore, special mechanisms must be provided for handling of events such as closed doors, obstructed or unavailable destinations, lack of resources, e.g. battery power, or upcoming new and more urgent tasks.

Although obstacle avoidance is a highly investigated area in past robotics research [1], handling of task relevant exceptions is still a rarely studied field. Some authors have focused attention to failure diagnosis and rule based recovery schemes for robot assembly operations [4,9]. In this type of application only a comparatively small amount of knowledge on the environment is needed. Other authors implemented rule based exception handling schemes for mobile robots in rather constrained situations, where again a small number of rules suffices for recovery [10].

However, for more versatile service robot applications a purely rule based system would need a large number of Horn clauses to cope with the above mentioned types of more complex exceptions. On the other hand, large numbers of rules have a detrimental effect on rule base maintainability: Knowledge is represented in a rather unstructured

* This work was supported by *Deutsche Forschungsgemeinschaft* within the *Sonderforschungsbereich 331, "Informationsverarbeitung in autonomen, mobilen Handhabungssystemen"*, project L7

way and modifications might cause unexpected side-effects. To achieve an orthogonal structure of the knowledge, we propose to translate the rules into tables. Subsequently these tables can be modified by relational algebra, a well-known formalism in the area of databases [5]. This transformation leads to a structured knowledge representation, which can be easier maintained and which provides more flexibility when autonomous mobile robots are applied to more challenging service tasks.

CLASSIFICATION OF EXCEPTIONS

Exceptions for service robots can be distinguished by *sources* and *severity*. *Sources* for unexpected events considered are:

- the supervising control system, which transmits a new or modified task to the service robot,
- other intelligent systems, which demand communication or cooperation,
- internal (sub-) systems, e.g. sensors, actors and their data processing units,
- discrepancies between real environment and its internal representation.

Exceptions of different *severity* are:

- obstacles, which may be avoided by path modification,
- technical failures, which can be fixed by redundant systems,
- task-relevant exceptions, which require rescheduling of operations.

Obstacles or technical failures might result into task-relevant exceptions, if they cannot be fixed at the subsystem level.

Assuming that the actual situation has been correctly identified, this paper focuses on task-relevant exceptions and the related *therapy*, i.e. the procedure that defines how to react in such a case and how to return into a nominal state.

An analysis of the required therapies shows that only in a few cases a complete replanning of the ongoing task(s) is required, sometimes just the insertion of one or two tasks suffices for recovery, in other cases the additional performance of single operations helps to resume nominal task execution. Beside these measures, low-level commands are needed, which have no specific relation to the ongoing process and which are associated to the system itself, e.g. *repeat previous action* or *clear operation buffer*.

METHODOLOGY

The knowledge how to react to the above-mentioned exceptions might be represented by a set of rules. However, with increasing number of rules the knowledge becomes less structured, and by each modification the risk of unexpected side-effects increases. [9] proposes the representation of knowledge by ordered trees. This may however lead to redundancy or alternatively vast interconnections of the branches. This paper proposes a different approach, which is basically the transformation of the rules into tables or *relations*. For manipulation of such tables relational algebra offers an appropriate mathematical background [5]. Corresponding definitions might be found in [6,8].

The fundamental intent of the algebra is to allow the setup of expressions, which serve as a high-level and symbolic representation of the user's intent and as a convenient basis for normalisation of the relations. The related process can be characterised as the sequential reduction of a given collection of relations to a more appropriate form. This procedure is reversible, which means that no information is lost in the normalisation process. With progressing normalisation clarity of knowledge increases and insertion, deletion and update dependencies are avoided. A drawback of this approach is the effort needed for concatenation of normalised relations for queries. Consequently, the selection of the key attributes along with the separation of relations needs careful study.

The relations are queried by the *exception handler*. Its inputs are the identified unexpected events and information on their current context. After inference, the resulting tasks, operations and low-level-commands have to be enforced by appropriate modules. Here a hierarchical control system architecture with decentralised agents is considered, with the exception handler located at the supervisory level, Fig. 1.

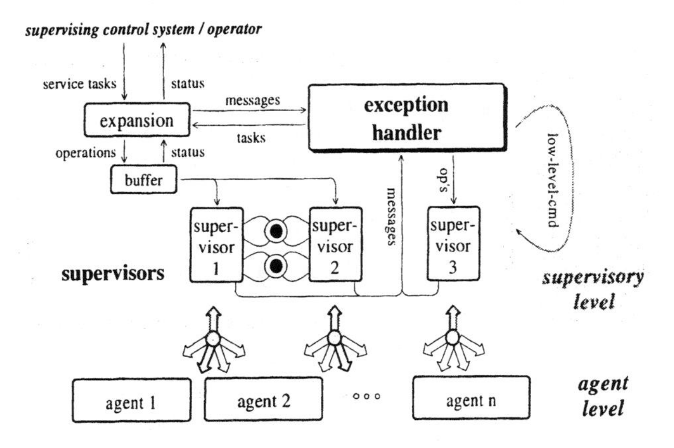

Figure 1. Overview of supervisory and agent level

A service task desired from the supervising control system or the operator is broken down into operations by the expansion unit. Upon request, a supervisor removes one of the operations from the buffer and "feeds" an appropriate agent with it, adding information such as map updates or the expected final state of the previous operation. For information on the status of the agents and for bumpless transfer of control at least two supervisors are required. Their control activity is represented by Petri Nets [12], Fig. 2.

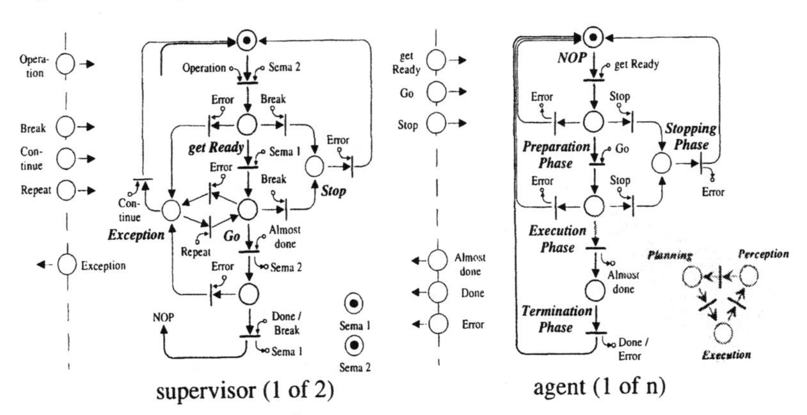

Figure 2. Petri Net of supervisor and agent

Whenever an exception at the subsystem level occurs which cannot be fixed by the agent currently in charge, the exception handler is notified through one of the supervisors. A third independent supervisor is reserved.for enforcing single operations generated by the exception handler.

PROTOTYPE IMPLEMENTATION

Especially in the area of autonomous mobile robots the therapy of task-relevant exceptions is characterised by certain basic approaches with many variations depending on the current context. The agents of such a system have to perform specific operations, such as local manoeuvring [3] or long distance motion [1]. We call such an agent *expert*, if its kernel consists of a perception - planning - execution - cycle performed in realtime. More detailed information about this system architecture can be found in [2].

After normalisation of the exception handling knowledge, the capabilities of the experts are summarised in relation D, Fig. 3. Information about locations is put into relations E, F and H. The known exceptions can be found in relation A, with their basic recovery approaches in relation B. C and G represent the available low-level commands and tasks, respectively.

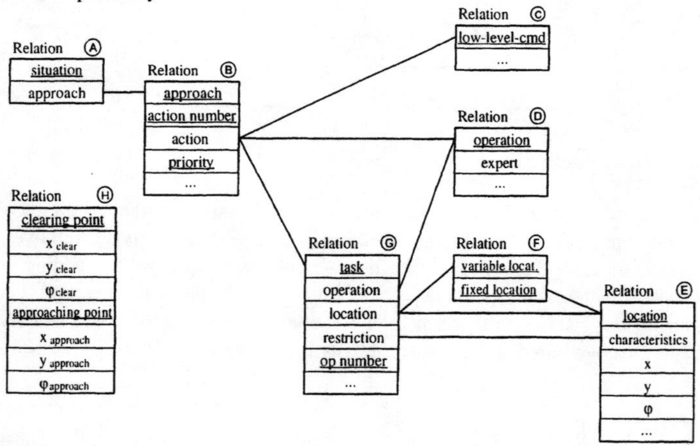

Figure 3. Normalised relations with relationships

Except of relation H, all relations are decomposed into the 5th normal form and are as such free of dependencies. Instead of referring many rules, only one query is required for providing at least one therapy:

$$\sigma_{expression} (A \Diamond\Diamond B \Diamond\Diamond (\Pi_w(C) \cup \Pi_w(D) \cup \Pi_w(G))) \quad ** \qquad (1a)$$

with "expression": (*identified situation*) = `A.Situation` (1b)

Whenever a task is generated by query (1), it is expanded by an additional query (2):

** ◊◊ symbolises the natural join

$$\sigma_{expression}\,(\, G \times D \times \underbrace{(\, E \cup \Pi_E\,(E \lozenge\lozenge F\,))}_{\hat{E}}) \tag{2a}$$

with "expression"

$$(task) = \texttt{G.task} \wedge \\ \texttt{G.operation = D.operation} \wedge \\ \texttt{G.location} = \hat{E}\texttt{.location} \wedge \\ \texttt{G.restriction} = \hat{E}\texttt{.characteristic} \tag{2b}$$

EXPERIMENTAL RESULTS

The roughly described system including the new exception handling approach was implemented and successfully tested with the multifunctional robot MACROBE. The agents applied in the experiment are the experts for local manoeuvring, for long distance motion and for local interaction. Other experts, such as for floor coverage [7] or for fork lift truck operations were not used in this experiment. Fig. 4 shows the track of the vehicle performing a transportation task from area ① to ⑤.

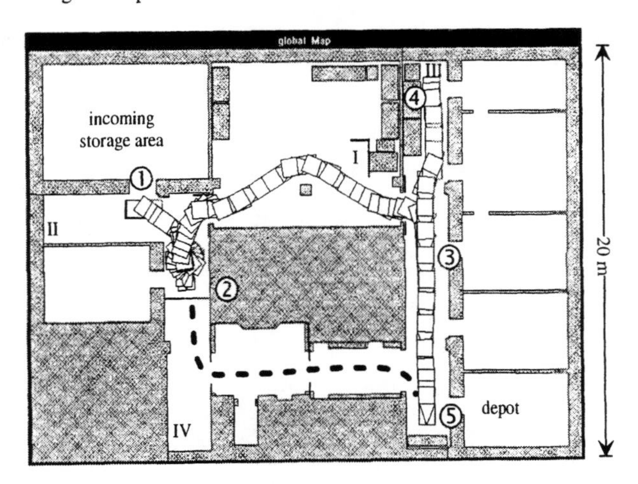

Figure 4. Track of the full-scale mobile robot MACROBE

Starting at ①, the expert for long distance motion selects the shortest path to destination point ⑤ (sketched by dots). During path execution, the multisensor system detects a closed door at ②. Due to the narrow environment, the expert in charge gets stuck and the concerned supervisor alarms the exception handler. The application of query (1) generates a therapy, consisting of one operation (*turn vehicle*) and a low-level-command (*repeat previous operation*). The vehicle turns, selects an alternative route to destination and continues long distance motion. Later, at ③, a malfunction of a subsystem is detected. A query determines the next available service station and puts it into relation F, such that (1) results into two tasks: Move to docking station III (④) and wait for acknowledgement; return to interruption point (a variable location) and resume nominal execution. Application of query (2) twice expands both tasks, taking into account the current context, such that after execution of the therapy the vehicle reaches the desired destination point ⑤.

CONCLUSIONS

The transformation of the exception handling rules into relations leads to a more structured representation of exception handling knowledge for mobile service robot applications. Detailed information is put into normalised relations that are connected according to demand. Consequently the knowledge remains clear and balanced so that future extensions concerning functionality, applications, and new layouts can be added omitting undesired side-effects. The implementation by use of a relational database allows comfortable cooperation with other databases and on-line modifications through two-phase commit.

A requirement for this approach is a deliberate database-design: The selection of relevant attributes (*keys*) is a strategic consideration. However, its appropriate selection allows higher degree of flexibility while avoiding unnecessary data overhead.

The database Rdb/ELN used in the experiments runs onboard the vehicle on a μVAX 3300. The database maintenance is performed externally through wireless Ethernet on a VAX cluster.

REFERENCES

1. Azarm, K.; Schmidt, G., *Integrated Mobile Robot Path Planning and Execution in Changing Indoor Environments*, Proceedings of the IEEE International Conference On Intelligent Robots and Systems (IROS), München, (1994), 298-305
2. Azarm, K.; Bott, W.; Freyberger, F.; Glüer, D.; Horn, J.; Schmidt, G., *Autonomiebausteine eines mobilen Roboterfahrzeugs für Innenraumumgebungen*, Informationstechnik und Technische Informatik 36, (1)(1994), 5-11
3. Bott, W.; Freyberger, F.; Schmidt, G., *Automatic Local Manoeuvre Planning and Execution for Car-like Mobile Robots*, Proceedings of the IFAC Workshop on Motion Control, Munich, (1995)
4. Cheng, X., *An On-line Planning System For A Mobile Two-Arm Servicing Robot In A Manufacturing Environment*, Rembold, U. et. al. [Hrsg.], Intelligent Autonomous Systems IAS-4, Karlsruhe, (1995), 101-108
5. Codd, E. F., *A Relational Model of Data for Large Shared Data Banks*, Commun. of the ACM, Vol. 13, No. 6, (1970), 377-387
6. Date, C. J., *An Introduction to Database Systems*, Addison-Wesley Publishing Company, 1995
7. Hofner, C.; Schmidt, G., *Path planning and guidance techniques for an autonomous mobile cleaning robot*, Special Issue "Research on Autonomous Mobile Systems", Robotics and Autonomous Systems, (14), (1995), 199-212
8. Maier, D., *The Theory of Relational Databases*, Computer Science Press, Rockville, (1983)
9. Meijer, R. M., *Autonomous Shopfloor Systems - A Study into Exception handling for Robot Control*, PhD thesis, University of Amsterdam, (1991)
10. Noreils, F. R.; Chatila, R. G., *Plan Execution Monitoring and Control Architecture for Mobile Robots*, IEEE Transactions on Robotics and Automation, (11), (1995), 255-266
11. Wang, F. W.; Saridis, G. N. et.al., *A Petri-Net Coordination Model for an Intelligent Mobile Robot*, IEEE Transactions on Systems, Man, and Cybernetics, (4), (1991), 777-789
12. Holvoet, T., *Agents and Petri-Nets*, Petri Net Newsletter, (49), (1995), 3-8

A Stereotactic Microscope For Robot Assisted Neurosurgery

Cesare Giorgi M.D., Davide Sebastiano Casolino M.S.
1st. Division of Neurosurgery, Istituto Neurologico. Milano, Italy.

Howard Eisenberg M.D.
Division of Neurosurgery, University of Maryland Medical System. Baltimore, Maryland, U.S.A.

Ettore Gallo M.S., Giovanni Garibotto M.S.
Telerobot. Genova, Italy.

ABSTRACT

In this paper we describe a mechanic antropomorphic arm connected to a neurosurgical operative microscope through a force-feedback sensor. The arm and the microscope are passively moved by the surgeon within the neurosurgical field, which is "framed" in a reference system Morphological and functional data relative to the patient under scrutiny are acquired preoperatively in the same reference system that is used to calibrate the position of the microscope at operation.

KEYWORDS: image-guided surgery, stereotaxy, robotic neurosurgery.

INTRODUCTION

This project involves the merging of an instrument that has contributed to the development of minimally invasive neurosurgery--the operative microscope--with a navigation technique that describes the position of each volume element of the brain within a surgical reference system-- stereotaxy.

Stereotaxy consists of the fixation of an external reference system (the stereotactic frame) to the patient's head, prior to the acquisition of pertinent data for the morphological description of the operative site (lesion and surrounding healthy structures). MRI, CT, angiograms, as well as SPECT and PET, can be acquired in stereotactic conditions: this information can then be made available in the metrics of a surgical reference system.

Contemporary computer technology offers an ideal representation of these data, providing very effective surface or volume rendering, merging of different image modalities, transparencies, and reformatting of images along orthogonal, or arbitrary (surgical) planes. Surgical instruments mounted or calibrated on the stereotactic reference can be displayed together with the anatomy of the patient : trajectories of approach can be selected according to the relationship of the lesion with the surrounding eloquent structures.

Stereotaxy originated as a method to identify and reach target points within the brain by means of needles and electrodes mounted on protractors that are hinged on the frame itself. While still valid to perform brain biopsies or procedures for disfunctions like epilepsy or Parkinsonism that can be improved by lesioning or stimulating very small cerebral areas, this procedure shows its limitations when it is employed for removal of lesions of more substantial volume. Graduated arcs and linear rulers that identify single trajectories of approach, even in the most straightforward case of an "isocentric" system, are difficult to set so as to identify a sequence of trajectories, necessary in removing a cerebral lesion. The possibility of contamination and of errors in setting the linear coordinates makes this procedure awkward and calls for the development of surgical instruments that provide the benefit of intraoperative morphological information without such hindrances.

Fig.1. Graphic simulation of the surgical frame, and trajectory of approach. In the LL insert, reformatted image at the probe tip, orthogonal to the selected trajectory

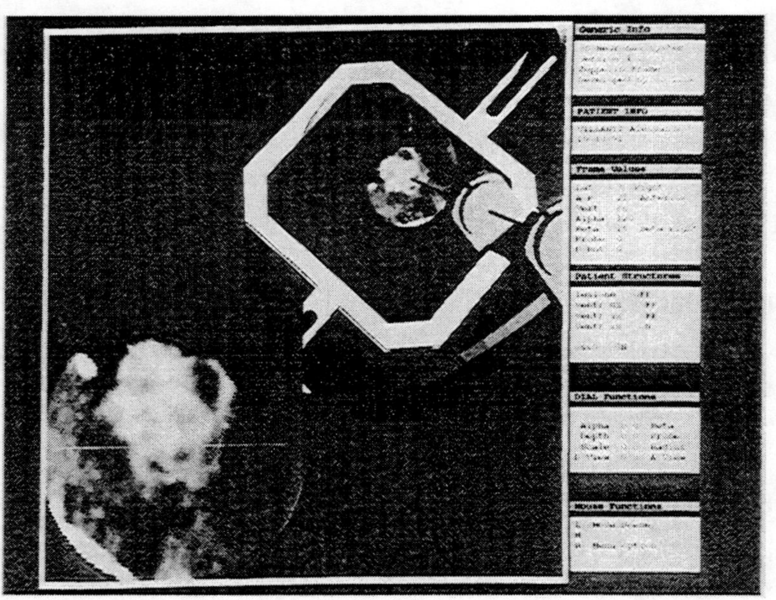

Image processing subsystem

The graphic system is based on a SGI platform, and allows for the surface rendering of structures of interest that are manually or automatically outlined on the original images. These images can be "reformatted" along arbitrary planes to display sequential sections orthogonal to the chosen trajectory of approach (Fig 1). Images from the microscope digital camera, matching those reformatted sections, can then be displayed since the optical trajectory is known. Multimodal images can be "fused" to integrate information; SPECT and PET functional data can also be incorporated in the same reference system.

Since preoperative images lose their value during surgery due to the opening of the skull, the drainage of cerebrospinal fluid, and to the progressive removal of the lesion, intraoperative ultrasound digital images, acquired in stereotactic conditions, can be calibrated and displayed with reference to the optical trajectory.

The stereotactic robot

This project represents the evolution of solutions that have been explored by our group other authors in the field of image-guided neurosurgery. Industrial robots (4, 1) or custom designed motorized tool carriers (5) have been used to perform brain biopsies; passive articulated arms have been manufactured to function as 3-D digitizers in the stereotactic frame of reference (3, 6).

Our development has been directed along these baselines.

1) The motorized tools should only assist the surgeon during surgery by providing additional data.

2) The new instrument should be incorporated in the surgical field, rather than simply "added" to it.

Following these criteria, some solutions have been proposed based on optical tracking, ultrasound, or magnetic field sensors, with the aim of minimizing the intrusion of the added technology. In our opinion, non-mechanical designs generate concern about "unobstructed visual pathways," or potential inaccuracy caused by suboptimal, and undetectable, wave propagation measurement or magnetic field interference.

We have adopted a more traditional, but reliable, solution based on a mechanical 6 degrees of freedom anthropomorphic arm, the base of which is rigidly connected to the operative table, and the end effector connected to the microscope. The core of the proposed solution is represented by a 3-D sensor, mounted at the end effector, that allows the orientation of the microscope with respect to the surgical reference frame. Insofar the head of the patient is also rigidly connected to the bed, an initial calibration of the system allows the surgeon to establish a unique coordinate transformation between the two systems and to track the movements of the microscope within the reconstructed volumetric model of the brain.

Input from the 3-D sensor activates the motors of the joints, so that the arm can be back-driven by the surgeon without effort. Compensation for the arm's weight and inertia has been made, thereby allowing the surgeon to move the microscope easily.

The following is a brief description of the main characteristic of this prototype.

Dimensions and kinematics: A surgical robot must not interfere with the standard equipment setup. To find the best solution for the arm configuration, we installed a mockup with adjustable links in the operative room. As a result af this test, 2 degrees of freedom have been introduced at the shoulder, 1 at the elbow, and 3 at the wrist (roll-pitch-roll, with coaxial shafts). The critical measurements of 500 mm forearm length, 450 mm forearm, and 70mm forearm diameter have been obtained by a special design of the wrist: all actuators and sensors for the wrist have been mounted near the elbow, and the motion of the three wrist axis is transmitted by a group of three concentric shafts connected to the final joint by bevel gears. Backlash reduction has been achieved by an accurate adjustment of the position of the primitive cones of the bevel gears. The robot base is also adjustable in height for different table positions.

High dexterity: Because the arm must accommodate various patient positions, it must be capable of wide adjustment. Such demands necessitated the design of joints with a stroke greater than that normally used for industrial applications: turret=±180°, shoulder pitch= ±120°, elbow=±150°, wrist first roll=±180°, wrist pitch=±120°, and wrist second roll=±180°.

Velocity and accelerations: The robot and the effector must move smoothly and slowly (50mm/sec), a speed that allows for the use of motors smaller than those used in industrial robots, thus contributing to the overall reduction of dimensions.

Accuracy: In image-guided surgery the precision requirements are set by the dimensions of the image voxels, and it is generally agreed that it should be kept around ±1mm. The arm is mechanically very accurate, but the precision at operation is largely determined by the calibration procedure. Our system uses the following calibration procedure:
the arm end effector mounts on a conical pin during calibration and is passively moved to touch a series of eight points, whose coordinates are known in the surgical frame of reference. At each point, the six angular values of the joints are stored. Through the direct kinematic algorithm, the coordinates of the same points are calculated in the arm's reference system. A first transformation matrix gives a "rough estimate" of both the relative position and orientation of the arm with respect to the surgical reference systems, and of the position of the measured points in this same reference. An "errror function" is therefore defined for the differences between the actual and the calculated measurements. Through a minimization algorithm, the values of the transformation matrix that minimize the discrepancy are found and introduced into the final transformation matrix. At this point, the microscope is positioned and oriented in the surgical frame of reference.

Sterilization: The arm's body is finished with smooth, watertight surfaces; all cables are routed within the structure. Contact sterilization is therefore possible, although, for practical purposes the forearm and end effector's measurements allow for simple wrap-up solutions.

Safety: Each robot axis is equipped with a "coarse" resolver mounted on the output shaft, and a "fine" resolver mounted on the motor shaft. The latter is used to close the position loop, and a consistency check is performed by the controller at each reading of the sensors. A warning message is displayed in case of inconsistent position values.

Fail-safe brakes are mounted at each axis, and are released together with the miscroscope stand brakes in cases when the robot is linked to a traditional operative microscope.

Space constraints can be implemented to limit the robot's degree of freedom and avoid interference with the patient's body, even in cases of operation in a "passive" mode.

The control system: Forces and torque applied by the surgeon to intraoperatively orient the microscope in the surgical field must be interpreted by the robot control system that will activate the motors accordingly (Fig.2). In order to obtain this precision, a commercial force/ torque sensor capable of measuring three forces and three torque components is interposed between the robot and the microscope handles. When the surgeon applies a force to the handle, a reaction force and /or torque is sensed and transmitted through a parallel line to the robot control system. At the same time, the control system acquires data from the robot encoders. It is then possible to calculate the Transpose Jacobian, thereby obtaining the torques that should be exerted by the robot joints to balance the force/torque applied to the end effector.

Since the desired motion of the arm must follow the "surgeon's intention" unhindered by the static forces generated by the balance of the arm, the torque values are transformed through experimentally determined coefficients into increments of joint velocity. These values are generated by the PID axis controllers, whose output is connected to the motor drivers.

Other parameters can be introduced to obtain a "viscous dumping" of the joint motion, to reduce the vibration, and to obtain a smooth movement of the arm. The time necessary to execute the loop is approx 5 msec.

From the acquisition of the joint position sensors and from the direct kinematic algorithm, the position and orientation of the microscope axis within the arm reference system is computed. By means of a transformation matrix, the parameters of which have been computed in the calibration phase, the position and orientation of the optical axis are computed in the patient's frame of reference. These values are sent to the graphic workstation to obtain the superimposition of the optical image on the corresponding reformatted image from the preoperative data.

DISCUSSION

The prototype robot for neurosurgery described in this paper has been tested at the Istituto Neurologico di Milano for a period of 6 months, starting in September 1994. In its original configuration, the arm was to be connected to a standard operative microscope after the calibration procedure. In practice we found that , due to its elasticity, the microscope stand could not be safely linked to the force/torque sensor without exerting low frequency oscillations that were amplified by the sensor's response.

This problem, theoretically solvable with the increase of the "viscous" character of the arm response, led us to abandon the original configuration, in favor of a more practical solution consisting of a microscope head mounted "on board."

Preliminary experience gathered with a digital camera mounted on the end effector have been promising (2), although the final assessment will only be possible when the optical microscope head, specifically designed for this application, is available.

Fig.2. The Robot arm, connected to an operative microscope.

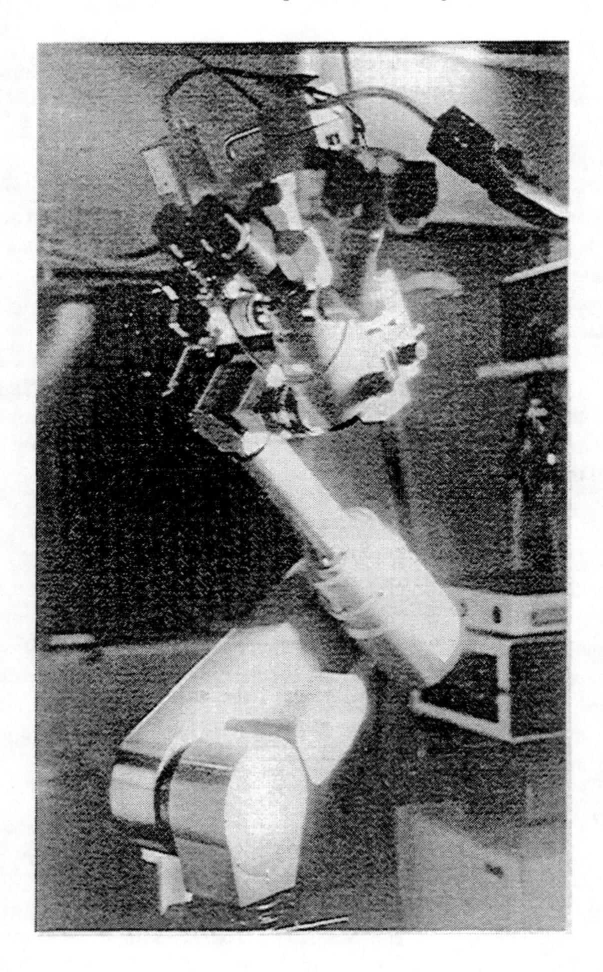

ACKNOWLEDGEMENTS

This project has been supported by the Italian Research Council CNR/PFR. (Progetto Finalizzato Robotica, Director Professor U. Cugini)

REFERENCES

1. Benabid AL, Cinquin P, Lavallée S, Le Bas JF, Demongeot J, de Rougemont J (1987) Computer-driven robot for stereotactic surgery connected to CT scan and magnetic resonance imaging. Appl Neurophysiol 50:153-154.
2. Giorgi C (1994) Intraoperative fusion of field images with CT / MRI data by means of a stereotactic mechanical arm. Min Invas Neurosurg 37:53-55
3. Giorgi C,Luzzara M, Casolino SD, Ongania E (1993) A computer controlled stereotactic arm: virtual reality in neurosurgical procedures. Acta Neurochir Suppl 58:75-76
4. Kwoh YS, Hou J, Jonkeere EA, Hayati S (1988) A robot with improved absolute positioning accuracy for CT guided stereotactic brain surgery. IEEE Trans Biomed Eng 35:153-160
5. Villorte N, Glauser D,Flury P, Burckhardt CW (1992)Conception of stereotactic instruments for the neurosurgical robot Minerva. Paper presented at the 14th Annual International Conference of the IEEE Engineers in Medicine and Biology Society.
6. Watanabe E,Watanabe T, Manaka S, Mayanagi Y, Takakura K (1987) New equipment for computer tomography guided neurosurgery. Surg. Neurol 27: 543-547.

EVOLUTIONARY ROBOTICS AT SUSSEX

I. Harvey, P. Husbands, D. Cliff, A. Thompson, N. Jakobi
School of Cognitive and Computing Sciences
University of Sussex, Brighton BN1 9QH, UK
inmanh, philh, davec, adrianth, nickja@cogs.susx.ac.uk

ABSTRACT

We give an overview of evolutionary robotics research at Sussex. We explain and justify our distinctive approaches to (artificial) evolution, and to the nature of robot control systems that are evolved. We illustrate by presenting results from research with evolved controllers for autonomous mobile robots; simulated robots, coevolved animats, real robots with software controllers or with a controller directly evolved in hardware.

KEYWORDS: Evolutionary Robotics, Artificial Evolution

WHY EVOLUTIONARY ROBOTICS?

When designing a control system for a robot, there are at least three major problems:
- It is not clear *how* a robot control system should be decomposed.
- Interactions between separate sub-systems are not limited to directly visible connecting links, but also include interactions mediated *via the environment*.
- As system complexity grows, the number of potential interactions between sub-parts of the system grows *exponentially*.

Classical approaches to robotics have often assumed a primary decomposition into Perception, Planning and Action modules. Many people now see this as a basic error [2]. Brooks acknowledges the latter two problems above in his subsumption architecture approach. This advocates slow and careful building up of a robot control system layer by layer, in an approach that is explicitly claimed to be inspired by natural evolution — though each new layer of behaviour is wired in by hand design.

An obvious alternative approach is to abandon hand design and explicitly use evolutionary techniques to incrementally evolve increasingly complex robot control systems. Unanticipated interactions between sub-systems need not directly bother an evolutionary process where the only benchmark is the behaviour of the whole system. Other individuals and groups have taken a similar evolutionary approach, such as [1][5][12]; here we concentrate on an overview of work at Sussex. We start with theoretical questions of what artificial evolutionary techniques and classes of control system are appropriate for evolutionary design. We discuss the relationship between robot simulations and reality, and the problem of evaluation within a noisy and uncertain environment. Sussex projects in this area are described, with both simulations and real robots, including hardware evolution.

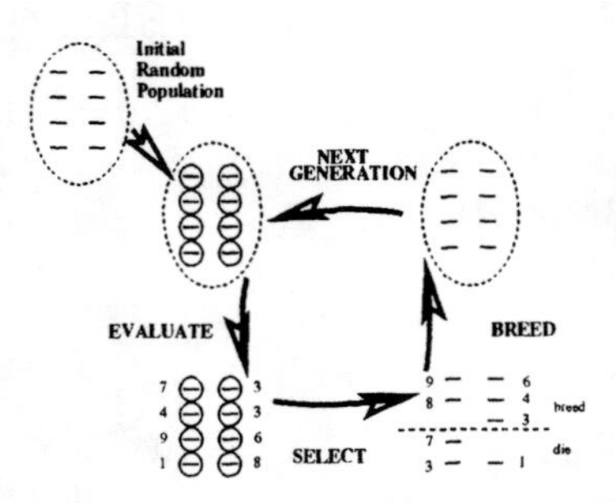

Figure 1: The GA Cycle.

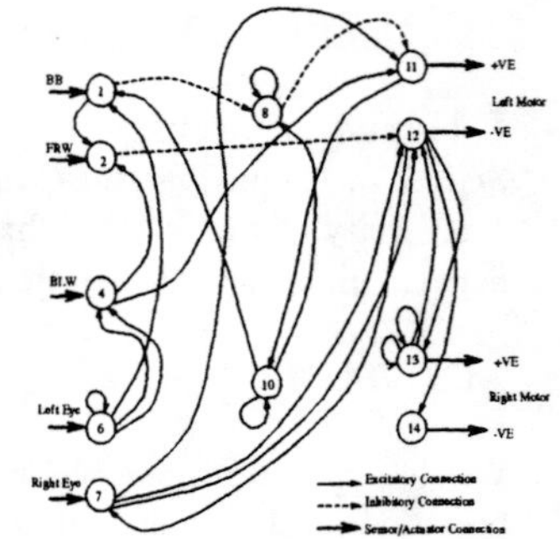

Figure 2: Network without redundant units.

ARTIFICIAL EVOLUTION FOR ROBOTS

Genetic Algorithms (GAs) are the most commonly used evolutionary algorithm for optimisation. Evolutionary Robotics (ER) typically needs adaptive improvement techniques [8] rather than optimisation techniques — a critical distinction.

Optimisation problems can be seen as search problems in some high-dimensional search space, of known size. Each dimension corresponds to a parameter that needs to be set, coded for on a small section of the genotype; in robotics, a genotype specifies the characteristics of a control system. Using a population of such genotypes (often initially random), each is evaluated on how good is the potential solution that it encodes. Fitter genotypes are preferentially selected to be parents of the next generation; offspring inherit genetic material from parents, and also undergo random mutations. This cycle of selection, reproduction with inheritance of genetic material, and variation, is repeated over many generations (Fig. 1).

A GA optimisation approach typically starts with a population of random points crudely sampling the whole search space. Successive cycles focus the population of sample points towards fitter regions of the space. However, some domains — including much of evolutionary robotics — do not always fall into this convenient picture of a fixed-dimensional search space. Standard GA theory does not necessarily then apply.

SAGA — Species Adaptation Genetic Algorithms

In ER a genotype will specify the control system of a robot which is expected to produce appropriate behaviours. The number of components required may be unknown *a priori*; and when using incremental evolution, through successively more difficult tasks, the number of components needed will increase over time. Such incremental evolution calls for *GAs as adaptive improvers* rather than *GAs as optimisers.*

Species Adaptation Genetic Algorithms (SAGA) were developed for this purpose [6]. It was shown that progress through such a genotype space of increasing complexity will only be feasible through gradual increases in genotype length; this implies the evolution of a *species* — the population is largely genetically converged. With successive generations, selection is a force which tends to move such a population up hills on a fitness landscape, and keep it centred around a local optimum; whereas mutation explores outwards from the current population. For a given selection pressure, there is a maximum rate of mutation which simultaneously allows the population to retain a hold on its current hill-top, whilst maximising search along relatively high ridges in the landscape, potentially towards higher peaks. In SAGA, this means that rank-based selection should be used to maintain a constant selective pressure, and mutation rates should be of the order of 1 mutation per genotype [6].

What building blocks for a control system?

We must choose appropriate building blocks for evolution to work with. Primitives manipulated by the evolutionary process should be at the lowest level possible. Any high level semantic groupings inevitably incorporate the human designer's prejudices. We agree with Brooks [2] in dismissing the classical Perception, Planning, Action decomposition of robot control systems. Instead we see the robot as a whole — body, sensors, motors and 'nervous system' — as a dynamical system coupled (via sensors and motors) with a dynamic environment [1]. Hence the genotype should encode at the level of the primitives of a dynamical system.

One such system is a dynamic recurrent neural net (DRNN), with genetic specification of connections and of the timescales of internal feedback. These DRNNs can in principle simulate the temporal behaviour of any finite dynamical system, and are equivalent (with trivial transformations) to Brooks' subsumption architectures. We also deliberately introduce internal noise at the nodes of DRNNs, with two effects. First, it makes possible new types of feedback dynamics, such as self-bootstrapping feedback loops and oscillator loops. Second, it helps to make more smooth the fitness landscape on which the GA is operating.

ER IN SIMULATION

Experiments at Sussex have used a round two-wheeled mobile robot performing navigational tasks. Initial experiments [3] used simulations of such a robot with touch sensors and two visual inputs — simulated photoreceptors, with genetically specified fields of view. The robot task was to reach the centre of a circular arena, with white walls and black floor and ceiling. Grey-level visual inputs to each photoreceptor were calculated by ray-tracing. Robot motion was modelled carefully, including collisions and noisy motor properties, using measurements from a real robot.

The genetically specified DRNNs used had input nodes for each sensor, output nodes for each motor, and an arbitrary number of 'hidden' nodes. All nodes were noisy linear threshold devices. Connections were *excitatory* (weighted link joining the output of one unit to the input of another) or *veto* (an infinitely inhibitory link between two units). The task is set implicitly by the evaluation function, and robots were rated on the basis of how much time they spent at or near the centre of the arena; they always started near the perimeter, facing in a random direction.

Robots with successful evolved control systems make a smooth approach towards the centre of the arena, and circle there. Success was also achieved when the height of the wall was allowed to vary over one order of magnitude, each robot being given 10 trials with differing wall-heights across the full range; for robustness, the evaluation was based on the *worst* score it obtained across its trials.

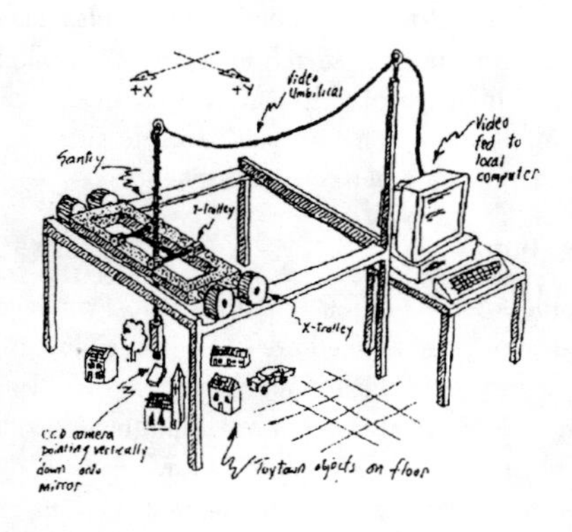

Figure 3: A cartoon sketch of the Gantry.

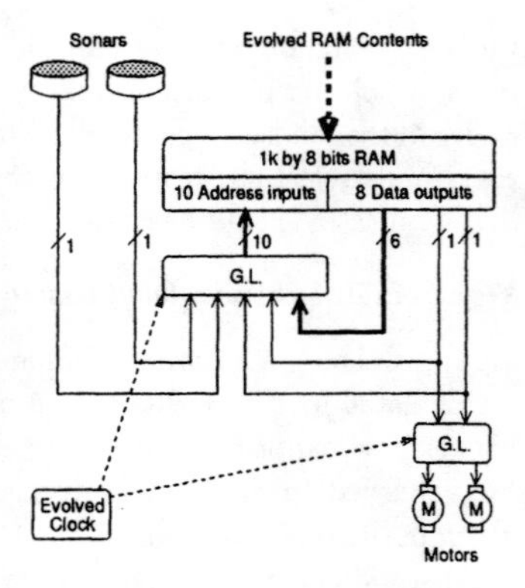

Figure 4: The evolved DSM.

Analysis of an evolved network starts with identification of redundant units and connections. Since early stages of evolution allowed visual signals to prevent the robot from bumping into the walls, the touch-sensors are unused, and their nodes can be recruited as extra 'hidden' nodes. The results of eliminating redundant nodes from a successful network are shown in Fig. 2. An analysis of the attractors of the dynamics of the robot with such a control system has been made [9].

Coevolution

At Sussex further work in simulation has involved exploring the dynamics of coevolution in pursuit-evasion contests [4]. One species of pursuing animats have their fitnesses determined by the current strategies of another species of evaders, and *vice versa.* Such a coevolutionary 'Arms Race' may have implications for incremental evolution of robots, as a method of automatically increasing task complexity whilst taking humans out of the loop.

THE GANTRY

Ray-tracing in simulation is computationally expensive. For dynamic real-world domains with noisy lighting conditions it is necessary to use a real robot. Evolution requires the evaluation of many trials, which should be automated. We developed a specialised piece of visuo-robotic equipment for this — the gantry-robot.

The robot is cylindrical, 150mm in diameter, and moves in a real environment. Instead of using wheels, the robot is suspended from the gantry-frame with stepper motors that allow translational movement in X and Y directions (Fig. 3), corresponding to that which would be produced by left and right wheels. The visual input is from a CCD camera pointing down at a mirror inclined at 45°, which can be rotated about a vertical axis so as to 'see' along the direction the 'robot' is facing. The CCD image is subsampled into 3 or more genetically specified virtual photoreceptors, or receptive fields — we are using minimal bandwidth vision.

We used the same networks and genetic encoding schemes as before. Tasks were navigating towards white paper targets, in a predominantly dark arena. Using an incremental evolutionary methodology, simple visual environments were used initially, moving on to more complex ones in this sequence of tasks [7]:

(1) Forward movement **(2)** Movement towards large target
(3) Movement towards small target **(4)** Distinguishing triangle from rectangle

An initial random population of 30 needed about 10 generations to achieve success at each stage, which each had appropriate evaluation functions. Control systems capable of reaching the small target were found to generalise to following a moving target of similar size. For the final task, two white targets were fixed to one wall, one triangular and one rectangular. The robots were given trials with differing start positions, not biased towards either target. The evaluation function added a bonus for getting close to the triangle, but subtracted a penalty for nearing the rectangle.

The successful networks were of a similar complexity to that of Fig. 2. The networks evolved such that robots rotated on the spot when visual inputs were both low or both high; but moved in a straight line when only one was high. The visual morphology evolved such that the visual inputs changed in unison when crossing a vertical dark/light edge, and only differed significantly at an oblique edge. Thus the control system was an 'oblique dark/light boundary detector' rather than a 'triangle detector'. In the context, it performed the required task of detecting the triangle, and rejecting the square.

EVOLVABLE HARDWARE

The robot control systems for the experiments above, though conceptualised as dynamical systems, have been implemented in software. They can also be implemented directly in hardware [13], using *intrinsic* hardware evolution, where each genetically specified piece of hardware is tested for real *in situ*. The low-level physics of the hardware can be utilised, and the components can behave at their natural timescales, without the necessity of global clocking or other design constraints.

Thompson used artificial evolution to design a hardware controller, a *Dynamic State Machine* (DSM), for a mobile robot using sonars to avoid walls in a corridor. Success was achieved with a DSM of just 32 bits of RAM and 3 flip-flops (excluding clock generation) which took sonar echo pulses directly, without pre-processing, and output appropriate pulses direct to the motors. The genetic specification of the DSM (Fig. 4) determined whether each signal was synchronised by a clock; and if so, the frequency of that clock. The DSM was intimately coupled to the real-time dynamics of its sensorimotor environment.

In very recent work, to be published, Thompson has applied these techniques to a Field Programmable Gate Array (FPGA) from the forthcoming Xilinx XC6200 family. Circuits on an unclocked FPGA can be evolved to generate desired output frequencies over a wide range, from 10Hz to 1MHz.

EVOLUTION WITH *KHEPERA*

When using simulations it is an important to decide just how realistic the model should be, and how noise should be handled. Jakobi [11] built a simulator, *Khepsim*, for the *Khepera* robot from EPFL in Lausanne. This was based on a spatially continuous, two dimensional model of the underlying real world physics, using a profile derived from the motors and sensors of a real *Khepera.*

IR and ambient light values were calculated by ray-tracing. Runs were performed in simulation with different noise levels — zero, observed noise, double observed noise — and tested on a real robot, for obstacle-avoiding and light-seeking tasks. It was concluded that simulated noise levels should be similar to real levels for systems evolved in simulation to transfer properly. If there is a significant difference in noise levels (too high *or* too low), then whole different classes of behaviours become available which, while acquiring high fitness scores in simulation, fail to work in reality.

SUMMARY

We have discussed the use of SAGA for incremental evolution through a space of dynamical robot control systems. Other relevant aspects are also being researched at Sussex*, such as artificial morphogenesis [10], the design of fitness functions to 'shape' evolution towards desired goals, interactions between learning and evolution.

Evolutionary Robotics is a research area in its infancy; the tests for all newborn AI philosophies are whether they can grow up into the real world, and scale up with increasing complexity. In the evolutionary experiments at Sussex we have started to demonstrate the possibilities in simulation, on real robots, and directly in silicon.

REFERENCES

1. R. Beer and J. Gallagher "Evolving dynamic neural networks for adaptive behavior". *Adap. Beh.* 1(1):91–122, 1992.
2. R. Brooks "A robust layered control system for a mobile robot". *IEEE J. Rob. Autom.*, 2:14-23, 1986.
3. D. Cliff, I. Harvey, and P. Husbands. "Explorations in evolutionary robotics". *Adap. Beh.*, 2(1):71–104, 1993.
4. D. Cliff and G. Miller. "Tracking the Red Queen" In F. Morán et. al., eds., *Advances in Artificial Life: Proceedings of the 3rd ECAL*, pp. 200–218. Springer-Verlag, 1995.
5. D. Floreano and F. Mondada. "Automatic creation of an autonomous agent", In D. Cliff et. al., eds., *From Animals to Animats 3*, MIT Press/Bradford Books, 1994.
6. I. Harvey. "Evolutionary robotics and SAGA: the case for hill crawling and tournament selection". In C. Langton, ed., *Artificial Life III*, pp. 299–326. Addison Wesley, 1993.
7. I. Harvey, P. Husbands, and D. Cliff. "Seeing the light: Artificial evolution, real vision". In D. Cliff et. al. eds., *From Animals to Animats 3*, MIT Press/Bradford Books, 1994.
8. J. Holland. *Adaptation in Natural and Artificial Systems*. Univ. Mich. Press, Ann Arbor, 1975.
9. P. Husbands, I. Harvey, D. Cliff. "Circle in the round", *J. Rob. and Aut. Sys.* 15:83-106, 1995.
10. P. Husbands, I. Harvey, D. Cliff, and G. Miller. "The use of genetic algorithms for the development of sensorimotor control systems". In P. Gaussier and J.-D. Nicoud, eds., *From Perception to Action*, pages 110–121, Los Alamitos, CA, 1994. IEEE Computer Society Press.
11. N. Jakobi, P. Husbands, and I. Harvey. "Noise and the reality gap". In F. Morán et. al., eds., *Advances in Artificial Life: Proceedings of the 3rd ECAL*, pp. 704–720. Springer-Verlag, 1995.
12. S. Nolfi, D. Floreano, O. Miglino, and F. Mondada. "How to evolve autonomous robots". In R. Brooks and P. Maes, eds., *Artificial Life IV*, pages 190–197. MIT Press/Bradford Books, 1994.
13. A. Thompson, I. Harvey, and P. Husbands. "Unconstrained evolution and hard consequences". In E. Sanchez and M. Tomassini, eds., *Towards Evolvable Hardware*. Springer-Verlag, 1996.

* More information is available via WWW on http://www.cogs.susx.ac.uk/lab/adapt/

A Kinematic Formulation of Path Tracking Problem for Low Speed Articulated Vehicles

A. Hemami and V. Polotski
Dept. of Electrical and Computer Engineering
Ecole Polytechnique, Montreal, Que., Canada

ABSTRACT

The paper is about the problem of path tracking control in the class of vehicles made up of two single axis units pivoted together. The reason for this articulation is twofold, to give more manoeuvrability to the vehicle for curve negotiation and to serve for steering. Examples of such a vehicle are used for transfer of ore inside the relatively narrow and restricted underground mine galleries. Steering is performed by a linear actuator that changes the angle between the two units. Path following problem consists of making the vehicle follow a desired path based on the sensory information about the position and orientation errors at each instant of time. A feedback system governs the motion by an appropriate steering action. This paper reports the analysis of the path following control problem and the results of experimentation on a laboratory model.

KEYWORDS: Path tracking, Mobile robot, Articulated Vehicle

INTRODUCTION

The type of vehicle this paper is concerned consists of two single axis units pivoted together such that the entire vehicle has four non-steerable wheels. The steering of this vehicle is carried out by changing the pivot angle between the two units. This class of vehicles is widely used in underground mining for transportation of ore. The reason for this particular structure is that it gives more manoeuvrability for turning in the rather restricted areas of underground mine galleries. The steering action is performed by changing the length of a (or a pair of) hydraulic cylinder(s) installed between the two units. In operation, this vehicle moves in both forward and backward direction. For this reason the seat of the driver is often sidewise, so that he can easily see whichever side the vehicle moves. A schematic of a typical sample of this vehicle, usually called LHD (for Load-Haul-Dump) is shown in figure 1. Automation of the navigation of this vehicle implies that by using natural guidelines like the mine walls, or installed guidelines such as reflective tape, electric wires and so on it can follow a required path.

Automation of wheeled vehicles has been the subject of research for long time, starting from Automated Guided Vehicles (AGV). Still there are unsolved problems that need to be investigated, like the subject of this paper. From a vehicle dynamics point of view, the stability of a vehicle in negotiating a corner and in lane change, and the ride comfort for passenger vehicles are the main objectives. On the other hand, for certain classes of vehicles, such as AGV and LHD, because of the relatively slow speed of operation the vehicle stability and passenger comfort are not important issues. Moreover, for the same reason the effect of wind and similar matters do not carry much importance, unlike the case of high speed vehicles. On the other hand, for an automated vehicle the steering (path tracking control), stability of path tracking, obstacle detection and avoidance are the issues to be investigated.

Figure 1- Schematic of a LHD

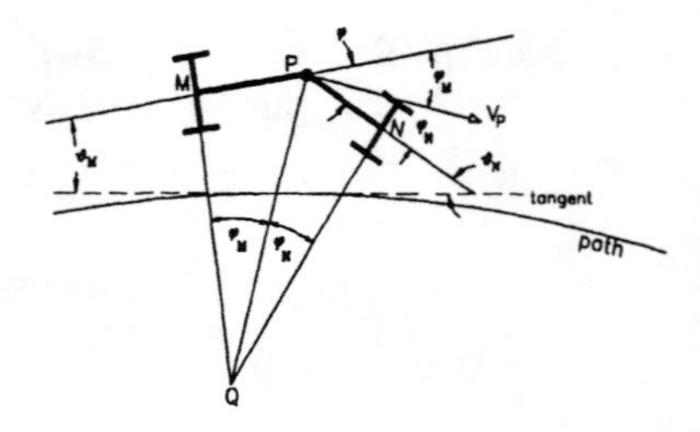

Figure 2 - Definition of parameters

As far as path tracking automation is concerned the equations of motion must be developed. All the various wheeled systems share a common property. Their motion is subject to nonholonomic constraints. It has been shown ,[7], that the control system representing the path tracking dynamics consists of slow and fast modes. The fast modes are associated with the dynamics of the vehicle and the slow modes are associated with the kinematics of motion. The fast modes are, however, stable but the slow modes are unstable. The problem of stabilization of the system, thus, can be formulated based on the kinematics of motion [9]. This greatly simplifies the synthesis of a control policy.

The list of contributors to the involved research in dynamics of wheeled vehicles and automation of their navigation is exhaustive. A few number of references are cited here to exemplify the kind of work and the variety of approaches for the analysis of problem. Vehicle dynamics are discussed in various papers; we mention only two books by Ellis [5] and Wong [22]. As for the dynamics and kinematics of motion the works of Fenton et al [6], Larcombe [14], Borenstein and Koren [3], Alexander and Maddocks, Muir and Newman [17], Saha and Angeles [20] and Hemami et al [7] must be mentioned. These deal only with single steering vehicles. Recently, systems equipped with double steering have been introduced, where both the front and rear wheels are steerable. On this subject a reader may consult the work by Nisonger and Wormley [18], Lee [15], Mehrabi [16], Ackermann and Sienel [2] and Hemami [8]. Finally, on the class of vehicles under study the works of DeSantis [5], Hurteau et al [11], Juneau [12], Juneau and Hurteau [13], Polotski et al [19] and St-Amant et al [21] can be cited.

CONTROL PROBLEM FORMULATION

Figure 2 illustrates certain parameters of a vehicle and its general configuration with respect to a desired path. The first step, in terms of path tracking, is to define the error in position and error in orientation. For a single unit vehicle these errors are readily defined. For the vehicle under study, however, this is not straightforward. There are, nevertheless, a variety of options for the choice of these errors for the vehicle under study. One can take the position and orientation errors of the front unit (depending on the direction of motion the sense of front and rear change) as the errors for the complete system. In this paper the same definitions as in reference [10] are adopted. These are

$$\varepsilon_d = \varepsilon_N + \left(\frac{l_N}{l_M}\right)\varepsilon_M \quad , \quad \varepsilon_\theta = \phi_N + \theta_N \tag{1}$$

The position error ε_d comprises a linear function of the position errors of the individual points M and N, the mid-axis points of the rear and front units, respectively (See figure 2). The orientation error ε_θ is based on the angle between the velocity vector of point P for a momentarily fixed articulation angle and the tangent to the path at the corresponding point of the path at each moment. ϕ is the angle between the two units and θ is the angle the front unit makes with the (tangent to the) path. The effect of curvature must be superimposed to both errors. In practice, it can be regarded as a disturbance for the controlled system, and can be rejected by the addition of necessary integral action to the controller [10]. Moreover, the following relationship exists between ϕ_M and ϕ_N

$$\phi = \phi_M + \phi_N \tag{2}$$

as can be seen by inspection from figure 2. The sign notation for ϕ and the two errors is that a deviation to right is negative and a deviation to left is positive. ϕ_M and ϕ_N have the same sign as ϕ. When ϕ is zero both ϕ_M and ϕ_N are zero, and vice versa. For a zero value of ϕ the orientation error as defined reduces to that of an ordinary vehicle.

The equations representing the dynamics of path tracking, after linearization of trigonometric terms ($\sin\phi = \phi$ and $\cos\phi = 1$, assuming small articulation angle), can be expressed in the following state space form

$$\begin{bmatrix} \dot{\varepsilon}_d \\ \dot{\varepsilon}_\theta \\ \phi \end{bmatrix} = \begin{bmatrix} 0 & aV & 0 \\ 0 & 0 & bV \\ 0 & 0 & 0 \end{bmatrix} \begin{bmatrix} \varepsilon_d \\ \varepsilon_\theta \\ \phi \end{bmatrix} + \begin{bmatrix} 0 \\ 1 \\ 1 \end{bmatrix} \dot{\phi} + \begin{bmatrix} d_1 \\ d_2 \\ 0 \end{bmatrix} \tag{3}$$

where V is the speed.

A smaller size vehicle of this type is developed for experimentations. Each wheel is separately powered by a permanent magnet electric DC motor. An electric cylinder (DC motor plus worm-gear) is used between the two units for steering. Each motor has a separate driver which at the present time consists of a digital to analog convertor and a voltage-to-current amplifier. Values of rotations of each wheel and the articulation angle are measured by encoders for feedback purposes.

The above formulation implies that first, in order to take care of the unmeasurable disturbances integral control must be added to the controller and, then that a linear feedback of the measurable states ε_d, ε_θ and ϕ can stabilize the open-loop unstable system. For the first issue, a closer look at the structure of the system indicates that it is not necessary to include the integral control for the third state, since the element in the disturbance vector corresponding to ϕ is identically zero. Moreover, because the position error ε_d is proportional to the integral of the orientation error ε_θ there is no need to add the same term once more. As a result, it would be sufficient to add only the integral of the position error and the augmented system assumes the form

$$\begin{bmatrix} \dot{\varepsilon}_d \\ \dot{\varepsilon}_\theta \\ \phi \\ \varepsilon_d \end{bmatrix} = \begin{bmatrix} 0 & aV & 0 & 0 \\ 0 & 0 & bV & 0 \\ 0 & 0 & 0 & 0 \\ 1 & 0 & 0 & 0 \end{bmatrix} \begin{bmatrix} \varepsilon_d \\ \varepsilon_\theta \\ \phi \\ \int \varepsilon_d \end{bmatrix} + \begin{bmatrix} 0 \\ 1 \\ 1 \\ 0 \end{bmatrix} \dot{\phi} \tag{4}$$

As can be seen the speed of the vehicle appears in the plant matrix and, consequently, influences the location of the closed-loop poles and the performance of the controlled system. This is true and the faster the speed the faster is the response of the system in reducing the tracking errors (in the range of speeds that do not violate the validity of equation). Alternatively, the problem can be formulated in terms of the distance of travel rather than the time, and then the closed-loop poles imply the rate of reduction of errors in terms of metres and not seconds. For example, if errors decay to 5% within 4.5 meters of displacement (instead of 3 seconds for a speed of 1.5 m/sec.) it will be 4.5 metres whatever the speed is. The augmented system in this latter case is

$$
\begin{bmatrix} d\varepsilon_d/ds \\ d\varepsilon_\theta/ds \\ d\phi/ds \\ dg/ds \end{bmatrix} = \begin{bmatrix} 0 & a & 0 & 0 \\ 0 & 0 & b & 0 \\ 0 & 0 & 0 & 0 \\ 1 & 0 & 0 & 0 \end{bmatrix} \begin{bmatrix} \varepsilon_d \\ \varepsilon_\theta \\ \phi \\ g \end{bmatrix} + \frac{1}{V} \begin{bmatrix} 0 \\ 1 \\ 1 \\ 0 \end{bmatrix} \dot{\phi} \tag{5}
$$

where a change of variables in the following form of is made

$$
ds = Vdt \quad , \quad g = \int \varepsilon_d \, ds \tag{6}
$$

ANALYSIS OF THE FEEDBACK AND CONTROL

No matter whether the system in equation (4) or that in (5) is used the rate of change of the articulation angle is the only input to the system. A control law has the general form (without loss of generality only case of equation 4 is shown)

$$
\dot{\phi} = Kx = k_1 \, \varepsilon_d + k_2 \, \varepsilon_\theta + k_3 \, \phi + k_4 \int \varepsilon_d \tag{7}
$$

K is the feedback matrix and x denotes the states. In this section details of the effects of the hardware structure and the results of experiments are discussed. The mechanical system under consideration has two sets of actuators. The first set consists of the wheel driver motors and the second is the articulation angle drive system. In a real size vehicle normally the left and right wheel are interconnected by a differential; it compensates for the variation of the force/torque due to the ground reaction, and adjustment of the speeds when the vehicle turns. In the laboratory model, however, each side of each axle is equipped with a DC motor and a gear/belt reducer. The adjustment of the speeds must be done by means of the command to the motor controller through software, by taking into account the kinematic relationships, otherwise some of the wheels skid. The effect of changing the input to the wheel motors is only changing the speed of motion. In this sense, although there are four inputs to the four motors they can be regarded as one category of input. A second input to the system is the steering actuator(s) whose effect is governing the path following. Although speed appears in the path tracking equations it is not normally used to govern the tracking performance. Thus, it can be said that the two inputs are decoupled, that is, each input is used for a separate function.

Our concentration in what follows is on the steering function, as governed by the aforementioned equations. The reason why two outputs, ε_d and ε_θ, can be controlled by one input lies in the fact that one is (proportional to the) derivative of the other. The third state, on the other hand, is the integral of the input. The entire system consists of, in fact, two interlocked second order systems. A closed-loop system is formed by substituting from equation (7) in equation (4). At this point, however, it lacks the effect of the dynamics of feedback component.

The schematic block diagram of the control loop is shown in figure 3. The output from the control law is forwarded to an amplifier with a constant α, which drives the DC motor. The latter, in turn, changes the length of the stem of a linear actuator through a worm gear. The length change is proportional to the motor displacement by a factor β. Two types of nonlinearities exist in the control loop, as depicted in figure 3. The first one is due to the friction forces in the actuator, which introduces a dead zone for the actuating signal. The second is due to the geometry; it stems from the variation of angle ϕ which is not linear with respect to the variation of the actuator active length. Compensation for both of these nonlinearities must be provided by a controller.

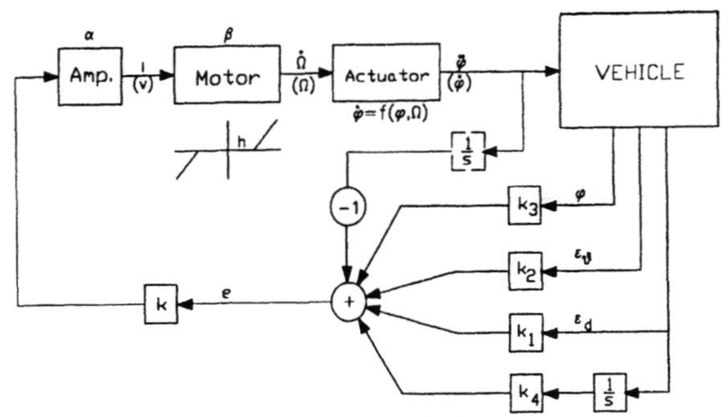

Figure 3 - control system

In the experimentations carried out a low frequency oscillation was noticed in the controlled motion. This oscillation persisted to exist despite changes of certain parameters such as speed, sampling time, and the like. The source of this vibratory behaviour was not evident. A closer study of the dynamics of the closed-loop system, thus, was to be done.

The role of the amplifier is to deliver a signal which is proportional to the input signal and, from electrical characteristics, is at the level that can be used by the motor. In the present configuration the amplifier behaves as a voltage to current convertor. An input voltage between of v volts gives rise to an output current maintained at αv amperes. This enforces an angular motion of the motor with an acceleration proportional to αv, which after the gear effect would appear as an acceleration of $\alpha\beta v$ for the linear actuator. If the nonlinearity between the length of the piston and angle ϕ is ignored for the sake of simplicity of discussion, the result is in fact in the form of

$$\ddot{\phi} = \alpha\beta v = \alpha\beta k e \quad , \quad e = \dot{\phi}_{desired} - \dot{\phi}_{measured} = Kx - \dot{\phi}_m \tag{8}$$

where α, β, k and e are as shown in figure 3. K and x denote the feedback matrix and the states. $\Delta\dot{\phi}$ represents the required change in the speed of articulation. When the vehicle is on the track e is zero and so is $\Delta\dot{\phi}$, meaning that the input remains constant. Moreover, since at the present time an external means of measurement of the tracking errors is not at hand, measurement of the errors is according to an estimation based on the measure of the articulation angle ϕ [19]. In this sense, in one way or another ϕ appears in the measure of all the states. That is to say, the closed-loop plant contains terms of the form

$$e \cong A\phi \quad , \quad \ddot{\phi} \cong \alpha\beta k A\phi \tag{9}$$

Whereas for the vehicle speed control the same type of loop with identical amplifier has been used and works well it seems that combined with the hardware, as described above, some unnecessary oscillations are introduced to the system. The remedy of this unpleasant malfunction, as it seems, would be one of the three options:

a) To use an amplifier in the voltage-voltage mode, so that an input voltage maintains a constant output voltage. In this regard, the angular velocity and not the acceleration of the motor is controlled.
b) An independent system is used to measure the tracking errors. This is more costly, but it is the proper way of measurement of errors.
c) A control law which takes into consideration the entire closed-loop system be used. The It must include the dynamics that stem from the behaviour of the amplifier.

At the present time work is being carried out on the three options. The results are not obtained yet. The tendency is to compare the outcome of all the three options.

CONCLUSION

The control of the steering for a particular class of vehicles, used mainly in underground mines, is discussed. The implementation of a linear feedback control based on the presented formulation of the problem has shown to introduce an undesirable oscillatory behaviour to a laboratory model of the vehicle. The source of this behaviour seems to lie with the elements of the control loop. A complete analysis of the closed-loop plant, taking into account the dynamics of each individual component is underway.

REFERENCES

1. Alexander, J.C. and J.H. Maddocks, "On the Kinematics of Wheeled Mobile Robots", Int. J. Robotic Research, Vol. 8(5),(1989), 15-27.
2. Ackermann J. and Sienel, W., "Robust Yaw Dumping of Cars with Front and Rear Wheel Steering', IEEE Trans. Control Systems Technology, Vol. 1(1),(1993), 15-20.
3. Borenstein, J. and Y. Koren, "Motion Control Analysis of a Mobile Robot", ASME Journal of Dynamic Systems, Measurement and Control, 109,(1987), 73-79.
4. DeSantis, R.M., "Modelling and Path Tracking for a Load-Haul-Dump Vehicle, ASME J. Dynamic Systems, Measurement and Control, (1996) in press.
5. Ellis, J.R., Vehicle Dynamics, London Business Books Ltd, London, UK, (1969)
6. Fenton, R.E., Melocik, G.C. and K.W. Olson, "On the Steering of Automated Vehicles: Theory and Experiment", IEEE Trans. Automatic Control, Vol. AC-21, (1976), 306-314.
7. Hemami, A., M.G. Mehrabi and R.M.H.Cheng, "Synthesis of an Optimal Control law for Path Tracking in Mobile Robots", Automatica, Vol. 28(2),(1992), 383-387.
8. Hemami, A., "A Control Scheme for Low Speed Automated Vehicles with Double Steering", IEEE Control and Decision Conf., Vista, FL, (1994), 2452-2454.
9. Hemami, A., "Steering Control Problem Formulation of a Tricycle-Model Vehicle", Int. J. Control, Vol. 61(4),(1995), 783-790.
10. Hemami, A., "Path Tracking Automation of Articulated Vehicles", Proc. of 13th World Congress IFAC'96, San Francisco,(1996), in press.
11. Hurteau, R., St-Amant, M., Laperriere, Y. and Chevrette, G., "Optical Guidance System for Underground Mine Vehicles", IEEE Int. Conf. on Robotics and Automation, Nice, (1992), 639-644.
12. Juneau, L., "Etude d'un système de guidage automatique pour des environments structurés", MSc Thesis, Ecole Polytechnique, Montreal, (1993), (Study of a Guidance System for structured environments)
13. Juneau, L. and R. Hurteau, "Automatic Guidance of a Mining Vehicle Using Laser Range Data: Simulation and Preliminary Experimental Results", Proc. 5th Symp. on Robotics and Manufacturing, Hawai, (1994), 545-550.
14. Larcombe, M.H.E., "Tracking Stability of Wire Guided Vehicles", Proc. 1st Int. Conf. on Automated Guided Vehicle Systems, Stratford-upon-Avon, U.K.,(1981), 137-144.
15. Lee, A.Y., "A preview Steering Autopilot Control Algorithm for Four-Wheel-Steering for Passenger Vehicles", J. Dynamic Systems, Measu. and Control, Vol. 114,(1992), 401-408.
16. Mehrabi,M.G., "Path Tracking Control of Automated Vehicles: Theory and Experiments", PhD Thesis, Concordia University, Montreal, (1993).
17. Muir, P.F. and C.P. Neuman, "Kinematic Modelling of Wheeled Mobile Robots", J. of Robotic Systems, 4(2), (1987), 281-340.
18. Nisonger R.L. and Wormley, D.N., "Dynamic Performance of Automated Guideway Transit Vehicles with Dual Axle Steering", IEEE Trans. Vehicular Technology, Vol. VT-28, (1979), 88-94.
19. Polotski, V., Boisgontier, X. and Lopez, P., "estimation de position d'un robot mobile articulé", Proc. Canadian Conf. on Electrical and Computer Engg., (1995), 489-493 (Positiion Estimation of an Articulated Mobile Robot).
20. Saha, S.K. and J. Angeles, "Kinematics and Dynamics of a Three-Wheeled 2-DOF AGV", IEEE Int. Conf. on Robotics and Automation, Scottsdale, Arizona, (1989), 1572-1577.
21. St-Amant, M. Laperriére, Y Hurteau, R. and Chevrette, G., "A Simple Robust Vision System for Underground Vehicle Guidance", Proc. 1st Int Symp. on Mine Mechanization and Automation, Golden CO, (1991), 6-1 to 6-20.
22. Wong, J.Y., "Theory of Ground Vehicles", John Willey & Sons, New York, (1978).

Using Projective Geometry to Derive Constraints for Calibration-Free Visual Servo Control

SETH HUTCHINSON

Dept. of Electrical and Computer Engineering
The Beckman Institute, University of Illinois 61801

ABSTRACT

Most approaches to visual servo control require calibration of the hand/eye system, or use adaptive methods to estimate the calibration on-line. In this paper, we review recent results from computer vision that show how projective geometry can be used to derive visual constraints that are invariant with respect to the hand/eye configuration. Such constraints can be used to specify goal positions for the robot that are invariant with respect to the camera calibration.

KEYWORDS: visual servo control

INTRODUCTION

Many approaches to visual servo control require calibration of the hand/eye system (e.g., [3] [6] [4] [2] [11]) or use adaptive methods to estimate the calibration on-line (e.g., [16] [10]). In many cases, this calibration can be tedious for the user and is prone to error. Thus, an important next step in visual servo control research is the development of calibration free approaches, i.e., approaches which either do not require explicit calibration of the camera system, or which are robust in the face of calibration errors [8][9][17].

In visual servo systems, calibration information is used for two purposes: (1) to define an error function, either directly in terms of the image observables, or in terms of the reconstructed 3D structure of the robot workspace, and (2) to define the parameters of a control system that can reduce the error function to zero. In this paper, we address the first of these problems, namely, specifying goal positions using coordinate systems that are invariant with respect to camera calibration. To this end, we review recent results from computer vision that show how projective geometry can be used to derive visual constraints that are invariant with respect to the hand/eye configuration.

We discuss several scenarios. First, we describe the case of using a single uncalibrated camera to control robot motion that is constrained (using, e.g., force feedback) to lie in a plane in the workspace. Following this, we consider the case of motion of a point in three-space, observed by a pair of cameras whose image planes are co-planar. Then, we consider the case for which the relative position and orientation of two stereo cameras is known. Finally, we consider the case in which the relative position and orientation of two stereo cameras is unknown. We begin by reviewing some basic concepts from projective geometry.

GENERAL DEFINITIONS

In this section we review concepts results from projective geometry. More comprehensive introductions to this material can be found in a variety of texts, including [14] [15] [13].

In the one dimensional case, the projective space $\mathcal{P}_1$ is referred to as the projective line. To each point $p \in \mathcal{P}_1$ we may assign a set of homogeneous coordinates (x_0, x_1). All coordinates $\lambda(x_0, x_1)$ correspond to the same point in $\mathcal{P}_1$, for $\lambda \in \Re$. We define the *projective parameter* of this point by $\theta = x_0/x_1$. The *cross ratio* of four points is defined as follows.

Definition 1 *If P_1, P_2, P_3, P_4 are four points in $\mathcal{P}_1$, their cross ratio is defined as $\zeta(\theta_1, \theta_2, \theta_3, \theta_4)$, where θ_i is the projective parameter for P_i in some coordinate system for $\mathcal{P}_1$.*

Definition 2 *The cross ratio of four elements, $\theta_1, \theta_2, \theta_3, \theta_4$, of a field is given by:*

$$\zeta(\theta_1, \theta_2, \theta_3, \theta_4) = \frac{(\theta_1 - \theta_3)(\theta_2 - \theta_4)}{(\theta_2 - \theta_3)(\theta_1 - \theta_4)} \tag{1}$$

If we consider the embedding of $\mathcal{P}_1$ in the plane $\Re^2$, using the standard basis we obtain the following, equivalent, definition for the cross ratio of four collinear points:

Definition 3 *The cross ratio of four collinear points, P_1, P_2, P_3, P_4 is defined to be the cross ratio of the pairwise distances between the points, in particular*

$$\zeta(P_1, P_2, P_3, P_4) = \frac{(|P_1 - P_3|)(|P_2 - P_4|)}{(|P_2 - P_3|)(|P_1 - P_4|)} \tag{2}$$

At times, we shall wish to consider a pencil of lines (i.e., a set of lines that pass through a common point) defined by five points in the plane (this will be discussed in more detail below).

Definition 4 *The cross ratio, $\zeta(l_1, l_2, l_3, l_4)$, of a pencil of four lines l_1, l_2, l_3, l_4 passing through some point P, is defined to be the cross ratio of the intersections of these lines with any other line not incident on P.*

Finally, when we consider the case of $\mathcal{P}_3$, we will require the following definition for the cross ratio of a pencil of planes.

Definition 5 *The cross ratio of a pencil of four planes intersecting in the line l is defined to be the cross ratio of the intersections of these planes with any other plane not containing l (i.e., the cross ratio of the pencil of lines formed by the intersection of these four planes with any other plane that does not contain l).*

A *projective basis* for $\mathcal{P}_1$ can be defined by three points with homogeneous coordinates given by $P_0 = (1,0)^T$ $P_1 = (0,1)^T$ $P_2 = (1,1)^T$. This basis is referred to as the standard basis, with the points P_0 and P_1 as the vertices and P_2 as the unit point. We may use the cross ratio to define a non-homogeneous coordinate system for $\mathcal{P}_1$ by assigning $k = \zeta(P_0, P_1, P_2, P)$ as a non-homogeneous coordinate for the point P.

The basis for much of the research using projective invariants is the following well-known theorem from projective geometry [13]:

Theorem 1 *Let D and D' be projective lines, with the points* P_1, P_2, P_3, P_4 *on D and* P_1', P_2', P_3', P_4' *on* D'. *There exists a projective transformation* $u : D \to D'$ *taking the points* P_1, P_2, P_3, P_4 *to* P_1', P_2', P_3', P_4', *respectively, if and only if the cross ratios* $\zeta(P_1, P_2, P_3, P_4)$ *and* $\zeta(P_1', P_2', P_3', P_4')$ *are equal.*

Thus, since the cross ratio is invariant with respect to arbitrary projective transformations, this non-homogeneous coordinate, $k = \zeta(P_0, P_1, P_2, P)$, for P is an invariant.

MOTION IN THE PLANE

Constrained motion is useful for a variety of robotic applications. For example, we might wish for the robot to trace out some path on a surface while maintaining a constant contact force with that surface. Previously, we have implemented a hybrid controller that uses force feedback to control the orientation of the manipulator's tool as well as the contact forces, while using a visual control loop to control the motions of the tool along the unconstrained directions of the contact surface [7]. In particular, orientation is controlled to keep the tool's approach, or Z axis, aligned normal to the contact surface. By orienting the gripper in this manner, control of the contact forces are limited to a single Cartesian degree of freedom in the tool's coordinate frame. The two remaining translational degrees of freedom, those tangent to the contact surface, are controlled using the vision system. The ability to specify the desired force and visual position provides a convenient method for describing many automation tasks, such as sanding and grinding, in which the vision system can control where on the surface to operate while the force control loop determines how hard to press the tool onto the surface.

In this paper, we are concerned with the specification of goal trajectories in the plane of robot motion, such that this specification is invariant with respect to camera calibration. We begin by defining a coordinate system for $\mathcal{P}_2$.

We can define a basis for $\mathcal{P}_2$ by the four points with coordinates $P_0 = (1,0,0)^T$, $P_1 = (0,1,0)^T$, $P_2 = (0,0,1)^T$, $P_3 = (1,1,1)^T$. This particular choice of basis is referred to as the standard basis for $\mathcal{P}_2$. The following theorem [14], concerning collineations of $\mathcal{P}_2$, implies that, given any four points $P_i \in \mathcal{P}_2$, we may always assign coordinates in this way (provided an independence condition is satisfied).

Theorem 2 *There is a unique collineation which transforms four given points, no three of which are collinear, into four given points, no three of which are collinear.*

As shown in [15], we may express the ratios of homogeneous coordinates in this basis as cross ratios. In particular, let l_{ij} be the line defined by the points P_i and P_j; then for a point P_j, with homogeneous coordinates (x_1, x_2, x_3), we define a pair of non-homogeneous coordinates (k_1, k_2) as follows [15]:

$$\begin{cases} k_1 = x_1/x_3 = \zeta(l_{10}, l_{12}, l_{13}, l_{1j}) \\ k_2 = x_2/x_3 = \zeta(l_{01}, l_{02}, l_{03}, l_{0j}). \end{cases} \tag{3}$$

Given Theorem 2, and the fact that both the image plane and the plane defining the workspace are isomorphic to $\mathcal{P}_2$, it is evident that using standard basis for $\mathcal{P}_2$ to define a coordinate system for the workspace plane in terms of the points P_0, P_1, P_2, P_3 (as described above) we establish a collineation between the workspace plane and the image plane, given the correspondences between P_i and the image points p_i. Thus, the following theorem establishes the invariance of the system of non-homogeneous coordinates (k_1, k_2) described above.

Theorem 3 *The cross ratio of a pencil of lines is invariant with respect to collineations of the projective plane.*

Therefore, we may use the non-homogeneous coordinates k_1, k_2 as a set of coordinates that are invariant with respect to perspective projection. In particular, if the goal location of a point P is expressed in these coordinates, we can determine immediately the image plane location p to which P projects, given that we have identified the projections of the four basis points in the image.

STEREO CAMERAS WITH PARALLEL IMAGE PLANES

The special case of two cameras, positioned so that their image planes are parallel, and their $x-$axes coincide has been considered in [1] [8].

Let the point P_i have world coordinates $(x, y, z)^T$. The projection of P_i onto the two image planes gives two image points, whose left and right image coordinates are given by $p = (u, v)^T, p' = (u'v')^T$, respectively, with $v = v'$ (since the $x-$axes of the image planes coincide).

Define the determinants D_{ijkl} and V_{ijkl} as

$$D_{ijkl} = \det \begin{bmatrix} p_j - p_i & p_k - p_i & p_l - p_i \\ q_j - q_i & q_k - q_i & q_l - q_i \\ u'_j - u'_i & u'_k - u'_i & u'_l - u'_i \end{bmatrix} \tag{4}$$

$$V_{ijkl} = \det \begin{bmatrix} x_j - x_i & x_k - x_i & x_l - x_i \\ y_j - y_i & y_k - y_i & y_l - y_i \\ z_j - z_i & z_k - z_i & z_l - z_i \end{bmatrix} \tag{5}$$

Then, as shown in [1][8],

$$\frac{D_{0123}D_{0145}}{D_{0124}D_{0135}} = \frac{V_{0123}V_{0145}}{V_{0124}V_{0135}}$$
$$\frac{D_{0123}D_{0245}}{D_{0124}D_{0235}} = \frac{V_{0123}V_{0245}}{V_{0124}V_{0235}}$$
$$\frac{D_{0123}D_{1245}}{D_{0124}D_{1235}} = \frac{V_{0123}V_{1245}}{V_{0124}V_{1235}}.$$

These final three equations can thus be used to define invariant coordinates for the point P_5 in the projective basis defined by the points $P_0 = (1, 0, 0, 0)^T$, $P_1 = (0, 1, 0, 0)^T$, $P_2 = (0, 0, 1, 0)^T$, $P_3 = (0, 0, 0, 1)^T$, $P_4 = (1, 1, 1, 1)^T$.

STEREO CAMERAS IN KNOWN RELATIVE CONFIGURATIONS

Consider the case of a stereo camera system, positioned so the coordinate frame of the right camera is related to the coordinate frame of the left camera by a rotation matrix, $\mathbf{R}$, and a translation vector $\mathbf{t}$. Again, let the projection of a point, P_i onto the two image planes, have left and right image coordinates given by $p_i = (u_i, v_i)^T, p'_i = (u'_i, v'_i)^T$, respectively.

Assuming that the image formation process can be modeled by perspective projection, then a set of homogeneous image coordinates $(u, v, u', s)^T$ can be obtained from the homogeneous coordinates of the point P by a non-singular linear transformation (since this image formation model is projective linear [5]).

For the case of $\mathcal{P}_3$, we may use the cross ratio of a pencil of planes to define the non-homogeneous projective coordinates (see, e.g., [12]). In particular, let the points $P_0 = (1,0,0,0)^T$ $P_1 = (0,1,0,0)^T$ $P_2 = (0,0,1,0)^T$ $P_3 = (0,0,0,1)^T$ $P_4 = (1,1,1,1)^T$ define a projective basis for $\mathcal{P}_3$. Then the non-homogeneous projective coordinates for the point P are given by

$$\begin{cases} k_1 = x_1/x_4 = \zeta(P_1P_2P_0, P_1P_2P_3, P_1P_2P_4, P_1P_2P) \\ k_2 = x_2/x_4 = \zeta(P_2P_0P_1, P_2P_0P_3, P_2P_0P_4, P_2P_0P) \\ k_3 = x_3/x_4 = \zeta(P_0P_1P_2, P_0P_1P_3, P_0P_1P_4, P_0P_1P) \end{cases} \tag{6}$$

where $P_iP_jP_k$ denotes the plane defined by these three points.

Since the coordinates $(u, v, u', s)^T$ are obtained from the homogeneous coordinates for P by a projective transformation that is a function of the position and orientation of the camera system, these coordinates are invariant with respect to the position and orientation of the camera system.

STEREO CAMERAS IN UNKNOWN RELATIVE CONFIGURATION

One of the most significant results in applying projective geometry to uncalibrated vision systems has been given by Faugeras [5]. Space permits only a brief summary that result here. More detail about the application of this result to visual servo control may be found in [8]. Suppose that five points P_0, P_1, P_2, P_3, P_4 are used to establish the standard projective basis for $\mathcal{P}_3$, and that the correspondences between the images of these five points and at least three other points in the left and right cameras are known. Then it is possible to reconstruct these three other points, or any other point whose projection can be located in each image, in the projective coordinate system defined by the projective basis given by P_0, P_1, P_2, P_3, P_4. Furthermore, this reconstruction is unique, to within a projective transformation of the environment.

CONCLUSIONS

We have reviewed several results from the computer vision literature on using projective geometry to establish invariants in images. We have described, for a number of special cases, how these results may be used to establish goal coordinates for robot tasks that are invariant with respect to camera calibration parameters.

References

[1] E. B. Barrett, P. M. Payton, N. N. Haag, and M. H. Brill. General methods for determining projective invariants in imagery. *Comp. Vision, Graphics, and Image Proc.: Image Understanding*, 53(1):46–65, January 1991.

[2] A. Castano and S. A. Hutchinson. Visual compliance: Task-directed visual servo control. *IEEE Trans. on Robotics and Automation*, 10(3):334–342, June 1994.

[3] P.I. Corke. Visual control of robot manipulators — a review. In K. Hashimoto, editor, *Visual Servoing*, volume 7 of *Robotics and Automated Systems*, pages 1–31. World Scientific, 1993.

[4] B. Espiau, F. Chaumette, and P. Rives. A New Approach to Visual Servoing in Robotics. *IEEE Transactions on Robotics and Automation*, 8:313–326, 1992.

[5] O.D. Faugeras. What can be seen in three dimensions with an uncalibrated stereo rig? In G. Sandini, editor, *Proc. European Conference on Computer Vision*, volume 588 of *Lecture Notes in Computer Science*, pages 563–578, Santa Margherita, Italy, 1992. Springer-Verlag.

[6] J.T. Feddema and O.R. Mitchell. Vision-guided servoing with feature-based trajectory generation. *IEEE Trans. on Robotics and Automation*, 5(5):691–700, October 1989.

[7] F. Geiger. Hybrid force/vision control of robotic manipulations. Master's thesis, University of Illinois at Urbana-Champaign, Dept. of Electrical and Computer Engineering, 1995.

[8] G. D. Hager. Calibration-free visual control using projective invariance. DCS RR-1046, Yale University, New Haven, CT, December 1994. To appear Proc. ICCV '95.

[9] Nicholas Hollinghurst and Roberto Cipolla. Uncalibrated stereo hand eye coordination. *Image and Vision Computing*, 12(3):187–192, 1994.

[10] N. P. Papanikolopoulos and P. K. Khosla. Adaptive Robot Visual Tracking: Theory and Experiments. *IEEE Transactions on Automatic Control*, 38(3):429–445, 1993.

[11] N. P. Papanikolopoulos, P. K. Khosla, and T. Kanade. Visual Tracking of a Moving Target by a Camera Mounted on a Robot: A Combination of Vision and Control. *IEEE Transactions on Robotics and Automation*, 9(1):14–35, 1993.

[12] J. Ponce and Y. Genc. Epipolar geometry and linear subspace methods: A new approach to weak calibration. In *Proc. IEEE Conf. on Comp. Vision and Patt. Recog.*, page to appear, 1996.

[13] P. Samuel. *Projective Geometry*. Springer-Verlag, New York, New York, 1986.

[14] J. G. Semple and G. T. Kneebone. *Algebraic Projective Geometry*. Oxford University Press, Oxford, 1952.

[15] C. E. Springer. *Geometry and the Analysis of Projective Spaces*. W. H. Freeman and Co., San Francisco, 1964.

[16] W.J. Wilson. Visual servo control of robots using kalman filter estimates of robot pose relative to work-pieces. In K. Hashimoto, editor, *Visual Servoing*, pages 71–104. World Scientific, 1994.

[17] B. Yoshimi and P. K. Allen. Active, uncalibrated visual servoing. In *Proc. IEEE Int'l Conference on Robotics and Automation*, pages 156–161, San Diego, CA, May 1994.

OmniNav: Obstacle Avoidance for Large, Non-circular, Omnidirectional Mobile Robots

B. Holt[1], J. Borenstein[1], Y. Koren[1], and D. Wehe[2]

The *University of Michigan*

1) Department of Mechanical Engineering and Applied Mechanics

2) Department of Nuclear Engineering and Radiological Science

Ann Arbor, MI 48109-2110, USA

ABSTRACT

This paper describes an obstacle avoidance method designed specifically for large, non-circular robots with omnidirectional motion capabilities. The method, called OmniNav, is reflexive, meaning that it is able to make navigation decisions without having to discern details of the environment, such as edges. The OmniNav method combines elements of the previously developed Virtual Force Field (VFF) and Vector Field Histogram (VFH) methods to protect arbitrarily shaped mobile platforms from collisions. To this end the method computes a preliminary direction of travel based on the VFH method, which assumes the robot was point-sized. Then, highly localized potential fields are applied to "act-on points" around the robot's real perimeter. A transfer function uses these "virtual forces" to alter the preliminary direction of motion and to add rotation as necessary to avoid collisions.

Keywords: Mobile robots, Obstacle avoidance, Omnidirectional, Ultrasonic sensors.

1. INTRODUCTION

The specific problem addressed by our system is how to guide an irregularly shaped three-degree-of freedom vehicle around obstacles. Our method, called *OmniNav*, has three unique features:

- **OmniNav works in uncertain environments.** Previously, an effective way of dealing with large-robot navigation has been to represent obstacles in the workspace of the robot as configurations of the robot that are not allowed. This approach, called the configuration space approach, was first described by Lozano-Perez [1987]. In this approach, it is necessary to accurately locate the edges of all obstacles in the workspace, and then either change their location or search for robot configurations which intersect these edges. BecauseIn contrast, the OmniNav method uses a *histogramic map* representation of the environment (see Section 2), which does not require knowledge of specific obstacle features.
- **OmniNav is strictly reflexive and can operate independently of any higher-level navigation architecture.** By using global path planning, many researchers have

developed methods that optimize the orientation of an irregular vehicle with respect to the obstacles in the environment. This is usually done by constructing a Voronoi Diagram and applying potential fields to multiple act-on points on the perimeter of the robot [Hague et al., 1989; Barraquand et al., 1992]. All of these systems operate in an open loop, i.e., they compute the path once prior to moving and not again. Our method, on the other hand, operates in a closed loop, updating the control commands at a rate of approximately 25 Hz. Thus, OmniNav is able to continually sample its environment and upgrade its map.

- **OmniNav controls three degrees of freedom.** As discussed by Borenstein and Raschke [1992], a similar method has been successfully applied to two degree-of-freedom vehicles, such as the University of Michigan's CARMEL. OmniNav is able to control not only the translation of the vehicle, but also its rotation, independently. The additional degree-of-freedom allows it to pass through areas that would not be traversable with only two degrees-of-freedom.

The following two sections discuss in detail the OmniNav method. Section 2 explains how the VFF method is applied to act-on points around the vehicle. In Section 3 the VFH method, used to initially find open pathways, is explained. Section 4 discusses the transfer function that adjusts the direction of travel as selected by VFH and optimizes its orientation in response to the VFF forces. Section 5 gives the results in simulation for OmniNav and describes our current effort to employ OmniNav on a real mobile robot.

2. THE VIRTUAL FORCE FIELD (VFF) METHOD

The VFF method is specifically designed to accommodate and compensate for inaccurate range readings from ultrasonic or other sensors [Borenstein and Koren, 1989]. To do so, it uses a two-dimensional Cartesian grid, called the *histogram grid C*, to represent data from ultrasonic (or other) range sensors. Each cell *(i,j)* in the *histogram grid* holds a *certainty value (CV)* $c_{i,j}$ that represents the confidence of the algorithm in the existence of an obstacle at that location. This representation was derived from the certainty grid concept that was originally developed by Moravec and Elfes, [1985]. In the *histogram grid*, *CVs* are incremented when the range reading from an ultrasonic sensor indicates the presence of an object at that cell.

Combining the *histogram grid* with the potential field concept, the VFF method allows the immediate use of real-time sensor information to generate repulsive force fields. Fig. 1 illustrates this approach: As the vehi-

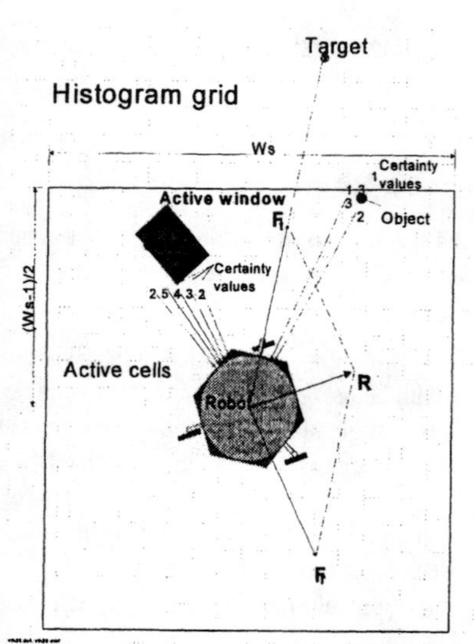

Figure 1. The Virtual Force Field (VFF) concept: Occupied cells exert repulsive forces onto the robot, while the target applies an attractive force.

cle moves, a square "*window*" accompanies it, overlying a region of C. We call this region the "*active region*" (denoted as C^*), and cells that momentarily belong to the *active region* are called "*active cells*" (denoted as $c^*_{i,j}$). The window is always centered about the robot's position.

Each active cell exerts a virtual repulsive force $F_{i,j}$ toward the robot. The magnitude of this force is proportional to $c^*_{i,j}$ and inversely proportional to d^n, where d is the distance between the cell and the center of the vehicle, and n is a positive number (usually, $n = 2$). All virtual repulsive forces add up to yield the resultant repulsive force $\boldsymbol{F}_r$. Simultaneously, a virtual attractive force $\boldsymbol{F}_t$ of constant magnitude is applied to the vehicle, "pulling" it toward the target. Summation of $\boldsymbol{F}_r$ and $\boldsymbol{F}_t$ yields the resultant force vector $\boldsymbol{R}$. The direction of $\boldsymbol{R}$ is used as the reference for the robot's steering command.

In the course of ealier experimental work with the VFF algorithm it was found that a robot guided by this method alone tended to develop oscillations when in cluttered environments or traveling in down narrow passages. Koren and Borenstein, [1991] introduced a mathematical analysis of these inherent problems. As this analysis shows, one main contributing factor to instability of motion with potential field methods (PFMs) is the non-linear force function of the repulsive forces. However, non-linearity is employed by most PFMs to assure that the repulsive force is strong enough to be effective at a certain distance from the obstacle, so that the robot can begin an avoidance maneuver in time.

3. THE VECTOR FIELD HISTOGRAM (VFH)

To overcome these problems, Borenstein and Koren [1991] developed a new obstacle avoidance method called the *vector field histogram* (VFH). The VFH method builds the *histogram grid* the same way the VFF method does. However, the VFH method then introduces an intermediate data representation called the *polar histogram*. The *polar histogram* retains the statistical information of the *histogram grid* (to compensate for the inaccuracies of the ultrasonic sensors), but reduces the amount of data that needs to be handled in real-time. This way, the VFH algorithm produces a sufficiently detailed spatial representation of the robot's environment for travel among densely cluttered obstacles, without compromising the system's real-time performance.

The *polar histogram* H is an array comprising 72 elements; each element represents a 5°-sector of the robot's surroundings. During each sampling interval, the active region of the *histogram grid* C* is mapped onto H as shown in Fig. 2, resulting in 72 values that can be interpreted as the instantaneous *polar obstacle density* around the robot.

After the *polar histogram* has been constructed, the VFH algorithm computes the required steering direction for the robot, Θ. A polar histogram typically has peaks (sectors with high *obstacle density*), and valleys (sectors with low obstacle density). Any valley with obstacle densities below a selected threshold level is a candidate for travel. Since there are usually several candidate-valleys, the algorithm selects the one that most closely matches the direction to the target. Note that the width of a valley, W_V, can be measured in terms of the number of consecutive sectors that are below threshold. A small W_V indicates a narrow passageway or corridor. For narrow valleys, i.e., closely spaced obstacles, Θ is chosen to be the center of the valley.

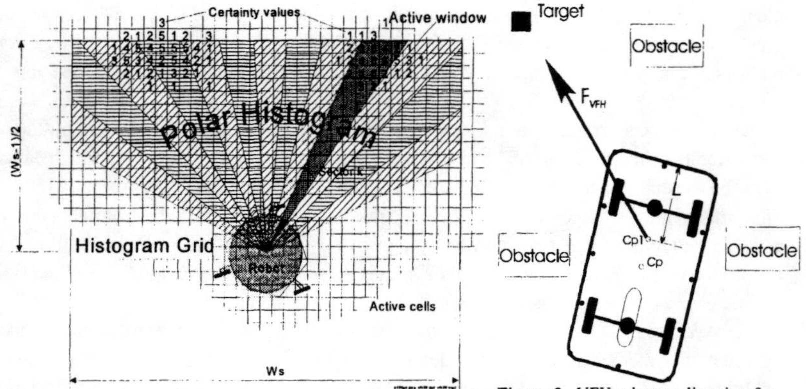

Figure 2. Mapping the *histogram grid* onto the *polar histogram*.

Figure 3. VFH selects a direction for travel.

4. OMNINAV FOR NON-CIRCULAR, OMNIDIRECTIONAL ROBOTS

While both VFF and VFH methods were originally designed for *point-sized* robots, a combination of these methods can be used to efficiently guide non-point mobile robots through densely cluttered obstacle courses. The combination of the two methods allows us to use each one to its advantage, while avoiding the disadvantages. For example, since stable motion and better spatial resolution are the strength of the VFH method, it is used to determine what we call the "*principal steering direction*." The VFF algorithm, on the other hand, can provide local corrective measures to account for the shape of the vehicle. Limiting the potential fields-based VFF method to a corrective function, we can use a *steep* force profile with only *short-range* effects. This way we reduce the oscillatory tendency of PFM-based obstacle avoidance. The following discussion explains in more detail how the OmniNav system works.

a) Computing the principle steering direction Θ with the VFH method (Figure 3)
The task of the VFH component is to find the openings through which the robot should travel to reach the goal. To do so, the VFH algorithm is applied at a point CP_1 to determine the principal steering direction Θ. A vector $\boldsymbol{F}_{VFH}$ is computed and applied in that direction. CP_1 is located on the longitudinal axis of the vehicle, but its optimal location (in terms of distance L from the front of the vehicle) differs for different vehicles.

b) Local protection with the VFF method (Figure 4)
The VFF method is applied to each one of the n act-on points A_n on the periphery of the robot. Technically, this is done for each act-on point A_n by adding up all individual repulsive forces $\boldsymbol{F}_{i,j}$ from filled cells in the histogram grid, yielding n repulsive forces $\boldsymbol{F}$, (note that Fig. 4 does not show the histogram grid explicitly). The force fields that act on each act-on point are very steep; in our application they are generated by $\boldsymbol{F}_{i,j} \propto 1/d^5$. This way, the effective range is very short. Consequently, $\boldsymbol{F}_k \neq 0$ only when an act-on point is very close to an obstacle. To avoid having the repulsive forces impact the speed of the vehicle, the force component that is parallel to the direction of travel is eliminated.

c) Extracting the relevant information

As can be seen in Figure 5, there are times when two adjacent act-on points straddle an obstacle and may thus generate conflicting signals. In the first part of Figure 5 the robot is traveling directly to the right, approaching the small obstacle. After the repuslive forces are projected perpendicular to the direction of travel, the two remaining force vectors tend to cancel each other. To reduce this problem, the repulsive forces felt by each act-on point are always oriented toward the center of the robot. This way, the repulsive forces protect the whole robot, instead of each act-on point individually.

In calculating the moments on the robot, however, a different treatment is needed. At the corners, the moments are found by computing the cross product of the VFF-generated force with the arm from the center point of the robot to the act-on point. In the third illustration of Figure 5 it can be seen that forces along the side of the vehicle, if computed in the same manner, will actually rotate it toward an obstacle. Thus, at the side act-on points, the components of force along the length of the robot are eliminated from the moment calculation.

d) Manipulating the correction information

The kinematic control program for the robot and the simulation is designed to accept translation velocity and orientation commands [Borenstein, 1995]. The translation command begins with the summation $\mathbf{V} = a\boldsymbol{F}_{VFH} + b\boldsymbol{F}_{VFF}$, where $\boldsymbol{F}_{VFH}$ is the vector found by the VFH method, and $\boldsymbol{F}_{VFF}$ is the sum of the repulsion forces found by the VFF method (see Fig. 6). The coefficient a is determined as $a = 1/W_V$, i.e., the inverse of the width of the candidate-valley selected by the VFH method (see Sect. 3.2). The effect of this latter scaling factor is as follows: when the robot circumnavigates a single obstacle, the candidate-valley will be wide and a will be small. Consequently, the correctional effect of $\boldsymbol{F}_{VFF}$ will be quite significant. In a narrow corridor, on the other hand, the robot is forced to be close to the walls and very large repulsive forces develop (because of $\boldsymbol{F}_{i,j} \propto 1/n^5$). Furthermore, small diversions of the robot from the centerline result in dramatic fluctuations of the forces and consequently in oscillatory motion. However, a will be relatively large and will dominate the resultant vector $\boldsymbol{F}_S$. This way, the oscillatory behavior usually associated with potential field control is avoided.

The orientation command begins with a sum of moments $W = aW_{VFH} + bW_{VFF}$. W_{VFF} is completed from the forces on each act-on point, and W_{VFH} is the angular difference between the vehicle's current orientation and the VFH. The coefficient b indicates the tendency of the robot to align itself in the free direction. This helps avoid any situation where the robot straddles an opening and generates conflicting moments from the act-on points to either side (see Fig. 5).

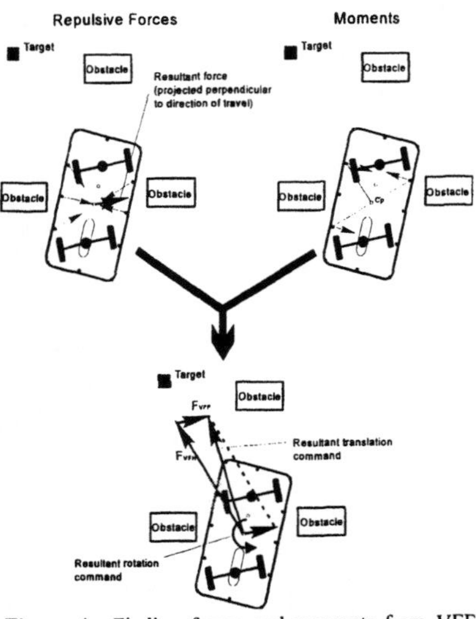

Figure 4. Finding forces and moments from VFF applied to the act-on points. Summation with the VFH-based direction yields the final direction of travel.

After summing the moments and forces from VFH and VFF, the two commands are then normalized with respect to each other as follows:

$$|V| = \frac{d \cdot V}{\sqrt{V^2 + W^2}} \qquad |W| = \frac{e \cdot W}{\sqrt{V^2 + W^2}}$$

Where e and d are weighting coefficients.

The effect of this normalization is to slow the robot's translation down when experiencing large moments, giving it time to adjust its orientation as necessary before proceeding. It also provides a means of keeping the commands to levels that are controllable by the kinematic control algorithm [Borenstein 1995].

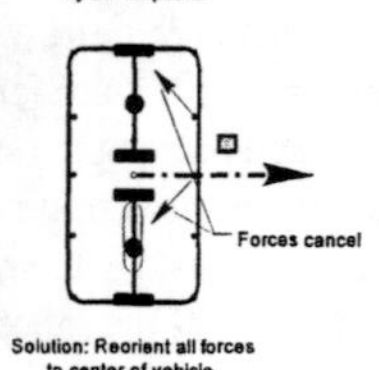

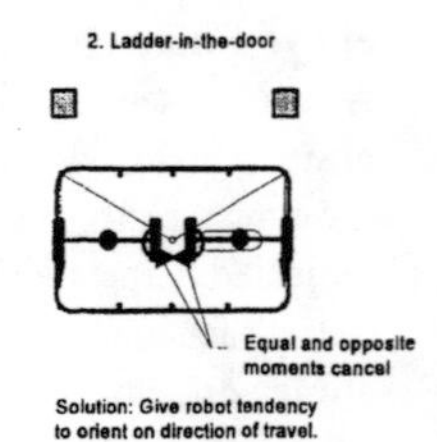

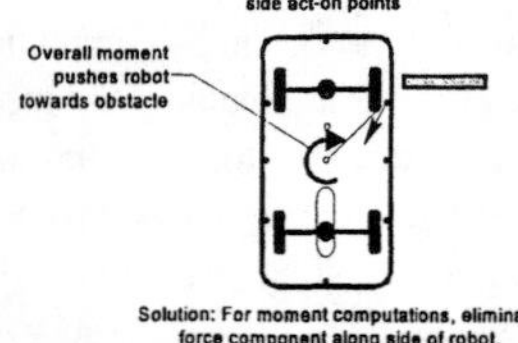

Figure 5. Special cases in the application of the OmniNav method and their heuristic solutions.

5. SIMULATION RESULTS AND FUTURE TESTBED

To verify the performance of the OmniNav method, a multi-degree-of-freedom robot, called the Compliant Linkage Autonomous Platform with Position Error Recovery (CLAPPER), will be employed as the platform. The CLAPPER is a four-degree-of-freedom robot, consisting of two differential-drive, TRC LabMate robots in tandem, connected by a compliant linkage. The controller of the CLAPPER coordinates the motion of the two TRC trucks and assures smooth, omnidirectional motion [Borenstein 1995]. Figure 6 shows a photograph of the CLAPPER, which is currently being instrumented with 32 ultrasonic sensors. The results presented in this section, however, are simulation results only. We expect to have actual performance results from the real robot available for presentation at the conference.

5.1 Simulation Setup

The OmniNav method was simulated by adapting the simulator used to develop the CLAPPER. In this simulation, an obstacle course can be created and is then converted into a histogramic map, from which the VFH and VFF commands can be generated. The simulated robot uses odometry and OmniNav obstacle avoidance to move from an arbitrary starting point to an arbitrary target. Figure 7 shows the type of virtual obstacle used to simulate the program. The square poles are 100mm on a side, and are spaced such the most optimal paths among them would have at most 25 cm of combined clearance

Figure 6: The University of Michigan's CLAPPER platform is currently being equipped with 32 ultrasonic sensors for testing the OmniNav method.

on both sides of the robot. This minimum spacing is consistent with the ultrasonic sensor array to be used on the real robot, in which the transducers have a minimum range of 27 cm.

5.2 Simulation Results

A sample run through the previously described obstacle course is shown in Figure 7. With the OmniNav obstacle avoidance algorithm the simulated robot did maintain a minimal clearance of 10 cm from all obstacles at all times. On a 486/50 MHz PC, the average sample time was less than 20 ms, and the simulated velocity of the robot averaged 230 mm/s.

6. CONCLUSIONS

In this paper we have introduced the OmniNav reflexive obstacle avoidance method. The OmniNav method combines the previously developed VFH and VFF methods and can be applied to robots of arbitrary shape to guide them through densely cluttered obstacle courses. The system has been verified by simulation and is currently being installed on a real test platform.

Acknowledgments

This research was funded by the Department of Engergy grant DE-FG02-86NE37969. The authors wish to thank Brian Costanza, Liquang Feng, Chris Minekime, and Iwan Ulrich for their many useful suggestions.

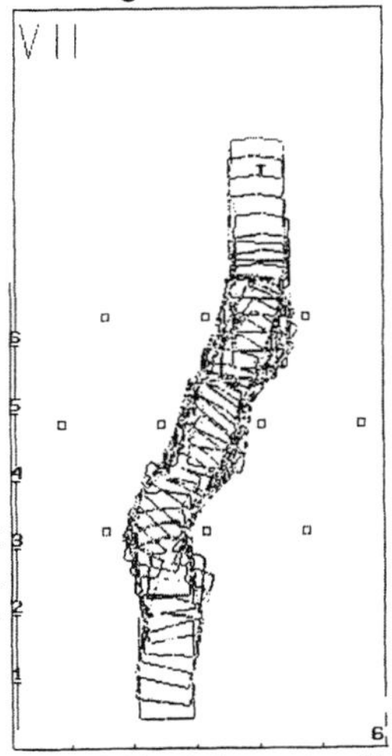

Figure 7: Simulation results with the OmniNav method.

7. REFERENCES

Barraquand, J.,Langloi, B., and Latombe, J.C., 1992, "Numerical Potential Field Techniques for Robot Path Planning." *IEEE Transactions on Systems, Man, and Cybernetics*, Vol. 22, No. 2, pp. 224-241.

Borenstein, J. and Koren, Y., 1989, "Real-time Obstacle Avoidance for Fast Mobile Robots." *IEEE Transactions on Systems, Man, and Cybernetics*, Vol. 19, No. 5, Sept/Oct, pp. 1179-1187.

Borenstein, J, and Koren, Y., 1990, "Real-time Obstacle Avoidance for Fast Mobile Robots in Cluttered Environments." *Proceedings of the IEEE Conference on Robotics and Automation*, Cincinnati, Ohio, May 13-18, pp. 572-577.

Borenstein, J. and Koren, Y., 1991, "Potential Field Methods and Their Inherent Limitations for Mobile Robot Navigation." Proceedings of the *IEEE Conference on Robotics and Automation,* Sacramento, California, April 7-12, pp. 1398-1404.

Borenstein, J. and Koren, Y., 1991, "The Vector Field Histogram—Fast Obstacle-Avoidance for Mobile Robots." *IEEE Journal of Robotics and Automation*, Vol.7, No.3, pp. 278-288.

Borenstein, J. and Raschke, U., 1992 "Real-time Obstacle Avoidance for Non-Point Mobile Robots." *The Fourth World Conference on Robotics Research.*

Borenstein, J. 1995, "Control and Kinematic Design of Multi-Degree-of-Freedom Mobile Robots with Compliant Linkage" *IEEE Transactions on Robotics and Automation*, Vol. 11, No. 1, February, pp. 21-35.

Hague, T., Brady, M., and Cameron, S., 1990, "Using Moments to Plan Paths for the Oxford AGV." *Proceedings of the 1990 IEEE International Conference on Robotics and Automation*, Cincinnati, Ohio, May 13-18, pp. 210-215.

Lozano-Perez, T., 1987, "A Simple Motion-Planning Algorithm for General Robot Manipulators." *IEEE Journal of Robotics and Automation*, Vol. RA-3, No. 3, pp. 224-238.

Moravec, H.P. and Elfes, A., 1985, "High Resolution Maps from Wide Angle Sonar." *IEEE Conference on Robotics and Automation*, Wahington D.C., pp. 116-121.

Problems of adaptive positioning systems with pneumatic cylinders

Krzysztof. B. Janiszowski, Mariusz Olszewski
Institute of Automatics and Robotics, Warsaw University of Technology
02-525 Warsaw, ul. Narbutta 87, Poland, email iap@mp.pw.edu.pl

Abstract

Problems of application of new ideas in control of pneumatic positioning systems are described. A model of system and its local presentation are introduced together with used control systems for point-to-point (P-P) positioning and tracking positioning tasks. Identification of parameters of model and problems of states reconstruction are discussed. Some results of laboratory experiments of tracking positioning and point-to-point positioning are presented.

Keywords: adaptive control, identification, pneumatic cylinder, tracking control.

1. Introduction

A proportional servo-valve and pneumatic cylinder integrated with a fine measurement and signal processor control unit create together a modern positioning system. Such system can be fast, light, simple and robust as pneumatic elements, contemporary precise and programmable as electrical servomechanisms. A modern control in hydraulic positioning systems has longer tradition. In case of pneumatic cylinders, compressed flows together with a strong nonlinear friction of piston, induce problems in dynamics of closed-loop control system. The PD or PDD controllers, common in hydraulic control, were not useful for pneumatic - oscillatory systems and introduce big overshoots in closed-loop [11,12]. More suitable seem to be state space controllers with pole-placement design [1,11,12].

Manufacturers of positioning systems try to use a gain matrix approach or calculate physical parameters of mechanic systems for determination of controller settings. In both cases, beside displacement, controller processor has to know many information, like e.g. position of cylinder, direction of movement, pressure of supply air, kind of sealing, friction coefficients or a mass of load. These approaches for controller tuning are not adaptive.

The control theory approach, based on parametric identification or adaptive learning is not common. A tuning of controller is not easy one, but most important is a question how to model and handle with variable, not measurable friction forces. An interesting approach, based on observer of not measurable disturbances [3], did not yield sufficient quality of control. Nevertheless an application of modern control seems to be the only way for creating full adaptive, pneumatic positioning systems.

2. Pneumatic cylinder as control plant

A dynamics of a loaded pneumatic cylinder, Fig. 1a, can be locally presented as a system of a mass suspended on two air springs and subjected to a friction, Fig. 1b and in result a simplified dynamic presentation is equivalent to an oscillatory linear system, Fig 1c.

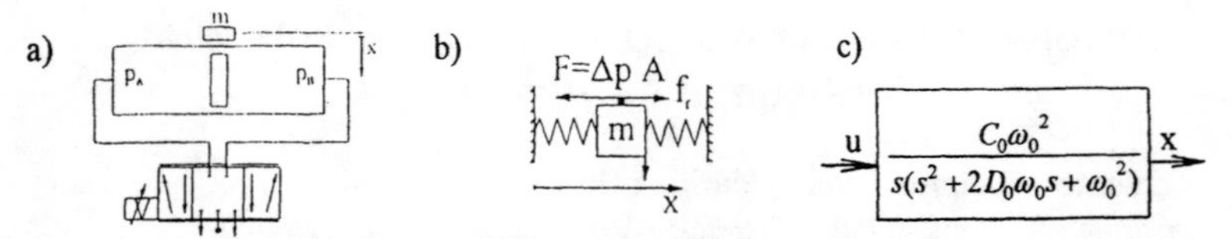

Fig. 1. a) loaded pneumatic cylinder, b) local presentation of system structure, c) linear approximation of dynamic of system transfer function

A space state presentation of system dynamics can be presented by equations

$$\frac{d}{dt}\begin{bmatrix} x \\ v \\ a \end{bmatrix} = \begin{bmatrix} 0, & 1, & 0 \\ 0, & 0, & 1 \\ 0, & -\omega_0^2, & -2D_0\omega_0 \end{bmatrix}\begin{bmatrix} x \\ v \\ a \end{bmatrix} + \begin{bmatrix} 0 \\ 0 \\ C_0\omega_0^2 \end{bmatrix} u \tag{1}$$

$$y = [1, \; 0, \; 0][x, \; v, \; a]'$$

where x, v, a are piston position, velocity and acceleration and u denotes control of proportional servo-valve. A transformation from flow and balance equations to the form (1) is described in e.g. [5,11]. Relations between dynamic parameters: gain C_0, eigenfrequency ω_0 and dumping factor D_0 and physical parameters are following

$$C_0\omega_0^2 = \frac{RA\theta}{m}\left[\frac{n_a k_M}{V_a} + \frac{n_b k_M}{V_b}\right], \qquad 2D_0\omega_0 = \frac{k_F}{m}, \qquad \omega_0^2 = \frac{A^2}{m}\left[\frac{n_a p_a}{V_a} + \frac{n_b p_b}{V_b}\right] \tag{2}$$

where A is a surface of piston, V_a, V_b and p_a , p_b are air volumes and pressures in cylinder, θ is an air temperature, m is a mass, factors k_F and k_M express influences of friction and flow characteristic of valve. n_a, n_b presents a polytropic gas transformation and R is the gas constant. These relations reveal a local character of presentation (1). The values of dynamic system parameters are usually within ranges $C_0 \in$ <0.15, 4.5 m/s/V> (at $\pm$10V valve input), $\omega_0 \in$ <7-100 rd/s> and $D_0 \in$ <0.1, 2.5> but the dynamic parameters for the same valve-cylinder system can vary from values determined for a centre of cylinder with factors: (0.4, 3). These variations motivate looking after adaptive procedures in control algorithms.

3. Control in pneumatic positioning systems

The used structure of pneumatic positioning systems is presented on. Fig. 2a. A control is based on a position feedback with states controller. The complete control system structure can be presented as on. Fig. 2b. A feedback [K_x, K_v, K_a] determines dynamic parameters (C, ω, D) of a closed-loop system, Fig. 2c. .

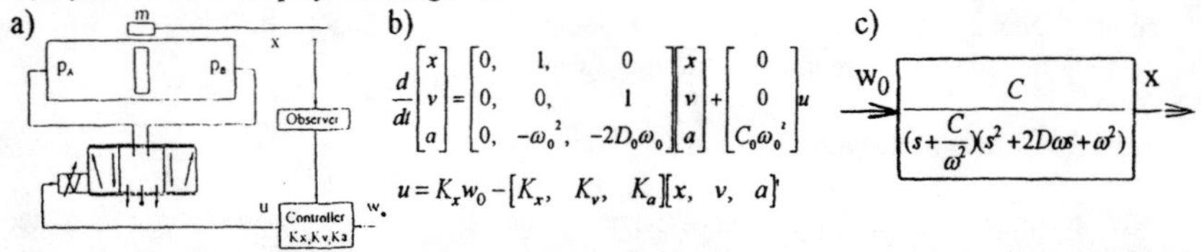

Fig. 2. State variable control of pneumatic positioning system: a) structure of system, b) control algorithm, c) transform function of the closed-loop system

Small values of positioning error can be achieved by a large coefficient K_x. Other used approach, e.g. an integral action in controller algorithm will induce stick-slip effects in control. The design of controller feedbacks is usually made with pole-placement technique for predefined closed-loop system dynamics, e.g. [11,12]. Experimental rules for tuning of closed-loop system to the dynamic parameters (C_0, ω_0, D_0) are presented in [6]

$$C = \mu\omega_0^{\,2} / 2\varepsilon_m, \qquad \omega \cong 1.45\omega_0, \qquad D = 0.05 \div 0.4 \tag{3}$$

where ε_m is acceptable error, μ is estimated valve hysteresis and value of D depends on character of transients at positioning. Small overshoots values can be achieved for $D \cong 0.05$ but then the control time will be long and valve action is excessed, Fig. 3c. On the Fig. 3b are presented standard valve action and positioning transients. Fig. 3a presents best transients - time minimal positioning. An adaptive control with iterative model estimation demands rather safe control rules, hence transients like on Fig 3c are common.

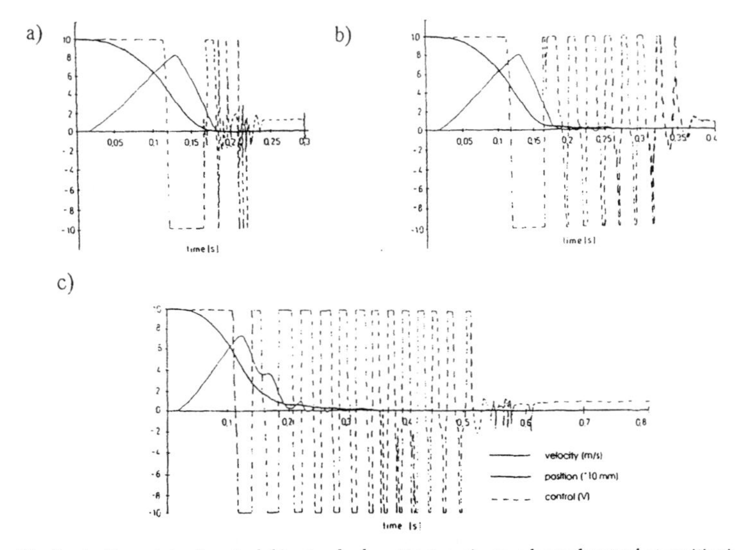

Fig. 3: a) - time minimal control, b) - standard positioning, c) - overdumped control at positioning

The classic state-space control in P-P point positioning is performed without influence on transients, Fig. 3. Hence valve action is to strong and piston accelerations can be to big.

Other algorithm of control can be developed with application of Dynamic Matrix Control (DMC) approach. A successful application of predictive control idea with state-feedback was derived in [7]. This way of positioning can be defined as follows

$$\begin{aligned}
&1^o\ x(t) \rightarrow x_0(t),\ v(t) \rightarrow v_0(t),\ a(t) \rightarrow a_0(t)\ for\ (t_p,\ t_k)\\
&2^o\ x(t_p) = x_0(t_p),\ x(t_k) = x_0(t_k),\ v(t_p) = 0,\ v(t_k) = 0,\ a(t_p) = 0,\ a(t_k) = 0\\
&3^o\ v_0(t) = \int a_0(\tau)\,d\tau,\ x_0(t) = \int v_0(\tau)\,d\tau\ for\ (t_p,\ t_k).
\end{aligned} \tag{4}$$

The piston has to follow predefined trajectories of position $x_0(t)$, velocity $v_0(t)$ and acceleration $a_0(t)$ with boundary conditions for start (t_p) and stop (t_k). The controller algorithm fits the actual action u(t) of controller to a task of tracking the trajectories $x_0(t)$, $v_0(t)$ and $a_0(t)$ for an interval (t, t+H T), where H is a prediction horizon and T is a sampling interval. As in DMC algorithm are calculated demanded states trajectories: $\underline{x}_0 = [x_0(t+T), ..., x_0(t+HT)]'$, $\underline{v}_0 = [v_0(t+T),..., v_0(t+HT)]'$ and $\underline{a}_0 = [a_0(t+T),..., a_0(t+HT)]'$ and predicted state vectors: $\underline{x}_{pr} = [x_{pr}(t+T),..., x_{pr}(t+HT)]'$, $\underline{v}_{pr} = [v_{pr}(t+T),..., v_{pr}(t+HT)]'$ and $\underline{a}_{pr} = [a_{pr}(t+T), ..., a_{pr}(t+HT)]'$. The control u has to minimize a performance index

$$I=\left[\underline{x}_0-\underline{x}_{pr}\right]^T \underline{Q}_x\left[\underline{x}_0-\underline{x}_{pr}\right]+\left[\underline{v}_0-\underline{v}_{pr}\right]^T \underline{Q}_v\left[\underline{v}_0-\underline{v}_{pr}\right]+\left[\underline{a}_0-\underline{a}_{pr}\right]^T \underline{Q}_a\left[\underline{a}_0-\underline{a}_{pr}\right]+\underline{u}^T \underline{P}\underline{u} \quad (5)$$

where $\underline{Q}_x$, $\underline{Q}_v$, $\underline{Q}_a \in R^{H \times H}$ are positive defined matrices and $\underline{P}$ is a matrix of control effort. This algorithm yields quiet positioning with controlled transients, e.g. Fig 4,

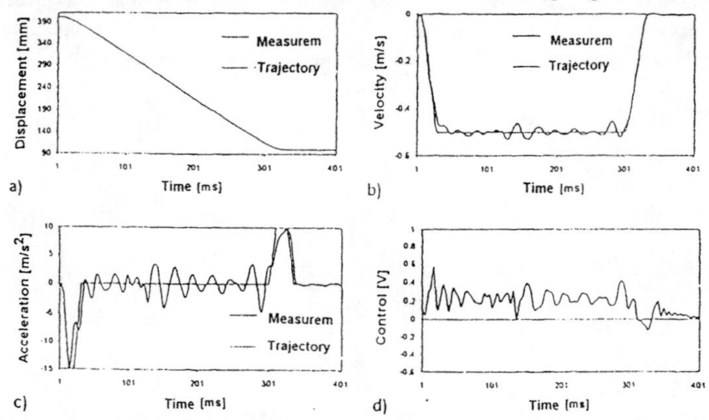

Fig. 4. Transients at predictive control in pneumatic positioning system: a) piston position, b) piston velocity, c) piston acceleration, d) valve activity (continuous lines denote demanded states trajectories)

The error in position control is under 1% of demanded displacement and errors of other states are limited too. Most remarkable is very smooth valve action. The positioning time it short, however the maximal piston velocity is not so high as at Fig.3. It is result of decreasing of interval fine control, which was long Fig. 3b, 3c. This way of control is very convenient for the applications with limited or demanded profile of the piston velocity.

4. State variable reconstruction

All state variable algorithms need information of actual states of the system. There are different ways for evaluation of the state for this particular application, e.g. [2]. The state observers need a good knowledge of model parameters (1) or (2). Hence an observer, determined for a medium value of piston velocity, will be not efficient at very low piston velocity, when the impact of adhesive friction forces is much more visible on the piston behaviour. At the noiseless measurement systems one can are apply simple formulas

$$v(t)=\frac{x(t)-x(t-\Delta)}{\Delta}, \qquad a(t)=\frac{v(t)-v(t-\Delta)}{\Delta} \quad (6)$$

of numerical derivation with a derivation interval Δ equal to 4÷8 T. This way of reconstruction is simple but in case of fast cylinders, a phase shift introduced by the derivation interval Δ is remarkable and induces oscillations in a closed-loop system. For removing of this effect an extrapolation algorithm (7) was derived and tested [2]

$$v(t) = \frac{217x(t) - 62x(t-T) - 145x(t-2T) - 102x(t-3T) - 3x(t-4T) + 82x(t-5T) + 83x(t-6T) - 70x(t-7T)}{252T} \tag{7}$$

Similar formula have been used for calculation of acceleration.

5. Identification of system dynamics

An effective identification of the local system dynamics is most important problem in adaptive control based on parametric model of the system. An application of classic iterative schemes [9] to this problem was not robust because of limited interval of identification, see. e.g. Fig. 3. The interval of the piston acceleration is usually equal to 80 - 150 ms and so is the braking interval (when the controller has already to know good estimation of system parameters!). At sampling interval T equal 1 ms, possible data for identification are based on 50 - 200 samples of signals, what is not enough to perform robust parametric identification. There was derived a new approach to the identification problem, based on application of state space model of the system [4]. For the considered system the investigated parameters (C_0, ω_0, D_0) can be estimated from a discrete-time model

$$\begin{bmatrix} x(k+1) \\ v(k+1) \\ a(k+1) \end{bmatrix} = \begin{bmatrix} 1, & T, & 0 \\ 0, & 1-\alpha T, & \beta T \\ 0, & -2\alpha\beta, & 1-\alpha T-2\beta(1-\beta) \end{bmatrix} \begin{bmatrix} x(k) \\ v(k) \\ a(k) \end{bmatrix} + \begin{bmatrix} 0 \\ C_0 T\alpha \\ 2C_0\alpha\beta \end{bmatrix} u(k) \tag{8}$$

with $\alpha = 0.5\omega_0 \mathrm{T}$ *and* $\beta = 1 - D_0\omega_0\mathrm{T}$.

The estimation of (C_0, ω_0, D_0) with application of the above model is very efficient in case of adaptive control system and allows robust detect changes of system parameters due to e.g. change of load mass of moving parts. A typical transients of estimated parameters shows Fig. 5

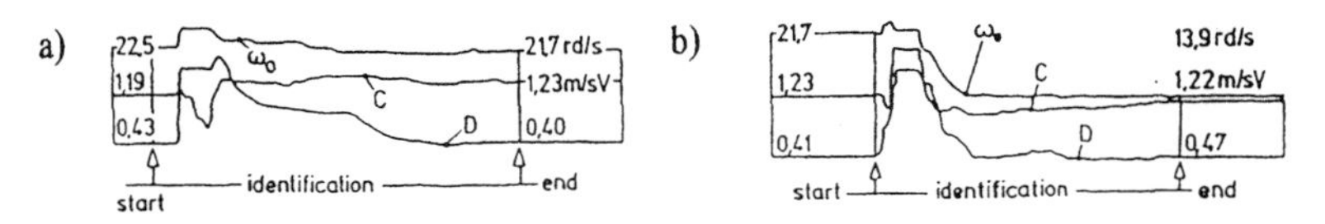

Fig. 5. Estimation of system parameters: a) no variation load, b) increase up to 250 % of load.

It can be observed an interval (appr. 40-50 samples) of transients in estimation of the dynamic parameters (C_0, ω_0, D_0) and the rest of estimation did not change of parameters.

6. Adaptive pneumatic positioning systems - concluding remarks

The presented results of adaptive control in pneumatic positioning systems are only a short presentation of variety of problems observed in these systems. Very important, here not explicit mentioned problem, is a delay in control system and observed nonsymmetry, very important in at vertically mounted, loaded cylinders. Development of estimation procedures for structured model of valve-cylinder system is now performed.

The present state of adaptive P-P positioning (with identification, controller design and states reconstruction) can be considered as robust way of control. Only for small piston displacements (0.1 - 5 mm), when estimation interval is to short, the identification is not robust. Then a model from introductory identification experiment is used. The parameters of P-P positioning: error appr. 20 μm, overshoot not greater 100 μm at displacements from 1 mm up 90% cylinder length, for standard quality pneumatic cylinders with teflon sealing and 5 μm resolution measurement systems are acceptable. The tracking positioning is more comfortable but it induces problems at parameter estimation - actuation abilities of input signal (see Fig. 4d) are limited and poor in comparison of P-P control (see Fig. 3a, b, c).

The other methods of determination of the controller settings - like fuzzy approach are more easy but not so fast in adaptation as described space controllers. The neural networks approach for modelling nonlinear phenomena in pneumatic systems is now intensively investigated. The first results are not very optimistic due to long calculation time of model of system but the future can provide new faster algorithms.

7. Acknowledgements

This results were obtained under financial support of Polish Scientific Research Committee, grant Nr S5/8/8506/92/03.

8. References

1. Ackermann,J.(1985): Sampled Data Control Systems, Springer, New York.
2. Chudzik, Z., Olszewski M.(1994): Problems of state reconstruction in positioning systems with pneumatic drive. 4th Intern. Symp.on Fluid Control, Measurem. and Visual., Toulouse, 1994.
3. Isermann,R. Keller,H.(1992): Intelligent actuators, IFAC Workshop on Motion Control..., Perugia, Italy
4. Janiszowski, K. (1995): Identification of Parameters of dynamic Processes for discrete-time Control Systems, SACCS'95 Conf. Iasi, pp 234-240
5. Janiszowski,K., Olszewski M. (1994): Modelling and identification for determination of dynamics of pneumatic positioning system, MMAR 94, Conf., Poland, pp. 231-237,
6. Janiszowski,K., Olszewski M. (1994): State Space Adaptive Control for Nonlinear Systems, IFAC Symp. on Advances in Control Education, Tokio, Japan,
7. Janiszowski,K., Olszewski M. (1995): Multitrajectory position control of servopneumatic drives, Fluid Power Conf. , Olomuc, pp 1-11.
8. Klein, A.(1993): Einsatz der Fuzzy-Logik zur Adaption der Positionsregelung fluidtechnischer Zylinderantriebe, PD Thesis, Aachen,Germany
9. Ljung, L. Soderström,T. (1983): Theory and practice of recursive identification, Cambridge Mass. MIT Press
10. Olszewski M., Janiszowski, K. (1993): Automatische Inbetriebnahmeidentifikation an servopneumatischen Zylinderantrieben für Lageregelungen. 9.Fachtagung "Hydraulik und Pneumatik", Dresden 1993, 441-452.
11. Olszewski,M.(1991): Konzept der Zustandregelung fur schwachgedämpfte Fluidantriebe. Ölhydraulik und Pneumatik, 35, ss. 932-941
12. Scholz,D. (1990): Auslegung servopneumatischer Antriebssysteme, PhD Thesis, Aachen, Germany

DYNAMIC OPTIMIZATION OF GRASPING MOBILE OBJECTS BY ROBOT MANIPULATORS

Anne-Dominique Jutard-Malinge, Guy Bessonnet
Laboratoire de mécanique des solides, URA CNRS, Université de Poitiers S.P.2M.I., BP 179, 86960 Futuroscope Cedex, France

ABSTRACT

The study deals with a grasping in motion of objects by robot manipulators. This method allows dynamic optimization of the grasp operation, i.e. the place and the time, during the manipulator transfer. The purpose of this study is to determine the conditions of an optimal grasp between the manipulator configuration and the object position. This resulting transfer problem with its final state characterized by local constraints is solved by applying Pontryagin's Maximum Principle, which yields new optimality conditions. An example of grasping an object placed on a moving belt is illustrated in this study to show the practice of our theory.

KEYWORDS : Robotics, Grasping, Optimization, Dynamics

INTRODUCTION

In manufacturing industry, robots are mainly used for tasks of object handling, such as the loading of tool machines, the transfer of objects from a moving belt for loading palettes... The handling of objects is generally made by a robot of manipulator arm type at fixed work station if the work zone is limited. However, if the transfer has to be completed over a distance that goes beyond the range of the manipulator or over a long distance, one may use a robot with a mobile base or a mobile robot.

In general, grasp and placement sites are fixed and perfectly defined. The transfer cycle can be analyzed in two phases. The approach phase (phase I) corresponds to the movement of the robot from an initial position to the unknown place where the object is grasped. Moving the object to the placement makes up the removal phase (phase II).

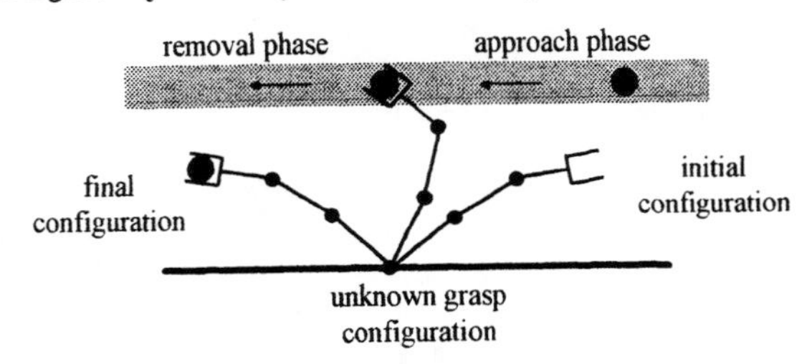

Figure 1. Grasp in flight of a mobile object.

Using current applications often leads one to place too much importance on the control of the kinematics of the system during this operation. Nevertheless, these types of approaches will not allow to apprehend the dynamics. Others studies have been devoted to the dynamic aspect in order to improve mechanical conditions of the object transfer [1-4]. Problems of optimal transfer concern robot manipulators marked by a movement executed between two completely specified configurations [5-7]. These initial and final configurations are always reached to zero velocity. These transfer problems of fixed objects imply that the place of the grasp is known, as well as the configuration of the manipulator robot at the time of the grasp.

Another point of view is based on not only the removal phase (phase II), but also the approach phase of the manipulator. The initial configuration of the manipulator and the configuration of the placement are known. Grasping of the object then becomes an intermediate operation. The localization of this grasp and/or the configuration of the manipulator at this time are not fixed a priori. This intermediate state is not completely specified, but is characterized by conditions that will define the grasp. It is thus possible to continue toward a global optimization of the movement, including phases I and II, and also to minimize the selected performance criterion on the operation's complete cycle. In the case of an object grasped in movement, placed on a moving belt for example, it is a priori better not to choose the place of the grasp and the manipulator configuration, in order to optimize this operation. Working with these choices, a slight or great economy of time and energy may be achieved.

We will deal with this intermediate operation which is the grasping of an object, considering the relative velocity between the manipulator and the object in movement. It will therefore concern a dynamic grasping of the object. The approach presented here allows one to optimize the dynamics of the grasp and transfer operation by determining the optimal place and the optimal grasp time without interrupting the dynamics of the manipulator. Local constraints characterizing the grasp in flight are formulated in the first part of our study. They define the forms of the grasp by linking the manipulator and object kinematics, i.e. the configuration of the manipulator to the position of the object. The problem of dynamic optimization is dealt with trough the optimal control theory by using Pontryagin's Maximum Principle. The constrained intermediate state implies new necessary conditions of optimality that is developed in the second part. The proposed approach is illustrated by the study of the grasping of an object placed on a moving belt by a 3R planar manipulator.

TRANSFER PROBLEM FORMULATION

The approach phase (phase I) leads to the study of the dynamic problem with the initial state entirely defined (initial configuration) and with the final state characterized by local constraints defined at the time the object is grasped. The removal phase (phase II) corresponds to a dynamic problem with initial state entirely defined, namely the final state of the phase I, which is obtained after finding the solution of the first problem. The final state of the phase II, specified by the placement, is completely defined. These two problems have to be successively solved to reach an optimal solution of the complete grasp cycle. Phase II corresponds to a problem already solved [7],consequently in this study we will only develop the new problem defined by phase I.

Movement equations

The particularity of the problem developed here is due to the final conditions of the phase I. However, the formulation of the dynamic model is identical to a problem marked by the completely defined initial and final states. For a manipulator with n degrees of freedom, described by n joint parameters $q_i = q_i\,(t)$, the direct dynamic model, using the Hamiltonian form, is :

$$t \in [t_0, t_1]\ ,\ \dot{x}(t) = f\big(x(t), u(t)\big) \tag{1}$$

with : $x = (x_1,...,x_{2n})^T$ phase variables (n joint parameters and n conjugate momentum)

$\dot{x} = (\dot{x}_1, ... \dot{x}_{2n})^T$ phase velocities

$u = (u_1,...,u_n)^T$ control variables

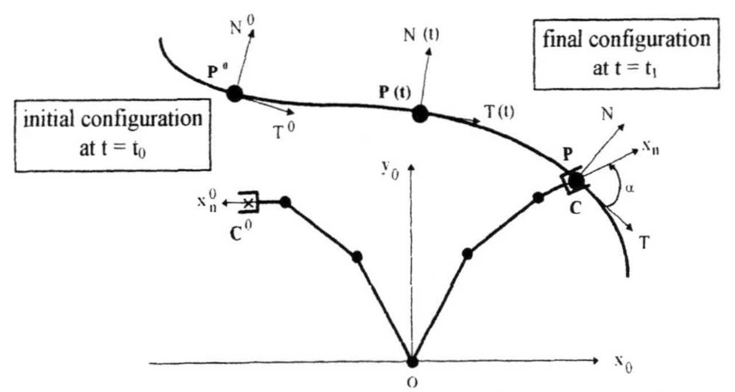

Figure 2. Case of an object picked up in flight by a manipulator.

Boundary conditions

The conditions specifying the initial state of the manipulator are :

$$x\,(0) = x^0 \tag{2}$$

The state of the system at the grasp time t_1 is characterized by next local relationships :

- position constraint : $\quad \mathbf{OC}\,(t_1) = \mathbf{OP}\,(t_1) \quad (3)$

- velocity constraint : $\quad \mathbf{V}\,(C(t_1)) = \mathbf{V}\,(P(t_1)) \quad (4)$

- orientation constraint: $\quad (\mathbf{T}\,(t_1), \mathbf{x}_n\,(t_1)) = \alpha \quad (5)$

- rotation velocity constraint : $\quad \mathbf{\Omega}\,(S_n\,(t_1)) = \mathbf{\Omega}\,(\mathbf{T}(t_1),\mathbf{N}(t_1)) \quad (6)$

with : $\mathbf{V}\,(A)$ velocity of a point A in the reference frame $(0\ ;\ \mathbf{x}_0, \mathbf{y}_0)$

α specified orientation angle

$\mathbf{\Omega}\,(S_n)$ rate of rotation of S_n in the referential $(O, \mathbf{x}_0, \mathbf{y}_0)$

$\mathbf{\Omega}\,(\mathbf{T}, \mathbf{N})$ rate of rotation of the local base in P to the path

These constraints can be recast in the next generic form, at grasp time t_1 :

$$\Psi_i\,(x(t_1), t_1) = 0\,,\ \ i = 1,...,k\,,\ \ k \leq 2n \tag{7}$$

OPTIMAL TRANSFER CONDITIONS

Phase and control constraints, performance criterion

As in [8-11],the unknowns of the optimal transfer problem are the phase trajectory x(t), the control vector u(t) and the final time t_1. The solution has to respect unilateral constraints introduced on actuating inputs, joint parameters and velocities in order to respect geometrical and technological limitations :

$$\forall t \in \left[t_0, t_1\right] \quad \begin{cases} \left|u_i(t)\right| - u_i^{max} \leq 0 \\ h_i(x(t)) = x_i^{min} - x_i(t) \leq 0 \\ h_{n+i}(x(t)) = x_i(t) - x_i^{max} \leq 0 \\ h_{2n+i}(x(t)) = \dot{x}_i^{min} - \dot{x}_i(t) \leq 0 \\ h_{3n+i}(x(t)) = \dot{x}_i(t) - \dot{x}_i^{max} \leq 0 \end{cases} \quad i = 1,.., n \tag{8}$$

The phase constraints $h_j(x(t))$, $j \leq 4n$ are accounted for in the penalized mixed performance criterion [10] :

$$J = \int_{t_0}^{t_1} L(x,u)\, dt \tag{9}$$

with : $L(x,u) = 1 - \mu + \frac{\mu}{2}\sum_{i=1}^{n}\left(u_i / u_i^{max}\right)^2 + \frac{r}{2}\sum_{j=1}^{4n}\left(h_j^+(x(t))\right)^2$

$\mu \in [0\,,1]$: weight factor $\qquad$ $r \geq 0$: penalty multiplier

$h_j^+(x(t)) = \underset{t \in [t_0,t_1]}{Max}(h_j(x(t)),0)$ $\qquad$ $h^+ = (h_1^+,\ldots,h_{4n}^+)^T$

μ defines the relative weight between the optimization of the traveling time and the reduced quadratic control. r is intended to tend towards infinity so as to minimize any excess of the authorized values, when the constraint is infringed. Moreover, the researched solution has to satisfy initial conditions (2) at $t = t_0$ and local constraints (7) at $t = t_1$.

Necessary conditions of optimality

To solve the stated problem of optimal control, let us introduce the Hamiltonian function :

$$H(x(t), u(t), w) = w^T\, f(x(t), u(t)) - L(x(t), u(t))\ ,\ \ w \in \Re^{2n} \tag{10}$$

Since the k constraints (7) have to be verified only at the last instant, k associated constant multipliers are introduced.

By applying Pontryagin's Maximum Principle, one obtains the following necessary conditions of optimality [12-14] :

- there exists a 2n-vector costate function $t \rightarrow w(t)$ solution of the adjoint equation :

$$\forall t \in [t_0, t_1] \qquad \dot{w}(t) = -H_{,x}^T(x(t), u(t), w(t)) \tag{11}$$

and which verifies the transversality condition :

$$\text{at } t = t_1 \qquad w(t_1) = -\Psi_{,x}^T(x(t_1), t_1)\lambda \tag{12}$$

- the Hamiltonian satisfies the maximality condition :

$$\forall t \in [t_0, t_1] \qquad H(x(t), u(t), w(t)) = \max_{v \in U} H(x(t), v, w(t)), \tag{13}$$

and the final condition at $t = t_1$:

$$H(x(t_1), u(t_1), w(t_1)) = \lambda^T \Psi_{,t}(x(t_1), t_1) \tag{14}$$

Condition (12) is due to the fact that the final state is not specified but characterized by constraints. For the same reason, the condition (14) is no longer the usual condition $H(x(t_1), u(t_1), w(t_1)) = 0$, since the vector constraint Ψ depends explicitly on t.

As in [10], the stated problem has been resolved by means of a finite difference scheme. The technique consists in reformulating all previous equations and conditions on the constant interval [0,1], by re-scaling the running time t into the new variable $\tau = t/T$. We introduce k+1 additional phase variables which satisfy the k+1 trivial differential equations $\dot{T} = 0$ and $\dot{\lambda}_i = 0$, $i = 1,...,k$. The problem is reduced to the resolution of a (4n+k+1)-dimensional ordinary differential system submitted to 4n+k+1 boundary conditions. This ordinary two-point boundary value problem has been numerically solved by using the routine D02RAF of the Fortran Nag Library which implements a finite difference method.

APPLICATION

The method presented in this study is illustrated by the grasp of a mobile object by a 3R planar manipulator (Fig. 3). The movement of the object is rectilinear, the velocity of the moving belt is constant.

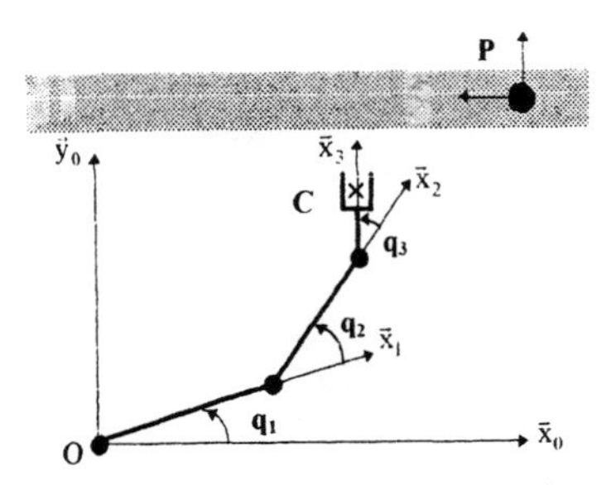

Figure 3. Grasp of mobile object by a 3R planar robot manipulator.

For the phase I, initial conditions at $t = t_0$ are :

- *Object :* $P^0/x_0 = 1.2$ m $P^0/y_0 = 1$ m
  $\alpha = 10°$ $V = 1$ m/s $\Omega = 0$ rad/s

- *Manipulator :* $q_1^0 = 0°$ $q_2^0 = 20°$ $q_3^0 = 0°$
  $\dot{q}_1^0 = \dot{q}_2^0 = \dot{q}_3^0 = 0$ rad/s

Constraints characterizing the final state of the phase I, six in number for a planar problem, are the following, at $t = t_1$:

$$\Psi_1 = r_1 \cos(q_1) + r_2 \cos(q_1 + q_2) + r_c \cos(q_1 + q_2 + q_3) - (Vt_1 + X_0)$$

$$\Psi_2 = r_1 \sin(q_1) + r_2 \sin(q_1 + q_2) + r_c \sin(q_1 + q_2 + q_3) - Y_0$$

$$\Psi_3 = -(q_1 + q_2 + q_3) + \alpha + \pi$$

$$\Psi_4 = -r_1\dot{q}_1 \sin(q_1) - r_2(\dot{q}_1 + \dot{q}_2)\sin(q_1 + q_2) - r_c(\dot{q}_1 + \dot{q}_2 + \dot{q}_3)\sin(q_1 + q_2 + q_3) - V$$

$$\Psi_5 = r_1\dot{q}_1 \cos(q_1) + r_2(\dot{q}_1 + \dot{q}_2)\cos(q_1 + q_2) + r_c(\dot{q}_1 + \dot{q}_2 + \dot{q}_3)\cos(q_1 + q_2 + q_3)$$

$$\Psi_6 = \dot{q}_1 + \dot{q}_2 + \dot{q}_3$$

For the phase II, initial conditions at $t = t_1$ are equal to the final conditions of the preceding phase, obtained after solving the approach problem. Selected final conditions are, at the unknown time $t = t_2$: $q_1^2 = 150°$ $q_2^2 = 60°$ $q_3^2 = 30°$
$\dot{q}_1^2 = \dot{q}_2^2 = \dot{q}_3^2 = 0$ rd/s.

Figure 4 presents the evolution of the system during the approach phase as well as the transfer phase of the object. The complete optimal travel time is equal to 2.96s, which represents a rapid movement considering its amplitudes. At the moment of grasp time t_1, resulting from the optimization process observed in the phase I, the manipulator configuration and the object position on the moving belt verify the six relationships characterizing the grasp. The actuator limitations and the geometrical limitations of joint movement are not surpassed, the obtained solution respects the introduced unilateral constraints (8). For each phase, one can observe two changes of concavity between the first link and the second. The action of bending the arm allows a reduction to occur in the actuating torques during the transfer.

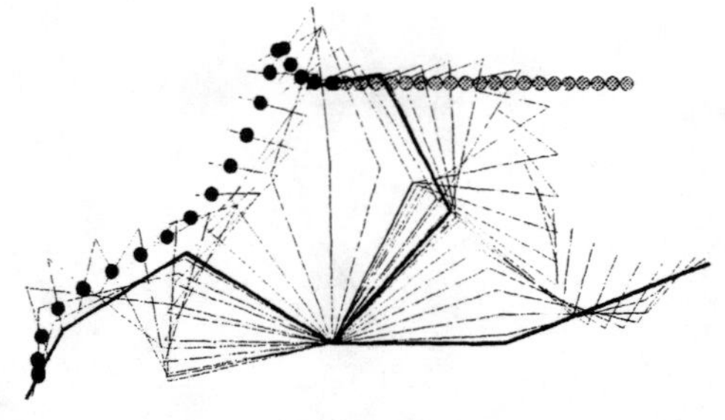

Figure 4. Optimal motion example with the approach phase and removal phase.

Figure 5 shows the continuity of velocities within the change from phase I to phase II. Actuating inputs are not saturated as shown in figure 6. One notices furthermore certain discontinuities in the change from one phase to another. The modification of mass characteristics in the moving system at the time of the object grasp (additional mass) explains this phenomenon.

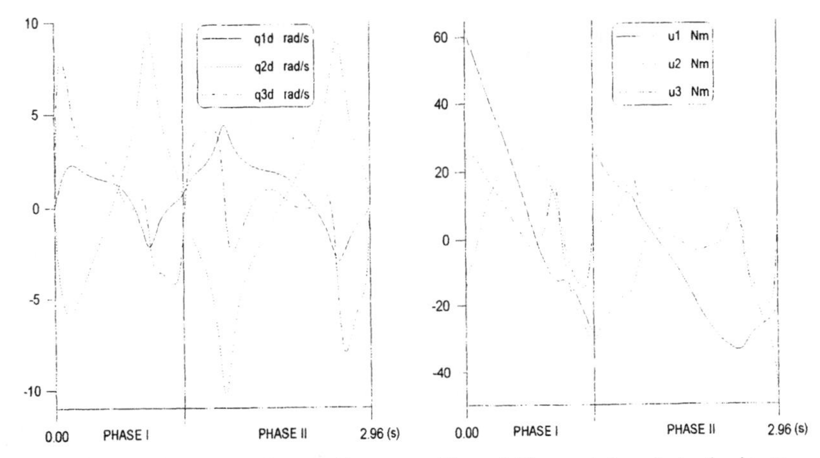

Figure 5. Time variation of joint velocities.

Figure 6. Time variation of actuating inputs.

CONCLUSION

Our study presents a dynamic optimization method to transfer objects by a robot manipulator. Here, we are primarily interested in the approach phase of grasping an object. The particularity of this study is shown by the fact that the grasp time, the location of grasp concerning mobile objects and the configuration of the manipulator at grasp time are not known. The purpose is to determine the conditions of an optimal junction between the object and the manipulator in order to execute a rapid transfer performing a grasp in motion. The resolution of this optimal transfer problem, marked by a partially defined final state and characterized by local constraints, allows for optimizing the movement of the manipulator during the approach phase until the moment when the object is grasped.

The example presented of grasping a moving object by a fixed manipulator is one of the possible applications under consideration here. Experiments are in progress to extend this study to a manipulator with a mobile base. As a general rule, every problem with an effective relative movement between the manipulator and the object to be transferred could be solved. This method will also be applied to manipulators with certain degrees of mobility that could be submitted to specified movement laws. This application would most notably allow one to take into account, in the process of optimization, any movement of the manipulator base as well as kinematic constraints concerning the end-effector, adapted to special features of the handled object.

REFERENCES

1. Dissanayake M. W. M. G., Goh C. J. and Phan-Thien N. "Time-Optimal Trajectories for Manipulators." *Robotica*, Vol. 9, pp. 131-138 (1991).
2. Shiller Z. "Time-Energy Optimal Control of Articulated Systems with Geometric Path Constraints." *IEEE Int. Conf. on Robotics and Automation*, pp. 2680-2685, San Diego (1994).
3. Shin K. G. and McKay N. D. "Minimum Cost Trajectory Planning for Industrial Robots." *Control and Dynamic Systems*, Vol. 39, pp. 345-403 (1991).
4. Singh S., Leu M.C. "Optimal Trajectory Generation for Robotic Manipulators Using Dynamic Programming." *Journal of Dynamic Systems, Measurement and Control*, Vol. 109, pp. 88-96 (1987).
5. Kim B. K., Shin K. G. "Minimum-Time Path Planning for Robot Arms and their Dynamics." *IEEE Transactions on Systems, Man and Cybernetics*, Vol. SMC 15, n°3, pp. 213-223 (1985).
6. Renaud M., Fourquet J. Y. "Time-Optimal Motions of Robot Manipulators including Dynamics." *Robotics Review* 2, PP. 225-259 (1992).
7. Bessonnet G. "Optimisation Dynamique des Mouvements Point à Point de Robots Manipulateurs." *Thesis*, Poitiers University (1992).
8. Bessonnet G. and Lallemand J.P. "Optimal Motions of Robot Manipulators with Bounded Actuator Powers." *Eighth World Cong. IFTOMM*, pp 413-416, Prague (1991).
9. Bessonnet G. and Lallemand J.P. "Planning of Optimal Free Paths of Robotic Manipulators with Bounds on Dynamic Forces" *IEEE Int. Conf. on Rob. and Automation*, pp. 270-275,Atlanta (1993).
10. Bessonnet G. and Jutard-Malinge A.D. "Optimal Free Path Planning of Robot Arms Submitted to Phase Constraints" *IASTED Int. Conf. on Robotic and Manufacturing*, pp. 51-56, Cancun (1995).
11. Bessonnet G. and Lallemand J.P. "Optimal Trajectories of Robot Arms Minimizing Constrained Actuators and Traveling Time" *IEEE Int. Conf. on Robotics and Automation*, pp. 112-117 (1990).
12. Lewis F. L. "Optimal Control" *John Wiley and Sons* (1986).
13. Pontryagin L., Boltiansky V., Gamkrelidze A., Mishchenko E. "The Mathematical Theory of Optimal Processes" *Wiley Interscience*, New-York (1962).
14. Athans M., Falb P.F. "Optimal Control" *Mc Graw Hill book company* (1966).

DATA

Manipulator :	$m_1 = 15$ kg	$m_2 = 10$ kg	$m_3 = 5$ kg	link masses
	$r_1 = 0.7$ m	$r_2 = 0.6$ m	$r_3 = 0.3$ m	link lengths
	$a_1 = 0.35$ m	$a_2 = 0.3$ m	$a_3 = 0.2$ m	gravity center positions
	$r_c = 0.2$ m			position of grasp point C
	$I_1 = 2.45$ kg m^2	$I_2 = 1.2$ kg m^2	$I_3 = 0.25$ kg m2	moments of inertia/Z_0
Object :	M = 5 kg			object mass

Bounds for actuating inputs : $u_1^{max} = 80$ Nm $u_2^{max} = 50$ Nm $u_3^{max} = 30$ Nm

Bounds for phase variables : $q_3^{max} = 135°$

Omnidirectional Planetary Wheel-type Mobile Robot

Byung-Soo Kim, Seungho Kim
Advanced Robotics Team / Korea Atomic Energy Research Institute
P.O.Box 105 Yusong, Taejon 305-600, Korea

Seung-Hong Hong
Dept. of Electronics Engineering / Inha University
253 Nam-ku, Incheon 402-020, Korea

ABSTRACT

This thesis focuses on the design and construction of Omnidirectional Planetary Wheel-type Mobile Robot (OPWMR) and the development of locomotion algorithm enabling OPWMR to be operated remotely without collision by giving an operator a sense of force that is proportional to the distance between the robot and the obstacles. The Omnidirectional Planetary Wheel (OPW) has a configuration of triangular arm in which 3 omnidirectional wheels are attached to the end of it.

The results of experiments on the floor showed that, even though there are some differences in human dexterity, the proposed force reflected algorithm is more efficient than that of general teleoperation.

KEYWORD : planetary wheel, omnidirectional wheel, force reflection, mobile robot, teleoperation

INTRODUCTION

The teleoperated mobile robot is an important device for decontamination, inspection, maintenance, and so on, because of high radiation inside a nuclear power plant [1,2]. It should be able to go over obstacles, climb up and down the stairs, and turn the narrow corner for applying in the nuclear facilities. There are 3 approaches in designing the mechanism of the mobile unit such as wheel, leg, and crawler types. Wheel mechanism has been commonly used since the ancient time due to its simple shape, high efficiency, and high speed. However, it can't go over obstacles and stairs. Other types, including crawler, have good mobility over unterrain, whereas they have a lot of problems on the floor. The planetaty wheel type is a very effective mechnism for the above application because it has both merit of wheel and leg type. The mobile robots, such as TO-ROVER, AMOOTHY,

AIMARS, and KAEROT employed the planetary wheel mechanism showed good performance on the stairs [3,4], but they have a weak point on the steering function. OPW is newly designed to improve the above defect. It has a configuration of a triangular arm in which 3 omnidirectional wheels are attached to the end of it.

The kinematic solution of OPWMR is proposed. It is based on the Muir's kinematic solution [5] except the variation of the distance between the center of robot body and each omnidirectional wheel. To evaluate the effectiveness of the designed mechanism the force reflecting algorithm is proposed such that the measured distance between the robot and the obstacles can be converted into the corresponding force and transmitted to the operator through the joystick, giving him a sense of contact with remote task environments.

In the following section, we will describe the design of OPWMR, the kinematic analysis, locomotion algorithm, and experimental results.

DESIGN of OPWMR

There are some stairs, obstacles, and narrow hall way and right angle edge in nuclear power plant. The mobile robot should be able to climb up and down the stairs and turn in narrow space. OPW is newly designed for the above functions. It enables a mobile robot to run freely in any direction on the floor by omnidirectional wheels, and to go up and down the stairs by triangular arms of planetary wheel. However, the omnidirectional wheel that can be attached to the planetary wheel mechnism is limited to 4 driving wheel mechanism. That is to say, 3 driving wheel mechanism that many omni-directional mobile robots generally adopt can't be used for OPWMR. We designed OPW of which structure is similar to the swedish wheel except the number of roller and the attachement between the wheel hub and the roller. The two modifications are considered; the number of roller is reduced 6 from 12 (swedish wheel), the attachment place of the roller is changed to the center of wheel hub from both side of it. The first modification can improve the mobility of mobile robot because the condition number of omnidirectional wheel is reduced about 50 % [6]. And the second one can increase the weight of mobile robot. Figure 1 and figure 2 show the appearance of OPW and OPWMR, respectively.

KINEMATIC ANALYSIS

The designed mobile robot has 5 DOF; x, y, and z axis translation and rotation around x and z axis, and 8 actuation joints; 4 wheels and 4 triangular arms. In this mechanism the forward and inverse kinematic solution is very difficult to solve. Therefore we devided robot kinematic model into planar and side case. The kinematic model of side case is develped to climb up and down the stairs [4]. In this paper 2D planar kinematic model is proposed assuming the triangular arm of OPW is rotated very slowly. In this case, kinematic structure of OPWMR is similar to Uanus [6] except the point that the distance between the

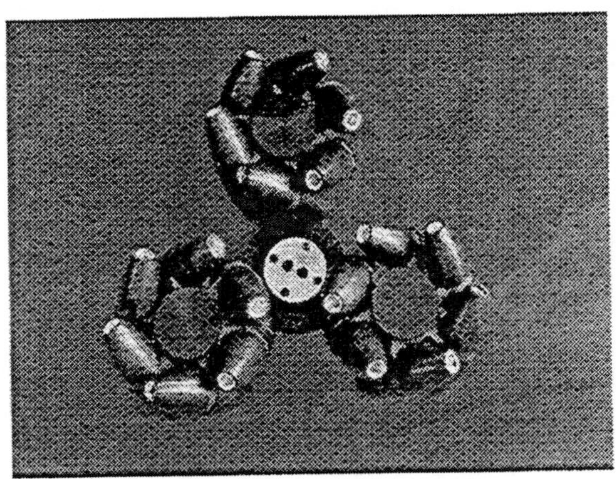

Figure 1. The appearance of OPW.

Figure 2. The appearance of OPWMR

center of robot body and omnidirectional wheel is changable during moving on the floor. Figure 3 shows 2D planar model of OPWMR. The forward and inverse kinematic solution is given as follows;

$$\dot{\theta}_w = J\dot{P} \tag{1}$$

where

$$\dot{P} = \left[{}^{\bar{B}}V_{Bx} \; {}^{\bar{B}}V_{By} \; {}^{\bar{B}}\omega_{Bz} \right]^T$$

$$\dot{\theta}_{\omega} = \left[{}^{A_1}\omega_{W_{1x}} \; {}^{A_2}\omega_{W_{2x}} \; {}^{A_3}\omega_{W_{3x}} \; {}^{A_4}\omega_{W_{4x}} \right]^T$$

$$J = \frac{1}{R}\begin{bmatrix} -1 & 1 & l_a + l_{bf} \\ 1 & 1 & -(l_a + l_{bf}) \\ -1 & 1 & -(l_a + l_{bb}) \\ 1 & 1 & l_a + l_{bb} \end{bmatrix}$$

$$\dot{P} = J^{+}\dot{\theta}_{\omega}, \quad J^{+} = (J^T J)^{-1} J^T. \tag{2}$$

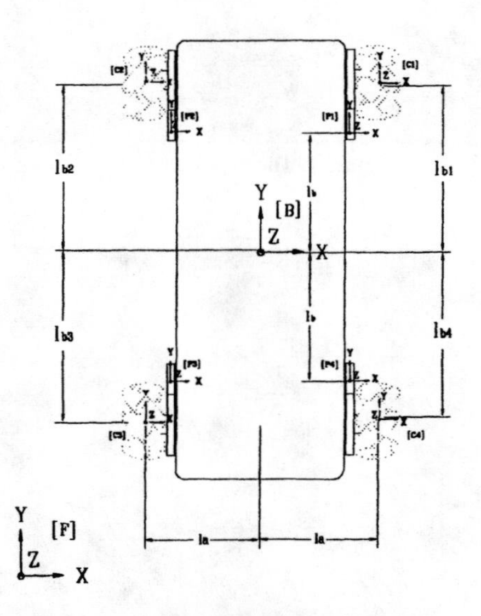

Figure 3. Kinematic modelling of OPWMR

And the length of Y-axis between robot body and omnidirectional wheel in the above equations is described

$$l_{bi} = l_b \cos\theta_a + L\cos(\theta_{ti} + \theta_a) \tag{3}$$

where

$$l_{bf} = l_{b1} = l_{b2}, l_{bb} = l_{b3} = l_{b4}.$$

LOCOMOTION ALGORITHM

Force reflecting locomotion algorithm is proposed such that, during navigation at remote site the distance between the robot and obstacles can be converted into the corresponding force by utilizing ultrasonic sensors mounted on the body of the robot, and such that the operator at remote site could control the robot successfully using the motorized joystick, allowing him to feel spatial sense with remote task environments. The force reflecting method is presented as follows. If the distances from left and right obstacles is d_1, d_2 respectively, the torque of motor attached to joystick, τ_j, is

$$\tau_j = K_D \left(\frac{1}{d_1} - \frac{1}{d_2}\right)^2 \quad (4)$$

The force reflecting gain plays great role in determining the performance of the robot. The operational parameters such as force range and optimal force reflecting gain were determined through experiments.

EXPERIMENTAL RESULTS

To evaluate the effectiveness of the designed mechanism, OPWMR, and the usefulness of force reflecting algorithm simulations and experiments were conducted on both virtual and real hallway. The experimental results, using force reflected joystick, on virtual hallway showed that the minimum torque to be sensed by the operator is 2.7 Ncm, and that the mobility of the robot is remarkably improved with the force reflected gain of 17 in figure 4. In figure 5 we compared the motion of joystick with feedback torque.

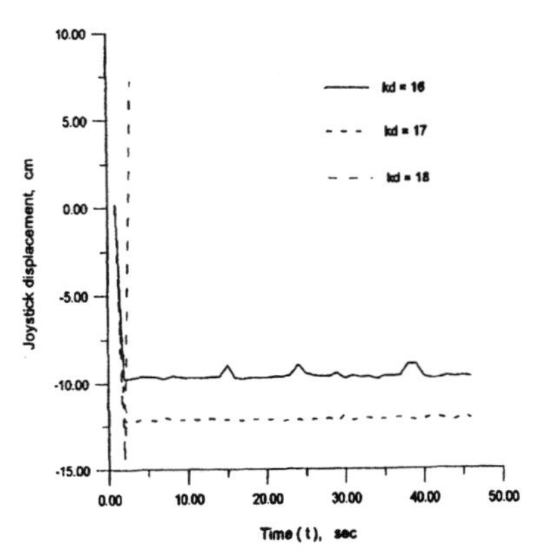

Figure 4. Experimental result for force reflection gain

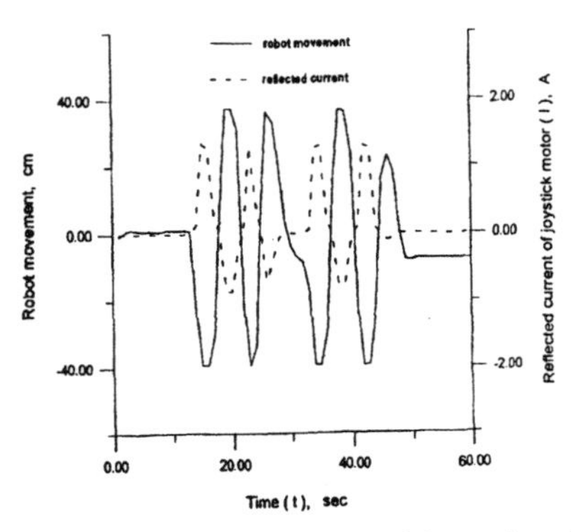

Figure 5. The comparison of the motion of the joystick and feedback torque

CONCLUSION

The importance of development in this work is in providing a teleoperated mobile robot that can be operated at a remote site to perform inspection taskes in nuclear facilities without endangering human operators. The experimental results obtained from the floor test shows that, even though there are some difference in human dexterity, the proposed force reflected locomotion algorithm is more efficient than that of general teleoperation.

Further study is being given to the improvement of OPW mechanism and the development of force reflection method including robot dynamics.

REFERENCES

1. J.R.White, Evaluation of Robotic Inspection Systems at Nuclear Power Plant, NUREG/CR-3717 (1984).

2. J.R.White, Demonstration Testing of a Surveillance Robot at Brown Ferry Nuclear Plant, NUREG/CR-4815 (1987).

3. T. Arai and et al., " A Stair-Climbing Robot for Maintenance : AMOOTHY, " Conf. on Robotics and Remote Handling in Nuclear Industry (1984), 444-458.

4. B.S.Kim and et al., " Teleoperated Mobile Robot (KAEROT) for Inspection in Nuclear Facilities, " Proc. of 2nd Specialist's Meeting on AIR, JAPAN (1994), 369-379.

5. Patrick Fred Muir, Modelling and Control of Wheeled Mobile Robots, Ph.D. Dissertation, Dept. of Electrical and Computer Science, The Robotics Institute CMU (1988).

6. B.S.Kim, Locomotion Algorithm of Omnidirectional Planetary Wheel Type Mobile Robot, Ph.D. Dissertation, Dept. of Electronics Engineering, Inha University (1995).

DECLARATIVE DESIGN BASED ON FUNCTIONAL AND TECHNICAL FEATURES

Yuji KOYAMA and André CROSNIER

LIRMM, 161 rue Ada, 34392 Montpellier Cedex 5, France
Phone: (+33) 67 41 85 56, Fax: (+33) 67 41 85 00
e–mail: koyama@lirmm.fr, crosnier@lirmm.fr

ABSTRACT

The paper suggests an approach that aims to define complementary representations used as links between the functional aspects of the design and the geometrical aspects of the product. This approach is based on the declaration of the functional choices adopted by the designer. It uses functional features dedicated to the representation of working principles. The framework of the study is concerned with the assembly of mechanical parts including the motion and the contact between components.

KEYWORDS: Design Process, Functional Features, Declarative Design

INTRODUCTION

The design process can be defined as a resolution process associated with the activity of searching a path through a space representing all the possible solutions that the designer knows [1 , 6]. Each step of this process involves the manipulation of the following elements: the requirements, the functions that fulfill the requirements, the alternatives that can be adopted by the designer, the constraints to take into account, and finally the solutions retained from the resolution of the constraints. The current method for designing a product results from the functional analysis [3]. This method consists in a first step to specify the requirements leading to the definition of design problems. In a second step, the method proceeds to the decomposition of each design problem into sub–problems, more simple to resolve. The decomposition process is carried out as far as working principles can be associated with each sub–problem. However, the CAD/CAM systems which play a crucial role do not offer facilities for modeling the different concepts manipulated by the designer during the design process. Especially, the proper functionalities of the product defined during the conceptual design step can not be expressed, explicitly. First of all, this situation prevents to easily capture the intents of the designer. And on the other hand, it prevents to establish and to maintain the links between the "functional" and the "geometrical". In [4], Shah suggests the Feature Based Modeling approach that aims to procedurally define the product model as a combination of form features instantiated from generic features placed in a library. On the other hand in [5], a declarative approach that aims to explicitly state the spatial relationships between geometric entities constituting a feature is suggested. However, the features have been created at the beginning in order to associate an engineering significance with the geometric entities. Consequently they represent topological structures lacking of functional significance.

The paper suggests an approach that aims to define complementary representations used as links between the functional aspects of the design and the geometrical aspects of the product. This approach is based on the declaration of the functional choices adopted by the designer. It uses functional features dedicated to the representation of working principles. The framework of the study is concerned with the assembly of mechanical parts including the motion and the contact between components.

APPROACH

Design process

A first definition of the product coming from the Value Analysis can be given: "The product has to fulfill the requirements through a set of functions associated with it ". Our approach is based on the logic resulting from the problem solving process (Requirements – Functions – Product) which starts by the analysis of the requirements leading to the functions to be performed by the product. Then the different components of product appear. The concept of functionality plays a key role during the design process. This last one can be decomposed into four phases (Figure 1) [6]: the specification step leads to elaborate a requirements list, the conceptual design step in which the overall function is decomposed into sub–functions leading to determine the function structures and the principle solutions, the preliminary design step leads to the specification of layout of a technical system (structure construction), and finally the detailed design step in which the arrangement, the geometry and the dimensions of individual parts are laid down. As the designer advances in his task, the representation of the product becomes more concrete.

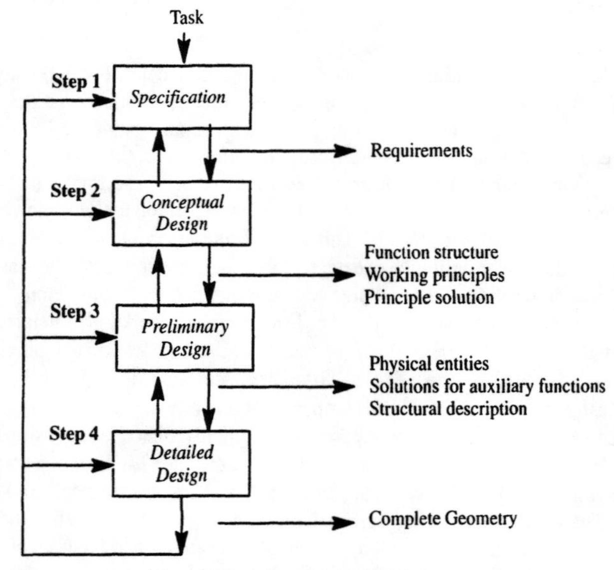

Figure 1. Basic framework of the design process.

Because the design process is a step–by–step analysis, the link between two representations of the product associated with two different steps which are not in order is not kept up. For instance, the link between the functions identified during the conceptual design step and the geometric entities elaborated during the detailed design is not explicitly maintained. Our approach aims to associate with the conceptual design step a description of the working principles retained by the designer.

Example

We present in this section an example used as support in the paper for explaining the different concepts of the approach. The geometric model of the product is represented on Figure 2. The overall function that can be associated with this product is expressed by the following sentence: "The piece *A* must be temporally immobilized with respect to the piece *B* in order to be cut out". Some assignments are presented in Table 1 concerning the environment.

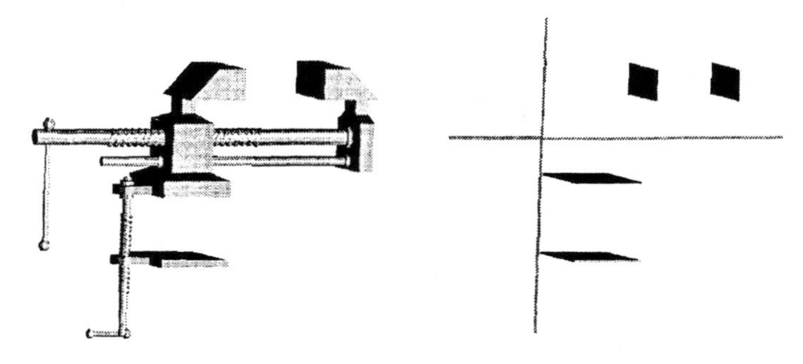

Figure 2. Geometric and functional models of the product.

Piece *A*	Shape	Cylinder of revolution Max. radius: 30mm Max. length: 200mm
	Material	Steel
	Weight	2kg
Piece *B*	Shape	Parallelepiped Height: 40mm
	Material	Wood

Table 1. Assignments relating to the environment.

The specification step leads to the requirements list as the examples presented in Table 2.

Requirements:
1 – The cut plane must be orthogonal to the axis of revolution of *A* 2 – The trajectory of the cut tool is contained into the plane yz 3 – The power is supplied by a human operator

Table 2. Table of requirements.

DECLARATIVE DESIGN

Functional features

According to [4], a feature can be defined as a set of geometric entities associated with a shape, behavior and engineering significance. From this basic definition, we define the concept of functional feature as a structured set of entities involved in the description of the working geometry that fulfills a specific function. This definition can be illustrated by the example presented on Figure 3. With the function "Clamp the piece *A* with respect to *R0*" is associated one working principle based on the motion characterizing the relative translation of two plane contact surfaces. This principle is illustrated by a working geometry in which the motion is represented by the axis of translation. From the working geometry, it is possible to identify a set of working entities: the contact surfaces (*S1*, *S2*), the axis of translation (*AX1*). These entities represent the functional invariant of the working principle.

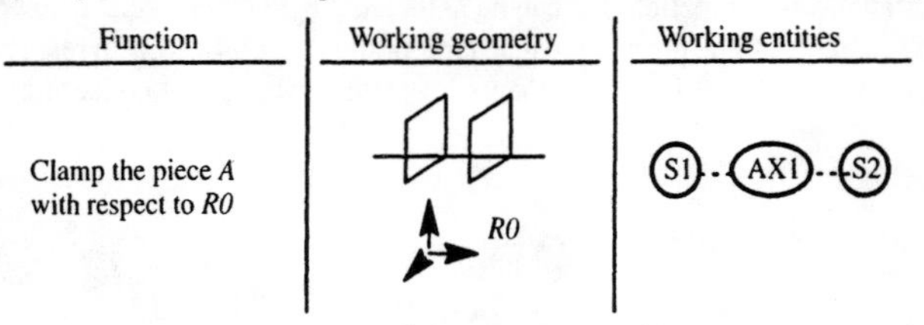

Figure 3. Working entities.

The requirements are used in a first time to realize the choice of a working principle, and in a second time, they lead to establish a set of constraints that must be taken into account in order to define the product from the working principle. The constraints induce geometrical relationships between the working entities explicitly stated through a undirected connected graph whose the nodes are the initial working entities and whose the links are the relationships. On the other hand, the working principle is schematized as a 3D skeleton referring to the graph. For the function presented on Figure 3, the set of constraints leading to the graph and the 3D skeleton are illustrated on Figure 4.

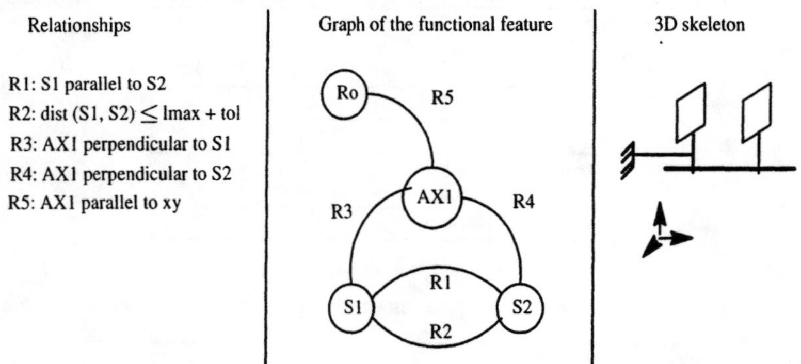

Figure 4. Representation of a functional feature.

Declarative design

We define the declarative design as an approach that aims to explicitly represent the function and the working principle associated with it. The principle solutions being chosen, it is necessary to establish the structural representation of the product which is mainly based on entities associated with a physical significance and basically representing one component. Some physical entities are directly identified and characterized from the initial working entities that interact with elements of the environment, like *J1* (resp. *J2*) which refers to the node *S1* (resp. *S2*) and inherits the properties from *S1* (resp. *S2*) (Figure 5). Moreover, the working principles inducing a motion of the working entities require the definition of auxiliary functions fulfilling the corresponding requirement. Different working principles may be associated with one auxiliary function (i.e. a translational motion can be realized by either a translation or the combination of a translation with a rotation). Depending on the choice, new physical entities (like the worm *W1*) which permit to realize the function may emerge. Then, the knowledge of the requirement, on one hand, and the working principle, on the other hand, is sufficient to define the nature of the links between the different physical entities.

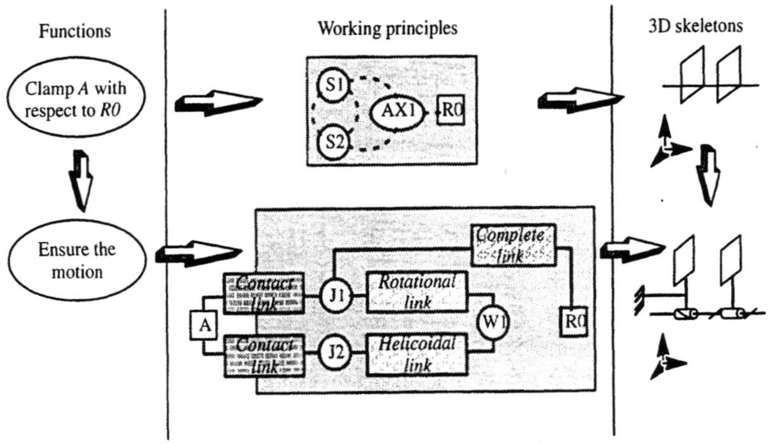

Figure 5. Framework of the declarative design

REPRESENTATION OF THE PRODUCT

During the design, different complementary viewpoints associated with the product are manipulated by the designer. Each viewpoint can be represented by a particular model. We distinguish three models. The functional model is composed of the overall function that must be fulfilled, the sub–functions resulting from the decomposition process with which the working principles represented as functional features are associated. The physical model contains the components resulting from technical solutions and the description of the structure representing their assembly. Finally the geometric model contains the geometric entities representing the components of the physical model and the set of constraints used to parameterize the geometry. Each model contains a set of constraints used either to specify its elements or to define relationships between the elements.

Each model of the product can be more or less associated with a different step of the design process (Figure 6): the functional model with the conceptual design step, the physical model with the preliminary design step and the geometrical model with the detailed design step. The declarative design offers facilities for adding to the functions complementary representations (functional features, 3D skeleton) which are used as links between the functional model and the physical model. The 3D skeleton representation resulting from the conceptual design step is progressively completed during the design process until ending in the detailed geometric model. This representation may be used either during the top–down or the bottom–up phases.

CONCLUSION

This paper addresses an approach dedicated to the declarative design that aims to represent the working principles adopted by the designer. This representation based on the concept of functional features can be used to establish links between the geometric model and the functional model. Moreover this representation can be used to report the intents of designer into the product. The future work concerns the representation of the design process. Indeed the design activity should be represented by the sequence of decisions taken by the designer.

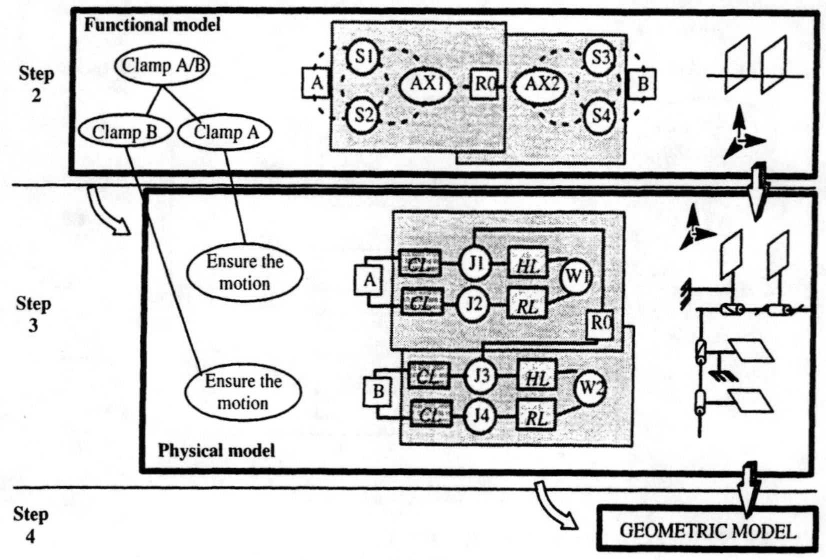

Figure 6. Models of the product

ACKNOWLEDGEMENT

The paper presents a part of the work developed in relation with the laboratories (*) involved in a French project financially supported by the Technical and Scientific Mission (DSPT8) of the French Ministry of Education and Research (MENESR).

REFERENCES

1. Bañares-Alcántara, R. "Representing the engineering design process: two hypotheses." *Computer Aided Design Journal* 23 (9) (November, 1991), 595–603.

2. Chou, Y.C.,Srinivas, R.A., Saraf, S. "Automatic Design of Machining Fixtures: Conceptual Design" *Int. J. of Advanced Manufacturing Technology* 9 (1994), 3–12.

3. Feru, F., Vat, C.,Cocquebert, E., Deneux, D., Rouchon, C. "Computer-Aided Functional design-for-manufacturing" *Int. J. of CADCAM and Computer Graphics* 9 (5) (1994), 665–683.

4. Shah, J.J. "Assessment of features technology" *Computer Aided Design Journal* 23 (5) (June, 1991), 331–343.

5. Shah, J.J., Balakrishnan, G., Rogers, M.T., Urban, S.D. "Comparative study of procedural and declarative feature based geometric modeling" *Proc. of the IFIP International Conference*, Vol.2, Valenciennes (May, 1994), 647–672.

6. Pahl, G., Beitz, W. *"Engineering design: A systematic Approach"* Springer Verlag (1996).

---

* CRAN: Centre de Recherche en Automatique de Nancy
LAMIH: Laboratoire d'Automatique et de Mécanique Industrielles et Humaines de Valenciennes
LAN: Laboratoire d'Automatique de Nantes
LIRMM: Laboratoire d'Informatique, de Robotique et de Microélectronique de Montpellier

Cooperative Tasks in Multiple Mobile Manipulation Systems

O. Khatib, K. Yokoi, K. Chang, D. Ruspini, R. Holmberg, and A. Casal
Robotics Laboratory, Computer Science, Stanford University

ABSTRACT

Mobile manipulation capabilities are key to new applications of robotics in space, underwater, construction, and the service environments. The work presented in this paper builds on models and methodologies developed for fixed-base manipulation: the Operational Space Formulation; the Dynamic Coordination of Macro/Mini structures; the Augmented Object Model; and the Virtual Linkage Model. We present the extension of these methodologies to mobile manipulation systems and propose a new decentralized control structure for cooperative tasks. Experimental results obtained with two holonomic mobile manipulation platforms built at Stanford University are also discussed.

KEYWORDS: vehicle/arm coordination, cooperation, decentralization

INTRODUCTION

A central issue in the development of mobile manipulation systems is vehicle/arm coordination. Our approach to controlling redundant systems is based on two models: an *end-effector dynamic model* [1] obtained by projecting the mechanism dynamics into the operational space, and a *dynamically consistent force/torque relationship* [2] that controls joint motions in the null space associated with the redundant mechanism.

Another important issue in mobile manipulation concerns the cooperation between multiple vehicle/arm systems [3][4]. Our approach is based on the integration of two models: The *augmented object* [5] and *virtual linkage* [6]. The *augmented object* describes the system's closed-chain dynamics, while the *virtual linkage* characterizes object internal forces. These models have been successfully used in cooperative control of two and three PUMA arms. We present here their extension to multiple vehicle/arm systems and describe a new strategy for decentralized cooperative operations.

VEHICLE/ARM COORDINATION

In our approach, a mobile manipulator is treated as a macro/mini redundant system: the "macro" mechanism with coarse, slow, dynamic responses (the mobile base), and the relatively fast and accurate "mini" device (the manipulator). With a holonomic mobile base, the joint space equations of motion of a mobile manipulator are

$$A(\mathbf{q})\ddot{\mathbf{q}} + \mathbf{b}(\mathbf{q}, \dot{\mathbf{q}}) + \mathbf{g}(\mathbf{q}) = \mathbf{\Gamma}; \tag{1}$$

where $\mathbf{q}$ is the n joint coordinates and $A(\mathbf{q})$ is the $n \times n$ kinetic energy matrix. $\mathbf{b}(\mathbf{q}, \dot{\mathbf{q}})$ is the vector of centrifugal and Coriolis joint-forces and $\mathbf{g}(\mathbf{q})$ is the gravity joint-force vector. $\mathbf{\Gamma}$ is the vector of generalized joint-forces. The operational space dynamics for a redundant manipulator are obtained by the projection of the joint-space equations of motion (1), by the *dynamically consistent* generalized inverse $\overline{J}^T(\mathbf{q})$ [2],

$$\overline{J}(\mathbf{q}) = A^{-1}(\mathbf{q})J^T(\mathbf{q})\Lambda(\mathbf{q}); \tag{2}$$

where $J(\mathbf{q})$ is the Jacobian matrix. The end-effector equations of motion are

$$\overline{J}^T(\mathbf{q})\left[A(\mathbf{q})\ddot{\mathbf{q}} + \mathbf{b}(\mathbf{q},\dot{\mathbf{q}}) + \mathbf{g}(\mathbf{q}) = \mathbf{\Gamma}\right] \quad \Longrightarrow \quad \Lambda(\mathbf{q})\ddot{\mathbf{x}} + \mu(\mathbf{q},\dot{\mathbf{q}}) + \mathbf{p}(\mathbf{q}) = \mathbf{F}; \tag{3}$$

where $\mathbf{x}$, is the vector of the m operational coordinates describing the position and orientation of the effector, $\Lambda(\mathbf{q})$ is the $m \times m$ pseudo kinetic energy matrix associated with the operational space. $\mu(\mathbf{q},\dot{\mathbf{q}})$, $\mathbf{p}(\mathbf{q})$, and $\mathbf{F}$ are respectively the centrifugal and Coriolis force vector, gravity force vector, and generalized force vector acting in operational space. The above property also applies to non-redundant manipulators, where the matrix $\overline{J}^T(\mathbf{q})$ reduces to $J^{-T}(\mathbf{q})$. The dynamic coordination and control of the mobile manipulator rely on the dynamically consistent relationship between joint torques and operational forces for redundant systems.

$$\mathbf{\Gamma} = J^T(\mathbf{q})\mathbf{F} + \left[I - J^T(\mathbf{q})\overline{J}^T(\mathbf{q})\right]\mathbf{\Gamma}_{\text{coordination}}. \tag{4}$$

This relationship provides a decomposition of joint forces into two dynamically decoupled control vectors: joint forces corresponding to forces acting at the end effector ($J^T\mathbf{F}$); and joint forces that only affect internal motions, $[I - J^T(\mathbf{q})\overline{J}^T(\mathbf{q})]\mathbf{\Gamma}_{\text{coordination}}$.

COOPERATIVE MANIPULATION

The *augmented object* model provides a description of the dynamics at the operational point for a multi-arm robot system. The simplicity of these equations is the result of an additive property that allows us to obtain the system equations of motion from the dynamics of the individual mobile manipulators. The *augmented object* model is

$$\Lambda_\oplus(\mathbf{x})\ddot{\mathbf{x}} + \mu_\oplus(\mathbf{x},\dot{\mathbf{x}}) + \mathbf{p}_\oplus(\mathbf{x}) = \mathbf{F}_\oplus \quad \text{with} \quad \Lambda_\oplus(\mathbf{x}) = \Lambda_{\mathcal{L}}(\mathbf{x}) + \sum \Lambda_i(\mathbf{x}); \tag{5}$$

where $\Lambda_{\mathcal{L}}(\mathbf{x})$ and $\Lambda_i(\mathbf{x})$ are the kinetic energy matrices associated with the object and the ith effector, respectively. $\mu_\oplus(\mathbf{x},\dot{\mathbf{x}})$, $\mathbf{p}_\oplus(\mathbf{x})$, and $\mathbf{F}_\oplus$ also have the additive property. Object manipulation requires accurate control of internal forces. We have proposed the *virtual linkage* [6], as a model of internal forces associated with multi-grasp manipulation. In this model, grasp points are connected by a closed, non-intersecting set of virtual links (Figure 1.) For an N-grasp manipulation task, the *virtual linkage* model is a $6(N-1)$ degree of freedom mechanism that has $3(N-2)$ linearly actuated members and N spherically actuated joints. By applying forces and moments at the grasp points we can independently specify internal forces in the $3(N-2)$ members, along with $3N$ internal moments at the spherical joints. Internal forces in the object are then characterized by these forces and torques in a physically meaningful way. The relationship between applied forces, their resultant and internal forces is

$$\begin{bmatrix} \mathbf{F}_{res} \\ \mathbf{F}_{int} \end{bmatrix} = \mathbf{G} \begin{bmatrix} \mathbf{f}_1 \\ \vdots \\ \mathbf{f}_N \end{bmatrix}; \tag{6}$$

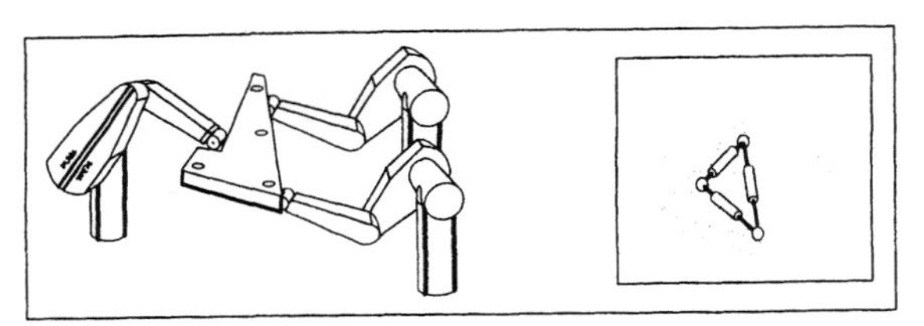

Figure 1: The Virtual Linkage

where $\mathbf{F}_{res}$ represents the resultant forces at the operational point, $\mathbf{F}_{int}$ the internal forces and $\mathbf{f}_i$ the forces applied at the grasp point i. $\mathbf{G}$ is called the grasp description matrix, and relates forces applied at each grasp to the resultant and internal forces in the object. Furthermore, $\mathbf{G}$ can be written as

$$\mathbf{G} = [\ \mathbf{G}_1 \mathbf{G}_2 \ldots \mathbf{G}_N\ ] \quad \text{with} \quad \mathbf{G}_i = \begin{bmatrix} \mathbf{G}_{res,i} \\ \mathbf{G}_{int,i} \end{bmatrix};$$

where each $\mathbf{G}_i$ represents the contribution of the ith grasp to the resultant and internal forces felt by the object and where $\mathbf{G}_{res,i}$ is the contribution of $\mathbf{G}_i$ to the resultant forces in the object and $\mathbf{G}_{int,i}$ to the internal ones. $\mathbf{G}^{-1}$ provides the forces required at the grasp points to produce the resultant and internal forces acting at the object.

$$\begin{bmatrix} \mathbf{f}_1 \\ \vdots \\ \mathbf{f}_N \end{bmatrix} = \mathbf{G}^{-1} \begin{bmatrix} \mathbf{F}_{res} \\ \mathbf{F}_{int} \end{bmatrix} \quad \text{with} \quad \mathbf{G}^{-1} = \begin{bmatrix} \overline{\mathbf{G}}_1 \\ \vdots \\ \overline{\mathbf{G}}_N \end{bmatrix} \quad \text{and} \quad \overline{\mathbf{G}}_i = [\ \overline{\mathbf{G}}_{res,i}\ \ \overline{\mathbf{G}}_{int,i}\ ]; \qquad (7)$$

where $\overline{\mathbf{G}}_{res,i}$ represents the part of $\overline{\mathbf{G}}_i$ that correspond to the resultant forces at the object; and the matrix $\overline{\mathbf{G}}_{int,i}$ represents the part corresponding to the internal forces.

Centralized Control Structure

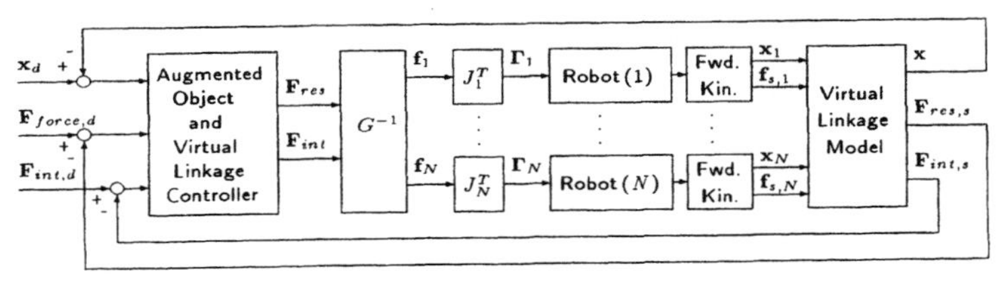

Figure 2: Centralized Control Structure

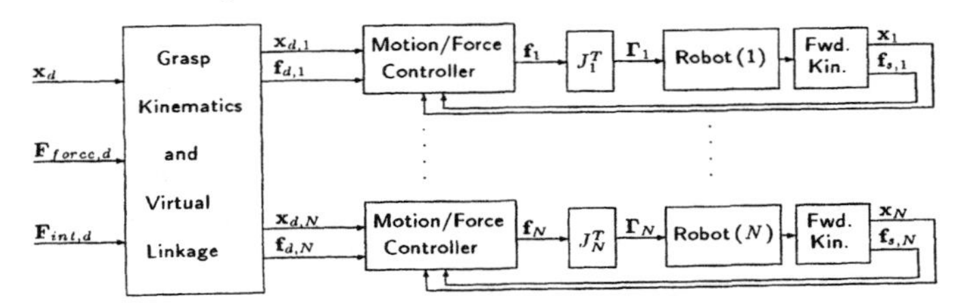

Figure 3: Decentralized Control Structure

For fixed base manipulation, the *augmented object* and *virtual linkage* have been implemented in a multiprocessor system using a centralized control structure. However, this type of control is not suited for autonomous mobile manipulation platforms. Before presenting the decentralized implementation, we begin with a brief summary of the centralized control structure. The overall structure of the centralized implementation is shown in Figure 2. The force sensed at the grasp point of each robot, $\mathbf{f}_{s,i}$, is transformed, via $\mathbf{G}$, to sensed resultant forces, $\mathbf{F}_{res,s}$, and sensed internal forces, $\mathbf{F}_{int,s}$, at the operational point, using equation (6)

$$\begin{bmatrix} \mathbf{F}_{res,s} \\ \mathbf{F}_{int,s} \end{bmatrix} = \mathbf{G} \begin{bmatrix} \mathbf{f}_{s,1} \\ \vdots \\ \mathbf{f}_{s,N} \end{bmatrix}.$$

The centralized control strategy consists of (i) a unified motion and force control structure for the *augmented object*, $\mathbf{F}_{res}$; and (ii) $\mathbf{F}_{int}$, corresponding to the internal forces acting on the *virtual linkage*. These are

$$\mathbf{F}_{res} = \mathbf{F}_{motion} + \mathbf{F}_{contact}; \tag{8}$$

where

$$\mathbf{F}_{motion} = \hat{\Lambda}_{\oplus}\Omega\mathbf{F}^*_{motion} + \hat{\mu}_{\oplus} + \hat{p}_{\oplus} \quad \text{and} \quad \mathbf{F}_{contact} = \hat{\Lambda}_{\oplus}\overline{\Omega}\mathbf{F}^*_{contact} + \mathbf{F}_{contact,s}. \tag{9}$$

$\hat{\Lambda}_{\oplus}$, $\hat{\mu}_{\oplus}$, and $\hat{p}_{\oplus}$ represent the estimates of $\Lambda_{\oplus}$, $\mu_{\oplus}$, and $p_{\oplus}$. The vector $\mathbf{F}^*_{motion}$ and $\mathbf{F}^*_{contact}$ represent the inputs to the decoupled system. Ω is the *generalized selection matrix* associated with motion control and $\overline{\Omega}$, its complement, is associated with force control. The control structure for internal forces is

$$\mathbf{F}_{int} = \hat{\Lambda}_{\oplus}\mathbf{F}^*_{int} + \mathbf{F}_{int,s}; \tag{10}$$

where the vector $\mathbf{F}^*_{int}$ represents the inputs to the decoupled system. A suitable control law can be selected to obtain $\mathbf{F}^*_{motion}$, $\mathbf{F}^*_{contact}$ and $\mathbf{F}^*_{int}$. The control forces of the individual mobile manipulator, $\mathbf{f}_i$, are given by using equation (7),

$$\begin{bmatrix} \mathbf{f}_1 \\ \vdots \\ \mathbf{f}_N \end{bmatrix} = \mathbf{G}^{-1} \begin{bmatrix} \mathbf{F}_{res} \\ \mathbf{F}_{int} \end{bmatrix}.$$

Decentralized Control Structure

In the proposed decentralized control structure, the object level specifications of the task are transformed into individual tasks for each of the cooperative robots. Local feedback control loops are then developed at each grasp point. The task transformation and the design of the local controllers are accomplished in consistency with the *augmented object* and *virtual linkage* models. The overall structure is shown in Figure 3. The local control structure at the ith grasp point is

$$\mathbf{f}_i = \mathbf{f}_{motion,i} + \mathbf{f}_{force,i}. \tag{11}$$

The control vectors, $\mathbf{f}_{motion,i}$, are designed so that the combined motion of the various i^{th} grasp points results in the desired motion at the object operational point. On the other hand, the vectors $\mathbf{f}_{force,i}$ create forces at the grasp points, whose combined action produces the desired contact and internal forces on the object. The motion control at the ith grasp point is

$$\mathbf{f}_{motion,i} = \hat{\Lambda}_{\oslash,i}\Omega\mathbf{f}^{*}_{motion,i} + \hat{\mu}_{\oslash,i} + \hat{p}_{\oslash,i}; \tag{12}$$

with

$$\hat{\Lambda}_{\oslash,i} = \hat{\Lambda}_{g,i} + \overline{\mathbf{G}}_{res,i}\hat{\Lambda}_{\mathcal{L}}\overline{\mathbf{G}}^{T}_{res,i}; \tag{13}$$

where $\hat{\Lambda}_{g,i}$ is the kinetic energy matrix associated with the ith effector at the grasp point. The second term of equation (13) represents the part of $\hat{\Lambda}_{\mathcal{L}}$ assigned to the ith robot and described at its grasp point. The vector, $\hat{\mu}_{\oslash,i}$, of centrifugal and Coriolis forces associated with the ith effector is

$$\hat{\mu}_{\oslash,i} = \hat{\mu}_{g,i} + \overline{\mathbf{G}}_{res,i}\hat{\mu}_{\mathcal{L}}; \tag{14}$$

where $\hat{\mu}_{g,i}$ is the centrifugal and Coriolis vector of the ith robot alone at the grasp point. Similarly, the gravity vector is

$$\hat{p}_{\oslash,i} = \hat{p}_{g,i} + \overline{\mathbf{G}}_{res,i}\hat{p}_{\mathcal{L}}; \tag{15}$$

where $\hat{p}_{g,i}$ is the gravity vector associated with the ith end effector at the grasp point. The sensed forces at the ith grasp point, $\mathbf{f}_{s,i}$, combine the contact and internal forces felt at the ith grasp point, together with the acceleration force acting at the object. The sensed forces associated with the contact and internal forces alone, $\mathbf{f}_{\bar{s},i}$, are therefore obtained by subtracting the acceleration effect from the total sensed forces

$$\mathbf{f}_{\bar{s},i} = \mathbf{f}_{s,i} - \overline{\mathbf{G}}_{res,i}\left(\hat{\Lambda}_{\mathcal{L}}\ddot{\mathbf{x}}_{d} + \hat{\mu}_{\mathcal{L}} + \hat{p}_{\mathcal{L}}\right). \tag{16}$$

Here, the object desired acceleration has been used instead of the actual acceleration, which would be difficult to evaluate. The force control part of equation (11) is

$$\mathbf{f}_{force,i} = \hat{\Lambda}_{\oslash i}\mathbf{f}^{*}_{force,i} + \mathbf{f}_{\bar{s},i}; \tag{17}$$

where $\mathbf{f}^{*}_{force,i}$ represents the input to the decoupled system associated with the contact forces and internal forces.

EXPERIMENTAL PLATFORMS

We have built two autonomous mobile manipulation platforms (Figure 4). These platforms were developed in collaboration with Oak Ridge National Laboratories and Nomadic Technologies. Each platform consists of a PUMA 560 arm mounted on a holonomic mobile base. The arm is equipped with a 6-axis wrist force sensor and an electric two fingered gripper. The base consists of three "lateral" orthogonal universal-wheel assemblies which allow the base to translate and rotate holonomically [7]. The base houses two PCs, motor amplifiers, and batteries. The above control strategies have been successfully implemented in the two platforms. Motion and force control performance are comparable to results obtained with fixed base PUMA arms.

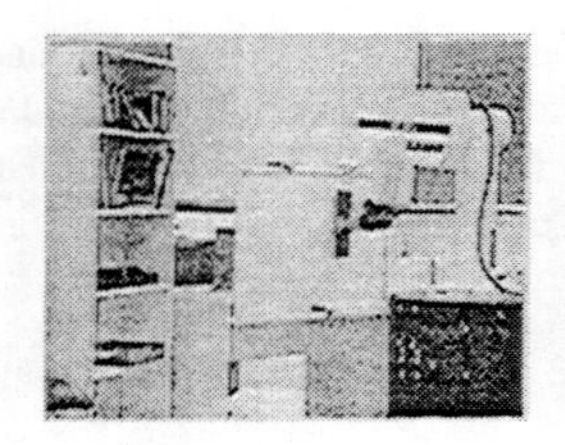

Figure 4: Stanford Mobile Platforms

Erasing a whiteboard, cooperating in carrying a basket, and sweeping a desk are examples of tasks demonstrated with the Stanford Mobile Platforms [8].

CONCLUSION

We have presented extensions of various operational space methodologies for fixed-base manipulators to mobile manipulation systems. A redundant vehicle/arm platform is treated as a macro/mini structure and is controlled using a dynamic coordination strategy. For cooperative operations, we have developed a new decentralized control structure based on the *augmented object* and *virtual linkage* models that is better suited for mobile manipulator systems. Vehicle/arm coordination and cooperative operations have been successfully implemented on two mobile manipulator platforms developed at Stanford University.

ACKNOWLEDGMENTS

The financial support of Boeing, General Motors, Hitachi Construction Machinery, and NSF (grants IRI-9320017 and CAD-9320419) is gratefully acknowledged. Many thanks to Andreas.Baader, Alan Bowling, Oliver Brock, Francois Pin, James Slater, John Slater, Stef Sonck and Dave Williams who have made significant contributions to this project.

REFERENCES

1. Khatib, O., "A Unified Approach to Motion and Force Control of Robot Manipulators: The Operational Space Formulation," *IEEE J. Rob. and Auto.*, 3(1)(1987), 43-53.
2. Khatib, O., "Inertial Properties in Robotics Manipulation: An Object-Level Framework," *Int. J. Robotics Research,* 14(1)(1995), 19-36.
3. Uchiyama, M. and and Dauchez, P., "A symmetric Hybrid Position/Force Control Scheme for the coordination of Two Robots," *Proc. ICRA*, (1988), 350-356.
4. Adams, J. A., Bajcsy, R., Kosecka, J., Kumar, V., Mandelbaum, R., Mintz, M., Paul, R., Wang, C., Yamamoto, Y., and Yun, X.,"Cooperative Material Handling by Human and Robotic Agents: Module Development and System Synthesis," *Proc. IROS* (1995), 200-205.
5. Khatib, O.,"Object Manipulation in a Multi-Effector Robot System", *Robotics Research 4*, R. Bolles and B. Roth, eds., MIT Press: Cambridge (1988), 137-144,
6. Williams, D. and Khatib, O., "The Virtual Linkage: A Model for Internal Forces in Multi-Grasp Manipulation," *Proc. ICRA* (1993), 1025-1030.
7. Pin, F.G. and Killough S.M., "A New Family of Omnidirectional and Holonomic Wheeled Platforms for Mobile Robots," *Oak Ridge National Lab. Report*, (1994).
8. Khatib, O., K. Yokoi, K. Chang, D. Ruspini, R. Holmberg, A. Casal, and A. Baader. "The Robotic Assistant," *IEEE Int. Conf. Rob. and Auto. Video Proc.*, (1996).

An Underwater Mobile Robotic System for Reactor Vessel Inspection in Nuclear Power Plants

Jae H. Kim, Jae K. Lee, Heung S. Eom
Korea Atomic Energy Research Institute, Duckjindong 150, Yousung, Taejon, Korea

Jong W. Choe
LG Electric Equipment R&D Center, Songjungdong 1, Cheongju, Korea

ABSTRACT

Reactor pressure vessels of the nuclear power plants should be inspected periodically. The conventional inspection machine with a big structural sturdy column is so bulky and heavy that installation and maintenance of this machine are extremely difficult. To solve this problem we have developed an advanced robotic inspection system by adopting recent robotic technology. In this paper, the design of this system and the brief test results using the pressure vessel mockup are described.

KEYWORDS: underwater, mobile, robot, inspection, reactor vessel

INTRODUCTION

The reactor pressure vessel is one of the important equipments in the nuclear power plant in view of its function and safety. The reactor pressure vessel is usually constructed by welding large rolled plates, forged sections or nozzle pipes together. In order to assure the integrity of the vessel, these welds should be periodically inspected using ultrasonic transducers. Ultrasonic testing is performed by scanning the weld area of interest precisely.

This examination has been performed using bulky inspection machine which is commercialized by several companies worldwide. The conventional inspection machine with a big structural sturdy column is so bulky and heavy that maintenance and handling of the machine are extremely difficult. Most of the machine have over 5 meters high column and big supports which should be mounted on the upper flange surface of the reactor pressure vessel. The machine must be operated under water to minimize exposure to the radioactively contaminated vessel walls.

Our institute, KAERI, has also used such machine to perform vessel examinations since 1985. It was a good and efficient system to inspect the reactor pressure vessel those times, however, it is so huge and heavy that it requires much efforts to transport the system to the site and also requires continuous use of the utility's polar crane to move the manipulator into the building and then onto the vessel. Setup beside the vessel requires a large volume of work preparation area and several shifts to complete.

In order to resolve these problems we have concentrated on the development of an advanced robotic machine by adopting recent robotic technology. We have finished developing the prototype system and in the near future we will use this system practically. The system will reduce dramatically the critical path process for preservice inspection of the pressurized water reactors. By adopting this compact system, overall inspection time can be greatly reduced by deploying two robots simultaneously in the vessel. In this paper, we introduce our newly developed robotic inspection system.

SYSTEM DESCRIPTION

As shown in Fig. 1, the reactor pressure vessel in pressurized water reactor has cylindrical shape. It has the inlet and outlet nozzles around upper shell. It has many welds such as circumferential seam, the weld of nozzle to upper shell, the weld of flange to upper shell and so on. When inspecting the welds of the vessel wall, the reactor head is removed to the other place and the reactor internals are also moved to the next canal so that the inspection can be performed efficiently. The reactor vessel is filled with water up to the top of the canal to reduce the radiation exposure during inspection. So the inspection machine should be operated under water.

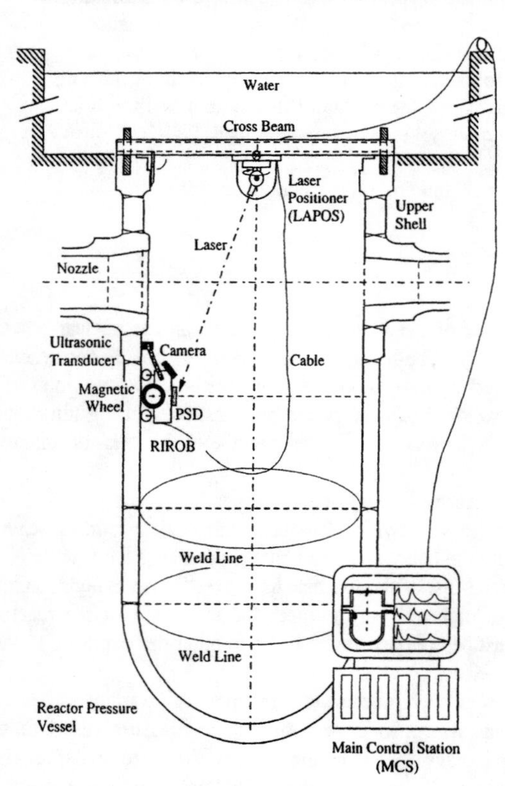

Figure 1. Configuration of the reactor vessel inspection system (RISYS).

Our reactor inspection system (RISYS) consists of the reactor inspection robot (RIROB), laser positioner (LAPOS), main control station (MCS), and so on.

Reactor Inspection Robot

RIROB is a submarine type mobile robot whose weight is approximately 30 kg in the air and zero in the water by the aid of floats. Most of the reactor pressure vessel in PWR is composed of carbon steel and clothed with austenitic stainless steel inside.

In order to climb on the vertical wall of the vessel, RIROB has four magnetic wheels. The magnetic material of the wheel is neodymium with 12.9 Kgauss of residual induction and 318 KJ/m^3 of maximum energy product. The ring shaped magnet has N and S poles on each side of the magnet. The circular pure steel plates are attached on each side of the magnet to maximize the attraction force to the vertical wall. Smooth rubber is clothed around the magnet to prevent the slippage on the vertical wall.

RIROB has four magnetic wheels: two are caster wheels and the other two are driven by the DC servo motor so that the robot can move in any direction on the vertical inner wall of the reactor vessel. The robot can control the linear velocity and angular velocity by the sum and difference of the velocities of the left and right driving wheels. Both the front and rear caster wheel are mounted on the parallelogram links with the robot body plate, as shown in Fig. 2. It always makes the robot body parallel to the wall even though the wall is cylindrical.

The robot has also the light and long manipulator, and the ultrasonic probes are attached to its end-effector. The manipulator has three degrees of freedom which are translation, rotation and 4 consecutive translations, as shown in Fig. 3. The manipulator can reach up to 100 cm using 4 consecutive translation links. It is not so easy to design the long reach manipulator kinematically, because it has the constraints to be light and not bulky to be mounted on a small mobile robot.

The camera and lamp are mounted on the robot and the visual image from the camera is transmitted to the main control station. The robot has an inclinometer to measure the inclination of the mobile robot and to control the robot posture. The depth sensor is also mounted on the robot body to measure the water pressure and to calculate the current vertical depth of the robot.

The robot has the position sensitive detector on its back and the laser positioner induce the robot to the next position by pointing the position using the laser beam.

Laser Positioner (LAPOS)

The robot is induced by the laser positioner (LAPOS) which is fixed in the middle of the crossbeam across the reactor upper flange. The laser positioner emits the laser beam to the next position for the robot to move. The robot has the position sensitive detector on its back, detects the deviation of the laser beam spot from the center of the position sensitive detector, and moves in the direction to make this deviation zero.

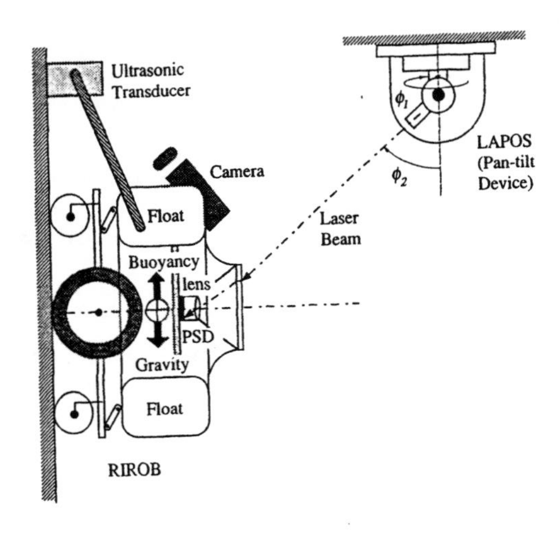

Figure 2. Schematic diagram of RIROB and LAPOS.

The laser positioner is a kind of pan-tilt device on which the diode laser is mounted. The device is accurately driven by the micro stepping motors of which resolution is less than 0.01

deg/step. The positioner induces the robot to the next position by emitting the laser beam. The laser positioner is covered by hemispherical shaped plastic cap to prevent the deflection of the laser beam and water penetration.

Main Control Station

The main control station has the function to control RIROB, the laser positioner and the sonic data acquisition subsystem. It is PC based control station with operating software and interfaces. It has the geometric information of all reactor vessels operating in Korea, so we can plan the inspections and simulate it on the 3D graphic display.

During inspection, the main control system generates the scan path for RIROB to move and the motion command for the manipulator links to move. Simultaneously, the current posture of the robot is displayed graphically and the image captured by the camera on the robot is also displayed. After inspection the examination reports are generated using the stored data. The system can also be operated in manual mode against the malfunction of computer control.

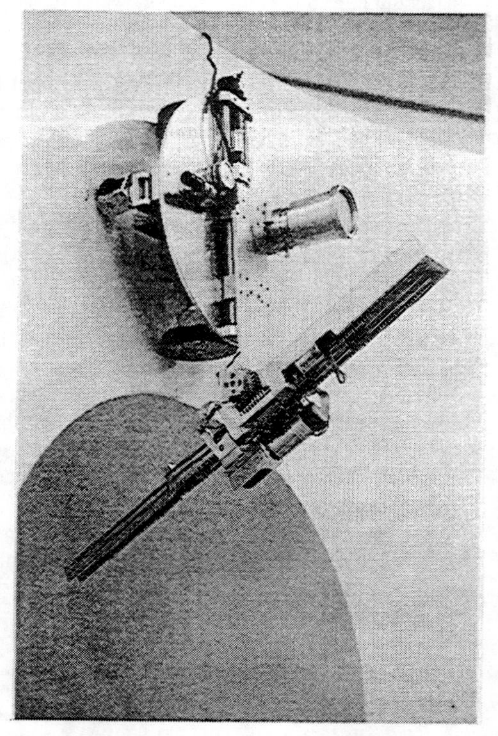

Figure 3. RIROB working on the vessel wall (mockup).

The other systems such as sonic data acquisition subsystem and the data evaluation subsystem are also under development. The data acquisition subsystem drives the ultrasonic sensor, collects the reflected signal data and then displays and stores these.

MOTION CONTROL

As shown in Fig. 2 and 3, the position sensitive detector (PSD) is mounted on the RIROB body plate. When the laser beam points a position P on the PSD surface (Fig. 4), the sensor generates the currents corresponding to the deviation (e_x, e_y) of the laser spot with respect to the center of the PSD.

The control objective is to drive the RIROB in such a way that

$$e_x = 0, \quad e_y = 0 \tag{1}$$

By considering the fact that the linear velocity of the robot center, v_c, has the relationship with the y-directional deviation, e_y, while the angular velocity of the robot center, $\dot{\Phi}$, is strongly related to the x-directional deviation, e_x. So we propose a control law as

$$
\begin{aligned}
v_c &= K_{py}\, e_y + K_{dy}\, \dot{e}_y \\
\dot{\Phi} &= (K_{px}\, e_x + K_{dx}\, \dot{e}_x) / L
\end{aligned}
\tag{2}
$$

where K_{ij} is the corresponding control gain and L is the length between the robot center and each wheel. Finally the linear velocity and angular velocity of the robot is implemented through two driving wheels, so the velocity of the left wheel, v_l, and that of the right wheel, v_r, become

$$
\begin{aligned}
v_l &= v_c + \dot{\Phi} L \\
&= K_{py} e_y + K_{dy} e_y + K_{px} \dot{e}_x + K_{dx} \dot{e}_x \\
v_r &= v_c - \dot{\Phi} L \\
&= K_{py} e_y + K_{dy} \dot{e}_y - K_{px} e_x - K_{dx} \dot{e}_x
\end{aligned}
\tag{3}
$$

Stability of the above control law was proven using Liapunov function and its control performance has been investigated through a series of experiments [1-2].

EXPERIMENTS

We performed the path tracking experiments on the reactor vessel mockup as shown in Fig. 5. The mockup is half cylindrically shaped reactor vessel, whose dimension is 2 meters high, and 4 meters in diameter. Prior to underwater experiments, we performed the experiment in the air circumstance.

The rotation angle of the laser pan-tilt device is calculated according to the predetermined path of the robot. The laser pan-tilt device is actuated and the RIROB is driven by actuation signals which are determined by the use of the control law explained in the previous section.

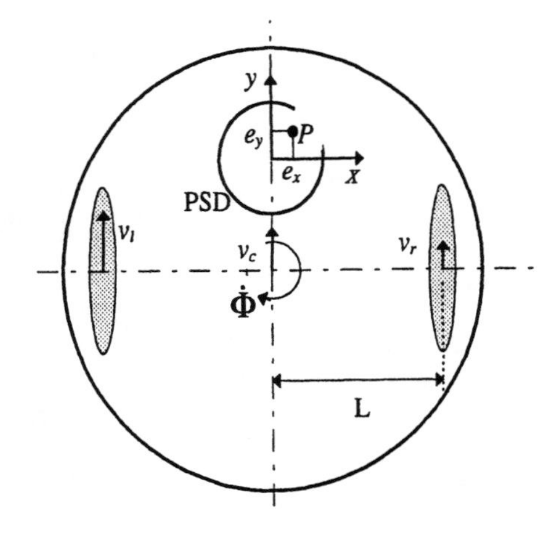

Figure 4. Position sensitive detector (PSD) and robot motion control.

The results show that RIROB moves smoothly and precisely along the given path. It is noted that the robot wheel slips in the vertical direction due to the gravity. It would be compensated when experimenting underwater by the aid of the

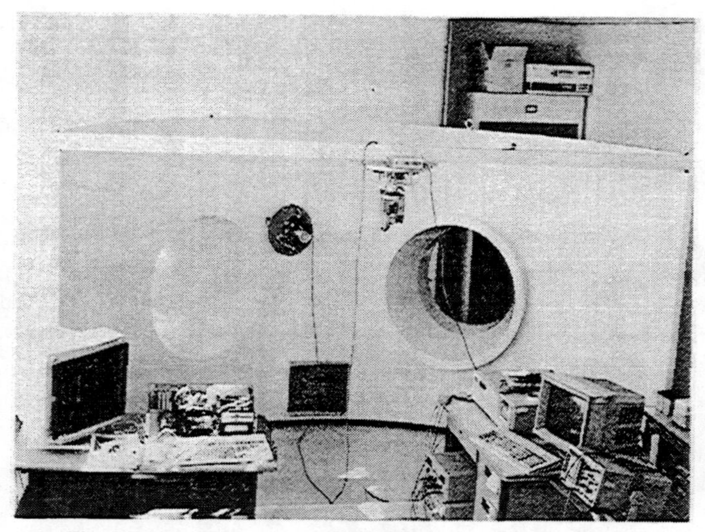

Figure 5. Overview of the experiments on the reactor vessel mockup.

floats installed on the robot for buoyancy force.

SUMMARY

In order to improve the reactor vessel inspection system, we have developed a new robotic inspection system. We completed the laser induced control of the mobile robot, and the method is thought to be applied to other industries. When our system is practically used for reactor vessel inspection instead of the conventional machine, a lot of benefits is expected such as reduction in the critical path process and improvement of handling safety, examination reliability and positioning accuracy and so on.

REFERENCES

1. Kim J. H. "Robot based Reactor Vessel Inspection Technology at KAERI." *Proceedings of 3rd KAIF/FAF Round table Conference*, Seoul, Korea (July, 1995).
2. Lee J. C. and Kim J. H. "Path Tracking Control of a Wheeled Mobile Robot using Position Sensitive Detector." *Proceedings of the Korean Automatic Control Conference*, (1) Seoul, Korea (October, 1994), 340-344.
3. Fallon J. B., Shooter S. B. Reiholtz S. W. and Glass S. W. "URSULA: Design of an Underwater Robot for Nuclear Reactor Vessel Inspection." *Proceeding of American Society of Civil Engineers, Specialty conference on robots for challenging environments*, Albuquerque, NM, (Feb. 1994).

High-Speed Image Processing Method for the Line Mark Painting Robot in Outdoor Environments

Shinji Kotani and Hideo Mori

Department of Electrical Engineering and Computer Science,
Yamanashi University, 4-3-11, Takeda, Kofu, Yamanashi 400, Japan
E-mail kotani@kki.esi.yamanashi.ac.jp

ABSTRACT

We have been developing a Line Mark Painting Robot since 1992. The robot detects and follows a half-faded line mark on an asphalt road using image processing.
In this paper we describe the system configuration of the testbed robot and we propose the image processing method. It ensures reliable detection and high-speed following in ordinary outdoor environments. The image processing method is based on a template matching with a "Line Mark Model". For the shadow problems we use discriminant method.
Experimental results showed the robustness of our proposed image processing method and motion control method. The image processing time for the following mode was about 100 [msec] and the maximum deviation from the center of the half-faded line mark was less than 10 [mm].

KEYWORDS: image processing, template matching, robot, threshold value, discriminant method, half-faded line mark

1 Introduction

In Japan line marks on an asphalt road are painted or repainted by civil engineering corporations. Seven or eight workers make a team to paint a line.

The members of the team share the following process: (1) Controlling traffic around the target line mark, (2) Pouring a heated paint from the paint tank on the truck to the paint pot of the machine, (3) Sweeping away fallen leaves and the sand from the line mark, (4) Painting a guide line called "Kegaki" in Japanese, (5) Sprinkling a liquid primer, (6) Painting or repainting a line mark and sprinkling beads for reflecting car light. Controlling traffic process requires two workers. Others processes require one worker for each process. The operation has not progressed for a long time. There are many problems of the line mark painting processes using conventional machines. Examples of them are dangerness, unclean environment, distress and shortage of workers. One of the solutions for the above problems is to make a robot in stead of the conventional machines. We can substitute processes (3), (5) and (6) by the robot. The robot requires a operator, so the robot save two workers.

We have been developing a Line Mark Painting Robot since 1992. The robot detects and follows a half-faded line mark on an asphalt road using image processing. However in an outdoor environment there are many difficult problems, for example, the noise, the shadow problems and variability of line marks associated with dynamic worlds.

Line Mark detection methods are studied by many researchers. For instance, in German, Graefe developed a fully autonomous road vehicle, which tested in real-world scenes on the German Autobahn [1]. In Japan, FUJITSU Co. and NISSAN Co. developed Personal Vehicle System PVS [2], SUBARU Co. developed a stereoscopy image recognition system and recognized road boundaries and obstacles [3]. Nouson developed pipelined image processing system and estimated road structure and camera position from continuous road images [4]. However in these studies following a line mark with accuracy is not so important. And the line mark is not half-faded but clear.

In this paper we describe the system configuration and the image processing method of the testbed robot. It ensures reliable detection and high-speed following in an ordinary outdoor environment. The image processing method is based on a template matching with a "Line Mark Model". For the shadow problems we use discriminant method. Experimental results showed the robustness of our proposed image processing method.

2 Design of a Line Mark Painting Robot

2.1 System Configuration

The photograph of the testbed robot is shown in Figure 1. Hardware specification of the robot is shown in Table1.

Figure 1: Testbed robot: Line Mark Painting Robot

Table 1: Specification of the hardware

item	specification
Dimension	100(W)170(L)90(H)(cm)
Weight	450Kg
Servo Motor	300W*2
E-Generator	100V-20A-50Hz
CPU(Image)	MC-68040. 25MHz
CPU(Control)	i80486, 20MHz
CPU(Interface)	V25(NEC Co.), 16MHz
CPU(Start/Stop)	V25(NEC Co.), 16MHz
Communications	RS-232C 9600bps
Image Memory	512*480 8bit
Video Camera	CCD f=11mm AutoIris

2.2 Image Processing

There are two modes of the image processing method. One is a "Detection Mode" and the other is a "Following Mode". The former is carried out before running, the latter is carried out while running.

2.3 Line Mark Model

We assume that the slope of the road is constant and the robot should be set up as follows:

- The center of the video camera lens, optic axis and the center of the paint outlet of the robot are on the same plane (See Figure 2).
- There is no vibration when taken image.
- The robot does not lean on either side (Rolling).
- There is not very much change of the depression angle of the video camera (Pitching).

In this case, a "Line Mark Model" is defined as a triangle which is surrounded by the Vanishing Point, and the Left and the Right sides of the line mark (see Figure 3).

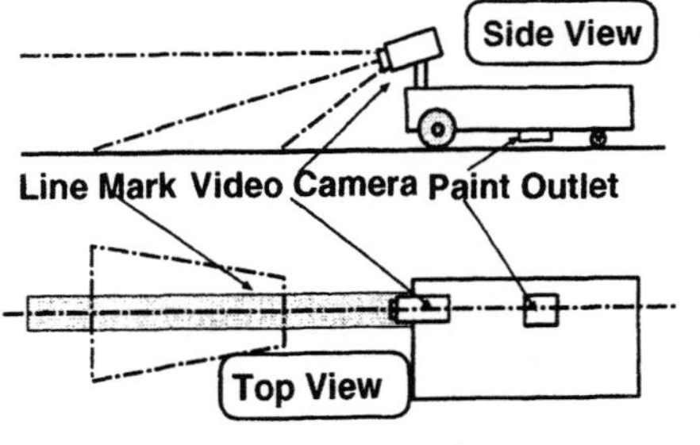

Figure 2: Relationship between the robot and video camera

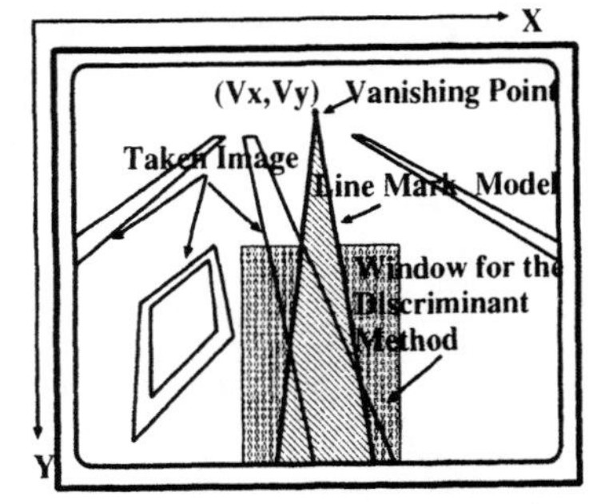

Figure 3: Line Mark Model, Taken image, Window for the Discriminant Method

We calculate the difference between the line mark model and a taken image and then get the deviation of the robot of the lateral direction (mm) and longitude direction ($degree$) [6].

2.4 Deciding the threshold value

There are many binarization methods [7], for example, Mode method, P-Tile method, Discriminant method and so on. We use the discriminant method [8] for getting the threshold value because a brightness is the important factor. The pattern of the line mark and road surface are variable. Because we must consider shadow problems(see Figure 4). So we check the value of the between the inside group and the outside group variance.

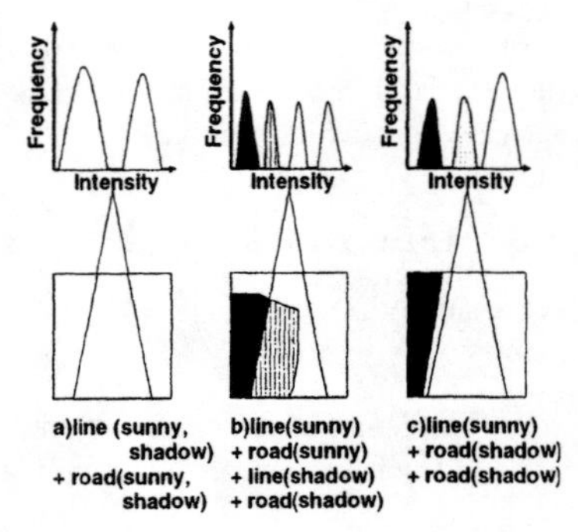

Figure 4: Examples of the discriminant method for shadow problems

2.5 Template matching

We set up the temporary vanishing point (Vx, Vy) from initial condition of the video camera. We use template matching algorithm for detecting the line mark. The method gets the value of matching with a line mark model which moves and rotates (see Figure 5). The matching procedures are as follows:

1) movement of Vy
2) movement of Vx
3) rotation
4) matching

The detected line mark has the highest value of the sum of "white" points after binarization.

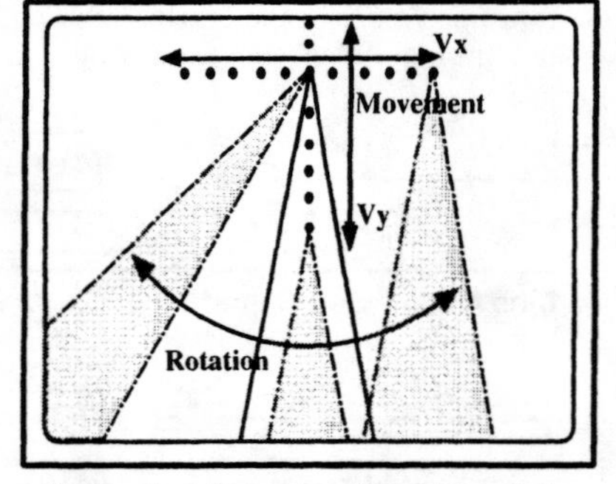

Figure 5: Movement, Rotation of the Template

2.6 Speed-Up

We must carry out the image processing as fast as possible at the following mode, so we apply speed-up techniques as follows:

1. **SSDA Method:** At first we use estimating vanishing point (Vx, Vy) and the degree of the direction, and then we move and rotate template. If the value decreased three times continuously, the direction of moving or rotation will be terminated.

2. **Rough to Minute Matching:** If we need precise difference, the degree of the rotation must be smaller, but it is costly. So we rotate it roughly and we get the degree which is the highest value of the sum of "white" points. Then we rotate it minutely around that degree.

3. **Estimating a temporary vanishing point:** We can estimate the temporary vanishing point because we know the previous vanishing point and encoder output of the left and right wheels. The encoder output is reliable if the interval is short.

3 Experimental Results

3.1 Binarization

A original image is shown in Figure 6(b). There are sunny part and shadow part on a road and a line mark surface in the image. The intensity histogram is shown in Figure 6(a). If we use simple binarization method, for example assumed Figure 4(a) only, the binarized result may be fault. The example of using simple binarized method is shown in Figure 6(c). The example of using our method is shown in Figure 6(d).

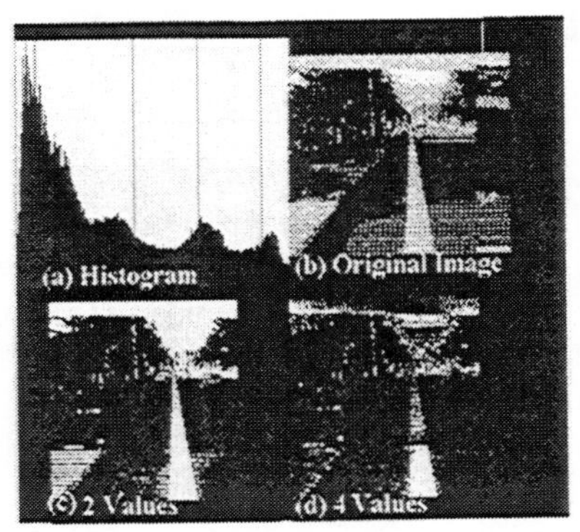

Figure 6: Results of the discriminant method

3.2 Experimental Setup

The robot has the instrument system which measures deviation of the robot and the postures (Roll, Pitch, Yaw) of the robot. The system configuration is shown in Figure 7. The specification of their sensors are shown in Table2 (for Deviation), 3, (for Pitch, for Roll) and 4 (for Yaw).

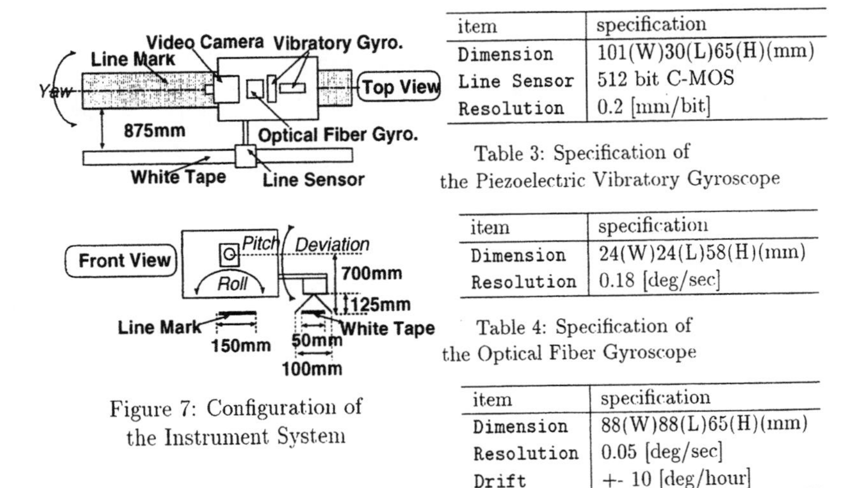

Figure 7: Configuration of the Instrument System

Table 2: Specification of the Line Sensor

item	specification
Dimension	101(W)30(L)65(H)(mm)
Line Sensor	512 bit C-MOS
Resolution	0.2 [mm/bit]

Table 3: Specification of the Piezoelectric Vibratory Gyroscope

item	specification
Dimension	24(W)24(L)58(H)(mm)
Resolution	0.18 [deg/sec]

Table 4: Specification of the Optical Fiber Gyroscope

item	specification
Dimension	88(W)88(L)65(H)(mm)
Resolution	0.05 [deg/sec]
Drift	+- 10 [deg/hour]

3.3 Run Test

The robot runs at 600 [mm/sec] on an asphalt road with the target line mark 150 [mm] in width. The trajectories and postures of our robot are shown in Figure 8 and Figure 9.

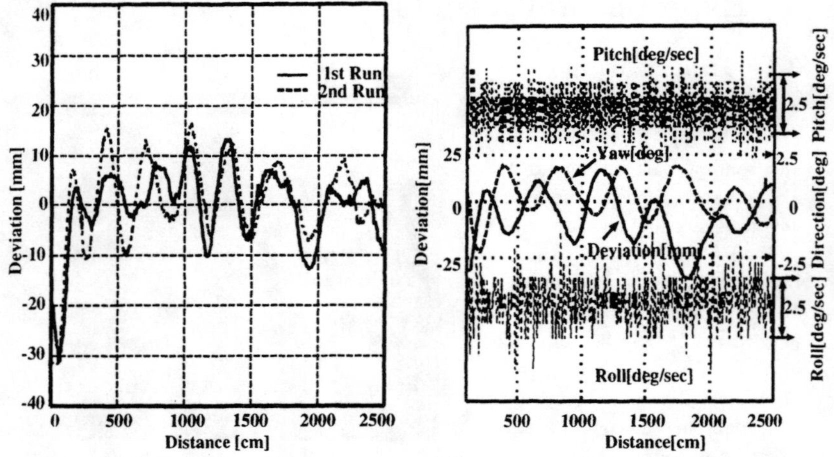

Figure 8: measured trajectories with visual feedback

Figure 9: Measured trajectory and postures

The solid line and the dashed line (Figure 8) show the trajectory with feedback in image processing. The dashed lines (Figure 9) show the postures of the robot.

The position estimation interval by dead reckoning was 20 [$msec$]. The feedback interval of the image processing was 200 [$msec$]. The instrument interval of the deviation and postures was every 30 [$msec$]. We can estimate the posture angles of our robot to integrate (trapezoid formula) the pitch and roll value (see Figure 10). The maximum roll and pitch angles were 0.5 [deg]. When we calculate the position of our robot using inverse perspective transformation, the influence of their posture angles is considerably: especially the roll angle is very big. Judging from Figure 10, the influence of their postures was less than 1 [mm] at the instrument system and 6 [mm] at the vision system.

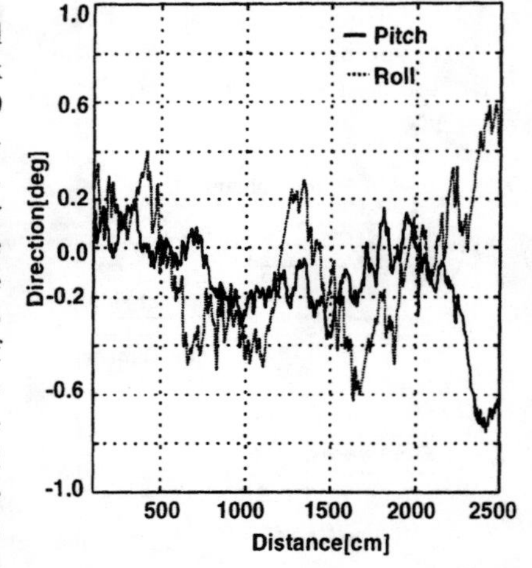

Figure 10: Calculated Posture angles(Roll and Pitch)

We confirm that the Line Mark Detection, the Line Mark Following and Motion Control System are effected by these results. The deviation from the center of the line mark is less than 10 [mm].

The causes for the deviation error in this experiment are 1) noises at the image processing, 2) parameters of our PID control, 3) delay time and dead zone, 4) assumption of the rolling is zero. If we need more precise following of a half-faded lane mark, we must consider these problems.

4 Conclusion

We developed the testbed system of the line mark painting robot. We proposed the method for detecting and following a half-faded line mark in actual outdoor environments and implemented the method to the robot. We confirmed the availability of our image processing method. Furthermore, using the speeding up method, we can follow a half-faded line mark at a very high speed.

Now we have been developing an image processing algorithm for a curved line mark with the new image processing system using IP-X(HITACHI Co.Ltd.) on an IBM PC/AT system. The point is a vision system with a conventional video camera. The camera is insufficient for real outdoor scenes. So we need wide dynamic range vision sensor [9].

References

[1] Volker Graefe., "Vision for Intelligent Road Vehicles", In Intelligent Vehicles '93 Symposium, Tokyo, pp135-140, July, 1993.

[2] M.Ohzora et al., "Video-Rate Image Processing System for an Autonomous Personal Vehicle System", IAPR Workshop on Machine Vision Applications, MVA'90, pp.389-392

[3] Keiji Saneyoshi et al., "3-D Image Recognition System for Drive Assist", In Intelligent Vehicles '93 Symposium, Tokyo, pp60-65, July, 1993.

[4] K.Nohsoh, S.Ozawa, "A Simultaneous Estimation of Road Structure and Camera Position from Continuous Road Images"(in Japanese), Trans.IEICE, D-2 Vol.J76-D-2, No.3, pp514-523, 1993

[5] S.Kotani, H.Mori, et al., "Image Processing and Motion Control of a Lane Mark Drawing Robot", in Proc. of IEEE/RSJ Int. Conf. IROS'93, Japan, 1993, pp.1035-1041

[6] T.Kobayashi, S.Ozawa, et al., "On-line Estimation of Road Parameters from Road Images"(in Japanese), Trans.IEICE, D-2 Vol.J75-D-2, No.1, pp67-75, 1992

[7] A.Rosenfeld, A.C.Kak, "Digital Picture Processing", Academic Press, Inc., 1976

[8] N.Ots, "An automatic threshold selection method based on discriminant and least squares criteria"(in Japanese), Trans.IEICE, Vol.J63-D, No.4, pp349-356. 1980

[9] K.Yamada, T.Nakano and S.Yamamoto, "Wide Dynamic Range Vision Sensor for Autonomous Vehicles", in Proc. of Int. Conf. Robotics and Automation, pp770-775, 1995

NEW DESIGN OF A REDUNDANT SPHERICAL MANIPULATOR

Sylvie LEGUAY-DURAND*, Claude REBOULET

CERT-ONERA – SUP'AERO – Département d'Automatique 2 avenue Edouard Belin 31055 TOULOUSE cedex – FRANCE*

E-mail : Leguay@cert.fr – Reboulet@cert.fr

ABSTRACT

A new kinematic design of a parallel spherical wrist with actuator redundancy is presented. A special feature of this parallel manipulator is the arrangement of actuators with co-axial axes which allows unlimited rotation around any axis inside a given cone-shaped workspace. Detailed kinematic analysis has shown that actuator redundancy not only removes some singularities but increases workspace. The structure optimization has been performed with a global dexterity criterion coupled with constraints on usable workspace.

KEYWORDS : Spherical wrist, parallel manipulator, redundancy, optimization, dexterity.

1 INTRODUCTION

A wrist is intended to modify end-effector orientation around any direction in space. In the ideal case, performance will be constant and independent of its configuration. This property, known as isotropy, is not always guaranteed for many existing mechanisms. Their performances are bad if, in some configurations, actuators have to carry out large-scale movements to change end-effector orientation. However, this isotropy property is necessary in many implementations as same capacities (in terms of acceleration, dexterity, accuracy, wrench ...) are desirable around any direction, whatever the configuration of the manipulator is.

Solving this problem firstly requires to choose an appropriate mechanical concept. Secondly, the design parameters of the structure have to be optimized. A dexterity measure is also needed to perform this task.
For mechanical concept, it is possible to choose between serial and parallel kinematic chains. The serial chains, in this case three consecutive rotational joints, are characterized by a large usable workspace but a bad isotropy. They also have singular points for which end-effector cannot be rotated around a particular direction. An additional joint can be used to avoid this singularity problem. This is the case of kinematic redundancy.

The main advantages of parallel kinematic chains are lightness and rigidity. The use of these chains in a robotic context is of recent date. Their lightness, mainly due to the ability of bringing the actuators as close as possible to the fixed base is interesting in many applications requiring high rate of acceleration (Speed-R-Man [11]) or effort control as teleoperation with a force-reflecting controller.

Many authors have analysed the dexterity of manipulators and compared different measures ([7], [10]). One of the first papers to consider dexterity is [13], introducing the condition number of Jacobian J $\left(k = \|J\|\|J^{-1}\|\right)$ which is simply the ratio of the radii of the largest and smallest principal axes of the manipulability ellipsoid described in [14]. A direct physical significance of this local measure has also been shown ([1],[7],[8]). Other measures taking into account inertial properties have been defined (generalized inertia ellipsoid in [1], dynamic manipulability measure in [15]). As they characterize the dexterity of a robot at a given configuration, the above measures are local. But for design optimization, a global measure may be more desirable. In [4] and [8], global measures are defined by integrating local dexterity indices over the workspace.

Another feature of the structure is the actuator redundancy. Studied particularly in [8], actuator redundancy is dual to the kinematic redundancy in serial chains. It means that the mechanism has more actuators than necessary, without increasing mobility. Actuator rates are uniquely determined by a given trajectory but actuator torques are undetermined. It is used to increase dexterity and eliminate certain type of singularities as it will be shown.

The purpose of this article is to present this new design of redundant spherical parallel wrist. To this end, the inverse kinematic problem and the Jacobian matrix are reviewed. Then, the problem of singularities of different types is discussed. Finally, a global dexterity measure is introduced. This measure, coupled with constraints on workspace volume, is used to optimize the geometrical parameters of the structure.

2 DESCRIPTION OF THE MECHANISM

The mechanism is composed of two pairs of sub-arms, the first pair (1) and (2) attached to point $P_{1,2}$ of the moving platform, the second pair (3) and (4) to point $P_{3,4}$. Each sub-arm consists of two spherical links. All the joints are of revolute type and their axes (active and passive joints) are concentric to point O, defining the rotation center of the spherical wrist. As it is shown on figure **1a**, a particularity of this structure is to have the axes of the four actuators collinear along the z-axes.

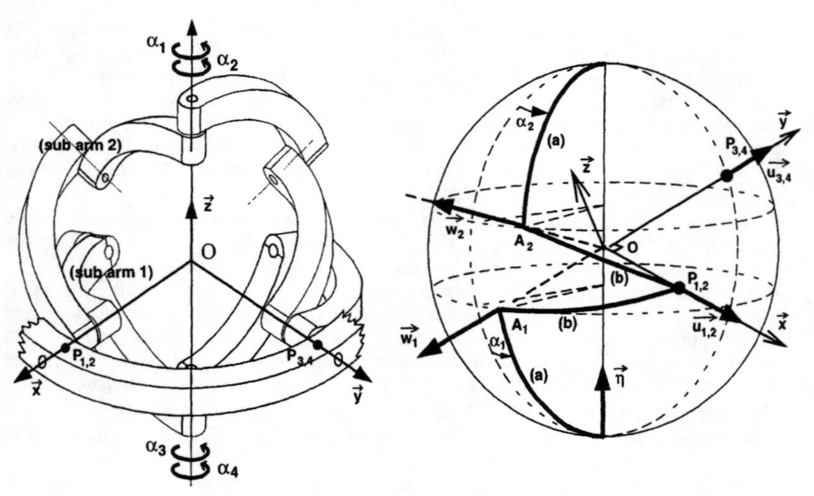

Figure 1a,1b. The spherical parallel wrist and its geometric parameters

This mechanism which is close to non-redundant structures studied in [2], [3] and [5], differs in its conception as there are only two attachment points on the moving platform. In its nominal configuration (fig **1a**), the structure is symmetrical about the xy plane. An angle of 90° between the two attachment points has been chosen to simplify the design but it corresponds also to the best conditioning as it can be demonstrated.

Due to the four collinear actuators, unlimited rotation is possible around any axes inside a cone. Since it is interesting to have an as wide open cone as possible, a constraint on the opening angle of the cone has been introduced. Moreover, actuator redundancy permits to eliminate parallel-type singularities and to standardize the dexterity in the workspace.

3 KINEMATIC ANALYSIS

3.1 Inverse Kinematics

The inverse kinematic problem for this manipulator consists in finding the joint variables α_i, $(i = 1, ..4)$ corresponding to a given orientation of the moving platform. As two configurations exist for each sub-arm, it leads to 16 solutions. But in practice, a unique solution is feasible because of the risk of collisions between the sub-arms.
Orientation of the moving platform is specified by three Euler angles ψ, θ, φ, where ψ is the rotation around **z**-axis, θ is the rotation around new axis **x** and φ is the rotation around new axis **z**. The rotation matrix with respect to the base coordinates is then introduced as : $R = Rot_z(\psi).Rot_x(\theta).Rot_z(\varphi)$.

Let us consider the geometric parameters of the manipulator on fig. **1b**. Let us denote by **a** and **b** the link angles of the four sub-arms and define $\vec{\mathbf{u}}_i$ as the unit vector along the axis of the revolute joint connecting the moving platform and the adjacent link (OP_i). In the chosen layout, the points P_1 and P_2 merge in $P_{1,2}$ and P_3, P_4 in $P_{3,4}$. $[x_i, y_i, z_i]^T$ the components of $\vec{\mathbf{u}}_i$ in the fixed base, are functions of the three Euler angles ψ, θ, φ.
Moreover, $\vec{\mathbf{w}}_i$ is defined as the unit vector along the axis of the intermediate revolute pair of each sub-arm. The components of $\vec{\mathbf{w}}_i$ are given by :

$$\vec{\mathbf{w}}_i = \begin{bmatrix} \sin a \cos \alpha_i \\ \sin a \sin \alpha_i \\ (-1)^i \cos a \end{bmatrix} \tag{1}$$

The solution to the inverse kinematic problem is then obtained by solving the equation :

$$\vec{\mathbf{u}}_i.\vec{\mathbf{w}}_i = \cos b \tag{2}$$

For each sub-arm, it leads to :

$$X_i \cos \alpha_i + Y_i \sin \alpha_i = Z_i \qquad \text{where} \qquad \begin{aligned} X_i &= x_i \sin a \\ Y_i &= y_i \sin a \\ Z_i &= \cos b - (-1)^i z_i \cos a \end{aligned} \tag{3}$$

This classic form equation gives the solutions to the kinematic inverse problem :

$$\alpha_i = \beta_i \pm \arccos\left(\frac{Z_i}{d_i}\right) \qquad \text{with} \qquad \begin{aligned} d_i &= \sqrt{X_i^2 + Y_i^2} \\ \beta_i &= \arctan 2\left(\frac{Y_i}{X_i}\right) \end{aligned} \tag{4}$$

The feasible solution is selected by testing the sign of $\det|\vec{\mathbf{w}}_i, \vec{\mathbf{u}}_i, \vec{\eta}|$, with $\vec{\eta}$, unit vector along the axis of the actuators.

3.2 Jacobian matrix

The Jacobian matrix can be found by differentiation of equation (2) as shown in [5], which leads to :

$$\dot{\vec{\mathbf{u}}}_i.\vec{\mathbf{w}}_i + \vec{\mathbf{u}}_i.\dot{\vec{\mathbf{w}}}_i = 0 \tag{5}$$

where $\dot{\vec{\mathbf{u}}}_i = \vec{\omega} \wedge \vec{\mathbf{u}}_i$, $\vec{\omega}$ being the angular velocity of the moving platform. Equation (5) can be rewritten as :

$$b_i \dot{\alpha}_i + (\vec{\mathbf{u}}_i \wedge \vec{\mathbf{w}}_i).\vec{\omega} = 0 \qquad \text{with} \qquad b_i = \vec{\eta}.(\vec{\mathbf{w}}_i \wedge \vec{\mathbf{u}}_i) \tag{6}$$

which gives the 4x3 inverse of the Jacobian matrix J^{-1} :

$$\dot{\vec{\alpha}} = \begin{bmatrix} \cdots \\ -\frac{(\vec{\mathbf{u}}_i \wedge \vec{\mathbf{w}}_i)^T}{b_i} \\ \cdots \end{bmatrix} \vec{\omega} = J^{-1}\vec{\omega} \tag{7}$$

$\dot{\vec{\alpha}}$ being the 4x1 vector of joint rates.
It is also interesting to note that the motor axes being all collinear along z-axis, the choice of ψ, rotation around the same z-axis, allows to deal only with angles θ and φ. The inverse of Jacobian matrix J^{-1} does actually not depend on angle ψ, which simplifies the kinematic analysis and the optimization process.

3.3 Discussion on singularities

As shown in [6], equation (7) can be written as :

$$\begin{bmatrix} b_1 & & 0 \\ & \ddots & \\ 0 & & b_4 \end{bmatrix} \dot{\vec{\alpha}} = \begin{bmatrix} \cdots \\ -(\vec{u}_i \wedge \vec{w}_i)^T \\ \cdots \\ \cdots \end{bmatrix} \vec{\omega} \qquad B\dot{\vec{\alpha}} = A\vec{\omega} \qquad \text{with } J^{-1} = B^{-1}A \qquad (8)$$

• The singularities of the first kind [6] (which can be called serial type singularities), appear when $\det(J) = 0$ which corresponds to $\det(B) = 0$. They consist of the set of points where different branches of the inverse kinematic problem meet and are known to lie on the boundary of the workspace. As matrix B is of diagonal form, this happens if one of the b_i is equal to zero, ie :

$$\vec{\eta}.(\vec{w}_i \wedge \vec{u}_i) = 0 \qquad i = 1,..,4 \qquad (9)$$

This equation states that the two links of a sub-arm are coplanar, i.e. the corresponding sub-arm is totally unfolded or folded, which can be expressed as :

$$\begin{matrix} \vec{u}_i.\vec{\eta} = (-1)^i \cos(a \pm b) \\ i = 1,..,4 \end{matrix} \qquad \text{which leads to} \qquad \begin{matrix} \sin\theta \sin\varphi = \pm\cos(a \pm b) \\ \sin\theta \cos\varphi = \pm\cos(a \pm b) \end{matrix} \qquad (10)$$

It is also possible to plot the locus of first type singularities in the θ–φ plane.

• The second kind of singularities [6] (or parallel type singularities) corresponds to configurations in which the gripper is locally movable even when all the motors are locked. As opposed to the first one, this kind of singularity lies inside the workspace and corresponds to a set of points where different branches of the direct kinematic problem meet. These singularities appear when $\det(J) \to \infty$, i.e. $\det(A) = 0$. Since $\vec{u}_1 = \vec{u}_2$ and $\vec{u}_3 = \vec{u}_4$, the matrix A can be written :

$$A = -[(\vec{u}_1 \wedge \vec{w}_1) \;\; (\vec{u}_1 \wedge \vec{w}_2) \;\; (\vec{u}_3 \wedge \vec{w}_3) \;\; (\vec{u}_3 \wedge \vec{w}_4)]^T \qquad (11)$$

A is singular if its rank becomes less than two, i.e. if three of its line vectors are collinear or if its four line vectors are coplanar. As sub-arm (1) is coupled with sub-arm (2) (respectively (3) with (4)), the first two line vectors being collinear implies : $\vec{w}_1 = \pm\vec{w}_2$

– $\vec{w}_1 = \vec{w}_2$ is impossible as $0^\circ < a < 90^\circ$

– $\vec{w}_1 = -\vec{w}_2$ is impossible without crossing a singularity of the first kind, as the chosen layout gives :

$$\det(\vec{w}_1, \vec{u}_1, \vec{\eta}) > 0 \quad \text{and} \quad \det(\vec{w}_2, \vec{u}_1, \vec{\eta}) > 0 \qquad (12)$$

From above, the first two line vectors of A define a plane perpendicular to $\vec{u}_1$ and the last two line vectors a second plane perpendicular to $\vec{u}_3$. Since angle $(\widehat{\vec{u}_1, \vec{u}_3})$ is constant and equal to 90°, it is impossible for the two planes to merge. No singularity of the second kind exists in the workspace.

4 DEXTERITY MEASURE AND OPTIMAL DESIGN

It is difficult to define optimality as it first depends on the manipulator application and secondly because it is almost impossible to take into account for example geometrical, inertial and dynamic aspects in a single measure. Another goal of design is generally the maximization of the workspace, which is frequently a drawback of parallel mechanisms and often gives results opposed to structure optimization. In the case of the spherical wrist, it is possible to choose parameters **a** and **b** to have isotropy at nominal configuration. Unfortunately, the usable workspace is then limited to a cone less than 20°. Therefore it may be interesting to accept to be further from isotropy to obtain a larger workspace.

In the case of three-dof parallel manipulators with all identical actuators, it can be shown that Jacobian matrix J is involved in kinematic, force, inertial and dynamic relationships. The condition number of J (or its inverse called dexterity) seems therefore appropriate to perform the structure optimization.

It is defined by : $k(J) = \frac{\sigma_{\max}}{\sigma_{\min}}$ σ_i singular values of J (13)

As this measure has only a local character, a solution proposed by [4] and [8] is to integrate the dexterity over the workspace and to normalize by the volume N_w of the workspace :

$$D_g = \frac{1}{N_w} \int_W \frac{1}{k(J)} dw \qquad N_w = \int_W dw \tag{14}$$

This measure is also approximated by a discrete sum : $D_g = \frac{1}{N_w} \sum_W \frac{1}{k(J)}$ (15)

For the sum to approximate the integral, the orientation **w** should be uniformly distributed across the workspace. A uniform sampling of the Euler angles will obviously not be appropriate. Finite rotations can be conveniently interpolated using **quaternions** coordinates q_0, q_1, q_2, q_3 defined by : $q_0 = \cos\left(\frac{\alpha}{2}\right)$ $[q_1\ q_2\ q_3]^T = \sin\left(\frac{\alpha}{2}\right)\vec{u}$ where $\vec{u}$ is the axis of the rotation and α the angle. Any rotation is represented by a quaternion of unit magnitude, thus by a point of the unit hypersphere in the four-dimensional quaternion space. The rotation defined by Euler angles ψ, θ, φ corresponds to the quaternion :

$$q = \left[\cos\left(\frac{\theta}{2}\right)\cos\left(\frac{\psi+\varphi}{2}\right),\ \sin\left(\frac{\theta}{2}\right)\cos\left(\frac{\psi-\varphi}{2}\right),\ \sin\left(\frac{\theta}{2}\right)\sin\left(\frac{\psi-\varphi}{2}\right),\ \cos\left(\frac{\theta}{2}\right)\sin\left(\frac{\psi+\varphi}{2}\right)\right]^T$$

An element of volume on the unit hypersphere is then calculated using the Euler angles, for which the limits of the workspace are well-known, as : $dV = \sin\theta\, d\psi\, d\theta\, d\varphi$

The new dexterity measure used to optimize the mechanism is simple to formulate and to compute as it only depends on θ and φ as stated before ; It takes the form :

$$D_g = \sum_{\theta,\varphi} \frac{1}{k(J)} \sin\theta \Big/ \sum_{\theta,\varphi} \sin\theta \tag{16}$$

Figure **2a** shows the value of this measure with variation of geometric parameters **a** and **b**. However, this single measure is not sufficient to determine optimal values of **a** and **b**. It is interesting to add a constraint on the volume of the workspace, in fact to specify a minimal opening for the cone within which unlimited rotation around any axis is possible. The limitations on the workspace, given by singularities of the first kind when two links are totally folded or unfolded, can be expressed formally as :

$$a + b \geq \lambda + \frac{\pi}{2} \qquad a - b \geq \lambda - \frac{\pi}{2} \tag{17}$$

Figure **2b** shows the contour lines of the dexterity measure and the constraint on the workspace (17) with λ fixed at 60° :

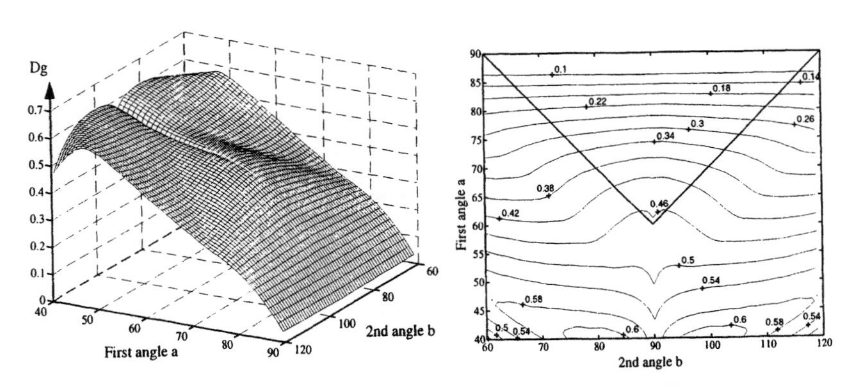

Figure 2a, 2b. Global dexterity measure for redundant spherical wrist and 60° cone

The previous study allows to choose optimal values for parameters **a** and **b**. Thus it is interesting to plot the inverse of the condition number of J as a function of euler angles θ and φ on figure **3**. As a larger workspace requires to be further from isotropy at nominal configuration ($\theta = 0$), it can be seen that, as soon as the opening of the workspace cone is larger than about 30°, the nominal configuration becomes a local minimum, giving a mean to estimate the performances. For the chosen values of **a** and **b** (figure 3) the dexterity measure is greater than 1/3 over the largest part of the workspace.

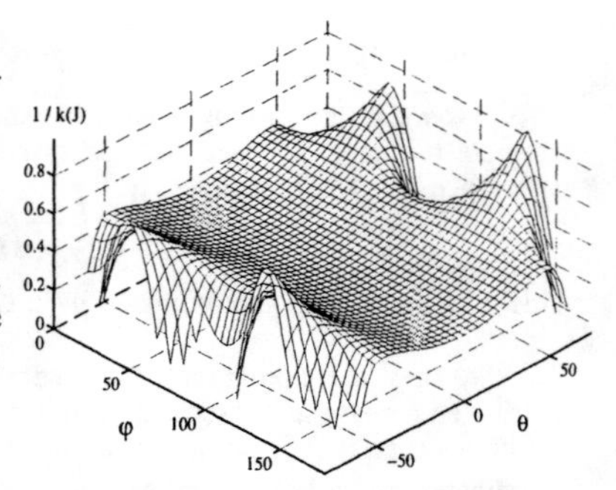

Figure 3. Local dexterity measure for spherical wrist

5 CONCLUSION

A new kinematic design of a parallel spherical wrist with actuator redundancy has been presented. The kinematic analysis has shown that actuator redundancy removes the second type singularities. The optimization is based on a global measure whose formulation and implementation are quite simple. An additional constraint on the workspace volume has been introduced, which allows unlimited rotation around any axis situated within a 60° cone (a property which is not common in most known parallel mechanisms). Comparisons between this structure and other equivalent non-redundant mechanisms are presented in [12], showing a notable improvement in performance in terms of dexterity and workspace volume for the redundant wrist.

REFERENCES

1. H. Asada. A geometrical representation of manipulator dynamics and its application to arm design. *Journal of Dynamic Systems, Meas. and Control (ASME)*, 105:131–135, 1983.
2. H. Asada and J.A. Cro Granito. Kinematic and static characterization of wrist joints and their optimal design. In *IEEE Int. Conf. on Robotics and Automation*, pages 244–250, 1985.
3. D.J. Cox and D. Tesar. The dynamic model of a three degree of freedom parallel robotic shoulder module. In *4th Int. Conf. on Advanced Robotics*, pages 13–15, 1989.
4. C. Gosselin and J. Angeles. A global performance index for the kinematic optimization of robotic manipulators. *Journal of Mechanical Design (ASME)*, 113:220–226, 1991.
5. C.M. Gosselin and J. Angeles. The optimum kinematic design of a spherical 3 dof parallel manipulator. *J. of Mechanisms, Trans. and Automation in Design*, 111:202–207, 1989.
6. C.M. Gosselin and J. Angeles. Singularity analysis of closed-loop kinematic chains. *IEEE Trans. on Robotics and Automation*, 6(3):281–290, 1990.
7. C.A. Klein and B.E. Blaho. Dexterity measures for the design and control of kinematically redundant manipulators. *The Int. Journal of Robotics Research*, 6(2):72–83, 1987.
8. R. Kurtz and V. Hayward. Multiple-goal kinematic optimization of a parallel spherical mechanism with actuator redundancy. *IEEE Trans. on Robotics and Automation*, 8(5), 1992.
9. J.P. Merlet. *Les robots paralleles*. Editions HERMES - Paris, 1990.
10. F.C. Park and R.W. Brockett. Kinematic dexterity of robotic mechanisms. *The Int. Journal of Robotics Research*, 13(1):1–15, 1994.
11. C. Reboulet, C. Lambert, and N. Nombrail. A parallel redundant manipulator : Speed-r-man and its control. In *4th ISRAM*, pages 285–291, 1992.
12. C. Reboulet and S. Leguay. The interest of redundancy for the design of a spherical parallel manipulator. In *Advances in Robot Kinematics (To appear)*, 1996.
13. J.K. Salisbury and J.Craig. Articulated hands : Force control and kinematic issues. *Int. Journal on Robotics Research*, 1(1):4–17, 1982.
14. T. Yoshikawa. Analysis and control of robot manipulators with redundancy. In MIT press, editor, *1st Int. Symp. on Robotics Research*, pages 735–747, 1984.
15. T. Yoshikawa. Analysis and design of articulated robot arms from the viewpoint of dynamic manipulability. In *3th Int. Symp. of Robotics Research*, pages 273–279, 1985.

The Next Generation Munitions Handler Critical Technologies and Dual-Use Potential

M. B. LEAHY, JR.
Headquarters Air Force Materiel Command Science & Technology
4375 Chidlaw Rd, Suite 6
Wright-Patterson AFB, OH 45433-5006
m.leahy@ieee.org

ABSTRACT

The Armed Services will improve the quality of the aircraft munitions loading process by fielding a new generation of munition handling equipment incorporating emerging telerobotics technology. An active program is underway to develop a Next Generation Munitions Handler (NGMH) Advanced Telerobotics Technology Demonstrator (ATTD). This research used air campaign planning principals to address the development of the technology roadmap and dual use business case study required to transition the ATTD into a full-scale prototype. Analysis of the munitions handling process revealed a tentative list of telerobotics technologies required to enhance operator performance. The maturity level and validity of that list was investigated through an intelligence preparation operation that supports the selection of nine specific technology targets. Courses of action to bring those technologies to commercial-off-the-shelf availability were explored. Scenarios for technology application in a range of alternative military and commercial applications lay the groundwork for development of a dual use business case. Civilian industry coalition partners were identified. Creation of a full scale NGMH prototype acquisition campaign is now possible.

KEYWORDS: telerobotics, robotics, military, human-machine interface, hydraulics

INTRODUCTION

The United States Air Force and Navy have an ongoing program to determine the feasibility of significantly enhancing the capabilities of munitions handling equipment by incorporating emerging telerobotic technologies into a new system design. The Air Force Material Command (AFMC) Robotics and Automation Center of Excellence (RACE) is providing technical direction for the design and development of the Next Generation Munitions Handler (NGMH) Advanced Telerobotics Technology Demonstrator (ATTD) [1]. The ATTD will enable the first realistic evaluation of emerging telerobotics technologies for flightline applications. Successful testing will pave the way for a full scale acquisition program. The objective of this research was the critical technology development roadmap and business case analysis necessary to support an acquisition campaign whose end-state is a commercial-off-the-shelf (COTS) telerobotic system which provides the mobility and dexterous manipulation necessary to reduce combat aircraft munitions load crew size to two individuals.

The aim of this paper is highlight the salient features of that research and encourage the reader to examine the full report [2]. The NGMH ATTD and the munitions loading problem are reviewed in [1] and discussed in detail in [2]. This paper jumps right into a summary of the extensive literature search. Section three identifies the nine specific critical technology challenges and corresponding courses of action. Section four summarizes the results of a top level analysis of the military and dual-use business cases. Conclusions and future directions finish the paper.

INTELLIGENCE PREPARATION OF THE TECHNOLOGY BATTLEFIELD

A critical technology target is defined as a set of capabilities that directly support a specific system performance requirement and are not commercially available. The end-state of the intelligence preparation is the comprehensive understanding necessary to define the proper critical technology targets and recommend courses of action to attack those targets. To reach that end-state, a comprehensive picture of the industrial and research communities' ability to support development of the NGMH prototype was gained through a review of the current commercial state-of-the-art product specifications and recent research publications. The goal of that examination was twofold. First, determine if the critical system performance requirements are achievable with existing commercial components. Then, if the results of the commercial search are negative, seek out existing laboratory research that has the potential to meet the requirements. The examination focused on the key system performance requirements provided by the NGMH ATTD project and was subdivided into five major technology areas: control systems, hydraulic actuation systems, sensors, omnidirectional mobility platforms, and human-machine interfaces. A summary of those results is presented. Consult the full report for the over 100 references which support these findings [2].

Control. The Unified Telerobotic Architecture Program (UTAP) provides an open architecture compatible with emerging standards. Current program plans will lead to a commercial product within five years. Techniques to mask the weight and inertia of a heavy payload are embedded in commercial control systems for electric drive robots. A single manufacturer of hydraulic robots provides gravity compensation as a standard control system feature. Force control, again on electric drive robots, has been extensively evaluated in the laboratory, and should emerge in commercial products within the next several years. But once again, hydraulic system implementation and evaluation results are scarce. Impact control is an immature technology. Numerous techniques have been proposed, but experimental evaluations are restricted to simple 2 DOF electromechanical devices. No hydraulic experiments were found. High precision chamferless part mating has not progressed beyond the basic research stage. Researchers are currently working to solve 2 DOF planar assembly problems with simple parts and fixtures.

Current industrial robot control systems are not designed to support real-time switching of operator and program inputs. However, the UTAP specification supports that capability. No commercial control currently supports redundancy resolution for greater than 7 DOF. Redundancy resolution is a popular research topic, but experimental evaluations for greater than 7 DOF are scarce. Real-time criteria and constraint switching has not been experimentally validated. Using redundancy to enable obstacle avoidance schemes has been demonstrated, and the concept of using redundancy to absorb impact forces was suggested. Two techniques appear suitable for NGMH application. However, no relative comparison results are in print. Insufficient experimental data is available to select a single methodology for the NGMH.

Hydraulic Actuation Systems. The requirement for high payload and small size is achievable with COTS equipment. Hydraulic robots with up to a 350 lb. payload capacity have demonstrated human bandwidth at low speed. High precision at low speed remains a significant basic research challenge. Developments in high performance servovalves show great potential and commercial vendors can provide low friction cylinders at moderate cost. However, high payload high precision actuator performance sufficient for missile loading has not been laboratory demonstrated.

Sensors. A commercial vendor can provide a force/torque sensor that meets the NGMH ATTD performance specifications. However, that one-off design will push the technology envelope and research on alternative approaches is very limited. A whole arm obstacle avoidance concept is being transferred from the Department of Energy (DOE) laboratories to a small business. That system will meet the basic NGMH requirements. The basic technologies necessary to sense an obstacle in a Chemical Nuclear Biological (CNB) environment have been identified by other DOD projects.

Omnidirectional Mobility Platform. Three basic techniques for designing omnidirectional vehicles were discovered. The all steerable wheel concept is commercially available, but lacks the payload capacity and simultaneous rotation and translation ability necessary for the NGMH. Vehicles based on the universal wheel and orthogonal wheel concepts have demonstrate sufficient potential to warrant further investigation. Once again, a comprehensive relative comparison between concepts has not been published. Insufficient intelligence information exists to support advocating selection of a single methodology.

Human-Machine Interface (HMI). The standard industrial operator interface, and a wealth of laboratory telerobotic systems, are all designed to control the manipulator from a distance. Intelligence on HMI for near-the-weapon control is limited to the NGMH design reviews. A pure come-along control mode, where the operator directly moves the munition, is impractical. The proposed near-the-weapon HMI retains the critical performance aspects of hands-on control while negating the drawbacks. Several telerobotics architectures, including UTAP, support on-line degree of autonomy selection.

CRITICAL TECHNOLOGIES & COURSES OF ACTION

Analysis of the information gathered from the literature search reduces the initial list of NGMH technology voids to nine explicit critical targets.

1. A UTAP compliant telerobotics control system specification
2. Force sensor with heavy payload and high sensitivity and overload protection
    - Fz=5000 lbs, Fxy=1500 lbs, Txyz=10,000 lbs-in
    - linear force resolution: Fz=5 lb., Fxy=1 lb.
    - moment resolution = 7 lbs-in
    - overload protection, factor of 10 for force and factor of 5 for moment
    - mechanical stop force protection
    - physical dimensions: diameter 8.5 in. with less than 4 in. thickness

3. Hydraulically powered omnidirectional platforms for large payloads
    - 4000 lb. payload capability
    - simultaneous translation and rotation
    - capable of traversing the flightline
4. Whole arm obstacle avoidance system
    - hierarchical sensing network
    - compact modular sensors
    - completely passive, no infra-red or electronic emissions
    - function in a CNB flightline environment
5. Redundant motion planning algorithm for 10 DOF mobile manipulator system
    - full time joint limit avoidance
    - sensor driven simultaneous multiple link obstacle avoidance
    - loop-rate switchable task optimization criteria and constraints
6. Human-machine interface for human augmentation tasks
    - near-the-weapon operator control
    - operator selectable control modes
7. High precision and performance heavy payload hydraulic actuation systems
    - very high motion resolution with a 3000 lb. payload
        - less than 1 mm measured at the NGMH end-effector
        - less than 0.0001 radian measured at the joint
    - end-effector velocity of 1 ft/sec with 3000 lb. payload
    - human-like acceleration at low speeds
8. Active impact control for large payload redundant hydraulic robots
    - compensate for impacts up to the force sensor overload limits
9. Munition installation aids
    - simultaneous insertion of two rectangular pegs into two chamferless rectangular slots
    - realistic peg, slot surfaces, and sensor resolution
    - insertion direction perpendicular to gravity vector
    - less than 3 mm clearance in all dimensions

The list is prioritized based on degree of difficulty. The first item is estimated to require less resources and time to commercialize then the last. The variation in difficulty is not linear.

The remaining task was determining a strategy for maturing the critical technologies into commercial-of-the-shelf (COTS) capabilities and/or products. The strategy was defined by a set of concurrent Courses of Action (COAs). Those COAs separated the necessary actions into technology transfer and applied and basic research categories. Realistic experimental evaluation was an overriding theme. Opportunities to achieve desired effects by leveraging existing programs were exposed. Coalitions of laboratories and companies best suited to accomplish the mission were suggested. A quick synopsis of the COAs follows. Complete details are in [2].

A UTAP compliant telerobotics control system specification is a pure leverage opportunity. All the major bugs in the interface specification should be discovered during the current phase two activity. Implementing the UTAP specification during the ATTD development should answer any NGMH unanticipated requirements prior to prototype development. The only action required is continued political support and collaboration.

The roadmap for force/torque sensor systems focuses on applied research. The technology exists, but has not been applied on the scale and with the mechanical overload features required for the NGMH. A commercial version of the prototype being developed for the ATTD should satisfy this requirement.

The roadmap for platform mobility systems focuses on applied research and experimental evaluation of two omnidirectional locomotion methods. While the intelligence preparation identified three fundamental approaches, the all-steerable wheel method is not suitable for further evaluation. A full scale comparative experimental evaluation of the orthogonal wheel and universal wheel platform designs should be incorporated into the ATTD.

The whole arm obstacle avoidance system development shall be heavily leveraged against the existing DOE technology transfer activity. DOE has to develop an obstacle avoidance system within the next five years to meet their hazardous waste remediation project goals. The architecture they are currently helping to commercialize is sufficient.

The COA required to convert redundant motion planning technology into a commercial product calls for a combination of basic and applied research prior to technology transfer. The Full Space Parameterization approach clearly has tremendous potential. Just as clearly, the Configuration Control Approach was successfully applied in a realistic scenario. However, neither approach has been rigorously evaluated for a 10 DOF combined mobility manipulation human augmentation application

The HMI COA is again a combination of basic and applied research. Past research provides numerous insights into how to design the driver's station platform interface, but the field of near-the-weapon HMI for human augmentation is wide open.

The high precision and performance requirements of the redundant manipulator can be achieved by advanced hydraulic actuation system designs, use of advanced control software, or some combination of the two. Therefore, both alternatives shall be evaluated in a coordinated fashion. Both evaluations involve a degree of basic research

Active impact control for redundant heavy payload hydraulic manipulators requires a COA stretching from basic research through full scale testing. Completion of full scale testing is dependent on the availability of prototypes of the previously discussed high performance high pressure hydraulic systems.

The path to providing the operator with installation aids is again dominated by basic level research. This target is the toughest to attack due to the immaturity of the field. Installing a missile on a launcher is a long difficult journey from planar assembly with simple friction

BUSINESS CASES

To secure technology investment, one must demonstrate significant return on investment. Business plan development started with a verification of the munitions handling economics. Estimates showed that improving the quality of the munitions handling process has a DOD wide manpower reduction potential of over $75M per year [2]. Those findings certify the purely military justification for an NGMH critical technologies research and development program. However, the DOD alone will never generate the sales volume necessary to foster commercialization of the critical technologies. To achieve the goal of an NGMH based on COTS, the benefits of the relevant technologies must extend to alternative military and industrial high payoff applications.

The key NGMH capabilities empowering alternative applications are: omnidirectional transport of heavy payloads, heavy lift capability with minimal operator exertion, high fidelity manipulation, operator rides on the platform, totally self contained portable heavy lift manipulator system. Based on those capabilities, three generic employment configuration were postulated. Configuration variations centered on replacing the munitions end-effector with: a general purpose gripping device or a tooling device, forklift tines, and/or an operator work platform. Employing those key capabilities and possible reconfigurations as a search space filter, resulted in identification of two additional military applications. Development of the NGMH critical technology suite would reduce manpower requirements in civilian and military heavy lifting and hazardous tasks. NGMH systems would have a payback period of 2-3 years for each individual removed from those dual-use non-munitions applications.

A search for indirect commercial applications of NGMH systems and critical technologies was conducted. Emphasis was on identifying industrial sectors, not individual applications. The objective was to locate industries with a high potential payoff from augmenting their existing human workforce with NGMH type systems. The construction, manufacturing, security, and commercial cleaning industries emerged at the top of that list. Notional scenarios and resultant cost analysis utilizing an NGMH compatible system were conducted [2].

Potential construction industry manpower cost savings, from utilization of NGMH based human augmentation systems, are in the trillions of dollars per year. The manufacturing sector, especially automotive, will also significantly reduce manpower costs and enhance product quality by incorporation of telerobotic systems. Security system, and commercial cleaning applications could exploit the improved performance and lower cost systems enabled by NGMH critical technologies to increased market penetration and thereby reduce manpower costs by tens of millions of dollars per year. Clearly, the commercial application potential warrents more detailed exploration.

CONCLUSION

A comprehensive picture of the industrial and research communities ability to support development of the NGMH prototype was developed through a review of the current commercial state-of-the-art and recent research publications. A list of critical technologies was identified. Course of action necessary to mature those technologies into COTS components were developed. Top level investment justification is in place for a range of military and industrial service robot applications. Creation of a full scale NGMH prototype acquisition campaign is now possible.

REFERENCES

1. Deeter, T. E., et. al. "The Next Generation Munitions Handler Advanced Telerobotics Technology Demonstrator Program." *Proc of ISRAM*, May 1996.
2. Leahy, M. B. Jr., and N. Hammel "The Next Generation Munitions Handler Prototype Acquisition Campaign: Targets and Courses of Action." *Air Command and Staff College Report 205*, May 1995.

Towards Error Recovery in Sequential Control Applications

Peter Loborg **Anders Törne**

Dept. of Computer and Information Science, Linköping University, Sweden

ABSTRACT

In this paper we describe how to generate a large set of restart points for a sequential control program, thus enabeling fast restart of a plant after a fault and subsequent repair. This is work in progress, and will recapitulate the main idea as developped in the Aramis project, and discuss its applicability to the programming languages defined by IEC1131-3.

KEYWORDS: error recovery, sequential control, task-level programming

INTRODUCTION

The languages used to instruct the production equipment at a shop floor are primarily designed for describing the normal flow of activities — sequential control in combination with continuous control. The handling of an abnormal or unforeseen situation is generally not supported. If such situations must be dealt with, this must be explicitly encoded using the same language elements as for the normal control.

There exist several attempts to use different form of planning formalisms to synthesise the program, given a specification of goals or intended behaviour. In these formalisms, the problem of exception due to simple, manageable faults such as misaligned parts are handled either by generating special code for catching the problem and retry the operation [1], to plan for a set of actions to overcome the problem and then return to the original program [3][5], or to perform a replanning step and extend the original program with this new plan [11][12].

We are focusing on the situation when a piece of equipment is to be restarted after a failure and subsequent repair. The machinery is almost guaranteed to be in a state in the middle of its activities (it seldom breaks in its 'home state'), and the state of the controlling computer/PLC is often inconsistent with respect to the controlled machinery. The task is to resynchronise and restart with minimal loss (in time and material) from an almost arbitrary state. The resulting restart state may be a state in the same sequence of states as was passed when the error occurred, or it might be a state in an alternative sequence of states.

The tool we use for this is a world model, containing an abstract view of the state space and additional information about possible operations on each device of the equipment. At compile time, we analyse the program and construct a database with information of all possible restart points. The problem of exponential blowup of the state space is avoided by exploiting the parallelism inherent in the application and by using the abstraction of the world model. If the machinery to restart is not in a precomputed restart point, a graph search can be adopted to find such a point 'close' to the current state.

As a byproduct, the analysis phase will also reveal programming errors such as an attempt to apply an operation or start an activity in a state where it is not applicable, and if timing information for all primitive operations is provided, it is also possible to obtain cycle time estimates at compile time, as well as to reschedule the default behaviour of the system to reduce the cycle time.

THE PROBLEM OF RESYNCHRONISATION

To resynchronise the controller with the plant, we need to have more knowledge about the plant available in the controller than what is customary. The task of resynchronisation can informally be described as *given a current observable state of the plant, find a controller state such that if reached during normal execution it would have produced the observable plant state.*

By *observed plant state* we mean a set of values for all sensors and actuators of the plant (not the actual plant state). By *controller state* we mean an observed plant state *and* the internal description of the program counter, the stack and all argument bindings of the controller, at some suitable level of abstraction.

Using this definition, it is clear that there can be several controller states matching an observed plant state, since a program can pass the same observed plant state multiple times but with a different trace of observable plant states each time, or since there can be several programs passing this observed plant state. It is also clear that we prior to the execution need a translation table that given an observed plant state returns a set of possible controller states. If such information is available, the task of resynchronisation is reduced to the task of picking the proper controller state and continue execution. In current practice, this information is generally not available, although a small subset of the information can be hand coded, i.e. implementing a set of restart points in the program. Although it is not realistic or even possible to hand code all possible restart points, it is possible to generate them as a part of the compilation of a program — provided there exists a more detailed description of the data structures and devices used in the program as compared to the current practice.

EXTENDED DESCRIPTION OF A DEVICE

In ordinary computers such as IBM-PC's, Unix machines etc., a device interface consists of a set of functions and procedures (interface routines) designed to control the device. For each device there is also a set of rules describing the intended usage of the interface routines, often presented to the programmer as a manual for the device in question.

As an example, consider the device *read-only file*. This device is controlled using three functions and a procedure:

- *open(filename)* will try to open a file identified by *filename*, and if it succeeds it will return a reference to the open file, a descriptor *file*
- *read(file)* will read and return the next data item of *file* if there is one
- *eof(file)* is a predicate function for testing if there exist more data items to read or not
- *close(file)* is a procedure that will close and forget the object referenced by descriptor *file*

The rules for when these interface routines are applicable can be described using a state transition graph (Fig. 1).

If this description is available to the compiler, the compiler can employ abstract interpretation to analyse the program, and it will be able to catch most of the errors resulting from

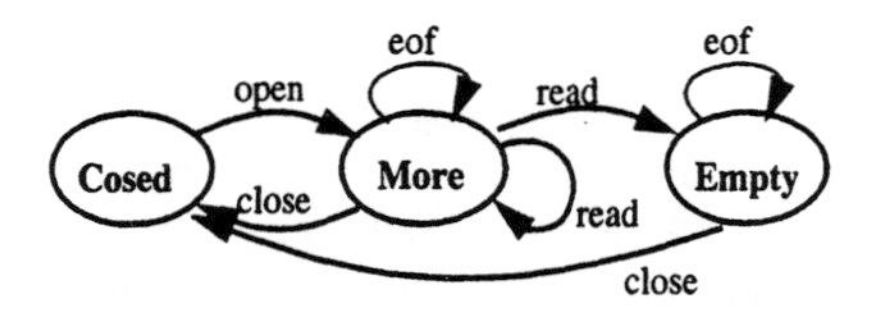

Figure 1. A state transition diagram showing the possible states for a read-only file and allowed operations in each state — normally not available to the compiler, only to the programmer.

miss-usage of the device.

In a programmable controller (for controlling a plant) implementing the standard IEC 1131 [4], the information available about each device or component in the plant is represented by a set of variables of a known, finite domain. Some of the variables represent sensors, and other variables represent actuators. The controller operates by reading input variables and changing output variables, and if the change is consistent with the intended usage of the device the execution will proceed. If not, the device will not respond as expected. This situation should not be misconceived with the situation when the change of values for the output variables is correct but the device fails to deliver the expected response due to other circumstances such as a broken actuator or sensor. In the latter case the problem can be foreseen, and an exception handler can be implemented to look for and react to such situations.

If all devices in the plant where modelled by both the variables and a state transition diagram showing the intended usage of the device (Fig. 2), the problem of misusage of the device can be caught by the compiler.

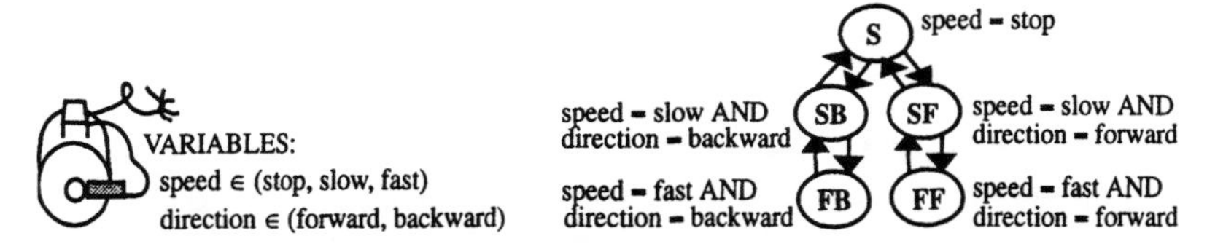

Figure 2. An electrical, bidirectional two speed engine is modeled by two variables, direction and speed (left hand part). The behaviour of this device is modeled by a state transition diagram with five abstract states, each representing a combination of variable values according to its constraint. The arcs between the states describe legal transitions, rouling out the posibility to reverse the engine without requesting it to stop in between.

The analysis technique to be used, abstract interpretation [2], operates by exploiting all possible execution paths of a program in parallel. All variables in the program are assumed to have any value consistent with their domain when the algorithm starts, and the set of possible values is reduced and modified during the analysis. The algorithm continues until it reaches a fix point, i.e. the set of possible values for the variables does not change any more. During the analysis the algorithm constructs a mapping between different (control) states of the program and possible values for the set of variables in the program.

Our goal is to construct a reversed mapping, i.e. for each state (as defined by the variables) a set of controller states are stored.

ARAMIS

The device model presented above has been developed in the scope of Aramis — A Robot And Manufacturing Instruction System [7][8][9][10][13]. In this system, the device interface is a part of a world model, which acts as an abstraction barrier between the task programming level on top of the world model and the control level below the world model. A program executed at the task level operates by reading and writing values from and into the variables defined by each device. The change of an input variable of a device is regarded as a request to the device, a request to change the internal state *iff* this change is consistent with the state transition diagram for the device. The device will then execute a control algorithm at the control level, an algorithm responsible for changing the plant into a state that is consistent with the requested state of the device model. When the control algorithm succeeds, the request is completed and the calling task level program returns from the call. Using this architecture we achieve the following characteristics and functionality:

- a well-defined abstraction of the actual state space at the task level, defined by the actuators and senors available in the plant
- a decoupling of task specification and device implementation
- a possibility to execute a task level program on a simulated system, i.e. to experiment with tasks and task level programs during the construction or reconstruction of the plant
- a guarantee against violations of the implemented abstraction layer during the life cycle of the plant
- reusable, hardware independent task level programs

We will return to these characteristics and compare them with a system based on the standard IEC1131 in the next section. In the remainder of this section we will describe the analysis algorithm in more detail.

Analysis of a task level program in Aramis

The algorithm for analysing a task level program is not a trivial algorithm, and in the general case using an arbitrary task level language it will either have an unacceptable complexity or will not compute at all. The following is a list of the prerequisites needed for the algorithm to be useful:

- the task level program must be compileable into a safe petri net, implying that it cannot include a goto statement or equivalent. If constructs such as procedures exist, they must also be compileable into safe petri nets.
- only top level programs or processes may have global references to devices and other data objects, procedures or other subprograms must only reference objects passed to them as parameters
- iterations in the application must be bounded and the bound must be known at compile time. This is normally not a problem in the manufacturing context, since the only unbounded iteration is the repeated production of a product, and we restrict the analysis to one such cycle.
- in the definition of the state transition diagram in a device, the set of nodes of that diagram (possibly extended with an implicit error node) must form an equivalence relation over the state space defined by the device variables. The rationale is that each possible combination of variable values in a device must match at most one node in the diagram.

Given these prerequisites, a task level program can be analysed. The algorithm consists of the following basic steps:

- Translate each named program entity, process or procedure, into a safe petri net. This translation can be done at definition time, i.e. each time a process or procedure is edited and saved. The symbol table generated for that entity should also be saved along with the petri net.
- For the petri net under analysis (PNuA): translate each condition in the petri net to the smallest set of abstract device states that will satisfy the condition. An abstract device state is a node in the state transition diagram of the device. This static translation will fail in cases where run time values are needed, in which case the translation is postponed until needed. However, all conditions consisting of a boolean combination of expressions on the form $v\rho c$ (where v is a variable, c is a constant in the domain of v and ρ is a relation defined over that domain) can be translated initially.
- based on the symbol table information of the top level process, all devices and other objects referenced are known. Form an *initial state set* (ss_i) for the PNuA as a set of possible device states, where a possible device state in turn is a set of all abstract device states for that device. As an example, consider a process referencing a single electrical engine (fig X) and a single file object (fig Y), the initial state set will be the set: {{FB, SB, S, SF, FF}, {Closed, More, Empty}}
- using the initial state set, perform an abstract interpretation of the PNuA by incrementally generating the graph of reachable and valid markings of the PNuA. A marking is valid *iff* all elements of the *current state set* (ss_c) for that marking are non-empty. For each transition with a condition, let the ss_c for the marking after that transition be the intersection of the previous ss_c and the state set representing the condition. For each place representing a state changing request for a device d into a new abstract state a, verify that the request is applicable, i.e. that the element d of the ss_c (ss_{cd}) is a subset of the set of abstract states in d from which a is reachable with one transition. If not, an error due to missusage of the device is found. Otherwise, the new ss_c is formed by replacing ss_{cd} with $\{a\}$.
- If a place in the PNuA represents a call to a sub-program, that place is replaced by the petri net representing that sub-program after substituting all local variables with conflicting names with a set of new unique variables. Thereafter the algorithm continues as described above.

As a result of this process, all reachable and valid markings are found, and for each such marking there is one or several state sets associated. Thus it is possible to construct a mapping from state sets to markings, to be used if there should be a need for resynchronisation of the controller and the plant.

The complexity of this algorithm is estimated as in the same size as the complexity of generating all reachable markings for a safe petri net. The fact that the conditions on transitions are used to restrict the size of the reachability graph is favourable. However, a marking can be reached more then once and with different current state sets, and thus a sub tree of the reachability graph can be searched multiple times. This indicates a slightly worse complexity. However, since the algorithm will be used at compile time, i.e. offline, the complexity is not a major issue.

THE LANGUAGES IN IEC 1131-3

In the international standard IEC1131-3 a set of five different languages for programming a controller is defined, namely: *instruction language* (IL), *ladder diagram* (LD), *function block diagram* (FBD), *sequential function chart* (SFC) and *structured text* (ST). Two of the languages are textual languages (IL and ST), and the other three are graphical languages.

Two of the more important improvements made by the standard is the introduction of a rigorous type system and the specification of how programs written in these languages can interact with each other. Any new device or program, regardless of implementation language, which can be provided in the controller using a function block as interface is an allowed extension of the controller.

Applicability of the analysis algorithm to IEC 1131-3 programs

Since all programs (POU — *program organisation unit* in the standard) can be viewed as either a mathematical, side-effect free function or as a function block, the following discussion will only cover these language elements.

In order for the analysis algorithm to operate there is a need for a layer of device descriptors, a layer that defines a border between a higher and a lower level of the system. Such a layer *can* be implemented using the FBD language, using a set of function blocks with the implementation part described as an SFC. However, the petri net notation is primarily a graph of activities, while a state transition diagram is primarily a graphical notation for a state changing system where the state is defined by the variables in the system*. There is no support for making it mandatory to use the intermediate device layer. Each programmer is free to call other parts in the lower layer directly from within the upper layer. This problem can, however, be handled by introducing these restrictions in the editors used for writing programs in the system.

Another requirement is that it should be possible to compile each program into a safe petri net. Although this is also the recommended practice when writing control programs [6], this is not the case for an arbitrary IEC 1131 program since a goto statement or equivalent can be expressed in all languages except FBD. Thus, this requirement imposes additional restrictions on the languages to be verified at compile time.

In Aramis, the possibility to select a suitable data abstraction is used extensively to reduce the state space. The new type system introduced in the standard provides the possibility for a similar data abstraction in the IEC1131 languages, but again, it is not mandatory to define an abstract version of a type used at the lower level of the system in order to use it in the upper level.

Concluding from the arguments above, it seems as if the analysis algorithm would apply given that a set of programming rules where strictly followed. There are, however, more details in the IEC 1131 languages that must be investigated in order to relax the restrictions imposed as much as possible, and thereby retaining a flexible programming system.

CONCLUSIONS

We have presented a method for generating all possible restart points of a program, given an extended representation of the devices used in the program. The essential properties of this extended representation and the programming language used has been discussed, and the applicability of this method on a program written in one of the IEC 1131-3 languages

---

* In the context of languages, by the set of strings consumed or generated up to the current node.

has been investigated. We have found that this method is applicable, given that a set of programming rules are followed. This is a preliminary result, and the details are yet to be investigated.

REFERENCES

[1] Cao, T. and Sanderson, A.C. "Sensor-based Error Recovery for Robotic Task Sequences Using Fuzzy Petri Nets". In *Proceedings of IEEE International Conference on Robotics and Automation*, pages 1063-9, 1992.

[2] Cousot, P and Cousot, R. *Abstract Interpretation: A Unified Lattice Model for Static Analysis of Programs by Construction or Approximation of Fixpoints* Conf. Record of Fourth ACM Symposium on POPL, pp. 238-252, Los Angeles, 1977

[3] Gini, M. "Recovering from Failures: A New Chalenge for Industrial Robotics". In *Proceedings of the 25'th IEEE Computer Society International Conference (COMPCON-83)*. pages 220-227, Arlington 1983.

[4] IEC 1131-3 International Standard, Programmable Controllers — part 3

[5] Lee, M. H., Barnes, D. P. and Hardy, N. W. "Knowledge Based Error Recovery in Industrial Robots". *In Proceedings of the International Joint Conference on Artificial Intelligence*. pages 824-826, 1983

[6] Lewis, R.W. *Programming industrial control systes using IEC 1131-3*. IEE Control Engineering Series 50. The Institution of Electrical Engineers, London (1995)

[7] P. Loborg and A. Törne. "A Layered Architecture for Real-Time Applications". *The 7:th Euromicro Workshop on Real-Time Systems*, Odense, Denamrk 1995

[8] P. Loborg. *Error Recovery Support in Manufacturing Control Systems*. Licentiate theses no. 440, Linköping University, Sweden, 1994.

[9] Loborg, P., Holmbom, P., Sköld, M. and Törne, A. "A Model for the Execution of Task Specifikations for Intelligent and Flexible Manufacturing Systems." *Integrated Computer-Aided Engineering*, p.185-194, no. 1(3), 1994

[10] Loborg, P. and Törne, A. "A Hybrid Language for the Control of Multimachine Environments". *Proceedings of EIA/AIE-91*, Hawaii, June 1991.

[11] Noreils, F. R. "Integrating error recovery in a mobile robot control system". *Proceedings 1990 IEEE International Conference on Robotics and Automation*, pages 396-401 vol.1, 1990.

[12] Noreils, F.R. and Chatila, R. G. "Control of mobile robot actions". *Proceedings of 1989 IEEE International Conference on Robotics and Automation*. pages 701-7 vol.2, 1989

[13] Törne, A. "The Instruction and Control of Multi-Machine Environments". *Applications of Artificial Intelligence in Engineering V, vol. 2, pp. 137-152, proc. of the 5th Int. Conf.* in Boston July 90, Springer-Verlag, 1990.

PLANNING AUTONOMOUS VEHICLE DISPLACEMENT MISSION IN CLOSED LOOP FORM

Alain Lambert, Olivier Lévêque
Nadine Lefort-Piat, Dominique Meizel
UTC/HEUDIASYC.URA CNRS 817
BP 529, 60205 Compiègne, FRANCE
fax : (33) 3-44-23-44-77
E-mail : {Alain.Lambert, ..., Dominique.Meizel}@utc.fr

ABSTRACT
This paper presents a Robot-Task Planner for displacement missions of mobile robots in 2D environments. The particularity of such a planner consists in the definition of the mission as a sequence of closed-loop controls, which means that observations as well as actions are specified. The produced plans are then Petri nets whose places are elementary feedback (exceptionally open-loop) motions and whose transitions are sensor-based events. The vehicle's observation ability considered in this paper is restricted to a cooperation of contact sensors with dead-reckoning.

KEYWORDS: Robot-Task, Planning under Uncertainty, Mobile Robots.

INTRODUCTION

Planning is an archetypal activity in robotics as soon as the missions to be performed did not resume to mimic a human operator; this question has been first solved by geometric planners which compute a collision free path between two configurations that optimize a performance criterion (typically minimum length) [3]. They produce nominal controls which results in the desired path assuming that the environment and the robot are accurately modelled. Since this hypothese is unrealistic, several ways have been explored to robustify the plan execution with respect to deviation between reality and its simplified model used in the planner.
A top-down approach resulted in the use of preimages in planning with uncertainty. This technique has been used by [4] to plan the motion of a punctual robot with a contact sensor in a 2D polygonal world under the assumption that the motions are piecewise linear with an uncertain orientation. Another top-down approach consists to state that the unavoidable unrealistic features of the plan will be adapted to the reality by an execution control module [6] [7] that will take into account small deviations as trajectory inflexions caused by an obstacle as more serious failures such like a closed

door, which causes replanning. In the latter case, sensing the environnement is quite decoupled from planning whereas in the first reference [4], the specification of a contact sensor seems to be strongly induced by the geometric characteristics of the planner; the adaptation of the method to a robot equipped with a real sensor like a telemeter, for instance, doesn't seem straightforward.
The method developped in this communication is a bottom-up approach: it consists in stating all motions wich should be performed by closed loop schemes in order to intrisically adapt the plan execution to the map-reality mismatch.

This implies that the observation as well as the actions must be specified along the path performing the displacement mission. The present work is an extension of the one initiated in [8]. The robot-mission is still a displacement of a car like vehicle in a 2D world with a polygonal map and the observation consists to localize the vehicle into landmarks based reference frames called *local maps*. In the previous work, a static localization algorithm using a discretized range scanner was considered. In the present one, a dynamic localization procedure involving both dead reckoning and contact measurements is considered in the same framework. The paper is organised as follows. The next section details the prerequisite and the objective of the planner. After that, a measurement of the localization unprecision is introduced. This induces a potential like function wich is used for planning. Results are finally presented.

PROBLEM STATEMENT

The achievement of a mission, whatever its application's context, introduces at least two "components": the monitor and the executor. The monitor (here the human operator) defines the objective of the mission, the executor (here the robot) must achieves this objective. For example, if we consider indoor applications like cleaning rooms or material transport in a hospital , the operator will indicate the objective of the mission by specifying the office to reach or the wall or room to clean by the robot. In each case, the mission is specified by indicating a precised element of the environment and the task to realize. In order to perform the mission, it is necessary to have an "enriched" map in which it is easy for the operator to specify the mission: the designated element, the task to realize and the initial and the goal configuration. These two last informations must be defined by a check-list of data measurable at the execution step. The robot is then able during the plan execution to decide whether the mission is finished or not. This map is composed of the mission's relevant elements. Conversely, the mission should be referenced with respect to beacons and landmarks in the environment.
The objective of the mission must be planned in this context. It consists in defining the mission by a sequence of elementary tasks to be performed by the robot so that each of these tasks can be easily controlled and checked at execution time. The planning results must be compatible and understandable by the execution control.
With this aim in mind, observations needed to control a task wich should be planned at the same time as actions. In other words this consists in defining closed loop tasks. To generate the sequence of close loop tasks corresponding to a specific mission, the planning method is based on several models. First of all, a model of the indoor environment is given. This model is not limited to a geometrical description but integrates, for each map primitive, perception possibilities. Secondly, models of real sensors are needed in which technical data like range, emitted beam, ..., must be precised.

These two models characterize the interactions between robot sensors and elements of the environment. These interactions form the observation part of the task, allowing its execution control. At last, geometric and kinematic features must be precised, in order to define the motion. The result of the planner is a sequence of closed loop tasks corresponding to a specific mission. The observations correspond to a list of couples associating map primitives and sensors.
Closed loop task is the essential underlying idea in the Robot-Task concept introduced by Espiau and exposed for instance, in [5]. Each of these tasks is modeled with both a continuous aspect (control law) and a discret one (reactions upon events). The basic idea of the robot-task concept, is to combine these two aspects into a single object called a *robot-task*. Practically, any robot-task represents an elementary action to be performed by the robot and is composed of four elements that are *the preconditions, the postconditions, the observers and the control procedure.*
The control procedure pilots the robot, the pre(post)-conditions trigger in (off) the control procedure and observers can be termed as a receptivity to some interruptions. Finally, as shown in 1(a) the result of the planner is a Petri Net where at each place is associated the control law of an action and at each transition the detection of the primitives composing the local map.

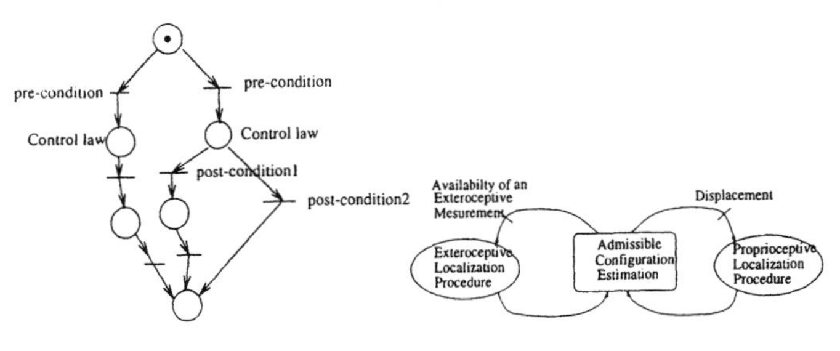

(a) The result of planning by means of robot-task.

(b) Cooperation of proprioceptive and exteroceptive processes.

Figure 1.

LOCALIZATION ABILITY AND CONFIGURATION UNCERTAINTY POTENTIAL

Localization consists in specifying geometric relationships between a mobile and a set of beacons and landmarks which define a reference frame. There are two paradigms to determine a current localization : the former is a static one and consists to derive it from exteroceptive measurements (i.e from interactions between on-board sensors and the environment), the latter induces it from both the previous estimation and the self-knowledge of the vehicle motion.

In more formal terms, the configuration estimation $\hat{q}$ is obtained in the static case as the solution of a system where each individual equation is an exteroceptive measurement : $\{z_i = h_i(\mathcal{S}_{j(i)}, q)\}$. In this expression, z_i denotes the raw measurement,

h_i is the measurement principle, $\mathcal{S}_{j(i)}$ terms the beacon or landmark with which the ith sensor interacts and $q \in \mathcal{C}$ the vehicle configuration.

Admissible Localization Domain Statement

Practically, the obvious fact that each individual measurement is corrupted by noise implies that the estimation unprecision should be precised. Formally, it consists in considering the disturbed equation system $\{z_i = h_i(\mathcal{S}_{j(i)}, q) + w_i\, ;\, i = 1, \ldots, N(i)\}$ (w_i terms the noise) and to derive the admissible subset $\hat{\mathcal{Q}} \subset \mathcal{C}$ that satisfies the noise characteristics. Such a set can either be defined as a confidence set as in [1] when w_i is expressed in statistical terms or merely as a feasible set when only the bounds of the noise are known ($w_i \in [-\beta_i, -\beta_i]$) [2]. For example, Fig.2(a) displays a situation where a robot gets unaccurate telemetric measurements and Fig.2(b) represents the corresponding admissible localization domain. For computational ease, the admissible domain $\hat{\mathcal{Q}}$ is often replaced by its prismatic or ellipsoidal envelop with a simpler analytical expression.

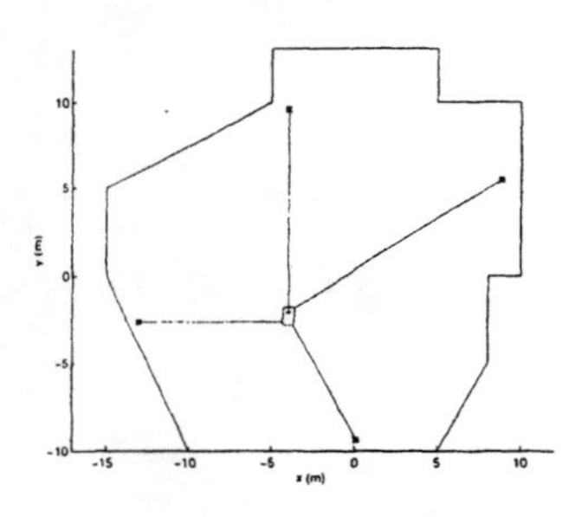

(a) Unaccurate telemetric measurements.

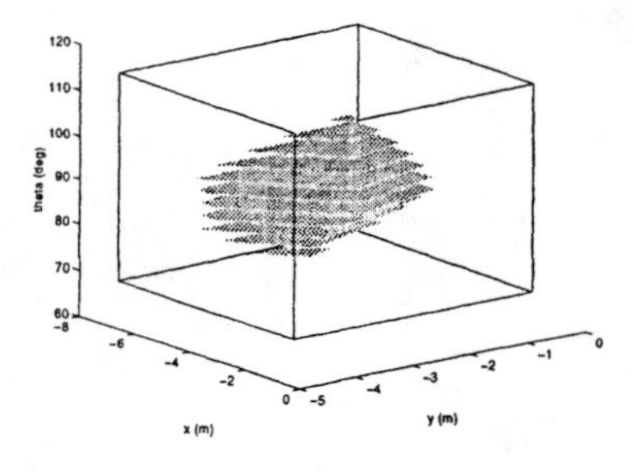

(b) Configuration domain.

Figure 2. Admissible localization domain.

Updating the Admissible Localization Domain

Computing the admissible domain is generally performed in a recursive way : successive measurements will iteratively precise a rough estimation of the domain which can either be the initial one or the set deduced from the previous one by integrating the motion self-perception through a displacement model (typicallly a kinematical one). Such a prediction can be expressed as $\hat{\mathcal{Q}}^+ = f(\hat{\mathcal{Q}}, \delta\hat{\mathcal{P}})$ where $\delta\hat{\mathcal{P}}$ is the path followed by the vehicle between the instants where $\hat{\mathcal{Q}}^+$ and $\hat{\mathcal{Q}}$ are computed. If there are available exteroceptive measurements, they will subsequently precise the admissible domain as shown in Fig.1(b)

Configuration Uncertainty Potential For planning purposes, two elements of the vehicle self localization procedure are emphasized

1. the former is a positive scalar function $size(\hat{\mathcal{Q}}(q))$ defined for instance as the surface of the mean $x - y$ section or the volume of its cubic envelop,

2. the latter is the list $\mathcal{L}(q)$ of beacons or landmarks used by the vehicle for localizing itself.

Notice that the viewpoints of the planner concerning self localization are twofolds : since we consider the robot at the planning phase, the localization q is perfectly known and is used for the simultaneous computation of the sets $\hat{\mathcal{Q}}$ and $\mathcal{L}(q)$. These items are relevant at the execution step for the statement of the robot self-localization ability (when the real q is unaccurately known). The previous items are used for planning by definition of an iterative potential function composed of three terms.

$$\begin{aligned} U(q,q^-) &= \lambda.size(\hat{\mathcal{Q}}(q,q^-)) & (1)\\ &+ \mu.\max(0, Card(\mathcal{L}(q) - \mathcal{L}(q^-)) \\ &+ \nu.Rep(q) \\ size(\hat{\mathcal{Q}}(q,q^-)) &= \min\{size(\hat{\mathcal{Q}}(q)), (1+\beta.length(\delta\hat{\mathcal{P}})).size(\hat{\mathcal{Q}}(q^-))\} & (2)\end{aligned}$$

In this expression, q^- denotes the node preceding q when this one is examined in the graph-search and $size(\hat{\mathcal{Q}}(q,q^-))$ is computed by a combination of dead reckoning and static localization (Fig.1(b)) by a simple formula (2).

λ, μ, ν, β are non-negative scaling constants,

$size(\mathcal{Q}(q,q^-))$ penalizes the unprecision of the robot self localization when it takes place around the configuration q.

The term $Card(\mathcal{L}(q) - \mathcal{L}(q^-))$ penalizes the addition of new beacons and landmarks with the aim to simplify the localization procedure which is the most difficult part of the vehicle's control algorithm.

The term $Rep(q)$ is a repulsion potential from obstacles which are not landmarks. It is classically defined by (3) in which d is the distance to the closest undetectable obstacle, R is the robot's radius, $\alpha.R$ is an influence distance and ε is a "small" distance w.r.t R.

$$Rep(q) = \begin{cases} \frac{1}{\varepsilon} - \frac{1}{R(\alpha-1)} & if \quad d < R+\varepsilon \\ \frac{1}{d-R} - \frac{1}{R(\alpha-1)} & if \quad R+\varepsilon \le d < \alpha.R \\ 0 & if \quad \alpha.R \le d \end{cases} \quad (3)$$

This potential function (1) being defined planning a path consists in finding the path $\mathcal{P}^*$ from q_{init} to q_{goal} which minimizes the functional $\mathcal{J}(\mathcal{P})$ (4)

$$\mathcal{J}(\mathcal{P}) = \int_{\mathcal{P}} U(q,\dot{q}).ds \quad (4)$$

Finding such an optimal path is practically performed by

1. discretizing the configuration space
2. find the optimal path by use of an A^* algorithm.

The result of this planning step is a sequence $\{q_i, \mathcal{L}(q_i)\,;\, i = 1,\ldots,N\,;\, q_1 = q_{init}\,,\, q_N = q_{goal}\}$ where each configuration q is associated with the set $\mathcal{L}(q)$ of beacons and landmarks used for an optimal self localization around q. This is performed, as detailed in [8] by agregation of the data in the sequence $\{\mathcal{L}(q_i)\,;\, i = 1,\ldots N\}$ by means of a dynamic clustering algorithm. Finally, the planning results in a sequence of local maps

$\{\mathcal{M}_0, \mathcal{M}_1, \ldots \mathcal{M}_F\}$ in which sequence of passing configurations are defined in a local reference frame $\mathcal{F}_i$ defined by using one or more element of $\mathcal{M}_i$. Tasks to be performed at the execution phase are then path following or parking mechanisms defined in the successive local frames and maps.

Planning this displacement missions is then completed by the definition of a sequence of local maps in which the successive displacements will be referenced. Passing from the task $\mathcal{T}_i$ defined in the local map $\mathcal{M}_i$ to the successive one can be triggered when the localization by using elements of $\mathcal{M}_{i+1}$ is more accurate than when it is performed with reference to $\mathcal{M}_i$. Those events are monitored during the execution phase but the fact is that they are defined by the planner, the specification of a mission by means of Robot-Tasks is then complete. In the sequel we propose to discuss a significative set of results.

RESULTS

We consider displacement missions of a non-holonomic cart-like mobile robot. The considered sensors in this paper are a belt of contact sensors used in combination with odometry. The case of telemetric sensors (without odometry) has already been presented in [8].

Displacement in a Hall

The first result displayed on fig.3(a) shows a classical path from the initial configuration to the goal one under non-holonomy constraints. The initial configuration is considered to be known with precision and the dead-reckoning process is assumed to be perfect. The localization is then fully accurate along the path. The result exhibits then minimum length characteristics.

In the second example represented on fig.3(b), the odometry is no more so accurate and the vehicle tends to follow the walls where it gets localization information by its contact sensors. In this experiment, the repulsion potential of the non-detectable obstacles is not used ($\nu = 0$ in eq.1). The overall behaviour of the vehicle consists then to follow detectable obstacles (due to the localization part $size(\mathcal{Q}(q, q^-))$ of the potential function), which may be seen as a good thing but this implies that the vehicles follows non-detectable obstacles too, (due to the minimum-length expression $\int_{\mathcal{P}} (.)ds$ of the criterion $\mathcal{J}(\mathcal{P})$) which is dangerous in any way. This dangerous behaviour is circumvented by using the repulsion part ($\nu \neq 0$ in eq.1) of the potential function. The result is shown on fig.3(c) In this result, the vehicle follows the detectable walls $d_1 \& d_2$ until the wall d_1 stops. When this event occurs, localization is fully accurate. Then it follows the wall d_2 on a given distance and next a path in open loop until it finds the detectable corner of walls $d_3 \& d_4$. Here again its localization is fully accurate as this event occurs. It follows then an open loop path away from the dangerous zone caused by non-detectable obstacles until it finally reaches wall d_5 and achieves its mission.

A Turn-left Manoeuver

Fig.4(a) displays a simple turning left manoeuver. q_{init} and q_{goal} are first considered to be perfectly known and the dead-reckoning is assumed to be perfect : one obtains the typical minimum-length planned path of Fig.4(a).

In Fig.4(b), one considers the fact that the x coordinate of q_{init} is poorly defined. It implies that the vehicle will first get close to d_1 and later on reach the corner of walls

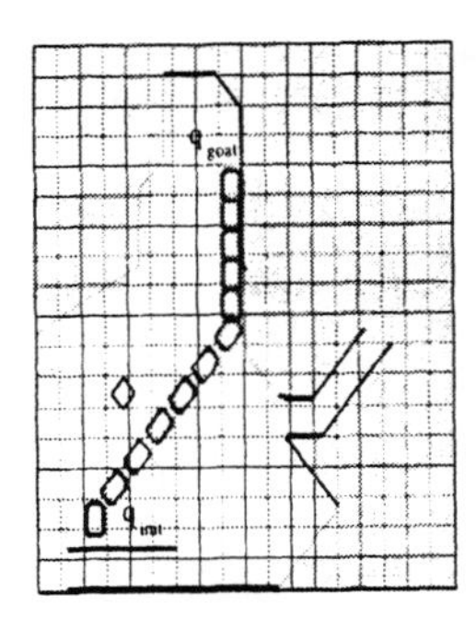

(a) Planning assuming accurate localization. The size of the room is $9m\,x\,9m$.

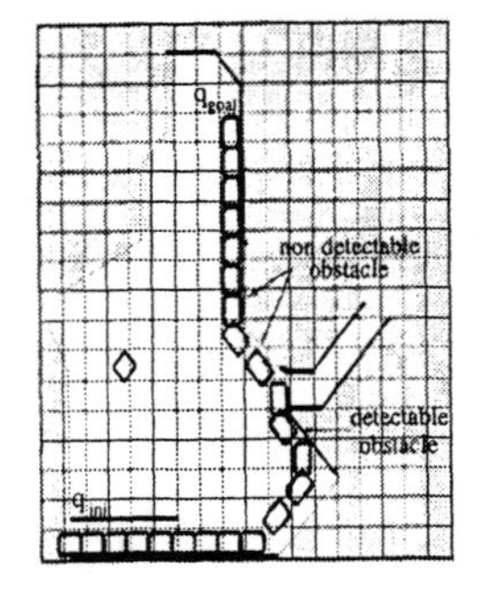

(b) Planning with localization uncertainty and no repulsion of non detectable obstacles.

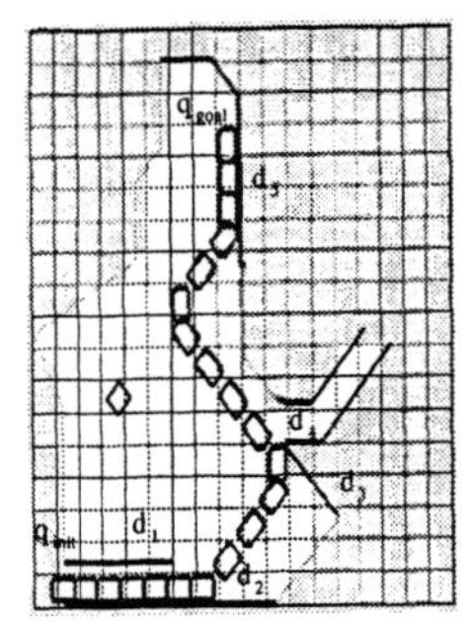

(c) Planning with localization uncertainty and repulsion of non detectable obstacles.

Figure 3. Different plannings for a displacement mission.

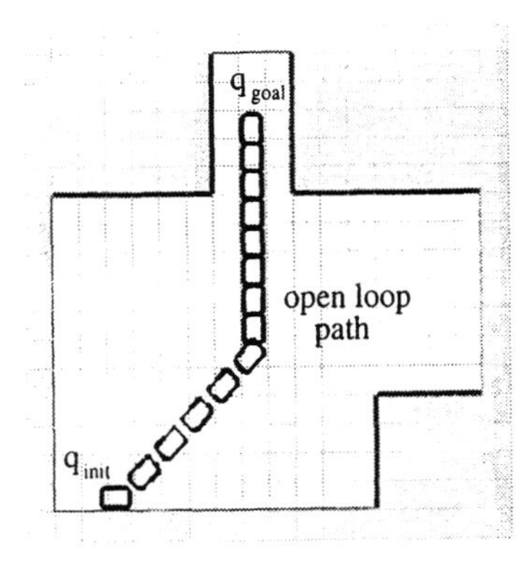

(a) Planning a turn left assuming accurate initial localization.

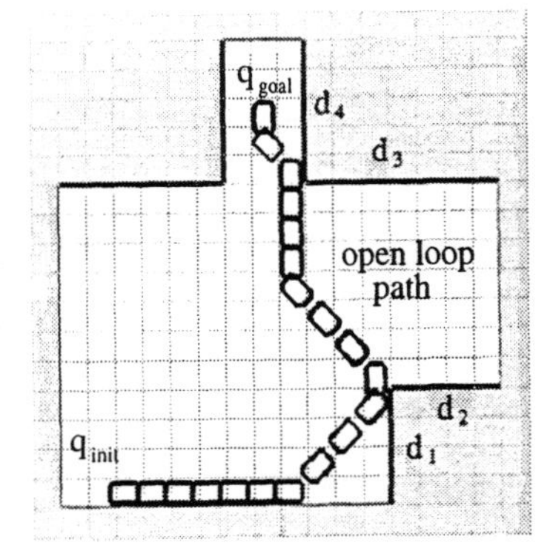

(b) Planning a turn left assuming unaccurate initial localization.

Figure 4. Turn left manoeuver.

$d_1\&d_2$ to get a good localization (like in [4]). It can then cross the horizontal road in order to reach the corner of walls $d_3\&d_4$ and achieves the mission.

CONCLUSION

This paper has demonstrated the adaptability of the concept introduced in [8]. The main idea consists in planning observation at the same time as actions in order to obtain tasks easily executable in closed loop form and thus robust with model reality mismatch. The concept of local map introduced with telemetric measurement is presently very restricted here because of the simplicity of the contact sensor which as been considered and which generally only permit to interact with one single landmark. Anyway, although trivial in this case, the local map concept is still valid. Considering other robot-task than path following and parking is now an extension in progress. Another significative improvement would be to consider now the difficulty of the environnement recognition (i.e the complexity of the matching step in the localisation process) in the planning.

REFERENCES

[1] J. J. Leonard and H. Durrant-Whyte. *"Mobile Robot Localization by Tracking Geometric Beacons."* IEEE Transactions on Robotics & Automation, 7(3), 376-382, (1991).

[2] A. Preciado and D. Meizel and A. Segovia and M. Rombaut. *"Fusion of Multi-Sensor Data: a Geometric Approach."* Proceedings of the IEEE Int. Conf. Robotics and Automation, 2806-2811, (1991).

[3] J.C. Latombe. *Robot Motion Planning.* Kluwer Academic Publishers, (1991).

[4] R. Alami and T. Simeon. *"Planning Robust Motion Strategies for a Mobile Robot."* Proceedings of the IEEE Int. Conf. Robotics and Automation, 1312-1318, San Diego USA (1994).

[5] E. Coste-Manière and B. Espiau and D. Simon. *"Reactive Objects in a Task Level Open Controller."* Proceedings of the IEEE Int. Conf. Robotics and Automation, 2732-2737, Nice France (1992).

[6] F. R. Noreils. *"An Architecture for Cooperative and Autonomous Mobile Robots."* Proceedings of the IEEE Int. Conf. Robotics and Automation, 2703-2710, Nice France (1992).

[7] M. Hassoun and C. Laugier. *"Reactive motion planning for an intelligent vehicle."* Proceedings of the Intelligent Vehicles'92 Symposium, 259-264, Detroit USA (1992).

[8] I. Collin and D. Meizel and N. LeFort and G. Govaert. *"Local maps and Task-Function Planning for Mobile Robots."* Proceedings of the IEEE/RSJ/GI Int. Conf. on Intelligent Robots and Systems, 273-280, Munchen Germany (1994).

Aknowledgements

Olivier Lévêque is funded by the Picardie Regional Council within the "DIVA" project.

INTEGRATED DESIGN FOR A COMPUTER INFORMATION MANAGEMENT (CIM) PROJECT : DISTRIBUTED HYPERMEDIA INFORMATION SUPPORT SYSTEMS

M. LOMBARD

CRAN-GGP, Université de Nancy I
B.P. 239 - 54506 Vandoeuvre-lès-Nancy Cedex (France)
phone. +33-83.91.21.50 fax. +33-83.91.23.90
E-mail. lombard@cran.u-nancy.fr

ABSTRACT

The geographical partner distribution of a project as well as the manipulated data complexity need the utilization of information super-highway. To go more far than a simple management of file, we propose to couple to the classic server an object database. Thus, we present an architecture coupling P.C.T.E. to a WWW server. Finally, we will show that the WWW server can be seen as a layer of integration in the sense of C.I.M..

KEYWORDS : Concurrent Engineering, WWW (World-Wide Web), PCTE (Portable Common Tool Environment)

1. INTRODUCTION

Regrouping various Production Engineering competencies in a research-laboratories consortium plays a fundamental role in rapid circulation of results, in mutual understanding between different partners and in the establishment of a cooperation spirit.

This is the reason why, our project named "Scenario for a Concurrent Engineering in Manufacturing Integrated Systems", supported by the Technical and Scientific Mission (DSPT8) of the French Ministry of University Education and Research, aims at putting in evidence what a Concurrent Engineering should be, with the implementation of a real semantic communication between the different partners involved in the consortium*.

In fact, successful design requires the inclusion and integration of downstream issues in order to achieve feasible and affordable design. Distributed information support systems have a role to play in such an integrated, concurrent view of design.

---

* **CRAN** Centre de Recherche en Automatique de Nancy
LAMIH Laboratoire d'Automatique et de Mécanique Industrielles et Humaines de Valenciennes
LAN Laboratoire d'Automatique de Nantes
LIRMM Laboratoire d'Informatique, de Robotique et de Microélectronique de Montpellier

To resume, our problem is to design and realize an architecture which allows the integration of different points of view related to specific skills, taking into account heterogeneous organisations, as well as distributed and cooperative ones.

Our experimental framework is an industrial problem of design and realization issued from an assembly application dealing with the riveting of DASSAULT airplane sub-sets.
The first part of our proposal deals with a presentation of different communication supports as well as their limits, as there is a lot of information to be exchanged in this project.

Then, and because traditional communication supports do not allow to share and to save information, we present a distributed hypermedia architecture which allows to capitalize all information as well as taken decisions during the whole project, using the multimedia documentation capabilities offered by the Internet's World-Wide Web (WWW).
Moreover, accessing information or documentation is not enough in such an architecture as the one specified before.
Actually, taking into account the geographic distribution of the different members of the consortium, as well as the specific use of softwares, it is necessary to allow, through the WWW server, a remote access to specific softwares, in a transparent way for the user. Firstly, and because of information protection problems, the access to the information is limited in a consultation mode.

The second part of this paper deals with the interfacing process between WWW and PCTE (Portable Common Tool Environment), on the one hand in order to present a structured data management and not a file management anymore, and on the other hand in order to present the integration repository offered by PCTE.
The modeling allowing the implementation of the database is also the same as the one used in order to realize the architecture of the WWW server.

As a conclusion, we present the state of progress regarding the different implementation phases, as well as the development prospects that are considered.

2. SYSTEM OF COMMUNICATION

In this type of project, where partners are geographically distributed it is necessary to envisage some supports of communication allowing to transmit pictures and information in general. All intervening stages in the life of an engineering project, therefore of a product, are concerned : meetings of work from a distance in phase of project definition, writing of cost notebook, in phase of systems analysis, in phase of design within multi skills teams in concurrent engineering, and even in phase of production
as envisaged in the french project " Potential Enterprise" for the communication exchanges beetween Manufacturing Inside-institution Workshops.

To solve the problem of the geographical distribution of the different actors, it is necessary to lean on :

- "philosophical" works, such that concepts of concurrent engineering, that attach especially to instal communication beetween actors within an even disciplines or between different interest disciplines. Concepts of retro-engineering that correspond to ascend the know-how of the manufacturer to the designer tend also to improve trades of information within the enterprise.

- "technical" works as EDI (Computerized Data Exchange), considered as an universal language for un-partition off the data-processing, have been initiated to facilitate relationships between customers, subcontractors, etc ... The implementation of EDI has to strengthen not only trades with external partners, but also to coordinate the administrative view with other views.

The first stage to put in place a system of communication has been to choose the utilization of the electronic messaging (email) for on the one hand the exchange general information as for the organization and to the functioning of the project, and on the other hand for the exchange of un-formal data since not responding to no structure and preliminary procedure.
Following a logic of project, these exchanges are addressed to a common mailbox to which all members of the project are connected since declared in a common distribution list allowing the distribution of messages to all partners.

On the other hand, after utilization, the analysis of communications of the project exposed in (1) on 150 exchanges by email shows how without a minimum of rigor to the level of the identification of speakers, the classification of messages , the identification of modifications, . .. it is not possible to extract some necessary relevant information. More, the work to several on a same document has shown how it was difficult to identify modified points and their authors, it is similarly for the validation. This is why, in parallel to management of un-fomal data it is appeared essential on the one hand to consolidate existent data with formal manner and on the other hand to propose a new organization of work.

Thus, it is appeared necessary to better manage exchanged information and to preserve this historical. The globalization and the virtual geographical distribution of the information is rendered possible thanks to of systems such that the World Wide Web on the Internet network. Indeed, the geographical cover (international), services such that the environment HTML (Hyper Text Markup Language) allow easily to make cooperate the totality of actors. To make this, we have leaned on an interface MOSAIC commonly used in this environment.
But, the main difficulty to develop an HTML server concerns the structuration and the navigation between pages which compose the hypertext document. At this present time, it does not exist methods to model, to generate and to validate HTML document structures. Thus, we propose to develop these hypertext documents by using entity-relationship modelling as in (2). Thus these models serve on the one hand to organize the structure of pages and on the other hand to the structure of the database as explained in what follows.

3. DATA MANAGEMENT FOR COLLABORATIVE ENGINEERING

The second stage in the installation of the communicating infrastructure has focused on the contribution of supplementary functionalities for the assistance with the idea, namely :

- management of cooperative technical data existent in the project and common to all partners in the manner to capitalize results in view of their futur re-use,
- the assistance on-line to the user,

- the possibility to preserve the utilization of the electronic messaging but with stuctured manner,
- the utilization of modelized and design tools a distance.

Thereby, an integrated environment such that specified previously has to be adaptable, flexible allowing thus the different partners and actors of the project the dedicated tool manipulation and distribute geographically coherent manner. The utilization of P.C.T.E. (Portable Common Tool Environment) (3) as platform of experimentation for the architecture D.M.M.S. (Design Management and Manufacturing System) (4) has served as reference as for the modelisation of technical data. Indeed, the OMS model (Object Management System) of P.C.T.E. plays a major role since it provides necessary functions for the modelisation and to the manipulation of objects of the environment as well as to the stocking of their authorities. The object model of P.C.T.E. is an entity-relationship model extended by the relationship of inheritance. It draws one's inspiration from "object-oriented" models by allowing the definition of classify objects, classe instances, simple attribut inheritance, and views on objects by the notion of totality of partial diagrams or SDS (Schema Definition Set) juxtaposed in a space of work (Working Schema). P.C.T.E. as repository structure for the integration of software tools has a similar approach that CALS (5).

More, to have a real hypermedia integrated dynamic environment, the access to the P.C.T.E. database been made via a WWW server (World Wide Web) on the Internet network using the HTTP protocol (Hyper Text Protocol Transmission). A document provides by a such server is at the HTML format, it is interpreted by the tool of consultation : MOSAIC (Figure 1).

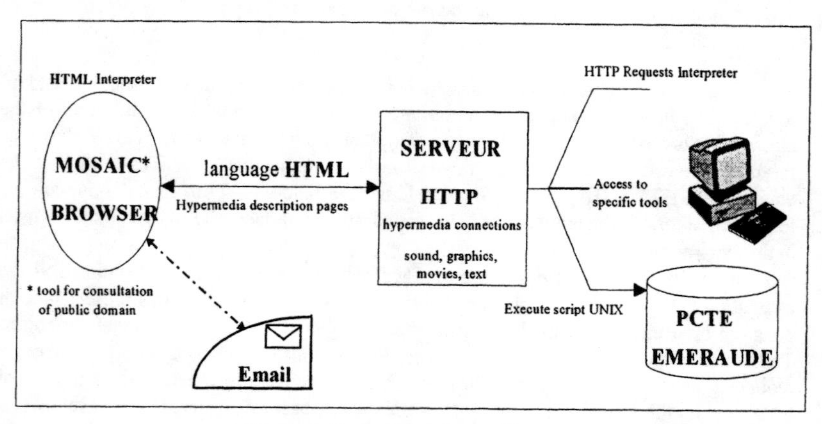

Figure 1. Communication architecture

In our project, the utilization of architecture connected P.C.T.E. with the WWW server allows us to have classic services for the user assistance, but also to access directly to dedicated tools and to data, instances of objects in the database that are anchors in the hypertext sense. It is thus possible to consult, to instance, or to destroy data following rights of access of the user.

Finally, un-formal communications by email are made now via MOSAIC but of more structured manner since it is thus possible to address a message to the responsible person for the HTML page, therefore of data and present tools and that the subject is then well identified.

4. PROTOTYPE

The execution of an external program by a WWW server been made in two stages :

- it is necessary first to have an HTML document containing forms allowing the user to enter parameters,
- it is necessary obviously to have a capable program to process these parameters.

```
<HTML>
...
<FORM METHOD='GET' ACTION='http://aipsun11.cran.u-nancy.fr/program'>
<LI>Enter information : <INPUT NAME = 'name' SIZE='10'>
...
To validate your choice : <INPUT TYPE='SUBMIT' VALUE = 'validate'><P>
To unvalidate your choice : <INPUT TYPE='RESET' VALUE = 'unvalidate'><P>
</FORM>
...
</HTML>
```

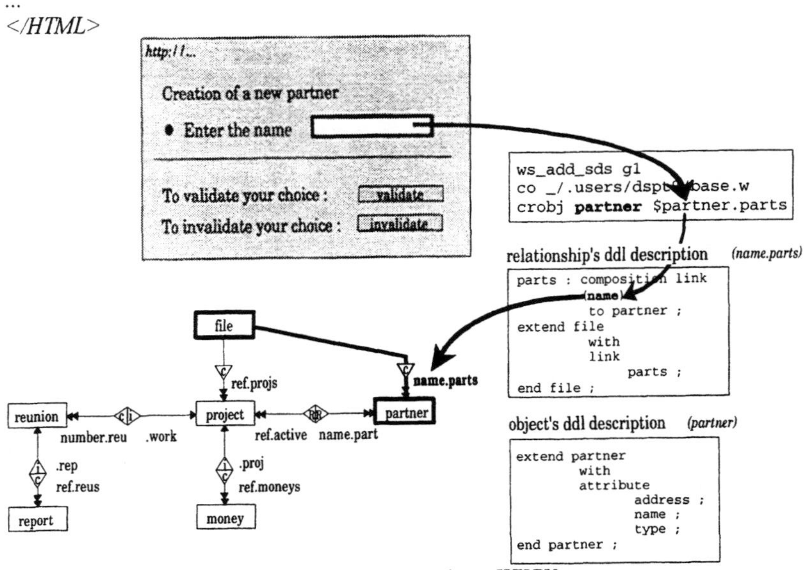

Figure 2. P.C.T.E. connected to a WWW server

The role of the external program is :

- to recuperate parameters acquired by the intermediary of forms,
- to process them,
- to provide among others a result in exit that will be sent to the client program that has submitted the request.

5. CONCLUSION

Our prototype focuses on the interfaçage of a WWW server and Emeraude (software supporting P.C.T.E.) for the consultation and the instanciation of objects of the database. This first implementation has allowed already to release rules of passage between an entity/relationship model and structures of pages in html format. Once these identified rules, it is possible to implemente them and to envisage then a quasi automatic generation of the server. In this case, the work of the designer consists in render convivial the environment of work by addition of document, text, picture, video . . . to optain a sort of guide to the user.
This work can be widened to the interfaçage with others data base. It can be assisted as with Oracle 5 that has primitive of accesses to the WWW server, or developed by leaning on the work presented. Thus the operator since its position of work can access transparent manner to data base geographically distributed. This new manner to envisage the integration gives to the acronym C.I.M. feel it to Computer Information Management.

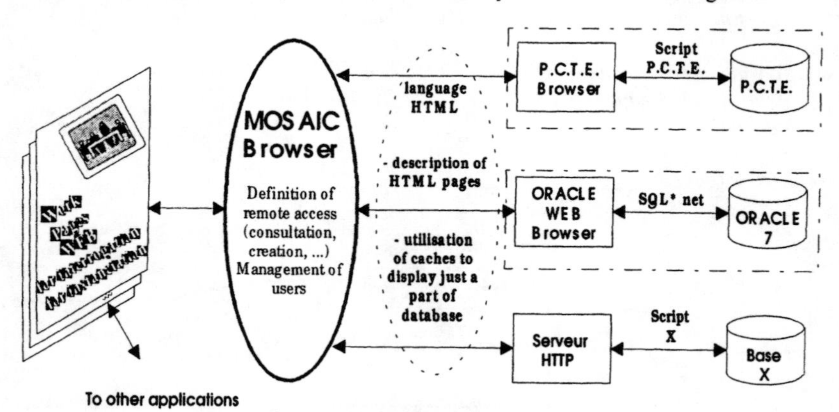

Figure 3. Multi database integration

6. REFERENCES

1. Petiot, J.F., SOMMER, J.L., Jacquet, L., Koyama, Y., *Analysis of the communications in concurrent engineering design.* IDMME'96 (Integrated Design and Manufacturing in Mechanical Engineering), Nantes (France), April 15-17,1996.
2. Pinon, J.M.,Laurini, R., *La documentation multimédia dans les organisations.* Ed. Hermès, Paris (France), 1990.
3. *P.C.T.E. a basis for Portable Common Tool Environment.* Functional Specification. Version 1.5, Commission of the European Communities - Brussels (Belgique), 1988.
4. LOMBARD-GREGORI, M., *Contribution au Génie Productique : Prototypage d'une Architecture d'Ingénierie Concourante des Systèmes Intégrés de Fabrication Manufacturière.* Thèse de Doctorat de l'Université H. Poincaré Nancy I - 1994
5. Chevalier, P., CALS et les systèmes d'informations électroniques. Ed Hermès, Paris (France), 1993.

Development and Operation of Modular Robots

P.M. Lourtie, J.L. Sequeira and J.P. Rente
IDMEC-Instituto de Engenharia Mecânica, IST
Av Rovisco Pais, 1096 LISBOA CODEX , Portugal
E-mail: {lourtie, jluis, rente}@gcar.ist.utl.pt

J. Esteves
CAUTL Centro de Automática da Univ. Técnica de Lisboa, IST
Av Rovisco Pais, 1096 LISBOA CODEX , Portugal
E-mail: estevesj@alfa.ist.utl.pt

ABSTRACT

Robotics is a technologic domain with a strong development in research and in industrial applications. Nowadays robotics is an educational topic at high school as well as at university levels and also in technical education. At a university level a robotics kit allows for the application of the knowledge from different scientific domains. For instance the study of kinematic and dynamic behaviour, control, task or trajectory planning, allowing for easy implementation of small projects in robotics and automation. It is also possible to interconnect with projects in areas like instrumentation, artificial vision, and so on. This paper is an introduction to a project taking place at Instituto Superior Técnico, with the objective of building a didactic robot to be used in robotics and automation courses.

KEYWORDS: modular robot, educational robot, robot design

INTRODUCTION

Nowadays, the didactic robots available in the market, have fixed configurations with restricted dynamic specifications, disallowing for the use in the study of dynamic behaviour of robots.

The didactic robot developed has a modular concept allowing for parts of the full system to be replaced and/or reconfigured, in order to fulfil some objective of a specific area of teaching and research. The modular concept is not restricted to geometric aspects. As the overall control architecture is open, e.g., the axis controller is based on personal computer cards, the modularity can be achieved at other levels. This is a main contribution to the robotics educational and research domains.

The robotic modules that have been developed, have actuators that are powerful enough to explore the dynamic capabilities of the robots and to be use by research teams as a working tool that may be developed.

The idea of building a modular robot resulted from the fact, when it was decided to buy robotic equipment for the laboratory of the Control, Robotics and Automation Group, this type of equipment was not commercially available. Either there were fixed configuration robots or modular didactic robots very limited from the point of view of performance, capabilities and size. The fixed configuration robots, not only are they fixed from a point of view of the mechanical arm, but also from the point of view of the control algorithms that may be used and of the possibility of easily replacing the standard controller by a different one.

The mechanical modularity means that the available components may be assembled in a number of configurations, namely producing an articulated arm, with or without offset, SCARA, cylindrical or spherical robots (Figure 1). This mechanical modularity could only be implemented because modularity was achieved in power and signal cabling.

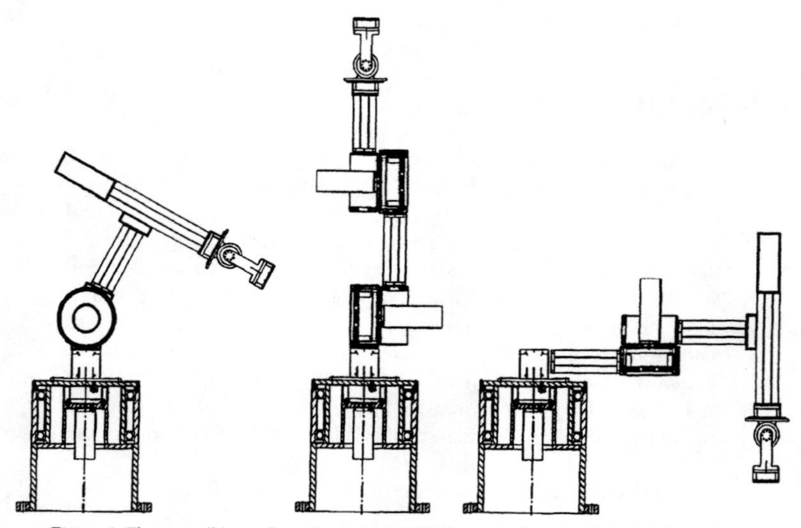

Figure 1. Three possible configurations with 3 DOF plus wrist (Spheric, Articulated and Scara)

The controller was designed independently from the actuators with access by the user to the actuators control signals. This makes it possible for the user to replace the controller by a different one of his choice or design. The modularity of the controller allows the implementation of different control algorithms.

The robot was conceived as an educational and research aid that may be used in higher education, but also at secondary education or technical training. For this aim, the implementation of commercial programming languages, on top of the DORM own language, is being studied.

DESIGN OF MECHANICAL MODULES

The fundamental idea in the design of the kit elements was modularity, which is in conflict with the objective of optimizing the dynamic behaviour of a given configuration. The first step was to develop an articulated arm (RRR, three rotary axes corresponding to waist, shoulder and elbow) as a basis for the project, as this is the most demanding case in what gravity forces are concerned.

The overall indicative specifications used for the design of the mechanical modules, joints, links, base and wrist, were the following:

- Articulated arm with Euler (spherical) wrist;
- Horizontal reach 1200mm;
- Achievable angular of the joints of 90°/s;
- Achievable angular acceleration of $180°/s^2$;
- Precision of 0.5mm.

From these specifications and a preliminary estimation of the masses involved, in particular of the motors, an estimate of the required torques was produced and the joints, base and links were designed.

An universal flange was designed, in such a way that joints and links may be assembled in a number of configurations. This flange has a rectangular shape of 100x80mm and is held together by four screws.

A concession to modularity consists of having a base element that is structurally different from the other rotary joints. A second, minor, concession to modularity consisted of having two different motoring alternatives for the shoulder and elbow type joints. In this case, the structural elements are the same and there are attachments for the two different sizes of motors and of harmonic drives. In any case the motor and harmonic drive used in the base are the same as those used in the elbow type joint.

This is a minor concession, as it is possible to work with two elbow type joints, as part of an articulated arm, if dynamic performance specifications are relaxed and/or the shorter link elements are used, reducing the horizontal reach.

The mechanical elements that have, so far, been produced are:

1. One base element;
2. Two joints, one of each of the possible motoring alternatives;
3. Two long and two short link elements;
4. One three axes wrist, with three alternative configurations (see below).

In may be seen, from the specifications referred above, that these modules do not fall in the same class of the available didactic robots. Its power and dynamic behaviour opens the possibility for a number of applications beyond the ordinary classroom demonstrations.

The structural elements are made mainly of an aluminium alloy, except for the base that is made of steel. The overall weight of the articulated arm configuration with the Euler wrist is approximately 140Kg.

The stiffness and natural frequencies of the reference configuration were studied using a finite element program (ALGOR). The results of the natural frequency calculations were confirmed by shock physical tests. Both results concur to the fact that the lowest natural frequency is around 47Hz.

Static deflection was calculated using the same program, giving results within the specifications for a carried load of 5Kg. The results have shown, however, that the point where stress and deflections were more important was the wrist, due to its particular design.

JOINT DESIGN

The joints design are powered by electric motors that drive the joints through harmonic drives. The gearing ration of the harmonic drives used is 1:160.

The motors used are of the brushless type, with brake and resolver incorporated. These brushless motors are synchronous ones with $SmCo_5$ permanent magnets in order to achieve a good torque/weight ratio. The maximum torque of the motors used for the shoulder and elbow type joint are respectively 1.9 and 1.0Nm.

In the design of the structural elements there were several objectives:

1. Simplicity of production;
2. Robustness to manipulation, for instance during reconfiguration tasks;
3. Low cost.

The option was for a cast aluminium alloy, available in the market, and the machining of the areas of contact, for assembly. The joint has a closed shape, so that an open version of the harmonic drive might be used, also protecting from intrusion of objects or the hand during manipulation.

The overall shape is of two cylinders with the same axis, moving in relation to each other, each having a flat flange for assembly with another module.

LINK DESIGN

The links are made of the same alloy as the structural elements of the joints and have a very simple structural design. They consist of a tube with a cross section that is approximately rectangular. At each end there is a standard flange to assemble it together with any other element, a joint but eventually another link.

The length of the link to satisfy the original reference specification is 245mm, but shorter links were produced with 105mm.

BASE DESIGN

The base is of cylindrical shape, the top half rotating with respect to the bottom one. On the upper face of the base element, there is a plate with a dice that allows for the connection to a standard flange, both horizontally and vertically.

It is built of steel and the rotating parts are supported by large bearings that ensure a high stiffness and resistance to torques resulting from offset loads.

WRIST DESIGN

The full wrist designed is made of three joints (Roll, Pitch and Roll) and is built of the same aluminium alloy. Part of the joints may be removed, resulting in either a Pitch-Roll or a simple Roll wrist. This is achieved by replacing the first roll with a fixed element, for a Pitch-Roll wrist, or by removing the two final joints, for a Roll wrist.

The motoring of the wrist is obtained by the same type of brushless motors and harmonic drives, but smaller, all the joints of the wrist using the same model. The motor has a maximum torque of 0.06 Nm, the gearing ratio of the harmonic drives being also 1:160.

The design of the wrist gave special consideration to ease of assembly and disassembly. The result was a design that, from the point of view of stiffness is not ideal, although within the initial specifications. In future editions, after the present configuration has been fully tested, this point will be given some attention.

The wrist is equipped with an universal gripper mounting, allowing for commercially available grippers to be attached.

CABLING MODULARITY

The mechanic modularity is only possible because modularity was achieved in power and signal cabling. The cabling modularity is based in two general buses indexed by joint. These buses go through the joints and links till the wrist. Both buses are not ended and the connections are made at joint level with T connectors and each link is cabled. These T connectors are specific for each joint and therefore the indexation.

In the first bus, power bus, travels the power for motor and brakes. The second bus, signal bus, is used for resolvers, temperature sensors and limit switches signals. This bus has some spare wires for the inclusion of different sensors, e.g., force sensors at the wrist or strain gauges for flexible link.

Both buses stop at wrist level because the wrist has only a small (three) number of configuration and the interchangeability between modules is not possible. The only possibility is to connect or not each module of the wrist

CONTROL UNIT

The control unit is divided in two levels. The upper level is responsible for generating the reference signal in speed (± 10V) to the servo ampliflier unit. The upper level of the control unit is mounted in a rack multibus with an Intel 486 based board and two Galil DMC-1004 motion control cards.

At the lower level, the servo amplifiers unit powers the brushless motor with a PWM signal using at the control inner loop signals from the resolvers. As the upper level uses encoder signal, and not resolver, the conversion is also made at this level.

The controller was designed independently from the actuators with access by the user to the actuators control signals. This makes it possible for the user to replace the upper level by a different one of his choice or design. The modularity of the upper level allows the implementation of different control algorithms. As the angular sensors are resolvers, needed for the inner loop of the brushless motor, the user can use them directly or the encoder signal that is generated in the servo amplifier unit.

SOFTWARE

The software is composed by two packages, the modeller package and the simulator package.

In the first package the user can build/define, in a 3D graphical environment, the configuration of is choice. As the frames are placed in an automatic procedure, using Denavit-Hartenberg algorithm, the user can see the homogeneous transformation matrix.

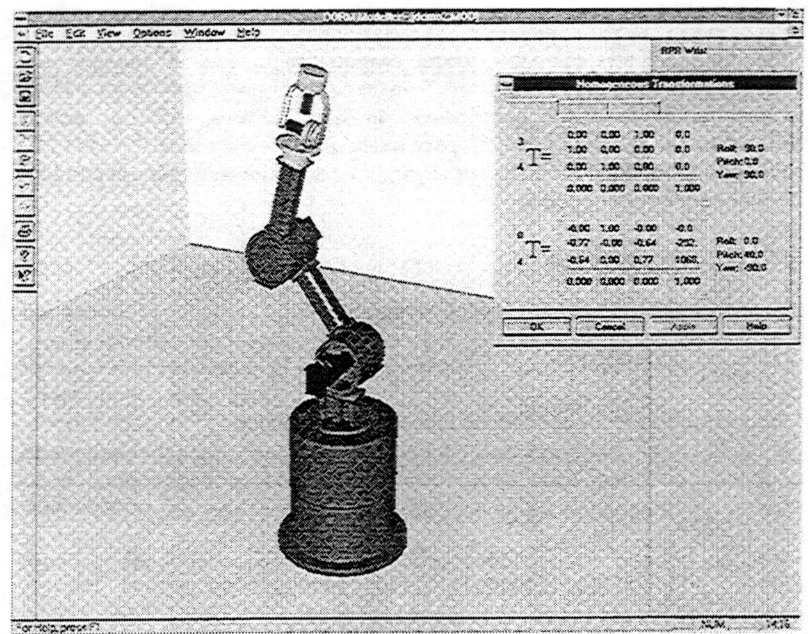

Figure 2. Screen shot of the Modeller package with an articulated configuration with a RPR wrist

Using direct kinematics, the position of the tool central point is produced by the program. Figure 2 shows a screen of the modeller package.

In the simulator package the user can simulate and program the robot. It is also a 3D graphical environment and uses the model created in the previous package. The two activities ,simulation and programming of the robot, can be perform in a teach-by-doing mode or on DORM own language.

REFERENCES

1. Cohen, R., Lipton, M.G., Dai, M.Q. and Benhabib, B. , "Conceptual Design of Modular Robot", *ASME J. Mechanical Design* 114 (1992) 117-125.
2. Matsumaru, T., "Design and Control of the Modular Robot System:TOMMS.", *IEEE International Conference on Robotics and Automation* (1995) 2125-2131.
3. Klafter, R.D., Chmielewski, T.A., Negin, M., *Robotic Engineering, an Integrated Approach*, Prentice-Hall, Inc. (1989).
4. Denavit, J., Hartenberg, R.S., "A Kinematic Notation for Lower-Pair Mechanisms Based on Matrices.", *J.App. Mech* 77 (1955), 215-221.

COOPERATION AMONG DISTRIBUTED CONTROLLED ROBOTS BY LOCAL INTERACTION PROTOCOLS

TIM C. LUETH
FZI Research Center, Haid-und-Neu-Str. 10-14, D-76131 Karlsruhe, Germany, email: t.lueth@ieee.org

RONALD GRASMAN, THOMAS LAENGLE
IPR Institute for Real-Time Computer Systems and Robotics, U. of Karlsruhe, D-76128 Karlsruhe, Germany, email: laengle@ira.uka.de

JING WANG
College of Engineering, U. of California, Riverside, CA 92521, U.S.A

ABSTRACT

Inter-robot cooperation is useful on the symbolic and physical levels. It requires the dynamic coupling of planning systems but also the coupling of real-time control systems. This paper describes the basic cooperation mechanisms. Furthermore, the use of a cooperative robot operation system CAIC and the FZI's local infrared communication system for the Khepera robots during inter-robot cooperation is described.

KEYWORDS: Inter-robot cooperation, coupled cooperative robot control, local communication

COOPERATION DURING INDEPENDENT TASK EXECUTION

Typically, robots are designed in a first step without considering multi robot environments. These robots receive their tasks from one or more *clients* and operate as a *server*. During task execution the robots elaborate the *symbolic task description* depending on the current environment, *map* the elaborated task to a *network of control loops* [1], and *physically change* the environment. For instance, they transport or manipulate assembly parts. This means, besides the symbolic task description there is an actual physical robot-environment-interaction.

If several mobile robots are active as servers in the same environment and receive independent tasks from the same clients, they physically have to share resources, i.e., at least their work space. To master arising conflicts, passive or active interaction levels can be chosen. For each interaction level, a different execution-control architecture is required. The levels can roughly be classified in:

1 The individual robot classifies an other robot as an *uncooperative* robot. It delays, accelerates, or adapts its own task execution (i.e., individual conflict compensation).

2 The individual robot classifies an other robot as an equal environment *participant* that obeys a common set of passive physical interaction rules such as using traffic lanes or access rules (CSMA/CA/CD). Both robots minimize the conflict potential together [2].

3 The individual robot classifies an other robot as an equal environment *participant*. It expects *broadcasts* of the participant's resource-requirement schedule automatically or

on request. The exchange of symbolic description avoids misinterpretation by individual estimation of other participants' future resource-requirements. The information exchange requires communication capability and a protocol [3].

4 The individual robot classifies an other robot as an environment *partner*. Both robots exchange their resource schedules and optimize the joint schedules based on individual (negotiation) or common needs [4].

In all four *interaction levels*, the individual tasks are performed individually and independently even if communication, coordination, and symbolic cooperation are used for optimization. Each robot has to classify an unknown robot as *uncooperative*, *participant*, *communicating participant*, or *partner*. It has to select the required interaction protocol for optimal individual behavior. Optimal means overall-cost efficiency for the individual robot. Overall-costs depend on delays in physically task execution but also on delays by communication, negotiation and planning and in addition on the availability of communication and information processing capacity. Therefor, an individual robot can decide not to behave like a partner even if it could so, since from its point of view it's inefficient. Further considerations belong to *cooperative game theory*.

The number of possible conflict relations among two individual robots is proportional to the square of the number of robots. If the number of robots is increasing, the time between two conflict situations will decrease correspondingly. In a similar manner, the available time to master a conflict situation is decreasing and the need for efficient conflict avoidance becomes evident. The need of an individual robot to behave like a partner will automatically arise.

Cooperation during independent task execution is caused by resource conflicts and requires coordination (delay, acceleration) of alternating execution. Therefore, the following steps must be performed in each robot:

1 Decision on the maximal offered individual interaction level.
2 Negotiation (Offering, accepting, refusing, ignoring) on the common interaction level.
3 Using the common understood interaction level.

In contrast to that, cooperation by coupled distributed controlled system is caused by the need for joint physical environment changes [5]. It requires simultaneously and timely executions.

COOPERATION BY COUPLED DISTRIBUTED CONTROL SYSTEMS

If a task cannot be physically performed by only one robot it must be given to a team of robots that are able to perform the task by physical cooperation. Consequently, the symbolic task description must be elaborated in terms of the team's configuration and mapped to a network of control loops. Then, the network is separated into smaller control sub-networks in terms of the available capacity (robots). Afterwards, the *subnets* are distributed to the individual robot-control systems [6].

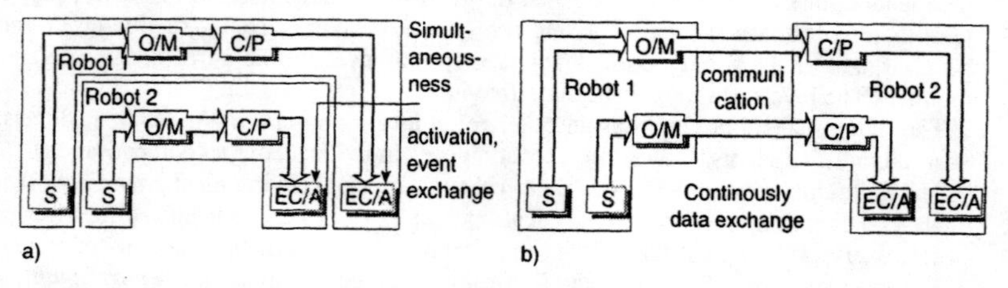

Figure 1. a) Coupled activation of control networks. b) Coupling within control networks

Often, closed subnets of the network are distributable completely to one robot (Fig. 1a). Then, there is only a need for *event-based synchronizing the activation* of local subnets with subnets running on other robots.

Sometimes, it is necessary to distribute even single closed-control loops to different robots by *separating* the control loops into linked loop modules for *sensing* (S), *observation/modeling* (O/M), *control/planning* (C/P), and *execution control* (EC/A) (Fig. 1b). Therefore, *continuous information flow* (*time-based* or *on request*) must be established between two robots and also the module activation must be synchronized [7].

LOCAL COMMUNICATION FOR COOPERATIVE ROBOTS

Dynamic distribution of control-network modules and dynamic distribution of events and control information between the modules require a *robot operating systems*. For this purpose, the *CAIC* system (*Cooperative Architecture for Intelligent Control*) has been developed in Karlsruhe [7,8]. CAIC is an extension for existing real-time operating systems. It allows the flexible linkage of encapsulated control modules by supporting different module activation (time-based, event-based) and information exchange (continuously, on request, on event). Furthermore, it allows to initiate the execution of control modules on other robots.

To support robot operating systems such as CAIC and to allow the dynamic coupling of real-time control loops operating on different robots, a local communication system is required. It should have the following features:

- No central communication backbone even if wireless (Fig. 2a),
- real-time data exchange, i.e., little and known transfer delay,
- selection of communication partners based on physical conditions (Fig. 2b), and
- limited channels width to avoid disturbance by other teams (Fig. 2c).

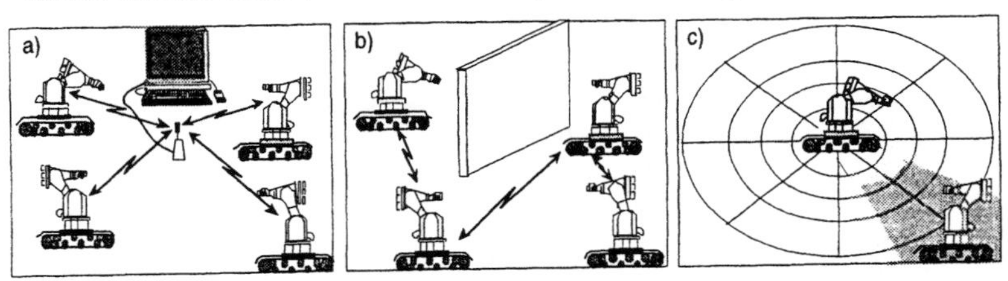

Figure 2. a) Central backbone. b) Visible communication partner selection. c) Limited consumption of the communication medium.

THE ROBOTS AND THEIR COMMUNICATION SYSTEM

For our experiments, small mobile manipulators (100 g, Ø 55 mm) of the Khepera family [9] have been used. For mobility, two servo motors equipped with encoders and gearboxes (25:1) are used. The incremental encoder resolution is 12 tics/mm. For each motor a PID control loop is used for controlling the speed. The PID-coefficients are programmable and the internal counters (I-term) are readable.

Eight infrared proximity sensors (sender and receiver) are able to measure back light with a resolution of 10 Bit. The sender emits light with a frequency of 2 or 10 Hz depending on the BIOS version. Therefore, the sensors are able to measure distances of the robot to other objects in the range between 10 and 60 mm depending on the objects' reflection.

The robot is controlled by a 32 Bit micro controller (MC68331) with 256 K ROM and 256 K RAM running with 16 MHz. The BIOS supports a simple multi-tasking operating system for 15 concurrent tasks. Task switching is performed by a 5 ms round-robin mechanism. After a reset, the BIOS boots the system through an optional serial line with 9.6 or 38.4 KBit/s. A serial line can be used for power supply and communication with a work station too.

The gripper turret on top of the robot's basis turret is equipped with a 8 Bit micro controller (68HC11) and an EPROM for the BIOS extensions. The turrets are connected physically and electrically via a bus connector system. The gripper has two DoF: jaw opening width and gripper angle. Both are controlled by stepper motors. A photosensitive sensor can detect objects between the gripper jaws.

To support local robot interaction without a central communication workstation/server an infrared communication (IRC) turret (Fig. 3) has been developed at FZI/IPR [10]. The IRC-turret is mounted on top of the robot's I/O-turret.

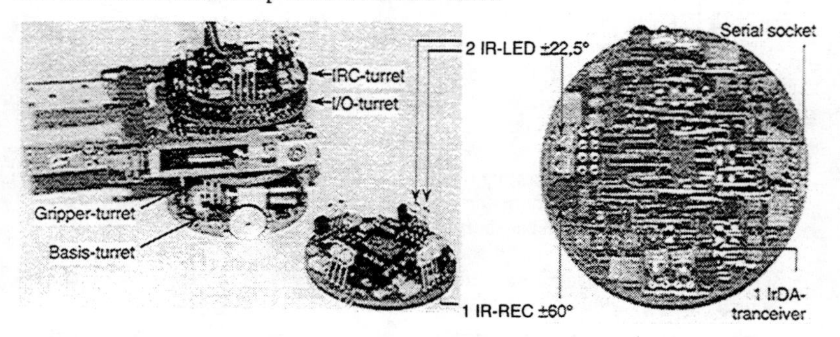

Figure 3. The FZI's inter-robot infrared communication turret for the Khepera robots

The IRC-turret for inter-robot communication switches the serial port (UART) of the MC68331 either to the serial cable connector of the turret or to one of four independent infrared communication transceiver modules of the turret. Each module is based on the IrDA standard [11, 12] (half-duplex) and uses two Ir-senders for emitting the information with 9.6 KBit/s within an approximately 90° degree area. The Ir-receiver of each module is able to receive of an approximately 120° degree area (Fig. 4a).

One feature of the FZI's IRC-turret is the programmable sender resistor and receiver sensitivity. By changing the sender's resistor, the maximal communication range (100% reliable communication) can be increased up to 100 cm. By changing the receiver's sensitivity (5-Bit), the communication range can be modified as shown in Fig. 4b.

To broadcast a message into all directions, it is necessary to sequentially select the four modules and send the message. Since the communication rate is only 9.6 KBit/s, the micro controller is able to send a Byte in all directions in less than 1 ms. On the other hand, it is not possible to receive concurrently data from all four module by switching. Nevertheless, by a special circuitry it is possible to detect, which modules are receiving data. Then, the program can select which module will be used for listening.

To support the use of the IRC turret, a set of real-time communication procedures and processes (tasks for the multi-tasking BIOS) are available. The communication procedures are used for establishing [13] basic exchange of events (signal/wait), for synchronization, and for continuously data (processed sensor information, execution commands) exchange. For each 90°-sender module, a cycle send-buffer exists that is sent and emptied with a defined frequency or on demand. For the selected 120°-receiver module, a cycle receive-buffer is filled with a frequency of 1 ms. All control modules/tasks can write messages into the send-buffer. The content of the receive-buffer is interpreted independently by several control modules/tasks that wait for external information.

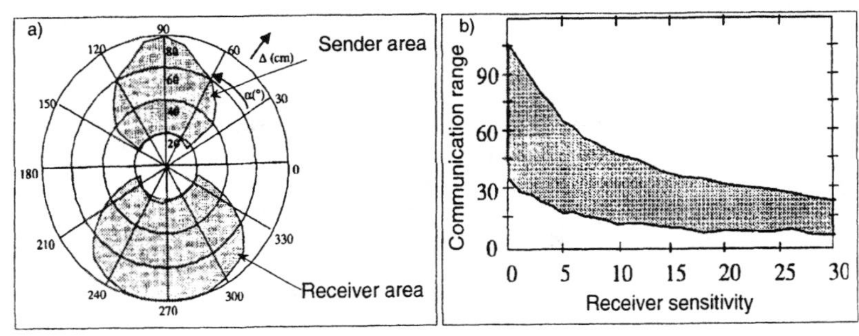

Figure 4. a) Sender and receiver range field for 100% reliable communication. b) Reliable communication range in terms of 5 Bit receiver sensitivity [10]

INTERACTION PROTOCOLS AND EXPERIMENTS

Using the IRC-turret and the CAIC operating system, inter-robot cooperation is achieved by the following strategy. In CAIC, the *control-state* of a control-subnet describes which control loops are active simultaneously. If two robots want to synchronize their states, they exchange their readiness for changing to a new state: If a robot itself is ready for a change and interprets an external request for a change, it broadcasts that it will now change its state. Afterwards, both robots switch to the new state in which, typically, a different interpreter and communication protocol are used for processing the received information. In Fig. 5. an example is shown for using this method in transporting an object in a closed kinematic chain. Doing this in a master-slave configuration requires the continuous transfer of data from the master to the slave.

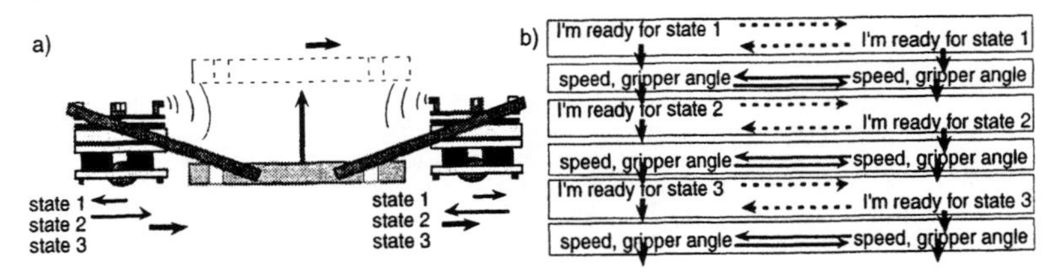

Figure 5. a) Cooperative movement. b) Interaction protocol

After achieving the correct position for grasping and closing the gripper, both robots signalize each other the readiness for lifting the object upwards: state 1. During state 1, both robots have to lift the object and move slowly outwards. The arbitrarily chosen master sends its wheel speed and gripper angle to the slave robot and announces the need to switch to state 2. During state 2, both robots have to lift the object further but to move slowly inside. Again, an arbitrarily chosen master sends its wheel speed and gripper angle to the slave robot and announces the need to switch to state 3. During state 3, both robots move in the same direction to transport the object.

Previous experiments [6] have shown that an individual robot can be used even for assembly operations (grasp, transport, insert) of the *Cranfield-Assembly-Benchmark.* By using the CAIC system, the IRC-turret, and the interaction protocols, two robots are able to grasp and transport cooperatively big and heavy parts. Ongoing research deals with cooperative execution of assembly tasks such as insert by two robots.

CONCLUSION

Cooperation among distributed controlled robots is more than plan synchronization, coordination, and cooperative planning. For physical cooperation, a common control network must be distributed to several robots that communicate to synchronize the activation of control modules and to transfer control loop information between the robots' control systems. To achieve a dynamic coupling of robot control systems, a cooperative robot control system (for instance CAIC [7,8]) must be supported by an adequate local communication system. The FZI's IRC-turret for the Khepera robot is such a system. Interaction protocols that are grounded inside encapsulated control modules are the basis for dynamically coupled control systems.

ACKNOWLEDGMENT

This research is being performed at FZI • Forschungszentrum Informatik, Division TE&R • Robotics, and at IPR • Institute for Real-Time Computer Systems and Robotics (Prof. Dr.-Ing. U. Rembold, Prof. Dr.-Ing. R. Dillmann), University of Karlsruhe. The project is being partially funded by the NATO Project CRG 950340 of Prof. J. Wang and Dr. T. Lueth. Thanks to Prof. Alex Zelinsky (U. of Wollongong, Australia) for his support and Johan Hellqvist (Univ. of Lund LTH, Sweden) for his work.

REFERENCES

1. Lueth, T.C., U. Rembold, T. Ogasawara: Task Specification, Scheduling, Execution, and Monitoring for Centralized and Decentralized Intelligent Robots. R&A IEEE Int. Conf. on Robotics and Automation, Nagoya, Japan, May, 1995, WP2 Task Driven Intelligent Robotics.
2. Parker, L.: Heterogeneous Multi-Robot Cooperation, Doctoral dissertation, MIT AI Lab technical report 1465, MIT Press, 1994.
3. Noreils, F.R.: Coordinated Protocols: An Appraoch to Formalize Coodination Between Mobile Robots. IROS Int'l. Conf. Intelligent Robots and Systems, Raleigh, NC, July, 1992, pp. 717-724.
4. Le Pape, C.: A Combination of Centralized and Distributed Methods for Multi-Agent Planning and Scheduling. IEEE Int'l. Conf. Robotics and Automation, Cincinatti, U.S.A, May, 1990, pp. 488-493.
5. Zimmermann, M.: Concurrent Behavior Control - A System's Thinking Approach to Intelligent Behavior, Swiss Federal Institute of Technology, Dissertation, 1993.
6. Lueth, T.; Th. Laengle, J. Heinzman: Dynamic Task Mapping for Real-Time Controller of Distributed Cooperative Robot Systems. IFAC WS Distributed Computer Control Systems, Toulouse-Blagnac, France, September, 1995, pp. 37-42.
7. Lüth, T.; Th. Längle, J. Hellqvist (1995): Distributed Robot Control by Structural Adaptive Control (in German). In Dillmann, Rembold, Lüth (Ed.), Autonome Mobile Systeme , Springer, pp. 250-258.
8. Hellqvist, J.: Design and Implementation of Real-Time Robot Operation System (in German), University of Karlsruhe, Master Thesis, Faculty of Informatics, February, 1996.
9. K-Team: Khepera User Manual, Ver. 3.0, LAMI-EPFL, Lausanne, Swiss, 1994.
10. Grasman, R.: Local Communication and Coupled Control for Small Mobile Manipulators (in German), University of Karlsruhe, Master Thesis, Faculty for Electrical Engineering, February, 1996.
11. IrDA: Press Release, IrDA Walnut Creek, CA, USA, 1995.
12. TEMIC: Design Guide: IrDA-Compatible Transmission, Telefunken Semiconductors Issue 05.95
13. Premvuti, S.; J. Wang: A Medium Access Protocol (CSMA/CD-W) Supporting Wireless Inter-Robot Communication. In Distributed Autonomous Robot Systems, Springer-Verlag, 1994.

System-based Methodology for Concurrent Engineering

M. LOMBARD, J.L. SOMMER, E. GETE, F. MAYER

CRAN-GGP, Université de Nancy I
B.P. 239 - 54506 Vandoeuvre-lès-Nancy Cedex (France)
phone. +33-83.91.21.50 - fax. +33-83.91.23.90
E-mail. <name>@cran.u-nancy.fr

ABSTRACT
Analysis methods are needed by Concurrent Engineering in mechanical design. In the other hand, functional approach and flows analysis might give support to this concept. In this paper, we deal with two examples extracting from the approach based on Value Analysis of the French project entitled « Scenario for a Concurrent Engineering in Manufacturing Integrated Systems » supported by the Technical and Scientific Mission (DSPT8) of the French Ministry of University Education and Research. Next we show how disadvantages of functional approach are pointed out by the flows analysis (systemic approach). Last, we suggest a partial method joining this two methods of analysis.

KEY WORDS : Concurrent Engineering, flows analysis, value analysis

1. INTRODUCTION

The set up of a Concurrent Engineering reduces ***Time to Market*** and improves both design and product quality in contrast to the classical linear approach. Following this point of view, the French project entitled *« Scenario for a Concurrent Engineering in Manufacturing Integrated Systems »* supported by the Technical and Scientific Mission (DSPT8) of the French Ministry of University Education and Research, aims at putting in evidence what should be a Concurrent Engineering, with the implementation of a real semantic communication between the different partners involved in the consortium*.

Our problematic is situated in the implementation of multidisciplinary exchanges related to mechanics, automatics, quality management, ..., insofar as each of these domains consists of activities sub-set.

These « co-operating » exchanges to be co-ordinated are stemmed from various sources of expertise, which are research-laboratories works.

The purpose of this co-operation in a communicating infrastructure is, on one hand the establishment of semantic links between different trades, and on the other hand in the

---

*
CRAN	Centre de Recherche en Automatique de Nancy
LAMIH	Laboratoire d'Automatique et de Mécanique Industrilles et Humaines de Valenciennes
LAN	Laboratoire d'Automatique de Nantes
LIRMM	Laboratoire d'Informatique, de Robotique et de Microélectronique de Montpellier

emergence of a common methodology for the construction of a manufacturing system based on the product that has to be realised.

Our support of experimentation is an industrial problem of design and realisation issued of an assembly application by riveting of DASSAULT airplane sub-sets.

Modelling methods used by the consortium are therefore methods called « systemic » ones (system-based), in the sense that they correspond to modelling rules stemmed from the general system theory and from its extensions [1].

Many modelling methods, characterised by analytical methods, perceive complex systems as being complicated, namely reducible to models, themselves being complicated yet capable of simplification and potentially furnishing a basis for their automatisation [1]. However, the inadequacy of those systems' models can be established at the time of application for the complex phenomenon's' representation because it only describes their internal structure and not their finality.

Presently, our solution to our industrial problem is to apply another **complementary but not contrary** approach based on flows analysis to the Value Analysis propounding to model complex systems in a way to propose a new modelling method.

2. VALUE ANALYSIS

2.1. Concepts used for the DSPT8 project

No matter what kind of design activity is developed, between product design or manufacturing process design, it is possible and even necessary to carry out a functional analysis of the element to be designed. An approach based on the Value Analysis process [2][3] has allowed us to identify the needs that can be registered regarding a product, or a manufacturing process.

Expressing a need, regarding an object, justifies the fact that it has to be designed (or at least studied). Therefore, each object in the design phase originates from the need expressed by one ore more users. The first thing that has to be done is therefore to clearly identify the users of the product to be designed. By studying the different steps of the product life cycle, one can identify its potential users, and therefore define its ***Use Cycle*** (figure 1).

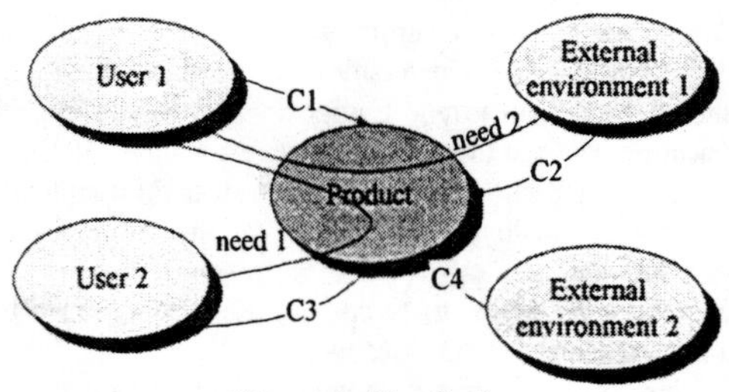

Figure 1. *Use Cycle*

In this case, no existing methodology allows us to clearly define this use cycle. Therefore, we have decided to base our process on the *Use Cycle* recommended by [4].

After all users are identified, one has to isolate the external environments that are related to the object to be designed. For each step of the product *Use Cycle*, one or several users may express a need and elements of the external environment that constraint our product.

In order to propose an answer regarding the need expressed by the user, the product has to bring solutions in the form of ***service functions***. Moreover, the reply to the constraints imposed by its environment (external environment) obliges it to have ***constraint functions***. Aiming at correctly define what we mean about these functions, we had to document the external environments of our product. This step corresponds to the development of a detailed glossary allowing to precise the role and the parameters that distinguish them.

The problem that appears at this is that all external environments have been correctly identified. This is even more embarrassing when one knows it can be a cause of customer's dissatisfaction.

On the other hand, the expression of service functions and constraints must not be interpretable. It therefore necessary appears that one has to formalise the syntax of the sentences describing each type of function :

- *the product must allow '**user**' of '**action verb**', '**external environment**' (service function),*
- *the product must '**action verb**' '**external environment**' (constraint function).*

Even if this syntax seems to resolve several ambiguities, it does not allow the exhaustive definition of the service functions.

The next step consists in documenting each function and in attaching to them several evaluation criteria.

On this base, it will be possible, after a phase of technical solution research, to evaluate the adequation between the proposed solution and the service function needed.

At this level, it is important to classify the importance of the different functions associated to the product. Actually, this classification aims at allowing the assurance of a certain designers adaptability regarding their technological choices.

2.2. Examples

In these examples, we shown two types of problem coming from this analysis. First, we are not absolutely sure that external environment have been correctly identified : « *Has 'Downstream stock' been in direct environment ?* » (figure 2). In the other hand external « Control system » is not plainly defined and we are not able to say « Who control What » : « *Have 'Riveting system', 'Upstream stock', 'Transport out system' been control through system to design ?* » (figure 3).

In this kind of analysis, we can plainly answer to the needs express by the customers with a group of functions. Yet, we are not sure that these needs are well represented by this group because of the definition of external environments.

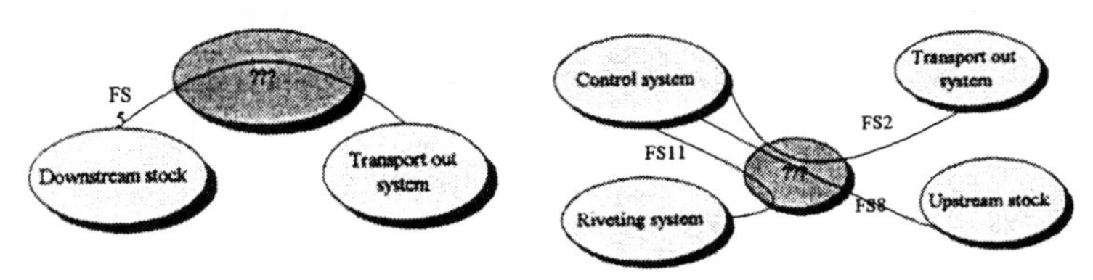

Figure 2. *External environment* **Figure 3.** *Control architecture*

3. SYSTEM ANALYSIS

3.1. Concepts

The Systemic theory gives the only foundation for concept formalization of systems. So process, which can be easily linked to the notion of action, is defined when, during the ***time***, there is a modification of attitude, in a referential *<Space-Shape>*, of a lot of products identifiable by their ***Shape***. In this way [1] proposes to identify process in a referential ***<Time-Space-Shape>*** allowing to define process' canonical model.

In our point of view, usage of the term operator with the sense of holding of the different operations existing in this referential. In the case of works, [5] introduces a fourth operator named *« Nature »* and an associated Function, ***« Control »***, in a way which represents controlled action. The role attributed to this operator is to direct a special transformation, a transmutation, namely a change of the nature of something. In fact, we think that each **Nature** relationship between a given **Manufacturing Function** (Mechanical point of view) and **Control Function** (Automation point of view) may be considered as a base for a concurrent engineering gateway definition. It is fitting to take into account the ***Nature operator*** with which the Control systemic function is associated, to link any flows and finally to obtain a complete system.

The various stages in the development of a scenario, according to the systemic methodology, show how different skill stations work in concurrence [6]. To do this, each operator is associated to a **Manufacturing Function**, namely for the component Shape, ***« Transform »*** (milling, turning, assembling, ...), for the component Space, ***« Transport »*** (moving, convoying, ...) and for the component Time, ***« Stock »*** or to a **Control Function**, respectively named ***« Process »***, ***« Communicate »***, ***« Store »***.

With a view to obtain a realistic architecture which is used by all the designers in a Concurrent Engineering context, it is necessary to process functions (Figure 4) along each axis proposed by [7].

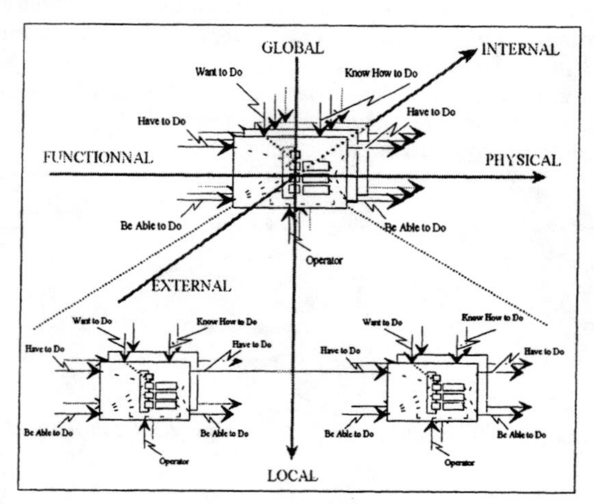

Figure 4. *Skill processings in a Concurrent Engineering context*

In practice, the organisation of activities must follow a syntagmatic scheme, that is to say along the axis of a sequence of words which correspond to activities considered. In fact, according to this organisation, it is not possible to have two successive activities of the same nature. In this case, it is probable that one activity has been forgotten or one is more complex and requires a decomposition.

From a practical point of view and regarding to our industrial problem, our systemic concurrent engineering starts with the Assembly Activity. To describe interactions between functions, [8] proposes a modal typology which can be applied to our model with four flows inducing partial behaviours to come closest to explain *« what it has to do »*, *« what it knows how to do »*, *« what it is able to do »*, and *« what it want to do »*.

3.2. Example

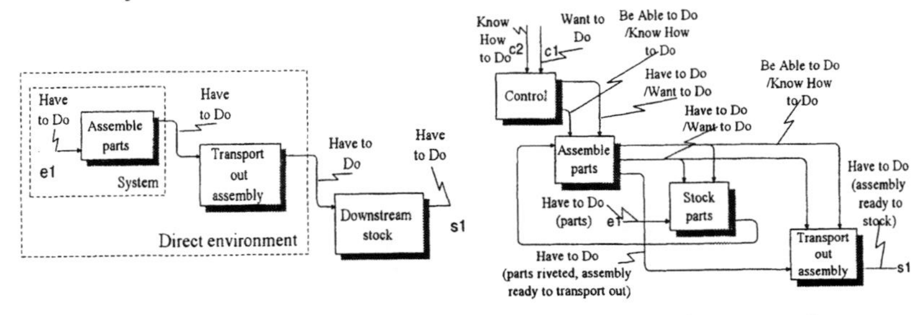

Figure 5. *External environment definition* **Figure 6.** *Master and servant architecture*

Thus, *« Have to do »* flow historical (figure 5) shows that *'Downstream stock'* is not in the direct environment this resolve the problem of figure 2. Furthermore, figure 6 shows that figure 3 analysis translate a *« master and servant »* architecture which is not asked in specifications.

4.CONCLUSION - PARTIAL METHODOLOGY

The proposed partial methodology is stemming from Value Analysis and Flows Analysis. Thus retained concepts are :

. the clear expression of the function.

. functions and flows typologies in view to identify and to validate without ambiguity the different elements of the direct external environment.

In fact, for the identification of an external environment one attaches to explore a flow in a diagram called *« context diagram »*. Thus, different skills are emerged by system' decomposition in sub-functions : automation corporation for the control function studied and mechanical corporation (designers) for the physical function studied namely *'Transport in parts'*, *'Position parts'*, *'Rivet parts'* (figure 7).

Each transformational function of the part flow in "time" "space" and "shape" is realised by an *« atomic sub-system »* interfacing the Informational process of the control system with the physical process via some action and observation process according to [9] applied by [8].

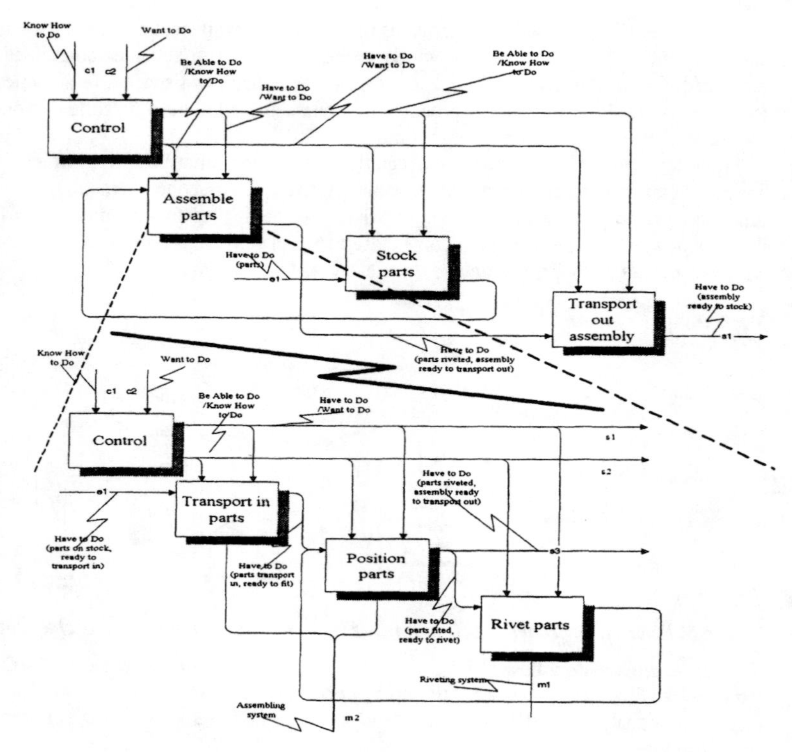

Figure 7. *System' decomposition along « Global/Local » axis*

5. BIBLIOGRAPHY

1 LE MOIGNE J.L., Modelisation of Complex Systems, AFCET Systems, Dunod, Paris (France), 1990, (in French).

2 AFNOR Standards NFX 50 - 150......152, 1989 - 93 (in French).

3 JOUINEAU C., Value Analysis, Method - Preparing - Applications , ESF Editions 1982 ISBN 2 7101 0370 (in French).

4 CHAVALIER J., Product and Value Analysis, CEPADUES Editions 1989 ISBN 2 85428 189 6 (in French).

5 LE GALLOU F., Graphical representation and global modelisation, Proceedings of the 2nd *European Systemic Congress*, Prague (Czecho-Slovak), 4-8/10/93 (in French).

6 LOMBARD-GREGORI M., Contribution to discrete part manufacturing engineering : prototyping a concurrent engineering architecture for manufacturing integrated systems, Thesis of the University of Nancy I, 16 February 1994, (in French).

7 LOMBARD-GREGORI M., MAYER F., System-based concurrent methodology for discrete part manufacturing engineering, IFIP, Proceedings of feature modeling & recognition in advanced CAD/CAM systems, Valenciennes (France), 24-26/05/1994.

8 MAYER F., Contribution to Manufacturing Engineering ; application to Education Engineering at the Lorraine Manufacturing Inside - institution workshop, Thesis of the University of Nancy I, 1995, (in French).

9 MOREL G., LHOSTE P., IUNG B., Towards intelligent actuation and measurement system, Advance Summer Institute'94 (ASI'94) Laboratory for automation and robotics, Vol. 1, pp 121 - 126, Patras (Greece), 26/06-1/07/1994.

Near-Optimal Commutation Configuration of a Mobile Manipulator for Point-to-Point Tasks using Genetic Algorithms

Jae-Kyung Lee

Advanced Robotics Team
Korea Atomic Energy Research Institute
P.O.Box 105, Yusong, Taejon 305-600, KOREA
Tel : +82-42-868-8838/ Fax : +82-42-868-8833/ e-mail : jklee@nanum.kaeri.re.kr

Hyung Suck Cho

Dept. of Mechanical Engineering
Korea Advanced Institute of Science and Technology
373-1 Kusong-dong, Yusong-ku, Taejon 305-701, KOREA
Tel : +82-42-869-3213/ Fax : +82-42-869-3210/ e-mail : hscho@sorak.kaist.ac.kr

ABSTRACT

In this paper, a method for finding the near-optimal commutation configurations which are essentially required for a mobile manipulator to execute a multiple task which consists of a sequence of point-to-point subtasks is proposed. We formulate this problem as a global non-linear optimization problem taking into account the motion trajectories of the mobile manipulator between subtasks. Because the solution spaces for this problem are generally unconnected and nonconvex, the genetic algorithms are adopted as a solution method. Computer simulations are performed to verify the effectiveness of the proposed method.

KEYWORDS: commutation configuration, mobile manipulator, multiple point-to-point task, genetic algorithms.

INTRODUCTION

A mobile manipulator has a mobility function in addition to the basic manipulation function. By utilizing this mobility function, a mobile manipulator can perform various tasks over a large space which exceeds the workspace of a fixed-base manipulator. When a mobile manipulator performs multiple tasks which consist of a sequence of subtasks, the final configuration of each subtask becomes the initial configuration of the subsequent subtask. This configuration is called as a commutation configuration, the concept of which was first suggested in [1]. Because the commutation configurations become the boundary conditions for each subtask, they affect the efficiency of the multiple task significantly. For efficient motion for a mobile manipulator from one subtask to the subsequent subtask, it is essentially required to determine the commutation configuration with consideration of the motion feasibility between subtasks.

Yamamoto and Yun[2] proposed a coordination algorithm based on the concept of preferred operation region and Seraji[3] proposed a coordinated motion control method using the configuration control treating the mobile manipulator as a redundant manipulator. Pin and Culioli[1][4] obtained the optimal commutation configuration by using the Newton and Tunneling algorithms and the minimax approach. Carriker et al. [5][6] formulated the coordination of mobility and manipulation as a non-linear optimization problem and solved by using a simulated annealing method. Recently, Zhao et al.[7] obtained the solution by using genetic algorithms. However, no authors addressed the approach which takes into account the motion trajectories between subtasks and the commutaiton configurtions simultaneously.

In this paper, a new method for finding the near-optimal commutation configurations which are essentially required for a ÿmobile manipulator to execute a multiple task which consists of a sequence of point-to-point subtasks efficiently is proposed. We formulate this problem as a global non-linear optimization problem by considering the motion trajectories and commutation configurations simultaneously. Because the solution spaces for this problem are generally unconnected and nonconvex, local optimal solution is expected by conventional gradient descent method. Therefore, the genetic algorithms are adopted as a solution method, which is robust stochastic searching algorithm for various optimization problems[8].

PROBLEM FORMULATION

In order that a mobile manipulator can perform the desired task, the end-effector configuration should be equal to the desired task vector at each subtask, that is, following equality constraints should be satisfied.

$$\overline{X}_{d,i} = F(\overline{Q}_i), i = 1,\ldots,N \tag{1}$$

In the above equation, $\overline{X}_d$ is the desired task vector and $\overline{Q} = \{\overline{P}^T : \overline{\Theta}^T\}^T$ is the coordinated variable vector of the mobile manipulator where $\overline{P}$ is a posture vector of the mobile platform and $\overline{\Theta} = \{\theta_1, \theta_2, \cdots, \theta_n\}^T$ is the manipulator joint variable vector.

In addition to the constraints given by Eq. (1), inequality constraints given by the joint limit, joint torque limit or obstacle avoidance should be added. Porvided that above constraints are satisfied for each of the N subtasks, in order to determine the most desirable motion trajectories between subtasks, we introduce a cost function that measures the motion trajectories for the completion of the multiple tasks as follows:

$$C = \sum_{i=1}^{N-1} \rho_i \int_{t_{i0}}^{t_{if}} G_i(\overline{Q}_i, \dot{\overline{Q}}_i)\, dt \tag{2}$$

where ρ_i is a weighting factor and G_i is the criteria function.

Then, the problem can be summarized as follows:

$$\text{Find } \{\overline{P}_1, \overline{P}_2, \ldots, \overline{P}_N\}$$

$$\text{which minimize } C = \sum_{i=1}^{N-1} \rho_i \int_{t_{i0}}^{t_{if}} G_i(\overline{Q}_i, \dot{\overline{Q}}_i)\, dt \tag{3}$$

Subject to $\overline{X}_{d,i} = F(\overline{Q}_i)$, $i = 1,2,\ldots,N$

$\overline{\Theta}_{\min} \le \overline{\Theta}_i \le \overline{\Theta}_{\max}$, $i = 1,2,\ldots,N$

$\overline{\tau}_{\min} \le \overline{\tau}_i \le \overline{\tau}_{\max}$, $i = 1,2,\ldots,N$.

MOTION TRAJECTORY GENERATION BETWEEN SUBTASKS

Let us consider the i-th motion trajectory. The necessary conditions are written by the Lagrange equations as follows:

$$\frac{\partial L_i}{\partial \overline{Q}_i} - \frac{d}{dt}\left(\frac{\partial L_i}{\partial \dot{\overline{Q}}_i}\right) = 0. \tag{4}$$

As a reasonable choice, we adopt the criteria function of the following form,

$$G_i(\overline{Q}_i, \dot{\overline{Q}}_i) = \frac{1}{2}\dot{\overline{Q}}_i^T W_i \dot{\overline{Q}}_i + g_i(\overline{Q}_i) \tag{5}$$

where W_i is a diagonal matrix whose elements reflect the relative cost and $g_i(\overline{Q}_i)$ is the function of the configuration such as manipulability measure or distance to some obstacle. Substituting Eq.(5) into Eq.(4), we get second order kinematic differential equation as follows:

$$W_i \ddot{\overline{Q}}_i - \nabla g_i = 0 \tag{6}$$

where g_i is the gradient vector of $g_i(\overline{Q}_i)$.

Introducing $\mathbf{X} = \{\overline{X}_1^T : \overline{X}_2^T\}^T = \{\overline{Q}^T : \dot{\overline{Q}}^T\}^T$, then the order of Eq.(6) can be reduced to first order as following form:

$$\dot{\mathbf{X}} = \left\{\overline{X}_2^T : \nabla g(\overline{X}_1)\right\}^T \tag{7}$$

OPTIMAL COMMUTATION CONFIGURATION BY USING GENETIC ALGORITHMS

In this section, we address the method for determining the optimal commutation configuration by using genetic algorithms. Genetic algorithms are robust stochastic searching algorithms based on the mechanics of the natural selection and natural genetics[8]. This algorithm finds an optimal solution that is represented by the fittest individual from the evolution of population. The population is initially created to be a set of individulas that are randomly generated. Based on the fitness of each individual, population evolves to the next generation through genetic operations such as crossover and mutation.

Because the manipulator configuration can be obtained through inverse kinematics of the manipulator if the platform posture is determined, we choose the sequence of platform postures for coding. Each individual is assigned a fitness value which means the probabilty that the individual survives or creates offsprings to the next generation. In this paper, the fitness function is defined by

$$f = K / C. \tag{8}$$

In order to remove the candidates which violate the constraints, we use penalty function K as follows:

$$K = \begin{cases} 0 : \text{in case of violation occurs} \\ a : \text{otherwise} \end{cases} \tag{9}$$

where a is a positive constant. Figure 1 shows the flow chart of the proposed method.

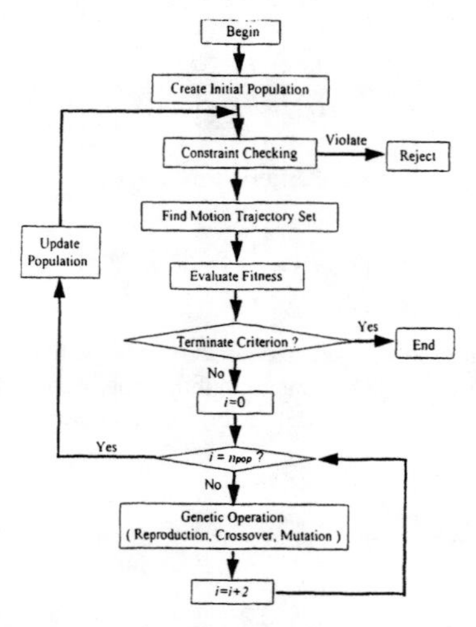

Figure 1. Flow chart of the proposed method.

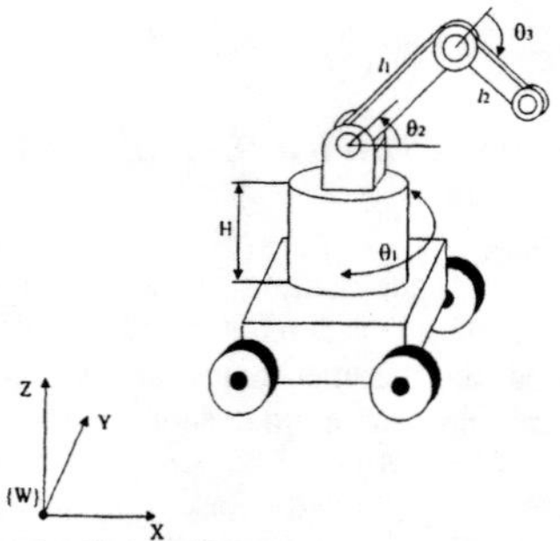

Figure 2. A mobile manipulator which consists of a 3-DOF manipulator and an omni-directional mobile platform.

SIMULATION

We performed simulation for a mobile manipulator which consists of an omni-directional mobile platform and 3-DOF manipulator which is shown in Figure 2.

The kinematic relations for this mobile manipulator system are given as follows:

$$\begin{aligned} X_e &= x + \{l_1\cos(\theta_2) + l_2\cos(\theta_2+\theta_3)\}\cos(\theta_1) \\ Y_e &= y + \{l_1\cos(\theta_2) + l_2\cos(\theta_2+\theta_3)\}\sin(\theta_1) \\ Z_e &= H + \{l_1\sin(\theta_2) + l_2\sin(\theta_2+\theta_3)\}. \end{aligned} \tag{10}$$

In the simulation, we set $l_1 = 1.5\ m, l_2 = 1.0\ m, H = 0.5\ m$.
The desired task vectors and the force vector applied to the end-effector are given by

$$X_{d,1} = \{\,3,\ 3,\ 1.25\,\}^T,\quad X_{d,2} = \{\,9,\ 3,\ 1.0\,\}^T, \tag{11}$$

$$F_{d,1} = \{\,2,\ 2,\ 0\,\}^T,\quad F_{d,2} = \{\,0,\ 2,\ -2\,\}^T. \tag{12}$$

The joint limits of the manipulator are as follows:

$$-150^\circ \le \theta_1 \le 150^\circ,\ \ 0^\circ \le \theta_2 \le 120^\circ, -120^\circ \le \theta_3 \le 90^\circ. \tag{13}$$

The Jacobian matrix can be obtained from Eq.(11) as follows:

$$J = \begin{bmatrix} -s_1(l_1c_2 + l_2c_{23}) & -c_1(l_1s_2 + l_2s_{23}) & -c_1l_2s_{23} \\ c_1(l_1c_2 + l_2c_{23}) & -s_1(l_1s_2 + l_2s_{23}) & -s_1l_2s_{23} \\ 0 & l_1c_2 + l_2c_{23} & l_2c_{23} \end{bmatrix}. \tag{14}$$

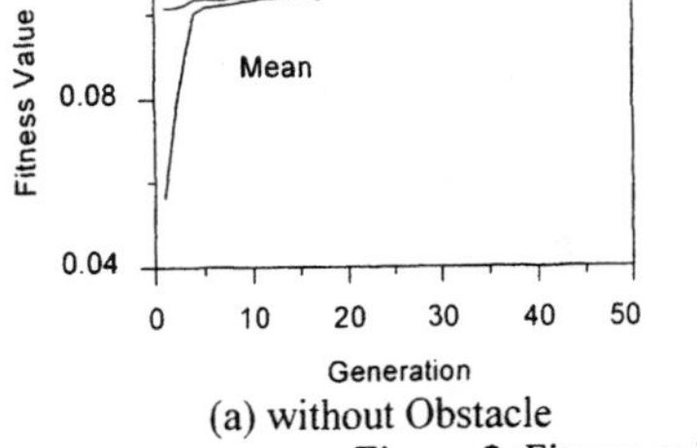

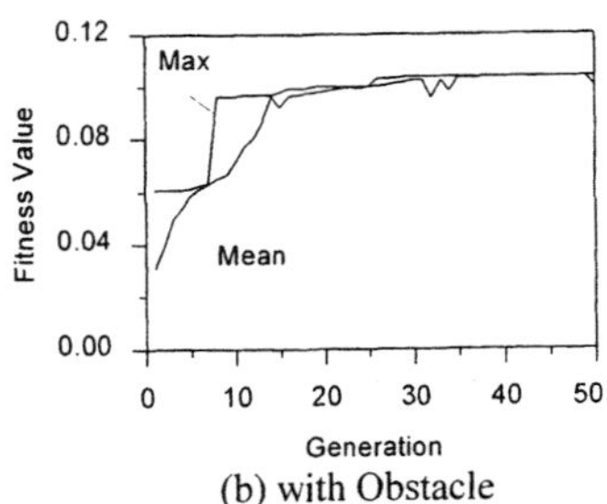

(a) without Obstacle (b) with Obstacle

Figure 3. Fitness value versus generation.

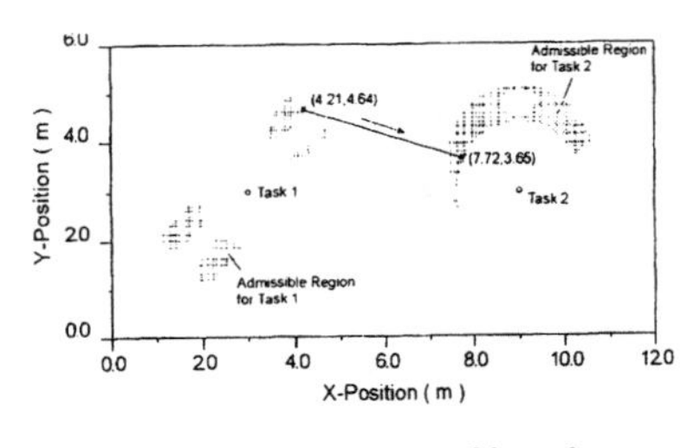

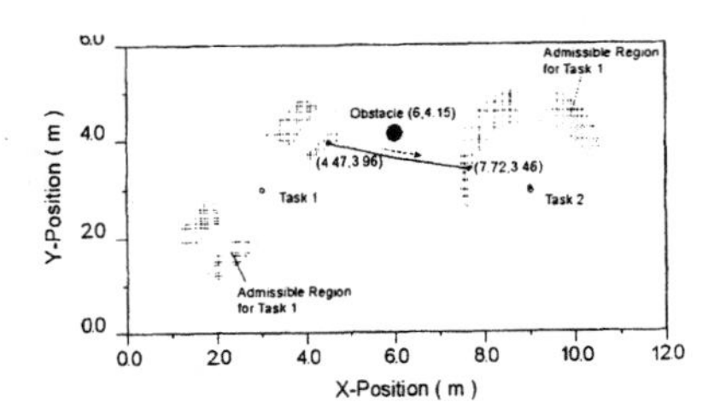

(a) without Obstacle (b) With Obstacle

Figure 4. Mobile platform path for two task problem.

The actuator torques can calculated by using the relationship of $\bar{\tau} = J^T \bar{F}$. We set the torque limit is 2.1 N-m. As a first simulation, we performed simulation for two-task problem in case of there is not obstacle and next, we performed simulation in case that there is a point obstacle at (6,4.15). Figure 3 (a) shows the maximum and average fitness value versus generation plot in case of without obstacle and (b) in case of with obstacle, respectively. Figure 4 (a) shows the mobile platform path in case of without obstacle. As can be seen in the figure, the path is straight line. Figure 4 (b) shows the mobile platform path in case of with obstacle. As can be seen in the figure, the path avoids the obstacle.

CONCLUSIONS

In this paper, a new method for finding the near-optimal commutation configurations for a mobile manipulator to execute a multiple task which consists of a sequence of point-to-point subtasks efficiently was proposed. We formulated the problem as a global non-linear optimization problem by incorporating the motion trajectories in the cost function. Because the solution spaces for this problem are unconnected and nonconvex, solution obtained by any gradient descent method is liable to be local optimal. In order to overcome such difficulties, the genetic algorithms which are robust stochastic searching algorithm for various optimization problems is adopted as a solution method.

A series of simulations were performed to verify the effectiveness of the proposed method. The simulation results showed that the proposed method can effectively be applied in finding the near-optimal commutation configurations for a mobile manipulator to execute mutiple point-to-point tasks.

REFERENCES

1. Pin,F.G. and Culioli,J-C."Optimal Positioning of Combined Mobile Platform-Manipulator Systems for Material Handling Tasks." ÿ*Journal of Intelligent and Robotic Systems* 6, (1992), 165-182.
2. Yamamoto,Y. and Yun,X. "Coordinating Locomotion and Manipulation of a Mobile Manipulator." *Proc. 31st IEEE Conf. on Decision and Control*, Tucson (December, 1992), 2643-2648.
3. Seraji,H. "An On-Line Approach to Coordinated Mobility and Manipulation." *Proc. IEEE Intern. Conf. on Robotics and Automation*, Atlanta (May 1993), 28-35.
4. Pin,F.G. and Culioli,J-C. and Reister,D.B. "Using Minimax Approaches to Plan Optimal Task Commutation Configurations for Combined Mobile Platform-Manipulator Systems." *IEEE Trans. on Robotics and Automation*, Vol.10, No.1, (1994), 44-54.
5. Carriker,W.F., Khosla,P.K. and Krogh,B.H."The Use of Simulated Annealing to Solve the Mobile Manipulator Path Planning Problem." *Proc. IEEE Intern. Conf. on Robotics and Automation*, Cincinnati (May, 1990), 204-209.
6. Carriker,W.F., Khosla,P.K. and Krogh,B.H. "Path Planning for Mobile Manipulators for Multiple Task Execution." *IEEE Trans. on Robotics and Automation*, Vol.7, No.3, (1991), 403-408.
7. Zhao,M., Ansari,N. and Hou,E.S.H. "Mobile Manipulator Path Planning by a Genetic Algorithm," *Journal of Robotic Systems* 11(3), (1994), 43-153.
8. Goldberg,D.E. *Genetic Algorithms in Search, Optimization and Machine Learning.* Addison-Wesley, MA (1989).

SOME NEW ASPECTS IN WORKSPACE OPTIMISATION

JADRAN LENARČIČ
The "J. Stefan" Institute, University of Ljubljana,
Jamova 39, Ljubljana, SLOVENIA

ABSTRACT

Herein, we present a concise review of some new workspace optimisation criteria that we propose in the design of robot kinematic structures, in planning, and programming of robot applications. These criteria are related to the robot motion abilities termed kinematic flexibility, to the force and velocity characteristics, and to the distribution of the positional accuracy.

KEYWORDS: robot workspace, optimisation, performance

INTRODUCTION

The analysis of the robot workspace has been an interesting research subject ever since the first days of robotics. The majority of reported contributions, however, were dedicated to the development of numerical and analytical methods that enable an efficient determination of different types of the robot workspaces [1,2]. In fact, the calculation of the workspace for a mechanism with an arbitrary kinematic structure with a high number of degrees of freedom is time-consuming and there is a need to visualise the obtained result with a sophisticated computer graphics technique [3,4].

The workspace geometry workspace appears to be the most widely used criterion in the development of the mechanism's structure. A common approach has been to maximise the volume of the workspace associated with the total lengths of the links [5], and in combination to optimise the form of the workspace [6]. However, such an approach to evaluate the robot motion abilities is questionable. For instance, the reachable workspace represents the region of the space where the end effector reaches every position without regard to the orientation. In a practical implementation, a programmer is not informed if the robot is able to grip the object in a specified (although reachable) point in the requested orientation and if the robot is able to provide enough driving force to lift the object or move it with the specified accuracy and velocity. It appears that the reachable workspace is associated only with possible collisions of the tip of the robot with the objects placed nearby. It even doesn't give an information what happens with other parts of the linkage.

Hence, we trust that it is more (or at least equally) important to investigate the insight characteristics of the robot workspace, for instance, the distribution of the velocity capability. The objective of the present work is to introduce some novel criteria to quantify the insight of the robot workspace which, for our opinion, should be taken into account in the process of the design of the mechanisms' kinematic structures, as well as in the procedures of planning and programming of robot applications.

MULTIPLE CONFIGURATIONS, KINEMATIC FLEXIBILITY

This rather unknown criterion was first proposed by the author in [7,8]. It is directly connected to the number of the inverse kinematics solutions (number of configurations) in a given point of the workspace. Suppose that the mechanism in Fig. 1 possesses two kinematic configurations that correspond to a position of the tip. Hypothetically, if there is an obstacle placed in the workspace and if (because of possible collisions) the given position cannot be achieved with one of the configurations, there is a possibility that it is achieved with the other one.

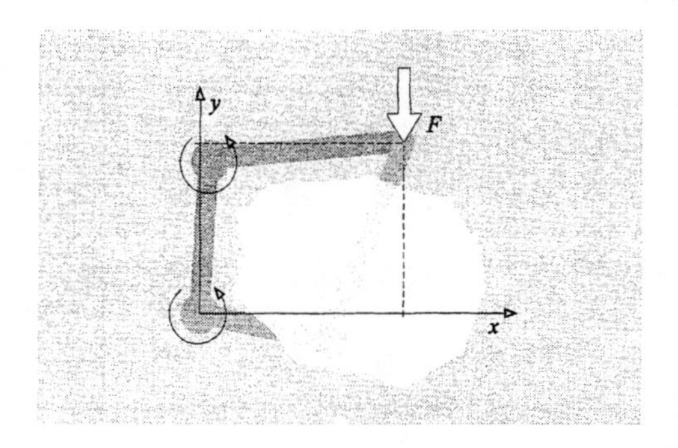

Figure 1: A 2-R mechanism

Additionally, assume that a force is applied to the end effector and the task is to resist to this force. In one configuration the necessary joint torques exceed the permissible maximum. It may be, however, that in the other configuration the resulting joint torques are consistently smaller. Consequently, the mechanism that has more inverse kinematics solutions enables to program the task in different ways and is, therefore, kinematically more flexible.

According to [7,8], the kinematic flexibility π is referred to as the number of the inverse kinematics solutions (configurations of the mechanism) in a Cartesian point of the end effector. In serial mechanisms with unlimited joint angles with perpendicular or parallel joint axes, the kinematic flexibility $\pi = 1,2,4,8,...$ depending on the structure of the mechanism and on the number of degrees of freedom *n*. If there are limits on joint angles or if the joint axes are not parallel or perpendicular, we will find regions in the workspace with $\pi = 1,2,3,4,5,...$ The goal of an optimisation procedure is to maximise those regions with the maximum kinematic flexibility. The relative kinematic flexibility is defined as

$$\pi_{rel} = \sum_r \frac{V^r}{V}\,, \tag{1}$$

where V is the volume of the mechanism's workspace, and V^r is the volume of the subspace with $\pi = r$. Clearly, the relative flexibility depends on the number of degrees of freedom, on the arrangement of joints, as well as on the proportions of the links and on the ranges of joint coordinates. The optimum design problem is then determined as the maximisation of the relative flexibility for a given kinematic structure of the mechanism where the optimisation parameters are the link lengths and the ranges of joint coordinates.

This is valid for any non-redundant serial linkage, however, the most simple and illustrative is the *2-R* planar mechanism that is found in many industrial manipulators, for instance, in the widely used scara-type structure. Here, we deal with two link lengths, d_1 and d_2, and the ranges of two joint angles, h_1 and h_2. Assume that in the first case the link lengths are $d_1 = 0.77$, $d_2 = 0.63$ and the ranges of motion $h_1 = 264^\circ$, $h_2 = 216^\circ$ and in the second case $d_1 = 0.42$, $d_2 = 0.98$ and $h_1 = 144^\circ$, $h_2 = 336^\circ$. In both cases the sum of the link lengths is $d_1 + d_2 = 1.4$ and the sum of joint ranges is $h_1 + h_2 = 480^\circ$. It will be observed that the kinematic flexibility in the first case is *2* almost in the whole workspace, while in the second case the major part of the workspace has the kinematic flexibility *1*. If we study the relative kinematic flexibility as a function of the ratio between the link lengths and the ratio between the ranges of motion, we will discover that the best proportions of the mechanism are when the first link is much longer than the second link, and when the first range of motion is much larger of the second one. The relative kinematic flexibility reaches the theoretical maximum when the first range of motion is greater or equal to 360°. These principles can be well used in robot design.

CARTESIAN AND JOINT VELOCITIES

As is well known, the Cartesian velocities $\dot{\boldsymbol{p}}$ of the end effector and the joint velocities $\dot{\boldsymbol{q}}$ are connected through the Jacobian matrix $\boldsymbol{J}$

$$\dot{\boldsymbol{p}} = \boldsymbol{J}\dot{\boldsymbol{q}} \ . \tag{2}$$

The distribution of the velocity capabilities inside of the workspace illustrate the manipulability (velocity) ellipsoids [9]. The shapes of these ellipsoids (rather than their volume) can be represented in terms of the ratio between the minimum and the maximum principal axes, that is by the conditioning of $\boldsymbol{J}$. The shortest principal axis is the minimum attainable velocity of the end effector in the given point of the workspace, and the longest principle axis is the maximum attainable velocity. In a singularity configuration (singularity of $\boldsymbol{J}$), the ellipsoids become flat in the direction in which the mechanism cannot provide any Cartesian velocity. Inside the workspace, the manipulability ellipsoids change their shape and volume depending on the position and on the values of the kinematic parameters, such as the link lengths. The distribution of the manipulability ellipses along the radial line in the workspace of a *2R* mechanism is presented in Fig. 2.

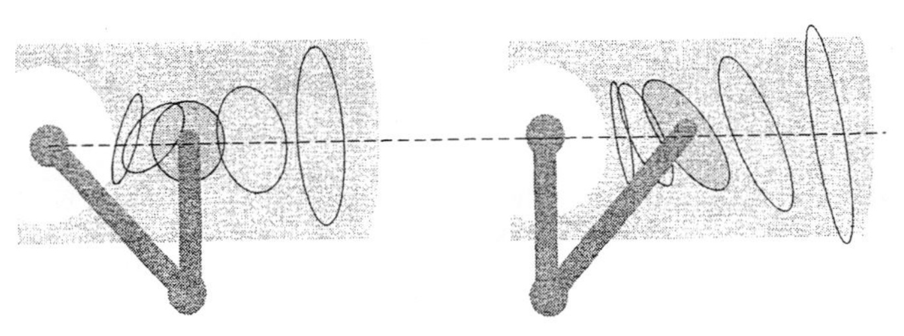

Figure 2: Manipulability ellipses of a 2R mechanism

Here, the ellipses collapse in lines at the borders of the workspace (at the inner and the outer singularity) and get more round inside the workspace. By changing the ratio between the link lengths, we can produce more round or more flat ellipses. Hence, from the viewpoint of robot control, the kinematic singularities hamper the velocity characteristics of the manipulator and a general objective is, therefore, to design a mechanism that produces uniform Cartesian velocities in all directions, conversely, a mechanism that possesses more round ellipses in a major part of the workspace. The two mechanisms in Fig. 2 have identical workspace geometries. However, the ellipses of the first one are more round. In one point, the so-called isotropic configuration ($\det\{\boldsymbol{J}\} = 1$), the ellipse becomes a circle. In this point it is able to achieve the same Cartesian velocity in all directions. A Cartesian-type mechanism is isotropic in the whole workspace. In [10] and related previous reports, the objective is to develop other spatial mechanisms with the isotropic property in order to increase the so-called kinematic dexterity.

EXTERNAL FORCES, JOINT TORQUES, ACCURACY AND COMPLIANCE

In contrast, velocity limitations and disturbances in the mechanism's motion caused by singularities and ill-conditioned configurations of a robot mechanism can offer extremes in terms of applicable forces and can, additionally, increase accuracy in certain directions [11]. Namely, the manipulability ellipsoids (representing the vicinity to the singularity configurations) can be changed to more flat (rather than round if isotropy is given as the main criterion) so that the robot can exert or resist large forces with relatively weak actuation in a wider region of the workspace. This is relevant in a variety of applications of robot manipulators, such as in lifting or pushing of heavy loads, or in applications where extreme positional accuracy in a given direction is required. This also calls for a development of more sophisticated control approaches that are able to deal with kinematic singularities.

Since the external force $\boldsymbol{F}$ and the vector of the joint torques τ are connected by the transpose of the Jacobian

$$\tau = \boldsymbol{J}^T \boldsymbol{F} \ , \tag{3}$$

the force-torque ellipsoids possess inverted lengths of the principal axes in the same directions of the manipulability ellipsoids. Consequently, the force-torque properties can also be explained by the manipulability ellipsoids. When a force is applied to the end effector in the direction of the shortest principal axis of the ellipse, the joint torques will be small. If the same force is applied in the direction of the longest principal axes, the joint torques will be much larger. If the force-torque characteristics of the mechanism are taken into account, the objective of the optimum design will be to obtain flat manipulability ellipsoids deeply inside the workspace. The right-hand side mechanism in Fig. 2 can exert higher force inside the workspace in the radial direction than the mechanism with the isotropic point.

Similarly, it is possible to develop a criterion that favours the kinematic singularities (maximises the conditioning of $\boldsymbol{J}$) when the objective is to produce an extremely accurate motion in a given direction. Near kinematic singularities, a large motion in joint coordinates results in a small motion in Cartesian space in at least one direction. In this direction the manipulator is able to produce a very precise movement. On the contrary, in the opposite direction the accuracy is usually much worse and, as is generally recognised, in an absolute sense the accuracy near kinematic singularity is disturbed (also in relation to the control difficulties).

As is well known, the compliance matrix is related to the Jacobian as follows

$$\boldsymbol{C} = \boldsymbol{J}\boldsymbol{K}^{-1}\boldsymbol{J}^T \ , \tag{4}$$

where $\boldsymbol{K}$ is the diagonal matrix of the joint spring constants. If the spring constants are equal (or very similar) for all joints, the compliance matrix is proportional to $\boldsymbol{JJ}^T$. Hence, a manipulator with flat manipulability ellipsoids is more compliant in the direction of the longest principal axis and is more rigid in the opposite direction. In analogy to the velocity and force-torque capabilities, a designer is interested either to build a mechanism with more round manipulability ellipsoids in order to obtain uniform compliance and accuracy or to build a mechanism with more flat manipulability ellipsoids to achieve extremes. Since both can be done with the same volume and form of the workspace (see Fig. 2), one can find a compromise in an iterative procedure in which he alternates between the optimisation of the geometrical characteristics of the workspace, such as the volume and the form, and the optimisation of its insight properties.

ON HUMAN ARM WORKSPACE

The human arm is a complex mechanical structure of non-rigid bodies. In order to study the kinematics it is necessary to introduce ponderous approximations. A unique study of the human arm reachable workspace is reported in [12], where the kinematics of the arm is modelled of six revolute joints (for the major linkage without the wrist). The workspace for a healthy left human arm is shown in Fig. 3 - the coordinate frame is in the centre of the sternoclavicular joint, the anterior hemisphere and the posterior void are seen. In later investigations two additional dependent translational joints were included. From the positioning point of view the human arm is redundant. In a given point in the workspace, the mechanism can produce an infinity of configurations and the manipulability ellipsoids can change their form and size in the same position of the end effector. In the analysis of the human arm workspace, we discovered a series of interesting design peculiarities related to its kinematic structure and workspace characteristics.

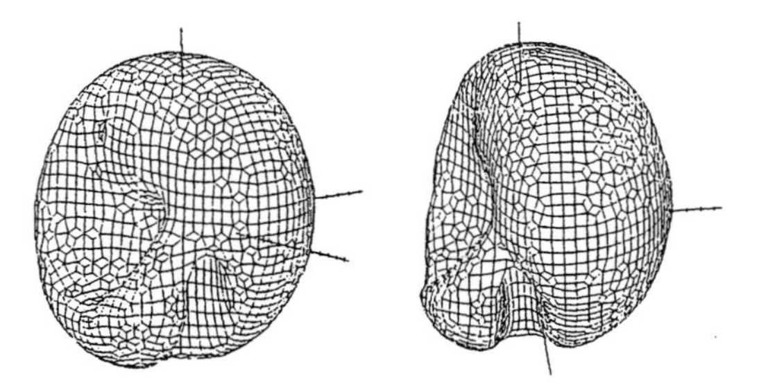

Figure 3: The workspace of a healthy left human arm in different views

When we optimise of the workspace volume and compactness it is typical for an industrial-type mechanism [6] that the maximum volume and the maximum compactness are obtained with consistently different ratios between the link lengths. In an anthropomorphic mechanism [12], they are obtained simultaneously with the same ratio between the lengths of the upper arm and the forearm. It is also interesting that the proportions of the human arm and the ranges of motion in joints (although dependent on each other) do not enable collisions between the links of the forearm, upper arm, and clavicle. It seems that each joint has its own task. The elbow is the only joint that provides the volume to the workspace. If the sternoclavicular joint is fixed, the workspace will be unexpectedly small (despite that it has very small ranges of motion). Obviously it is because the sternoclavicular joint helps to avoid collisions between the arm and the body.

Surprising is also a small self-motion of the arm that is equally distributed throughout the workspace, even very close to the limits. Because of the redundancy, the manipulability ellipsoids can change in every point of the workspace. An interesting fact is that in the region where the hand is close to the head, the ellipsoids can be very round and very flat in the same point. For example, if the elbow is down, the ellipsoids are flat and in such a configuration heavy weights can be caught with less fatigue. If the elbow is up, the ellipsoids are round and the motion is more uniform. The arm is able of to produce precise movements.

CONCLUSIONS

In the present work, we propose a series of criteria related to the robot motion static abilities that should be used in the optimum design. All these criteria are referred to as the insight characteristics of the robot workspace and are introduced as an alternative to the classical optimisation of the workspace that is usually associated only with its volume and (even rarely) its form. Throughout the article, the background and practical significance is described for the kinematic flexibility, the velocity capability, the force and torque capability, accuracy, and compliance. Finally, some comments on the human arm reachable workspace are presented that may have an important impact to the design of anthropomorphic robots in the future.

REFERENCES

1. A. Kumar, K.J. Waldron, "The workspaces of a mechanical manipulator", *ASME J. Mech. Design*, Vol. 103, 1981.
2. K.C. Gupta, B. Roth, "Design considerations for manipulator workspace", *ASME J. Mech. Design*, Vol. 104, 1982.
3. X. Chen, K.C. Gupta, "The geometry, structure, and visualisation of manipulator workspaces", *Laboratory Robotics and Automation*, Vol. 4, No. 2/3, 1992.
4. M. Ceccarelli, "Determination of the workspace boundary of a general n-revolute manipulator, *Advances in Robot Kinematics and Computational Geometry*, J. Lenarčič and Ravani, Eds., Kluwer Academic Publishers, Dordrecht, 1994.
5. D.C. Yang, T.W. Lee, "Heuristic combinatorial optimisation in the design of manipulator workspace", *IEEE Trans. on Syst. Man and Cybern.*, Vol. 14, No. 4, 1984.
6. J. Lenarčič, P. Oblak, U. Stanič, "Some kinematic cnsiderations for the design of robot manipulators", *Robotics & CIM*, Vol. 5, No. 2/3, 1989.
7. J. Lenarčič, A. Košutnik, U. Stanič, P. Oblak, "Performance optimisation of hyper-redundant snake-like robot manipulator", *22nd Int. Symp. on Ind. Robots*, Detroit (USA), 1991.
8. J. Lenarčič, U. Stanič, P. Oblak, "Study of kinematic flexibility of standard welding robots", *23rd Int. Symp. on Ind. Robots*, Barcelona, (Spain) 1992.
9. T. Yoshikawa, "Manipulability of robotic mechanisms", *J. Robot. Res*, Vol. 4, No. 2, 1985.
10. K.E. Zanganeh, J. Angeles, "On the isotropic design of general six-degree-of-freedom parallel manipulators", *Computational Kinematics*, J.P. Merlet, B. Ravani, Eds., Kluwer Academic Publishers, Dordrecht, 1995.
11. J. Lenarčič, "Improvement of velocity and force capability of robot manipulators", *Laboratory Robotics and Automation*, Vol. 6, No. 6, 1994.
12. J. Lenarčič, A. Umek, "Simple model of human arm reachable workspace", *IEEE Trans. on Syst. Man and Cybern.*, Vol. 24, No. 8, 1994.

CALIBRATION OF A PARALLEL TOPOLOGY ROBOT

A. B. LINTOTT and G. R. DUNLOP

Department of Mechanical Engineering
University of Canterbury
Christchurch, New Zealand
Email: R.Dunlop@canterbury.ac.nz

ABSTRACT

A calibration method for a Stewart platform has been developed as part of a project aimed at developing a calibration method for a Delta robot. The Delta has 3 degrees of freedom (DOF) but is more complex than the Stewart platform for calibration purposes because an extra link is inserted in the kinematic chain between the base and the Nacelle member.

INTRODUCTION

Parallel topology robots have not yet made a significant impact in industrial applications and this is probably why there has been very little work on the calibration of parallel manipulators. The Stewart platform[1] was originally proposed as a flight simulator platform, and is widely used for this today. McCallion and Pham[2] used the mechanism as a 6 DOF platform for robotic assembly, while Afzulpurkar and Dunlop[3] applied the mechanism to satellite tracking. The Stewart platform has also been considered for use as a machine tool([4], [5]).

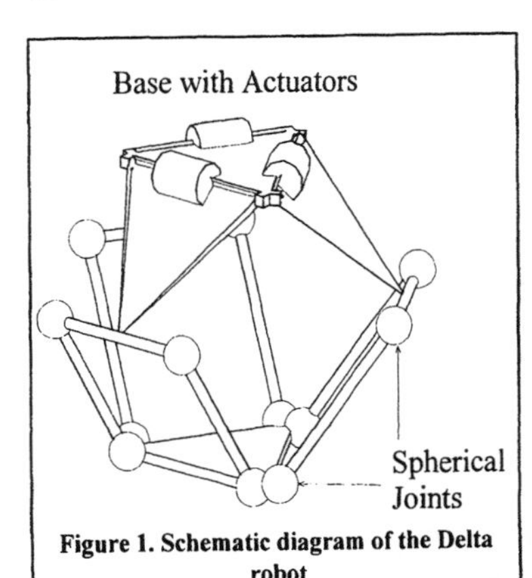

Figure 1. Schematic diagram of the Delta robot

The Delta robot was first proposed by Clavel([6], [7]). The robot is a simplification of a 6 DOF mechanism first proposed by Hunt[8] which, in turn, bears a resemblance to the Stewart platform. The Delta robot has 3 translational degrees of freedom and is actuated by 3 kinematic chains consisting of a rotary arm in series with 2 ball-jointed parallel arms. A number of prototypes have been constructed at the Swiss Federal University of Lausanne (EPFL) and a production version now exists.

The work presented in this paper is at an early stage of development. The goal is to develop an effective calibration model for the full Delta robot. Thus far, work has focused on developing a calibration model for the Stewart platform because it is similar in form to a sub-structure of the Delta robot. In this paper, the model for a variant of the Stewart platform with fixed leg lengths and mobile base joints is developed as an intermediate step to the Delta structure.

THE FIXED-LEG STEWART PLATFORM

This variant of the Stewart platform structure is not proposed as a practical design. It serves as a simplified structure with similar properties to part of the Delta robot.

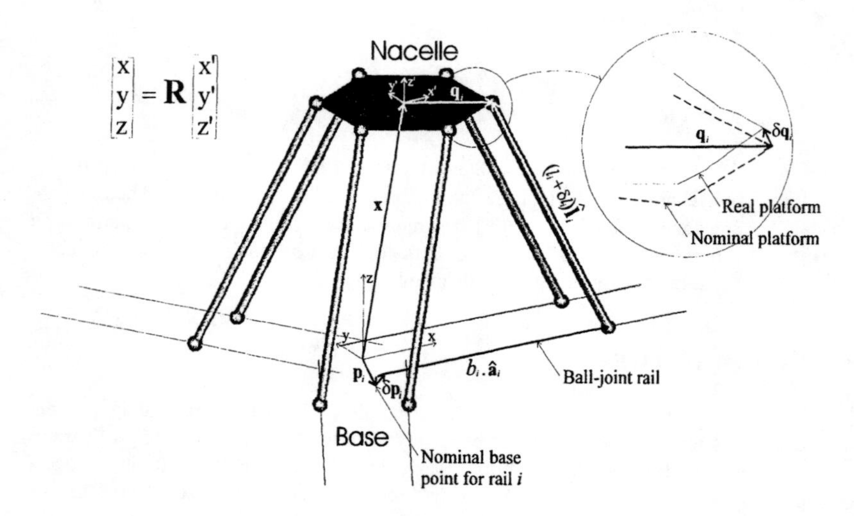

Figure 2. The fixed-leg Stewart Platform.

In this design, the Stewart platform has 6 inextensible legs that are connected by ball joints to the nacelle and the base (fig. 2). The nacelle ball joints are fixed in position relative to the nacelle coordinate frame while the base ball joints are allowed to move along straight rails. The robot is actuated by sliding the base ball joints along these rails a distance b_i from the base point of the rail. In the Delta robot model, the 6 straight rails will be replaced by 3 paired circular arcs representing the loci of the 3 actuator arms.

The error model allows for deviations in the ball joint positions and in the lengths of the legs, giving 42 error parameters in total. For this simplified model, the rails are assumed to be perfectly straight and accurately located.

Modelling

For the fixed-leg Stewart platform, the loop closure equation is as follows (see fig. 2):

$$\mathbf{x} + \mathbf{R} \cdot (\mathbf{q}_i + \underline{\delta \mathbf{q}}_i) - (\mathbf{p}_i + \underline{\delta \mathbf{p}}_i + b_i \cdot \hat{\mathbf{a}}_i) - (l_i + \delta l_i)\hat{\mathbf{l}}_i = 0 \tag{1}$$

The *inverse geometric solution* expresses the actuator coordinates as a function of the endpoint position. If the nacelle is fixed in space and leg i is allowed to swing from the nacelle, then the locus of the base ball joint traces out a sphere. The sphere intersects the rail in 2 locations and the convention is adopted that the point of intersection furthest from base platform centre is taken to be the attachment point. Given **x**, the centre of the spherical locus is located in the base coordinate frame by vector **c** The sphere is described by (3) and the line is described by (4)

$$\mathbf{c} = \left\{ \mathbf{x} + \mathbf{R} \cdot \left(\mathbf{q}_i + \underline{\delta \mathbf{q}}_i \right) \right\} \tag{2}$$

$$\left(s_x - c_x\right)^2 + \left(s_y - c_y\right)^2 + \left(s_z - c_z\right)^2 = \left(l_i + \delta l_i\right)^2 \tag{3}$$

$$\mathbf{s} = \mathbf{p}_i + \delta \mathbf{p}_i + b_i \hat{\mathbf{a}}$$

$$\begin{bmatrix} s_x \\ s_y \\ s_z \end{bmatrix} = \begin{bmatrix} p_x + \delta p_x \\ p_y + \delta p_y \\ p_z + \delta p_z \end{bmatrix} + b_i \cdot \begin{bmatrix} a_x \\ a_y \\ a_z \end{bmatrix} \tag{4}$$

Substituting (4) into (3), a quadratic equation (5) is generated in terms of b_i. Provided that **x** is within the workspace of the robot, (5) will generally have 2 real roots (6). Following the convention mentioned above, the largest root is the one that is selected.

$$\left(b_i \cdot a_x - h_x\right)^2 + \left(b_i \cdot a_y - h_y\right)^2 + \left(b_i \cdot a_z - h_z\right)^2 = \left(l_i + \delta l_i\right)^2 \tag{5}$$

$$b_i = \frac{E \pm \sqrt{E^2 - 4EF}}{2} \tag{6}$$

where:

$$\begin{aligned} h_x &= c_x - p_x - \delta p_x & E &= 2 \cdot \left(a_x h_x + a_y h_y + a_z h_z \right) \\ h_y &= c_y - p_y - \delta p_y & F &= h_x^2 + h_y^2 + h_z^2 - \left(l_i + \delta l_i \right)^2 \\ h_z &= c_z - p_z - \delta p_z \end{aligned} \tag{7}$$

The *direct geometric solution* expresses the endpoint position as a function of the actuator coordinates, robot geometry, and error parameters. While the direct geometric solution has been solved analytically[9], it is much more computationally efficient to use a numerical solution algorithm such as the Newton-Raphson algorithm[10].

The *differential solution* expresses rates of change in the endpoint coordinates as a function of changes in the error parameters. If the error parameters are grouped into a single 42 by 1 partitioned vector as in (8) then the Jacobian matrix is defined as in (9).

$$\underline{\delta} = \left[\delta l_1 \ \cdots \ \delta l_6 \ \ \underline{\delta \mathbf{q}}_1 \ \cdots \ \underline{\delta \mathbf{q}}_{18} \ \ \underline{\delta \mathbf{p}}_1 \ \cdots \ \underline{\delta \mathbf{p}}_{18} \right]^{\mathrm{T}} \tag{8}$$

$$\mathbf{J} \equiv \frac{\partial \mathbf{x}}{\partial \underline{\delta}} = \left[\frac{\partial \mathbf{x}}{\partial (\delta l_i)} \ \ \frac{\partial \mathbf{x}}{\partial \left(\underline{\delta \mathbf{q}}_j\right)} \ \ \frac{\partial \mathbf{x}}{\partial \left(\underline{\delta \mathbf{p}}_j\right)} \right], \qquad \begin{aligned} 1 \le i \le 6, \\ 1 \le j \le 18, \end{aligned} \tag{9}$$

Measurement and Identification

For simplicity, the measurement device in the example is assumed to be capable of measuring all 6 components of the endpoint position expressed in the world frame. A set of measurement points in the workspace is chosen and readings are taken from the measurement device when the robot is moved to these points. It is appropriate to consider which set of measurement points in the robot's workspace guarantees a complete solution for the error parameters $\underline{\delta}$. This is a question of *observability*.

Menq and Borm[11] present an observability index based on the singular value decomposition of the *identification Jacobian* **M** (11). It is normally advantageous to choose measurement points that maximise the observability index. Borm and Menq[12] use a numerical optimisation procedure to search for optimum measurement positions. For the example presented here, a similar numerical optimisation was performed using a steepest ascent method. It was observed that the measurement points generally migrate towards the edges of the workspace and toward singular configurations as the optimisation progresses. The near-singular points caused difficulties during the identification phase.

The data for this example were generated numerically. A set of values was chosen for the $\underline{\delta}$ error parameters. The object was to recover these values from the data using the identification technique.

Once the measurement data has been obtained, the values of the error parameters $\underline{\delta}$ must be determined. This is a model fitting exercise. The world-frame components of the measurement points are concatenated into a single vector (10). In the following expressions, the superscript represents the ordinal number of the measurement points.

$$\mathbf{X}_{nom} = \left[\mathbf{x}^1_{nom} \quad \mathbf{x}^2_{nom} \quad \cdots \quad \mathbf{x}^n_{nom}\right]^{\mathrm{T}} \tag{10}$$

Each nominal measurement position, $\mathbf{x}^i_{nom}$, is associated with a Jacobian matrix $\mathbf{J}^i$ evaluated at $\mathbf{x}^i_{nom}$. The identification Jacobian **M** is defined as

$$\mathbf{M} = \left[\mathbf{J}^1 \quad \mathbf{J}^2 \quad \cdots \quad \mathbf{J}^n\right]^{\mathrm{T}} \tag{11}$$

The simplest approach to solving for $\underline{\delta}$ is to use a linear least squares solution. The endpoint deviations are detected by a measuring device that returns the measured endpoint position $\mathbf{x}^i_{act}$ plus a small measurement noise component $\tilde{\theta}^i$. The readings are then concatenated into a set of data **B**.

$$\mathbf{B} = \mathbf{X}_{act} + \tilde{\Theta} \tag{12}$$

where

$$\mathbf{X}_{act} = \left[\mathbf{x}^1_{act} \quad \mathbf{x}^2_{act} \quad \cdots \quad \mathbf{x}^n_{act}\right]^{\mathrm{T}}, \qquad \tilde{\Theta} = \left[\tilde{\theta}^1 \quad \tilde{\theta}^2 \quad \cdots \quad \tilde{\theta}^n\right]^{\mathrm{T}} \tag{13}$$

The noise component is generally assumed to be zero-mean, normally distributed noise. The measurement data **B** may be approximated by using a first order Taylor expansion about $\mathbf{X}_{nom}$:

$$\mathbf{B} = \mathbf{X}_{nom} + \mathbf{M}\underline{\delta} \tag{14}$$

which is rearranged to give:

$$\underline{\Delta} = \mathbf{M}\underline{\delta} \qquad \text{where } \underline{\Delta} = \mathbf{B} - \mathbf{X}_{nom} \tag{15}$$

The linear least squares estimate for $\underline{\delta}$ is as shown in (16).

$$\tilde{\delta} = \left(\mathbf{M}^{\mathrm{T}}\mathbf{M}\right)^{-1}\mathbf{M}^{\mathrm{T}}\underline{\Delta} \tag{16}$$

Linear least squares estimation works satisfactorily if the identification Jacobian is well conditioned, the observability is relatively high, and the calibration model is sufficiently linear. For situations where the condition number is poor, singular value decomposition[13] is often used. For improved performance[14], a weighted least squares solution (17) is possible which weights the contribution of each data point to the result according to the inverse of its uncertainty by means of a weighting matrix **W**.

$$\tilde{\delta} = \left(\mathbf{M}^T\mathbf{W}\mathbf{M}\right)^{-1}\mathbf{M}^T\mathbf{W}\underline{\Delta} \tag{17}$$

Non-linear least squares solutions are generally more computationally demanding but they often give a more accurate result. A practical technique is the Levenberg-Marquardt algorithm[13]. The technique involves minimising a cost function (18)

$$\chi^2 = \mathbf{X}_{act} - \vec{\mathbf{G}}(\mathbf{X}_{nom}, \tilde{\delta}) \tag{18}$$

where $\vec{\mathbf{G}}(\mathbf{X}_{nom}, \tilde{\delta})$ is the direct geometric solution.

For this example, the simulated data were processed using MatLab[15]. The results for noisy and noise free data analysed by both linear and non-linear methods are presented in table I.

RESULTS

To evaluate the improvement in accuracy due to calibration, the RMS error in the endpoint position may be evaluated before and after calibration. The tabulated results were obtained using a set of randomly distributed points within the workspace, not equal to the points used for measurement. Because the units of the error parameter vector $\tilde{\delta}$ are homogeneous, the concept of distance in the 42-dimensional parameter space has meaning. A good measure of the success of a result in simulation may be gained from evaluating the norm of the difference between the estimated and actual error parameter vector (19). These results are also tabulated to aid comparison.

$$\lambda = \left\| \tilde{\delta} - \underline{\delta} \right\| \tag{19}$$

The iterative linear least squares results were generally poor because the calibration model is significantly non-linear. The Levenberg-Marquardt result for noise free data converges to a result that is very close indeed to the actual error parameter vector, while the same method applied to noised data reduces the RMS position error in world space to about 2% of the original uncalibrated RMS error.

When measurements that have been optimised for Menq-and-Borm observability are used, a problem is encountered in that some of the measurement points are close to

Solution method	$\underline{\delta}$ Not calibrated	$\tilde{\delta}$ Least Sqr. No noise	$\tilde{\delta}$ Least Sqr. Noise	$\tilde{\delta}$ Lev.-Marq. No noise	$\tilde{\delta}$ Lev.-Marq. Noise	$\tilde{\delta}$ Lev.-Mar. No noise Optimised	$\tilde{\delta}$ Lev.-Mar. Noise Optimised
$\left\|\tilde{\delta} - \underline{\delta}\right\|$ (mm)	—	0.196	1.65	1.30×10^{-3}	0.760	1.39×10^{-4}	0.154
RMS Error Position (mm) Orientation (rad)	 4.33 0.0546	 0.0774 2.80×10^{-4}	 0.175 1.13×10^{-3}	 1.38×10^{-5} 5.87×10^{-8}	 0.082 1.36×10^{-4}	 3.65×10^{-5} 1.92×10^{-7}	 0.086 1.31×10^{-4}

TABLE I . A set of typical results from simulated calibration procedures.

physical singularities of the robot. Near singularities, the iterative direct geometric solution methods are less likely to converge. Apart from that, the repeatability of the physical robot would suffer near singularities, adding noise to the data. It was necessary to move these points away from the singularities in order to achieve convergence. The optimised measurements generally provide better parameter estimates than the non-optimised measurements.

CONCLUSION

A method for robot calibration has been applied successfully to the simulated calibration of a parallel manipulator. The high degree of non-linearity in the calibration model strongly favours the use of non-linear solutions in the identification phase.

Problems were encountered when the observability criterion of Menq and Borm was applied to optimise the measurement positions. Methods for ensuring convergence of the direct geometric solution by constraining the optimised points to avoid areas of poor conditioning need to be applied.

Future work is aimed at expanding the fixed-leg Stewart platform model so that it may be applied to calibration of the Delta robot.

REFERENCES

1. Stewart, D. 1965–66. *A Platform with Six Degrees of Freedom*. Proc. IMechE (London), Vol. 180, Pt. 1, No. 15.
2. McCallion, H. and Pham, D.T. 1979. *The analysis of a six degree of freedom work station for mechanized assembly*. 5th World Congress on Theory of Machines and Mechanisms, pp. 611-616.
3. Afzulpurkar N. V., and Dunlop G. R., 1989. *A novel antenna mount for orbital satellite tracking and marine communications*. Fifth National Space Engineering Symposium, Canberra, Australia, Nov.27-Dec.1. The Institute of Engineers Australia Publication No.89/20, pp 92-96.
4. Stix, G. 1995. *Nice Legs*. Scientific American, December 1995 issue, pp. 24.
5. Rathbun G., 1985. *A Stewart platform six-axis milling machine development*. ME thesis, University of Canterbury, Christchurch, New Zealand.
6. Clavel, R. 1988. *DELTA, a Fast Robot with Parallel Geometry*. Proc. Int. Symposium on Industrial Robots, pp. 91–100, April.
7. Clavel, R. 1991. *Conception d'un Robot Parallèle Rapide à 4 Degrés de Liberté*. Thèse No. 925, Institut de Microtechnique, Ecole Polytechnique Fédérale de Lausanne..
8. Hunt, K. H. 1983. *Structural Kinematics of In-Parallel-Actuated Robot-Arms*. ASME J. of Mechanisms, Transmissions, and Automation in Design, Vol. 105, pp. 705–712, December.
9. Dasgupta, B. and Mruthyunjaya, T. S. 1994. *A Canonical Formulation of the Direct Position Kinematics Problem for a General 6–6 Stewart Platform*. Mech. Mach. Theory, Vol. 29, No. 6, pp. 819–827.
10. Merlet, J. P. 1993. *Direct Kinematics of Parallel Manipulators*. IEEE Trans. on Robotics & Automation, Vol. 9, No. 6, pp. 842-846, December.
11. Menq, C. H. and Borm, J. H., Lai, J. Z. 1988. *Estimation and Observability Measure of Parameter Errors in a Robot Kinematic Model*. Proc. 2nd USA–Japan Symp. on Flexible Autom., pp. 73–79.
12. Borm, J. H. and Menq, C. H. 1989. *Experimental Study of Observability of Parameter Errors in Robot Calibration*. Proc. IEEE Int. Conf. on Robotics & Automation, Vol. 1, pp. 587–92.
13. Press, W. H., et al., 1992. *Numerical Recipes in C: The Art of Scientific Computing. 2nd ed.* Cambridge University Press, Cambridge & New York.
14. Mooring, B. W., Roth, Z. S., Driels, M. R. 1991. *Fundamentals of Manipulator Calibration*. John Wiley & Sons, New York.
15. MathWorks Inc., 1994. *MatLab Version 4.2c.1*. Numerical analysis software package.

HYBRID CONTROLLERS FOR ROBOTICS

BUD MISHRA[1]
Robotics Laboratory, Courant Institute, New York University

ABSTRACT

In this paper, we survey various theoretical and algorithmic questions related to a compiler that takes a specification of the desired behavior of a hybrid system (i.e., combining both discrete and continuous operations) and builds a hybrid supervisory controller for the system. A version of the compiler has been constructed at NYU with Marco Antoniotti and has been used in such areas as robotics and manufacturing. We briefly describe how the system works with an example of a controller for bipedal locomotion.

KEYWORDS: Controller, Temporal Logic, Walking Machine.

INTRODUCTION

We describe a research project (with collaborators from NYU, Rutgers and ICSI Berkeley) for rapid prototype construction of discrete controllers based on a practical approach that we have developed and implemented on small experimental robotics systems successfully. The research described here involves two components.
Theoretical Work: This involves formalizing the specification, synthesis, verification, simulation and query-processing of a very powerful hybrid controller (involving both continuous time and discrete event structure) for large scale robotics and manufacturing systems.
Experimental Work: This involves implementing a large distributed software system that automatically generates a controller (either directly or in a two step process, with the intermediate structure described by GRAFCETS) and the associated visualization and interface tools.

THEORETICAL FRAMEWORK

A crucial component of our controller synthesis algorithm is based on the *modal control* approach to the construction of supervisory controller for a discrete event system. However, we differ in one significant way: namely, by the specification formalism based on the classical propositional temporal logic.

---

[1]Address: 251 Mercer Street, New York, NY, 10003, U.S.A. E-mail: `mishra@nyu.edu`
We are grateful to many of our colleagues for their help, advice and comments: Dr. Marco Antoniotti of ICSI Berkeley who contributed to all the ideas described here, Prof. Mohsen Jafari of Rutgers, Prof. E.M. Clarke of CMU, Prof. P.K. Khosla of CMU, Dr. R. Kurshan of Bell Labs, Prof. R. Wallace of Lehigh, Mr. F. Hansen of NYU, and Dr. M. Teichmann also of NYU.

Discrete Event System

We start with a discussion of the original work of Ramadge and Wonham [9].

For our purpose, a *discrete event system* stands for any dynamical system that can be represented in terms of finite state automata. Thus our discrete event system is represented as an automaton $G = (U, X, f, x_0)$ where U = set of inputs/controls[2], X = set of states, $f: U \times X \to X$ = the state transition functions and $x_0 \in X$ represents the initial state of the system. We also allow each state to be labeled by a set of propositional formulas that hold true in that state. Classically, such a labeled finite state automaton is called a Kripke structure. The inputs, U is assumed to be a disjoint union of U_c = the set of controllable events and $U_u = U \setminus U_c$ = the set of uncontrollable events.

The modal controller works by simply disabling some subset of controllable events and the choice of this subset is based solely upon the current state of the dynamical system (i.e., plant). Equivalently, we may base the choice of the disabled controllable events upon some propositional (possibly modal/temporal) formula that holds in the current state. Note that, the controller has *no control* over the uncontrollable events and they cannot be disabled.

A controller for the given dynamical system is then simply specified by a map: $\eta : X \to \Gamma : x \mapsto \gamma_x$, where $U_u \subseteq \gamma_x \subseteq U$, and γ_x represents all the inputs and controls that have *not been disabled*. Such a γ_x is called an *admissible control* and $\Gamma \subseteq 2^U$ is the class of admissible controls.

Given a plant as defined earlier, we can describe the language generated by the plant (assuming it is a trim automaton with every state being a final state) by $M \subseteq U^*$. The controller's job is then to modify the plant's behavior by restricting the accepted language to some subset $L \subseteq M$, where L is to be specified by the user.

The discussion above should serve well to expose one of the major pragmatic problems with Ramadge-Wonham formalism. Namely, the task of accurately modeling the desired behavior by the language L is rather cumbersome. The problem is two folds: There may be more than one such language that should be acceptable to the designer. However, the choice is somewhat arbitrary. Furthermore, great care has to be taken in order not to restrict the language too strongly. Secondly, a good specification of L depends heavily on the designer's understanding of the plant model and the associated language M. In a hierarchically organized system, the designer makes relatively poor usage of the underlying structure. Furthermore, a minor change in any component can significantly alter the choice of the language L. In order to alleviate this problem, we propose using only a very high level temporal logic specification of the desired behavior. Starting with fairly general description of the plant model (e.g., in SADT and IDEFO), we construct automatically both the language M and a sub-language $L \subseteq M$ satisfying the temporal properties.

The second problem with the approach explained in this subsection, is that the language L may not be controllable, in the sense that it may not allow a supervisory controller achieving the language L. Ramadge and Wonham in their seminal work[9] solve this problem elegantly by proposing the concept of a *supremal controllable* sub-

[2]These represent the very primitive events in the system.

language $L^\dagger \subseteq L \subseteq M$, for which there is always a supervisory controller. The supremal controllable sub-language $L^\dagger$ is simply the largest controllable sub-language of L, and since the controllable sub-languages of L is nonempty (the empty-set $\emptyset \subseteq L$ is controllable) and is closed under set union operation, it can be easily defined as

$$L^\dagger = \bigcup\{K : (K \subseteq L) \wedge (\overline{K}U_u \cap M \subseteq \overline{K})\}.$$

This also leads to a fixed point characterization of $L^\dagger$ which allows it to be computed fairly efficiently. However, in the approach we propose, the controllable language $L \subseteq M$ satisfying the temporal properties can be computed directly and the synthesis of supervisor is then fairly straightforward.

Temporal Logic

The logic that we use to specify is a propositional temporal logic of branching time, called *CTL* (*Computation Tree Logic*)[7]. The syntax for CTL is as follows: Let $\mathcal{P}$ be the set of all the atomic propositions in a given language, $\mathcal{L}$. Then (1) Every atomic proposition P in $\mathcal{P}$ is a formula in CTL. (2) If f_1 and f_2 are CTL formulas, then so are $\neg f_1$, $f_1 \wedge f_2$, $\mathbf{AX}\ f_1$, $\mathbf{EX}\ f_1$, $\mathbf{A}[f_1\ \mathbf{U}\ f_2]$ and $\mathbf{E}[f_1\ \mathbf{U}\ f_2]$.

In this logic the propositional connectives $\neg$ and $\wedge$ have their usual meanings of *negation* and *conjunction*, respectively. The temporal operators **AX** and **EX** are the unary *nexttime* operators. The intuitive meaning of $\mathbf{AX}\ f_1$ (respectively, $\mathbf{EX}\ f_1$) is that f_1 holds in every (respectively, in some) immediate successor state of the current state. The temporal operators **AU** and **EU** are the binary (*strong*) *until* operators. For the sake of readability, we write $\mathbf{A}[f_1\ \mathbf{U}\ f_2]$ and $\mathbf{E}[f_1\ \mathbf{U}\ f_2]$ instead of the customary $\mathbf{AU}(f_1, f_2)$ and $\mathbf{EU}(f_1, f_2)$. The intuitive meaning of $\mathbf{A}[f_1\ \mathbf{U}\ f_2]$ (respectively, $\mathbf{E}[f_1\ \mathbf{U}\ f_2]$) is that for every computation path (respectively, for some computation path), there exists an initial prefix of the path such that f_2 holds at the last state of the prefix and f_1 holds at all other states along the prefix.

We also use the following syntactic abbreviations:

$f_1 \vee f_2 \equiv \neg(\neg f_1 \wedge \neg f_2)$, $f_1 \rightarrow f_2 \equiv \neg f_1 \vee f_2$, **and** $f_1 \leftrightarrow f_2 \equiv (f_1 \rightarrow f_2) \wedge (f_2 \rightarrow f_1)$.

$\mathbf{AF}\ f_1 \equiv \mathbf{A}[\text{True}\ \mathbf{U}\ f_1]$ **and** $\mathbf{EF}\ f_1 \equiv \mathbf{E}[\text{True}\ \mathbf{U}\ f_1]$

$\mathbf{AG}\ f_1 \equiv \neg\,\mathbf{EF}\ \neg f_1$ **and** $\mathbf{EG}\ f_1 \equiv \neg\,\mathbf{AF}\ \neg f_1$

The semantics of a CTL formula is as follows. A CTL structure is a triple $\mathcal{M} = (S, R, \Pi)$ where (1) S *is a finite set of states*, (2) R *is a total binary relation on* S *($R \subseteq S \times S$) and denotes the possible transitions between states*, and (3) Π *is an assignment of atomic proposition to states, i.e.* $\Pi : S \rightarrow 2^{\mathcal{P}}$.

A *computation path* is an infinite sequence of states $(s_0, s_1, s_2, \ldots)$ such that $\forall_i[\langle s_i, s_{i+1}\rangle \in R]$. For any structure $\mathcal{M} = (S, R, \Pi)$ and state $s_0 \in S$, there is an *infinite computation tree* with root labeled s_0 such that $s \rightarrow t$ is a directed edge in the tree *if and only if* $\langle s, t\rangle \in R$.

The truth in a structure is expressed by $\mathcal{M}, s_0 \models f$, meaning that the temporal formula f is satisfied in the structure $\mathcal{M}$ at state s_0. The semantics of temporal formulas is defined inductively as follows:

$$\begin{array}{lll}
s_0 \models P & \textit{iff} & P \in \Pi(s_0). \\
s_0 \models \neg f & \textit{iff} & s_0 \not\models f. \\
s_0 \models f_1 \wedge f_2 & \textit{iff} & s_0 \models f_1 \text{ and } s_0 \models f_2. \\
s_0 \models \mathbf{AX}\ f_1 & \textit{iff} & \text{for all states } t \text{ such that } \langle s_0, t\rangle \in R, t \models f_1. \\
s_0 \models \mathbf{EX}\ f_1 & \textit{iff} & \text{for some state } t \text{ such that } \langle s_0, t\rangle \in R, t \models f_1. \\
s_0 \models \mathbf{A}[f_1\ \mathbf{U}\ f_2] & \textit{iff} & \text{for all computation paths } (s_0, s_1, s_2, \ldots), \\
 & & \quad \exists_{i \geq 0}[s_i \models f_2 \wedge \forall_{0 \leq j < i}[s_j \models f_1]]. \\
s_0 \models \mathbf{E}[f_1\ \mathbf{U}\ f_2] & \textit{iff} & \text{for some computation path } (s_0, s_1, s_2, \ldots), \\
 & & \quad \exists_{i \geq 0}[s_i \models f_2 \wedge \forall_{0 \leq j < i}[s_j \models f_1]].
\end{array}$$

The Model Checker for CTL can now be thought of as an algorithm that determines the satisfiability of a given temporal formula f_1 in a model $\mathcal{M}$. Thus, given a plant description with the language $M \subseteq U^*$, our first goal is to find a sub-language $L \subseteq M$ such that if the computation paths of the plant's computation tree is restricted to the strings of L, then the resulting "restricted" computation tree is a model for the CTL specification, denoting the desired overall behavior.

However, in the general setting, the associated operators are non-monotonic; the synthesis problem may not have any or unique solution and may not be computationally tractable. In a recent paper, we have shown that in fact the problem in its full generality ("Unrestricted CTL Supervisory Synthesis Problem") is NP-complete [5]. In particular, we have shown that disjunctions of arbitrary CTL formulas, do not admit an efficient algorithm for the synthesis of a supervisor. The resulting implementation in the Control-D system [1], therefore either allowed only disjunctions with only one *path* disjunct, or used some heuristics to synthesize the supervisor.

CONTROLLER FOR A WALKING MACHINE

Using Control-D, a hybrid controller for the walking machine was implemented in Common Lisp with Motif-based graphical animation and was extensively tested. The desired behavior of the system was specified in CTL.

Each leg was modeled by a finite state machine with five states corresponding to the following activities of an individual leg. Start (`s`): The leg is in a rest position. Unload (`u`): The leg begins not to support any weight anymore. Recover (`r`): The leg is brought forward in a "flying motion". Load (`l`): The leg starts to bear weight. Drive (`d`): The leg thrusts forward the hip exerting a force on the ground. Slipping (`sl`): The leg was not able to firmly stand on the ground.

The events `es`, `eu`, `er`, `el`, `ed` and `esl` (corresponding respectively to 'end-of-start,' 'end-of-unload,' 'end-of-recover,' 'end-of-load,' 'end-of-drive,' and 'end-of-slipping') are all *controllable*, slip ('begin-of-slipping') is in contrast *uncontrollable*. For our testbed, we chose to introduce an uncontrollable event, slip, which indicates a condition where the leg has somehow lost the necessary "stance" on the ground.

Since we want to model a system with many legs we proceed in standard way by taking the *interleaving product* of the machines for each leg. This yields a "combined machine" that effectively represents the discrete behavior of the comprehensive, unregulated system.

Now, we need to ensure that the bipedal system (when augmented with a controller) guarantees various desirable properties. For example:

<u>*Safety Condition:*</u> $\neg \mathbf{EF}(\mathrm{r}_i \wedge \mathrm{r}_j)$, $i,j \in \{1,2\}, i \neq j$.

> It is not the case that for some computation path, there exists a state on the path where both legs are recovering simultaneously. In other words, we wish to avoid situations where both legs are "up in the air."

<u>*Liveness Condition:*</u> $\mathbf{AG}[\mathrm{d}_i \rightarrow \mathbf{AF}(\mathrm{d}_j)]$ $i,j \in \{1,2\}, i \neq j$.

> For each state in which one of the legs drives, it must be the case that in every computation path starting there, for some state on the path the other leg drives. Thus, we wish to have each leg contact the ground infinitely often.

<u>*"Rear-to-Front" Wave:*</u> $\mathbf{AG}[(\mathrm{d}_i \wedge \mathrm{r}_j) \rightarrow \neg \mathbf{EX}(\mathrm{u}_i \wedge \mathrm{r}_j)]$ $i,j \in \{1,2\}, i \neq j$.

> Both legs alternate in driving in a wave like manner.

In order to give a flavor of the current usage of our system, we present an excerpt of a session where we consider the behavior of one train of legs (i.e. a *front* (1) and a *rear* (2) leg). Figure 1 shows the state machine representing the behavior of one leg and the machine representing the interleaving of the legs (i.e. `legs` is the `shuffle` of the two machines (`leg1` and `leg2`)).

```
(define-state-machine leg2
    :states (s2 r2 l2 d2 u2 sl2)
    :start s2
    :alphabet (es2 er2 el2 ed2 eu2 esl2 slip2)
    :uncontrollable (slip2)
    :delta ((s2 es2 u2) (r2 er2 l2)
            (l2 el2 d2) (d2 ed2 u2)
            (u2 eu2 r2)
            (d2 slip2 u2) (sl2 esl2 u2)
            )
    :final-states (s2 r2 l2 d2 u2 sl2)
    )
```

(a)

```
(define-state-machine legs
    :op (shuffle leg1 leg2))
```

(b)

```
>> (omega-op K legs
       uncontrollable-events)
;; Debugging deleted...
>> OMEGA(0):
       removable states =
        ((R1 D2) (D1 R2)
         (D1 U2) (D1 SL2)
         (U1 D2) (SL1 D2))
----
;; Debugging deleted...
>> OMEGA(1):
       removable states =
         NIL
#<Representation for
  the approximation to K>
>>
```

(c)

Figure 1: (a) State machine for one leg. (b) Combined state machine. (c) Results derived by Control-D.

We build a representation for the desired behavior K by removing from the machine `legs` those states which are "inconsistent" with our aims. As an example, we are not interested in those situations when both legs are recovering: this state is represented as `(r1 r2)`. The resulting language K is not controllable, hence we need to build an approximation for it. In this case the approximation algorithm terminates after two iterations. The results are shown in (c) of Figure 1. We can now immediately infer that the gait of the train of legs will be completely "stable", since the states `(r1 d2)` and `(d1 r2)` (which represent states of the train where the system is "unbalancing" in order to proceed) have been removed from the supervisor.

FINAL REMARKS

While there is a tremendous push in robotics and manufacturing to simplify the very basic building blocks (e.g, RISC: Reduced Intricacies in Sensing and Control, Reactive Robotics, or Reconfigurable Robotics), there is a complementary drive to be able to automatically combine a large number of such building blocks to obtain a huge gain in functionality (e.g., Swarm Robotics or Collective Robotics). We believe that in this setting, our techniques offer many fundamental contributions.

References

[1] M. Antoniotti. *Synthesis and Verification of Controllers for Robotics and Manufacturing Devices with Temporal Logic and the Control-D System.* Ph.D. Thesis, N.Y.U., NY 1995.

[2] M. Antoniotti, M. Jafari and B. Mishra. "Applying Temporal Logic Verification and Synthesis to Manufacturing Systems." *Proc. IEEE SMAC*, Vancouver, 1995.

[3] M. Antoniotti and B. Mishra. "Automatic Synthesis Algorithms for Supervisory Controllers." In *Proceedings of the Fourth International Conference on Computer Integrated Manufacturing and Automation Technology*, Troy, NY, October 10–12, 1994.

[4] M. Antoniotti and B. Mishra. "Discrete Event Models + Temporal Logic = Supervisory Controller: Automatic Synthesis of Locomotion Controllers," *Proc. ICRA'95*, Nagoya, 1995.

[5] M. Antoniotti and B. Mishra. *The Supervisor Synthesis Problem for unrestricted* CTL *is* $\mathcal{NP}$-*complete*, Tech Report, NYU No. 707, No. `ICSI TR-95-062`, Nov 1995.

[6] E.M. Clarke, E.A. Emerson, and A.P. Sistla. "Automatic Verification of Finite-State Concurrent Systems Using Temporal Logic Specifications." *ACM Transactions on Programming Languages and Systems*, 8(2):244–263, 1986.

[7] E.A. Emerson and E.M. Clarke. "Characterizing Properties of Parallel Programs as Fixpoints." *Proc. ICALP'85*, LNCS 85, Springer 1981.

[8] B. Mishra and E.M. Clarke. "Hierarchical Verification of Asynchronous Circuits using Temporal Logic." *Theoretical Comp. Sc.*, 38:269–291, 1985.

[9] P.R. Ramadge and W.M. Wonham. "Supervisory Control of a Class of Discrete-Event Processes." *SIAM J. Cont. Opt.*, 25:206–230, 1987.

[10] P.R. Ramadge and W.M. Wonham. "Modular Feedback Logic for Discrete Event Systems." *SIAM J. Cont. Opt.*, 25, 1987.

[11] W.M. Wonham and P.R. Ramadge. "On the Supremal Controllable Sublanguage of a Given Language." *SIAM J. Control Optim.*, 25:637–659, 1987.

Designing a Parallel Manipulator for a Specific Workspace

J-P. Merlet
INRIA Sophia-Antipolis

Abstract:

We present an algorithm to determine all the possible geometries of Gough-type 6 d.o.f. parallel manipulators whose workspace has to include a desired workspace. This desired workspace is described as a set of geometric objects, limited here to segments, which describe the desired locations of the center of the moving platform. This algorithm takes into account the leg length limits, the mechanical limits on the passive joints and legs interference.

Keywords:parallel manipulators,workspace,design

Introduction

In this paper we consider a 6 d.o.f. Gough-type parallel manipulator constituted by a fixed base plate and a mobile plate connected by 6 extensible legs (figure 1). For a parallel manipulator, workspace limits are due to the bounded range of linear actuators, mechanical limits on passive joints and interference between legs. One important step in the design of a parallel manipulator is to define its geometry according to the desired workspace. Various geometrical algorithms for computing the workspace boundary when the platform's orientation is kept constant have been described by Gosselin [3] and Merlet [5].

The problem we address in this paper is to find all the possible locations of the centers of the passive joints such that the robot workspace includes the desired workspace. [1]

This research area has been addressed by very few authors. Stoughton [7] has proposed a numerical method for optimizing the workspace of a specific parallel manipulator whose length limits are known. Liu [4] has characterized some extremal positions of the robot as a function of its geometry and maximal leg lengths. Gosselin [2] has established design rules for the spherical 3 d.o.f parallel manipulator type and has studied the optimization of the workspace of planar three d.o.f parallel manipulator [1].

Let A_i, B_i denote the attachment points of the leg on the base and on the platform (figure 1). For a pair of A_i, B_i we attach a reference frame O, x, y, z to the base such that the z coordinate of A_i is equal to 0. In the same manner we attach to the platform a mobile frame C, x_r, y_r, z_r such that the z_r coordinate of B_i is equal to 0 (in other words we assume that we know the planes in which are located the A_i, B_i, which can however be different for each joint). A subscript $_r$ will denote a point or a vector whose coordinates are written in the mobile frame. The rotation matrix between the mobile frame and the reference frame will be denoted Rot. Let α_i be the angle between the Ox axis and OA_i and let β_i be the angle between the Cx_r axis and CB_{i_r}. We denote by R_1 the distance between O, A_i and r_1 the distance between C, B_i.

[1] The proposed algorithms have been implemented in C on a workstation under the X-windows system and every drawings appearing in this paper is a result of this program. This program is available via anonymous `ftp` on `zenon.inria.fr`

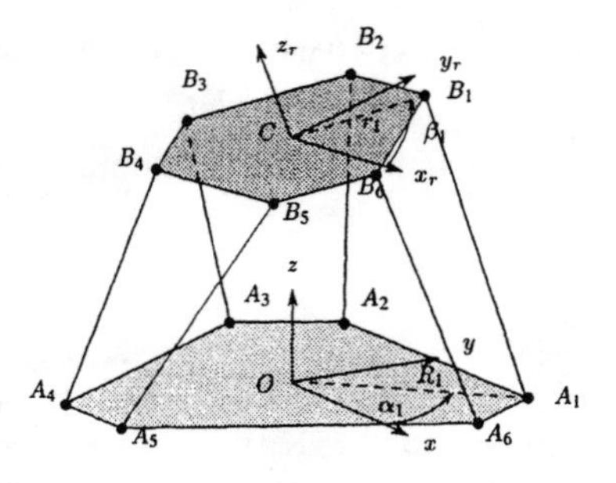

Figure 1: Notation

Consequently we have:

$$\mathbf{A_iO} = \begin{pmatrix} -R_1 \cos \alpha_i \\ -R_1 \sin \alpha_i \\ 0 \end{pmatrix} = R_1 \mathbf{u_i} \quad \mathbf{CB_{ir}} = \begin{pmatrix} r_1 \cos \beta_i \\ r_1 \sin \beta_i \\ 0 \end{pmatrix} = r_1 \mathbf{v_i} \tag{1}$$

where $\mathbf{u_i}, \mathbf{v_i}$ are unit vectors. As soon as the values of R_1, r_1 are fixed for each pair of (A_i, B_i) the geometry of the robot is completely determined. The purpose of this paper is to determine all the possible values of R_1, r_1 for each pair of (A_i, B_i) such that the workspace of the corresponding robot includes a given workspace under the following assumptions: 1) the minimum and maximum value of the leg lengths are known and 2) the angles α_i, β_i are known.

We attach a plane $\mathcal{P}_i$ to each pair of A_i, B_i., with a frame whose axes will correspond to the value of the R_1, r_1 parameters for the pair A_i, B_i. The algorithm presented in the sequel will compute the boundary of the regions of this plane such that any point in the regions defines a robot whose workspace includes the specified workspace. The regions will therefore be called the *allowable region* for the pair (A_i, B_i).

As for the desired workspace we will assume that it is defined as a set of segments describing the location of the point C of the moving platform, the orientation of the moving platform being fixed for each object. Note that extension to other types of geometric objects is possible.

Remark: When studying the effects of the workspace constraints on one leg, the desired workspace will be specified in the reference frame attached to this particular leg. The allowable region will be expressed in the plane $\mathcal{P}_i$ attached to this particular reference frame which could be different for each leg. Consequently there is no assumption on the location of the joint centers A_i, B_i which may be in a general position.

Segments workspace

In the following sections the desired workspace is described by a set of pairs (segment, *Rot*). For each pair point C should be able to describe the segment while the moving platform has the orientation defined in the pair.

Remark: To deal with the orientation aspect it is possible to specify in the set various segments which differ only by their orientation components.

Allowable regions for the length constraints

Sets of maximal and minimal ellipses

In this section we will assume that the constraints limiting the workspace are only the leg lengths. A trajectory segment is defined by two points M_1, M_2 and for any C belonging to the trajectory we may write: $\mathbf{OC} = \mathbf{OM_1} + \lambda \mathbf{M_1M_2}$ where λ is a scalar parameter in the range [0,1]. Remember that the leg length ρ is the norm of the vector $\mathbf{AB}$. For a point C on the segment we have:

$$\rho^2 = \mathbf{AO}\cdot\mathbf{AO}+\mathbf{CB}\cdot\mathbf{CB}+2(\mathbf{AO}+Rot\,\mathbf{CB_r}).\mathbf{OC}+2Rot(\mathbf{CB_r})\cdot\mathbf{AO}+\mathbf{OC}\cdot\mathbf{OC} \quad (2)$$

This equation may be rewritten as:

$$\begin{aligned}\rho^2 \; = \; & R_1\mathbf{u}\cdot(\mathbf{OM1}+\lambda\mathbf{M_1M_2})+r_1 Rot(\mathbf{v})\cdot(\mathbf{OM_1}+\lambda\mathbf{M_1M_2})+R_1^2+r_1^2+ \\ & \lambda^2\mathbf{M_1M_2}\cdot\mathbf{M_1M_2}+\mathbf{OM_1}\cdot\mathbf{OM_1}+2\lambda\mathbf{OM_1}\cdot\mathbf{M_1M_2}+R_1r_1\,Rot(\mathbf{v})\cdot\mathbf{u}\end{aligned}$$

or in various other forms:

$$\mathcal{E}(R_1,r_1,\lambda,\rho)=0 \quad F(\lambda)=A_2\lambda^2+A_1\lambda+A_0=0 \quad (3)$$

The structure of equation (3) leads to the following theorem: *For a given set of ρ, λ values, equation (3) defines an ellipse in the R_1-r_1 plane.*

Consider now the function $\mathcal{E}_{max}(\lambda)=\mathcal{E}(R_1,r_1,\lambda,\rho_{max})$ with λ in the range [0,1]. This function defines a set of maximal ellipses. If $\mathcal{E}(R_1,r_1,\lambda,\rho_{max})\leq 0$ for any λ in the range [0,1], then the leg length is less than of equal to ρ_{max} for any posture of the moving platform on the trajectory. Consequently, the allowable region for the maximum length constraint is the set of points $M(R_1,r_1)$ such that $\mathcal{E}_{max}(\lambda)\leq 0$ for any λ in [0,1]. Hence any such M point must lie inside all the maximal ellipses of the set and, therefore the allowable region with respect to the constraint $\rho \leq \rho_{max}$ is the intersection $\mathcal{I}$ of all the maximal ellipses. We denote $\mathcal{E}_{max}(0)$ and $\mathcal{E}_{max}(1)$ the two ellipses obtained for $\lambda=0$ and $\lambda=1$. The following theorems hold (due to the lack of space proofs are not presented but can be found in [6]):

Theorem 1 *As λ varies the center of the corresponding maximal ellipse lies on a segment which in some cases may be reduced to a point. The angles between the main axes of the ellipses and the R_1, r_1 axes are $\pi/4$.*

Theorem 2 *If the maximal ellipses $\mathcal{E}_{max}(0)$, $\mathcal{E}_{max}(1)$ exists, then the ellipse exist for any value of λ in the range [0,1]. The intersection $\mathcal{I}$ of all the ellipses in the set is equal to $\mathcal{E}_{max}(0)\cap\mathcal{E}_{max}(1)$. Therefore the allowable region is simply the intersection of the ellipses computed for the extremal points of the trajectory.*

Let $\mathcal{E}_{min}(\lambda)$ be the function $\mathcal{E}(R_1,r_1,\lambda,\rho_{min})$. This function defines a set of minimal ellipses. If $\mathcal{E}_{min}(\lambda)>0$ for a given point M and a given λ, then the corresponding leg length is greater than ρ_{min}. Therefore for any point belonging to the allowable region this relation has to be verified for all λ in [0,1]. Consequently, any point of the allowable region must lie outside the region $\mathcal{U}$ defined by the union of the minimal ellipses. Hence the allowable region $\mathcal{A}_l$ for the leg length constraint is:

$$\mathcal{A}_l = (\mathcal{E}_{max}(0) \cap \mathcal{E}_{max}(1)) - \mathcal{U} \tag{4}$$

Computing this union is a more complex than computing the previous intersection but an efficient method has been developed.

For a desired workspace defined by a set of segments we compute the allowable region for each pair and then compute the intersection of all these regions. In our implementation, the boundary of all the regions are approximated by polygons in order to simplify the intersection and the Boolean operations on the regions.

An example of the result of the algorithm is shown on figure 2: three segments have been defined and it is assumed that all the attachment points lie on the same circle (the allowable region is therefore the intersection of the allowable regions of each leg).

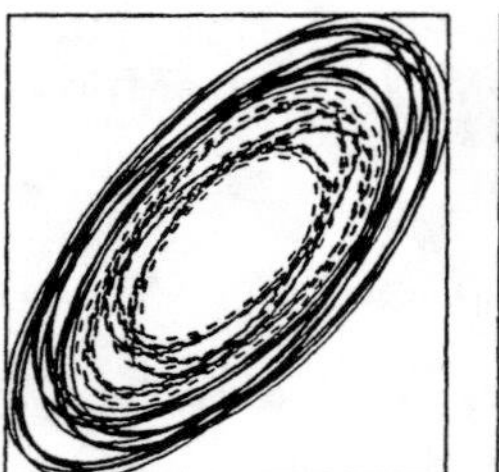

Figure 2: On the left the minimal ellipses (dashed lines) and the maximal ellipses (in thin lines) for a workspace defined by three segments. On the right the allowable region.

Allowable region for the mechanical limits on the joints

In this section we will use a classical model for the mechanical limits: it is assumed that the operator is able to define a pyramid with planar faces and apex at A_i such that if the joint constraints are satisfied, then the leg A_iB_i lies inside the pyramid. The normal to the faces of the pyramid are denoted n_j^i.

For a given posture of the moving platform the leg A_iB_i will lie inside the pyramid (which means that the mechanical limits of the joint are not violated) if

$$\mathbf{A_iB_i} \cdot \mathbf{n_j^i} \leq 0 \quad \forall j \in [1, k]$$

This inequality may be rewritten for a specific leg and a specific face of a pyramid as:

$$R_1\mathbf{u} \cdot \mathbf{n} + r_1 R\mathbf{v} \cdot \mathbf{n} + \lambda \mathbf{M_1M_2} \cdot \mathbf{n} + \mathbf{OM_1} \cdot \mathbf{n} \leq 0 \tag{5}$$

Let $\mathcal{L}(R_1, r_1, \lambda)$ denote the left side of this inequality. The equation $\mathcal{L}(R_1, r_1, \lambda) = 0$ defines a set of lines with identical slope in the R_1-r_1 plane which span a region in the plane whose boundary is constituted of the line $\mathcal{L}(R_1, r_1, 0) = L_0$, $\mathcal{L}(R_1, r_1, 1) = L_1$. One of the two lines L_0, L_1 separates the plane in two half-planes $\mathcal{P}_{+1}, \mathcal{P}_{-1}$ such that $\mathcal{L}(R_1, r_1, \lambda) \leq 0$ for any λ in [0,1] if $M(R_1, r_1)$ belongs to $\mathcal{P}_{-1}$ and $\mathcal{L}(R_1, r_1, \lambda) > 0$ for some values of λ in [0,1] if M belongs to $\mathcal{P}_{+1}$. Therefore, $\mathcal{P}_{-1}$ defines a half-plane which is the allowable region for this face of the pyramid.

The process is repeated for each face of the pyramid, thereby leading to a set of half-planes. The intersection of these half-planes is the allowable region with respect to the mechanical limits on the joint: it leads to a polygonal region.

We may now compute the allowable region for a set of segments as the intersection of the polygonal region with the region computed for the leg lengths constraints

Conclusion

The algorithm presented in this paper enables to compute the location of the attachment points of all the robots whose workspace contains a desired workspace under the assumption that the minimum and maximum leg lengths are known and that the general direction of the lines on which the attachment points lie are also known. Due to a lack of space we have not considered interference between the legs but this problem has been addressed in [6].

Afterward various criteria may be used to determine an "optimal" robot using a numerical algorithm, whose search domain is now reduced. These possible criteria might be, for example the maximal dexterity over the workspace, the absence of singularities in the workspace, or to minimize the maximum of the positioning errors for the platform for a given error of the sensors measuring the leg length. Finding procedures for efficiently evaluating some of these points are still open problems and will constitute the object of our future research.

As an example, the maximum accuracy criterion has been used for designing the robot of the European Synchrotron Radiation Facility (ESRF) presented in figure 3. At the start of this project the workspace was specified as a cube of 0.02m side, the required accuracy was 1 μm for a nominal load of 500 kg and the range of the actuators was $\pm$ 0.04m. As a symmetrical manipulator was desired this information was included in the definition of the α_i, β_i angles. The allowable region was determined and a regular grid was defined on the region. At each node of the grid (i.e., for a given robot geometry) the minimal sensor accuracy for obtaining the desired positioning accuracy was determined and a node leading to a sensor accuracy of 0.87 μm was found. During the process it was noted that for another robot whose workspace was satisfactory the sensor accuracy was 0.0435 μm; hence the performances of robots whose workspace includes a given workspace may largely vary and the design has to be carefully studied.

Extension of these algorithms to other type of workspace description can be made. The case where the desired workspace is described by a set of spheres as been treated in [6]. The algorithm described in this paper was illustrated on the Gough-type parallel manipulator, but a similar algorithm may be used for other type of parallel robots.

References

[1] Gosselin C. *Kinematic analysis optimization and programming of parallel robotic*

Figure 3: The ESRF robot

manipulators. Ph.D. Thesis, McGill University, Montréal, June, 15, 1988.

[2] Gosselin C. and Angeles J. The optimum kinematic design of a planar three-degree-of-freedom parallel manipulator. *J. of Mechanisms, Transmissions and Automation in Design*, 110(1):35–41, March 1988.

[3] Gosselin C., Lavoie E., and Toutant P. An efficient algorithm for the graphical representation of the three-dimensional workspace of parallel manipulators. In *22nd Biennial Mechanisms Conf.*, pages 323–328, Scottsdale, September, 13-16, 1992.

[4] Liu K., Fitzgerald M.K., and Lewis F. Some issues about modeling of the Stewart platform. In *2nd Int. Symp. on Implicit and Robust systems*, Warsaw, 1991.

[5] Merlet J-P. Détermination de l'espace de travail d'un robot parallèle pour une orientation constante. *Mechanism and Machine Theory*, 29(8):1099–1113, November 1994.

[6] Merlet J-P. Designing a parallel robot for a specific workspace. Research Report 2527, INRIA, April 1995.

[7] Stoughton R. and Kokkinis T. Some properties of a new kinematic structure for robot manipulators. In *ASME Design Automation Conf.*, pages 73–79, Boston, June, 28, 1987.

A REGISTRATION AND CALIBRATION PROCEDURE FOR A PARALLEL ROBOT

P. MAURINE, E. DOMBRE
L.I.R.M.M., Université Montpellier II, France

ABSTRACT

A two stage calibration method for the parallel robot Delta 4 is presented. It allows one to identify the offsets on the three first joints and the absolute location of the robot base. It involves a low cost displacement sensor and dedicated targets which can be easily moved on the work area. Intensive simulations show the robustness of the protocol and experimental results validate this procedure.

KEYWORDS : parallel robot, calibration

INTRODUCTION

Many comprehensive studies and works have been done in the area of parallel robots [13], [17], [16]. This is due to their interesting features by comparaison with serial robots: great dynamic capabilities and rigidity, a high positioning repeatability, and so a high positioning accuracy if the actual parameter values are known.
The loss of accuracy of such structures is mainly due to the joint offsets, the manufacturing tolerances and the errors of the robot registration in the environment. The solution to compensate this loss of accuracy is known as robot calibration. The identification of the real robot kinematic parameters allows one to compensate the nominal kinematic model.
Several papers can be found about the calibration of serial robots [8], [19], [14]. Nevertheless, few papers have been published about the calibration of parallel robots. Bennett et al., [1] use the approach they have developed about the autonomous calibration of single closed loop kinematic chain to calibrate the RSI-6DOF wrist. The offsets and the gain of this structure are identified. Using the same approach Nahvi et al. [15] identify the joint offsets and three other kinematic parameters of a 3-degree-of-freedom (dof) platform. Like in the previous work, experimental results are given and compared to those obtained using an external calibration device.
Other works have been carried out on the calibration of the Stewart platform. Zhuang et al. [21] propose a new solution but they conclude that the main drawback of their method is that the parameters can not be identified globally. Masory et al. [10] develop a more robust method to identify the platform parameters. No experimental studies are performed but extensive simulations considering measurement noise show that the positioning error of the platform can be reduced by at least one order of magnitude. Geng et al. [7] simulate a two stage calibration procedure of the Stewart platform.
None of these methods takes into account the registration of the robot in the environment. So the location errors of the robot base which occur when the robot is first installed or is moved on its work area cannot be compensated. Moreover no experimental results are given about the three last methods.
We have developed a method to calibrate the robot Delta 4. This method is efficient and can be easily carried out in a real environment. It allows one to identify the robot location in the work area and the joint offsets. These parameters are subject to change due to maintenance operation and when the robot is moved on the line.

The other kinematic parameters whose influence on the robot accuracy is less important, are not taken into account. Therefore, in our approach the robot is supposed to have been previously calibrated.
The measurement device we use is a low cost laser displacement sensor. It operates in a range finder mode or in a detection mode in both cases on dedicated targets. These targets are easily placed and moved in the work area due to their small dimensions.
The first section of this paper describes the delta robot. The specifications of the procedure are then developed. Discussion about the modeling approximations and simulations are given in the second section. Experimental validation on the robot Delta 4 is presented in the last section.

THE ROBOT DELTA 4

The Delta 4 is a very fast 4-dof parallel robot suitable to pick and place tasks. It can move and place low weighted objects at high speed along trajectories about 200 mm long. Its fully parallel structure [18] complies with this kind of applications. Mechanism specifications are given in [4][5].
The robot consists of a base plate, a traveling plate and three identical kinematics chains made of two parts (figures 1, 2) :

- the arm actuated by one of the three motors attached to the top plate and distributed on a circle at 120 degrees mutually.
- the lower parallelogram, which drives the traveling plate.

Thus the traveling plate always remains parallel to the base plate and the translational motions result from the combined motions of the three actuators. The end effector is attached to the mobile plate and it is connected to a fourth actuator attached either on the top plate or directly on the traveling plate.
Several approaches have been developed to establish direct and inverse kinematics models [20] [6]. We will use here the modelization proposed by Pierrot [18].

CALIBRATION PROCEDURES

As in our previous works [3] [11] [12], the idea is to identify a restricted set of parameters instead of identifying the whole set of parameters simultaneously. In this section we present the procedures used to identify the three offsets θ_{1off}, θ_{2off}, θ_{3off} of the three first joints, the position errors d_x, d_y, d_z and the orientation errors δ_x, δ_y, δ_z of the robot base with respect to the environment. These errors are assumed to be small. Two procedures allow the identification of two sets of parameters: (1) δ_x, δ_y, d_z, θ_{1off}, θ_{2off}, θ_{3off} and (2) d_x, d_y, δ_z. Note that the offsets can also be identified using the second set of parameters. The offset on the fourth joint θ_{4off} is not identified since it has no influence on the end effector position and a negligible influence on its orientation along z_4 axis.
The frames used in the procedures are the following (figures 1, 2):

R_w: the reference frame associated to the work cell,
R_b: the base plate frame tied to its center O_b,
R_t: the traveling plate frame tied to its center O_t,
R_c: the sensor frame associated to the point O_c, the z_c axis is along the sensor optical axis.

Using an homogeneous matrix, ${}^iT_j$, [2] to describe a frame R_j in a frame R_i we can define ${}^wT_b$, ${}^bT_t$, ${}^tT_c$:

$$
{}^wT_b = \begin{bmatrix} 1 & -d_z & d_y & X_0+d_x \\ d_z & 1 & -d_x & Y_0+d_y \\ -d_y & d_x & 1 & Z_0+d_z \\ 0 & 0 & 0 & 1 \end{bmatrix}, \quad {}^bT_t = \begin{bmatrix} 1 & 0 & 0 & X_t \\ 0 & 1 & 0 & Y_t \\ 0 & 0 & 1 & Z_t \\ 0 & 0 & 0 & 1 \end{bmatrix},
$$

where $[X_0\ Y_0\ Z_0]^T$ are the nominal coordinates of O_b in R_w. The coordinates of $[X_t\ Y_t\ Z_t]^T$ are computed using the direct kinematic equations (DKE) taking into account the offsets θ_{1off}, θ_{2off},

θ_{3off}:

$$[X_t \quad Y_t \quad Z_t]^T = DKE(\theta_1 + \theta_{1off}, \theta_2 + \theta_{2off}, \theta_3 + \theta_{3off}, \theta_4)$$

where θ_i is the ith joint variable. If $[X_c\ Y_c\ Z_c]^T$ are the O_c coordinates in R_t then:

$$^tT_c = \begin{bmatrix} 1 & 0 & 0 & X_c \\ 0 & 1 & 0 & Y_c \\ 0 & 0 & 1 & Z_c \\ 0 & 0 & 0 & 1 \end{bmatrix}$$

where $X_c=0$, $Y_c=0$, Z_c are the nominal coordinates of O_c; the sensor optical axis is aligned with the z_t axis of R_t.

Identification of δ_x, δ_y, d_z, θ_{1off}, θ_{2off}, θ_{3off}

Let P be an horizontal support plane accurately positioned in the work area. Its height in R_w is h_P. The robot is driven to reach the knots N_i of a virtual horizontal pattern grid whose heigth in R_w is h_μ (figure 1). For each knot N_i the distance d_i between the sensor head O_c and the plane P is recorded.

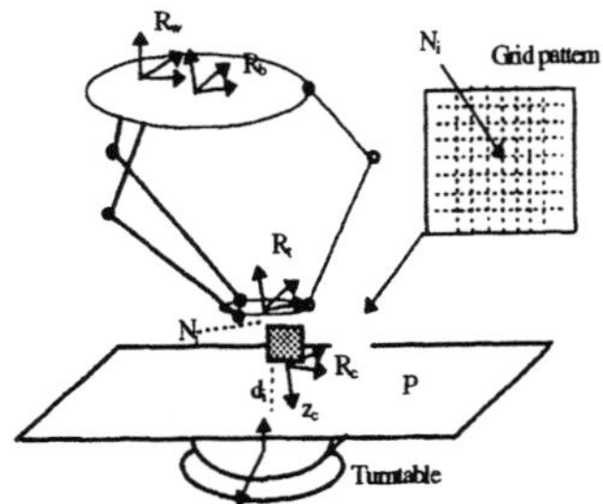

Fig. 1: Identification of δ_x, δ_y, d_z, θ_{1off}, θ_{2off}, θ_{3off}

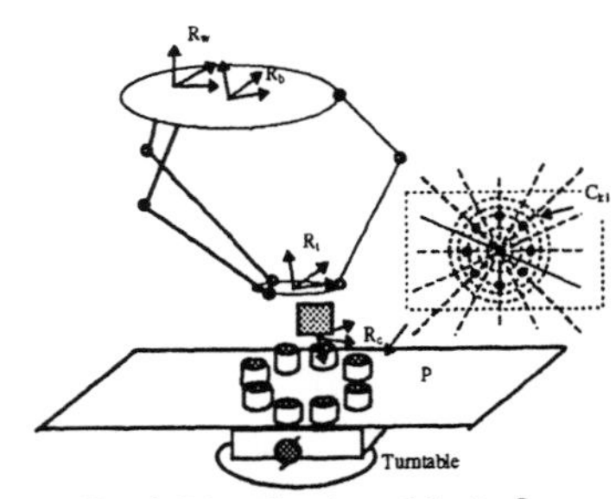

Fig. 2: Identification of d_x, d_y, δ_z

Using the previous definitions of ^wT_b, ^bT_t, ^tT_c, the matrix ^wT_t is computed:

$$^wT_t = \begin{bmatrix} 1 & -d_z & d_y & X_0 + X_t + d_y \cdot (Z_t + Z_c) - d_z \cdot Y_t + d_x \\ d_z & 1 & -d_x & Y_0 + Y_t - d_x \cdot (Z_t + Z_c) + d_z \cdot X_t + d_y \\ -d_y & d_x & 1 & Z_0 + Z_t + Z_c + d_x \cdot Y_t - d_y \cdot X_t + d_z \\ 0 & 0 & 0 & 1 \end{bmatrix} = \begin{bmatrix} 1 & -d_z & d_y & X' \\ d_z & 1 & -d_x & Y' \\ -d_y & d_x & 1 & Z' \\ 0 & 0 & 0 & 1 \end{bmatrix}$$

If the sensor optical axis is aligned with the z_c axis of R_c frame, the theoretical distance d_i' between O_c $[X'\ Y'\ Z']^T$ and the plane P along the z_c direction whose components in R_w are $[\delta_y\ -\delta_x\ 1]^T$ can be calculated by :

$d_i' = (Z' - h_P) \cdot \sqrt{(d_y^2 + (-d_x)^2 + 1^2)}$, leading $d_i' = (Z' - h_P)$ when neglecting the second order terms.

The vector of parameters $p_1=[\delta_x\ \delta_y\ d_z\ \theta_{1off}\ \theta_{2off}\ \theta_{3off}]^T$ being identified is the one that minimizes the following criterium :

$$\sum_{i=1}^{M} e_i^2 = \sum_{i=1}^{M} (d_i' - d_i)^2 \ ; \ \varepsilon_i = [Z_0 + Z_{ti} + Z_c + d_x \cdot Y_{ti} - d_y \cdot X_{ti} + d_z - h_P] - d_i$$

where M is the number of measurements N_i and :

$$[X_{ti} \quad Y_{ti} \quad Z_{ti}]^T = DKE(\theta_{1i} + \theta_{1off}, \theta_{2i} + \theta_{2off}, \theta_{3i} + \theta_{3off}, \theta_{4i})$$

$[\theta_{1i}, \theta_{2i}, \theta_{3i}, \theta_{4i}]^T$ is the nominal configuration given to the robot controller to reach N_i (no location errors and no offsets are taken into account). The non linear minimization problem is solved using MATLAB library.

Identification of d_x, d_y, δ_z, (θ_{1off}, θ_{2off}, θ_{3off})

A number I of cylinders are plugged on the plane P. The positions are uniformely distributed on a circle C_k (figure 2). The position and the orientation of each cylinder are accurately known in R_w. The robot is moved so the center of the traveling plate O_t describes a circle C_k' above the cylinders. The laser sensor attached to the mobile plate detects the first edge B_{1j} of the cylinder j. Then the robot moves to detect the second one B_{2j} and so on for each cylinder. The configurations θ_{1ij}, θ_{2ij}, θ_{3ij} (i=1, 2 ; j=1, ..., I) are stored. Using the definition of ${}^wT_c$, B_{ij} coordinates are given by:

$$X_{ij} = X_0 + X_{tij} + d_y \cdot \left(Z_{tij} + Z_c\right) - d_z \cdot Y_{tij} + d_x \; ; \; Y_{ij} = Y_0 + Y_{tij} + d_x \cdot \left(Z_{tij} + Z_c\right) - d_z \cdot X_{tij} + d_y$$

where $\left[X_{ti} \quad Y_{ti} \quad Z_{ti}\right]^T = DKE\left(\theta_{1i} + \theta_{1off}, \theta_{2i} + \theta_{2off}, \theta_{3i} + \theta_{3off}, \theta_{4i}\right)$

Using δ_x, δ_y, d_z, θ_{1off}, θ_{2off}, θ_{3off} values identified by the first procedure, the parameter vector $p_2=[d_x \; d_y \; \delta_z]^T$ to be identified is the one that minimizes for all cylinders:

$$\sum_{i,j}\left(r_i - \sqrt{\left(X_{ij} - x_{ci}\right)^2 + \left(Y_{ij} - y_{ci}\right)^2}\right) \text{(i=1, ..., I ; j=1,2)}$$

where x_{ci}, y_{ci} are the coordinates of the ith cylinder center in R_w. As in the previous procedure, this non linear problem is solved using MATLAB library. Note that the offsets can also be included in the vector $p_2=[d_x \; d_y \; \delta_z \; \theta_{1off} \; \theta_{2off} \; \theta_{3off}]^T$ to be identified with d_x, d_y, δ_z. This allows one to verify the results given by the first procedure.

DISCUSSION AND SIMULATIONS

In the previous section, the location errors of the sensor on the mobile plate are not taken into account. Practically the laser sensor is roughly attached to the plate in order to save time and to reduce the cost of fixturing. Thus, the sensor head is placed with d_{xc}, d_{yc}, d_{zc} position errors and with δ_{xc}, δ_{yc}, δ_{zc} orientation errors. This introduces a bias on the coordinates of the measured points on P and on the cylinder edge coordinates.

To compensate these errors [11], for each point N_i and for each edge B_{ij} two measurements are performed respectively at the orientation θ_4 and $\theta_4+\pi$ of the mobile plate with respect to z_4 axis. The value of d_i and θ_{1off}, θ_{2off}, θ_{3off} used in the minimization algorithm is the mean value of the measurements performed respectively for θ_4 and $\theta_4+\pi$.

Intensive simulations have been done to validate these hypotheses [12]. They show that the identification procedure is robust to location errors of the sensor on the end effector. The parameter sensitivity to the measurement noise and the choice of the number of measurements are also discussed.

EXPERIMENTAL VALIDATION

Experimental set up

The experimental setup is shown in figures 1 and 2. A rigid plate is used as the measurement plane P. Making use of drilled holes whose coordinates are precisely known in the frame associated to the plate, several cylinders can be accurately plugged on different circles C_k. The plate is attached to a precision turntable providing O_x, O_y fine translational motions along x and y axis and O_x, O_y, O_z fine rotational motions about x, y and z axis. So the location errors are directly given to the plane P rather than to the robot base. The offsets are introduced on the robot controller. The sensor is a Keyence LB-12 laser displacement meter (resolution 2 µm, linearity 0.5%).

Sensor calibration

The sensor is attached to the end effector in such a way that the location errors are minimized. For both procedures distance between the sensor head and the targets is close to 40 mm. For the identification of d_x, d_y, δ_z, θ_{1off}, θ_{2off}, θ_{3off}, it comes that the sensor has to be oriented in such a way that the plane formed by the emitted and received beams is orthogonal to the displacement direction [11]. Furthermore, to increase the sensitivity of the sensor, for the procedure 1 the plane is painted in white whereas for the procedure 2 the top of cylinders are painted in white and a black support is put on plane P.

Experimental results

Since the absolute orientation of the support plane and the absolute location of the cylinders on the plane are unknown in R_w, a first series of measurements is done in order to identify a reference orientation and position of the plane and a reference location of the cylinders.

Identification of δ_x, δ_y, θ_{1off}, θ_{2off}, θ_{3off}

When using the precision turntable the rotation angles δ_{xP} and δ_{yP} are given to the plane P. The turntable does not allow us to add d_{zP} errors which makes this parameter unidentifiable. The procedure is carried out in less than 8 minutes. The identified parameters are shown in table 1.
These results show that with the proposed procedure, δ_x, δ_y, θ_{1off}, θ_{2off}, θ_{3off} may be accurately identified. The worst parameter accuracy is $\pm 0.17°$ for δ_x, δ_y and $\pm 0.31°$ for the offsets.

δ_{xP} ° nominal	δ_{yP} ° nominal	θ_{1off} ° nominal	θ_{2off} ° nominal	θ_{3off} ° nominal	δ_{xP} ° identified	δ_{yP} ° identified	θ_{1off} ° identified	θ_{2off} ° identified	θ_{3off} ° identified
1	0	0	0	0	1.12	0.12	0.16	-0.14	0.28
0	2	0	0	0	-0.17	1.91	0.12	-0.17	0.23
-0.5	1	0	0	0	-0.41	1.07	0.09	0.04	0.22
0	0	2	-1.5	-2	-0.07	0.12	2.18	-1.39	-1.84
0	0	2	2	0	0.11	0.07	1.76	1.82	0.12

Table 1: Experimental results for δ_x, δ_y, θ_{1off}, θ_{2off}, θ_{3off}

Identification of d_x, d_y, δ_z, (θ_{1off}, θ_{2off}, θ_{3off})

The plane P is oriented in such a way that $\delta_{xP}=\delta_{yP}=0$; no offsets are introduced in the robot controller. Using the first procedure, the real values of the robot parameters δ_x, δ_y, d_z, θ_{1off}, θ_{2off}, θ_{3off} are computed. Then d_{xP}, d_{yP}, δ_{zP} are given to the plane P where 8 cylinders are plugged uniformly on the circle C_k whose radius is 200 mm. The procedure is run in less than 15 mn, then using δ_x, δ_y, θ_{1off}, θ_{2off}, θ_{3off} identified with the first procedure the parameters d_x, d_y, δ_z are computed. Experimental results are shown in table 2. Once again, they validate the proposed procedure.

d_{xP} mm nominal	d_{yP} mm nominal	δ_{zP}° nominal	d_{xP} mm identified	d_{yP} mm identified	δ_{zP} ° identified
2	-5	-1	2.07	-4.87	-0.96
-2	-3	0	-1.86	-3.09	0.02
-3	4	1	-3.07	3.97	1.16
5	3	3	4.95	2.78	2.82
1	-1	1	0.82	-1.12	1.01

Table 2 : Experimental results for d_x, d_y, δ_z

CONCLUSION

We have developed a two stage calibration method to identify on one hand the offsets of the three first joints of the parallel robot Delta 4 and, on the other hand the robot registration in the environment. This procedure is well suited for Delta 4 robot and it could be used for other structures. It is easy to implement on the shop floor. It involves a low cost displacement sensor.
Intensive simulations have been performed to evaluate the sensitivity with respect to the sensor location errors, the measurement number and the measurement noise. They show that identification

procedure is robust with respect to these variations. Experimental results validate these procedures. Current works concern the optimisation of these procedures in term of accuracy.

REFERENCES

[1] D.J. Bennett, J.M. Hollerbach, "Autonomous calibration of single loop closed kinematics chain formed by manipulators with passive endpoint constraints", *IEEE transactions on Robotics and Automation*, Vol 7, N 5, pp 597-605, 1989.

[2] E. Dombre, W. Khalil, "Modélisation et commande des robots", Edition Hermès, Paris, 1988.

[3] E. Dombre, A. Fournier, "Yet another calibration technique to reduce the gap between CAD world and real world", *Proc. Fifth International Symposium on Robotics and Manufacturing, ISRAM'94*, Maui, Hawai; August 1994.

[4] R. Clavel, "Delta, a very fast robot with parallel geometry", *Proc. International Symposium on Industrial Robots*, pp 91-100, 18th, April 1988.

[5] R. Clavel, "Une nouvelle structure de manipulateur parallèle pour la robotique légère", R.A.I.R.O. APII, Vol 23, N6, 1989.

[6] R. Clavel, "Conception d'un robot parallèle rapide à 4 degrés de liberté", Thèse de doctorat, Ecole polytechnique de Lausanne, 1991.

[7] Z.J. Geng, L.S. Haynes, "An effective kinematics calibration method for Stewart Platforms", *Proc. WAC intelligent autonomous soft computing*, Hawaï, USA, Vol 2, pp 87-92, 1994.

[8] J.M. Hollerbach, "A survey of kinematics calibration", *In Robotics Review*, Mit Press, Vol 1, pp 207-242, 1989.

[9] J.M. Hollerbach, D.H. Lokhorst "Closed loop kinematics calibration of the RSI-6DOF hand controller", *Proc IEEE International Conference on Robotics and Automation*, Atlanta, Georgia, USA, Vol 2, pp 142-148, 1993.

[10] O. Masory, J. Wang, H. Zhuang, "On the accuracy of a Stewart platform", *Proc IEEE International Conference on Robotics and Automation*, Atlanta, Georgia, USA, Vol 1, pp 725-731, 1993.

[11] P. Maurine, E. Dombre, A Fournier, «An integrated low cost registration procedure for robots manipulators", *2nd Japan-France Congress on Mecatronics*, Takamatsu, Kagawa, Japan, November 1-3, 1994.

[12] P. Maurine, E. Dombre, "A calibration procedure for the parallel robot Delta 4", *Proc IEEE International Conference on Robotics and Automation*, Minneapolis, Minnesota, April 1996 (to be published).

[13] J.P. Merlet, "Les Robots Parallèles", Traité des Nouvelles Technologies, Série Robotique, Edition HERMES Eds., 1990.

[14] B.W. Mooring, Z.S. Roth, M.R. Driels, "Fundamentals of manipulator calibration", John Wiley & Sons, 1991.

[15] A.N. Nahvi, J.M. Hollerbach, V. Hayward, "Calibration of a parallel robot using multiple kinematics closed loops", *Proc. IEEE International Conference on Robotics and Automation*, San Diego, Californie, USA, pp 407-412, 1994.

[16] N. Nombrail, "Analyse et commande d'une famille de robots manipulateurs à structure parallele et redondante", Thèse de Doctorat, Toulouse, France, 2 Décembre 1993

[17] S. P. Patarinski, "Parallel robots: a review", Technical Report. Dept. Mechatronics and Precision Eng., Tohoku university, Sendai, Japon, pp. 1-30, 1993.

[18] F. Pierrot, "Robots pleinement parallèles légers: conception, modélisation et commande", Thèse de Doctorat, Montpellier, France, 24 Avril 1991.

[19] Z. S. Roth, "An overview of robot calibration", *IEEE Journal of Robotics and Automation*, Vol 3, N 5, pp 377-385, 1987.

[20] F. Sternheim, "Computation of the direct and inverse geometric models of the delta parallel robot", *Robotersysteme*, pp 199-203, 1987

[21] H. Zhuang and Z. Roth, "A method for kinematics calibration of Stewart Platforms", *Proc. ASME Annual winter meeting*, Atlanta, GA, pp 43-48, 1991

Learning robot behaviours and their coordination

Jean-Pierre Müller*
IIIA - University of Neuchâtel
rue Emile Argand 11, 2007 Neuchâtel - Switzerland
Email: Muller@info.unine.ch

INTRODUCTION

It is important to methodologically distinguish between the environment dynamics, the agent dynamics and their coupling. In order to make this distinction clearer, let us define the following vocabulary. We will call a *sensory-motor loop* the internal mechanism linking perception to command whether fixed or adaptive and however sophisticated the percepts and commands can be for the sake of generality. We will call *behaviour* the externally observed behaviour (in the intuitive sense) produced by the execution of such a sensory-motor loop in a particular context. For example, the sensory-motor loop computing some command in answer to some light intensity variations can produce the behaviour of going toward a light source (or going away).

The *design* of a suitable behaviour amounts to designing a sensory-motor loop which produces the desired result by interaction with the environment. In order to do that, a detailed analysis of the environment dynamics, control loop dynamics and their interactions has to be made. Most of the time in behaviour-based robotics, this design is made intuitively, but see [11] for a formal account.

The most common approach to the *learning* of a suitable behaviour consists in defining a behaviour as optimising a function to carefully specify and letting the sensory-motor loop learn the best command to choose in each state regarding this function. The contribution (positive or negative) of each command to the function to optimise is given by an instantaneous reinforcement value. The goal of the learning process is to take at each step the command which maximises the discounted sum of reinforcement values over time. This kind of learning, called *reinforcement learning* is related to adaptive optimal control [1] and will be explored in this paper.

We will first present a general framework for modelling both the agent, the environment and their related couplings including through reinforcement.

Secondly, we will apply this framework to a possible design of learning sensory-motor loops (actually a set of these).

Finally, complex tasks requires the global sensory-motor loop to be decomposed into sensory-motor sub-loops. If these loops control the same motors, they have to be selected one at a time or weighted. We present the case of selection and show that it is a variant of the above mentioned learning case allowing for a hierarchical structure of the learning process.

*This work is financed by the Swiss National Foundation grant no 21-40936.94

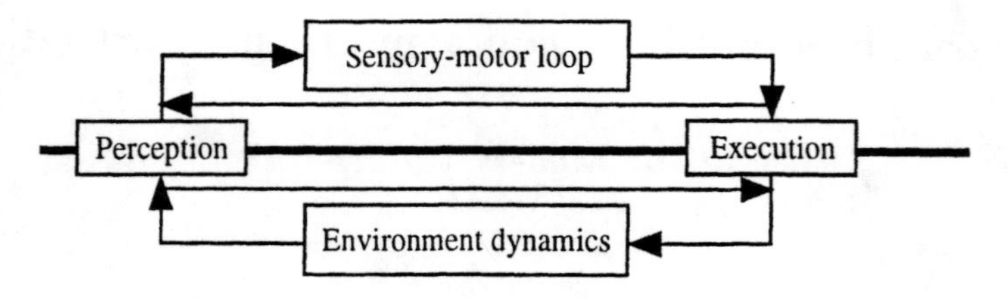

Figure 1: The global model

THE MODEL

The model that we propose has to take into account the distinction that we want to make between the environment and the agent. We have therefore to define separately the model of the environment in which the agent is seen as a black box acting on the environment and the model of the agent encapsulating the decision process on the basis of its perceptions.

This distinction is illustrated by the figure 1. The former puts equally the distinction between the state of the environment and the perception that the agent has as well as between the decision of the agent and the effect on the environment. This figure suggests equally that there is a coupling between two dynamics and not a cycle passing through the agent and the environment. Therefore it is necessary also to describe this coupling if one wants to be complete. We will use the notation coming from control theory [2].

The environment

To model the environment, we need to take into account its *state* comprising the agent itself (the agent being situated and then part of the environment) and its *dynamics* under the influence of the agent action u. The dynamics is a function from state to state under the effect of the action: $x_{t+1} = F(x_t, u_t)$. We will assume a stationary dynamics, i.e. F does not change over time and a discrete formulation.

The agent

The agent is characterised by its perception, its commands and the decision process we will describe successively.

One must clearly distinguish between the *environment state* (x) and the agent *perception* (written $p \in P$ in the following). This perception is not only local but moreover a simple set of measures. For example, the agent may have an obstacle in front of him (from a designer point of view and defined as a qualification of the neighbourhood relationship) while an agent with a distance sensor only gets a measure known by the designer (but a priori not by the robot) as correlated with the distance to the obstacle.

In the same way, one must distinguish between the *command* executed by the agent ($c \in C$) and the *action* realised in the environment ($u \in U$). For example,

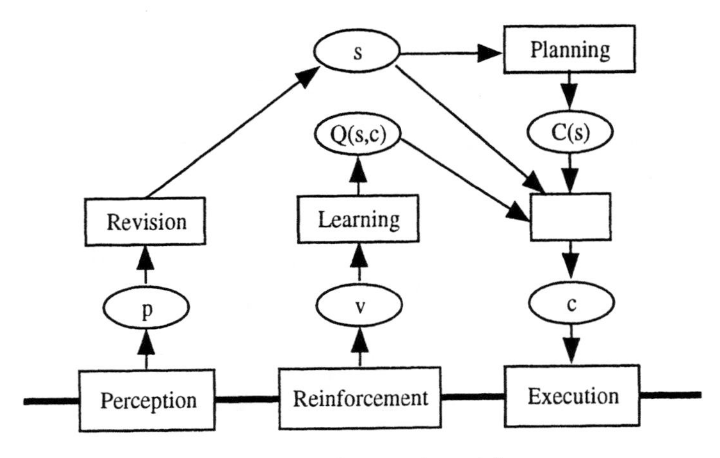

Figure 2: The agent's model

an agent may decide of a certain motor speed or number of wheel turns (the command) realising a movement in the environment (the action) which is not necessarily correlated and not perceived as such by the robot.

With perception only, one assumes that the instantaneous perception is sufficient to decide meaningfully of the best next command to execute or to predict the next perception: the decision process, respectively the internal dynamics, is said to be markovian. This is seldom the case, the reason why we introduce the notion of an internal state $s \in S$ which is assumed markovian and a revision process: $P \times S \rightarrow S$. The revision process can range from the identity function (when the perception is directly markovian) up to the most sophisticated knowledge revision process.

For each internal state s corresponds the set $C(s)$ of possible commands. The process which builds $C(s)$ will be called *planning* but may be a simple table lookup up to an arbitrary action planner.

In control theory, decision process is modelled by a function $\pi : S \rightarrow C$. Moreover, the function π is based on an evaluation function $Q(s,c)$. Finally Q can be given or computed through various decision theories including learning. Q can be learned *directly* or *indirectly*. It is learned directly when it is updated at each step t through the received reinforcement r_t, the new internal state s_{t+1}, for example by Q-learning ($Q_{t+1}(s_t, c_t) = Q_t(s_t, c_t) - \nu(Q_t(s_t, c_t) - (r_t + \gamma \max_{c'} Q_t(s_{t+1}, c')))$ where ν is the learning rate and γ the discounted factor, see [1, 5]). It is learned indirectly if the internal dynamics is learned, i.e. the probability to be in s_{t+1} after having performed c_t in s_t. In this latter case, any computation of Q, for example through dynamic programming, can be performed once the model is acquired.

The resulting agent model is illustrated by the figure 2 which is sufficiently general to capture reactive agents up to cognitive ones.

The couplings

In the different figures, we can see three boxes which have to be further elicitated when designing the sensory-motor loop: perception, execution and reinforcement.

The perception is a function of the interaction of the agent with the environment modelled as a function $Perception : X \times C \rightarrow P$.

The distinction we made between the command ($c \in C$) and the action in the environment (u$\in U$) becomes evident when we observe that u has to be a function of both c and the current state of the environment. We have $Execution : X \times C \rightarrow U$.

However, these couplings can remain black boxes if we turn to learning and only the reinforcement one has to be elicitated. This amounts to:

1. defining the optimality measure of the desired behaviour;
2. formulating how each action contributes (positively or negatively) to this optimality measure;
3. finding out how this contribution can be internalised in the agent dynamics or, alternatively, finding out such a formulation as a function in terms of its commands and internal states.

LEARNING BEHAVIOURS

In order to illustrate the design of learning sensory-motor loops, we will consider the behaviours of following corridors in some absolute orientations.

To follow a corridor, we can consider maintaining the robot in the middle of the free space. Let us define $d_{lw}(t)$ and $d_{rw}(t)$ the distances to the left wall, respectively to the right wall, at time t. The more the robot is in the middle, the smaller is the difference between these two distances. In order to avoid oscillations, we will consider the absolute value. Finally, following a corridor indefinitely is any sequence of actions (designer's point of view) minimising $\sum_t |d_{lw}(t) - d_{rw}(t)|$. By the same reasoning, follow an absolute direction dir minimises $\sum_t |or(t) - dir|$ where $or(t)$ is the orientation of the robot at time t. The resulting behaviour must minimise the weighted sum. The ratio of the weights rates the importance of keeping the heading against keeping the middle of the corridor. The instantaneous reinforcement from the agent point of view is straight forwardly computable from left and right sonar readings, respectively from odometry or a digital compass (to avoid drift).

The set of commands consists in orientation changes, the translation speed remaining constant (or controlled by another sensory-motor loop). In the set of perceptions, we decided to essentially get the sonar readings at front left and front right to have the commands chosen on an anticipation of the left and right wall positions.

For the following, we still need an applicability condition which is satisfied when we actually get readings from front left and front right and there is nothing directly in front.

LEARNING COORDINATION

In this section we will apply the above described model to learning coordination. It is a reformulation of what has been described in [8] in terms of indirect learning of Q.

Perception and commands

First of all, we need a formalism to talk about the sensory-motor capabilities. In our case the perception is based on sensory-motor loop applicability. Conversely the commands are the selections of one of the sensory-motor loop.

Given $L = \{l_1, \ldots, l_m\}$ the set of *sensory-motor loops* and the operator $\mathcal{P}(X)$ defined as the power set of a set X, we introduce $P = \mathcal{P}(L)$ the *perception space*(the set of all the possible configurations of applicable sensory-motor loops), $p \in P$ a particular *perception*[1], $C = L$ the *command space*, $C(p) = p$ the set of the possible commands when p is perceived which is nothing but the set of applicable sensory-motor loops.

The sensory-motor loops allowing to go to North, South, East and West respectively (applicable when the route is free) can be represented by:

- $C = L = \{n, s, e, w\}$
- $P = \mathcal{P}(L) = \{\{\}, \{n\}, \{s\}, \ldots, \{n, s, e, w\}\}$

The resulting loops and states are represented in the figure 3.

The internal state space

Internally, the effect of selecting a sensory-motor loop can only be a change of perception. Therefore, every cycle of the cognitive loop starts with the observation of a new perception state p ($p \in P(F)$ which determines $C(p)$, the set of stimulated loops in state p) and gives rise to the selection of a particular sensory-motor loop.

We define the revision process as the cognitive ability to recognise and structure the regularities of the sensory-motor interaction[2].

The perception alone is in general not markovian. Most of the time, the same sensory-motor loop enabled when being in the same sensory state at different times can produce different results. From an designer's point of view, it can be either because different actual situations correspond to the same sensory state (the sensory state is ambiguous) or because the sensory-motor loop can produce non-deterministic results. However, from the robot's point of view, these two possibilities cannot be distinguished and can only be interpreted as non-determinism. The notion of *context* is introduced to reduce or eliminate this non-determinism. Formally, we define a

[1]In order to avoid vegetating running phases, we consider that the agent has always a sensory-motor loop which is applicable, for example to produce a random movement.

[2]See work of Maja Mataric [7] or U. Nehmzow and T. Smithers [10] for other examples of robots learning sensory-motor sequences. Gary Drescher [3] is inspired by the schema mechanism proposed by Piaget.

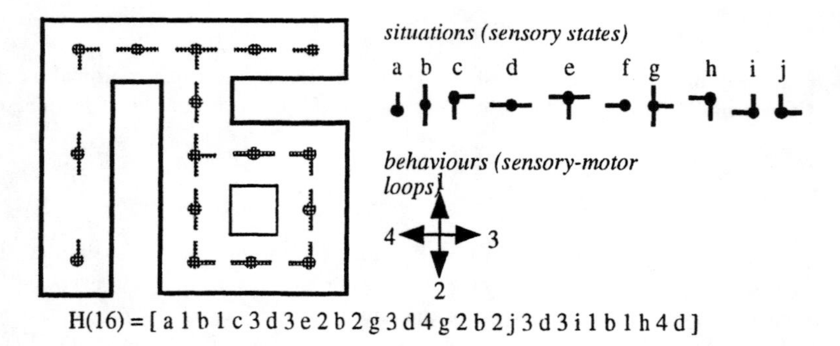

Figure 3: The maze, the behaviours and sensory states and one resulting history

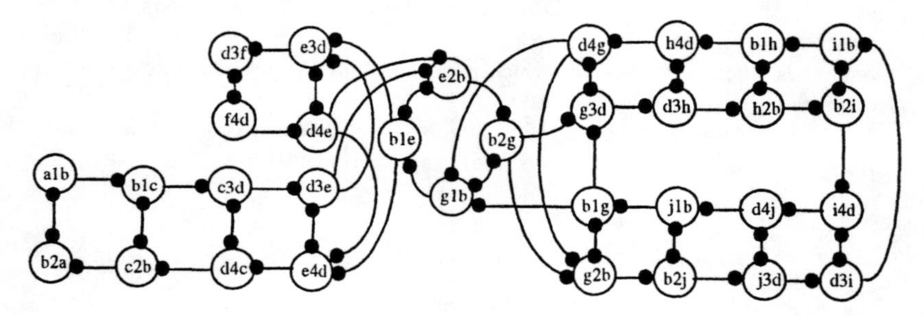

Figure 4: A global sensory-motor graph

context c_k as a sub-history of length $2k-1$:

$$c_k = [p_1\ c_1\ p_2\ c_2 \ldots c_{k-1}\ p_k]$$

where $k \geq 1, p_i \in P(F)$ and $c_i \in L$. A context of length 1 is reduced to a perception state. A sufficient value of k is assumed to favour deterministic anticipations.

The resulting regularities are represented in a *graph* $G(t) = (V(t), E(t))$. The vertices of this graph (elements of $V(t)$) correspond to particular *contexts* the agent can be in. The edges (elements of $E(t)$) correspond to sensory-motor loops which represent the *transition* from a source context to a destination context.

Figure 4 gives the graph made of unique contexts and behavioural transitions corresponding to the (expected) interaction of our agent in the environment of figure 3.

Therefore, the internal state is made of the graph and the current context. The revision process is in charge of building the graph, i.e. of learning the internal state dynamics for computing Q when necessary. We are using a simple algorithm with contexts of constant length. Our experiments in the frame described in figure 3 resulted in the construction of graphes which were very close to the expected graph (see figure 4).

The decision process

The decision process can be explored simultaneously or separately into three modes: the *localisation mode* when the current context does not correspond to any context of the graph, the *learning mode* when the objective is simply to acquire the graph as efficiently as possible, the *planning mode* when the agent must go to a certain place. In the following we will describe how Q should look like in each case. Note that, unlike the basic behaviours, Q is simple to express when the dynamics is known.

In localisation mode, the strategy depends on a trade off with learning mode. If no learning occurs, the problem is to find the shortest sequence of commands forming a known context. Given the confidence in the current perception, the first command of the most rapidly disambiguating sequence will be chosen. Note that this formulation is recurrent at each step allowing for an incremental and noise-recovery (when the next perception is still unexpected) process. In learning context, the problem is to decide between being lost or having discovered a new context.

In learning mode, the strategy consists in building the graph in the fewer possible steps ideally by passing by each context no more than there are different sensory-motor loops to try in this context.

Finally, in planning mode, a place to go is given by the user. The first problem is to translate a place in the external world to a context in the agent internal state. Therefore, the goal must be given in terms of contexts or subcontexts. Q reflects the shortest path in the graph and is computed by a spreading activation technique (see [6, 9] for other examples) inspired from the equations proposed by Huberman and Hogg in [4]. This system attributes to each vertex an activation level which is a function of the distance to and density of the goal contexts. This mechanism allows to always know the best transition to follow wherever the agent is in the graph. In particular, it is insensitive to the loss of localisation.

CONCLUSION

We have presented a general model of environment/agent interaction carefully distinguishing environment dynamics from the internal agent dynamics. This methodological distinction allows for designing in a clear and principled way behaviour-based systems. In particular, the proposed model expresses in a single formalism:

- the perception/environment state and command/action distinctions;
- the perception/internal state distinction to take into account revision processes and the hidden state problem[12];
- the decision process suitably expressed for both control theory and reinforcement learning.

This formalism is intended to root a behaviour-based system language.

This paper has focused on designing learning systems where learning is not a way out of design but a simpler way to do it, especially when the environment/agent interaction becomes too intricate to model. We have presented its applications to designing learning sensory-motor loops respectively to learning a sensory-motor loop selection system. The architecture being based on closed-loop control is equally suited for dynamic environments.

We have not tackled the problem of learning in hierarchical behaviour-based systems but we guess that our approach should help understand learning interaction between levels. It is the focus of our on-going work.

REFERENCES

1. A. Barto, S.Bradtke et S. Singh "Learning to act using real-time dynamic programming" University of Massachussets, Dept of CS, Tech report, 1994.
2. Dean et Wellman, *Planning and Control*, Morgan Kaufmann, 1991.
3. Gary L. Drescher *Made-Up Minds, A Constructivist Approach to Artificial Intelligence* The MIT Press (1991).
4. Bernardo A. Huberman, Tad Hogg "Phase Transition in Artificial Intelligence Systems" *Artificial Intelligence*, vol. 33, pp. 155-171, 1987.
5. L. Kaelbling, *Reinforcement learning in embedded systems*, MIT Press, 1995.
6. Pattie Maes "A Spreading Activation Network for Action Sélection" *Intelligent Autonomous Systems*, pp. 875-885, T. Kanade, F.C.A. Groen, L.O. Hertzberger (editors), Amsterdam, 1989.
7. M. J. Mataric "Environment learning using a distributed representation" *IEEE International Conference on Robotics and Automation*, 1990.
8. J.P.Müller et M.Rodriguez "Representation and Planning for Behavior-based Robots" *Environment Modelling and Motion Planning for Autonomous Robots* H.Bunke, T.Kanade (Eds), World scientific Pub., 1995.
9. David W. Payton "Internalised Plans: A Representation for Action Ressources" *Designing Autonomous Agents*, pp. 89-103, P. Maes (editor), MIT Press, 1990.
10. Ulrich Nehmzow, Tim Smithers "Mapbuilding using Self-Organising Networks in "Really Useful Robots"" *ECAL'91*, pp. 152–159, Francisco J. Varela, Paul Bourgine (editors), MIT Press, 1991.
11. Ulrich Nehmzow, Tim Smithers "On the Role of Dynamics and Representation in Adaptive Behaviour and Cognition" *Third International Workshop on Artificial Life and Artificial Intelligence*, San Sebastian, 1994.
12. S.D. Whitehead - L.-J. Lin "Reinforcement learning in non-markovian decision processes" *Artificial Intelligence* vol 73, pp 271-306, 1995

Fault Detection in the Delft Intelligent Assembly Cell

Bart R. Meijer MSc.
Delft University of Technology, CIM Centrum Delft *
Faculty of Mechanical Engineering and Marine Technology
Landbergstraat 3 NL 2628 CD Delft, The Netherlands
email: B.R.Meijer@WbMT.TUDelft.NL

ABSTRACT

As quality cannot be inspected into a product, fault-tolerance cannot be added to a system. Like a good quality product is the result of both the product design and the production process that made it, a fault tolerant system is the result of its design in combination with operational guidelines. This is particularly true for fault detection in control hardware. Timing problems in control hardware usually go by unnoticed, unless special detection circuits have been built in. Yet these timing problems can cause noticeable motion disturbances. This paper will discuss design principles and the error management approach taken in the Delft Intelligent Assembly Cell project. The design considerations of transputer based interface boards for a robot controller are presented here to illustrate the error detection and damage confinement problem.

KEYWORDS : fault tolerance, error management, robot control, flexible assembly

INTRODUCTION

There exists a strong parallel between quality and fault tolerance. As quality cannot be inspected into a product, fault-tolerance cannot be added to a system. Like a good quality product is the result of both the product design and the production process that made it, a fault tolerant system is the result of its design in combination with operational guidelines.

Most "construction" principles that are applied to achieve fault tolerance are not particularly new. The use of symmetry to achieve self-compensation is well known in mechanical construction as well as in electronics. The principle of redundancy is very common in computer technology and in information transmission. Feedback control can be considered as a design principle against a limited class of internal disturbances. These principles have been discussed extensively by other researchers in fault tolerance and error management [1,2].

A "well designed" fault tolerant system will be relatively insensitive towards internal and external disturbances. This means the system is capable of detecting the failure mode that is caused by the disturbance and the system knows a cure to escape from this mode.

---

* This research was done when the author was with the Production Engineering Laboratory

However these systems will not change their behaviour based on experience gained from failure modes. To change future behaviour, in order to further decrease the sensitivity to internal and external disturbances, is the goal of error management.

In a generic error management approach eight different phases can be distinguished that have to be incorporated into the design and the operations of a system. These eight phases are: exception detection, damage confinement, damage assessment, recovery planning, fault/error diagnosis, fault/error treatment, fault documentation and maintenance [3]. The first four phases give us a fault tolerant system. The last four are needed to improve future response to disturbances.

The layout of this paper is as follows. First the Delft Intelligent Assembly Cell and our approach to fault tolerance and error management are presented. The remainder of the paper will discuss the impact of fault detection and damage confinement on the design of control hardware. This discussion is illustrated with an example from the DIAC project, the interface boards for a transputer based robot controller.

DELFT INTELLIGENT ASSEMBLY CELL

The Delft Intelligent Assembly Cell (DIAC) project is a government (SPIN_FLAIR) sponsored multi-disciplinary research project at Delft University of Technology aimed at the development, implementation and integration of the technology that makes up a flexible assembly cell [4,5].

Some characteristics of the design of DIAC are summed below:

- DIAC is built around two different robots with overlapping workspace. One robot is a 4 DOF Scara type robot (Bosch SR800 Turboscara) that is especially good for assembly actions in vertical directions. The second robot is a 6 DOF Anthropomorphic type of robot (ABB IRB2000). This robot can cover any assembly direction and offers a large workspace.
- The cell can exchange product specific tools, such as clamping units and product supports automatically. Both part supply trays and product specific tools are brought to and from the cell using an AGV. In order to prevent unnecessary transfers, part and tool carriers inside the cell en outside the cell are integrated. Both make use of the same transport frame, a bottomless euroframe of size (400*300*94.5 mm3).
- Within the cell transport frames are handled by a TTT handling system that serves both robots, a 24 position random access buffer store and two AGV docking stations. This solution was taken because it uses less space and because it can offer uncoupled serving of both robots.
- Parts are supplied in a semi-ordered fashion. Parts are supplied on universal trays that have been customised through the use of small number of pegs. These pegs ensure that the parts don't touch or overlap with other parts on this tray after transportation. At the feed position 2D and 2.5D model based vision is used to verify the part identity and to measure its 3D position and orientation.
- Robot endeffectors are smart tools that can be exchanged automatically. As part of the DIAC project a smart gripper as well as a 'handheld' laser triangulation camera have been developed. Because the number of electrical connections supported by

the tool exchange system is limited, transputer link technology is adopted as the data interface between the tools and the robot. Both the gripper and the camera have a transputer inside that is not only used for data-exchange, but also for control. The control software can be downloaded over the 20 Mbit serial transputer link.

The number of different sensors and actuators present in DIAC offer the necessary redundancy for experimental work on fault tolerance and error management. Moreover the number of sensors and actuators and the level of autonomy in these systems, make the inclusion of error management in the design a necessity. It is not possible to pre-program responses for all possible system states during the design.

FAULT TOLERANCE AND ERROR MANAGEMENT

For a general error management system a total of eight basic functions may be distinguished.

- *exception monitoring/detection.*
- *damage confinement*, errors will be prevented to cause more damage through error propagation.
- *damage assessment.*
- *recovery planning.*
- *fault/error diagnosis*, to establish the cause of an exception.
- *fault/error treatment*, to prevent an exception from happening again.
- *fault/error documentation.*
- *maintenance.*

In error recovery planning there are two basic approaches: repair (forward error recovery) or abandoning some or all of the results of recent activity in the system (backward error recovery). Fault/error documentation and maintenance are two meta functions. These functions are called meta functions because they are operating above the level of the detected error. The fault/error documentation function serves to find fault/error patterns in order to improve the system performance through either prevention or faster diagnosis of these kinds of faults. Maintenance serves to keep up system reliability and fault tolerance for continued service.

Error management and detection and damage confinement are decentralised reflex like functions. An example will illustrate this. A robot that is trying to move outside its workspace boundary is switched off automatically. A robot link passes a limit switch (detection) and as a result the power is switched off automatically (damage confinement). Finally some message is sent to a higher level of control, indicating that the robot has stopped because of a workspace boundary problem. This message is not to be misunderstood as diagnosis; this message only serves as damage assessment from which recovery may or may not be possible. Diagnosis means finding the fault in the robot program or robot model that caused the robot to move outside its workspace boundary.

Modelling exception detection and damage confinement as a reflex like decentralised function does not only apply for actuator systems. It applies to all systems (virtual as well as physical) present in a production system. The reason for this is that exception detection involves specialised knowledge that is (only) available with the people that design or build these systems. This knowledge may vary from sensor values that may not be exceeded to statistical models used for detection of system degradation. In the case of a complex system composed of sub-systems, error detection and damage confinement can only be done if sufficient knowledge is available about the interaction of sub-systems.

Based on damage assessment (defining the system state after an exception has occurred) recovery planning can be done. Both forward and backward recovery planning rely heavily on the assumption that a goal is still available. This stresses the importance of a task oriented process plan from which achievable goals can be derived. The choice between forward and backward recovery strategies depends on the outcome of damage assessment and the costs associated with both strategies to get back in business.

Although diagnosis may not be possible it is still possible to treat the system to prevent the occurrence of these exceptions in future occasions. A mechanism that can be used for this purpose influences the cost-estimates of the strategies that were chosen to carry out a certain operation. Unreliable strategies can be ruled out in future plans this way. The same accounts for fault/error documentation and maintenance. The occurrence of error patterns or a change of frequency of certain faults can be an indication that maintenance becomes necessary.

CONTROL PROCESSES

As a reference for our control processes the state model of figure I was used .

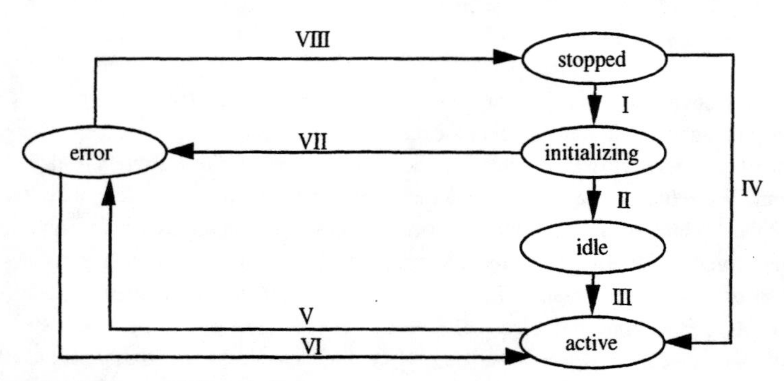

Figure I State model for control processes.

The bubbles in figure I represent the states. The states themselves are handled by processes that may show the same internal structure with the same internal states. The arrows represent state transitions that can only take place upon either external events or internal events from within the states. After a cold start a process is installed and remains in the state stopped. Upon an initialisation request (I) the initialisation process starts. From within this process two events can be generated. The initialisation may fail (VII). This is a fatal error from which automatic recovery is not possible. The initialisation process may succeed (II) and the state changes from initialising to idle. The idle state is introduced because the active state cannot be enabled before all initialisation processes in the system are finished. Therefore active-enable (III) is an external event. In the active state the normal tasks of this control process are carried out. There are two events that end the active-state. The process may be stopped (IV) or an error may occur (V) upon which a transition to the error state takes place. If a local recovery is possible (VI) the active state is resumed. If recovery is not possible the process is stopped (VIII). The error process itself takes care of signalling other error processes that the process has been stopped for some reason. The primary task of this error process is damage confinement and damage assessment.

DESIGN CONSIDERATIONS FOR INTERFACE HARDWARE FOR ROBOT CONTROL

The state model for a control process shows how error detection and damage assessment and damage confinement are integrated into the design of a reference model for building control processes. However, the reference state model only addresses the problem of a synchronised start of all processes. Keeping these processes synchronised once they are in active state cannot be solved this way. Keeping these processes synchronised is vital for distributed actuator controllers performing synchronised motions. In the DIAC project Twente LINX transputer technology, developed at the University of Twente, has been adopted for this purpose[6, 7]. One of the features of Twente LINX technology is a hardware timing signal, available to all control boards in the system, that can be used to trigger all the DA-converters for a synchronised release of new set points. The same signal is used to trigger circuits that take feedback values. Another feature of Twente LINX is a global error signal. If one of the control boards fails, the error signal can be set to stop all other controllers in the system.

Interface hardware for control systems should also be capable of supporting various types of controllers. The calculation time for a new set point for a PID or an impedance controller is constant. The implementation of the algorithm has to be fast enough to ensure the set point is always ready before the next control cycle is to start. This is not the case for controllers where the number of calculations depends on the system state. In order to notice these timing irregularities it is necessary to detect underrun and overrun conditions. Underrun means, the calculation of the set point is to slow to keep up with the timer (hardware or software) that controls the cycle time. Overrun means the software is fast enough but attempts to release two set point values within one cycle. If overrun occurs there is usually a software bug in the controller. The damage confinement after an underrun situation depends on the controller. With a PID controller, underrun for one cycle is usually acceptable and active state can be resumed (recoverable). Yet it is wise to monitor the frequency of underrun occurrence and compare this with the cycle frequency of the controller.

In order to be safe, interface hardware for control systems is also equipped with circuitry to detect feedback problems such as loose sensor cables and quadrature failures for optical encoders. Furthermore there has to be a hardware watchdog that switches the outputs to ground if the controller crashes and the last set point remains on the output terminals. Grounding the outputs will cause a velocity controlled servo to stop with maximum effort. However it is not the preferred solution for torque controlled systems.

CONCLUSIONS AND FUTURE WORK

In this paper the error management approach of the DIAC project has been presented. To illustrate that fault tolerance and error management have a profound impact on the design of control systems a reference model for building control processes has been presented and the design considerations for hardware interfaces for a robot controller have been discussed.

Most sensor and actuator systems in DIAC have now become reliable tools. This opens a perspective for experiments with error management on the assembly planning level. On this level the problems of fault/error diagnosis and treatment are the most challenging.

ACKNOWLEDGEMENTS

The Delft Intelligent Assembly Cell project is a multi-disciplinary research project at Delft University of Technology. In this project 8 research groups from the department of Mechanical Engineering, the department of Applied Physics, the department of Electrical Engineering and the department of Mathematics and Computer Science co-operated. The cell has been built in the CIM Centre Delft. Three national research institutes. the Institute of Applied Physics (TPD-TNO), the Institute for research on Plastics and Rubber (KRI-TNO) and the Metal Research Institute (MI-TNO) are involved as sponsor and/or subcontractor. The author wishes to express his thanks to all colleagues in the above mentioned institutes for their discussions and valuable contributions to this project. Finally he DIAC project would not have been possible without a government grant from the national SPIN-FLAIR program.

LITERATURE

1. Adlemo A., "On dependability in distributed computerised manufacturing systems", PhD. thesis, Chalmers University, Gothenburg, Sweden, 1993, ISBN 91-7032-789-0
2. Stigter J.O., "Error management or how a robot can bear murphy's law", PhD. thesis, Delft University of Technology, 1994, ISBN 90-407-1032-5
3. Meijer B.R., Stigter J.O., "Integration of control and error management for a flexible assembly cell using a cost function",Preprints of 1992 IFAC/IFIP/IMACS Int.Symp. on Artificial Intelligence and Automatic Control, Delft, The Netherlands, 1992, pp.251-256.
4. Meijer B.R., Jonker P.P., "The architecture and philosophy of the DIAC (Delft Intelligent assembly Cell)", Proceedings of the 1991 IEEE International conference on Robotics and Automation, Sacramento California, April 1991, Vol.3 pp. 2218-2223.
5. Meijer B.R., "The Delft Intelligent assembly cell; technology that enables flexible assembly",Proceedings of the 10th ISPE/IFAC International Conference on CAD/CAM, Robotics and Factories of the Future, Ottawa, Canada, 1994, ISBN 1-895634-06-7, pp.960-965
6. Klomp C., "Sensor based fine motion control", PhD. thesis, Delft University of Technology, 1994, ISBN 90-9007162-8
7. Schwirtz M.H., Wijbrans K.C.J., Bakkers A.W.P., Hoogzaat E.P., Bruis R., "The Twente LINX backplane", Proceedings of the NATUG-5 conference, Baltimore VA, USA, April 6-7, 1992

Enhancing CIMOSA with Exception Handling

S. Messina, P. Pleinevaux

Swiss Federal Institute of Technology, Lausanne, Computer Engineering Department, EPFL-DI-LIT,CH-1015 Lausanne, Switzerland

e-mail: messina@di.epfl.ch, fax +41.21.693.47.01

ABSTRACT:

CIMOSA (Open System Architecture for CIM) [2], an architecture for the modelling of manufacturing applications, does not provide a facility for exception definition and handling. Exceptions, traditionally associated to programming language and operating systems, are necessary in all types of languages, including specification languages. Our contribution consists of the enhancement of the CIMOSA model with a complete facility and methodology for the specification of the system behaviour in case of exception.

KEYWORDS: enterprise modelling, exception, exception handling, requirement specification, integrity constraint.

INTRODUCTION

CIMOSA (Open System Architecture for CIM) [2] defines a framework for the definition, development and continuous maintenance of a consistent model of manufacturing applications. We are interested in studying the Requirement Definition Model of the CIMOSA methodology. During this model definition the end-user expresses his view of the business needs and describes entirely the important aspects of the manufacturing system. Among these needs, the definition of the exceptions that may occur in the enterprise represents a fundamental part of the requirement specification. The current version of CIMOSA does not provide exception handling.

Our goal is to enhance the CIMOSA Requirement Definition Model with a facility for the definition and handling of exceptions.

We first present a brief overview of the main concepts of exception handling. Then we give a justification for the introduction of the exceptions into the specification phase of CIMOSA. The second part of the paper starts with a brief presentation of the CIMOSA modelling concepts. A possible way to introduce exceptions in CIMOSA is then proposed and discussed. This solution is based on Borgida's idea [3] of defining integrity constraints on the enterprise objects and raising an exception whenever these constraints are violated. Finally, a real example of a manufacturing application is proposed in order to show how exception handling is integrated in the CIMOSA model.

BASIC CONCEPTS OF EXCEPTION HANDLING

The terms exception and exception handling have been defined by many authors in several different ways. We give here the definitions proposed by Goodenough and Wirth in [5] and [6].

Exceptions are unusual conditions detected while attempting to perform some operation and brought to the attention of the operation's invoker.

This definition is very general and applies to all phases of system development from user requirements specification to implementation. It is also important to note that exceptions are not necessarily error conditions as generally considered.

Bringing the exception condition to the invoker's attention is called **raising an exception**.

The special processing required to react to the exception, or the invoker's response, is the **handling** of the exception.

Exception handling mechanisms have been proposed in order to cope with system failures [4].

'*Exceptions and exception handling mechanisms are not needed just to deal with errors. They are needed as means of interleaving actions belonging to different levels of abstraction*' [5]. The user defines the exception conditions and the exception handling actions, while at a lower level of abstraction, the exception mechanism takes care of detecting and raising the exceptions.

The classical exception model is based on the definition of pre- and post-conditions defined on each operation, that must always be satisfied before and after each operation execution. Borgida [3] is the first to integrate the notions of exceptions and integrity constraints: he defines exceptions as constraints violations. His exception model is based on an object representation of the system entities. The objects are not only described in terms of operations and attributes, but also in terms of constraints that must be checked. These constraints may be defined on single object, or they may involve several objects. The constraints must be checked each time the objects are modified.

JUSTIFICATION FOR EXCEPTIONS IN SPECIFICATIONS

Exceptions are traditionally associated to programming languages, particularly since the introduction of Ada. An exception facility allows to develop robust programs, that are resistant to anticipated problems.

Exceptions can also be introduced in specifications. The reason is simple. In a system development project that must satisfy robustness requirements, it is essential that faults be anticipated in the specification and design phases so that the behaviour of the system can be analysed and predicted before its installation.

The specified behaviour of an enterprise functionality may be interrupted, modified or it may fail and be signalled through an exception. The user must have a tool for expressing the behaviour of the enterprise, even in case of exceptions. Especially in critical applications, where a failure or interruption of the activity is fatal, the user feels the need to describe how the system should react to an exception.

Defining exceptions at an early phase of the enterprise modelling allows to specify the system behaviour in all anticipated unusual conditions, to analyse it, to prevent possible inconsistency and to integrate it in the "usual" system behaviour. Instead of waiting for a later definition at implementation time (for instance, using a programming language that supports exception handling), if the model includes from the beginning an exception mechanism, the subsequent design and implementation phases are easier to realise, give a more coherent result and require less modifications.

Briefly, we can say that exceptions are necessary in all types of languages, whether they are programming, design or specification languages, to notify and react to system's exceptional conditions.

CIMOSA OVERVIEW

The CIMOSA Modelling Framework [2] provides the necessary guidance to enable end users to model the enterprise and its associated CIM system in a coherent way. The CIMOSA modelling approach is based on a Reference Architecture composed of reusable generic building blocks, which are aggregated to describe the enterprise model.

The CIMOSA model development is composed of three phases, where the first one is the Requirement Definition Modelling phase [1]. This model is described by the end-user that provides his view of the business needs. He gives his knowledge about the function, information and resources of the system. The next phases are the Design and the Implementation. The example given below concentrates on the Requirement Definition Level, and we develop our proposal referring to the Requirement Definition Model of the enterprise.

The first step in the Requirement Definition Model development consists in the definition of the Domain to be modelled, its objective and constraints.

The **Domain** describes a part of the enterprise relevant for achieving a defined set of business objectives. Example of domains of activity in a real scenario may be the Engineering Department or the Flexible Manufacturing System [8], [7]. Domains communicate with one another through events and describe the enterprise activities through Enterprise Objects and processes acting on them.

An **Enterprise Object** is a generic entity of the enterprise that can be described by many Object Views. One Enterprise Object may be viewed from different points of view, thus it may correspond to several Object Views.

The functionalities and the behaviour of a Domain are defined through **Domain Processes**. A Domain Process is a stand-alone process triggered by events and governing the execution of Enterprise Activities (the basic functionalities) according to the so called Procedural Rules.

Each Domain Process is decomposed into Business Processes and/or Enterprise Activities. That is the Domain Process is decomposed into hierarchically structured functions, that are elementary functions.

A set of **Procedural Rules** define the sequence of activation of Business Processes and/or Enterprise Activities. Let us consider the following example that will be used later to introduce the exception handling facility. Within a real manufacturing scenario, we take into consideration a Flexible Manufacturing System (FMS) Domain. Inside this domain, we consider the Transport Domain Process, which deals with transportation of palletised tools and raw material. This Domain Process is composed of the following Business Process and Enterprise Activities:

```
Domain Process: Transport
  Business Process: Transport Pallet from A to B
    Enterprise Activity: Load Pallet from A
    Enterprise Activity: Transport Pallet from A to B
    Enterprise Activity: Unload Pallet into B
```

The Business Process 'Transport Pallet from A to B' transports the pallet from the starting point A to the ending point B (A and B may be a warehouse cell or one of the system components) by means of a cart. The component Enterprise Activities decompose the transport from A to B into more elementary steps: load the pallet on the cart, transport it and then unload it from the cart.

An Enterprise Activity and a Business Process are detailed by describing their functionality, composed of several components, some of which are:

- the Function Input (FI): set of Object Views to be processed and transformed;

- the Function Output(FO): set of Object Views produced or returned by the activity or process;
- the Ending Status (ES): non-empty set of the possible termination states of the activity or process;
- the Required Capability: is the possibly empty set of resources needed for the execution of the activity or process;
- Where Used: is the set of next upper level structures (Domain or Business Processes) using this activity or process.

The Ending Status of an Enterprise Activity is a value describing one of the possible termination states of that Enterprise Activity. The Ending Status of a Business Process is the Ending Status of the last employed Enterprise Activity. The Ending Status equal to DONE indicates that the Enterprise Activity has been successfully completed. The Ending Status is needed for a Procedural Rule to determine which Enterprise Activity should be further triggered to continue the execution of the Business Process or Domain Process.

Here is an example of Enterprise Activity (EA) functionality description:

```
Enterprise Activity: Load Pallet from A

  FI: A, Pallet
  FO: A, Pallet
  ES: DONE
```

The Business Process (BP) functionality description is as follows:

```
Business Process: Transport Pallet from A to B

  FI: A, B, Pallet, Cart
  FO: A, B, Pallet
  ES: DONE

  Procedural Rules:
    WHEN Start DO EA Load Pallet from A
    WHEN ES(EA Load Pallet from A)=DONE DO EA Transport Pallet from A to
    B
    WHEN ES(EA Transport Pallet form A to B)=DONE DO EA Unload Pallet
    into B
    WHEN ES(EA Unload Pallet into B)=DONE DO FINISH
```

OUR PROPOSAL

The approach that we consider to detect and then to treat exceptions in CIMOSA is to signal them through the use of the Ending Status. Our solution suggests to define a particular Ending Status, the 'Exception' status, for any Business Process and Enterprise Activity of a Domain Process. This status is returned whenever an exception occurs during the current Business Process or Enterprise Activity. In such a case, the current Business Process or Enterprise Activity is suspended and the Business Process or Enterprise Activity that handles the exception is executed. This approach allows to define specific exception handling actions for each particular exception condition.

Another alternative that we have studied proposes to use events to signal the occurrence of an exception and to trigger a Domain Process 'Exception Handler'. Because of space limitation, we cannot present here this second proposal and the motivations of our preference for the first one[1]. We just affirm that our choice was motivated by a higher flexibility and the better propagation provided by the Ending Status proposal.

We extend our model by integrating not only the classical exception model based on the satisfaction of pre-conditions or post-conditions before or after the execution of an instruction, but also Borgida's model [3], which is focused on the integrity constraint violations on modified objects.

The pre- and post-conditions can express several constraints:

- object integrity constraints, defined on the attributes of the objects accessed during the Enterprise Activity;
- availability of physical or logical resources needed for the execution of the Enterprise Activities or Business Processes;
- synchronisation constraints with another Business Process or Enterprise Activity defined in another Domain Process, through an event.

For instance, two physical resources, an AGV (Automatic Guided Vehicle) and a machining centre, must synchronise themselves in order to perform the 'unload pallet' Enterprise Activity from the AGV to the pallet store of the machine. The pre-condition of the Enterprise Activity is defined to check that the AGV is present at the machine location and that the pallet store has free space for the pallet.

In CIMOSA we define exceptions as class occurrences, in other words as objects. The exception is an object, just like any other enterprise object. It is created when an exception is raised and it belongs to a class. It has attributes such as the Object Views involved in the exception, their attributes, plus other attributes proper to the exception object, i.e., context of occurrence, seriousness, and so on (see Figure 1). The exception object provides information about the occurrence to the handler. It may also be recorded in a log for later analysis. Grouping exceptions into classes has the advantage of treating uniformly exceptions of the same type.

We describe CIMOSA processes, from Domain Process to Enterprise Activity, according to Borgida's model: their definition includes the objects on which the transaction acts, the pre-conditions and post-conditions that must be satisfied before and after the transaction and the actions to be performed. When an Enterprise Activity is executed, it manipulates and accesses a set of objects (or more precisely, Object Views). In order to ensure the satisfaction of integrity rules on the attributes of these objects, the pre-condition is tested before the execution. If the rules are satisfied, the action is executed. After this, the post-condition is tested to check if the action has not affected the integrity of the objects. If the constraints are not satisfied, an exception is raised and a handler is called to handle it. In our proposal, an exception handler is an enterprise function, just like any Enterprise Activity. It is pre-defined or it is defined by the user. It acts upon an exception object, of which it accesses the attributes, and on the Object

1. The reader interested in more details may contact the authors to get an extended version of the paper.

Views, of which it may modify the attributes. The exception handler, like any EA, is composed of pre-condition, action and post-condition (see Figure 1).

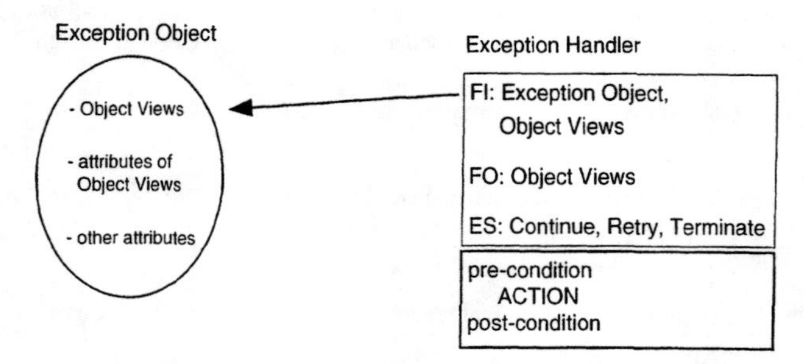

Figure 1 Exception object and handler definition

EXAMPLE

We refer here to the FMS Domain example introduced in the CIMOSA Overview Section. We introduce the exception handling in the Business Process 'Transport Pallet'. In order to do that, we define a possible exception that can occur during the Business Process and an Enterprise Activity devoted to the handling of the exception.

The Enterprise Activity 'Load Pallet from A', is enhanced with the definition of a pre-condition, that consists of the state of the cart equal to empty, the state of A full and the pallet present in point A, and a post-condition, that consists of the pallet loaded in the cart, the state of A empty and the Ending Status of the Enterprise Activity equal to DONE. An exception Exc1 is defined as occurring in case the pre- or post-conditions are not satisfied. For example, if the pallet is not located at point A the exception Exc1 is raised.

```
Enterprise Activity: Load Pallet from A

   FI: A, Pallet
   FO: A, Pallet, Excl
   ES: DONE, Excl

   Pre-condition: Pallet in A & cart empty & A full
        ACTION: load
   Post-condition: A empty & Pallet in cart & ES=DONE
   Exception: Excl
```

The Enterprise Activity 'Exception Handler' is executed when an exception of type Exc1 occurs. It has a pre-condition as well, which is the contrary condition of the Enterprise Activity 'Load Pallet from A': the point A is empty or the pallet is not in A or the cart is not empty. The post-condition and the exceptions that can occur during the handling are not defined for reasons of simplicity of the example.

```
Enterprise Activity: Exception Excl Handler

   FI: Excl, A, Pallet
   FO: Excl
   ES: Terminate
```

```
Pre-condition: Pallet not in A OR cart not empty OR A empty
     ACTION: write exception message: `exception 1 occurred'
Post-condition: /
Exception: /
```

This simple example shows an handling activity which is limited to displaying the message exception and terminating the interrupted action. A different approach could define several exception types, one for each violated condition. This would indicate which component of the pre-condition has not been satisfied and which corrective action must be taken (e.g., retry or modify the system state). For instance, we could have defined an exception ExcP in case the location of the pallet is wrong, and exception ExcC in case the state of the cart is incorrect and an Exception ExcA in case the state of the point A is empty. With this solution, we would then define an Exception Handler for each exception type.

We also define a pre-condition for the Business Process, that coincides with the pre-condition of the first contained Enterprise Activity and we define a post-condition, that corresponds to the post-condition of the last Enterprise Activity.

```
Business Process: Transport Pallet from A to B
   FI: A, B, Pallet, Cart
   FO: A, B, Pallet, Excl
   ES: DONE, Excl

   Pre-condition: Pallet in A & cart empty & A full
   ACTION:  WHEN Start     DO    EA Load Pallet from A
        WHEN ES(EA Load Pallet from A)=DONE  DO  EA Transport Pallet from A
to B
        WHEN ES(EA Load Pallet from A)=Excl  DO  EA Exception Excl Handler
        WHEN ES(EA Exception Excl Handler)= Terminate DO  FINISH
        WHEN ...
   Post-condition: Pallet in B & ES=DONE
   Exception: Excl
```

The Procedural Rules of the Business Process define its behaviour, even in case an exception of type Exc1 occurs. For the sake of simplicity we define only Exc1 as the possible exception occurring in the Business Process.

CONCLUSION

The goal of this work is to study the notion of exception and exception handling and their introduction into CIMOSA. Motivation of our work is the fact that in a system development project that must satisfy robustness requirements, it is essential that faults be anticipated in the specification phase so that the behaviour of the system can be analysed and predicted before its installation.

The notion of exception is defined in a vague and confusing way in CIMOSA. Furthermore, no exception handler definition and exception propagation mechanisms are proposed by CIMOSA. Our contribution is essentially the proposal of a complete exception handling mechanism that provides to the user a tool for defining the system behaviour in case of exceptions, like the unavailability of resources, the interruption of an activity or the inaccessibility of a process.

REFERENCES

1. ESPRIT Consortium AMICE, "Integrated manufacturing - a challenge for the 1990s", *Computing and Control Engineering Journal*, pp. 101-125, May 1991, Special Issue.

2. ESPRIT Consortium AMICE, "*CIMOSA: Open System Architecture for CIM*", Springer Heidelberg, 1994.

3. A. Borgida, "Language features for flexible handling of exceptions in information systems", *ACM Trans. on Database Systems*, 10(4):565-603, 1985.

4. F. Cristian, "Exception handling and software fault-tolerance", *IEEE Trans. on Computers*, 31(6):531-540, June 1982.

5. J. B. Goodenough, "Exception handling: Issues and a proposed notation", *Communications of the ACM*, 18(12):683-696, Dec. 1975.

6. Robert W. Sebesta, "*Concepts of Programming Languages*", The Benjamin/Cummings Publishing, 1993.

7. Siemens, "*CIM: a management perspective*", Siemens Aktiengesellschaft, 1990.

8. A. Storr, U. Rembold, B. O. Nnaji, "*Computer Integrated Manufacturing and Engineering*", Addison Wesley, 1993.

Two New Algorithms for Forward and Inverse Kinematics under Degenerate Conditions

Ezio MALIS **Lionel MORIN** **Sylvie BOUDET**

Direction des Etudes et Recherches d'Electricité De France
6, Quai Watier - 78401 Chatou - FRANCE.

ABSTRACT

We propose a fast algorithm for the computation of any order derivatives of robots' forward kinematics. We calculated second derivatives to show its application in real-time robotics. Then, an algorithm for solving the inverse kinematics problem at singularities under degenerate conditions is proposed. The algorithm was successfully tested on a 7 degree of freedom robot.

KEYWORDS: Kinematics, Derivatives, Degenerate Conditions, Serial Manipulators, Real-time Control.

INTRODUCTION

To solve the inverse kinematics problem we need to find the joint displacement that moves the end-effector of the robot to a specified position. The closed form solution can be obtained only for non-redundant robots with special geometry. For a general serial manipulator different iterative methods are used. The majority of the methods in the literature use the information provided by the Jacobian matrix: the Jacobian transpose method [1], the Jacobian pseudo-inverse method [2], the damped least-squares Jacobian inverse method [3], [4]. Real-time control of robots is done without difficulty with one of these methods to find joint displacements from Cartesian commands as long as the Jacobian is not singular.

Suppose the robot is at a singularity and the desired motion is in a degenerate direction. In this case the robot may get stuck. This condition does not mean that the displacement along the degenerate direction cannot be performed. If the movement is possible and all the points along this direction are in the workspace, there is at least one joint position for each of these points. Nielsen [5] outlined that this solution can be found by using the second derivatives of forward kinematics (i.e. Hessian matrices) the Jacobian matrix being only the first term in a Taylor expansion of the forward kinematics of a robot. If the first term is rank deficient, other terms (i.e. the higher order derivatives) give the information needed to accomplish the desired motion. In order to perform the Taylor expansion of forward kinematics, we propose in this paper an algorithm for the real-time computation of any order derivatives of forward kinematics of any serial robot (n joints, rotary or prismatic).

Egeland [6] has shown that the Cartesian acceleration term provides the information required to exit the singularity. Thus, he proposed to move the robot inside the joint null space. This is easy if the dimension of the null space is 1. But in general case for a n joints robot, its dimension is n-r (where r is the rank of the Jacobian matrix). The direction of the null space in which we have to move is difficult to know, and it is possible that a null space motion doesn't take a redundant robot out of the singularity. In order to solve the inverse kinematics problem at singularities, when the desired motion is in a degenerate direction, a new algorithm for a general serial robot is proposed in the paper.

This paper is organized as follows. In the first section we introduce a notation for orientation coordinates which facilitates the mathematics of the algorithms presented here. In the second section we illustrate the algorithm to calculate any order derivatives of forward kinematics. The final result is the real-time Hessian matrices computation. In the third section we illustrate the algorithm for inverse kinematics at singularities under degenerate conditions. The experimental results are given in the fourth section.

FORWARD KINEMATICS

To represent the kinematics of a robot, we will use the Denavit-Hartemberg modelisation modified by Dombre and Khalil [7]. The transformation matrix between the end-effector frame R_n and the reference frame R_0 is ${}^0\mathbf{T}_n$. This representation is redundant because the orientation matrix has 9 parameters, when only 3 are needed. A rotation of θ around a unitary vector $\mathbf{u}$ can be described with the following four parameters:

$$\mu_1 = \cos(\theta), \quad \mu_2 = u_x \sin(\theta), \quad \mu_3 = u_y \sin(\theta), \quad \mu_4 = u_z \sin(\theta) \tag{1}$$

If we calculate these parameters in function of the elements of the matrix ${}^0\mathbf{T}_n$ we obtain:

$$\mu_1 = \frac{1}{2}(s_x + n_y + a_z - 1),\ \mu_2 = \frac{1}{2}(n_z - a_y),\ \mu_3 = \frac{1}{2}(a_x - s_z),\ \mu_4 = \frac{1}{2}(s_y - n_x) \tag{2}$$

where $\mathbf{s} = \begin{bmatrix} s_x & s_y & s_z \end{bmatrix}^T$, $\mathbf{n} = \begin{bmatrix} n_x & n_y & n_z \end{bmatrix}^T$ and $\mathbf{a} = \begin{bmatrix} a_x & a_y & a_z \end{bmatrix}^T$ are the columns of the orientation matrix. Euler's parameters are not used in this algorithm because they are non-linear functions of the elements of the transformation matrix ${}^0\mathbf{T}_n$. The notation uses the advantages of the homogeneous matrix for transformation multiplication and those of linear vector expressions for deriving easily forward kinematics. Furthermore, if we make the hypothesis that $-\pi/2 \le \theta \le \pi/2$, the first parameter can be discarded and the representation of the rotation is minimal. The commanded motions on a robot in one sampling time are usually not large. If the rotation is larger than π, it can be split into two rotations. The vectorial form of the forward kinematics of a robot is:

$$\mathbf{x} = \mathbf{f}(\mathbf{q}) = \begin{bmatrix} p_x & p_y & p_z & \frac{1}{2}(n_z - a_y) & \frac{1}{2}(a_x - s_z) & \frac{1}{2}(s_y - n_x) \end{bmatrix}^T \tag{3}$$

this is a set of linear functions of the elements of the transformation matrix ${}^0\mathbf{T}_n$.

DIFFERENTIAL KINEMATICS

We take the p order Taylor development of the transformation matrix ${}^0\mathbf{T}_n$ calculated in $\mathbf{q}$. Let $\tilde{\mathbf{q}} = \mathbf{q} + \delta\mathbf{q}$ so that:

$$ {}^0\mathbf{T}_n(\tilde{\mathbf{q}}) = {}^0\mathbf{T}_n(\mathbf{q}) + \sum_{j=1}^{n} \frac{\partial\left[{}^0\mathbf{T}_n(\mathbf{q})\right]}{\partial q_j} \delta q_j + \frac{1}{2} \sum_{i=1}^{n} \sum_{j=1}^{n} \frac{\partial^2\left[{}^0\mathbf{T}_n(\mathbf{q})\right]}{\partial q_i \partial q_j} \delta q_i \delta q_j + \ldots + \mathbf{R}_p(\mathbf{q}) \tag{4} $$

To know the general derivative of the matrix ${}^0\mathbf{T}_n$ we use the following theorem:

Theorem: If the reference frame is the initial end-effector frame:

$$ {}^0\mathbf{T}_n(\tilde{\mathbf{q}})\Big|_{\tilde{\mathbf{q}}=\mathbf{q}} = {}^0\mathbf{T}_n(\mathbf{q}) = \mathbf{I} \tag{5} $$

the general derivative of the forward kinematics may be calculated as follows:

$$ \left.\frac{\partial^p\left[{}^0\mathbf{T}_n(\tilde{\mathbf{q}})\right]}{\partial \tilde{q}_1^{p_1} \partial \tilde{q}_2^{p_2} \cdots \partial \tilde{q}_n^{p_n}}\right|_{\tilde{\mathbf{q}}=\mathbf{q}} = \left(\left.\frac{\partial\left[{}^0\mathbf{T}_n(\tilde{\mathbf{q}})\right]}{\partial \tilde{q}_1}\right|_{\tilde{\mathbf{q}}=\mathbf{q}} \right)^{p_1} \left(\left.\frac{\partial\left[{}^0\mathbf{T}_n(\tilde{\mathbf{q}})\right]}{\partial \tilde{q}_2}\right|_{\tilde{\mathbf{q}}=\mathbf{q}} \right)^{p_2} \cdots \left(\left.\frac{\partial\left[{}^0\mathbf{T}_n(\tilde{\mathbf{q}})\right]}{\partial \tilde{q}_n}\right|_{\tilde{\mathbf{q}}=\mathbf{q}} \right)^{p_n} \tag{6} $$

where $p = p_1 + p_2 + \ldots + p_n$ and p_j is a positive or null integer.

This theorem show that all derivatives can be calculated only by calculating the first derivatives.

First derivatives: the Jacobian matrix

The Jacobian matrix is calculated by deriving the forward kinematics. We will note with j the j-th columns of the Jacobian matrix. The subscript "p" indicates the position and the subscript "o" the orientation. The forward kinematics defined by equation (3) is a set of linear functions of the elements of the transformation matrix ${}^0\mathbf{T}_n$. If the reference frame is the initial end-effector frame, the elements of the j-th column of the Jacobian matrix are obtained directly by using equation (6) for p=1. The first derivative of the matrix ${}^0\mathbf{T}_n$ can be written as follows:

$$ \left.\frac{\partial\left[{}^0\mathbf{T}_n(\tilde{\mathbf{q}})\right]}{\partial \tilde{q}_j}\right|_{\tilde{\mathbf{q}}=\mathbf{q}} = \begin{bmatrix} {}^j\hat{\mathbf{g}}_o & {}^j\mathbf{g}_p \\ \mathbf{0} & 0 \end{bmatrix} \tag{7} $$

where ${}^j\hat{\mathbf{g}}_o$ is the skew matrix associated to the vector ${}^j\mathbf{g}_o$. The terms of the Jacobian matrix in this vector are the derivatives with respect to q_j of rotation around axis x, y and z. Equation (7) is available for both prismatic and rotary joint. The difference between rotary and prismatic joints is made when the Jacobian matrix elements are calculated.

Second derivatives: the Hessian matrices

Let $h^k_{i,j}$ be the (i,j) element of the k-th Hessian matrix $\mathbf{H}_k$ ($\{k = 1, 2, \ldots, 6\}$):

$$h^k_{i,j} = \frac{\partial f_k(\mathbf{q})}{\partial q_i \partial q_j} \tag{8}$$

We calculate only elements with $1 \le i \le n$ and $i \le j \le n$ because the transformation matrix is a continuous function of **q** and so by using the Schwartz theorem, we have $h^k_{j,i} = h^k_{i,j}$. As for first derivatives, the (i,j) element of the Hessian matrix are obtained directly from equation (6) for p=2. By using equation (7), the second derivatives can be written in function of the elements of the Jacobian matrix:

$$\left.\frac{\partial^2\left[{}^0\mathbf{T}_n(\tilde{\mathbf{q}})\right]}{\partial \tilde{q}_i \partial \tilde{q}_j}\right|_{\tilde{\mathbf{q}}=\mathbf{q}} = \begin{bmatrix} {}^i\hat{\mathbf{g}}_o & {}^i\mathbf{g}_p \\ \mathbf{0} & 0 \end{bmatrix} \begin{bmatrix} {}^j\hat{\mathbf{g}}_o & {}^j\mathbf{g}_p \\ \mathbf{0} & 0 \end{bmatrix} \tag{9}$$

This equation is available for both prismatic and rotary joint. The difference between rotary and prismatic joints is made when the elements Jacobian matrix are calculated.

INVERSE KINEMATICS

The algorithm proposed in this paper minimizes the following function:

$$\min_{\delta\mathbf{q}} \quad \Phi = \|\delta\mathbf{q}\|^2 \tag{10}$$

subject to:

$$\min_{\delta\mathbf{q}} \quad \Psi = \|\delta\mathbf{x} - (\mathbf{f}(\mathbf{q} + \delta\mathbf{q}) - \mathbf{f}(\mathbf{q}))\|^2 \tag{11}$$

Our idea is to take a second order Taylor expansion of the norm of the Cartesian error:

$$\hat{\Psi} = \delta\mathbf{x}^T\delta\mathbf{x} - 2\delta\mathbf{x}^T\mathbf{J}(\mathbf{q})\delta\mathbf{q} + \frac{1}{2}\delta\mathbf{q}^T\left[2\mathbf{J}^T(\mathbf{q})\mathbf{J}(\mathbf{q}) - \sum_{i=1}^{m}\delta x_i \mathbf{H}_i(\mathbf{q})\right]\delta\mathbf{q} \tag{12}$$

We define C, **B** and **E** in the following way:

$$C = \delta\mathbf{x}^T\delta\mathbf{x}, \quad \mathbf{B} = -2\delta\mathbf{x}^T\mathbf{J}(\mathbf{q}), \quad \mathbf{E} = 2\mathbf{J}^T(\mathbf{q})\mathbf{J}(\mathbf{q}) - \sum_{i=1}^{m}\delta x_i \mathbf{H}_i(\mathbf{q}) \tag{13}$$

Equation (12) gives a better approximation of the Cartesian error than other methods, thanks to the term using the Hessian matrices. We test the vector **B** of equation (13). If $\mathbf{B} \neq \mathbf{0}$, we can use any of the existing algorithms. If $\mathbf{B} = \mathbf{0}$ and $\delta\mathbf{x} \neq \mathbf{0}$, $\delta\mathbf{x}$ is a degenerate vector, we cannot use those algorithms, and we have to solve the set of equations (10) and (11) written hereafter. The minimum value of Ψ is 0, thus our problem is equivalent to minimize equation (10) with the following constraint:

$$\hat{\Psi} = C + \frac{1}{2}\delta\mathbf{q}^T\mathbf{E}\delta\mathbf{q} = 0 \tag{14}$$

The Lagrange operator of equations (10) and (14) is:

$$L = \delta\mathbf{q}^T\delta\mathbf{q} + \lambda\left(C + \frac{1}{2}\delta\mathbf{q}^T\mathbf{E}\delta\mathbf{q}\right) \tag{15}$$

The necessary conditions for solving this set of equations are:

$$\frac{\partial L}{\partial \delta\mathbf{q}} = \delta\mathbf{q} + \lambda\mathbf{E}\delta\mathbf{q} = 0 \tag{16}$$

$$\frac{\partial L}{\partial \lambda} = C + \frac{1}{2}\delta\mathbf{q}^T\mathbf{E}\delta\mathbf{q} = 0 \tag{17}$$

If $\lambda = 0$, equation (16) implies $\delta\mathbf{q} = \mathbf{0}$ and equation (17) doesn't hold when $\delta\mathbf{x} \neq \mathbf{0}$. So λ is not null and μ is defined as $\mu = -1/\lambda$. Equation (16) can be expressed as follow: $\mathbf{E}\delta\mathbf{q} = \mu\delta\mathbf{q}$. Therefore $\delta\mathbf{q}$ is an eigen vector of $\mathbf{E}$ and μ is its eigen value. We choose a normalized vector $\mathbf{v}$ and a scalar α so that $\delta\mathbf{q} = \alpha\mathbf{v}$. By plugging this equation into equation (17), we get:

$$C + \frac{1}{2}\alpha^2\mu = 0 \tag{18}$$

C is positive, therefore we have to choose a negative eigen value of $\mathbf{E}$ to verify equation (18). The solution to equation (18) is:

$$\alpha = \pm\sqrt{-\frac{2C}{\mu}} \tag{19}$$

The bigger is the norm of the eigen value μ, the smaller is α. Since $\mathbf{v}$ is normalized the norm of α is the norm of $\delta\mathbf{q}$. We want the smallest norm of $\delta\mathbf{q}$ to satisfy equation (10), so we have to choose the lowest eigen value of $\mathbf{E}$. The sign of α determines the configuration of the robot when it exits the singularity.

EXPERIMENTAL RESULTS

The algorithm was tested on the PA-10 from Mitsubishi, a 7 degree of freedom robot. The Singular Value Decomposition (S.V.D.) algorithm was used to calculate the eigen values and vectors of $\mathbf{E}$. The pseudo-inverse algorithm was used to inverse the Jacobian matrix.The computation period is 20 milliseconds on a MVME 167 processor card (VME bus).

Suppose now the robot in a singularity. We want to move the end-effector of the robot along a degenerate direction. When we use the pseudo-inverse algorithm the robot doesn't move because we filter noise on joint measurements. If we don't filter the joint measurements, it is possible that robot moves but is not possible to control when movement occurs nor the configuration of the robot. On the other hand when we use also Hessian information the Cartesian error norm is small and the movement start immediately. It is therefore possible to move away from a singular configuration in a degenerate direction with little tracking error. The lower the initial velocity the lower the initial Cartesian error norm.

We have carried out several tests at every singularity with similar results: we can reduce the Cartesian error as we wish if we accept a lower initial velocity and we can choose the initial configuration of the robot. If there are no negative eigen values, the Taylor expansion needs to be taken to a higher order, and a new algorithm is needed. In the case of the seven degree of freedom Mitsubishi robot arm, our experimental results confirmed that there is at least one negative eigen value at each singularity.

CONCLUSIONS

The first algorithm described in this paper allows recursive computation of all derivatives of the robot kinematics from their first derivatives. In this way, the Hessian matrices are easily obtained from the Jacobian matrix. To calculate all Hessian matrices of an n joints serial robot, we need $6n^2$ multiplication and $3n^2$ additions. With this algorithm, it possible to calculate in real-time, the Hessians from closed form equations, with significant improvement of performance of robots around singularities.

The second algorithm described in this paper solves the problem of moving a general manipulator in a degenerate direction from a singularity of its workspace. The basic idea is based on taking a Taylor expansion of the norm of the function to minimize. To illustrate the algorithm, a second order expansion was performed in this paper. There may be robots for which higher orders are needed. Using the first algorithm given in the paper we can calculate the necessary matrices used for solving the problem at any order.

REFERENCES

1. Chiaverini S., Sciavicco L., Siciliano B.: "Control of Robotic Systems through Singularities", *Proc. of the Int. Workshop on Nonlinear and Adaptive Contr.: Issues in Robotics*, Grenoble, France (Nov. 1990), 285-294.
2. Withney D.E.: "The Mathematics of Coordinated Control of Prosthetic Arms and Manipulators", *Trans. ASME J. Dynamic Systems, Measurement and Control*, 94 (Dec. 1972), 303-309.
3. Nakamura Y., Hanafusa H.: "Inverse kinematics solutions with singularity robustness for robot manipulator control", *Trans. ASME, J. Dynamic Syst., Meas. and Contr.*, 108 (Sept. 1986), 163-171.
4. Wampler C.W.: "Manipulator Inverse Kinematics Solution Based on Damped Least-squares Solutions", *IEEE Trans. Systems, Man and Cybernetics*, 16(1) (Jan. 1986), 93-101.
5. Nielsen L., Canudas de Wit C., Hagander P.: "Controllability Issues of Robots near Singular Configurations", *Proc. of the Int. Workshop on Nonlinear and Adaptive Control: Issues in Robotics,* Grenoble, France, (Nov. 1990), 307-314.
6. Egeland O., Spangelo I.: "Manipulator Control in Singular Configurations - Motion in Degenerate Directions", *Proc. of the Int. Workshop on Nonlinear and Adaptive Control: Issues in Robotics,* Grenoble, France, (Nov. 1990), 296-306.
7. Dombre E., Khalil W.: "Modélisation et commande des robots", *Série Robotique* Hermès, (1986).

THE ASSEMBLY OF ROUND PARTS USING DISCRETE EVENT CONTROL

BRENAN McCARRAGHER

Department of Engineering
The Australian National University
E-mail: brenan@faceng.anu.edu.au

ABSTRACT

The assembly of round parts is one of the most common industrial assemblies. This paper presents a method for the modelling and assembly of round parts using discrete event control. The assembly task is abstracted according to the possible discrete states of contact. A force sensor is used to detect any changes in the discrete states of contact. The force sensor is also used to automatically determine the proposed direction of motion. The discrete event controller then issues position or velocity commands to direct the system to the next desired contact state. The method has been implemented in an industrial setting, successfully assembling the gear mechanism of a starter motor.

KEYWORDS: Robotic Assembly, Discrete Event Control

1 Introduction

The focus of this paper is a new approach to the task-level control of the robotic assembly of round parts. The industrially-based task is the assembly of a gear mechanism for a starter motor, as shown in Figure 1. One of the goals of the work is to reduce the costs of the grippers and fixturings through a control methodology that can accomodate and react to relatively large uncertainties. Thus, simple grippers were used with an uncertainty of $\pm 2.5mm$. However, the minimum tolerance for the assembly is $0.075mm$. Due to the large uncertainty-to-tolerance ratio, impedance control was unable to reliably execute the insertion task. However, a discrete event controller with force sensing was able to reliably execute the task.

Two primary components are developed. First, a discrete event model of the gear assembly is derived. The model exploits the axial symmetry of the workpiece using a planar projection of the workpiece. The projection is determined according to the sensed force of contact. Second, a discrete event controller is derived consisting of velocity commands which maintain desired contacts, gain new contacts, and avoid unwanted contacts. These velocity commands are then integrated to give appropriate position commands. The interaction of these components produces a powerful means of intelligent control of robotic assembly, and this method can be applied within the constraints of an industrial setting.

Many people have looked at the issue of modelling robotic assembly. Hirai and Asada [7] use polyhedral convex cones to describe the kinematics of the assembly. Cao and Sanderson [5] use fuzzy Petri nets to analyze task sequences for assembly. Tsuda and Takahashi [18] propose the use of local, planar models for multi-agent assembly. A method to align the peg and hole before insertion is given by Bruyninckx *et al.* [4]. The work presented in this paper is based on [10, 11, 12] where the initial discrete event modelling and control synthesis of robotic assembly is presented for polygonal objects in 6 degrees-of-freedom.

Most of the research work in the area of discrete event systems is concerned with developing and proving control-theoretic ideas for specific classes of systems. Ostroff

Figure 1. Gear Mechanism for a Starter Motor

and Wonham [13] provide a powerful and general framework (TTM/RTTL) for modelling and analyzing real-time discrete event systems. Holloway and Krogh [8] use cyclic controlled marked graphs (CMG's) to model discrete event systems, allowing for the synthesis of state feedback logic. Some additional methods include formal languages [15], finite state machines [17], and Petri nets [14]. Brockett [3], Stiver and Antsaklis [16], and Gollu and Varaiya [6] have given more general formulations for the modelling and analysis of hybrid dynamical systems. Recently, Astuti and McCarragher [1, 2] presented a method for the optimal synthesis of a hybrid dynamic controller for constrained motion systems which guarantees convergence to the final desired discrete state. These papers have tried to capture a wide range of system attributes in simple models to allow for tractable analysis and control optimization.

2 Discrete Event Model of Assembly

We treat this industrial assembly as a discrete event dynamic system modelled with Petri nets. The actual gear assembly is shown in Figure 1. The abstraction to discrete event modelling highlights the necessary transitions for successful assembly. These discrete changes in state are the focal point of the control strategy since they indicate the significant changes in the system dynamics and control. The changes in both the dynamics and the control of assembly processes precisely at the point when the discrete state of contact changes serve as the motivation for the Petri net modelling of assembly since these are highlighted with Petri nets. Additionally, Petri nets allow for a compact mathematical description of the geometric contraints and admissible transitions for an assembly task.

2.1 Planar Projection of Round Parts

The actual task for this research is a round workpiece into a round hole. To date, the Petri net model of assembly has only been applied to 6 degree-of-freedom polygonal objects [12]. As such, direct application to the gear mechanism was not possible. Moreover, other methods for round parts, such as [4], are not directly applicable due to the more complicated structure of the gear mechanism and the lack of control options, see Figure 1.

Instead, we sought to exploit the axial symmetry of the workpiece and maintain the Petri net modelling framework by using a planar projection of the workpiece as shown in Figure 2. In projection, the workpiece resembles a "two-tiered" peg-in-hole problem. Note that by incorporating symmetry, only one side of the hole need be

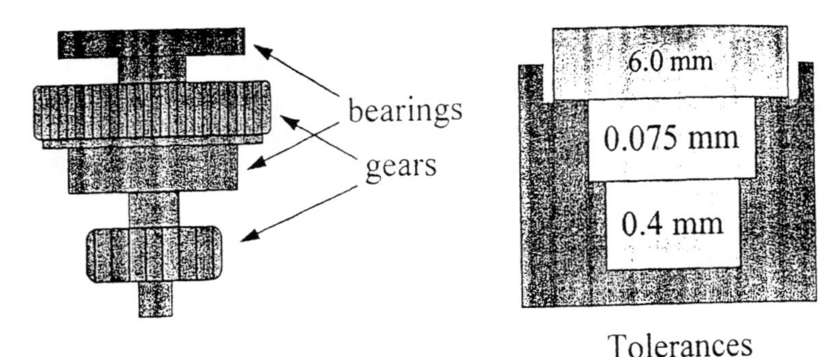

Figure 2. Planar Projection of Gear Mechanism

considered, significantly reducing the size of the Petri net.

The main issue in using a planar projection is determining the plane of projection. That is, onto which plane should the workpiece be projected? This decision is based on the state of force contact indicating the most promising directions of movement. For each control command, a single planar projection is determined.

We use force sensing to determine the projection plane. Since the axial symmetry is in the z−direction, all that is needed is an angle defined in the $x - y$ plane. Thus, the projection plane is defined by an angle θ from the x-axis according to the following equation.

$$\theta = \arctan \frac{\Delta M(y)}{\Delta M(x)} \tag{1}$$

where $\Delta M(\cdot)$ represents the change in moment in the $(\cdot)$ direction occurring at a discrete event. In essence, the angle θ indicates the direction in which motion will be least inhibited. It reasonably assumes an initial single point contact and relatively clean force/moment signals.

2.2 Petri Net Model of Assembly

A standard Petri net is composed of four parts [14]: a set of places $\mathcal{P}$, a set of transitions $\mathcal{T}$, an input function $\mathcal{I}$, and an output function $\mathcal{O}$. The input function $\mathcal{I}$ is a mapping from the places to the transitions, while the output function $\mathcal{O}$ is a mapping from the transitions to the places. The number of transitions, is denoted c. The number of places is p. Places can be thought of as conditions of the current state that enable the transitions to occur, or fire.

For planar models, the edges, or vertices, of both the workpiece and the constraint are denoted generally as ϕ. The facets, or surfaces, of both the workpiece and the constraint are denoted generally as ψ. In modelling the assembly process with a Petri net, then, we define a place to represent one contact pair (ϕ, ψ): either a surface of the workpiece in contact with an edge of the fixture or an edge of the workpiece with a surface of the fixture. Essentially, this means that each place represents only one constraint equation. Combinations of places can be used to describe cases of two-point contact where two constraints are simultaneously active. To make the description complete there is also a place modelling the condition of no-contact, that is, the null constraint equation.

Consistent with the definition of a place, we define a transition as the gaining or losing of a single contact pair, or constraint. Therefore, the occurrence of a transition is a discrete event, or change in contact state. The input function defines the places that must be active for a given transition to fire; that is, the contact pairs necessary for a given change of contact to occur. When the place conditions for a given transition are met, the transition is said to be enabled. The output function defines the contact pairs resulting from a discrete event.

A marking γ of a Petri net is a (px1) vector assignment of tokens to the places of the net, where p indicates the number of places; that is, the total number of possible contact pairs. Tokens can be thought of residing in the places of the net. A token residing in a place indicates that the constraint represented by that place is currently active. For our purposes here, a place can only have either zero or one token.

The execution of the Petri net is controlled by the distribution of the tokens. If a token exists in each of the input places to a transition, that transition is said to be enabled. A transition that is enabled may fire. A transition fires by removing the token from each of the input places and establishing a token in each of the output places. Note that a place may be both an input and an output of a single transition. The firing of a transition results in a next desired marking γ_d, describing the new state of contact represented by the distribution of tokens. By enabling various transitions and executing the net, we can direct the system through a series of discrete contact changes (events) to the desired final state.

The full potential of exploiting axial symmetry in the Petri net modelling of assembly has not yet been realized. Rigorous methods for the automated generation of Petri net models using axial symmetry are currently under development. For this project, the Petri net model and the symmetry pruning of the model were conducted by hand.

3 Discrete Event Control

As was pointed out earlier, during assembly the state of contact changes. Each contact represents a constraint on the rigid body dynamics describing the motion of the manipulator and the workpiece. The changes in contact, therefore, result in different constraint equations. Changing constraint equations result in a varying number of motion degrees of freedom and a varying number of equations of motion. Using Petri net modelling, we will develop a condition on the admissible velocity commands that will maintain a current contact, called the maintaining condition. Additionally, for nominal part locations, we will derive a necessary condition for a discrete event to occur, called the enabling condition, and we will derive a sufficient condition for a discrete event not to occur, called the disabling condition.

The generalized coordinates for planar rigid body motion are the x and y position in inertial space and the orientation α. The generalized coordinates will be expressed as a vector $\mathbf{q} = [x, y, \alpha]^T$. We denote the position vector from the inertial origin to an arbitrary vertex ϕ as $\mathbf{d}_\phi$. A surface ψ, on the other hand, is represented by the unit normal vector $\mathbf{n}_\psi$ and the position vector $\mathbf{d}_\psi$. For the sake of convenience, we define the normal vector to be an outward vector, and the position vector is one of the two vertices that bound the surface.

Let $h_{\phi\psi}$ be the distance between edge ϕ and surface ψ given by

$$h_{\phi\psi} = (\mathbf{d}_\phi - \mathbf{d}_\psi)^T \mathbf{n}_\psi \tag{2}$$

The contact between the edge and the surface is described simply by

$$h_{\phi\psi} = 0 \tag{3}$$

To derive admissible velocities that satisfy the geometric constraints, we simply differentiate equation (3).

$$\frac{d}{d\mathbf{q}} \left[(\mathbf{d}_\phi - \mathbf{d}_\psi)^T \mathbf{n}_\psi \right] \frac{d\mathbf{q}}{dt} = 0 \tag{4}$$

Equation (4) describes the velocity vector that allows the system to move without violating the constraint. We can rewrite equation (4) to yield

$$\mathbf{a}_i \dot{\mathbf{q}} = 0 \tag{5}$$

where $\mathbf{a}_i$ is a 1x3 row vector. Equation (5) is our maintaining condition.

In addition to determining motion that maintains a contact, it is desired to determine the motion such that the workpiece makes the next discrete change. Since we only allow one contact to be either lost or gained at any one instant, the next constraint to become active or inactive can be determined uniquely. We again denote this general constraint with a contact pair as (ϕ, ψ), where ϕ is the edge of contact and ψ is the surface of contact. We now describe two cases, one for a gain of contact and one for a loss of contact.

Gain of Contact – Since the desired contact is not yet active, the current distance between edge ϕ and surface ψ is positive. To gain contact, the distance must decrease:

$$\frac{d}{dt}\left[h_{\phi\psi}(\mathbf{q})\right] = \mathbf{a}_i\dot{\mathbf{q}} < 0 \tag{6}$$

The change of marking for a gain of contact is given by

$$\Delta_i = 1 \tag{7}$$

Note that Δ_i is a scalar.

Loss of Contact – The second case is when the desired transition results in a loss of contact. In this instance, the distance between the edge and surface of the contact pair must increase. Therefore,

$$\frac{d}{dt}\left[h_{\phi\psi}(\mathbf{q})\right] = \mathbf{a}_i\dot{\mathbf{q}} > 0 \tag{8}$$

The change in marking for a loss of contact is then

$$\Delta_i = -1 \tag{9}$$

Combining equations (6) through (9) yields

$$\Delta_i\mathbf{a}_i\dot{\mathbf{q}} < 0 \tag{10}$$

Equation (10) stipulates that any commanded velocity should cause the distance between the current position and the desired position to decrease. It is our enabling condition. It is a necessary equation for the transition specified by Δ_i to fire.

Lastly, it is desired to determine a condition for a discrete event to not occur. Since (10) is a necessary condition for a transition to occur, we can change the inequality sign to determine a sufficient condition for a transition not to occur. Hence, our disabling condition for a transition specified by Δ_j is

$$\Delta_j\mathbf{a}_j\dot{\mathbf{q}} \geq 0 \tag{11}$$

Now that the maintaining, enabling and disabling conditions are derived, we can apply them to determine the velocities which effect a desired transition in a predetermined event trajectory. If the desired transition is a gain of contact we need apply (5) to maintain contact, and we need to apply (10) to the velocity command for the desired transition to occur. If the transition is a loss of contact, we need only apply (10) for the desired transition to occur. Additionally, we wish to disable all other transitions and so apply (11) to each of these transitions. The problem has now been reduced to solving a set of simultaneous linear inequalities (10) and (11) with, perhaps, one equality constraint (5). The solution to these simultaneous inequalities may not be unique. Any continuous control command that satisfies the inequality constraints may be used. We find the velocity that maximizes the minimum distance to each constraint for a robust solution.

4 Experiments

A significant number of experiments were run using a six degree-of-freedom robot built by Nippondenso. The robot was belt driven and was position controlled with the Nippondenso Robot Language (NDRL) using VX-Works. There were two servo loops. The low level servo loop implemented independent joint control at a rate of 1ms. The higher level servo loop interpreted NDRL commands/positions and calculated the joint path at a rate of 32ms. The robot was necessarily equipped with a six-axis force sensor. The force sensor was sampled at 600 Hz. In addition to the uncertainty of the grippers, there was additional uncertainty due to the belts. The belt drives also generated significant noise on the force signal. Moreover, the robot arm was mounted on a mobile base. At high speeds, there was noticable vibrations due to the mobile base.

The goal of this series of experiments was to demonstrate the effectiveness of discrete event control and qualitative reasoning [9] as a stand-alone control method. This series also shows the effectiveness of the approach for error detection and recovery. For comparison, an impedance controller was also implemented. The impedance controller proved to be successful approximately 50% of the time. The discrete event controller proved to be highly successful, completing insertion approximately 90% of the time. An example force signal and the corresponding position trajectory are shown in Figure 3.

5 Conclusions

This project represents a highly significant advance in robotic assembly. First, a complex assembly involving a two-tiered round workpiece and round hole was successfully completed. Planar projection of the model and on-line force sensing were necessary for the implementation. Second, this project demonstrates the effective industrial impelementation of discrete event control and qualitative reasoning. Using these tools, a very powerful assembly system has been developed and implemented. Third, this system allows for the introduction of very inexpensive grippers, and hence, very inaccurate grippers. The position accuracy of the gripper was greater than $\pm 2.5mm$. However, the smallest tolerance of the assembly was only $0.075mm$. Despite this great difference, the system successfully assembles the gear mechanism for a automobile starter motor. This project has demonstrated great potential savings through advanced control of the assembly process, allowing for a significantly less controlled environment. This advanced ability is one of the most important strengths of the Discrete Event Systems approach to assembly. This project is currenty progressing toward implementation on a fully automated assembly line.

References

[1] Astuti, P. and B. McCarragher, "Controller Synthesis of Manipulation Hybrid Dynamic Systems", *International Journal of Control, submitted 1995.*

[2] Astuti, P. and B. McCarragher "The Stability of Discrete Event Systems Using Markov Processes", *International Journal of Control, to appear.*

[3] R. Brockett, "Hybrid Models for Motion Control Systems", in *Essays on Control: Perspectives in the Theory and its Applications*, H.L. Trentelman and J.C. Willems *eds.*, Birkhauser Publishers, Boston 1993.

[4] H. Bruyninckx, S. Dutre and J. De Schutter, "Peg-on-Hole: A Model Based Solution to Peg and Hole Alignment", *International Conference on Robotics and Automation*, pp 1919-1924, 1995.

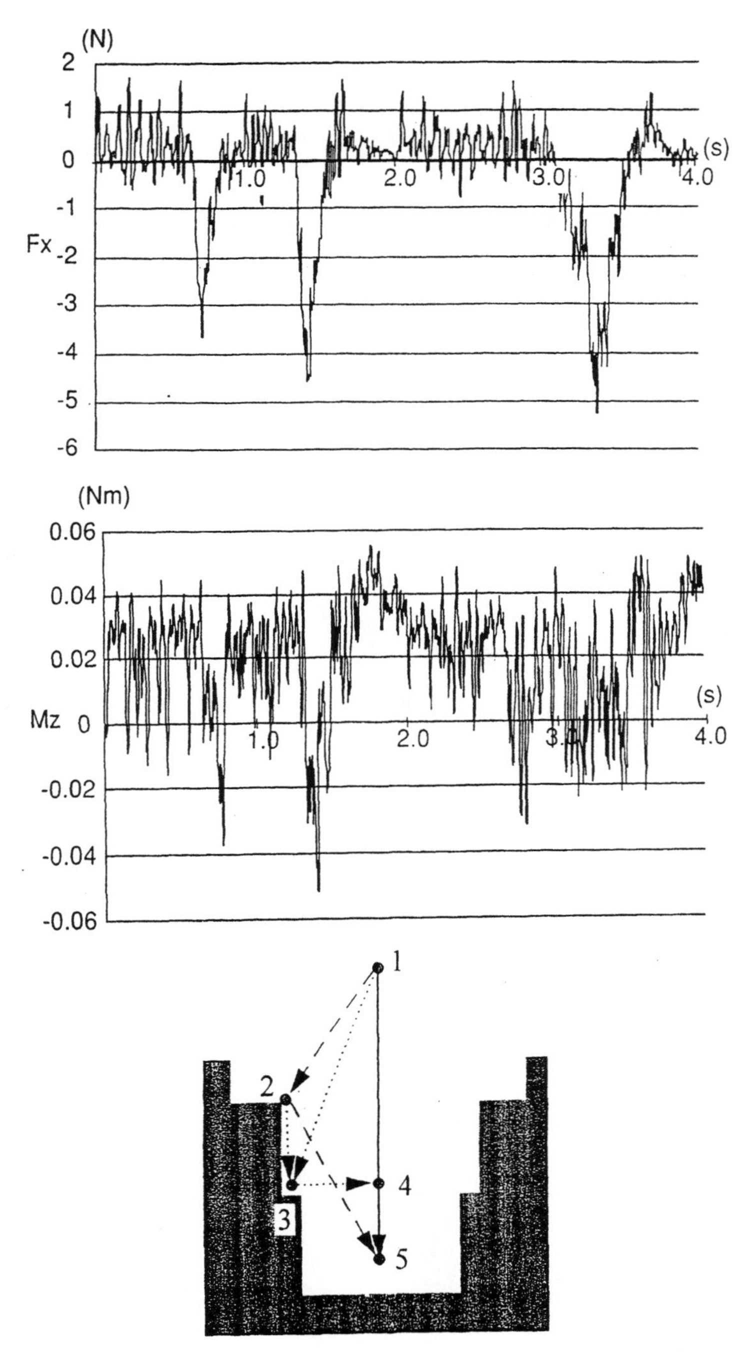

Figure 3. Force (F_x) and Moment (M_z) Signals and Position Trajectory - Both Desired Commands and Actual Trajectory - for a Successful Assembly

[5] T. Cao and A.C. Sanderson, "Sensor-Based Error Recovery for Robotic Task Sequences Using Fuzzy Petri Nets", *IEEE Conference on Robotics and Automation*, pp. 1063-1069, May 1992.

[6] A. Gollu and P. Varaiya, "Hybrid Dynamical Systems", *28th Conference on Decision and Control*, December 1989.

[7] S. Hirai and H. Asada, "A Model-Based Approach to the Recognition of Assembly Process States Using the Theory of Polyhedral Convex Cones", *Proc. of 3rd Japan-USA Symposium on Flexible Automation*, 1990.

[8] L.E. Holloway and B.H. Krogh, "Synthesis of Feedback Control Logic for a Class of Controlled Petri Nets." *IEEE Trans. on Automatic Control*, vol 35, No 5,May 1990.

[9] McCarragher, B. and H. Asada, "In-Process Monitoring for Robotic Assembly using Dynamic Process Models", *ASME Journal of Dynamic Systems Measurement and Control*, 115(2A), pp. 261-275, June 1993.

[10] McCarragher, B. and H. Asada, "The Discrete Event Modelling and Planning of Robotic Assembly Tasks", *ASME Journal of Dynamic Systems Measurement and Control*, vol 117 no 3, p394-400, September 1995.

[11] McCarragher, B. and H. Asada, "The Discrete Event Control of Robotic Assembly Tasks", *ASME Journal of Dynamic Systems Measurement and Control*, vol 117 no 3, p 384-393, September 1995.

[12] McCarragher, B. "Task Primitives for the Discrete Event Modelling and Control of 6-DOF Assembly Tasks", *IEEE Transactions on Robotics and Automation*, to appear April 1996.

[13] J.S. Ostroff and W.M. Wonham, "A Framework for Real-Time Discrete Event Control", *IEEE Transactions of Automatic Control*, Vol 35, No 4, April 1990.

[14] James L. Peterson, *Petri Net Theory and the Modeling of Systems*, Prentice Hall, Inc., 1981.

[15] Peter J. G. Ramadge, "Some Tractable Supervisory Control Problems for Discrete-Event Systems Modeled by Büchi Automata", *IEEE Transactions on Automatic Control*, Vol 34, No 1, January 1989.

[16] J. Stiver and P Antsaklis, "Modelling and Analysis of Hybrid Control Systems", *Proc. 31st Conference on Decision and Control*, December 1992.

[17] Gilead Tadmor and Oded Maimon, "Control of Large Discrete Event Systems: Constructive Algorithms", *IEEE Transactions on Automatic Control*, Vol 34, No 11, November 1989.

[18] M. Tsuda and T. Takahashi, "A Method for Changing Contact States for Robotic Assembly by using Some Local Models in a Multi-Agent System", *IEEE International Conference on Robotics and Automation*, pp. 2713-2719, 1995.

The Spatial Peg–in–Hole Problem

TH. MEITINGER and F. PFEIFFER
TU München, Lehrstuhl B für Mechanik, Germany

ABSTRACT

In this paper the modeling and simulation of a rectangular peg–in–hole insertion carried out by a six axis manipulator is presented. The dynamic model of the robot coupled with the assembly process model is established, where all possible contact points between the peg and the environment are monitored. Every occurring contact point represents a geometrical constraint on the robots dynamic which is then characterized by closed loops. The contact laws are formulated as a Linear Complementarity Problem (LCP), which allows the determination of the contact forces for dependent constraints. No assumptions about the sequence of the contact points have to be made in advance, rather the sequence results from the time evolution of the system, which depends on the initial conditions. Through the simulation we are able to predict the behavior of the manipulator during the task as well as the load on the mating parts. The comparison with measurements show that this approach is capable of describing the regarded process rather realistically.

KEYWORDS: Peg–in–Hole, Robot Dynamics, Contact Mechanics

INTRODUCTION

Small tolerances between the mating parts are the often characteristic for mounting tasks. During the automatic assembly with a robot, the parts will contact each other due to uncertainties in the manipulators position and in the parts geometry. Undesireable high strains on the workpieces or even the unfeasibility of the task, e. g. because of jamming, may result. A well known example for parts mating is the peg–in–hole problem. Many assembly processes can be reduced to this example. Thus the effects mentioned above can be studied.

There are three different approaches to handle these problems. One solution is the development of special passive compliant mechanisms, based on the Remote Center of Compliance (RCC). Through this measure the area, where no problems during the assembly might occur, is enlarged. Such mechanisms were first developed for the peg–in–hole problem in two dimensions. In [20] the analysis for designing the RCC was made quasistatically, in [1] the dynamics of the peg and the supporting mechanism were also taken into account. An extension to three dimensions can be found in [17]. The same problem is regarded in [18], where the principle of the Spatial Remote Center of Compliance (SRCC) is used for the analysis.

A second solution is the additional use of sensor information. Nearly all authors follow the same principle: First an initial contact state has to be identified. A new method for this step was introduced in [5]. Different contact topologies in the presence of sensing and control uncertainties are tested. The method succeeds, even if the the contact forces are statically indefinite. Then the peg has to be moved towards the hole. This phase can be called peg–on–hole [3]. Afterwards the peg is aligned and inserted into the hole. Thereby, especially two point contact, which might lead to jamming, has to be avoided. For this phase in [19] an optimized controller is presented. If the operation fails due to sensing, model or control errors, the method of error detection and recovery is applied in order to complete the given task [9, 8]. In [10] all possible contact sequences are modeled as a discrete event system using Petri nets. As new control methods appeared, they were also applied to the peg–in–hole problem, like fuzzy control [13] or neural networks [12]. In newer investigations [2], additional sensors are utilized for not only detecting the contact state, but additionally finding out manufacturing defects like burrs and ruts.

A third approach is theoretical investigation. The insertion task is described by geometrical and mechanical models, where uncertainties can also be taken into account. If only the geometry is considered one talks of fine motion planning [6]. More insight into the peg–in–hole insertion is gained when additionally the stiffness of the supporting mechanism is considered. In [20] for the two dimensional case the area is calculated, where jamming and wedging might occur. In [4] similar considerations are made for the same problem including three dimensions. The dynamics of the complete parts mating process including a complete dynamic model of the manipulator is presented in [15]. We have already presented dynamic simulations of a peg–in–hole insertion combined with compliant mating parts [11]. By numerical simulation problems which might occur during the mating task can be recognized. Thus the feasibility of the task can be investigated in advance.

In this paper we present the spatial modeling and simulation of the insertion of a rectangular peg into a hole executed by a PUMA 560. In the next section we will first show the dynamic model of the manipulator. Then we will describe the geometric constraints arising between the peg and the hole in case of contact. The contact laws (nonpenetrability in normal direction and Coulomb's frictional law in tangential direction) are formulated as a Linear Complementarity Problem (LCP). Our approach is verified through a comparison of the calculated and measured forces acting on the robots gripper during the operation.

ROBOT MODEL

The industrial robot, here a PUMA 560 (Fig.1a), posseses six axis $(\gamma_{A1}, \cdots, \gamma_{A6})$ and is modeled as a tree-like multibody system with rigid links. Since the natural frequency of oscillations due to the stiffness in the first three joints are in the range of interest, an elastic joint model is introduced. A link–joint unit consists of two bodies, the drive and the arm segment (Fig.1b). They are coupled by a gear model which is composed of the physical elements stiffness c and damping d (1). Thus three additional degrees of freedom are introduced between motor shafts and arm segments: $\gamma_{M1}, \gamma_{M2}, \gamma_{M3}$. In the remaining links no joint model is necessary, because there the stiffness is high compared to the acting forces and the elasticities of these joints have no effect on the system dynamics under consideration. During the assembly task only small distances are covered, so that the robot's dynamics is linearized around a working point γ_0. The vector $\boldsymbol{q} = \gamma - \gamma_0 = (q_{M1}, q_{M2}, q_{M3}, q_1, q_2, q_3, q_4, q_5, q_6)^T$ denotes the deviation from this working point. We obtain the following equations of motion:

$$\boldsymbol{M}\ddot{\boldsymbol{q}} + \boldsymbol{P}\dot{\boldsymbol{q}} + \boldsymbol{Q}\boldsymbol{q} = \boldsymbol{B}\boldsymbol{u} \in \mathbb{R}^9, \qquad (2)$$

with the inertia matrix $\boldsymbol{M}$, the damping matrix $\boldsymbol{P}$ and the stiffness matrix $\boldsymbol{Q}$. $\boldsymbol{Bu}$

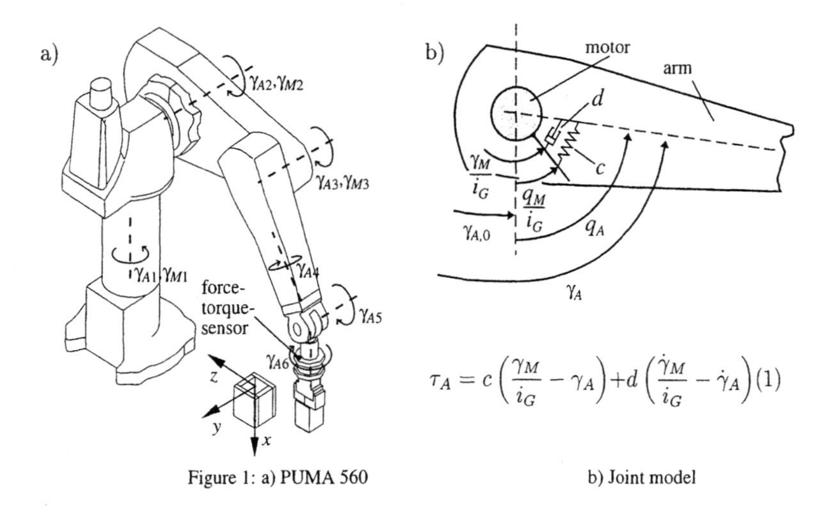

$$\tau_A = c\left(\frac{\gamma_M}{i_G} - \gamma_A\right) + d\left(\frac{\dot{\gamma}_M}{i_G} - \dot{\gamma}_A\right) \quad (1)$$

Figure 1: a) PUMA 560 b) Joint model

regards the influence of the controller. For PD position control we have the typical form of the vector $\boldsymbol{u}$:

$$\boldsymbol{u} = -\boldsymbol{K}_p \boldsymbol{B}^T (\boldsymbol{q} - \boldsymbol{q}_D) - \boldsymbol{K}_D \boldsymbol{B}^T (\dot{\boldsymbol{q}} - \dot{\boldsymbol{q}}_D). \quad (3)$$

The matrices $\boldsymbol{K}_P$ and $\boldsymbol{K}_D$ contain the positional and velocity feedback gains. Through changing the desired positions and velocities $\boldsymbol{q}_D$ and $\dot{\boldsymbol{q}}_D$ the motion of the manipulator along a trajectory is realized.

CONTACT KINEMATICS

The peg with the chamfer and the hole (Fig.2a) are regarded as rigid bodies. Ev-

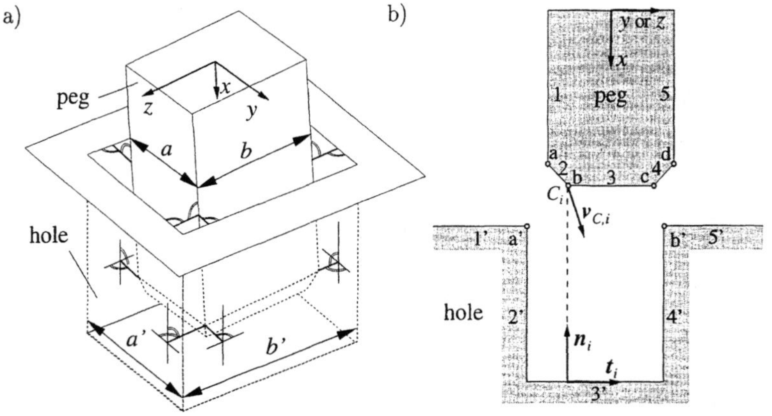

Figure 2: Contact kinematics of the peg and the hole

ery contact point closes a kinematical loop between the robot and the environment. Mathematically every contact point represents one or more additional constraints on the system, depending on whether sliding or sticking occurs. The constraints reduce the degree of freedom of the manipulator. The spatial peg consists of four sides. Ten possible contact points exist between one side of the peg and the hole (Fig.2b). Two sides are situated within the xy–plane, and two in the xz–plane. There are two different types of contact points: point–plane and edge–edge. Let the letters a, b, c, d in Fig.2b denote points and the numbers $1', 2', 3', 4', 5'$ denote planes. Then there are four possible contact points of the type point–plane: $a-2'$, $b-1'$, $c-5'$, $d-4'$. For the other type edge–edge the numbers $1, 2, 3, 4, 5$ denote edges on the peg and the letters a', b' denote edges of the hole. There exist six potential contacts of this type: $1-a'$, $2-a'$, $3-a'$, $3-b'$, $4-b'$, $5-b'$. Thus we have altogether 40 potential constraints between the peg and the hole with the sketched geometry in the spatial case. In case of contact the mathematical condition for a normal or tangential constraint is, that the velocity of a potential contact point projected in the normal or tangential direction has to be zero:

$$\dot{g}_{N,i} = \boldsymbol{n}_i^T \boldsymbol{v}_{C,i} = \boldsymbol{n}_i^T \boldsymbol{J}_{C,i} \dot{\boldsymbol{q}} = 0, \qquad (4) \qquad \dot{g}_{T,i} = \boldsymbol{t}_i^T \boldsymbol{v}_{C,i} = \boldsymbol{t}_i^T \boldsymbol{J}_{C,i} \dot{\boldsymbol{q}} = 0, \qquad (5)$$

$(i = 1, \ldots, 40)$ where $\boldsymbol{J}_{C,i}$ is the Jacobian of translation with respect to the i-th contact point. An active constraint in normal direction means, that the rigid bodies do not penetrate each other. In tangential direction an active constraint means stiction.

CONSTRAINT DYNAMICS

In order to combine the geometrical constraints with the dynamics of the robot, (4) and (5) are differentiated with resepct to time:

$$\ddot{g}_{N,i} = \underbrace{\boldsymbol{n}_i^T \boldsymbol{J}_{C,i}}_{\boldsymbol{w}_{N,i}} \ddot{\boldsymbol{q}} + \underbrace{\left(\boldsymbol{n}_i^T \boldsymbol{J}_{C,i}\right)^{\cdot} \dot{\boldsymbol{q}}}_{\widetilde{w}_{N,i}} = 0, \quad (6) \quad \ddot{g}_{T,i} = \underbrace{\boldsymbol{t}_i^T \boldsymbol{J}_{C,i}}_{\boldsymbol{w}_{T,i}} \ddot{\boldsymbol{q}} + \underbrace{\left(\boldsymbol{t}_i^T \boldsymbol{J}_{C,i}\right)^{\cdot} \dot{\boldsymbol{q}}}_{\widetilde{w}_{T,i}} = 0. \quad (7)$$

However only some of all 40 potential constraints are active. We define n_N and n_T as the number of constraints in normal and tangential direction, which are active at any one moment. Through the active constraints additional contact forces act on the system which are considered in the differential equations of motion (2) with the aid of the Lagrangian multiplier method:

$$\boldsymbol{M}\ddot{\boldsymbol{q}} + \boldsymbol{P}\dot{\boldsymbol{q}} + \boldsymbol{Q}\boldsymbol{q} = \boldsymbol{B}\boldsymbol{u} + \boldsymbol{W}_N\boldsymbol{\lambda}_N + \boldsymbol{W}_T\boldsymbol{\lambda}_T + \boldsymbol{W}_R\boldsymbol{\lambda}_N \qquad (8)$$

$$\boldsymbol{W}_N := \{\boldsymbol{w}_{N,i}\}\,;\ i = 1, \ldots, n_N \quad \boldsymbol{W}_T := \{\boldsymbol{w}_{T,i}\}\,;\ i = 1, \ldots, n_T$$

$$\boldsymbol{W}_R := \{\boldsymbol{w}_{R,i}\}\,;\ \boldsymbol{w}_{R,i} := \begin{cases} \boldsymbol{0} & ;\ i = 1, \ldots, n_T \text{ (sticking)} \\ -\mu \boldsymbol{w}_{T,i} \text{sign}(\dot{g}_{T,i}) & ;\ i = n_N - n_T, \ldots, n_N \text{ (sliding)}\,. \end{cases}$$

The components of the vectors $\boldsymbol{\lambda}_N = (\lambda_{N1}, ..., \lambda_{Nn_N})^T$ and $\boldsymbol{\lambda}_T = (\lambda_{T1}, ..., \lambda_{Tn_T})^T$ correspond to the unknown constraint forces normal and tangential to the respective tangent plane. The term $\boldsymbol{W}_R \boldsymbol{\lambda}_N$ considers frictional forces in all contact points where sliding occurs. Besides the system dynamics (8), the constraint conditions (6, 7) have to be fulfilled. Eliminating the acceleration $\ddot{\boldsymbol{q}}$ from Eq. (6, 7) by inserting Eq. (8) leads to a set of linear equations for the unknown constraint forces $\boldsymbol{\lambda} = (\boldsymbol{\lambda}_N^T\ \boldsymbol{\lambda}_T^T)^T$ and the relative accelerations $\ddot{\boldsymbol{g}} = (\ddot{\boldsymbol{g}}_N^T\ \ddot{\boldsymbol{g}}_T^T)^T$. This method however is only valid for a set of active constraints. Due to the time varying structure of the system, additional conditions have to be fulfilled in order to check the validity of the present constraints:

- Penetration of the bodies is not allowed ($\ddot{g}_{N,i} \geq 0$). No negative forces are not acting ($\lambda_{N,i} \geq 0$). Thus result the complementary conditions for a normal constraint: $\ddot{g}_{N,i} \geq 0$, $\lambda_{N,i} \geq 0$, $\ddot{g}_{N,i}\lambda_{N,i} = 0$.

- For contact points with vanishing relative velocity $\dot{g}_{T,i} = 0$, the friction force is limited by the friction saturation $\Delta F_{T,i} = \mu_0\,\lambda_{N,i} - |\lambda_{T,i}| \geq 0$ which models the Coulomb friction cone. If stiction is present $\ddot{g}_{T,i} = 0$ and $\Delta F_{T,i} \geq 0$ must be satisfied, whereas at a transition to sliding the friction force $\lambda_{T,i}$ opposes the starting relative motion, hence the condition $\ddot{g}_{T,i}\,\lambda_{T,i} \leq 0$ must hold for both cases sliding $\ddot{g}_{T,i}\,\lambda_{T,i} < 0$ and rolling $\ddot{g}_{T,i}\,\lambda_{T,i} = 0$. If we furthermore assume a friction characteristic that is steady for $v = |\dot{g}_{T,i}| = 0$, ($\lim_{v\to 0}\,\mu(v) = \mu_0$, μ_0: static friction coefficient) the magnitude of the sliding friction force is specified by $\Delta F_{T,i} \geq 0$.

All requirements which the constraint dynamics (8) must satisfy can thus be summarized by a set of inequalities and complementary conditions:

$$\left.\begin{array}{l} \lambda_{N,i} \geq 0 \quad \ddot{g}_{N,i} \geq 0 \quad \lambda_{N,i}\,\ddot{g}_{N,i} = 0 \quad i = 1, ..., n_N \\ \Delta F_{T,i} = \mu_0\,\lambda_N - |\lambda_{T,i}| \geq 0 \quad \ddot{g}_{T,i}^T\,\lambda_{T,i} \leq 0 \quad \Delta F_{T,i}\,\ddot{g}_{T,i} = 0 \quad i = 1, ..., n_T \end{array}\right\} \tag{9}$$

A solution $(\boldsymbol{\lambda}_N^T, \boldsymbol{\lambda}_T^T, \ddot{\boldsymbol{g}}_N^T, \ddot{\boldsymbol{g}}_T^T, \Delta\boldsymbol{F}_T^T)^T$, which satisfies (8) and the conditions (9) is only valid for the actual structure of the system. A change of the structure because of a change of constraints causes the velocity, acceleration and constraint force in the specific contact point to be unsteady, especially if stick–slip phenomena or impacts occur. Up to now, the solution that is consistent with the dynamic constraints had to be found by a combinatorial search, where all contact combinations had to be tested. The combinatorial problem, however, can be avoided if we transform the contact laws into a linear complementary problem (LCP). The unilateral constraints in normal direction were already derived in standard form. For the bilateral tangential constraints we introduce auxiliary, unilaterally bounded variables ($\lambda_{T,i}^{(+)}$, $\lambda_{T,i}^{(-)}$) and ($\ddot{g}_{T,i}^{(+)}$, $\ddot{g}_{T,i}^{(-)}$) which act in two opposing directions and replace the terms $\lambda_{T,i}$ and $\ddot{g}_{T,i}$. Thus two complementary conditions can be formulated for the tangential constraints. The equations of motion (8) are then rewritten:

$$\boldsymbol{M}\ddot{\boldsymbol{q}} + \boldsymbol{P}\dot{\boldsymbol{q}} + \boldsymbol{Q}\boldsymbol{q} = (\boldsymbol{W}_N + \boldsymbol{W}_R)\,\boldsymbol{\lambda}_N + \boldsymbol{W}_T\,(\boldsymbol{\lambda}_T^{(+)} - \boldsymbol{\lambda}_T^{(-)}). \tag{10}$$

The reformulated contact laws define a Linear Complementary Problem (LCP) in standard form:

$$\boldsymbol{x} = \widehat{\boldsymbol{A}}\,\boldsymbol{y} + \widehat{\boldsymbol{b}} \quad \boldsymbol{x}^T\,\boldsymbol{y} = 0 \quad \boldsymbol{x} \geq 0 \quad \boldsymbol{y} \geq 0 \quad \widehat{\boldsymbol{A}} \in \mathbb{R}^{n_N+2n_T, n_N+2n_T} \quad \widehat{\boldsymbol{b}}, \boldsymbol{x}, \boldsymbol{y} \in \mathbb{R}^{n_N+2n_T},$$

where $\boldsymbol{x}$ and $\boldsymbol{y}$ represent the constraint forces and the relative accelerations, the matrix $\widehat{\boldsymbol{A}}$ contains the characteristic matrix of the constraint dynamics and the definition of the friction cone. The vector $\widehat{\boldsymbol{b}}$ represents the violation of the system dynamics with respect to the constraint directions. The derivation and the structure of the above LCP can by found in detail in [7] and [16]. The solution of this problem is provided by a modified Simplex algorithm, e.g. method of Lembke, and indicates which of the possible constraints are currently active.

RESULTS

Measurements were conducted with a PUMA 560 manipulator inserting a rectangular peg with a chamfer into a rectangular hole. The starting position of the manipulator was $\gamma_0 = (4.6^\circ, -157.2^\circ, 27.5^\circ, 0.0^\circ, -50.3^\circ, 4.6^\circ)^T$, which is similar to the configuration shown in Fig.1. The equations of motion of the robot were linearized around this working point. The mating parts can be seen in Fig.2, where the peg had the measures $a = 45.2mm, b = 45.4mm$ with a chamfer $45^\circ \times 4mm$ and the hole had the dimensions $a' = 46.0mm, b' = 45.8mm$. The robot's path during the mating task was $80mm$ in positive x direction. We show here the results of four experiments compared to numerical simulations using the approach described above. The initial lateral displacement between the peg and the hole was set to $\pm 4mm$ in the two cartesian directions y and z.

Displacement in y direction

Let us first regard the experiments, where the displacement was $\Delta y = \pm 4mm$. In

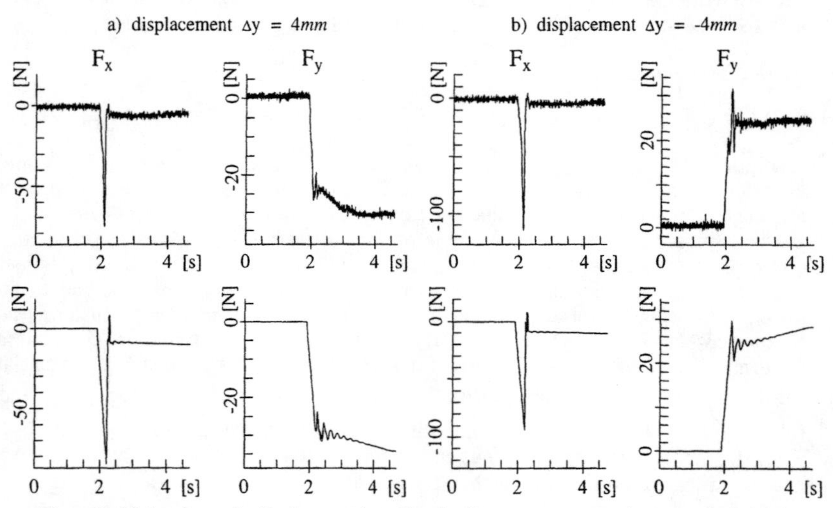

Figure 3: Mating forces for displacement in y direction (top: measurements, bottom: calculations)

Fig.3 the gripper forces during insertion F_x and F_y are shown. The upper plots are measurements, the lower plots are the calculated results for the same starting configuration. In both cases there is a peak of F_x versus the manipulator motion, when the chamfer of the peg comes in contact with the upper edge of the hole (see Fig.2b, contact points of type $4 - b'$ in case of pos. or $2 - a'$ in case of neg. displacement). After having passed the edge, it is sliding downwards, having contact with one side of the hole (see Fig.2b, contact points of type $5 - b'$ in case of pos. or $1 - a'$ in case of neg. displacement). The force F_y due to this contact acts towards the center of the hole.

Displacment in z direction

More interesting are the experiments, where the displacement was varied in the z direction: a) $\Delta z = +4mm$, b) $\Delta z = -4mm$. Here the behavior of the manipulator is different for both cases, see Fig.4 (top: measurement, bottom: calculation). If there is displacement $\Delta z = +4mm$, there is again a force peak in F_x at the first contact (contact points of type $4 - b'$), when the chamfer slides at the upper edge of the hole. The peg is then sliding into the hole, having contact with the upper edge $(5 - b')$,

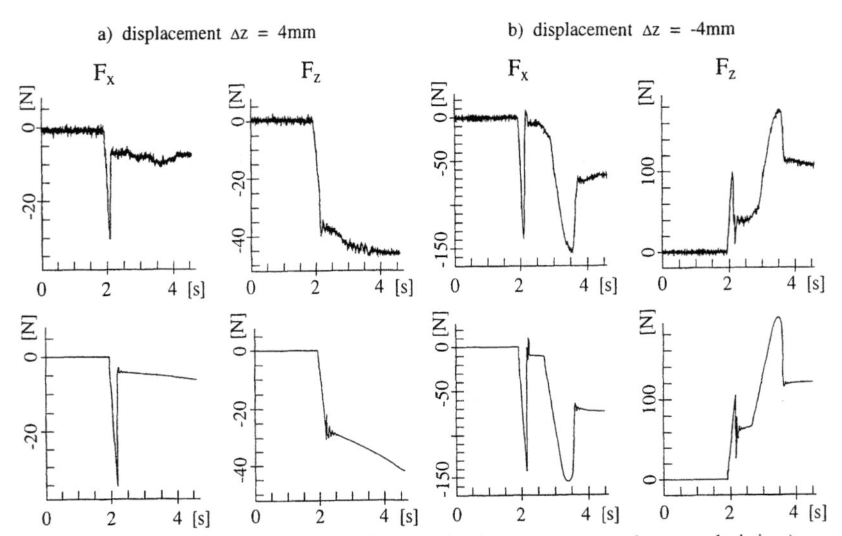

Figure 4: Mating forces for displacement in z direction (top: measurements, bottom: calculations)

similar to the first two experiments. Completely different behavior can be observed, when the lateral displacement is $\Delta z = -4mm$. Here only the beginning of the insertion is similar to the other cases ($2 - a'$ and $1 - a'$). But as the peg proceeds deeper into the hole, there are additional contact points (of type $d - 4'$) inside the hole–after about $2.7s$. The contact forces and thus the mating forces F_x and F_z become very large, because jamming occurs. The insertion finally succeeds, because drive torques are increased by the controller.

The reason for the unsymmetric behavior of the robot in cases a) and b) can be found in the robots starting configuration. If a force in negative x direction is applied, the manipulator is not only displaced in the same direction ($-x$), but also in negative z direction because of couplings in the stiffness matrix. This means for the example with initial displacement $\Delta z = +4mm$ (Fig.4a), that the gripper is moved towards the center of the hole, when the peg is in contact with the hole. Therefore the contact forces are reduced. The opposite happens, if the lateral displacement is $\Delta z = -4mm$. As mating forces act on the gripper, the gripper moves away from the hole, whereas the mating forces additionally increase.

CONCLUSION

A method was presented to model a robot performing an assembly process, where both the robot dynamics and the contact mechanics between the mating parts are considered. The contact mechanics during insertion are described using an LCP approach [7]. The comparison between the measured and the calculated results show, that our method is capable of describing mating tasks with rigid bodies executed by a robot. All important effects that show up in the experiments are covered by the simulation. As an interesting result we found, that the forces during the mating operation depend not only on the magnitude of the initial displacement, but also on its sign, which shows, whether the position error is directed towards the robots basis or away from it. In the future we will use our simulations to find optimal positions and controller gains for the manipulator in order to reduce the strains and stresses on the mating parts and on the robot as well.

ACKNOWLEDGEMENT

The research work presented in this paper is partially supported by a contract with the DFG (Deutsche Forschungsgemeinschaft, SFB 336).

REFERENCES

1. Asada,H., Kakumoto,Y.: "The Dynamic RCC Hand for High–Speed Assembly." *Proc. IEEE Conf. on Robotics & Automation*, 1988, 120-125
2. Bergqvist,C., Söderquist,B., Wernerson,Å: "On Combining Accelerometers, Force/Torque-sensors, and Electrical Sensing for Detecting Contact Errors during Assembly." *Proc. of IEEE/RSJ Int. Conf. on Intelligent Robots & Systems*, 1994, 1736-1743
3. Bruyninckx,H., Dutré,S., De Schutter,J.: "Peg–on–hole: A Model Based Solution to Peg and Hole Alignment." *Proc. IEEE Conf. on Robotics & Automation*, 1995, 1919-1924
4. Caine,M.E., Lozano-Pérez,T., Seering,W.P.: "Assembly Strategies for Chamferless Parts." *Proc. IEEE Conf. on Robotics & Automation*, 1989, 472-477
5. Farahat,A.O., Graves,B.S., Trinkle,J.C.: "Identifying Contact Formations in the Presence of Uncertainty." *Proc. IEEE/RSJ Conf. on Intelligent Robots & Systems*, 1995, Vol.III,59-64
6. Lozano-Pérez,T., Mason,M.T., Taylor,R.H.: "Automatic Synthesis of Fine–motion Strategies for Robots." *Int. J. Robotics Research*, 1984, Vol.3(1)
7. Glocker,C., Pfeiffer,F.: "Stick–Slip Phenomena and Application." *Nonlinearity and Chaos in Engineering Dynamics*, John Wiley & Sons Ltd, 1994, 103-113
8. Gottschlich,S.N., Kak,A.C.: "A Dynamic Approach to High–Precision Parts Mating." *Proc. IEEE Conf. on Robotics & Automation*, 1988, 1246-1253
9. Jennings,J., Donald,B., Campbell,D.: "Towards Experimental Verification of an Automated Compliant Motion Planner Based on a Geometric Theory of Error Detection and Recovery." *Proc. IEEE Conf. on Robotics & Automation*, 1989, 632-637
10. McCarragher,B., Asada,H.: "A Discrete Event Controller Using Petri Nets Applied to Assembly." *Proc. of IEEE/RSJ Int. Conf. on Intelligent Robots & Systems*, 1992, 2087-2094
11. Meitinger,T., Pfeiffer,F.: "Dynamic Simulation of Assembly Processes." *Proc. IEEE/RSJ Conf. on Intelligent Robots & Systems*, 1995, Vol.II, 298-304
12. Park,Y.K., Cho,H.S.: "A Neural Network–based Assembly Algorithm for Chamferless Parts Mating." *Proc. IEEE Symp. on Assembly and Task Planning*, 1995, 381-386
13. Park,Y.K., Cho,H.S., Park,J.O.: "A Fast Searching Method for Precision Parts Mating Based upon Fuzzy Logic Approach." *Proc. of IEEE/RSJ Int. Conf. on Intelligent Robots & Systems*, 1992, 1319-1323
14. Pfeiffer,F.: "Dynamical Systems with Impact and Stick-Slip Phenomena." *CSME Mechanical Engineering Forum 1990*, Toronto, Canada
15. Seyfferth,W., Pfeiffer,F.: "Dynamics of Assembly Processes with a Manipulator." *Proc. of IEEE/RSJ Int. Conf. on Intelligent Robots & Systems*, 1992, 1303-1310
16. Seyfferth,W.: "Modellierung unstetiger Montageprozesse mit Robotern." *Fortschrittberichte VDI*, Reihe 11, Nr. 199, 1993 (in German)
17. Strip,D.: "A Passive Mechanism for Insertion of Convex Pegs." *Proc. IEEE Conf. on Robotics & Automation*, 1989, 242-248
18. Sturges,R.H., Laowattana,S.: "Passive Assembly of Non-Axisymmetric Rigid Parts." *Proc. of IEEE/RSJ Int. Conf. on Intelligent Robots & Systems*, 1994, 1218-1225
19. Wapenhans,H., Seyfferth,W., Pfeiffer,F.: "Hybrid Position and Force Control with Unsteady Assembly Dynamics." *IMACS/SICE Int. Symp. RMS^2*, 1992, 1279-1284
20. Whitney, D.E.: "Quasi-Static Assembly of Compliantly Supported Rigid Parts." *Trans. of the ASME Journal of Dynamic Systems*, Measurement & Control, Vol.104, March 1982, 65-77

Modelling Ultrasonic Sensing for Mobile Robots

Phillip McKerrow, David Crook and Janos Tsakiris
Department of Computer Science, University of Wollongong, Northfields Avenue, Wollongong, NSW, 2522. Phone: (042) 21 3771, Fax: (042) 21 4870, Email: phillip@cs.uow.edu.au

ABSTRACT

Ultrasonic sensing is used by mobile robots for collision avoidance, localisation, wall following, feature tracking, mapping and object recognition. In order to improve the performance of ultrasonic sensors researchers are developing models of the sensing process.

KEYWORDS: ultrasonic sensors, mobile robots, models, transducers

INTRODUCTION

The problems with ultrasonic sensing systems are the result of the physics of backscatter, the geometry of the sensors, fluctuations in the propagation medium, clutter in signals due to echo overlap, limited lateral resolution, and lack of information processing algorithms. The speed of sensing is limited by the speed of sound. When multiple sensors are used they can interfere with one another. Each of these problems can be modelled and algorithms developed from these models to overcome them

The aim of modelling is to provide a theoretical basis for developing algorithms for processing the information in the echo. Several models have proved to be useful. However, to achieve the long term goal of recognising objects by analysing the radiation scattered off them, we need models that capture the geometry of the object, are computable within a reasonable time, work in the far field, and have the potential to be inverted.

SENSING MODEL

Ideally, a sensing system should be able to locate an object in 3 dimensions: range, bearing and elevation. Also, it should be able to locate all the objects relevant to the current task. Currently, the most common arrangement for ultrasonic sensors is a ring. Range readings are usually recorded on a horizontal plane through the sensors and the resultant map is assumed to represent the floor plan of the environment. With a ring the robot is unable to distinguish between echoes from the floor, the walls and the ceiling.

While a robot moves on a plane, it has height, and the sensing plane is above floor level. A 3D sensing system will enable the robot to determine if the echo is coming from an object on the floor, such as an electrical chord, a overhead object, such as a cupboard door, or an object it could run into. The size of the field of insonification and the height of the sensors above the floor determine how much of the floor and ceiling are detected [13].

TRANSDUCER MODELS

An ultrasonic transducer produces a spherical wave front where the pressure amplitude in the far field is a function of angle to the axis of the transducer and the applied voltage. Thus the directivity function of a round ultrasonic transducer is a conical beam, which is modelled with an impulse model [9], a Gaussian function [7] or a Bessel function [11].

The Bessel function is the radiating plane piston model of the transducer and models the side lobes as well as the main lobe. This model is reasonably accurate for pulsed signals. The Gaussian function models the main lobe accurately for transducers with a wide bandwidth but does not model the side lobes.

PROPAGATION MODELS

In echolocation, variations in the characteristics of the air that the pulse and echo propagate through affect both the time of flight and the amplitude of the echo. Variations in air temperature, pressure and humidity can be compensated for by measuring them and applying appropriate equations to the results. The equation for amplitude variation with environment conditions is quite complex [2].

Temperature measurements can be used to calibrate time of flight. But, local temperature gradients that occur within one metre of the ground and gusts of wind create local variations that are difficult to measure. In a practical system, regular measurement of an object at a known location relative to the sensor provides a simple method of calibration without measuring environment parameters.

ECHO DETECTION MODELS

All sensing systems rely heavily on measuring the time of flight. Accurate times of flight can be obtained by matching either the echo with the transmitted pulse or the envelope of the echo with the envelope of the pulse [15]. Kleeman and Kuc [8] model the shape of the pulse in order to obtain accurate time of flight measurements when correlating the echo with the transmitted pulse.

ENVIRONMENT MODELS

Building a 3D model of the world with ultrasonic sensors requires robust algorithms for detecting 3D acoustic features: geometric primitives that are robustly detectable. These features can be classified as flat surfaces (planes), lines (intersections of 2 planes), points (intersections of 3 or more planes), curved surfaces and intersections of curved surfaces with planes. Complex features consist of a number of these in close proximity. Akbarally and Kleeman [1] identify nine acoustic features that occur in a room where all the planes are vertical or horizontal and meet at right angles.

IMPULSE MODEL

Kuc and Siegel [9] model a receiver by dividing it up into elemental areas and summing their inputs to find the impulse response of a receiver. As the transmitter directivity function is similar to that of the receiver, the impulse response of the transmitter/receiver pair is found by self convolution. The result is a general equation that gives the impulse response in terms of transducer radius and inclination angle. A similar approach is used to model the impulse response of walls, corners and edges.

Kleeman and Kuc [8] take this approach further to accurately measure of time of flight, by matching the received echo with the transmitted echo. They measure the impulse responses of the transmitter, the receiver, the air and a plane reflector. From these they obtain a model of the wave shape of the echo. By cross-correlating an echo with this modelled wave shape they can find the exact time of flight of a pulse.

ARC MODEL

Significant improvements in ultrasonic sensing have been made with simple models. The development of the arc model [11] explained many of the errors that people had reported with ultrasonic sensing. A circular single-element transducer has a conical field of insonification and the wave front is a circular section of a spherical shell. The edges of the field of insonification are set at the angle where the pressure amplitude is n dB down on the axial pressure, where n is determined by the task and the characteristics of the transducer.

In 2 dimensions this cone maps into a sector and the wave front into an arc. The pressure on the arc varies with angle, as discussed above, but in a simple arc model, the pressure on the arc is assumed to be constant with angle. A range reading is the distance to an object lying on that arc. It is a one dimensional value and contains no information about the bearing or elevation of the object except that it lies within the field of insonification. If the surface is specular the object is at an orientation nearly orthogonal (within the beam angle) to the axis of the transducer.

From this model, McKerrow [12] developed a robust algorithm for extracting surface segments by fusing ultrasonic range data with motion to produce outline segment maps of a room. Further research [14] has shown that this feature extractor breaks down in a consistent way at corners and edges enabling their identification.

REGIONS OF CONSTANT DEPTH MODEL

While the regions of constant depth (RCD) maps produced by Leonard and Durrant-Whyte [10] look similar to the arc maps produced by McKerrow they represent different data. An RCD is the sweep angle over which a panning sensor at a fixed location will measure the same range. Hence, the plot of the range readings taken while panning through this angle is an arc, and consequently an RCD map drawn from pans at several locations looks like the arc map.

RCDs occur on surfaces and at corners because the multi-path reflections from a corner produce the same range readings as a surface located at the corner. They take longer to measure than the arc model because several readings are required to construct each RCD. However, scanning enables the RCD to give an indication of the specularity and hence detectability of a surface, making an RCD a useful feature for localisation.

VIRTUAL SOURCE MODEL

Replacing the individual sensors with binaural sensors enables the sensor to accurately measure bearing as well as range. The virtual source model [3] is fundamental to thinking about binaural sensing. In it, geometry is used to model sets of sensors [5, 8, 17]. Some of these models are complex and approximations are used to reduce computation time. All rely on accurate measurement of the time of flight to obtain high resolution, so thresholding systems are inadequate.

STEREO CORRESPONDENCE MODEL

To achieve the goal of finding the range and bearing of multiple objects, binaural systems must handle multiple echoes. Binaural sensing introduces the stereo correspondence problem [16]. Like vision, there is a trade off between resolution and matching accuracy when selecting the distance between transducers.

Increasing the separation between the receivers, increases the accuracy of the bearing calculation because the difference in the times of flight is increased. But the ability to resolve between two objects is decreased. Reducing the separation between the receivers makes the matching of the echoes easier because the range of time of flight value over which matching is possible is reduced.

ECHO FORMATION MODEL

When a pulse of ultrasonic energy strikes an object it is scattered. The directions in which it is scattered and the amplitudes of the scattered pulses are determined by the geometry and acoustic characteristics of the surface. The process of echo formation is complex and includes direct scattering, multiple scattering and diffraction.

Freedman [6] developed a model of direct scattering for underwater sonar that predicts the shape and amplitude of echoes that return to a receiver, coincident with the transmitter, from the geometry of the object. This model assumes that the wavelength is smaller than the dimensions and radii of curvature of the object. As the wavelength at 50 KHz is 6.8 mm, this model has the potential to apply to many mobile robot situations.

Crook and McKerrow [4] transferred this model to air and added a radiating plane piston model of the transducer. The model (Equation 1) calculates the voltage *E(g)* produced at the receiver by echoes from the surface of the object at range *g*. The receiver voltage is the pressure of the echoes multiplied by the directivity function of the receiver for a transmitted pulse with source pressure A_0.

$$E(g) = \frac{-A_0}{\lambda r_m^2} e^{j(\omega t - 2kr_g)} e^{-\alpha 2 r_g} \sum_{n=0}^{\infty} \frac{D(A_W, g, n)}{(j2k)^n} \tag{1}$$

The right hand side of Equation 1 calculates the echo amplitude as a function of the incident pressure and the object geometry. The incident pressure is the pressure of the transmitted sinusoid $A_0 e^{j(wt - 2kr_g)}$ at range r_g and time t attenuated by the spreading loss $1/r_m$ and the attenuation in air $(e^{-a2r_g})/r_m$ where r_m is the mean range to the object and a is the attenuation constant of air.

This model calculates the effect of object geometry on the echo. Freedman showed that the pressure of each echo component that originates from range r_g is proportional to the sum of the discontinuities in the derivatives of the projected area of the object at that range. The projected area A_W at range g is the area of intersection of a plane orthogonal to the axis of the transducer with the object at range g, as shown for a sphere in Figure 1. For convex objects this area is a minimum at the point nearest to the transducer and increases to a maximum at the equator of the object - that is the plane furtherest from the transducer that is insonified. Thus it is a curve with a slope greater than or equal to zero.

When this curve is differentiated, discontinuities occur at points where there are instantaneous changes in the slope. Echoes are initiated at planes through the object where these discontinuities in the derivatives of the projected area occur. The amplitude of the echoes is proportional to the sum of the magnitudes of these discontinuities.

The summation includes wave number k raised to the power n in the denominator, where n is the order of the derivative. Thus, for large values of k, the significance of the higher order derivatives decreases rapidly. For 50KHz waves in air $k = 2\pi/\lambda = 1{,}000$, so the contribution of each succeeding derivative is 60 dB down on the previous one. Thus, the third derivative is 120dB down on the first and, hence, unlikely to make any measurable contribution to the echo.

In Figure 1, $A(r)$ denotes the projected area at range r and $d^nA(r)/dr^n$ denotes the nth derivative of the projected area with respect to r. For scattering from a sphere, discontinuities occur at the front face of the sphere (range r_1) and at the equator (range r_2) and the values for each of the derivatives of the projected area are:

$A(r) = \pi y^2 = \pi(a^2 - (r_1+a-r_2)^2)$ $\qquad$ $A^1(r) = 2\pi(r_1+a-r_2)$

$A^2(r) = -2\pi$ $\qquad$ $A^3(r) = 0$

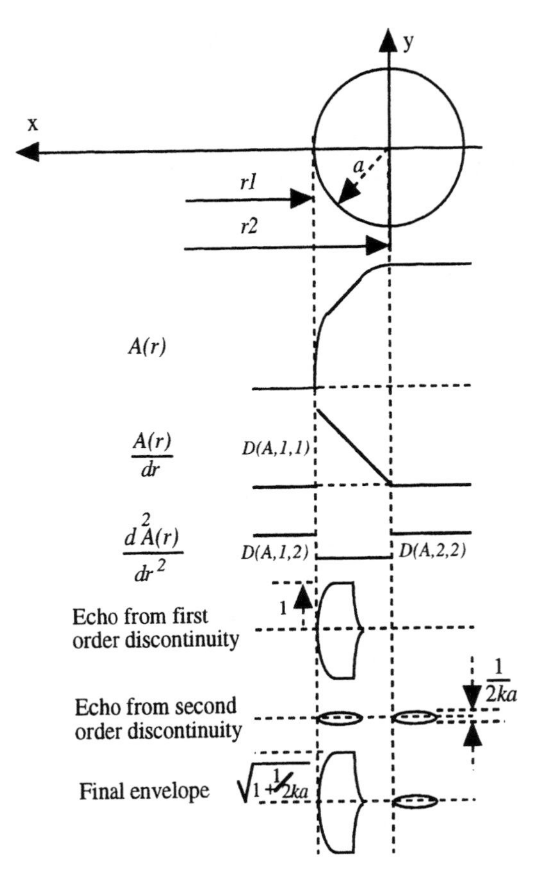

Figure 1. Pictorial representation of echo components generated by a sphere

Therefore, the magnitude at the discontinuities r_1 and r_2 are :

$D(A,1,0) = 0$	$D(A,1,1) = -2\pi a$	$D(A,1,2) = 2\pi$
$D(A,2,0) = 0$	$D(A,2,1) = 0$	$D(A,2,2) = -2\pi$

By substituting these values into Equation 1 the echo components and the final echo envelope can be determined as shown graphically in Figure 1. Crook and McKerrow [4] validated this model for symmetrical convex objects with diameters from 10λ to 40λ over the range 60λ to 140λ.

CONCLUSION

In order to improve the performance of ultrasonic sensors researchers are formulating a theory of ultrasonics sensing. This theory consists of a set of models and algorithms based on these models. This work will continue until a set of simple robust models is obtained, as many current models are either too computationally expensive or give inadequate results in some common situations. As each new model is developed, it is applied to different tasks to determine if robust algorithms can be derived from it.

Freedman's model is useful for predicting the echo from objects with simple shape. Echo components occur when there are discontinuities in the first two derivatives of projected area. The resultant echo is the superposition of these components. We are experimenting with a range of simple shapes, that are found in indoor environments, to see if we can classify these based on the echo predicted by Freedman's model.

REFERENCES

1. Akbarally, H. and Kleeman, L. "A Sonar Sensor for Accurate 3D Target Localisation and Classification", *Proceedings IEEE International Conference on Robotics and Automation*, Nagoya, (1995) 3003-3008.
2. American National Standard, *Method for the Calculation of the Absorption of Sound by the Atmoshpere*, Acoustical Society of America, ANSI S1.26-1978 (1978).
3. Barshan, B. and Kuc, R. "Differentiating Sonar Reflections from Corners and Planes by Employing an Intelligent Sensor", *PAMI*, 12 (6), (1990) 560-569.
4. Crook, D. and McKerrow, P.J. "Models of ultrasonic sensing", *Proceedings of the National Conference of the Australian Robot Association,* ISBN 0 9587583 0 1, Melbourne, (1995) 158 - 165.
5. D'Sousza, R.P. and McKerrow, P.J. "A Model for Ultrasonic Binaural Sensing", *Proceedings of the National Conference of the Australian Robot Association,* ISBN 0 9587583 0 1, Melbourne, (1995) 166-177.
6. Freedman, A. "A Mechanism of Acoustic Echo Formation", *Acustica*, 12, (1962) 10-21.
7. Hong, M.L. and Kleeman, L. "A Low Sample Rate 3D Sonar Sensor for Mobile Robots", *Proceedings IEEE International Conference on Robotics and Automation*, Nagoya, (1995) 3015-3020.
8. Kleeman, L. and Kuc, R. "Mobile Robot Sonar for Target Locaslization and Classification", *IJRR*, 14 (4), (1995) 295-318.
9. Kuc, R. and Siegel, M.W. "Physically Based Simulation Model for Acoustic Sensor Robot Navigation", *PAMI*, 9, (6), November, (1989) 766-778.
10. Leonard, J.J., Durrant-Whyte, H.F. "Mobile Robot Localisation by Tracking Geometric Beacons", *IEEE Trans on R and A,* 7 (3)(1991), 376-382.
11. McKerrow, P.J. and Hallam, J.C.T. "An Introduction to the Physics of Echolocation", *Proceedings of the Third National Conference on Robotics, Melbourne*, ARA, (1990) 198-209.
12. McKerrow, P.J. "Echolocation - From Range to Outline Segments", *Robotics and Autonomous Systems*, Elsevier, 11, (4), (1993) 205-211.
13. McKerrow, P.J. "Progress in Ultrasonic Sensing for Service Robots", *Proceedings of First workshop on Robotics ior the Service Industries* , IARP, invited, ISBN 0 646 24178 8, Sydney, May, (1995) 44-52.
14. McKerrow, P.J. "Robot Perception with Ultrasonic Sensors using Data Fusion", Proceedings IEEE SMC Conference, October, Vancouver, (1995) 1380-1385.
15. Peremans, H., Audenaert, K. and Van Campenhout, J.M. "A High-Resolution Sensor Based on Tri-aural Perception", *IEEE Transactions on Robotics and Automation*, 9, (1), 36-48 (1993).
16. Peremans, H. "A maximum likelihood algorithm for solving the correspondence problem in tri-aural perception", *Proceedings IEEE International Conference on Multisensor Fusion and Integration for Intelligent Systems*, (1993) 485-492.
17. Sabatini, A.M. and Benedetto, O.D. "Towards a Robust Methodology for Mobile Robot Localisation Using Sonar", *Proceedings IEEE Robotics and Automation Conference*, (1994) 3142-3147.

TASK-ORIENTED DYNAMIC MODELING OF TWO COOPERATING ROBOTS

R. Mattone, A. De Luca

Dipartimento di Informatica e Sistemistica (DIS)
Università di Roma "La Sapienza"
Via Eudossiana 18, 00184 Roma, Italy

mattone@labrob.ing.uniroma1.it adeluca@giannutri.caspur.it

ABSTRACT

We present a general formalism for deriving a control-oriented model of cooperating tasks for multi-robot systems and its specific application to a cell with two different robots involved in a mechanical finishing operation. The modeling approach is flexible enough so as to cover different possible interaction types between the end-effector of each robot and the common payload and can be easily applied even to complex cooperating tasks, as in the considered example. Since the interesting aspects and quantities needed for the correct execution of the task are intrinsically characterized, this description of robotic tasks is useful for control purposes.

KEYWORDS: Cooperating robots, Modeling, Force-motion control.

1. INTRODUCTION

Modeling and control of cooperating robots have been considered extensively in the literature (see, e.g., [1–3]). However, most papers are restricted by a common set of assumptions. In particular: the payload is a single rigid object; all robots have similar mobility; each robot rigidly grasps the payload (power grasp); the contact points are fixed in the object frame; no other environmental constraints are imposed on the payload. With these assumptions, the overall task description is obtained by coupling, through the exchanged contact forces, the dynamic model of the object with the dynamics of the multiple robots. From the control point of view, hybrid tasks are then defined in a direct way in terms of tracking the object motion and regulating the internal forces. The force/load distribution among the robots can be formulated as a optimization problem [4–5].

Indeed, the above modeling assumptions can be relaxed or generalized. Handling of a flexible object is analyzed in [6], while the case of payloads consisting of multiple rigid bodies is considered in [7]. Different grasping types have been classified using grasp matrices in [8]. In particular, point contacts are assumed in [9], where the coordination of robots pushing against a common object is studied. Much research is devoted to the case of rolling contacts (a nonholonomic constraint) in multi-fingered dextrous hands [10]. Whenever the grasp of a single robot does not determine completely the object pose, a model of the friction at the contact is also required. The analysis of this effect is usually performed in quasi-static conditions.

In the above works, specific robotic systems have been considered without using a unified modeling approach that could cover them altogether. This prevents a systematic generation of the governing equations and the general analysis of control aspects. On the other hand, in [11] we have started developing a task-based modeling approach of cooperating robots that complies with all the above open issues. The key point is a parametric description of the given task, with the introduction of suitable sets of task variables. As an example, general grasping conditions are explicitly characterized by means of kinematic variables describing the feasible relative motion in the robot-payload interaction. In this way, even moving contact points on the payload surface can be taken into consideration. Furthermore, specific variables can be defined along directions where friction (of any type) acts during dynamic operation.

Although the formalism can be extended so as to model the presence of external constraints, of rolling contacts, or of payload with extra degrees of freedom (see [7]), we limit here our analysis to the physical conditions occurring in the robotic cell used as a case study.

In the next section we briefly outline the modeling steps and the associated definition of kinematic, static, and dynamic quantities. An automatic decomposition of contact forces is provided, based on energy-transfer (and thus, unit invariant) arguments. Then, we apply our approach to a robotic cell constituted by a 4-dof SCARA robot arm and a 6-dof parallel platform (SmartEE).

2. COOPERATING ROBOTS-ENVIRONMENT MODEL

2.1 Kinematics

Consider a system of $m = 2$ robots with n_i joints ($i = 1, 2$) that cooperate in the execution of a task defined on a common payload, that may possibly be subject also to other environmental constraints. We assume that the contact between each robot and the payload is never lost and that, during the whole task execution, the contact points are either fixed or mobile. For the mobile contacts, the associated friction effects are considered explicitly.

Each manipulator configuration is identified by the joint variables vector $\mathbf{q}_i \in \mathbb{R}^{n_i}$, while its end-effector pose is given by a vector $\mathbf{p}_i \in \mathbb{R}^{g_i}$, being g_i the number of degrees of freedom (dofs) of motion of the ith robot end-effector. The end-effector pose collects the position $\mathbf{r}_i \in \mathbb{R}^{g_{r_i}}$ and a minimal representation of the orientation $\mathbf{o}_i \in \mathbb{R}^{g_{o_i}}$ (e.g., Euler angles if $g_{o_i} = 3$), where g_{r_i} and g_{o_i} respectively are the number of linear and rotational dofs of the ith robot end-effector and $g_i = g_{r_i} + g_{o_i}$. The end-effector velocity is $\mathbf{v}_i = (\dot{\mathbf{r}}_i, \omega_i) \in \mathbb{R}^{g_i}$, related to the time derivative $\dot{\mathbf{p}}_i$ of the pose by a matrix transformation depending in general on $\mathbf{o}_i$. Then, the end-effector kinematics is described *from the* i*th robot side* by

$$\mathbf{p}_i = \mathbf{k}_i(\mathbf{q}_i), \qquad \mathbf{v}_i = \mathbf{J}_i(\mathbf{q}_i)\dot{\mathbf{q}}_i, \qquad i = 1, 2, \tag{1}$$

where $\mathbf{J}_i(\mathbf{q}_i)$ is the geometric Jacobian of the ith robot.

Let l be the total number of parameters describing the payload dynamics and e the number of *extra dofs* that describe the payload dynamics when the contact points are kept fixed. Thus, $d = l - e$ will be the number of dynamic coordinates needed to characterize the grasp. Then, the payload configuration is identified by a set of generalized coordinates $\mathbf{s}_L \in \mathbb{R}^l$ (needed to describe the *environment* dynamics). Vector $\mathbf{s}_L$ splits in general in two vectors of generalized coordinates: $\mathbf{s}_{LE} \in \mathbb{R}^e$ for the extra dofs and $\mathbf{s}_{LD} \in \mathbb{R}^d$ that completes the description of the payload configuration and is necessary to characterize the grasp. From now on, we suppose that there are no extra dofs in the payload ($e = 0$). Depending on the type of contact between each robot and the object, other variables are in general necessary, in addition to $\mathbf{s}_{LD}$, to complete the description of the pose of the ith robot end-effector *from the environment side*. These variables describe the relative motion between the robots and the payload and may be partitioned in those on which the generalized friction forces at the contact perform work, $\mathbf{s}_{F_i} \in \mathbb{R}^{f_i}$ (*friction* variables), and those on which no work is performed by any force, $\mathbf{s}_{K_i} \in \mathbb{R}^{k_i}$ (*kinematic* variables). Then, we can write:

$$\mathbf{p}_i = \mathbf{\Gamma}_i(\mathbf{s}_{LD}, \mathbf{s}_{F_i}, \mathbf{s}_{K_i}), \qquad \mathbf{v}_i = \mathbf{T}_i(\mathbf{s}_{LD}, \mathbf{s}_{F_i}, \mathbf{s}_{K_i}) \begin{bmatrix} \mathbf{s}_{LD} \\ \mathbf{s}_{F_i} \\ \mathbf{s}_{K_i} \end{bmatrix}, \qquad i = 1, \ldots, m. \quad (2)$$

Matrix $\mathbf{T}_i$ can be partitioned as $\mathbf{T}_i = [\mathbf{T}_{D_i} \;\; \mathbf{T}_{F_i} \;\; \mathbf{T}_{K_i}]$, expliciting the contributions respectively due to *dynamic*, *friction*, and *kinematic* degrees of freedom in the robot-environment interaction. This matrix is assumed to be full *column* rank, at least in the region of interest for the task execution. Its columns are generalized directions used as a basis for the vector space of the admissible end-effector velocities.

Equations (1) and (2), written for each robot separately, can be rewritten for the whole robotic system as

$$\mathbf{p} = \mathbf{k}(\mathbf{q}) = \mathbf{\Gamma}(\mathbf{s}), \qquad \mathbf{v} = \mathbf{J}(\mathbf{q})\dot{\mathbf{q}} = \mathbf{T}(\mathbf{s})\dot{\mathbf{s}}, \quad (3)$$

where $\mathbf{q} = (\mathbf{q}_1, \mathbf{q}_2)$, $\mathbf{p} = (\mathbf{p}_1, \mathbf{p}_2)$, $\mathbf{v} = (\mathbf{v}_1, \mathbf{v}_2)$, $\mathbf{s} = (\mathbf{s}_{LD}, \mathbf{s}_{F_1}, \mathbf{s}_{F_2}, \mathbf{s}_{K_1}, \mathbf{s}_{K_2})$, $\mathbf{k} = (\mathbf{k}_1, \mathbf{k}_2)$, $\mathbf{\Gamma} = (\mathbf{\Gamma}_1, \mathbf{\Gamma}_2)$, and

$$\mathbf{J} = \begin{bmatrix} \mathbf{J}_1 & \mathbf{O} \\ \mathbf{O} & \mathbf{J}_2 \end{bmatrix}, \quad \mathbf{T} = \begin{bmatrix} \mathbf{T}_{D_1} & \mathbf{T}_{F_1} & \mathbf{O} & \mathbf{T}_{K_1} & \mathbf{O} \\ \mathbf{T}_{D_2} & \mathbf{O} & \mathbf{T}_{F_2} & \mathbf{O} & \mathbf{T}_{K_2} \end{bmatrix} = [\mathbf{T}_D \, \mathbf{T}_F \, \mathbf{T}_K]. \quad (4)$$

2.2 Statics

The generalized forces exchanged between the ith end-effector and the payload are collected in a g_i-dimensional vector $\mathbf{F}_i$ of forces and torques. In a dual way with respect to (3), the vector of all contact forces $\mathbf{F} = (\mathbf{F}_1, \mathbf{F}_2)$ can be parameterized as $\mathbf{F} = \mathbf{Y}(\mathbf{s})\lambda$. The columns of matrix $\mathbf{Y}$ are generalized directions used as a basis for the vector space of the contact forces. They are such that $\mathbf{T}_K^{\mathrm{T}}\mathbf{Y} = \mathbf{0}$, since contact forces do not perform work on kinematic degrees of freedom. Matrix $\mathbf{Y}$ is assumed to be full rank, and thus the dimension of the vector of force parameters λ will be $g - k$ (with $g = g_1 + g_2$, $k = k_1 + k_2$), namely the difference between the total number of dofs of the end-effectors and the total number of kinematic dofs.

At each $\mathbf{s}$, the vector space of contact forces, *span*$[\mathbf{Y}(\mathbf{s})]$, can be decomposed into two subspaces. The subspace *span*$[\mathbf{Y}_R(\mathbf{s})]$ of reaction forces, i.e., those forces which do not cause payload motion, is such that $[\mathbf{T}_D \, \mathbf{T}_K]^{\mathrm{T}} \mathbf{Y}_R = \mathbf{0}$ and has dimension $g - k - d$. Note that the generalized forces in *span*$[\mathbf{Y}_R(\mathbf{s})]$ may perform work on the $\mathbf{s}_F$ variables, hence $\mathbf{T}_F^{\mathrm{T}}\mathbf{Y}_R \neq \mathbf{0}$ in general. The subspace *span*$[\mathbf{Y}_A(\mathbf{s})]$, defined as the complement of $\mathbf{Y}_R$ in $\mathbf{Y}$, is the d-dimensional subspace of active forces responsible for payload motion. Any matrix $\mathbf{Y}_A$, whose columns are used as a basis for this subspace, is such that $\mathbf{T}_D^{\mathrm{T}}\mathbf{Y}_A$ is nonsingular. Following the decomposition of $\mathbf{Y}(\mathbf{s})$, $\mathbf{Y} = [\mathbf{Y}_A \, \mathbf{Y}_R]$, a partition is induced on the parameter vector $\lambda = (\lambda_A, \lambda_R)$. We call in the sequel *dynamic directions* those in the span of the columns of either $\mathbf{T}_D$ or $\mathbf{Y}_A$.

Any choice satisfying the above properties is generally admissible for matrix $\mathbf{Y}$ and, correspondingly, for matrices $\mathbf{Y}_R$ and $\mathbf{Y}_A$. On the other hand, the available freedom in this choice can be used to achieve a physical correspondence of the components of λ with either forces or torques. In particular, λ can be normalized so that the components of λ_R represent the so-called *internal forces* in the robot-payload interaction, while λ_A are the net active forces producing motion of the payload. We recognize that different choices of $\mathbf{Y}$ correspond, for a given task, to different load distributions among the manipulators. From a control point of view, choosing one of the admissible forms for $\mathbf{Y}$ is equivalent to imposing a given load distribution among the robots.

Some further considerations are needed in the presence of friction. In the generalized directions along which friction is effective, a general relation of the form $\Phi(\mathbf{s}_F, \dot{\mathbf{s}}_F, \lambda) = \mathbf{0}$ holds between the generalized coordinates $\mathbf{s}_F$, their derivatives and the parametrization of the generalized contact forces λ. For simplicity, suppose that the grasp is capable to compensate the friction forces arising at the moving contact points (*force closure*). In this case, $\mathbf{Y}$ can be chosen so that all generalized friction forces are parametrized by some components of the subvector λ_R alone. This natural choice implies that friction will not be used for achieving payload motion. Then, we model the friction at the contact points through the following f-dimensional equation (with $f = f_1 + f_2$):

$$\lambda_{R_F} = \mu_F(\mathbf{s}_F, \lambda_{R_P}, \lambda_A) + \nu_F(\mathbf{s}_F, \lambda_{R_P}, \lambda_A)\dot{\mathbf{s}}_F, \tag{5}$$

where $\lambda_R = (\lambda_{R_F}, \lambda_{R_P})$ has been partitioned so that the f-dimensional subvector λ_{R_F} parametrizes the generalized friction forces. The functions μ_F and ν_F model *dry* and *viscous* friction, respectively. Following the decomposition of λ_R, a partition is induced on $\mathbf{Y}_R = [\mathbf{Y}_{R_F}\ \mathbf{Y}_{R_P}]$. As a result, *span*$[\mathbf{Y}_{R_P}(\mathbf{s})]$ is the subspace of generalized contact forces that do not perform work on any degree of freedom, while forces in *span*$[\mathbf{Y}_{R_F}(\mathbf{s})]$ perform work only on the $\mathbf{s}_F$ variables. Thus, matrix $\mathbf{Y}_{R_F}$ will be such that $\mathbf{T}_F^{\mathrm{T}}\mathbf{Y}_{R_F}$ is square and nonsingular.

2.3 Dynamics

Following the Lagrangian approach, the dynamic model of the overall system of two robots and payload can be written in the standard form as

$$\mathbf{B}(\mathbf{q})\ddot{\mathbf{q}} + \mathbf{n}(\mathbf{q}, \dot{\mathbf{q}}) = \mathbf{u} - \mathbf{J}^{\mathrm{T}}(\mathbf{q})\mathbf{F}, \tag{6}$$

$$\mathbf{B}_L(\mathbf{s}_{LD})\ddot{\mathbf{s}}_{LD} + \mathbf{n}_L(\mathbf{s}_{LD}, \dot{\mathbf{s}}_{LD}) = \mathbf{T}_D^{\mathrm{T}}(\mathbf{s})\mathbf{F}, \tag{7}$$

with forces $\mathbf{F} = \mathbf{Y}(\mathbf{s})\lambda$ acting from the robots to the object and where

$$\mathbf{B}(\mathbf{q}) = \begin{bmatrix} \mathbf{B}_1 & \mathbf{O} \\ \mathbf{O} & \mathbf{B}_2 \end{bmatrix}, \quad \mathbf{n}(\mathbf{q}, \dot{\mathbf{q}}) = \begin{bmatrix} \mathbf{n}_1 \\ \mathbf{n}_2 \end{bmatrix}, \quad \mathbf{u} = \begin{bmatrix} \mathbf{u}_1 \\ \mathbf{u}_2 \end{bmatrix},$$

with the usual definitions of inertia, Coriolis, centrifugal, and gravity terms.

The couplings between the dynamics of the robots and of the payload are given by the constraints (3) on the end-effectors pose and velocity. In order to obtain a more compact dynamic model that is useful for control purposes, the joint accelerations $\ddot{\mathbf{q}}$ can be explicited from the robots dynamic model (6), and substituted in the differentiated expression of the contact constraints (3). Two alternatives are then available, corresponding to the elimination of the accelerations $\ddot{\mathbf{s}}_{LD}$ or, respectively, of the active forces λ_A from the model equations. In fact, we can explicit one of these two sets of motion and force parameters in terms of the other (and of $\mathbf{s}_K$ and $\mathbf{s}_F$), using the dynamic model of the payload (7). Due to the definition of $\mathbf{s}_{LD}$ and λ_A, both choices are always feasible. When friction is present at the contact, the friction model (5) can be used to further eliminate the explicit appearance of λ_{R_F} in the dynamic equations.

As a result, we obtain the following alternative forms for system description:

$$\mathbf{Q}(\mathbf{q},\mathbf{s})\begin{bmatrix}\lambda_A\\ \lambda_{R_P}\\ \ddot{\mathbf{s}}_F\\ \ddot{\mathbf{s}}_K\end{bmatrix} = \mathbf{m}(\mathbf{q},\dot{\mathbf{q}},\mathbf{s},\dot{\mathbf{s}}) + \mathbf{J}\mathbf{B}^{-1}\mathbf{u}, \tag{8}$$

$$\widehat{\mathbf{Q}}(\mathbf{q},\mathbf{s})\begin{bmatrix}\ddot{\mathbf{s}}_{LD}\\ \lambda_{R_P}\\ \ddot{\mathbf{s}}_F\\ \ddot{\mathbf{s}}_K\end{bmatrix} = \widehat{\mathbf{m}}(\mathbf{q},\dot{\mathbf{q}},\mathbf{s},\dot{\mathbf{s}}) + \mathbf{J}\mathbf{B}^{-1}\mathbf{u}. \tag{9}$$

All the involved matrices and vectors depend on the dynamic models of the robots and the payload, and on the chosen parametrization of the system motion. In particular, $\mathbf{m}$ and $\widehat{\mathbf{m}}$ depend also on the friction model through μ_F and ν_F. Examples of the actual form of the terms in eqs. (8) and (9) can be found in [7]. It can be shown that, under mild hypotheses, both $\mathbf{Q}$ and $\widehat{\mathbf{Q}}$ are nonsingular.

Equations (8) and (9) serve as a basis for the design of model-based hybrid motion/force controllers. In particular, by means of an input-output nonlinear control law, we can impose desired and independent evolutions either to the set $(\lambda_A, \lambda_{R_P}, \mathbf{s}_F, \mathbf{s}_K)$ or to the set $(\mathbf{s}_{LD}, \lambda_{R_P}, \mathbf{s}_F, \mathbf{s}_K)$ (see also [11]).

3. APPLICATION TO AN EXPERIMENTAL ROBOTIC CELL

In this section the described formalism is applied to the modeling of a cell with two cooperating robots, available at DIS for experiments in mechanical finishing. The cell consists of a 6-dof parallel platform (SmartEE Hughes) and a 4-dof SCARA-type serial manipulator (IBM 7545), both equipped with an ATI force/torque sensor at their end. The platform can be arbitrarily positioned and oriented in its workspace, so $g_{r_1} = g_{o_1} = 3$ and $g_1 = 6$. Its end-effector is assumed coincident with the center of the top plate. The SCARA end-effector can reach any point in its cartesian workspace, while it can only rotate about the fixed Z axis (normal to the basement plane). Hence, $g_{r_2} = 3$, $g_{o_2} = 1$, and $g_2 = 4$. A total of $g = 10$ end-effector dofs are involved. A CAD picture of the cell is shown in Fig. 1.

The typical task can be described as follows. The payload is a rigid object (*workpiece*) of general but smooth form carried by the SmartEE platform and rigidly attached to it. The platform can be arbitrarily positioned and oriented in its workspace. The SCARA end-effector is equipped with a milling tool that moves in contact with the object surface. The orientation of the tool is constrained to a single direction because of the limited degrees of freedom of the SCARA arm. The interaction between tool and workpiece can be modeled as a point contact. In this interaction normal and tangential forces arise, the latter being caused by the existing friction in the sliding motion. Forces and torques acting on the workpiece can be measured at the contacts with the SCARA and with the SmartEE respectively, through the two 6D-sensors. The joint positions of the two robots are measured with encoders. The mechanical finishing operation requires a prescribed motion trajectory of the tool on the workpiece surface while controlling the exerted normal force.

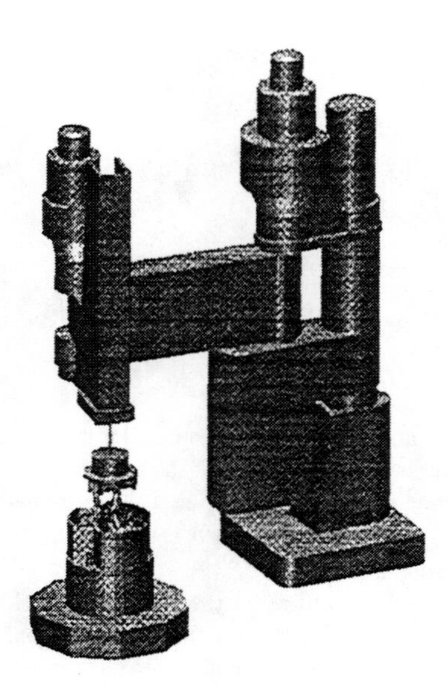

Fig. 1: CAD picture of the experimental robotic cell at DIS

The object can move in the free space within the described grasp configuration, so that $l = 6$. The vector $\mathbf{s}_L$ can be chosen as

$$\mathbf{s}_L = (\mathbf{r}_O, \mathbf{e}_O), \tag{10}$$

being $\mathbf{r}_O$ the absolute position of the workpiece center of mass and $\mathbf{e}_O$ a minimal representation of the orientation of a frame attached to it. The payload has no extra dofs (its pose is completely determined by the positions of the contact points with the two robots), so that $\mathbf{s}_L = \mathbf{s}_{LD}$ and $l = d$.

As the workpiece is rigidly grasped by the platform, the dynamic variables $\mathbf{s}_{LD}$ uniquely determine the pose $\mathbf{p}_1$ of the platform, yielding $f_1 = k_1 = 0$. Three other variables are needed instead to identify the SCARA end-effector pose, as its position on the object surface and its orientation vary during the task. In particular, we need two local coordinates defined on the object surface and the absolute orientation of the SCARA end-effector about the fixed Z axis. Friction forces act tangentially to the contact surface, while we assume that the friction torque about the normal direction is negligible. Thus, the two surface coordinates are collected in the vector $\mathbf{s}_F$ of friction variables, while the orientation angle is the single kinematic variable s_K. Hence, $f = f_2 = 2$, $k = k_2 = 1$, $\mathbf{s} = (\mathbf{s}_{LD}, \mathbf{s}_F, s_K) \in I\!R^9$, and

$$\mathbf{p}_1 = \begin{bmatrix} \mathbf{r}_O + {}^B\mathbf{R}_O(\mathbf{e}_O)^O\mathbf{r}_{c_1} \\ \mathbf{e}\left({}^B\mathbf{R}_O(\mathbf{e}_O)^O\mathbf{R}_{R_1}\right) \end{bmatrix} = \mathbf{\Gamma}_1(\mathbf{s}), \tag{11}$$

$$\mathbf{p}_2 = \begin{bmatrix} \mathbf{r}_O + {}^B\mathbf{R}_O(\mathbf{e}_O)^O\mathbf{r}_{c_2}(\mathbf{s}_F) \\ s_K \end{bmatrix} = \mathbf{\Gamma}_2(\mathbf{s}), \tag{12}$$

where ${}^B\mathbf{R}_O(\mathbf{e}_O)$ is the rotation matrix representing the orientation of the object with respect to the base frame, ${}^O\mathbf{R}_{R_1}$ defines the *constant* relative orientation between the frame attached to the object and the platform end-effector, ${}^O\mathbf{r}_{c_i}$ is the vector locating the ith contact point in the object frame, and $\mathbf{e}(\mathbf{R})$ is a 3-dimensional vector that extracts a minimal representation of the orientation from the rotation matrix $\mathbf{R}$. From (11) and (12), it follows

$$\mathbf{v}_1 = \begin{bmatrix} \mathbf{I}_{3\times3} & \frac{\partial}{\partial \mathbf{e}_O}\left[{}^B\mathbf{R}_O(\mathbf{e}_O)^O\mathbf{r}_{c_1}\right] \\ \mathbf{O}_{3\times3} & {}^B\mathbf{R}_O(\mathbf{e}_O)^O\mathbf{K}(\mathbf{e}_O) \end{bmatrix} \begin{bmatrix} \dot{\mathbf{r}}_O \\ \dot{\mathbf{e}}_O \end{bmatrix} = \mathbf{T}_{D_1}(\mathbf{s}_{LD})\dot{\mathbf{s}}_{LD}, \tag{13}$$

$$\begin{aligned} \mathbf{v}_2 &= \begin{bmatrix} \mathbf{I}_{3\times3} & \frac{\partial}{\partial \mathbf{e}_O}\left[{}^B\mathbf{R}_O(\mathbf{e}_O)^O\mathbf{r}_{c_2}(\mathbf{s}_F)\right] \\ \mathbf{0}_{1\times3} & \mathbf{0}_{1\times3} \end{bmatrix} \begin{bmatrix} \dot{\mathbf{r}}_O \\ \dot{\mathbf{e}}_O \end{bmatrix} + \begin{bmatrix} {}^B\mathbf{R}_O(\mathbf{e}_O)\frac{\partial^O\mathbf{r}_{c_2}(\mathbf{s}_F)}{\partial \mathbf{s}_F} & \mathbf{0}_{3\times1} \\ \mathbf{0}_{1\times2} & 1 \end{bmatrix} \begin{bmatrix} \dot{\mathbf{s}}_F \\ \dot{s}_K \end{bmatrix} \\ &= \mathbf{T}_{D_2}\dot{\mathbf{s}}_{LD} + \mathbf{T}_{F_2}\dot{\mathbf{s}}_F + \mathbf{T}_{K_2}\dot{s}_K. \end{aligned} \tag{14}$$

where ${}^O\mathbf{K}(\mathbf{e}_O)\dot{\mathbf{e}}_O$ is the angular velocity of the object (expressed in the object frame).

Being $\mathbf{T}_D$, $\mathbf{T}_F$, and $\mathbf{T}_K$ defined by the above equations, feasible choices for the (10×3) matrix $\mathbf{Y}_R$ and the (10×6) matrix $\mathbf{Y}_A$ in the contact force parametrization $\mathbf{F} = \mathbf{Y}_R(\mathbf{s})\lambda_R + \mathbf{Y}_R(\mathbf{s})\lambda_R$ are given by:

$$\mathbf{Y}_R = \begin{bmatrix} -{}^B\mathbf{R}_{sur}(\mathbf{s}_{LD}, \mathbf{s}_F) \\ {}^B\mathbf{R}_O(\mathbf{e}_O)\mathbf{S}[\mathbf{r}_{c_1} - \mathbf{r}_{c_2}(\mathbf{s}_F)]{}^B\mathbf{R}_O^{\mathrm{T}}(\mathbf{e}_O){}^B\mathbf{R}_{sur}(\mathbf{s}_{LD}, \mathbf{s}_F) \\ {}^B\mathbf{R}_{sur}(\mathbf{s}_{LD}, \mathbf{s}_F) \\ \mathbf{0}_{1\times3} \end{bmatrix}, \tag{15}$$

$$\mathbf{Y}_A = \begin{bmatrix} \mathbf{I}_{3\times3} & \mathbf{O}_{3\times3} \\ -{}^B\mathbf{R}_O(\mathbf{e}_O)\mathbf{S}[\mathbf{r}_{c_1}]{}^B\mathbf{R}_O^{\mathrm{T}}(\mathbf{e}_O) & \mathbf{I}_{3\times3} \\ \mathbf{O}_{4\times3} & \mathbf{O}_{4\times3} \end{bmatrix}. \tag{16}$$

Here, ${}^B\mathbf{R}_{sur}(\mathbf{s}_{LD}, \mathbf{s}_F) = {}^B\mathbf{R}_O(\mathbf{s}_{LD})^O\mathbf{R}_{sur}(\mathbf{s}_F)$ defines the relative orientation between the base frame and a frame attached to the object surface, having its origin at the point identified by $\mathbf{s}_F$, with the $X_{sur} - Y_{sur}$ coordinate plane tangent to the contact surface, and with the Z_{sur} axis along the (ingoing) normal to the surface. Moreover, $\mathbf{S}[\mathbf{r}]$ is the skew-symmetric matrix generated by a vector $\mathbf{r}$.

With the above choices of $\mathbf{Y}_R$ and $\mathbf{Y}_A$, $\lambda_A \in I\!R^6$ is the parametrization of the active forces on the object expressed in the base frame, while $\lambda_R \in I\!R^3$ —the parametrization of the internal forces— coincides with the contact force exerted by the SCARA robot expressed in the frame characterized by ${}^O\mathbf{R}_{sur}$ at the contact point (*task frame*). In particular, the first two components of λ_R parametrize the contact force acting tangentially to the object surface (i.e., the friction force), while the third component represents the force exerted by the SCARA robot along the normal to the surface. Hence, $\lambda_{R_F} = (\lambda_{R_1}, \lambda_{R_2})$ and $\lambda_{R_P} = \lambda_{R_3}$, according to the definitions in Sect. 2.2.

As friction is a relevant phenomenon in the task we are interested in, a reliable friction model should be adopted. In [12], an experimentally validated description of friction in deburring tasks is proposed. Using the described formalism, it can be expressed as

$$\lambda_{R_F} = \lambda_{R_P} \mathbf{K}_F \dot{\mathbf{s}}_F, \tag{17}$$

where $\mathbf{K}_F$ is a diagonal matrix of friction coefficients (constant in the region of task execution). As previously indicated, eq. (17) can be used to eliminate the appearance of λ_{R_F} in the model equations.

As a result of the modeling phase, we may use e.g. the description (9) to define a hybrid controller of the ten quantities $\mathbf{s}_{LD}$, $\mathbf{s}_F$, s_K (nine motion parameters), and λ_{R_P} (one force parameter).

4. CONCLUSION

We have presented a task-oriented modeling approach for general cooperating robot systems and its specific application to a cell with two different robots involved in a mechanical finishing operation. The proposed formalism is useful for control purposes because it leads to a system description in terms of the kinematic and dynamic quantities that characterize the task. Thus, generalized hybrid motion/force controllers can be directly designed (see, e.g., [7] and [11]) starting from the obtained model of the system.

REFERENCES

[1] J.Y.S. Luh and Y.F. Zheng, "Constrained relations between two coordinated industrial robots for motion control," *Int. J. of Robotics Research*, Vol. 6, No. 3, pp. 60–70, 1987.

[2] T. Yoshikawa and X. Zheng, "Coordinated dynamic hybrid position-force control for multiple robot manipulators handling one constrained object," *1990 Int. Conf. on Robotics and Automation*, Cincinnati, OH, pp. 1178–1183, 1990.

[3] T.J. Tarn, A.K. Bejczy, and X. Yun, "New nonlinear control algorithms for multiple robot arms," *IEEE Trans. on Aerospace and Electronic Systems*, Vol. 24, No. 5, pp. 571–583, 1988.

[4] F.-T. Cheng and D.E. Orin, "Efficient algorithm for optimal force distribution—The compact-dual LP method," *IEEE Trans. on Robotics and Automation*, Vol. 6, No. 2, pp.178–187, 1990.

[5] H. Bruhm, J. Deisenroth, and P. Schädler, "On the design and simulation-based validation of an active compliance law for multi-arm robots," *Robotics and Autonomous Systems*, Vol. 5, pp. 307–321, 1989.

[6] C. von Albrichsfeld, M. Svinin, and H. Tolle, "Learning approach to the active compliance control of multi-arm robots coupled through a flexible object," *3rd European Control Conf.*, Roma, I, pp. 1900–1905, 1995.

[7] A. De Luca and R. Mattone, "Modeling and control for cooperating robots handling payloads with extra degrees of freedom," *3rd European Control Conf.*, Roma, I, pp. 1923–1930, 1995.

[8] R.M. Murray, Z. Li, and S.S. Sastry, *A Mathematical Introduction to Robotic Manipulation,* CRC Press, Boca Raton, 1994.

[9] E. Paljug, X. Yun, and V. Kumar, "Control of rolling contacts in multi-arm manipulation," *IEEE Trans. on Robotics and Automation*, Vol. 10, No. 4, pp. 441–452, 1994.

[10] A. Bicchi and R. Sorrentino, "Dexterous manipulation through rolling," *1995 IEEE Int. Conf. on Robotics and Automation*, Nagoya, J, pp. 452–457, 1995.

[11] A. De Luca and R. Mattone, "Modeling and control alternatives for robots in dynamic cooperation," *1995 IEEE Int. Conf. on Robotics and Automation*, Nagoya, J, pp. 138–145, 1995.

[12] A. Fedele, A. Fioretti, and G. Ulivi, "Experiments in robotic deburring with on-line surface identification," *2nd IEEE Mediterranean Symp. on New Directions in Control and Automation*, Chania, GR, pp. 120–126, 1994.

ROBODOC® System for Cementless Total Hip Arthroplasty Clinical Results and System Enhancements

Brent D. Mittelstadt, Peter Kazanzides

Integrated Surgical Systems, Inc.
829 W. Stadium Lane, Sacramento, CA , 95834 USA
Tel: (916) 646-3487 Fax: (916) 646-4075 e-mail: iss@netcom.com

ABSTRACT

The ROBODOC® System brings the accuracy of computer-based planning and computer-controlled machining to surgical procedures. This paper presents the clinical results of using the first generation ROBODOC System for over 300 cementless total hip replacement surgeries in the United States and Germany. This clinical experience has also identified enhancements that are necessary for widespread commercial acceptance of the system.

KEYWORDS: Medical Robotics, Computer Assisted Surgery, Orthopedics, Hip Replacement, Clinical Results

INTRODUCTION

The ROBODOC® System has been developed to address an existing void in cementless total hip replacement (THR) surgery. In THR surgery, the surgeon replaces the articulating surfaces of the ball and socket hip joint using an acetabular cup and femoral stem. Clinical success of the procedure is dependent on achieving long term fixation of the artificial components to the patient's host bone. There are two main approaches for attaching the implants to the bone. These approaches differ depending on whether or not bone cement is used to fix the components. Implant designs which do not require the use of cement (cementless) have evolved in an attempt to create a biologic interface between the implant and the host bone that will, in theory, last for the life of the patient. Current cementless femoral implant designs are often based on large human CT and x-ray databases to create a geometry which optimally fits the internal structure of the proximal femur. The implant designs are then accurately fabricated using advanced computer aided manufacturing technologies. The deficiency or void in the overall process is related to the conventional tools that are used to plan and carry out the surgical procedure. Conventional planning uses plane x-rays and two dimensional overlays to determine component size and placement, and conventional bone preparation uses hand held broaches and reamers to prepare the implant cavity.

Our belief is that this void in technology can be eliminated using image-guided computer-assisted surgery techniques where CT data are used to accurately plan the case and a precise surgical robot is used to prepare the femur for the implant. The primary goal of this development has been to significantly improve implant selection and sizing, improve positioning accuracy of the implant within the bone and improve the accuracy of preparation of the bone cavity to accept the implant. Research indicates that these improvements may result in a consistently improved success rate and an increased useful life of the implant/bone aggregate.

The first generation of the ROBODOC System has been clinically used to assist with hip surgery since 1992. This system was designed with an emphasis on ensuring that accuracy and safety requirements were reliably satisfied. Because this was one of the first active robot systems to operate on human subjects, the primary focus of the system development was to fully characterize system performance, develop and test redundant safety mechanisms and to develop a simple man/machine interface. The first generation system and its evolution has been the subject of numerous publications [1-5].

The development approach for ROBODOC has been to complete the implementation and testing for all safety and performance components (without completely addressing OR time and ease of use) and then install these systems in selected hospitals where they can be closely monitored. This approach has allowed for the installation of systems in highly competent centers and then the collection of clinical use feedback. This information is then used to refine the system for more widespread use. Another advantage to this development approach is that it allows for an early start of long term clinical studies that can lead to documented differences between conventional and robotic surgical techniques.

This paper describes the clinical results with the first generation system in addition to the enhancements that are being developed to support widespread acceptance of the ROBODOC system. To aid with system understanding, a short overview of the surgical protocol is included in the following section.

ROBODOC THR PROTOCOL

For the ROBODOC surgical protocol, patients undergo a separate procedure where three titanium locator pins are implanted in the affected femur under local anesthesia. These pins (two at the knee and one at the proximal femur) are used for registering preoperative planning coordinates with intraoperative coordinates during the robotic procedure. After the pins are installed, a CT scan of the femur is taken and loaded into the ORTHODOC presurgical planning workstation (ORTHODOC). Using ORTHODOC, the surgeon manipulates three dimensional implant models relative to reconstructed CT data of the femur to determine the optimum implant size and position. The planning data, along with pin locations, are then transferred to the surgical robot system (ROBODOC).

During surgery, after manual preparation of the acetabular cup, the femur is secured in a stabilization device (the femoral fixator) and the three locator pins are exposed. Bone movement during the robotic procedure is sensed using a bone motion detection device. This device measures bone movement and causes an interrupt if the displacement is outside a defined limit. The next step involves using a sterile ball probe, attached to the robot, to determine the robot (surgical) coordinates for each locator pin. These data, along with the corresponding ORTHODOC data, are used to compute a rigid body transformation which can relate pre-surgical plan coordinates to intraoperative surgical coordinates. Following registration, a high speed, pneumatically-powered cutting tool is installed and the robot machines the cavity in the femur. Following cavity preparation, the surgeon disconnects the robot, installs the implant, and completes the surgery in the standard fashion. More detailed descriptions of the procedure, user interface and the safety systems have been previously published [6-8].

CLINICAL RESULTS

To date, over 300 patients have been successfully operated on using the ROBODOC system. These surgeries have been done in the United States and Germany.

United States Clinical Trial

In 1992, the first patient was operated on using the ROBODOC system as part of an FDA (Food and Drug Administration) authorized ten patient clinical pilot study. The goal of this single-center clinical trial was to collect data to justify that the system was safe to operate on humans. In 1993, following successful completion of this study, a randomized multi-center clinical trial was started. For this study, any patient meeting the enrollment criteria is randomly assigned to either receive a hip replacement using the ROBODOC system or one using conventional instrumentation (control group). The purpose of this study is to collect data which could be used to evaluate differences between conventional and robot cases. Three centers have contributed to this study, where each center had 2-4 surgeons participating. As of January 1, 1996, 116 patients with 131 hips have been enrolled in the study and 127 hips (67 ROBODOC and 60 Control) have completed a minimum 3 month follow-up. Two different implants systems have been used so far in the study, the OSTEOLOCK (Howmedica) and the AML (DePuy).

The patients were evaluated using the modified Harris Hip Scale, the Hip Society Rating System, and the SF 36 Health Status Questionnaire. Surgical time, blood loss, length of stay and complications were also recorded. Three month post-operative radiographs were reviewed in a blinded fashion using a strict objective criteria to grade fit, size selection, implant position and reaming defects.

The results for these patients are the following. When the Harris Hip Scale results were broken into total score and pain score, there were no significant differences at 3 months. However, of the patients who had one year follow up results, statistically more patients from the ROBODOC group fell into the "no pain" category. Average length of stay was not significantly different but surgical time and the associated blood loss was significantly greater for the ROBODOC cases. The post-operative complication rate was not significantly different, but the control group included 3 acute intra-operative femoral fractures (cracks) versus 0 for the ROBODOC group.

The most significant differences were seen in the analysis of the post operative x-rays. For both implant designs, the ROBODOC hips had consistently improved alignment and positioning. For the OSTEOLOCK cases, where there was a greater sample size, the implant to bone fit, implant alignment and size selections were all significantly better for the ROBODOC cases.

German Clinical Use

In Germany, regulations required the TÜV testing of the System, which, contrary to the FDA procedure, did not challenge the efficacy of the system, but put emphasis on its technical safety. Regarding the System as a tool in the surgeons control, its use and evaluation of its benefits were left to the surgeon's judgment.

Following development of system modifications necessary to support the surgical approach used in Germany, the first successful surgery was completed in August 1994 at the Berufsgenossenschaftliche Unfallklinik in Frankfurt, Germany. At this site, the ROBODOC system is not being used in a regulated clinical study, but instead as indicated by the surgeons. As a result, there are no control data that can be used for direct comparison. Results can, however, be evaluated based on previous series of patients done with conventional techniques.

The first 120 cases were successfully completed by one surgical team. Results for this series are the following. The OR time for the first 15 cases averaged 180 minutes, whereas for the remaining 105 cases, the average time was 120 minutes with the fastest time being 99 minutes. The overall complication rate was lower than in comparable cementless series previously published.

As of February 1996, a total of 235 patients have been successfully operated on using the ROBODOC System. A number of clinical observations have been noted. There have been no acute intra-operative fractures in any of the patients. This is compared to reported rates that range from 10 - 25%. All post operative x-rays revealed correct (as planned) implant position and orientation. The post operative recovery was significantly improved and allowed for full weight bearing in 3-5 days as compared to the common practice of 4-6 weeks. This occurred in spite of persistent complaints of knee pain due to pin implantation. The knee pain, also common in the United States patients, is not severe and dissipates within 4-6 weeks.

SYSTEM ENHANCEMENTS

During the early part of the above clinical testing, a number of system enhancement were identified. These enhancements were not related to accuracy or safety but were primarily focused on refining the system for widespread acceptance where OR time and ease of use would be critical factors. The clinical feedback identified the following problem areas:

- The time required to set-up and perform the diagnostics was too long.
- Bone fixturing during surgery was time consuming and difficult for the occasional user.
- Implant cavity machining time, which ranged from 25 - 50 minutes depending on implant size, was too long.
- Pin implantation and robotic location was not desirable due to patient discomfort and the required additional surgical procedure.

The system enhancements to address these and other concerns are being implemented and tested in two separate phases.

Phase One Enhancements

The first phase of improvements focus on reducing set-up and cutting time yet still require implantation of three locator pins. For safety reasons, the ROBODOC system runs extensive start-up diagnostics to verify that all safety and performance related components are functioning within specified tolerances. After detailed analysis of the data collected during the early clinical cases, we were able to significantly reduce the time required to perform these tests (from 25 to 10 minutes) without any compromise in safety.

Another item that was addressed in the first phase of enhancements was the time required for the robot to machine the cavity. The first generation system machined the cavity in 25 - 50 minutes, depending on implant size. This time was reduced to 15 - 30 minutes by optimizing the tool paths and changing the tool design. One aspect of this change was to use cutters with larger diameters when possible. This allowed for reduction in time for the some of the larger stems.

The final area addressed in this phase was the improvement of the bone fixturing approach. The previous design utilized a proximal clamp attached to a frame that used sharp pins at the distal end to penetrate the soft tissue and secure the distal femur. This distal fixation was often difficult to install, and required incisions that would not normally be done in a conventional THR surgery. The new design corrects these problems by using a soft tissue constraint applied below the knee to stabilize the distal femur, coupled with an improved proximal bone clamp which increases stabilization.

Phase Two Enhancements

The second phase enhancements which are currently under development involve the removal of the need for locator pins, additional reduction of robotic cavity machining time, and an improved design of the ROBODOC surgical component to address manufacturability, serviceability, and European regulatory requirements for commercial products (CE marking).

A new method for registering the pre-operative data and the surgical robot has been developed and is currently being clinically tested and compared to the pin-based method. This approach involves collecting intra-operative canal center point data and femur neck surface data during surgery and then using optimization techniques to determine the transformation which causes agreement with similar data collected from the ORTHODOC CT-based pre-surgical planning system. Laboratory testing and early clinical results have shown that this new method of registration can produce results that are comparable to the pin-based technique. Additional work is underway to evaluate the repeatability and robustness of this approach.

The second phase reduction in robot cavity preparation time has focused on the implementation of a force control cutting strategy which continuously adjusts the feed rate of the tool based on the sensed cutting force [6]. Previously, the feed rate for each portion of the cavity was set at a fixed value based on the worst-case assumption that the entire cavity was being machined in dense bone. This was done to ensure high dimensional accuracy in all cases. In reality, there is a significant amount of implant volume which is machined in the softer trabecular bone. The new algorithm is designed to take advantage of this by increasing the feed rate when there is reduced resistance. This method should reduce the cavity machining time to approximately 10 - 20 minutes.

The final major development for this phase is the redesign of the OR component to address the European regulatory standards that have recently been enacted; in particular the standards for electromagnetic emissions and susceptibility. The new design also represents a significant improvement to the packaging of the system, including a reduction from three to two physical units, a reduction in the number and size of the interconnecting cables, and improvements to the maneuverability of the components. A more detailed description of the commercial version of the system is presented in [9].

CONCLUSION

The successful linking of the fields of imaging and robotics has resulted in a new tool for precision in surgery. The original application of this device to cementless total hip replacement has shown the technology to be of benefit. It is through the development of additional applications that this new category of surgical "smart tools" will allow the operating room of the future to provide improved outcomes that make this technology cost effective.

REFERENCES

1. Mittelstadt, B.D., Paul, H.A., Musits, B.L., Hayes, D., Taylor, R.H., "Accuracy of Surgical Technique in Femoral Canal Preparation in Cementless Total Hip Replacement," *Proc. American Academy of Orthopaedic Surgeons Annual Meeting*, New Orleans, LA, February 1990.
2. Paul, H.A., Bargar, W.L., Mittelstadt, B., Musits, B., Taylor, R.H., Kazanzides, P., Zuhars, J., Williamson, B., Hanson, W.: "Development of a Surgical Robot For Cementless Total Hip Arthroplasty," *Clinical Orthopaedics*, Vol. 285; pp. 57 - 66, December 1992.
3. Mittelstadt, B., Paul, H., Kazanzides, P., Zuhars, J., Williamson, B., Pettitt, R., Cain, P., Kloth, D., Rose, L., and Musits, B.: "Development of a surgical robot for cementless total hip replacement," *Robotica,* Vol. 11, pp. 553-560, 1993.
4. Taylor, R.H., Mittelstadt, B.D., Paul, H.A., Hanson, W., Kazanzides, P., Zuhars, J.F., Williamson, B., Musits, B.L., Glassman, E., Bargar, W.: "An Image-directed Robotic System for Precise Orthopaedic Surgery," *Computer Integrated Medicine*, MIT Press, Boston, MA, 1995.
5. B. D. Mittelstadt, P. Kazanzides, J. Zuhars, Bill Williamson, P. Cain, F. Smith and W. Bargar, "The Evolution of A Surgical Robot From Prototype to Human Clinical Use", *Computer Integrated Medicine*, MIT Press, Boston, MA, 1995.
6. Kazanzides, P., Zuhars, J., Mittelstadt, B.D., Taylor, R.H.: "Force Sensing and Control for a Surgical Robot," *Proc. IEEE Conf. on Robotics & Automation*, pp. 612-616, Nice, France, May 1992.
7. Cain, P., Kazanzides, P., Zuhars, J., Mittelstadt, B., Paul, H. "Safety Considerations in a Surgical Robot," *Biomedical Sciences Instrumentation*, Vol. 29, pp. 291-294, San Antonio, Texas, April 1993.
8. Bargar, W.L., Taylor, J.K., Leathers, M., Carbone, E.J., "Preoperative Planning and Surgical Technique for Cementless Femoral Components Using 3-D Imaging and Robotics: Report of Human Pilot Study," *Proc. American Academy of Orthopaedic Surgeons Annual Meeting*, New Orleans, LA, February 1994.
9. Kazanzides, P., Cain, P.W., Wasti, H.A., "Distributed Architecture for a Fail-Safe Robot System," *Proc. Signal Processing Applications Conf. at DSP^x '96*, San Jose, CA, March 1996.

A Multi-Agents Approach for Autonomous Mobile Minirobotics

MOSTEFAÏ Nadir
Laboratoire d'Automatique de Besançon
CNRS, URA 1785
25 rue Alain Savary 25000 Besançon - France
Phone : 81 40 27 92 Fax : 81 40 28 09
e-mail : mostefai@ens2m.fr

BOURJAULT Alain
Laboratoire d'Automatique de Besançon
CNRS, URA 1785
25 rue Alain Savary 25000 Besançon - France
Phone : 81 40 28 01 Fax : 81 40 28 09
e-mail : abourjau@ens2m.fr

ABSTRACT

This paper deals with a multi-agents approach applied to minirobotics. Each agent is a physical entity that consists of an autonomous mobile minirobot. By using lots of relatively simple minirobots in place of one large complicated and expensive robot, we can do works in the environment at a fraction of the time and cost. Such multi-agents systems can be matched to many real world tasks and may solve problems in better. The multi-agents context suggests a cooperation between different entities. This cooperation is not limited to avoiding interference with others but can ensure an exchange of messages between the agents involved in the system. This idea is illustrated by scenarios inspired from ethology. Indeed, the recruitment mechanisms used in ants societies are good examples of cooperative works.

KEYWORDS: multi-agents, hybrid approach, cooperation, ethology, recruitment, minirobotics, control.

INTRODUCTION

In multirobots systems [1], two main approaches are known : the cognitive approach and the reactive one. In the cognitive approach, the functioning mode for each robot is in the form of perceive--process--act. The processing is itself a composition of functions that are: identification, planification and decision. These functions use knowledge bases which contain domain knowledge (purposes, plans, environmental knowledge) and control knowledge (methods, strategies, their chain). A robot can have an explicit representation of the environment including other robots and may take into account the past activities.

An application illustrating the latter approach has been developed for multirobot cooperation. It is based on a paradigm called plan-merging[2], where the robots incrementally merge their plans into a set of already coordinated plans. This is done through exchange of information about their current state and their future actions.

However, in practice, such cognitive systems are so far feasible only with a small number of robots. Furthermore, we cannot argue for a reduction of time between mission conception and implementation.

The reactive approach is very interesting. It privileges the behavior notion upon the knowledge one. The functioning mode is rather in the form of perceive -- act. The subsumption concept [3][4] is the most illustrative methodology in building intelligent reactive control. It consists on a behaviors hierarchy where a behavior from a superior level subsume a behavior of lower levels. For instance, if we have two behaviors for a robot that are "turn left" and "go forward", with the first behavior subsuming the second one, this means since its preconditions are true, it happens. Consequently, it suspends and can even suppress the second behavior if its preconditions become unsatisfied. This concept emphasizes that there would be no traditional notion of planning and the notion of world modeling is impractical and unnecessary.

The idea of scaling down the size of the robots inspires missions which will capitalise on several agents. The "emergent functionality" concept [5] means that the system function is not specified at first but emerges from the elementary interactions of agents. This concept leads to an approach which is indeed flexible and robust but sets the problem of relationship between control and emergence; if the degree of a behavior specification decreases, it becomes more difficult to explicit the control that generates it. So one way to solve this inconsistency is to focus the interest on the environmental conditions leading to the desired behavior rather than on the robots individual behaviors in their environment.Then, one should not think of agents swarms as machines which are told what to do but rather as autonomous individuals when turned on, do what is in their intelligent control programs.

Our research deals with a cognito-reactive approach. The idea with hybridizing both previous approaches is to let a reactive behavior-based system take care of the real time issues involved in world interactions while a more traditional artificial intelligence system (based on a learning cognitive process) sits on top, making longer term executive decisions as in following a memorized path.

MULTI-AGENTS TAXONOMIES

The cooperation can be defined in terms of different concepts such as organization, communication, coordination, negotiation or global coherence. In model A (see figure1), the organization and communication are from a practical point of view two complementary ways that lead to cooperation [6] while in model B (figure 1), the problem distributed solving notion is introduced instead of cooperation notion [7]. Another taxonomy related to DAI (Distributed Artificial Intelligence) is defined in model C (figure 1) where systems are considered from a social angle [8].

The basic question for the organization is : which agent does what ?. The communication can be accomplished in a traditional manner by exchanging messages or by environment. The stigmergy notion stipulates, in order to coordinate activities of a group, the interactions between agents and their environment are sufficient. This is particularly true in insects societies such as ants. Indeed, they communicate thanks to chemical trails.
The coordination manages the different activities for acheiving a goal. It may be either a coordination by plans using conventions in stable and known environments or a coordination by retroaction using communication in variable and unknown environments.
When pursuing different goals, the agents may conflict. So, with negotiation, any problem can be solved. To converge as soon as possible to the global objective, a system needs ensuring coherence which can be assimilated to a long term coordination.

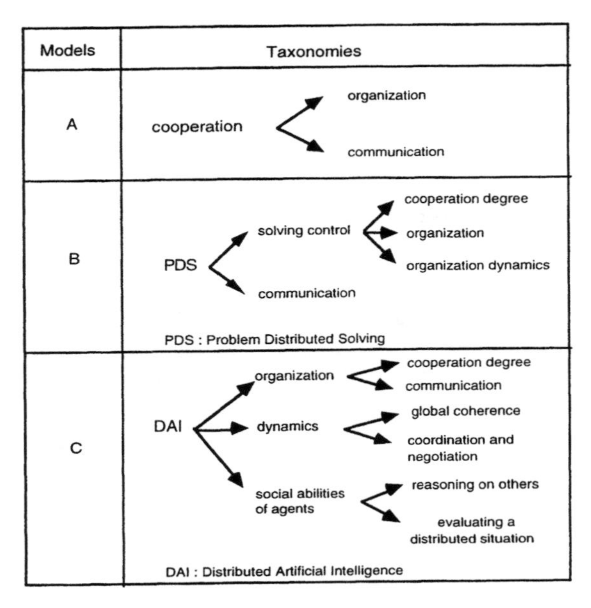

Figure 1. Multi-agents taxonomies

The main problem for any taxonomy is then to integrate the concepts it defines in a given multi-agents system, in order to reach a high level of performance.

SOME COOPERATION MODELS

The ethology leads to the analysis of mechanisms involved in insects social organization. When we consider the behaviours of the workers in ants societies, we fall into one of the three following foraging categories [9][10]:

- purely individual foraging;
- individual foraging + recruitment;
- individual foraging + recruitment + permanent trail.

In the second category, the workers forage individually, but if one encounters an important food source, it recruits nestmates which initiate the food source's exploitation. Different recruitment mechanisms exist. In mass recruitment, a scout (the recruiter) discovers the food source and returns to the nest laying a chemical trail. In the nest, some of its nestmates (the recruited) detect the trail and follow it to the food source. There they ingest food and return to the nest reinforcing the trail. In tandem recruitment, the scout invites ants at the nest to accompany her back to the food. One recruit succeeds in following the leader, the two animals being in close contact. In group recruitment, the recruiter leads a group of recruits to the food source by means of a short-range chemical attractant. After food ingestion, in each recruitment type, the recruited become recruiters.

A learning ability should be a central element in the functioning of ants societies. Two memories related to spatial foraging activities are known for ants. A short-term memory is used when an ant that has discovered some food will return to the same neighbourhood while the long-term memory (a longer time scale) is involved for route fidelity to permanent food sources. Obviously, for an agent like a robot, this learning function would be a path memorization from the base until the target. This allows finding easily the way when coming back to the base.

An analogy between a bio-system (ants society) and a multi-minirobots system is shown below in figure 2 :

agent / concept	ant	minirobot
organization	- scout - worker	-explorator -exploiter
communication	stigmergy	radiocommunication or optical communication
coordination	specialization and interactions	goal achievement and synchronization
global coherence	learning	long-term coordination
negotiation	?	none so far

Figure 2. Bio-system and miniobotics system analogy.

In following section, we define cooperative scenarios based upon the tandem and the mass recruitment mechanisms.

COOPERATIVE SCENARIOS

First scenario

Let us assume we have an environment whose dimensions are 200/200 meters. This environment include four bases where minirobots parks. We consider a population of eight minirobots : four explorators and four exploiters. In a base, one robot from each category takes place (see figure 3).

The scenario is based on a tandem recruitment mechanism since once an explorator finds the target, it comes back to the base by a memorized path in order to recruit and lead an exploiter until the target.

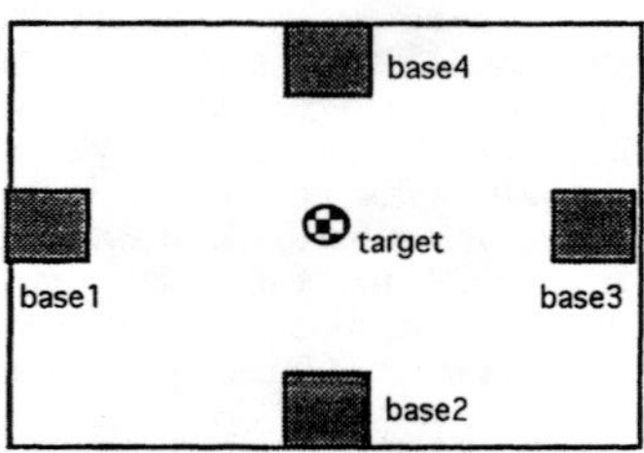

Figure 3. Environment in first scenario

A simulation has been done on a SiliconGraphics workstation, using Automod programming language. We have defined for this purpose some parameters:
v_i as the velocity of the minirobot r_i, r_i as the minirobot taking place in base i, zc as a zone capacity and represents the admitted number of minirobots near the target at the same time and perturbation (x,y) means any problem that may affect a minirobot with a periodicity of

N.B : the curves have been obtained considering an updating time of thirty (30) seconds.
The limitation of the zone capacity to zc =2 make the transition in figure 4 less abrupt than for zc = 4 because an intermediate value corresponding to a number of minirobots equal two is considered. The nil plateau is due to the exploration duration where none exploiter has reached yet the target.
In figure 5, a perturbation is introduced. The minirobot r_1 that has not been affected by this perturbation spent on average a longer time near the target. This should be interpreted by the fact when detecting any perturbation, a minirobot must escape it.

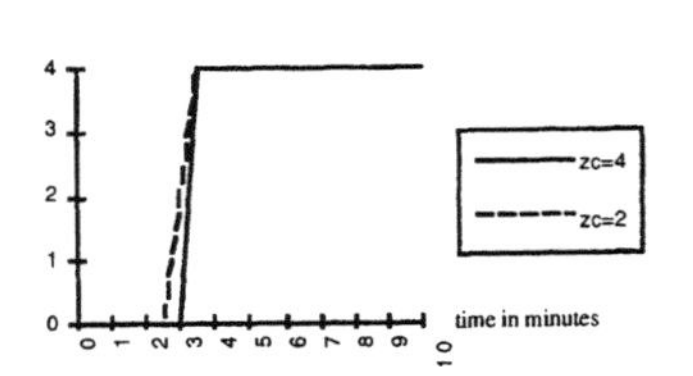

Figure 4. Total of minirobots that reached the target (v_i = 3m/s).

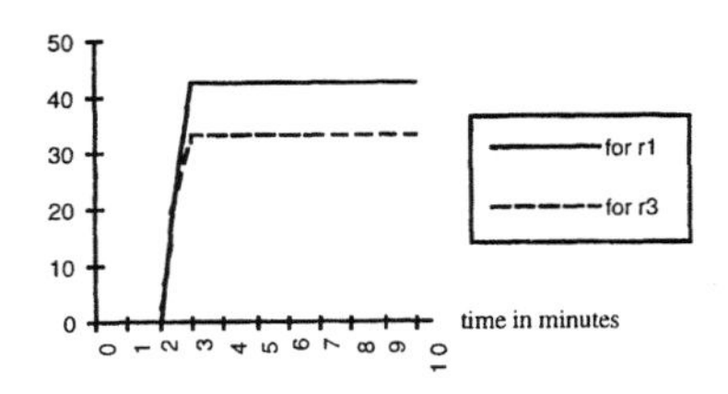

Figure 5. Average time in seconds spent near the target with a perturbation (3min, 40sec) affecting r3.

Second scenario

The environment still include four bases but contains two targets. We consider a population of twenty minirobots where five of them take place in each base (see figure 6). The scenario is based on a combination between two recruitment mechanisms : the mass recruitment because the minirobots move partially on stable paths joining the base to the target neighbourhood and a pseudo-tandem recruitment since only one recruit in the same base succeeds in detecting the target discovery message and go to exploit it.

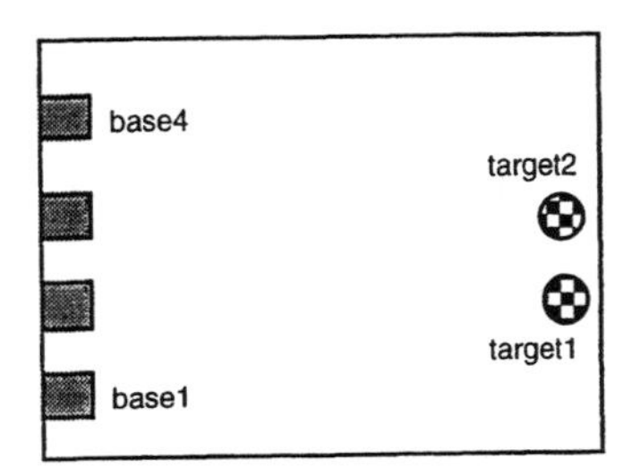

Figure 6. Environment in second scenario

For this scenario, we consider the parameters d_1 and d_2 where d_1 is the minirobots number at target1 (density d_1) and d_2 is the minirobots number at target2 (density d_2). So, when a minirobot has to choose which target to exploit, it can decide according to criterions:

Density balancing criterion (the default criterion).
If $d_1 < d_2$ then target1
If $d_2 < d_1$ then target2

Agregation criterion.
If $d_1 > d_2$ then target1
If $d_2 > d_1$ then target2

As results, we note that most of time, the total number of minirobots reaching target2 is greater than for target1 as shown in figure 7. The agregation criterion privileges the target1

In case of figure 9, the minirobot r_4 spent on average more time at target1 when it is affected by perturbations.This may be due to a conjunction of circumstances such as environmental configuration and the occurence of perturbations. A weak influence of perturbations on the exploitation process of target2 is shown by figure 10.
Consequently, according to such results, a control strategy for the minirobotics system can be adopted by assigning for example adequat values to velocity parameters.

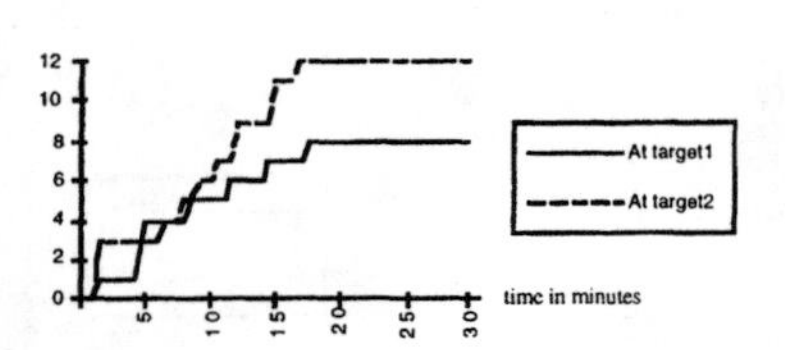

Figure 7. Total number of minirobots that reached the target (zc =4, v_i =3m/s).

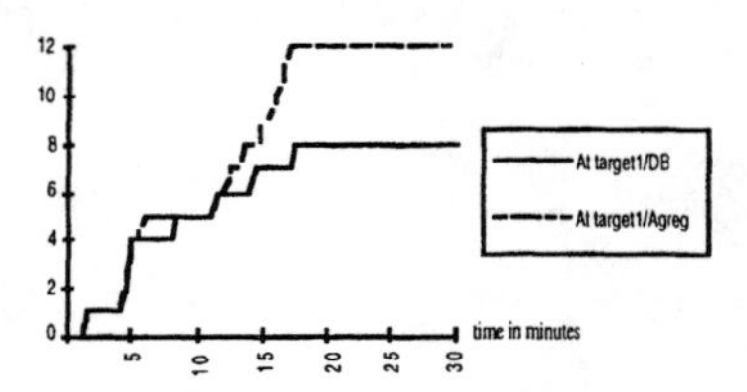

Figure 8. Comparison related to the criterion used (Density Balancing : DB or Agregation : Agreg) of minirobots that have already reached target1.

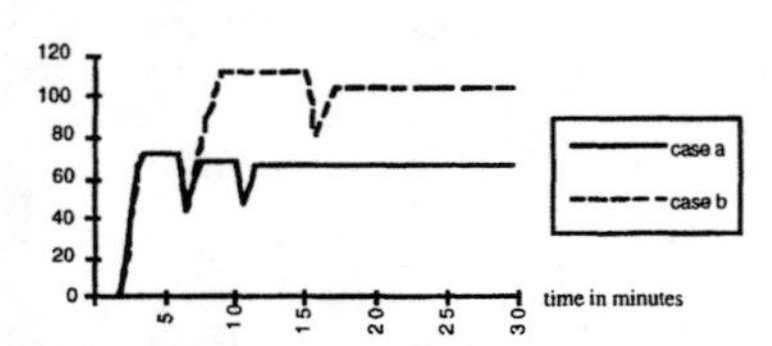

Figure 9. Average time (sec) spent by r4 at target1 (zc = 2, v_i = 3m/s). In case a, there is no perturbation. In case b, there is a perturbation (5min, 3min).

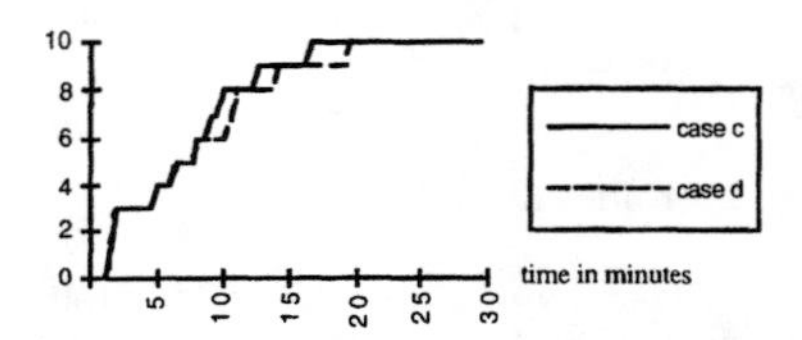

Figure 10. Total number of minirobots at target2. (zc=2, v_1=2m/s, v_2=3m/s, v_3=4m/s, v_4=5m/s). In case c, there is no perturbation affecting target2 zone. In case d, there is a perturbation (5min, 3min).

CONCLUSION

We have presented an outline for a multi-agents minirobotics system used in ethology inspired scenarios. The study of some animals that have been considered so far less social than ants or bees interests more and more researchers. We think particularly to fishes shoals or birds squadrons. One should raise a question : why multi-agents systems when many problems are still unsolved for a single agent ? We can then argue by robustness, flexibility, faster development time and costs reductions.

REFERENCES

1. Premvuti, S. and Yuta, S. "Consideration on the cooperation of muliple autonomous mobile robots." IEEE Int Workshop on Intelligent Robots and Systems, pp 59 - 63, vol-1, Ibaraki-Japan, July 1990.
2. Alami, R. and al "A paradigm for plan-merging and its use for multi-robots cooperation." IEEE Int Conference on systems, man and cybernetics, San Antonio, October 1994, pp 612 - 617.
3. Brooks, R. "A robust layered control system for a mobile robot." IEEE Journal of Robotics and Automation, vol RA-2, pp 14 - 23, March 1986.
4. Brooks, R. and al "Lunar base construction robots." IEEE Int Workshop on Intelligent Robots and Systems proceedings, Tokyo-Japan, July 1990.
5. Steels, L. "Towards a theory of emegent functionality." Int Conference on Simulation of Adaptive Behaviors, from Animals to Animats, pp 451 - 461, 1991.
6. Buron, T. "Structures de communication et d'organisation pour la cooperation dans un univers multi-agents." Thèse de l'Université Paris 6, 17 Novembre 1992.
7. Keith, S.D. "Distributed Problem Solving Techniques : A Survey." IEEE Transactions on Systems, Man and Cybernetics, vol SMC-17, N°-5, September / October 1987.
8. Chaïb-Draa, B. and al "Trends in Distributed Artificial Intelligence." Artificial Intelligence Review 6, pp 35 - 66, 1992.
9. Pasteels, J. and al "Self-Organization mechanisms in ant societies (I) : trail recruitment to newly discovered food sources." Behavior in social insects, vol - 54, 1987.
10. Deneubourg, J.L. and al "Random behavior, amplification process and number of participants : how

Integrating assembly planning with compliant control

J. Nájera*, C. Laugier
LIFIA - INRIA Rhône Alpes
46, avenue Felix Viallet
38031 Grenoble cedex, France

W. Witvrouw†, J. De Schutter
Katholieke Universiteit Leuven, PMA
Celestijnenlaan 300B
3001 Heverlee, Belgium

Abstract

This paper presents an approach to the integration of planning and compliant control for robotic assembly. Departing from the desired final assembly configuration of an object, a fine motion planner derives an assembly graph by inverting the actions that lead to a fully unconstrained situation. Contacts are used to reduce the uncertainties which exist between the nominal world and the real world. In each contact situation the planner generates a set of potential motions which either use the existing contacts to lead the manipulated object closer to its goal, or which switch to another contact situation if necessary. The planned assembly sequences are then transformed into compliant motion task specifications, which a force controlled robot needs to execute the plan. The most interesting feature of the presented controller is its textual interface with the planner, allowing the specification of all kinds of tasks, including tracking of uncertainties. The integration between planner and controller is made possible by the fact that both use similar modeling primitives, and that the planner consults an extensible database of heuristic rules, which are based on experience with the real world execution of compliant motions.

Keywords: Fine Motion Planning, Compliant Control, Force Control.

Introduction

The automation of mechanical assemblies is an open problem which has received a lot of attention during the past two decades. Unfortunately, the level of intelligence necessary for a robot to automatically perform an assembly task is far from being attained. One of the main reasons for this (besides the complexity of the problem), is the lack of integration between planning and control. We present an approach to plan robust motions for the workpieces involved in an assembly task, and explain how these can be transformed into a compliant motion specification language that can be interpreted by a force controlled robot. A simple task concerning the assembly of a revolute joint is planned and executed by the robot using the proposed approach.

1 Planning

The problem of fine motion planning concerns the generation of a robust strategy (a plan) to place a manipulated object in a particular assembly configuration despite the position, sensing, and control uncertainties. To solve this problem, the contacts between the manipulated object $\mathcal{A}$, and the obstacles forming the assembly environment $\mathcal{B}$, can be used as a "geometric guide" to direct the robot to the target assembly configuration while reducing its position uncertainty [4]. A popular approach for fine motion planning is applying a *backward chaining* strategy to determine the different ways in which an assembly can be dismantled, then reversing the resulting sequences to obtain the corresponding assembly plans [6]. Following this approach, our fine motion planner derives an assembly graph by inverting the actions that lead to a fully unconstrained situation of $\mathcal{A}$. In each contact situation, the planner generates a set of potential motions which either use the existing contacts to lead the manipulated object closer to its goal, or which switch to another contact situation if necessary. We illustrate the planning process with an example.

*e-mail: Jose.Najera@imag.fr
†Wim.Witvrouw@mech.kuleuven.ac.be

1.1 Planning a sequence

Consider the assembly of a simple revolute joint consisting of six pieces and seven screws (figure 1a). In order to join the pieces forming each link, it is necessary to find a way of aligning them before screwing. This can be done by using a base support forming a corner where the individual pieces can be aligned. The full assembly problem can be then decomposed into thirteen sub-assemblies (figure 1b).

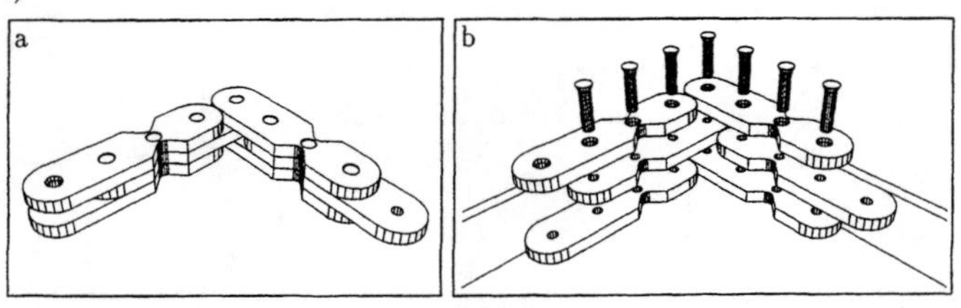

Figure 1: (a) A revolute joint. (b) The assembly sequence necessary to assemble the revolute joint.

This decomposition can be derived by a *sequence* planner [9]. From this ordered sequence, each sub-assembly can be defined as the assembly of a mobile object $\mathcal{A}$, and a static environment $\mathcal{B}$ formed by the base support and the previous sub-assemblies. To illustrate the planning process, let's consider the first sub-assembly consisting on the alignment of one of the pieces in the corner. Departing from the final configuration of assembly, we can determine the local translational freedom of $\mathcal{A}$. This is done by intersecting the half-spaces denoting the unconstrained translation directions derived from the *face-X* contacts (where X stands for surface, edge or vertex). The resulting motion constrains can be mapped onto the unitary sphere $SO(1)$ resulting in what is called the *local translational freedom cone* [9] (figure 2).

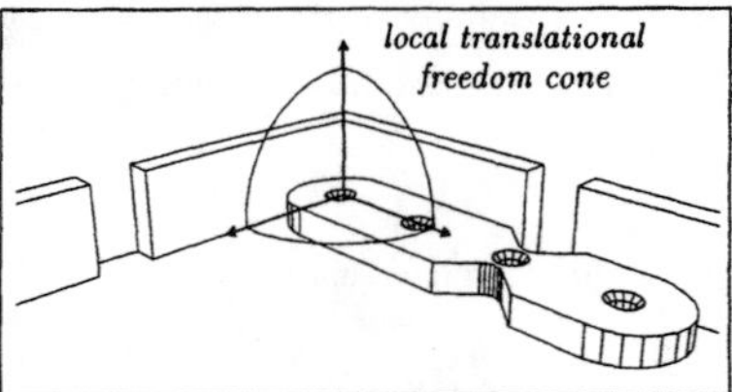

Figure 2: The local translational freedom cone.

The planner analyzes this cone and finds that there are three motion directions maintaining a maximum number of two contacts: those corresponding to the edges delimiting the cone. Each one of these motion directions breaks one *face-X* contact, maintaining two face contacts while the selected motion is applied. The global translational freedom (collision detection in the direction of motion) associated to each potential motion direction is validated using a closest-features distance algorithm [5]. Some heuristics are applied to determine the range of each motion with regard to the position uncertainty of $\mathcal{A}$ in the new contact configurations (maintaining the face contacts while leaving enough space to manipulate the object). In the following steps, the algorithm finds two robust motion directions for each of the three newly generated contact configurations. This process is recursively applied until the manipulated object is completely unconstrained. The result is a directed graph containing a set of disassembly sequences which can be reversed to obtain the associated assembly plans (figure 3). Each node in the graph represents a contact configuration, and each link, the geometric transformation necessary to pass from one contact situation to the next (rotation directions are not discussed here but they are already implemented in the planner). The user is responsible for selecting a plan from the graph.

Because of the position and control uncertainties, a geometric description of the actions is not enough to execute the task with the robot. Sensing information must be included in the plans in order to give the robot the capacity of "feeling" its environment during execution. For this reason, the nominal actions in the graph must be transformed into control commands which clearly specify the velocities and reaction forces to maintain during execution. Also, force-based termination conditions must be specified in order identify the new contact configurations and to stop the motions. In the next section we describe a robust compliant motion control system which has been used to execute the planned actions.

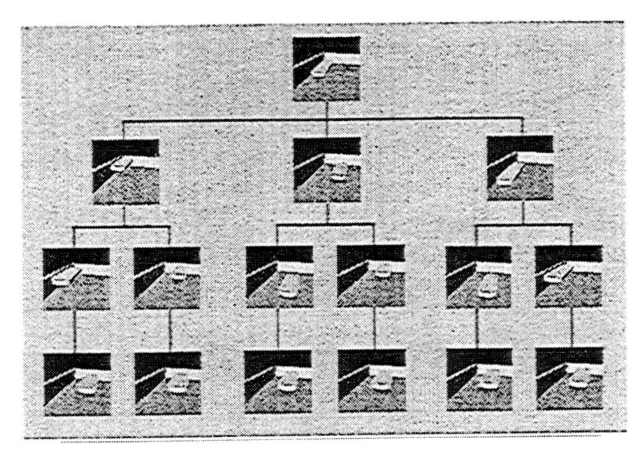

Figure 3: The disassembly graph produced by the planner.

2 A compliant motion control system

Compliant control deals with robotic tasks involving contacts between the manipulator and its environment. These tasks require the use of position and force sensing as a source of feedback during execution [8, 3]. The use of force signals enhances the quality of the task execution, making it more robust while offering an alternative control approach [7]. To prevent *ad-hoc* compliant motion programming, a general development framework is needed. COMRADE [10] (COmpliant Motion Research And Development Environment) has been developed with this purpose in mind. This system offers the user a modular portable software which can be applied to solve a wide variety of generic compliant motion tasks. The control architecture of COMRADE is based on the concept of hybrid position/force control. The force and tracking control loops are closed around a motion control system which is connected to the standard robot controller [2].

2.1 The task specification language

A compliant task is programmed using a robot-independent task specification language. The user specifies the commands in an English-like language, making the task description clear and easy to understand. Three types of commands are available:

- **Free space motion commands:** these are used to specify the robot position in both, Cartesian and joint coordinates.
- **Compliant motion commands:** to separate the task geometry from the motion specification, the user selects a *task frame* whose axes can be controlled as velocities, forces biases (to maintain), or tracking directions [1]. The values of all of the measurable parameters can be used as termination conditions to determine the duration of motions (figure 4).
- **Miscellaneous commands:** that do not fall under the previous categories and are usually application dependent (opening/closing a gripper, taking a snapshot with a camera, screwing, etc) can be easily integrated into the specification language.

Clearly, this control language can be applied to solve the alignment and insertion operations necessary for the revolute joint assembly (and other similar tasks). The next section presents an algorithm to automatically transform the nominal plans produced by the planner into the compliant control specification language.

3 Transforming a plan into a control program

The integration between the planner and the controller is made possible by the fact that both use similar modeling primitives, and that the planner consults an extensible database of heuristic rules, which are based on experience with the real world execution of compliant motions. The following rules describe the transformation of a nominal plan into a compliant control motion program:

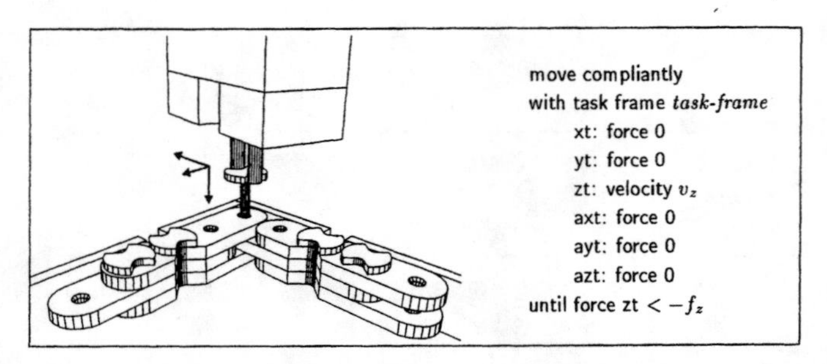

Figure 4: Each one of the axes (xt, yt, zt, axt, ayt, azt), denotes a velocity, a force or a tracking direction to maintain. The boolean expression at the end of the control specification determine the termination conditions for the motion.

- **Placing the task frame.** This is maybe the most delicate part in a compliant motion specification (it often needs the intervention of the user). A badly placed task frame may result in a failing execution of the task. It is important to remind that the planner eliminates the contacts progressively. For the case of translations, we can tentatively center a task frame F in the geometric feature of $\mathcal{A}$ (polygonal face, edge, vertex) which is supposed to establish a new contact (new contacts are detected by analyzing consecutive nodes in the graph). The F_z axis of the task frame must point in the direction of motion. For the case of rotations, this axis is aligned with the rotation axis using a right-hand rule convention. This task frame setting may be overridden by a heuristic rule derived from expert knowledge (e.g. placing the task frame in the common edge of a double face contact to maintain).

- **Termination conditions.** Because F_z axis is pointing in the direction of motion, the termination condition will normally be a force threshold zt $< -f_z$. For rotations, a negative moment threshold will stop the motion azt $< -t_z$.

- **Velocities.** Because the F_z is pointing in the direction of motion, for translations there will always be a linear velocity v_z to control. For rotations it will always be an angular velocity ω_z.

- **Forces to maintain.** The contacts which are maintained while passing from one contact situation to the next denote the directions of the force bias to maintain. These force directions are associated to the face normals of the obstacles in contact, and must be expressed with respect to the task frame F. Let's denote A as $\mathcal{A}$'s reference frame, and ${}^AT_F$ as the geometric transformation between A and F. Each face normal must be first expressed with respect to A, and then transformed to F using ${}^AT_F$.

After the user selects a plan from the graph, the previous rules are applied iteratively traversing the nodes and the links denoting the plan. The result is a compliant control program that can be used to execute the task. Figure 5 shows one of the plans in the graph and its corresponding control program.

The first control instruction establishes a face contact between the side of $\mathcal{A}$ and the wall (notice that the task frame is placed in the middle of the face of $\mathcal{A}$ which is supposed to be aligned). By setting zero moments, the alignment of the two faces is guaranteed, at least for small rotational uncertainties. The second instruction is similar except that a force bias is specified in the $-f_x$ direction. This compliant motion establishes a double face contact between $\mathcal{A}$ and the support table. The last motion aligns the manipulated object in the corner (a heuristic was used to re-position the task frame aligning its F_z axis with the edge of $\mathcal{A}$ orthogonal to the force biases). The rest of the objects can be aligned with similar control programs. The screwing is performed with an adapted control primitive (the position of the holes is determined using on-line information and the geometric model of the task).

4 Conclusion

We have presented a simple to approach to integrate fine motion planning with compliant control. The integration between our planner and the compliant control formalism is made possible by the fact that both use similar modeling primitives. A simple assembly task (figure 1) has been successfully planned and executed applying this approach (human intervention has been needed for selecting the

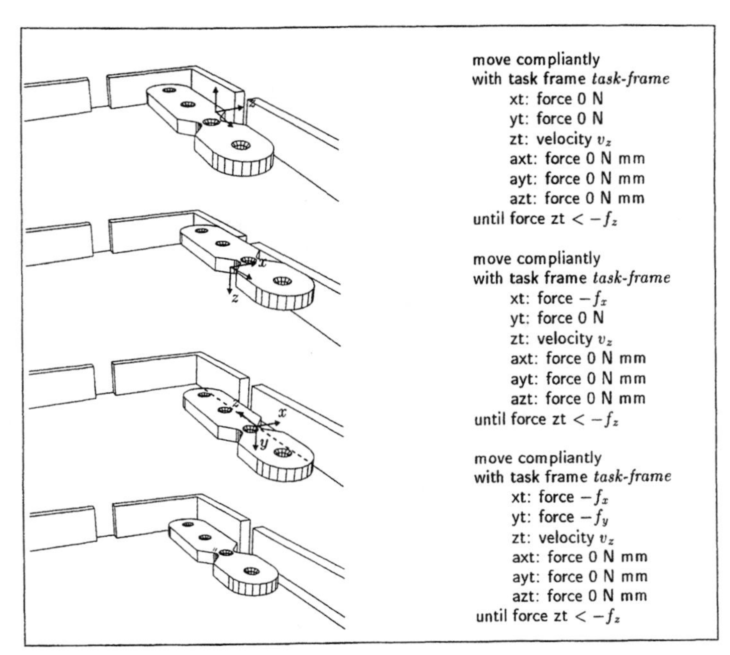

Figure 5: An assembly sequence and it corresponding control program.

magnitudes of the control parameters). It was executed using a Kuka IR361 robot equipped with a Schunk force sensor (figure 6).

References

[1] J. De Schutter and H. Van Brussel. Compliant robot motion: I. a formalism for specifying compliant motion tasks. *International Journal of Robotics and Automation*, 7(4):3 – 17, August 1988.

[2] J. De Schutter and H. Van Brussel. Compliant robot motion: II. a control approach based on external control loops. *International Journal of Robotics and Automation*, 7(4):18 – 33, August 1988.

[3] Oussama Khatib. A Unified Approach for Motion and Force Control of Robot Manipulators: The Operational Space Formulation. *IEEE Journal on Robotics and Automation*, 3:43–53, 1987.

[4] Christian Laugier. Planning fine motion strategies by reasoning in the contact space. In *IEEE International Conference on Robotics and Automation*, pages 653 – 659, Scottsdale, AZ, 1989.

[5] Ming Lin and John Canny. Efficient collision detection for animation. In *Proceedings of the Third Eurographics Workshop on Animation and Simulation*, Cambridge, England, 1991.

[6] Matthew T. Mason Tomás Lozano-Perez and Russel H. Taylor. Automatic synthesis of fine-motion strategies for robots. *International Journal of Robotics Research*, 3(1):3–23, 1984.

[7] Mattew T. Mason. Compliance and force control for computer controlled manipulators. In *Proceedings of the International Conference on Robotics and Automation*, June 1981.

[8] M.H. Raibert and J. Craig. Hybrid Position/Force Control of Manipulators. *ASME Journal on Dynamic Systems, Measurements and Control*, pages 126–133, 1981.

[9] Randall H. Wilson. *On Geometric Assembly Planning*. PhD thesis, Department of Computer Science, Stanford University, Stanford, California 94305, March 1992.

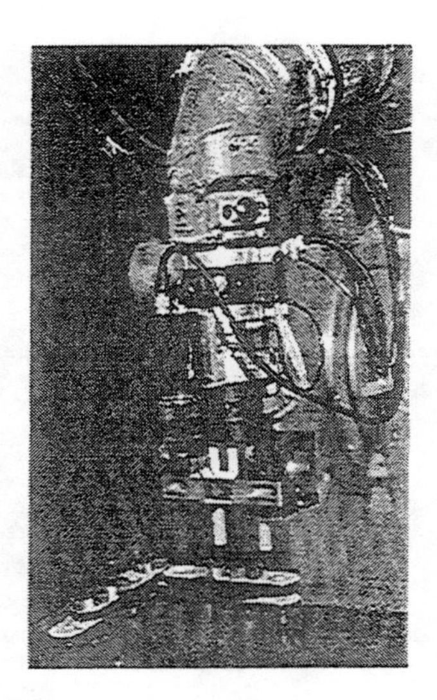

Figure 6: A Kuka IR361 robot equipped with a Schunk force sensor executing the revolute joint assembly task.

[10] Wim Witvrouw, Peter Van De Poel and Joris De Schutter. COMRADE: COmpliant Motion Research And Development Environment. In *3rd IFAC/IFIP workshop on Algorithms and Architectures for Real-Time Control (AARTC'95)*, pages 871–88, Ostend, Belgium, May 1995.

Reactive collision-free motion control for a mobile two-arm system

U.M. Nassal
Institute for Real-Time Computer Systems and Robotics (IPR)
University of Karlsruhe
e-mail: nassal@ira.uka.de

ABSTRACT

Mobile manipulators are a promising direction for the development of more flexible robot systems. However, most of the mobile manipulators considered in current research are equipped with only one manipulator. Two-arm systems provide extended manipulation capabilities without assistance of a second stationary or mobile manipulator. This paper presents a new scheme for motion coordination and control, which allows to consider a two-arm manipulator system on a single mobile platform. Furthermore, a collision avoidance scheme is presented, which can easily be integrated in the motion control loop.

1 INTRODUCTION

Since mobile manipulators are kinematically redundant systems, most of the research concerned with control of mobile manipulators uses schemes which were successfully applied to redundant manipulators. In [1] a thorough discussion of these different schemes is presented. These methods, however, are not directly applicable to mobile two-arm systems, since the mobile platform is now part of two different kinematic chains. The known methods have in common, that they are mapping the end-effector position and orientation to the *complete* configuration of the mobile manipulator. In [2] a new coordination scheme has been introduced, which allows for multiple manipulators on a mobile platform.

In most papers concerning mobile manipulators, the presence of obstacles is neglected. Some of the work which takes this into account shall be mentioned here: In [3], obstacles for the manipulator are considered using a superquadric representation of an object. This superquadric is used to construct a potential field which exerts a repulsive force towards the manipulator. Another approach is shown in [4], where virtual forces are exerted on all joints and limbs, including the mobile platform. However, the problem of collision avoidance in a complex environment with non-convex obstacles has not been adressed, yet. In this paper, a model-based approach is chosen, where a configuration space model for the degrees of freedom of the mobile platform is set up. The varying geometry of the manipulators is considered to avoid collisions of the whole mobile manipulator, too.

The paper is organized as follows: in Section 2, the new coordination scheme which is called transparent coordination is outlined. Section 3 presents the control concept for the mobile manipulator. In Section 4 the collision avoidance scheme is presented and in Section 5 experimental results are shown. The paper ends with the conclusions.

2 TRANSPARENT COORDINATION

The idea of the transparent coordination scheme is the distribution of different responsibilities to the different mechanisms: The manipulators are responsible for the end-effector motion, which can be controlled in different ways (position/force). The task of the mobile platform is the surveillance of the manipulators configuration. This separation of end-effector motion and manipulator configuration can be carried out due to the kinematic redundancy. To eliminate the influences of the mobile platform to the end-effector position and orientation, a kinematic decoupling scheme has to be used. In the following, this principle is discussed in the velocity domain.

The end-effector velocity in world coordinates can be written as:

$$\dot{\mathbf{x}}_{e/w} = \mathbf{T}_1\left(\mathbf{x}_{p/w}\right)\dot{\mathbf{x}}_{e/p} + \mathbf{T}_2\dot{\mathbf{x}}_{p/w}, \tag{1}$$

where $\mathbf{T}_1\left(\mathbf{x}_{p/w}\right)$ is a transformation matrix to map the end-effector motion from a platform-fixed coordinate system to world coordinates, which depends on the current configuration of the platform $\mathbf{x}_{p/w}$. All the vectors $\mathbf{x}$ have a dimension of 6, since they contain position and orientation. $\mathbf{T}_2$ denotes the transformation matrix to transform the platform motion into end-effector motion. Since the end-effector in general is not located in the centre of the mobile platform, this matrix represents the translational velocity of the end-effector which is induced by the platform rotation. Rewriting Eqn. 1 in matrix form and introducing the manipulator Jacobian $\mathbf{J}_m$ yields:

$$\dot{x}_{e/w} = \underbrace{\begin{bmatrix} T_1(x_{p/w})J_m & T_2 \end{bmatrix}}_{\mathbf{J}} \begin{bmatrix} \dot{\theta}_m \\ \dot{x}_{p/w} \end{bmatrix}. \tag{2}$$

The transparent cooperation approach considered here is different from the approaches described in [1] since the platform is merely used to control the internal motion and has no immediate effect on the external motion:

$$\dot{\theta}_m = J_m^{-1}\, T_1^{-1}\left[\dot{x}_{e/w} - T_2\, \dot{x}_{p/w}\right], \tag{3}$$

$$\dot{x}_{p/w} = f(\theta_m, \dot{\theta}_m, \text{obstacles}). \tag{4}$$

Equation (3) outlines the principle of decoupled motion which allows to move the vehicle without affecting the end-effector motion. The control of the mobile platform is expressed by Equation (4), where a control function can be designed that complies with different requirements (controlling the manipulator configuration, avoiding obstacles etc.). Rewriting these equations in matrix form yields

$$\begin{bmatrix} \dot{\theta}_m \\ \dot{x}_{p/w} \end{bmatrix} = \begin{bmatrix} J_m^{-1}\, T_1^{-1} \\ 0 \end{bmatrix} \dot{x}_{e/w} + \begin{bmatrix} J_m^{-1}\, T_1^{-1} T_2\, \dot{x}_{p/w} \\ f(\theta_m, \dot{\theta}_m, \text{obstacles}) \end{bmatrix} \tag{5}$$

where a decomposition into *effective* and *self-motion* can be recognized.

The *effective motion* of the end-effector is used to control the manipulator joints only. The self-motion of the mobile manipulator is controlled by the vehicle motion in order to perform local optimization for the configuration of the manipulator while avoiding obstacles for the mobile manipulator. Since the self-motion allows an optimal configuration to be left without resulting in a positioning error for the end-effector, it is reasonable to assign this task to the mechanism with the slower dynamic response.

3 MOTION CONTROLLER FOR MOBILE MANIPULATION

As shown in section 2, the mobile manipulation controller has to compute a platform motion which allows all manipulators on the platform to keep away from their workspace boundary and which avoids collisions between the platform and obstacles (equ. 4). This task is performed in two steps: at first a suitable vehicle acceleration is computed to optimize the manipulators configuration and then this acceleration is evaluated whether a collision risk occurs or not. In this section, the focus is on the first part, .i.e. the control of the vehicle to support the manipulators.

3.1 VALUATING THE MANIPULATOR CONFIGURATION

To enable the mobile manipulation controller to assess the present state of the arms, each manipulator configuration is valued by a scalar cost-function $c(\theta)$. For designing this cost-function it has to be considered that the workspace is restricted by joint limits and singularities. Joint limits can be measured by

$$c_l = \sum_{i=1}^{dof} \left(\frac{1}{\theta_i - \theta_{i,\min}} + \frac{1}{\theta_{i,\max} - \theta_i} \right) \tag{6}$$

where dof specifies the number of joints and $\theta_{i,\min}$, $\theta_{i,\max}$ are the upper and lower joint limits.

Usually, the manipulability measure $w = |\det(J(\theta))|$ defined by Yoshikawa [5] is used for singularity-avoidance, where $J(\theta)$ is the Jacobian matrix. Yet, for mobile manipulation, where the vehicle motion can merely control the distance of the end-effector to its singularities in the plane of the vehicle motion, a measure is needed which increases consistently and steady when the end-effector moves towards a singular configuration with constant Cartesian speed. For that purpose a cost-function is established which evaluates the distance to the singularities in the Cartesian space. Eqn. (7) shows this cost-function for a PUMA manipulator, where $d_s(\theta)$ and $d_e(\theta)$ are the distances to the shoulder and elbow singularity of the PUMA respectively:

$$c_s = \frac{1}{d_s(\theta)} + \frac{1}{d_e(\theta)} \tag{7}$$

These singularities are the most relevant ones for the workspace boundary. The resulting cost function is the sum of the two functions c_s and c_l.

3.2 MOTION CONTROL

The concept of transparent coordination induces that the manipulator can be controlled by directly mapping the end-effector motion to the joint-motion, since this is no longer a kinematically redundant system. The focus of this paper is the control of the mobile platform, which is explained in this section.

In equ. (3) it is outlined that vehicle motions modify the joint angles of each manipulator and thereby also affect its cost-function. The derivation

$$\mathbf{g} = \frac{\partial c(\theta)}{\partial \mathbf{x}_{p/w}} \tag{8}$$

can be used as a measure which defines how the position of the vehicle has to be changed in order to reduce the cost of the respective manipulator. By using the chain rule this derivation can be calculated in three steps: First, the costs have to be derived with respect to the joint-angles. The complexity of this computation depends on the specific cost function. Then, the derivation of these angles with respect to the end-effector coordinate in Cartesian space can be computed by using the inverse Jacobian. Finally, the derivation with respect to the vehicle-position in world coordinates can be computed by regarding the current velocity transformation between end-effector frame and vehicle frame. So the dimension of **g** is the number of degrees of freedom of the platform.

Since it is necessary to take all manipulators on the platform into consideration, the different derivations $\mathbf{g}_i$ for each manipulator are summed up. When all elements of **g** are identical to zero, the sum of all cost-functions is minimal and the vehicle is located in an optimal position. Therefore, this gradient can be regarded as an error signal, which can be processed by a control algorithm. In [6] a Fuzzy controller is described, which controls the mobile platform to optimize the manipulator configuration by means of this gradient.

4 COLLISION AVOIDANCE

Collision avoidance for the mobile two-arm systems is accomplished in a simular manner as the actual motion coordination. There is also a separation of different responsibilities. Three different types of collision risks can be distinguished (see Figure 1):

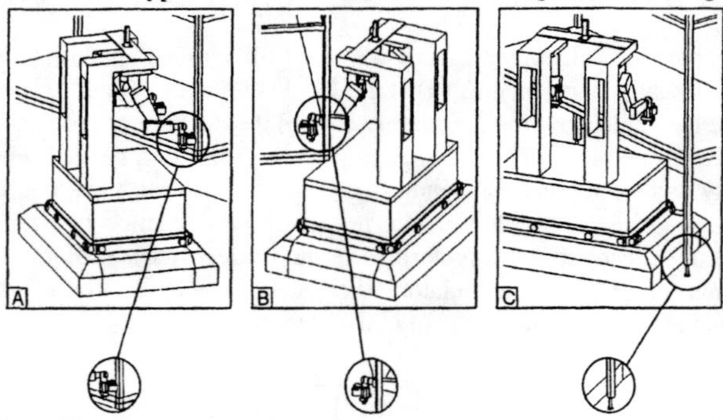

Figure 1: Distinction between different collision risks

- *Collisions of the end-effector:* Since the mobile platform cannot affect the end-effector position, the manipulator controller is responsible for collisions of the end-effector.
- *Collisions of the arm:* The manipulator configuration is changed due to the effective motion and the internal motion. Therefore, both subsystems, manipulator controller and mobile platform controller are responsible for collisions of the arm.
- *Collisions of the mobile platform:* This is completely in the responsibility of the platform controller.

In this paper, the controller of the mobile platform is considered. Therefore, the collision risks for the arm and the mobile platform have to be regarded for the collision avoidance scheme. This is accomplished by extending the motion controller described in section 3. In Figure 2 the structure of the extended motion controller is shown.

The acceleration $\mathbf{a}_d$ which is computed by the mobile manipulation controller is evaluated, regarding the current velocity **v** and the current local environment which is looked up in the configuration space model. Then, the acceleration is reduced if necessary to avoid collisions. This is accomplished by a second controller which is designed to keep the distance to the configuration space obstacles above a given limit. The configuration space used in this scheme has three dimensions for the degrees of freedom of the mobile platform. Using the current manipulator joint angles, the configuration space model can be adapted to the changing shape of the mobile manipulator. Thus, the mobile platform can also regard those collision risks, which are due to the self-motion of the manipulator.

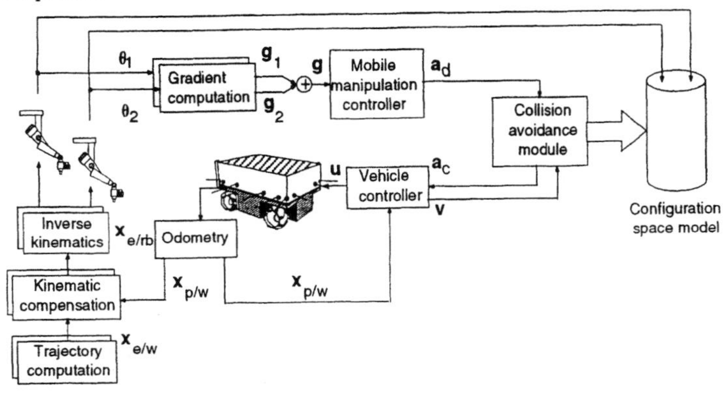

Figure 2: Structure of the extended motion controller

5 EXPERIMENTAL RESULTS

In order to evaluate the developed approach, the control scheme has been applied to a real mobile two-arm system, the mobile robot KAMRO. It consists of a mobile platform with an omnidirectional drive system using MECANUM wheels and two PUMA manipulators with 6 degrees of freedom which are mounted in a hanging configuration. For the evaluation of the motion controller, three different cases are considered:

a) Both manipulators are moving along parallel trajectories.

b) One manipulator is moving along a trajectory , while the other is not active (this corresponds to the case of a mobile manipulator with a single arm).

c) Both manipulators are active. One of them is moving along a trajectory while the other one is resting in world coordinates.

Figure 3 shows different phases of the coordinated motion in the three cases. In case a) it can be recognized that the mobile platform performs a translational motion which is approximately parallel to the end-effector motion. The resulting rotational component of the gradient is zero since the gradients of the single manipulators have an opposite rotational component. In case b) the mobile platform chooses the motion which is optimal to improve the configuration of the active manipulator. This is a superposition of rotation and translation. In c) one of the manipulators is resting in world coordinates. Therefore the mobile platform performs a movement which is close to a circular motion around the end-effector of the resting manipulator.

In principle, the mobile platform can support an arbitrary number of manipulators. In practical applications however, two arm systems already have considerably increased manipulation capabilities compared to single manipulator systems. Three or more manipulators can seldom be usefully applied. Furthermore, each manipulator introduces additional restrictions to the platform motion, so there is a practical limit for the number of manipulators.

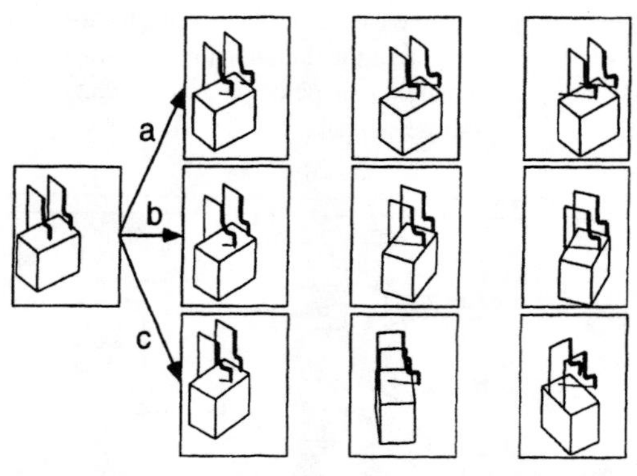

Figure 3: a) Two manipulators moving along parallel trajectories b) Only one manipulator is active c) Both manipulators are active; one of them is resting in world coordinates

CONCLUSIONS

In this paper, a new approach for motion control for mobile manipulation has been presented, which allows geometric obstacles to be considered and multiple manipulators to be used on a single platform. The main principle is the distinction between different responsibilities for the mobile platform and the manipulator system. As far as motion coordination is concerned the transparent coordination scheme was presented, which regards the mobile platform as a means to control the manipulator configuration. The collision avoidance scheme extends this control concept by considering constraints imposed by the geometry of the environment.

ACKNOWLEDGEMENTS

This work was performed at the Institute for Real-Time Computer Systems and Robotics, IPR (Prof. Dr.-Ing. U. Rembold, Prof. Dr.-Ing. R. Dillmann) at the University of Karlsruhe and is supported by the German Research Foundation (DFG) in the national research program SFB 314 "Artificial Intelligence-Knowledge Based Systems".

References

[1] D.R. Baker, C.W. Wampler: *Some Facts Concerning the Inverse Kinematics of Redundant Manipulators.* ICRA IEEE Int. Conf. on Robotics and Automation, 1987, pp. 604-609.

[2] U.M. Nassal et al.: *Mobile Manipulation—Cooperation of Manipulators And a Mobile Platform for an Autonomous Robot.* Fifth World Conference on Robotics Research, Cambridge, Massachussetts, September, 27-29, 1994, pp. 11.11-11.24.

[3] Y. Yamamoto, X. Yun: *Coordinated Obstacle Avoidance of a Mobile Manipulator.* ICRA IEEE Int. Conf. on Robotics and Automation, Nagoya, Japan, 1995, pp. 2255-2260.

[4] Cameron, J. et al : *Reactive Control for Mobile Manipulation.* IEEE Int. Conf. on Robotics and Automation, Atlanta, May 1993, pp. 228 - 235.

[5] T. Yoshikawa: *Analysis and Control of Robot Manipulators with Redundancy.* In M. Brady and R. Paul (Ed.), Robotics Research: The First Int. Symp., MIT Press, Cambridge, Mass., 1984, pp. 735-748.

[6] Nassal, U.M, Junge R.: *Fuzzy Control for Mobile Manipulation.* To appear in the Proceedings of IEEE Int. Conf. on Robotics and Automation, 1996, Minneapolis.

PARALLEL ROBOTS AND MICROROBOTICS

PERNETTE Eric, CLAVEL Reymond
Swiss Federal Institute of Technology, EPFL
1015 LAUSANNE, SWITZERLAND
Tél.: (+41) 21 693 59 59 / 38 21 - Fax: (+41) 21 693 38 66

ABSTRACT

During the past few years, there has been an increasing demand in the field of precision engineering for fine motion of multi-degrees of freedom. This motivated the development of a new robotics application field, microrobotics. In this paper, are presented the requirements of microrobotics as well as a state of the art in order to show the different advantages of parallel robots in very high precision applications. Parallel micromanipulators using elastic joints are introduced as well as manufactured structures from one single solid. Metallic bellows are also used and some solutions are suggested to obtain important working space.

KEYWORDS: Parallel micromanipulator, precision, stiffness, elastic joint, monolithic structure, metal bellows, working-space, modularity.

INTRODUCTION

During the past few years, important improvements in the fields of microstructures, micromechanic devices as well as in the fields of microelectronics and optics have allowed the development of microsystems and of integrated optical elements. These small size, high technology products require robots capable of manipulating and assembling micro-components with the highest precision, (typically 0.1 μm). These microrobots also have to be flexible enough to adapt to the different microassembly tasks and other applications.

In order to obtain a submicronic accuracy, the structure of these micromanipulators must be constituted of a high-precision mechanic, without backlash, nor friction or hysteresis. It must also have a high structural frequency, be rigid, compact, light and have a minimum length of the line of force [1]. One is thus very quickly limited by the classical serial structures. If those are frequently used in macroscopic applications, they do not seem quite adapted to very high precision applications. The arrangement in series of segments or of discrete stages, each giving a degree of freedom, produces an accumulation and an amplification of errors [2-4]. These devices may be modular, but they rapidly become voluminous. Moreover, the major inadequacy of serial architectures as far as their utilization in microrobotics is their lack of stiffness. This is seen in the unmeasured and important deflections in the structure of the robot [5, 6]. Increasing the stiffness by increasing the mass is a solution to limit this problem but other difficulties arise. Many others ideas have been discussed such as: an active stiffness control [7], to take into account these deflections at the kinematic modelling level [5], and the local support concept [6] or the bracing strategies [8]. These solutions are rather delicate, they complicate the control, and reduce the flexibility in certain cases. Finally, another solution aiming to increase the stiffness of the structure without increasing its mass, is the utilization of parallel robots. They allow the obtaining of rigid architectures and also the limiting of errors and the obtaining of mobilities of six degrees of freedom in a very compact manner. Moreover, actuators may be directly fixed on the ground. These robots, already used in different applications such as assembly, the transfer of light objects and the machine tool [9-12], are more and more solicited in micromanipulation. Many authors present concepts of parallel micromanipulators which offer great possibilities as far as the different tasks of micromanipulation are concerned, whether it be at the level of mobility or flexibility.

A state of the art is carried out in this article in order to show, through the literature, the advantages offered by parallel structures within microrobotics and the important part they are playing in this rapidly expanding sphere.

PARALLEL ROBOTS AND MICROROBOTICS

The first applications and realizations of parallel robots are approximately dated of the 40s, by Pollard in 1938, to paint cars [13], and Gough in 1947, to test tyres [14]. Then, Stewart in 1965 [14] uses a parallel structure to realize a flight simulator slightly different from figure 1. This architecture is by the way close to that used by Gough, but it is often reported in literature as the "Stewart platform". Since then, with the important development and the decreasing of the cost of the computation and control systems, this structure has given birth to numerous 6 or 3 DOF parallel robot concepts [15]. The choice of parallel structures for very high precision applications is justified by numerous advantages.

- **Stiffness and Structural frequency:** stiffness is important because several kinematic chains join the fixed part "base" and the mobile part "platform". Moreover, the bending solicitations of the under mechanisms are reduced. Thus, the specific frequencies are high, giving way to few repeatability errors due to the uncontrolled oscillation of the mobile structure.
- **Precision:** the parallel disposition of the segments prevents the accumulation and the amplification of errors and also limits the effect of the manufacturing tolerances. We obtain a great volumetric precision and not only a good repeatability as with a lot of industrial robots; this great volumetric precision is necessary for microrobotics applications.
- **Mobility and compactness:** for limited working volume applications, parallel structures are more compact than the serial structures, no matter what their number of degrees of freedom (3 or 6...).
- **Fixed actuators:** the fixing of the actuators directly on the ground allows the mechanical structure of the manipulator to be independent from the used actuators and the mobile mass is reduced [9, 16-17]. The fact of moving the actuators away from the working zone is also an important point in microassembly (work in clean atmosphere). We also reduce the problems of precision due to thermal dilatation caused by the actuators (heating, friction...).

Other criteria can be determining in certain microassembly tasks:

- the sensivity to temperature, alleviated by the symmetry of structures and by the number of kinematic chains.
- the uniform distribution of the load on the different chains and actuators in nominal position.
- the possibility of realizing compliant systems (insertion of pieces, adaptation of 2 surfaces in contact), of determining the torque/force acting on the platform (contact detection, active control of the force) [15, 16].

These parallel manipulators have as main disadvantages a limited working space and a coupling of the DOF between the translations and the orientations. Some solutions are suggested in this article to remedy these problems for certain applications in microrobotics.

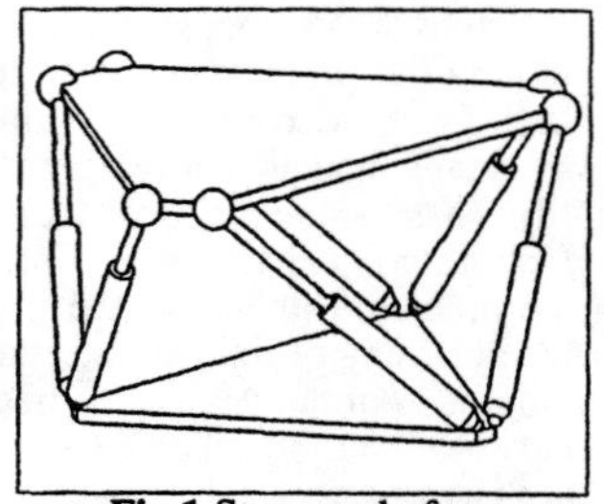

Fig.1 Stewart platform

PARALLEL MICROMANIPULATORS

Ellis, as early as 1962, suggests the realization of a piezoelectric micromanipulator by organizing the ceramics in parallel rather than in series [18]. Since then, numerous prototypes of parallel micromanipulators have been developed for different applications, such as biotechnologies and micro-surgery [19-22].

The utilization of parallel architectures as microrobots for the microassembly requires articulations and transmission elements without backlash, friction, nor hysteresis. For that purpose, articulations without relative motion (rolling or sliding motion between different elements) are used. They are articulations that are bent elastically. Thus are realized, by simple manufacturing, different types of elastic joints [23]: revolute, prismatic, universal and spherical, that can be integrated within the level of parallel structures. The rigidity of the whole is thus increased.

The machining of these elements must be very careful. Different structures which use these flexible articulations are reported in literature.

Stoughton [24] introduces a micromanipulation concept composed of 2 parallel structures commanding a chopsticks set. He uses, for each chopsticks a stewart platform activated by 6 piezoelectric elements *(fig.2)*. Each segment is connected to the base and to the platform by flexible links, type ball joint. The emphasis is on compactness, rigidity, on the whole of identical arms and on fabrication simplicity.

Hudgens and Tesar [25] also propose a fully parallel micromanipulator with 6 DOF, which is a variation of the stewart platform *(fig.3)*. This architecture comes in the second type of parallel robots [15], where we do not find a robot with segments of variable lengths, but a robot with variable positions of the articulation points. In that case, the low articulations of the segments linked to the mobile part describe a circle. The actuators, in this instance the motors, are therefore fixed on the base, which gives the possibility to use a four-bar linkage transmission device with an eccentric to increase the mechanical advantages and amplify the movements of the motors. The whole of the transmission system is realized from only one solid.

Hara [26] presents a micromanipulator which uses an architecture suggested by Artigue *(fig.4)* [11]. It corresponds to an identical structure to the Stewart platform but with a different spatial disposition. The kinematic chains are divided in 3 groups. Thus, for small movements, the different degrees of freedom are uncoupled. The actuators are of a piezoelectric type, associated with linear elastic transmission systems. A resolution of 0.01μm and a movement of 10 μm are thus reached.

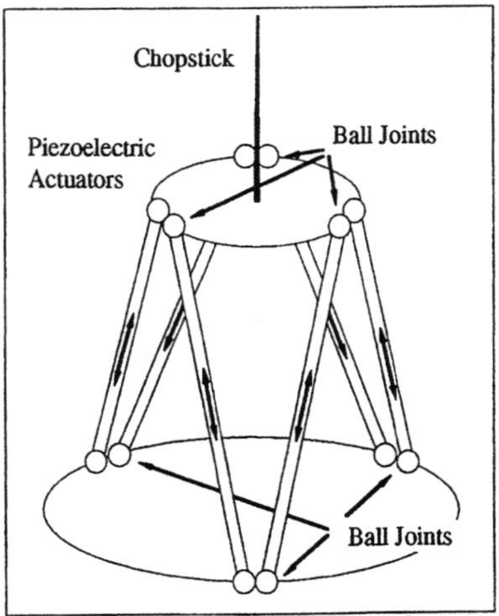

Fig.2 Stoughton Micromanipulator

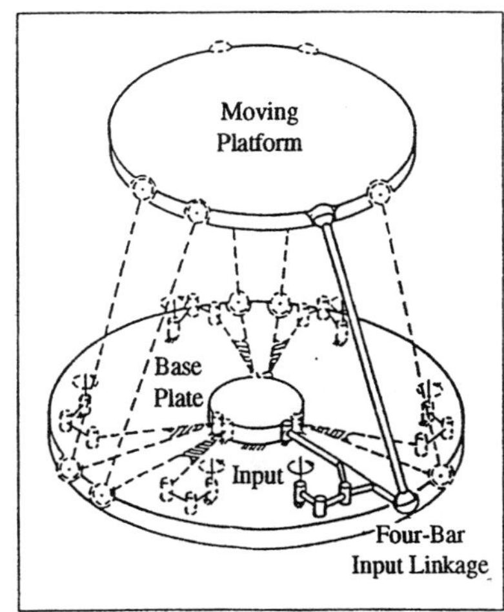

Fig.3 A 6 DOF Micromanipulator

Hemni [27] uses another distribution of the actuators, distribution which we find also with Artigue at the macroscopic level [11]. The kinematic chains are disposed in pairs this time, following 3 perpendicular planes in order to obtain a minimum and identical number of actuators for each DOF. The uncoupling works as in the previous example for small movements.

Lee [28] realizes a 3 DOF structure *(fig.5)*, of which one is a vertical translation and tow are rotations (θx, θy). He uses another type of a flexible spherical joint. A lever allows the amplification of the piezoelectric actuators.

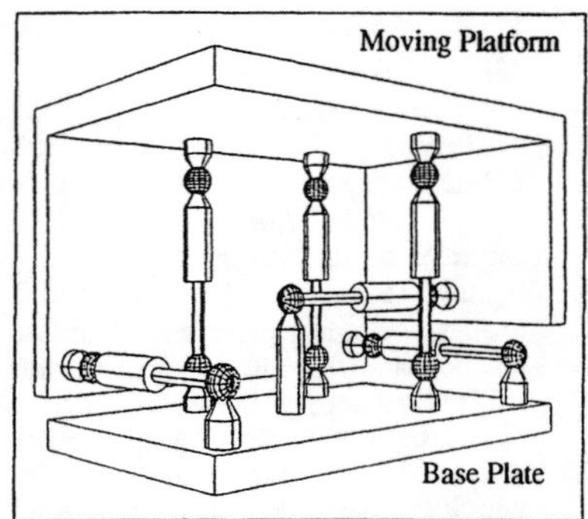

Fig.4 Artigue Platform

MONOLITHIC STRUCTURES

One realizes that it is possible to conceive by electro-discharging, the whole of a kinematic chain from only one piece, by a combination of simple elastic articulations. However, certain structures can even be fully manufactured from one same solid. These are the "monoblock structures". They allow a total freeing from the operations and tolerances of the assembly components. Rigidity is therefore maximal and the precision excellent.

Several planar microrobots with 3 DOF (X,Y, θz) are elaborated following this principle *(fig.6)* [3, 23, 26]. They are exempt from any assembly of pieces, except the actuators.

The current methods of machining also allow the production of "volumic monoblock structures". Thus, Magnani [23] realizes a 3 DOF microrobot in one single solid, "Orion" *(fig.7)*, which uses the same configuration as that of Lee's *(fig.5)*. The motorization of this structure is insured by 3 actuators of the Inch-Worm type, with piezoelectric elements. A different distribution of the articulations in the kinematic chains (2 parallel universal joints instead of a spherical joint + 1 rotate joint) allows the obtaining of a monolithic structure of a micro-Delta: it is a 3 translation-micromanipulator. The particularity of this robot is that the orientation of its mobile part is always constant, in a passive manner.

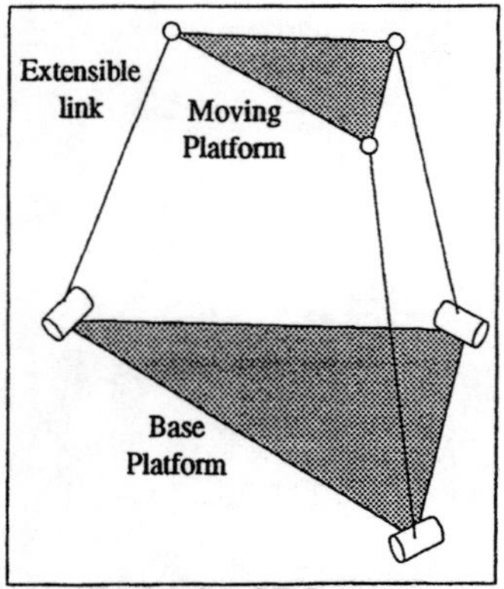

Fig.5 A 3 DOF Structure

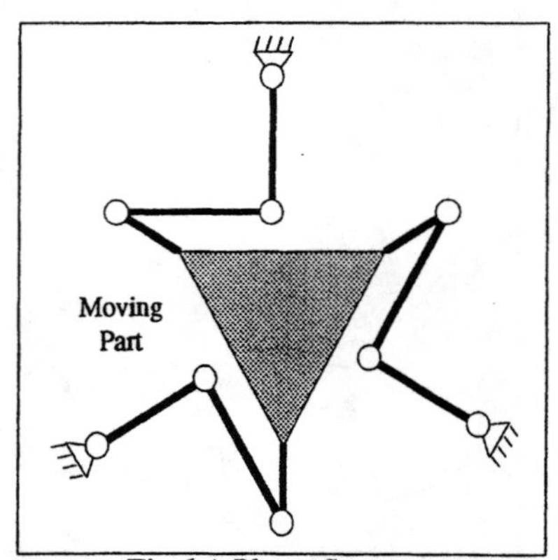

Fig.6 A Planar Structure

The advantage of these monolithic robots is very important, in their realizations as well as in their characteristics. This concept fully benefits from high precision manufacturing new techniques, but also and above all from the advantages of parallel robots, in which the structure is actuator-free. However, one has to know that articulations using material elastic deformation have a few disadvantages: weak displacements, often complex simulation with finite elements and linearity problems; change in strain according to the joint position and non fixed rotation centre joints.

STUCTURES WITH METALLIC BELLOWS

Parikian [29] suggests a new microrobot concept, close to that of the Delta Concept *(fig.8)*, one that utilizes metallic bellows. These elements allow movements in all directions, except the rotation following the bellows axis. They also have the advantages of elastical structures. If we dispose the bellows following 3 perpendicular planes and if we activate the mobile part with linear actuators from the inside of the bellows, we obtain the principle of the Delta Robot: 3 DOF of translation. Each bellows prevents a rotation. The rigidity of the whole is directly linked to the bellows (materials, diameter, number of undulations...). It is possible to add an extra DOF at the level of each kinematic chain, an active rotation at the base and following the axis of the bellows, in order to obtain a modular microrobot with 6 DOF. The rotations are transmitted to the mobile part through the intermediary of the bellows. Thus the DOF of translation and of rotation are perfectly uncoupled as well as the different rotations. Moreover, the number of DOF may be adapted according to the task. The advantage of this principle is that it can be combine flexible elements and rigid ones, and it can work in any environment: hermetic, aggressive environment. When in movement, it is possible to compensate the deformation strains of the bellows if working on the pressure inside the bellows.

Salcudean [30] also proposes a 6 DOF architecture with 6 pairs of externally pressurized bellows. In this case, the bellows are not used as passive elements in linear guides, but directly as pneumatic actuators. The movements are of about +- 6.5mm and +-5° and the precision, as for now, is of 5μm.

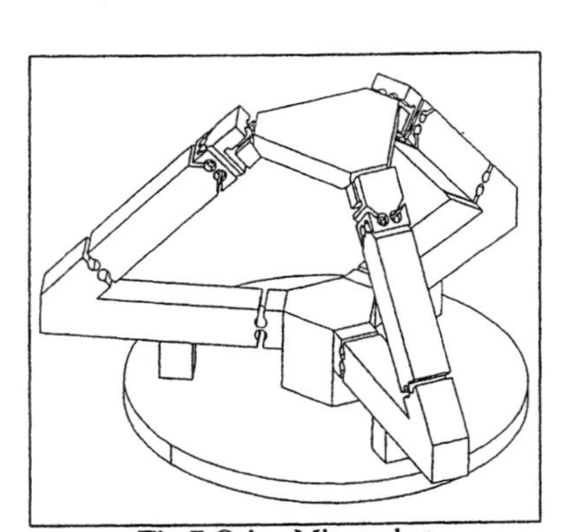

Fig.7 Orion Microrobot

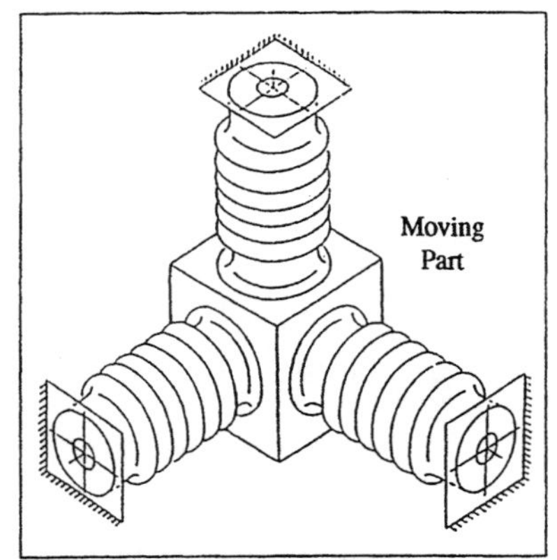

Fig.8 A Bellows Structure

INCREASE OF WORKING SPACE

It is noted that one of the current concerns of microrobotics is precisely to reconcile precision (0.1μm) with the working space (>1cm3): one is often obtained to the detriment of the other, as it is here the case with the utilization of parallel robots with elastic joints. In fact, in this case the working zone stays essentially limited by the elastic articulations, the number of kinematic chains linking the base and the platform, and sometimes the interferences. Indeed, the possibility of fixing the actuators on the base, allows, with an appropriate choice [31], to free ourselves from their limitation. However, several solutions are suggested to cumulate precision and working volume.

The simplest way consists in the utilization of a parallel micromanipulator combined with classic-type robots, as repositioning modules. Two approaches are possible according to that principle: the first consists in the using of the precise manipulator as the left hand [15, 32], and the second as the active wrist (terminal tool) [6, 8, 33-35]. The parallel micromanipulator thus compensates the static errors of the sequential robot with serial structure. This principle is often designated by the name "Macro-Micro displacement" or "Coarse-Fine robot".

It is also possible to use hybrid structures. One can make work in series, or one facing the other, 2 parallel architectures with complementary DOF in order to obtain the desired number of DOF. In choosing architectures with a limited number of DOF, we diminish the possible interferences and the number of chains linking the base and the mobile part [23]. Moreover, this principle has the advantage of providing modular systems.

Stoughton [24] suggests the addition of redundant actuators under the base of the parallel micromanipulator in order to increase the movements.

Several architectures of parallel robots which have been little solicited in microrobotics may in part alleviate the problem of limited volume. We find as an architecture the one suggested by Merlet *(fig.9)* [17, 22]. It is a Stewart platform with fixed actuators, and variable position articulation points following a vertical line. In this case, the movement following the vertical axis is theoretically unlimited, being only dependent upon actuators stroke. The delta robot variant with linear actuators *(fig.10)* [23, 36] also allows us to privilege an axis, following the same principle.

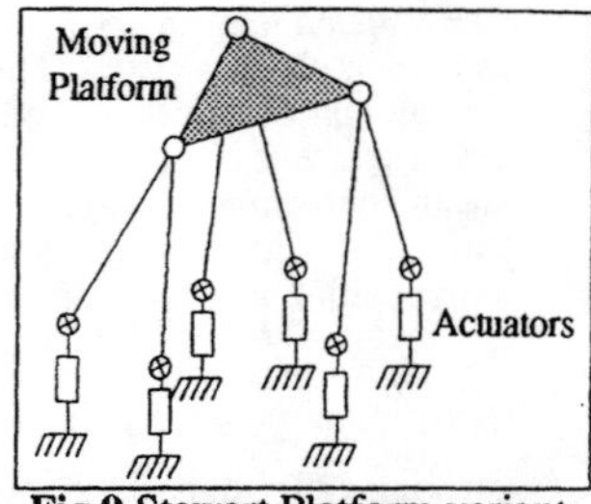

Fig.9 Stewart Platform variant

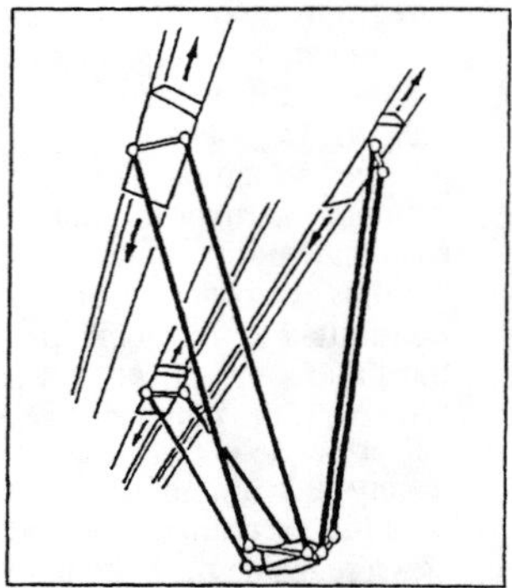
Fig.10 Linear Delta

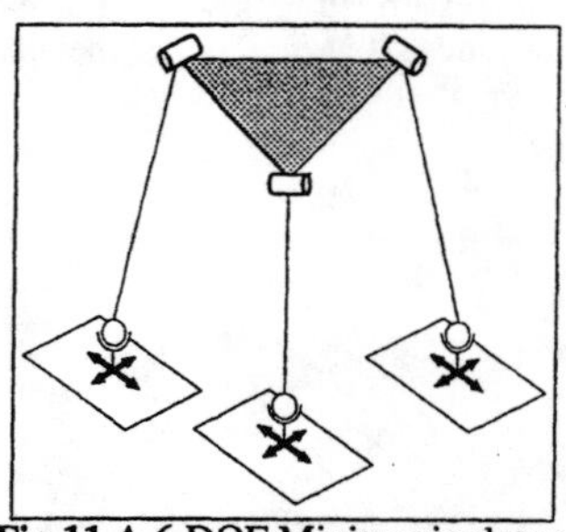
Fig.11 A 6 DOF Minimanipulator

The architecture which Tsai [37] suggests also allows us to obtain a working zone theoretically unlimited, essentially in the direction of the plane defined by the 2 DOF actuators *(fig.11)*. A very particular attention is to be brought at the level of the articulations and of the vertical movement. By driving this structure with micro-translators with 2 DOF (Multi-DOF micro-actuators) [38], one can acquire a good precision.

Behi's architecture [39], using 2 actuators per chain can also be interesting. However, the fact of multiplying the number of actuators per chain to increase the working volume might be done to the detriment of precision.

CONCLUSION

The great possibilities of parallel robots have been presented in this article. This allowed us to show that these architectures are well adapted to micromanipulation: stiffness, precision, compactness, fixed actuators. Architectures using elastic articulations, directly cut into the mass have been discussed as well as their advantages. Certain concepts of microrobots with metallic bellows have also been shown, as well as certain solutions to increase the working volume and limit the coupling of the DOF of parallel structures. Thus, high precision microrobots with an important working space can be developed. The modularity aspect has also been emphasized so as to facilitate the adaptation to different tasks of microrobotics.

Microrobotics is a quickly developing field which presently conditions the development of numerous elements such as microsystems, integrated optics elements, as well as certain applications such biotechnologies and microsurgery. We have thus been able to show that parallel robots have an important role to play in microrobotics even if they come in the second place in the field of the robotics in general.

REFERENCES

1. Nakazawa, H. "The principle of the compliance Minimizing the length of the line of force." *Principles of the Precision Engineering,* Oxford Science publications, 95-96.
2. Unetami, Y. "Piezo-Electric Micro-Manipulator in Multi-Degrees of Freedom with tactile sensibility." *10 ISIR* , Milan (1980), 571-579.
3. "Product for micropositioning." *Physik Instrument catalog,* Germany (1995).
4. "Scientific of Laboratory Products." *The 1994 Newport catalog.*
5. Yao, J. "Accuracy Improvement: Modelling of elastic deflections." *Robotics, Vol. 9.,* (1991), 327-333.
6. Sharon, A. "Enhancement of Robot accuracy using endpoint Feedback and a Macro-Micro manipulator System." *Proc. America Control Conf.,* (1989), 1836-1842
7. Zalucky, A. "Active control of Robot, Structure Deflections." *ASME Journ. of Dynamic Syst. Measur. and Control,* (March 1984).
8. Hollis, R. L., Hammer, R. "Real and Virtual Coarse-Fine Robot Bracing Strategies for Precision Assembly" *Proc. of the Int. Conf. and Robotics and Automation,* Nice (May 1992), 767-774.
9. Pierrot, F. "Manipulations robotiques à haute vitesse: une solution pleinement parallèle." *RAIROAutom. prod. Inf. Ind., Vol. 26.,* (1992), 4-14.
10. Callion, H. Mc "The analysis of a six-Degree-of-freedom workstation for mechanised assembly." *Proc. of the Fifth World Congress on Theory of Machines and Mechanisms,* (1979), 611-617.
11. Artigue, F. "Analyse cinématique de Systèmes de repositionnement et de mesures tridimensionnels pour l'assemblage automatique." *Thèse de Doctorat d'Etat* Université Pierre et Marie Curie, Paris 6 (1984).
12. Goedetics, "Hexapods come of age" *Geodetic Technology Inter. Holding NV: pamphlet,* Genèva.
13. Pollard, W. L. V. "Position Controlling Apparatus." *Brevet USA 1942,* priorité Avril 1938.
14. Stewart, D. "A platform with six degrees of freedom." *Proc. Inst. Mech. Engrs., Vol. 180. N°5,* (1965), 371-378.
15. Merlet, J. P. "Les robots parallèles."*Traité des Nouvelles Technologies série Robotique,* Edition Hermes (1990).
16. Magnani, I., Clavel, R. "Les microrobots: de nouveaux outils pour de nouveaux produits." *ASMT - Journées de microtechnique 94,* EPFL, 29-34.
17. Merlet, J. P.,Gosselin, C. "Nouvelle Architecture pour un manipulateur parallèle à 6 degrés de liberté." *Mech. Mach. Theory, Vol. 26(1),* (1991), 77-90.
18. Ellis, G. W. "Piezoelectric Micromanipulators." *Science Instruments and Techniques, Vol. 138,* (Oct. 1962), 84-91.
19. Fukuda, T., Fujiyoshi, M. "Design and dextrous control of Micromanipulator with 6 D.O.F." *IEEE ICAR,* (1991), 343-348.
20. Hunter, I. W. "Manipulation and Dynamic Mechanical testing of Microscopic Objects Using a Tele-Micro-Robot System." *Proc. of the IEEE Int. Conf. on Robotics and Automation,* (1989), 1553-1558.
21. Arai, T. "Micro Hand Module using Parallel Link Mechanism." *Japan/USA Symp. on Flexible Automation, Vol. 1,* ASME, (1992), 163-167.

22. Grace, K. "A six Degree of Freedom Micromanipulator for ophthalmic surgery." *Int. Conf. on Robotics and Automation, Vol. 2*, (1993), 630-635.
23. Magnani, I. "New Designs for micro-robots." *Int. Precision Engineering Seminar*, Compiègne (1994), 537-540.
24. Stoughton, R. "Kinematic Optimization of a chopsticks-type micromanipulator." *Japan/USA Symp. on Flexible Automation, Vol. 1*, ASME, (1992), 151-157.
25. Hudgens, J. C., Tesar, D. "A Fully-parallel six degree-of-freedom micromanipulator: Kinematic Analysis and Dynamic Model." *Proc. 20 th Biennal ASME, Mechanisms Conf. Trends and Development in Mechanisms Machines and Robotics, Vol. 15(3)*, (1988), 29-37.
26. Hara, A., Sugimoto, K. "Micromanipulator with Multi-Degrees of Freedom." *20 th ISIR*, 505-512.
27. Henmi, N. "A six-degrees of freedom Fine Motion Mechanism." *Mechatronics, Vol. 2(5)*, (1992), 445-457.
28. Lee, K. M., Arjunan, S. "A three degree of Freedom Micro-Motion-In-Parallel Actuated-Manipulator." *Proc. of the IEEE Int. Conf. on Robotics and Automation*, (1989), 1698-1703.
29. Parikian, T. "Kinematic Analysis and Design of Parallel Robots." *Technical Report N° 95-02*, Swiss Federal Institute of Technology (1995).
30. Salcudean, S. E. "A six Degree of Freedom Wrist with Pneumatic Suspension." *Proc. of the IEEE Int. Conf. on Robotics and Automation*, (1994), 2444-2450.
31. Pernette, E. "Ultra-Accurate Actuator with long Travel." (à paraître).
32. Kasai, M. "Trainable assembly System with an active sensory table possessing 6 axes." *11 ISIR*, Tokyo (1981), 393-404.
33. Reboulet, C. "Hybrid control of a 6-DOF In-parallel actuated Micro-manipulator mounted on a scara Robot." *Int. Journ. of Robotics and Automation, Vol. 7(1)*, (1992), 10-14.
34. Berthomieu, T. "Etude d'un micro-manipulateur parallèle et de son couplage avec un robot porteur." *Thèse de Doctrorat*, ESAE (1989).
35. Baartman, J. P., Brennemann, A. E. "Placing Surface Mount Components Using Coarse/Fine Positioning and Vision" *IEEE Transactions on Components Hybrids and Manufacturing Technology, Vol. 13(3)*, (Sept. 1990), 559-564.
36. Stevens, B., Clavel, R. "The Delta parallel robot, its Future in Industry." *ISRAM 94*, Hawaï (1994).
37. Tsai, L. W., Tahmasebi, F. "Design and Analysis of a New six-Degree-of-Freedom Parallel Minimanipulator." *Proc. of the 6 Int. Conf. on CAD / CAM Robotics and Factories of the Future*, 568-575.
38. Ferreira, A., Minotti, P. "New multi degrees of freedom piezoelectric micromotor aiming at micromanipulator applications" *IEEE Ultrasonics Symp.*, Seattle (1995).
39. Behi, F. "Kinematic Analysis for a six-Degree-of-Freedom 3-PRPS Parallel Mechanism." *IEEE Jour. of Robotics Automation, Vol. 4(5)*, (1988), 561-565.

OUTPUT FEEDBACK FORCE/POSITION REGULATION OF RIGID JOINT ROBOTS VIA BOUNDED CONTROLS

ELENA PANTELEY
Institute of Problems of Mech. Engg., Academy of Sciences of Rusia St. Petersbourg, RUSIA elena@ccs.ipme.ru

ANTONIO LORIA and ROMEO ORTEGA
Université de Technologie de Compiègne, URA C.N.R.S. 817, BP 529, 60205 Compiègne, FRANCE, e-mail : aloria(rortega)@hds.univ-compiegne.fr

ABSTRACT

In this note we deal with the problem of force/position control of constrained manipulators subject to input constraints, with only position measurements. We assume that the manipulator interacts with an infinitely stiff environment, hence that the constraints are modeled by a singular algebraic equations. In order to cope with these singularities, we use a reduced order model for which the coordinates space is restricted to a subset where the constraints Jacobian is non singular. Our contribution consists of a controller which ensures that if the generalized positions start in this subset, they never escape of it. Moroever, we prove asymptotic stability even with bounded control inputs.

KEYWORDS: saturation functions, holonomic constraints, force/position robot control.

1. INTRODUCTION

During last decade different approaches to control manipulators interacting with their environment have been proposed in the literature, they can be classified according to the control objective and the adopted model of contact force, for instance Whitney [24] identified 6 different approaches for force control which may be divided in two groups: in the first one the manipulator position and exerted force are simulteniously controlled in a non conflicting way while control approaches belonging to the second group focus on controlling the relationship between the manipulator position and the interaction force.

A popular example lying in the first group is hybrid control (introduced in [18]) which aims at controlling force in the directions constrained by the environment and position in the others. Later it was extended to dynamic hybrid control [26] and to an operational space formulation in [11]. See also [14], [4], [16], [5], [25], and [13].

Some examples of the second group are active stiffness control [10], compliant control [21], [22] and impedance control [7], [6].

On the other hand, concerning the adopted model, mainly two types of contact force models are used: elastic and infinitely stiff envirnoment based. In the first case the interaction with the complaint environment is considered to be elastic so the compliant force is supposed to be proportional to the deformation of the environment. When considering an infinitely stiff environment, starting with [26], [14], [12] the interaction is modelled by holonomic (algebraic) constraints imposed to the manipulator's motion. Unfortunately these equations are singular, then in order to cope with this difficulty, various techniques for deriving so-called *reduced order models* have been proposed based on the projection of the dynamic robot equations onto a submanifold described by the algebraic equation of constraints.

An alternative approach is based on the *principle of orthogonolization* [2] whose key feature is the introduction of a projection matrix that projects velocity and position error signals to a plane tangent to the constraint surface (in joint space) in order to distinguish force and position signals.

The regulation problem has been considered among others in [23], [1], [5], [13]; in the first two references authors consider that the end-effector interacts with an infinitelly stiff environment, while in the last two interaction with an elastically compliant environment was considered. In [23] a PD feedback + compensation for both gravity and contact force at the desired position was proposed and Lyapunov's direct method was used to guarantee local stability. Based on the principle of ortogonalization assymptotically stable PID control scheme was used in [1] and some passivity properties of constrained manipulator were used.

Nevertheless, the above cited solutions (except from [13]) for constrained manipulators present two major disadvantages; the first is the need of velocity meausurements which are often contaminated with noise [3], the second concerns the assumption about non-singularity of constrained Jacobian, i.e. it is supposed that the real trajectory lies always close to the set-point and hence, that the Jacobian is always non-singular.

To the best of our knowledge the force/position control problem whithout velocity measurements has only been treated in [8], [17], and [13]; in [8] the (open loop) observer design of constrained robots is studied, more recently in [17] authors proposed a nonlinear observer and proved for the first time closed loop asymptotic stability cinsidering an infinitely stiff environment. Finally, in

[13] authors proposed the first focre/position controller without velocity measurements and with uncertain gravity knowledge considering an eleastical environment. The latter uses a linear filter in order to avoid measuring velocities, its simplicity allowed to prove semiglobal asymptotic stability.

On the other hand, in the majority of control schemes used for stabilization of constrained manipulators assumption about global solvability of constraint equation is used. In the papers devoted to regulation problems it is assumed either that the constraint Jacobian is non singular in the whole state space or that the actual trajectory is contained in the neibourhood of the desired set-point. However, it should be noted that these assumptions are quite restrictive, for instance the first one can be not fulfilled even in very simple cases like for a planar two-link revolute-joint manipulator whose endpoint is constrained to a plane.

In this note we consider that the manipulator's endtool interacts with an infinitely stiff environment. To cope with the singularities of the constraints equation we use the reduced order model proposed in [16] whose coordinates space is reduced to a submanifold defined by the constraints equation and where it can be ensured that the constraint Jacobian is non singular. Our contribution consists of a controller which ensures that for every initial conditions strictly contained in this subset, the generalized positions remain in it. Based on this, asymptotic stability of the closed loop system is also proved. Moreover, we achieve our control objective assuming that only position measurements are available and using bounded controls. As far as we know this result is the first in its kind.

This note is organized as follows: In next section we introduce the reduced order model originally proposed in [16] and stress some important properties of it, in section 3, we present our main result. In last section we conclude with some remarks.

2. DYNAMIC MODEL

2.1 Preliminaries

We consider in this note the standard model of a rigid revolute joint robot manipulator with potential energy $U(q)$ and kinetic energy $T(q,\dot{q}) = \frac{1}{2}\dot{q}^\top D(q)\dot{q}$ where $D(q) = D^\top(q) > 0$ is the robot inertia matrix. Then using the Euler-Lagrange equations we derive the dynamic model [21]

$$D(q)\ddot{q} + C(q,\dot{q})\dot{q} + g(q) = u - f \tag{1}$$

where $g(q) \triangleq \frac{\partial U}{\partial q}$, is the vector of gravitational forces and $C(q,\dot{q})\dot{q}$ iss the Coriolis and centrifugal forces matrix; u are the applied torques and f is the reaction forces vector. In order to introcuce our notation, we recall that there exist some positive constants k_g, d_m d_M and k_v such that $\forall\ q \in \Re^n$

P1. $d_m I \leq D(q) \leq d_M I$

P2. $\|\frac{\partial g(q)}{\partial q}\| \leq k_g$

P3. $\|g(q)\| \leq k_v$

We assume that the manipulator's end-effector interacts with an infinitely stiff environment hence, its motion is constrained to a smooth $(n-m)$–dimentional submanifold Φ, defined by

$$\phi(q) = 0 \tag{2}$$

where the function $\phi : \Re^n \longrightarrow \Re^m$ is at least twice continuously differentiable.

Assumption **A1** below, concerns the solvability of the constraints equation (2) and will be used in the sequel in order to derive a reduced order model of consrained motion as well as in the design of our controller.

A1 We assume that there exists an operating region $\Omega \subset \Re^n$ ($\Omega \triangleq \Omega_1 \times \Omega_2$, where Ω_1 is a convex subset of $\Re^{n-m}$, Ω_2 is an open subset of $\Re^m$) and a function $k \in \mathcal{C}^2(\Omega_1 \rightarrow \Re^m)$, such that $\phi(q^1, k(q^1)) = 0$ for all $q^1 \in \Omega_1$. Then the vector q^2 can be uniquely defined by the vector q^1 such that $q^2 = k(q^1)$ for all $q^1 \in \Omega_1$.

1 Remark In words, assumption **A1** guarantees the solvability of (2) only in the set Ω, which is equivalent to assume that the Jacobian $J(q)$ is non singular only $\forall q \in \Omega$. This is the main difference with other similar reduced order models used in the literature where it is assumed that $J(q)$ is full rank in the entire state space. See for instance [15].

2 Remark Notice that under this assumption the matrix $J(q) = \partial\phi(q)/\partial q$ can be partitioned as

$$J(q) = [J_1(q), J_2(q)], \tag{3}$$

where $J_1(q) = \partial\phi(q)/\partial q^1$, $J_2(q) = \partial\phi(q)/\partial q^2$ and the Jacobian matrix $J_2(q)$ never degenerates in the set Ω. Necessary and sufficient conditions of global solvability are presented in [17], laying on results of Sandberg [19], [20].

Thus, without loss of generality we can assume that for all $q \in \Omega$ there exist positive constants β_1, β_2 such that

$$\beta_1 \leq \|J_2(q)\| \leq \|J(q)\| \leq \beta_2 \tag{4}$$

To this point we introduce below an assumption concerning the Hessian of the constraints equation

A2 We assume that for all $q \in \Omega$ there exists a constant β_3 such that for all $i \in [1, \ldots, n]$ the following bound holds true

$$\left\| \frac{\partial J(q)}{\partial q_i} \right\| \leq \beta_3 \tag{5}$$

Considering that the end-effector motion is constrained to the submanifold Φ, the above definition of the set Ω allows a parametrization of the generalized coordinates vector in which only $n - m$ *independent* coordinates need to be controlled [14], [9]. Thus, without loss of generality let us choose q^1 and q^2 as independent and dependent coordinates respectively and we can write

$$\dot{q} = H(q)\dot{q}^1 \tag{6}$$

where

$$H(q) = \begin{pmatrix} I_{n-m} \\ -J_2^{-1}(q)J_1(q) \end{pmatrix} \tag{7}$$

Similarly to the parametrization of the vector $\dot{q}$, the generalized reaction forces f which do not deliver power on admissible velocities, i.e.

$$\dot{q}^\top f = 0 \tag{8}$$

can be parameterized by the vector of Lagrange multipliers $\lambda \in \Re^m$ as

$$f = J^\top(q)\lambda \tag{9}$$

To this point we have defined a new coordinates space, now we are ready to introduce our reduced order model.

2.2 Reduced order model and its properties

In this section we define our reduced order model which was introduced in [16] and we stress some similar properties to **P1** – **P3** useful for further analysis. First notice that using equations (6), (7) and (9) in (1) we get

$$D_*(q)\ddot{q}^1 + C_*(q,\dot{q})\dot{q}^1 + g_*(q) = H^\top(q)u \tag{10}$$

$$\lambda = Z(q)\left(C_\lambda(q,\dot{q}^1)\dot{q}^1 + g(q) - u\right) \tag{11}$$

where we have used the identity $JH = 0$ and we have defined

$$\begin{aligned} D_*(q) &= H^\top(q)D(q)H(q) \\ C_\lambda(q,\dot{q}^1) &= D(q)\dot{H}(q) + C(q,\dot{q}^1)H(q) \\ C_*(q,\dot{q}) &= H^\top(q)C_\lambda(q,\dot{q}) \\ g_*(q) &= H^\top(q)g(q) \\ Z(q) &= \left(J(q)D^{-1}(q)J^\top(q)\right)^{-1} J(q)D^{-1}(q) \end{aligned} \tag{12}$$

Next, similarly to [14] we introduce a decoupled control scheme which allows to control generalized positions and constraint forces separately, thus consider the control input of the form

$$u = {H^+}^\top(q)u_a + J^\top(q)u_b \tag{13}$$

where $u_a \in \Re^{n-m}$, $u_b \in \Re^m$ and $H^+(q) = H^\top(q)\left(H^\top(q)H(q)\right)^{-1}$ which exists under the conditions of assumption **A1**. Then the *decoupled* reduced order dynamic model for robot-manipulators with holonomic constraints can be rewritten in the form [16]

$$D_*(q)\ddot{q}^1 + C_*(q,\dot{q})\dot{q}^1 + g_*(q) = u_a \tag{14}$$

$$\lambda = Z(q)\left(C_\lambda(q,\dot{q}^1)\dot{q}^1 + g(q) - H^{+^\top}(q)u_a\right) - u_b \tag{15}$$

$$\phi(q) = 0 \tag{16}$$

Notice that now we are able to design two separate position and force control laws u_a and u_b respectively.

To finish with this section we formulate some useful properties of the reduced order model (14) - (16) which will be exploited in the sequel to design our control algorithm.

For all $q^1 \in \Omega_1$ the reduced order model (14) - (16) has the following properties.

P4 The matrix $D_*(q) = D_*^\top(q) > 0$ has full rank for all $q \in \Re^n$, furthermore, $z^\top D_*(q)z \geq d_m z^\top z$ for all $q \in \Re^n$ and any $z \in \Re^{n-m}$.

P5 A suitable choice of $C_*(q,\dot{q})$ implies that for all $q \in \Re^n$ the matrix $\dot{D}_*(q) - C_*(q,\dot{q})$ is skew-symmetric.

P6 There exist constants β_1, β_2, such that

$$1 \leq \|H(q)\| \leq 1 + \beta_2/\beta_1 \tag{17}$$

P7 There exist some positive finite constants k_{g_*}, k_{v_*} such that

$$\left\|\frac{\partial g_*(q)}{\partial q}\right\| \leq k_{g_*} \tag{18}$$

$$\|g_*(q)\| \leq k_{v_*} \tag{19}$$

P8 There exist a constant β_4, such that

$$\|H(q)^+\| \leq \beta_4 \tag{20}$$

Due to lack of space we do not provide a proof of the properties above. Interested readers are kindly requested to contact the second author. Next we state our control problem and propose our main result as solution.

3. PROBLEM STATEMENT AND ITS SOLUTION

Assume that only position and force measurements are available, and that the manipulator is subject to the *input constraint*

$$|u_i| \leq u_i^{\max}, \quad i = 1, \ldots, n \tag{21}$$

then, design a smooth control law such that

$$\lim_{t\to\infty} \tilde{q}(t) \triangleq \lim_{t\to\infty}[q(t) - q_d] = 0$$

$$\lim_{t\to\infty} \tilde{f}(t) \triangleq \lim_{t\to\infty}[f(t) - f_d] = 0$$

where f_d and q_d stand for the constant desired values of force and position respectively which are supposed to satisfy the following restrictions:

A3 The desired position lies within the set Ω, i.e. $q_d \in \Omega$. Furthermore, it belongs to the interior of Ω with some distance from the margin of the set, i.e.

$$\min_{q_m^1 \in \mathrm{fr}\Omega_1} \|q_d^1 - q_m^1\|^2 > \epsilon, \tag{22}$$

where $\mathrm{fr}\Omega_1$ denotes the margin of the set Ω_1 and ϵ is a small positive constant whose choice may depend on the initial conditions.

A4 For desired position $q_d \in \Omega$ the constraint equation $\phi(q_d) = 0$ is satisfied. Furthermore, we can express $f_d = J^\top(q_d)\lambda_d$ for all $q_d \in \Omega$.

3.1 Main result

In this section we propose a solution to the above formulated problem, this constitutes our main result. First, let us stress some useful additional properties of the reduced order model (14) – (16).

Proposition 3.1 Assume **A1** – **A4** hold and that $\|\tilde{q}(0)\| < \epsilon$, consider the model (14) – (16) in closed loop with the control law (13) where for the $i-th$ coordinate we define

$$
\begin{aligned}
u_{a_i} &= -k_{p_i}\tanh(\tilde{q}_i^1) - k_{d_i}\tanh(\vartheta_i) + g_*(q_d)_i &(23)\\
\dot{q}_{c_i} &= -a_i \tanh(q_{c_i} + b_i q_i^1) &(24)\\
\vartheta_i &= q_{c_i} + b_i q_i^1 &(25)\\
u_{b_i} &= \lambda_{d_i} - k_{p_i}\tanh(\tilde{\lambda}_i) &(26)
\end{aligned}
$$

where we suppose that

$$k_p^{\min} \triangleq \min_{i \le n-m} k_{p_i} \ge \frac{4k_{g_*}}{\operatorname{sech}^2\left(\frac{2k_{v_*}}{k_{g_*}}\right)} \tag{27}$$

then the equilibrium point $x = \operatorname{col}(\tilde{q}, \dot{q}, \vartheta, \tilde{f}) = \operatorname{col}(0,0,0,0)$ is asymptoticaly stable inside the set $\Omega_x \times \Re^n$ where

$$\Omega_x \triangleq \{x \in \Re^{3n} \mid q^1 \in \Omega_1, q^2 \in \Omega_2, \ \operatorname{col}(\dot{q}, \vartheta) \in \Re^{2n}\}$$

Furthermore, if

$$u_i^{\max} > \beta_4(k_{v_*} + k_p^{\min}) + \beta_2(\|\lambda_d\| + k_p^{\min})$$

then the input constraints (21) are satisfied. □

Due to lack of space we do not provide a proof of the proposition above, interested readers are kindly requested to contact the second author.

4. CONCLUSIONS

As far as we know we provided in this note the first solution to the practically interesting problem of output feedback force/position regulation with bounded control inputs.

Our approach considers a reduced order model whose coordinates are defined in a subset Ω of the entire coordinate space. The main property of this model is that it is ensured that, if the generalized positions start and remain in this set, the constraints Jacobian is non singular.

We have proposed a controller which makes use only of position and force measurements to keep the generalized positions bounded in the set Ω. Moroever, it was proved that the closed loop system is asymptotically stable.

Our scheme uses bounded controls, hence we have proved that it is posible to achieve asymptotic stability in the set Ω respecting some given input constraints.

Acknowledgements

This work was carried out while the second author was visiting the Institute of Problems of Mechanical Engineering of the Academy of Sciences of Rusia, he gratefully acknowledges the hospitality of Prof. A. Fradkov. This work was supported in part by the INTAS colaboration project and by CONACyT, Mexico.

References

[1] S. Arimoto. State-of-the-art and future research directions of robot control. In *Proc. 4th. Symposium on Robot Control*, pages 3–14, Capri, Italy, 1994.

[2] S. Arimoto, Y.H. Liu, and T. Naniwa. Principle of orthogonalization for hybrid control of robot arms. In *Proc. 12th. IFAC World Congress*, Sydney, Australia, 1993.

[3] P. Bélanger. Estimation of angular velocity and acceleration from shaft encoder measurements. In *Proc. IEEE Conf. Robotics Automat.*, volume 1, pages 585–592, Nice, France, 1992.

[4] R. Carelli and R. Kelly. An adaptive impedance/force controller for robot manipulators. *IEEE Trans. Automat. Contr.*, 36:967–971, 1991.

[5] S. Chiaverini and L. Sciavicco. The parallel approach to force/position control manipulators. *IEEE Trans. Robotics Automat.*, 9:289–293, 1993.

[6] D.M. Dawson, F.L. Lewis, and J.F. Dorsey. Robust force control of a robot manipulator. *Int. J. Rob. Research*, 11(4):312–319, 1993.

[7] N. Hogan. Impedance control: An approach to manipulation. Parts I–III,. *ASME J. Dyn. Syst. Meas. Contr.*, 107:1–24, 1985.

[8] H.P. Huang and W.L. Tseng. Asymptotic observer design for constrained robot systems. *IEE Proceedings-D*, 138:211–216, 1991.

[9] Mills J.K. and Goldenberg A.A. Force and position control of manipulators during constrained motion tasks. *IEEE Trans. Robotics Automat.*, 5(1):30–46, 1989.

[10] Salisbury J.K. Active stiffness control of a manipulator in cartesian coordinates. In *Proc. 19th. IEEE Conf. Decision Contr.*, pages 95–100, Albuqnerque, NM, 1980.

[11] O. Khatib. A unified approach for motion and force control of robot manipulators: The operational space formulation,. *IEEE J. Robotics Automat.*, RA-3:43–53, 1987.

[12] H.N. Koivo and R.K. Kankaanrantaes. Dynamics and simulation of compliant motion of a manipulator. In *Proc. IEEE Conf. Robotics Automat.*, volume 2, pages 163–173, 1988.

[13] A. Loria and R. Ortega. Force/position regulation for robot manipulators with unmeasurable velocities and uncertain gravity. *Automatica*, 1996. to appear.

[14] H. McClamroch and D. Wang. Feedback stabilization and tracking of constrained robots. *IEEE Trans. Automat. Contr.*, 33:419–426, 1988.

[15] N.H. McClamroch. Singular systems of differential equations as dynamical models for constrained robot systems. In *Proc. IEEE Conf. Robotics Automat.*, San Francisco, CA, 1986.

[16] E. Panteley and A. Stotsky. Adaptive trajectory/force control scheme for constrained robot maqnipulators. *Int. J. Adapt. Control Signal Process.*, 7(6):489–496, 1993.

[17] E. Panteley and A. Stotsky. Asymptotic stability of constrained robot motion observer based control schemes. In *Proc. 2nd. European Contr. Conf.*, Groningen, The Netherlands, 1993.

[18] M. Raibert and J. Craig. Hybrid position/force control of manipulators. *ASME J. Dyn. Syst. Meas. Contr.*, 103:126–133, 1981.

[19] I. Sandberg. Global inverse function theorems. *IEEE Trans. Circuits and Systems*, 27:998–1004, 1980.

[20] I. Sandberg. Global implicit function theorems. *IEEE Trans. Circuits and Systems*, 28:145–149, 1981.

[21] M. Spong and M. Vidyasagar. *Robot Dynamics and Control*. John Wiley & Sons, New York, 1989.

[22] B.J. Waibel and H. Kazerooni. Theory and experiments on the strability of robot compliance control. *IEEE Trans. Robotics Automat.*, 7(1):95–104, 1991.

[23] D. Wang and H. McClamroch. Position force control for constrained manipulator motion: Lyapunov's direct method. *IEEE Trans. Robotics Automat.*, 9:308–313, 1993.

[24] D.E. Whitney. Force feedback control of manipulator fine motion. *ASME J. Dyn. Syst. Meas. Contr.*, pages 91–98, 1977.

[25] B. Yao, S.P. Chan, and D. Wang. Robust motion and force control of robot manipulators in the presence of environmental constraint uncertainties. In *Proc. 31st. IEEE Conf. Decision Contr.*, pages 1875–1880, Tucson, Arizona, 1992.

[26] T. Yoshikawa. Dynamics hybrid position/force control of robot manipulators–description of hand constraints and calculation of joints driving forces. In *Proc. IEEE Conf. Robotics Automat.*, pages 1393–1398, San Francisco, CA, 1986.

UNIFIED TELEROBOTIC ARCHITECTURE FOR AIR FORCE APPLICATIONS

S. B. PETROSKI, T. E. DEETER
AFMC Robotics and Automation Center of Excellence
Technology & Industrial Support Dir., Robotics and Automation Branch
San Antonio Air Logistics Center, Kelly AFB, TX 78241-6435
ti-race@sadis05.kelly.af.mil

M. B. LEAHY, JR.
Headquarters Air Force Materiel Command
Science & Technology
4375 Chidlaw Rd, Suite 6
Wright-Patterson AFB, OH 45433-5006
m.leahy@ieee.org

ABSTRACT

The United States Air Force is leading the way toward automating industrial processes previously beyond the capability of classical robotic technologies. Telerobotics will open the door on our ability to automate processes characterized by small batch sizes, feature uncertainty and varying workload. The Air Force Materiel Command's Robotics and Automation Center of Excellence (RACE) is providing the framework needed to bring telerobotics to the shop floor in support of fleet maintenance operations with the Unified Telerobotic Architecture Project (UTAP). The objective of this paper is to present the case for telerobotics, highlight the need for a cohesive approach, and provide an overview of the UTAP.

KEYWORDS: robotics, telerobotics, open systems, remanufacturing, depot, military, aircraft, maintenance

INTRODUCTION

The United States Air Force currently has five Air Logistic Centers (ALCs), or depots, which monitor, maintain, and upgrade its weapon systems as long as they remain in the inventory. The industrial processes performed at the ALCs differ from typical original equipment manufacturers (OEM) processes in several fundamental ways which conspire to make ALC automation difficult, if not impossible, to achieve. The main difference between ALC and OEM processes is the level of uncertainty in the operation. System integrators strive to engineer uncertainty out of OEM processes when designing robotic workcells, while ALC processes are mainly repair oriented and therefore full of uncertainty by their nature. Because no two wear patterns are exactly alike, nor do any two bird strikes occur at exactly the same place or magnitude, any attempt to automate the ALC process must be much more robust and flexible. The robotics and artificial intelligence necessary to provide that robustness and flexibility are currently beyond our grasp technically and economically. Other factors peculiar to ALCs that exacerbate the problem are small batch sizes (as low as one per quarter) and high variability in workload. Add to the technical challenges a workforce that is resistant to and fearful of complete automation, a management structure disappointed by the unfulfilled promise of past robotics projects, and an austere funding climate and the future of robotics in the ALCs begins to look particularly bleak.

Having made the case that ALC processes are difficult to automate, it must be pointed out that the need for robotic based solutions is growing. New processes that are environmentally safe but too demanding for human operators, the need for increased process consistency with lower tolerances, and competition with private industry are just a few of the factors demanding the judicious application of advanced robotics technologies. Telerobotics is the means to bridge the cultural and technical gap between manual operation and full automation.

MOTIVATIONS FOR TELEROBOTICS

We broadly define telerobotics as *the technologies and systems that permit a human operator to direct and/or supervise the operation of a remote robotic effector mechanism* [1]. Telerobotics encompasses a spectrum of operation modes, ranging from direct mapping of a human operator's inputs to the robotic manipulator (telepresence) to nearly autonomous operation with occasional input from the operator (supervisory control). The telerobotic approach to automation is an attractive solution to the problems described previously. The telerobotic system is not a replacement for the human worker, it is an augmented tool that improves his quality of life while allowing his cognitive skills to solve the uncertainties of the process. Small batch sizes and workloads that vary over time no longer create tremendous problems because the system does not rely on specific programs or motion routines for each distinct workpiece. The programming required in a telerobotic system is generic to the application, such as maintaining stand-off distance, normality and tool velocity in a telerobotic painting application. Operators accept the system as an innovative tool rather than a threat to job security; and while operator acceptance is not a guarantee of a system's success, lack of it will guarantee failure.

A short list of candidate ALC processes that require or could benefit from the implementation of telerobotics include aircraft skin repair [2], fuel tank sealing [3], corrosion control and surface finishing [4]. Additional applications recently targeted for telerobotic implementations include munitions handling and non-destructive inspection of aircraft and engines.

Unfortunately, telerobotic systems are not available off the shelf. Existing telerobotic systems have been geared toward extremely hazardous tasks or remote operations such as undersea, space, or nuclear material handling. Because these applications inherently require robotic solutions, there has never been the competitive pressure to drive costs down relative to other alternatives. Air Force depots represent the first market for telerobotics where it is an option, but not the only option. In this case, telerobotics must compete successfully with strictly manual operation, which is cheap, low risk, and the accepted practice. It is doubtful that any individual telerobotic application could compete with the manual process. Any gains in productivity, quality or safety would be hard pressed to offset the initial system acquisition cost. In order to bring telerobotics to the shop floor, we must drive down system prices by spreading the development costs over multiple applications.

A cohesive, unified approach to developing a common architecture for multiple telerobotic applications fulfills that need, and provides the benefits of upgradability, supportability and reliability while lowering the insertion cost of each individual implementation. For stand-alone custom solutions, software development and system integration constitute 60% or more of the system cost. By creating a common framework for operator interfaces and basic commands for movement, gripping, obstacle avoidance, etc., we reduced the software development task to writing the specific code peculiar to the new implementation. The unified architecture approach also allows easy upgrading or replacement of the individual components of the system. This is accomplished by specifying the architecture standards at the interface level. After modifying a desired joystick (for example) to conform to its interface specification, swapping it into the system is no more complicated than switching printers on a computer. The unified architecture also maximizes the amount of similarity between the components and operator interfaces of different applications, reducing training requirements and maintenance support costs. Finally, the commonality between systems allows for the selection of proven hardware components and minimizes the amount of new code that must be written for each implementation, resulting in greater reliability [4].

RACE has seized this opportunity to champion the development of this common framework with the Unified Telerobotic Architecture Project (UTAP).

UNIFIED TELEROBOTIC ARCHITECTURE PROJECT

The UTAP began in FY93 as an engineering study performed for RACE by NASA's Jet Propulsion Laboratory (JPL) to define a telerobotic system capable of accomplishing a wide range of depot applications. JPL began by distilling a representative set of processes into a global set of functional requirements sufficient to span the needs of depot applications. A comparison of these requirements and commercially available and near-available technology resulted in an architecture describing the components and connectivity needed to solve the original set of tasks [1]. In FY94, RACE then partnered with the National Institute of Standards and Technology (NIST) to define the interfaces between the functional blocks of the architecture. The interfaces are in the object oriented computer language C++, and define the messages that may be passed between the functional blocks of the architecture. The document containing the interface messages and their descriptions is available from NIST via "anonymous ftp" [5]. These interfaces were then validated by a system integrator, Advanced Cybernetics Group (ACG), who implemented a system using UTAP analogous routines on an Adept motion control platform. ACG performed a validation test set consisting of several combinations of the functional requirements defined by JPL in the initial phase. One example of the validation test the system performed was autonomous regulation of tool stand-off distance while a human operator controlled the other two cartesian coordinates via joystick. The remainder of the validation tests are described in [4]. A video of the system accomplishing these validation tests was produced and is available from ACG or RACE [6]. For FY95, efforts continued toward refining the specifications. NIST produced an updated version of the UTAP specification, incorporating lessons learned from the validation implementation, suggested modifications from industry and national lab reviewers, and insights gained from work on their own Enhanced Machine Controller. ACG was again on contract, this time to perform an implementation/validation of telerobotic surface finishing. The implementation demonstrated the capability of a UTAP based system to autonomously regulate the contact forces of a surface finishing process while the other two tangential cartesian coordinates are regulated via joystick. This validation of the architecture is particularly significant because telerobotic surface finishing was one of the original target applications and will become one of the early shop floor implementations. The remainder of this section addresses current UTAP activities.

Next Generation Munitions Handler

Telerobotics also has a place beyond the shop floor, and RACE is working with Oak Ridge National Lab (ORNL) to introduce telerobotics to the flight line for the first time in the Next Generation Munitions Handler (NGMH) project. The NGMH is a replacement technology for the current jammer vehicles the Air Force uses to load bombs and missiles onto fighter aircraft. The jammers are essentially diesel powered specialized forklift-like support equipment with hydraulic actuators for loading the munitions. The actuators are crude, resembling farm or earth moving equipment, with separate manual controls for each joint of the manipulator. The NGMH project will replace this equipment with an omnidirectional vehicle carrying a high precision, high payload hydraulic manipulator with coordinated axes which are operated in a telerobotic fashion. The UTAP based system controls will employ gravity compensation and force feedback to achieve a man-amplification effect. The human operators will be able to lift and finely position a 2500 pound bomb while exerting twenty-five pounds of effort. More detailed descriptions of the NGMH project may be found in [7] and [8].

The NGMH represents the first "real world" test of the UTAP architecture. From the beginning of the NGMH program, the two efforts have been linked and are expected to converge in the NGMH prototype phase. ORNL contributed to the refinement of the original UTAP document and has maintained a cross-feed of information with NIST and ACG. As the first implementation of the architecture, a good deal of latitude with respect to compliance is acceptable and expected. The UTAP is intended as a growing, evolving specification that enhances the chances of successful implementations of telerobotics, not a rigid set of restrictions. One factor that will greatly influence the amount of conformity of the NGMH prototype is the level of openness in available commercial controllers. As greater levels of openness are available, the opportunity to utilize more of the UTAP features increases. A successful NGMH prototype will be a significant milestone in the validation of the UTAP architecture.

Telerobotic NDI for Aircraft

Decreasing Air Force expenditures for new aircraft pushes the service life of the existing airframes in the inventory beyond their originally intended lifetimes. Extending their service life increases demand for non-destructive inspection capacity and methods. RACE is working with JPL to meet these requirements. As with the original UTAP study, the initial phase of this effort is the identification of characteristic applications for which generalized solutions may be developed. The applications identified so far fall into three general classes. The first is large area inspection of flat or gently curved surfaces, typically searching for newly forming cracks or areas of delamination. Second is inspection of large constrained or obstacle filled spaces, either for similar structural defects or the presence of dangerous conditions such as leaking fluids or gasses. The third class is inspection of interior regions of aircraft or their engines which are accessible only through small openings such as through fastener holes or engine access ports. The likely solution to the large area inspection class is a mobility platform capable of crawling on the exterior of aircraft surfaces attached by suction cup feet. The large constrained space inspections will likely be accomplished by a more sophisticated walking robot capable of negotiating over or around obstacles. The solution to the last class will rely on hyper-redundant or "snake" robot technology.

As NASA's lead center in the areas of robotics (both mobile and hyper-redundant) and miniaturization (necessary for both the sensor technology and the snake robots themselves), not to mention telerobotics, JPL is uniquely capable of performing this study. Development in these areas will add to our telerobotic "toolkit" capabilities involving mobility, obstacle avoidance and methods of presenting remotely gathered data to operators in the most meaningful ways.

OAC Testbed

The main obstacle to implementing telerobotic systems at a reasonable cost today is the closed nature of commercially available controllers. Even those considered relatively "open", such as the Adept platform on which ACG performed the validation test set and surface finishing systems, make algorithms utilizing force control or direct manipulation of P-I-D gains impossible or prohibitively difficult. However, the push to bring telerobotics to the shop floor economically dictates the use of commercial, off the shelf equipment. Toward this goal, RACE is creating a testbed for comparison of emerging commercial controllers which possess greater levels of openness and show promise as potential controllers for telerobotic systems.

RACE has acquired two Cincinnati Miliacron T3-7X6 robots on which we will implement telerobotic systems using these emerging control platforms. Current offerings from Trellis and Cimetrix have been acquired or are on order. The work to lobotomize the T3s and to integrate the new controllers will begin soon at Kelly AFB. The T3 represents a large pool of older industrial robots that may be mechanically sound, but are obsolete because of the limitations of their original controllers. By demonstrating the feasibility of retrofitting these older manipulators with today's more capable controllers, we reap the benefit of increased utility from the old hardware and simultaneously increase the potential market size for these new controllers - promoting and supporting the move toward open systems. Any resultant positive impact this has on the trend toward openness only enhances the opportunity for successful implementation of telerobotic systems.

CONCLUSION

The Air Force depots face numerous challenges. Dramatically reduced funding and manpower levels, stricter environmental and safety legislation, increased quality demands, and the need for tighter control over process parameters due to newer, more high-tech weapon systems all indicate the need for increased use of robotic technologies. However, the unique requirements of depot remanufacturing and repair workloads present significant technical, cultural and economic hurdles. RACE is championing the development of telerobotics as the means for clearing those hurdles. The groundwork has been laid and the first implementations are within sight. We believe the current UTAP activities will produce the initial success stories required to propel telerobotics into widely accepted use throughout the military and private sector manufacturing and remanufacturing industrial bases.

REFERENCES

1. *Generic Telerobotic Architecture for C-5 Industrial Processes*, JPL Final Report, Aug 93.
2. M. B. Leahy, Jr. and B. K. Cassiday, "RACE Pulls for Shared Control", *Proceedings of SPIE Conference on Cooperative Intelligent Robotics in Space III*, Boston, MA, Nov 16-18, 1992.
3. M. B. Leahy, Jr. and S. B. Petroski, "Telerobotics for Depot Applications", *Proceedings of NAECON 93*, Dayton, OH, May 1993.
4. M. B. Leahy, Jr. and S. B. Petroski, "Telerobotics for Depot Modernization", *Proceedings of AIAA Conf. on Intelligent Robotics in Field, Factory, Service and Space*, Houston, TX, Mar. 20-24, 1994.
5. R. Lumia, et al, *Unified Telerobotic Architecture Project (UTAP) Interface Document*, NIST, available for anonymous ftp at giskard.cme.nist.gov (directory = pub/utap)
6. UTAP Phase 2 Validation (NIST Contract 50SBN4C8083), ACG Video, Advanced Cybernetics Group, 234 E. Caribbean Dr, Sunnyvale, CA, 94089.
7. T. E. Deeter, G. J. Koury and M. B. Leahy, Jr., "The Next Generation Munitions Handler Advanced Telerobotics Technology Demonstrator Program", *Proceedings of International Symposium on Robotics and Manufacturing*, Montpellier, France, May 27-30, 1996.
8. M. B. Leahy, Jr. and Neil Hammel, *The Next Generation Munitions Handler Prototype Acquisition Campaign: Targets and Courses of Action*, USAF Air Command and Staff College research, May 1995.

APPLICATION OF WORKSPACE ANALYSIS TO ROBOT LOCATIONING PROBLEM

Young Soo Park, Jisup Yoon

Nuclear Environment Management Center
Korea Atomic Energy Research Institute
P.O.Box 105, Yusung, Taejon, Republic of Korea

ABSTRACT

This paper presents a computational platform for integrating the workspace analysis into the problem of finding a continuous trajectory feasible location of a mobile manipulator in cluttered environment. In this algorithm, task feasibility is tested in operation space over the octree coding of task trajectory. Subsequently, the task feasible location of robot base is determined through a direct search type of optimization process by adopting the task feasibility as cost function. The result is coherent with existing workspace analysis methods and complete in that it provides the full characterization of task capability of workspace as well as the optimal base locations of the manipulator.

KEYWORDS: task capability, t-connectivity, octree, direct optimization, mobile manipulator, operation space

INTRODUCTION

In teleoperation, an immediate responsibility of the human operator is to adequately place mobile base so that the manipulator has full access to the specified task trajectories. Such a decision making, hereafter called base locationing, requires the determination of the boundaries and the task capabilities of various topological subspaces of manipulator's workspace in the presence of obstacles. Due to the complexity of robot kinematics, however, trial and error based approaches relying solely on operator's intuition is practically infeasible. Thus, an efficient off-line planner must be provided.

To this end, the problem addressed here is a so called *find space problem* whose goal is to find a collision free space for a manipulator among known set of obstacles. Used as a decision making criteria is the robot's ability to reach and move along prescribed task trajectories and/or subspaces in the workspace. Naturally, the study incorporates the extension and application of existing works on topological workspace analysis. Several authors have early proposed topological characterization of the workspace of robots based on motion capabilities. Borrel and Legeois [1] have introduced the notion of aspects and revealed that they characterize trajectory following capability of most industrial robots in the absence of obstacles. Wenger and Chedmail extended the scope of such characterization to robots with generic kinematic structures and in the presence of obstacles [2]. In the application side, many researchers have looked into the automatic positioning of robot manipulators [3][4], but little have taken into account the global task capability of the workspace. Among the few, Wenger [5] has proposed an optimal method based on workspace coverage, but no consideration is given to the detailed task capability of the workspace. Recently, Reynier et. al [6] has proposed an optimal method which directly searches the task feasibility of a trajectory segment in operation space. This method, however, adopts a mathematical description of task trajectory which is less coherent to workspace analysis.

Complementing such drawbacks of existing methods, presented in this paper is a computationally efficient base locationing method which performs the path feasibility test in the operation space. In this regard, research efforts are given in the followings: 1) the provision of quantitative enumeration of trajectory capabilities, 2) Adaptation of a direct optimization technique for the base locationing problem, 3) Improvement of computational efficiency of the base locationing methods. Computational complexity is reduced by restricting the search space to the task trajectory, replacing the costly construction of configuration free spece with simple collision test.

DESCRIPTION OF ROBOT AND ENVIRONMENT

The foregoing method for continuous trajectory feasibility analysis is applicable for all non-redundant revolute joint manipulators whose inverse kinematic model is analytically known. In the illustration of the method, the body of the robot is approximated by serially connected links with infinitesimally small thickness. It is claimed that such an assumption does not cause loss of generality of the method because the addressed work is made independent of the geometric modelling of robot manipulators.

Both the environment and test trajectory are assumed to be known. Obstacles are describe with three dimensional polygons, whose shapes and locations are specified with the coordinate vectors of the vertices. Task is described as a continuous trajectory in operation space. An operation space of degree $m<=6$ is spanned by the location and orientation vectors of the end effector. Thus, each point on the trajectory is specified by the position and orientation. In implementation, the trajectory is scanned in operation space and stored into discrete elements in octree [7]. Resolution of octree is made sufficiently small so that each increment in octree corresponds approximately to one pixel of video display. In the subsequent part of the algorithm, this octree of unit pixel resolutions is useful for performing connectivity tests.

IDENTIFICATION OF TRAJECTORY FEASIBILITY

A continuous trajectory is feasible, or equivalently 't-connected', if the end-effector is able to follow it without changing posture and without colliding into obstacles. It is known that, for type 1 nonredundant manipulators, a connected portion of the trajectory in configuration space is also a connected component in operation space under geometric operator. Thus, trajectory capability can be tested in operation space[6]. However, for nonredundant type 2 manipulators, the kinematic mapping is not one-to-one, and it may seem insufficient to characterize the feasibility of a trajectory in operation space.

Therefore, for generality, it is necessary to refer to the connectivity analysis in configuration space. It is worth mentioning at this point that the inverse kinematic operator is a multivalued function from operation space to configuration space. However, it should also be noted that the inverse kinematic operator is a collection of many one-to-one mapping from $\underline{x}$ to each unique joint angles $\underline{\theta}^i$, where the superscript i indicates respective uniqueness sets. The t-connected subspaces are then identified as obstacle free portions of the uniqueness domains. The uniqueness domains of type 2 robots are the maximal subspaces constructed based on one-to-one correspondence of the inverse kinematic operator. Therefore, it is plausible that the C-image of the task trajectory pertaining to a uniqueness domain can be constructed by identifying and separately storing the single valued inverse kinematics solutions for all trajectory points. Subsequently, the t-connected portion of the trajectories can be obtained by eliminating from each uniqueness groups based on collision test and subsequently identifying the connected components of the remaining elements. Specifically, the following procedure is implemented to identify trajectory feasibility of workspace:

1) Code the trajectory into octree
2) Increment along the octree cells of the task trajectory and proceed with the following steps:
   - compute joint angles corresponding to the trajectory point by inverse kinematics routine
   - for each joint angle sets, perform collision test
   - if no collision occurs, store the trajectory point in respective uniqueness group
3) Code the trajectory points in each uniqueness group into octree entities.
4) Perform connected component labeling for the octree entity of each uniqueness group to identify the t-connected portion of the trajectory

This process is illustrated in Figure 1. Thus obtained t-connected trajectories are denoted in this paper as *WUtraj* and its octree representation is denoted as *Oct(WUtraj)*, where *Oct(.)* is an operator for conversion to octree element.

DESCRIPTION OF ALGORITHM

The base locationing method consists of finding a task capable location of the robot base and the associated posture. The terms of the problem are as follows:

Given data: the kinematic parameters of the robot, its joint limits,
the task trajectory, the environmental obstacles

Goal : To find the base location, $\underline{X}_d$.

Here, the base location $\underline{X}_d$, is a vector defined as

$$\underline{X}_d = (x, y, z, \alpha) \tag{1}$$

where x, y, z are respectively the position of the robot base in world frame, and α is the rotation angle around the vertical axis of the world frame.

We propose to implement an optimal searching method for the resolution of this problem. The optimization routine tends to minimize a criteria which gives a good estimation of the robot's ability to follow the continuous trajectory. Thus, used as a cost function is the maximum attainable length of the task feasible trajectory, as

$$F_{opt}(\underline{X}_d) = \underset{j,k}{Max}\ Area[CC_k(Oct(WUtraj_j(\underline{X}_d))]\ . \qquad (2)$$

Here, the functions $Area[\ ]$, $CC_k(\ )$ are volume of octree entity and operator for connected component labeling, respectively. Also, $WUtraj_j(\underline{X}_d)$ denotes the jth unique inverse kinematics solution set of the trajectory at base location $\underline{X}_d$.

Given the cost function as in the above, a proper optimization scheme must be adopted to determine task feasible base location of the robot. However, due to the nonanalytic nature of the workspace characterization process, a direct search type of method should be adopted. Among many direct search methods, simplex method is adopted in this work because it makes virtually no assumption on the nature of the cost function, not even continuity, and provides robust operation [8].

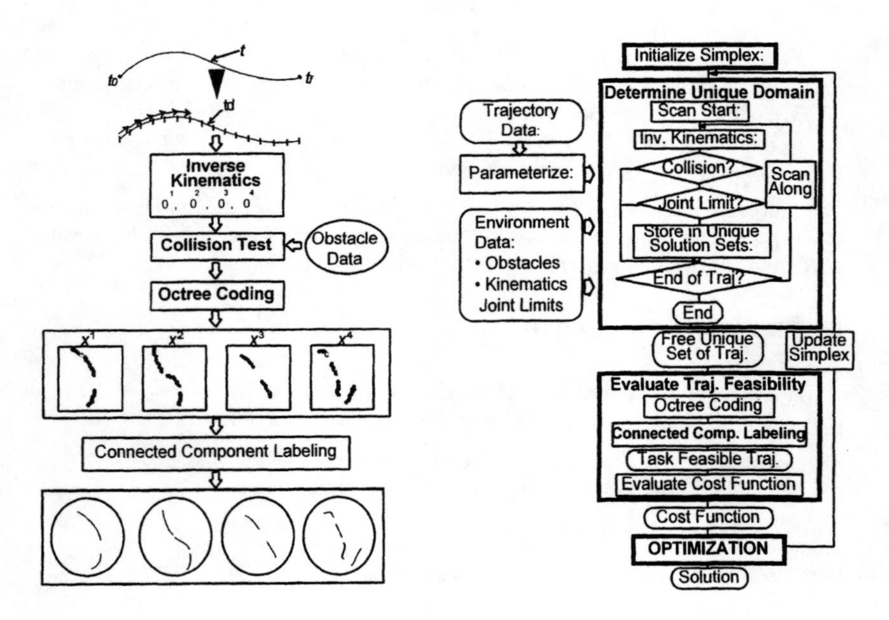

Figure 1. Identification of Trajectory Feasibility

Figure 2. Optimization Procedure

APPLICATION EXAMPLES

The trajectory is described with straight lines, each point being referred to by a vector representing the location of the specified position and orientation of the end effector. A 6-DOF manipulator, Unimation PUMA, is used as shown in Figure 3(a). Its kinematic parameters can be found in many texts. The obstacle environment consists of eight obstacles, and the trajectory is a straight line segment as shown in Figure. 3(b). The orientation of the hand is constrained in such a manner that the approach vector of the hand is normal to and pointing into the nearest surface of the obstacle. The total lengths of the trajectory is 8.49m, and collision detection accuracy is 3.4cm.

The parameters of optimization are decided to be $\underline{X} = (x, y, z, \alpha)$. The termination criteria is defined as the magnitude fraction of change in function values to the sum of functional values to be less than a certain tolerance value, namely *FTOL*. Specifically, it is given as

$$FTOL > \frac{2 * \left| F_{opt}(ihigh) - F_{opt}(ilow) \right|}{\left| F_{opt}(ihigh) + F_{opt}(ilow) \right|} \tag{3}$$

where *ihigh* and *ilow* are the highest and the lowest value of cost function evaluated at the current simplex points. *FTOL* is chosen to be 0.001, and no joint limit is considered. All results have been obtained on a 80486 PC.

The optimization process ended after 69 calls to the cost function evaluations as shown in Fig. 4(a). The total time required was 5 minutes and 34sec. The result gives the optimal base location of (x = 0.5728m, y =-0.5664 m, z=2.199m, α =54.57^{o}) where 92.5% coverage of the trajectory is achieved. The corresponding posture is revealed to be (LEFT ARM, ELBOW DOWN, NO FLIPWRIST). Fig. 4(b) shows the t-connected portions of the trajectory, shown as dark lines overlaid on the horizontal trajectory, for four different postures at the selected base location. In all test conditions, the calculation time remain in the order of minutes, which is in the same order of most efficient algorithms of similar kinds. Therefore, it is claimed that the presented method is computationally comparable while incorporating more practical characterization of the task capability.

CONCLUSION

In this work, a computational platform for optimization method is presented for selecting optimum location of mobile manipulator. Computational efficiency is improved by performing the task feasibility analysis along octree representation of task trajectory in operation space. The method may provide an essential functionality of operator aid in teleoperation, where frequent repositioning of mobile base is required during task performance.

REFERENCES

1. Borel, P., Liegeois, A., 1986, "A study of multiple manipulator kinematic solution with application to trajectory planning and workspace determination," *Proceedings of the 1986 IEEE International Conference on Robotics and Automation*, pp.1180-1185

2. Wenger, P., Chedmail, P., 1991, "On the connectivity of manipulator free workspace," *Journal of Robotic Systems*, Vol. 8(6), pp. 767-799
3. Nelson, Pederson, Donath, 1987, "Locating assembly tasks in a manipulator's workspace," *Proc. IEEE Rob. Aut. 1987*, pp. 1367-1372
4. Shiller, 1989, "Interactive time optimal robot motion planning and work-cell design," *Proc. IEEE Rob. Aut. 1989*, pp. 964-969
5. Chedmail, P., Wenger, P., 1989, "Design and positioning of robot in an environment with obstacles using optimal research," *IEEE Conf. on Robotics and Automations*, pp. 1069-1074
6. Reynier, F., Chedmail, P.,Wenger, P., 1992, "Automatic positioning of robots, continuous trajectories feasibility among obstacles," *IEEE Int. Conf. on Syst. Man and Cybern.* '92, Chicage, USA, pp.189-194
7. Shaffer, C.A., Samet, H., 1987, "Optimal Quadtree Construction Algorithms," *Computer Vision, Graphics, and Image Processing,* Vol. 37, pp. 402-419
8. Beightler, C.S., Phillips, D.T., Wilde, D.J., 1979, *Foundation of Optimization,* Prentice-Hall Inc., New Jersey

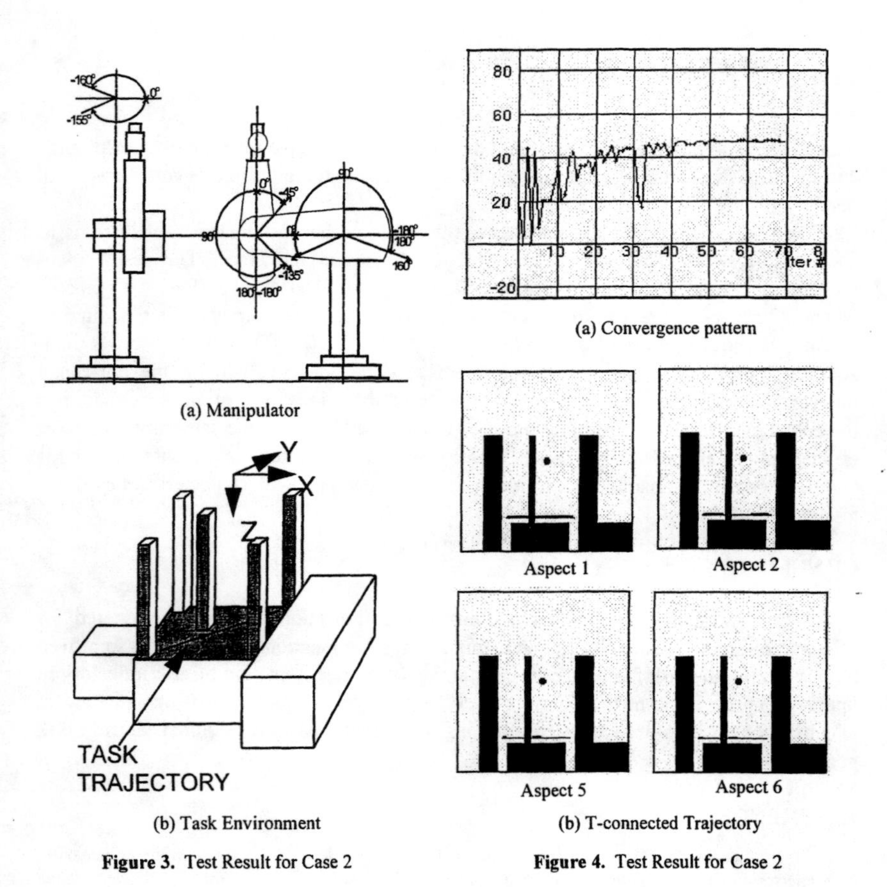

(a) Manipulator

(b) Task Environment

Figure 3. Test Result for Case 2

(a) Convergence pattern

(b) T-connected Trajectory

Figure 4. Test Result for Case 2

INTEREST OF HYBRID POSITION-FORCE CONTROL FOR TELEOPERATION

Yann PLIHON*, Claude REBOULET

CERT-ONERA – SUP'AERO – Département d'Automatique 2 avenue Edouard Belin 31055 TOULOUSE cedex – FRANCE*

E-mail : Plihon@cert.fr – Reboulet@cert.fr

ABSTRACT

This paper deals with the problem of teleoperation with time delays. These delays render direct force feedback teleoperation infeasible. We are proposing a novel teleoperation methodology based on hybrid position-force control and "virtual reality". At the beginning of the task, the operator sets hybrid parameters of the two robots (the master and the slave robot), thanks to a 3D man machine interface coupled with a graphical model of the remote environment. Then, the operator can exert forces and perform displacements from the master to the slave along the selectioned degrees of freedom. The other degrees of freedom are automatically controlled at the remote site. The experimental testbed is described in the final part.

KEYWORDS

teleoperation, telerobotic, hybrid position-force control.

1 INTRODUCTION

Many applications necessitate a capability to perform work in remote environment, where human being can not go: radioactive objects manipulation, underwater ressource extraction industries, pipeline maintenance or satellites manipulation. As defined by [1], we can classify this remote operating systems according to the rather low or high level of human's intervention. With "low level" systems (teleoperation), the physical ability of a human operator is needed to control motions and forces of the remote robot. In "high level" systems (telerobotics), the human operator gives "high level" orders (like: "go there", "turn this crank", ...). Of course, most of advanced systems combine high level and low level interventions of the operator. In this paper, we restrict remote robot control to teleoperation.

A teleoperation system is composed by two sites: the operator's site and the remote site. In the operator's site, the human operator is equipped with a master robot. The slave robot, at the remote site, generally duplicates the motion generated by the operator handling the master robot. The first teleoperation systems, for nuclear plants applications, used mechanically coupled robots [2]. Later, more complex systems used electrically coupled robots and computer capabilities that allowed remote teleoperation and the use of kinematically different robots. Those early systems are called "position bilateral control systems". The two robots are position controlled and the references for position control loops are the positions measured at the other site. With such a device, the force felt by the operator is due to the positional errors in the slave's control loop. In an ideal case, the operator should only sense the forces due to external forces applied on the slave robot. However, the position control loop generates unavoidable position errors which give to the operator an impression of manipulating a damped system. The systems with force feedback overcome this problem by using an external sensor: the force felt by the operator is the one measured by the force sensor at the remote site. This scheme generally increases the global performances.

For orbital or underwater applications, communication delays can reach several seconds, and render direct teleoperation infeasible [3]. Many solutions were proposed to stabilize those systems [4], [5],[6]. Unfortunely, all these solutions reduce the bandwith of the master-slave loop as time delays increase. In presence of a great time delay, the force feedback becomes useless since the reaction becomes too slow. Other solutions are based on "predictive display" and "shared control" [7], [8], [9]. The systems based on predictive displays make use of graphical models of the remote environment to allow for immediate visual feedback to the operator. With shared control systems, the operator manages only few degrees of freedom and the others are automatically controlled. Funda proposed the teleprogramming solution for time delays problem [10]: the operator interacts with a graphical simulation of the remote environment. The system monitors the operator's activity and generates a corresponding stream of robot instructions, which are sent to the remote site for execution.

In this paper, we present another solution for time-delayed remote control of robotic systems. We first present a new way to use hybrid position-force control in teleoperation. Then we proposed a new teleoperation methodology based on this scheme. Finally, we give our first experimental results, and present a conclusion and future researches.

2 HYBRID POSITION-FORCE CONTROL IN TELEOPERATION

2.1 Definitions

Hybrid position-force control, first proposed by Raibert [11], consists in divinding the controlled workspace into two complementary and orthogonal subspaces. One subspace is position controlled and the other is force controlled. There are several approaches to describe an "hybrid task". We present here the simplest one. First, we must define two frames (figure 1):

— Fo: the object frame. This frame is defined relatively to the object hold by the robot. The origin of this frame is O.

— Fe: the environment frame. This frame is defined relatively to the environment. The origin of this frame is E.

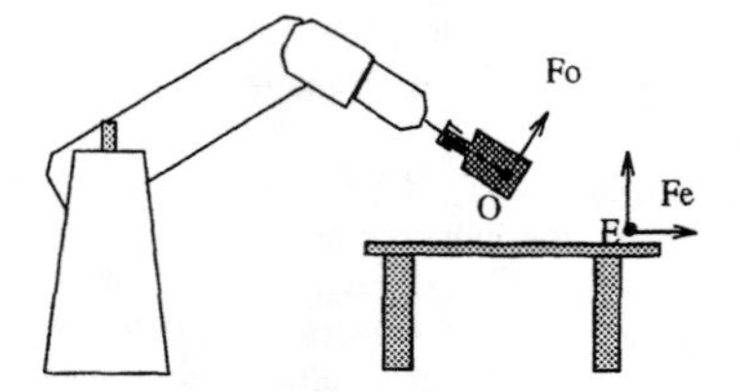

Figure 1 The different frames used

Then, the space partition is defined in the environment frame Fe by six values s[i], which are either 0 (in that case, the corresponding axis is force controlled) or 1 (in that case, the corresponding axis is position controlled). Finally, we defined the desired position Xd and the desired forces Fd. Xd is a six dimensional vector defining the position and the orientation of Fo relatively to Fe, only in the position controlled subspace. Fd is composed by the forces and the torques applied by the robot on the environment, only in the force controlled subspace. Forces and torques are written in Fe and the origin for torque measurements is O, the origin of Fo. Thus O is the center of rotation.

To sum up, the behavior of a robot under hybrid position/force control is completely defined by setting few parameters:

— Fo (the origin of Fo is the center of rotation).

— Fe (the axes of Fe are used to define the position axes and the forces axes).

— S, the space partition.

— Fd, Xd, the desired positions and forces.

There are several ways to set this parameters. Let us consider one basic example: the operator wants to control the direction and the intensity of a force vector.Two ways are possibles:

— Firstly, the three Fe axes x,y,z can be set as force controlled axes. Thus, we can control the force direction and the force intensity by setting Fx, Fy, Fz.

— Secondly, way is to choose one force controlled axis (x for instance) can be chosen. Thus, the intensity is controlled by setting Fx and we choose the direction of the force by setting the orientation of the environment frame Fe.

2.2 The dual hybrid position-force scheme

Yves Brière proposed to use hybrid position-force control for the master robot [12], [13], [14]. Thus, one can choose a space partition composed by two orthogonal and complementary subspaces. In one subspace, the robot is position controlled, and in the other one, it is force controlled. In the first subspace, the position is kept constant and the operator's force is measured. In the second subspace, the force is set to null value. Thus, if the operator exerts a force, he will encounter almost no resistance. Thanks to this property, the operator can move the master device anywhere he wants in the force controlled subspace. For a teleoperation application, such a device can be called a operator's position-force sensor. This hybrid control master robot can be coupled with a hybrid control slave robot. Let us consider again the two dual examples described above. In the first case, the master robot is force controlled and is used as an operator's position sensor. The corresponding position controlled subspace of the slave robot can duplicate the operator's displacements. In the second case, the master robot is position controlled and used as an operator's forces sensor. The corresponding force controlled subspace of the slave can duplicate the operator's force. This scheme is called "dual hybrid position force control" since the master's position controlled subspace is connected to the slave's force controlled subspace and vice-versa. Let us give an example: the task depicted in figure 2 consists in wiping out a board with a parallelepipedic eraser.

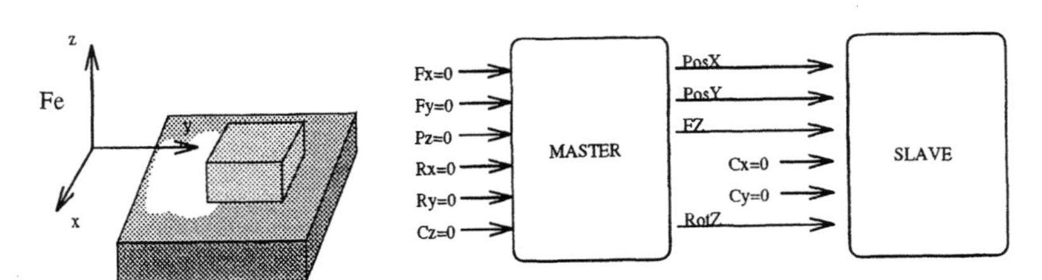

Figure 2 Wiping out a board

The torques Cx and Cy are set to zero in order to constraint the eraser to remain in plane on plane contact with the board. The operator only deals with four degrees of freedom. He manages the x and y axes position and the z axis rotation of the eraser. He also manages the z axis force.

2.3 Interest of this scheme for teleoperation

We can divide the tasks that the operator tries to achieve in teleoperation in two groups.

Firstly, let us consider the positionning task. It is well known that an operator can hardly deal with six degrees of freedom in the same time. The classical adopted strategy is to split the task into subtasks. To reach a desired position/orientation for instance, the operator firstly achieves the translation and then the rotation. An important property of the dual hybrid teleoperation scheme: the implementation of shared control is facilitated. Some degrees of freedom of the master robot can remain fixed (position controlled) but can not be interpreted as force commands. In the corresponding slave's subspace, a constant position is commanded. For instance, if the x axis is only connected, the operator manages the position of the slave robot along a straight line defined by the x axis of the slave Fe frame. If the x and y axes are connected, the operator moves the master hand device (and so the slave robot) as he were on a plane.

Secondly, let us consider contact task, as peg-on-hole or grasping task. To achieve this task, the operator must manage the force and torque between the object and the environment. By setting the space partition, the operator can choose the contact he wants. Then he only manages a few degrees of freedom, and the other degrees of freedom are automatically controlled. All the contacts between two polyhedral objects can be modeled with the hybrid control. Thus, this teleoperation scheme allows the operator to manage any type of contacts.

2.4 A new solution for teleoperation with time delays

In this section, we propose a new teleoperation methodology, based on dual hybrid position force control. A high level view of this approach is shown in figure 3. We see that this system is based on shared control and predictive display.

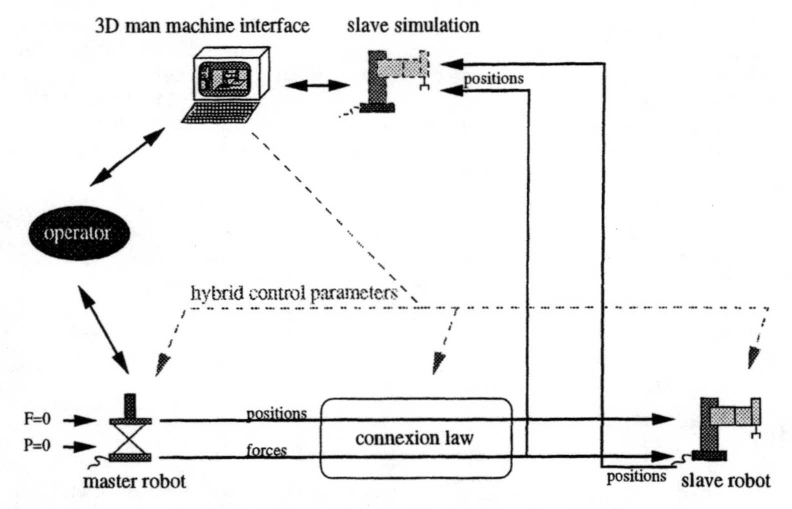

Figure 3 A high level view of the "teleinteracting system"

The positionning and contact tasks necessitate an accurate setting of the hybrid control parameters. Fe axes define force and position controlled directions. If the operator wants to go in one direction and manages a force in another direction, he must set accurately Fe axes relatively to the remote environment. This problem can be solved by using a graphical simulation of the environment coupled with a 3D man machine interface. The operator defines the slave Fe axes by positionning a "drawing" of this frame in the 3D virtual word. So he previews the directions of the position and force controlled axes of the slave robot. As communication delays exist, this graphical simulation is also used to preview the slave robot displacements, when the two robots are coupled.

Figure 3 shows that the system is divided into five components:

— The slave robot is hybrid position force controlled as defined above. At the beginning of the task, the operator defines the hybrid control parameters (Fe, Fo, S), thanks to the 3D man machine interface.

— The master robot space partition is dual of the one of the slave. The environment frame Fe is set at the center of the master workspace, and the frame Fo is set along the master handle.

— The graphical simulation of the remote environment is only used to visualize the hybrid parameters of the two robots and to preview the slave robot displacements. Thus, any model structure can be taken. As we use a high level 3D graphical software (Open Inventor), we do not take care of model structure chosen by the software.

— A connexion law manages the data transfer between the two robots and the slave simulation. In the force controlled subspace of the master robot, the measured positions are first scaled. Then, they are sent to the slave robot and to the slave robot simulation. In the other master subspace (the position controlled subspace), the measured forces are scaled and then sent to the slave robot. In the same slave subspace, the measured slave positions are sent back to the slave simulation.

— The 3D man machine interface is coupled with the graphical slave environment simulation. We create a set of 3D widgets which allow the operator to define accurately all the hybrid parameters. The two slave frames Fo and Fe are modeled by three orthogonal cylinders. The operator can translate and rotate these two virtual frames on the graphical display thanks to the computer mouse. To choose the slave space partition, the operator selects with the mouse the axes he wants to configurate.

The operator must first split the task into substasks to achieve a complete task. He must take into place the following strategy for each substask:

— Firstly, he must choose the movement direction by setting the slave Fe axes. This action is done by positionning the Fe virtual frame in the graphical display.

— Secondly, he must choose the slave frame Fo which essentially defines rotation center of the slave object.

— The operator chooses the force and position controlled axes by selecting the axes on the display with the mouse.

— He must choose the force and position scales between the master and the slave robot (the master space partition is dual to the one of the slave).

— Finally, he switches on the connexion between the two robots and he manages both the positions and the forces along the chosen axes. When this substask is finished, he switches off the connexion and begins the next one.

Time is lost by the accurate setting of the hybrid parameters with the man machine interface. But the more accurate the parameters will be set, the shorter the task execution will last.

3 EXPERIMENTAL SET-UP

We developed an experimental teleoperation set-up composed of two robots, each of them having six degrees of freedom. The slave robot is RCE1, a prototype developed at the CERT-DERA between 1984-1986. RCE1 is a macro/mini device: the macro robot is a SCARA type robot and the mini is a fully parallel six degrees of freedom wrist. The interest of this prototype comes from the hybrid position-force control scheme based on the macro/mini architecture, which combines the advantage of both a serial robot (extented workspace) and a parallel robot (lightness and high quality force control).

The master robot is a fully parallel device having six degrees of freedom. Operator interface and remote environment simulation are done on an INDY graphical workstation.

4 CONCLUSION

A new teleoperation methodology has been proposed. With this solution, the operator commands a remote hybrid slave robot with a hybrid master robot in a dual configuration. The hybrid parameters of the two robots can be set accurately thanks to a 3D man machine interface coupled with a graphical simulation of the remote environment. A general system has been described and designed. As this system allows the operator to manage accurately the remote interactions between the slave robot and its environment, we have chosen to call it "***teleinteracting system***".

REFERENCES

1. Thomas B. Sheridan. Telerobotics. *automatica*, 25(04):487–507, 1989.
2. Jean Vertut and Philippe Coiffet. *téléopération:évolution des technologies*, volume 3a of *Les robots*. HERMES, 1984.
3. William R. Ferrel. Remote manipulation with transmission delay. *IEEE Transactions on Robotics and Automation*, HFE-6(1):24–32, 1965.
4. R.J. Anderson and M.W. Spong. Asymptotic stability for force reflecting teleoperators with time delay. *The international Journal of Robotics Research*, 11(2), apr 1992.
5. Blake Hannaford. A design framework for teleoperators with kinesthetic feedback. *IEEE Transactions on Robotics and Automation*, 5(4):426–434, aug 1989.
6. G. Raju, G.C. Verghese, and T.B. Sheridan. Design issues in 2-post network models of bilateral remote manipulation. In *IEEE International Conference on Robotics and Automation*, pages 1316–1321, 1989.
7. A.K. Bejczy, W.S. Kim, and S.C. Venema. The phantom robot: Predictive displays for teleoperation with time delays. In *Proceedings of the IEEE International Conference on Robotics and Automation*, pages 546–551, 1990.
8. L. Conway, R.A. Volz, and M.W. Walker. Teleautonomous systems: Projecting and coordinating intelligent action at a distance. *IEEE Transactions on Robotics and Automation*, 6(2):146–158, apr 1990.
9. G. Hirzinger. Rotex- the first robot in space. In *Sixth International Conference on Advanced Robotics-'93*, 1993.
10. Janez Funda. Teleprogramming: Toward delay-invariant remote manipulation. Master's thesis, University of Pennsylvania, 1991.
11. M. H. Raibert and J. J. Craig. Hybrid position/force control of manipulators. *Trans. ASME J. Dyn. Syst. Meas. Contr.*, (102):126–133, 1981.
12. Yves Brière. Téléopération en présence de retard: le concept de téléopération hybride duale. Master's thesis, Ecole Nationale Supérieure de l'Aéronautique et de l'Espace, Toulouse, France, 1994.
13. Y. Briere, Y. Plihon, and C. Reboulet. The dual hybrid position-force concept for teleoperation. In *Euriscon 94*, august 1994.
14. C. Reboulet, Y. Plihon, and Y. Briere. Interest of the dual hybrid control scheme for teleoperation with time delays. In *Fourth International Symposium on Experimental Robotics, ISER'95*, july 1995.

A NOVEL APPROACH TO COLLISION AVOIDANCE OF ROBOTS USING IMAGE PLANES

G. QUICK and P. C. MÜLLER
University of Wuppertal, Safety Control Engineering, Gaußstraße 20, D-42097 Wuppertal, FRG

ABSTRACT

A lot of presentations in the field of collision avoidance for robots which appeared in the past are related to both collision free path planning and particular joint control for collision avoidance. On the other hand there are differences between on-line-techniques, off-line-techniques, and the way to manipulate the robots path. With these various approaches there are several specific solutions for the collision problem. Usually the procedures assume the existence of an ideal sensor system and that there is no problem in obstacle detection, because possible obstacles are known. Thus position and orientation of obstacles in the space can be calculated for the used algorithms. However for a universal insert unknown obstacles must be taken into consideration an so the present work tries to describe a control for collision avoidance using image planes with special considerations to the sensors.

KEYWORDS: CCD-camera, collision avoidance, image processing, potential field, robot control, robot safety

INTRODUCTION

Collision avoidance for robots is a large subject with much publication. For an additional presentation in this field it is necessary to show the relationship to these other publications. An essential criterion is the application to fixed or mobile robots. Another distinction is given by local procedures or algorithms for multi robot systems working on hierarchical structures. For these systems much computing power is required. This claim can be guaranteed easier for local algorithms on the whole particular with the view on extensions to further robots. Real time computing and processing is another demand which can be rated in the context of the method to get real information about efficiency. First it is important whether the collision avoidance algorithm is for off line or on line use. Another difference is the aim of manipulating the robots motion. This can be a path planning with optimization or on the other hand a system for safety aspects which is realized as a rule by a distance control to obstacles. As a standard the algorithms are working either with cartesian coordinates or robot coordinates in the configuration space of the robot. Among other procedures potential fields are often used as a device. On the other hand algorithms are based on graphic or analytic methods. The stated different features make it possible to have many combinations with almost unlimited solutions. As an example Khatib [1] and Kuntze [2] have suggested procedures with potential fields as a distance control with effect to moments of the robot drives. In this case there is passably consideration of dynamic even though the algorithms are based on path planning with slow movement and without discussion of the dynamics. Potential field algorithms are also used for path planning [3]. There is a changing potential from the starting point up to the destination position and

special potential edges on representation of obstacles for pushing the robot to a collision free path. Another method for all multi robot systems,taking into consideration the dynamics, are presented by Hoyer and Freund [4]. In this application a preview is taken to the next position of the robots with a check whether there is a possible collision with another robot and obstacle or not. But the behaviour of the obstacle from the time of preview up to the real situation might be unpredictable.
The attempt for a different method on collision avoidance has the aim to get an integrated and complete solution. This method must be usable for cooperating robots and robots with obstacles in their environment. Here the view on safety and the avoidance of destruction, damage, etc. is primary and not some path planning aspects. But it is important that the procedure works in relation to all classes of risk. This requirement can be realized on principle by combination of methods demonstrated as before.However there is assumed in general a settlement for the sensor problems on most known methods. For the class of robot to robot the necessary information is given by the robot controllers but for the more general case of moving obstacles the real position is mostly supposed. Furthermore there is an idealization to well-known obstacles although unknown objects have to be especially detected. With image systems for viewing obstacles the delay by computation are rarely remarked. These delay is in a range about some seconds and thus it too long for robots with high dynamic. Another problem results from the necessary signal processing for information like position, orientation, and type of recognized objects. In most cases there is a replacement by approximated geometrical objects. Transformations of coordinates are also required and so the range for disposal time is very much reduced. Because of a need for an integrated solution, the next chapter discusses sensors for collision avoidance on robots.

COMPARISON AND SELECTION OF SENSORS

On closer inspection of different sensors there is a special view on fixed robots again, because evaluation of sensors would be different on use for mobile robots. Standard applications for safety like electronic fences are not considered, but they could be integrated into the following conception. The closer inspection takes care under the aspect of collision sensors within a realistic insert on the robot environment and without relations to the particular process. Switches, pressure sensitive stuffing for robots, and protective clothing as well as sensors by capacity and induction are impractical for implementation on realistic environments. The often mentioned ultrasonic sensors are difficult in use if disturbing noise is present. An insert is of interest for applications on mobile robots only. In the field of optical distance sensors there must be distinguished between two methods. First there is a costly procedure using run time measurement for laser light. The accuracy is higher even for longer distances than on applications for collision avoidance necessary. The time for a single measurement of $10\mu s$ seems to be fast enough. On the other hand there is a more simple procedure based on triangulation. Realization of these kind of sensors by position sensitive detectors (PSD) makes possible a high relative resolution. The absolute resolution is given by presence of an optical lens. The measurement rate is about some kHz. However both measurement systems are only determining the distance to an object respectively obstacle and they determine only a very small measurement point. Such kind of systems must be movable at least for one degree of freedom to supervise a defined area. This means that the measurement system must be equipped with a drive for changing the measurement direction periodically. Combinations of these technical modules, called scanners, are available with drives for deflection on one or two degrees of freedom. Systems with one degree of freedom are applied mostly on mobile robots. Equipment with two degrees of freedom make it possible to get a snap like a photo with information about additional distances. Finally there are video sensors realized as usual by cameras. A CCD matrix represents the light sensitive elements so that a two dimensional image can be snapped. These image processing systems have an adaptable measurement range given by the lens, a resolution about 512 x 512 point respectively pixels, and usually a rate of 25 Hz. In contrast to the optical distance meters there is a two dimensional measurement plane without any movement of components. For a serious collision avoidance system the mentioned optical sensors are of interest only so that these sensors must be compared to each other. Both sensors are initiating a local discretization of the supervised area with the result of segments of pyramid stumps coming into existence. For every segment there is a distance measurement by laser scanner and on the other hand a brightness measurement by the CCD camera to the next object with a view from the working site of the sensor. At first sight the laser scanner seems to be the more qualified sensor because of the distance measurement so that there is a 3D information at all but common to the image processing, which can detect contours, it is possible to measure in front of objects but not behind the obstacles. On the other hand the range and the resulting resolution must be discussed. On CCD cameras the rate and bandwidth consequently are characterized by commercial TV technique. On building up a camera with a standard CCD matrix, such as TH7864A [5] with relations 3 by 4, the sensitive area becomes 75% of the whole supervised area and this value depends not on the range adaption by lenses. On laser scanners the rate is limited by the laser measurement system and the scanned area. The area of operation respectively the scanned angles in space is determined by rotating mirrors and the resolution of a scanned image, given by lines and points each line, depends on the periodical movement of the scanner drives. For example a structure with polygonal mirrors is mentioned in [6]. On measuring angles about 60 degrees respectively 51 degrees and a resolution about 41 lines and 81 points every line, there is a rate of 5 images per second. This corresponds to a laser measuring frequency of 17 kHz. Increasing this value up to 100 kHz makes it possible to have a rate about 25 Hz similar to the CCD camera but without such high resolution. Supplementary the laser beam supervises only small points of the whole area and there are variations on the selected range. Furthermore consideration on accuracy in construction for laser scanners are more costly than CCD cameras with corresponding accuracy. So for application on a decent collision avoidance sensor system with some sensors in an area it is more advantageous to use CCD cameras instead of laser scanners. Additionally for technology it is more simple to increase performance of CCD cameras by higher rates than improvements of laser scanners for similar results.

SENSOR DESCRIPTION AND REPRESENTATION OF OBSTACLES

For evaluation of the quality of information which is available by the sensor system it is necessary to make some close inspections to the camera itself. Also signal processing must be taken into consideration.

The camera consists of optical lenses, sometimes apertures, and the sensitive CCD chip. And so errors are resulting from the used optical elements and mechanical arrangement to each other. Linearity errors especially are initiating mistakes on the coordinates of obstacles which are detected by the following signal processing. However these errors are small in contrast to extra charges for safety and approximated objects on large scale in exchange for the obstacles of previous procedures. So it is possible to have system errors as a factor on extra charges for safety as well. Thus the sensor projects all objects which are existing in the true angle range to the image plane which is given by the CCD-matrix. This corresponds to perspective projection with correct projection of lines and without correct projection of angles. The following signal processing based on three kinds of discretization. Discretization by time has to be discussed in connection with computing power. Local discretization is based on the CCD-matrix and the resulting errors are called linearity

errors.Linearity errors on discretization of light intensity are given by the pixels and the AD-converter of the image processing. These kind of errors are insignificant on closer inspection because of the only interest in a binary image. The image signals are passing an automatic gain control (preprocessing) and filters for image correction. In the following the snapped real scene has to be compared to the nominal scene. Because of the characterization of the sensor by position P and orientation O in a world coordinates system (x, y, z) as well as the scale given by the lens, all objects from the operating space of the robot can be transformed in the image plane of the CCD-matrix. So a calculation of the difference about the images by the binary value of one bit emphasis make it possible to detect all unknown objects. The projected position of the robot has to be calculated to the nominal image, so that there are no results in the binary image based on the robot itself. Additionally there can be information about lighting of the scene and references from a knowledge base. So every discretized element of the binary image contains the information whether there is an obstacle in the subspace represented by the pixel or not. It must be annotated, that on one hand known obstacles, which are taken into consideration by the path planning, cannot be recognized but on the other hand sudden changes in position of these obstacles can be recognized as well. This is helpful on sudden changes in the environement.
The necessary computing power and the resulting speed of response base on the CCD-matrix and on the following signal processing. The transfer of pixel information inside the CCD-matrix occurs such as every 40ms, which effects in a delay. After that it starts reading of the pixel information and transportation to the image processing. Additionally the necessary existence of pixel information of neighboring image lines for filtering initiate a delay. Further processing steps like difference calculation extend the processing time again but a continuous data flow can be guaranteed by real time system. So with a realization by DSPs there is constant information for the collision avoidance algorithm which are not delayed more than 50ms. Finally it must be noted, that this kind of representation makes it easy to take inadmissible spaces of the robots configuration space into consideration. The projection of these prohibited spaces have to be used with a logical "or"-combination to the prior calculated binary image, so that a new binary image comes into existence.

JOINT CONTROL FOR COLLISION AVOIDANCE

First the aim is to have a protection against collisions with the most flexible part of the robot. In most cases this is the here considered tool centre point (TCP). Collisions to other parts of the robot within its kinematic linkage are taken into consideration later on. Further the description refers to a local collision avoidance, because there is a supposed global path planning. The joint controller of the standard robot control structure is receiving the set values from in front of an inverse coordinate transformation. A camera is recording the current scene and the following image processing calculates the data so that there is a disposal about a binary image. By the known position $P_c(x, y, z)$, orientation $O_c(\theta, \phi, \psi)$, and the focal length of the lens f_c a (u, v)-plane for the collision avoidance is characterized and by the dimension of the used CCD- matrix the boundaries in this plane are set. The current position of the TCP, given by q, has to be transformed respectively projected into this (u, v)-plane. Together with a well known nominal im age of the current scene there are different algorithms which can be used for the collision avoidance. Among other possibilities a potential field method is preferably selected. The specific procedure bases on repulsive forces. First there is an interest in the pixel with the projected position of the TCP. Around this TCP-pixel a circle line with average radius r selects some particular discretized image elements. These selected pixels are verified about the state if they are filled or not. In case of exclusive free pixels there is no collision avoidance algorithm necessary. However if the movement of the TCP is as near to the pixels which are filled by the projected obstacle, the algorithm must work with effect of movement to the TCP. This movement has to be calculated and can be described by a virtual force. This force results as a superposition from single forces given by reciprocal relation to the distance from the TCP to the filled pixel.

$$F \sim \sum \begin{pmatrix} a \\ b \end{pmatrix} \frac{1}{a^2 + b^2}, \quad \sqrt{(a^2 + b^2)} \approx r \tag{1}$$

The Jacobian J^T transforms the resulting force F into torques Γ for the drives. Different velocities in movement and extra charges for safety are regarded by the radius r. The resulting velocity of image flow is given by the bandwidth of the sensor system. So from an image of a sample t_k up to a new image of sample t_{k+1} a relative movement of maximum one pixel related to the image, respectively the resolution of discretization, is permitted. The special kind of switching included in these algorithms may be a disadvantage. Acting a low pass filter on the binary image makes possible a really potential field. So the existing levels are a function of distance to the obstacle and the pixels on the ring around the TCP can be interpreted as forces directly.

EXTENSION FOR THREE DIMENSIONAL SPACE

An extension for collision avoidance in the space by using image processing makes it necessary to operate with at least two cameras. In the so called 3D image processing systems the two cameras are used with an arrangement as in human eyes. With a special image processing and computing system the obstacles are namely detectable, but nevertheless a view behind the obstacles is not possible and there is no information about this space. Other arrangements of two cameras also cannot avoid the shadow effects based on the obstacles. It is not possible to have an efficient collision avoidance in the space by using two cameras. This can be concluded from the case of collision avoidance in the plane also, because there is no efficient procedure by using one camera. Finally there is always a three dimensional space in reality so that a supposed 2D collision avoidance by using only one camera cannot realize a decent collision avoidance at all. Reasons for this are the restricted refuge space of the current scene particular for situations on which the projected obstacle nearly fills up the whole image plane. Study of these findings make it clear that effectiveness of collision avoidance depends on the quantity of used sensors respectivily cameras. This context can be understood by another point of view. In the described procedure here, there is a primary interest in areas without obstacles and not in the areas of the image plane where the obstacles are projected in. On conventional procedures using two or more cameras, the aim is to recognize a special obstacle by detection of features in the images and to determinate position and orientation of the obstacle by comparison of the scenes. The idea of collision avoidance in the new approach using image planes makes it possible to have algorithms processing parallel for the different image planes and combine the results like the superposition theorem. Another possibility of combination can be given by a hierarchical structure with an automatic dynamic switching between the image planes depending on the availability of refuge space. For the realization of both possibilities there is a multiple arrangement of cameras, image processing and resulting superpositioning of the several correction torques as a disturbance feedforward added to the set torque given by the position controller.
On this procedure every image plane represents pyramid stumps of square and oblique pyramids. If there is a termination with "1" of busy pixels given by the projected obstacle and a termination by "0" for refuge pixels, the calculation of busy space elements is possible by special transformations. But the operations for calculation are very costly and a substitution by another mathematical body as a representative obstacle would be necessary. The

view to the refuge space is more simple and the whole refuge space can be described by

$$\overline{Q} = \overline{Q(u,v)_1} \cup \overline{Q(u,v)_2} \cup \ldots \cup \overline{Q(u,v)_n} \tag{2}$$

recently. In this equation Q represents the subspace of interest and $Q(u,v)_i$ are projections of this subspace in different image planes i.
Another advantage is found in the case of failure. Standard image processing systems cannot work when there are disturbances with one or more cameras. If there is a failure with one camera this effects only in a reduced flexibility of collision avoidance because the algorithm loses one image plane.

CONCLUSION

In this presentation collision avoidance is handled with a special view on sensors. The sensor system is based on standard cameras, but with a different kind of image processing. The actual collision avoidance algorithm and control was carried out in the coordinate system of the image plane. These advantages are clear with regard to the practical realization. First collision avoidance was described to a special point on the robot, in this special case the TCP. Blowing this point up to a cover like a sphere or taking a safety area in consideration makes it possible to take other parts of the robot into account. However for a wise extension it is necessary to have a separate inspection on the robot links for collision avoidance in the future. For this the links can be divided or summarized into parts. Another problem for collision avoidance algorithms based on distance control is the possibility of a stopping robot as a result of the different distance forces based on different obstacles. This problem is tolerated because the whole collision avoidance is built on safety aspects. The system has to be optimized with regard to the disposal computing power and on the other hand to the optimum use of the refuge space.

REFERENCES

1. Khatib, O. "Real-Time Obstacle Avoidance for Manipulators and Mobile Robots". *The International Journal of Robotics Research*, Vol. 5, No. 1, 1986, pp. 90-98.

2. Kuntze, H.-B., Schill, W. "Maßnahmen zur Optimierung der Zuverlässigkeit und Sicherheit von Handhabungsgeräten." *Fraunhofer-Institut für Informations- und Datenverarbeitung IITB* Niederschrift eines Vortrages vom 3.2.1981.

3. Adolphs, P., Nafziger, D. "Schnelle kollisionsvermeidende Bahnplanung im Konfigurationsraum mit Entfernungsfeldern." *Robotersysteme* , Vol. 6, pp. 236-244, 1990.

4. Freund, E., Hoyer, H. "Real-Time Pathfinding in Multirobot Systems Including Obstacle Avoidance." *The International Journal of Robotics Research*, Vol. 7, No. 1, 1988, pp. 42-70.

5. Databook: *CCD Products.* Thomson composants militaires et spatiaux.

6. Karl, G. *Eine 3-D Laserentfernungskamera zur Bewegungsführung mobiler Roboter.* Ph. D. Dissertation, Technical University Munich, Germany 1990.

BEHAVIOR COOPERATION BASED ON MARKERS AND WEIGHTED SIGNALS

JUKKA RIEKKI*

University of Oulu
90570 Oulu Finland
email: jpr@ee.oulu.fi

YASUO KUNIYOSHI

Autonomous Systems Section
Electrotechnical Laboratory
Tsukuba, Japan
email: kuniyosh@etl.go.jp

ABSTRACT

In this paper we propose *markers* for coordinating behavior cooperation. Markers ground task-related data on sensor data flow. Behaviors command markers by specifying weights for the different possible command parameter values. Cooperation is achieved by combining the commands sent to the same marker. We discuss also multi-agent cooperation and propose a scheme for learning cooperative behaviors. Markers have an important role in multi-agent cooperation and learning. We present experiments on a control system that enables a real robot to help other robots in transferring objects.

KEYWORDS: behavior-based, markers, behavior coordination, multi-agent cooperation

INTRODUCTION

The behavior-based architecture has been applied successfully to mobile robot control [1,2]. However, the tasks performed by these robots have been rather simple. The behavior-based architecture does not scale well to complex tasks because they require more powerful behavior coordination methods than the behavior-based architecture offers. The scaling problem has been discussed in more detail by Tsotsos [3].

We propose in this paper *Samba*, a behavior-based architecture with powerful coordination methods. The key idea is to use *markers*, which describe tasks and ground task-related data on sensor data flow. Behaviors execute tasks by commanding markers. Commands contain *weights* for the different possible values of command parameters. These weights are used in arbitrating behaviors. Markers are applied in coordinating cooperating behaviors - also in multi-agent cooperation - and in learning cooperative behaviors.

The architecture was inspired by the work of Brooks [1] and Chapman [4]. We have extended Chapman's work by attentive behaviors for the 3D domain and by integrating image space data with mobile space data. Also Brill et al. [5] have reported a marker-based architecture. The major difference between these architectures and ours is that in our system, markers have arbitration functionality.

* This research was done at the Electrotechnical Laboratory. The research was supported by the Science and Technology Agency program of the Japanese government.

The arbitration method was inspired by the distributed command arbitration method reported by Payton et al. [6]. We have generalized the method for more complex commands, for arbitrating markers, and for the image space. In recent work also Rosenblatt & Thorpe [7] extend the distributed command arbitration method for more complex commands.

In the following chapters we describe our architecture and discuss cooperation. We also describe the experiments done so far.

THE SAMBA ARCHITECTURE

The Samba architecture contains *Sensor, Actuator, Marker, Behavior,* and *Arbiter* modules. The architecture is presented in Figure 1. The control system is connected to the external world through sensor and actuator modules. Markers connect task-related environment features to behaviors. Actuators and markers are the resources of the control system. Behaviors execute tasks by sending commands to the markers and actuators. Arbiters solve the conflicts among the behaviors commanding the same resource.

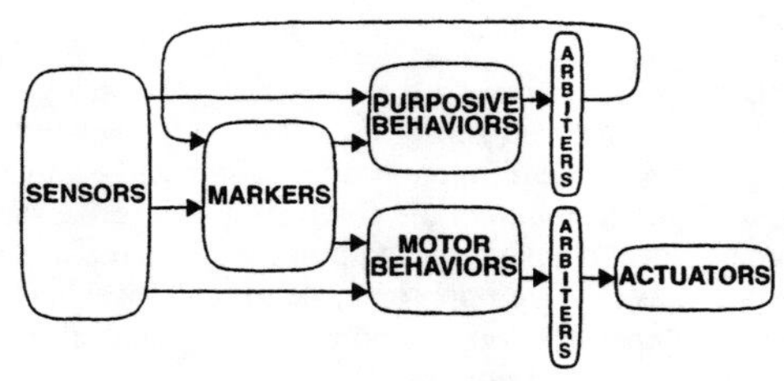

Figure 1. The Samba architecture.

Markers coordinate behaviors at several stages of task execution. First, markers activate the behaviors needed to perform the task. Secondly, markers share data among the behaviors and focus their attention on the important environment features. Thirdly, markers arbitrate behaviors. Each marker has an arbiter, that combines the commands sent by the behaviors. Actuators have similar arbiters. Finally, markers control cooperating behaviors. When several tasks can be executed in parallel, the commands sent to a marker are combined.

Signals

Modules communicate by sending signals to each other. A signal specifies weights for the different possible values of some data fields. The number and type of the data fields is not restricted. A weight is a real number in a range [-1.0, 1.0]. The data fields can be grouped and weights can be specified for these groups separately.

In the simplest case, a signal contains one set of values. The interpretation is that these values have the maximum weight and the rest of the possible values have zero weights. Or, the one set of values can have a weight in a range [0.0, 1.0]. This weight can be given by a filter producing the values, or as a function of a parameter that has some relation to the values. Such a weight function is illustrated in Figure 2. The weight increases linearly from 0.0 to 1.0 when the parameter value increases from the minimum value to the maximum value. The parameter can be, for example, camera speed for a signal describing an image processing result.

In the general case, the weights are specified for all possible sets of data field values. For a position in two-dimensional space, the weights form a three-dimensional surface. Figure 3 shows a simple weight surface.

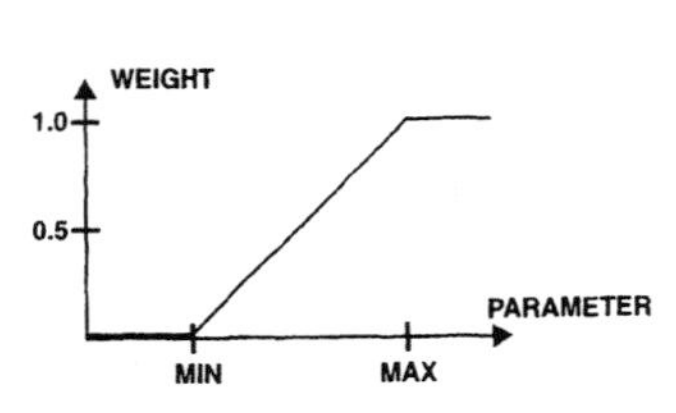

Figure 2. A weight function.

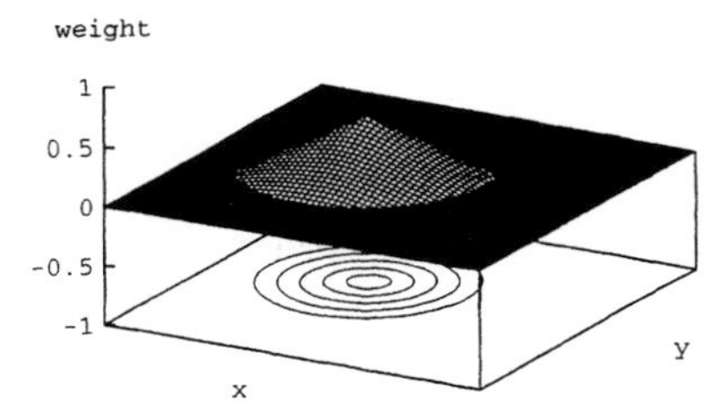

Figure 3. A weight surface.

Markers

A marker connects a task-related environment feature to motor actions. A marker is bound to a feature either by a behavior or automatically based on sensor data. It can be interpreted as binding task parameters to environment features. Binding activates the behavior executing the task.

A feature is indexed by its position in the environment. The feature position can be described either in the egocentric coordinate system (mobile space markers) or in the image coordinate system (image space markers). The feature position is updated automatically based on observations on the feature. Mobile space markers are updated also based on ego-motion.

Markers are also used to describe goals for the agent as positions related to the feature position. Further, feedback data describes the state of the task. Feedback is sent by the behavior executing the task.

A marker specifies weights for the different possible feature positions. The weights are initialized when the marker is bound. After that, the marker updates the weights based on the observations on the feature. The weights decay over time. The maximum weight is the activation level of the marker. When it decreases below a threshold, the marker deactivates itself. A marker specifies weights also for the goal positions.

Behaviors

A behavior transforms input signals received from sensors and markers into commands to actuators and markers (that is, to resources). A behavior calculates first its activation level based on the input weights and the previous activation level. If the activation level exceeds a threshold, the behavior transforms input signals into output signals. The activation level describes the importance of the behavior. The weights of the output reflect the activation level. A behavior also reports the progress of the task by sending a state signal to the corresponding marker.

The system contains two types of behaviors: motor behaviors and purposive behaviors. Motor behaviors control actuators based on signals received from sensors and markers.

Purposive behaviors execute tasks by controlling markers. When task conditions are fulfilled, a purposive behavior binds the markers needed in task execution with environment features. It also fills in task-specific data related to the features. For each marker there is either a motor behavior that controls an actuator based on the marker data or a lower-level purposive behavior that decomposes the task described by the marker further and binds the corresponding markers. A purposive behavior monitors the execution of the task by analyzing sensor data and the marker feedback data. Figure 4 illustrates task execution.

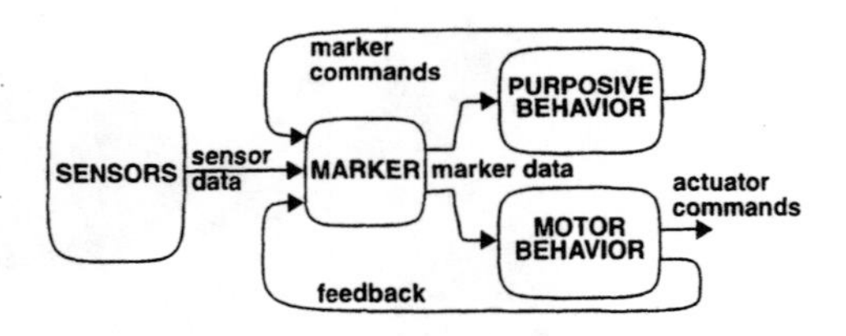

Figure 4. Task execution.

Arbiters

An arbiter combines the commands that behaviors send to a resource. Each input command has a gain that can be modified dynamically. The arbiter multiplies the weights of the commands by the gains, sums the weights, and selects the command parameter values having the maximum weights. Then, the arbiter sets the weights of the combined command based on the maximum weight and its neighborhood.

For an actuator arbiter we specify a second element, that sums the commands from stabilizing behaviors to the arbitrated command. The stabilizing behaviors take other actuators into account. For example, the cameras can be stabilized by subtracting the amount of agent rotation from the camera commands.

COOPERATION

Single Agent

As behaviors are capable of executing only simple tasks, many behaviors must cooperate to perform a complex task. Cooperation is implemented using markers and weighted signals. The signal weights specify how much each behavior affects the resulting command. The behaviors produce separate weights for each marker position. The arbiter combines the positions separately.

Multiple Agents

Cooperation among multiple agents is based on markers initialized by visual observations. A marker triggers the behaviors performing the cooperative task. Thus, cooperating multiple agents corresponds to the process of controlling an agent's own behaviors. This cooperation by observation is discussed in more detail by Kuniyoshi et al. [8].

Cooperation among multiple agents can also be learned. An agent can learn tasks by imitating other agents. Imitation consists of seeing, understanding, and doing [9]. The agent observes another agent when it performs the task and represents the observed actions by its own actions. The agent represents also preconditions for the actions by its own sensor and marker values. After this, it can start to execute the task together with the other agents.

Commands for markers and actuators form a natural representation for observed actions. Once a set of actions has been observed, the creation of a new behavior is straightforward. The behavior outputs the observed set of actions. Further, sensor and marker values are a natural choice for representing the conditions, as these values are produced also when the agent executes the task itself. The new behavior receives the sensor and marker signals that are used in the conditions.

The commands for markers and actuators form the action tree of the agent. A node in the structure describes one action known by the agent, a command to either a marker or to an actuator. Nodes describing actuator commands are leafs in the tree. A node describing a marker command has child nodes for commands produced by the behavior that executes the task described by the marker.

When the agent observes another agent, it first represents the other agent's actions by commands to its own actuators. Then it replaces sequences of commands by a command to a marker, until there are no more sequences of commands in the observed set that match a set of child nodes in the action tree. The remaining set of actions is produced by the new behavior when the conditions for the task are satisfied.

The learned behaviors can also be generalized. When several behaviors produce the same sequence of actions, a new behavior executing the common sequence is created. Also a new marker is created. The marker activates the new behavior. The common sequence of actions is replaced in each behavior by activation of the new marker.

The complexity of the learning problem is managed by learning gradually. At each step, the agent learns a new way to combine the existing actions. After learning a new action, the action is added to the action tree. When possible, behaviors are generalized. The agent builds incrementally new layers of behaviors and markers in top of existing ones.

EXPERIMENTS

We have tested the Samba architecture in the application of Posing, Unblocking, and Receiving. The Pure control system controls an agent to help other agents in their transfer tasks. The agent is equipped with a stereo gaze platform. The gaze platform has 2 DOF vergence control. Zero Disparity Filter (Zdf) extracts from the image data the features that are at the center of both cameras, that is, the features that the cameras are fixated on. Refer to papers [10,11] for more details on the Pure system. Lately we have improved the Pure system by adding weights for signals, activation levels for behaviors and arbiters for resources. So far we have tested the posing and unblocking tasks.

Posing and Unblocking Separately

In the posing experiment, the agent followed another agent successfully for several minutes. This experiment demonstrates the basic characteristics of our architecture. The active modules are shown in Figure 5. Pose behavior initializes the Moveto marker, which contains a point bound to an object and a goal point for the agent. The Moveto behavior moves the agent to the goal point and turns it towards the object point. The Moveto marker updates the internal representation of the points continuously based on ego-motion. Pursuit behavior controls the cameras towards the posed object based on the output of Zdf module, which is described by the Zdfcent marker. The Moveto marker updates its points based on the fixation point of the cameras.

In the unblocking experiment, the agent helped another agent by pushing away an obstacle blocking that agent. This is an example of multi-agent cooperation. When there is an obstacle in the trajectory of another agent, the Unblock behavior controls the Moveto marker in such a way that the agent heads towards the obstacle at a predefined distance and pushes the obstacle away.

As an example of weight calculation, see the weight of the Zdfcent marker in Figure 6. The weight is calculated based on the size of the Zdf region (in pixels) and the speeds of the cameras. The other agent was tracked until 245 seconds (approximately). The small valleys in the weight before that moment are caused by variations in the Zdf region size. When an obstacle is searched the weight goes to zero. When the obstacle is found, the weight increases quickly. At the end the weight varies considerably, as the obstacle fills the images.

Cooperation of Posing and Unblocking

We tested behavior cooperation also by executing posing and unblocking tasks in par-

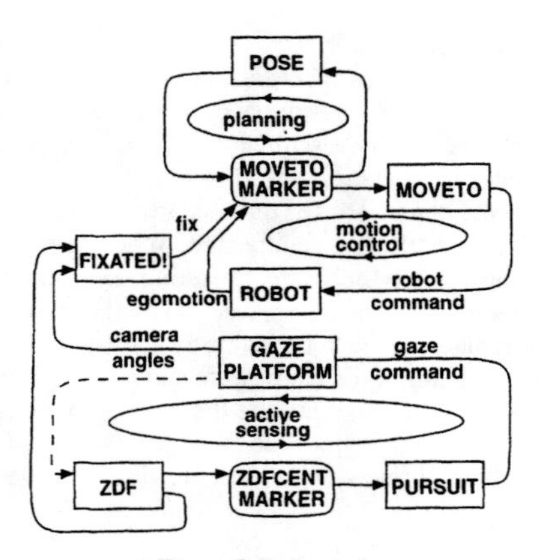

Figure 5. Posing task.

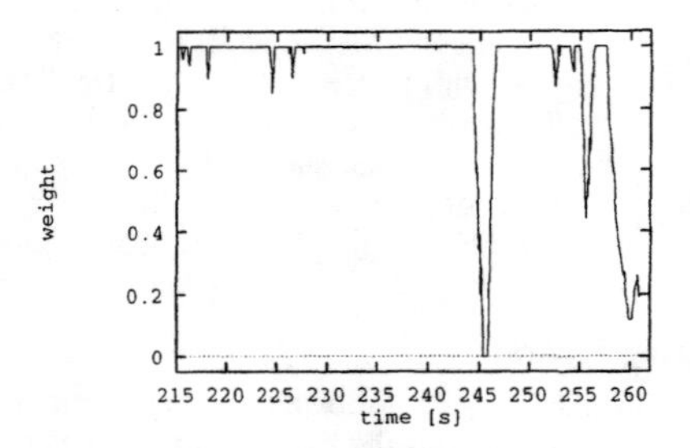

Figure 6. Zdfcent weight.

allel. Both behaviors sent commands to the Moveto marker. This is an example of cooperating behaviors of a single agent. The behaviors cooperate to unblock other agents. Pose keeps the agent near the other agent, so the area in front of the other agent can be checked periodically and Unblock can remove the obstacles.

In the first version Unblock simply subsumed Pose when an obstacle was found. The behaviors did not cooperate very well. The reason was that during posing the agent tends to drive behind the other agent, which makes unblocking difficult.

We are currently implementing a new version in which Pose and Unblock send weighted signals to the Moveto marker. The signals are combined by an arbiter. The maximum weights of the combined signal specify the goal and target positions of the Moveto marker. This approach enables Unblock to control the agent towards a good unblocking position while Pose keeps the agent near the other agent. When the area in front of the other agent is known to be free, Unblock sends small weights and thus does not have big effect on the Moveto marker. But when the known free area shortens, Unblock increases the weights and the goal position shifts toward the good unblocking position.

Figure 7 illustrates the combining of the goal positions. In the signal sent by Pose, weights are large at a constant distance from the target. Unblock specifies large weights at a constant distance from the future trajectory of the target. The maximum weight of the combined signal specifies the goal position of the Moveto marker.

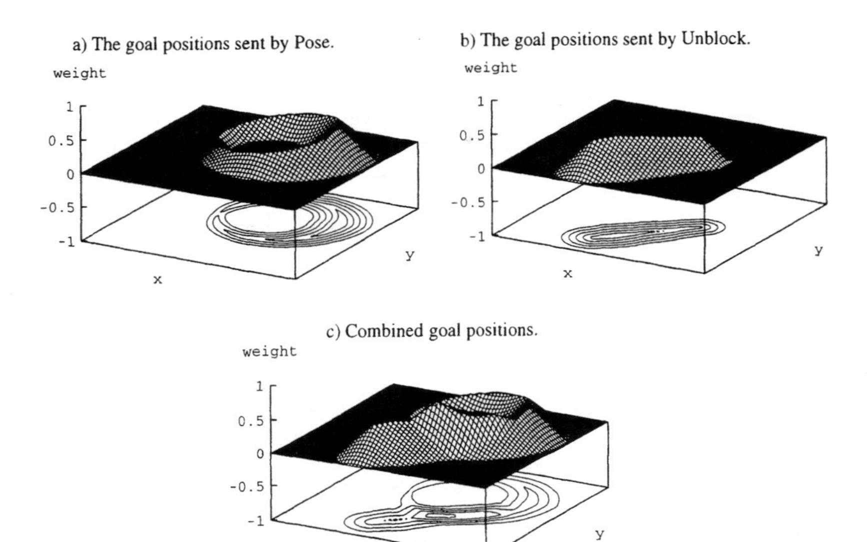

Figure 7. An example of combining goal positions.

DISCUSSION

The lack of powerful behavior coordination methods prevents applying the behavior-based architecture to complex tasks. To solve this problem, we proposed behavior coordination methods based on markers and weighted signals. The weights enable a continuous shift from one behavior to another. Further, weights enable the anticipation of tasks, as behaviors can send commands with small weights when the task conditions have not yet been fully satisfied.

We improved the behavior-based architecture bottom-up, without introducing a symbolic model or reasoning based on the model. As all coordination methods are local, the system is robust and scalable.

These coordination methods can also be used to cooperate behaviors of multiple agents. We proposed a method for learning a common task by imitating other agents. As there is no explicit communication between agents, the multi-agent system is robust and extensible. Further, there is no need for transformation between the representations of the agents.

The arbitration approach was chosen as it is an open mechanism, that can easily be modified by other modules such as learning modules. Cooperation can be adapted based on experiments. Weighting each command enables dynamic arbitration based on the agent and environment state. The use of gains allows further tuning based on additional information.

When we specified the arbiter, we considered first weighted averaging. In this approach the resulting command is constrained by the commands sent by the behaviors. In many cases this is too loose a constraint. For example, when triangulation is used to calculate object positions, the cameras must always be turned towards an object. Also, a marker must be bound to one feature, not between several features. The chosen arbitration method produces commands that are suggested by at least one behavior. The disadvantage of the chosen arbitration approach is that it is computationally expensive. We are currently investigating methods to implement it effectively.

In the near future, we will finish the experiments of unblocking. We expect these experiments to demonstrate how the arbitration method enables a smooth transition from one behavior to another. After that we will integrate the components based on the optical flow in the Pure system.

We will continue to develop the Samba architecture. More complex systems are needed to test the arbitration method. The proposed learning scheme has to be specified in more detail and implemented. Although the learning scheme contains challenging problems, we believe that it also has a great potential as an approach to building complex control systems.

CONCLUSION

In this paper we discussed behavior cooperation. We described how markers and weighted signals can be used to coordinate cooperating behaviors. We discussed the cooperating behaviors for both single and multiple agents. We proposed a method for learning cooperative behaviors. We tested the architecture with real robots.

ACKNOWLEDGMENTS

The authors would like to thank colleagues at the Electrotechnical Laboratory for their assistance in this research project. The discussions with Paul Bakker, Polly Pook, and Alex Zelinsky were valuable. Special thanks to Kenji Konaka, who helped in the implementation. Further, we are grateful to Nobuyuki Kita and Sebastien Rougeaux, who contributed to the original image processing and robot control programs.

REFERENCES

1. Brooks, R.A. (1986) A robust layered control system for a mobile robot. IEEE Journal of Robotics and Automation, RA-2(1):14-23.
2. Brooks, R.A. (1990) Elephants don't play chess. Robotics and Autonomous Systems 6(1-2):3-15.
3. Tsotsos, J.K. (1995) Behaviorist intelligence and the scaling problem. Artificial Intelligence 75:135-160.
4. Chapman, D. (1991) Vision, instruction, and action, MIT Press.
5. Brill, F.Z., Martin, W.N. & Olson, T.J. (1995) Markers elucidated and applied in local 3-space. International Symposium on Computer Vision, Coral Gables, Florida, November 21-23, 1995. pages 49-54.
6. Payton, D.W., Rosenblatt, J.K. & Keirsey, D.M. (1990) Plan guided reaction. IEEE Transactions on Systems, Man, and Cybernetics, 20(6):1370-1382.
7. Rosenblatt, J.K. & Thorpe, C.E. (1995) Combining goals in a behavior-based architecture. IEEE/RSJ International Conference on Intelligent Robots and Systems (IROS'95), Pittsburgh, USA, August 1995, pp 136-141.
8. Kuniyoshi, Y., Kita, N., Rougeaux, S., Sakane, S., Ishii, M. & Kakikura, M. (1994) Cooperation by observation - The framework and basic task patterns. IEEE International Conference on Robotics and Automation, pp. 767-774.
9. Kuniyoshi, Y., Inaba, M. & Inoue, H. (1992) Seeing, understanding and doing human task. 1992 IEEE International Conference on Robotics and Automation, May 1992, Nice, France.
10. Kuniyoshi, Y., Riekki, J., Ishii, M., Rougeaux, S., Kita, N., Sakane, S. & Kakikura, M. (1994) Vision-based behaviors for multi-robot cooperation. IEEE/RSJ/GI International Conference on Intelligent Robots and Systems (IROS'94), Munchen, Germany, September 1994, pp 925-932.
11. Riekki, J. & Kuniyoshi, Y. (1995) Architecture for vision-based purposive behaviors. IEEE/RSJ International Conf. on Intelligent Robots and Systems (IROS'95), Pittsburgh, USA, August 1995, pp 82-89.

SENSOR-BASED TASKS : FROM THE SPECIFICATION TO THE CONTROL ASPECTS

P. RIVES, R. PISSARD-GIBOLLET, L. PELLETIER

INRIA-centre de Sophia Antipolis, 2004 Route des Lucioles,
06565 Valbonne, France.
fax: (+33) 93 65 78 45
e-mail Patrick.Rives@sophia.inria.fr

ABSTRACT

In this paper, we discuss both theoretical and experimental issues allowing to realize sensor-based tasks efficiently. The aim of this work is, on the first hand, to establish a coherent framework from the task specification level to the control level and, on the other hand, to develop programming tools which easily allow to implement and validate sensor-based tasks. An important requirement was to be able to support different types of sensors in a similar way. Both theoretical and experimental issues are presented.

KEYWORDS: interaction screw, virtual linkage, continuous closed control loop, reactivity to logical events, canonic sensor-based tasks..

SENSOR-BASED TASK FRAMEWORK

In this framework, we claim that, in many cases, a robotic application can be successfully performed by sequencing elementary sensor-based tasks. At each elementary task is associated a servoing law, continuous on a time interval $[T_{begin}, T_{end}]$, and built from the sensor's outputs. The transitions between the elementary tasks are scheduled by the events occuring during the task (reactive approach) [11, 3]. Our purpose is to provide the end-user with a library of well-conditionned elementary sensor-based tasks which could be assembled to realize more complex robotic actions. The keypoint of our framework is lying on a generic model of the interactions between the robot and its local environment [4], [7], [1].

Modeling the interaction robot-environment

Modeling the 3D environment: Firstly, we will assume that we deal with an *a priori* known geometric 3D scene, composed of a set of objects described by their surface. We choose implicit equations (i.e. $\mathcal{F}(x, y, z) = 0$) to represent the 3D surfaces in the scene. Thereby, to compute the normal vector at a point T of a 3D feature, we have only to apply the gradient of the fonction $\mathcal{F}$ at this point T :

$$\vec{n_T} = k \cdot \nabla\mathcal{F}(x_T, y_T, z_T) \text{ where } k \text{ is a normalisation factor.}$$

Modeling the sensor output: Let us consider a sensor S rigidly fixed to a robot, with associated frame F_S. This sensor perceives an object T in the environment, with associated frame F_T.

As output, it provides a vectorial signal $s = (s_1, \ldots, s_k)^T$. We will assume later on that each component s_i depends only on the relative location of F_T with respect to F_S. This means that if the sensor is motionless with respect to the target, then the signal remains constant. This assumption is valid in practice for a large class of sensors such as force, vision (geometric features) or range sensors.

The relative pose (position and orientation) of F_T with respect to F_S can be represented by an element $\bar{r}$ of the Lie group of displacements SE_3. We may thus write:

$$s_i(\bar{r}) = s_i(F_S, F_T) \tag{1}$$

where s_i is a C^2 function defined over SE_3, and valued in R. Moreover, se_3, the Lie Algebra of SE_3 (which is its tangent space at the identity), can be assimilated to the space of screws. The differential of s_i is a linear mapping from se_3 to R, it can be represented by a screw $H_i = (H_i(S), u)$ so that :

$$\dot{s}_i = \frac{\partial s_i}{\partial \bar{r}} \dot{\bar{r}} = \frac{\partial s_i}{\partial \bar{r}} T_{ST} = (H_{i,j}(S), u) \bullet (V_S, \omega_{F_S/F_T}) \tag{2}$$

where the product of screws is the scalar $\langle H_i(S) \,|\, \omega_{F_S/F_T} \rangle + \langle V_S \,|\, u \rangle$.

We can note that H_i represents uniquely the interaction between the sensor and the robot's environment, so it is called *Interaction Screw*. The matricial representation, called *Interaction Matrix*, is given by the relation :

$$\dot{s}_i = L_i^T . T_{ST} \qquad \text{with} \qquad L_i^T = H_i . \begin{pmatrix} 0 & I_3 \\ I_3 & 0 \end{pmatrix} \tag{3}$$

In a rough manner, L_i^T can be viewed as a sort of *Jacobian* relating the variation of the elementary sensor output s_i to the relative displacement between the sensor and the environment.

For example, let us consider a pin-point range finder. We assume a purely geometric model which delivers an output signal $s = f(\delta)$ where δ is the distance between the sensor and the target. Explicit computation of the corresponding interaction matrix can be found in [10] and provides :

$$\dot{s} = L^T \cdot T_{ST} = -\frac{\partial f}{\partial \delta} \cdot \frac{1}{\langle n_T \,|\, n_S \rangle} \left[n_T^T \;\; , \;\; \delta \cdot (n_T \times n_S)^T \right] \cdot \begin{bmatrix} V_S \\ \omega_{F_S/F_T} \end{bmatrix} \tag{4}$$

where

- n_S is the line of sight of the sensor and n_T the normal to the surface of the target at the impact point
- $\langle \cdot \,|\, \cdot \rangle$ and $(\cdot \times \cdot)$ respectively represent scalar and cross products
- $\frac{\partial f}{\partial \delta}$ is a known constant defined by the sensor technology used.

Let us now suppose that this sensor interacts with a sphere of center $\mathcal{O} = (x_0, y_0, z_0)$ and with radius R. Its implicit equation will be :

$$\mathcal{F}(x, y, z) = (x - x_0)^2 + (y - y_0)^2 + (z - z_0)^2 - R^2$$

and the normal vector will be :

$$\nabla \mathcal{F}(x_T, y_T, z_T) = 2 \begin{pmatrix} x_T - x_0 \\ y_T - y_0 \\ z_T - z_0 \end{pmatrix}$$

So we can compute the interaction matrix in a straightforward manner. Since T is the intersection point of the sensor axis and of the feature, its coordinates are given by $ST = \delta n_S = (\delta \; 0 \; 0)^T$ and we obtain : $n_T = K \cdot (\delta - x_0 \;\; -y_0 \;\; -z_0)^T$. Then, the interaction matrix is given by :

$$L^T = K \cdot \frac{\partial f}{\partial \delta} \cdot \begin{pmatrix} -1 & -\frac{y_0}{x_0 - \delta} & -\frac{z_0}{x_0 - \delta} & 0 & -\frac{\delta z_0}{x_0 - \delta} & \frac{\delta y_0}{x_0 - \delta} \end{pmatrix} \tag{5}$$

By assembling pin-point sensors, we can build models of more complex sensors like laser scanning device or 3D camera and compute their corresponding interaction matrices. For example, the laser scanning device will be defined by its plane of sight and its scanning step angle and it will provide a vector signal $\underline{s} = (s_1 .... s_i .... s_K)^T$ such that $s_i = f(\delta_i(\vec{n_{S_i}}, \vec{n_{T_i}}))$ where $\vec{n_{S_i}} = (\cos\theta_i \;\sin\theta_i\; 0)^T$ is a parametrization of the sweep angle.

In previous papers, we show that a closed form of the interaction matrix can be explicitely computed for a large class of sensors and 3D primitives. Examples of such computations for visions sensors can be found in [4, 2]. More recently [6], closed forms of interaction matrix was obtained for range data provided by thin-field sounder, laser scanning device and 3D camera interacting with elementary 3D primitives described by implicit equations like plane, ridge, sphere or cylinder.

Sensor-based task : Specification aspects

At the end-user level, it is important to deal with a description of the tasks which can be easily handled. A description based on specification of geometric and kinematic constraints between frames linked to the robot and frames linked to the local environment seems to be well-adapted. Such a description is already used to design mechanisms and known as the formalism of *mechanical linkages*. This formalism, lying on the properties of screws, allows to specify contacts between solids in terms of free or constrained motions between frames attached to each solid. Due to the nature of the interaction matrix, it exists a strong analogy between sensor-based tasks and motions with contact. Eq. 2 shows that the differential of the sensor signal s_i is a screw product. The nullspace S^* of this differential (defined by $\dot{s}_i = L^T \cdot T^*{}_{ST} = 0$) can be interpreted as a set of displacements $T^*{}_{ST}$ leaving the sensor signal invariant for a given configuration $\bar{r}_d$. A basis of S^* is easily obtained by computing $Ker(L^T)$. The subspace S^* represents the "free motion space" corresponding to a *virtual contact* between the sensor S and the target T. The underlying idea is the following: imposing $\dot{\underline{s}} = 0$ is equivalent to introduce constraints in the configuration space SE_3, as if the target and the sensor were linked geometrically. This set of compatible constraints $\underline{s}(\bar{r}) - \underline{s}_d = 0$ determines a *virtual linkage* between the sensor S and the target T. For instance, in the case of a pin-point range finder, one can illustrate the constraint $ST = \delta_d$ by imaginating a tactile finger of constant length δ_d rigidly linked S. Then, the free motions will be those keeping the tip of the finger in contact with the surface of the object. Moreover, translating these sensor constraints in the configuration space allows us to define the notion of *virtual* $\mathcal{C}$-surfaces in a sense analog to those used by Mason [5] in compliant motion planning. A *virtual* $\mathcal{C}$-surface is a differential submanifold of R^6 and has the following properties : (i) a couple sensor/3D geometric primitive is characterized by its $\mathcal{C}$-surface, (ii) its dimension represents the number of dof unconstrained by the linkage and is equal to the $rank(L^T)$ (iii) a closed form of its equation may be computed for 3D primitives described by implicit equations.

Sensor-based task : Control aspects

Here, we present only a brief overview of the framework used at the control level. For more details, we refer the reader to papers previously published [4, 9, 7] and to the Samson's reference book [10].

Using the general formalism introduced by Samson , we can express a robot task in term of the regulation of an output function $\underline{e}(\bar{r}, t)$. We consider the task is perfectly achieved during the time interval $[0, T]$ iff : $\underline{e}(\bar{r}, t) = 0, \; \forall t \in [0, T]$

Applying this formalism to the sensory based task, we showed in previous papers [4] that a generic form of $\underline{e}(\bar{r}, t)$ can be built using the sensor output $\underline{s} = [s_1 \dots s_k]^T$ such that :

$$e(\bar{r},t) = \begin{pmatrix} s_1(\bar{r},t) - s_1^d(t) \\ \vdots \\ s_k(\bar{r},t) - s_k^d(t) \end{pmatrix}$$

It was also shown that a very simple *gradient based approach* embedded in a velocity control scheme is sufficient to ensure an exponential convergence to 0 of the task function $\underline{e}(\bar{r},t)$ by using the following desired velocity screw T^d as control input :

$$T^d = -\lambda . L^{T+}_{\underline{s}=\underline{s}^d} . e(\bar{r},t) \quad (6)$$

where λ is a positive gain and $L^{T+}_{\underline{s}=\underline{s}^d}$ is the pseudo inverse of the interaction matrix computed at the desired value $\underline{s}^d$.

Exemples of sensor-based control laws have been validated both on a hand-eye arm and a mobile robot using vision and ultrasonic belt for different tasks like wall-following or positionning with regard to a target [1, 7, 9, 8].

Towards a library of sensor-based tasks

Using this framework, we have defined a catalog of elementary sensor-based primitives using range and vision sensors interacting with 3D geometric features like line, plane, sphere, cylinder.... Each elementary task is defined by :(i) the type of the sensor signals used (ii) the symbolic form of the associated interaction matrix (iii) the type of virtual linkage which allows to realized : prismatic, revolute, planar, ball-and-socket....(iv) the degrees of freedom constrained by the linkage. Thanks to *Maple*, the interaction matrix, the corresponding *virtual linkage* and the value of the desired signal s_d can be computed automatically. The characteristics of the elementary sensor-based primitives are summarized in a tabular form :

Type of interaction : range scanning device / Cylinder

Virtual Linkage	Class T	Class R	3 D primitive	Sensor signals	Free space basis
Line/cylinder contact	1	1	a cylinder of radius R with axis $(a,b,c)^T$ passing by (x_0,y_0,z_0)	$\begin{pmatrix} \delta_1 \\ \vdots \\ \delta_i \\ \vdots \\ \delta_N \end{pmatrix}$	$\begin{pmatrix} a & c\,y_0 - b\,z_0 \\ b & a\,z_0 - c\,x_0 \\ c & b\,x_0 - a\,y_0 \\ 0 & a \\ 0 & b \\ 0 & c \end{pmatrix}$

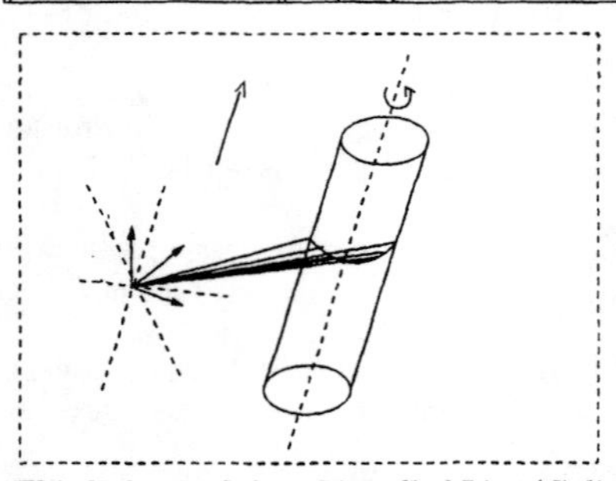

Interpreting these data shows that the free motions are:

- the translation parallel to the axis of the cylinder
- the rotation around the axis of the cylinder

This linkage of class 2 is called Line/Cylinder linkage and can be used, for instance, in an underwater pipe following task. At the end-user level, these sensor-based primitives are used to specify elementary tasks like surface following, positionning with regard to an object or tracking a moving target.

Associated sensor-based control laws computed by using 6 characterize, in continuous time, the physical motion of the robot during the $[0, T]$ interval of their validity. Nevertheless, when we want to execute this motion in a realistic environment we need to take into account and react in time to various situations, at least for ensuring the integrity of the robot. These two tightly coupled aspects of a robotic action are coherently handled by the *Robot-task* (RT) concept [11]. Let's remind that a RT models an elementary robotic action. It is formaly defined as the entire parametrized specification of a control law, and a logical behavior associated with a set of events which may occur just before, during and just after the task execution.

The behavior of the system is handled by the framework of reactive systems theory: it consists of the legal sequences of input/output signals received/emitted by the system. Its specification is methodic; events are typed in pre-conditions, exceptions with three types of reaction and post-conditions. The control-law activation starts at the instant that all pre-conditions are satisfied. During its execution exceptions are monitored; they are handled either localy changing in-line a control parameter or globaly asking from the application to interrupt the current RT and activate a recovery program or imposing the total application interruption driving the robot in a safe position. Finaly the action stops when the set of post-conditions is satisfied.

EXPERIMENTAL RESULTS

We have developped a versatile testbed in order to validate sensor-based control approaches in real experiments. This robotic system uses a nonholonomic wheeled cart carrying a two d.o.f head with a CCD camera and recently equiped with a belt of eight sounders. Different *Robot-tasks* has been already validated with success and discussed in previous paper [8]. We just present here an example of *Visual wall following task*. To cross the room the robot has to follow a wall. The parallel lines corresponding to a skirting board at the bottom of a wall 1) are used for its control. This visual-servoing task is handled by the sensor-based control previously described. For this task *WallFollowing*, thanks to the vision system, line parameters are extracted at video rate 2 for the visual servoing control loop and events are detected as the *End-of-wall* (post-condition) or the exception *Target-lost*.

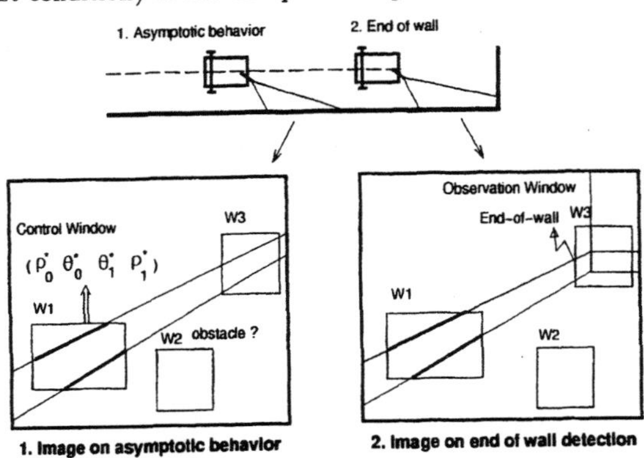

Figure 1: The wall following task

CONCLUSION

The framework developped in this paper allows to handle elementary robotic actions tasks at different levels : from the specification up to implementation issues. Each robotic action is constituted by a continuous sensor-based control law and a logical behaviour. Using such a framework, we expect to provide the end-user with a library of canonic sensor-based tasks which can be assembled to perform a complex robotic mission.

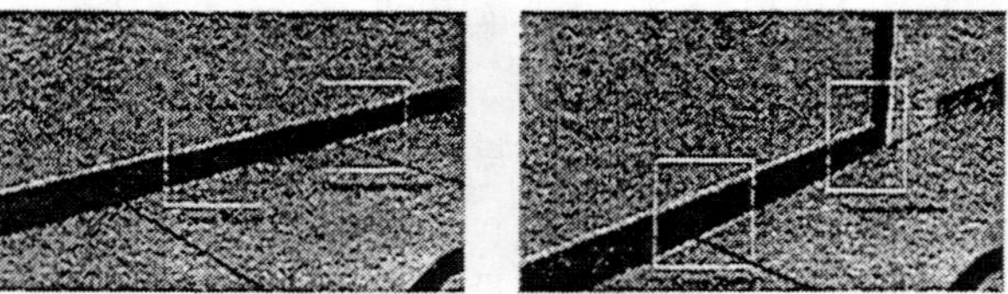

Figure 2: Real time image processing

References

[1] F. Chaumette. *La Relation Vision-Commande: Théorie et Application à des Tâches Robotiques.* PhD thesis, Université de Rennes I, July 1990.

[2] F. Chaumette, P. Rives, and B. Espiau. Classification and realization od the different vision-based tasks. In *Visual Servoing - Automatic Control of Mechanical Systems with Visual Sensors; World Scientific Series in Robotics and Automated Systems Vol 7.* K. Hashimoto (ed.), London, 1993.

[3] E. Coste-Manière, B. Espiau, and D. Simon. Reactive objects in a task level open controller. In *IEEE Conference on Robotics and Automation*, Nice, France, May 1992.

[4] B. Espiau, F. Chaumette, and P. Rives. A new approach to vcisual servoing in robotics. *IEEE Transactions on Robotics and Automation*, 8(3):313–326, 1992.

[5] M. T. Mason. Compliance and force control for computer controlled mainpulators. *IEEE Transactions on Systems, Man, and Cybernetics*, June 1981.

[6] L. Pelletier. Specification of sensor-based tasks with a view to an application to motion planning. Master's thesis, Université de Nice Sophia Antipolis, 1995. Rapport de stage de DEA Robotique et Vision.

[7] R. Pissard-Gibollet and P. Rives. Applying visual servoing techniques to control a mobile hand-eye system. In *IEEE Int. Conference on Robotics and Automation*, Nagoya, Japan, May 21-27 1995.

[8] R. Pissard-Gibollet, P. Rives, K. Kapellos, and J. Borrelly. Real-time programming of a mobile robot actions using advanced control technics. In *4th Int. Symposium on Experimental Robotics*, Stanford, California, June 30 - July 2 1995.

[9] P. Rives and H. Michel. Visual servoing based on ellipse features. In *SPIE*, Boston, MA, USA, September 1993.

[10] C. Samson, B. Espiau, and M. Leborgne. Robot control: The task function approach. In *Oxford Engineering Sciences Series 22.* Oxford University, 1991.

[11] D. Simon, B. Espiau, E. Castillo, and K. Kapellos. Computer-aided design of a generic robot controller handling reactivity and real-time control issues. *IEEE Trans. on Control Systems Technology*, 1(4), 1993.

ROBOT NAVIGATION UNDER APPROXIMATE SELF-LOCALIZATION

ALESSANDRO SAFFIOTTI*
IRIDIA, Université Libre de Bruxelles
50 av. F. Roosevelt, CP 194/6, Brussels, Belgium
E-mail: asaffio@ulb.ac.be, URL: http://iridia.ulb.ac.be/saffiotti/

ABSTRACT

The ability to pursue specific goals is essential for purposeful autonomous navigation. Goals usually refer to objects in the environment whose position is stored in a map. To effectively use this information, the robot needs to translate absolute positions in the map into internal coordinates, based on an estimate of its own position in the map. We describe a technique to incorporate map information into a fuzzy controller when this estimate is approximate, and it is represented by a fuzzy set. We illustrate this technique by showing an example on a mobile robot.

KEYWORDS: Mobile robots, navigation, fuzzy sets, fuzzy control, maps, position estimation, uncertainty.

INTRODUCTION

Autonomous robot navigation requires the ability to pursue specific goals, like crossing a door or reaching a given location, while maintaining reactivity. Goals are often generated at higher levels of abstractions, through the use of symbolic planners or topological maps, and then have to be considered at the control level. The problem of how to integrate high-level and low-level representations and processes is one of the central issues in mobile robotics.

In previous works [4, 3], we have proposed a technique to bring abstract goals into a fuzzy controller based on the use of *object descriptors*: the relevant properties of the object of an action (e.g., a door to cross) are extracted from the map and wrapped into a data structure — the descriptor. This is then incorporated in the local state of the controller so that its properties can be used as an input to the controller. For example, consider the goal of going down a given corridor for which we have some information in the map. From this information, we build the descriptor shown in Fig. 1 (left), including the current position of the corridor *relative to the robot*, i.e., expressed in the robot's local frame R. The fuzzy rules for corridor following (right) access the fields of this descriptor to evaluate their antecedents; these fields are then continuously updated whenever the robot's sensors perceive the actual corridor to achieve a closed-loop response. The principal advantages of using an object descriptor are that: (a) it

*Work supported by the BELON project, founded by the *Communauté Française de Belgique*.

NAME: Corr1 POSITION: (1200 500 −20°) WIDTH: 2000 LENGTH: 14000	IF left-wall-too-close THEN turn-right IF right-wall-too-close THEN turn-left IF close-to-midline AND heading-right THEN turn-left IF close-to-midline AND heading-left THEN turn-right

Figure 1: A descriptor for a corridor, and the fuzzy rules to follow it.

provides default values for properties that are not currently sensed; (b) updating the descriptor's properties is easier than updating a global map; and (c) the control rules are easier to write and to evaluate, as all the geometric properties of the descriptor are expressed in the robot's frame.

To build a descriptor for an object, we need to convert its absolute position, stored in the map, into a robot-relative position (in R) to be stored in the descriptor, using the robot's current *self-location* in the map. Self-localization, however, is rarely precise, and the robot's self-location is in general affected by uncertainty. Although some solution has been proposed for navigating using approximate maps (e.g., [2]), these do not deal with the case in which self-location is itself approximate. In the rest of this paper, we extend the descriptor technique to take this case into account.

APPROXIMATE SELF-LOCALIZATION

We represent approximate locations by fuzzy subsets of a given space, read under a possibilistic interpretation [7, 8]: if L_o is a fuzzy set representing the approximate location of object o, then we read the value of $L_o(\vec{x}) \in [0,1]$ as the *degree of possibility* that o be located at $\vec{x}$. This representation allows us to model different aspects of locational uncertainty. Figure 2 shows four approximate locations in one dimension: in (a), the object is located at approximately 5 ("vagueness"); in (b), it can possibly be anywhere between 5 and 10 ("imprecision"); in (c), it can be either at 5 or at 10 ("ambiguity"); in (d), we are told that the object is at 5, but the source may be wrong, and there is a small possibility that it be located just anywhere ("unreliability"). In what follows, we assume that all locations in the map are fuzzy, including the robot's estimate of its own position. (A single point is a special case of a fuzzy set.)

In [5] we proposed a recursive algorithm for self-localization based on the above representation. The outline of the algorithm is very simple. At each time-step t the robot has an approximate hypothesis of its own location in the map, represented by a fuzzy subset H_t of the map frame. During navigation, the robot's perceptual apparatus recognizes relevant features, and searches the map for matching objects using a fuzzy measure of similarity. Each matching pair is used to build a *fuzzy localizer*: a fuzzy set representing the approximate location in the map where the robot *should be* in order to see the object where the feature has been observed. So, each localizer provides one imprecise source of information about the actual position of the robot. All these

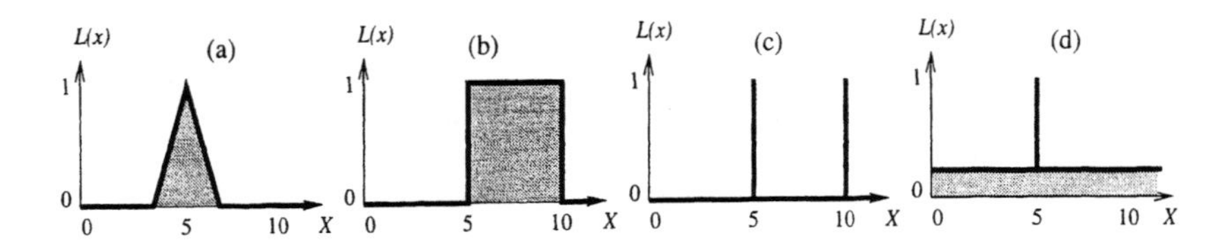

Figure 2: Representing different types of locational uncertainty by fuzzy sets.

localizers, plus odometric information, are combined by fuzzy intersection to produce the new hypothesis H_{t+1}, and the cycle repeats.

BUILDING APPROXIMATE DESCRIPTORS

Suppose we are given a goal involving an object o, whose position in the map is given by the fuzzy set O. We need to build descriptor for o in the controller, and to initialize its position to the correspondent of O in the robot's frame R. The pivot for this correspondence is the current position of the robot in the map. Here, we are interested in the case when the robot's estimate about its position is represented by a fuzzy set, like the H_t produced by the algorithm above.

We assume to have a function $T_r : R \longrightarrow M$ that maps locations from the robot frame R to the map frame M, given that the robot is located at $\vec{r} \in M$. (When both M and R are Cartesian frames, T_r is simply a rotation plus a translation.) We then define the *expected position* of o in the R frame at time t by:

$$Exp_t[o](\vec{x}) = \sup_{\vec{r} \in M} \left(H_t(\vec{r}) \wedge O(T_r(\vec{x})) \right), \tag{1}$$

where $\wedge$ denotes the minimum. Intuitively, we expect to find object o at $\vec{x} \in R$ if, and to the extent by which, there exists a position $\vec{r} \in M$ such that: (a) $\vec{r}$ is a possible location for the robot, as measured by the current hypothesis H_t; and (b) given $\vec{r}$, $\vec{x}$ corresponds (via T_r) to a map location that is possible for o. It is easy to see that (1) reduces to simple forms when either O or H_t (or both) are single points.

We shall use the value of $Exp_t[o]$ to set the position of the descriptor for o. We have two alternatives: to use this value as it is; or to defuzzify it, say by a center of gravity computation

$$\hat{o} = \frac{\int_R \vec{x}\, Exp_t[o](\vec{x})\, \mathrm{d}\vec{x}}{\int_R Exp_t[o](\vec{x})\, \mathrm{d}\vec{x}}, \tag{2}$$

and set the descriptor position to $\hat{o}$. In the former case the position is a fuzzy subset of R, rather than a single point; correspondingly, all the quantities that depend on it (e.g., the distance from the robot) are fuzzy quantities, and must be considered as such in the evaluation of the control rules. In the next section, we present an example of this evaluation, and show that there are good reasons not to use (2).

In practice, the evaluation of (1) can be computationally prohibitive, and some simplifications have to be made. We have written a sample implementation where fuzzy locations are restricted to be triangular — i.e., the projection on each axis of M has the shape shown in Fig. 2 (a) — and (1) is approximated by simple operations on the extreme points of the triangles. With these assumptions, the self-localization and prediction algorithms can be evaluated in a few milliseconds.

EXPERIMENTS

Figure 4 shows an experimental run of the above technique on a small mobile robot. Fuzzy locations are drawn as ellipsoids with radiuses proportional to the width of the corresponding component; the width of the θ component (orientation) is represented by an arc oriented as θ. The right windows show the robot's internal map of the environment, including the robot's self-location hypothesis H_t. The cross indicates the real position of the robot, measured by hand. The left windows show the state of the controller as a Cartesian representation of the space around the robot (in top view, facing up); the small dots indicate readings from the sonar sensors: the real walls can be easily seen from these readings.

In (a), the robot believes that it is approaching the end of one corridor, although it is actually further ahead, and is starting to turn left to enter the next one. The right window shows that the H_t hypothesis is reasonably sharp (and correct) along the X and Θ axes, but pretty vague (and off) along Y: this is because previous self-localization steps used the corridor walls to update the position along X and Θ, but could not use any clue to correct the Y coordinate. On the left, the two ellipsoids indicate the positions of the descriptors of the walls of the next corridor (centered at the wall end), built from their positions in the map (which here is crisp) and from H_t using an approximation of (1). The positions are extremely vague in the Y component as a consequence of the uncertainty in H_t. Correspondingly, the robot's estimate $dist_R$ of its distance from the mid-line of the next corridor is also vague. Figure 3 plots the fuzzy variable **midline-dist**, together with the **close-to-midline** predicate used in the first two rules in Fig. 1. By intersecting these two fuzzy sets, we obtain a truth value of 0.4 for the antecedent of the first rule, which makes the robot turn smoothly left. Note that the center of **midline-dist**, $\hat{d}$, lays outside from the **close-to-midline** predicate: if we had used (2) to set the descriptor position, the robot would have not started the turning maneuver now, and it would have not been able to engage the new corridor.

In (b), the robot has entered the new corridor, and uses the wall descriptors to compute the values of the fuzzy variables **left-wall-dist** and **right-wall-dist**. Given these

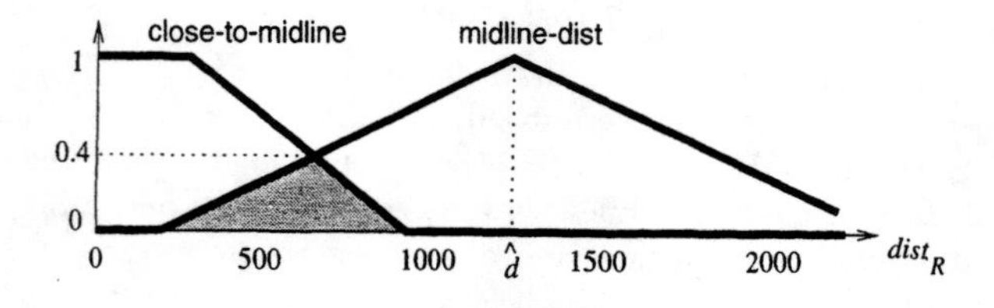

Figure 3: Evaluation of the **close-to-midline** fuzzy predicate.

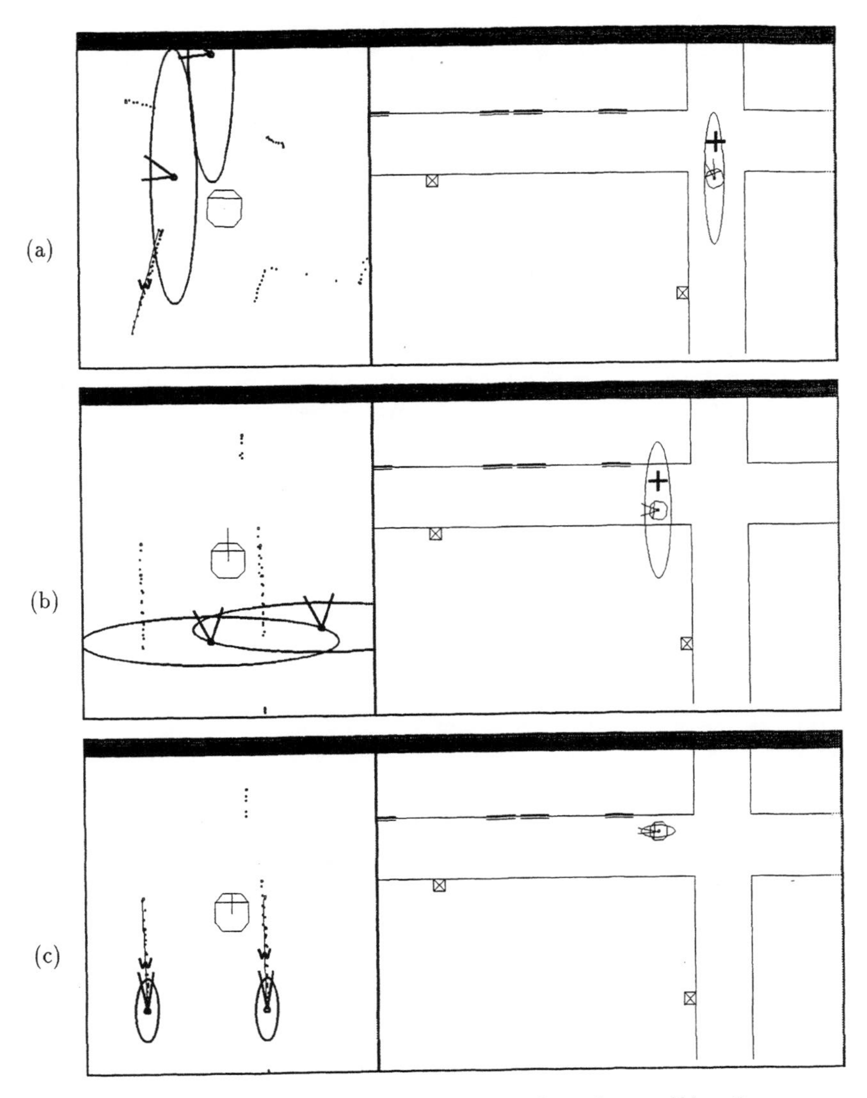

Figure 4: An example of corridor switching using a fuzzy self-location.

values, the corridor following rules generate a command to turn smoothly right. These rules are combined with the obstacle avoidance ones, that try to keep the robot far from the sonar returns produced by the right wall, resulting in a straight trajectory close to the right wall. Finally (c), the corridor walls are recognized ("W") and used to update the position of the wall descriptors (left). The new observation is also used by the localization algorithm to update the self-location hypothesis H_t (right).

CONCLUSIONS

We have presented a technique for incorporating map information into a fuzzy controller to support real-time indoor robot navigation. This technique can be used when the robot's estimate of its location is approximate. We represent locational uncertainty by fuzzy sets; this contrasts with most current approaches to self-localization which are probability-based [1, 6]. Probabilistic techniques typically assume that errors are normally distributed, and that they are small enough to justify linear approximations; recent results suggest that fuzzy techniques can be effectively used in situations when these assumptions cannot be guaranteed [2], [5].

Our technique is still preliminary, and several points remain for future work. The most urgent one is probably the development of a less restrictive, but still computationally tractable, representation for fuzzy locations. Considering locations which are not necessarily triangular brings about the challenging problem of how to deal with situations in which H_t, is ambiguous, i.e., in which several disjoint locations are possible for the robot — which makes the expected locations ambiguous as well.

REFERENCES

1. P. Moutalier and R. Chatila. Acquisition of beliefs through perception. In *Stochastic multisensory data fusion for mobile robot location and environment modeling*, pages 207–216, Tokyo, Japan, 1989.

2. G. Oriolo, G. Ulivi, and M. Vendittelli. Potential-based motion planning on fuzzy maps. In *Second EUFIT Congress*, pages 731–735, Aachen, D, 1994.

3. A. Saffiotti, K. Konolige, and E. H. Ruspini. A multivalued-logic approach to integrating planning and control. *Artificial Intelligence*, 76(1-2):481–526, 1995.

4. A. Saffiotti, E. H. Ruspini, and K. Konolige. Blending reactivity and goal-directedness in a fuzzy controller. In *Procs. of the 2nd Fuzzy-IEEE Conf.*, pages 134–139, San Francisco, CA, 1993.

5. A. Saffiotti and L.P. Wesley. Perception-based self-localization using fuzzy locations. In M. van Lambalgen, editor, *Reasoning with Uncertainty in Robotics*. LNCS, Springer-Verlag, Berlin, Germany, 1996. To appear.

6. R. Smith, M. Self, and P. Cheeseman. Estimating uncertain spatial relationships in robotics. In I. Cox and G. Wilfong., editors, *Autonomous Robot Vehicles*. Springer-Verlag, 1990.

7. L. A. Zadeh. Fuzzy sets. *Information and Control*, 8:338–353, 1965.

8. L. A. Zadeh. Fuzzy sets as a basis for a theory of possibility. *Fuzzy Sets and Systems*, 1:3–28, 1978.

A MODEL OF THE FUTURE ENTERPRISE - MULTI-FUNCTIONS INTEGRATED FACTORY

LIANGXIN SI and P. HERBERT OSANNA
Institute of production engineering(IFT), vienna university of technology
Karlsplatz 13/3113, A-1040 Vienna, Austria

ABSTRACT

Multi-functions integrated factory (MFIF) is an innovative concept and a new model for future enterprises developed to meet demand for cost-effective and customer-driven multi-functional product(MFP). In this paper, an overview about models, conceptual architecture and working mechanism of the ICAx systems in MFIF which have the features of concurrent, interactive, collaborative, modular, integrative, learning, autonomous, self optimising and self organising functions, are given. Internet, distributed computing environment (DCE) and parallel-processing technologies and STEP standard which have been partially used and developed are bases to realize MFIF.

KEYWORDS: MFIF, MFP, IMS, enterprise automation, enterprise integration, advanced data exchange

INTRODUCTION

To meet high-level demands for comfortable daily life in the future, manufacturing enterprises must be flexible and agile enough to quickly respond to product demand changes, and new models and configurations for future manufacturing systems and enterprises in which IMSs are usually applied need to be investigated.

Multi-functions integrated factory (MFIF) is an innovative concept and model for future enterprises and is initiated with the aim to provide cost-effective, agile and optimum ways to produce customer-driven multi-functional products (MFPs) in near future, based on intelligent production technology, information highway, distributed computing environment (DCE) technology, parallel-processing computing and advanced engineering data exchange techniques.

MFIF (MULTI-FUNCTIONS INTEGRATED FACTORY)

By means of information highway (internet), DCE and parallel-processsing technologies and advanced product data representation and exchange model STEP, factories which produce cars, aircrafts and ships respectively, for instance, could be linked to each other to form a new kind of factory with all three functions (Figure 1) according to needs. The product - MFP would be so produced that the different function tasks of the product should be manufactured in adequate function factory, and then integrated together to realize the combination of the functions. The factory works by using its advantages of multi-functions, and high efficiently and agilely produces low cost customer-driven multi-functional products.

Such MFIF has the potential to improve industrial competitiveness, fully manufacturing automation and optimally to manufacture the customer-driven MFP worldwide. Intelligent manufacturing systems (IMS) are the basis for realization of MFIF. In MFIF individual functional enterprises are functionally and configurationally integrated with other functional enterprises located in world wide for producing corresponding MFPs.

MFIF will be exist and stepwise realize because of the following reasons:

- In order to survive on the future market and to meet demands for comfortable future daily life, the manufacturers and factories would like to integrate together to meet customer-driven production and to enhance the market competition.
- Internet application such as World Wide Web has been utilized worldwide in past few years. DCE environment that enables networked computers to share data and services efficiently and securely has been developed [4] and partially used, and will be widely applied in engineering automaton.
- Significant process has been made in integrating parallel-processing computers into PC- and workstation-based engineering networks, and efforts to rewrite software codes are beginning to take off [1].
- In near future enterprises or concerns will be not located in one contentrated place, but distributed worldwide according to production functions and needs for world marketing.
- The different enterprises which before produce cars, planes and boats respectively, for example, could be easily integrated by means of powerful international networks or information highway, using DCE technology and new product data representation and exchange model - STEP.
- Mutimedia, telecommunication and concurrent engineering technology will make simultaneous interactive on-line information exchange of all production processes possible and comfortable, being used with applications of internet and DCE technology.
- Intelligent machines, cells and systems will be widely used in manufacturing production environments [2][6][7][8][9][10][11].

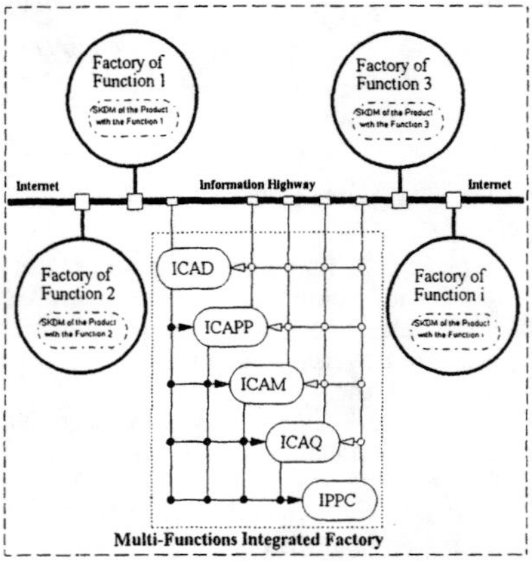

Factory of Function 1
SKDM of the Product with the Function 1
Factory of Function 3
SKDM of the Product with the Function 3
Internet
Information Highway
Internet
Factory of Function 2
SKDM of the Product with the Function 2
ICAD
ICAPP
ICAM
ICAQ
IPPC
Factory of Function i
SKDM of the Product with the Function i
Multi-Functions Integrated Factory

Figure 1. Model of Product Data Exchange between ICAx Systems in MFIF

Features of MFIF

One feature of MFIF is the use of cross-functional design and manufacturing production teams, in which engineering staffs with different skills and expertise work concurrently, collaboratively and interactively together on a MFP project.

MFIF is based on the assumption that it works only under the condition that the each single-functional factory has a possible full-scale IMS working environment and is an integration of intelligent manufacturing machines, cells and systems. Concurrent, interactive, collaborative, modular, integrative, learning, autonomous, self optimising and self organising functions are the main features of MFIF.

The production information sources and systematization knowledge for design and manufacturing (SKDM) of each factory in MFIF could be provided to the other companies. But the function-oriented intelligent manufacturing systems and components can not easily be exchanged, so that different function parts of the product should be produced in identical function skill factories. It is hoped that the benefits of individual function skills employed by the companies could be optimally utilized and integrated.

The factories are reconfigurable to take advantages of agile manufacturing production for the MFPs. The MFIF provides a function-business-shared feature to create new customer-driven markets.

Failurefree concurrent exchange of production information and data, concurrent processing and executing of production processes through distributed computing environment DCE, STEP or new standard, and learning of the all collaborative production processes are also the features of MFIF.

MODEL OF INTELLIGENT DESIGN AND MANUFACTURING SYSTEMS

MFIF is controlled and arranged by collaborative activities between the individual factories. Cooperative activities between all units of the factories can also be concurrently on-line carried aut. Learning is carried out step to step by using the methods of evolution, and used to optimize process control. It possible that all the systematization knowledges for design and manufacturing (SKDM) of each factory in MFIF and all production information between ICAx systems of MFIF can be simultaneously on-line exchanged (as shown in Figure 1).

ICAD (Intelligent CAD)

In MFIF, the design tasks of MFP can intelligently, interactively and collaboratively be carried out. Distributed design system [5] is used for multi-functional product design. Using DCE and parallel-processing technologies, the Engineers can work on parts of a design task, but the content of the design as a whole is a corporate resource to be managed and secured. Each group has only the right to modify the relevant parts that they have responsibility. This system makes it possible that the product designers of different function factories can parallelly work on the all subtasks of the product (Figure 2).

Collaborative design working method in MFIF has the goals to realize not only electronic data exchange function but also an interactive working function on-line, and to work at the different places and at heterogeneous systems out the same product model. Transmission of words, figures and sound by means of multimedia will also be integrated in such iinteractive CAD systems which could recognize handdraft drawings, learn the design process of the product, even understand the natural language instruction for the design, and optimize the process.

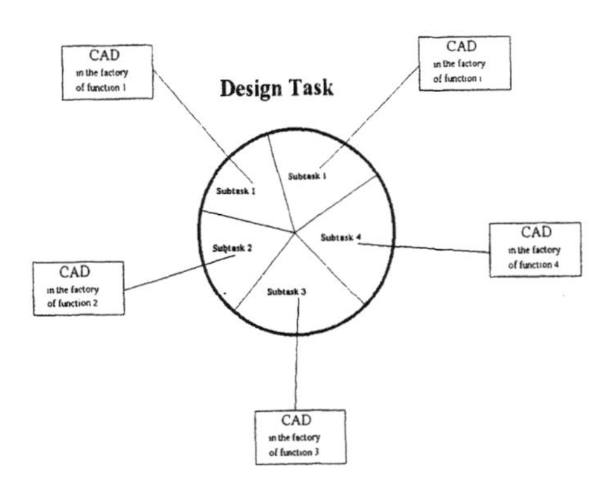

Figure 2. Distributed Design Model of MFP

The unique data structure of the CAD systems in MFIF is necessary. It provides full associativity among all engineering disciplines, tying together the entire design-through-manufacturing of a MFP. MFIF design system could be linked with the customers and suppliers through information highway, in order to achieve whole interactive design process of the product with the outside programs.

An effective use of analysis, simulation, and visuallization tools takes much advantages for MFP design. In this system the designers from the different single-functional enterprises typically use parallel-processing, virtual reality and virtual prototyping technologies, to design and simulate the customer-driven multi-faunctional products and their systematical function activities as well as to create a fully digital MFP production and entire manufacturing process programs.

ICAPP and ICAM (Intelligent CAPP and Intelligent CAM)

The ICAPP system operates collaboratively between process planning systems of all the factories and is a generative process planning system using feature-based workpiece models, and generates concurrently and collaboratively the process plans which are used for manufacturing processes of the adequate function parts of the product in the factories.

The collaborative process planning deals mainly with exchanges of production information, knowledge data, actual process planning status, integration of the plans, process planning of interface-part tasks and hybrid-functional process planning etc.. In order to get automation of the collaborative process planning, and to optimise the process planning through itself, learning method for process planning should be used. The ICAPP components of the factories in MFIF link process steps to the design model and other systems, so manufacturing informations are automatically regenerated when the design is changed.

In the ICAM system, the manufacturing process informations and programms would also be on-line concurrently, interactively and collaboratively exchanged and modified. It runs autonomously according to the adequate functional task and organizes all manufacturing activities and units optimally in adequate factories. Learning the processes from the processes, the ICAM units improve the all manufacturing process parameters continously. Intelligent flexible manufacturing system in MFIF is an autonomous FMS. Industry robot will be an autonomy [6], and computer aided manufacturing system will be an autonomous production system.

Assembly. This process is a final manufacturing step for MFPs in MFIF. The coupling areas of the different function parts of MFP are produced for ease-assembling, and they could be very easily combined and integrated together in one of the individual factories which is near to the customer.

ICAQ (Intelligent CAQ)

The quality assurance process would be used in all the product production processes - from the design to the assembly. There would be no final quality assurance process for assembled MFPs.

In order to realize the automatic quality assurance and to deal with the complex, variable and dynamic quality assurance problems of the production processes in MFIF, quality assurance system will be greatly enhanced through self-optimising processes thrust in the design system and all manufacturing production processes in MFIF. Learning with self improving ability makes the way to "Zero Error" production. A method has been developed enabling the supervision of quality in the process chain as well as its optimization by means of a knowlwdge and neuronal-network-based learning management system, and a self-learning system with neuronal networks has been realized [10]. The method can be used in MFIF and permits to learn stepwise from deviations and to improve the processes continuously.

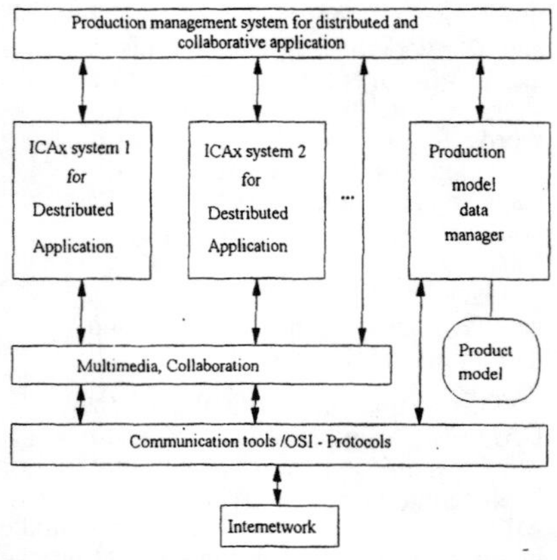

Figure 3. Model of Distributed and Collaborative ICAx Systems

Data Comunication in MFIF

Internet, DCE technology and advanced STEP model for product data representation and exchange make it possible to establish a global information highway for simultaneous on-line exchange of production data, for collaboration and interaction on the design and all the production processes, and for concurrent communication of all the systematization knowledges for design and manufacturing of each factory in MFIF, and to interact with the factory's suppliers and customers in world wide. Mutimedia and telecommunication make the interactive transmission of words, Figures, sound and wireless communication in MFIF possible and comfortable.

STEP. By means of STEP standard or its new version failurefree communication and concurrent exchange of production data would be realised in MFIF. A common product data representation and exchange model and common product database at any stage of the product life cycle, from its design, manufacture, use, maintenance to disposal are provided by STEP.

Parallel-processing computing. High-performance parallel-processsing computers function well as application servers and database servers, and they can run multiple programs or support X terminals that require additional processors [1]. The system enables engineers to model physical phenomena and to simulate such phenomena far more realistically than has ever been possible.

DCE technology. Furthermore, the level of cooperation and the depth of information exchange have to be defined to compile such a MFIF, software tools are required to support this model of cooperation among units of one or more single-functional enterprises. This software tools are:

- distributed computer environment and networks, to allow data exchange among the components.
- distributed data and knowledge bases, to share data and knowledge for cooperative design and production processes.

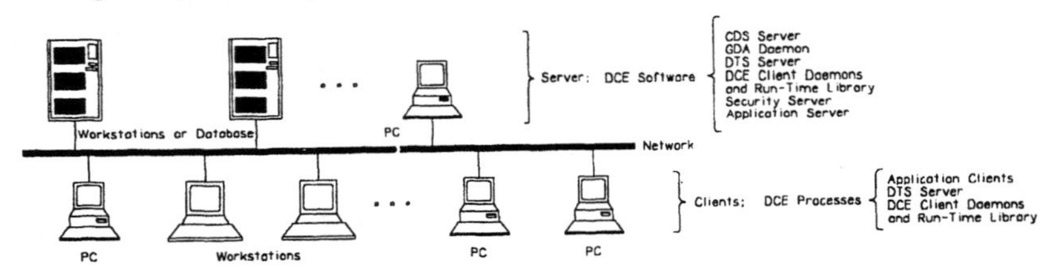

Figure 4. A Single Functional Enterprise (DCE Cell) Configuration for Data Communication in MFIF

DCE technology was developed to fill the need for a standardized approach to creating and executing secure client/server applications in complex, highly networked environments[4]. DCE is based on the client/server model in which an application's functionality is divided between clients and servers. This technology can be used to realize the complex collaborative and interactive communication, resource sharing and all inter-enterprise business activities and security. Applications developed using the DCE software system are portable and interoperable over a wide range of computers and networks.

A single-functional enterprise - a DCE cell. In a DCE environment, there might be several thousands host systems, some of which might be from different vendors, and many different categories of users and applications. To deal with this heterogeneous and diverse environment, DCE defines a basic unit of operation and administration called a cell, which allows users, systems, and resources to be grouped together according to their needs and mutual interests. In MFIF, a single-functional factory can be seen as a typical DCE cell. A typical DCE cell can span several systems and networks.

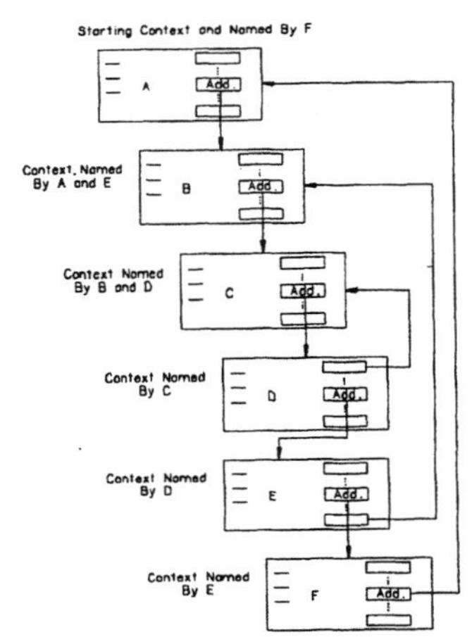

Figure 5. Model of Inter-Enterprise Global Naming System in MFIF

Global naming system. To find users, files, devices, and resources inside and outside the single-functional enterpriseas requires a naming system that allows each enterprise and the objects contained inside it to have unique names and a directory service that can cope with different naming systems. A global naming system model used in MFIF is illustrated in Figure 5.

The major technologies and features in DCE are included [4]:

- Threads;
- Remote Procedure Call (RPC);
- Security;
- Cell Directory Services (CDS);
- Global Directory Service (GDS);
- Distributed Time Service (DTS).

Although there exist several DCE products developed
for DCE environment and they are suitable for the MFIF application, but the advanced DCE application products have to be developed to fulfil the future needs for MFIF.

Fuzzy Logic as A Kernel in MFIF

In MFIF there are many inter-functional or cross-factorial problems and tasks to be solved. The design and all manufacturing production tasks of interface-parts of MFP can not clearly and easily be represented and distributed through precise statements or numerical values and other methods. Fuzzy logic could be used to model such gray areas. It enables to define functional system structure of MFP and to distribute functional tasks for interface-parts of MFPs. Fuzzy logic can also be used for distribution of benefits obtained through the integration of the factories, for collaborative activities among the factories, and other inter-functional and inter-factorial behaviors.

CONCLUSIONS

MFIF is a innovative concept and a new model for future enterprises developed to meet demand for cost-effective customer-driven MFP design and manufacturing, to realize agile and optimal manufacturing production. An overview about models and conceptual architecture of the ICAx systems in MFIF which have the features of concurrent, interactive, collaborative, modular, integrative, learning, autonomous, self optimising and self organising functions, are given.

Internet, DCE and parallel-processing technologies and STEP standard which have been partially used and developed are bases to realize MFIF. Fuzzy logic will be as a kernel for the task distribution, collaborative task execution, obtained benenfit distribution and intelligent autonomous manufacturing units in MFIF. It should be pointed out that MFIF will be stepwise realized in near future.

REFERENCES

1. Deitz, D. "Parallel Processors for The Work Patch." *Mechanical Engineering* (Oct.. 1995), 58-65.
2. Groumpos, P.P. "The Challenge of Intelligent Manufacturing Systems (IMS): The European IMS Information Event." *J. Intelligent Manufacturing* (June, 1995), 67-77.
3. Gupta, M.M. and Rao, D.H. "On The Principles of Fuzzy Neural Networks." *Fuzzy Sets and Systems* 61 (1994), 1-18.
4. Kong, M.M. "DCE: An Environment for Secure Client/Server Computing." *Hewlett-Packard Journal* (Dec., 1995), 6-15.
5. Krause, F.-L., Kiesewetter, T., and Kramer, S. "Distributed Product Design." *Annals of the CIRP* 43(1)(1994), 149-152.
6. Milberg , J. and Koch, M. R. "Autonome Fertigungssyteme." *VDI-Z* 135 (Nov./Dec., 1993), 63-69.
7. Si, L. and Osanna, P. H. "Intelligent Manufacturing Systems 2000." *Preprints of 2nd international CIRP workshop Learning in IMS*, Edited by L. Monostori. Budapest (1995).
8. Si, L. and Osanna, P. H. "Multi-Functions Integrated Factory." *Proceedings of 11th ISPE/IEE/IFAC international conference on CARS&FOF'95*, Edited by H. Bera. Colombia (1995).
9. Warnecke, H. J. und Hüser, M. "Selbstorganisation im Produktionsbetrieb." *ZWF* 90 (Jan./Feb., 1995) 12-15.
10. Westkämper, E. "Zero-Defect manufacturing by Means of A Learning Supervision of Process Chains." *Annals of the CIRP* 43(1)(1994), 405-408.
11. Yoshikawa, H. "Systematization of Design Knowledge." *Annals of the CIRP* 42(1)(1993), 131- 134.

HYBRID DYNAMIC MODELLING AND ADAPTIVE KALMAN FILTERING: APPLICATION TO MANEUVERING TARGET TRACKING

C. Sarrut and B. Jouvencel
Laboratoire d'Informatique, de Robotique
et de Microélectronique de Montpellier
UM 9928 CNRS/UMII
161 rue Ada , 34392 Montpellier Cedex 5, France
Phone: (+33) 67 41 85 60, Fax:(+33) 67 41 85 00
e-mail: sarrut@lirmm.fr, jouvence@lirmm.fr

ABSTRACT

In target tracking, an estimator is used to determine the kinematic components of a target (position and velocity) from measurements. Generally, the estimator is based on Kalman filtering. In mobile robotics, estimation processes are used in sensor based robot control, for relative location, obstacle avoidance or tracking. But in both topics, the unpredictable aspects of mobile dynamics require the adaptivity of estimators. In target tracking the most important problem to solve is caused by maneuvers, corresponding to acceleration occurence. In mobile robotics the estimation problems are different but we believe, that solutions developed in target tracking research could be transposed. This paper presents the methods classically used in target tracking to avoid estimation divergence. An approach based on hybrid modelling is presented in more details. We discuss about the interest of this approach. A non linear process of estimation by neural network is also proposed.

KEYWORDS: Mobile robotics, Hybrid dynamic, Kalman filtering, Neural network

INTRODUCTION

In mobile robotics, when the movements take place in dynamic environments, many tasks like obstacle avoidance need the knowledge of relative movement between the mobile robot and the object. The movement components (position, velocity) are given by a state estimation process. Then these components are used by a robot sensor based control. In such situations, the estimation process is the same as the one used in target tracking.

Two principal types of problems are then encountered. First, the target is non-cooperative and its dynamics is unknown. Second, the mobile robot movements are neither well-known not well-controled (slidings, skiddings).The difficulty of relative movement modelling is then evident, especially when there is no hypothesis on the linearity or stationnarity of this movement.

Usually, estimators are based on Kalman filtering which use the linear Gaussian markovian formalism (LGM).The system is described with discrete-time equations like:

$$x_{k+1} = Fx_k + w_k \tag{1}$$
$$y_k = Cx_k + v_k \tag{2}$$

where x_k represents the state vector at time k, w_k and v_k are independant zero-mean white Gaussian noises and F and C are respectively state evolution and measurement matrices. However, the estimation accuracy depends on the reliability of the a priori chosen evolution model F. When the movement variations of the target causes the model disability, the bandwidth of the filter must be enlarged to prevent errors from increasing. In other words, it is necessary to adapt the filter. In case of movement in a plane X,Y, and with a state vector x :

$$x = (posX , speedX, \quad posY, speedY)' \tag{3}$$

the movement corresponds to a constant-velocity kinematic model. The disturbance w_k allows to compensate for slight changes the trajectory. But, if the variations are too large, it is convenient to associate the phenomena to an acceleration appearance. In order to avoid estimation divergence, many methods have been proposed. The most known and used ones are presented below.

Additive stochastic disturbance

The most classical and simplest method consist of adding a stochastic disturbance to the evolution model. This approach assumes that all modelling errors can be compensated by means of an overestimated noise.The increase of the noise covariance generates high filter gains The essential drawback of this method is the bad accuracy of estimates when there is no large changes in trajectory. Adjustable white noise level can be used to avoid this phenomenon. A simple detection process of the occurence of sudden changes on the innovations permits to increase the noise level via a multiplicative coefficient. However, in single-model filtering, the most used solution has been presented in [5] and is based on acceleration modelling with colored noise.

Multiple-model filtering

The evolution model reliability has also been treated with multiple-model approach. The method presented in [1] consists of assuming two or more levels of process noise. Several models of the form (2) are postulated, and a filter is set up for each one. With likelihood fonctions, the probability for each model to be correct is calculated. The second approach, also presented in [1], considers that modelling the acceleration with noise is not sufficient. These different models must be coherent with the real movement. So, the idea is to have two filters with different orders.The 2nd-order filter (constant-velocity model) is used during permanent regime, and when an acceleration occurs, the estimator switches to the 3rd-order filter (constant-acceleration model) with the state vector:

$$x = (posX , speedX, accX \ posY, speedY, accY)' \tag{4}$$

The evolution of target performance and the appearance of new types of sensors have generated new approaches. In the maneuvering target tracking domain, the method proposed in [2], [3] and [4] considers the target maneuvers as discrete events. and uses the hybrid dynamics modelling to make a fast detection of this event. The employed formalism and the method are rapidly exposed below.

Hybrid dynamics modelling - Interactive Multiple Model algorithm.

The common hybrid system equation is:

$$dx_t = Ax_t \, dt + B_\phi(x_t) \, d\Phi_t + dv_t \tag{5}$$

where $B_\phi(x_t) \, d\Phi_t$ represents the discret term (maneuvers), Φ_t is called regime or state indicator, with $d\Phi_t =1$ if a maneuver occurs and 0 otherwise.
The target trajectory is considered as a succession of sequences corresponding to the possible movements of the target. These movements, in finite number, are called regimes and depend on target capabilities (2D motions for mobile robots). Each regime represents the "continuous" component of the model, and constitutes the finite states of a Markov

chain..The jumps between regimes are discrete events, appearing when a target maneuver occurs.The dynamics of these transitions are modelled with the equation:

$$d\Phi_t = \Pi' \Phi_t + dz_t \tag{6}$$

with $\Pi = (\pi_{ij})_{i,j=1 \text{ to } m}$, the transition rate matrix, m the number of regimes, and dz_t a martingale.

The resulting system is made of two evolution equations , one for the target state , the other concerning the regimes transitions. In [4], the author introduces an imaging sensor to detect discrete transitions with a regime filter. The m interacting Kalman filters on state are driven by the regime filter.

APPROACH

In order to transpose these results to mobile robotics, we consider the problem differently. The process we want to estimate, relative movement between the mobile robot and a target, can be non linear.

Motion estimation

For a given regime, the evolution of the process is described in discrete-time with:

$$X_{k+1} = \mathbf{F}(X_k) \tag{7}$$

where **F** is a non-linear fonction of the state vector X. To estimate the process X with the Kalman filter formalism, it is necessary to use the extended Kalman filter. It consists of linearizing the state equation around the current estimate with a Taylor development. This operation amounts to use the jacobian $\mathbf{J_F}$ to make a local linearization. Since $\mathbf{J_F}$ must vary in time to be coherent with the real values of **F**. It is essential to detect variations occurences in order to modify $\mathbf{J_F}$. Actually, the knowledge of $\mathbf{J_F}$ is impossible. The only solution is then, to choose an a priori model for the evolution called F and to make a local approximation with it.

Maneuvers estimation

A constant velocity kinematic model, common in target tracking, remains localy reliable until a too large variation in the evolution appears, generating an estimation error. The models of the different Kalman filters used in IMM algorithm are specific. In [4], these are two pure rotation movements (depending on known planes capabilities) and a constant-velocity model. In our case, we simply choose two different types of constant velocity models for the state filters and the regime filter.

As presented before, there are two state evolution state. The measurements being hybrid (detection and state), the observation process is decomposed in two equations. The final system is:

$$\begin{cases} xi_{k+1} = Fi\, x_k + B_\phi(x_k)\, \Phi_k + v_k \\ \Phi_{k+1} = \Pi'\, \Phi_k + z_k \end{cases} \tag{8}$$

$$\begin{cases} y_{xk} = C_x\, x_k + w_{xk} \\ y_{\Phi k} = C_\Phi \Phi_k + w_{\Phi k} \end{cases} \tag{9}$$

with i, the number of the Kalman filter considered, C_x and C_Φ respectively the state and regime measurement matrices and w_{xk}, $w_{\Phi k}$ independent martingales

The estimates resulting from the regime filtering process are the probabilities of each regime to be correct. The coefficients in C_Φ are determined to the means of the different state variables in each regime. If $R^{\phi\phi}{}_{k/k}$ is the error covariance on regime estimate, and $R^{w\phi}$ the variance of w_ϕ, the discrete-time regime estimation algorithm is given by:

1) prediction $$\Phi_{k+1/k} = \Pi' \, \Phi_{k/k} \tag{10}$$

2) correction $$\Phi_{k+1/k+1} = \Phi_{k+1/k} + G_{k+1} \, (y_{\Phi k+1} - C_\Phi \, \Phi_{k+1/k}) \tag{11}$$

with the gain G_{k+1}

$$\begin{aligned} G_{k+1} = [\, \mathrm{diag} \, (\Phi_{k+1/k}) - \Phi_{k+1/k} \, \Phi'_{k+1/k} \,] \, x \\ C'_\Phi \, [\, R^{w\phi} + \ C_\Phi \, \mathrm{diag} \, (\Phi_{k+1/k} \,) \, x \\ C'_\Phi - C_\Phi \, \Phi_{k+1/k} \, \Phi'_{k+1/k} \, C'_\Phi \,]^{-1} \end{aligned} \tag{12}$$

and the covariance of the error estimation

$$R^{\phi\phi}{}_{k+1/k+1} = \mathrm{diag} \, (\Phi_{k+1/k+1} \,) - \Phi_{k+1/k+1} \ \Phi'_{k+1/k+1} \tag{13}$$

Motion estimation updating

In state estimation, regime probabilities are used at two steps. After the prediction step, the state estimates and the error covariance of each filter are modified.

$$x^+{}_{k/k} \, (i) = \sum_{j=1}^{m} \pi_{ji} \ (\Phi_{k/k} \, (j) \, / \, \Phi_{k+1/k} \, (i) \,) \, x_{k/k} \, (j) \tag{14}$$

$$\begin{aligned} R^{xx+}{}_{k/k} \, (i) = \sum_{j=1}^{m} \pi_{ji} \ [\, (\Phi_{k/k} \, (j) \, / \, \Phi_{k+1/k} \, (i)] \ [\, R^{xx}{}_{k/k} \, (j) + \\ (x_{k/k} \, (j) - x^+{}_{k/k} \, (i) \,)^2] \end{aligned} \tag{15}$$

The prediction and correction steps for the state Kalman filters are similar to those of the extended Kalman filter ones. The final state estimate is obtained with:

$$x_{k+1/k+1} = \sum_{i=1}^{m} \Phi(i) \ x_{k+1/k+1}(i) \tag{17}$$

The structure of the entire estimation process is summarized below in figure 1.

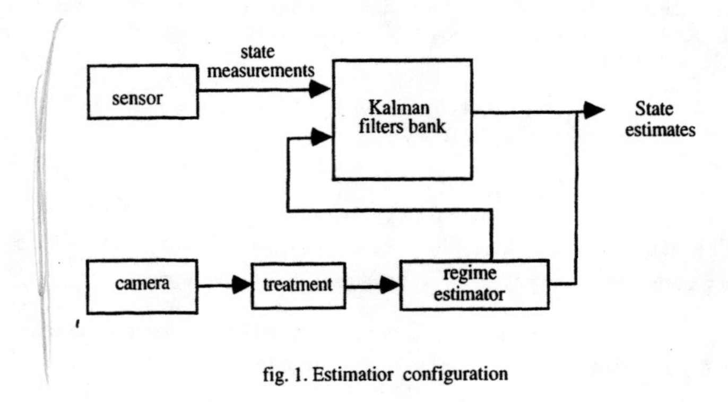

fig. 1. Estimatior configuration

Jacobian estimation

We want to underline the importance we give to the determination of the jacobian J_F of the real evolution fonction **F**. Our goal is to solve the problem encountered with the lack of knowledge on J_F. This problem is commonly avoided with the choice of an a priori model which has a time limited reliability. The method we want to introduce consists in a non-linear estimation of the jacobian with a neural network. This leads to an adaptive non-linear Kalman filter with neural learning of the model.
If consider the ith state equation in the bank of filters in the figure 1, the figure 2 the neural network scheme, allowing the learning of J_F , and applied to each Kalman filter.

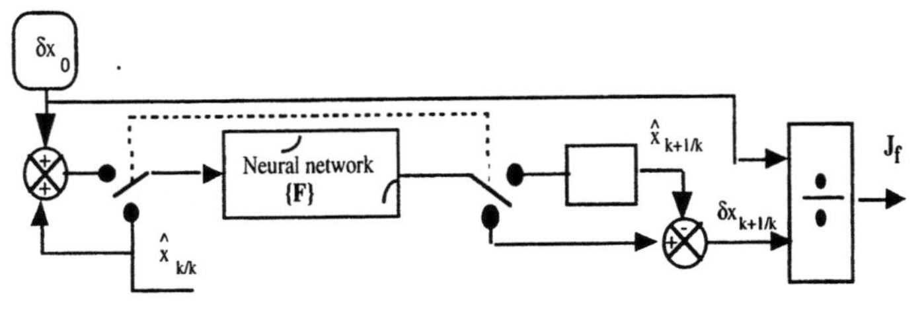

fig.2 Jacobian learning scheme

The figure 3 presents the entire Kalman filter with neural learning.

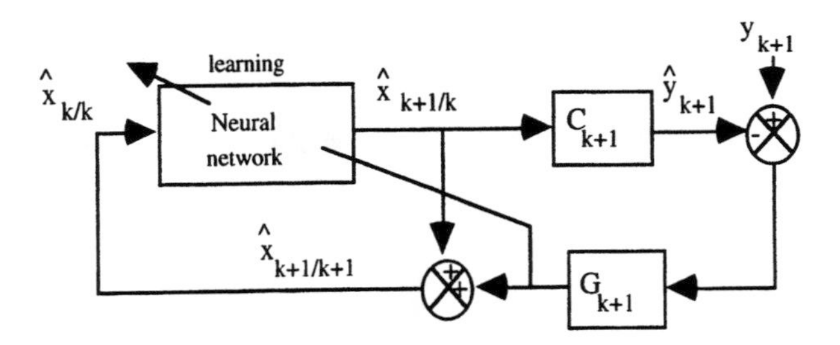

fig. 3 Structure of the proposed Kalman filter

EXPERIMENTAL SITE

The algorithms have been implemented on a transputer™ network and validations are in progress. The connection with the sensorial system constituted of a laser telemeter and a fast integrated camera image processing will be made in future.The telemeter is motorized to make azimuthal measurements. The camera is also equipped with two actuators which axes coincide with the vertical and horizontal image axes. These two sensors are mounted on a six wheel drive robot. The control of the sensorial system and the estimation process will be made on the transputer™ network. Our aim is to implement on the robot an algorithm allowing relative movement estimation. The camera will be dedicated to detect sudden changes (discrete events). The telemeter will give measurements on range and azimuth to obtain state estimates.
The figure 4 presents, first results obtained with a Kalman based on neural learning.

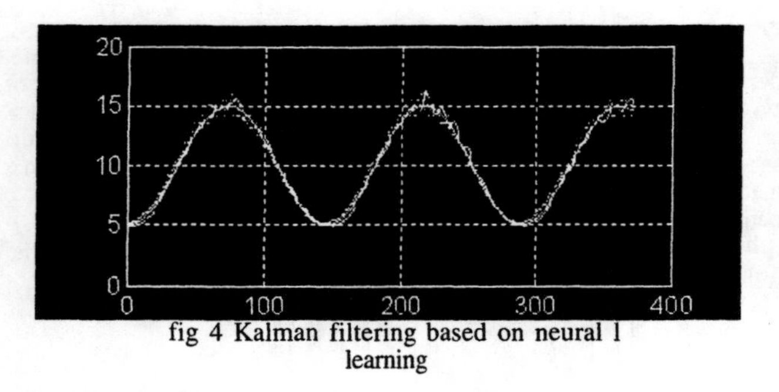

fig 4 Kalman filtering based on neural l learning

CONCLUSION

We have described in this paper the approach of the IMM algorithm adapted to our mobile robotics problems. We have also discussed the interest of using neural network to estimate the real evolution fonction. A direction to be explored, is the association of hybrid approach detection with an adapted Kalman filter based on neural learning. The detection of jumps into the real movements allowing to correct the estimated jacobian. Future research will consist of validating this algorithm on a six wheel drive robot.

REFERENCES

1 . Y.Bar-Shalom and T. E.Fortmann, *Tracking and Data Association,* Academic Press, Orlando, 1988

2 . H. A. P. Blom, *A Sophisticated Tracking Algorithm for ATC Surveillance Data,* Proceedings of the International Radar Conference, Paris, May 1984.

3 . H. A. P. Blom, *An Efficient Filter for Abruptly Changing Systems*, Proceedings of 23rd Conference on Decision and Control, Las Vegas, December 1984.

4 . M. Mariton, *Maneuvering Target Tracking: Overview and the Role of Imaging Sensors,* Treitement du Signal, Vol.10, n°2, pp 110-137, 1993.

5 . R. A. Singer, Estimating Optimal Tracking Filter Performance for Manned Maneuvering Targets, IEEE Transactions on Aerospace and Electronic Systems, Vol. AES-6, n°4, pp. 473-483, July 1970.

ARE SONAR RANGE ERRORS A SOURCE OF INFORMATION FOR OBJECT IDENTIFICATION?

ANGELO M. SABATINI
ARTS–Lab, Scuola Superiore S. Anna
V. Carducci, 40
56127 Pisa, ITALY

ABSTRACT

A method is proposed for classifying a number of naturally occurring structures, e.g.,walls, corners, edges, and cylinders using an inexpensive multi–aural linear array of sonar sensors. We intend to exploit the fact that the threshold–based range estimation method used in our sensor system is intrinsically biased, and the bias is somewhat deterministic in nature. The influence of the measurement offsets on the results of a simple "goodness–of–fit" test in the course of a sonar scan can be used in the attempt to disambiguate the hypotheses concerning some of the geometrical features of interest. The experimental results discussed in this paper offer promise for considering measurement offsets as a potentially useful acoustic cue in the process of object classification.

KEYWORDS: acoustical imaging, statistical pattern classification

INTRODUCTION

In–air sonar sensing techniques and related methods of signal processing are currently investigated for several applications of interest in the field of Robotics; among them, obstacle detection [1], positioning from sensing known environmental features [2], mapping of unknown environments for robot navigation [3]. A serious shortcoming of sonar sensing is that most robots have to accomplish their tasks in open air, a medium which severely affects the propagation of acoustic waves in many respects. Investigation of signal processing and filtering methods for effective sonar data interpretation are therefore mandatory for advancing the state–of–the–art in the field.

So far, most researchers have attempted to perform some limited imaging tasks in environments where the specularity is prevalent [3]-[7]. Using physically based models of sonar sensors allows to design multitransducer configurations, where a number of spatially distributed transmitters and receivers cooperate for the purpose of discriminating a few selected objects of interest on the grounds of their geometrical properties. In accordance with a geometric feature model approach, we propose to exploit a quite unusual source of information for object classification, which involves one type of range errors that sonar sensors using threshold–based range estimation methods necessarily perform. In fact, threshold–based range estimators are intrinsically biased, and the corresponding measurement offsets are proven to change with the operating conditions of the sensors; consequently, the biases of the range estimates are somewhat "deterministic" in nature, in contrast to the stochastic nature of the range errors due to the thermal noise in the receiver front end. We develop a method for evaluating the influence of the measurement offsets on the results of a "goodness–of–fit" test, which is executed in the attempt to discover the identity of selected reflecting objects, such as walls, corners, and cylinders with finite or null radii. The classification system is built for the operation of a multi–aural configuration of sonar sensors (multi–aural sensing head, or MSH) [8].

DESCRIPTION OF THE EXPERIMENTAL SYSTEM

The multi–aural sonar device consists of a linear array of three sonar sensors, with center-to-center spacing d (d=15 cm, in the current version), see Fig. 1.

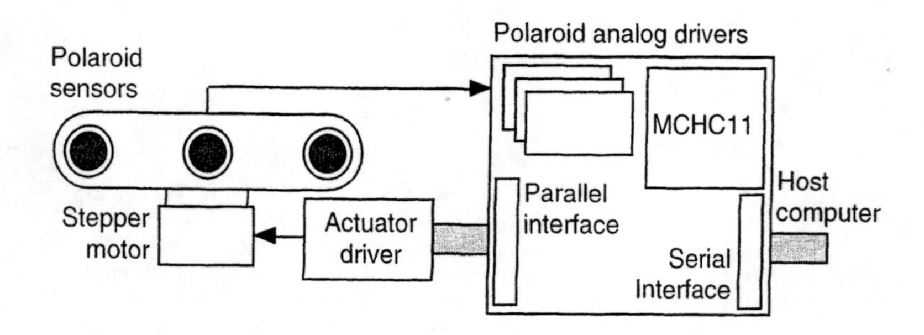

Fig. 1: A block diagram of the developed MSH, a servo–mounted linear array of Polaroid sensors.

The array is servoed by a stepper motor under computer control which can rotate the array with minimal scanning steps of about 0.6°. The sonar sensors are sequentially fired Polaroid electrostatic transducers; in our system they are used without any modification of the circuit board the sensor manufacturer provides together with the sensing element [9]. The Time-of-Flight (TOF) measurements are performed using a low–cost microcontroller of the Motorola HC11 family, connected to a PC486 via an RS232 interface [7].

An important point we would like to stress concerns the fact that in our implementation the complexity of the MSH is reduced to a minimum, at the expense of some interesting functionalities, e.g., multiple ranging capabilities, and exploitation of the Polaroid sensors as pure receivers are not considered here.

A FEATURE BASED GEOMETRIC MODEL

A sector scan, e.g., the collection of range readings that are obtained from a single rotating sensor placed at differently oriented sensing positions, takes approximately the form of a circular arc in case of reflections from a smooth planar surface of sizable extent; such an arc is sometimes called the region of constant depth (RCD) [2]. Sector scans of corner, cylinders, and edges give rise to RCDs as well, [4]. Refer to Fig. 2 for gaining a better understanding of the MSH operation in the case of reflections from a cylindrical beacon of radius R. Given that the three transducers fire in sequence, a whole measurement cycle consists of a vector of three range readings, the **r**–vector. In noiseless conditions, simple trigonometrical equations applied to the triangles *ABD* and *DBC* allows to relate the components of an **r**–vector to the polar coordinates of the point P (the signal reflection point for the central transducer) with respect to a sensor–based reference frame (B is the centre of the cylindrical reflector under examination):

$$\begin{aligned}(r_2 + R)^2 &= (\rho + R)^2 - 2\,d\,(\rho + R)\sin\theta + d^2 \\ (r_3 + R)^2 &= (\rho + R)^2 + 2\,d\,(\rho + R)\sin\theta + d^2\end{aligned} \qquad (1)$$

where $\rho = r_1$, and the angle θ give the distance and the orientation of a given feature. It is worthy noting that Eqs.(1) are valid for cylinders with finite radius R, for walls, e.g., cylinders with infinite radius, and for edges, e.g., cylinders with zero radius. Albeit the corner reflecting properties and the related sector scan appearance are explained in terms of an equivalent wall [4], Eqs.(1) are also valid for corners yielding R = 0 [8]. Of course, corners and knife edges are indistinguishable with the described MSH device on the basis of purely geometrical considerations from Eqs.(1).

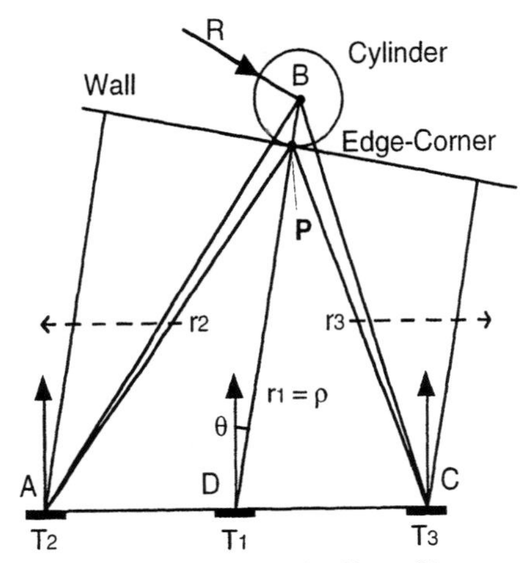

Fig. 2: Geometric model of ultrasonic reflections from the objects of interest.

In practice, the range measurements are noisy:

$$\mathbf{z} = \mathbf{r} + \mathbf{n} \tag{2}$$

In the sequel we assume that the noise is modeled as a gaussian random vector:

$$\mathbf{n} = \mathcal{N}(\mathbf{b}, C) \tag{3}$$

Here **b** is the measurement offset vector, and $C = \sigma^2 I$ is the covariance matrix (I is the 3 x 3 identity matrix). Of particular interest in this paper are those errors of threshold–based range estimation methods which produce range biased estimates, e.g., the measurement uncertainties that are due to the return signal exceeding the threshold at a time instant that actually depends on the echo amplitude and shape. For low–amplitude signals, the detection takes place with some delay; in the worst case, the delay is equal to the transmit pulse width (1 msec of pulse width corresponds to a range error of approximately 30 cm).

Because of the narrow beamwidth of typical sonar sensors, the RCD is limited in its angular extent [10]. When the rangefinder fails to detect the reflecting object, it returns its maximum sensing range. In practice, when the angle of incidence to the feature of interest is large, the Signal–to–Noise ratio (SNR) decreases, and the measurement offset thus determines an anomalous elongation of the range readings; the RCDs tend therefore to be more similar to straight line segments than to true circular arcs. This trend is particularly evident for those reflecting structures that are characterised by the lowest target strengths.

METHOD OF TARGET CLASSIFICATION

The first step of our method of target classification consists of performing a Least–Squares (LS) interpolation, in order to fit a line equation to the three range readings of a **z**–vector. A good criterion for the quality of the line fit is provided by the root mean square (RMS) error e. It is straightforward to demonstrate that:

$$e = \frac{|z_2 + z_3 - 2\,z_1|}{\sqrt{18}} \tag{4}$$

Substituting the expressions for the components of the **r**–vector of Eqs.(1) into Eq.(4) leads to the following expressions of the interpolation error e, which are proven to be (approximately) valid for noiseless and unbiased range measurements:

$$\begin{aligned} e(\theta) &= 0 && \text{for walls} \\ e(\theta) &\approx d^2/\rho && \text{for corners and edges} \\ e(\theta) &\approx d^2/(\rho + R) && \text{for cylinders with radius R} \end{aligned} \tag{5}$$

While no simplification is introduced in Eq.(5a), the other expressions are to be considered valid for small center spacing–range ratios [8]. Since a fundamental condition for accomplishing successful object classification will be shown to require the execution of appropriate sector scans, we prefer to explicitly indicate that the interpolation error is a function of the orientation angle θ, henceforth called the e–θ curve. Note that the interpolation errors are almost constant over the whole extent of the sector scan in the ideal case of unbiased and noiseless range measurements. As shown in Eqs.(4), the changes of e with the orientation angle θ, if any, are negligible if the measurement offsets are exactly the same for all sensors. Since we can assume that the return signal amplitude is approximately independent of range, due to the Automatic Gain Control (AGC) action built in the Polaroid receiver, the main factors affecting the bias turn out to be: a) the orientation angle, and b) the target strength. While performing the sector scan, the viewing angles from which the array sensors look at the feature in their field of view change. For walls, the viewing angles are equal for all sensors at any sensing position. The consequence is that the SNR–dependent offsets should be the same for the three sensors, with no changes regarding the interpolation error. This does not hold for the other objects, since they present different incidence angles to the three sensors. Also, we have to consider that edges and cylinders suffer from a further SNR reduction due to their lower target strengths, as compared to corners or walls. In consequence of the changes in the viewing angles, the e–θ curves appear to be roughly bowl–shaped functions. The concavity and the minimum of an e–θ curve are somewhat related to the radius value; however, the concavity of an e–θ curve for a corner is similar to that of a cylinder with infinite radius (wall).

The method of classification requires that a sector scan is performed. At each sensing position, a check of data consistency regarding the estimate of the orientation angle is made using Eqs.(1); we assume that a **z**–vector is validated when the corresponding azimuth estimate is $\theta \leq 9°$. The LS interpolation is then carried out for each validated **z**–vector, leading to the corresponding e–θ curve. The e–θ curve is then subjected to a further LS interpolation using a second–order polynomial; the feature upon which the identification process is based concerns the minimum and the curvature of the polynomial fitted to the interpolation error which results in turn from a linear interpolation on the data.

EXPERIMENTAL RESULTS

The reflecting objects that are considered for validating the theory developed in the previous Sections are: one cardbox planar surface, one cardbox corner, and a set of three cylinders (cyl_1: R=5 cm, cyl_2: R=10 cm, cyl_3: R=12.5 cm). Each object is isolated in the field of view of the sonar array at a distance of about D=1.7 m; a sector scan of extent $\Delta\psi = 24°$ is performed, starting from the position where the linear array is inclined by an angle $\psi = -12°$ w.r.t. the reflecting object; the selected scanning step is $\Delta\beta = 1.2°$, thus each sector scan is composed of M=21 sensing positions. Each sector scan is repeated a number of times (N=10). Fig.3a depicts a typical e–θ curve for both the planar surface and the corner; Figs.3b–3e show the e–θ curves with regard to one sector scan involving the cylindrical reflectors.Superimposed on the experimental points, the second–order interpolating curves are also reported. Note that the portion of the whole scan which leads to estimated orientation angles $\theta \leq 9°$ is of lesser extent for cylindrical reflectors, as expected. In Fig.4 we report the constellation of the feature points that are used to perform object classification in their 2D space.

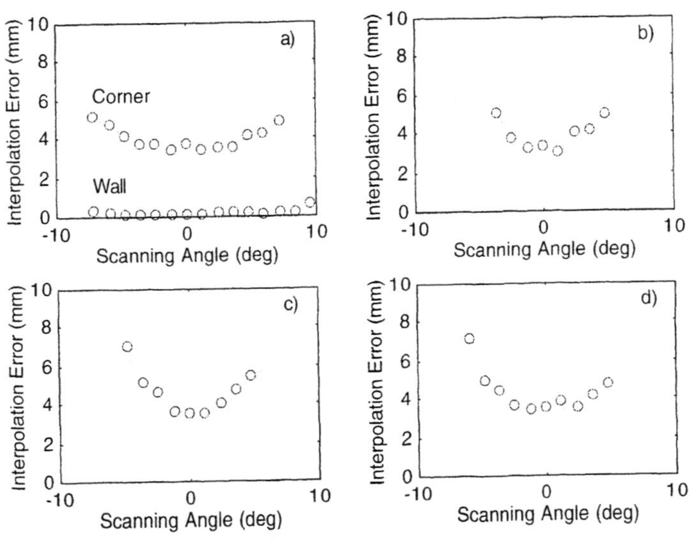

Fig. 3: Examples of e–θ curves for the objects explained in the text. a) corner – wall, b) cyl_1, c) cyl_2, and d) cyl_3.

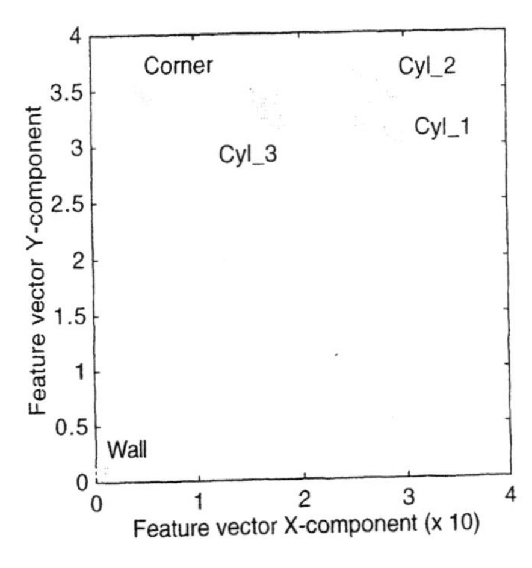

Fig. 4: Feature space for the set of objects to be classified (experimental data).

The X–coordinate of any feature point is the value of the curvature of the second–order interpolating polynomial. Since the extent of the validated scan tends to decrease with decreasing values of the radius of cylindrical reflectors, the X–coordinate is further multiplied by the ratio between the extent of the validated scan and the extent M of the actual scan. This device tends to increase the separation distance between the various clusters in the feature space. The Y–coordinate of any feature point is the minimum value of the second–order interpolating polynomial. Note that the "clouds" concerning the cylindrical targets are spread out quite a lot in consequence of the greater amount of noise, as compared to the clouds related to corners and walls. It should be pointed out that proper 2D decision regions are easily found for achieving successful object classification in all the experiments we have performed.

CONCLUDING REMARKS

Contrary to a common belief, sonar sensors are sufficiently stable and repeatable, so that their geometrical information can be successfully predicted in several operating conditions. In this paper we have shown that the behaviour of sonar sensors can be predictable as for some of the errors that they necessarily perform as well.

Some errors that arise in consequence of the used range estimation method are indeed somewhat "deterministic" in nature. The influence of the measurement offsets on the results of a simple "goodness–of–fit" test in the course of a sonar scan allows to disambiguate the hypotheses concerning some of the geometrical features to be classified.

The implementation of a very simple system for identification of selected environmental structures is demonstrated using a multi–aural configuration of sonar sensors.

ACKNOWLEDGMENTS

Dr. O. Di Benedetto is gratefully acknowledged for his help in performing the experiments described in this paper. This work is supported in part by funds from the Italian Minister of University and of Scientific and Technological Research (MURST 60%).

REFERENCES

1. Elfes, A. "Sonar–Based Real–World Mapping and Navigation." *J. Robotics Automat.*, 3 (3) (1987), 249–265.
2. Leonard, J. J., Durrant–Whyte H. F. "Mobile Robot Localization by Tracking Geometric Beacons." *IEEE Trans. Robotics Automat.*, 7 (3) (1991), 376–382.
3. Peremans, H., Audenaert K., Van Campenhout J. "A High–Resolution Sensor Based on Tri–Aural Perception." *IEEE Trans. Robotics Automat.*, 9, (1) (1993), 36--48.
4. Kuc R., Siegel M. W. "Physically Based Simulation Model for Acoustic Sensor Robot Navigation." *IEEE Trans. Patt. Analys. Mach. Intell.*, 9 (6) (1987), 776--788.
5. Barshan B., Kuc R. "Differentiating Sonar Reflections from Corners and Planes by Employing an Intelligent Sensor." *IEEE Trans. Patt. Anal. Mach. Intell.*, 12 (6) (1990), 560--569
6. Bozma Ö, Kuc R. "Building a Sonar Map in a Specular Environment Using a Single Mobile Sensor." *IEEE Trans. Pattern Anal. Mach. Intell.*, 13 (12) (1991), 1260--1269.
7. Sabatini A. M., Di Benedetto O.: "Spatial Localisation Using Sonar as a Control Point Matching Task." *Proc. 3rd Int. Symp. Intell. Rob. Syst.*, Pisa, Italy, (1995), 11--23.
8. Sabatini A. M. "A Statistical Estimation Method for Segmentation of Sonar Range Data." *Autonomous Robots*, 1 (2) (1995) , 167--178.
9. Polaroid Corporation. "*Polaroid US Ranging Systems Handbook: Application Notes/Technical Papers.*" (1984).
10. Holenstein A. A., Müller M. A., Badreddin E. "Mobile Robot Localization in a Structured Environment Cluttered with Obstacles." *Proc. IEEE Conf. Robotics Automat.*, Nice, France, (1992), 2576–2581.

Enabling Open Control Systems - An Introduction to the OSACA System Platform

Wolfgang Sperling, Peter Lutz
FISW GmbH, Rosenbergstr. 28, D-70174 Stuttgart, Germany

Abstract

The basis for any open control systems is formed by a standardised system platform. The ESPRIT III project OSACA (Open System Architecture for Controls within Automation systems) has worked out the necessary specifications and has jointly developed the software modules for the system platform. The platform is based upon object-oriented principles and has three major parts : operating system, communication system, and configuration system. After a general introduction in open control systems the basic functionality of the system platform is described and a sample configuration is presented.

Keywords: open systems, numerical controls, automation, system platform

INTRODUCTION

In the past and even today numeric controls (NC) are usually vendor specific solutions. They are offered as complete packages where the user has hardly a possibility to integrate his own software solutions. Users therefore heavily depend on the willingness of control vendors to implement required software extensions.

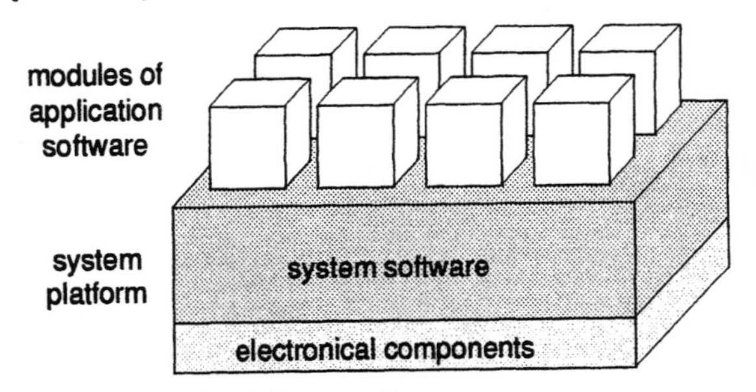

Figure I. Basic structure of modular control systems

Today we can observe a great variety and flexibility of machine tools, the upcoming of new technologies like laser cutting and new machine kinematics like Hexapod structures, and the wish to unite NC and RC activities. This forced control vendors to implement a modular software structure in their controls. The basic architecture for all modular control systems consists of a system platform containing the hardware and system software like the operating system and a set of modules of application software which contain the control specific functionality (Figure I).

Users need to integrate software themselves into control systems and want to profit from cost saving effects like reuse of software and use of standard-hardware. To achieve this modular control systems have to be transformed into open control systems.

DEFINITION OF AN OPEN SYSTEM

Today control vendors claim very different products to be open systems. This requires a clear definition of open control systems. From the user's point of view openness is focused on capabilities to integrate, extend and reuse software modules in control systems (Figure II). The required capabilities have to be supplied mainly by the system platform of the control.

capability of module	meaning
portable	a module can run in different control systems
extendable	the functionality of modules can be extended
exchangeable	a module can be replaced by one with comparable functionality
scalable	multiple instances of modules are possible to increase perfomance
interoperable	modules cooperate (exchange data)

Figure II. Five capabilities of modules in open control systems

SYSTEM PLATFORM

The functionality of the system platform can be directly derived from the requirements imposed on the control system by the openness for modules (Figure III).

capability of module	requirements to platform
portable	uniform application program interface (API) of platform
extendable	application independent platform hosting any module
exchangeable	configuration system(binding of modules)
scalable	configuration system (multiple instantation of modules)
interoperable	communication system (standardized protocol on application layer)

Figure III. Requirements to system platform for open control systems

Three basic elements have to be found in a system platform for open control system:

- The *operating system* guaranties an independent, quasi parallel execution of modules and ensures independence from the specific hardware.
- The *communication system* ensures the co-operation of modules in a standardised way.
- The *configuration system* serves to build up a software topology of the set of available modules by instantiating and connecting the different modules.

These three elements, integrated into the system platform, are accessible through a common application program interface (API) (Figure IV). The API also allows vendors to choose optimized solutions for their systems without violating the criteria for open systems. The API has to be vendor-neutral to allow the portability of application modules onto systems of different vendors. Because of their object-oriented nature, including encapsulation and multiple-instantiation, the application modules shall be called architecture objects (AO).

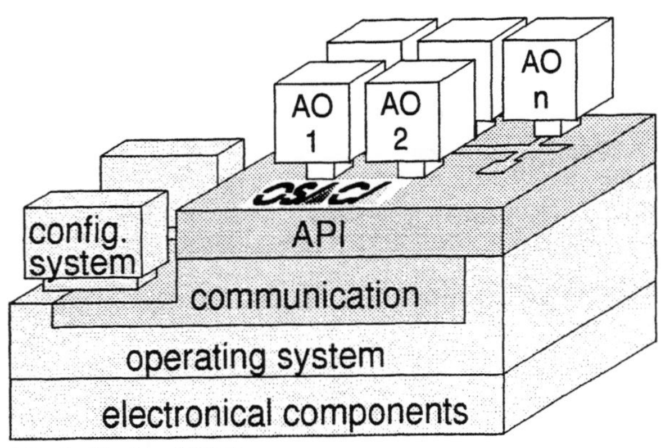

Figure IV. System architecture for open control systems

Hardware and Operating System

Basis for any system platform is the hardware consisting of processor-boards, I/O-boards and other peripheral equipment. Because of the much shorter innovation cycles for hardware compared to software the system platform has to be independent of specific hardware-requirements. It must be possible at any time to easily integrate the most suitable and cost efficient hardware, which is available.

Numerous operating systems in different flavours for different purposes are available on the market. To profit from latest innovations it is not desirable to select a specific product as a standard. Instead the API of the operating system has to be standardized. As POSIX is the only well established standard for operating systems which also includes real-time definitions it shall be selected.

Communication System

The communication system is the only mean of the system platform for information interchange between AO's. It must support both, the exchange of information between AO's located on the same processor-board as well as also between AO's located on different boards connected through a bus-system. A standardised protocol has to be defined to ensure uniform data formats and a fixed set of messages.

The chosen protocol architecture is derived from the OSI base reference model. It comprises however only two layers (Figure V): a message transport system (MTS) equivalent to the layers 1 to 4 of the OSI base reference model and an application services system (ASS) equivalent to the layers 5 to 7 of it.

The MTS offers connection oriented services for a transparent transport of arbitrary messages between AO's. It can be adapted to use any kind of existing mechanisms for information exchange, e.g. operating system services like message queues, LAN-protocols like TCP/IP.

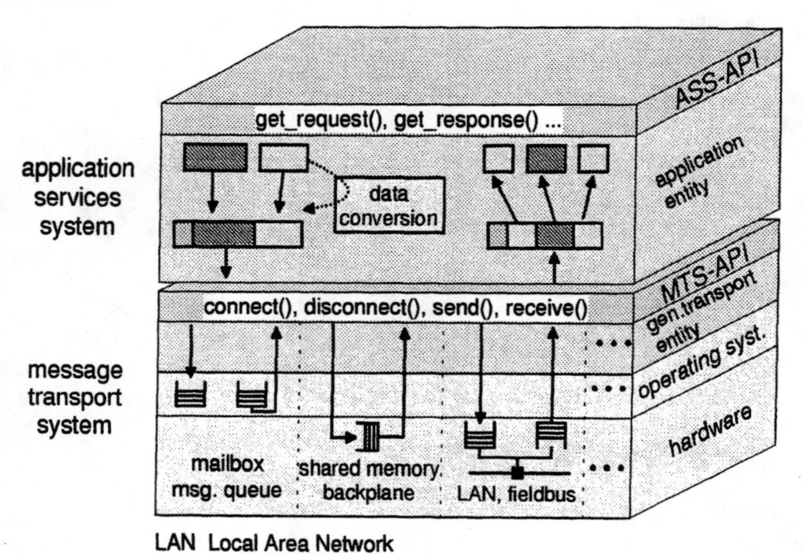

Figure V. Communication System

The ASS has the task to handle the application protocol. This includes connection management, the assembly/disassembly of messages and data conversion between different data representations like Little and Big Endian.

The application protocol is realized on a client/server basis using object oriented principles. In a server AO every information, data or services, that shall be accessible from external is mapped to a communication object (Figure VI. From a client's point of view a server is a set of communication objects which can be accessed by using ASS services to send and receive messages. AO's can be servers and clients at same time.

There is a fixed set of classes for communication objects, the most important ones are the class *variable* for reading and writing data and the class *process* to trigger actions in state machines. Additionally a class *event* is used for unsolicited sending of events and reports.

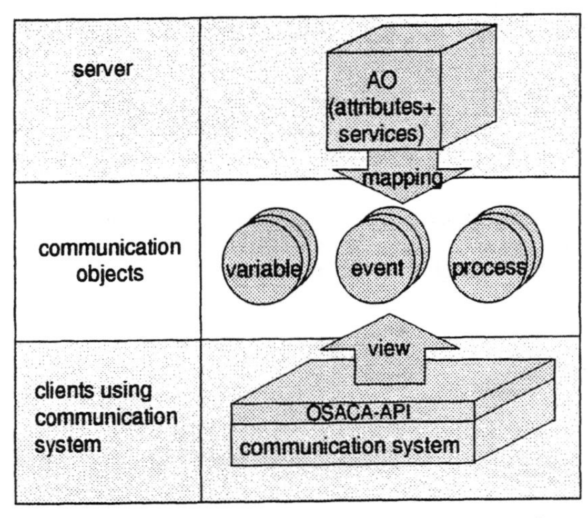

Figure VI. Communication Objects: Basis for the Application Protocol

Structure of Architecture Objects

Architecture Objects (AO) contain the specific application software. In order to relieve application programmers from solving communication specific problems a uniform software layer was introduced to manage the communication objects of the AO. This communication object manager (COM) holds lists of the available communication objects and performs requested services on them.

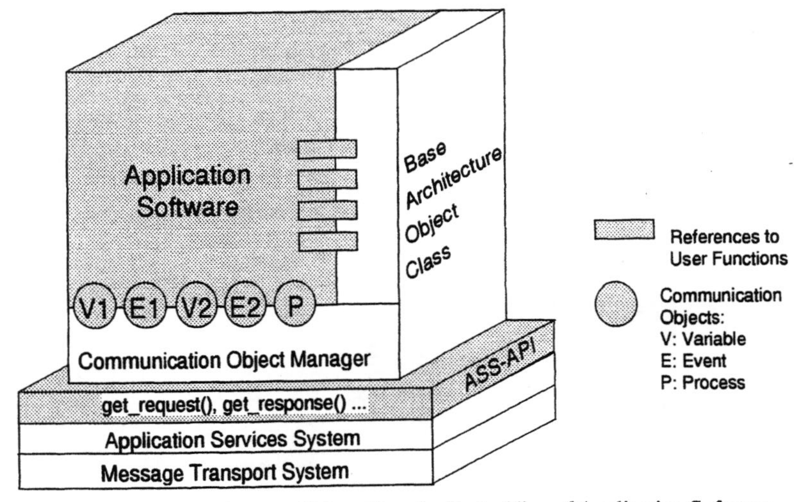

Figure VII. Base Architecture Object Class for Embedding of Application Software

The COM is embedded into a Base Architecture Object Class (BAOC) which allows to completely manage an AO. The BAOC contains predefined links to user specific functions for initalizing, configuring and resetting an AO. The application programmer can fully focus on solving his control specific problem. He only has to define the list of communication objects he wants to offer and supply the necessary links into his application software.

In order to achieve interoperability between AO's from different vendors the functionality and external behaviour of AO's has to be specified. This is achieved by defining a characteristic set of communication objects for every AO. Thus a reference architecture for application software is defined consisting of a set of specific AO's with their characteristic list of communication objects.

Configuration System

Today the usual proceeding for realizing a specific configuration for a control system is of a static nature. Before run-time the software is completely compiled and linked and forms one executable which is downloaded into the control. This solution is however inflexible and requires high efforts if modifications and additions are necessary after delivery of the control.

To overcome this situation a fully modular system is needed where the actual topology of the software is generated at the boot-up of the system. For this, the system platform has to contain a configuration system which is able to handle a library of AO classes. At boot-up AO's of the different classes are instantiated and communication connections between AO'S are established (Figure VIII).

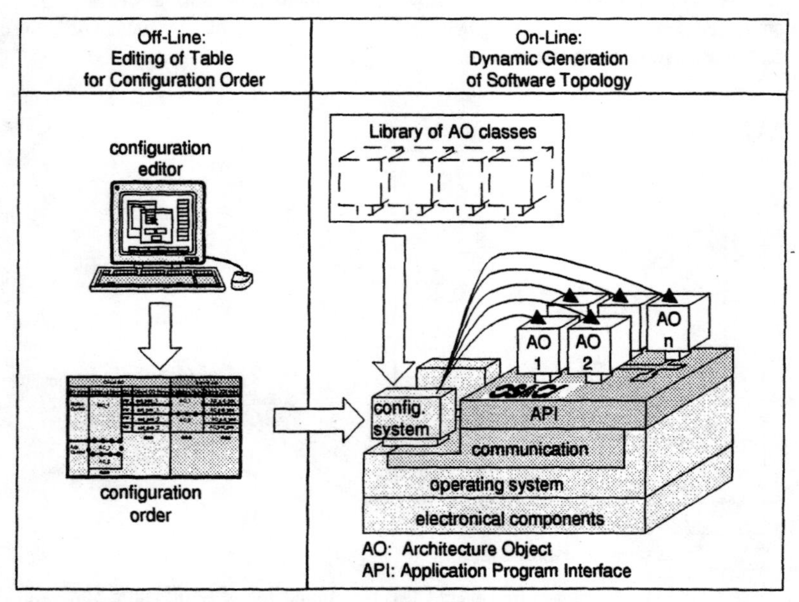

Figure VIII. Elements for Dynamic Configuration of Control Systems

The actual topology of the control system is described in an externally generated configuration order which is interpreted by the configuration system. A graphical configuration editor can be used to define the configuration order in a manner like CAD-systems are used to define the layout of a processor-board.

The configuration order contains a list of all AO instances that will be found in a specific control system (Figure IX). It is possible that there are several instances of the same AO class, e.g. class Axis Control. For every AO the client/server relationships are defined. Every relationship between a client and a server AO consists of a list of communication objects of the server which will be used under specific names in the client.

Client AO				Server AO	
AO class	Instance Name	COC	Client CO Name	Instance Name	Server CO Name
Motion Control	MC_1	Var.	set_pos_1	AC_1	AC_act_pos
		Var.	act_pos_1		AC_set_pos
		Var.	set_pos_2	AC_2	AC_act_pos
		Var.	act_pos_2		AC_set_pos
			•••	•••	•••
Axis Control	AC_1				
	AC_2				
	•••				

CO: Communication Object COC: Communication Object Class Var: Variable

Figure IX. Configuration Order

EXAMPLE

A typical configuration for control systems is the use of a PC-compatible computer for the operator's panel and VME-bus based components for the motion-control and I/O (Figure X). Both parts are connected using a serial bus, in this case Ethernet with TCP/IP. Such a configuration allows cost efficient solutions using commercially available hardware and software components.

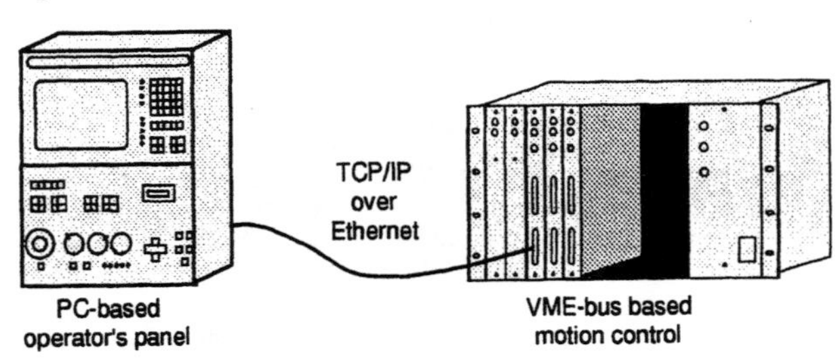

Figure X. Sample Configuration: PC for operator's panel and VME-Bus System for Motion Control

The resulting open control system consists of two part-platforms: the PC-based part under Windows 95 and the VME-bus based part using a processor board and VxWorks as real-time operating system. The communication system allows the AO's to communicate transparently no matter whether the communication partners are located on the same board or on different ones.

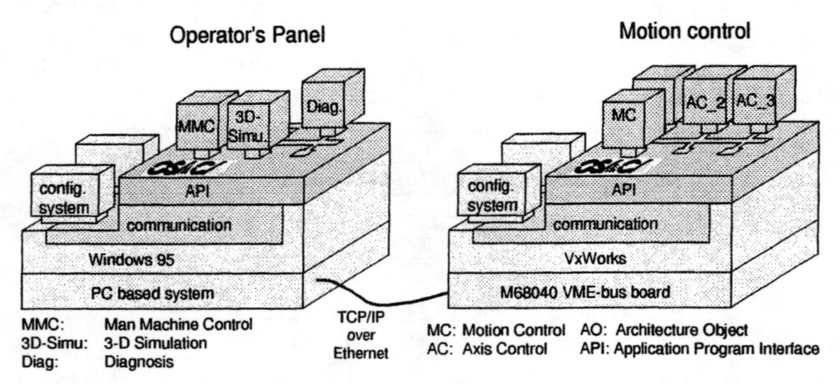

Figure XI. System Platform for Sample Configuration

CONCLUSION

The presented system platform allows to realize open control systems based on vendor-neutral specifications. This enables users to realize portable, re-usable applications runnable on controls of a vendor complying to the specifications.

The specifications for this platform were defined within two projects of the ESPRIT framework: OSACA (Open System Architecture for Controls within Automation Systems) phase I and II. The consortium that works on these specifications comprises five major European control vendors (Bosch, Fagor, Grundig-Atek, Num, Siemens), machine tool builders (Comau, Huron, Index) and research institutes (INTEC, FISW, WZL). Work started in 1992 and the consortium is now in the possession of a basic set of stable specifications. Currently further national and European projects are set up which will apply and improve the OSACA applications.

REFERENCES

1. ESPRIT III Project 6379 OSACA: Final Report. Stuttgart: FISW GmbH, 1995.
2. Pritschow, G.; et.al.: Open System Controllers - A Challenge for the Future of the Machine Tool Industry. In: Annals of the CIRP Vol. 42/1/993. Bern, Stuttgart: Verlag Technische Rundschau, 1993.
3. Pritschow, G.et.al.:Information Interchange in Open Control Systems. Production Engineering Vol. II/1 (1994)
4. Lutz, P.; et.al.: Communication System for Open Control Systems. In: K.-R. von Barisani et al. [Hrsg.]. Opening Productive Partnerships. Proceedings of the Conference on Integration in Manufacturing, Vienna 13-15 September 1995. Amsterdam etc., IOS Press, 1995.

VARIABLE STRUCTURE CONTROLLER DESIGN FOR FLEXIBLE ONE-LINK MANIPULATOR

SUSY THOMAS
Dept.of Elect.Engg., Calicut Reg.Engg.College, India

B.BANDYOPADHYAY
Faculty of Elect.Engg.,Ruhr-Univ. of Bochum,Germany

H.UNBEHAUEN
Faculty of Elect.Engg.,Ruhr-Univ. of Bochum,Germany

ABSTRACT

A variable structure controller is designed for the tip position control of a flexible one-link manipulator.The controller drives the system's error and it's derivatives (error signal is chosen to be that between the plant states and ideal desired states) referred to as the system's representative point(RP) to a hyper surface which is designed to yield an asymptotically stable system in sliding mode(SM).The discontinuous control law designed to satisfy the existence condition of SM, then sets up SM motion on the hypersurface whereby the RP slides to the origin.

KEYWORDS: flexible one-link manipulator, sliding mode(SM), switching vector, variable structure controller(VSC)

INTRODUCTION

Various advantages of flexible links make them desirable to rigid links[1].But the flexibility of such links make the position controller design difficult because one has to control not only the rigid mode but also the highly vibratory modes.Theoretically, a flexible arm is an infinite order system.Designing a controller for a high order system and implementing it may not be practically feasible. Hence a reduced order system is a pre-requisite for the design of any practical controller.Owing to the fact that actuators and sensors cannot operate in the high frequency range, the flexible arm is approximated by finite models that consist of a finite number of modes. In this paper a VSC is designed for the tip position control considering a finite model.This controller ensures insensitivity to parameter variations and disturbance effects [2, 3].

FLEXIBLE ONE-LINK MANIPULATOR

Fig2.1 shows the schematic of a flexible one-link manipulator.The controller is a motor at the rotating joint.One end of the link is clamped on a rigid hub mounted directly on the vertical shaft of the motor.The link is free to move in the horizontal plane

but not in the vertical plane or torsion. Notations employed are:
I_b moment of inertia of the beam,
I_h moment of inertia of the motor hub,
L length of the arm,
OX referenceaxis,
$q_i(t)$ generalised coordinate of the system,
T torque applied by the motor,
$w(x,t)$ elastic deformation at x,
$y(x,t)$ net movement of the point at x from the hub,
$\theta(t)$ angular rotation of the beam,
ω_i natural frequency of ith vibrational mode,
ξ_i damping ratio of ith vibrational mode.

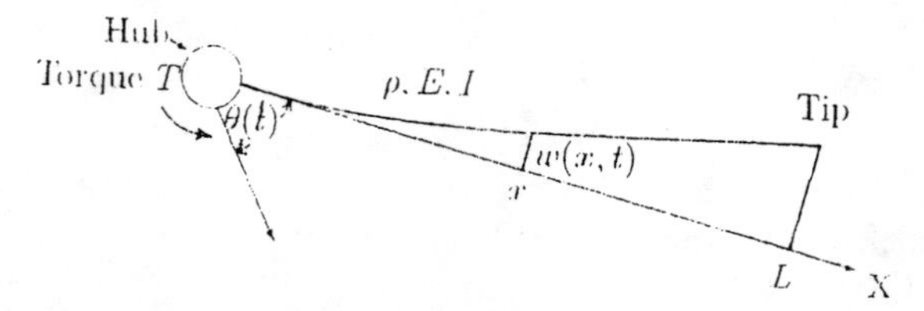

Figure 1. Schematic diagram of a flexible one-link manipulator

The modelling process is not described since it is available in [4]. In [4] the state space representation is obtained as:

$$\left.\begin{aligned} \dot{\boldsymbol{x}} &= \boldsymbol{A}\boldsymbol{x} + \boldsymbol{b}u \\ y &= \boldsymbol{c}^T\boldsymbol{x} \end{aligned}\right\} \tag{1}$$

where

$$\boldsymbol{A} = \begin{bmatrix} 0 & 1 & & & & & \\ 0 & 0 & & & & & \\ & & 0 & 1 & & & \\ & & -\omega_1^2 & -2\xi_1\omega_1 & & & \\ & & & & \ddots & & \\ & & & & & 0 & 1 \\ & & & & & -\omega_N^2 & -2\xi_N\omega_N \end{bmatrix}, \boldsymbol{b} = \frac{1}{I_b + I_h}\begin{bmatrix} 0 \\ 1 \\ 0 \\ \phi_1'(0) \\ \vdots \\ 0 \\ \phi_N'(0) \end{bmatrix} \tag{2}$$

$$\boldsymbol{c}^T = \begin{bmatrix} L & 0 & \phi_1(L) & \dots & \phi_N(L) & 0 \end{bmatrix} \tag{3}$$

$$\boldsymbol{x}^T = [q_0 \ \dot{q}_0 \ q_1 \ \dot{q}_1 \ \dots \ q_N \ \dot{q}_N]^T = [x_1 \ x_2 \ x_3 \ x_4 \ \dots \ x_{2N}]^T \tag{4}$$

$u = T$ and $y \equiv$ tip position. N can be truncated to obtain a low-order state space description on which the controller design can be based.

VARIABLE STRUCTURE CONTROLLER DESIGN

A truncated sixth order state space model is considered, maintaining the fact that the design can include any number of desired modes without any additional complexity.Table I gives the model parameters for the first two vibrational modes.These correspond to the experimental flex-arm used in [4].The physical parameters of this arm are given in table II. We make use of these parameters in the design.

TABLE I. Model parameters of the vibrational modes

Mode i	ω_i	ξ_i	$\phi_i'(0)$	$\phi_i(L)$
1	55.89	0.0015	2.886	-0.931
2	131.75	0.0015	-2.345	-1.027

TABLE II. Physical parameters of the flex-arm

Physical parameters	Symbol	Numerical value
Length	L	1.0 m
Width	B	3.4 mm
Height	H	2.54 cm
Modulus of elasticity	E	$6.9 \times 10^{10}\ N/m^2$
C.S.A. Moment of inertia	I	$8.39134 \times 10^{-11}\ m^4$
Linear density	ρ	$0.233172\ kg/m$
M.I. of the beam	I_b	$7.7724 \times 10^{-2}\ kg - m^2$
Hub inertia	I_h	$5.176 \times 10^{-3}\ kg - m^2$

For simplicity of notations $\boldsymbol{b}$ in eq.(2) is written as:

$$\boldsymbol{b} = \begin{bmatrix} 0 & b_0 & 0 & b_1 & 0 & b_2 \end{bmatrix}^T, \tag{5}$$

where b_0, b_1, b_2 are given appropriate values from eq.(2). The error vector is defined as:

$$\boldsymbol{e} = \boldsymbol{x} - \boldsymbol{x}_d = \begin{bmatrix} e_1 & e_2 & e_3 & e_4 & e_5 & e_6 \end{bmatrix}^T \tag{6}$$

where $\boldsymbol{x}_d$ represents the desired state.Without loss of generality it is assumed that $\boldsymbol{x}_d = \boldsymbol{0}$.Hence,

$$\dot{\boldsymbol{e}} = \dot{\boldsymbol{x}} = \boldsymbol{A}\boldsymbol{x} + \boldsymbol{b}u. \tag{7}$$

The hypersurface is chosen as:

$$s = \boldsymbol{g}^T\boldsymbol{e} = \boldsymbol{g}^T\boldsymbol{x} = 0 \quad \text{where} \quad \boldsymbol{g}^T = \begin{bmatrix} g_1 & g_2 & g_3 & g_4 & g_5 & g_6. \end{bmatrix} \tag{8}$$

Now

$$\dot{s} = g_1\dot{x}_1 + g_2\dot{x}_2 + g_3\dot{x}_3 + g_4\dot{x}_4 + g_5\dot{x}_5 + g_6\dot{x}_6. \tag{9}$$

The control function is chosen as :

$$u = \boldsymbol{k}^T\boldsymbol{e} = \boldsymbol{k}^T\boldsymbol{x} \quad \text{where} \quad \boldsymbol{k}^T = \begin{bmatrix} k_1 & k_2 & k_3 & k_4 & k_5 & k_6 \end{bmatrix}. \tag{10}$$

$\dot{x}_i$ is substituted in eq.(9) from eq.(1) and u is substituted from eq.(10).Both sides of the emerging equation is then multiplied by s.The right hand side of the resulting equation will have to be less than zero in order to satisfy the existence condition for SM, i.e., to satisfy $s\dot{s} < 0$ [5]. This leads to the inequality:

$$\begin{aligned} & sx_1(g_2b_0k_1 + g_4b_1k_1 + g_6b_2k_1) \\ & +sx_2(g_1 + g_2b_0k_2 + g_4b_1k_2 + g_6b_2k_2) \\ & +sx_3(g_2b_0k_3 + g_4b_1k_3 + g_6b_2k_3 - g_4\omega_1^2) \\ & +sx_4(g_3 + g_2b_0k_4 + g_4b_1k_4 + g_6b_2k_4 - 2g_4\xi_1\omega_1) \\ & +sx_5(g_2b_0k_5 + g_4b_1k_5 + g_6b_2k_5 - g_6\omega_2^2) \\ & +sx_6(g_5 + g_2b_0k_6 + g_4b_1k_6 + g_6b_2k_6 - 2g_6\xi_2\omega_2) < 0. \end{aligned} \tag{11}$$

The controller gains can be solved for by letting each term in eq.(11) to be seperately less than zero. But the first step towards solving eq.(11) is to design an asymptotically stable sliding surface s [6], i.e., $\boldsymbol{g}$ of eq.(8) has to be selected such that the system in SM is asymptotically stable. What follows serve as guidelines for the selection of the switching vector $\boldsymbol{g}$.

Synthesis of sliding surface

Consider the system of eq.(1).The following transformation of states is made:

$$\boldsymbol{x}' = \boldsymbol{M}\boldsymbol{x} = \begin{bmatrix} \boldsymbol{x}_1' \\ x_2' \end{bmatrix} \quad \text{where} \quad \boldsymbol{M} = \begin{bmatrix} \boldsymbol{M}_1 \\ \boldsymbol{m}_2^T \end{bmatrix}. \tag{12}$$

$\boldsymbol{M}$ is chosen such that:

$$\boldsymbol{M}_1\boldsymbol{b} = 0 \quad \text{and} \quad \boldsymbol{m}_2^T\boldsymbol{b} = b_N \tag{13}$$

where b_N is a non-zero scalar. we have

$$\dot{\boldsymbol{x}}' = \boldsymbol{MAM}^{-1}\boldsymbol{x}' + \boldsymbol{Mb}u. \tag{14}$$

Let

$$\boldsymbol{MAM}^{-1} = \begin{bmatrix} \boldsymbol{A}_{11} & \boldsymbol{A}_{12} \\ \boldsymbol{A}_{21} & \boldsymbol{A}_{22} \end{bmatrix} \tag{15}$$

We now have

$$s = \boldsymbol{g}^T\boldsymbol{x} = \boldsymbol{g}^T\boldsymbol{M}^{-1}\boldsymbol{x}' = 0. \tag{16}$$

Let

$$\boldsymbol{g}^T\boldsymbol{M}^{-1} = \begin{bmatrix} \boldsymbol{g}_1^T & g_2 \end{bmatrix}. \tag{17}$$

Therefore

$$s = \boldsymbol{g}_1^T\boldsymbol{x}_1' + g_2x_2' = 0 \tag{18}$$

where $\boldsymbol{x}_1' \in R^5$, $x_2' \in R^1$ and $\boldsymbol{g}_1^T$, g_2 are respectively, 1×5 and 1×1 subvectors of $\boldsymbol{g}^T\boldsymbol{M}^{-1}$. In SM

$$x_2' = -g_2^{-1}\boldsymbol{g}_1^T\boldsymbol{x}_1'. \tag{19}$$

Using eqs. (15) and (19) the equilavent system in SM can be represented by:

$$\dot{\boldsymbol{x}}_1' = (\boldsymbol{A}_{11} - \boldsymbol{A}_{12}g_2^{-1}\boldsymbol{g}_1^T)\boldsymbol{x}_1' \tag{20}$$

Now the problem at hand is to select $\boldsymbol{g}^T$ such that the eigen values of the equivalent system in eq.(20) be placed at desired(stable) locations.

Using the above design strategy, $\boldsymbol{g}$ was selected such that $\boldsymbol{g}_1^T$ and g_2 as in eq.(17) make the equivalent system in eq.(20) asymptotically stable. The switching vector was chosen as:

$$\boldsymbol{g}^T = \begin{bmatrix} 10 & 3.7 & 55.7 & -0.02 & -167 & 0.6 \end{bmatrix}. \tag{21}$$

And the transformation matrix considered to check on the stability of eq.(20) was :

$$\boldsymbol{M} = \begin{bmatrix} 1 & 0 & 0 & 0 & 0 & 0 \\ 0 & 0 & 1 & 0 & 0 & 0 \\ 0 & 0 & 0 & 0 & 1 & 0 \\ 0 & 2.89 & 0 & -1 & 0 & 0 \\ 0 & 2.345 & 0 & 0 & 0 & 1 \\ 0 & 0 & 0 & 0 & 0 & -1 \end{bmatrix} \tag{22}$$

SIMULATION RESULTS

Using the values in eq.(21), the inequality of eq.(11) was solved to obtain the controller gains $\boldsymbol{k}^T$.The gain values obtained are:

$$k_5 = 386.2, \; k_6 = 6.2$$

$$\begin{aligned} k_1 &= -2, \quad if \quad sx_1 > 0 & \qquad k_2 &= -2, \quad if \quad sx_2 > 0 \\ &= 2, \quad if \quad sx_1 < 0 & &= 2, \quad if \quad sx_2 < 0 \end{aligned}$$

$$\begin{aligned} k_3 &= -4, \quad if \quad sx_3 > 0 & \qquad k_4 &= -4, \quad if \quad sx_4 > 0 \\ &= 1, \quad if \quad sx_3 < 0 & &= 1, \quad if \quad sx_4 < 0 \end{aligned}$$

Simulation studies of the VSC were made using the software package MATLAB. Fig.2 shows the tip position and control torque thus obtained.Fig.3 presents the vibrational mode responses of the flexible link.It is noted that the vibrational modes are damped out quickly and the tip is positioned at the set point. The robustness property of the designed controller was tested

by introducing a decrease of 45% in ω_1 and an increase of 40% in $\phi_1(L)$.These bounds were taken into consideration while solving eq.(11).The simulation results obtained showed that the system remained insensitive to these variations.

CONCLUSION

A variable structure controller has been designed for the tip position control of a flexible one-link manipulator. The control algorithm presented is simple and easy to implement.The simulation results show that the controller succeeds in damping out the vibrational modes very quickly and achieving the desired tip position response.The system also remains insensitive to parameter variations and disturbance effects.

Acknowledgment: The second author acknowledges the financial support from Alexander von Humboldt Foundation,Germany.

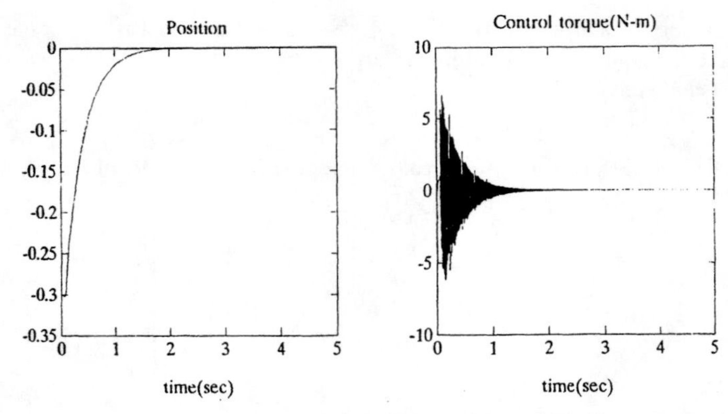

Figure 2. Tip position and Control Torque

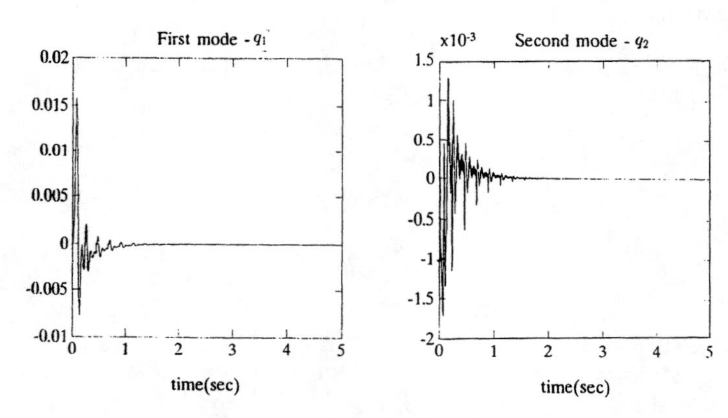

Figure 3. Vibrational mode responses

References

[1] Cannon and Schmitz, "*Initial experiments on the end point control of flexible one link robot*", Int.J.Robotic Res., Vol.3(1984), pp.62-75.

[2] B.Drazenovic, "*The invariance conditions in variable structure systems*", Automatica., Vol.5(1969), pp.287-295.

[3] U.Itkis, "*Control systems of variable structure*", New York, Halsted(1976).

[4] H.Krishnan and Vidyasagar, "*Control of a single limb flexible beam using Hankel norm based reduced order model*", In Proc. IEEE. Int. Conf. Robotics. Contr. Vol. 1(1988), pp. 9-14.

[5] V.I.Utkin, "*Sliding modes and their applications in variable structures*", Mir publishers, Moskow(1978).

[6] K.K.D.Young, "*Design of variable structure model following control systems*", IEEE.Trans. AC., Vol.23(1978), pp.1079-1085.

WORKSPACE SENSING BY FUSION OF OPTICAL AND ACOUSTIC RANGE DATA FOR UNDERWATER ROBOTICS

M. UMASUTHAN, J. CLARK and M. J. CHANTLER
Dept of Computing and Electrical Engineering
Heriot-Watt University
Riccarton, Edinburgh, EH14 4AS, United Kingdom
suthan, jclark and mjc@cee.hw.ac.uk

ABSTRACT

Current generation Remotely Operated Vehicles (ROVs) fitted with manipulators are controlled by an operator whose short range view of an underwater scene is provided by a video camera. The operator has no perception of depth and the image can become degraded in a turbid environment. In this work we describe preliminary results of a sensor system providing depth data from an underwater scene using a laser triangulation sensor in conjunction with a high frequency pencil beam sonar. The triangulation sensor provides high resolution data at a speed orders of magnitude greater than the sonar, but the sonar has the advantage of being able to operate in turbid environments. Data from the triangulation sensor is used in a fusion process to improve the quality of the sonar data. The operation and fusion of the sensors are described, and initial results presented and discussed.

KEYWORDS: triangulation, sonar, fusion, depth sensing

1 INTRODUCTION

Sensory deprivation is recognized as a major factor that limits the capability and productivity of robotic systems [1]. This is particularly relevant to current generation underwater Remotely Operated Vehicles (ROVs) due to the extreme conditions in which they operate [2, 3]. Both the successful automation of offshore inspection, repair and maintenance (IRM) tasks and improvements in operator feedback require accurate and reliable sensing.

Currently ROVs rely heavily on video cameras and forward looking sonar. The sonars provide long range data and the video cameras allow for short range inspection. This combination is effective for inspection but severely limits the capabilities and productivity of the operator during manipulator tasks.

The manipulators are operated in a master-slave configuration by an operator on the surface vessel. The movement of the smaller master arm is replicated by the larger slave arm and they form an approximately spatially correspondent system. The operator's perception of the underwater scene from the video camera is two dimensional and so there is no depth cue [2, 4]. This image can often be of poor quality and the already limited visibility can degrade dramatically in a turbid environment[5, 6], such as that which results from debris removal and sea-bed operations. The presentation to the operator of three-dimensional data of the underwater scene would increase the likelihood of completing the specified task, and reduce the chance of collisions between the manipulator and environment [7, 8].

In this paper we describe the acquisition and fusion of range data acquired from optical and acoustic sensors. A short range full frame triangulation device [9] is used in conjunction with a fully steerable high frequency pencil beam sonar [10]. The respective depth data are then fused to give an improved representation of the workspace.

2 SYSTEM DESCRIPTION

The laser depth sensor (**Figure** 1(a)) employs the well known technique [11] of laser triangulation to measure the three dimensional coordinates of points in an underwater scene with respect to a global reference frame. The ranging system consists of a line illumination source (laser), a rotating mirror, and a two dimensional light sensing element (CCD camera); whose optical axis is positioned at some finite disparity angle from the plane of the source. The plane of light is swept across the scene by the rotating mirror and the resulting 3-D curve (stripe) is observed through a calibrated camera. The camera is arranged so that the image of the stripe intersects each image row (or column) at most once and the range can be linked directly to one image coordinate.

The sonar, a dual axis Mesotech 971 (**Figure** 1(b)), has a high frequency (2.25 MHz) sonar head producing a pencil beam sub-tending approximately one degree, and providing high resolution over short ranges of up to ten metres. The first axis of rotation is provided by movement of the transducer with respect to the sonar head (producing a typical sector scan). The second axis is actuated by a drive unit which rotates the whole head round an axis perpendicular to that of the transducer rotation.

3 SYSTEM CALIBRATION

The calibration of the system is a three stage process in which we individually calibrate the camera, laser plane, and the pencil beam sonar.

A standard non-coplanar camera calibration technique [12, 13] is used to recover both the cameras intrinsic parameters (focal length, image centre, scale factor, and lens distortion), and the extrinsic parameters (the cameras rotation and translation relative to a known world co-ordinate system). A flat target, consisting of a series of black dots on a white background is imaged by the camera in several locations. The target is constrained to move in only one direction. The images are then processed in order to obtain pairs of image and world co-ordinates, for subsequent processing by the calibration program.

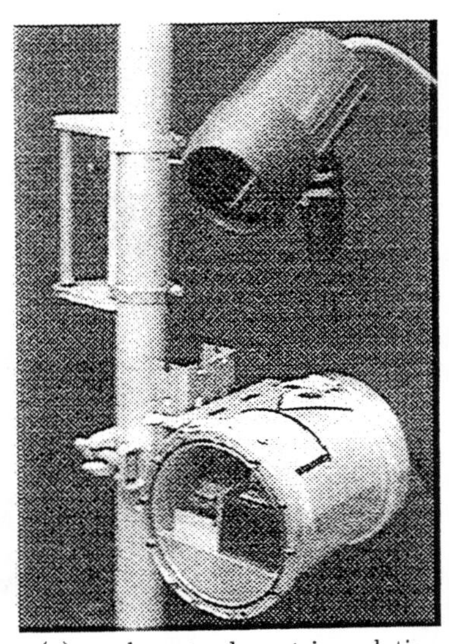

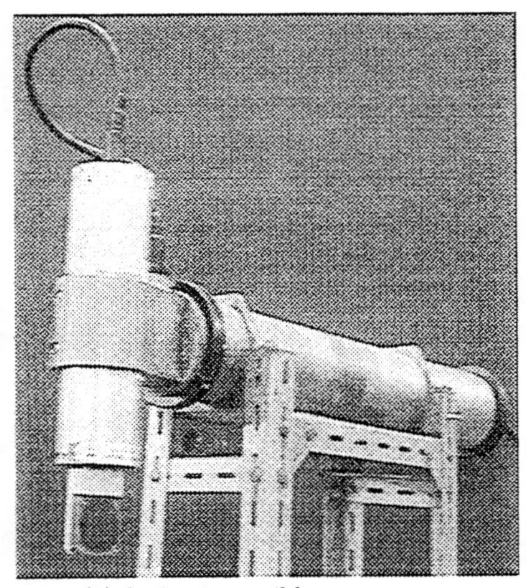

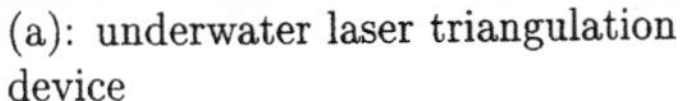

(a): underwater laser triangulation device

(b):dual axis pencil beam sonar

Figure 1: The optical and acoustic sensors.

An accurate knowledge of the laser plane parameters (**Figure** 2) i.e. disparity angle (θ), tilt angle (ϕ), and baseline distance, are required for the calculation of depth. The laser stripe is scanned across a flat plane by the galvanometer scanner. At each angle of rotation a stripe image is acquired. This process is repeated for several planes displaced from the base plane by a known z value. The parameters are obtained using robust least squares line fitting of image points extracted from images of the laser stripe.

The sonar head calibration provides calibration parameters relating to beam elevation angle, beam width, and intensity quantization. The Mesotech 971 displays depth data as an intensity image which is stored by a computer based framestore. Suspended point targets and wires are used to calibrate beam width, the elevation angle is determined in a similar manner, but with sonar operated in side scan mode to fix the azimuth angle. The intensity quantization is calibrated by acquiring multiple sonar images and applying a statistical process to reduce the risk of depth misclassification.

4 DEPTH CALCULATION BY TRIANGULATION

The coordinates of a point in the workspace of the sensor can be derived as a function of the image plane coordinates (x_i, y_i) and camera calibration parameters. The system geometry is shown in **Figure** 2.

The coordinates of an (x, y, z) point are given by

$$z = \frac{z_0}{1 + \frac{tan(\phi)}{cos(\theta)} \frac{y_i}{f} + tan(\theta) \frac{x_i}{f}}$$

$$x = \frac{\frac{x_i}{f} z_0}{1 + \frac{tan(\phi)}{cos(\theta)} \frac{y_i}{f} + tan(\theta) \frac{x_i}{f}}$$

$$y = \frac{\frac{y_i}{f} z_0}{1 + \frac{tan(\phi)}{cos(\theta)} \frac{y_i}{f} + tan(\theta) \frac{x_i}{f}},$$

where $z_0 = x_{BL} tan(\theta)$ and x_{BL} is the baseline distance; θ the disparity angle; ϕ the tilt angle; and x_i, y_i the point on the image plane.

The accuracy of the depth data was determined experimentally using a manual translation stage; flat planes were placed in known z positions and then measured by the triangulation device. The sensor was found to have an average accuracy of $\pm 1.5mm$ over a distance of 1 to $2m$. Several system parameters affect the resolution principally; baseline separation; distance of camera from object being measured (stand-off); camera focal length; and the level of sub-pixel precision in the peak detection algorithm. An analysis of the system geometry [14] has shown that we can increase the resolution of our system, to approximately $\pm 0.25mm$, by proper lens selection and increasing the baseline separation. However, increasing the baseline will increase the risk of occlusion. In occluded areas the viewing system is unable to observe the projected laser stripe and no depth data will be gathered.

5 FUSION OF SONAR AND LASER DATA

Methodologies for the fusion of data from sonar range sensors in the form of two dimensional Cartesian maps have been developed by Moravac [15] with Stewart [16] extending the probabilistic approach to underwater environments. The emphasis in the present work is to improve the accuracy of the sonar range data by fusing it with laser range finder data. Sonar range sensors are preferred in underwater environments as they give direct measurement of depth, require very little processing to produce the three dimensional location of the targets, can be used in turbid water and do not require an extensive calibration to produce a three dimensional depth map.

The problem of fusion is exactly dual to the problem of motion. The motion estimation (or registration) problem estimates the transformation between two visual maps, in our case sonar and laser data of a calibrated object. This process makes the data from each sensor commensurate in its spatial dimension.

Let (P, Q) be the set of primitives extracted from the sonar and laser data and (p, q) be the corresponding parameters of the primitives. The transformation between the primitives of the sets (P, Q) are defined as follows.

- First by the geometric displacement d which transforms them from coordinate system R to coordinate system R',

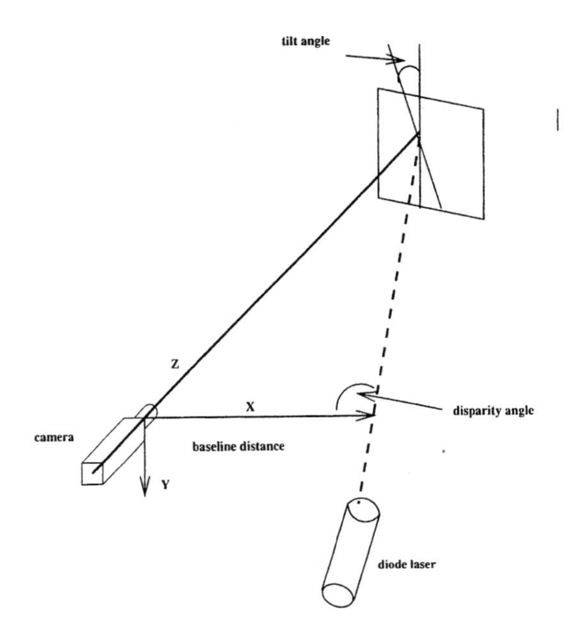

Figure 2: Triangulation geometry.

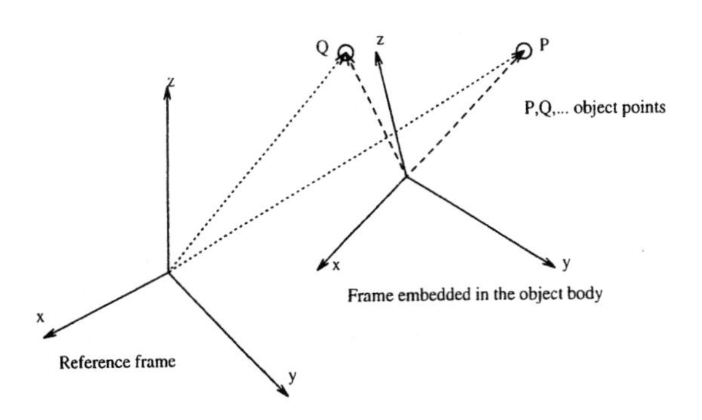

Figure 3: Co-ordinate systems.

- Then by the rotation transformation Rot between the laser and sonar coordinate systems.

The above relations can be written in the form of a vector equation

$$f(p, q, M) = 0, \text{ where } M = (Rot, d) \tag{1}$$

In the above equation the transformation M is applied to the parameters p. This transformation is recovered using a singular value decomposition method.

After recovering the transformation, the fusion stage computes a better estimate of parameters (p, q). Let the state vector s and the measurement vector m be defined as follows.

$$s = (p, q) \quad m = M = (Rot, d)$$

These vectors also satisfy equation 1. An extended Kalman filter is applied to compute the new state vector $\hat{s}$, which improves the original estimate s.

The set of primitives used in the above formulation consists of sets of planes which are extracted from the depth points. If we define the state vector as the sets of original raw depth points then the fusion stage improves the depth estimates. This is a very important advantage of our method since this improves the sonar depth data acquisition stage in real time.

6 RESULTS

Initial registration experiments were performed by sensing four planes, orientated at arbitrary angles, using both sensors and recovering the transform from the parameters of the fitted planes. Both the plane fitting and the minimization of the transformation error function were performed using least squares minimization.

The inaccuracy of the recovered transform (using least squares minimization) can be seen in **Figure 4**; in this example, of data acquired from a brick, accurate registration has not been achieved. The laser data is represented by the white lines on the top half of the brick.

The experiments were repeated for different data sets; however, the recovered transformation estimates did not improve. In order to eliminate the possibility that the error may be due to the least squares minimization method being used, the singular value decomposition method and the quaternion method were also implemented; however, the transformation estimates recovered from these methods still did not provide precise registration. This suggested that the error in the transformation is due to the error in the input data.

The accuracy of the sonar was measured experimentally and found to be approximately $10mm$ over a range of $1m$, which is lower than expected. Upon further investigation it was found that the low accuracy was due to the framestore being used to record the sonar data. The framestore, a Sun VideoPix board, had been designed to operate with both the NTSC (640 pixels per line) and PAL (768 pixels per line) video standards. To accomplish this the framestore produces the same number of pixels per line for NTSC and PAL video; 720 pixels per line. The framestore then compensates for this in software by converting the 720 pixels to 640 for NTSC and

Figure 4: Superimposed sonar and laser data.

768 for pal; this is achieved by dropping pixels in the NTSC case and replicating pixels in the PAL case. This problem first became apparent when we were unable to calibrate the camera using the same framestore. The problem has since been rectified in the case of the camera but not for image acquisition from the sonar and is therefore a major source of error for the sonar device.

7 DISCUSSION

We have described an underwater 3D sensing system employing a laser triangulation device and scanning pencil beam sonar. A three stage method for calibrating the system was presented; and a process for improving the resolution of the sonar data by fusion with the laser triangulation data was given.

Initial experimental results demonstrate that full registration between the the sonar and optical sensor has not been achieved. Further experimentation and investigation has shown that the error in registration is due not to the fusion process but rather the erroneous nature of the sonar data; the error in the sonar data has been identified as a hardware problem, principally the framestore used to store the sonar data.

Development of the sensor continues. Some re-engineering of the system components is necessary to address problems of system calibration. We wish to improve the quality of of the sonar data by removing the error introduced by the framestore; and improve the sonar calibration method. We would also like to fully evaluate the system performance over a variety of working conditions, especially in turbid media.

8 ACKNOWLEDGMENTS

This work was partly funded by the European Union under ESPRIT BRA Project No. 8972, UNderwater Intelligent Operation and Navigation (UN.I.O.N).

References

[1] R. Bajcsy, "Active Perception.", *Proc. IEEE*, vol. 76, No. 8, pp 996-1005, 1988

[2] D. McKeown, "The visual imaging requirements of the next generation of ROVs.", *Underwater Technology*, vol. 15 pp11-15, 1989

[3] A. R. Henderson, "Light and lasers underwater." *Proc of Intervention* 1988

[4] D. Maddalena, W. Prendin, M. Zampato, " Innovations on underwater stereoscopy: the new developments of the tv-trackmeter." in Proc. of *Oceans 94*, vol 2, pp150-156, Brest, 1994

[5] S. W. Tetlow, "Use of laser light stripes to reduce backscatter in an underwater viewing system", PhD Thesis, Cranfield Inst. of Technology, U.K. 1993

[6] J. S. Fox, "Structured light imaging in turbid water." *In Proc SPIE, Underwater Imaging*, Vol. 980, pp66-71, 1988

[7] D. J. Wenzel, S. B. Seida, and V. R. Sturdivant, "Telerobot control using enhanced stereo viewing." in Proc SPIE *Telemanipulator and Telepresence Technologies* Vol. 2351, pp233-249, 1994

[8] R. E. Cole, J. O. Merrit, D. Fore, and P. Lester, "Remote manipulator tasks impossible without stereo TV." in Proc SPIE *Stereoscopic Displays and Applications* Vol. 1256, pp255-265, 1990

[9] M. J. Chantler, D. Lindsay, C. S. Reid and V. J. C. Wright, "Optical and acoustic range sensing for underwater robotics." in Proc. of *Oceans 94* Vol. 1, pp205-210, Brest, 1994

[10] M. J. Chantler, C. S. Reid and V. J. C. Wright, "Probabilistic sensing for underwater robotics." in Proc. *2nd Int. Con. Intelligent Systems Engineering*, Hamburg, 1994

[11] P. Besl, *Active, Optical Imaging Sensors*, Machine Vision and Applications, pp127-152, 1988

[12] R. Y. Tsai, "A versatile camera calibration technique for high accuracy 3D machine vision metrology using off-the-shelf TV cameras and lenses." IEEE Transactions on Pattern Analysis and Machine Intelligence, 3(4), pp323-344, 1987.

[13] R. G. Wilson, "Modeling and calibration of automated zoom lenses." PhD Thesis, Carnegie Mellon University, Pittsburgh, U.S.A, 1994

[14] J. Clark and A. M. Wallace, "Depth sensing by variable baseline triangulation." Proceedings of the Sixth British Machine Vision Conference, Birmingham, September 1995, pp227-236

[15] H. P. Moravec, "Sensor fusion in certainty grids for mobile robots." AI Magazine, Summer, pp61-74, 1988

[16] W. K. Stewart, "Multisensor modeling underwater with uncertain information.", Ph.D Thesis, Woods Hole Oceanographics Inst. July 1988

Equivalent Open-Loop Kinematic Calibration Avoiding Expensive Measurement Systems

G. Volpi[1]**, R. Cammoun, P. O. Vandanjon**
CEA - Service de Téléopération et Robotique - C.E.R.E.M. D.P.S.A.
B.P. 6 - 92265, Fontenay-aux-Roses cedex, FRANCE
Email: cammoun@cyborg.cea.fr

W. Khalil
Laboratoire d'Automatique de Nantes - URA C.N.R.S. 823 - Ecole Centrale de Nantes - 1, rue de la Noë - 44072, Nantes cedex, FRANCE
Email : khalil@lan.ec-nantes.fr

ABSTRACT

Most of kinematic calibration methods use expensive measurement equipment. To avoid external sensors, it is possible to use a calibration method based on a closed kinematic chain, formed between the robot base and the robot tool end-point constrained to the workspace. In this paper we propose a closed-loop approach which allows, recording only joint variables, to identify the same number of robot kinematic parameters as in the open-loop approach using an external measurement system. Experimental tests on a telemanipulator have demonstrated the method reliability.

KEYWORDS : manipulators, kinematic calibration

INTRODUCTION

Kinematic calibration is a process which improves robot positioning accuracy through adjusting robot geometric parameters. This means to identify a more accurate relationship between joint sensor data and actual workspace position of robot end-effector. Calibration method is divided into two main categories: the open-loop methods based on external pose measurement systems and the closed-loop methods based on constraining the robot end-effector.

Open-loop methods could be contact-less or contact methods according to the measurement system. Among contact-less methods, the main systems are based on theodolites [18] [3], which present the drawback of slowness, and on cameras [19]. Other measure systems are based on laser tracking [11] and on time of flight devices [15]. Among contact methods, we can find methods based on coordinate measurement machines [2], on precalibrated plate [17], and on three cables attached to the end-effector [10].

Closed-loop methods could be classified by the type of constraint. [16] use a plane as constraint. Line constraints has been achieved by a fixture [13] or by a laser beam [12].

---

[1]Currently funded by University of Trieste, Italy

Point constraint has been used for robots which can achieve the same end-effector pose by more than one configuration. This constraint has been accomplished by a support on which a pointed tool is leant or by a special tool with a terminal spherical joint fixed to the workspace [9] or a hinge link [1]. Point constraint can be also carried out by grasping an extensible ball-bar with a terminal spherical joint [5].

The method proposed in [9] is very promising because it does not need external sensors and it represents a low cost approach. Its algorithms are very robust but they do not allow to identify the same number of robot kinematic parameters as in the open-loop method. This paper presents an improvement to this approach which overcomes the problem. Moreover joint gain identification is introduced and a method to select the identifiable parameters is presented.

PROBLEM APPROACH

Parameter calibration requires three levels, the **joint level calibration,** the **geometric level calibration** and the **nonkinematic level calibration** [14].

The joint level consists of identifying the parameters which characterise the relationship between the signals $\mathbf{q}_{read}$ produced by the joint sensor and the actual joint displacements $\mathbf{q}$. We assume the joint relationship linear, such as $\mathbf{q} = \mathbf{k}\ \mathbf{q}_{read} - \mathbf{q}_{off}$, where $\mathbf{q}_{off}$ represents the joint offsets and $\mathbf{k}$ the transmission gains. Non-linear effects (nonkinematic level calibration) as joint compliance or backlash are not taken into account in this paper.

The geometric level consists of identifying the geometric model parameters which define the transformations between all joint frames, with the assumption that the robot is composed of rigid links.

Calibration process could be divided in the following successive steps. The first step is the **modelling step** based on choosing the modelling notation and building the geometric model. The **selection step** consists of eliminating the unidentifiable parameters and selecting a set of identifiable parameters. The **measurement step** means to collect data from sensors. The **identification step** is based on a mathematical process using the collected data and the calibration algorithm to identify all the selected parameters.

GEOMETRIC MODEL

In this paper the modelling step has been achieved using the modified Denavit-Hartenberg notation to describe the link frames [7]. In this notation, a robot is composed of (N+1) links, link **0** is the base, link **N** is the terminal link, the joint **i** connects link **i-1** and link **i**. The direct geometric model

$$^{0}T_{N+1} = {}^{0}T_{1}\ {}^{1}T_{2} \ldots {}^{N}T_{N+1} \tag{1}$$

represents the tool pose, it also represents the transformation matrix which defines the tool frame **N+1** with respect to the base frame **0**.

Definition of Robot Geometric Parameters

A coordinate frame R_i is defined fixed with respect to link **i**. The axis of joint **i** is supposed along $\mathbf{z}_i$ while the $\mathbf{x}_i$ axis is defined as the common perpendicular to $\mathbf{z}_i$ and $\mathbf{z}_{i+1}$.

The definition of a joint is still carried out by means of 4 parameters as in the original Denavit-Hartenberg notation.

The four parameters (Figure. 1) which describe the transformation between the links are: α_i (angle between $\mathbf{z}_{i-1}$ and $\mathbf{z}_i$ around $\mathbf{x}_{i-1}$), θ_i (angle between $\mathbf{x}_{i-1}$ and $\mathbf{x}_i$ around $\mathbf{z}_i$), d_i (distance between $\mathbf{O}_{i-1}$ and $\mathbf{z}_i$ along $\mathbf{x}_{i-1}$), r_i (distance between $\mathbf{x}_{i-1}$ and $\mathbf{O}_i$ along $\mathbf{z}_i$). For a rotational joint $\theta_i = q_i$, for a prismatic joint $r_i = q_i$.

The transformation ${}^{i-1}T_i$, which defines the tool frame **i** with respect to the base frame **i-1**, is equal to:

$$ {}^{i-1}T_i = Rot(\mathbf{x},\alpha_i)\ Trans(\mathbf{x},d_i)\ Rot(\mathbf{z},\theta_i)\ Trans(\mathbf{z},r_i) \qquad (2) $$

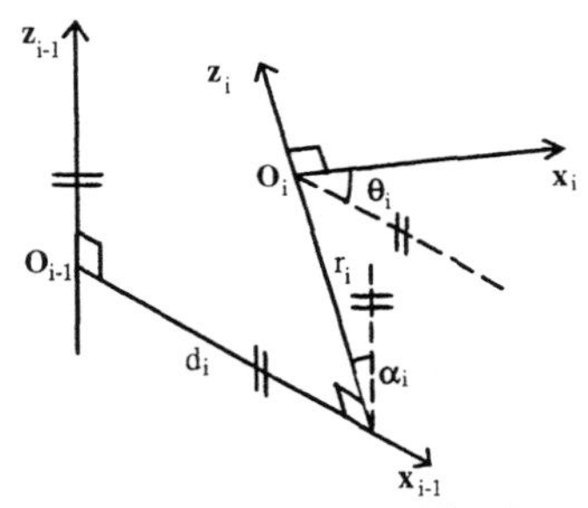

Figure. 1. Definition of link frames

Definition of Consecutive Parallel Axis

When two consecutive axes are parallel, for slight misalignments the geometric parameter r_i can approach infinity. To solve this problem it is possible to use the previous model (2) for all axes except for consecutive parallel axes [4]. In this case it is reasonable to add a parameter β which represents a rotation around the $\mathbf{y}_{i-1}$ axis. The matrix ${}^{i-1}T_i$ becomes equal to:

$$ {}^{i-1}T_i = Rot(\mathbf{y},\beta_i)\ Rot(\mathbf{x},\alpha_i)\ Trans(\mathbf{x},d_i)\ Rot(\mathbf{z},\theta_i)\ Trans(\mathbf{z},r_i). \qquad (3) $$

Definition of Tool Parameters

The tool definition is based on six parameters to define arbitrarily the tool frame. The tool is defined by:

$$ {}^{N}T_{N+1} = Rot(\mathbf{z},\gamma_{N+1})\ Trans(\mathbf{z},b_{N+1})\ Rot(\mathbf{x},\alpha_{N+1})\ Trans(\mathbf{x},d_{N+1})\ Rot(\mathbf{z},\theta_{N+1})\ Trans(\mathbf{z},r_{N+1}). \qquad (4) $$

This transformation can be reduced and the parameters γ_{N+1} and b_{N+1} are brought back and added respectively to θ_N and r_N. Finally we have $\theta'_N = \theta_N + \gamma_{N+1}$ and $r'_N = r_N + b_{N+1}$ in transformation ${}^{N-1}T_N(\alpha_N, d_N, \theta'_N, r'_N)$ and the tool transformation ${}^{N}T_{N+1}(\alpha_{N+1}, d_{N+1}, \theta_{N+1}, r_{N+1})$ will be built as in (2). Both θ_{N+1} and r_{N+1} are geometric parameters.

PARAMETER IDENTIFICATION

With the previous model the robot parameters consist of the offsets $\mathbf{q}_{off}$, the gains $\mathbf{k}$, the geometric parameters $\boldsymbol{\alpha}$, $\mathbf{d}$, $\boldsymbol{\theta}$, $\mathbf{r}$, the parameters for the parallel axes $\boldsymbol{\beta}$.

Open-Loop Classic Calibration

In classic approach an external measurement system is required to collect tool positions. During measurement step the real tool positions $\mathbf{y}_r$ and joint variables $\mathbf{q}$ are recorded. After collecting M configuration data, we proceed to parameter identification.

Using a first order Taylor development of parameter error it is possible to solve the identification problem, which is a non-linear estimation problem. The linearization gives the relation

$$\Delta\mathbf{y} = J_{\mathbf{X}}(\mathbf{q})\, \Delta\mathbf{X}, \tag{5}$$

where $\mathbf{y} = \mathbf{y}_r - \mathbf{y}_c$ represents the error between the measured real tool position $\mathbf{y}_r$ and the computed position $\mathbf{y}_c$, $J_{\mathbf{X}}(\mathbf{q})$ represents the (3×K) matrix generic Jacobian computed with respect to all the K kinematic parameters $\mathbf{X} = [X_1, X_2, ..., X_K]$, $\Delta\mathbf{X}$ represents error between the geometric parameters $\mathbf{X}_r$ and the nominal values $\mathbf{X}$.

To identify the kinematic parameters, equation (5) is applied to M configurations. It results in

$$\Delta\mathbf{Y} = \mathbf{W}\, \Delta\mathbf{X}, \tag{6}$$

$$\text{where } \Delta\mathbf{Y} = \begin{bmatrix} \Delta\mathbf{y}_1 \\ \dots \\ \Delta\mathbf{y}_M \end{bmatrix}, \mathbf{W} = \begin{bmatrix} J_{\mathbf{X}}(\mathbf{q}_1) \\ \dots \\ J_{\mathbf{X}}(\mathbf{q}_M) \end{bmatrix} \tag{7}$$

The tool position $\mathbf{y}_c$ is computed from direct model ${}^0\mathbf{T}_{N+1} = \begin{bmatrix} {}^0\mathbf{s}_{N+1} & {}^0\mathbf{n}_{N+1} & {}^0\mathbf{a}_{N+1} & {}^0\mathbf{P}_{N+1} \\ 0 & 0 & 0 & 1 \end{bmatrix}$ as $\mathbf{y}_c = {}^0\mathbf{P}_{N+1}$. The least-square solution to system (6) is given by

$$\Delta\mathbf{X} = (\mathbf{W}^T\, \mathbf{W})^{-1}\, \mathbf{W}^T\, \Delta\mathbf{Y}. \tag{8}$$

Once calculated $\Delta\mathbf{X}$, the new parameters $\mathbf{X}'$ are computed as

$$\mathbf{X}' = \mathbf{X} + \Delta\mathbf{X}, \tag{9}$$

then the matrix $\mathbf{W}$ is updated and the solution (8) is iterated again until we obtain a satisfactory precision.

Point Constraint Closed-loop Calibration Methods

The point constraint approach avoids the use of an external measurement system. The tool position is constrained on a point, while the orientation is free. The position of the constraint point is not known. By subtracting each member of two equations (5), computed on the configurations $\mathbf{q}_a$ and $\mathbf{q}_b$, the real position $\mathbf{y}_r$ is eliminated [9] and we obtain

$$\Delta\mathbf{y}^* = [\, J_X(\mathbf{q}_b) - J_X(\mathbf{q}_a) \,]\, \Delta\mathbf{X}, \tag{10}$$

where $\Delta\mathbf{y}^*$ represents the error between two computed tool position ${}^0\mathbf{P}_{N+1}(\mathbf{q}_a,\mathbf{X})$ - ${}^0\mathbf{P}_{N+1}(\mathbf{q}_b,\mathbf{X})$. This method is very robust for big errors, but it does not allow to identify the same number of robot kinematic parameters as in the open-loop approach.

A solution to this drawback is to use a system of equation (5) like the open-loop classic method, with an external measurement system, replacing the real position $\mathbf{y}_r$ by the average value $\mathbf{y}_m$, estimated on all computed tool positions as

$$\mathbf{y}_m = \sum_{i=1}^{M} {}^0\mathbf{P}_{N+1}(\mathbf{q}_i, \mathbf{X}) / M, \tag{11}$$

where $\mathbf{X}$ are the nominal parameters. Thus the error vector $\Delta\mathbf{y}$ in equation (5) is replaced by

$$\mathbf{y}_m - {}^0\mathbf{P}_{N+1}(\mathbf{q},\mathbf{X}) \tag{12}$$

Each iteration $\mathbf{y}_m$ and $J_X(\mathbf{q})$ are updated. The algorithm convergence is slow in comparison to (10), but the system dimension is smaller. The identifiable parameters are the same as those of classic calibration using an external sensor. To achieve a kinematic calibration not connected to the workspace point where we constrain the tool, each set of measures should be realised with the tool constrained on different points of workspace.

PARAMETER SELECTION

After modelling step the second calibration step consists of selecting the set of parameters which will be calibrated. The main problem is that the identification algorithm does not work if in the parameter set $\mathbf{X}$, there is any dependent parameters. The selection step can be carried out only by simulation. The set of identifiable parameters can be selected before measurement step.

Simulation of Matrix W

The procedure consists of building the matrix $\mathbf{W}$ (7) on a simulated set of configurations. For open-loop classic calibration the set of configurations can be casually computed, on the contrary in point constraint closed-loop calibration, the simulated configurations must not give any tool position error. To compute this configuration set we need an inverse geometric model, a good method consists of building an algorithm which generates a new configuration, starting from a random configuration close to the initial configuration and using an iterative least-square process. Once achieved a configuration, the procedure restarts until all configuration set is computed. The dimension of the configuration set must be K/3, where K is the number of kinematic parameters, to have a complete system. Since the objective is to compute the set of identifiable parameters and not the parameters themselves, the values of kinematic parameters are not significant in parameter selection and we can use the non-calibrated values.

Identifiable Parameters

The linearly independent parameters can be selected by analysing the diagonal of matrix R computed by QR decomposition of observation matrix **W**. All non-zero diagonal values of matrix R correspond to linearly independent parameters. The other parameters are divided in unidentifiable parameters and linearly dependent parameters.

The unidentifiable parameters have not effect on the model and the relative Jacobian column is zero.

The linearly dependent parameters can be calibrated. They depend by other kinematic parameters, thus each calibration process can identify only one parameter in a group of connected parameters. The Jacobian columns of a dependent parameter are a linear combination of some other columns. The relation between dependent parameters can be computed [8].

Selection Approach

To select a parameter in a set of dependent parameters, we take into account the following remarks. The joint offsets are the largest contribute to position error, accounting for almost 90% of root mean squared (RMS) value of error and link length errors explains only about 5% of the RMS errors [6]. Moreover it is more practical to identify the parameters which can be updated easily in robot control system without changing the model. Consequently the joint offset, the non-zero lengths and the tool parameters must be firstly selected.

Finally, in calibration algorithm, the matrix **W** will be composed by the columns related to the selected parameters.

EXPERIMENTAL RESULTS

The method has been tested on the slave manipulator RD500, used in teleoperation application at C.E.A. laboratories. The robot is a six axis robot characterised by a closed-loop structure as a parallelogram mechanism. The joint sensors are placed on motor axes.

After robot modelling to obtain an equivalent open-loop robot, we collect all the joint configurations while the robot grasps a link with a terminal spherical joint. During measurements, the spherical joint has been fixed to five different points of robot workspace.

Parameter selection allows to verify that it is possible to identify the same parameter as in open-loop method. The number of identifiable parameters is 26. All parameters are identifiable except r_6 which represents firstly a zero tool dimension (brought back from tool transformation) and a set of unessential parameters: the parameters r_2, α_7, θ_7, which are initially zeros. The first joint parameters have been deselected because an absolute calibration of base frame is not taken into account in this paper.

The following step is the parameter identification. Carried out in 15 iterations the algorithm leads to a new set of calibrated parameters with a final RMS error of 1.2 mm (maximum error 3.8 mm) while the initial RMS position error was 5.4 mm (maximum error 10 mm). TABLE I and II give respectively the initial parameters and the correct values. Big errors on robot lengths, which exceed the machine tolerance, are justified because we have not taken into account the joint compliance.

TABLE I. Parameters of RD500 and its tool (mm and degrees)

Joint	$\theta_{off\,i}$	k_i	α_i	d_i	r_i	β_i
1	0.8709	-3.1416	0	0	0	
2	-92.4066	1.5708	-90	0	0	
3	-1.4267	3.1416	0	575	0	0
4	-13.6937	3.1416	90	0	825	
5	58.4875	1.2566	-90	0	0	
6	0	3.1416	90	0	0	
tool	$\theta_7 = 0$		0	-155.8	337.8	

TABLE II. Calibrated parameters (mm and degrees)

Joint	$\theta_{off\,i}$	k_i	α_i	d_i	r_i	β_i
1	0.8709	-3.1416	0	0	0	
2	-91.8533	1.5694	-89.94	1.65	0	
3	-1.8855	3.1242	0.089	575.6	2.6	0.22
4	-14.5125	3.1398	89.71	-1.49	831	
5	58.5879	1.2710	-90.07	1.88	-0.41	
6	0.1974	3.1358	90.02	-0.03	0	
tool	$\theta_7 = 0$		0	-156.8	335.7	

CONCLUSION

This paper presents a methodical approach to robot kinematic calibration. An open-loop calibration needs expensive external sensor to compute the tool position. A closed-loop method, based on point constraint, does not need this equipment, but it only requires joint variable data. Before parameter calibration a selection of identifiable parameters must be performed. The calibration result and the easiness of realisation show the reliability of the method.

REFERENCES

1. D.J. Bennett and J.M. Hollerbach. "Autonomous calibration of single-loop closed kinematic chains formed by manipulators with passive endpoint constraints". *IEEE Trans. Robotics and Automation, vol. 7, no. 5*, 1991.
2. J.-H. Borm and C.-H. Menq. "Experimental study of observability of parameter errors in robot calibration". *Proc. IEEE, Robotics and Automation Conf., pp. 587-592*, 1989.
3. J. Chen and L.M. Chao."Position in error analysis for robot manipulators with all rotary joints". *Proc. IEEE, Robotics and Automation Conf., pp. 1011-1016,* 1986.
4. S.A. Hayati. "Robot arm geometric link calibration". *Proc. IEEE, Decision and Control Conf. , pp. 798-800,* 1985.

5. A. Goswami, A. Quaid and M. Peshkin. "Complete parameter identification of a robot from partial pose information". *Proc. IEEE, Robotics and Automation Conf., pp. 168-173*, 1993.
6. P. Judd and B. Knasinski. "A technique to calibrate industrial robots with experimental verification". *Proc. IEEE, Robotics and Automation Conf., pp. 351-357*, 1987.
7. W. Khalil and J.F. Kleinfinger. "A new geometric notation for open and closed-loop robots". *Proc. IEEE, Robotics And Automation Conf., pp. 1174-1180,* 1986.
8. W. Khalil and G. Gautier. "Calculation of the identifiable parameters for robots calibration". *9th IFAC/IFORS Symp. Identification and System Parameter Estimation, pp. 888-892,* 1991
9. W. Khalil, G. Garcia and J.-F. Delagarde. "Calibration of the geometric parameters of robots without external sensors". *Proc. IEEE, Robotics and Automation Conf., pp. 3039-3044*, 1995.
10. D. Payannet, M.J. Aldon and A. Liégeois. "Identification and compensation of mechanical errors for industrial robot". *Proc. 15th ISIR, Tokyo, pp. 857-864,* 1985.
11. J. Prenninger, M. Vincze and H. Gander. "Measuring dynamic robot movements in 6 D.O.F. and real time". *Robot Calibration,* ed. R. Bernhardt and S. L. Albright, London, Chapman & Hall, 1993
12. W.S. Newman and D.W. Osborn. "A new method for kinematic parameter calibration via laser line tracking". *Proc. IEEE, Robotics and Automation Conf., pp. 160-165*, 1993.
13. J.-M. Renders, E. Rossignol, M. Becquet and R. Hanus. "Kinematic calibration and geometrical parameter identification for robots". *IEEE Trans. Robotics and Automation, vol. 7, no. 6, pp. 721-731*, 1991.
14. Z. S. Roth, B.W. Mooring and B. Ravani. "An overview of robot calibration". *IEEE J. Robotics and Automation, vol. RA- 3, no. 5, pp. 377-385,* 1987.
15. W. Stone, A.C. Sanderson and C.P. Neuman. "Arm signature identification". *Proc. IEEE, Robotics and Automation Conf., pp. 41-48,* 1986.
16. G.-R. Tang and L.-S. Liu. "A comparison of calibration methods based on flat surfaces". *Int. Conf. Automation, Robotics and Computer Vision,* RO-14.3.3, 1991.
17. W.K. Veitschegger and C. Wu. "A method for calibrating and compensating robot kinematic errors". *Proc. IEEE, Robotics and Automation Conf., pp. 39-44*, 1987.
18. D.E. Whitney, C.A. Lozinsky and J.M. Rourke. "Industrial robot forward calibration method and result", *J. Dynamics Systems, Measurements and Control, vol. 108, pp. 1-8,* ASME,1986.
19. H. Zuang, L. K. Wang and Z. Roth. "Error-model-based robot calibration using a modified CPC model". *Robotics & Computer-Integrated Manufacturing, vol. 10, no. 4, pp. 287-299*, 1993.

ROBOTICS FOR MINE COUNTERMEASURES

JOHN P. WETZEL
ALLEN D. NEASE
Wright Laboratory, Air Base Technology Branch
Tyndall AFB, FL 32404

ABSTRACT

This paper describes a program currently underway to develop an assault beach minefield breaching system. The key component of this system is an unmanned ground vehicle that carries all of the various subsystems used in the clearing process. The vehicle, a D7G tractor, is teleoperated and equipped with a GPS-based navigation and mapping system for accurate positioning and tracking.

KEYWORDS: robotics, navigation, mine countermeasures, teleoperation

BACKGROUND

The US Air Force/Wright Laboratory/Air Base Technology Branch, Tyndall Air Force Base, Florida (WL/FIVC), performs civil engineering research in the areas of fire protection and crash rescue, pavements and structures, energy systems, and construction automation and robotics. The construction automation and robotics group, which is also part of the US Office of Secretary of Defense Joint Robotics Project (OSD/JRP), supports agencies in developing and fielding unmanned vehicles for use in unexploded ordnance remediation, mine countermeasures, and other hazardous missions.

The Joint Amphibious Mine Countermeasures (JAMC) Program is aimed at providing a capability for assault beach minefield breaching using unmanned vehicles as prime movers. The program will develop and demonstrate near-to-midterm techniques and equipment to neutralize landmines and light obstacles that hinder vehicle-mounted assaults from the sea. The system will clear assault lanes from the beach/water transition zone to the craft landing zone and will provide on-the-beach mine clearance for rapid follow-up force projection.

WL/FIVC is developing, integrating, and testing the subsystems on the prime mover, including teleoperation and navigation and mapping, for use in JAMC. The current focus of the program is on participation in the 1997 Countermine Advanced Concept Technology Demonstration sponsored by the US Office of Secretary of Defense.

JAMC SYSTEM

The basic requirement for the JAMC System is to neutralize mines and defeat light obstacles in a 46 m by 46 m (150 ft by 150 ft) area in 1.5 hours. JAMC is intended for use on beaches with limited direct fire and no observed indirect fire.

The JAMC System consists of a suite of tools which are delivered to the beach in an amphibious landing craft. The suite of tools is intended to permit the clearing operation to be tailored to specific beach conditions. To outline operations of the JAMC System, a basic concept of employment was developed which describes the events involved in a clearance mission. Employment of the JAMC System is consistent with the emerging operational maneuver from the sea doctrine.

The JAMC System uniquely combines mechanical, electromagnetic, and explosive mine countermeasures on a teleremotely operated platform to perform the beach minefield neutralization and marking mission. The integrated JAMC system is shown in figure 1. To meet the objective of a near to mid-term MCM system, the JAMC Program is maximizing the use of non-developmental items (NDI) and streamlined research and development. Previous testing and integration of MCM technologies is being leveraged to the maximum extent possible.

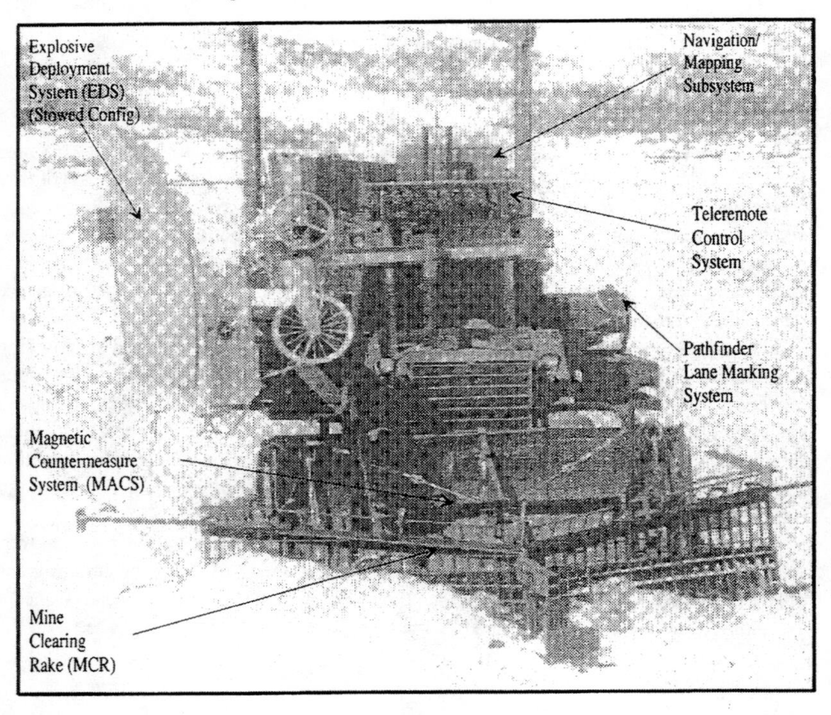

Figure 1. Integrated JAMC System

The Caterpillar D7G tractor was chosen as the mobility platform for the JAMC System. The D7G is a common stock item in the United States Marine Corps inventory, and is therefore an available asset for use for such a system. With a weight of nearly 50,000 lbs, the D7G is capable of employing all the MCM equipment while still maintaining a low enough weight to be delivered to the beach on an amphibious landing craft.

To allow manned operation on the D7G, a protective armor kit is integrated on the tractor. The D7G fording kit allows a fording capability of up to 60 inches of water. This allows the system to be deployed in the surf zone from an amphibious landing craft near the beach. A mine rake is used for mechanical clearance of the initial transit assault breach lanes.

A chain array is being developed as another means for mechanical clearance. The chain array will be dragged between a pair of D7G tractors to clear light obstacles such as concertina wire and engineers stakes. In addition, the chain array will assist in the initiation of tilt rod fuze and anti-handling device mines. Redundancy is incorporated in the chain array design by providing a chain mesh along the length of the array to allow for potential mine detonations which will sever the chain. The chain array also contains a magnetic segment to assist in neutralizing magnetic fuzed mines. The magnetic segment is attached to the rear of the chain segment of the array.

The Magnetic Countermine System (MACS) will be integrated on the front of the JAMC System. The MACS is designed to initiate magnetic fuzed mines in front of the vehicle to enhance survivability. As a future part of the USMC inventory, MACS is considered NDI and GFE for the JAMC System. This electromagnetic capability has been implemented on other USMC vehicles, including the Amphibious Assault Vehicle (AAV).

An explosive net array containing shape-charged munitions will be mechanically deployed to the side of the D7G tractor. The overall net dimensions are 24 feet wide by 150 feet long. The net contains two fuzes, one on each end for redundancy. The fuzes are used to remotely arm and detonate the net array. The net array has been designed to defeat all mine types, regardless of fuzing, to an overburden depth of 5 inches. This mine neutralization is accomplished with shape-charged munitions. The jet formed from the shape charges damages or detonates the mine through kinetic energy.

To physically mark the cleared area for assault troops, a marking system will be mounted to the D7G. The marking system will deploy marker poles along the perimeter of the cleared area for subsequent troop maneuvers through the area.

TELEOPERATION

A teleoperation subsystem (TS) is being developed for use on the D7G tractor. This system is part of an overall effort to develop a standardized teleoperation system for use on a variety of military ground vehicles, including the D7G tractor, M1A1 Tank, and HMMWV. The system is being designed for versatility and commonality to facilitate integration on these vehicles. The D7G is the first vehicle in which this standardized teleoperation system was integrated.

A vehicle-mounted computer provides the intelligence, and a set of antennas and radios provide the necessary communications. A set of actuators and cables are used by the TS to control the vehicle. Video cameras are provided for non-line-of-sight operations. A portable operator control unit (OCU) provides the necessary controls for an operator to command vehicle motion and subsystem functions. Through the TS, an operator has full control of the vehicle and all subsystems. Figure 2 shows a picture of the OCU. A single joystick is used to control vehicle throttle and steering. A second joystick is used to control the blade lift/lower and tilt. Switches are used to control vehicle discrete functions, including engine start blade float. A panel of switched is included to provide control of all required subsystem functions. The monitor directly above the control joysticks and switches displays the vehicle status guages, such as oil pressure and fuel level. A video display is housed in the cover of the OCU; a single camera view, or a quad view containing all four camera views, may be chosen for display. A joystick is also provided to control the pan/tilt camera motion.

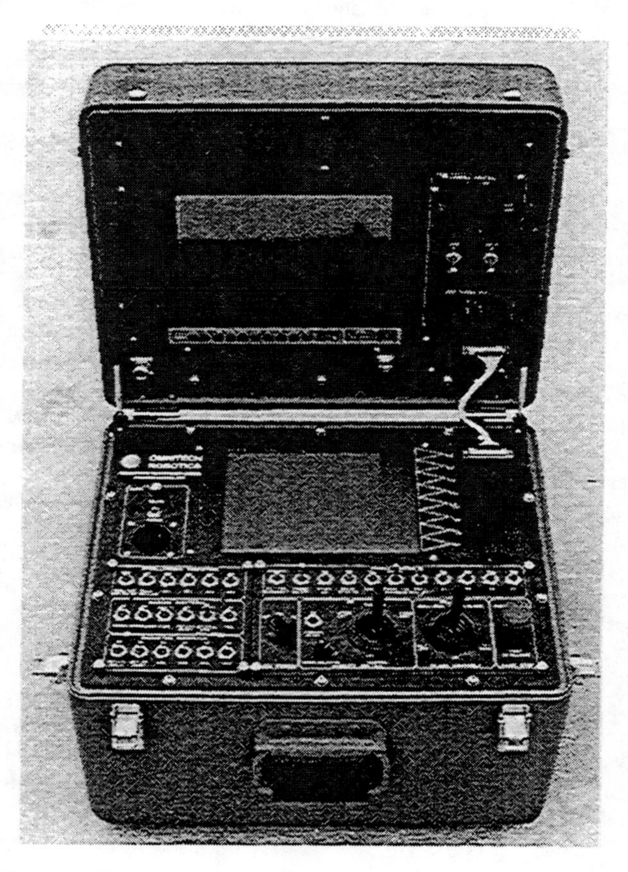

Figure 2. Teleoperation Subsystem Operator Control Unit (OCU)

NAVIGATION/MAPPING

To assist in achieving the navigational accuracy required to deploy the explosive net arrays, a navigation and mapping subsystem (NMS) is being developed. The first function of the NMS is to graphically display the D7G dozer position and orientation to aid the remote operator in controlling the dozer. The second function of the NMS is to allow the dozer to be navigated autonomously. The operator will input high level commands such as deploy an explosive array or return to a resupply point. The dozer will then be controlled autonomously by on-board computers to accomplish this navigation task. Electronic navigation information will be provided to the naval assault operation through the NMS.

The NMS was developed by integrating an Inertial Navigation System (INS) and Global Positioning System (GPS). The INS system is capable of outputting vehicle position at a rate of ten hertz relative from where the system is initialized. The drift error of the unit is approximately two feet after four minutes of operation. The GPS is employed in a differential correction mode. One of the GPS units always remains stationary at a known location. Error correction data from this base station unit is communicated to the GPS unit on board the moving vehicle. Tests have shown that the GPS units can provide accurate real time (one hertz rate) measurement of position to within a standard deviation of approximately 1.5 inches.

INTEGRATION AND TEST

Each of the JAMC subsystems described above was integrated on the D7G tractor. All integration of the JAMC subsystems and testing of the JAMC System was conducted at WL/FIVCF, Tyndall AFB. Developmental testing was performed on each of the prototype MCM subsystems integrated on the D7G tractor independently. Following these tests, an overall prototype system test was conducted. Results of these tests are being used in the next phase of the program to modify subsystem hardware to enhance JAMC System performance.

Following the first stage of developmental testing conducted in 1995, a single, fully-integrated prototype JAMC System was demonstrated on 2 November 1995. A Marine Corps Combat Engineer, with less than seven days of training, teleremotely conducted the operation (figure 3). The purpose of the JAMC Demonstration was to demonstrate the abilities of the individual and integrated technical MCM concepts to neutralize landmines and light obstacles from a designated beach area. The Demonstration scenario was representative of the JAMC mission, and was successful in demonstrating the combined MCM technologies.

Figure 3. Marine Corps Combat Engineer OCU Operations

ACKNOWLEDGMENTS

This work is being sponsored by the Marine Corps Systems Command (MARCORSYSCOM) Amphibious Warfare Technology Directorate (AW) in order to satisfy a mine countermeasures requirement for a near to mid-term mine field/light obstacle breaching capability. The authors would like to thank LtCol Walt Hamm and Rick Ellis for their program oversight and management. The authors would also like to acknowledge the dedicated work of the hardware development, integration, and test support team.

EVALUATION OF AN INTEGRATED INERTIAL NAVIGATION SYSTEM AND GLOBAL POSITIONING SYSTEM UNDER LESS THAN OPTIMAL CONDITIONS

Jeffrey S. Wit, Carl D. Crane III, David G. Armstrong II
Center for Intelligent Machines and Robotics
University of Florida, Gainesville, Florida 32611

ABSTRACT

Accurate real time position data is required for an outdoor autonomous ground vehicle to successfully navigate a pre-planned path. Two commonly used positioning systems for the navigation of an outdoor autonomous ground vehicle are an Inertial Navigation System (INS) and a Global Positioning System (GPS). The INS offers position data at a 10 Hz rate but tends to drift over time. The GPS offers position data without drift but at a 1 Hz rate. Through the use of an external Kalman filter, the integration of these two independent positioning systems offers accurate real time position data at a 10 Hz rate. Although the INS data is continually drifting, the Kalman filter is able to maintain an accurate navigation solution by using the GPS data. Since the Kalman filter depends on GPS to maintain its accuracy, the loss of or inaccurate GPS data will result in a less accurate filter solution. The focus of this paper is to evaluate the accuracy of the Kalman filter solution during these circumstances. Test results have shown that the Kalman filter was able to maintain position accuracies of approximately two meters or less for a period of five minutes after the loss of GPS data and was able to smooth through momentary inaccurate GPS data .

KEYWORDS: Global Positioning, Inertial Navigation, Kalman Filter, Autonomous Navigation, Unexploded Ordnance.

INTRODUCTION

An engineering program has been established at Wright Laboratory, Tyndall Air Force Base, Florida. The objective of this program is the autonomous cleanup of various Department of Defense facilities containing buried unexploded ordnance (UXO). Autonomous operation removes the human operator from a potentially hazardous environment and offers quality control by insuring that 100% of the site is efficiently searched.[1]

Removing buried munitions is a two step process. First, the buried munitions must be located. This is accomplished by an Autonomous Survey Vehicle (ASV) towing a sensor package across 100% of the area to be surveyed. As the ASV navigates, it collects and stores time-tagged position data as well as data form the sensor package. This data is then post-processed to determine the locations of the buried ordnance. Second, the buried munitions must be uncovered. This is accomplished by an autonomous excavator. The autonomous excavator navigates to each identified ordnance and removes it.[2]

The Center for Intelligent Machines and Robotics (CIMAR), at the University of Florida, has been contracted to develop the navigation system for these vehicles required to locate and remove the buried ordnance. A Kawasaki Mule 500 all-terrain vehicle was modified for computer control in order to be used as a Navigation Test Vehicle (NTV) (Fig. 1). Computer control was realized by mounting actuators and encoders on the vehicle's steering wheel, throttle, brake and transmission. Closed loop control of each actuator was then obtained by means of a VME based computer running under the VxWorks operating system. Autonomous navigation was accomplished by first calculating a survey path based on the coordinates of the survey site boundaries.[3] Figure 2 shows a typical path generated for the NTV to navigate. The vehicle's position relative to this path is then determined by utilizing an external Kalman filter which integrates the position data from an Inertial Navigation System (INS) and a Global Positioning System (GPS).[4]

Figure 1. Navigation Test Vehicle.

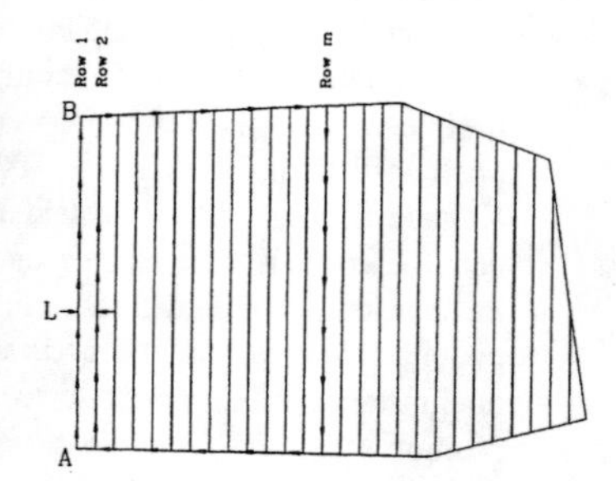

Figure 2. Planning of Survey Path

The INS used on the NTV is the H-726 Modular Azimuth Positioning System (MAPS) marketed by Honeywell Inc., Clearwater, Florida. The MAPS is a completely self-contained and strapped down system. It requires only an initial position at the start of operation. It then uses three ring laser gyros and three accelerometers to calculate a position solution relative to the initial position. This solution is made available to the controlling computer at a rate of 10 Hz.[5]

A problem encountered with INS is that the position solution tends to drift over time from the actual position due to measurement and system errors. In order to reduce the amount of drift, the MAPS can make use of zero-velocity updates (ZUPT) at a determined interval, typically every four minutes. This is not an acceptable solution since stopping the vehicle every four minutes is undesirable. Figure 3 shows the results of a dynamic test of the MAPS position data. The error in the MAPS position was calculated by comparing it to a post processed GPS position.[6]

The GPS used by the NTV consists of two Z-12 receivers marketed by Ashtech Inc.,Sunnyvale, California. The Z-12 receiver is designed to make full use of the Navstar Global Positioning System. It has twelve independent channels and can track all of the

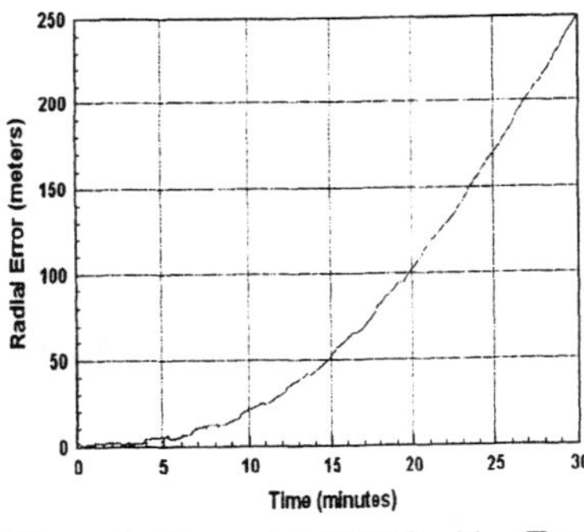

Figure 3. Dynamic MAPS Position Test

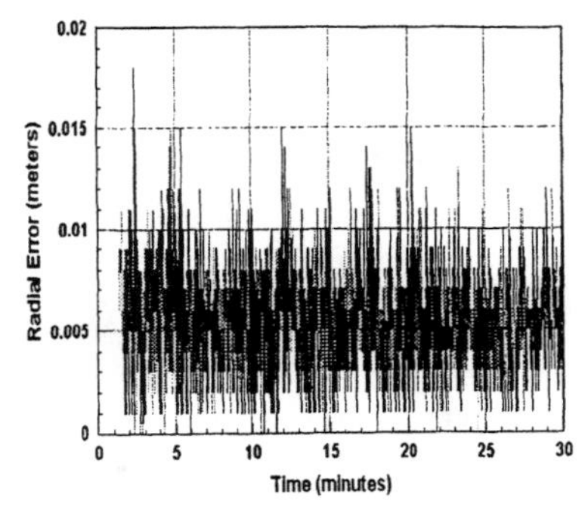

Figure 4. Dynamic GPS Position Test

satellites in view automatically. Data from the Z-12 is made available to the controlling computer at a rate of 1 Hz.[6]

The two GPS receivers operate in a carrier phase differential mode. In this mode, position can be calculated with centimeter level accuracy. Figure 4 shows a dynamic test of the real time differential GPS position data. The error in the real time differential GPS position was calculated by comparing it to a post processed GPS position.[6]

An external Kalman filter is used to integrate the INS and GPS position data to calculate a navigation solution. The filter includes nine states for the INS's position, position rate and tilt errors. Since the filter is being used for a ground vehicle, it is implemented in a local level geodetic frame. As a result, other INS errors can be included as process noise terms in the filter as tuning parameters. The filter processes the Differential GPS data by alternating between position and position change.[4]

The filter assumes that the position data from the GPS is accurate to within two-tenths of a meter. It uses this data to build a model of the INS position, position rate and tilt errors. By building this model of the INS errors, the filter has the ability to check the quality of the GPS data before it processes the data and it can continue to calculate an accurate position in the event of the loss of GPS data.

Under normal circumstances, GPS data is accurate within the two-tenths of a meter accuracy requirement. However, there are circumstances where GPS is unable to meet these requirements. First, GPS requires line of sight with at least four orbiting satellites in order to calculate a position. Tall trees and buildings are two examples where the NTV may not have line-of-sight with at least four satellites. Second, the accuracy of the calculated position depends on the changing geometry of the satellites in view. Third, the accuracy of the calculated position also depends on the amount of multi-path (reflected signals) which results from the surrounding environment.

The focus of this paper is an evaluation of the external Kalman filter during circumstances where there is a loss of GPS data and where there is inaccurate GPS data. The following sections will discuss the tests done in order to evaluate the Kalman filter navigation solution and the results derived from these tests.

TEMPORARY LOSS OF GPS DATA

Temporary loss of GPS data can occur when the NTV is in a place which does not allow a direct line-of-sight with at least four orbiting satellites. In order to test the filter's capability to maintain an accurate position solution during a temporary loss of GPS data, the following test was performed. Three areas within a survey site were marked off and the location of these areas were measured using GPS. The NTV was then allowed to autonomously navigate the survey site. Any time the vehicle entered any of the three marked areas, the GPS data was withheld from the Kalman filter.

Figure 5 shows the results of this test, where CALC is the calculated filter solution. In order for comparison, Figure 6 shows the same survey without any loss of GPS data. Table I shows the statistical results of these two surveys. From this test, it has been shown that the temporary loss of GPS data does not reduce the accuracy of the Kalman filter navigation solution.

TABLE I. Deviation of Real Time Data from Post Processed GPS Data.

	Temporary GPS loss		No GPS loss	
	CALC	GPS	CALC	GPS
Average Deviation (m)	0.23	0.02	0.23	0.03
Maximum Deviation (m)	1.19	1.12	1.54	1.86
Standard Deviation (m)	0.17	0.09	0.18	0.13

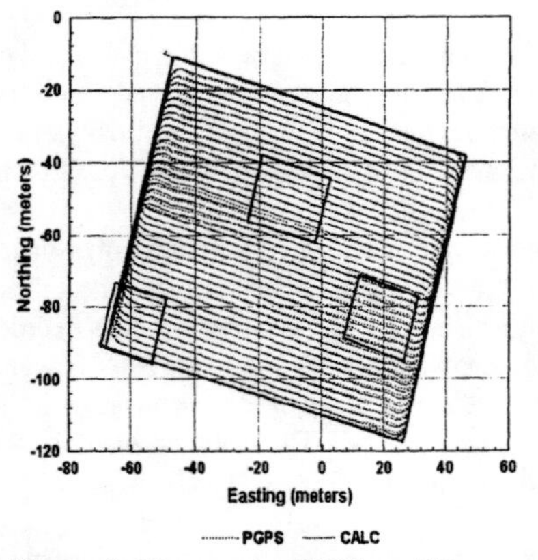

Figure 5. Temporary GPS Loss Test

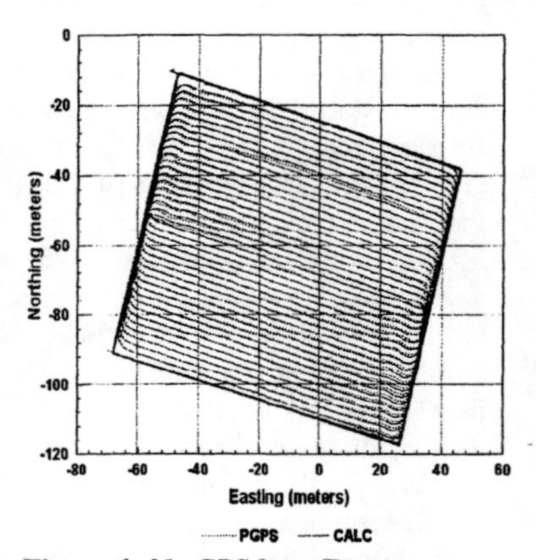

Figure 6. No GPS Loss Test

COMPLETE LOSS OF GPS DATA

Complete loss of GPS data is less likely to occur than a temporary loss. A complete loss of GPS data could occur when communication between the remote receiver and the base receiver fails. In order to test the filter's ability to maintain position accuracy after the loss of GPS data, the following tests were performed. The NTV was allowed to navigate autonomously for fifteen minutes. During this time, the filter constructed a model of the INS errors. After fifteen minutes, the GPS data was withheld from the filter. The NTV was allowed to continue to navigate for an additional ten minutes. For the next two tests, the NTV was allowed to autonomously navigate and construct its INS error model for thirty and forty-five minutes, respectively, before the GPS data was withheld from the filter.

The results of these tests are shown in Figure 7. In each of the tests, the filter was able to maintain an accuracy of four meters or less for two minutes after the loss of GPS data. Also from figure 7, it is apparent that the longer the filter has to construct its INS error model, the longer its navigation solution remains accurate after the loss of GPS data.

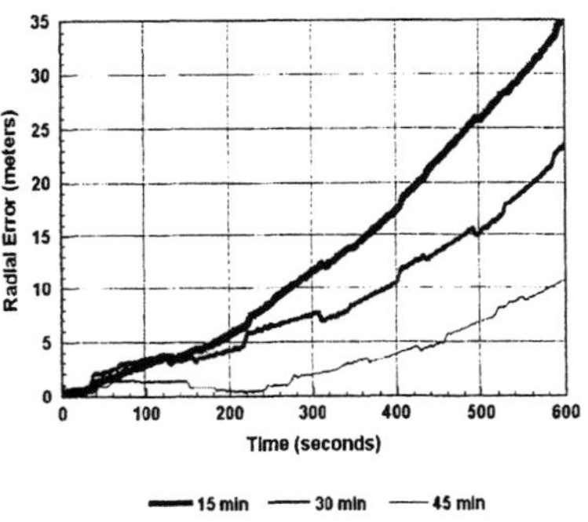

Figure 7. Complete GPS Loss Test

INACCURATE GPS DATA

Inaccurate GPS data could occur due to the changing geometry of the orbiting satellites. It could also occur in a high multi-path environment. In order to test the Kalman filter's ability to reject inaccurate GPS data, the following test was performed. The NTV was allowed to navigate a survey site autonomously. Approximately every two minutes error was introduced, by means of software, into one of the GPS positions. No error was introduced to the GPS data before or after this position. The amount of error was slowly increased from one-half meter to nine meters.

The results of this test are shown in figure 8. The filter was able to smooth through the inaccurate GPS position data. However, the filter did not reject all of the inaccurate GPS position data. The filter may require some additional fine tuning of its parameters so that all of the inaccurate GPS position data will be rejected.

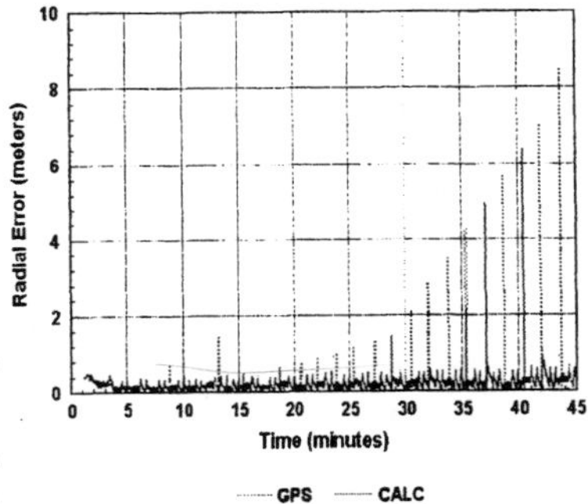

Figure 8. Inaccurate GPS Test

CONCLUSION

MAPS and GPS are two positioning systems which can be used to navigate an outdoor autonomous ground vehicle. Both positioning systems have inherent weaknesses that make it difficult to use one of them independent of another positioning system for the purpose of autonomous navigation. MAPS position data has been shown to drift over time due to measurement and system errors. GPS data is available at a rate of only 1 Hz. Also, GPS requires line-of-sight with at least four satellites and its accuracy depends on the satellite geometry and the amount of multi-path it experiences.

The integration of the MAPS and GPS position data, through the use of a Kalman filter, helps overcome the weaknesses of the two systems operating independently. The Kalman filter constructs a model of the INS error from the GPS position data. From the use of this error model and new MAPS position data, the filter is able to provide accurate position in between the GPS position data. With this model it is also able to provide an accurate position in the event of the loss of GPS data. It has been shown that the filter is able to provide an accurate position when there is a temporary loss of GPS data. It has also been shown that it is able to provide a position accurate to within four meters for two minutes after a complete loss of GPS. Finally it has been shown that the filter has the ability to smooth through inaccurate GPS data due to multi-path or poor satellite geometry.

The Kalman filter's integration of MAPS and GPS position data provides an accurate real time navigation solution. This navigation solution makes it possible to effectively control the NTV autonomously.

ACKNOWLEDGMENTS

The authors wish to acknowledge the large contribution of Robert M. Rogers who developed the Kalman filter software used in this study. Also the authors wish to acknowledge the support of Wright Laboratory, Tyndall Air Force Base, Florida, and the Navy EOD Tech. Division, Indian Head, Indiana.

REFERENCES

1. Rankin, A.L., Crane, C.D., and Armstrong, D.G., "Navigation of an Autonomous Robot Vehicle," ASCE Conference on Robotics for Challenging Environments, Albuquerque, February 1994, 44-51.
2. Crane, C.D., Armstrong, D.G., and Rankin, A.L., "Autonomous Navigation of Heavy Construction Equipment," Microcomputers in Civil Engineering, 10, 1995, 357-370.
3. A. L. Rankin, C. D. Crane, D.G. Armstrong, A. D. Nease, and H. E. Brown, "Autonomous Path Planning Navigation System used for Site Characterization," To be published in: Proceedings of SPIE, Vol. 2738, 1996.
4. Robert M. Rogers, Jeffrey S. Wit, Carl D. Crane III, David G. Armstrong II, "Integrated INU/DGPS for Autonomous Vehicle Navigation," To be published in: Proceedings of IEEE PLANS 96 Symposium, Atlanta, April 1996
5. Bye, C.T. and Dahlin, T., "An RGL Implementation for the Army MAPS," Proceedings of the Workshop on Automation and Robotics for Military Applications, GACIAC PR-86-02, October 1886, 173-182.
6. Crane, C.D., Rankin, A.L., Armstrong, D.G., Wit, J.S., and Novick, D.K., "An Evaluation of INS and GPS for Autonomous Navigation," Proceedings of the 2nd IFAC Conference on Intelligent Autonomous Vehicles, Espoo, Finland (1995), 208-213.

A SENSOR–BASED APPROACH OF THE COLLISION AVOIDANCE PROCESS OF AUTONOMOUS ROBOTS

R. ZAPATA, P. LÉPINAY
Laboratoire d'Informatique, de Robotique et de Microélectronique
LIRMM – UM CNRS C9928 – Université de Montpellier II,
161 rue Ada, 34392 Montpellier cedex 5, FRANCE
Tel:(33) 67.41.85.60. Fax:(33) 67.41.85.00. E–mail: zapata@lirmm.fr

ABSTRACT

This work deals with a sub–problem (reflex actions) of the very general Motion Planning problem for mobile robots [1] [2] [3] [4] [5] [6] [7] [8] .
We have developed a method for controlling the reactive behaviors of wheeled robots, which is based on the DVZ concept first described in [9] and improved in [10]. This problem can be seen as a member of the Differential Game Theory class of problems [11] and hence, be solved by using optimal control laws. Several other methods have been developed in this sense [12], [13]. This method has been implemented on 3 wheeled mobile robots, a manipulator and a 6–leg walking machine.

KEYWORDS: Mobile Robot, Motion Planing, Reflex Action

INTRODUCTION: THE DVZ THEORY

For a mobile robot, we define the *Deformable Virtual Zone* as a state function, denoted by Ξ, representing a deformable zone whose geometry characterizes the interaction between the robot and its environment. The DVZ is the sum of 2 terms:

$$\Xi = \Xi_h + \Delta, \qquad (1)$$

where Ξ_h is an undeformed protecting zone and Δ represents a deformation of Ξ_h due to the intrusion of information in the robot space (denoted by I and sensed by proximity sensors). This intrusion of proximity information deforms the manifold R representing the range of the proximity sensors into B (Figure 1).
The undeformed zone depends on a vector π characterizing the motion capabilities of the robot (its translational and rotational velocities for instance).
The complete evolution of this function Ξ is modeled by a differential equation of its deformation :

$$\dot{\Delta} = A(\pi,I)\ \phi + B(\pi,I)\ \psi \qquad (2)$$

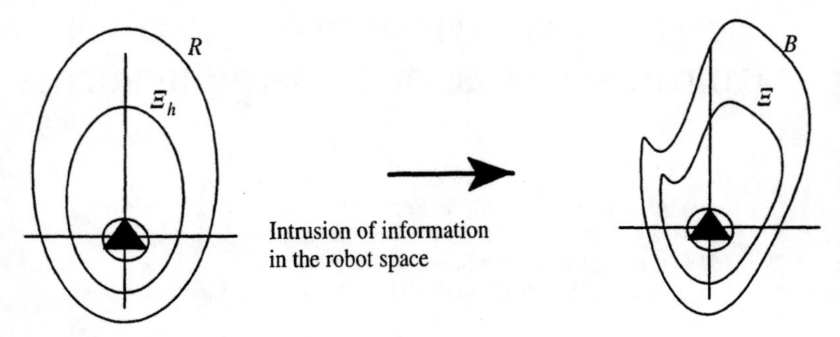

*Figure 1:*Deformation of the DVZ

driven by a 2–fold input vector $u = (\phi \quad \psi)^T$. The first control vector $\phi = \dot{\pi}$ tends to minimize the deformation of the DVZ. The second one, $\psi = \dot{I}$, is induced by the environment itself :

$$\Xi = \Xi_h + \Delta$$

$$\Xi_h = \varrho_{\Xi}(\pi) \qquad\qquad \Delta = \alpha(\Xi_h, I)$$

$$\dot{\Delta} = \left[\frac{\partial\alpha}{\partial\Xi_h}(\Xi_h, I) \times \varrho_{\Xi}'(\pi)\ (\pi)\right]\ \phi + \frac{\partial\alpha}{\partial I}(\Xi_h, I)\ \psi$$

This problem can be seen as a member of the Differential Game Theory class of problems. Figure 2 illustrates this concept.

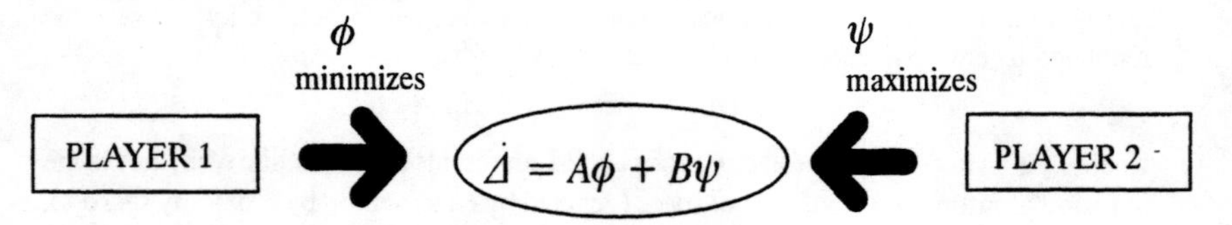

*Figure 2:*Playing with the DVZ deformation

IMPLEMENTATION

This method has been implemented on 3 wheeled mobile robots (SNAKE, RAT and RATMOBILE).

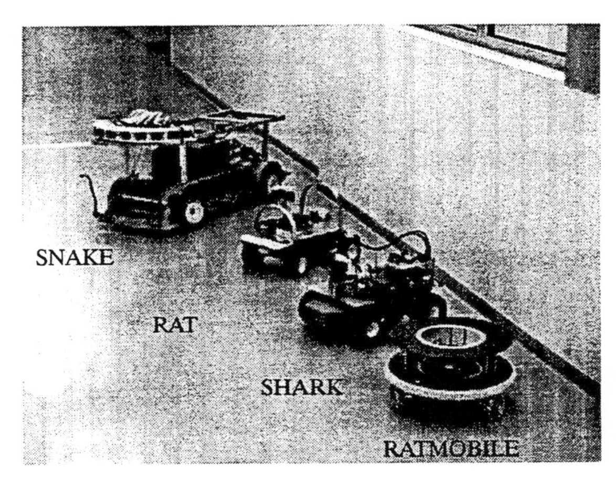

Figure 3: Mobile robots

SNAKE

SNAKE (Sensor–based Autonomous Kinetic Expert) was our first robot using this approach in a simplified way (no continuous minimisation of Δ but a direct control of Ξ_h through π). SNAKE (a 1/4 scale car–like autonomous vehicle equipped with an internal combustion engine) moved in unstructured and dynamic environments, avoiding randomly positioned obstacles at medium speed (up to 2m/s) without any model of the world, and single obstacles approximately up to 7m/s.

– The robot dynamics π is given by its translational velocity v and its rotational velocity $\dot{\Theta}$. These ones are directly related to the inter–axle distance L and to the turn angle δ of the robot by : $\dot{\Theta} = \frac{v}{L} \times \tan(\delta)$.

– The range manifold R is an arc of circle and can be deformed along the 7 directions into the information boundary manifold B.

– Two circular interaction zones (Ξ_{Stop} for the Emergency Stops procedure and Ξ_{Avoid} for the Dynamic Collision Avoidance procedure) are sampled by using the proximity data provided by the 7 ultrasonic sensors. The sizes of Ξ_{Stop} and Ξ_{Avoid} are proportional to the translational velocity v. For Ξ_{Avoid} we have: $R_{\Xi_{Avoid}} = k \times v + R_0$

where $R_{\Xi_{Avoid}}$ is the radius of Ξ_{Avoid}. The parameters k and R_0 (0–velocity risk radius) have been experimentally identified. These coefficients highly depends on the characteristics of the ground and on the reaction times of the servo–motors and of the ultrasonic system.

– The sampled interaction components $\hat{\Xi}_{Stop}$ and $\hat{\Xi}_{Avoid}$ are computed as the intersections of homothetic transformations of **R** and 2 limiting δ–oriented angular zones working as blinkers. These virtual blinkers make the robot look the direction it will move on. These two zones are deformed when **B** is smaller than Ξ. Otherwise, there is no reaction (Figure 4).

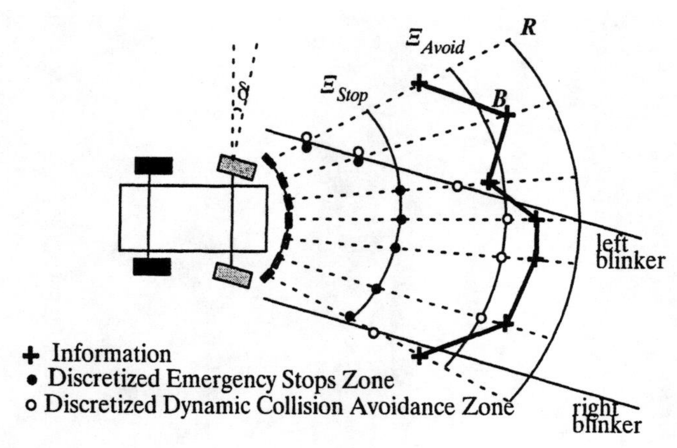

Figure 4: Implementation of the DVZ on SNAKE

RAT

RAT (Autonomous Robot with Transputer) has almost the same design than SNAKE, except that this electric 1/10–scale car–like robot can also move backward and therefore has increased dynamic capabilities. Three ultrasonic sensors protect the robot front space while a single ultrasonic sensor protects its back. Artificial whiskers are used to detect close obstacles. The DVZ was implemented in a neural architecture.

WAAL

The DVZ concept can equally be applied to the problem of stability maintaining of walking machines. In this case, the DVZ can be seen as a generalization of the support polygon. Here, the external cause of deformation is the displacement of the center of gravity of the moving body. *WAAL* (Walking Autonomous Artificial Locust) was a 6–leg, 6 degrees of freedom robot with only forward motion capabilities (from 0.5 to 1 m/s). Its purpose was to demonstrate indoor safe motion in presence of unknown and dynamic obstacles and to maintain its postural stability. Because of the poor dynamical capabilities of the robot actuators, the maintaining of the postural stability was only validated in simulation and not on the real robot.
In Figure 5 we can see the simulated avoidance of a moving ball by a 3–leg robot (the initial velocity of the ball is 30m/s). In this case, the machine is virtually protected by a 1–meter radius cylinder (DVZ) which is deformed by the intrusion of the ball. When this deformation is detected, the robot tries to avoid the collision by lowering its center of mass.

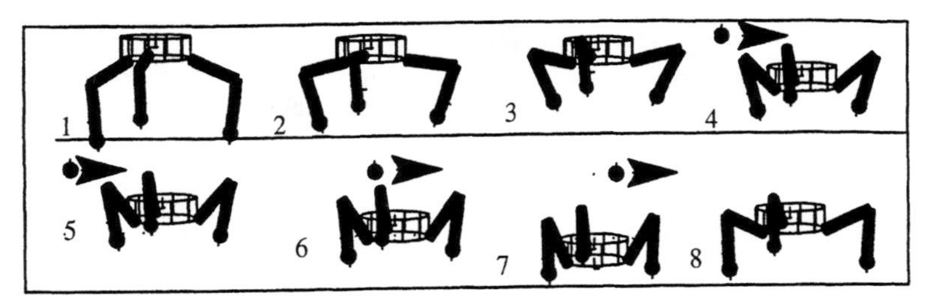

*Figure 5:*Avoiding an impact

A PUMA 560

A PUMA robot was able to avoid another moving manipulator, whose position was known. This application is an example of a 3D implementation of the DVZ approach.

RATMOBILE

RATMOBILE (a mobile quasi–holonomic robot moving in the plane) was developed in order first to implement the DVZ concept on a omni–directional robot and second to implement the concept of dynamic and 'egocentric memory': along a direction in which there is no sensor, the robot has to 'imagine' how varies the distance of the closest object in this direction. If this direction was previously sensed, the robot must keep alive this information and make it evolve in this memory of proximity. This principle is commonly used by living beings, able to move toward goals without seeing them (except at an initial time), only by measuring their own ego–motions (vestibular system). This basic principle was previously implemented to simulate the dynamic memory in vision systems [14].
RATMOBILE has only three sensors. The robot uses its proper rotation to refresh its dynamic memory. This memory is used as virtual proximity sensors (Figure 6) and allows to create the deformation of the DVZ. Therefore, the robot can avoid imagined obstacles.

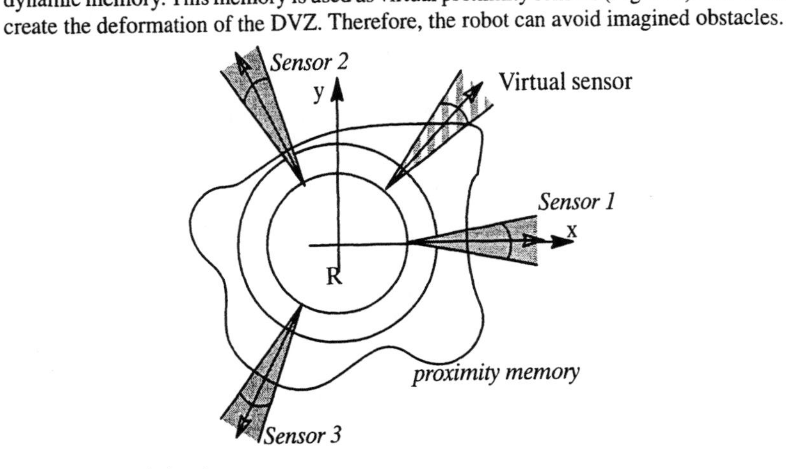

*Figure 6:*Real and virtual sensors

As we have seen in the introduction, the evolution of the DVZ deformation is described by a differential equation with 2 inputs : $\dot{\Delta} = A(\pi,I)\ \phi + B(\pi,I)\ \psi$, where f is controled by the robot. The control algorithm consists in choosing the desired evolution of this deformation $\nabla = \dot{\Delta}_{desired}$. A simple and efficient control law implemented on RATMOBILE is to choose this desired deformation asproportional to the real deformation and its derivative :

$$\nabla = -M\Delta - N\dot{\Delta} \quad (3)$$

The control vector is then defined by :

$$\phi_{best=}(A^T A)^{-1} A^T \quad \nabla \quad (4)$$

This control law tends to minimize the deformation and its velocity.

SHARK

Our next work would mainly consist in an extension of the DVZ approach to the problem of target following in unstructured environment. SHARK (Sensor–based Highly Autonomous Robot Killer) has been developed to emulate the Predator/Prey system (both with RAT). This 1/8–scale 6–wheel–drive robot is equipped with an ultrasonic system and a CCD camera. One of the main problems we will have to solve will be the cohabitation between avoidance DVZ and catching DVZ. This cohabitation would probably enlarge the formalization of the DVZ approach.

REFERENCES

1. R. Chatila "Système de navigation pour un robot mobile autonome : modélisation et processus décisionnel", Thèse de docteur ingénieur, Université Paul Sabatier, Toulouse, 1981.
2. R.A. Brooks "Robot Beings", International Workshop on Intelligent Robots and Systems '89, Sep 1989, Tsukuba, Japon .
3. O. Khatib "Real–time obstacle avoidance for manipulators and mobile robots", Proc. IEEE Int. Conf. on Rob. and Aut., St Louis, March 1985.
4. J. Angeles, S.K. Saha, J. Darcovich "The Kinematic Design of a 3–dof Isotropic Mobile Robot" International Conférence on Robotics and Automation, 1993, Atlanta, USA.
5. R.C. Arkin "Motor scheme based navigation for mobile robot : An approach to programming by behavior" IEEE International Conference on Robotics and Automation, pp 264-271 Avril 1987. Raleigh, USA.
6. J.D. Boissonnat, C. Laugier, J.P. Laumond "Planification de mouvements pour un véhicule planétaire" Bulletin de liaison de la recherche en Informatique et en Automatique, 1992.
7. R. Chatila, J-P. Laumond, T. Siméon, M. Taïx "Motion planning for a planetary rover" First IARP workshop on Robotics in Space, 17-18 juin 1991, Pise, Italy.
8. J.C. Latombe "Robot Motion Planning" Kluwer Academic Publishers, Boston.
9. R.Zapata "Quelques aspects topologiques de la Planification de Mouvements et des Actions Reflexes en Robitique Mobile", These d'etat, Université Montpellier II, July 91, France.
10. R.Zapata, P.Lépinay, P. Thompson "Reactive Behaviors of Fast Mobile Robots", Journal of Robotic Systems, January 1994
11. R. Isaac "Differential Games", John Wiley and Sons, Inc., New York 1965.
12. P. Lépinay, R. Zapata, B.Jouvencel "Sensor–based Control of the Reactive Behaviors of Walking Machines", International Conference IECON'93, Maui, Hawaii, December 1993
13. P.Deplanques, C. Novales, R. Zapata, B. Jouvencel "Sensor–based control versus neural network technology for the control of fast mobile robots behaviors in unstructured environments" IECON'92, 92, San Diego.
14. J. Droulez, A. Berthoz "Concept of Dynamic Memory in Sensorimotor Control", Motor Control : concepts and Issues, John Wiley & Sons Ltd

FIXTURELESS ASSEMBLY: MULTI-ROBOT MANIPULATION OF FLEXIBLE PAYLOADS

Wai Nguyen and James K. Mills

Laboratory For Nonlinear Systems Control
Department of Mechanical Engineering
University of Toronto
5 King's College Rd.
Toronto, Ontario
Canada M5S 3G8

ABSTRACT

This paper addresses methods of control of a multi-robot system designed for the assembly of flexible sheet metal parts. The work begins with the dynamic modeling of two robots and their sheet metal payloads. Recognizing that the flexible sheet metal states are practically unobservable, since none of the sheet metal parts will be instrumented, a practical control algorithm is proposed based on the rigid body dynamics of the robot and payloads. Stability analysis is carried out using the theory of singularly perturbed dynamic systems to determine the effect of the "rigid" feedback on the flexible system. Experimental results are given to illustrate the dynamic response of the two robot system, while carrying out the assembly operation with the two robots under force control. The experiments are performed with two 6-DOF commercial industrial robots.

1. INTRODUCTION

A modern automotive body is comprised of from 300 to 500 pieces of bent sheet metal, Brooke et al. (1991). Using current manufacturing technology within the automotive industry, these sheet metal pieces, each stamped from rolls of steel, are oriented and held firmly in complex clamping fixtures prior to being spot welded together, to form sub-assemblies. Sub-assemblies are oriented and clamped in fixtures, and then welded together to form major sub-assemblies, i.e. engine compartment, passenger compartment, and the rear trunk. In turn, these major sub-assemblies are oriented and clamped in other fixtures before being welded together by welding robots, to form an entire auto body.

In order to ensure that the position tolerances of subassemblies that must mate are met, and the dimension tolerances of the entire auto body are met, each individual sheet metal piece is required to be positioned within its fixture to within an accuracy on the order of 0.5 mm in three Cartesian axes. This position specification is achieved through the use of complex fixtures. These fixtures are very costly, ranging from \$75 million to \$150million (U.S.) to design and manufacture. The fixtures, which require from three to four years to design and manufacture, are not designed to be reconfigurable or reusable.

In order to update the automotive body assembly manufacturing technology, a number of innovations are under development. One such technology is referred to as "flexible fixtureless assembly". The first published work addressing technical aspects of this assembly approach are due to Mills (1992). The basis of this technology is the replacement of inflexible hardware fixtures with robot technology. Briefly, flexible fixtureless assembly is comprised of the following sequence of operations. Two robots each grasp one piece of sheet metal from a crude fixture. Using vision sensors and proximity sensors, the robots are servoed such that the sheet metal parts can be mated for welding by a third robot. Once the first two pieces are welded together, additional sheet metal parts can be mated sequentially to the welded subassembly.

Although a significant challenge, the benefits of such a technology are tremendous. The flexible fixtureless assembly workcell can be reprogrammed to assemble new parts. With this innovation, the high cost of the design, development and installation of new hardware fixtures each year can be avoided. With reprogrammable fixtureless assembly, more than one automobile body could be assembled in a single plant.

This paper discusses the problem of dynamics and control of a multi-robot system during the assembly of two flexible sheet metal parts, prior to welding. A number of issues arise when considering the dynamics and control of robots bringing two flexible payloads into contact. Since sheet metal parts are considerably more compliant than the links and transmissions of industrial robots, the payload flexibility cannot, in all likelihood, be ignored. Additionally, the sheet metal parts cannot be arbitrarily deformed during assembly, since this may lead to permanent deformation of the sheet metal parts. To ensure that the sheet metal partsare not deformed during the mating phase of the assembly process, the forces applied by the assembly robots must be precisely controlled. Hence, control of forces during the process is an important issue. To control the forces and achieve required position tolerances during mating, the robots must be properly coordinated.

This paper describes the theoretical development and experimental evaluation of a multi-robot control method to achieve stable robot position and force trajectory following while mating two flexible sheet metal parts. This control must also provide a smooth transition from noncontact to contact motion without generating large impact forces. Furthermore, the controller must be able to maintain the position and orientation of the sheet metal parts in the presence of disturbance torques and forces caused when spot welding electrodes come into contact with the sheet metal.

The outline of this paper is as follows. Section 2 gives a brief review of relevant literature. Section 3 gives an overview of the development of the dynamic equations of motion of the two robot system and the flexible sheet metal parts. Control laws are formulated in Section 4. Section 5 gives details of the experimental system. The experimental results are outlined in the next section and finally, Section 7 discusses the work done.

2. LITERATURE REVIEW

Coordinated control of multi-robot systems have been the subject of considerable study within the robotics research community for many years. There are two basic approaches to the solution of this problem, namely the master-slave and hybrid position-force approaches. In the master-slave approach, a group of robots is assigned the role of master, while the remaining robots act as slaves, and move in cooperation with the master, Arimoto et al. (1987). The hybrid position-force control method represents an alternative to the master-slave approach. This method is an extension of the hybrid control proposed by Raibert and Craig (1981) for single arm control. Numerous researchers have applied this methodology to the multi-robot control problem.

Many of the approaches to control of cooperating robots are similar in that they involve the manipulation of rigid objects. However, in contrast, the payloads involved in the fixtureless assembly task are flexible sheet metal parts. The joint coordinates are not subject to holonomic constraints. Hence, the two robots do not form a closed kinematic chain during contact of the sheet metal parts. So far, little work has been published addressing the subject of flexible payloads. Most work that does address this problem deals with a single payload of simple geometry. These objects have been modelled as a one dimensional spring, i.e. Bouffard-Vercelli (1992). Mills (1992) was the first to consider the control problem presented by the manipulation of flexible sheet metal parts. In this work a number of assumptions were made to simplify the control problem. A computed torque type of control was applied to the constrained robotic system, leading to position and force trajectory regulation, as observed in simulations. Subsequent work by Mills modelled the flexible sheet metal as continuous structures, and utilized the assumed modes method for structural modelling. A control which utilized feedback of only rigid body motions and contact forces was proposed. Mills and Ing (1995) went another step and modelled real sheet metal car parts using dynamic finite elements. Several different control strategies were developed and tested in numerical simulation.

3. DYNAMICS

In this section, we briefly outline the development of the dynamics of a two robot system, with each robot gripping a flexible sheet metal payload. We assume that i) the robot links and joints exhibit no compliance; ii) the payload deforms elastically; and iii) each robot grips the payload rigidly. This model has been reported in Mills and Ing (1995), hence only the final results are given. Using the method of finite element analysis, the payload can be discretized into finite elements with nodal degrees of freedom. We develop expressions for the kinetic, potential and strain energies of each finite element, and sum over the entire sheet metal part to obtain the Lagrangian for the entire sheet metal part. The equations of motion are then derived from evaluation of Lagrange's equations with respect to both the rigid body and flexible coordinates. Due to space constraints, the development of the equations of motion are not given, but can be found in Nguyen (1995). The final result is given below as:

$$M(\theta,q)\begin{Bmatrix}\ddot{\theta}\\ \ddot{q}\end{Bmatrix}+L(\theta,\dot{\theta},q,\dot{q})+\begin{Bmatrix}0\\ B\dot{q}\end{Bmatrix}+\begin{Bmatrix}0\\ Kq\end{Bmatrix}=\begin{Bmatrix}u\\ 0\end{Bmatrix}. \qquad (3.1)$$

where: $\theta \in R^n$ is the vector of the rigid robot joint coordinates, $q \in R^N$ is the vector of the flexible payload coordinates. $M(\theta,q) \in R^{(n+N)\times(n+N)}$ is the inertia matrix of the single robot payload system,

$L(\theta,\dot{\theta},q,\dot{q}) \in R^{(n+N)\times 1}$ is the vector of Coriolis, centrifugal forces, and the interaction between the rigid and flexible coordinates. $B \in R^{N\times N}$ is the payload damping matrix, $K \in R^{N\times N}$ is the payload stiffness matrix, $u \in R^{n}$ is the control input. Equation (3.1) represents the equation of motion of a robot rigidly grasping an elastically deformable payload. When the system is in contact with another robot and flexible payload. We treat the contact forces as static functions of the kinematics of the robot-payload system only. The underlying assumptions for such a treatment of contact forces are that the parts are forced together relatively slowly, and they do not undergo large accelerations while in contact. Inherent in our modelling of the two robot system in this manner, is that coupling between the rigid and flexible states of each robot is ignored. We now write the dynamics of the two robot system as follow, where the two robots are labeled "A" and "B" respectively:

$$\begin{bmatrix} M_A(\theta_A,q_A) & 0 \\ 0 & M_B(\theta_B,q_B) \end{bmatrix} \begin{Bmatrix} \ddot{\theta}_A \\ \ddot{q}_A \\ \ddot{\theta}_B \\ \ddot{q}_B \end{Bmatrix} + \begin{Bmatrix} L_A(\theta_A,\dot{\theta}_A,q_A,\dot{q}_A) \\ L_B(\theta_B,\dot{\theta}_B,q_B,\dot{q}_B) \end{Bmatrix} + \begin{Bmatrix} 0 \\ B_A\dot{q}_A \\ 0 \\ B_B\dot{q}_B \end{Bmatrix} + \begin{Bmatrix} 0 \\ K_A q_A \\ 0 \\ K_B q_B \end{Bmatrix} = \begin{Bmatrix} u_A \\ 0 \\ u_B \\ 0 \end{Bmatrix} - J^T(\theta_A,\theta_B,q_A q_B)F. \tag{3.2}$$

Without loss of generality, we can reorder the state vector of equation (3.2) and rewrite the above to give the following form:

$$M(\theta,q)\begin{Bmatrix} \ddot{\theta} \\ \ddot{q} \end{Bmatrix} + L(\theta,\dot{\theta},q,\dot{q}) + \begin{Bmatrix} 0 \\ B\dot{q} \end{Bmatrix} + \begin{Bmatrix} 0 \\ Kq \end{Bmatrix} = \begin{Bmatrix} u \\ 0 \end{Bmatrix} - J^{T}(\theta,q)F \tag{3.3}$$

where: $M(\theta,q) \in R^{2(n+N)\times 2(n+N)}$ is the reordered inertia matrix of the complete system consisting of the two robot-payload system, $L(\theta,\dot{\theta},q,\dot{q}) \in R^{2(n+N)\times 1}$ is the reordered vector of Coriolis, centrifugal forces, and the interaction between the rigid and flexible coordinates for the complete system, $B \in R^{2N\times 2N}$ is the combined payload damping matrix, $K \in R^{2N\times 2N}$ is the combined payload stiffness matrix, $u \in R^{2n}$ is the control input for the complete system. $J(\theta,q) \in R^{2(n+N)\times m}$ is the reordered Jacobian for the complete system and $F \in R^{m\times 1}$ is the vector of contact forces between the parts.

4. CONTROL OF MULTI-ROBOT SYSTEM

For a typical fixtureless assembly task, the control objectives are stated as follows:

i) to regulate contact forces between the sheet metal parts to desired values
ii) to cause the position of the sheet metal parts to follow prescribed trajectories
iii) to establish stable switching from non-contact to contact motion of the robot-payload system.

We propose a coordinated control leading to a control input which couples the states of the two robots during the contact phase of motion. This control scheme is based on the computed torque control. Application of the computed torque method of control to the coupled dynamic equations, derived in the last section, requires that the flexible payload states be measured and used in a linearizing, decoupling feedback. In an automotive assembly plant, measurement of the flexible modes of these payloads is not feasible. Hence, this reality leads to the use of a "rigid" control law in which the feedback is constructed from the robot joint states and the forces measured with wrist-mounted force sensors.

We begin the development of this control method with a transformation of the equations of motion, given by equation (3.3), to a set of task related coordinates. This transformation found in McClamroch and Wang (1988). Define a set of generalized coordinates for describing the task geometry and flexible states as follows:

$$\begin{Bmatrix} x \\ q \end{Bmatrix} = \begin{Bmatrix} x_1 \\ x_2 \\ q \end{Bmatrix} = \begin{Bmatrix} \phi_1(\theta,0) \\ \phi_2(\theta,0) \\ q \end{Bmatrix} \tag{4.1}$$

where $\phi_1(\theta,0)$ and $\phi_2(\theta,0)$ are the task geometries in the contact and non-contact directions respectively, and are evaluated at the zero deflection states of the flexible coordinates. Note that the functions $\phi_1(\theta,0)$ and $\phi_2(\theta,0)$ can be defines readily. However, if the functional dependency off these function on sheet metal deformation were included, it would be extremely cumbersom to explicitly write these functions. The coordinates, $x\ 1 \in R$ and $x\ 2\ \in R$ are defined as generalized coordinates corresponding to task related variables defined by the functions $\phi_1(\theta,0)$ and $\phi_2(\theta,0)$. Note that these new variables do not necessarily correspond to Cartesian coordinates. Under the assumption that the Jacobian of (4.1), J(x,0) has rank m, an inverse transformation is defined as:

$$\begin{Bmatrix}\theta\\q\end{Bmatrix}=\begin{Bmatrix}\theta_1\\\theta_2\\q\end{Bmatrix}=\begin{Bmatrix}\Omega_1(x_1,x_2,0)\\\Omega_2(x_1,x_2,0)\\q\end{Bmatrix} \tag{4.2}$$

where $\Omega_1(\bullet)$ and $\Omega_2(\bullet)$ are defined as in McClamroch and Wang. Note that the functions given in (4.2) would be cumbersome to derive for all but the simplest robot configurations. Denoting the Jacobian of (4.2) by T(x), the transformed dynamic equations of motion are given by:

$$\widetilde{M}(x,q)\begin{Bmatrix}\ddot{x}_1\\\ddot{x}_2\\\ddot{q}\end{Bmatrix}+\widetilde{L}(x,\dot{x},q,\dot{q})+\begin{Bmatrix}0\\0\\B\dot{q}\end{Bmatrix}+\begin{Bmatrix}0\\0\\Kq\end{Bmatrix}=\begin{Bmatrix}\widetilde{u}\\0_N\end{Bmatrix}-\begin{Bmatrix}F\\0\\J_q^T(x,q)F\end{Bmatrix} \tag{4.3}$$

where:

$$\widetilde{M}(x,q)=T^T(x,0)M(x,q)T(x,0),$$

$$\begin{Bmatrix}\widetilde{u}\\0_N\end{Bmatrix}=T^T(x,0)\begin{Bmatrix}u\\0_N\end{Bmatrix}.$$

Equation (4.3) describes the dynamics of the two robot-payload system with respect to the set of generalized coordinates given by equation (4.1). Transformation using T(x), leads to the form shown in (4.3) where the generalized contact forces appear as F. This is similar to a transformation in which a robot system is transformed to the task space, in which forces appear in Cartesian form. Details are found in McClamroch and Wang (1988). To better examine the dynamics of the multi-robot system, we can rewrite (4.3) utilizing the following identity matrix:

$$\begin{bmatrix}E_1^T & E_2^T & E_3^T\end{bmatrix}=\begin{bmatrix}I_m & 0 & 0\\0 & I_{2n-m} & 0\\0 & 0 & I_{2N}\end{bmatrix}. \tag{4.4}$$

With (4.4) we can split (4.3) into the following:

$$E_1\widetilde{M}(x,q)E_1^T\ddot{x}_1+E_1\widetilde{M}(x,q)E_2^T\ddot{x}_2+E_1\widetilde{M}(x,q)E_3^T\ddot{q}+E_1\widetilde{L}(x,\dot{x},q,\dot{q})=E_1\widetilde{u}-F \tag{4.5a}$$

$$E_2\widetilde{M}(x,q)E_1^T\ddot{x}_1+E_2\widetilde{M}(x,q)E_2^T\ddot{x}_2+E_2\widetilde{M}(x,q)E_3^T\ddot{q}+E_2\widetilde{L}(x,\dot{x},q,\dot{q})=E_2\widetilde{u} \tag{4.5b}$$

$$E_3\widetilde{M}(x,q)E_1^T\ddot{x}_1+E_3\widetilde{M}(x,q)E_2^T\ddot{x}_2+E_3\widetilde{M}(x,q)E_3^T\ddot{q}+E_3\widetilde{L}(x,\dot{x},q,\dot{q})+B\dot{q}+Kq=0 \tag{4.5c}$$

Equations (4.5a, b, c) represents the dynamics of the multi-robot system. We now apply a control to the system of dynamic equations, which for practical reasons, is not a function of the flexible states. While in principle, it is possible to instrument the sheet metal payloads, and to estimate the flexible states, it is not practical to suggest that such a procedure would be carried out in an automotive assembly plant. Hence, the proposed feedback is based only on the rigid body states of the dynamic equations of motion, given by (4.5a) - (4.5c). Additionally, although we want to control the forces of contact between the sheet metal pieces, the robot force sensors are mounted on the robot wrists, implying that we will regulate the wrist forces. The applied control, during contact, has the following form:

$$E_1\widetilde{u}=E_1\widetilde{M}(x,0)E_1^T\left[\ddot{x}_1^d+G_{v1}(\dot{x}_1^d-\dot{x}_1)+G_f(F^d-F)\right]+E_1\widetilde{M}(x,0)E_2^T\left[\ddot{x}_2^d+G_{v2}(\dot{x}_2^d-\dot{x}_2)+G_{p2}(x_2^d-x_2)\right]$$
$$+E_1\widetilde{L}(x,\dot{x},0,0)+F \tag{4.6a}$$

$$E_2\widetilde{u}=E_2\widetilde{M}(x,0)E_1^T\left[\ddot{x}_1^d+G_{v1}(\dot{x}_1^d-\dot{x}_1)+G_f(F^d-F)\right]+E_2\widetilde{M}(x,0)E_2^T\left[\ddot{x}_2^d+G_{v2}(\dot{x}_2^d-\dot{x}_2)+G_{p2}(x_2^d-x_2)\right]$$
$$+E_2\widetilde{L}(x,\dot{x},0,0). \tag{4.6b}$$

When the sheet metal payload is not in contact, the force terms are not in effect. For simplicity, the control law is not written in complete detail regarding this point. This control law is based mainly on the work of Lokhorst and Mills (1994). The closed-loop equations of motion of the resultant system are as follows:

$$0 = E_1\widetilde{M}(x,q)E_1^T\ddot{x}_1 + E_1\widetilde{M}(x,q)E_2^T\ddot{x}_2 + E_1\widetilde{M}(x,q)E_3^T\ddot{q} - E_1\widetilde{M}(x,0)E_1^T\left[\ddot{x}_1^d + G_{v1}(\dot{x}_1^d - \dot{x}_1) + G_f(F^d - F)\right]$$
$$-E_1\widetilde{M}(x,0)E_2^T\left[\ddot{x}_2^d + G_{v2}(\dot{x}_2^d - \dot{x}_2) + G_{p2}(x_2^d - x_2)\right]$$
$$+E_1\widetilde{L}(x,\dot{x},q,\dot{q}) - E_1\widetilde{L}(x,\dot{x},0,0) \qquad (4.7a)$$

$$0 = E_2\widetilde{M}(x,q)E_1^T\ddot{x}_1 + E_2\widetilde{M}(x,q)E_2^T\ddot{x}_2 + E_2\widetilde{M}(x,q)E_3^T\ddot{q} - E_2\widetilde{M}(x,0)E_1^T\left[\ddot{x}_1^d + G_{v1}(\dot{x}_1^d - \dot{x}_1) + G_f(F^d - F)\right]$$
$$-E_2\widetilde{M}(x,0)E_2^T\left[\ddot{x}_2^d + G_{v2}(\dot{x}_2^d - \dot{x}_2) + G_{p2}(x_2^d - x_2)\right]$$
$$+E_2\widetilde{L}(x,\dot{x},q,\dot{q}) - E_2\widetilde{L}(x,\dot{x},0,0) \qquad (4.7b)$$

$$-J_q^T(x,q)F = E_3\widetilde{M}(x,q)E_1^T\ddot{x}_1 + E_3\widetilde{L}(x,\dot{x},q,\dot{q}) + E_3\widetilde{M}(x,q)E_2^T\ddot{x}_2 + E_3\widetilde{M}(x,q)E_3^T\ddot{q} + B\dot{q} + Kq \qquad (4.7c)$$

The stability of the proposed "rigid" control law can be shown, utilizing the theory of singularly perturbed dynamic systems. Through a procedure that is detailed in Kokotovic et al. (1986), the dynamics of the two robot system and the flexible sheet metal payloads, during contact, can be separated into a multiple time-scale system. By showing that the perturbation dynamics of each of these subsystems is asymptotically stable, and also that the rigid system is asymptotically stable, then the overall system dynamics may be shown to be stable. For the two robot system with flexible sheet metal payloads, the details of this development, too lengthy to include here, are given in Nguyen (1995).

5. EXPERIMENTAL SET-UP

A two robot system suitable for testing and verification of fixtureless assembly control methods has been constructed. The two robots are commercially available 6-d.o.f. light industrial robots manufactured by CRS Robotics Corporation. The first three joints of each robot have been retrofitted in the laboratory with joint load torque control to permit the implementation of model based control methods The gripper is computer controlled through an interface connecting the VMS-2000 vacuum control system to the C500 robot controller. The robots are each controlled with a transputer based C500 controller, supplied by CRS. The multi-robot control algorithm introduced in the last section was executed at a servo rate of 170 Hz.

6. EXPERIMENTAL RESULTS

A large number of tests were conducted using the coordinated control of equations (4.6a,b).The test that is reported here involves three phases. In Phase I, the robots approach each other to bring the sheet metal into contact. In Phase II, the robots holding the sheet metal into contact, follow a prescribed trajectory. In Phase III, the robots hold the parts without motion. In Phase I, the robots are position controlled. In Phase II and III, the robots are under position and force control.

Here we give results for a typical test. In all test results shown, the parts come into contact at T= 6 sec.; the parts, while in contact translate until T= 14 sec.; and then for the remainder of the test the parts are stationary. Figure 2 illustrates Robot B force response obtained, while Figures 3 and 4 illustrate the gross motion of Robot A & B in the z direction respectively.

7. DISCUSSION

Our experimental results illustrate that the coordinated control, given by equation (4.6a,b), applied to the multi-robot fixtureless assembly tasks leads to acceptable performance. This performance is obtained in despite the very low control update rate, dictated by the significant computational loading of the control law. With a higher control update rate, the position errors could be reduced to within the allowable limits imposed by the automobile manufactures specifications. It has been shown theoretically, that the flexibility of the sheet metal parts will not destabilize the closed-loop system, Nguyen (1995). This has been confirmed in the experimental results obtained.

The test robots used each have a payload of only 3.0 kg., far too low to manipulate full sized auto parts. However, the results achieved here indicate the feasibility of using robots to assemble automobile sheet metal body parts. It is important to note that the robots have grasped the sheet metal parts in a manner which does not permit the precise coordinates of mating edges of the sheet metal to be known. This will lead to welding of the sheet metal parts to a final shape that is not within tolerances. To overcome this problem, a vision sensor must be used to augment the joint encoder information to servo the robots to permit correct mating of the sheet metal parts, prior to welding.

8. REFERENCES

1. Arimoto, S., Miyazaki, F., Kawamura, S. "Cooperative Motion Control of Multiple Robot Arms or Fingers", IEEE International Conference on Robotics and Automation, pp. 1407-1412, 1987.

2.Bouffard-Vercelli, Y., Dauchez, P., Degoulange, E., Delabarre, X., "Manipulation of Flexible Objects With a Two-arm Robot", Robotics and Flexible Manufacturing Systems, Elsevier Science Publishers B.V. North-Holland, pp. 109-118, 1992.

3. Brooke, L., Kobe, G., and McElroy, J., "The Case for Contour", Automotive Industries, pp 24-29, March, 1991.

4. Kokotovic, P.V., Khalil, H.K., O'Reilly, J., *Singular Perturbation Methods in Control: Analysis and Design*, Academic Press, Orlando, 1986.

5. Lokhorst, D.M., and Mills, J.K., "An Approach to Force and Position Control of Robots: Theory and Experiments", IEEE Transaction on Robotics and Automation, 1994.

6. McClamroch, N.H., and Wang, D., "Feedback Stabilization and Tracking of Constrained Robots", IEEE Transactions on Automatic Control, Vol.33, No. 5, pp.419-426, May 1988.

7. Mills, J.K., "Multi-Manipulator Control For Fixtureless Assembly of Elastically Deformable Parts", Proceedings, Japan-USA Symposium on Flexible Automation, pp. 1565-1572, 1992.

8. Nguyen, W. H., Fixtureless Assembly: *Control of Multiple Robotic Manipulators Handling Flexible Payloads*, M.A.Sc. Thesis, Dept. of Mechanical Eng., University of Toronto, 1995.

9. Raibert, M.H., Craig, J.J, "Hybrid Position/Force Control of Manipulators" Journal of Dynamic Systems, Measurement, and Control, Vol. 102, pp. 126-133, June, 1981.

Figure 1. Fixtureless Assembly using two robot manipulators

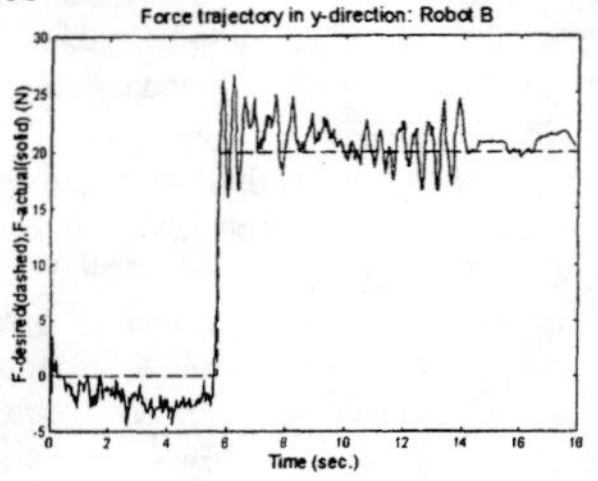

Force trajectory in y-direction: Robot B

Figure 2: Force measured from robot B under coordinated control

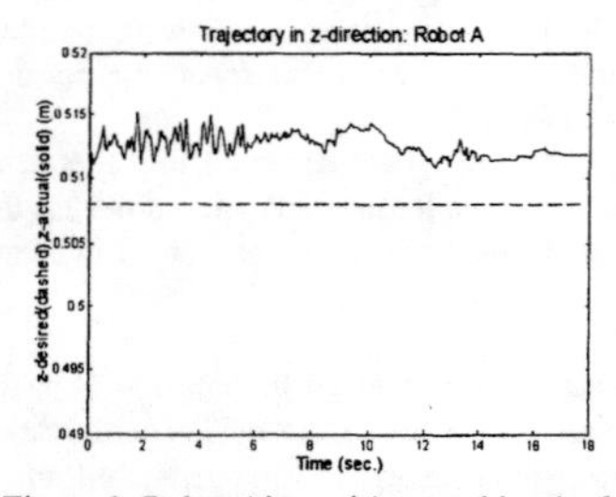

Figure 3: Robot A's position tracking in the z-direction under coordinated control

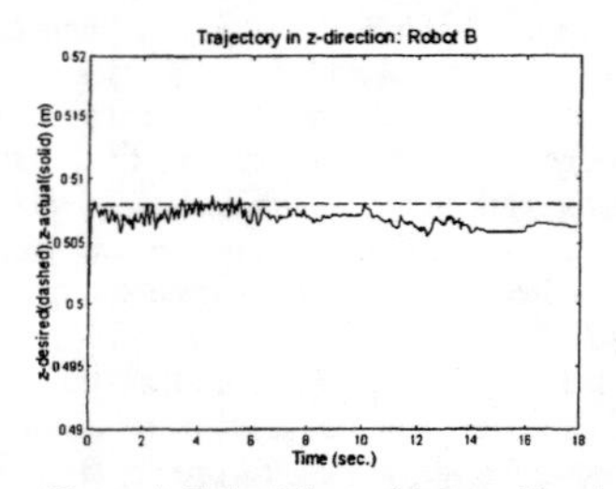

Figure 4: Robot B's position tracking in the z-direction under coordinated control

Controlling the Motions of an Autonomous Vehicle using a Local Navigator

Cyril Novales, Didier Pallard and Christian Laugier
INRIA Rhône-Alpes, 655 Av. de l'Europe, 38330 Montbonnot, France.
Ph: (33) 76.61.52.93. Fax: (33) 76.61.52.52. Email: Cyril.Novales@inria.fr

ABSTRACT

This paper addresses the problem of controlling the motion of an autonomous vehicle. This research is carried out within the framework of the French INRIA/INRETS Praxitele project on public transportation by individual electric cars. In this project the cars must have the ability to perform certain motions autonomously. For that purpose we have designed a specific control architecture based upon a local navigator,which projects in the robot evolution space admissible trajectories, called escape lines, on a short time horizon using a direct model of the mobile robot. After the removal of the escape lines that cross some obstacles, it chooses the best escape line that matches with the planned path and delivers it to the inputs of the servo controllers of the vehicle. The trajectories delivered by the local navigator respect the kinematic and dynamic constraints of the vehicle (as embedded in the direct model that is used). Since the local navigator only requires a proximity map obtained from sensory data rather than an exhaustive map of the vehicle's environment, its running time is low.

I. NAVIGATION PROBLEM

Our research work sits above the control level of an autonomous car like mobile robot which operates in a partially known environment. In this environment, the robot has to follow a predetermined path, given by a planner. But the robot is not necessary on this path and it must compute a trajectory to catch it up (figure 1). In case of unpredictive events occurrence (for example obstacles), this control level, called motion controller, must locally modify the path to take them into account. This functionality can be assumed by a reactive pilot, but generate oscillations due to this reactivity. To smooth these oscillations, we add to the reactive pilot a local navigator able to generate parts of trajectories on a few seconds. In sum-up, a motion controller must integrate two different kinds of constraints :

- The moving topology of the environment due to some unmapped static obstacles or unpredicted dynamical moves;
- Kinematic and dynamic constraints, particularly non-holonomic constraints, derived from the physical architecture of a car like mobile robot.

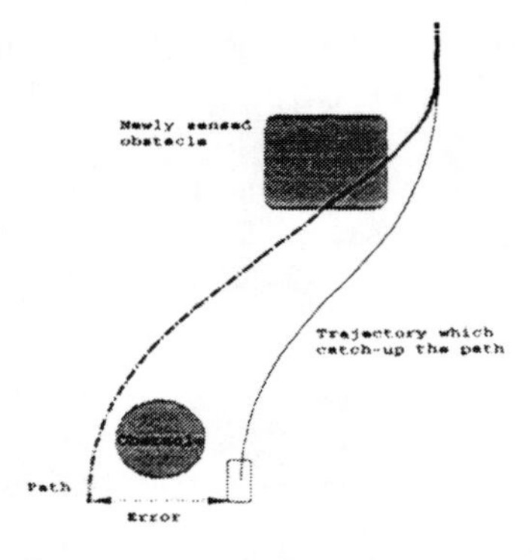

Figure 1: Plan following for an autonomous mobile robot

I.1. General Control Architecture

Our control architecture is composed of the three following layers: (figure 2)

- A long term layer which is called the ***"planner"***. This planner makes use of maps of the robot environment that includes static known obstacles, static detected obstacles and prediction of motion of mobile obstacles [1]. It delivers a plan to the next layer. We define a plan as a path with velocity profile;
- A short term layer which is called ***"local navigator"***. With a plan as input, this module delivers to the next layer a trajectory on a short time horizon. The purpose of the local navigator is to try to locally track the plan provided by the planner while taking into account the constraints of the mobile robot and the unpredictable events detected by the sensors (such new constraint map prevents the robot to accurately follow the plan given by the previous layer);
- An instantaneous layer which is called the ***"reactive pilot"***. This module converts the trajectory produced by the local navigator into controls for the robot actuators. It also reacts directly to sensor data in order to avoid collisions (e.g. emergency stop). [2].

To connect the layer of the *planner* which uses maps to the layer of the *reactive pilot* which reacts instantaneously to sensor data, we use a layer which works on short term: the *local navigator*. This *local navigator* and the *reactive pilot* form the ***motion controller***, which may be able to follow the plan computed by the planner in a reactive way.

I.2. Local Navigation

The short term layer must determine a trajectory on a short time horizon - a few seconds - and deliver it to the reactive pilot. There are two types of input for this navigator : proximity data and plan data, which is a path parameterised by linear velocity and given by the planner. The local navigator attempts to follow the nominal plan while satisfying the dynamic and kinematic constraints of the robot, and while taking into account newly sensed static or dynamic obstacles located on this plan

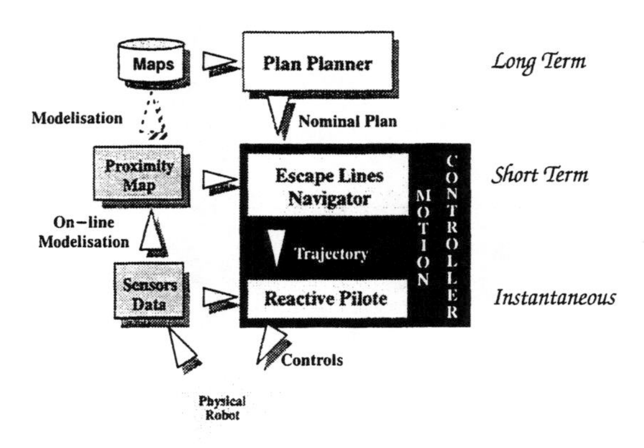

Figure 2: Control Architecture

(such obstacles will give rise to local modification of the executed trajectory). A proximity map of the environment built on-line from sensor data, is used to determine the required modification of the plan
This layer must therefore generate a trajectory that:

- attempts to follow the nominal plan as closely as it can whilst respecting constraints;
- avoids unmodelled or dynamic obstacles present on the plan.

II. LOCAL NAVIGATION BY ESCAPE LINES

II.1. Basic idea

To execute a given trajectory, one generally use an inverse model of the mobile robot [3] which transforms cartesian motions into differential controls. However, determining an inverse model for a given mobile robot is sometimes impossible because of non integrable constraints, due to the presence of non-holonomy [4]. An alternative to this approach is to use a **direct model** for predicting a restrictive set of possible pieces of trajectories for the mobile robot, before choosing one of these local solutions according to some given criteria [7]. Constructing a direct model of a given mobile robot is always possible. Moreover, the processing time to obtain a path from control inputs is lower than the previous method and dynamic and kinematic behaviours can be easily integrated. This is why we developed a navigator which makes use of a direct model of the robot. The basic idea is to "identify" the parameters of the small pieces of trajectories that the robot will execute when applying a given control law over a small time interval T, which is called **time horizon**, in a given set of known states of the robot.
Let us assume that we know the direct model of a mobile robot $\Sigma_{\mathcal{R}}$ in a given state X_0 at the time t_0. Let $\Omega_{[t_0, t_0+T]}$ be the set of admisible control laws which can be apply on $\Sigma_{\mathcal{R}}$ during time interval $[t_0, t_0+T]$. Applying all the control laws $\omega \in \Omega_{[t_0, t_0+T]}$

generates a set of small pieces of trajectories $\Lambda_{[t_0, t_0 + T]}$, that figure 3 illustrats. These

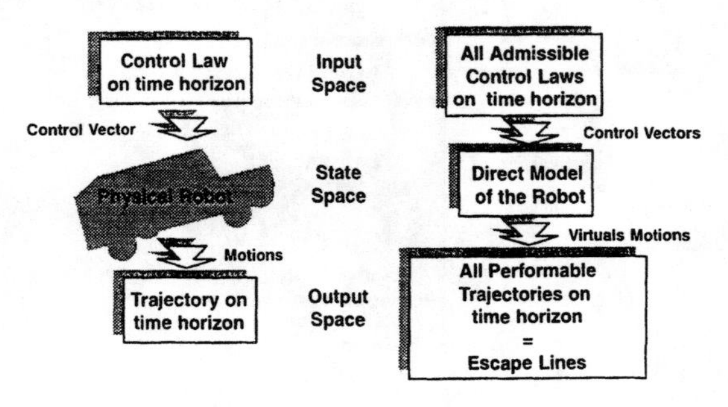

Figure 3: Escape Lines Generation

pieces of trajectories are called **Escape Lines**; they represents the local trajectories that the robot can follow from the current robot position. Then the basic idea is to construct at each time step the associated Escape Lines and to apply a given selection algorithm to choose in real time the more appropriate Escape Line to follow according to the nominal plan to track and to the detected obstacles. This method is called **Local Navigation using Escape Lines**.

II.2. Constructing Escape Lines

The general formalism of local navigation is derived from the control theory formalism of E.D. Sontag [6], and was developed in detail in previous papers ([5],[2]). In this formalism we make a difference between the internal state X of a mobile robot and its position/orientation e in the space. The internal state belongs to the state-space χ and the position/orientation belongs to the output space $\mathcal{E}$. The output function that links X to e is differential and since we keep as the input space the control space μ, the transition function between μ and χ remains differential.

We can model our electrical vehicle evolving on a plan ($\mathbb{R}^2$) without sliding as following: the input or control vector, $U = (\dot{v}, \dot{\delta})$, contains the linear acceleration and the steering angular speed; the state vector, $X = (v, \delta)$, is composed of the linear speed and the steering angle; and the output vector, $e = (x, y, \theta)$, is the Cartesian position and orientation of the robot.

Assume that a mobile robot is in the internal state X_0 at the time t_0. If we make the hypothesis of homogeneous floor (for wheel/floor contact) and assume an environment without obstacles, we can determine with a direct model of $\Sigma_{\mathcal{R}}$ the trajectory that it will follow if we apply a control law ω during T. To obtain the set of **Escape lines** $\Lambda_{[t_0, t_0 + T]}$ we apply to the model all the admissible control laws of $\Omega_{[t_0, t_0 + T]} \subset \mu^{[t_0, t_0 + T]}$. Actually this set $\Omega_{[t_0, t_0 + T]}$ depends on the state X_0 (*i.e. on the robot's abilities*) and also on the position/orientation e_0 at the time t_0 (*i.e. on the robot/environment relation*). With our hypothesis of homogeneous floor and environment without obstacles, we can determine $\Omega_{[t_0, t_0 + T]}$ with only the internal state X_0. Thus, we sever the internal state of a mobile robot from its position/orientation. To be admissible, each control law ω must only verify boundaries and dynamic constraints of the servoings. $\Lambda_{[t_0, t_0 + T]}$ on T gives all the trajectories that the robot will able to

Figure 4: Top view of forward Escape Lines for a car-like robot

perform from its position e_0 with its state X_0 on the time interval $[t_0, t_0 + T]$ if there is no obstacles.

For our car-like robot driving ahead with its steering wheels turned on the left at state X_0, the obtained Escape Lines are plotted on the figure 4.

The next step is to determine whether these Escape Lines are executable or not in the local environment of the robot (i.e. in the vicinity of the location e_0). For that purpose, the obtained Escape Lines are projected onto the output space $\mathcal{E}$ and they are matched with the proximity map constructing using sensor data. This allows us using a simple collision detection algorithm to construct a set of **Free Escape Lines** $\Lambda_{\mathcal{L}_{[t_0, t_0+T]}}$ (upper part of the figure 5).

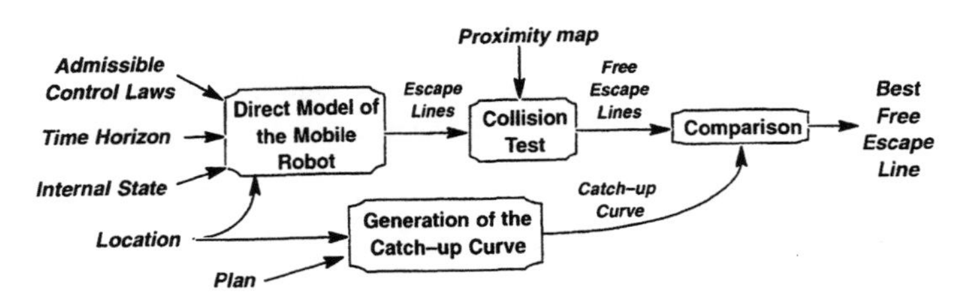

Figure 5: Algorithm of Local Navigation

II.3. Selecting an Escape Line

To choose a Free Escape Line, we use the plan given by the planner. But, if the mobile robot is not placed at the time t_0 on the right place of this plan, it need a catch-up curve to reach this plan. In that way, we define a geometrical curve $\mathcal{C}$, called **Catch-up Curve**, that joins e_0 to the position/orientation that the robot will be placed on the plan at $t_0 + T$ (lower part of the figure 5). This curve does not take into account robot constraints and obstacle places. It may only take into account initial and final robot velocities.

The best Free Escape Line is chosen by comparing the Catch-up Curve and the Free Escape Lines. The nearest Free Escape Line from the Catch-up Curve is delivered to the reactive pilot.

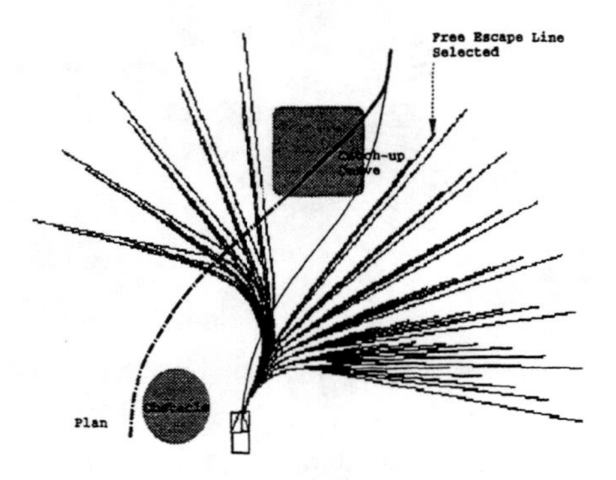

Figure 6: Free Escape Lines, Catch-up Curve and Plan

II.4. Sketch of the local navigator

Finally, the local navigator operates as follows:

- At each time step, the local navigator determines the set of Escape Lines $\Lambda_{[t_0, t_0 + T]}$ associate to X_0;
- The set $\Lambda_{[t_0, t_0 + T]}$ is pruned in order to remove the Escape Lines which generate a collision in the vicinity of e_0. This is done using the proximity map built with sensor data;
- A catch-up curve is computed using the plan delivered by the planner. This curve is realised without taking into account newly sensored obstacles,
- The Free Escape Line which is the nearest to the catch-up curve is chosen and delivered to the pilot.

The mobile robot does not follow the Best Free Escape Line to its end. The chosen escape line will be updated as the algorithm is recomputed.

III. IMPLEMENTATION

III.1. Initial experiments on small mobile robots

The first implementation of this local navigator was made on two small wheeled mobile robots in the LIRMM (Laboratory of computer scIence Robotics and Microelectronic of Montpellier): RAT and SNAKE II [5]. These robots drove at a maximum speed of 7m/s and had ultrasonic devices to perceive their environment. The local navigator ran on a single transputer and allowed the robot to move in a complete unknown environment with dynamic obstacles.

III.2. The Praxitele vehicle

Praxitele prototypes are electric car-like robots with a weight of 700kg. An asynchronous motor powered by lead Batteries allows to propel the car up to 70 km/h. Two persons can embark in each car and the vehicule gets an autonomy of about 80km. A Motorola VME162 with custom interfaces drives three servo-motors and a AC-motor controller. All proprioceptive data - like odometry or velocity - are computed on that device. Connected to ethernet, the computation is done using ORCCAD Software on a SUN workstation.
An ultra-sonic belt composed of fourteen ultrasonic sensors dispatched all around the vehicle has been constructed for sensing its local environment (Figure 7). Each of these sensors are synchronized and can detect obstacles in a range of eight meters with a resolution of one centimeter. It uses a measure of the flying time of ultrasonic waves to compute the distance between the vehicle and obstacles. This is done at a frequency of 50 ms (the time needed for the round trip of 8 meters of the wave) using a transputer network made of 4 transputers dynamically linked by a C004 circuit.

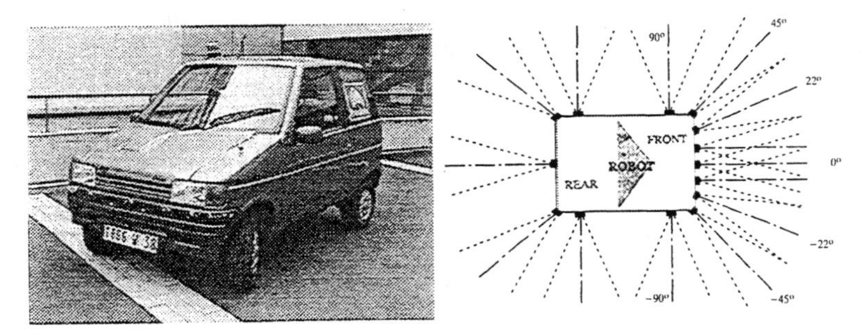

Figure 7: Praxitele prototype

Figure 8: Ultra-sonic sensors

III.3. Implementation on a transputer net and a VME board

The fourteen ultrasonic sensors are driven and synchronized by a custom board with a transputer and programmable integrated circuits. Data are computed in a second transputer which built a proximity map - a polar grid - of the local environment of the robot every 50ms, using the linear and the rotational instantaneous velocities. (Figure 9). This proximity map is sent to a third transputer which computes the escape lines. The best free escape line is sent to the fourth transputer which manages the data transfers with the VME board. The VME board is programmed using ORCCAD tool and it drives all the actuators and proprioceptive sensors. Two non-linear servo controls are implemented on it: one on the curvilinear velocity and one on the steering wheel angle. A fuzzy logic controller drives these 2 servo controls and takes place into this motion controller [2].

III.4 Synthesis

We have presented a local navigator for a mobile robot evolving in an unstructured and partially known environment. The data required by this navigator may come from

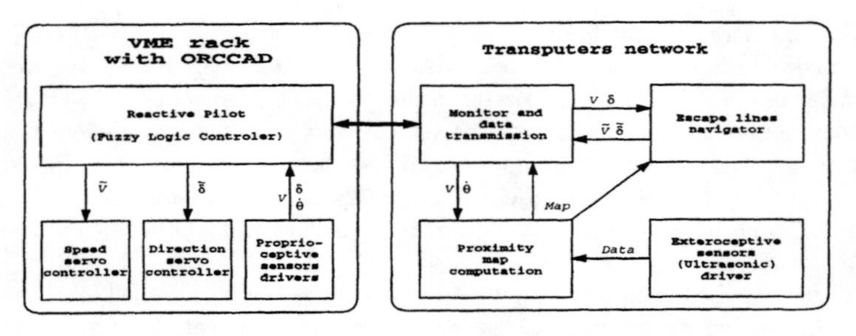

Figure 9: Motion controller implementation

non-accurate proximity sensors, like ultrasonic ones. Taking into account kinematic and dynamic constraints of the robot, as embedded in the direct model that it uses, the local navigator fits the computation abilities of the current technology. It was created to overpass classical oscillation problems due to the use of a reactive pilot. This local navigator forms with a reactive pilot a motion controller able to find a trajectory which takes into account constraints and unpredictive events, and follows the plan given by the planner.

References

[1] Th. Fraichard and A. Scheuer. Car-Like Robots and Moving Obstacles. In *icra*, volume 1, pages 64–69, San Diego, CA (USA), may 1994.

[2] Ph. Garnier, C. Novales, and Ch. Laugier. An Hybrid Motion Controller for a Real Car-Like Robot Evolving in a multi-Vehicle Environment. In *Intelligent Vehicles'95*, Detroit, USA, September 1995.

[3] Y. Kanayama, Y. Kimura, F. Miyazaki, and T. Noguchi. A Stable Tracking Control Method for a Non-Holonomic Mobile Robot. *IEEE/RSJ International Workshop on Intelligent Robots and Systems IROS'91, Osaka*, 1991.

[4] Jean-Claude Latombe. *Robot Motion Planning.* Kluwer Academic Publishers, Boston, 1991.

[5] C. Novales and R. Zapata. A Local Architecture for Controlling the Movments of Fast Mobile Robot. *International Symposium on Robotic And Manufacturing ISRAM'94, Hawai, USA*, 1994.

[6] E.D. Sontag. *Mathematical Control Theory, Deterministic and Finite dimensional System.* Springer Velag, 1990.

[7] B. Dacre-Wright T. Simeon. A practical motion planner for all-terrain mobile robots. *Proceeding of the International Conference on Intelligent Robots and Systems, Yokohama*, July 1993.

Evolving real-time behavioral modules for a robot with GP

Markus Olmer Peter Nordin Wolfgang Banzhaf
Fachbereich Informatik
Universität Dortmund
44221 Dortmund, GERMANY
email: olmer,nordin,banzhaf@ls11.informatik.uni-dortmund.de

Abstract

In this paper we demonstrate an efficient method which divides a control task into smaller sub–tasks. We use a Genetic Programming system that first learns the sub-tasks and then evolves a higher–level action selection strategy for deciding which of the evolved lower–level algorithms should be in control. The Swiss miniature robot Khepera is employed as the experimental platform. Results are presented which show that the robot is indeed able to evolve both the control algorithms for the different lower–level tasks and the strategy for the selection of tasks. The evolved solutions also show robust performance even if the robot is lifted and placed in a completely different environment or if obstacles are moved around.

INTRODUCTION

Robotics plays an increasingly important role in our society and its potential for the future is tremendous. Especially autonomous robots have a large potential in, for instance, medical applications, hostile environments, exploration tasks, or in the aiding of disabled people. An autonomous robot needs to be flexible enough to cope with an environment that is incompletely known and imperfectly modeled, when using sensors which might be noisy or even defective. Plainly, the number of possible errors in a real–life task is nearly infinite.

A useful strategy to create robust behavior is to divide complex actions into *action primitives* which only perform specialized tasks like avoiding obstacles or moving ahead, and then combine them again for the creation of higher *levels of competence* (Brooks [1]). Our approach uses a learning robot. The goal here is to have a robot that:

- can learn certain action primitives and a control structure and combine these to create complex behavior.
- is able to do this learning in real time, on–line.
- is fault-tolerant recovering after recognizing an error, always being able to act in a robust way in an otherwise unknown or only partially known environment.
- operates with an extendable system which adapts to changes, such as new action primitives.

Figure 1: The Swiss robot system KHEPERA. The diameter is 6 cm and its height is 4 cm, depending on how many expansion turrets are installed (none in this Figure).

For this purpose we implemented a variant of a genetic programming system called a compiling genetic programming system (CGPS) which uses different fitness functions for all its tasks. It creates binary machine code programs that fulfill the respective fitness criteria.

COMPILING GENETIC PROGRAMMING

The Compiling Genetic Programming System [5, 7, 8] is based on the Genetic Programming paradigm of John Koza [4]. The system generates machine code for SUN workstations by directly manipulating binary code with genetic operators. It has been shown to have several advantages when compared to another adaptive technique, neural networks. It is frequently faster, requires less memory and has good generalization capabilities [6]. It also produces output in a more symbolic form which is beneficial when trying to analyse why the robot behaves in a certain way. The system already previously proved to work for the complex task of a robot evolving obstacle avoiding behavior [9].

The method uses a stochastic sampling of the environment [9, 10] where randomly selected individuals compete against each other in a tournament selection procedure in *different* situations which could result in an "unfair" competition. However, in the long term the better performing individuals survive.

More precisely, the GP system randomly takes four individuals of the same population [1] and performs the following steps on every individual:

- Feed sensory values to one of the individuals in the tournament.
- Execute the individual and store the resulting value.
- Compute fitness.

After all steps the system computes the winners of the matches of individual 1 against individual 2 and of individual 3 against individual 4. Mutated and recombined versions of the winners then replace the losers as new individuals. This execution cycle is identical for all the action primitives and for the action selection mechanism. However, the action selection mechanism operates on a ten times slower time scale.

[1] We use separate populations of individuals with different fitness criteria for each of the tasks

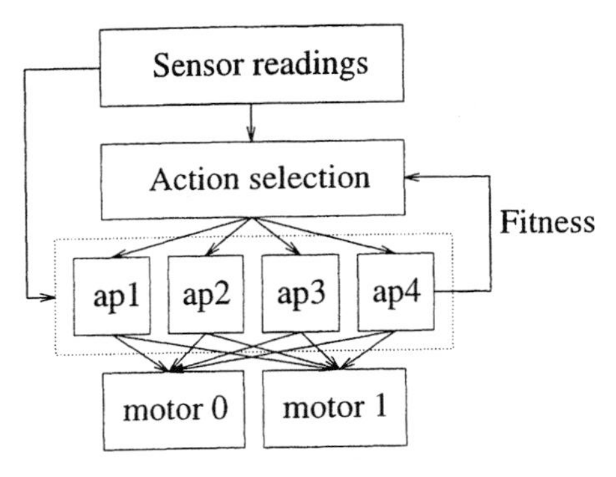

Figure 2: Data flow and architecture of the robot operating system.

THE ROBOT EXPERIMENT

The experimental environment consists of a black floor with moveable white walls so it can be changed easily to test the features and capabilities of the system. Walls can be rearranged and fixed with velcro fly. The walls provide good reflection for the robot's sensors. The GP system runs fully on the Khepera robot (see Figure 1, [3]) which has a simple multitasking operating system onboard.

The goal of the controlling GP system is to evolve several simple behaviors in a sense–think–act context. The controlling algorithm has a small population size, typically less than 40 individuals. The individuals of every species use selected values from the robot's sensors (8 ambient light, 8 reflected light, two measured motor speed, two distance traveled) as input and produce two motor speeds as output. The action selection population also uses the fitness values of the action primitives last computed.

The robot's GP system possesses five populations for the action primitives and the control structure:

GO AHEAD : the robot learns to move straight ahead at maximum speed.

AVOID OBSTACLE : the robot avoids obstacles at a learned maximum speed.

SEEK OBSTACLE : the inverse of *AVOID OBSTACLE*, useful to create wall following behavior. Wall following would be an oscillating task switching between obstacle avoidance and obstacle seeking.

FIND DARK : the robot searches for a dark gradient to head towards.

SELECT ACTION : this population contains functions which select one of the above action primitives.

The parameters for the GP system are shown in Table 1.

When the robot starts learning, the system feeds the required data and sensor readings into the input register space of the action selection mechanism population. Four individuals are selected for the action selection tournament. Every chosen individual selects one of the four action primitives. Again required values are copied into the GP register space and the tournament for an action primitive starts. The winners replace the losers and the genetic operators are used. This is done for all selected individuals from the action selection population. The genetic operators ensure that the offspring

Parameters	Values
Objective:	Evolve action primitives
Constants (Terminal set):	Integers form 0 to 8191
Instructions (Function set):	ADD, SUB, MUL, SHL, SHR, XOR, OR, AND
Input registers:	3 to 6
Output registers:	1 (sometimes interpreted as a vector)
Maximum population size:	36
Crossover probability:	100 percent
Mutation probability:	5 percent
Selection:	Tournament selection, size 4
Termination criterion:	None
Maximum number of generations:	infinite
Maximum number of instructions:	256

Table 1: Control parameters of the robot experiments with the CGPS system

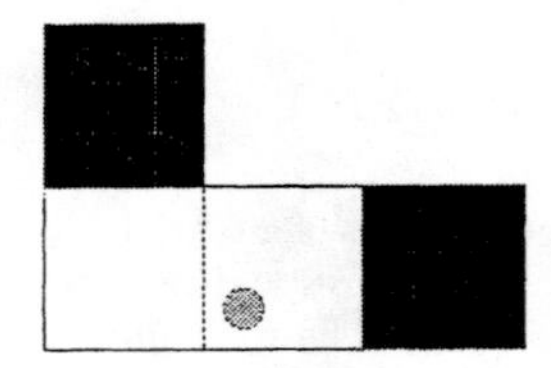

Figure 3: The experimental setup for the robot. The round spot is the starting point for the robot and the marked areas are dark places for the robot to hide in.

is syntactically correct and do not contain any unwanted machine code instructions or constants. Afterwards the GP system updates the selection population. The overall architecture is depicted in Figure 2. This mechanism continues forever to allow the robot to adapt continuously to new situations.

PERFORMANCE OF THE SYSTEM

Performance of the system has been measured in different experiments with the robot. The tasks were: collision avoidance, wall following and hiding in the dark. For every task six experiments were done, the best and the worst have not been counted. The performance of the evolved collision avoidance behavior and the action selection mechanism underlying it is shown in Table 2. The robot starts to show effective collision avoidance behavior after 800 tournament cycles. The second experiment dealt with the robot's capability to recognize dark areas in its environment and to stay there. Here, the performance measurement is the time the robot spends in the dark areas of its environment which is shown in Figure 3. Results are shown in Table 3.

The high values in the beginning of the evolution are caused by random movement of the robot in a dark area. The robot stays in the dark for a long time when it accidently keeps touching the wall while trying to evolve obstacle avoidance. In the end, the time spent in the dark constantly increases due to the success of the GP system. The next task for the robot is to follow the walls of the experimental environment. It comprises a combination of obstacle avoiding and obstacle seeking behavior. This time, we measure the number of times when the robot exits a narrow space (4 cm) between the robot and the wall.

Tournament cycles	Average Collisions	Best rate	Worst rate
0	30.25	22	38
100	17.75	9	20
200	13.25	4	24
300	14.25	6	20
400	12.75	5	12
500	7.50	6	9
600	10.25	7	16
700	8.00	2	11
800	6.00	3	8
900	5.75	3	11
1000	6.75	3	10

Table 2: The number of collisions during 100 tournament cycles, phase 0 only uses random programs. In two out of eight random tests the robot totally locked up in a corner. The learned behavior appears to be stable after 1000 cycles.

Tournament cycles	Average score
0	8
100	26
200	26.5
300	33.25
400	72.25
500	28.75
600	28
700	44.75
800	53.5
900	52.25
1000	68.25

Table 3: Robot scores during the finding–dark–behavior. The maximum score is 120 points where one point is scored for every second spent in the dark. Approximately 120 seconds is the reachable maximum but the time the robot needs to find the dark is between 10 and 15 seconds. Phase 0 only uses random programs.

RESULTS

After a short period of approximately 4-7 minutes, the robot has learned its action primitives and shows robust behavior in its test environment. When the robot is placed into a different environment, it takes about two additional minutes until it has adapted its behavior, e. g., by adjusting to new reflection properties of the walls. With the action primitives performing better the selection mechanism improves also. This requires, however, significantly larger amounts of time, because the fitness values of its actions are to be fed to the selection tournament: the selection mechanism can never get a good score without the actions receiving good fitness values in turn.

Our experiments have shown that GP can be a useful approach to robot control. It is, however, not error–free in every situation, since the robot sometimes touches a wall before converging again after a change in the environment. One of the advantages of this system is its property of easily adapting to new situations. A change in the environment does not cause the robot to need new instructions. Another advantage is the system's capability of generalizing what it has seen – the mechanisms emerging in its behavior would have taken a very long time to be modelled and implemented by a programmer. These results indicate the feasibility of GP in robotics, for reasons of speed, hardware needs, adapability and robustness.

FURTHER WORK

Presently, the robot is trained on tasks with low complexity. Our next steps will be to implement more action primitives in order to test the method more thoroughly. With more action primitives, more complex behavior is expected to emerge.

In the present version, the system does not use a representation or a model of the world, following the notion that "...the world is its own best model. It is always up to date. It always contains every detail there to be known. The trick is to sense it appropriately and often enough" ([2]). Whereas this is true for simple actions, Steels states: "Autonomous agents without internal models will always be severely limited" ([11]). For instance, a world model would be essential where path planning is needed. Also, by adding memory access to our GP system, in addition to the memory provided by the populations of programs, mapping of the world could become possible.

References

[1] Rodney A. Brooks. A robust layered control system for a mobile robot. *IEEE Journal of Robotics and Automation*, 2:14–23, 1986.

[2] Rodney A. Brooks. Why elephants don't play chess. *Robotics and Autonomous Systems*, 6:3–15, 1990.

[3] K-Team. *Khepera User Manual.* EPFL, Lausanne, 1994.

[4] John R. Koza. *Genetic Programming - On the Programming of Computers by the Means of Natural Selection.* MIT Press, Cambridge Mass., 1992.

[5] P. Nordin. Cgps - a compiling genetic programming system that directly manipulates the machine-code. In *Advances in Genetic Programming*, pages 311–331. MIT Press, Cambridge, Mass., 1994.

[6] P. Nordin. Comparison of a compiling genetic programming system versus a connectionist approach. In *Handbook of Evolutionary Computation.* Oxford University Press, to appear, 1997.

[7] P. Nordin and W. Banzhaf. Complexity compression and evolution. In L. Eshelman, editor, *Proceedings of the Sixth International Conference of Genetic Algorithms*, pages 310—317, San Mateo, CA, 1995. Morgan Kaufmann.

[8] P. Nordin and W. Banzhaf. Evolving turing-complete programs for a register machine with self-modifying code. In L. Eshelman, editor, *Proceedings of the Sixth International Conference of Genetic Algorithms*, pages 318—325, San Mateo, CA, 1995. Morgan Kaufmann.

[9] P. Nordin and W. Banzhaf. Genetic programming controlling a miniature robot in real time. Technical Report 4/95, Department of Computer Science, University of Dortmund, 1995.

[10] P. Nordin and W. Banzhaf. Real time evolution of behavior and a world model for a miniature robot using genetic programming. Technical Report 5/95, Department of Computer Science, University of Dortmund, 1995.

[11] Luc Steels. Exploiting analogical representations. *Robotics and Autonomous Systems*, 6:71–88, 1990.

DESIGN OF OPERATION PLANNING AND CONTROL SYSTEMS FOR FACTORIES ORGANISED AS FLEXIBLE NETWORK OF PROCESSORS

CLAUDE OLIVIER[1,2]

[1]*École de technologie supérieure, Université du Québec, Département de génie de la production automatisée, 4750 Henri-Julien ave, Montréal, (Québec), Canada, H2T 2C8, olivier@gpa.etsmtl.ca*

BENOIT MONTREUIL[2], PIERRE LEFRANÇOIS[2]

[2]*SORCIIER, Research Center on the International Competitiveness and the Engineering of the Network Entreprise, Université Laval, Ste-Foy, (Québec), Canada, G1K 7P4, benoit@osd.ulaval.ca, pierre@osd.ulaval.ca*

ABSTRACT

The new manufacturing paradigms, such as agility and mass customizing, are based on adaptability of products and production volumes to the customer needs. This paper presents a modeling approach, based on the concept of flexible networks of flexible processors, to design, plan and control these new manufacturing environments. The general context and the operating conditions of factories organized as flexible networks of processors are presented. Respecting these particular manufacturing conditions, a framework for the design of manufacturing resources planning and control systems is developed.

KEYWORDS: agile manufacturing, mass customization, manufacturing planning and control, network enterprise, flexibility.

INTRODUCTION

With the rapid changes taking place into the global market, it becomes clear that enterprises working on an agile manufacturing basis or using mass customization concepts will rapidly become leaders in the forthcoming economy. These new paradigms are no longer based on the stability of demand nor a narrow range of products. They are characterized by the adaptability of the products to the customer needs and a rapid variation of the production volume to follow the demand [1,2,3].

One of the ways to efficiently model the new manufacturing systems is through an approach based on the concept of *flexible* networks of *flexible* processors. In this concept, each resource of the manufacturing environment, (i.e. workstations, special crews, workers, machines, etc.), is to be considered as an independent processor or node of the network. This node is tied to other processors on a « as required » basis to perform value adding processes. Resources can then be combined for very short periods of time, a few minutes for example, or on a near-permanent basis. In a given network, many processors can have equivalent or yet very different technological capabilities. The links between processors are also kept as flexible as possible, they can be physically supported by automated or manual handling systems. Given the high flexibility of this kind of environment, many potential combinations of processors are conceivable to complete a particular task. In this context, the operating and marginal manufacturing costs of each processor and the handling costs, are key elements to consider in order to dynamically minimize the global manufacturing costs and maximize the global manufacturing performance.

In order to design efficient production systems and factories operating under the evolving paradigms, new manufacturing resources planning and control systems (MRPC) have to be developed. This paper synthetizes some of the research work done at the SORCIIER Center on the subject. First, the operating conditions of factories organized as flexible networks of processors are rapidly presented. Respecting these particular manufacturing conditions, a framework for the design of manufacturing resources planning and control systems is introduced, including five major phases from the *general configuration* management of the factory up to control management. Finally, the last section provides concluding remarks and presents some active research avenues.

OPERATING CONDITIONS AND GENERAL FRAMEWORK

As introduced briefly in the first section, the concept *of flexible networks of flexible processors* is a recent approach developed to efficiently model agile manufacturing systems [4]. In this approach, each manufacturing resource is considered as a processor. A processor could be, for example, a dedicated machine, a flexible workstation, a subcontractor a programmable machine tool, a specialized worker, an assembly workstation or system, a material handling system, etc. All these processors have specific technological characteristics, capabilities, capacities and availability. They may be located in the same physical area, such as within the manufacturing facility limits, but may be spread geographically. In the facility, processors having the same capabilities can be grouped in the same physical area, as in a functional layout, or they can be spread over the factory, as in a holographic layout [5].

All these processors are logically tied together using a *work-flow pattern* to perform specific value adding processes.Given the flexibility of the manufacturing environment, work-flow patterns are used, mainly in the planning phase, to estimate the processors load and the flow of items through the factory. The *work pattern* can be defined as the pattern of tasks *planned* to be accomplished by each processor in a given time slot. This pattern includes the specific groups of products, items or parts to be processed on each processor in this time period. The *flow pattern* expresses the *planned* routings of items between processors. The work-flow patterns respect the limitations imposed by the physical capacities of each processor, the material handling system and the physical storage spaces available for material, work in process or finished products. They respect also logistic constraints imposed by the managing team and by the dynamic conditions of this environment. For example, managers may want to spread the production over many processors to speed up the output, to partially insure the availability of some items in a case of an unpredictable breakdown or simply to uniformly distribute the work load on a given group of processors. The work flow patterns are determined with respect of the physical and logistic constraints and minimize the production and handling costs [4]. These processors and work-flow patterns can be totally internal to the manufacturing enterprise or can be linked to clients, subcontractors or suppliers. [6]

Many generic industries can be modeled by this approach. Particularly, enterprises producing sub-assembly or finished items, like OEM. These items are incorporated in larger products such as radios for motor vehicles, water pumps for washers or dishwashers, etc. In this context, large volume of sub-assemblies have to be produced. The demand for these products is tied to the finished product demands. But, even if these sub-assemblies are similar and their demand can be aggregated by *generic* products, the real demand for each of them is different and may vary from day to day, based on the clients preferences. The manufacturing environment required to produced in this context needs to be very flexible and efficient.

It is important to clearly define the general framework required to efficiently design, operate and control such a complex manufacturing environment. It can be synthetized by five management steps: 1) manufacturing configuration, 2) demand management, 3) global operation management, 4) operation deployment and, 5) operational process. Each is detailed in the next section.

MANAGEMENT PHASES

Manufacturing Configuration Management

The first step in manufacturing facility strategic planning and design is to determined, over a medium term planning horizon, the physical configuration of the manufacturing environment required. In agile manufacturing context, this environment should perform efficiently for a relatively stable and known aggregate demand but with a complex product mix, where real demand may vary from day to day. Dedicated manufacturing or assembly lines are not efficient in an operational context where products mix and lot sizes vary. In such an environment, it is important to design *adaptable* facilities, efficient under a given spectrum of uncertainty.

Traditional layout design considers products, flows or routings as static parameters. New approaches have to be developed in order to integrate the dynamic aspects of these parameters. Recent approaches, such as holographics layout design, open interesting ways to solve this complex question [6,7]. Moreover, the efficiency of the layouts generated by these approaches have to be evaluated by different criteria than the aggregate distance between groups of machines. Handling costs are still important, but operating costs, marginal processing costs and flow costs based on the physical and operational constraints also have to be considered. In this particular context, Olivier *et al.* [8] have proposed a model, based on a linear mathematical programming approach, to evaluate and compare layout efficiency in a flexible network environment. The approach proposed integrated the work-flow patterns previously introduced as an estimation of the real flows and work loads on the faciliy processors.

Demand Management

One of the most important data source for manufacturing planning is the demand pattern incoming from customers. Usuallty, demand does not have a totally random pattern. In fact, in this manufacturing context, most of customers are also producers. They have access to sophisticated production planning tools and are able to give good aggregate information on demand pattern. The vast majority of them are operating under just-in-time mode and expect a just-in-time delivery. In this context, it is compulsory to integrate into the production planning model all the information available on delivery schedules of items, types of transportation used, pickup schedules, delivery alternatives and marginal costs involved. But over this basic collaboration, the production planning should be done in a pro-active manner. Customers should be aware of the manufacturing facility possibilities and limits. They should try to help the producer to better exploit its facility but in a *win-win* manner. In the agreement between the producer and the customer, each party should know the limitations and the expectations of the other and should take them in account in the production plan determination wherever possible. In such an agreement, each party will be better serve by the other.

Global Operation Management

Given the flexibility of the manufacturing environment, multiplicity of routings, variation in order sizes, disparity of products, it is not possible to generate an optimal production plan in a very short period of time, even with an efficient mathematical model. It is essential to reduce the complexity of the problem by limiting the variables range. The first planning step is to produce *virtual manufacturing networks, VMN* [8]. VMN are determined based on the work-flow patterns but, this time, for a specific production day. VMNs specify, for a given demand and product mix, the specific processors to use, the volume of production for each product or item on each processor and the transit lot size between processors. For a given production day, the determination of VMN will minimize handling costs, operating costs and processing cost with respect to the manufacturing and logistical contraints. This phase limits the size of the problem to resolve and insure that capacity is available for a the production period investigated. The VMN are computed using a mathematical linear program.

Operations deployment

The next operation in the planning phase is to compute the production plan for each group of processors having equivalent technological capabilities but, this time, for a short period, an hour for example. In this phase, the sequence of operations is investigated and the availability of items is insured from product structure and temporal point of views. This second phase is also computed through a mathematical model but using the virtual manufacturing networks as input. The last procedure in the production plan generation, is to integrate the two phases of the aggregate planning into a detailed schedule. This schedule precisely shows, for each processor, the tasks to be done, starting time, expected finishing time, lot size, stockage area, location of components or raw material, etc.

Operational Process Management

Depending of the precision of the detailed production plan, operational process management could be two very different tasks. On one side, if lot of latitudes is given to processors groups managers, many short term planning decisions have to be taken in order to respect the schedule. If the detailed plan is very precise, managers do not take any planning decision.

But, either ways, it is essential to precisely monitor the production output to insure the delivery of finished products to the loading docks at the required time. This step can be done using sophisticated electronic equipment, or manually, depending of the production volume involved. In either cases, when a sufficiently important mismatch occurs between the production plan and the production itself, a new plan should be generated to take into account the actual network state. The mismatch can be evaluated through specific production differences introducing a raisonable uncertainty to not meet the delivery schedule within the allowed time limits.

CONCLUDING REMARKS

The ongoing research program on the design of operation planning and control systems for factories organised as flexible networks of processors have been briefly introduced, exposing some of the work done by the SORCIIER center. The space available in this paper does not allow to present the mathematical models involved and the real industrial applications. However, a large part of this work, such as layout evaluation, VMN determination or aggregate production planning is already operational and has been validated in an industrial context. The results obtained confirm that the direction taken is promising.

REFERENCES

1. Buzacott, J.,A., "A Perspective on New Paradigms in Manufacturing". Journal of Manufacturing Systems, Vol. 14, No. 2, pp. 118-125, (1995).
2. Golman, S.L., Nagel, R.N., and Canneth, P., *Agile Competitors and Virtual Organization,* Van Nostrand Reinhold, New York, (1995).
3. Pine, J., *Mass Customization: the New Frontier in Business Competition*, Harvard Business School, Boston, Ma. (1993).
4. Olivier, C., Montreuil, B., Lefrançois, P., Maley, J., "Evaluating Layouts for Factories Organized as Flexiblie Networks of Processors" Working Paper, faculté des sciences de l'administration, Université Laval, Québec, Cananda, (1995).
5. Marcotte, S. Montreuil, B., "Design of Holographic Layout for Agile Flow Shops", Proc. of the Int'l Industrial Eng. Conf., Montréal, Canada, pp. 929-938, (1995)
6. D'amours, S., Montreuil, B., Soumis, F., "Priced-Based planning and Scheduling of Multi-Product Orders in Symbiotic Manufacturing Networks",EJOR (1996) accepted.
7. Montreuil, B., Lefrançois, P., Marcotte, S., Venkadatri, U., "Layout for Chaos-Holographic Layout of Manufacturing Systems Operating in Highly Volatile Environments", Working paper, 93-53, Faculté des sciences de l'administration, Université Laval, Québec, Canada.
8. Olivier, C., Montreuil, B., Lefrançois, P., Maley, J., "Evaluating Layouts for Mass Customizing Factories", ISPE/IFAC Int'l Conf. on CAD/CAM, Robotics and Factories of the Futur. Ottawa, Canada, pp. 862-869. (1994)

A NAVIGATION SYSTEM FOR ADVANCED WHEELCHAIRS USING SONAR AND INFRARED SENSORS

LUCA ODETTI, ANGELO M. SABATINI
ARTS–Lab, Scuola Superiore S. Anna
Via Carducci, 40
56127 Pisa, ITALY

ABSTRACT
A navigation sensor system is developed for allowing a power wheelchair with omnidirectional steering capabilities to achieve a robust and safe reflex of obstacle avoidance. The sensor system is based on the integration of ultrasonic and infrared sensors in an array of so–called composite sensors. Robust and inexpensive analog signal processing techniques are proposed for improving the measurement abilities of the sensors; additionally, fast multisensor integration algorithms for environment analysis and map representation are illustrated.

KEYWORDS: power wheelchair, navigation aid, proximity sensors

INTRODUCTION

The operation of conventional power wheelchairs is difficult for many people with sensory, perceptual or motor impairments. Mainly, these difficulties concern the restricted functionality and limited mobility of conventional wheelchairs [1]. Several research projects are currently undergoing to stretch the state–of–the–art of power wheelchairs. Among them, the project *Office Wheelchair with High Manoeuvrability and Navigational Intelligence for People with Severe Handicaps* (OMNI) in the R&D–program TIDE (Technological Inititiative for Disabled and Elderly People) sponsored by the Euopean Community aims at the development of a generation of smart wheelchairs for vocational rehabilitation of severely disabled people [2].

The OMNI wheelchair adopts the omnidirectional driving concepts of the so–called Mecanum wheels, which provide the wheelchair with the ability to perform any sort of forward, sideward and rotational movements [3]. This feature helps to overcome one of the main limitations of the kinematic designs of conventional power wheelchairs. e.g., the need of performing prior rotations before changing direction. This limitation turns out be quite severe particularly when the user has to move within complex and densely crowded domestic environments, oftentimes a difficult activity even for experienced and skillful drivers. The functional augmentation that the omnidirectional driving concept gives to a power wheelchair can be fully exploited by the availability of a sensor–based navigation system. The aim of the sensor assistance to navigation concerns the safe generation of movements by the user. Our involvment in the project concerns the design and fabrication of such a navigation system module; in this paper we describe the approach we are pursuing towards this goal, and the main results achieved so far.

THE PROXIMITY SENSOR SYSTEM

Using several sensing techniques for detecting the presence of objects nearby the wheelchair is a key factor in an attempt to achieve robust and reliable behaviors of obstacle avoidance; the reason is that different sources of sensory information – sometimes redundant, oftentimes complementary – may help the system to build more robust environment representations. These representations, or maps, are vital to properly command in which direction and how fast the wheelchair would be safely driven to avoid collisions.

The complementary features of reflecting behaviors for ultrasonic waves and infrared radiation in an air medium make interesting the perspective to fuse together the information they supply. The sensor system is formed by a number of "composite" sensors; the sensors are deployed in a typical ring–like arrangement. With the term composite sensor we mean the physical integration within the same device of an ultrasonic rangefinder and an infrared detector, see Fig.1.

Fig. 1: Photograph of a composite sensor.

The ultrasonic rangefinder is composed of separate transmitting and receiving piezoelectric capsules. The infrared detector uses a light emitting diode and a pair of receivers, which are arranged in such a way to double the sensor sensitivity to the incoming radiation. Both ultrasonic and infrared sensing elements in each composite sensor are off–the–shelf, inexpensive devices which are typically found in remote control systems for household appliances, alarm systems and similar gadgets.

Our approach to the design of composite proximity sensor systems is based on the assumption that the use of expensive sensing elements and complex signal processing techniques are at odds with the desire to achieve acceptable cost–to–performance ratios for the navigation systems of smart wheelchairs. It should be pointed out, in fact, the importance of carefully considering the social and economical factors involved in the design of technological aids for the assistance to disabled people. Because one of the most attractive adavantages of using ultrasonic and infrared proximity sensors is their low cost and ease of installation, the analog drivers and the architecture of the control unit embedded in the navigation system are designed with great attention to the cost factor.

Signal Conditioning and Acquisition

We have investigated robust analog signal processing techniques for capturing the desired sensory information, e.g., the time–of–flight (TOF) as for the pulse–echo ultrasonic rangefinder, and the amplitude of the back–scattered radiation as for the pulsed–excitation infrared detector [4].

A novel threshold–based detection and estimation method for the TOF measurement is implemented in the ultrasonic analog drivers. A shortcoming of conventional threshold–based receiver is due to the fact that the time instant of threshold cross–over is actually dependent on the amplitude and shape of the return echo. Highly attenuated signals are thus responsible for anomalously elongated range readings. In our approach, a time–varying threshold is derived from the time integral of the return signal, in an attempt to make the TOF estimate somewhat insensitive to amplitude fluctuations. Our method is proven to yield unbiased TOF estimates, thus removing the noxious effect of signal amplitude on the performance of threshold–based receivers. The sensing range is between R_{min} = 5 cm and R_{max} = 200 cm.

The main feature of the infrared analog driver is the adoption of a time–windowed, band–pass filtering scheme. The synchronization between the transmitter and the receiver operation allows to "open" two temporal windows at the receiver for integrating the return signals; in the first window background noise and interfering radiations are integrated without the presence of the desired signal component, which is conversely integrated in the second window. If the statistical properties of the noise random processes do not change in the time interval between the two integrations, it is straightforward to remove most of the noise contamination by simply subtracting the first reading from the second one. This double reading scheme proves to be quite effective for attenuating the effects of some typical disturbances, e.g, the sunlight and the sources of artificial light within normal domestic environments. The sensing range is between R_{min} = 0 cm and R_{max} = 50 cm.

The Control Unit of the Sensor System

The same attention to the cost factor we devote to the choice of the sensing elements is also reflected in the selected architecture of the control unit of the sensor system, e.g., the unit which masters the sensor system, by deciding, for instance, which sensors and how many of them may be active at the same time. A single low–cost microcontroller of the Motorola HC11 family is embedded in the control unit; it performs the TOF measurements for the ultrasonic module of each composite sensor via its input capture functions (16 bit–resolution counters at the operating clock frequency f_o = 2 MHz). The 8 bit–resolution A/D converters integrated within the microcontroller are used to perform the measurements of the peak amplitude of the detected radiation for the infrared module of each composite sensor. Although, in principle, a distributed network of microcontrollers is conceivable so as to add how many composite sensors we would need in a specific application, only one microcontroller is presently used for managing a given number of composite sensors (N=16 for the OMNI wheelchair). The microcontroller communicates with the PC industrial board which provides the computational power for running the obstacle avoidance method via its asynchronous serial interface.

CONCEPT OF THE NAVIGATION SYSTEM

A smart wheelchair is functionally different from an autonomous robot. This means that, although advanced robotic technologies are to be used in the wheelchair, the wheelchair controller is relieved from handling tasks at a very high level of abstraction. Rather, the structure of the control system must be flexible and adaptable, and the control actions are to be always shared between the user and the control system itself. This approach allows to take the user in the active and psychologically rewarding role of master; the navigation system (NAVSENS) module looks like a sort of co–pilot, endowed with the ability to modify the intended user commands from the Human–Machine interface (HMI) module, see Fig. 2. In the wheelchair prototype described in [3], the High Level Control (HLC) module is at the interface between the HMI module and the MOTCON module. Trajectory back–tracing and play–back are handled at this level. In the present system the modification of user commands (direction of travel and velocity magnitude) critically depends on the environment perception by the proximity sensor system (environment analysis). The control reactive strategy of avoiding collisions with obstacles in the wheelchair surroundings is managed according to a variety of selectable operating modes.

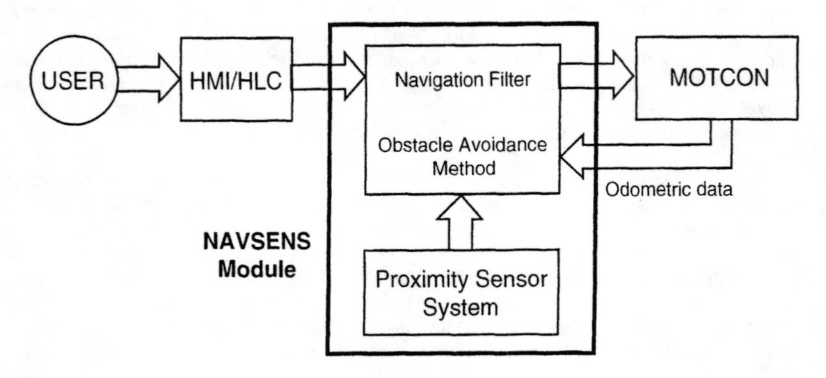

Fig. 2: System architecture.

A distinction is made in the OMNI system between: a) direct control of the wheelchair, b) sensor–supported control, and c) sensor–guided control. The term direct control concerns the possibility for the user to use a standard input device, such as a joystick, to control the chair motion, without any direct involvment of the navigation system. If needed, the sensor system may be switched on, so as to report the results of the environment analysis to the user. In this operating mode the navigation filter does not filter the user commands. The sensor–supported control allows the user to benefit from the navigation system in order to reduce the approaching velocity to the mapped obstacles. The sensor–guided control is a more sophisticated routine for avoiding collisions, where the navigation system is given the ability to change not only the velocity magnitude, but also the direction of travel. In this operating mode, interesting navigation routines are also defined and implemented, e.g., driving along a wall, and driving through a door.

The navigation computer is a PC industrial board, equipped with a math coprocessor in order to accelerate the most demanding operations implied by the method of obstacle avoidance. The communications among the navigation computer within the NAVSENS module and the other modules of the wheelchair, e.g., the Human–Machine Interface (HMI) module, the Motion Controller (MOTCON) module, and the proximity sensor system module are established using standard RS232 serial interfaces. The navigation computer drives sensorial acquisition, environment map building, according to the position data provided by the MOTCON module. Typical sensory frame intervals are in the order of about 200 msec. The main limitation in the achievable frame intervals depends on the worst–case latency due to the ultrasonic TOFs, rather than to the processing time involved in the method of obstacle avoidance (approximately 50 msec). The serial communication protocol captures messages coming from other modules, possibly re–routing them. The protocol recognizes velocity commands from the HMI module, and invokes the navigation procedure; such procedure modifies the command, depending on the current operating mode and the current status of the environment, as it is perceived by the method of obstacle avoidance.

METHOD OF OBSTACLE AVOIDANCE

The method of obstacle avoidance we intend to use is derived from the histogramic in–motion mapping for mobile robot and power wheelchair obstacle avoidance [5]–[6]. Since a single sonar reading is not able to convey reliable information about the position of any object around the robot for a number of reasons (multiple reflections, low spatial resolution, possibility of spurious readings, etc.), a grid representation of the world is built in an attempt to have a statistically significant view of the outer world. Although a single sonar reading has a small impact on the grid, fast firing of all the sonar sensors available on the robot while the robot itself is relentlessly moving allows to create a robust map for obstacle avoidance purposes. Cell updating can be very fast without requiring the computationally demanding sonar sensor models implied by the original method of occupancy grid proposed by Elfes [7]. In [5]–[6], a successive stage of data reduction is proposed, which corresponds to a projection of the data available on the Cartesian grid into a polar histogram. Such a polar histogram gives the probability of occupancy within each of a number of angular sectors. Safe direction of travel are identified among possibly wide valleys in between sharp peaks of the histogram profile. Additionally, the width of the valley and the height of the peaks defining the valley are used to change the velocity of approach in the appropriate steering direction.

Although the occupancy grid method lends itself quite easily to integrate other sensing modalities in the same framework, the histogramic in–motion mapping in its original implementation relies just upon sonar sensors. We propose to integrate the data from the infrared detectors of our composite sensors directly in the polar histogram representation. Although our infrared detectors are proportional devices, we are aware that too many physical factors – including target reflectivity, size, distance and orientation relative to the infrared sensor – have an effect on the measured infrared amplitude, thus making the rangefinding abilities of infrared sensors worse in comparison to those of their ultrasonic counterparts. Hence, it does not make too much sense to introduce infrared data directly in the occupancy grid. Infrared data are used in an on–off fashion to modify the histogram in the sense, for instance, to sharpen the peaks and to outline a valley in a region of space where sonar sensors are unable to properly work, due to their limited spatial resolution. The good edge detection capabilities of infrared sensors make them particularly suited for identifying door openings. For this reason, we suggest to structure the edges of any door within a domestic environment (the apartment of the disabled user) with small strips of highly reflective adhesive, placed at the height of the composite sensors deployed in the chair, so as to facilitate the process of door identification.

EXPERIMENTAL RESULTS

The navigation system has been tested as for its ability to correctly perform environment analysis in a number of interesting situations, including passing through a door, and following a long corridor. The reactive strategy has not yet been demonstrated in real–life experiments. At the time of writing this paper, the wheelchair prototype equipped with the sensor system is not available. We decided, therefore, to place a sensor frame with the same size as in the final implementation on a purposely built sensor carrier. The sensor carrier is simply a robot vehicle equipped with a dead–reckoning system, and remotely commanded by means of a standard joystick. The motion controller of the sensor carrier is based on a MC68HC11 microcontroller, similar to the one embedded in the NAVSENS module. The sensor data are collected while the sensor carrier is commanded to move by the remote operator; these data are then transferred from the NAVSENS module to a PC486DX, together with the odometric data from the motion controller of the sensor carrier. The purpose of the experiments carried on with the sensor carrier is to confirm the results of the simulations we preliminarly performed for deciding both the sensor allocation, and the strategy of spatio–temporal cell updating in the occupancy grid. An excellent qualitative agreement between them has been achieved, see Fig. 3. The polar histogram is augmented with the information coming from the infrared detectors of each composite sensor in the array. The sensor carrier experiments show that the sensor system has excellent abilities to accurately map the door openings. Ineffective door passage abilities of the original Vector Field Histogram are discussed in [6]; it is obvious, in our view, that the problem is not the method, but the properties of the proximity sensors that are used to implement the method.

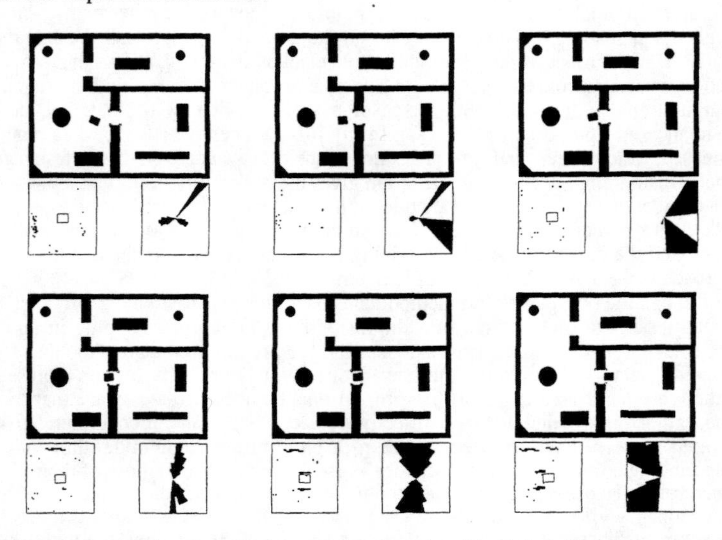

Fig. 3: Frames showing an approaching manoeuver of to a door (simulation). Each frame reports a) the room layout with the chair, b) the occupancy grid (bottom left), and c) the polar histogram (bottom right).

ACKNOWLEDGMENTS

This work is supported by funds from the R&D–program TIDE, project OMNI TP 1097. The other partners in the OMNI consortium are: the Chair of Process Control (PRT), FernUniversität Hagen, Germany, as coordinating partner; CALL–Centre, University of Edinburgh, U.K.; Forschungsinstitut Technologie–Behindertenhilfe (FTB), Evangelische Stiftung Volmarstein, Germany; Ortopedia, Kiel, Germany; Scientia Machinale s.r.l., Pisa, Italy.

REFERENCES

1. Cooper, R. A. "Intelligent Control of Power Wheelchair." *IEEE Eng. Med. Biol.*, 14 (4) (1995), 423–431.
2. Hoyer, H. "The OMNI Wheelchair." *Service Robot: An International Journal*, 1 (1) (1995), 26–29.
3. Hoyer, H., Hoelper R. "Intelligent Omnidirectional Wheelchair with a Flexible Configurable Functionality." *Proc. 17th RESNA '94 Annual Conf.*, Nashville TN, U.S.A., June 1994, 353–355.
4. Sabatini, A. M., Genovese, V., Guglielmelli, E., Mantuano, A., Ratti, G., Dario, P. "A Low-Cost, Composite Sensor Array Combining Ultrasonic and Infrared Sensors." *Proc. IEEE/RSJ Internat. Conf. Intell. Robots Syst.*, Pittsburgh PA, U.S.A., August 1995, 120–126.
5. Borenstein, J. "Histogramic In–Motion Mapping for Mobile Robot Obstacle Avoidance." *IEEE Trans. Robotics Automat.*, 7 (4) (1991), 535–539.
6. Bell, D. A., Levine, S. P., Koren, Y., Jaros, L. A., Borenstein J. "Design Criteria for Obstacle Avoidance in a Shared–Control System." *Proc. 17th RESNA '94 Annual Conf.*, Nashville TN, U.S.A., June 1994, 581–583.
7. Elfes, A. "Sonar–Based Real–World Mapping and Navigation." *J. Robotics Automat.*, 3 (3) (1987), 249–265.

ON THE DESIGN AND DEVELOPMENT OF MISSION CONTROL SYSTEMS FOR AUTONOMOUS UNDERWATER VEHICLES: AN APPLICATION TO THE MARIUS AUV *

P. Oliveira, A. Pascoal, V. Silva, C. Silvestre

Institute for Systems and Robotics, Instituto Superior Técnico
Av. Rovisco Pais, 1096 Lisboa Codex, Portugal
E-mail address: antonio@isr.isr.ist.utl.pt

ABSTRACT

This paper describes the design and development of a Mission Control System for the MARIUS AUV, and presents the results of sea tests for system design validation carried out in Sines, Portugal.

KEYWORDS: Autonomous Vehicles, Discrete Event Systems, Mission Control Systems, Underwater Robotics.

1 INTRODUCTION

Among the challenges that face the designers of underwater vehicle systems, the following is of the utmost importance: design a computer based *Mission Control System* that will *i)* enable an operator to define a vehicle mission in a high level language, and translate it into a mission plan, *ii)*provide adequate tools to convert a mission plan into a Mission Program that can be formally verified and executed in real-time, and *iii)* endow an operator with the capability to follow the state of progress of the Mission Program as it is executed, and modify it if required.

Meeting those objectives poses a formidable task to underwater system designers, who strive to develop vehicles that can be programmed and operated by end-users that are not necessarily familiarized with the most intricate details of underwater system technology. Identical problems face the designers of complex robotic systems in a number of areas that include advanced manipulators, industrial work cells, and autonomous air and land vehicles. The widespread interest of the scientific community in the design of Mission Control Systems for advanced robots is by now patent in a sizeable body of literature that covers a wide spectrum of research topics focusing on the interplay between event driven and time-driven dynamical systems. The former are within the realm of Discrete Event System Theory [4], whereas the latter can be tackled using well established theoretical tools from the field of Continuous and Discrete-Time Dynamical Systems [5].

Early references in this vast area include the pioneering work of K.S. Fu [6], Saridis [9] and Albus [1], which set the ground for the study of learning control systems, intelligent machine organization, and general architectures for autonomous undersea vehicles, respectively. For an overview of recent theoretical and applied work in the

*This work was supported by the Commission of the European Communities under contract MAS2-CT92-0021 of the MAST-II programme, and by the Portuguese PRAXIS programme under contract 3/3.1/TPR/23/94.

field, the reader is referred to [2, 10], which contain a number of papers on the design of advanced control systems for unmanned underwater vehicles, combined underwater vehicle and manipulator systems, intervention robots, and air vehicles.

As part of the international effort to develop advanced systems for underwater vehicle mission control, IST has designed a first version of a Mission Control System for the MARIUS AUV [3]. This paper provides a brief summary of the framework for design, analysis and implementation of the Mission Control System proposed, and reports the results of a series of sea tests for system validation conducted in Sines, Portugal. The work reported here has been influenced by the solid body of research carried out by INRIA/IFREMER in France, with applications to the VORTEX vehicle, and at NPS in the U.S. with applications to the PHOENIX vehicle, see [7] and the references therein.

The organization of the paper is as follows: Section 2 describes the basic framework adopted for Mission Control System design and implementation using the software programming environments CORAL and ATOL. Section 3 illustrates the basic steps involved in the design of a Mission Program for a simple mission example, describes the set-up for mission execution and mission follow-up from a shore station, and reports the results of running the mission described at sea.

2 MISSION CONTROL SYSTEM DESIGN

This section describes a framework for the design of Mission Control Systems for AUVs. Due to space limitations, only the key concepts will be presented here. For complete details the reader is referred to [8]. The framework proposed builds on the concepts of *System Task*, *Vehicle Primitive*, *Mission Procedure*, and *Mission Program*, that will be explained in the sequel.

System Task The concept of System Task arises naturally out of the need to organize into distinct, easily identifiable classes, the algorithms and procedures that are the fundamental building blocks of a complex Underwater Robotic System. For example, in the case of an AUV, it is convenient to group the set of all navigation algorithms to process motion sensor data into a Navigation Task that will be responsible for determining the attitude and position of the vehicle in space. A different task will be responsible for implementing the procedures for multi-rate motion sensor data acquisition. In practice, the number and type of classes adopted is dictated by the characteristics of the Robotic System under development, and by the organization of its basic functionalities, as judged appropriate by the Robotic System designer. These considerations lead naturally to the following definition:

A Vehicle System Task (abbv. *System Task - ST*) *is a parametrized specification of a class of algorithms or procedures that implement a basic functionality in an Underwater Robotic System.*

The implementation of a System Task requires the interplay of two modules: *i)* a *Functional module* that contains selected algorithms and procedures and exchanges data with other System Taks and physical devices, and *ii)* a logical *Command module*, embodied in a finite state automaton, that receives external commands, produces output messages, and controls the selection of algorithms, procedures, and data paths to and from the Functional module.

The design of the Functional module is carried out using well known tools from such diverse fields as navigation, guidance and control, instrumentation and measurements, communication theory, and computer science. The design of the Command module amounts to specifying a finite state automaton [4] that deals with the logical aspects of the System Task.

Vehicle Primitive The concept of Vehicle Primitive plays a central role in the general framework for Mission Control System design described in this paper. A Vehicle Primitive corresponds to an atomic, clearly identifiable action performed by an Underwater Robotic System, and constitutes the basic building block for the organization of complex robot missions. The following definition is offered:

A Vehicle Primitive (VP) is a parameterized specification of an elementary operation mode of an Underwater Robotic System. A Vehicle Primitive corresponds to the logical activation and synchronization of a number of System Tasks that lead to a structurally and logically invariant behavior of an underwater robot.

Associated with each Vehicle Primitive, there are sets of *pre-conditions* and *resource allocation* requirements that must be met in order for the Primitive to be activated, as well as a set of Vehicle Primitive *errors*. During operation, a Vehicle Primitive will generate messages that will trigger the execution of a number of System Tasks. The conditions that determine the occurrence of those events are dictated by the logical structure of the Vehicle Primitive itself, and by the types of message received from the underlying Vehicle System Tasks. The normal or abnormal termination of a Vehicle Primitive will generate a well defined set of *post-conditions* that are input to other Vehicle Primitives, and will release the resources that were appropriated during its execution.

By exploring the use Petri nets for the modeling of discrete event systems [4] it is possible to show that a Vehicle Primitive can be embodied in a Petri net structure defined by the five-tuple $(P_{VP}, T_{VP}, A_{VP}, w_{VP}, \mathbf{x}_{VP_0})$, where P_{VP}, T_{VP}, and A_{VP} denote sets of places, transitions, and arcs respectively, w_{VP} is a weight function, and $\mathbf{x}_{VP_0}$ is the initial Petri net marking. The set of places P_{VP} can further be decomposed as $P_{VP} = P_{pre} \cup P_{res} \cup P_{err} \cup P_{loc} \cup P_{pos}$, where $P_{pre}, P_{res}, P_{err}, P_{pos}$, and P_{loc} denote the subset of places that hold information related to the pre-conditions, resource allocation, errors, post-conditions, and the remaining state of the Petri net, respectively.

Based on the framework introduced, a Vehicle Primitive programming environment named CORAL has been developed. The left side of Figure 1 depicts the organization of the CORAL software tools that are available to *edit* and *generate* a *Library of Vehicle Primitives* which implement the complete set of atomic actions required for a specific Underwater Robotic System. Each Vehicle Primitive, embodied in its equivalent Petri net, can be input either graphically via a CORAL graphic input interface, or via a textual description using the declarative, LR1 synchronous language CORAL. A CORAL compiler/linker is in charge of accepting the vehicle primitive textual descriptions, and producing a Vehicle Primitive Library that is an archive containing the syntax and semantic descriptions of all Vehicle Primitives, as well as the data sets required for their execution.

In order to run the Vehicle Primitives described before, a CORAL Engine has been developed that accepts Vehicle Primitive descriptions and executes them in real-time.

Mission Procedure/Mission Program Given a mission to be performed by an Underwater Robotic System, the generation of the corresponding mission plan requires the availability of a set of entities aimed at specifying robot *Actions* at a number of abstraction levels. Those entities - henceforth referred to as *Mission Procedures* - allow for modular Mission Program generation, and simplify the task of defining new mission plans by modifying/expanding existing ones. The above introduction motivates the following definition:

A Mission Procedure is a parameterized specification of an Action of an Underwater Robotic System. A Mission Procedure corresponds to the logical and temporal chaining of Vehicle Primitives - and possibly other Mission Procedures - that concur the execution of the specified Action.

According to the definition, the execution of a robot mission entails the execution of a number of well defined Actions specified by Mission Procedures, which in turn synchronize the operation of Vehicle Primitives. In practice, the activation of Mission Procedures and Vehicle Primitives will be triggered by conditions imposed by the mission plan structure, and by messages received from the underlying Vehicle Primitives during the course of the mission.

In principle, simple Mission Programs could be embodied into - higher level - Petri nets that would implement the necessary Mission Procedure structures. However the analysis of even a simple mission plan programmed using that methodology will convince the reader that the complexity of the resulting Petri net structure can become unwieldy. See Section 3 for a detailed example. Furthermore, the approach described does not lend itself to capturing situations where the mission plan includes logical, as well as procedural statements (e.g., do loops for the repeated execution of Mission Procedures and Vehicle Primitives, etc.). These considerations motivated the need to define a specific environment for Mission Program/Mission Procedure design and implementation, named ATOL, which is currently being developed.

The framework for Mission Control System design and implementation proposed in this paper leads to the general structure of Figure 1 (right side), which captures the interaction among System Tasks, Vehicle Primitives, and Mission Program/Mission Procedures, at both programming and run-time. In the figure, the Human/Machine Interface provides the user with a text editor, and an on-line checking mechanism for the syntax and semantics of ATOL statements.

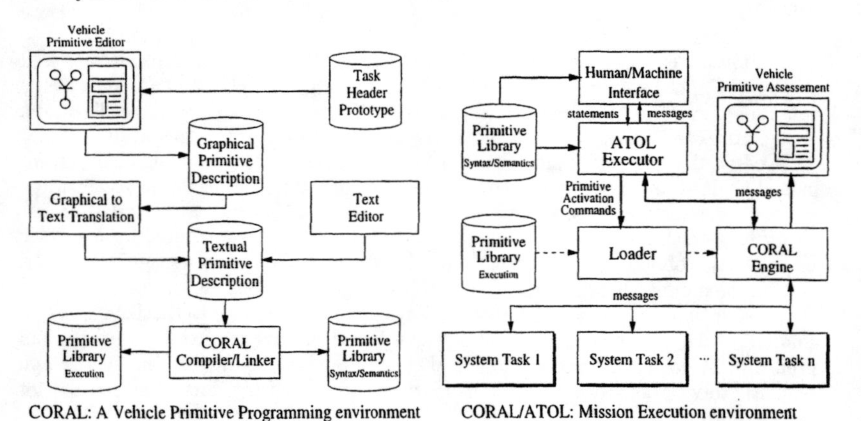

Figure 1: Mission Control System Organization.

From an execution point of view, the ATOL Executor - running an ATOL Mission Program - issues commands to the CORAL Loader, which transfers selected Vehicle Primitive descriptions from the Vehicle Primitive Library to the CORAL Engine. The Engine runs the Primitives selected by interacting with the System Tasks, and issues messages that condition the execution of the ATOL Mission Program. During mission execution, the status of any Vehicle Primitive can be displayed on a *Vehicle Primitive Assessment module* that allows visualizing the flow of markings on the corresponding Petri nets.

3 MISSION PROGRAM DESIGN USING CORAL. TESTS AT SEA

This section outlines the programming of a simple mission using CORAL, and presents the results of its execution using the MARIUS AUV [3]. The prototype vehicle is 4.5 m long, 1.1 m wide and 0.6 m high. It is equipped with two main back thrusters for cruising, four tunnel thrusters for station keeping maneuvers, and rudders, elevator and ailerons for vehicle steering in the vertical and horizontal planes. The vehicle has a dry weight of 1060 kg, a payload capacity of 50 kg, and a maximum operating depth of 600 m. Its maximum rated speed with respect to the water is 2.5 m/s. At the speed of 1.26 m/s, its expected mission duration and mission range are 18 h and 83 km, respectively.

The mission example requires that the AUV trace a square shaped trajectory, at constant depth and speed of 1.35 m and 2.0 m/s, respectively. The square maneuver is obtained by requesting the vehicle to change its heading by −90 deg every 40 seconds. The initial heading is 0 deg.

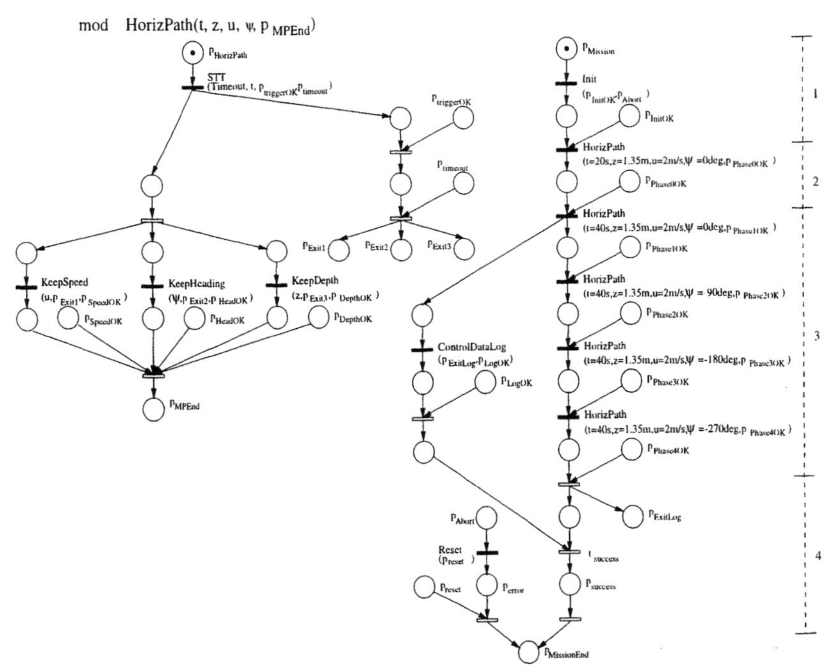

Figure 2: Mission Procedure and Mission Program in CORAL.

The design of the corresponding Mission Program involves a Mission Procedure named *HorizPath*, whose implementation using the CORAL programming environment is shown in Figure 2. This Mission Procedure parametrizes the action of keeping constant heading ψ, depth z, and speed u of the vehicle, for a period of time t.

The *HorizPath* Mission Procedure starts by setting a timer to generate a timeout after the required execution time has elapsed. This is done by issuing an timeout command with the required Mission Procedure duration time t. To perform the maneuver, three Vehicle Primitives are called in parallel: *KeepSpeed* with a velocity set-point u, *KeepDepth* with a depth set-point of z, and *KeepHeading* with a heading set-point of ψ. The generation of a timeout terminates the execution of *HorizPath* by exiting the three Vehicle Primitives.

The Mission Program can be explained with the help of Figure 2, which shows four distinct phases: in phase 1, all vehicle System Tasks are initialized by calling the *Init* Vehicle Primitive; in phase 2, the *HorizPath* Mission Procedure is called for a period

$t = 20$ s, with a velocity set-point of $u = 2$ m/s, a depth set-point of $z = 1.35$ m, and an heading set-point of $\psi = 0$ deg. At the end of this phase, the vehicle is headed north, and ready to start the required square maneuver; phase 3 calls the *HorizPath* Mission Procedure repeatedly, with heading set-points of 0 deg, −90 deg, −180 deg, and −270 deg, while maintaining the remaining input set-points equal to those in phase 2. The required duration of each Mission Procedure call is $t = 40$s. In parallel, the Vehicle Primitive *ControlDataLog* is called to start logging control loop data for later off-line analysis.

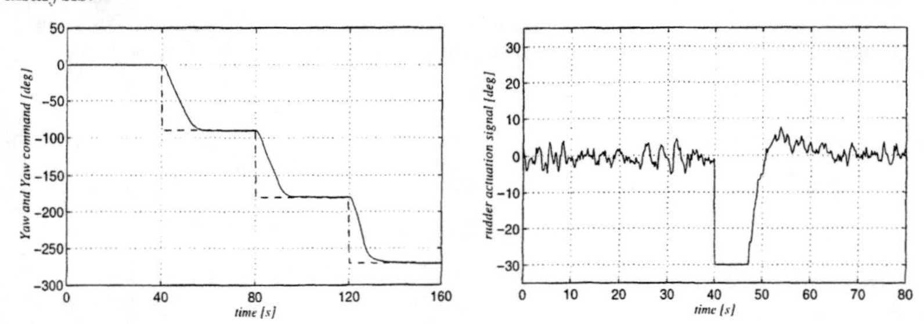

Figure 3: Left - Commanded and measured heading. Right - rudder deflection.

In order to assess the performance of the Mission Control System of MARIUS, a series of tests were conducted at sea in Sines, Portugal, in January 1996. The tests included programming and running the mission described above. Throughout the mission, the vehicle pulled a buoy with an antenna, thus enabling radio communications between the vehicle and a shore station. The software for Mission Control was run on the computer network installed on-board the AUV. Figure 3 shows the commanded and measured heading, and the rudder deflection, as examples of variables observed during the mission.

REFERENCES

[1] J. Albus, "System Description and Design Architecture for Multiple Autonomous Undersea Vehicles," National Institute of Standards and Technology, Technical Note 1251, September 1988.

[2] P. Antsaklis, K. Passino, *An Introduction to Intelligent and Autonomous Control*, Kluwer Academic Publishers, 1993.

[3] G. Ayela, A. Bjerrum, S. Bruun, A. Pascoal, F-L. Pereira, C. Petzelt, J-P. Pignon, " Development of a Self-Organizing Underwater Vehicle - SOUV, *Proceedings of the MAST-Days and Euromar Conference*, Sorrento, Italy, November 1995.

[4] C. Cassandras, *Discrete Event Systems. Modeling and Performance Analysis*,Aksen Associates Incorporated Publishers, 1993.

[5] G. Franklin, J. Powell, M. Workman, *Digital Control of Dynamic Systems*, Addison-Wesley, 1990.

[6] K.S. Fu,"Learning Control Systems-Review and Outlook," *IEEE Transactions on Automatic Control*, Vol.AC-15, No.2,1970

[7] M. Lee, R. McGhee, Editors, Proceedings of the IARP 2nd Workshop on Mobile Robots for Subsea Environments, Monterey, California, May 1994.

[8] P. Oliveira, A. Pascoal, V. Silva, C. Silvestre, "Design, Development, and Testing of a Mission Control System for the MARIUS AUV", *Proceedings of the 3rd Workshop on Mobile Robots for Subsea Environments*, Toulon, France, March 1996.

[9] G. Saridis, "Analytical Formulation of the Principle of Increasing Precision with Decreasing Intelligence for Intelligent Machines," *Automatica*, Vol. 25, pp. 461–467.

[10] K. Valavanis, G. Saridis, A. Pascoal, P. Lima, F-L. Pereira, editors. *Proc. of the Joint U.S./Portugal Workshop on Undersea Robotics and Intelligent Control*, Lisbon, Portugal, March 1995.

Experimental Evaluation of Remote Control Configurations: An Ergodynamics Approach

A. Pakbin, T.C. Chong, N. Sepehri and **V. Venda**

Department of Mechanical and Industrial Engineering
The University of Manitoba, Winnipeg, Canada R3T-2N2

ABSTRACT

This paper surveys literature and presents new experimental data on the influence of work factors on work efficiency. Particularly, it explores the first and second laws of ergodynamics stating the relation between work efficiency and the factor of mutual adaptation between the human work functional structure and work environment. Experimental subjects were operating an X-Y table in the labyrinth using two one-degree-of-freedom joysticks or a single two-degree-of-freedom joystick. The number of turns and the width of the labyrinths were changed as the work factors of efficiency-complexity. Bell shaped curves were obtained for both work functional structures allowing to choose single- or multi-joystick operations in different ranges of the work factors.

KEYWORDS: automation, manipulators, human-machine-environment mutual adaptation, task's complexity, ergodynamics, control.

INTRODUCTION

There exist many semi-automated engineering applications in which the operator controls a remote manipulator to perform various tasks. When using remote controls, however, the operator must be familiar with the manipulator configurations and capabilities. Sometimes, remote control configurations are not user friendly with respect to the ergonomics and logistics. This, however, depends on the application and the degree of involvement of the operator in the control loop that varies from direct control of active components to task planning [1]. In the past few years, primary industries have embarked a study examining the computerized motion control of heavy-duty machines such as forest harvesters and excavators. This was motivated by a possible reduction in learning time, reduction in adaptation time to different machines, reduction in fatigue, enhancement of safety, improvement in productivity and uniformity of production between operators with different motor coordination skills. The goal is to ultimately retrofit these machines into fully automated systems. The first step is to improve the operations by having a single joystick endpoint velocity control instead of the existing multi-joystick joint velocity control mode [2].

In this paper we discuss how much this step-by-step development benefits the human work efficiency and observe to what extend the transfer can be efficiently utilized. Specifically the objective of this paper is to explore the benefits of endpoint velocity control to the operator. We present, for the first time, a new approach and experimental data on influence of work factors on work efficiency for this problem. We implement and explore the laws of ergodynamics that relate the work efficiency to the factor of mutual adaptation between the human work functional structure and work environment [3]. The goal is to obtain meaningful characteristic curves, for both work functional structures, in order to choose between single- or multi-joystick operations in different ranges of the work factors.

The experimental setup was an X-Y table (see Figure 1) retrofitted as a system having its two axis controlled by joysticks. It can operate with either a pair of spring-centered one-degree-of-freedom (single-axis) joysticks or one two-degree-of-freedom (2D) joystick. The joystick translations actually control the motor speeds. The X-Y test-station allows the operator to simulate tasks such as pushing a log through a gate, pick and place or tracking. The operator can choose between two single-axis joysticks or a 2D-joystick. Experimental subjects operate the X-Y table in the labyrinths containing variation of turns and curves. The number of turns and acceptable widths of the labyrinths are changed as the work factors of efficiency-complexity. The preliminary results are encouraging in that they show that bell-shaped curves could be obtained for such work functional structures allowing to choose between single- or multi-joystick operations in different ranges of the work factors. Unlike morphological, kinematics, cognitive and other human performance structures, this approach displays a connection between the environment input and functional output of the human at work, and is expected to give an insight into the improvement of the ergonomics of existing hand controllers by using a computer to map the control from joint-space to coordinated-space.

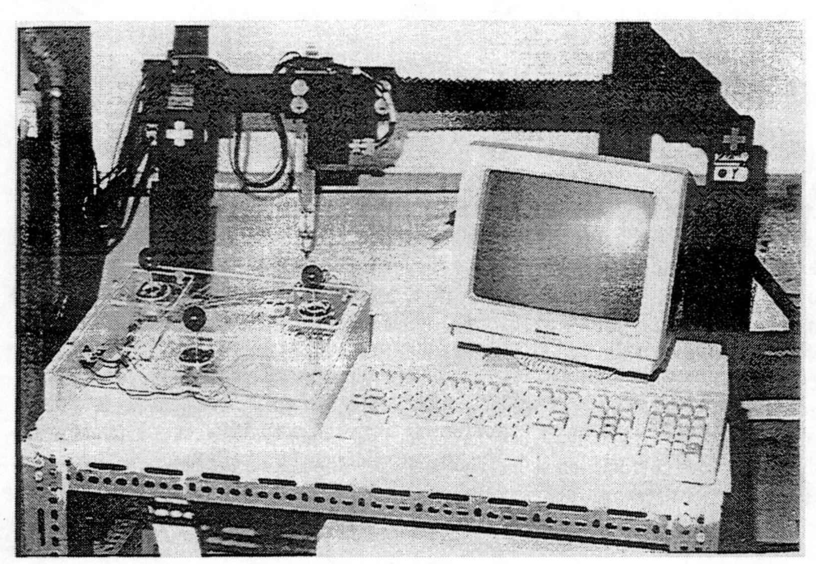

Figure 1. View of the X-Y table.

BACKGROUND

Experimental data on a bell-shaped influence of the factors of human-machine-environment mutual adaptation onto human work efficiency, have been presented since the beginning of this century. The first practical ergonomic study of optimal work conditions was done by Taylor in 1908 [4] who discovered an important ergonomic phenomenon: work productivity is a bell-shaped function of the work environment. Frank and Gilbreth [5] later studied the bell-shaped dependence of productivity on the ergonomic work factor, and displayed that human-environment interaction and mutual adaptation during training and work processes do not only depend on human skills, but also on the work organization.

In general any work output to be improved, is a criterion of a functional efficiency (Q). Efficiency is a positive measure of work and living processes. In an efficient system, the obstacles, difficulties, errors, and deviations from the optimal work processes and algorithms are minimal. We call these negative aspects, criteria of functional complexity (C). Functional efficiency and complexity criteria work against one another. The most typical criteria of complexity are extra time spent for the given work, the number of errors made, the portion of defective products, and the probability of failure.

Any model describing the influence of the environment, work organization or processes (F) onto work efficiency, has an optimal value; maximum for efficiency and minimum for complexity. The characteristic models of work functional structures, Q(F) or C(F), display a very general and important ergonomic law of human-environment mutual adaptation. Finding the essential, integral factor of interaction and mutual human-environment adaptation that influences the efficiency of human work is very important for successful ergonomic analysis. To reach maximum work efficiency, ergonomists should organize mutual adaptation between workers and work environment so that, from one side, the workers are trained in proper work skills, strength and motions and, from the other side, the work environment is adapted to the workers [6]. The three fundamental laws of ergodynamics are now described [7].

Ergodynamics Law I : Law of Mutual Adaptation- work efficiency is a bell shaped function of the factor of mutual adaptation between human work and environment. Thus, for every task there is an optimal efficiency value for an associated optimal factor of mutual adaptation. By simulating a working environment the efficiency at various factors can be measured and assuming that the productivity of these factors is independent from one another, or that a direct relationship between the factors can be found, one can establish the optimal working condition.

Ergodynamics Law II: Law of Work Structures Plurality- every work task can be done with different work structures. According to law I, the efficiency of each work structure is a specific bell-shaped function of the factor of mutual adaptation between the work structure and its environment. Thus, one may represent an individual's work structure as a family of respective characteristic curves.

Ergodynamics Law III: Law of Transformation- transformations between different work structures and interactions between different structures are maximally effective if they go through a state common, and equal for the structure. From laws I and II, every structure is modeled with a specific bell-shaped efficiency function. Thus the common states for the structures are modeled as intersect points of respective characteristic curves of the structures.

TESTS PROCEDURES

Male and female subjects, average 22 years of age, participated in this study. During the training period of approximately 15 minutes, they were given several labyrinths for the trial with both single-axis and 2D joysticks. The task was to drive a pen through the labyrinths as fast and as accurate as possible, i.e., having the lowest number of exiting the labyrinth boundary lines. The labyrinths, printed on 14×17*inches* papers, consisted of horizontal and vertical double lines with 90 degree turns. The complexity of the task was increased by adding turns to the labyrinths (see Figure 2). Two sets of labyrinths were used. The first set consisted of 23 labyrinths having turns varied from 3 to 63 with increments of 3. The second set consisted of 24 labyrinths having turns varied from 5 to 120 with increments of 5. The efficiency (also called here performance measure) was determined as the number of right turns crossed in a unit of time (turns per second).

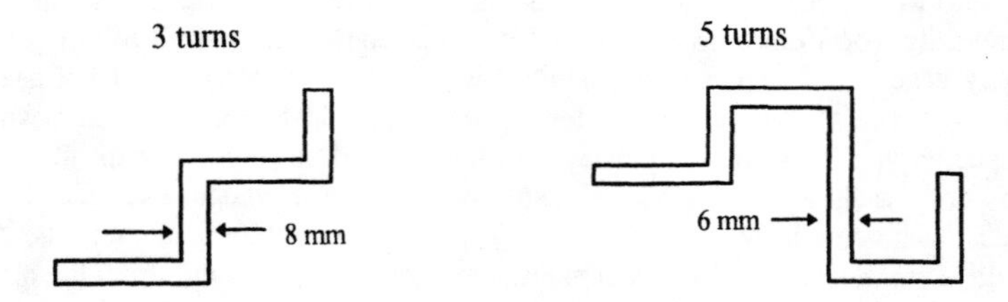

Figure 2. Three- and five- turns task.

RESULTS AND DISCUSSION

Figure 3 shows experimental data on the performance measure (Q) versus the overall number of turns in the labyrinth as a work factor (F). Maximum number of turns in the labyrinth in this experiment was 63, and the operator used the 2D joystick. The number of turns changed from 3 to 63, which was a factor of work efficiency-complexity. The result of this experiment clearly demonstrates the first law of the ergodynamics.

Figures 4 and 5 demonstrate the second law of the ergodynamics, i.e., every work functional structure may be presented as a certain bell-shaped curve Q(F). In the experiments, the operators were controlling the endpoint using either a 2D joystick, or two single-axis joysticks. The performance efficiency was measured for both sets of labyrinths -- 63-turns and 120-turns. Figure 4 shows the performance measure of operators for the 63-turns test. The bell-shaped curves for the two work structures (single- or multi-axis joysticks) are very close to each other implying that there is no statistically significant difference between the method of control. Work with the 120-turns

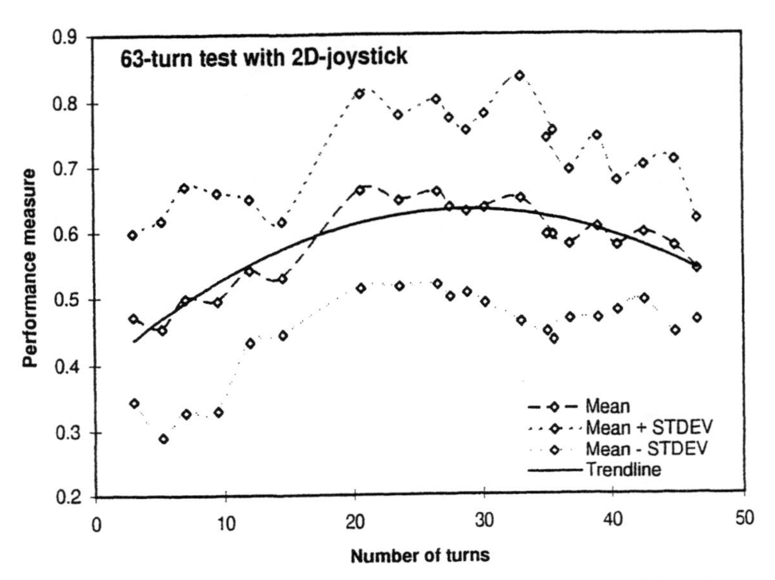

Figure 4. Typical performance measure versus number of turns.

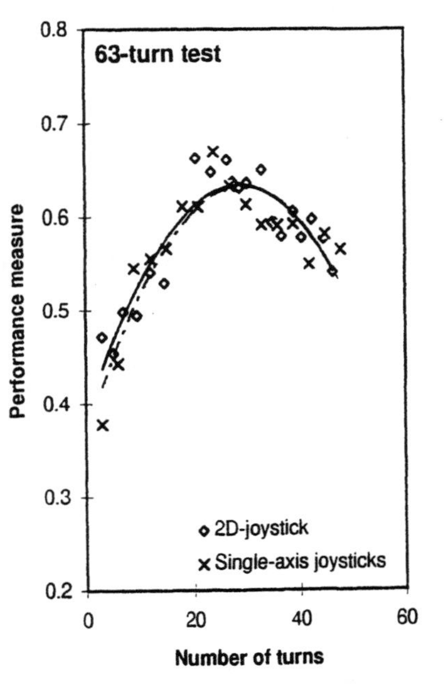

Figure 5. Performance; 63-turn test.

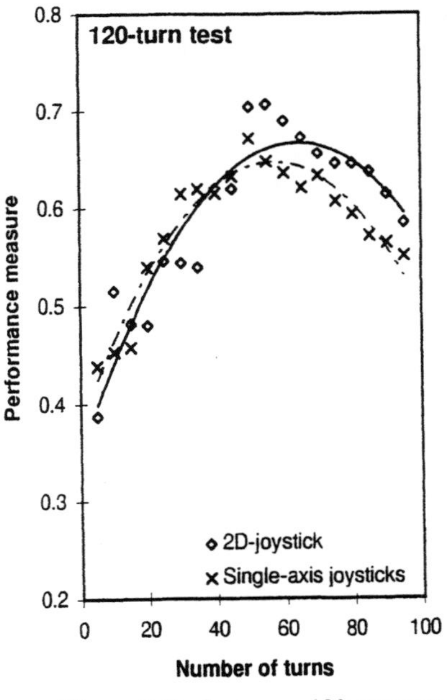

Figure 6. Performance; 120-turn test.

set, on the other hand, shows some difference in use of two types of the joysticks (see Figure 5). For work factor not exceeding 55 turns, the performance measures were almost the same with slight superiority of multi-joystick operation. For tasks with number of turns more than 55, work with a 2D joystick gave higher efficiency. This experiment adds an important information to the understanding of the second law of ergodynamics -- different work functional structures may have similar bell-shaped characteristic curves, Q(F), or different ones. If the curves are similar either work structure can be used. Different curves allow to distinguish the work functional structures by their efficiency. The one with higher efficiency is therefore recommended for the practical use in human-manipulator interaction.

The above experiments, although preliminary, have shown the significance of ergodynamics approach for modeling and prediction of the human behavior in a teleoperation environment. The initial study shows that different work conditions require different work functional structures in order to maintain high productivity. This means that instead of attempting to simply decide what type of manipulator operation is more effective, single-axis or 2D joysticks, one should first study the intervals of work conditions and tasks, then decide for a given work functional structure, on the type of joystick that leads to a maximal productivity and quality.

REFERENCES

1. Sheridan T.B. "Telerobotics, Automation and Human Supervisory Control." MIT Press, Cambridge, MA (1992).
2. Sepehri N., Sassani F., Wong D. and Lawrence P.D., "The Automation of Industrial Machines." *Proceedings International Conference on Engineering Application of Mechanics,* Tehran, Iran (1992) 380-387.
3. Venda V.F. "Individual Adaptation of Work Tools to the Operator." *Ergonomics*. 19(6) (1976).
4. Taylor F.W. "The Principles of Scientific Management." Harper and Row, N.Y. (1971).
5. Freivalds A. "The Ergonomics of Tools." *International Review in Ergonomics.* 1 (1987) 12-48.
6. Venda V.F. "Engineering Psychology and Design of Information Display Systems." Mashinostroenie, Moscow (in Russian) (1982).
7. Venda, V.F. and Venda, Y.V. "Dynamics in Ergonomics, Psychology, and Decisions: Introduction to Ergodynamics." Ablex: Norwood, N.J. (1995).

BUILDING A NETWORK MODEL FOR A MOBILE ROBOT USING SONAR SENSORS

Sollip Park
Seoul National University, Automatic Control Research Center

Hakyoung Chung
Seoul National Polytechnic University Dept. of Control & Instrumentation Eng.

Jang Gyu Lee
Seoul National University, Automatic Control Research Center

Hae Yong Yang
Seoul National University, Automatic Control Research Center

ABSTRACT

In an unknown environment, a mobile robot needs to use environmental information to perform its task. A grid model can be built with proximity sensors which are very sensitive to the environment but the path planning methods based on this model cannot give solutions to all pairs of locations. In a network model, path planning methods for a network give a path to every pair of nodes but this model cannot be built with proximity sensors. In this paper, a method to build a network model from a grid model is presented. Therefore, we can make a network model using proximity sensors. In converting a grid model to a network model, we use a quadtree model. A method to reduce the number of nodes is also proposed to reduce the computational time spent in searching for a path between every pair of nodes.

KEYWORDS: mobile robot, grid model, network model, proximity sensor, path planning, quadtree, node

INTRODUCTION

An autonomous mobile robot should be able to navigate by itself. If a robot has no information about its environment, it may be trapped due to peculiar shapes of obstacles or it may not find any optimal path. Therefore, environmental information is necessary for a mobile robot to perform its tasks. There are several methods using environmental models such as the configuration space method[1], the distance function method[2], the grid search method[3], and the network method[4]. These methods, except the grid search method, require accurate environmental models which cannot be given by proximity sensors commonly used with a mobile robot. The grid search method should rebuild a whole environmental model whenever it needs to give a path between other two nodes. Therefore, a method combining the grid search method and the network method was proposed[5]. This method plans a path using the network model which is computed by the grid search method. Based on this method, a grid model can cope with uncertainties and errors in sensor information and shortest path planning methods can give the shortest path between each pair of nodes in a network model. However, in a large workspace, the grid model needs too much memory to keep grid data.

In this paper, we propose a method to convert grid-type environmental information to a network model. To solve the memory problem in the grid model we use a quadtree in converting the grid information into the network model. The method to position nodes covering all the feasible paths in an environment is also proposed.

In the second section, the concept of a quadtree and the comparison with a grid model are presented and the proposed method of building a network model from a quadtree model is presented in the third section. Some experimental results of building a network model from sonar data are presented in the forth section.

QUADTREE MODEL

A grid model represents a workspace as a two-dimensional array of grids with the same size as shown in Fig. 1 and a grid has a certainty value which represents the possibility of the existence of obstacles. To build a precise grid model, accurate certainty values should be assigned. Therefore the method of utilizing neighboring sonar sensor readings at a time was introduced[6]. A disadvantage of a grid model is the inefficiency in memory. In a large workspace, this model needs too much memory to keep the grid data because this model divides a workspace into cells of the same size regardless of the clutter of an environment.

Using the quadtree data structure as an environment model is proposed by Samet[7]. The grid-like structure of the quadtree model allows sensor data to be readily included and the quadtree model overcomes the deficiencies of a grid model because the quadtree model is adaptive to the clutter of the workspace by subdividing arrays in a grid model into quadrants repeatedly until all the divided quadrants are either obstacles or vacant spaces. So the number of squares needed to represent the same workspace is decreased in the model as shown in Fig. 2. In Fig. 2, the quadtree model needs 28 squares to represent

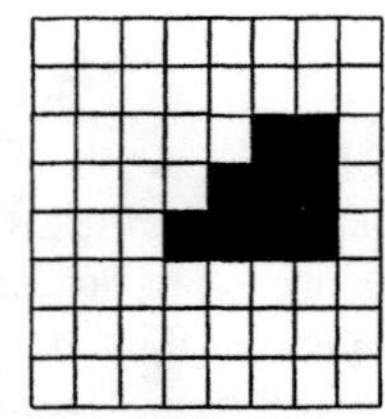

Figure 1. Grid model

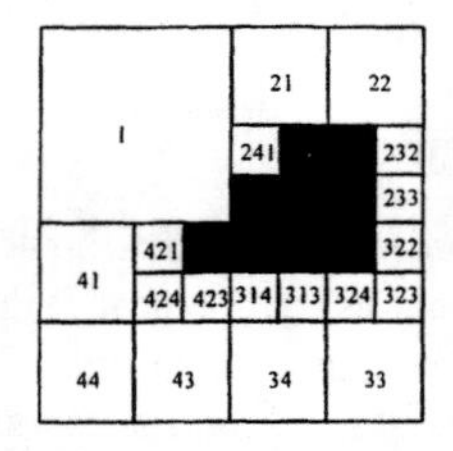

Figure 2. Quadtree model

an obstacle while the grid model needs 64 squares to in Fig 1.

NETWORK BUILDING ALGORITHM

Environments are made up of edges and nodes in a network model. A node is a point in a workspace on which a mobile robot can be located and an edge is a segment connecting two nodes without intersecting any obstacle. The cost of the edge is the distance or the running time between the two nodes which are connected by the edge.

Node establishment

In converting the quadtree model to a network model so as to use shortest path algorithms for networks, nodes should be established. First, the number and the positions of nodes are very important for path planning since too many nodes may decrease the speed of a path searching algorithm and ill-positioned nodes may not be able to represent all the feasible paths in a network model as shown in Fig 3. So it is necessary to represent all the feasible paths using small number of nodes to reduce the execution time of a path searching algorithm.

Our method sets a node at the center of every quadtree square and a new theorem is presented to prove that the nodes positioned at the center of every quadtree squares represent all the feasible paths in the quadtree model.

Theorem 1: For two adjacent squares in the quadtree model, the line between the two nodes set on the center of each square does not exit the two squares regardless of the sizes and the adjoining position of the two squares.

Proof) Theorem 1: In the Fig. 4, let A and B be the two adjacent squares in the quadtree model with nodes O and P at the center of the squares respectively. Let U and D be the end points of the line on the tangential line and let 2d be the length of the latus of square A. Because the squares A and B are quadtree squares, they are all regular squares. So, if we lay O at the origin of XY coordinate, the points and the line are defined as follows : P as (a,b), U as $(d,b+(a-d))$, D as $(d,b-(a-d))$, and OP as $y=\frac{b}{a}(x-a)+b$, where a, b and d are nonnegative and $a>d>b$ from the Fig. 4.

In order that a line connecting two nodes at the center of two adjacent squares does not exit the two squares regardless of the sizes and the adjoining position of the two squares, the point U should be located above the line OP and the point D under the line OP. So the equation(1) should be satisfied under the condition below.

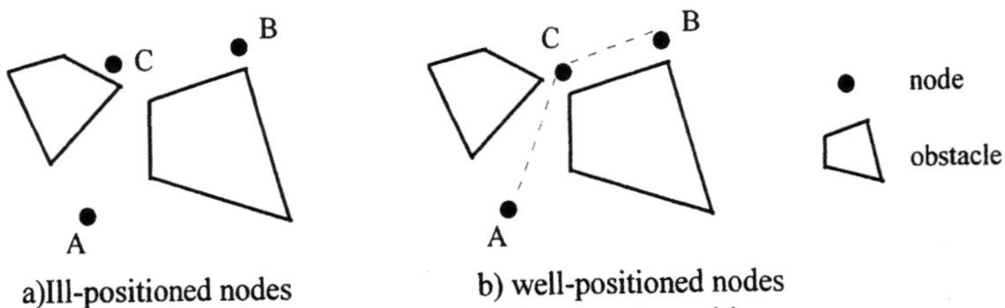

Figure 3. path according to node positions

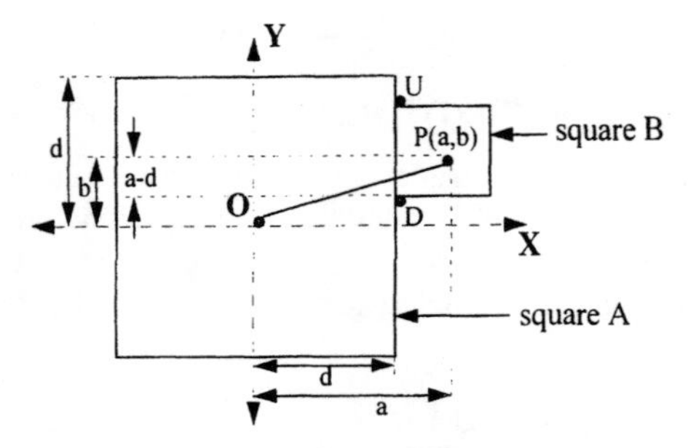

Figure 4. Two adjacent squares

$$b+(a-d) > \frac{b}{a}(d-a)+b > b-(a-d), \qquad a>d>b>0 \qquad\qquad \text{---}(1)$$

Solving the equation(1) yields two equations $(a-d)(a+b)>0$ and $(a-d)(a-b)>0$. These derived equations are true under the condition in the eq. (1). Therefore the line between any pair of two nodes set on the center of two adjoining squares does not exit the two squares regardless of the sizes and the adjoining position of the two squares.■

In the quadtree model, the path between every pair of squares can be represented as a connection of vacant squares. As the line connecting the centers of the squares does not exit the vacant squares by the Theorem 1, the path between every pair of nodes can be expressed by the chain of similarly directed lines connecting two adjacent nodes which are located at the center of the squares. Therefore, the nodes set at the centers of the squares don't omit any feasible path of the quadtree model.

Edge definition

After positioning nodes, we define edges between every pair of the nodes. The existence of an edge is easily determined by examining whether the line intersects with any obstacle quadtree square or not. The cost of the edge is the distance between the two nodes if the quadtree squares through which the edge passes are all vacant. Otherwise there is no edge between the two nodes.

Node diminution

Though the quadtree model increases the efficiency of memory adapting the sizes of squares to the clutter of the workspace, the number of squares may be determined differently by the gap between outlines of obstacles and dividing lines of the quadtree model. For example, Fig. 5 shows that the position of an obstacle makes difference in the number of squares representing the same workspace. In Fig. 5b), the number of squares is larger than that of Fig. 5a).

In a network model, because the number of nodes has much effect on the execution time of the path searching algorithm, extra nodes which are not necessary for the paths in the environments should be deleted. We propose a method to delete unnecessary nodes and omit the details for the simplicity.

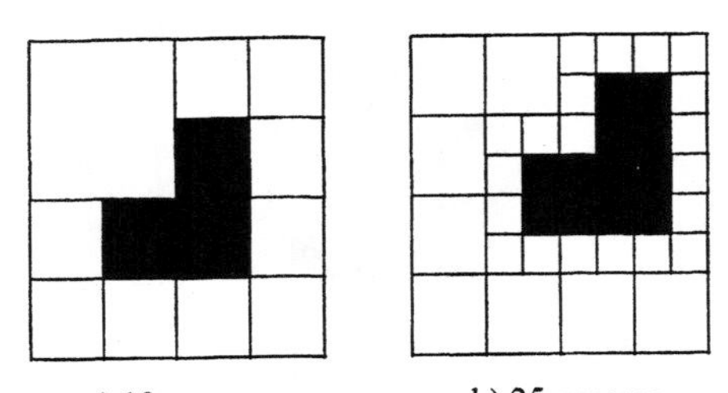

a) 10 squares b) 25 squares

Figure 5. The difference due to the position of an obstacle

When an outline of an obstacle does not coincide with a dividing line of the quadtree model, there is a horizontal or vertical succession of squares with a size through the gap between the two lines. As we set a node at the center of every square in the quadtree model, the succession of squares with a size has a line of nodes in itself. As a path composed of a succession of squares can be represented only by the two end nodes of the line, the nodes in a line except the end nodes are useless in the path planning between the two nodes. So we delete all the mid nodes in lines except two cases.

Exception 1: The first case is when two lines of nodes cross each other. In this case, if any latus of the quadrangle which has the two lines as its diagonal lines meets obstacles, a node should be located at the center of the crossing square of the two lines.

Exception 2: The second case is when a square is adjacent to a succession of squares. In this case, the path passing across the succession of squares from the adjacent square should be considered in path planning.

EXPERIMENTAL RESULTS

Our mobile robot explored the environment, 960cm × 320cm, making a grid model using proximity sensors and a network model was built from the grid model. In Fig. 6, a node is established at the center of every quadtree square. Using our methods, unnecessary mid nodes are eliminated and the network model is rebuilt as shown in Fig. 7. The number of nodes covering all the feasible paths in the environment is reduced to 43% of that of Fig. 6.

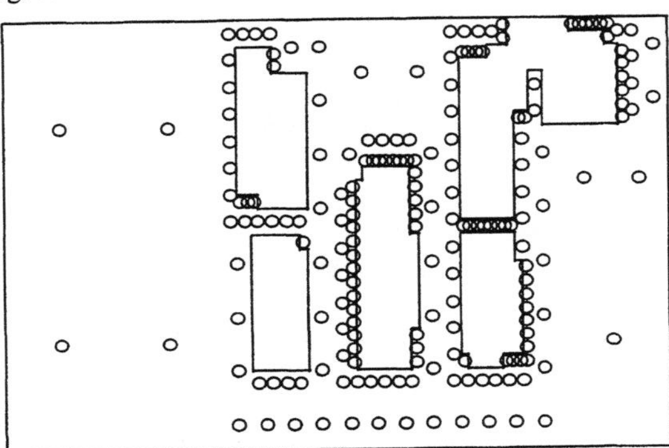

Figure 6. Locating nodes in all quadtree squares

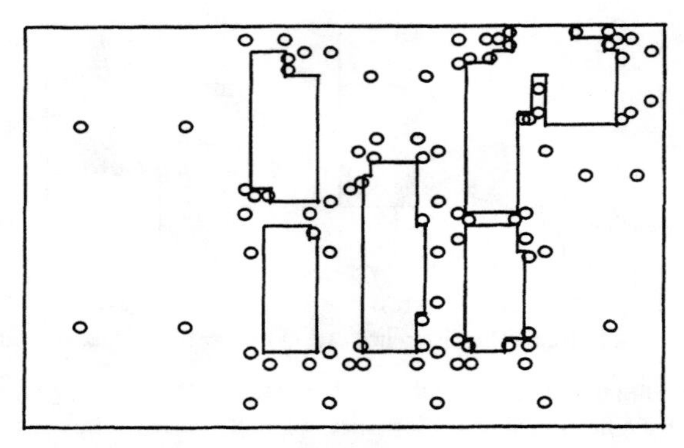

Figure 7. Eliminating the unnecessary nodes

DISCUSSIONS

A grid model is commonly used for a mobile robot because it is easily built using proximity sensors. But this model represents a workspace using grids of a size, so it needs much memory to contain precise environmental information in a large workspace. In this paper, a method to build a network model from a grid model was proposed for path planning between every pair of nodes. To solve the memory problem in a grid model, the quadtree was used and a method to reduce the number of nodes was proposed for an efficient path planning.

REFERENCES

[1] R A. Brooks, "Solving the find-path problem by good representation of free space," Trans. on System, Man and Cybernetics, vol. SMC-13, no. 3, pp. 290-297, 1983.

[2] E. G. Gilbert and D. W. Johnson, "Distance functions and their application to robot path planning in the presence of obstacle," IEEE Trans. Robotics Automat., vol. RA-1, no. 1, pp. 21-30, 1985.

[3] Alexander Zelinsky, "A mobile robot exploration algorithm," IEEE Trans. on Robotics Automat., vol.8, no. 6, Dec. 1992.

[4] Y. Tabourier, "All shortest distances in a graph: An improvement to Dantzig's inductive algorithm," Discrete Math., vol. 4, pp. 83-87, 1973.

[5] J. G. Lee and H. Chung, "Global path planning for mobile robot with grid-type world model," Robotics & Computer-Integrated Manufacturing, vol. 11, no. 1, pp. 13-21, 1994.

[6] H. P. Moravec, A. Elfes, "High resolution map from wide angle sonar," in Proc. IEEE Int. Conf. Robotics Automat., vol. 1, pp. 116-121, Mar. 1985

[7] Hanan Samet, "Distance transform for image represented by quadtrees," IEEE Trans. on Pattern Analysis and Machine Intelligence, volPAMI-4, no. 3, May 1982.

IMPROVING THE PERFORMANCE OF MODEL-BASED TARGET TRACKING THROUGH AUTOMATIC SELECTION OF CONTROL POINTS

I. PAVLIDIS M. J. SULLIVAN R. SINGH
N. P. PAPANIKOLOPOULOS

Artificial Intelligence, Robotics, and Vision Laboratory
Department of Computer Science
University of Minnesota
Minneapolis, MN 55455

ABSTRACT

In this paper, we further expand on work first presented elsewhere [9] regarding the automatic selection of control points for model-based target tracking. The shape descriptive qualities of the segmentation algorithm [9] proposed for the tracking task are tested experimentally. Comparative experiments are also presented for a model-based tracking scheme with and without the segmentation algorithm. The experiments highlight the positive features of the algorithm and verify the positive role the algorithm can play in a model-based tracker in terms of speed and quality of tracking.

KEYWORDS: curve segmentation, deformable-model-based trackers, rigid-model-based trackers, corners, key flat points.

1 INTRODUCTION

The need for target tracking arises in a number of different applications in robotics research. Characteristic examples include vision-based control of grasping and manipulation tasks [10, 13], and visual tracking of moving objects [3, 6]. Target tracking is also important in a number of other applications, like automated surveillance and traffic monitoring [12]. A spectrum of techniques has been developed for real-time visual tracking. Model-based tracking is a well-established and popular approach that involves the use of either deformable models [3, 4, 12, 13] or rigid models [5].

A necessary first step in the computation of certain models [3, 4, 5, 12, 13] is to determine a set of control points to approximate the tracked object's contour. Until recently, this was usually done by hand through a user-interface. However, the possibility of using a curve segmentation algorithm was often indicated. Picking control points manually, renders difficult the automation of the entire tracking task. In addition, since the user is picking the points randomly or at best by using some heuristic developed through his/her own experience, he/she tends to pick either too many or too few control points. On the other hand, using some classical curve segmentation algorithms [2, 7, 8] only half-automates the task since the performance of these algorithms depends upon the fine tuning of a number of parameters. Different object shapes may require different parameter settings or otherwise the segmentation algorithm will perform at times either excessive segmentation or sparse segmentation.

In [9], for the first time, a segmentation algorithm was proposed (named P & P), that filled out the existing gap in all the respects. Specifically, the proposed algorithm fully automates the selection of the control points since it does not depend on any parameters and works equally well for most kinds of shapes. Comparative experiments in [9] showed that the P & P algorithm comparatively to other curve segmentation algorithms, manages to select a small number of points that yet deliver a superior description of the original shape.

In this paper, the P & P algorithm is further analyzed and tested. It is also incorporated in a model-based tracker and its beneficial role in tracking in terms of speed and quality is verified experimentally. The organization of the paper is as follows: Section 2 refers to some model-based trackers that may benefit out of the proposed algorithm. Section 3 describes an experimental investigation of the algorithm's descriptive power. In Section 4, the performance of a model-based tracker with and without the algorithm is reported and discussed. Finally, in Section 5, the paper is summarized and conclusions are drawn.

2 MODEL-BASED TRACKERS

There are two major categories of model-based trackers: deformable-model-based trackers and rigid-model-based trackers. Some of them require the selection of control points along the contour of the target and may directly benefit from the P & P algorithm. As far as deformable-model-based trackers are concerned, Curwen *et al.* in [4] use a B-spline approximation to the original target contour. The control points of the B-spline could be appropriately placed by the P & P algorithm. The P & P algorithm is especially suitable for spline approximation of curves because it does not only locate high curvature points on the contour but also key in-between low curvature points. The latter helps in the reduction of the spline's approximating error at a small cost. Yoshimi *et al.* in [13] and Sullivan *et al.* in [11, 12] use a formulation of deformable models that involves an explicit placement of control points along the contour of the tracked object. This, and the fact that the computational cost of their methods is linear in the number of control points makes them ideal candidates for the testing of the proposed algorithm. In fact, Sullivan's implementation in [11] is the method we chose to highlight the potentially beneficial role of the P & P algorithm in model-based tracking (see Section 4). As far as rigid-model-based trackers are concerned, the algorithm can also be proved useful in automatically building a succinct and accurate model of a 2D object from its initial image.

3 EXPERIMENTAL INVESTIGATION OF THE ALGORITHM'S DESCRIPTIVE POWER

The P & P algorithm locates points of high curvature (corners) using a method similar to that in [2]. It also locates key in-between low curvature points (key flat points) by employing a procedure conjugate to that for locating corners. The P & P algorithm is described in detail in [9]. Here, only an interesting experimental investigation of the algorithm's shape approximating power is presented.

In order to get an indication of the goodness of the algorithmic selection of control points in terms of the accuracy of shape description, the following experiment was devised. Let a contour $\mathcal{C}$ of an arbitrary shape consist of N points ($\mathcal{C} = (\mathbf{P}_1, \mathbf{P}_2, \ldots, \mathbf{P}_N)$). Let the P & P algorithm select for the contour $\mathcal{C}$ a set $\mathcal{S}$ of m control points ($\mathcal{S} = (\mathbf{P}_{s1}, \mathbf{P}_{s2}, \cdots, \mathbf{P}_{sm})$). Let also a set $\mathcal{T}$ of m control points ($\mathcal{T} = (\mathbf{P}_{t1}, \mathbf{P}_{t2}, \cdots, \mathbf{P}_{tm})$) to be chosen in a way so that an error norm is driven to minimum (optimal polygonal fit). The norm chosen for the purposes of the particular experiment was the Euclidean distance error of the polygonal fit represented by the point set. The set $\mathcal{T}$ was determined after an exhaustive search of all the $\binom{N}{m}$ combinations for the contour $\mathcal{C}$. It is interesting to compare the set of control points given by the P & P algorithm with the optimal polygonal fit point set for a variety of shapes (see Figs. 1-4).

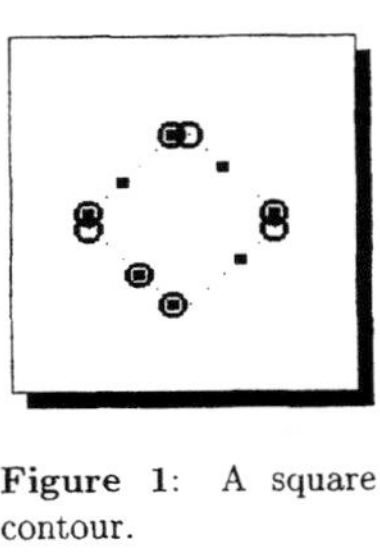

Figure 1: A square contour.

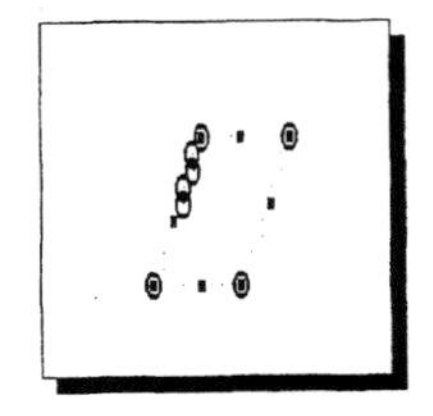

Figure 2: A parallelogram contour.

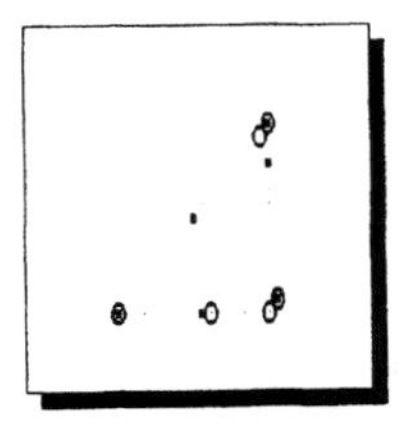

Figure 3: A triangular contour.

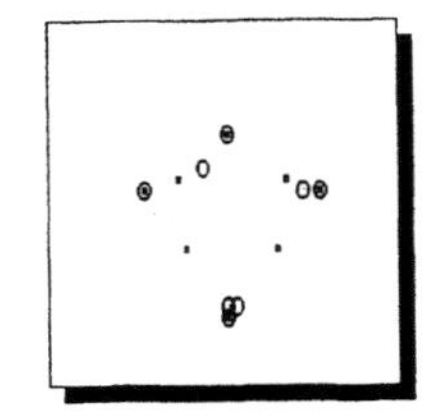

Figure 4: An irregular contour.

The small circles in the above figures represent the points of the optimal polygonal fit set while the points given by the P & P algorithm are represented by small squares. In all the shapes, the prominent corners are included in both the optimal polygonal fit set and the set of the P & P algorithm. Discrepancies arise only for the key flat points of the algorithm. The equivalent points of the optimal polygonal fit are mostly clustered in noisy areas of the shape. In contrast, the key flat points of the algorithm are uniformly distributed between the prominent corner points. This behavior is highly desirable, since the algorithm has not been designed specifically for a polygonal fit but for a more generic fit that may be even a spline fit. In fact, some model-based techniques use the control points for polygonal fits [11, 12, 13] and some others for spline fits [4]. The algorithm loses very little in terms of polygonal fit accuracy by placing the key flat points in a distributed instead of a clustered manner. For example, in the irregular contour case of Fig. 4, the error of the optimal fit is 0.8189 pixels while the error of the P & P fit is 2.1701 pixels. The error of an arbitrary polygonal fit for this shape could run as high as 42.8378 pixels. The small compromise the algorithm concedes in the polygonal fit case pays off in the spline fit case where a clustered distribution like the one favored by the optimal polygonal fit would give very poor results.

4 EXPERIMENTAL TRACKING RESULTS

Preliminary results of experiments incorporating the P & P algorithm for automatic control point selection in a model-based tracking scheme [11] suggest that this approach holds great promise. The P & P algorithm extends the previous system [11] in two important ways. It automates the selection of both the number and location of control points. In the previous implementation, the number of control points was preselected by the operator and their location was manually determined at run-time. By automating these tasks, the P & P algorithm makes the system more general and more independent of its operator. The system has been implemented on the Minnesota Robotic Visual Tracker ([1], see also Fig. 5).

Experiments were conducted in which a target was presented on a 27 inch monitor located one meter from the end-effector mounted camera. The target, a 7.3 cm tall square or triangle, moved around a rectangular path of 100 cm at approximately 8 cm/sec. The position commands sent to the robotic arm were collected and are graphically illustrated in Figs. 6 - 8. Previous results [11] (see Fig. 7) were compared to results using the P & P Algorithm (see Fig. 8).

The previous system used a *predetermined* number of control points irrespective of the target's shape. These points were *manually* placed near the object contour in a highly regular configuration. The generic constraints used by the tracking algorithm created a bias toward equidistant points and equal angles between edges. The new system uses the P & P algorithm to automatically select control points. Because the P & P algorithm does not choose equally spaced points, the constraints used during tracking were modified to reward configurations with angles close to the initial angles and distances close to the initial distances.

The model-based tracking scheme described in [11] worked well only when a small number of control points was selected and the points described the contour well. Since that system encouraged equidistance between control points and equal angles between edges, it performed best when the contour of the object being tracked could be approximated by an equilateral polygon (a highly regular shape) with as many vertices as the model had control points. For less regular shapes or control point configurations, performance degraded. For example, the system in [11] lost track of the square target after just one revolution when an eight-point model was used (see Fig. 7). The old system was not tested with the (non-equilateral) triangular target, since this target is not a highly regular shape.

The system using the P & P algorithm for automatic point selection performed substantially better. Ten trials were measured. In the first five, the arm tracked the moving square. In the second five, the triangular target was tracked. Results from the first trial with each target are presented in Figs. 6 and 8 respectively. The control point selection algorithm invariably selected ten points for the square and six points for the triangle that appropriately described the shapes. The tracker maintained tracking of the objects for several revolutions. In this experiment, the P & P tracker exhibited its ability to maintain tracking at fairly high speeds of different target shapes (square, triangle).

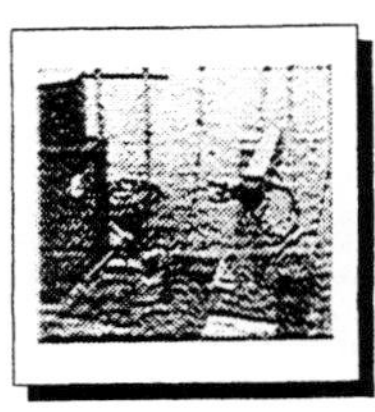

Figure 5: Experimental setup.

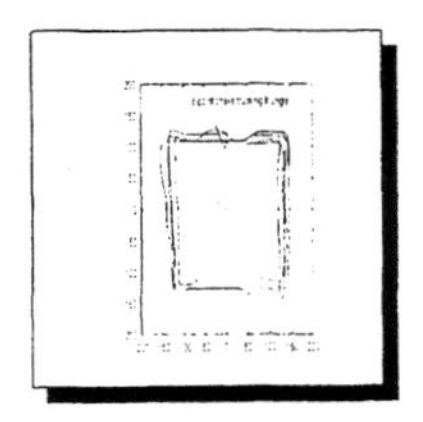

Figure 6: Tracking of a triangular target with the P & P algorithm.

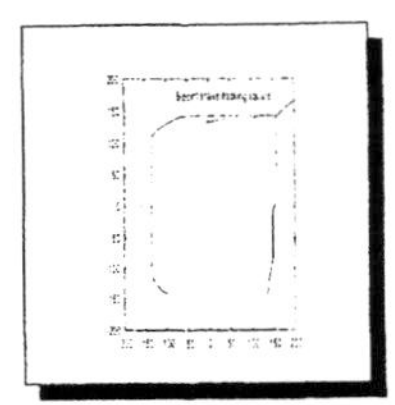

Figure 7: Tracking of a square target without the P & P algorithm. The target was lost after one revolution.

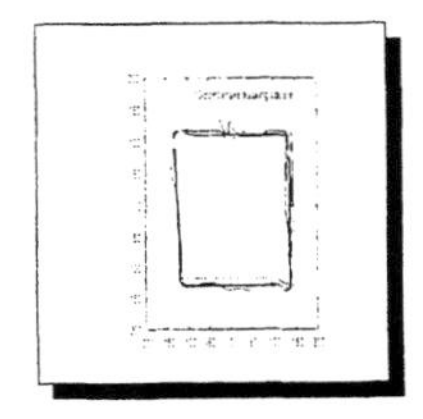

Figure 8: Tracking of a square target with the P & P algorithm.

5 SUMMARY

In this paper, further experimental investigation of the P & P algorithm, first appeared in [9], was reported. The P & P algorithm was designed to automate the selection of control points for certain model-based trackers. The algorithm was designed to perform satisfactorily for polygonal as well as spline fits, since both abound in model-based trackers. In the present work, the algorithm's output was compared with the corresponding point set that gave the optimal polygonal fit for a variety of shapes. The error of the algorithm's polygonal fit was very close to the error of the corresponding optimal fit. In particular, the corner points reported by the algorithm coincided with the corner points of the optimal set for every shape tested. Discrepancies between the algorithm's point set and the optimal polygonal fit set arose for some of the key flat points reported by the algorithm. These discrepancies cost a small approximation error to the polygonal fitness of the algorithm that is anticipated to pay off in the case of spline fits. Similar experiments for spline fits are under way and will be reported in the future.

The algorithm was also incorporated in a model-based tracker [11] and preliminary comparative experiments between the old and new systems highlight the beneficial role the P & P algorithm can play in model-based tracking. Further experiments with a greater variety of shapes and under a greater variety of conditions are under way and will be reported in the future.

ACKNOWLEDGEMENTS

This research was supported by the Department of Energy (Sandia Labs) through Contracts $\#AC-3752D$ and $\#AL-3021$, the National Science Foundation through Grants $\#IRI-9410003$ and $\#IRI-9502245$, the Minnesota McKnight Land-Grant Professorship Program, the Center for Transportation Studies at the University of Minnesota, and the Department of Computer Science at the University of Minnesota. The views and conclusions contained in this document are those of the authors and should not be interpreted as representing the official policies, either expressed or implied, of the funding agencies.

References

[1] S. A. Brandt, C. E. Smith, and N. P. Papanikolopoulos. "The Minnesota Robotic Visual Tracker: a Flexible Testbed for Vision-Guided Robotic Research." In *Proceedings the 1994 IEEE International Conference on Systems, Man and, Cybernetics*, pp. 1363-1368, 1994.

[2] J. J. Brault and R. Plamondon. "Segmenting Handwritten Signatures at Their Perceptually Important Points." *IEEE Transactions on Pattern Analysis and Machine Intelligence*, Vol. 15, pp. 953-957, 1993.

[3] P. A. Couvignou, N. P. Papanikolopoulos, and P. K. Khosla. "Hand-Eye Robotic Visual Servoing Around Moving Objects Using Active Deformable Models." In *Proceedings of the 1992 IEEE International Conference on Intelligent Robots and Systems*, pp. 1855-1862, 1992.

[4] R. Curwen and A. Blake. "Dynamic Contours: Real-Time Active Splines." In *Active Vision*, pp. 39-57, MIT Press, 1992.

[5] C. Harris. "Tracking with Rigid Models." In *Active Vision*, pp. 59-73, MIT Press, 1992.

[6] N. P. Papanikolopoulos and P. K. Khosla. "Feature-Based Robotic Visual Tracking of 3-D Translational Motion." In *Proceedings of the 1991 IEEE Conference on Decision and Control*, pp. 1877-1882, 1991.

[7] T. Pavlidis. *Structural Pattern Recognition*, pp. 168-184. Springer-Verlag, 1977.

[8] T. Pavlidis and S. T. Horowitz. "Segmentation of Plane Curves." *IEEE Transactions on Computers*, Vol. 23, pp. 860-870, 1974.

[9] I. Pavlidis and N. P. Papanikolopoulos. "Improving the Performance of Model-Based Target Tracking Through Automatic Selection of Control Points." To appear, *Proceedings of the 1996 IEEE International Conference on Robotics and Automation*, Minneapolis, Minnesota, April 22-28, 1996.

[10] C. E. Smith and N. P. Papanikolopoulos. "Grasping of Static and Moving Objects Using a Vision-Based Control Approach." In *Proceedings of the IEEE/RSJ International Conference on Intelligent Robots and Systems*, Vol. 1, pp. 329-334, 1995.

[11] M. J. Sullivan and N. P. Papanikolopoulos. "Using Active Deformable Models to Track Deformable Objects in Robotic Visual Servoing Experiments." To appear, *Proceedings of the 1996 IEEE International Conference on Robotics and Automation*, Minneapolis, Minnesota, April 22-28, 1996.

[12] M. J. Sullivan, C. A. Richards, C. E. Smith, O. Masoud, and N. P. Papanikolopoulos. "Pedestrian Tracking from a Stationary Camera Using Active Deformable Models." In *Proceedings of the Intelligent Vehicles '95 Symposium*, pp. 90-95, 1995.

[13] B. H. Yoshimi and P. K. Allen. "Visual Control of Grasping and Manipulation Tasks." In *Proceedings of the 1994 IEEE International Conference on Multisensor Fusion and Integration for Intelligent Systems*, pp. 575-582, 1994.

EXPERIMENTAL VALIDATION OF THE EXTERNAL CONTROL STRUCTURE FOR THE HYBRID COOPERATION OF TWO PUMA 560 ROBOTS

V. PERDEREAU - M. DROUIN
PARC Université Paris VI boite 164
4 Place Jussieu - 75252 Paris Cedex 05 - FRANCE

P. DAUCHEZ
LIRMM - UMR 9928 Université Montpellier II/CNRS
161 rue Ada - 34392 Montpellier Cedex 5 - FRANCE

ABSTRACT

For the successful coordination of two arms handling a common object in unstructured or ill-known environments, V. Perdereau and M. Drouin [1] proposed to implement at each arm level an efficient hybrid position/force controller where the force control loop is closed around the position loop. For now, the efficiency of this new hierarchical solution was only proven by simulation results. It was however suggested that real-time applications with industrial robots could be viable. This paper is devoted to reporting the validation of this method we have achieved in collaboration with P. Dauchez at the LIRMM in Montpellier on an experimental setup built around two PUMA 560 robots.

1. INTRODUCTION

When two robots operate in a complex environment and work on a same object interactively to achieve complicated and dexterous tasks, the object motion may be constrained in some directions due to interaction with external environment. It is then necessary to control the constraint force, i.e., the external force, in addition to the motion of the object and to the relative position/orientation of both end-effectors (or the reaction forces, i.e., the internal forces, between the arms). The control objective is therefore to realize the desired position and force profiles in a constrained coordinate frame located at the grasped object; controllers are supposed to explicitly use the forces sensed at the robot end-effectors.

One fundamental advantage of the master/slave approach [2] [3] is that the two arms are controlled independently allowing a distributed computer architecture and an easier implementation. However, both controllers do not share the same force and position errors, the force controller must react fast enough to changes in position to avoid dropping or damaging the object.

A better solution consists in giving an equal status to both arms. Hence, the cooperative multiple arm problem was often treated as a single multi-axis closed chain control problem.

In dynamic hybrid position/force control methods [4] [5] [6], equations of motion of the multi-arm system are derived regarding the object as part of each arm. The multi-robot system is then linearized and decoupled with respect to the object motion, the constraint force and the internal force by using the well known nonlinear feedback technique. The solution of inverse dynamics is time consuming and is not suitable for real time systems and limited to applications where the object properties and the contact conditions between the end-effectors and the object are well known in advance.

Other methods [7] [8] [9] are based on formulation of kinematics and statics for the two arms and the object considered as a unique system. The idea of hybrid position/force control is then extended to the cooperative control and schemes are very similar to the hybrid scheme originally proposed by Raibert and Craig [10] for a single arm robot. The authors [11] [12] however showed that this kind of structures implemented at the global level of the whole set arms-object has some drawbacks, mainly due to the many kinematic and geometric task configuration dependent transformations resulting in time varying gains in the control loops. These many on-line computations load the controller design and modify the dynamic position and force responses during task execution possibly leading to an unstable behavior.

Sustained by this analysis, V. Perdereau and M. Drouin proposed another structure [1] where each arm is controlled independently by a local hybrid position/force controller but where both arms play similar roles in using the same trajectory and force reference generator. In order to achieve any complex task, each local control scheme has to be completely independent of the task configuration. For this purpose, we developed an original scheme [13] for the simultaneous control of position and force of one robot manipulator on the basis of the one-dimensional external force control introduced by de Schutter and Van Brussel [14]. Thus, the force feedback loop is no more parallel to the position control loop as in other methods but rather closed around.

For now, the efficiency of this new hierarchical solution was only proven by simulation results. This paper is devoted to report the experimental validation we have achieved at the LIRMM in Montpellier in collaboration with P. Dauchez.

In the section that follows we shall recall the external control properties. Then after a description of the experimental set-up, we will present the results obtained when achieving a complex task with various objects while the position and force vectors have to be fully controlled. We will end by discussing several advantages of this method.

2. EXTERNAL CONTROL SCHEME PROPERTIES

For each manipulator, the position setpoint vector is compared to the actual end-effector position (Fig.1) so that the position controller computes appropriate arm commands. The robot arm is therefore first and always position controlled. However, the position setpoints of this internal position loop are not only the desired motion X_d in free space but a position correction term ΔX_d computed by integrating the force error along the task frame constrained directions is added. Hence, the force loop is hierarchically superimposed over the position loop.

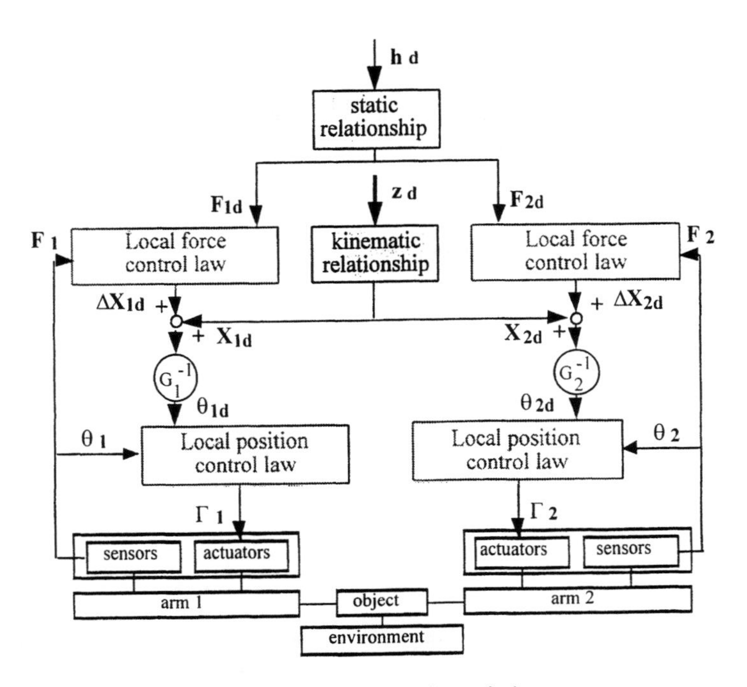

Figure 1. The cooperative external control scheme.

The task analysis is supposed to give appropriate position and force reference vectors but partial information about task configuration does not result in detrimental effects: a desired trajectory in constrained directions is viewed as a disturbance by the force controller and compensated for, and force input setpoints in unconstrained directions lead to a smooth end-effector motion with constant velocity. No selection matrices are necessary to avoid conflicts at actuator level since the position and force loops are added at an upper level. Hence, the position/force controller is no more implemented in the task frame but rather in the Cartesian reference frame. The transformations of both input vectors from the task frame are now outside the control loops. It follows that the dynamic behavior does not depend at all on the task configuration. This is the main advantage of this control scheme that we fully exploit afterwards.

Both robot arms are controlled independently by this external position/force controller. In fact, carrying out a coordinated motion only consists in determining the appropriate individual position and force input vectors in reference frame. For that, the task input setpoints described in the object frame are transformed into the reference frame and dispatched into the individual desired position and force vectors using static relationships at a specified object point. Since the force control loops are predominant on the position loops, any error in the object model is compensated for by the force controllers.

All the kinematic transformations are outside the control loops. This advantage makes the implementation easier and the system more flexible because the controller structure has not to be redesigned from task to task which is a common shortcoming of the existing position/force control methods. The dynamic position and force behaviors remain the same regardless of the configuration of the task or arm.

This solution permits a decentralized implementation. The computational time in a control cycle is therefore roughly equal to that in the case of controlling a single arm. Moreover, changing the number of robots in the system does not affect the overall data processing and computation structure. We can easily consider using it for more than two robots.

If the properties of this new hierarchical solution were long analyzed and proved by simulation results [12], its efficiency has to be verified on an experimental set-up. Next section is devoted to report this implementation.

3. HARDWARE AND SOFTWARE IMPLEMENTATION

3.1 The experimental set-up

The hybrid external coordination scheme was designed and implemented on two PUMA 560 robot arms (Fig.2). Each arm is equipped with a 6 axis force/torque analog sensor. The original Unimate controllers of these arms are only used for the power supply and the amplification of the commands to be sent to the 12 motors. A single real time controller from dSPACE company (Paderborn, Germany) proceeds powerful 32 bits computations for the software implementation. The CPU card is based on a Digital Signal Processor (DSP). Each sampling time, the program gathers information from the arms which includes optical encoders and force sensors signals and sends the actuator commands that have been computed. Hence, this controller also includes A/D and D/A converters to carry out the input/output communications via a double-access memory board and a Peripheral High Speed bus. Programs are developed in C language on a PC486 and then downloaded to the DSP. The PC is used as a terminal during the execution phase. This environment allows on-line visualization of the variables and modification of the parameters so that controller gains may be easily changed during the task to obtain an empirical adjustment. The powerful computation means of this controller leads to a very small sampling period of 3ms for 12 axis.

Figure 2. The experimental set-up.

3.2 The controller design

Each arm is controlled independently by an identical hybrid external controller. The position controller is a classical PID implemented in joint space. Gains are obtained for the particular mechanical structure but are not linked at all on the geometric configuration of the arm or of the task. The inverse dynamic model which is well known for this kind of robot manipulator alone might be easily used to decouple and linearize the system. We noted however that our classical controller was good enough to achieve a satisfactory tracking performance at the working velocities.

The joint setpoints θ_d are the result of the arm inverse kinematic model. The Cartesian input position vector, i.e., the combination of the desired end-effector trajectory X_d and the infinitesimal position variations along constraint directions ΔX_d, is obtained in the reference Cartesian space.

The force control loop is therefore implemented in this Cartesian space. A simple integrator gives a good dynamic behavior with no steady state error. Once the position controller gains are chosen to guarantee identical dynamic response in all space directions, then the force controller gain is identical for all force error components. This gain is proportional to the inverse environment stiffness as follows: $K_f K_e^{-1}$. This parameter is the only one depending on the task but is very easy to tune empirically.

The position controlled vector and the force controlled vector h_d are described in a frame attached to the object: the constraint frame [15]. The position vector merges the absolute and relative position/orientation of the end-effectors. These 6 component vectors are consistent respectively with the external force/moment vector of the total forces applied to the object that contribute to its motion, possibly in constrained directions, and the internal force/moment vector that produces the object deformation, both vectors forming vector. These task vectors are transformed to individual Cartesian space vectors. The end-effector force references F_{1d} and F_{2d} are linked to h_d by a static relationship and the desired end-effector positions X_{1d} and X_{2d} to z_d by the object kinematic model. More details about these transformations are given in [15].

4. EXPERIMENTAL RESULTS

This section is devoted to report several experiments we performed with the experimental set-up described previously with the aim to verify the efficiency of our method. Various task objectives have been achieved so that a large choice of complex tasks could be viable. The task frame is shown in Figure 3.

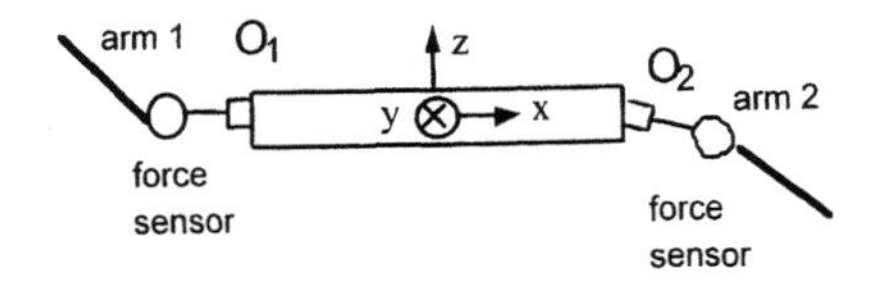

Figure 3. The task frame.

4.1 Absolute position response

To show the absolute position response, this experiment consists in moving a 2.5 kg mass grasped by both manipulators in free space. A fifth order polynomial is used for generating the trajectory of this object. It consists of a 10-centimeter translation along x axis and a 10-degree rotation along y axis. The velocity is about 30% of the given admissible velocity (0.25 m/s).

This desired motion is then dispatched to both controllers. The force control loop may be opened (giving no reference force) or maintained to avoid the object damage. We focus here however on the position response (Fig.4). The position error is very small although the object mass behaves as a disturbance in the position loop.

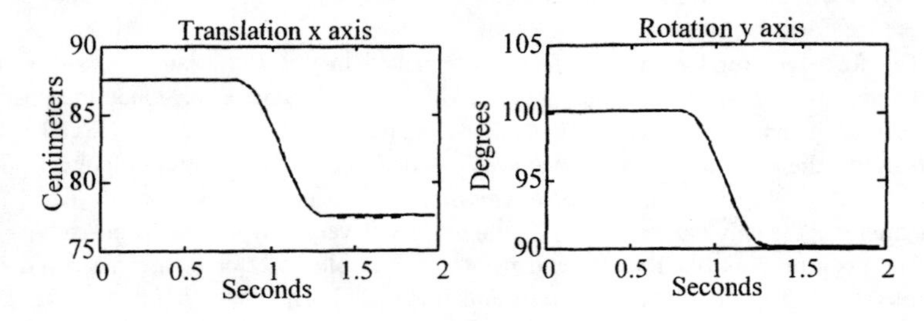

Figure 4. Absolute position response.

4.2 Internal force response

When both manipulators stretch a rope, the individual position setpoints are fixed so that the end-effector relative position roughly corresponds to the rope length. Since the absolute position of the center of this rope is supposed to be constant, the individual position setpoints remain unchanged during the task.

The stretching of the rope is only due to an internal force reference, here -20N, so that both effectors begin to move away with a constant velocity (Fig.5). When the rope is stretched, a force is detected and the force control loop operates. The desired force is reached in only 1s with no overshoot.

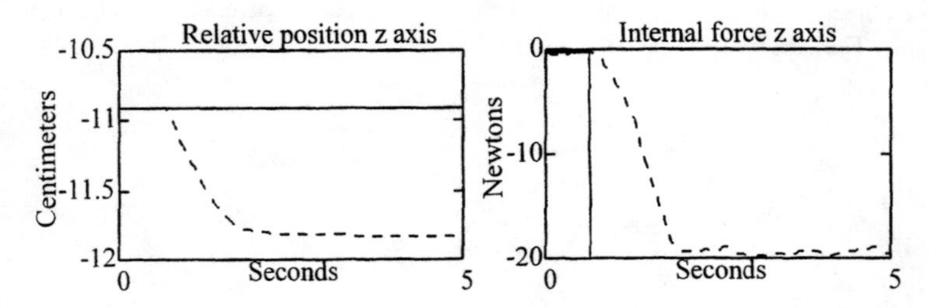

Figure 5. Internal force response.

4.3 External force response

Now the common manipulated object, a steel bar, is supposed to come into contact with an external environment. The arms go down, put the object on the table and press it to achieve a given -20N external force reference (Fig.6). It should be noticed that the force sensors behave as accelerometers at the beginning of the motion in free space. The force controller assures a continuous progression of the task and avoids collision effects.

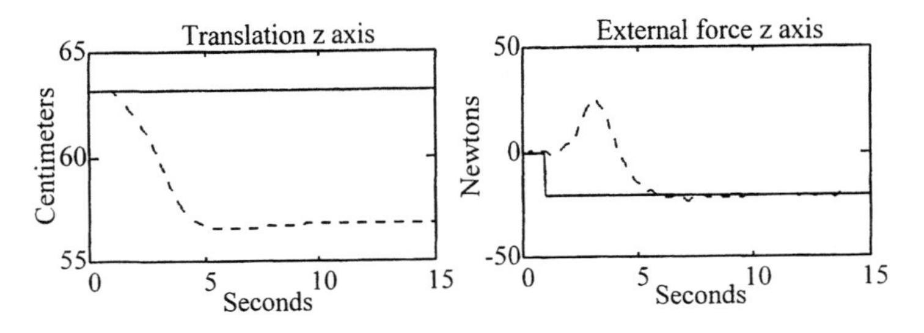

Figure 6. External force response.

4.4 Hybrid position/force response

The task consists in manipulating a cylindrical rigid object firmly held by the two robots. The system is commanded to move along the horizontal surface and the object exerts an external force on the surface while rolling on it (Fig.7). The velocity is kept small because the motion may disturb the force response.

The desired position and force are well realized. The surface is flexible. Its stiffness varies between 20N/cm and 200N/cm during the task while the force controller gains were tuned up for an average constant stiffness. Therefore we believe that the performance of our control structure for such a complex task is very good.

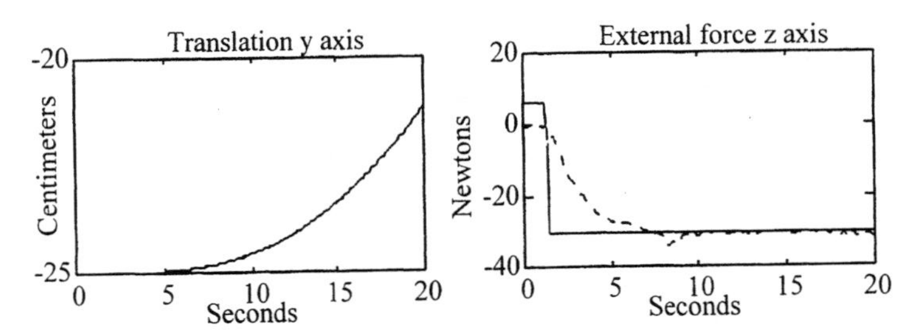

Figure 7. Hybrid position/force response.

CONCLUSION

Experimental results were carried out to validate the proposed method. Controller gains were tuned without looking for the best adjustment. Our aim was not to test advanced control algorithms but rather to show the capability of this new structure with a real experimental set-up. As will be shown by means of a video at the conference, all the tests we have conducted confirm that this external scheme is a valuable approach for industrial applications.

In addition to advantages already mentioned in this paper and in [1], its attractive hardware and software implementation must be emphasized. Indeed, industrial robots are often equipped with enclosed joint space control law, difficult to transform. Here no hardware modifications are needed. Software implementation is also made easy since few on-line computations are involved and almost no change is required in control laws regardless of the cooperative task. Only desired force and position have to be computed according to the task.

Other tests are planned in the future so as to obtain a more complete evaluation of the performance in regard to changes in the cooperative task.

REFERENCES

1. V. Perdereau & M. Drouin, "An hybrid external solution for two-robots cooperation.", 2nd France-Israel Symp. on Robotics, Saclay, France, April 1991, 1.15-1.22
2. T. Ishida, "Force control in coordination of two arms", 5th Int. Conf. on Art. Int., Aug. 1977, 717-722
3. S. Hayati, K. Tso & T. Lee, "Generalized Master/Slave coordination and control for a dual arm robotic system", 2nd Int. Symp. on Rob. and Man., Albuquerque, USA, Nov 1988, 421-430
4. S. Hayati, "Hybrid position/force control of multi-arm cooperating robots", IEEE Int. Conf. on Rob. and Aut., San Francisco, USA, April 1986, 82-89
5. T. Yoshikawa & X. Zheng, "Coordinated dynamic hybrid position/force control for multiple robot manipulators handling one constrained object", IEEE Int. Conf. on Rob. and Aut., Cincinnati, Ohio, USA, May 1990, 1178-1183
6. N. Xi, T. J. Tarn & A. K. Bejczy, "Event-based planning and control for multi-robot coordination", IEEE Int. Conf. on Rob. and Aut., Atlanta, USA, May 1993, 251-258
7. M. Uchiyama, N. Iwasawa & K. Hakomori, "Hybrid position/force control for coordination of a two-arm robot",IEEE Int. Conf. on Rob. and Aut., Raleigh, USA, April 1987, 1242-1247
8. M. Uchiyama & P. Dauchez, "A symmetric position/force control scheme for the coordination of two robots", IEEE Int. Conf. on Rob. and Aut., Philadelphia, USA, April 1988, 350-356
9. C. D. Kopf, "Dynamic two arm hybrid position/force control", Robotics and Autonomous Systems 5, 1989, 369-376
10. M. H. Raibert & J. J. Craig, "Hybrid position/force control of manipulators", Trans. of ASME, vol.102, June 1981, 126-133
11. V. Perdereau, "Contribution à la commande hybride force/position. Application à la coopération de deux robots.", Thèse de doctorat de l'université Paris VI, (February 18th, 1991) in French
12. V. Perdereau & M. Drouin, "Hybrid external control for two robots coordinated motion", Robotica, Cambridge University Press, to appear in 1996
13. V. Perdereau & M. Drouin, "A new scheme for hybrid force-position control", Robotica (1993), Cambridge University Press, volume 11, 453-464
14. J. de Schutter & H. Van Brussel, "Compliant robot motion II. A control approach based on external control loop", The Int. J. of Robotics Research, vol.4, n°4, August 1988, 18-33
15. P. Dauchez, "Task descriptions for the symmetric hybrid control of two-arm robot manipulator", LAMM Internal Report #90011, Montpellier, France, April 1990

VIRTUAL RECONSTRUCTION OF AN UNKNOWN REAL SPACE WITH AN ULTRASONIC SENSOR

Laurent PEYRODIE*,, Daniel JOLLY**, Anne-Marie DESODT****

* **Ecoles des Mines de Douai, Centre de recherche de Dorignies**
41 rue Charles Bourseul, B.P. 838
59508 DOUAI Cedex, FRANCE

** **Université des Sciences et Technologies de Lille**
Centre d'Automatique de Lille
59655 VILLENEUVE D'ASCQ Cedex, FRANCE

ABSTRACT : The aim of our work is to rebuild an environment completely unknown by using only a unique ultrasonic sensor which gives back a minimum of measures points, in order to obtain a rebuilding system, precise, available, and simple.

KEYWORDS : Fuzzy map, fuzzy aggregation, ultrasonic sensor

INTRODUCTION :

In a first part, we will explain a way to obtain, after rotations of the sensor, a possible environment or fuzzy map.

In a second part, we will define a method to aggregate some fuzzy maps of the same environment.

Each fuzzy map is obtained after an additional moving of the sensor (generally after a translation), this set of maps defines many points of view of the same environment. The aggregation method uses all the criteria, geometrical ones and fuzzy ones defined from a fuzzy map.

These results will be illustrated by a virtual reconstruction of real scenes.

1. FUZZY MAP OR POSSIBLE ENVIRONMENT

1.1 Ultrasonic problematic

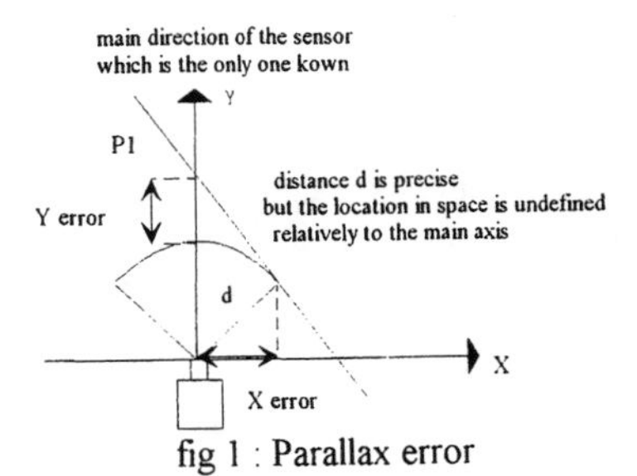

fig 1 : Parallax error

An ultrasonic sensor creates a circular wave due to this kind of wave, a precise location of an object in a space is impossible. On one distance measure derived from the sensor used during ours experimentations, we can make a parallax error of approximately 36° in a frame linked to the sensor (fig. 1) [WEI 94] [CAS 89].

We use a method based on possibility theory to deal with this parallax error [PEY 96]. We call this method a method of incidence research. The incidence angle is the angle builded by the main direction of the sensor and the normal to the surface.

1.2 Results of the algorithm of incidence research

The nature of computed results are geometrical ones (value of incidence, position in a Cartesian space) and mathematical ones (certainty level of the computing incidence). So a fuzzy map is defined in a Cartesian frame by some infinite lines to which we associate certainty levels.

We define mathematically a fuzzy map by a set of couples formed of the computed incidence and of the certainty level (computed at the measure point which is used like a support for the virtual rebuilded entity).The certainty level is equal for each point of the line.

The rebuilded lines are infinite ones because the impact on the target doesn't move in the circular trajectory of the sensor which is used to take the measures ; then, at this step of the algorithm the dimensions of the target are unknown. It is the reason why we must obtain many fuzzy maps of the same space if we want to rebuild all the entities with the good dimensions. We translate then the sensor to obtain the new fuzzy maps that is to say the new points of views. If we aggregate all the fuzzy maps in a same absolute frame, we can define more precisely all the entities and improve theirs positions in the reconstruction space.

2. AGGREGATION OF FUZZY MAPS

2.1 Aim of the aggregation

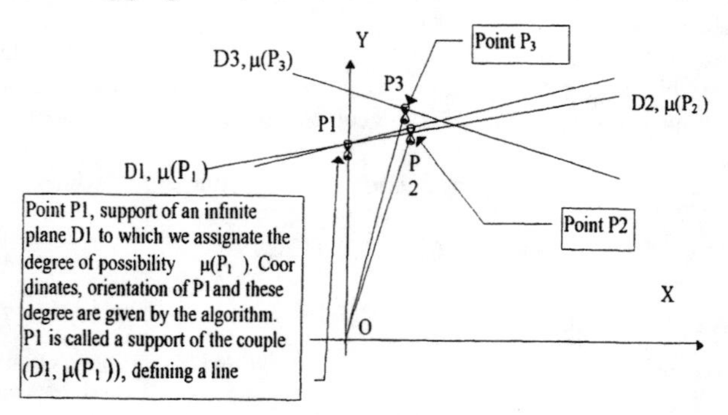

figure 2 : Example of a fuzzy map

We want to define at which conditions two lines figure the same entity.
Two lines are equal if their slope are neighbouring and their ordinates at zero point are neighbouring too. The slopes (linked to the computed incidence) and the ordinates at zero point are not certain data and are inaccurate data too, so we have to use a theory which overcomes those problems of imprecision and uncertainty.
So we must define a fuzzy distance between fuzzy sets which considers the geometrical (position, orientation) criteria and the certainty level associated to each entity.

That's why we take into account elementary entities (measure points) and no more infinite (line) entities ; and so we sum all the elementary entities to build a finite entity D. This sum aggregates all the geometrical and fuzzy criteria ; these criteria seem to be independent.

2.3 Definition of a simple element

Hypothesis 1 :

A finite element D is the union of "small " elements d_i $D = \bigcup_i d_i$

Hypothesis 2 :

The measures belonging to the same entity D are linked sequentially.

The second hypothesis allows us to link two elements d_i by using the notion of "transitivity".

Hypothesis 3 :

Two points (or simple elements) belong to the same entity if the incidences, that are associated to them, are neighbouring.

So we define D by using the hypothesis 1 and 2 to create some transitive links and we use the third hypothesis to unify the simple elements d_i.

2.3.1 Definition of a simple element d_i

A simple element is tailed in three parts : a surface of error S, an incidence i and a certainty level $\pi(i)$.

2.3.2 Definition of an error surface

To define this surface, the signification of a measure derived from the sensor must be understood : when the beam of the sensor comes back to the receptor, the littlest distance to an object, which is perpendicular to the main direction of the sensor is returned. The plane surfaces are located on circle tangents, the radius of this circle is equal to the measure produced by the sensor due to the aperture angle (fig 1). All this implies an error of position on the real plane surface in a Cartesian frame (Ox, Oy) of the space.

Because the points are given by our algorithm, these points have privileged values, that is why the error distance doesn't exceed ten percent of the original distance delivered by the sensor. In the same way, the error on the incidence angle doesn't exceed 20 degrees around the position of the point defined as the couple support line.

So, we create a error surface S_i around a measure point.

2.3.3 Definition of a simple element

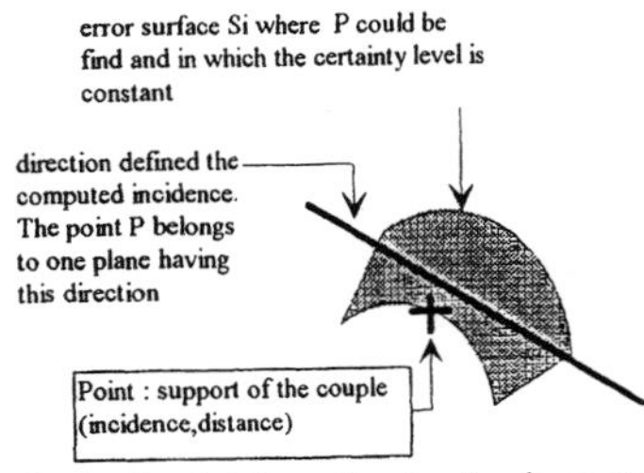

fig 3 : Definition of a simple element

At the error surface, we add the incidence (slope of line) and the certainty level computed. A simple element represents the locus point where we can find a measure point belonging to a line of slope computed with the algorithm and the certainty level is constant all over the error surface (fig. 5).

2.5. Aggregation method

This method takes into account the geometrical criteria, error surface and direction, and certainty levels too. We define aggregation rules applying on simple elements.

rule 1 :

We say that two elements are corresponding to one other if the intersection of their error surfaces isn't empty.

rule 2 :

Two simple elements are corresponding if the rule one is satisfied and if their incidence or direction are neighbouring.

rule 3 :

We associate at the new created element a new value of incidence by using an operator of aggregation Θ and a new spatial possibility distribution in R^2 obtained with the fusion method explained in [DUB 94] :

$$\Pi_S(P_1....P_n) = \max(\pi((P_1)...\otimes..\pi(P_n)) / h(\pi(P_1.....P_n)\ ,\ \min(1-h((P_1.....P_n),(\pi(P_1)..\vee...\pi(P_n))$$

Where S is the union of error surfaces which are defined around the points P_i or sources.

h is : $h(\pi(i_1),...\pi(i_n)) = \sup((\pi(i_1)\otimes\pi(i_n))$ where $\otimes$ is a conjunctive operator, we choose the operator minimum, h gives back the consensus level between sources.

The operator $\vee$, is an disjunctive operator, we choose the operator maximum.

1-h gives the possibility degree to be outside of the intersection support of sources for the studying parameter.

Dividing by h allows us to normalise the results.

This kind of aggregation enhances the consensus of the sources at the intersection of the sources that is to say in our case at the intersection of the error surfaces.

So, after the aggregation of certainty levels, we take care to the geometrical criteria, we add to the new created element a new value of incidence. We choice a barycentrical sum for computing the new incidence.

$$i = \Theta[(i1, \pi(i1)),\ ....,(i2, \pi(i2))] \quad (1)$$

These rules induce new features on the created element, but we don't modify the error surface for this new created element. We just create a new support point which is the union of the points verifying the rules one and two, we add to this new point the characteristics computed in the third rule.

Reconstruction algorithm of an entity of size D

- Definition of point sets :

  -1- Build a family $\Im$ of point sets by using the rules 1, 2 and 3

  $\Im = \{\mathcal{E}_j = \{P_j$ verifying the rules 1, 2 et 3$\}\}$ j=0 à n , n=card($\Im$ }

- We build then a new point set : $R=\{ P_i \}=\{ P_i \in \{E_j - (E_j \cap E_k)\}\}$

with E_j et $E_k \in \Im$ and $E_j \cap E_k \neq \emptyset$

  -2- Use of the second hypothesis

The second hypothesis allows us to apply rules 2 and 3 to each point of R and $\{ E_k - E_j \cap E_k \}$

then we obtain a set E_j' such as card(E_j')≥card(E_j) where E_j' is a set of points belonging to an entity D.

So, we begin again this algorithm at the step 2 with the sets E_j'.

-3- Obtaining the entity size

We have to remove the error surface around the measures points created in 2.

The points of the support E' define the support of a virtual line segment representing the real entity. But at this step of this algorithm, the error surface is not removed, it is still equal to the one illustrated at figure 6. The added knowledge resulting of the aggregation allows us to rebuild an entity D by using a linear regression applied on the points belonging to the support segment E'. This linear regression is applied not directly on the coordinates of the point but on corrected values. The knowledge of the direction of the support allows us to correct the coordinates if we suppose that the wave form emitted by an ultrasonic is circular, because these two facts define a constraint on the position of each point composing the segment.

2.6 Correction of the coordinates of measure points

All the points of E' belong to a line D which slope is given by the incidence i obtained by (5), this coefficient defines a constraint on the location of measure points in an absolute frame. If we consider the wave shape (circular arc) and the constraint, we are able to correct the coordinates of a measure point in the absolute frame. From this support point we search the intersection between the tangent to the circle (radius equal to the distance measured by the sensor in this direction, centre corresponding to the zero point of the frame) with the line which slope is given by the new incidence value obtained by (1).

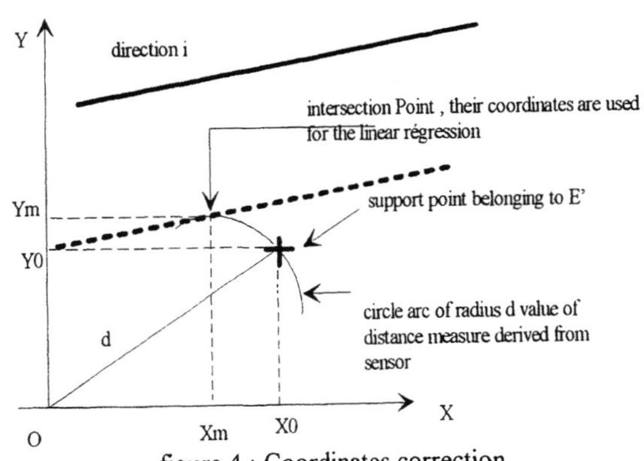

figure 4 : Coordinates correction

Those new computed coordinates give us the real location of the point impact on the target, so thanks to this correction we remove the error surface for each point of E'.

Remark : all the coordinates are expressed in an absolute frame.

Remark : On this segment, the certainty level proceeded from aggregation (5), could be displayed. So the operator in front of the virtual scene could define a trajectory going through the points which are the nearest from the most certain space sections.

3. EXPERIMENTAL RESULTS

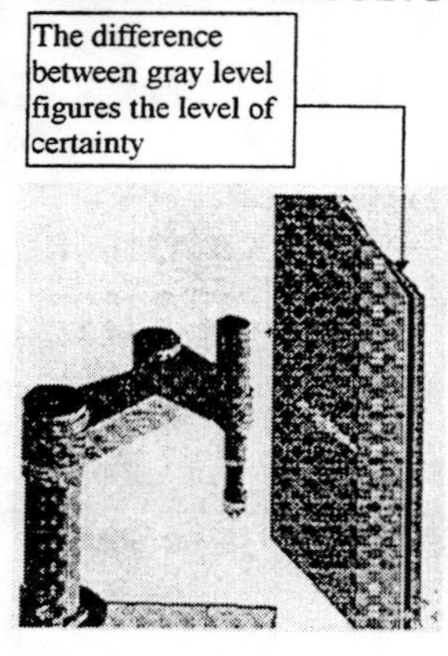

fig 5 : Set of fuzzy maps

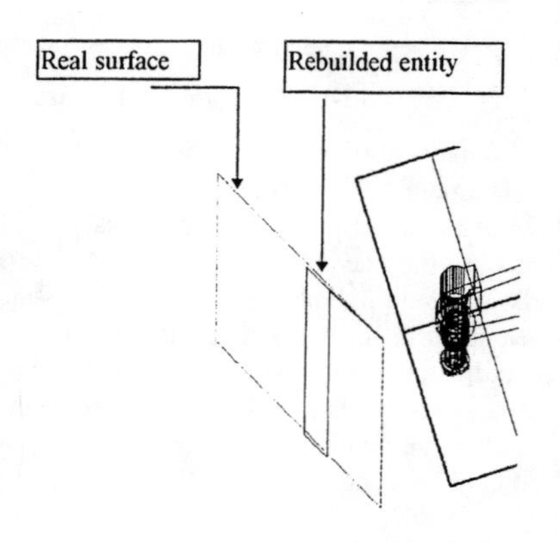

fig 6 : Aggregated environment

If the error on the slope of the rebuilded entity is null, the error on the localisation of the virtual surface is less than the error on the localisation of the real plane in the real space.

CONCLUSION

If we apply this algorithm of aggregation on many environments composed of plane surfaces and secant surfaces, the entities are well located, oriented and dimensioned.

If the objects possesses obtuse angle the informations related to the measures are so poor that the rebuilded line is rough. We must define a strategy of moving for the robot if we want to decrease the doubt on positions of these objects.

If we want to define more precisely dimensions of the rebuilded object, we could make the linear regression not only on the points belonging to E', but on all the measures data stored during experimentations too.

BIBLIOGRAPHY

[CAS 89] CASINIS, R. VENUTI, P. Sonar range data processing and enhancement, Intelligent Autonomous System, 1989, Vol. 1

[DUB 94] Dubois D, Prade H, La fusion d'informations imprécises, Rapport IRIT/94-44-R, Novembre 94

[PEY 96] Peyrodie L, Jolly D, Desodt AN, Modelisation of an ultrasonic sensor, AMPST 96, Bradford

[WEI 94] WEISBROD, J. Fuzzy exploration of an unknown environment, EUFIT, 1994, pp 1428-1433

A NEW LESS INVASIVE APPROACH TO KNEE SURGERY USING A VISION-GUIDED MANIPULATOR

M. ROTH, CH. BRACK, A. SCHWEIKARD

Bay. Forschungszentrum für wissensbasierte Systeme, Forschungsgruppe Kognitive Systeme (FG KS), Orleansstr. 34, 81667 Munich, Germany, e-mail: {rothm,brackc,schweika}@informatik.tu-muenchen.de

H. GÖTTE, J. MOCTEZUMA

TU München, Institut für Werkzeugmaschinen und Betriebswissenschaften (iwb), Karl-Hammerschmidt-Str. 39, 85609 Aschheim, Germany, e-mail: {gx,ma}@iwb.mw.tu-muenchen.de

F. GOSSÉ

Orthopädische Klinik, Medizinische Hochschule Hannover, Konstanty-Gutschow-Str. 8, 30625 Hannover, Germany

ABSTRACT

We describe new methods for on-line image processing and calibration in the context of a vision-guided robotics system for orthopaedic knee surgery: During knee surgery, e.g. insertion of knee implants, bone material of the femur and the tibia has to be removed. Conventionally, surgeons use templates to guide a hand-held saw. Therefore accuracy is limited by the surgeon's skills and dexterity. The proposed system uses a saw mounted on a guiding device fixed on a manipulator's hand. During surgery the hand is guided to preoperatively planned cutting planes. Hence the surgeon freely moves the saw to remove bony material, only limited by the guiding device keeping the movement within the cutting plane. This approach combines the robot's high accuracy with the surgeon's expertise and is therefore expected to provide safety, high overall accuracy and lower complication rates.
Preoperative planning is achieved interactively, based on 3D surface-oriented bone models derived from computer tomography (CT). Intraoperative usage of the system starts with calibration and registration of an X-ray system. For this matter a new calibration technique has been developed. It uses two permanently recognizable 3D calibration bodies and special image processing and recognition algorithms. Registration is based on error minimization between estimated and actual bone contours in the X-ray image.
This new method's advantages - compared to well-known techniques - are the achievement of highly accurate and robust calibration results, while constraints concerning the placement of landmarks relative to the X-ray system are avoided. Rigid patient immobilisation is not necessary, because small knee movements during surgery are allowed. This is due to the use of a real-time vision system for intraoperative bone tracking.

KEYWORDS: X-ray camera calibration, robotic knee surgery.

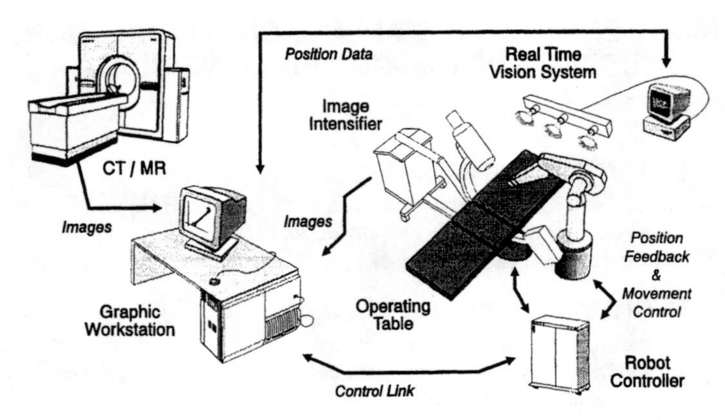

Figure 1. The System Components

SYSTEM OVERVIEW

The System Components

Besides a common X-ray image intensifier and a real time vision system the proposed system comprises a workstation for coordinating and controlling the process of treatment delivery, a surgical robot with tools and instruments specially developed for use in the operation theatre (these items being currently developed at the iwb) and an operating table with intelligent control (see fig. 1).

The Medical Application

The chosen medical application is the replacement of the knee joint with an artificial total knee prosthesis. Carried out in a conventional manner this treatment consists of a series of bone cuts delivered to femur and tibia by the surgeon with the help of a handheld powered saw guided by a set of templates. The lower end of the femur and the upper end of the tibia are shaped to fit the prosthesis in the best way. Afterwards the different parts of the prosthesis are mounted onto femur, tibia and patella. The size of the prosthesis and the different cutting planes are planned before the therapy using two X-ray images, transparent paper and a pencil. The accuracy of the whole planning and treatment delivery process is therefore limited by the surgeon's skills and dexterity.

The New Process Of Treatment Delivery

Our system for treatment delivery differs from the conventional method in several ways: Preoperative planning is achieved interactively, based on 3D surface-oriented bone models derived from computer tomography (CT) [7]. Instead of using templates to guide a handheld saw a special doublebladed saw will be positioned by the surgical robot to preoperatively planned cutting planes. This saw is a special development of iwb in cooperation with aesculap [10].

It is designed to reduce reaction forces and strain on the bone tissue. Sawing speed is enhanced and induction of thermal stress is reduced. It will be mounted on a guiding device fixed on a manipulator's hand. The surgeon can freely move the saw to remove bony material. He is only limited by the guiding device keeping the movement within the cutting plane (see fig 2). This enables him to perfectly control the direction of the

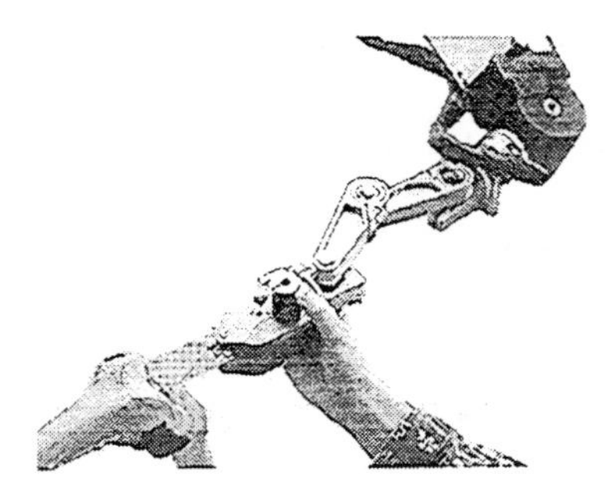

Figure 2. Robot hand, sawing device and saw

cut and the hereto applied force. The operating table is controlled by the workstation in order to achieve optimal positioning of the patient relative to the robot. This approach combines the robot's high accuracy with the surgeon's expertise and is therefore expected to provide high overall accuracy and lower complication rates and be safe in use. The intraoperative usage of the system starts with calibration and registration of the patient with a selfcalibrating X-ray system. For this matter a new calibration technique has been developed. It reduces the patients pain and stress by avoiding the preoperative insertion of markers into the bones [6]. Following steps will be performed before and during surgery in order to maintain precise localization and tracking of the patient:

1. Preoperative procedure:
   CT-scan *without* application of markers.

2. Intraoperative procedure:
   (a) Application of active markers (infrared LED's) on femur and tibia.
   (b) Initial selfcalibrating X-ray imaging of femur and tibia from at least two different angles with image intensifier.
   (c) Computing location of femur and tibia by correlation of CT-scan (1) and X-rays (2b).
   (d) Tracking of bone location with real time vision system.

Rigid patient immobilization is not necessary, because small knee movements during surgery are allowed. The background of the calibration technique is explained in detail further below. It uses two permanently recognisable 3D calibration bodies, special image processing and recognition algorithms.

CALIBRATION OF THE X-RAY SYSTEM

Motivation

In order to adequately model the previously unknown behaviour of the image intensifier a camera model with many degrees of freedom (DOF) has been chosen. The high correlation between the interior (distortion) and the exterior (position) camera parameters is a disadvantage related to the high DOF. With CCD-cameras this problem can be solved by taking more than one image from the calibration body, by moving the camera [4] or using stereo images [8].

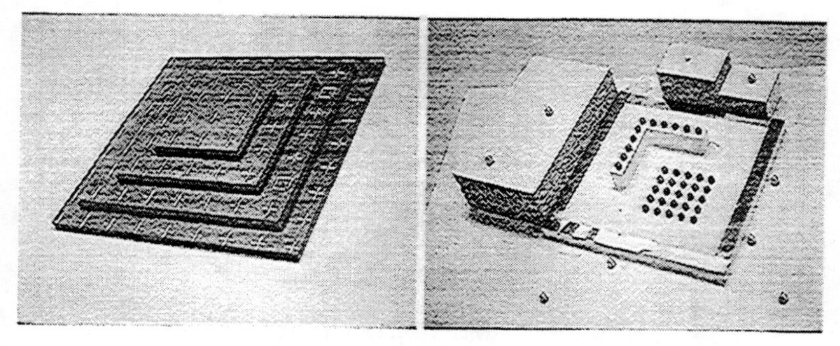

Figure 3. Calibration Bodies

Unfortunately mechanical and other instabilities of the X-ray system, e.g. mainly the detector's sensitivity towards electromagnetic fields, lead to unstable interior camera parameters over time and change in position. Therefore the above described methods fail to reduce the internal correlation of the X-ray system's camera parameters.

Another way to achieve a reduction of correlation of the camera parameters is the use of a 3D calibration body for multiplane calibration. A constraint of our application is that the calibration markers have to be recognizable from many different views. This creates a considerable risk of overlapping markers in the X-ray image.

Summarized, none of the existing methods for CCD-cameras can be fully applied to X-ray calibration. Therefore a new calibration technique was invented using two calibration bodies, both appearing in the image.

Calibration Environment

The first calibration body is a plate mounted rigidly on the operating table near the patient's leg (see fig. 4). It determines the actual reference coordinate system. The plate contains spherical metal landmarks, some of them just slightly hightened to avoid the above mentioned overlapping problem while still providing a weak 3-dimensionality.

The second body has a pyramidal design. On each level of the pyramid metal crosses have been placed in a concentrical manner. Some of the crosses have been replaced by rectangles to be able to distinguish different orientations of the symmetrically shaped pyramid (see fig. 3). This calibration object is mounted rigidly in front of the image intensifier's detector. Thus the resulting images of the patient's body are densly and entirely superimposed with cross-shaped regions (see fig. 5). These regions serve as evenly distributed markers in the images. Because their positions are constant despite any motion of the detector or the source of the beam they are suitable for calibration of the interior camera parameters (distortion).

This approach is not applicable to CCD-camera-calibration due to the rather small depth of focus intrinsic to the used optical systems. Objects which are placed directly in front of a X-ray-detector will always appear sharp edged and therefore be recognizable within the image. Mounting a similar but transparent pyramidal calibration body in front of the lens of a CCD-camera will produce fuzzy and unusable projections of the crosses.

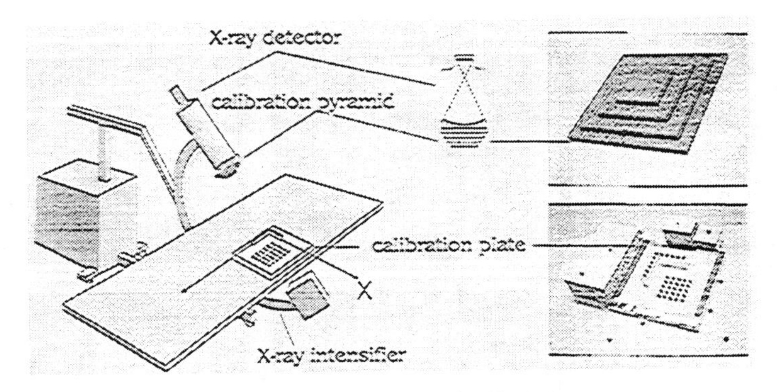

Figure 4. The Image Intensifier Environment

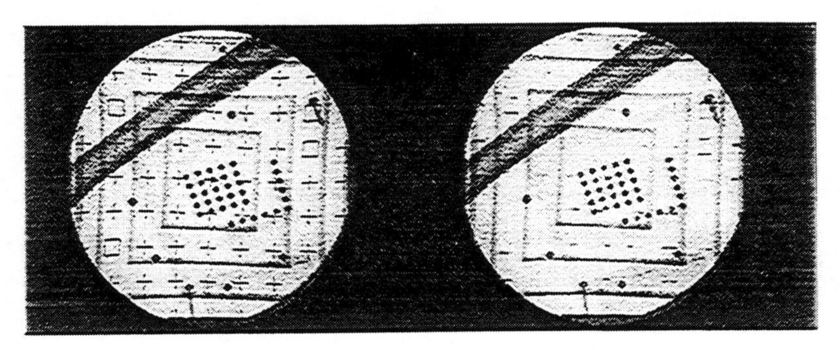

Figure 5. (a) Before, (b) After local smoothing

Calibration Techniques

The crucial condition for calibration is the correct assignment of image features and model-landmarks. According to the shape of the different markers on the two calibration bodies three different types of landmarks have to be extracted out of each image. The first step - well-known preprocessing (smoothing and dynamic thresholding) - yields a set of connected pixel-regions. The robust distinction between image-regions corresponding to the different types is based on characteristic shape-criteria:

- cross-shaped landmarks are detected by testing for perpendicular intersection of two lines, defined by means of connecting each region's horizontal and vertical extrema (minimum and maximum boundery points),

- rectangle-shaped landmarks by characteristic values of the region's run-length-encoding (i.e. average number of packed segments per line is expected next to a value of 2, due to the characteristic cavity within the rectangular shape) [2] and last, not least

- spherical landmarks by common geometrical features like circularity, compactness, convexity, area, lenght of contour etc. [2].

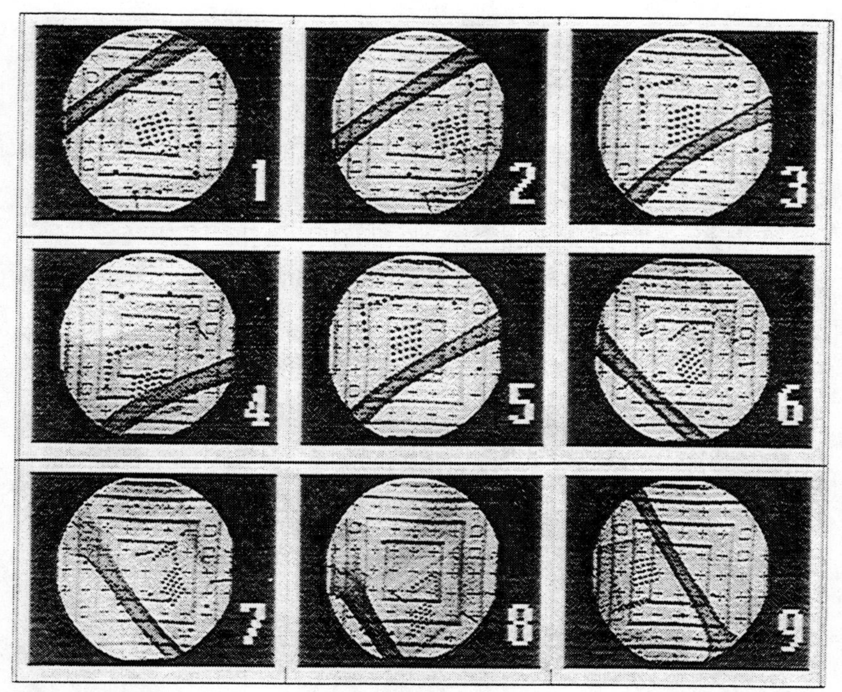

Figure 6. Set of images (showing both calibration bodies as well as part of femur), we used for error-analysis (see table I)

A robust algorithm for feature-assignment has been implemented. It copes with wrong detected as well as with missing image features. To ensure a unique solution for correct assignment it is essential to detect a sufficient number of features. Overlapping effects cannot be avoided and thus reducing the number of well-detectable features. In order to maximize this number the whole feature-detection process is split into two phases. First, a search for non-overlapping crosses and spheres is performed. The found features together with a-priori knowledge about the pyramidal calibration body are used to mask all crosses. Next, previously overlapped sphere-regions are reconstructed using morphological opening/closing-operations.

The calibration algorithm is based on error minimization between the detected image features and the estimated projection of the positioned model. We use a cubic camera model to represent a wide spectrum of distortion [11]:

$$\begin{pmatrix} u \\ v \end{pmatrix} = \begin{pmatrix} a_0 & a_1 & \cdots & a_9 \\ b_0 & b_1 & \cdots & b_9 \end{pmatrix} \begin{pmatrix} 1 \; x \; y \; xy \; x^2 \; y^2 \; yx^2 \; xy^2 \; x^3 \; y^3 \end{pmatrix}^T$$

where (u, v) represents the distorted and (x, y) the pinhole projection. Due to the missing initial values our calibration technique determines the linear camera parameters $(a_0, \ldots, a_2, b_0, \ldots, b_2)$ by applying the direct linear transformation method [3], [5] in the first step. The nonlinear parameters are set to zero. All variables are then estimated and improved by the Levenberg-Marquart (least squares) algorithm

[9]. After calibration a local smoothing-operation is applied to the X-ray image. This (see fig.5 (a)) permits the reconstruction of the image, no longer containing any markers of the pyramidal calibration body (see fig. 5 (b)). Thus we are able to minimize the perturbing effects of these markers on posterior image-processing steps e.g. bone-contour-extraction. The smoothing is based on both calibration results and model-based knowledge.

Results and Accuracy

For testing the quality and accuracy of our calibration method outside the operating theatre, the calibration plate is attached on a robust wooden board with additional spherical landmarks fixed on different levels of height around the plate (see fig. 3). Their exact position in space relative to the calibration plate has been measured beforehand with a coordinate measuring machine.

The goal of the test is to determine the positions of the additional landmarks with the X-ray system and to compare the results with the exact positions. Therefore a series of X-ray shots of the board are taken, each from a different angle (see fig. 6). They are calibrated in the described manner. Afterwards pairs of shots are randomly chosen and processed to triangulate the additional landmarks position in space.

To emphasize the improvement achieved by introducing the pyramidal body, the experiments results (average position error) with and without using the pyramid can be compared in table I. Accuracy of the whole method is better than 1 mm (average deviation between the beforehand determined and the triangulated positions of the landmarks).

pair of images	average error		pair of images	average error	
	plate only	**pyramid and plate**		plate only	**pyramid and plate**
5, 6	2.9619 mm	**0.5340 mm**	1, 6	5.3859 mm	**0.5489 mm**
8, 1	3.6071 mm	**0.9051 mm**	2, 5	2.1635 mm	**0.6452 mm**
7, 5	2.4420 mm	**0.6294 mm**	4, 1	2.0858 mm	**0.7093 mm**
3, 1	2.4649 mm	**0.5408 mm**	9, 5	1.9759 mm	**0.8001 mm**

TABLE I. Absolute Errors with and without using Pyramidal Calibration Body

Use for the Correlation in the Operation Theatre

A similar approach using the described calibration method is proposed in the operation theatre to locate the positions of the patient's femur and tibia. Initially after starting the surgery the patient's leg has to be opened for insertion of the active markers. They are applied to the tibia and the femur and tracked by the real time vision system. The calibration plate is rigidly mounted to the operating table close to the patients knee. Active markers have been fixed to the plate to be able to locate its absolute position in the vision systems coordinate system. Their positions on the plate have been measured with the coordinate measuring machine beforehand.

Pairs of calibrated X-ray shots of the bones are taken in the above described manner using both calibration bodies. Each time a shot is taken, the active markers positions are determined by the vision system in its coordinate system and memorized. The contours of the femur and the tibia are segmented from the X-rays in a known manner, based on an image processing and understanding system, called HORUS [2]. Then the positions of the bones are obtained by correlation of the CT-scan and the calibrated X-rays [1] in the coordinate system of the X-ray system. After this procedure each of the different coordinate systems can be transformed into each other. Thus the bones positions can be monitored just by tracking the active markers attached to them and applying simple coordinate transformations. From this time on the X-ray system is no longer needed to locate the positions of the femur and the tibia. They are monitored in real time just using the vision system and the active markers.

REFERENCES

1. J.L. Chen and other. Recovering and tracking pose of curved 3D objects from 2-D images. *Proceedings, CVPR '93, IEEE Computer Society Conference on Computer Vision and Pattern Recognition*, 1993.
2. W. Eckstein, G. Lohmann, and other. Benutzerfreundliche Bildanalyse mit HORUS: Architektur und Konzepte. *DAGM Symposium*, 15, 1993.
3. O.D. Faugeras and G. Toscani. Camera Calibration for 3D computer vision. *Proc. Int. Workshop on Industrial Application of Machine Vision and Machine Intelligence*, 1987.
4. S. Lanser, Ch. Zierl, and R. Beutlhauser. Multibildkalibrierung einer CCD-Kamera. *DAGM Symposium, Bielefeld*, 17, 1995.
5. T. Melen. *Geometric Modelling and Calibration of Video Cameras for Underwater Navigation.* PhD thesis, Institutt for teknisk kybernetikk Norges tekniske høgskole, 1994.
6. B.D. Mittelstadt and other. The Evolution of a Surgical Robot from Prototype to Human Clinical Use. In *First International Symposium on Medical Robotics and Computer Asssited Surgery (MRCAS'94)*, volume 1, pages 36–41, September 1994.
7. J.L. Moctezuma and other. A computer and robotic aided surgey system for accomplishing osteotomies. In *First International Symposium on Medical Robotics and Computer Asssited Surgery (MRCAS'94)*, volume 1, pages 31–35, September 1994.
8. G. Toscani O.D. Faugeras. The calibration problem for stereo. *IEEE CVPR, Miami Beach, FL, USA*, 1986.
9. Press and other. *Numerical Recipes in C.* Cambridge University Press, 1994.
10. H.-J. Schulz, T. Lutze, and H.P. Tümmler. Requirements of a planning and navigation device used in neuro-, ent- and maxillocraniofacial surgery. In *First International Symposium on Medical Robotics and Computer Asssited Surgery (MRCAS'94)*, volume 2, pages 272–276, September 1994.
11. P. Wunsch and K. Arbter. Kalibrierung eines nichtlinearen Hand-Auge Stereokamerasystems. *to be published*, 1994. DLR - Institut für Robotik und Systemdynamik.

MOTION PLANNING FOR MOBILE MANIPULATORS USING THE FSP (FULL SPACE PARAMETERIZATION) APPROACH

François G. Pin, Kristi A. Morgansen, Faithlyn A. Tulloch, and Charles J. Hacker
Robotics and Process Systems Division
Oak Ridge National Laboratory
P.O. Box 2008
Oak Ridge, TN 37831-6305

ABSTRACT

The efficient utilization of the motion capabilities of mobile manipulators, i.e., manipulators mounted on mobile platforms, requires the resolution of the kinematically redundant system formed by the addition of the degrees of freedom (d.o.f.) of the platform to those of the manipulator. At the velocity level, the linearized Jacobian equation for such a redundant system represents an underspecified system of algebraic equations. In addition, constraints such as obstacle avoidance or joint limits may appear at any time during the trajectory of the system. A method, which we named the FSP (Full Space Parameterization), has recently been developed to resolve such underspecified systems with constraints that may vary in time and in number during a single trajectory. In this paper, we review the principles of the FSP and give analytical solutions for the constrained motion case, with a general optimization criterion for resolving the redundancy. We then focus on a solution to the problem introduced by the combined use of prismatic and revolute joints (a common occurrence in practical mobile manipulators) which makes the dimensions of the joint displacement vector components non-homogeneous. Successful applications to the motion planning of several large-payload mobile manipulators with up to 11 d.o.f. are discussed. Sample trajectories involving combined motions of the platform and manipulator under the time-varying occurrence of obstacle and joint limit constraints are presented to illustrate the use and efficiency of the FSP approach in complex motion planning problems.

INTRODUCTION

For any robotic manipulator system, the forward kinematics are usually described by the equation

$$\bar{x} = F(\bar{q}) \tag{1}$$

where $\bar{x}$ is the location of a point (generally the end-effector) of the manipulator in the world coordinate system, $\bar{q}$ is the vector of joint angles measured in local coordinates, and $F()$ is the transformation function. In general, desired motions are expressed as trajectories in end-effector space. For loop-rate control, these trajectories are broken up into finite steps of length $\Delta\bar{x}$. The relationship between end-effector steps $\Delta\bar{x}$ and joint space steps $\Delta\bar{q}$ is found by differentiating and linearizing Eq. (1):

$$\frac{\Delta \bar{x}}{\Delta t} = J \frac{\Delta \bar{q}}{\Delta t} \tag{2}$$

where J is the linearized system Jacobian over the current time step Δt. The equation with which we will be working is then

$$\Delta \bar{x} = J \Delta \bar{q} \ . \tag{3}$$

In order to carry out trajectories, the robot must be given motions in terms of joint space variables. This task requires some type of inverse transformation to be made to convert from the known quantity $\Delta \bar{x}$ to the desired quantity $\Delta \bar{q}$. When the dimension of $\Delta \bar{q}$ (the number of joints in the system) is greater than that of $\Delta \bar{x}$, the system is kinematically redundant and Eq. (3) is underspecified. Several methods have been proposed for resolving underspecified systems of equations and [1] provides an excellent review of these methods for application to redundant manipulators. These methods, however, are quite varied and suffer from significant shortcomings (e.g., see discussions in [2] and [3]) when applied to real-time sensor-based systems. A novel approach, which we have named the Full Space Parameterization (FSP) method, has been recently developed [2], [3], [4], [5], [6] to remedy some of these shortcomings in cases where constraints and task criteria vary rapidly and unpredictably with time during a single trajectory.

OVERVIEW OF FULL SPACE PARAMETERIZATION

The FSP method has been specifically designed to optimally solve the inverse kinematics problem for redundant systems in the presence of applied constraints and behavioral criterion that may vary at loop rate [2], [3], [4], [5], [6]. For a redundant system, J will have fewer rows (n) than columns (m), and the number of vectors $\Delta \bar{q}$ which satisfy Eq. (3) will typically be infinite. This infinite set of solution vectors forms a subspace of the space spanned by $m-n+1$ linearly independent solution vectors $\bar{g}_k$, each of which satisfies the equation:

$$\Delta \bar{x} = J \bar{g}_k \ . \tag{4}$$

The vectors $\bar{g}_k$ can easily be found by inverting square submatrices J_k of the Jacobian J and inserting a 0 into the components corresponding to the columns of J that were removed to form J_k. The proof of existence and algorithms for the determination of the $m-n+1$ linearly independent solution vectors $\bar{g}_k$ can be found in [2], [4], and [5].

Once the $m-n+1$ solution vectors $\bar{g}_k$ have been found, any solution $\Delta \bar{q}$ can be written [2] as:

$$\Delta \bar{q} = \sum_{i=1}^{m-n+1} t_i \bar{g}_i \ , \quad \sum_{i=1}^{m-n+1} t_i = 1 \ , \tag{5}$$

where the parameters t_i, $i = 1, m-n+1$, can be found by minimizing the Lagrangian

$$L(t_i, \mu, \nu_j) = Q(t_i) + \mu \left(\sum_{i=1}^{m-n+1} t_i - 1 \right) + \sum_{j=1}^{r} \nu_j C^j(t_i) \tag{6}$$

in which Q is the optimization criterion to be satisfied by $\Delta \bar{q}$ with a set of r constraints C^j. The optimality conditions to be solved for t_i, $i = 1, m-n+1$, are:

$$\frac{\partial L}{\partial t_i}=0,\ i=[1,m-n+1];\ \frac{\partial L}{\partial \mu}=0;\ \frac{\partial L}{\partial v_j}=0,\ j=[1,r]\ . \tag{7}$$

As an example, assume that the criterion Q can be expressed as

$$Q=\left\|\Delta\bar{Z}(\bar{q},\Delta\bar{q})-\bar{Z}r\right\|^2 \tag{8}$$

with

$$\Delta\bar{Z}=B(\bar{q})\Delta\bar{q} \tag{9}$$

where $B(\bar{q})$ is a matrix; and the constraints C^j can be written [3] as:

$$\bar{\beta}^{j^T}\bar{t}-1=0;\ j=[1,r] \tag{10}$$

then the optimality conditions in Eq. (7) become:

$$\begin{cases} G\bar{t}+\bar{H}+\mu\bar{e}+\sum_{i=1}^{r}v_{i\bar{\beta}^{i=0}} \\ \bar{e}^T\bar{t}=1 \\ \bar{\beta}^{j^T}\bar{t}=1,j=[1,r] \end{cases} \tag{11}$$

with

$$\bar{H},H_k=\Delta\bar{Z}_r^T B\bar{g}_k;\ k=[1,m-n+1] \tag{12}$$

$$G,G_{ij}=\bar{g}_i^T B^T B\bar{g}_j;\ i,j=[1,m-n+1] \tag{13}$$

$$\bar{e},\ e_i=1;\ i=[1,m-n+1]\ . \tag{14}$$

Solving these equations gives (see [2] and [3])

$$\bar{v}=A^{-1}(a\bar{d}-\bar{b}(1+\bar{e}^T G^{-1}\bar{H})) \tag{15}$$

$$\mu=-(1+\bar{v}^T\bar{b}+\bar{e}^T G^{-1}\bar{H})/a \tag{16}$$

$$\bar{t}=-G^{-1}(\mu\bar{e}+\mathcal{B}\bar{v}+\bar{H}) \tag{17}$$

for non-nullspace motion. For nullspace motion Eqs. (15) and (16) are replaced by

$$\bar{v}=A^{-1}(a\bar{d}-\bar{b}\bar{e}^T G^{-1}\bar{H}) \tag{18}$$

$$\mu=-(\bar{e}^T G^{-1}\bar{H}+\bar{v}^T\bar{b})/a \tag{19}$$

where

$$a=\bar{e}^T G^{-1}\bar{e} \tag{20}$$

$$\bar{b},b_i=\bar{e}^T G^{-1}\bar{\beta}^i=\bar{\beta}^{i^T}G^{-1}\bar{e},\ i=[1,r] \tag{21}$$

$$\bar{d},d_i=1+\bar{\beta}^{i^T}G^{-1}\bar{H},\ i=[1,r] \tag{22}$$

$$A,\ A_{ij} = b_i b_j - a\overline{\beta}^{i^T} G^{-1} \overline{\beta}^j,\ i = [1, r],\ j = [1, r] \tag{23}$$

and $\mathcal{B}$ is a matrix whose columns are $\overline{\beta}^i$.

The approach for calculating the coefficient vectors $\overline{\beta}^i$ expressing the constraints has been described in detail in [3]. In particular, applications to redundant manipulators for the cases of joint limit and obstacle avoidance, and bounded joint accelerations were presented in [3] and [6], respectively. Figures 1 and 2 show two of the experimental testbeds on which implementations and tests of the FSP approach have been performed, initially using only the manipulator, then considering the complete mobile platform and manipulator system as discussed below. Figure 1 shows the ATLAS vehicle which includes a planar, 5 d.o.f. manipulator, mounted on an all-steerable wheeled platform. The two degrees of redundancy (d.o.r.) provide the system with added dexterity in the handling of various palletized cargo. Figure 2 shows the Next Generation Munition Handler (see paper in this conference) which utilizes advanced robotic technologies, including a fully omnidirectional platform (see [7]) and a 8 d.o.f. manipulator system to allow rapid aircraft reload turnaround. Including their platform, these systems involve 8 and 11 d.o.f., respectively. For a variety of tasks, some of these d.o.f. (i.e., joints) can be enabled or disabled, leading to systems with varying configurations and joint space dimensions (from 3 to 8 and 11, respectively). Also note that both systems include at least one prismatic joint.

Figure 1. The ATLAS mobile manipulator for palletized cargo handling.

Figure 2. The U.S. Air Force Next Generation Munition Handler for rapid aircraft turnaround.

APPLICATION TO MOBILE MANIPULATOR

For application to redundancy resolution for mobile manipulators under time-varying constraints, task requirements, and active configurations, unique advantages of the FSP method can be utilized. First, the FSP code [4], [5] has been developed to allow treatment of any dimension of the joint space or of the (Cartesian) control space; i.e., the Jacobian matrix can be an $n \times m$ matrix with any value for n (typically 3 or 6 for robot control applications) and m (the number of joints), and these values can change at every loop rate if necessary. Second, the FSP allows implementation of the most common constraints encountered in robotic motion planning (e.g., joint limits and obstacle avoidance as described in [3], non-holonomic constraints as described in [8]). The number and expression of these constraints (e.g., see Eq. (10) for one of the most common forms [3]) can vary at loop rate, i.e., can be based on sensor information in dynamic or *a priori* unknown environments. These aspects of the FSP have been treated in companion papers [2], [3], [4], [5], [8] and are only recalled here for completeness.

A particular aspect that derives from the formulation of Eqs. (8) and (9) is the capability to handle mixtures of revolute and prismatic joints in the system. This is particularly important since, as mentioned above, most practical mobile manipulators typically include prismatic joints in their "boom" or arm, but also because the platform motion can be seen as analogous to a combination of prismatic motions (with, of course, non-holonomic constraints between them, as appropriate [8]). The problem in Eq. (8) comes from the different dimensions (e.g., meters vs. radians) of the components of $\Delta\overline{q}$ for prismatic and revolute joints, making the *optimization* of the norm highly dependent on the choice of *relative* units in joint space. The solution involves using Eq. (9) to make the dimensions of the norm components uniform, or essentially dimensionless. The matrix B can therefore be expressed as $B = B_c B_d$, where B_c relates to the particular task criterion considered (e.g., if B is the identity matrix and the system only includes revolute joints, Eq. (8) with $\overline{Z}_r = \overline{0}$ provides the least norm of the joint displacements for comparison with the pseudo-inverse, as described in [2]) and B_d is a diagonal matrix used to "uniformalize" the dimensions of the norm components. This essentially corresponds to a *relative* weighing of the joint motions, and it is important to note that this weighing is not arbitrary but actually expresses a desired *relative behavior* between the prismatic and revolute joints. In the system shown in Figure 2, for example, the components of the diagonal matrix B_d were selected so that displacement of the prismatic joint (joint 3) over its entire range (.6 m) was "equivalent" to a motion of joint 1 over its entire range (1.52 rd). Thus, using units of meters and radians in the system, $B_{11} \times 1.52 = B_{33} \times .6$, or $B_{33} = 2.53 B_{11}$. Other schemes can, of course, be implemented and *changed at loop rate* (e.g., see [9]), as desired or required by the time-varying task requirements.

The various aspects of the FSP described above were implemented and tested on several systems consisting mainly of manipulator arms (e.g., see [2], [3], [9]). Several implementations and tests were also performed on mobile manipulators and Figure 3 shows examples of trajectories that were created using FSP for the HERMIES-III mobile manipulator. The HERMIES-III system [10] is composed of a three d.o.f. (in Cartesian Space) omnidirectional platform and a seven d.o.f. manipulator. The trajectories were created by specifying the start and finish location of the end-effector. Orientation control was not utilized to create these examples, i.e., only 3-D end-effector position is controlled, leading to a 3×10 Jacobian, or 7 degrees of redundancy. Both trajectories were specified

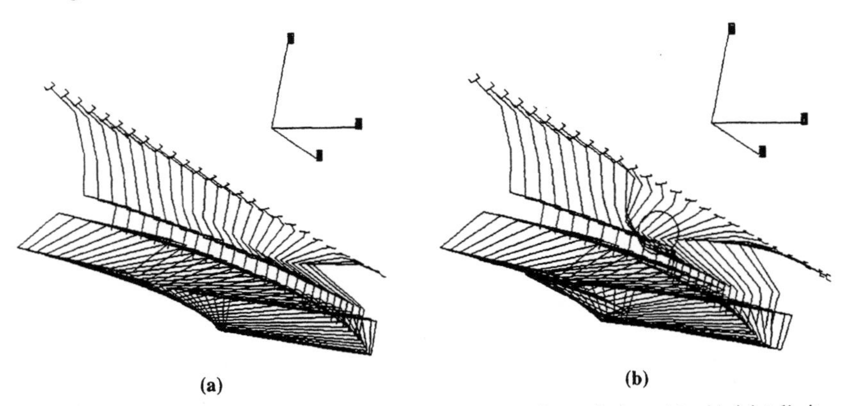

Figure 3. Examples of the FSP-produced motions for a mobile manipulator (a) with joint limits, (b) with joint limits and obstacle avoidance.

to use the least norm as the optimization criteria. As can be seen in Figure 3(a), the motions of both the platform and manipulator are simultaneous and smooth. The perspective from which the image was captured causes the dimensions of the platform to appear distorted as it rotates approximately 130 degrees in a clockwise direction from start to finish. In the second trajectory, a spherical obstacle was placed directly in the path that had been followed by the manipulator in the first example. As the manipulator moves to a location that would cause it to impact the object, the platform backs up slightly, and the links of the manipulator near the obstacle move into positions that prevent impact. The distortion effect due to the viewing perspective of the platform is present in the second image as it was in the first. Even with the sudden addition of an obstacle (and corresponding constraint) to the environment through which the robot moves, the motions of both the platform and the manipulator are smooth and simultaneous, and the end-effector reaches the desired position.

CONCLUSION

An approach to the motion planning of highly kinematically redundant mobile manipulators under time-varying constraints and task criterion has been presented. This approach is based on the use of the FSP method to resolve the constrained, underspecified, system of velocity equations. Emphasis in this paper has been placed on the treatment of particular features specific to practical mobile manipulators, in particular the handling of mixtures of prismatic and revolute joints. Sample trajectories for one of our mobile manipulator testbeds have also been presented to illustrate the use of the FSP approach in cases with time-varying obstacle and joint limit constraints.

REFERENCES

1. Siciliano, B., "Kinematic Control of Redundant Robot Manipulators: A Tutorial," *Journal of Intelligent and Robotic Systems* **3**, 201–212 (1990).
2. Pin, F. G., et al. "A New Solution Method for the Inverse Kinematic Joint Velocity Calculations of Redundant Manipulators," in *Proceedings of the IEEE Int. Conf. on Robotics and Automation*, San Diego, California, May 8–13, 1994, pp. 96–102.
3. Pin, F. G. and F. A. Tulloch, "Resolving Kinematic Redundancy with Constraints Using the FSP (Full Space Parameterization) Approach," in *Proceedings of the 1996 IEEE International Conference on Robotics and Automation*, Minneapolis, Minnesota, April 22–28, 1996.
4. Morgansen, K.A. and F. G. Pin, "Enhanced Code for the Full Space Parameterization Approach to Solving Underspecified Systems of Algebraic Equations," Oak Ridge National Laboratory Technical Report No. ORNL/TM-12816, 1995.
5. Fries, G. A., C. J. Hacker, and F. G. Pin, "FSP (Full Space Parameterization), Version 2.0" Oak Ridge National Laboratory Technical Report No. ORNL/TM-13021, 1995.
6. Morgansen, K. A. and F. G. Pin, "Impact Mitigation Using Kinematic Constraints and the Full Space Parameterization," in *Proceedings of the 1996 IEEE International Conference on Robotics and Automation*, Minneapolis, Minnesota, April 22–28, 1996.
7. Pin, F. G. and S. M. Killough, "A New Family of Omnidirectional and Holonomic Wheeled Platforms for Mobile Robots, *IEEE Transactions on Robotics and Automation* **10**(4), 480–489 (1994).
8. Pin, F. G., C. J. Hacker, and K. B. Gower, "Motion Planning of Mobile Manipulators Including Non-Holonomic Constraints Using the FSP Method," submitted to special issue of *Journal of Robotic Systems* (1996).
9. Bangs, A. L., F. G. Pin, and S. M. Killough, "An Implementation of Redundancy Resolution and Stability Monitoring for a Material Handling Vehicle" in *Proceedings of the IEEE/IES Intelligent Vehicles '92*, Detroit, Michigan, July 1–2, 1992.
10. Pin, F. G., M. Beckerman, P. F. Spelt, J. T. Robinson, and C. R. Weisbin, "Autonomous Mobile Robot Research Using the HERMIES-III Robot" in *Proceedings of the 1989 IROS International Workshop on Intelligent Robots and Systems*, "The Autonomous Mobile Robot and Its Application," Tsukuba, Japan, September 1989, 251–256.

ROGER : A MOBILE ROBOT FOR RESEARCH EXPERIMENTATIONS

Joaquim Salvi / Lluis Pacheco / Rafael Garcia-Campos
Computer Vision and Robotics Group. Department of Industrial Engineering, University of Girona. Avda Lluis Santaló s/n - 17003 GIRONA (SPAIN)
e-mail: qsalvi@ei.udg.es

ABSTRACT

This paper present an autonomous mobile robot able to operate in in-door structured environments and on some unevenness irregular out-door environments. The mobile is able to locate itself with a certain accuracy, and can do some basic operations of object manipulation with a 3-DOF arm. This sort of qualities allow the vehicle to be used in industrial and social applications.

KEYWORDS: Mobile robots, Colour segmentation, Real-Time, Fusion sensors.

INTRODUCTION

ROGER, a Generic Operational Robot for Research Experimentations, is an autonomous mobile robot made in order to be used like a working platform for the test of navigation systems and real-time images processing. The mobile robot has been designed for the navigation in structured in-door environments considering obstacles along its trajectory, and in out-door environments where the ground can have some important unevenness. Besides, its capacity to locate itself with a certain accuracy, and the incorporation of an arm with 3-DOF allow the mobile robot to do some basic operations of object manipulation. These sort of qualities allow the vehicle to be used in industrial and social environments.

A first prototype shown at Figure 1 has been made and tested in several applications like segmentation and tracking of objects, dinamical path planning in structured environment avoiding potential obstacles, and doing some object manipulation and transporting them. The mobile robot has taken part in the IAV 1995 Congress: The Competition of Intelligent Autonomous Vehicles, Otaniemi, Finland. In the following sections, we present its mechanical shape, the description of the sensorial system and the support hardware of the control system.

MECHANICAL SHAPE

The mechanical design is very significant, as it influences in an important way in the movement of the vehicle and its ability in path planning with a small error. Firstly the mobile robot shoud be able to work in grounds with some important irregularities and slopes. Trying to obtain these two purposes we chose for a mechanical traction like a military tank driven by two lateral independent rubber belts shown at Figure 2, that permit to turn round without any movement.

Figure 1. General view of ROGER.

Figure 2. View of the mechanical traction.

The mobile dimentions are: 1000 mm. long, 920 mm. wide and 620 mm. height. The vehicle has been built using a structure made of steel pipes, with three levels of platforms. the mobile weights about 60 Kgs. The supply batteries have been placed on the bottom level, with this we have been able to lower the gravity centre, so we get a better stability. The whole part of the engine control and the sensorial system centralized by a i486 computer have been placed on the middle level. We have placed the RGB camera and the 3-DOF arm on the top level.

Refering to the motricity specifications, we have chosen engines, that can give a initial acceleration of 0'5 m/s^2, and a maximum speed of 1 m/s, with a power traction that allow the mobile robot to climb slopes with an unvenness of 10º. The mobile can also save irregularities of the ground lower than 10 cm. In order to obtain this aim, we have chosen two engines of 24 V. cc. with reductor that can give up to 150 rpm. with a power of 200 W. Each engine has been equiped with an optic incremental encoder. The driven wheels impulse the mobile robot through two cogged rubber belts. Each rubber belt has a trapezoidal shape: the base is 600 mm. long the top side is 800 mm. long, with a global height of 500 mm. A wheel, with a diameter of 150 mm. has been placed in each apex of the shape. The driven wheels have been placed at the back upper apex.

The mobile robot has got a RGB camera located at he front part of the upper level, placed on a small platform that allows the camera a pitch movement of -40º up to 20º, so we can get the degree of freedom that the mobile robot traction system can't provide. The camera movement is achieved by means of a mechanical piston driven by a 12 V. cc. motor with a 50 W. of power. The position control is obtained by a resistor encoder.

The 3-DOF arm of the robot shown at Figure 3 and Figure 4 is made by three links: base, forearm and arm of 300 mm., 500 mm., and 720 mm. respectively, and three joints are driven by a 12 V. cc. motor, with a 50 W. of power, and controlled by resistor encoders. The open and close movement of the hand is driven by a servo-motor of 6 V. Although the accessibility of the arm is quite large, the base joint can revolve on 360°, the actual working area is rather limited by the scope of the camera lens.

Figure 3. The robot arm.

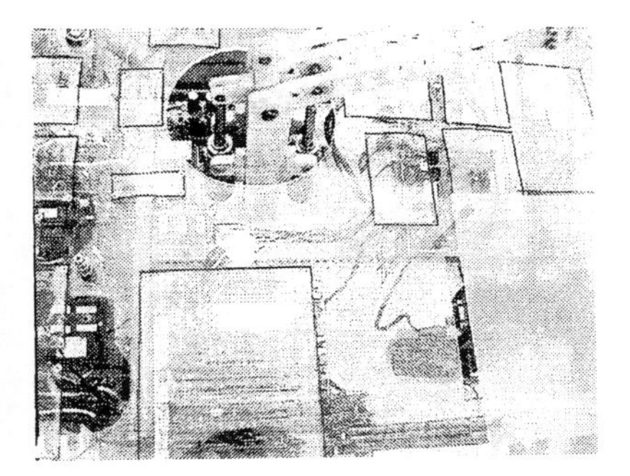

Figure 4. A view of the forearm joint.

DESCRIPTION OF THE SENSORIAL SYSTEM

As it is said by [3], there is fundamental common ground between the maximum performance of a mobile robot and the performance of the sensing hardware. For example, the maximum speed of a mobile robot is limited by the maximum range of the range sensor. So, when we place sensors on the mobile robot, we always have to answer the following questions: Is the path navigable?, Are there any obstacles?, What object is in front of me? The answers determine the selection of a sensor, the selection of an algorithm, the sensor aiming and data capture, and the interpretation of the results.

The ROGER vision system is composed of a RGB autofocus, autoiris, camera that provides a video signal. This signal is processed by a real-time hardware. The hardware is able to change from RGB space to HLS and obtain a segmented image using any margin of HLS components [2]. As the approximated colour features of the interesting objects, we want to recognize, are known, we can segment the input image, given an image that only contains the interesting objects, removing the rest of the image. Then, this video signal is processed by a PIP 1024B MATROX image processing hardware. We use digital filters to obtain a clear image and then some algorithms are applied to extract the 2D gravity centre and the apparent area of the segmented objects. Both of these magnitudes will be the main object features used in tracking.

When we want to manipulate any object we can usually settle on that those are motionless objects. To take an object or placed it on a known position in the working area of the robot arm, we have to know the *z* coordinate: the distance between the mobile and the object, or the distance of the spot where the object will be placed. As this spot can be also segmented, it has its own colour features, so we can assume both problems in one, trigonometry is applied to get an approximated distance as we are able to know the gravity centres of the same object obtained from two images with a small different pitch camera angle [6] [10].

In order to achieve a better robustness about the colour object segmentation, against potential light variations, another system in parallel work has been incorporated at the mobile. The light quantity present in the environment is given by a luxometer, so we can readjust the HLS components of the object segmentation dynamically.

For the structured indoor navigation, a infrared laser beam has been placed at the mobile. It generates a flat line that is useful to localize obstacles or apertures [5] [12].

The computer vision system and the navigation control algorithms sometimes may not respond as we wish, so one approach to building a robust system is to have a lower level that provides safety, even if the higher level systems fail. The vehicle will probably sense trouble and avoid it before it damages itself. This is the reason why ultrasonic sensors have been placed at the front and the sides parts of the mobile robot [4] [8] [9] [11] that can work in two different ways. The first way to work is based by interrumping the main computer periodically, giving information about the distance of the environment. On the other hand the sensorial system can work as an alarm system, so it interrupts the main computer only when the distance is shorter than a threshold. The ultrasonic sensors used have a range in depth up to 10 m. with 1% precision, at a frequency of 40 Khz. and a measuring cadence of 100 mseg.

SUPPORT HARDWARE OF THE CONTROL SYSTEM

The basis computing hardware shown at Figure 5 is formed by a i486 computer, with a PIP 1024B used for the object tracking, and another one to get the laser beam. We also have two 80552 micro-controllers to control the wheel motors and the robot arm [1].

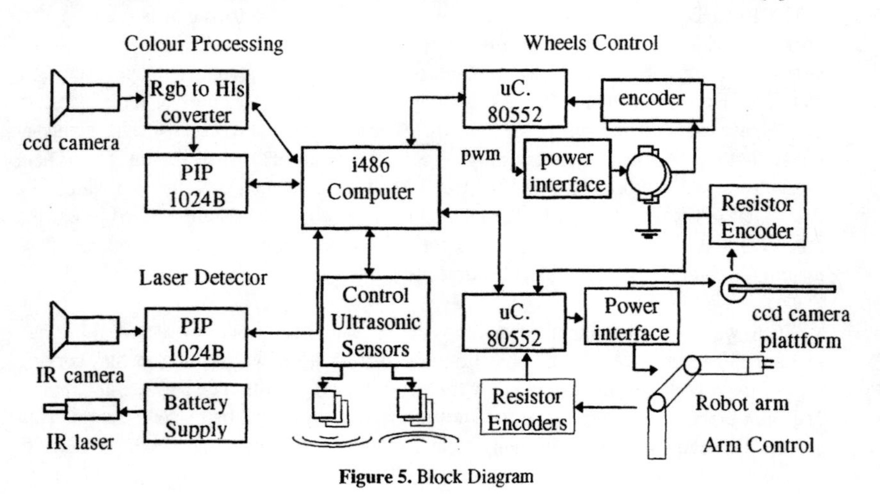

Figure 5. Block Diagram

The wheel motors system control

The driven wheels have got an incremental encoder with a reset position. Each encoder has got two photoemiters and two photodetectos separated about 90° ones from the others. With this method built in a PLA (Programmable Logic Array) allows one to detect the wheel direction without losing steps. The signals obtained by the two shaft encoders are being inputs of the 80552 micro-controllers. Then this micro-controller generates both PWM signals to drive the wheels depending on the commands sent by the PC using PID control algorithms.

The control system for arm positioning

The obsolute position of each joint of the arm is given by a resistor encoder. The readings of these encoders are converted to digital signals through the A/D converters of the 80552 micro-controller used. This micro-controller uses the order from the PC and the kinematical invert model to obtain the angles on each joint has to revolve. We also use PID controls to make sure that the hand receives the co-ordinates sent by the PC.

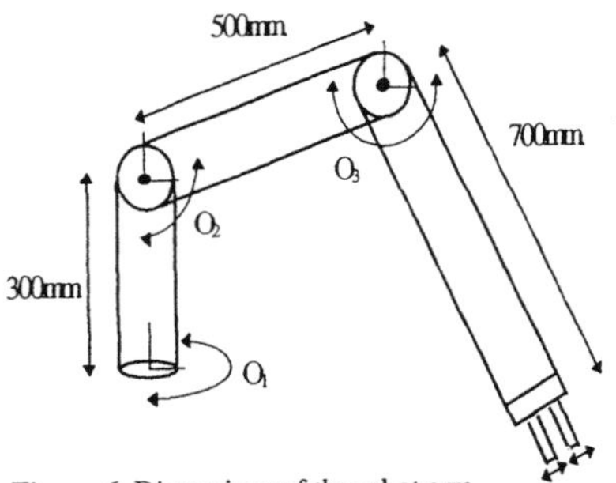

Figure 6. Dimentions of the robot arm..

TABLE I. Available angles for each joint.

Angle	Maximum	Minimum
O_1	360°	0°
O_2	45°	-20°
O_3	20°	-120°

EXPERIMENTATION AND FURTHER RESEARCH

The mobile robot has been tested in out-door environments which have some ground unevenness. The tasks that have been tested are the following: to locate an object in a scene, to follow it, to keep a safe distance, and pick it up if it stops moving. If the mobile can't find the objects of interest in the image, then the mobile will revolve on itself until the objects are located. It has also been tested in an in-door environment made up of walls, corridors and doors. The mobile is able to identify, more or less, the shape of the environment and therefore readjusting its path dynamically, without any collisions with the walls nor any potential obstacle. By means of a laser beam the mobile can identify the doors width, and only move through those that have a width bigger than a threshold.

We are designing a specific hardware that computes the gravity centre of the segmented objects in real-time. Thinking of the possibility to increase the autonomy of the mobile robot, we foresee the modification of its structure for a suitable implementation in electrogen equipment. Getting a better robustness in colour image processing, we are studying the possibility to readjust the colour features of the known objects to minimize the CCD saturation effects [7]. We are also thinking about the real navigation speed measured by using the Doppler effect of the mechanical waves, as well as the incorporation of DSP's to improve the control and positioning of the mobile.

CONCLUSIONS

In this paper we have presented the development of an autonomous prototype, with the purpose of making a mobile platform basically for the experimentation of navigation control and computer vision algorithms. Getting the mobile robot ready has not been an easy task, the integration of different systems always present co-ordination problems, especially in the case shown as it has a large number of tasks to carry out in parallel.

On the other hand, from the technological viewpoint, there are endless possibilities, at first hand they appear irrelevant, but we are challenged to carry out further necessary improvement and modification. So, for instance, an aspect so obvious as the insulation of the electronic noise among different devices may become a difficult problem. However the prototype is ready to use as a platform, especially for testing the control systems and robotics planning, as well as testing the tracking and navigation algorithms.

REFERENCES

1. Amat, J., Casals, A., "Processors for Road Images Segmentation from Textures" Workshop Computer Vision: Specialized Processors for Real-Time Image Analysis. Barcelona. Sept. 1991.
2. Benito, A., Batlle, J., Pacheco, Ll., "Preprocessador de Color en Tiempo Real" Revista Española de Electrónica 1995.
3. Thorpe, C.E., Herbert, M., "Mobile Robotics: Perspectives and Realities" ICAR 1995.
4. Charbonnier, L., Fournier, A., "Heading Guidance and Obstacles Localization for an Indoor Mobile Robot" ICAR 1995.
5. Aldon, M.J., Le Bris, L. "Mobile Robot Localization using a Light-Stripe Sensor" IV 1994.
6. Tan, TN., Baker, KD., Sullivan, GD., "3D Structure and Motion Estimation from 2D Image Sequence" IVC 1993.
7. Regincos, J., Batlle, J., "An approach at colour segmentation, minimizing the CCD saturation problems" Technical Report ref. 95/015-EI
8. Bozma, O., Kuc, R., "Single sensor sonar mapbuilding based on physical principles of reflexion" IEEE cat n° 91THO375-6
9. Wilkes, D., Dudek, G., Jenkin, M., "Multitransducer sonar interpretation" IEEE 1993 1050-4729/93
10. Chen, JL., Stockman, GC., Rao, K., "Recovering and tracking pose of curves 3D objects form 2D images" CVPR 1993.
11. Crowley, JL., "World modelling and position estimation for a mobile robot using ultrasonic ranging" IEEE 1989 International Conference on Robotics and Automation.
12. Motyl, G., Chaumette, F., Gallice, J., "Coupling a camera and laser stripe in sensor based control" Second International Symposium on Measurement and Control in Robotics, AIST Tsukuba Research Center, Japan 1992.
13. Fu, K.S., Gonzalez, R., Lee, C.S.G., "Robotic: Control, Sensing, Vison and Intellligence" Mc. Graw-Hill 1987.

MODEL UPDATE BY RADAR- AND VIDEO-BASED PERCEPTIONS OF ENVIRONMENTAL VARIATIONS

NORBERT O. STÖFFLER
TU München, Lehrstuhl für Prozeßrechner,
Prof. Dr.-Ing. G. Färber, 80290 München, Germany,
e-mail: stoffler@lpr.e-technik.tu-muenchen.de

THOMAS TROLL
TU München, Lehrstuhl für Hochfrequenztechnik,
Prof. Dr.-Ing. J. Detlefsen, 80290 München, Germany,
e-mail: troll@hfs.e-technik.tu-muenchen.de

ABSTRACT

The design of mobile robots which can cope with unexpected disturbances like obstacles or misplaced objects is an active field of research. Such an autonomous robot assesses the situation by comparing data from its sensors with an internal model of its environment. In this paper we present an object-oriented geometric model and an exemplaric model update: A new object is detected by a radar sensor and identified using a video sensor.

KEYWORDS: autonomous mobile robot, environmental modelling, microwave radar sensor

INTRODUCTION

This work is part of a research project[1] towards the development of autonomous mobile robots which can fulfil service and transport tasks in structured environments like office buildings and industrial plants. To be able to react to changes in its environment a robot has to survey its surroundings with one or more sensors. Based on the sensor data it updates an internal description (*environmental* or *world model*). This description includes the position of the robot itself, the positions of obstacles and the positions and states of other relevant objects. The model can also be considered as a set of hypotheses. These hypotheses can be tested and improved by comparison with the sensor data.

Raw sensor data is preprocessed to extract sensor-specific features. In case of video cameras these features are typically edges. In case of laser and microwave radar, which acquire 3-D range and velocity information, these features include intersection lines between object surfaces and the scanning sensor beam.

In our approach a common geometric model of the environment is used for all kinds of sensors. This avoids consistency problems between parallel sensor-specific

[1]This work was supported by *Deutsche Forschungsgemeinschaft* within the *Sonderforschungsbereich 331, "Informationsverarbeitung in autonomen, mobilen Handhabungssystemen"*, projects L4 and Q5.

descriptions. Sensor-specific features are calculated online corresponding to the sensor type and the hypothesis to be tested.

By comparing the two sets of features - extracted and predicted ones - the hypothesis can be modified until it is approved and finally fed back into the model.

RADAR SENSOR

Using microwave radar in the field of robotics seems to be very uncommon. Such sensors are generally associated with the detection of objects like airplanes or ships over far distances. Since the propagation of microwaves is nearly independent of atmospheric conditions and coherent signal processing allows high sensitivity, it is possible to cover distances up to 100 m at reasonable transmitted power. This feature combined with the direct access to the object's velocity lets a microwave radar usefully enhance the sensing capabilities of an autonomous system, as other self-illuminating sensors like laser scanners and ultrasonic sensors hardly exceed a detection range of 15 m.

The project L4 is engaged in the development and evaluation of an experimental 94 GHz sensor. It is designed to measure distances as well as velocities of objects. The distances are determined via the time of flight of single pulses. A very short pulse width of 1.7 ns results in a radial resolution of 25 cm. The distance of resolved objects can be measured with an accuracy of 5 mm. The velocity is directly accessible via the Doppler effect. This effect requires a coherent signal source to guarantee a fixed relation between transmitted and reflected signal. In our case an IMPATT oscillator, phase-locked by a Gunn oscillator, guarantees high phase stability at sufficient peak power. The carrier frequency of 94 GHz yields corresponding Doppler frequencies of 625 Hz per 1 m/s, which permit accurate velocity measurement within a short observation time. For the angular resolution of 1.5° the radar beam is focused by a Fresnel type lens of 170 mm diameter. As shown in figure 1, this sharp beam is deflected by a mirror for three-dimensional imaging of the surroundings. The implementation of the sensor is described in detail in [3], the most important system parameters are shown in table 1.

TABLE I

System Parameters of the Radar Sensor

Parameter	Value
Frequency	94 GHz
Wavelength	3.2 mm
Pulse peak power	10 mW
Radiated mean power	20 μW
Detection range	0 . . . 80 m
Radial resolution	25 cm
Distance accuracy	2 cm
Angular resolution	1.5°
Angular accuracy	0.15°
Velocity range	±8 m/s
Velocity accuracy	10 mm/s
Measurement rate	10,000 Voxels/s

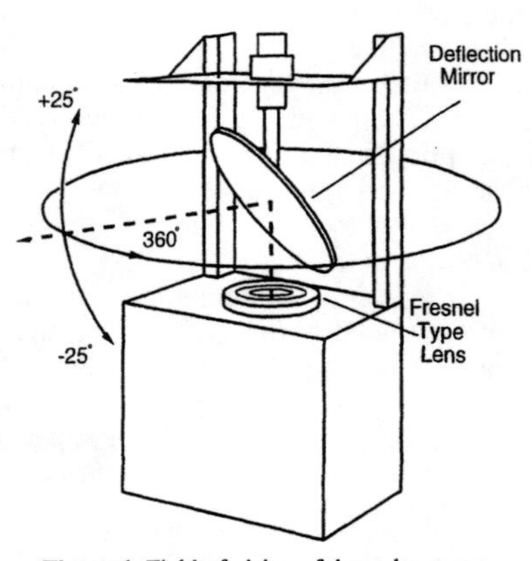

Figure 1. Field of vision of the radar sensor

Based on this system various applications like navigation and observation, which correspond to typical perception tasks of an autonomous robot, have been implemented in the last years [6].

VIDEO SENSOR

For the video sensor a standard industrial CCD camera is used. The projection of a 3D point in the scene into the corresponding 2D pixel of the video image is characterized by the model of a pinhole camera with radial distortions. The underlying camera parameters are determined by a specific calibration process (see [1] for a brief description). The image preprocessing extracting edges as image features is based on the image analysis system HORUS.

MODEL STRUCTURE

To permit sensor independent abstractions the model structure which is examined by the project Q5 is based on three-dimensional solid modelling techniques. Solid bodies are stored by a polyhedral boundary representation. If possible, sensor-specific features are calculated from this boundary representation using a corresponding sensor model. Intersections between e. g. a scanning radar beam and the boundaries can easily be calculated online. In the case of a video sensor an exact sensor model is difficult to obtain because the sensor data depends on various factors like surface properties and illumination. Furthermore an exact calculation would be very time-consuming. Therefore a simplified approach is used. The sensor model is reduced to perspective projection and edges of good visibility are represented separately by three dimensional line-segments, so-called *video-edges*. These video-edges are based on the same set of vertices as the boundary description but form only a subset of the boundary edges. This dualism reduces the prediction of video-features to a mere visibility test. It also allows a simplification of the boundary representation to exclusively convex polygons which additionally facilitates the visibility calculation.

To permit realtime access to the model appropriate index structures are necessary. In first experiments demonstrating localization in a static environment, a two-dimensional spatial-tree has proven to be an efficient index structure for otherwise unrelated model elements [5].

To allow more complex perception tasks and non-static environments, a hierarchic structure with additional symbolic information is currently examined [2] (figure 2a). Model elements are aggregated to *named objects*. Because it is neither possible nor necessary to describe the complete environment of the robot in terms of distinguishable, named objects, a pseudo-object called *background* is introduced. It encompasses all elements without special object assignment.

The description of a named object is built up recursively. An object can contain other objects which are termed *member-objects*. Object and member-object are connected by a joint which exhibits exactly one rotatory or translatoric degree of freedom, following the conventions used in manipulator kinematics. To deal with unknown states during a prediction the space potentially being occupied by a moving member-object is stored as an additional polyhedron, called *mask*. During the visibility test this polyhedron is used to literally mask out potentially hidden features.

Each branch in the object-tree carries its own boundary polygons and video-edges. Geometrically identical objects form an *object class*. The invariant parts of an object description are stored only once for each class; the objects (i. e. the instances of a class) differ in their individual positions and joint states.

MODEL UPDATE

In our approach the model update is divided into several widely independent perception tasks. They interact with the model on different levels of abstraction according to the parts of the model they regard as hypothetic (figure 2b).

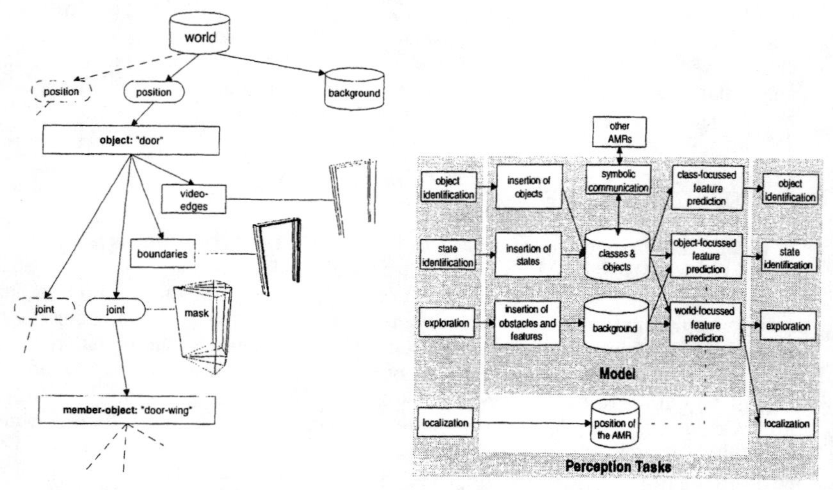

Figure 2. (a) Model structure, (b) Model update by parallel perception tasks

Each perception task is implemented by a separate client module which commands at least one sensor, extracts its own relevant features from the sensor data and simultanously requests predictions from the model server. Then it compares the two sets, interpretes the difference and updates the model accordingly. Interferences between the quasi-parallel model accesses of different tasks are avoided by private communication channels containing local copies of hypothetic model subsets.

A common parameter which is assumed to be known by several perception tasks is the position of the robot itself. Localizing with a microwave radar or a video sensor is accomplished by establishing correspondences between the sensor data and model information followed by estimating the position of the sensor using least square optimization [5]. In case of the radar sensor the two-dimensional echo map is matched with the intersection lines of the solid body boundaries and the scanning plane. In case of the video-sensor extracted edges are matched with the projected *video-edges* supplied by the model.

In consequence of the radar's far detection range a localization can be carried out with a coarse or without any a-priori hypothesis about the robot's position. In contrast to that, localization with the video-sensor already needs a good position hypothesis.

Once a valid position is found and has to be successively updated, a tracking approach which uses the spatio-temporal restrictions of the robot-position can be applied. The last estimated or dynamically extrapolated position is used as a position hypothesis for the next prediction. If the cycle-time is short enough in comparision with the speed of the robot, model information can be used to reduce measurement time by defining regions of interest. Both types of localization, the initial localization with the radar sensor and the successive localization with radar or video-sensor have been demonstrated in several environments [5, 7].

Severe mismatch of the features predicted for the current robot-position and the sensor data indicates a variation of the environment. Therefore a second task evaluates these mismatches and initiates object identification. If none of the known object classes can be matched, the object is inserted into the model as part of the background. This at least prevents collisions and further mismatches.

The applied algorithms for video-based object identification and localization are described e. g. in [1, 4]. Predictions corresponding to single object classes are requested from the model and matched with the sensor data.

In addition to the model-update done by the various perception tasks information can also be updated on a symbolic level. Independently operating robots communicate about environmental variations by exchanging object names and attributes, i. e. states and positions, via a symbolic communication medium.

EXPERIMENTAL EXAMPLE

The shown experiment was carried out at the experimental industrial plant of the *Institut für Werkzeugmaschinen und Betriebswissenschaften (iwb)* using the experimental mobile platform MAC1 (figure 3a). The platform is equipped with a radar and a video sensor and several computers connected by ethernet and TCP/IP. All modules, i. e. model and perception tasks, are implemented by RPC-servers respectively clients. The platform is moving along a clearance between the machine-tools which is expected not to be obstructed (figure 4a). The radar sensor is scanning an area of 5 m x 40° in front of the vehicle. To cover the space down to the floor the radar beam is inclined by 12.5°. In figure 3b radar echos which can be matched with predictions of the model are shown in light grey. Echos depicted in dark grey can not be matched and indicate the presence of a yet unknown object. Based on this coarse position hypothesis the video-based object identification consisting of an object recognition followed by a fine localization is initiated.

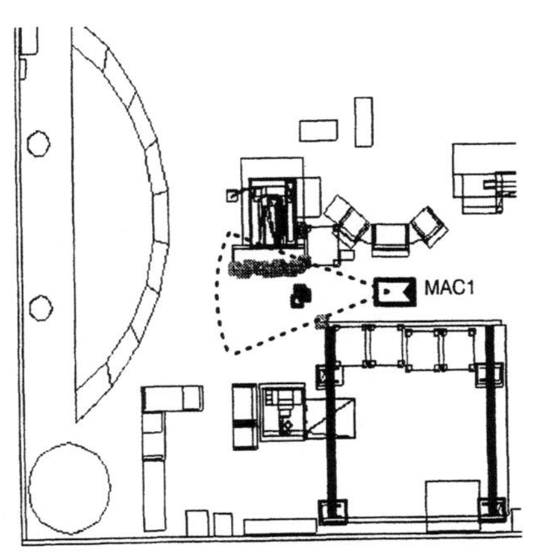

Figure 3. (a) Experimental platform MAC1, (b) correlation of radar echos and model

The video-image and a feature-prediction corresponding to the class "chair" and

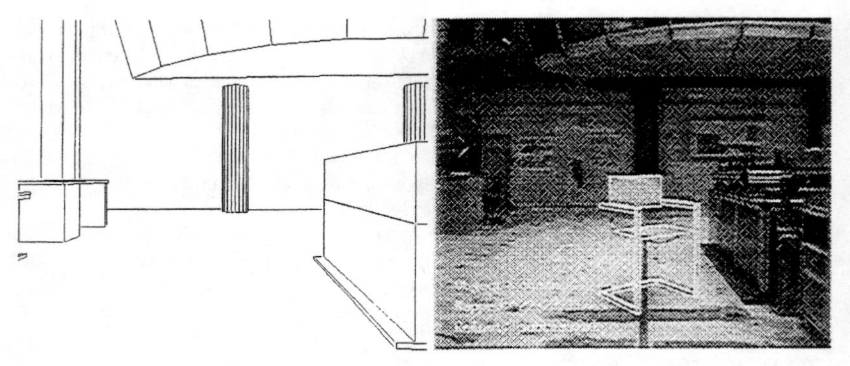

Figure 4. (a) Model before the update, (b) identified object

the final object position and are shown in figure 4b. Once the object is identified it is instantiated in the model.

CONCLUSION

The results of the experiments show the suitability of the described model structure to support concurrent perception tasks on a mobile robot. The division of model updates into several widely independent perception task facilitates combining the advantages of very different sensor types. The range data of the radar sensor is exploited in two ways to improve the performance of object identification. By defining a region of interest the computation time for the image preprocessing is typically reduced by the factor 3 and the initial position hypothesis yields additional constraints for object identification.

REFERENCES

1. S. Blessing, S. Lanser, and C. Zierl. Vision-based Handling with a Mobile Robot. In *6. ISRAM*, Montpellier, May 1996.
2. Alexa Hauck and Norbert O. Stöffler. A Hierarchic World Model Supporting Video-based Localisation, Exploration and Object Identification. In *2. Asian Conference on Computer Vision, Singapore, 5. - 8. Dec.*, 1995.
3. M. Lange and J. Detlefsen. 94 GHz Three-Dimensional Imaging Radar Sensor for Autonomous Vehicles. *IEEE Trans. on Microwave Theory Tech.*, 39(8):819–827, May 1991.
4. Stefan Lanser, Olaf Munkelt, and Christoph Zierl. Robust video-based object recognition using cad models. In U. Rembold, R. Dillmann, L.O. Hertzberger, and T. Kanade, editors, *Proc Conf. Intelligent Autonomous Systems*, pages 529–536. IOS Press, 1995.
5. Achim Ruß, Stefan Lanser, Olaf Munkelt, and Michael Rozmann. Kontinuierliche Lokalisation mit Video- und Radarsensorik unter Nutzung eines geometrisch-topologischen Umgebungsmodells. In *9. Fachgespräch "Autonome Mobile Systeme", München*, pages 313–327, 1993.
6. M. Rožmann and J. Detlefsen. Environmental exploration based on a three-dimensional imaging radar sensor. In *Proc. IROS*, pages 422–429, Rayleigh, NC, 1992.
7. M. Rožmann and J. Detlefsen. Standortbestimmung in Innenräumen mit einem hochauflösenden 94-GHz-Radarsensor. In *8. Radarsymposium der DGON*, pages 43–50, Köln, 1993. Verlag TÜV Rheinland.

Results, Problems, Future Trends of Research in and Application of Robot Calibration

Klaus Schröer, Michael Grethlein, Alexei Lisounkin
Fraunhofer-Institut für Produktionsanlagen und Konstruktionstechnik, Berlin

ABSTRACT

Taking into account the industrial needs to introduce robot calibration, the paper presents a survey on techniques, hints and advice to be regarded when implementing robot calibration procedures. It centres on kinematic and actuator modelling. Suggestions concerning future research topics in robot calibration are provided. Problems and requirements of applying calibration in industry are discussed.

KEYWORDS: robot calibration, kinematic model, actuator model, industrial needs.

1 INTRODUCTION

Calibration packages for use in industry are required, because of the demands big robot users have for calibrated robots, i.e. for robots showing a high absolute positioning accuracy, for robots being capable of executing off-line programmed robot task programs without manual re-teaching.

Many different solutions to calibrate a robot were developed in the past. Therefore, it is necessary to discuss their capabilities and advantages and to discuss practical experiences and requirements from applying robot calibration.

Robot calibration is a term applied to the procedures used in determining actual values which describe the geometric dimensions and mechanical characteristics of a robot or multi-body structure. A robot calibration system must consist of appropriate robot modelling techniques, accurate measurement equipment and reliable model parameter determination methods. For practical improvement of a robot's absolute accuracy, error compensation methods are required which use calibration results. Measurement systems and methods, although very important for practical implementation of calibration, are not a topic of this paper. A classification according the dimension of evaluated measurement information and on numerical methods for calibration can be found in the survey of Hollerbach and Wampler [Hol96], an overview on available measurement systems is included in the ISO document [Iso94], which is a result of the revision process of ISO 9283.

After discussing the scope of modelling being required for calibration (kinematic, actuator and elasticity model), section 2 centres on kinematic and actuator models.

Finally, section 3 discusses problems of industrial use of robot calibration and requirements of its industrial users.

2 MODELLING FOR CALIBRATION

Robot modelling for purposes of calibration has been investigated by several researchers. These investigations include the development of various robot modelling techniques for estimating the kinematic and sometimes mechanical features of the static robot. While not all calibration procedures developed include the identification of a robot's static-mechanical features, experience showed these features to have a significant impact on robot accuracy [Sro93a]. The kinematic geometry of the robot model used must therefore be extended to include effects of elastic deformations, at least of the rotary joints. In order to include elasticity, the reaction forces and torques induced by the robot's payload and its own body mass must be computed. Examples of how to provide the calibration system user with flexible options to model and identify joint, beam or plate elasticity are for example described in [Sro94]. Actuator parameters need be included, if the complete system from motor encoders to tool-centre point (TCP) has to be identified or checked. Due to practical reasons and user needs, a "target" model (specifying the TCP with respect to the robot flange) is suggested to be included in order to allow easy exchange between measurement target and different tools (e.g. for error compensation). A further practical reason is that integration of a target model allows for use of a calibration system with different types of measurement systems. This results in integrating the four models illustrated in Fig. 1.

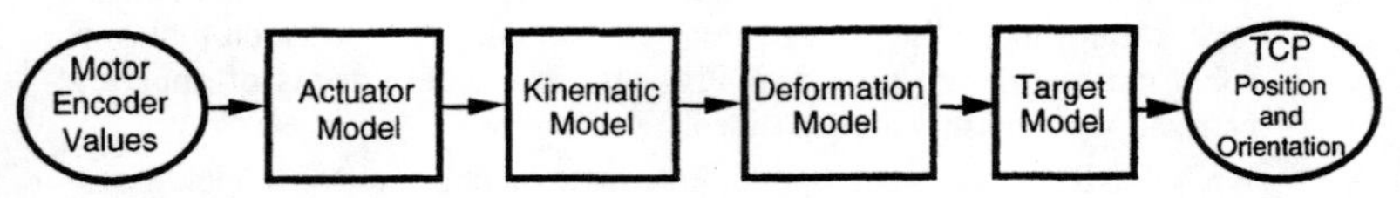

Figure 1. Robot model.

For calibration, this way of modelling requires so-called kinematic loop methods [Hol96], where the manipulator is placed into a number poses providing joint and TCP measurement information, and where all model parameters are identified simultaneously by exploitation of the differentiability of the system model as a function of its parameters. These are the calibration methods this paper refers to.

2.1 Kinematic Modelling

Important to robot calibration methods is an accurate kinematic model which has identifiable parameters. Many different alternatives to model the kinematic geometry have been suggested. The main statement of this section is, that a unique/single kinematic model being applicable for calibration of all arrangements of subsequent joints cannot exist.

Parameter identification demands that three basic requirements of the kinematic model be met:

Completeness: All spatial geometries of successive joints in an open kinematic chain must be describable.

Model-continuity: Small changes in the spatial geometry of joints must result in small changes of the describing parameters.

Minimality: The kinematic model must include only a minimal number of parameters.

Requirements of completeness and model-continuity are self evident [Eve87]. Particularly, model-continuity is an important prerequisite for non-linear optimisation procedures to work reliable. The requirement of minimality is plausible, because its violation results in redundancy of the model and rank deficiency during the numerical identification procedure.

Everett et al. [Eve88] have shown that a minimal and complete kinematic model has exactly

$$n = 4r + 2t + 6 \qquad (1)$$

model parameters, where r is the number of rotary joints and t the number of prismatic joints. The number n defined by (1) is also the maximum number of independent model parameters of a complete kinematic model.

It must be recognised that mathematically precise statements about minimality can only be made for the geometric relations defined by the kinematic model. Statements about minimality of the model cannot be made for non-geometric parts of the robot model such as the model of elastic deformations.

History of kinematic modelling of robots begins with the famous Denavit-Hartenberg convention which was used and modified by many authors [Kha86]:

$$T = R_z(\theta)\, T_z(d)\, T_x(a)\, R_x(\alpha), \quad \theta, d, a, \alpha \in \Re. \tag{2}$$

To overcome problems of model-continuity with parallel joints it was modified by Hayati [Hay85]:

$$T = R_z(\theta)\, T_x(a)\, R_x(\alpha)\, R_y(\beta)., \quad \theta, a, \alpha, \beta \in \Re. \tag{3}$$

Veitschegger [Vei86] suggested to use a five-parameter model:

$$T = R_z(\theta)\, T_z(d)\, T_x(a)\, R_x(\alpha)\, R_y(\beta), \tag{4}$$

and Stone/Sanderson [Sto87] presented the so-called S-model having 6 parameters:

$$T = R_z(\theta)\, T_z(d)\, T_x(a)\, R_x(\alpha)\, R_z(\gamma)\, T_z(b). \tag{5}$$

The CPC model developed by Zhuang and Roth [Zhu90] has 7 parameters and is different in as far as it directly uses the redundant representation of normalised, 3D vectors to model a joint-axis direction.

One motivation for these investigations was the objective to find a unique/single kinematic model being uniformly applicable for calibration of all arrangements of subsequent joints. But several years ago, it was proved that this single modelling convention cannot exist due to fundamental topological reasons concerning mappings from Euclidean vector spaces to spheres ([Got86], [Bak88], [Bak90]), known informally among topologists as "you can't comb the hair on a coconut."

The first consequence of this result is that a list of alternative modelling conventions for different types and configurations of joints has to be defined. Implicitly, the above models having five or more parameters also are lists, because some of the parameters are switched off for calibration depending on the geometric arrangement of subsequent joints. Therefore, why not explicitly define a list of alternatives suitable for the different situations? This was the objective of the modelling conventions derived by the authors.

The second consequence of the above topological result is that each parametrisation has singularities, e.g. joint configurations where the parametrisation is not model-continuous and thus cannot be used. The main innovation of the modelling conventions derived by the authors is that for each alternative the manifold of its singularities is determined by a mathematical procedure which guarantees that all singularities are found [Sro93a]. This is also made for prismatic joints which in the past have received little attention, possibly because of the obvious incompatibility of simultaneously meeting the model-minimality demand and including joint motion in the model.

Additional restrictions and demands on the kinematic model arise from its practical application for robot calibration:

a) If the current coordinate frame is not the robot base frame but a frame belonging to a joint, the parametrisation must begin with a R_z- or T_z-transformation in order to integrate the joint motion.
b) The transformations from last joint to TCP and from robot base to first joint have to be parametrised in a way different from ordinary joint transformations.
c) Integration of an elastic-deformation model shall be possible. This requires that the geometric locations of the joints described by the kinematic model coincide with their physical locations.

This requirements can only be met, if additional kinematic parameters (i.e. elementary transformations out of T_x, T_y, T_z, R_x, R_y, R_z) are inserted into parametrisations. But, it can be shown that it is possible to expand complete, minimal and model-continuous kinematic models by additional elementary transformations and that the following statements are valid [Sro93a]:

- The added elementary transformations are redundant and thus, are not permitted to be identified.
- An unexpanded, minimal model always exists which equivalently describes the same kinematic structure as the expanded model.

The authors have derived a set of 17 parametrisations being one example for a set of modelling conventions satisfying all above stated requirements [Sro93b]. The decision on the model used for a single joint transformation of a robot is made by a modelling program which requires a simple and redundant vector chain description of the robot as input.

2.2 Actuator Models

Most actuating elements (particularly gears) have a linear transmission and coupling characteristic, and therefore can simply be modelled by a matrix. But, in many industrial applications, robots can be found which use closed-loop mechanisms (KUKA IR663, Fanuc, ABB). They offer the advantage of increased stiffness and carrying capacity. In regards to modelling, closed-loop mechanisms in robots present a challenge. They introduce complex non-linear couplings between successive joints, and at times between actuating motors and joints, which do not exist in a purely open-loop manipulating robot.

First, two types of closed-loop structures have to be distinguished:

a) Fully parallel manipulators with multiple closed loops such as a Stewart platform or hexapod (see e.g. [Wam94]).
b) Manipulators with closed-loop actuating elements such as 4-bar linkages or slider-cranks which are used to actuate one or two rotary joints of a serial-manipulator's open kinematic chain [Eve89]. This is the most frequent way to utilise closed-loop mechanisms with industrial robots.

Thesis: Concerning manipulator models used for calibration, different solutions should be chosen for type a) (fully parallel manipulators) and type b) (serial manipulators with closed-loop actuating elements).

The first argument supporting this thesis is that fully parallel manipulators require more complex equations for forward kinematics (being used for calibration):

i) either distances (e.g. between unmeasured spherical wrist joints) have to be used for defining the forward-kinematic equations,
ii) or closure constraints have to be added explicitly to the forward-kinematics function (now called: constraint-function approach) which then require to solve a system of non-linear equations for their numerical solution.

If possible, both alternatives i) and ii) should be avoided, because they deteriorate the situation for the numerical calibration algorithms.

Both alternatives really can be avoided with serial manipulators having closed-loop actuating elements (type b), because these robot types use mechanisms such as 4-bar linkages (perhaps actuated by slider-crank mechanisms) with joint axes designed to be parallel. If these mechanisms are modelled to be planar mechanisms, then functions can be derived which explicitly compute joint positions of the serial manipulator from the closed-loop's input joint positions. Therefore, manipulators of type b) belong to the class of manipulators which comes with a forward kinematic function allowing for straight forward computation of TCP pose from actuator position values (in contrast to type a which comes with an inverse kinematic function).

For calibration, this means that the actuator model of the serial manipulator has to be expanded to include not only linear transmission and coupling characteristics, but also the models of planar 4-bar linkages and other closed-loop mechanisms.

Avoiding alternatives i) and ii) in this way results in a higher complexity of the model function itself, because the direct, analytic solution of the planar closed loop has to be integrated into the robot model. But this higher effort seems to be justified, because the lion's share of closed loops utilised in industrial manipulation systems is made up of those used in serial manipulators with closed-loop actuating elements (type b).

This section shall be concluded with a suggestion concerning research on modelling and calibration of fully parallel manipulators. As mentioned above, most approaches to model their kinematics use distances (e.g. between unmeasured spherical wrist joints) for defining forward kinematic equations and then compute the solution of the such-defined system of non-linear equations.

Another approach is to define a forward kinematic function for a serial part of the parallel manipulator which necessarily will include variables for unmeasured joints being unknown. If closure constraints are then added explicitly, they implicitly define the values of these unmeasured joints. Forward kinematics then requires to solve a system of non-linear equations. If this constraint-function approach is used to calibrate the system, the calibration problem numerically becomes an optimisation problem with a large number of non-linear equality constraints. The number of constraints is large, because for each robot pose (being measured for calibration purpose), values of the unmeasured joints must be determined implicitly. Numerical procedures for this type of problem are still subject of research in mathematics, but the field has come to a status of maturity that renders this approach worth to be investigated (see e.g. [Spe93], chap. 3.4).

This is in line with a tendency in the field of dynamic modelling of multi-body systems which tend to use no longer minimal coordinates (according to the system's degrees of freedom) and their derivatives to define the equations of motion. Instead, the so-called descriptor form is used where each body has 6 dof and the constraints introduced by joints are explicitly added to the system of ODEs as equality constraints (see e.g. [Eic92]).

3 INDUSTRIAL APPLICATION OF ROBOT CALIBRATION

Practical prerequisites for applying calibration techniques are availability of required measurement devices (i.e. their integration in the processes of production and production preparation) and a concept for handling measurement and calibration results. Both prerequisites are at the same time the most important obstacles to a broader industrial use of calibration techniques. The reason for this situation is that at first an investment in measurement technology and in organisational changes in the production process concerning the handling of measurement systems and the handling of data from measurement and calibration is required. Investment and organisational changes in the production process need decisions on the management level of a company.

Although it is evident

- that through on-going changes, a growing number of tasks in the field of planning, test, simulation and preparation of the production process move to the computer, and
- that a growing need thus exists for closing the gap between computer model and reality through applying measurement and calibration techniques,

company management mostly avoids making the required decisions, if the technical expert is not capable of guaranteeing return of investment within a quite short period. Therefore, only a very small number of the big robot users has already made the decision to introduce measurement and calibration.

To ease this process, the providers of measurement and calibration technology have to take account of industry's needs. Two of them are:

- Industry needs complete solutions for robot, (varying) tools and workcell. Complete solutions contain procedures and measurement systems.
- Industry needs graduated calibration capabilities for accuracy check, partial re-cali-

bration, full calibration, calibration for quality control in robot production. For partial re-calibration and accuracy check, simple and inexpensive measurement devices should be applied (like light beams or camera systems) which can permanently remain installed inside the robot workcell.

A further measure which should be taken to ease introduction of calibration in industry is standardisation of models and procedures. A first suggestion will be provided by the European project "Improvement of Robot Industrial Standardisation - IRIS" (Programme on Standards, Measurements and Testing, No. SMT4-CT95-2013) which began in Febr. 1996.

ACKNOWLEDGEMENT

Funding for parts of this work was provided by the Commission of the European Communities within ESPRIT project CAR-5220: Calibration Applied to Quality Control and Maintenance in Robot Production.

REFERENCES

[Bak88] Baker, Daniel R. and Ch.W. Wampler. "On the Inverse Kinematics of Redundant Manipulators". Int. J. of Robotics Research 7(2). (1988): 3-21.

[Bak90] Baker, Daniel R. "Some Topological Problems in Robotics". The Mathematical Intelligencer 12(1). (1990): 66-76.

[Eic92] Eich, Edda. Projizierende Mehrschrittverfahren zur numerischen Lösung von Bewegungsgleichungen technischer Mehrkörpersysteme mit Zwangsbedingungen und Unstetigkeiten. Diss. Uni. Augsburg 1991. Fortschrittsberichte VDI, Reihe 18, Nr. 109. Düsseldorf: VDI-Verlag, 1992.

[Eve87] Everett, Louis J., M. Driels and B.W. Mooring. "Kinematic Modelling for Robot Calibration". Proc. 1987 IEEE Int. Conf. on Robotics and Automation (1987): 183-189.

[Eve88] Everett, Louis J. and T.-W. Hsu. "The Theory of Kinematic Parameter Identification for Industrial Robots". Trans. ASME, Journ. of Dynamic Systems, Measurement and Control 110. (1988): 96-100.

[Eve89] Everett, Louis J., "Forward Calibration of Closed-Loop Jointed Manipulators," Int. J. of Robotics Research, 1989, vol. 8 no. 4, pp. 85-91.

[Got86] Gottlieb, Daniel H. "Robots and Topology". Proc. 1986 IEEE Int. Conf. on Robotics and Automation. (1986): 1689-1691.

[Hay85] Hayati, S.A. and M. Mirmirani. "Improving the Absolute Positioning Accuracy of Robot Manipulators". Journal of Robotic Systems 2(4). (1985): 397-413.

[Hol96] Hollerbach, J.M. and Ch.W. Wampler. "A Taxonomy for Robot Kinematic Calibration Methods". Int. J. of Robotics Research. 15(5) (Oct. 1996).

[Iso94] ISO / TC184 / SC2 / WG2 N291 Rev. 2. Manipulating Industrial Robots. On Overview of Test Equipment and Metrology Methods for Robot Performance Evaluation in Accordance with ISO 9283 (DTR 13309). Nov. 1994.

[Kha86] Khalil, W. and J.F. Kleinfinger. "A new geometric notation for open and closed loop robots." Proc. 1986 IEEE Int. Conf. on Robotics and Automation. (1986): 1174-1180.

[Spe93] Spellucci, P. Numerische Verfahren der nichtlinearen Optimierung. Basel,Boston,Berlin: Birkhäuser, 1993.

[Sro93a] Schröer, Klaus. Identifikation von Kalibrationsparametern kinematischer Ketten. Produktionstechnik - Berlin, Bd. 126. München: Hanser, 1993.

[Sro93b] Schröer, Klaus. "Theory of kinematic modelling and numerical procedures for robot calibration." Robot Calibration. eds: R. Bernhardt, S. Albright. London: Chapman & Hall, 1993, 157-196.

[Sro94] Schröer, K. (ed.) Calibration Applied to Quality Control and Maintenance in Robot Production - CAR 5220 - Report on Project Results. Berlin: Fraunhofer-IPK, 1994.

[Sto87] Stone, H.W. and A.C. Sanderson. "A Prototype Arm Signature Identification System". Proc. 1987 IEEE Int. Conf. on Robotics and Automation (1987) 175-182.

[Vei86] Veitschegger, W.K. and C.H. Wu. "Robot accuracy analysis based on kinematics." IEEE J. Robotics and Automation 2(3) (1986): 171-179.

[Wam94] Wampler Ch.W., J.M. Hollerbach, T. Arai. "An Implicit Loop Method for Kinematic Calibration and its Application to Closed-Chain Mechanisms". GM Publication R&D-8188, 1994.

[Zhu90] Zhuang, H., Roth, Z.S. and Hamano, F. "A Complete and Parametrically Continuous Kinematic Model for Robot Manipulators." Proc. 1990 IEEE Int. Conf. on Robotics and Automation (1990): 92-97.

ROBO-SHEPHERD: LEARNING COMPLEX ROBOTIC BEHAVIORS

Alan C. Schultz, John J. Grefenstette, and William Adams

Naval Research Laboratory, Washington, DC 20375-5337, U.S.A.
schultz@aic.nrl.navy.mil

ABSTRACT

This paper reports on recent results using genetic algorithms to learn decision rules for complex robot behaviors. The method involves evaluating hypothetical rule sets on a simulator and applying simulated evolution to evolve more effective rules. The main contributions of this paper are (1) the task learned is a complex behavior involving multiple mobile robots, and (2) the learned rules are verified through experiments on operational mobile robots. The case study involves a shepherding task in which one mobile robot attempts to guide another robot to a specified area.

KEYWORDS: Genetic Algorithms, Evolutionary Computation, Robot Learning

INTRODUCTION

In this study, a complex robotic behavior, embodied as set of stimulus-response rules, is learned by a robot. Learning initially takes place under simulation. This methodology reflects our belief that many robotic tasks are too expensive or dangerous to learn from experiences on a real robot in its real environment. This work demonstrates that non-trivial behaviors learned under simulation can transfer to an operational mobile robot, with similar behavior and performance levels being achieved in the real environment. The methodology depends on the assumption that important aspects of the real world are captured by the simulator. This assumption should not be viewed as a severe limitation on the approach. Our experience shows that even simulators with limited fidelity are sufficient for learning many interesting tasks for real-world robots.

As an example, we report on learning a shepherding task, in which one mobile robot seeks to guide another to a specified area. The behavior is learned under simulation. The resulting rules are then used to control an operational mobile robot. The resulting behavior and performance level is compared to that achieved under simulation. Adding to the difficulty of this task, all perception of the environment by each operational robot is via the actual sensors, and the robots must learn to avoid collisions with other objects in the world.

THE SHEPHERDING TASK

In this task, one robot is the shepherd and the other robot is the sheep. The performance task is for the shepherd to guide the sheep into a pasture within a limited amount of time. The sheep reacts to the presence of nearby objects by moving away from them. Otherwise, the sheep moves in a random walk. The shepherd must learn to control his own translation and steering to get the sheep to move into the pasture. Both robots are controlled by a reactive rule set that maps current sensors of each robot into appropriate motion commands. Only the shepherd's rules are learned.

THE LEARNING SYSTEM

The behaviors, which are represented as a collection of stimulus-response rules, are learned in the SAMUEL rule learning system. SAMUEL is a heuristic learning program that uses genetic algorithms and other competition-based heuristics to improve its decision-making rules. The system actively explores alternative behaviors in simulation, and modifies its rules based on this experience. SAMUEL is designed for problems in which payoff is delayed in the sense that payoff occurs only at the end of an episode that may span several decision steps. SAMUEL is particularly well-suited for learning strategies in competitive environments for tasks such as predator-prey problems, tracking problems, and other models involving multiple competing agents.

SAMUEL includes a competition-based production system interpreter, incremental strength updating procedures to measure the utility of rules, and genetic algorithms to modify strategies based on past performance. This system incorporates a convenient language for the expression of tactical decision rules, a graphical interface, and a number of heuristics for rule modification. The rule language in SAMUEL also makes it easier to incorporate existing knowledge, whether acquired from experts or by symbolic learning programs.

In SAMUEL, learning is driven by competition among knowledge structures. Competition is applied at two levels: Within a strategy composed of decision rules, rules compete with one another to influence the behavior of the system. At a higher level of granularity, entire strategies compete with one another using a genetic algorithm.

Previous papers on SAMUEL focused on the operations of the system in purely simulated environments [2][3]. We have also previously reported on using this method to learn simple robot behaviors such as navigation and collision avoidance [7][8].

Representation

Each stimulus-response rule consists of conditions that match against the current sensors of the robot, and an action that suggests a translation or steering velocity command to the robot based on the current situation. The condition and action values are described in more detail below. For example, a rule for the shepherd might be:

IF range = [35, 45] AND bearing = [340, 35] THEN SET turn = -24 (Strength 0.8)

During each decision cycle, all the rules that match the current state are identified. Conflicts are resolved in favor of rules with higher *strength*. Rule strengths are updated based on rewards received after each training episode. See [2] for further details.

LEARNING UNDER SIMULATION

The rule set for the shepherd is learned under simulation while the sheep's rule set is fixed. The learning task requires the shepherd to get the sheep to within a fixed range of the goal within a time limit, without hitting an obstacle (the enclosing walls of the simulated environment or the sheep). For learning, a differential performance score is given (the fitness) for each rule set in the population. The performance of an individual rule set is calculated by averaging over 20 episodes, in which each episode begins with the sheep and the shepherd placed in random initial positions and orientations, and end when the sheep enters the pasture, time expires, or a collision occurs. The performance of each episode is rated by an automated critic that gives full credit for getting the sheep into the pasture, partial credit for getting it close, and no credit for episodes that end due to collision with the sheep or with the surrounding walls. In these experiments, the pasture remains in a fixed location for all trials.

After each individual in the population is evaluated in simulation, genetic and other operators in the Samuel system are applied to the individuals to generate the next population of behaviors, presumable with higher fitness. This cycle is iterated until the desired fitness is achieved, or until a fixed upper limit of generations have elapsed.

TESTING ON THE ROBOTS

In order to verify the learned behaviors, the learned rules are used to control the actual shepherd robot. The shepherd robot and the sheep robot are placed in random locations and orientations within our laboratory environment, and the resulting performance is recorded.

The two robots used in this study are Nomad 200 systems manufactured by Nomadic Technologies, Inc., and include 16 sonars, 16 infrared sensors (not used in this study), tactile sensors, and a structured light range finder which is described below.

THE SHEPHERD ROBOT

For this task, five abstract sensors are defined for the shepherd: the range and bearing from it to the sheep, the sheep's heading with respect to the sheep's bearing, and the range and bearing to the center of the pasture to which it is herding the sheep. Conceptually, these abstract sensors are shown in figure 1. These abstract sensors are derived from information available through the actual sensors on the robots.

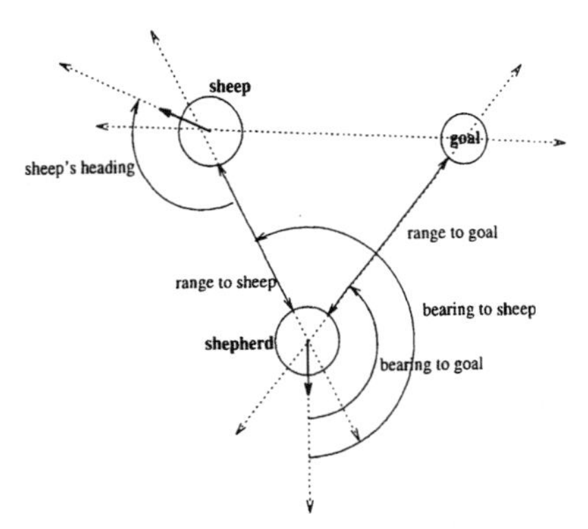

Figure 1: The meaning of the shepherd's sensors

The range, bearing and heading are discretized. The bearing values are partitioned into intervals of five degrees. The range is partitioned into intervals of five inches from 0 to 75 inches. The heading is partitioned into 45 degree segments.

The actual perception system, a structured light rangefinder, has a limited reliable detection range of approximately 8 feet, with the chance of losing the sheep increasing as the range increases. Under simulation, the true range and bearing are determined, and then are discretized as above. To simulate the shepherd's limited perception range its detection of the sheep is restricted to 6 feet.

In addition to the range, bearing and heading of the sheep returned by the shepherd's perception system, the shepherd also knows the coordinates of the pasture to which he wants to herd the sheep. In actual operation, the shepherd uses dead reckoning to determine this location. In the simulation, the true values are determined, and then discretized as above.

The learned actions are the velocity mode commands for controlling the translational rate and the steering rate. The translation is given as -4 to 16 inches/sec in 2 inch/sec intervals. The steering command is given in intervals of 4 degrees/sec from -24 to 24 degrees/sec.

Description of the shepherd's perception system

The range and bearing to the sheep, and the heading of the sheep are determined on the real robot by using a perception system mounted on the shepherd robot. Perception is performed by an on-board triangulation-based rangefinder. A laser on the robot projects a plane of laser light parallel to the floor and the intersection of that plane with any objects is detected by a video camera mounted above the laser. The positions of the illuminated pixels in the image are then mapped into Cartesian coordinates.

To provide both identification and pose information of the sheep robot, an irregular, extruded pentagon is placed on top of the sheep and a mathematical description of it is given to the shepherd. The pentagon vertices have five distinct angles. Software on the shepherd robot's processor continually fits lines to the points returned by its rangefinder, and measures the angles between the adjoining lines. Observation of a single vertex identifies the object as the sheep and, in conjunction with the poses of the adjacent lines, determines the relative orientation and position of the pentagon and therefore the sheep.

When the sheep is in rangefinder view, the shepherd attempts to center the visible vertex in its view. When not in view, the shepherd interpolates the sheep's relative position based on the sheep's last observed position and velocity and the shepherd's known position. An arc is defined encompassing the interpolated position, and the rangefinder is panned through the arc repeatedly. If the sheep is not observed again after a reasonable amount of time, it is considered lost and the shepherd spins the rangefinder in complete circles until the sheep is re-acquired.

THE SHEEP ROBOT

The sheep is controlled by a manually generated reactive behavior (rule-set). In general, its behavior is a random walk when no other objects are within a predetermined distance (150 inches in these experiments). Objects that are within this range tend to make the sheep move away from the object. The sheep makes its decisions based on two abstract sensors, namely, the range and bearing to the closest object within 150 inches. Outside of that distance, these sensors return the value UNKNOWN.

On the actual sheep robot, the 16 sonars are used to determine the range and bearing to the closest object. The minimum value is obtained from the sonars. If the value is less than the specified maximum range of 150 inches, then we determine which sonar returned that value, and then discretize the bearing and range. If the minimum range returned by all sonar sensors is over 150 inches, then we return the value UNKNOWN for the range and the bearing. As in the shepherd, the range is discretized by partitioning into intervals of five inches from 0 to 150 inches. The bearing is partitioned into intervals of five degrees. In the training simulator, the true bearing to the closest object is calculated, and then discretized as above.

	total	percent
Successful episodes	19	67%
Failure due to time limit	1	3%
Failure due to collision	6	20%
Failure due to lost perception	3	10%
Failure due to lost communication packets	1	3%

Table 1: Summary of results with real robots

The actions of the sheep are limited to a translational rate of -4 to 6 inches/sec and a steering rate of -24 to 24 degrees/sec.

RESULTS FROM LEARNING

In the simulation, the best shepherd behavior reached a success level of 86%. This rule set (from generation 250 in the first SAMUEL run) was then used to control the real shepherd robot. Using the operational robots, 27 episodes were performed. At the start of each episode, the shepherd and sheep robots were placed in random orientations and positions within a restricted portion of the room, similar to the simulation episodes. A success was counted if the shepherd guided the sheep to the pasture area, defined as a three foot radius area in another part of the room, within a fixed amount of time (60 secs). A failure was counted if the time was exceeded, if either the sheep or the shepherd collided with any other object in the environment, if the shepherd's perception system stopped tracking the sheep, or the robot's communication link with the host computer failed. In addition to the actual success rate, observations were subjectively made of the shepherd's behavior to determine if the behaviors were similar to those in the simulator.

The results are summarized in Table 1. The actual robot succeeds in 67% of the episodes. In these results, failures include the shepherd losing track of the sheep, as well as communication failures. However, in the simulation, it is not possible for either of these conditions to occur. If these cases are removed from consideration, then the observed success rate is 73%, much closer to the success rate obtained under simulation.

Subjective observations suggest that the same types of behaviors are exhibited by the real robot as in the simulation. For example, in both cases, the shepherd tends to maneuvers to the outside of the sheep and forces it toward the pasture. Videos of the operational robots will be shown at the Conference.

RELATED WORK

Other approaches to evolutionary robotics have been used before. The work of (Harvey, Cliff and Husbands) [4] has concentrated on bottom-up learning; for example, the perception system is learned as a neural network via evolutionary algorithms. Our approach differs in that we take an engineering view, and do not insist that all components must be learned. We start with designed parts, and with a specific decomposition of behaviors, and let the system learn the rules for each behavior.

Although similar in that robot behaviors are being learned, Dorigo [1] takes an entirely different architectural approach to evolution, using classifier systems to learn behaviors. Here the entire population of the genetic algorithm is taken as the behavior. In our work, each individual in the population is a behavior, and the population consists of competing behaviors.

In the system GA-Robot by Ram *Et. Al.* [6], a genetic algorithm with a floating point representation is used to optimize parameters that effect the behavior. In Samuel, the entire behavior is learned in a high-level stimulus-response language.

CONCLUSIONS AND FUTURE WORK

This paper reports on an approach for learning behaviors for mobile robots by evaluating hypothetical rule sets on a simulator and applying simulated evolution to evolve more effective rules. The resulting rules are then placed on an actual robotic system for testing. The learning system, SAMUEL, uses genetic algorithms applied to symbolic rules. We have previously reported on using this method to learn simple robot behaviors such as navigation and collision avoidance [7][8]. Here, we have demonstrated that the approach can be used to learn rules to perform a complex herding behavior involving multiple mobile robots, and that the learned rules can be verified through experiments on operational mobile robots.

Future work will aim at scaling these approaches to more complex multi-robot tasks, including cooperative mapping of an area, finding hidden objects and performing cooperative surveillance tasks.

References

[1] Dorigo, M. (1993). "Genetic and Non-Genetic Operators in Alecsys," *Evolutionary Computation*, 1(2): 151-164.

[2] Grefenstette, J. J., Ramsey, Connie L., and Schultz, Alan C., (1990). "Learning sequential decision rules using simulation models and competition," *Machine Learning*, **5(4)**, 355-381

[3] Grefenstette, J.J. (1991). "Lamarckian learning in multi-agent environments," *Proc. Fourth International Conference of Genetic Algorithms*, San Mateo, CA: Morgan Kaufmann, 303-310.

[4] Cliff, D., I. Harvey, and P. Husbands (1991). Cognitive Science Research Paper No. 318, School of Cognitive and Computer Science, University of Sussex.

[5] Holland, J. H. (1975). *Adaptation in Natural and Artificial Systems.* Univ. Michigan Press, Ann Arbor, 1975.

[6] Ram, Ashwin, R. Arkin, G. Boone, and M. Pearce (1994). "Using Genetic Algorithms to Learn Reactive Control Parameters for Autonomous Robotic Navigation," *Adaptive Behavior*, 2(3), 1994

[7] Schultz, Alan C. and Grefenstette, John J. (1992). "Using a genetic algorithm to learn behaviors for autonomous vehicles," *Proceedings of the of the AIAA Guidance, Navigation and Control Conference*, Hilton Head, SC, August 10-12, 1992.

[8] Schultz, Alan C. (1994). "Learning robot behaviors using genetic algorithms," *Intelligent Automation and Soft Computing: Trends in Research, Development, and Applications, v1*, Mohammad Jamshidi and Charles Nguyen, editors, Proceedings of the First World Automation Congress (WAC '94), 607-612, TSI Press: Albuquerque.

A Compact Cylinder-Cylinder Collision Avoidance Scheme for Redundant Manipulators

F. Shadpey

Department of Electrical & Computer Engineering, Concordia University, Montreal, Canada

F. Ranjbaran

Department of Mech. Eng. & Center for Intelligent Machines McGill Univ., Montreal, Canada

R. V. Patel

School of Electrical & Computer Engineering, Oklahoma State University, Stillwater, U.S.A.

J. Angeles

Department of Mech. Eng. & center for Intelligent Machines McGill Univ., Montreal, Canada

ABSTRACT

The development of a primitive-based collision and self-collision avoidance scheme for redundant manipulators is discussed. The method is based on modeling the arm and its environment by simple geometric primitives (cylinders and spheres). A compact method of detecting collisions between two cylinders is introduced. By resorting to the notions of *dual angles* and *dual vectors* for representing the axes of cylinders in space, a characterization of different types of collisions is discussed. The performance of the proposed scheme is demonstrated by computer simulation and graphical rendering.

KEYWORDS: collision avoidance, geometric modeling, redundancy resolution

INTRODUCTION

The need for compact and efficient collision-avoidance schemes is becoming increasingly apparent for the successful operations of robots in applications such as space, underwater, and hazardous environments. Much of the research work reported to date has dealt with obstacle avoidance as an off-line path planning effort to find a collision free path for the end-effector [1, 2] or by mapping the obstacles into joint space, and determining a collision free path in joint space [3]. In recent years, kinematic redundancy has been recognized as a major characteristic for operation of manipulators in cluttered environments e.g., see [4]. In order to implement an on-line collision avoidance scheme, three major areas are involved: redundancy resolution; geometric modeling of the robot and its environment; and distance calculation algorithms. Colbaugh et al. [5], used circles to model obstacles surrounded by a Surface of Influence (SOI), while modeling the links by straight lines. Shadpey et al., [6], extended the redundancy resolution scheme used in [5] to the three dimensional workspace of a 7-axis manipulator, where the manipulator links were represented by spheres and cylinders while the objects were modeled by spheres. The proposed method has been successfully implemented through hardware demonstration on REDIESTRO [7], (see Fig. 1). Although this method is convenient for spherical or bulky objects, it would result in major reduction of the workspace when dealing with long objects. One of the simplified geometric models for the links of industrial manipulators is the cylinder. Moreover, the cylinder is most suited for modeling objects in the workspace such as rods, mesh structures and openings without losing significant regions of the available workspace. In this paper, we present a compact method of detecting collisions between cylinders.

CYLINDER-CYLINDER COLLISION DETECTION

In order to determine the relative position of two cylinders, first the relative layout of their axes needs to be established. The axes of the cylinders being directed lines in three dimensional space, we resort to the notions of line geometry. Specifically, with the aid of dual unit vectors, (or line vectors), and the dual angles subtended by them, we will categorize the relative placement of cylinders and thus determine the possibility and the nature of collisions between the two cylinders in question.

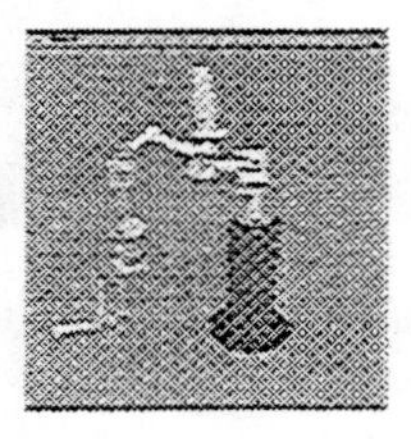

Figure 1: REDIESTRO with actual links

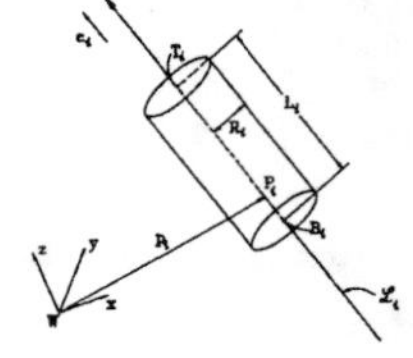

Figure 2: Basic notation

We consider each cylinder to be composed of three parts, the cylindrical surface plus the two circular disks as the top and the bottom bases of the cylinder. Four points along the axis $\mathcal{L}_i$ of each cylinder $\mathcal{C}_i$ are of interest, namely, P_i, B_i, T_i and H_i, as indicated in Fig.2. The point P_i specified by a position vector $\mathbf{p}_i$ is any point of reference on the line. The points B_i and T_i with position vectors $\mathbf{b}_i$ and $\mathbf{t}_i$ respectively, are the centers of the bottom and the top bases of the cylinder respectively, and H_i is the foot of the common normal H_iH_j of the two lines $\mathcal{L}_i$ and $\mathcal{L}_j$ on the line $\mathcal{L}_i$.

Review of Line Geometry and Dual Vectors

A line $\mathcal{L}$ can be defined via a dual unit vector, also called a line vector.

$$\hat{\mathbf{e}} = \mathbf{e} + \epsilon \mathbf{m} \tag{1}$$

where $\mathbf{e}$ defines the direction of $\mathcal{L}$, $\mathbf{m}$ bing the moment of $\mathcal{L}$ with respect to a predefined point , and ϵ being the *dual unity* has the property that $\epsilon^2 = 0$. Now, let $\mathcal{L}_i$ and $\mathcal{L}_j$ be two lines, their *dual angle* is defined as

$$\hat{\nu}_{ij} = \nu_{ij} + \epsilon h_{ij} \tag{2}$$

where ν_{ij} is the *projected angle* between $\mathbf{e}_i$ and $\mathbf{e}_j$ and h_{ij} is the *distance* between $\mathcal{L}_i$ and $\mathcal{L}_j$. More detailed discussion can be found in numerous papers, e.g., see [8].

Three different possibilities for the layout of two distinct lines $\mathcal{L}_i$ and $\mathcal{L}_j$ exist as explained below:
(**A**) Non-parallel and non-intersecting lines: $\hat{\nu}_{ij}$ is a *proper dual number*, i.e., $\nu_{ij} \neq k\pi$, with $k = 0, 1$ and $h_{ij} \neq 0$.
(**B**) Parallel lines: $\hat{\nu}_{ij}$ is a pure dual number, (its primal part is zero), i.e., $\nu_{ij} = k\pi$, with $k = 0, 1$ and $h_{ij} \neq 0$.
(**C**) Intersecting lines: $\hat{\nu}_{ij}$ is a real number, (its dual part is zero), i.e., $\nu_{ij} \neq k\pi$, with $k = 0, 1$ and $h_{ij} = 0$.
Now, for two cylinders $\mathcal{C}_i$ and $\mathcal{C}_j$ to collide, one of the three cases discussed below must occurs:
(**1**) Body-body collision: This situation—the most likely one—is shown in Fig. 3, where two cylindrical surfaces of the objects are intersecting
(**2**) Base-body Collision: The cylindrical body of one cylinder collides with one of the two circular disks of the other cylinder.
(**3**) Base-base collision: One of the circular disks of one cylinder collides with a circular disk of another cylinder.

(A) Cylinders with non-parallel and non-intersecting axes

In order to characterize the types of possible collisions for two cylinders whose major axes are represented by $\mathcal{L}_i$ and $\mathcal{L}_j$, that are non-parallel and non-intersecting, the following steps are taken:

- First we need to determine the location of the points H_i along $\mathcal{L}_i$, and H_j along $\mathcal{L}_j$, i.e., the feet of the common normal on the two lines. This can be done by determining the scalars h_i and h_j, as given below:

$$h_i = \frac{(\mathbf{p}_i - \mathbf{p}_j) \cdot (\mathbf{e}_j \cos \nu_{ij} - \mathbf{e}_i)}{\sin^2 \nu_{ij}} \tag{3}$$

$$h_j = \frac{(\mathbf{p}_j - \mathbf{p}_i) \cdot (\mathbf{e}_i \cos \nu_{ij} - \mathbf{e}_j)}{\sin^2 \nu_{ij}} \tag{4}$$

with, $\mathbf{h}_i = \mathbf{p}_i + h_i\, \mathbf{e}_i$, and $\mathbf{h}_j = \mathbf{p}_j + h_j\, \mathbf{e}_j$.

- Now, if $h_{ij} > (R_i + R_j)$, collision is not possible.
- If $h_{ij} \leq (R_i + R_j)$, collision is possible, as explained below:

(A-1) If $b_i \leq h_i \leq t_i$, and $b_j \leq h_j \leq t_j$, then we have a body-body collision, and the critical points (closest points on each axes which lies between top and bottom disks) P_i^c and P_j^c on the axes are H_i and H_j respectively, (Fig. 3), with critical directions $\mathbf{n}_{ij}$ and $-\mathbf{n}_{ij}$ for $\mathcal{C}_i$ and $\mathcal{C}_j$, respectively.

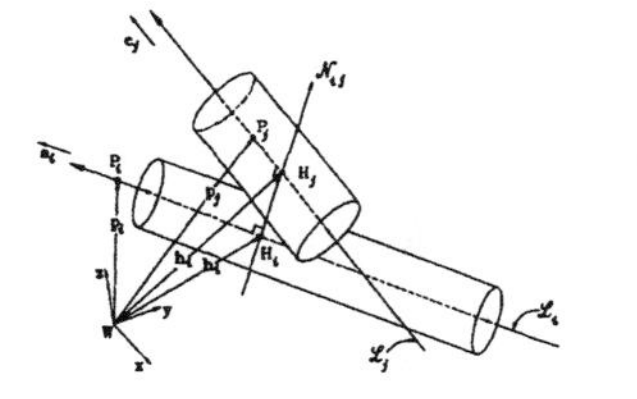

Figure 3: (A-1) Body-Body Collision (non-parallel and non-intersecting axes)

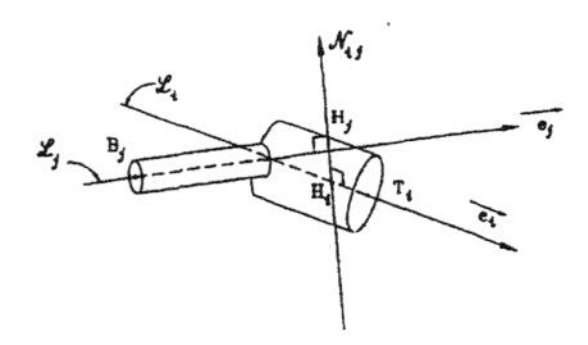

Figure 4: (A-2) Base-Body Collision (non-parallel and non-intersecting axes)

(A-2) If only one of H_i or H_j lies outside of its corresponding cylinder then, we may or may not have a collision. However, if the two cylinders collide, the collision has to be in the form of a base-body case only, (Fig. 4). As an example, in order to determine the critical points and the critical direction, we assume that H_i lies inside of $\mathcal{C}_i$ with H_j being outside of $\mathcal{C}_j$.

The critical point P_j^c of $\mathcal{C}_j$ will thus be one of the two points B_j or T_j, whichever lies closer to H_j. Moreover, the critical point P_i^c of the cylinder $\mathcal{C}_i$ is the orthogonal projection of P_j^c on $\mathcal{L}_i$. If $\mathbf{p}_j^c$ is the vector representing P_j^c, we will have

$$\mathbf{p}_i^c = \mathbf{p}_i + (\mathbf{p}_j'^c \cdot \mathbf{e}_i)\, \mathbf{e}_i. \tag{5}$$

where $\mathbf{p}_j'^c$ is the vector conecting P_j^c to P_i. We will thus consider that a collision occurs, whenever the following inequality is satisfied (Fig. 5):

$$\|\mathbf{p}_i^c - \mathbf{p}_j^c\| \leq (R_i + R_j) \tag{6}$$

The foregoing inequality gives a conservative prediction of collision between the base and the body of the two cylinders which assumes that the base of the cylinder is not a simple circular disk, but, a semi-sphere of the same radius.

(A-3) If both H_i and H_j lie outside of their corresponding cylinders, then we have a base-base collision, and the critical points and direction will be determined as explained below (Fig. 5):

Denote by $\{d_k\}_1^4$, the set of distances of B_i and T_i to B_j and T_j, i.e.,

$$\begin{aligned} d_1 &= \|\mathbf{b}_i - \mathbf{t}_j\|, \quad d_2 = \|\mathbf{b}_i - \mathbf{b}_j\| \\ d_3 &= \|\mathbf{t}_i - \mathbf{t}_j\|, \quad d_4 = \|\mathbf{t}_i - \mathbf{b}_j\| \end{aligned}$$

and $d_c \equiv \min\{d_1, d_2, d_3, d_4\}$, then we have a base-base collision if, $d_c \leq (R_i + R_j)$.

(B) Cylinders with intersecting axes

In order to characterize a collisions between two cylinders with intersecting axes, we first project the endpoints T_i and B_i of the cylinder $\mathcal{C}_i$ onto the line $\mathcal{L}_j$ and denote the projected points by T_j' and B_j'. Conversely, we project T_j and B_j of the cylinder $\mathcal{C}_j$ onto the line $\mathcal{L}_i$ and denote the projected points by T_i' and B_i'. The position vectors of the foregoing four points will take on the form

$$\begin{aligned} \mathbf{b}_i' &= \mathbf{p}_i + b_i' \mathbf{e}_i, \quad \mathbf{t}_i' = \mathbf{p}_i + t_i' \mathbf{e}_i \\ \mathbf{b}_j' &= \mathbf{p}_j + b_j' \mathbf{e}_j, \quad \mathbf{t}_j' = \mathbf{p}_j + t_j' \mathbf{e}_j \end{aligned}$$

with

$$\begin{aligned} b_i' &= -(\mathbf{p}_i - \mathbf{b}_j) \cdot \mathbf{e}_i, \quad t_i' = -(\mathbf{p}_i - \mathbf{t}_j) \cdot \mathbf{e}_i \\ b_j' &= -(\mathbf{p}_j - \mathbf{b}_i) \cdot \mathbf{e}_j, \quad t_j' = -(\mathbf{p}_j - \mathbf{t}_i) \cdot \mathbf{e}_j \end{aligned}$$

(B-2) If any one of the following four conditions holds, then we will have a base-body collision, and the critical direction will be a unit vector pointing along a vector joining the corresponding critical points, (Fig. 6),

$$\begin{aligned} b_i \leq b_i' &\leq t_i, \quad \text{and,} \quad \|\mathbf{b}_i' - \mathbf{b}_i\| \leq (R_i + R_j) \\ b_i \leq t_i' &\leq t_i, \quad \text{and,} \quad \|\mathbf{t}_i' - \mathbf{t}_i\| \leq (R_i + R_j) \\ b_j \leq b_j' &\leq t_j, \quad \text{and,} \quad \|\mathbf{b}_j' - \mathbf{b}_j\| \leq (R_i + R_j) \\ b_j \leq t_j' &\leq t_j, \quad \text{and,} \quad \|\mathbf{t}_j' - \mathbf{t}_j\| \leq (R_i + R_j) \end{aligned}$$

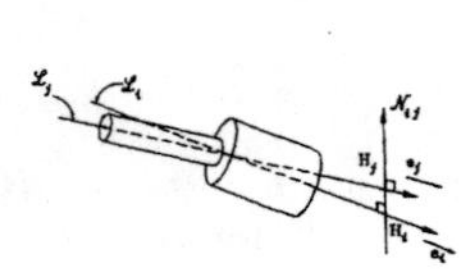

Figure 5: (A-3) Base-Base Collision

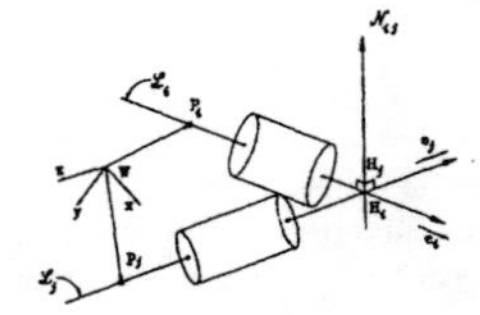

Figure 6: (B-2) Base-Body Collision

(B-3) If none of the foregoing conditions is satisfied, then we will not have a base-body collision. However, we may have a base-base collision. The procedure for base-base collision detection for a pair of intersecting lines is similar to that of case (A-3) explained earlier. (Fig. 7).

(C) Cylinders with parallel axes

For the special case of two parallel lines $\mathcal{L}_i$ and $\mathcal{L}_j$ for which an infinite number of common normals exist, we resort to a unique definition for one common normal lying closest to the origin [9]. If the line $\mathcal{N}_{ij}$ passes through the points H_i and H_j of $\mathcal{L}_i$ and $\mathcal{L}_j$ with H_i and H_j being the closest points of the two lines to the origin, then the dual representation of $\mathcal{N}_{ij}$ is given as

$$\hat{\mathbf{n}}_{ij} = \frac{\mathbf{h}_j - \mathbf{h}_i}{h_{ij}} + \epsilon \frac{\mathbf{h}_i \times \mathbf{h}_j}{h_{ij}} \tag{7}$$

where, $\mathbf{h}_i$ and $\mathbf{h}_j$ are the position vectors of the points H_i and H_j respectively and $h_{ij} \equiv \|\mathbf{h}_j - \mathbf{h}_i\|$ is the distance between the two lines.

If $h_{ij} > (R_i + R_j)$, then the two cylinders do not collide. However, if $h_{ij} \leq (R_i + R_j)$, then depending on the location of the cylinders along their axes relative to each other, two special cases of body-body (C-1) and base-base (C-3) collisions can occur.

SIMULATION AND GRAPHICAL RENDERING

A graphical rendering of REDIESTRO with its actual links and actuators is shown in Fig. 1, while Fig. 8 depicts the arm with each moving element of the arm enclosed in a cylindrical primitive. The links and the actuator units are modeled by 14 cylinders in total, the fourth link having the maximum number of 4 sub-links. The environment is modeled by three cylindrical objects form a triangular opening. The end-effector trajectory is defined as a straight line passing through this triangular opening. Fig. 9 shows successful operation of collision avoidance algorithm.

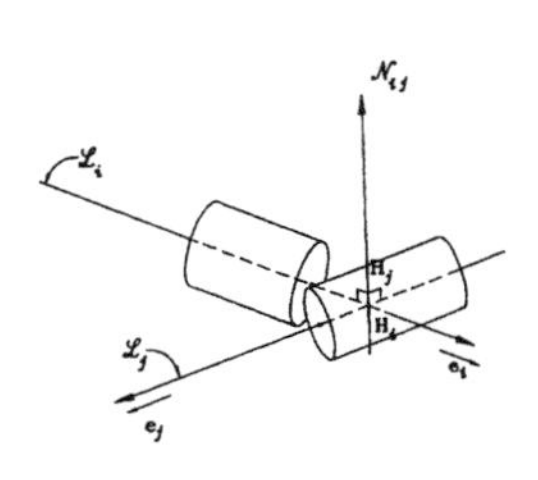

Figure 7: (B-3) Base-Base Collision

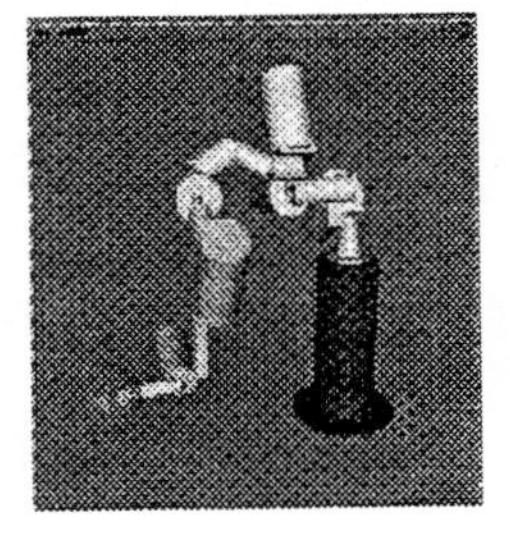

Figure 8: REDIESTRO represented by primitives

CONCLUSION

A cylinder-cylinder model for collision detection/avoidance applied to redundancy resolution of redundant manipulators has been presented. The notion of unit dual vectors was used to characterize the relative layout of the major axes of the two cylinders. Depending on this layout, different possible collisions between the two cylinders were examined. Having characterized and identified the critical points and the critical directions for each situation, we used the augmented Jacobian matrix to resolve redundancy in the manipulator while inhibiting the motion of the critical points along the critical directions. Graphical simulations of REDIESTRO, a redundant manipulator with a complex architecture were used to demonstrate the proposed method.

ACKNOWLEDGEMENTS

This research was supported by the Natural Sciences and Engineering Research Council of Canada under Grant OGP0001345, and by the Institute for Robotics and Intelligent Systems (IRIS).

Figure 9: Top row: collision avoidance inactive (a through c). Bottom row: collision avoidance active (d through f). Cylindrical Surfaces of Influence (SOI) shown transparent

References

[1] S. Borner, and R. B. Kelley, *IEEE Transaction on Syst., Man, and Cybernetics*, vol. 20, no. 6, pp. 1337-1351, 1990.

[2] T. Hasegawa, H. Terasaki, *IEEE Transaction on Syst., Man, and Cybernetics*, vol. 18, no. 3, pp. 337-347, 1988.

[3] P. Khosla and R. V. Volpe, *Proc. IEEE Int. Conf. on Robotics and Automation*, pp. 1778-1784, 1988.

[4] D. G. Hunter, *National Research Council Canada, NRCC no. 28817, Issue A, 7 April 1988.*

[5] R. Colbaugh, H. Seraji, and K. Glass, *J. Rob. Res.*, Vol. 6, pp. 721-744, 1989.

[6] F. Shadpey, C. Tessier, R.V. Patel, B. Langlois, and A. Robins, *AAS/AIAA American Astrodynamics Conference*, Aug. 1995, Halifax, Canada.

[7] F. Ranjbaran, J. Angeles, M. A. González-Palacios, and R. V. Patel, 1995, *Journal of Robotics and Intelligent Systems*, Vol. 13, pp. 1–21.

[8] G. R., Veldkamp, *Mechanism and Machine Theory*, vol. 11, pp. 141–156, 1976.

[9] M. A. González–Palacios, J. Angeles and F. Ranjbaran, *in Proc. IEEE Int. Conf. Robotics Automat.*, Atlanta, Georgia, 1993, vol. 1, pp. 450–455.

A LASER SCANNING SYSTEM FOR METROLOGY AND VIEWING IN ITER

P. T. SPAMPINATO, R. E. BARRY, M. M. MENON, J. N. HERNDON
Oak Ridge National Laboratory, P.O. Box 2008, Oak Ridge, TN, USA 37831-6304

M. A. DAGHER, J. E. MASLAKOWSKI
Rockwell Rocketdyne Division, Canoga Park, CA, USA

ABSTRACT

The construction and operation of a next-generation fusion reactor will require metrology to achieve and verify precise alignment of plasma-facing components and inspection in the reactor vessel. The system must be compatible with the vessel environment of high gamma radiation (10^4 Gy/h), ultra-high-vacuum (10^{-8} torr), and elevated temperature (200°C). The high radiation requires that the system be remotely deployed. A coherent frequency modulated laser radar-based system will be integrated with a remotely operated deployment mechanism to meet these requirements. The metrology/viewing system consists of a compact laser transceiver optics module which is linked through fiber optics to the laser source and imaging units that are located outside of a biological shield. The deployment mechanism will be a mast-like positioning system. Radiation-damage tests will be conducted on critical sensor components at Oak Ridge National Laboratory to determine threshold damage levels and effects on data transmission. This paper identifies the requirements for International Thermonuclear Experimental Reactor metrology and viewing and describes a remotely operated precision ranging and surface mapping system.

KEYWORDS: laser radar metrology, remote maintenance, gamma radiation environment, surface mapping

INTRODUCTION

The International Thermonuclear Experimental Reactor (ITER) is a fusion device planned to be built early in the next century. The performance and survival of plasma-facing components (PFC) located within the reactor's vacuum vessel depend on precise alignment and positioning with respect to the plasma edge. A remotely deployed and controlled three-dimensional metrology system is being developed to periodically verify the condition of in-vessel components in ITER. This metrology system has two basic functions: (1) frequent inspection to establish the dimensional status of in-vessel components and (2) extensive checking of in-vessel components and plasma-facing surfaces during scheduled maintenance shutdowns.

DESIGN REQUIREMENTS

The interior surface area of ITER is approximately 1500 m^2. In order to achieve acceptable mapping times, up to ten metrology systems will be required. Each system must be capable of acquiring in-vessel dimensional data accurately, under harsh environmental and radiological conditions. It must withstand gamma radiation levels that are expected to reach 3×10^4 Gy/h, while operating in a 200°C environment, in vacuum conditions. The undeployed system will experience a cyclic magnetic field that peaks at 0.15 T. In addition, an alternative requirement which is being investigated is the feasibility of deploying the system in a constant magnetic field of 6 T. The system must also function during scheduled maintenance activities when the vessel will be at atmospheric pressure. Because of the severity of the ITER environment, further development is required to adapt any commercially available range imaging systems. A survey of available mapping technologies[1] identified frequency modulated (FM) coherent laser radar (CLR) as a promising approach for ITER remote metrology.

THREE-DIMENSIONAL METROLOGY SYSTEM

The design concept for a precision measuring device for ITER is based on an FM CLR produced by Coleman Research Corporation.[2] A prototype model of the FM CLR has demonstrated submillimeter range accuracy at more than 10 m. The prototype showed very little sensitivity to surface type, color, and angle of incidence in tests performed at Oak Ridge National Laboratory (ORNL).

Range is determined by measuring the frequency difference between an FM laser beam reflected from a target surface and a local oscillator reference. The accuracy of the range is determined by the linearity of the FM over the counting interval. A prototype Coleman-CLR system that was developed for a joint Department of Energy/National Aeronautics and Space Administration project was tested at ORNL in June 1994. The results indicated that the basic technique is accurate enough to meet the ITER requirements. Since that time, the accuracy of the ranging system has been further improved by Coleman researchers.[3]

- Results show that the unit was able to measure absolute range from 4 to 12 m and accurately track 0.127-mm (0.005-in.) increments.

- Signal attenuation is well within the capability of the instrument's resolution.

- The instrument was able to obtain range data at near grazing angles on most surfaces, indicating low angular sensitivity.

In the ITER application, automated beam deflection techniques will be used to provide a scanning capability for fast mapping of the large surface areas of the vacuum vessel.

Figure 1 shows a schematic configuration with a one-dimensional acousto-optical (AO) beam deflector and a nodding mirror. This allows the radar to have a two-dimensional scanning capability with only one moving part.

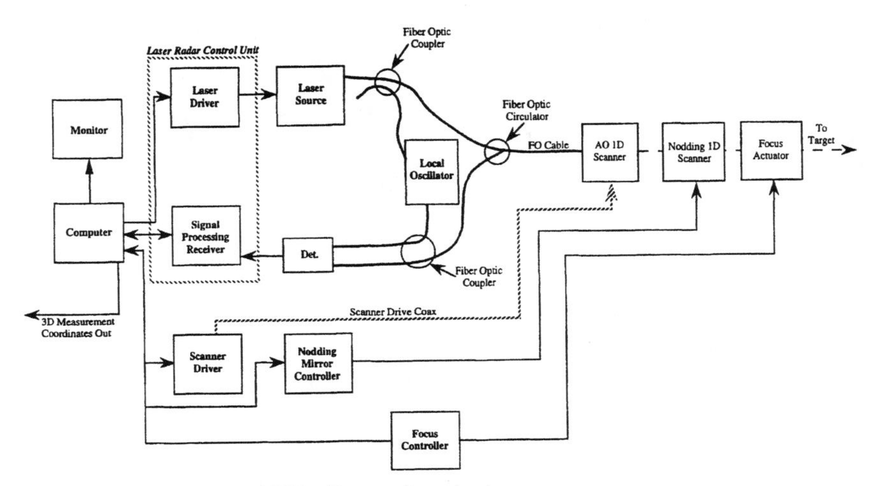

Figure 1. Schematic of the FM CLR with a two-dimensional scanner.

The “viewing” capability of the CLR system was also investigated.[3] A metrological map of a surface with features can be converted into a “picture quality” image that helps to make a visual assessment of surface patterns. The CLR can be used to acquire images in a wide range of formats and resolutions. The prototype version has a pointing accuracy of 0.01° and can be programmed to scan areas up to ±40° in azimuth and elevation. Scan patterns can be set up for raster-type, serpentine, or discrete point scans. The data may then be used to produce high-quality surface images. The image of a dime obtained by precision range scanning of its surface is shown in Figure 2.

IN-VESSEL DEPLOYMENT SYSTEM

A conceptual design effort is under way to develop a deployment mechanism mounted to a top port of the vacuum vessel. Two viewing/metrology deployment systems are being evaluated. A one-piece 150-mm-diam, 11-m-long mast to ensure positioning accuracy and minimize deflection and vibration would use a linear actuated drive (ball screw) and high-precision bearing guides to deploy it through the top port opening. The other system would use a telescopic mast that extends into the vacuum vessel. Both concepts will utilize an airlock housing mounted on top of the vacuum vessel. The amount of out-of-vessel storage space required by both systems represents an important design consideration. For example, in its retracted position the telescopic mast will require approximately one-third of the storage space needed by the single mast design. Figure 3 is a sketch of the deployment system mounted above the reactor.[2]

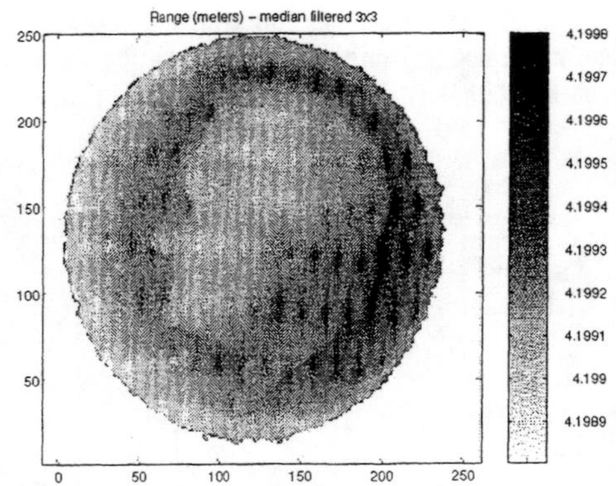

Figure 2. Image of dime developed from range measurements.

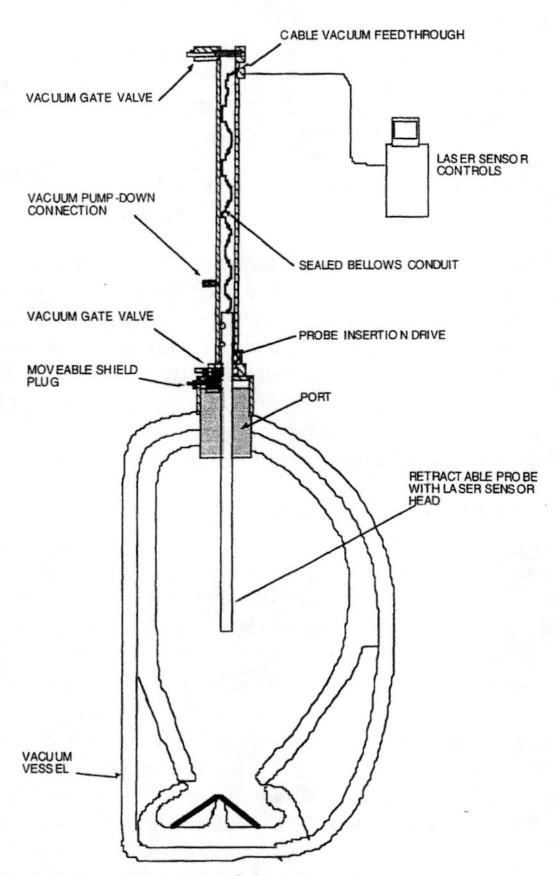

Figure 3. Viewing/metrology system mounted above the reactor.

The design concerns associated with the telescopic mast are also significant. For example, achieving the same extended stiffness and deflection resistance as the one-piece mast will be difficult. In addition, the stepped-down diameter will have a reduced cross-section of the mast at the lower end, which may limit the cabling and coolant lines that can feed through the system. Coolant lines will be needed to support a differential environment inside the mast to protect cables, motors, and sensors. For this configuration, special shielding at each sliding joint may have to be designed to maintain a temperature of <150°C inside the mast.

Either system must penetrate two boundaries: the vacuum vessel and the cryostat wall. This penetration scheme will allow quick access to the deployment system for removal and installation. Isolation valves will be used to seal the opening and maintain the boundary separation between the vacuum vessel and the cryostat. A movable neutron-gamma shield plug protects the stored system and provides access through the cryostat and vacuum vessel. A special feedthrough connection will be used for power and control cables, fiber-optic lines, and coolant lines. Both deployment system designs are presently envisioned to have 2 degrees of freedom (DOF), namely vertical travel and 360° rotation. All system motions will operate at constant velocity to meet the scanning and resolution requirements of the viewing/metrology sensor equipment.

To ensure quick deployment into the vacuum vessel, the viewing system will be permanently mounted to the top ports of the vacuum vessel at ten locations. Nuclear shielding will be required to prevent neutron and gamma radiation streaming. A special shielding mechanism mounted at the vacuum vessel port will allow deployment of the probe. The present concept is a key-lock mechanism that is being evaluated by the ITER central design group. The viewing/metrology mast and key-lock system will operate together as one mechanism.

RADIATION TESTING OF SENSOR COMPONENTS

The performance of two crucial components of the metrology system will be established under the radiation conditions anticipated in ITER. These components are (1) the polarization maintaining (PM) fiber and (2) the AO scanner. The latter is a complex device involving several active and passive components. The basic AO crystal planned to be used in the ITER metrology scheme is TeO_2.

The gamma radiation dose in proximity to the spent fuel rods in the High Flux Isotope Reactor at ORNL, soon after they are removed from the active reactor, is ~10^6 Gy/h. This dose decays with time, exhibiting a characteristic e-folding decay time of ~8 days. Thus, cumulative doses of ~10^7 Gy can be imparted to test objects if they can be introduced into the test region. The axial length of the test piece for which the radiation dose will be uniform within ±10% is about 0.2 m.

Plans are under way to test the effect of radiation on the transmission characteristics of a 1550-nm laser beam through the PM fiber and the TeO_2 crystal. If these tests prove promising, the scanning performance of an entire AO scanning device will be evaluated. Several meters of the PM fiber will be wound on a short spool (<0.2 m) and introduced to the test region. The physical size of the AO device (maximum dimension 45 mm) appears to fit in the experimental region although the details of the test itself are yet to be finalized.

CONTINUING WORK

An important aspect of the design work is to evaluate the time required for mapping PFCs. This is being done using three-dimensional modeling computer simulation. A model of the ITER vacuum vessel was created with access ports, blanket/shield components, and divertor components.[3] It included the sensor head with 2 DOF positioning the laser scanner device. This simulation is being updated and will be used to scan plasma-facing surfaces, particularly the divertors, from the top port locations. Modeling will be expanded to verify that ten ports are necessary for full surface coverage of the divertors, as well as other surfaces. Time and motion studies will be conducted to assess various operating scenarios. Preliminary modeling results have shown that a rough scan of the entire in-vessel surface can be completed in less than 8 h if all probes are utilized simultaneously. Metrology tests using the prototype sensor equipment will be repeated on PFC candidate material surfaces such as beryllium, tungsten, and carbon fiber composites.

CONCLUSIONS

A remotely deployed metrology system based on coherent FM laser radar is being designed for ITER in-vessel inspections. Experiments using a prototype system show that the technique has the capability to map complex surfaces to submillimeter accuracy from distances of ~20 m. The quality of the metrological image obtained by this technique also demonstrates that the system can be utilized for in-vessel viewing of reactor internals without the need for an illuminating source. The modifications necessary for adopting this system for use in the harsh ITER environment are being studied, along with development of a mast-like deployment system.

ACKNOWLEDGMENTS

The authors would like to acknowledge A. Slotwinski and R. Sebastian of Coleman Research Corporation for developing the metrology system design concept. This work is sponsored by the Office of Fusion Energy, U.S. Department of Energy, under contract DE-AC05-96OR22464 with Lockheed Martin Energy Research Corporation.

REFERENCES

1. Barry, R. E., et al., *An Analysis of Metrology Techniques Related to ITER Mapping Requirements*, ORNL/TM-13104, Oak Ridge Natl. Lab., Lockheed Martin Energy Systems Inc., September 1994.
2. *Remote Metrology System (RMS) Design Concept*, CDR/95-073, Coleman Research Corp. (Oct. 1995).
3. Barry, R. E., et. al., "A 3D Remote Metrology System for the International Thermonuclear Experimental Reactor," American Nuclear Society Winter Meeting, San Francisco, CA (Nov. 1995).

Prototyping a Three-link Robot Manipulator

Tarek M. Sobh, Mohamed Dekhil, Thomas C. Henderson,
and Anil Sabbavarapu

Department of Computer Science and Engineering
University of Bridgeport
Bridgeport, CT 06601, USA
and
Department of Computer Science
University of Utah
Salt Lake City, UT 84112, USA

Abstract

In this paper we will present the stages of designing and building a three-link robot manipulator prototype that was built as part of a research project for establishing a prototyping environment for robot manipulators. Building this robot enabled us determine the required subsystems and interfaces to build the prototyping environment, and provided hands-on experience for the real problems and difficulties that we would like to address and solve using this environment. Also, this robot is used as an educational tool in robotics and control classes.[1]

Keywords: Robotics, Prototyping, Control, Design.

1 Introduction

Teaching robotics in most engineering schools lakes the practical side and usually students end up taking lots of theoretical background and mathematical basis, and maybe writing some simulation programs, but they do not get the chance to apply and practice what they have learned on real robots. This is due to the fact that most of the robots available in the market are either too advanced, complicated, and expensive (e.g., specialized industrial robots), or toy-like robots which are too trivial and does not give the required level of depth or functionality needed to demonstrate the main concepts of robot design and control. One of our goals in this project, was to build a robot that is simple, flexible, and easy to use and connect to any workstation or PC, and at the same time, is capable of demonstrating some of the design and control concepts. We also tried to keep the cost as low as possible to make it available to any engineering school or industrial organization.

We consider the main contribution of this work to be building URK (Utah Robot Kit) which is a three-link robot prototype that has a small size and reasonable weight which is convenient for a small

[1]This work was supported in part by DARPA grant N00014-91-J-4123, NSF grant CDA 9024721, and a University of Utah Research Committee grant. All opinions, findings, conclusions or recommendations expressed in this document are those of the author and do not necessarily reflect the views of the sponsoring agencies.

lab or a class room. URK can be connected to any workstation or PC through the standard serial port with an RS232 cable, and can be controlled using a software controller with a graphical user interface. This software controller applies a simple PID control low for each link which does not require the knowledge of the robot parameters. Therfore, this software can be used to control any electro-mechanical system that can be controlled by a physical PID controller. The interface enables the user to change any of the control parameters and to monitor the behavior of the system with an on-line graphs and a 3-D view for the robot showing the current position of the robot.

2 Background and Related Work

Controlling and simulating a robot is a process that involves a large number of mathematical equations. To be able to deal with the required amount of computation, it is better to divide them into modules, in which each module accomplishes a certain task. The most important modules, as described in [2], are kinematics, inverse kinematics, dynamics, trajectory generation, and linear feedback control.

2.1 Robot Modules

There has been a lot of research to automate kinematic and inverse kinematic calculations. A software package called SRAST (Symbolic Robot Arm Solution Tool) that symbolically solves the forward and inverse kinematics for n-degree of freedom manipulators has been developed by Herrera-Bendezu, Mu, and Cain [5]. The input to this package is the Denavit-Hartenberg parameters, and the output is the direct and inverse kinematics solutions. Another method of finding symbolic solutions for the inverse kinematics problem was proposed in [11]. Kelmar and Khosla proposed a method for automatic generation of forward and inverse kinematics for a reconfigurable manipulator system [7].

Dynamics is the study of the forces required to cause the motion. There are some parallel algorithms to calculate the dynamics of a manipulator. Several approaches have been suggested in [8, 9, 10] based on a multiprocessor controller, and pipelined architectures to speed up the calculations.

Linear feedback control is used in most control systems for positioning and trajectory tracking. There are sensors at each joint to measure the joint angle and velocity, and there is an actuator at each joint to apply a torque on the neighboring link. The readings from the sensors will constitute the feedback of the control system. By choosing appropriate gains we can control the behavior of the output function representing the actual trajectory generated. Minimizing the error between the desired and actual trajectories is our main concern. Figure 1 shows a block diagram for the controller, and the role of each of the robot modules in the system.

2.2 Local PD feedback Control vs Robot Dynamic Equations

Most of the feedback algorithms used in current control systems are implementations of a proportional plus derivative (PD) control. In industrial robots, a local PD feedback control law is applied at each joint independently. Some ideas have been suggested to enhance the usability of the local PD feedback law for trajectory tracking. One idea is to add a lag-lead compensator using frequency response analysis [1]. Another method is to build an inner loop stabilizing controller using a multi-variable PD controller, and an outer loop tracking controller using a multi-variable PID controller [12]. In general, using a local PD feedback controller with high update rates can give an acceptable accuracy for trajectory tracking applications. It was proved that using a linear PD feedback law is useful for positioning and trajectory tracking [6].

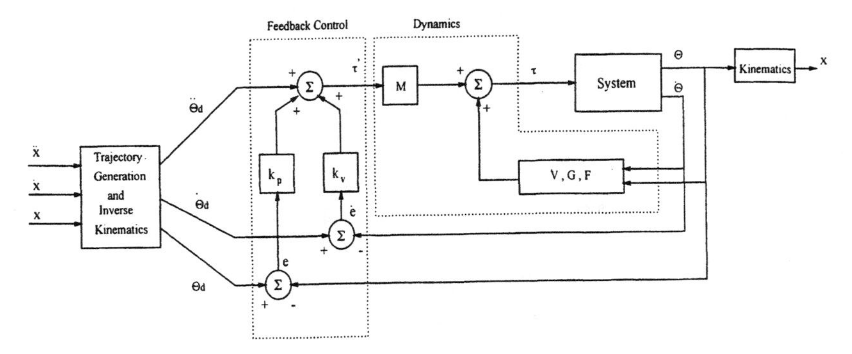

Figure 1: Block diagram of the Controller of a Robot Manipulator.

3 Prototyping a 3-Link Robot

3.1 Analysis Stage

This project was started with the study of a set of robot configurations and analyzed the type and amount of calculation involved in each of the robot controller modules (kinematics, inverse kinematics, dynamics, trajectory planning, feed-back control, and simulation). This phase was accomplished by working through a generic example for a three-link robot to compute symbolically the kinematics, inverse kinematics, dynamics, and trajectory planning; these were linked to a generic motor model and its control algorithm. This study enabled us to determine the specifications of the robot for performing various tasks, it also helped us decide which parts (algorithms) should be hardwired to achieve specific mechanical performances, and also how to supply the control signals efficiently and at what rates.

3.2 Controller Design

The first step in the design of a controller for a robot manipulator is to solve for its kinematics, inverse kinematics, dynamics, and the feedback control equation that will be used. Also the type of input and the user interface should be determined at this stage. We should also know the parameters of the robot, such as: link lengths, masses, inertia tensors, distances between joints, the configuration of the robot, and the type of each link (revolute or prismatic). To make a modular and flexible design, variable parameters are used that can be fed to the system at run-time, so that this controller can be used for different configurations without any changes.

The kinematics and the dynamics of the three models have been generated using some tools in the department called *genkin* and *gendyn* that take the configuration of the manipulator in a certain format and generate the corresponding kinematics and dynamics for that manipulator. For the trajectory generation, The cubic polynomials method was used. The error in position and velocity is calculated using the readings of the actual position and velocity from the sensors at each joint. Our control module simulated a PID controller to minimize that error. The error depends on several factors such as the frequency of update, the frequency of reading from the sensors, and the desired trajectory.

Figure 2: The physical three-link robot manipulator.

3.3 Simulation

A simulation program has been implemented to study the performance of each manipulator and the effect of varying the update frequency on the system. Also it helps to find approximate ranges for the required torque and/or voltage, and to determine the maximum velocity to know the necessary type of sensors and A/D. To make the benchmarks, as described in the next section, we did not use a graphical interface to the simulator, since the drawing routines are time consuming, and thus give misleading figures for the speed.

3.4 PID Controller Simulator

As mentioned in Section 2.2, a simple linear feedback control law can be used to control the robot manipulator for positioning and trajectory tracking. For this purpose, a PID controller simulator was developed to enable testing and analyzing the robot behavior using this control strategy. Using this control scheme helps us avoid the complex (and almost impossible) task of determining the robot parameters for our three-link prototype robot. One of the most complicated parameters is the inertia tensor matrix for each link, especially when the links are nonuniform and have complicated shapes.

3.5 Building the Robot

The assembly process of the mechanical and electrical parts was done in the Advanced Manufacturing Lab (AML) with the help of Mircea Cormos and Prof. Stanford Meek. In this design the last link is movable, so that different robot configurations can be used (see Figure 2).

There are three motors to drive the three links, and six sensors (three for position and three for velocity), to read the current position and velocity for each link to be used in the feedback control loop.

This robot can be controlled using analog control by interfacing it with an analog PID controller. Digital control can also be used by interfacing the robot with either a workstation (Sun, HP, etc.) or a PC via the standard RS232. This requires an A/D and D/A chip to be connected to the workstation (or the PC) and an amplifier that provides enough power to drive the motors. A summary of this design can be found in [3, 4].

4 Testing and Results

4.1 Simulator for three-link Robot

This simulator was used to give some rough estimates about the required design parameters such as link lengths, link masses, update rate, feedback gains, etc. It is also used in the benchmarking described earlier. This simulator uses an approximate dynamic model for the robot, and it allows any of the design parameters to be changed.

4.2 Software PID Controller

A software controller was implemented for the three-link robot. This controller uses a simple local PID control algorithm, and simulates three PID controllers; one for each link. Several experiments and tests have been conducted using this software to examine the effects of changing some of the control parameters on the performance of the robot.

The control parameters that can be changed in this program are:

- forward gain (k_g)
- proportional gain (k_p)
- differential gain (k_v)
- integral gain (k_i)
- input trajectory
- update rate

In these experiments, the program was executed on a Sun SPARCStation-10, and the A/D chip was connected to the serial port of the workstation. One problem we encountered with this workstation is the slow protocol for reading the sensor data, since it waits for an I/O buffer to be filled before it returns control to the program. We tried to change the buffer size or the time-out value that is used, but we had no success in that. This problem causes the update rate to be very low (about 30 times per second), and this affects the positional accuracy of the robot. We were able to solve this problem on an HP-700 machine, and we reached an update rate of 120 times per second which was good enough for our robot.

5 Conclusion

A prototype 3-link robot manipulator was built to determine the required components for a flexible prototyping environment for electro-mechanical systems in general, and for robot manipulators in particular. A local linear PD feedback law was used for controlling the robot for positioning and trajectory tracking. A graphical user interface was implemented for controlling and simulating the robot. This robot is intended to be an educational tool, therefore it was designed in such a way that makes it very easy to install and manipulate.

Acknowledgments

We would like to express our thanks to Mircea Cormos, Prof. Sanford Meek, and Prof. Beat Brüderlin for helping make this robot come to life.

References

[1] CHEN, Y. Frequency response of discrete-time robot systems - limitations of pd controllers and improvements by lag-lead compensation. In *IEEE Int. Conf. Robotics and Automation* (1987), pp. 464–472.

[2] CRAIG, J. *Introduction To Robotics*. Addison-Wesley, 1989.

[3] DEKHIL, M., SOBH, T. M., AND HENDERSON, T. C. URK: Utah Robot Kit - a 3-link robot manipulator prototype. In *IEEE Int. Conf. Robotics and Automation* (May 1994).

[4] DEKHIL, M., SOBH, T. M., HENDERSON, T. C., AND MECKLENBURG, R. Robotic prototyping environment (progress report). Tech. Rep. UUCS-94-004, University of Utah, Feb. 1994.

[5] HERRERA-BENDEZU, L. G., MU, E., AND CAIN, J. T. Symbolic computation of robot manipulator kinematics. In *IEEE Int. Conf. Robotics and Automation* (1988), pp. 993–998.

[6] KAWAMURA, S., MIYAZAKI, F., AND ARIMOTO, S. Is a local linear pd feedback control law effictive for trajectory tracking of robot motion? In *IEEE Int. Conf. Robotics and Automation* (1988), pp. 1335–1340.

[7] KELMAR, L., AND KHOSLA, P. K. Automatic generation of forward and inverse kinematics for a reconfigurable manipulator system. *Journal of Robotic Systems 7*, 4 (1990), pp. 599–619.

[8] LATHROP, R. H. Parallelism in manipulator dynamics. *Int. J. Robotics Research 4*, 2 (1985), pp. 80–102.

[9] LEE, C. S. G., AND CHANG, P. R. Efficient parallel algorithms for robot forward dynamics computation. In *IEEE Int. Conf. Robotics and Automation* (1987), pp. 654–659.

[10] NIGAM, R., AND LEE, C. S. G. A multiprocessor-based controller for mechanical manipulators. *IEEE Journal of Robotics and Automation 1*, 4 (1985), pp. 173–182.

[11] RIESELER, H., AND WAHL, F. M. Fast symbolic computation of the inverse kinematics of robots. In *IEEE Int. Conf. Robotics and Automation* (1990), pp. 462–467.

[12] TAROKH, M., AND SERAJI, H. A control scheme for trajectory tracking of robot manipulators. In *IEEE Int. Conf. Robotics and Automation* (1988), pp. 1192–1197.

Decentralized Control of Multiple Robots Moving in Formation

Wenzhong Tang and Hong Zhang
Department of Computing Science, University of Alberta
Edmonton, Alberta, Canada, T6G 2H1

ABSTRACT

We describe a single-neighbor following strategy for decentralized control of robot formation marching. Our robot group consists of a single leader robot and a collection of follower robots with identical mobility and sensing capability. The formation is maintained by each robot (except for the leader) sensing the position of and tracking one of its neighbors without central control or global information. The validity of the approach is demonstrated through simulation in which an arbitrarily specified formation involving a varying number of robots has been achieved accurately.

KEYWORDS: formation marching, multiple robots, decentralized

INTRODUCTION

In some multi-robot applications such as the mine-sweeping task, it is often desirable for the robots in the group to form and maintain a spatial relationship in order to perform the collective tasks efficiently and reliably. Two related but different research issues that have been considered are the *formation generation* problem [5], and the *formation marching* problem. The formation marching problem, with which this paper is concerned, assumes that the formation has already been created and that the objective is for the formation to move through a desired trajectory while maintaining the desired spatial configuration.

Centralized control strategies have been proposed for robot groups to maintain formation. Wang [7] proposed several distributed navigation strategies for movement in formation, namely, *nearest-neighbor tracking*, *multi-neighbor tracking*, and *mixed nearest-neighbor tracking and inertially referenced movements.* Each robot was treated as a point, and all the strategies assumed that a world coordinate frame (world frame) was known by all robots in the group (fleet), and that each robot knew the absolute coordinates of one or more nearest neighbors in the group and adjusted its movement accordingly. Chen and Luh [2] also investigated formation generation and formation marching of a small group of homogeneous mobile robots by imposing constraints. They assumed each robot has the knowledge of its current position and orientation with respect to the world coordinates. In addition, the leader robot of the group can broadcast its position and orientation with respect to the world coordinate frame to all other robots in the group. Approaches such as the ones described above where a substantial amount of global information is required are considered to be centralized control strategies.

In contrast, the formation marching problem has also been examined using strategies that are based on local interaction of robots in the group. Lynn Parker [4] investigated formation marching under a single leader. The author discussed how the performance of the robot group would be affected by the different proportion of global and local control information, and argued that in order to get satisfactory results, the following robots should have a proper balance between local and global control. In an unpublished study, Balch [1] discussed motor schema-based control for multi-agent robot formations. Four formations were considered for a group of four robots: *line*, *column*, *diamond*, and *wedge*. For each formation, each robot has a specific position and is assigned a unique ID. The *maintain-formation* motor schema generates a movement vector toward the desired formation position by using artificial potential fields. Three different formation position references are introduced: *Unit-center referenced*, *Leader-referenced*, and *Individual-referenced*; however, only the results of the unit-center reference were available.

In this paper, we will present our research in the decentralized control of multiple robots moving in formation. The desired formation is specified in terms of arbitrary desired relative position and orientation among the robots, and the formation is maintained using a neighbor following strategy. A robot acquires the position of its neighbor through sensing. The orientation is obtained by either local communication or estimation using consecutive positions. The problem of error propagation is handled through a method we call dynamic prediction. Kinematic and dynamic constraints are taken into account in order to produce realistic robot motion. Since the reliance on sensing makes our control strategies sensitive to sensor errors, we will examine how errors in sensory information affect the performance of formation marching controlled by different coordination strategies.

DECENTRALIZED FORMATION KEEPING STRATEGY

We propose a decentralized formation keeping algorithm, which we refer to as *single neighbor following (SNF)*. It is similar to *neighbor tracking* described in [7], except that in our case formation maintenance is based on sensing without a central reference frame. Neighbor following was also adopted by [4], but our emphasis will be placed on the generation of arbitrary formations, reduction of formation error through prediction, and orientation or heading estimation using consecutive positions. In addition, we will study the effect of inaccurate sensing on formation quality. We will specifically examine *single* neighbor following, in which each robot in the group except for the leader tracks the position of another robot nearby which is not necessarily the group leader.

Our robot model is based on the physical robot units that have been developed in our collective robotics research [3]. The robots are wheeled with two degrees of freedom. It occupies a circle of diameter d and has a mass of m. The robot is driven by two independent wheels placed on both the left and the right side of the robot. The speeds of the two wheels have an upper bound, which in turn define bounds on both the linear velocity and angular velocity of the robot. The mass of the robot is considered in the calculation of robot acceleration. Each robot has a number of different sensors that allow it to sense the position of nearby robots, obstacles, and goal position; it is assumed to be able to obtain its own orientation, use local communication to broadcast its own orientation and receive the orientations of other nearby robots.

Formation Keeping Algorithm

In the following strategy we propose, there is one group leader moving on its own and each follower robot tries to keep a pre-specified position and orientation with respect to its designated leading robot. Therefore, for a group of n robots, $n-1$ robots are follower robots and one is the group leader. Each follower robot has a *leading robot*, which is not necessarily the group leader, and there are $n-1$ leading robots.

The motion of each robot is driven by a potential function. Specifically, any robot

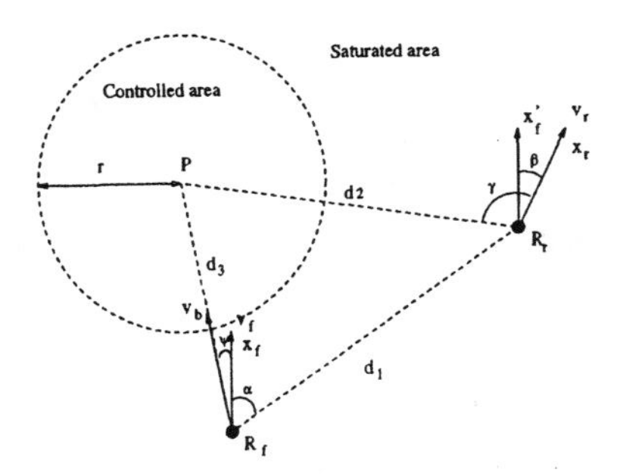

Figure 1: Calculate follow behavior of a follower robot

R_f in the group that is not the group leader calculates the position and heading of its leading robot R_l with respect to R_f's reference frame. It then generates an artificial potential with respect to R_l. The attractive potential in turn guides R_f to move to its desired position in the formation, subject to kinematic and dynamic constraints defined in the previous section.

Figure 1 illustrates how the keep formation behavior in 2-D space is calculated. At a given instant, R_f determines from the sensor information its reference robot R_r to be at a distance d_1 and a relative angle α. In addition, assume the angle between the moving direction of R_r and that of R_f to be β. According to the reference information known by the follower robot, the follower robot should be at point P of distance d_2 at an angle γ with respect to R_r. In order for R_f to reach to P, the keeping formation algorithm needs to know the distance d_3 between the current position of R_f and P, and the angle ψ between the current direction of R_f and the vector $\vec{v_b}$ from R_f to P.

.Using simple transformation matrix calculation, we can obtain the position of $P(x, y)$ in the follower robot frame as follows.

$$x = c\cos\beta - d\sin\beta + a \quad y = c\sin\beta + d\cos\beta + b$$

where $a = d_1 \cos\alpha$, $b = d_1 \sin\alpha$, $c = d_2 \cos\gamma$, and $d = d_2 \sin\gamma$. d_3 and ψ can then be defined in terms of x and y as: $d_3 = \sqrt{x^2 + y^2}$ and $\psi = \arctan\frac{y}{x}$.

Once the distance d_3 between the follower robot R_f and the desired position P and the angle ψ between them are known, the keep formation behavior generates a force vector $\vec{F_b}$ towards P. The magnitude of $\vec{F_b}$ depends on d_3. As the velocities and accelerations of the robot have upper limits, the force vector $\vec{F_b}$ must also have a upper limit. We define this upper limit as F_{max}. We define an area within radius r of the desired position P as the *controlled area*, and the area outside this area *saturated area*. When robot R_f is in the controlled area, the magnitude of $\vec{F_b}$ is in proportion of distance d_3. Otherwise, the magnitude of $\vec{F_b}$ is equal to F_{max}. The force function we use thus has the form

$$\|\vec{F_b}\| = \begin{cases} \frac{d_3}{r} F_{max} & 0 \le d_3 < r \\ F_{max} & d_3 \ge r \end{cases}$$

Direction Estimation Using Position Measurements

In the formation marching problem of multiple robots, each follower robot must know not only the relative position but also the relative heading direction of the robot being tracked. The latter can be obtained by sensing the heading directly or explicit inter-robot communication. However, the approximate heading of a robot can also be estimated from two or more consecutive measurements of its position This eliminates the need for inter-robot communication and that for additional sensors of instantaneous heading. Details of the estimation algorithm can be found in [6].

Error Reduction by Prediction

Static position errors in the formation are required to exist in order to activate the force function to move the robots. In a decentralized control scheme, if they are left uncompensated, these errors will propagate through the formation and degrade its quality, a phenomenon that is particularly pronounced in large robot groups. This problem can be partially overcome by moving the following robot at the same speed and direction as its leader, even in the absence of position errors. This can be accomplished by putting the center of the attractive force function of the follower robot somewhere ahead of its desired position.

Define d_c as the amount by which the new desired position is ahead of the robot's required position in the formation. The computation of d_c in our study has been based on empirical observations. First, d_c of a follower robot should be proportional to the speed of its leading robot v_l and the mass of the following robot m_f, and inversely proportional to the time interval between two consecutive positions Δt. Secondly, we should take into account the maximum force produced by the formation keeping algorithm F_{max}, and the maximum controlled distance d_{max}. Incorporating all these considerations, we calculate d_c by

$$d_c = K_c v_l \quad \text{where} \quad K_c = \frac{d_{max} m}{F_{max} \Delta t}$$

We call this method *dynamic prediction* and, with this method, we have been able to substantially reduce the static tracking errors between robots.

SIMULATION RESULTS

Simulation Results with Perfect Sensor Data

We have tested our formation marching algorithm in simulation. We either assume that the orientation of the leading robot is available or calculate it using positions. The effectiveness of dynamic prediction is also evaluated in simulation. Figure 2 illustrates three robots moving in a *row* formation where, when sensor data are assumed to be perfect, direction information of a leading robot can be accurately obtained from position information, and there is no difference in performance. However, without dynamic prediction, errors in the quality of the formation are noticeable. This is particularly clear in the right-most figure of the row formation where three robots maintain parallel to each other, unlike the middle or left figure. Tracking errors in the formation are also considerably less for the case using dynamic prediction. Finally note the smoothness of the trajectories, as a result of taking into account the kinematic and dynamic constraints. Figure 3 shows another example where a group of seven robots move in a V-formation. Other cases have also been tested and the results can be found in [6].

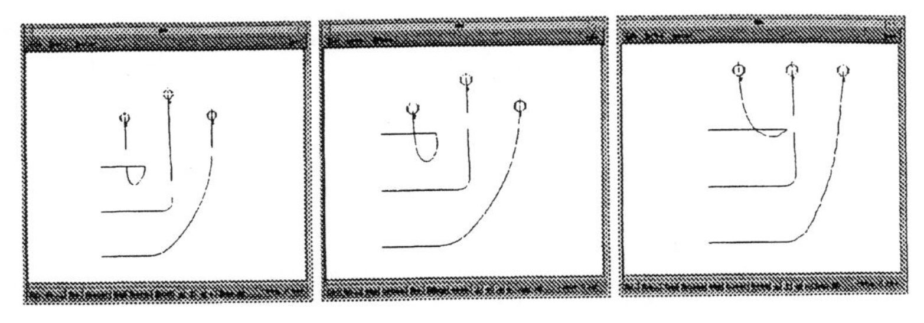

Figure 2: Three robots in a *row* formation. Left: SNF only; Middle: SNF with heading estimation; Right: SNF with heading estimation and dynamic prediction.

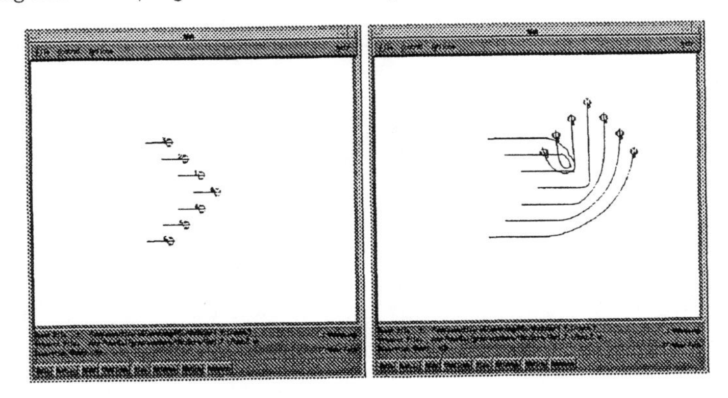

Figure 3: Seven robots in *V* formation before and after a right-angle turn

The Effect of Sensory Noise on Performance

Sensory errors affect the stability and quality of the formation. To simplify our study of the effect of sensor errors, we consider only errors in sensed position information. We use the formation error of all the robots averaged over the entire course of the mission as the metric to evaluate the quality of a formation. Two cases are examined here: (a) simple SNF, and (b) SNF with dynamic prediction. The formation errors for the two cases are shown in Figure 4, where the x-axis is the standard deviation δ of sensory errors along one of the principal axes of the R_l frame. A 2-D curve is sufficient if we set δ along the other axis to the same. The y-axis is the average formation error.

CONCLUSION

In this paper, we have discussed existing approaches to the formation marching problem, and presented our approach. Our study has shown that formation marching can be achieved in a decentralized fashion without an external supervisor, world reference frames, extensive world knowledge, or inter-robot communication. In addition, neighbor following strategy in general, and SNF in particular, are feasible in keeping formation for multiple robots. We have also found that direction information can be obtained by observing the change in the position of the robot being tracked, thus

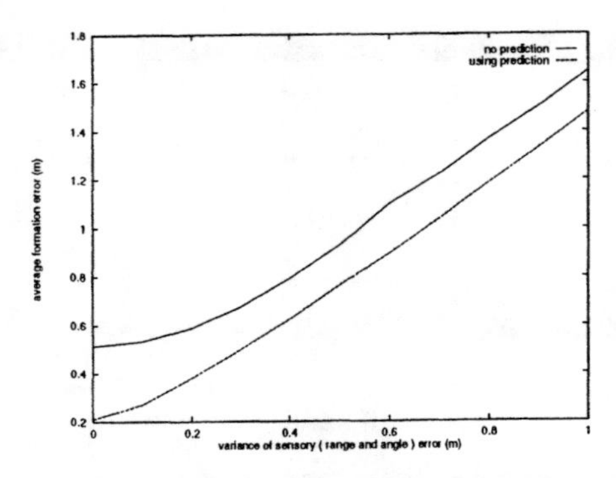

Figure 4: Formation error given leader's orientation with and without prediction

reducing the need for sensing. Static position error in formation can be eliminated by using prediction. The performance of formation algorithms under noisy sensory inputs heavily depends on the accuracy of the sensed relative direction of the robots being followed. When sensory information becomes unreliable, moderate communication can greatly enhance the robustness of formation marching. However communication must be employed with caution in order to avoid possible communication bottleneck. We are currently trying to implement the formation keeping strategies to a group of 10 physical mobile robots. Future work includes improving the tracking algorithm using multi-neighbor following strategies and the incorporation of other group behaviors.

References

[1] Tucker Balch. Motor schema-based control for multiagent robot formations. Working paper, Mobile Robot Laboratory, College of Computing, Georgia Intitute of Technology, June 1994.

[2] Q. Chen and J.Y.S. Luh. Coordination and control of a group of small mobile robots. In *Proceedings of IEEE International Conference on Robotics and Automation*, volume 3, pages 2315–2320, May 1994.

[3] C.R. Kube and H. Zhang. Collective robotics: From social insects to robots. *Adaptive Behavior*, 2(2):189–219, 1994.

[4] L.E. Parker. Designing control laws for cooperative agent teams. In *Proceedings of IEEE International Conferencence on Robotics and Automation*, volume 3, pages 582–587, 1993.

[5] K. Sugihara and I. Suzuki. Distributed motion coordination of multiple mobile robots. In *Proceedings of the 5th IEEE International Symposium on Intelligent Control*, volume 1, pages 138–143, September 1990.

[6] Wenzhong Tang. Decentralized control of multiple robots moving in formation. Master's thesis, University of Alberta, May 1995.

[7] P.K.C. Wang. Navigation strategies for multiple autonomous mobile robots moving in formation. *Journal of Robotic Systems*, 8(2):177–195, 1991.

FORCE CONTROL OF A MEDICAL ROBOT FOR ARTERIAL DISEASES DETECTION

Xavier THEROND, Eric DEGOULANGE, Etienne DOMBRE and François PIERROT
LIRMM - UMR 5506 - Université Montpellier II / CNRS
161, rue Ada - 34392 Montpellier Cedex 5 - FRANCE

ABSTRACT

This paper deals with the force control of a medical robot for arterial diseases detection. The robot is used to move probes on the patient's skin while exerting a given effort. First, the medical robotics application is described and the safety issues are summarized; the choices which have been made regarding the robot and the control scheme are justified. The so-called external force control scheme is described. Experimental results showing its efficiency are presented.

KEYWORDS : medical robotics, safety, redundant manipulator, force control

1- INTRODUCTION

Robots have first been used to replace operators in the realization of industrial tasks (painting, assembling, welding, ...) which are repetitive, tedious or dangerous tasks. Nowadays, robotics is more and more involved in servicing tasks. In these tasks, new scientific and technical challenges arise from the fact that the robot does not take the place of the operator anymore but has to closely cooperate with him.

The work presented in this paper is about such a cooperative robot for medical applications. We are working with a research team of the Cardiovascular Center of the Broussais[1] Hospital. In this center physicians are studying the evolution of patient arterial diseases. They analyze essentially the superficial arteries such as the right and left carotids, humeral and femoral arteries. Clinical observations have shown that the results could even be extended to others arteries such as coronaries. The aim of these medical procedures is to detect atheromatous plaques and to evaluate arterial elasticity. In that purpose, two kinds of probes are used: *ultrasound probes* to display arterial structures; *Doppler effect based probes* to measure arterial velocity profiles.

These probes are manually moved. However, a more reproducible resolution and a more reliable spatial location would be obtained if the probes were located and moved by a robot, which was the aim of the work presented in this paper.

The paper is divided in four parts. In the first part, the two medical procedures are described. In the second part we summarize the safety constraints. These constraints led

[1] Institut National de la Santé Et de la Recherche Médicale (INSERM U-28) - Hôpital Broussais, Paris

us to choose a redundant robot and a so-called external force control scheme. In the third part, we present the force control which has been implemented. Experimental results, both in free space and in constrained space, are given in the last part.

2- MEDICAL CONTEXT

In order to prevent infarctions, the team of the Cardiovascular Center of the Broussais Hospital studies the elasticity of patients' arteries. They are looking for atheromatous plaques filling up arteries. For this purpose, two parameters have to be evaluated: the thickness and the diameter of the artery.

Depending on the kind of probe, two different procedures (ultrasound-based and Doppler-based) are used to determine the artery dimensions. However, each procedure is run following two similar phases: probe placement, and measurements. These phases are described in [1], [2] and [3]:

- the *probe placement phase* consists in adjusting accurately both position and orientation of the probe until contact with the patient's skin is obtained; the probe is moved from any current location to a desired location on the skin,
- the *measurement phase* consists in moving the probe in contact with the skin and requires small displacements. Each measurement has to be synchronized with heartbeat.

In the ultrasound-based procedure, several images are taken when the probe is motionless: they provide information on the artery elasticity and on the atheromatous plaque shape (figure 1). In the Doppler-based procedure, small radial displacements are generated to provide a series of velocity curves.

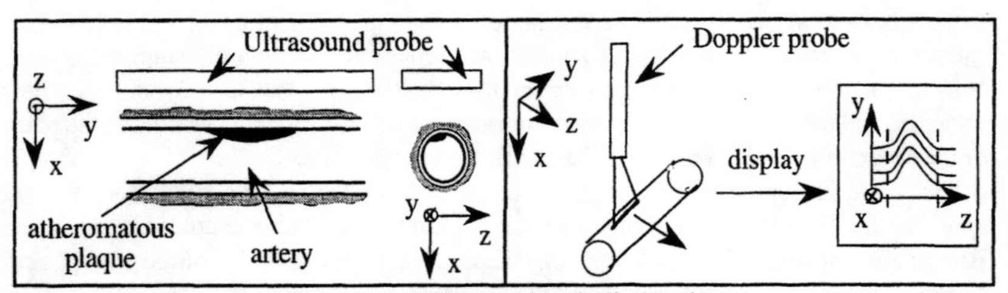

Figure 1: Carotid artery - Longitudinal section

During the measurements constant step-by-step displacements are required, ranging from 0.1 mm to 10mm depending on the probe and on the experimental procedure. It is clearly difficult to manually perform such small displacements in a reproducible way: they could be automatically and efficiently done with a robot.

3- SAFETY ISSUES

Safety is a very important aspect in medical robotics. Medical robots must operate in contact with physicians and patients whereas industrial robots are isolated in a cell with safety interlocks to prevent any robot motion when someone is inside the cell.

The safety constraint assessment is not very easy and depends on the application. The requirements are obviously different when the robot is used for instance as an assistant in neurosurgery, or when it has to move a non-invasive probe on the patient's skin. In the first case a very good accuracy is needed and the robot must stop instantaneously and hold the desired pose in case of failure. In the second case, the accuracy is less critical but the effort exerted by the probe on the skin has to be controlled and monitored in a safe manner.

Many authors have addressed the safety problem (see for instance [4], [5], [6], [7], [8], [9], [10]) and have proposed various solutions. Among these solutions, some of them are more relevant for our application:

- the robot behavior has to be controlled at any time. In particular, when a failure occurs, any uncontrolled motion must be prevented,
- when moving, only slow motions are allowed. This can be a build-in feature of the robot if the gear reduction ratios are large enough,
- it is better to make use of a Dead Man Switch (DMS) rather than an emergency stop button,
- the force applied by the robot on the patient's skin must be controlled,
- the working area of the robot must be restricted.

Regarding the implementation of these rules in a real system may lead to two different points of view. For the physician, the safety must be taken into account in the very beginning of the design of a new generation of intrinsically safe robot. This is probably the best solution but the more expensive one. For the engineer with a background in industrial robotics, it seems easier to modify a standard robot in order that it meets safety constraints. In our case, we have chosen an industrial redundant robot: PA-10 from Mitsubishi Heavy Industry [11]. This choice has been justified in terms of costs, performance and development constraints thanks to simulation studies presented in [1]. Redundancy is a key feature to avoid obstacles but it can also be exploited in order to give the physician the most convenient posture in terms of free space to operate.

In order to satisfy safety constraints, different features have been added:

- the gear reduction ratios have been multiplied by two with respect to the standard ones,
- in order to control the forces applied on the skin, which should range between 0.5 and 5N, we make use of a 6-axis force/torque sensor (Gamma model, 34N/3Nm from Assurance Technologies Inc.),
- during the measurement phase, automatic motion of the robot are monitored by the physician through a DMS,
- a hardware watchdog at the servo level must be added to stop the robot in case of communication failure between the servo level and the high level controller.

Figure 2 shows the workstation. A compatible personal computer (with 486 processor - 66MHz) allows us to control the servo-drivers through a LAN (ARCNET) connection with QNX, a real time and multitasking operating system.

4- FORCE CONTROL

The so-called external force control scheme [12], [13] presents two key features regarding safety: the joint position servo loop is always activated (this is not the case with the hybrid position/force control scheme), and it is possible to synthetize both position and force information on the same direction.

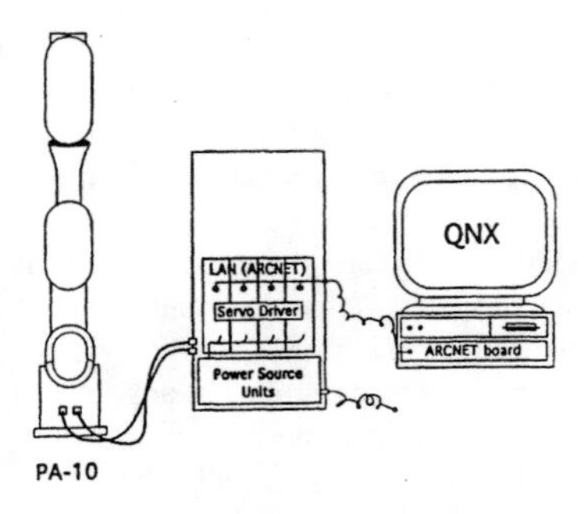

Figure 2: Workstation

The main idea is the following (figure 3): a variation of the robot cartesian pose $\Delta P = [\Delta X^T \ \Delta\phi^T]^T$ is calculated from the difference between the desired forces and moments $H_d = [F_d^T \ M_d^T]^T$ and the measured ones at the tool level $H_r = [F_r^T \ M_r^T]^T$. This increment is used to modify the command in the workspace: ΔP is added to the current pose $P_d = [X_d^T \ \phi_d^T]^T$ to obtain a new desired pose P.

To solve the inverse kinematics, a numerical method based on the pseudo-inverse has been implemented [14]. The inverse kinematics model transforms the desired cartesian pose into desired joint positions. At the joint level, a standard P.I.D. controller is implemented giving a very good stability[2].

In our application, we have two different phases. It was very important to use the same scheme for both. During the probe placement phase, H_d is kept equal to 0. During the measurement phase, the probe is moved in contact with the skin with small displacements (constrained space): H_d is obtained thanks to a cartesian trajectory generation.

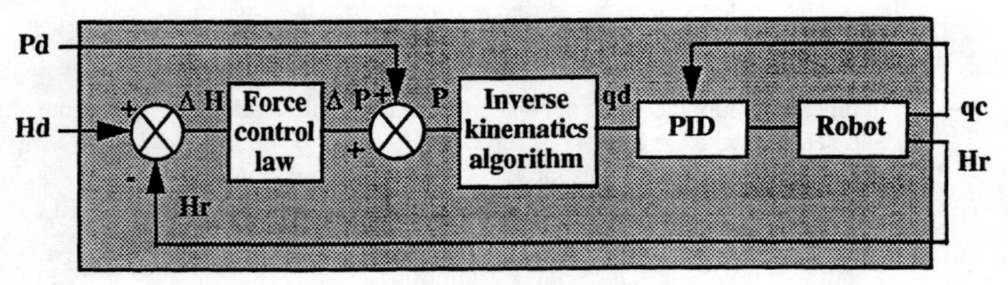

Figure 3: External force control

5- EXPERIMENTAL RESULTS

In this section, we present some experimental results obtained with the external force control scheme.

In the first series of curves (figures 4a and 5a), the robot is in the vicinity (0.6cm) of the leg of a dummy. The desired force along the z probe axis (orthogonal to the surface) is 3N. Figures 4a and 5a show the evolutions of the recorded force and of the corresponding

[2] It is not surprising when considering the large gear reduction ratios.

displacement along the z probe axis. While the probe is not in contact with the leg, the force is equal to 0. As soon as the probe is in contact with the leg, the force reaches a mean value of 3N. The displacements along z are computed from the joint encoders data.

For the second series of curves (figures 4b and 5b), the robot is in contact with the leg exerting a desired force of 3N along the z probe axis. The desired displacement along the x probe axis is equal to 10cm. Figure 4b shows that the force along the z probe axis can be maintained at a mean value of about 3N. Figure 5b presents the evolution of the displacement along x axis.

In both cases, the amplitude of the oscillations has a peak value less than 0.5N. They are due to the rather large sample period (10ms): it takes time for the robot to compensate for force errors. Better results would be obtained with a smaller one.

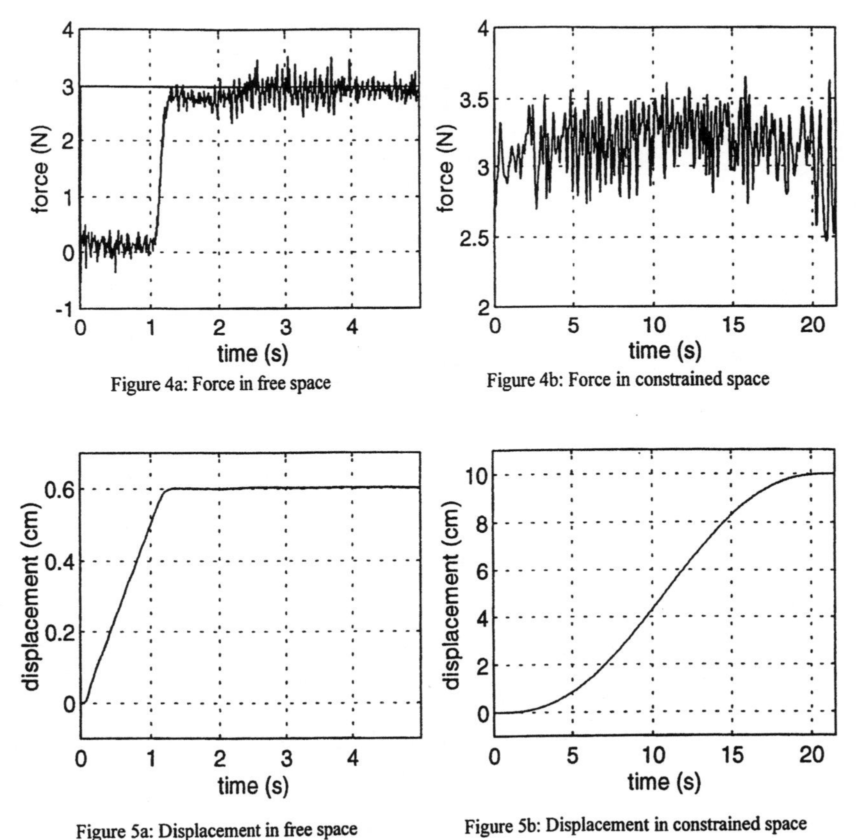

Figure 4a: Force in free space

Figure 4b: Force in constrained space

Figure 5a: Displacement in free space

Figure 5b: Displacement in constrained space

6- CONCLUSION

The work presented in this paper shows that it is possible to move probes in contact with a patient's skin in a safe manner with an external force control scheme involving a 6-axis force/torque sensor. This scheme makes the software developments easily portable on other applications requiring contact, generally speaking between a tool and a compliant surface, with a redundant or non redundant robot.

An interesting feature of our system is that it makes use of an on-the-shelves industrial robot and that the high level controller is a simple PC. It makes a cheap, robust, easy-to-use system for a physician, open for further improvements and evolutions.

Experimental results show the good behavior of the control scheme in both free space and constrained space. Clinical evaluations should be made in the very near future. For this purpose, the development of an user-friendly man-machine interface is in progress.

ACKNOWLEDGMENTS

We would like to thank the following institutions and individuals who have contributed to the work described here:

- Electricité De France/Direction des Etudes et Recherches (EDF/DER, Chatou) for its financial support and Jacques Pot,
- Alain Simon and Jaime Levenson, Broussais Hospital, Paris,
- Patrice Flaud and Jean-Louis Counord, Laboratoire de Biorhéologie et d'Hydrodynamique Physiologique (URA CNRS 343, Université Paris VII).

REFERENCES

1 X. Thérond, E. Dombre and F. Pierrot, "Arteries diseases detection: a cooperative-robot solution", *ICAR'95, Proc. 7th Int. Conf. on Advanced Robotics*, Vol. 1, pp 43-50, 1995.

2 J.-L. Counord and P. Flaud, "Device for positioning of U.S. probes with a rational method of exploration", *10th Annual Int. Conf. of the IEEE Engineering in Medicine and Biology Society*, New Orleans, Louisiane, Nov. 4-7, 1988.

3 P. Flaud, A. Bensalah, J.-L. Counord, J. Levenson and A. Simon, "A new geometric procedure for in vivo pulsed Doppler evaluation of velocity distribution inside the diametrical section of large arteries in humans", *Annuals of Biomedical Engineering*, Vol. 18, pp. 519-531, 1990.

4 R.O. Buckingham, "Safe active robotic devices for surgery", *Proc. IEEE Int. Conf. on Systems, Man and Cybernetics*, Vol. 5, pp. 355-358, Le Touquet, France, Oct. 17-20, 1993.

5 B.L. Davies, R.D. Hibberd, Ng WS, A.G. Timoney and J.E.A. Wickham, "A surgeon robot for prostatectomies", *Proc. Int. Conf. on Advanced Robotics*, Vol. 1, pp. 871-875, Pisa, Italy, June 19-22, 1991.

6 B. Davies, "Safety of medical robots", *Proc. Int. Conf. on Advanced Robotics*, pp. 311-317, Tokyo, Japan, Nov. 1-2, 1993.

7 T. Dohi, Y. Ohta, M. Tsuzuki, K. Miyata, D. Hashimoto and H. Iseki, "Robotics in computer aided surgery", *Proc. Int. Conf. on Advanced Robotics*, pp. 379-383, Tokyo, Japan, Nov. 1-2, 1993.

8 R. H. Taylor, H. A. Paul, P. Kazanzides, B. D. Mittelstadt, W. Hanson, J. Zuhars, B. Williamson, B. Musits, E. Glassman and W. L. Bargar, "Taming the bull: safety in a precise surgical robot", *Proc. Int. Conf. on Advanced Robotics*, pp. 865-870, Pisa, Italy, June 19-22, 1991.

9 J. Troccaz, S. Lavallée and E. Hellion, "A passive arm with dynamic constraints: a solution to safety problems in medical robotics?", *Proc. IEEE Int. Conf. on Systems, Man and Cybernetics*, Vol. 3, pp. 166-171, Le Touquet, France, Oct. 17-20, 1993.

10 J. Troccaz, S. Lavallée and E. Hellion, "PADyC: A passive arm with dynamic constraints", *Proc. Int. Conf. on Advanced Robotics*, pp. 361-366, Tokyo, Japan, Nov. 1-2, 1993.

11 K. Onishi, "New industrial world created by portable manipulator system", *Proc. 24th Int. Symp. on Industrial Robots*, pp. 815-820, Tokyo, Japan, Nov. 4-6, 1993.

12 V. Perdereau, "Contribution à la commande hybride force-position", *Thèse de Doctorat*, Université Paris VI, Feb. 1991.

13 E. Dégoulange, "Commande en effort d'un robot manipulateur à deux bras : application au contrôle de la déformation d'une chaîne cinématique fermée", *Thèse de Doctorat*, Université Montpellier II, Dec.1993.

14 E. Dombre, W. Khalil, "Modélisation et commande des robots", Edition Hermès, Paris, 1988.

A Simple Control Law for The Path Following Problem of a Wheeled Mobile Robot

A. Tayebi, M. Tadjine and A. Rachid

Laboratoire des Systèmes Automatiques
Université de Picardie - Jules Verne
7, Rue du Moulin-Neuf, 80000 Amiens, France.
E-Mail : tayebi@lsa.u-picardie.fr

ABSTRACT

This paper deals with the closed-loop path following problem of a unicycle-like mobile robot. The desired path is described by a fictitious reference mobile robot with the same kinematics constraints as the real one. A nonlinear control law based on partial feedback linearization and Lyapunov method is derived. The proposed controller ensures global asymptotic stabilization of the closed-loop system.

KEYWORDS: Wheeled mobile robot, Path following, Nonlinear feedback control design, Feedback linearization, Lyapunov method.

INTRODUCTION

A fundamental problem concerning wheeled mobile robots is the restricted mobility in the direction of the wheels axis which prevents the robot to move sideways. Therefore, the degrees of freedom are less than the configuration variables that are necessary to completely describe the kinematics behavior of such systems. Two main problems have extensively attracted the attention of many authors in the literature: the path following problem (see for example [2], [4-6], [10] and [12]), and the point-stabilization problem (see for example [1], [3], [7-8] and [11]). This paper deals with the first problem, where a kinematics model based on reference mobile robot tracking is used to derive a nonlinear control law which ensures global asymptotic stabilization of the closed loop-system. This controller is based on partial feedback linearization and the Lyapunov direct method.

The paper is organized as follows: In the next section, the kinematics model based on reference mobile robot tracking is derived. Thereafter, the tracking control law is proposed and the choice of control design parameters is discussed. Finally, some simulation results are presented to highlight the effectiveness of the proposed feedback control law, and some concluding remarks end the paper.

KINEMATICS MODEL

The mobile robot under consideration is a unicycle-like vehicle. The motion control of this vehicle can be achieved by dealing with the linear and rotational velocities (v, $\dot{\theta}$).

We assume that the vehicle rolls without slippage on horizontal ground, and the desired path is represented by a fictitious reference robot with the same nonholonomic constraints.

The configuration of the real robot (resp. reference robot) is described by its orientation θ (resp. θ_r) and the position of the point M (resp. M_r) located at mid-distance of the rear-wheels.

Let $(x_e, y_e, \theta_e)^T$ be the configuration-error vector, where (x_e, y_e) are the coordinates of the position-error vector $\overrightarrow{MM_r}$ in the basis of the mobile frame linked to the real mobile robot $(M, \vec{i}_1, \vec{j}_1)$, and θ_e is the orientation-error with respect to $\vec{i}_0$.

Let $(v, \dot{\theta})$ and $(v_r, \dot{\theta}_r)$ be respectively the linear and rotational velocities of the real and reference mobile robot.

In order to achieve a good tracking we must find an adequate feedback control law which allows to superpose the real robot and the reference one by vanishing the configuration-error $X_e = (x_e, y_e, \theta_e)^T$.

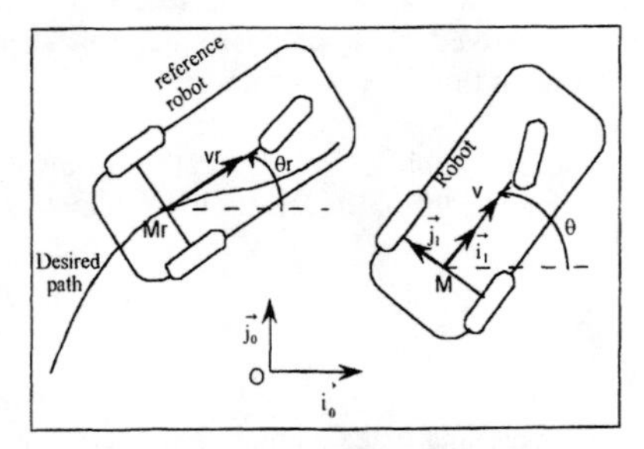

Figure 1. Reference mobile robot tracking.

In the mobile frame $(M, \vec{i}_1, \vec{j}_1)$ one has

$$\overrightarrow{MM_r} = x_e \vec{i}_1 + y_e \vec{j}_1, \tag{1}$$

$$\frac{d\,\overrightarrow{MM_r}}{dt} = \dot{x}_e \vec{i}_1 + \dot{y}_e \vec{j}_1 - y_e \dot{\theta}\, \vec{i}_1 + x_e \dot{\theta}\, \vec{j}_1. \tag{2}$$

The latter can also be written as follows

$$\frac{d\,\overrightarrow{MM_r}}{dt} = \frac{d\,\overrightarrow{OM_r}}{dt} - \frac{d\,\overrightarrow{OM}}{dt}, \tag{3}$$

where

$$\frac{d\,\overrightarrow{OM_r}}{dt} = v_r \cos\theta_e \vec{i}_1 - v_r \sin\theta_e \vec{j}_1, \tag{4}$$

and

$$\frac{d\vec{OM}}{dt} = v\vec{i}_1. \tag{5}$$

Substituting (4) and (5) in equation (3) and identifying with equation (2) yields

$$\begin{aligned} \dot{x}_e &= v_r \cos\theta_e + y_e\dot{\theta} - v \\ \dot{y}_e &= -v_r \sin\theta_e - x_e\dot{\theta} \end{aligned} \tag{6}$$

Let u_1 and u_2 be the control variables defined as:

$$\begin{aligned} u_1 &= v - v_r \\ u_2 &= \dot{\theta} - \dot{\theta}_r \end{aligned} \tag{7}$$

Substituting (7) in (6) we obtain the following global state representation

$$\begin{cases} \dot{x}_e = -u_1 + y_e u_2 + y_e\dot{\theta}_r + v_r(\cos\theta_e - 1) \\ \dot{y}_e = -x_e u_2 - x_e\dot{\theta}_r - v_r \sin\theta_e \\ \dot{\theta}_e = u_2 \end{cases} \tag{8}$$

CONTROLLER SYNTHESIS

Using the same kinematics model and a particular Lyapunov function depending on $(1-\cos\theta_e)$, Kanayama et al. [2] have proposed a non linear control law which stabilizes locally the equilibrium point $(0,0,0)$ for $v_r > 0$. However, this control law globally stabilizes the equilibrium points $(0,0,0\,mod(2\pi))$, it means that θ_e can converge to a multiple of 2π as it is shown in Figure 2.
Figure 3 shows the vehicle motion under the control law proposed in [2] for the following initial conditions:

$$x_{e0} = 2m, y_{e0} = 2m, \theta_0 = 3\frac{\pi}{4}rd.$$

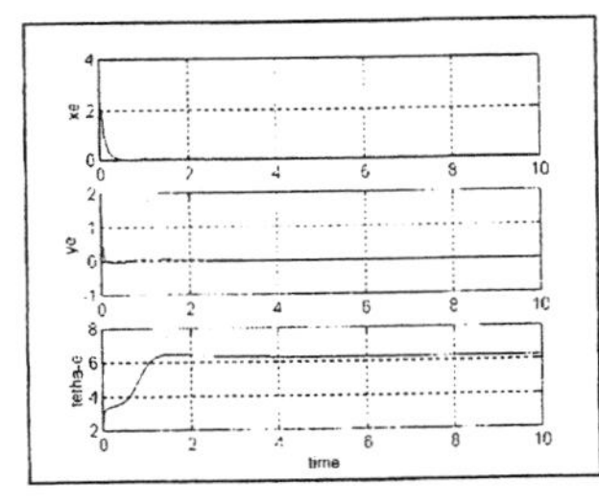

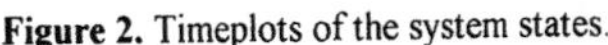

Figure 2. Timeplots of the system states.

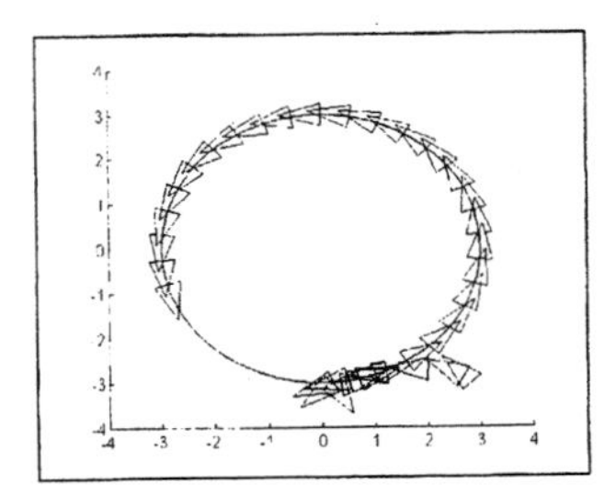

Figure 3. Circular path following.

In the sequel we present a nonlinear control law which globally stabilizes the equilibrium point $(0,0,0)$ if v_r is not equal to zero.

Proposition. Assume that $v_r \neq 0$, then the equilibrium point $(0,0,0)$ of system (8) is globally asymptotically stable under the following control law:

$$\begin{cases} u_1 = v_r(\cos\theta_e - 1) + y_e(\dot{\theta}_r + u_2) + k_1 x_e \\ u_2 = -k_2\theta_e + k_3 v_r y_e \dfrac{\sin\theta_e}{\theta_e} \end{cases} \tag{9}$$

where k_1, k_2 and k_3 are positive constant parameters. □

Proof. A global exponential stabilization of x_e to zero is obtained by using the linearizing feedback u_1 in the first equation of (8), which becomes

$$\dot{x}_e = -k_1 x_e \tag{10}$$

Since k_1 is a positive parameter, it is clear that x_e converges to zero as $exp(-k_1 t)$.
Let k_1 be chosen sufficiently large such that x_e converges to zero rapidly, and consider the following subsystem obtained from (8), when x_e is sufficiently small:

$$\begin{cases} \dot{y}_e = -v_r \sin\theta_e \\ \dot{\theta}_e = u_2 \end{cases} \tag{11}$$

Let us use the following Lyapunov function

$$V(X_e) = \frac{1}{2}(\theta_e^2 + k_3 y_e^2), \tag{12}$$

which is positive semi-definite and whose time derivative is

$$\dot{V}(X_e) = -k_3 v_r y_e \sin\theta_e + \theta_e u_2. \tag{13}$$

Substituting u_2 in (13) leads to

$$\dot{V}(X_e) = -k_2\theta_e^2. \tag{14}$$

Hence, $\dot{V}(X_e)$ is always negative semi-definite since k_2 is a positive parameter, therefore $V(X_e)$ decreases with respect to time and tends to a finite positive value when t tends to infinity, since k_3 is a positive parameter. Thus, $\dot{V}(X_e)$ tends to zero, and the convergence of θ_e to zero immediately follows.
Using La Salle theorem, the largest invariant set defined by $\dot{V}(X_e) = 0$ is restricted to $\{\ y_e = 0, \theta_e = 0\ \}$. Indeed, due to the convergence of θ_e it follows that $\dot{\theta}_e$ tends to zero when t tends to infinity. Due to this, from (11) u_2 tends to zero. Moreover, from (9) u_2 tends to $k_3 v_r y_e$. Finally, the convergence of y_e to zero immediately follows provided that $v_r \neq 0$. □

Above, we have demonstrated that the equilibrium point $(0,0,0)$ is globally asymptotically stable under the control law (9) for all positive values of k_1, k_2 and k_3.

However, since we want an optimal response we have to find some guidlines to the selection of these free design parameters. To this end, let us consider the case in which the reference robot is moving with a constant linear velocity in a neighborhood of the equilibrium point. The closed loop system becomes:

$$\begin{pmatrix} \dot{x}_e \\ \dot{y}_e \\ \dot{\theta}_e \end{pmatrix} = \begin{pmatrix} -k_1 & 0 & 0 \\ -\dot{\theta}_r & 0 & -v_r \\ 0 & k_3 v_r & -k_2 \end{pmatrix} \begin{pmatrix} x_e \\ y_e \\ \theta_e \end{pmatrix} \tag{15}$$

The closed-loop system is characterised by its eigenvalues which are solutions of the following equation:

$$(\lambda + k_1)(\lambda^2 + k_2 \lambda + k_3 v_r^2) = 0. \tag{16}$$

As shown before, $1/k_1$ corresponds to the time constant for the first order system which describe the global behavior of the coordinate x_e.
The local behavior of y_e and θ_e is described by the roots of:

$$\lambda^2 + k_2 \lambda + k_3 v_r^2 = 0 \tag{17}$$

Hence, we can choose k_2 and k_3 in accordance with the desired dynamics.

SIMULATION RESULTS

In this section we present simulation results for the path following problem when the real robot tracks a circular path ($v_r = 0.3m/s$ and $\dot{\theta}_r = 0.1rd/s$) and a rectilinear path ($v_r = 0.3m/s$ and $\dot{\theta}_r = 0$). The control parameters are chosen as follows:

$$k_1 = 1, k_3 = 10 \text{ and } k_2 = 2|v_r|\sqrt{k_3}.$$

Figure 4 shows the time evolution of the position and orientation errors, and Figure 5 shows the vehicle motion when the desired trajectory is circular, with the following initial conditions: $x_{e0} = 2m, y_{e0} = 2m, \theta_0 = 3\frac{\pi}{4}rd.$

In the case of rectilinear path tracking, the time evolution of the position and orientation errors are depicted in Figure 6 and the vehicle motion in Figure 7, for the following initial conditions: $x_{e0} = 1m, y_{e0} = 1m$ and $\theta_0 = \frac{\pi}{4}rd.$

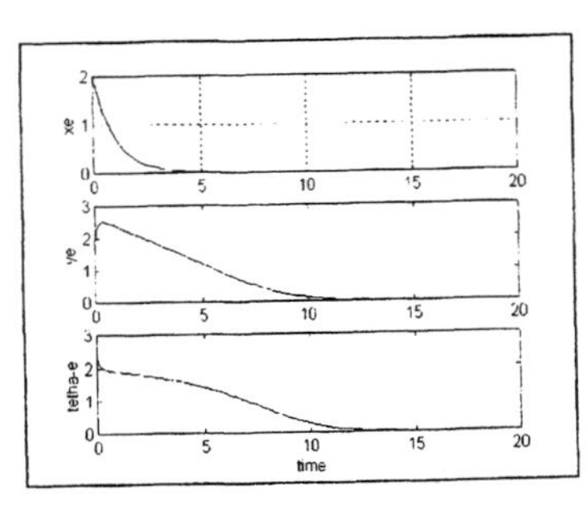

Figure 4. Timeplots of the system states.

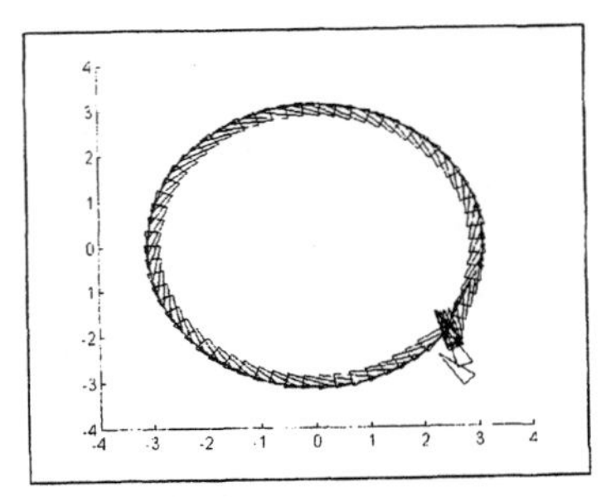
Figure 5. Circular path tracking

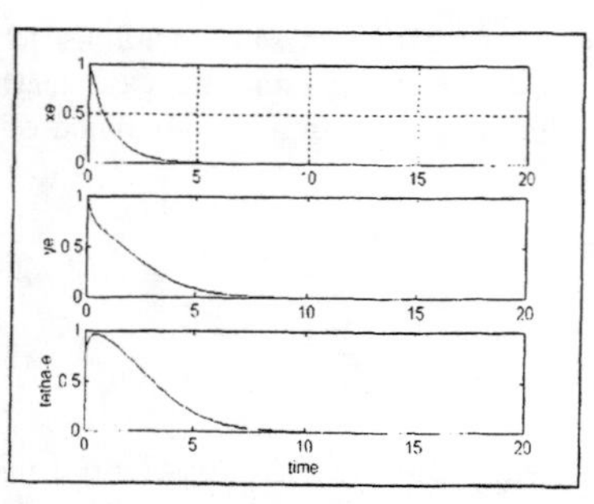

Figure 6. Timeplots of the system states

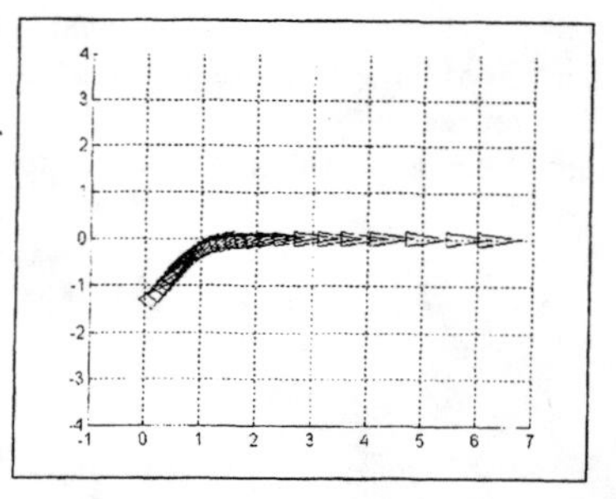

Figure 7. Rectilinear path tracking

CONCLUSION

In this paper, a nonlinear control law based on partial feedback linearization and Lyapunov method has been proposed for the closed-loop path following problem of unicycle-like mobile robots. This control law guarantees global asymptotic stabilization of the closed-loop system. Simulation results have been given to highlight the effectiveness of the proposed controller.

REFERENCES

1. C. Canudas de Wit and O. J. Sordalen, "Exponential stabilization of mobile robots with nonholonomic constraints," *IEEE Tran. On Autom. Control, Vol. 37, NO.11, November 1992, pp. 1791-1797.*
2. Y. Kanayama, Y. Kimura, F. Myazaki and T. Noguchi, " A stable tracking control method for a nonholonomic mobile robot, " *In Proc. IEEE Int. Workshop on Int. Robots and Systems IROS 91, Osaka, pp. 1236-1241.*
3. R. M. Murray and S. S. Sastry, "Nonholonomic motion planning: Steering using sinusoids, " *IEEE Trans. On Autom. and Control, vol.38, No.5, pp. 700-716, May 1993.*
4. M. Sampei, T. Tamura, T. Kobayashi and N. Shibui, "Arbitrary path tracking control of articulated vehicles using nonlinear control theory, " *IEEE Tran. On Cont. Syst. Thechnology, Vol.3, No.1, March 1995, pp.125-131.*
5. M. Sampei, T. Tamura, T. Itoh and M. Nakamichi, "Path tracking control of trailer-like mobile robot," *In Proc. IEEE Int. Workshop on Int. Robots and Systems IROS 91 Osaka., pp.193-198.*
6. C. Samson, "Mobile robot control, Part 1: Feedback control of a nonholonomic wheeled cart in cartesian space," *INRIA, Tech. Rep. 1288, Oct 1990.*
7. C. Samson and K. Ait-Abderrahim, "Feedback stabilization of a nonholonomic wheeled mobile robot," *In Proc. IEEE Int. Workshop on Int. Robots and Systems IROS 91 Osaka. pp. 1242-1247.*
8. C. Samson "Control of chained systems application to path following and time-varying point-stabilization of mobile robots," *IEEE Trans. on automatic control, vol.40 , pp. 64-77, January 1995.*
9. J. J. Slotine and W. Li, *Applied Nonlinear Control*, Prentice-Hall International, Ed. 1991.
10. O.J. Sordalen and C. Canudas de Wit, "Exponential Control Law for a Mobile Robot: Extension to path Following, " *In Proc. Int. Conf. On Robotics and Automation, Nice, France, May 1992, pp. 2158-2163.*
11. A. Tayebi and A. Rachid, " A time-varying-based robust control for the parking problem of a wheeled mobile robot, " *To appear in Proc. IEEE Int. Conf. On Robotics and Automation, Minneapolis, Minnesota, April 1996.*
12. A.Tayebi and A. Rachid "Robust tracking control for a wheeled mobile robot, " *To appear in Proc. of IEEE Mediterranean Electrotechnical Conference, Bari, ITALY, May 1996.*

POSITION CONTROL LAWS TAKING ADVANTAGE OF THE CYLINDER DYNAMIC POSSIBILITIES

Serge SCAVARDA
LAI - Institut National des Sciences Appliquées de LYON
20, avenue Albert Einstein
69621 VILLEURBANNE, Cedex

ABSTRACT
In most cases the control laws proposed for electropneumatic positioning systems do not take advantage of the posibilities offered by cylinder dynamic. This paper presents a series of linear or non linear position control laws which better use the local and or global pneumatic cylinder properties.

KEYWORDS : electropneumatic positioning systems, cylinder dynamic properties.

1.INTRODUCTION

During the last ten years a large amount of research deals with control of electropneumatic positioning systems. Usually the control law adopted is based on a state feedback scheme. The coefficients of the state feedback matrix were calculated by using a third order reduced linearised state model. This reduced model was obtained by a first order linearisation of the classical fourth order non linear state model initially proposed by Shearer [1], followed by a reduction order. This reduction is only possible if the Taylor series development is calculated by assuming both a complete servovalve symmetry, an equilibrium position at the central position and no force acting on the piston-load assembly at equilibrium Therefore the local dynamic behaviour is the desired behaviour only for the central piston position. Moreover, in most cases the control laws proposed do not take advantage of the cylinder dynamic possibilities. A group of researchers of the Laboratoire d'Automatique Industrielle of INSA of LYON [4], [5], [6], [7], [8], has tried to develop control laws which use better the local and or global cylinder dynamic possibilities..This paper resume and shows the different models used and the different contol laws developed for this purpose

2.THE SYSTEM UNDER CONSIDERATION

In order to have at ones disposal general system framework, we suppose that the electropneumatic system under consideration (figure 1) is composed of a rodless, double acting, linear pneumatic cylinder whose position in controlled by two three-way single stage electropneumatic servovalves.

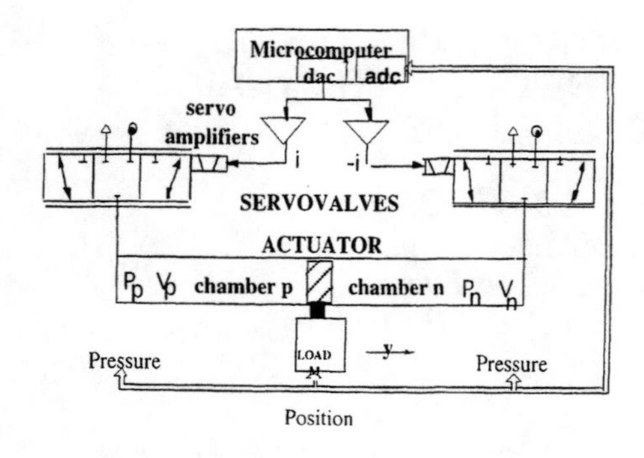

Figure 1: Electropneumatic positioning system

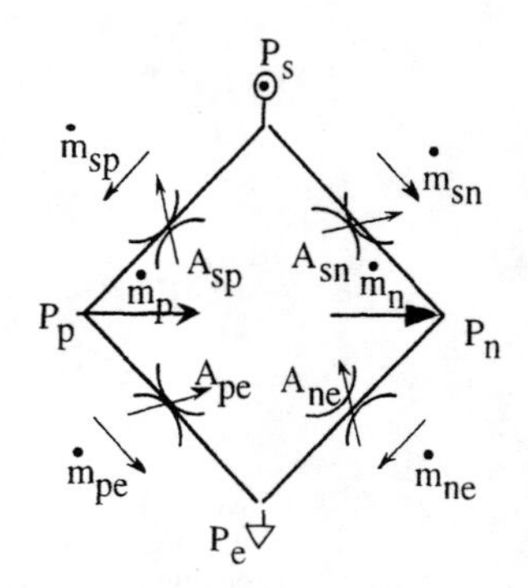

Figure 2: Flow stage

The pneumatic ram drives an inertial load
The system is controlled by a microprocessor
All sensors necessary are available
The two servovalve flow stages are schematized in figure 2.

3. NON LINEAR STATE MODEL AND BLOCK DIAGRAM

By adopting the classical modelling assumptions and by choosing as state variables the chamber pressures P_p and P_n, the piston velocity $\dot{y}$ and the piston position y, and, as input variable, the servovalve current i, we can obtain a non-linear state model under the form [1]:

$$\dot{\mathbf{x}} = \mathbf{f}(\mathbf{x},\mathrm{u}) \quad (1)$$

defined as below(2).

$$\frac{dP_p}{dt} = \frac{\gamma r T_s}{V_p(y)}\left[\dot{m}_{sp}(i,P_p) - \dot{m}_{pe}(i,P_p) - \frac{P_p A}{rT_s}\dot{y}\right]$$

$$\frac{dP_n}{dt} = \frac{\gamma r T_s}{V_n(y)}\left[\dot{m}_{sn}(i,P_p) - \dot{m}_{ne}(i,P_p) + \frac{P_n A}{rT_s}\dot{y}\right]$$

$$\frac{dy}{dt} = \dot{y}$$

$$\frac{d\dot{y}}{dt} = \frac{A}{M}P_p - \frac{A}{M}P_n - \frac{b_{vis}}{M}\dot{y} - \frac{F_{sec}}{M} - \frac{F_e}{M} \quad (2)$$

The mass flow rates $\dot{m}_{sp}$ $\dot{m}_{pe}$, $\dot{m}_{sn}$ and $\dot{m}_{ne}$ can be expressed as a product of a function of the servovalve input current by a function of the corresponding chamber pressure:

$$\dot{m}_{sp}(i,P_p)-\dot{m}_{pe}(i,P_p)=A_{sp}(i)D(P_s,P_p)-A_{pe}(i)D(P_p,P_e)$$

$$\dot{m}_{sn}(i,P_n)-\dot{m}_{ne}(i,P_n)=A_{sn}(i)D(P_s,P_n)-A_{ne}(i)D(P_n,P_e) \quad (3)$$

4.LINEARISED STATE MODEL AND LOCAL PROPERTIES.

4.1 Tangent Linearised State Model

For a non-linear system studying the properties of the tangent linearised state model is, in most cases, a way of understanding the dynamic behaviour better. This possibility is, and has been, largely used. The tangent linearised state model is obtained by a Taylor series development limited to the first order and calculated about an equilibrium state determined by neglecting the dry friction force.

By designating the equilibrium value in any variable by the subscript $_e$ and the changes in any variable by a superscript * added to the corresponding letter, the tangent linearised state model for a given piston position has the well known expression [1] [3] [4] given below:

$$\frac{d}{dt}\begin{bmatrix} P_p^* \\ P_n^* \\ y^* \\ \dot{y}^* \end{bmatrix} = \begin{bmatrix} -\dfrac{1}{\tau_p} & 0 & 0 & -\dfrac{\gamma P_{pe} A}{V_{pe}} \\ 0 & -\dfrac{1}{\tau_n} & 0 & \dfrac{\gamma P_{ne} A}{V_{ne}} \\ 0 & 0 & 0 & 1 \\ \dfrac{A}{M} & -\dfrac{A}{M} & 0 & -\dfrac{b_{vis}}{M} \end{bmatrix} \begin{bmatrix} P_p^* \\ P_n^* \\ y^* \\ \dot{y}^* \end{bmatrix} + \begin{bmatrix} \dfrac{\gamma r T_s}{V_{pe}} G_{ip} \\ -\dfrac{\gamma r T_s}{V_{ne}} G_{in} \\ 0 \\ 0 \end{bmatrix} i^* \qquad (4)$$

with:

$$C_{pp} = -\left[\left.\frac{\partial \dot{m}_{sp}}{\partial P_p}\right|_e - \left.\frac{\partial \dot{m}_{pe}}{\partial P_p}\right|_e \right] \qquad C_{pn} = -\left[\left.\frac{\partial \dot{m}_{sn}}{\partial P_n}\right|_e - \left.\frac{\partial \dot{m}_{ne}}{\partial P_n}\right|_e \right] \qquad \tau_p = \frac{V_{pe}}{\gamma r T_s C_{pp}}$$

$$G_{ip} = \left[\left.\frac{\partial \dot{m}_{sp}}{\partial i}\right|_e - \left.\frac{\partial \dot{m}_{pe}}{\partial i}\right|_e \right] \qquad G_{in} = -\left[\left.\frac{\partial \dot{m}_{sn}}{\partial i}\right|_e - \left.\frac{\partial \dot{m}_{ne}}{\partial i}\right|_e \right] \qquad \tau_n = \frac{V_{ne}}{\gamma r T_s C_{pn}}$$

(5)

One can easily deduce from the value of the state matrix coefficients the expression of the natural cylinder pulsation:

$$\omega = \sqrt{\frac{\gamma A^2}{M}\left[\frac{P_{pe}}{V_{pe}} + \frac{P_{ne}}{V_{ne}}\right]} \qquad (6)$$

with: $V_{pe}(y_e) = V_{pi} + A y_e \quad V_{ne}(y_e) = V_{ni} - A y_e$

These expressions clearly show that:

-the natural pulsation is minimal for the mid-stroke piston position.

-the time constants τ_p and τ_n are functions of the piston position.

4.2 Reduced Tangent Linearised State Model And Local Dynamic Properties

A state feedback control law can be easily deduced from the previous tangent linearised state model. However in most cases the state feedback is based on a reduced model. Many authors used the reduction proposed by Shearer [1] in the case of a symmetric ram, a symmetric four way servovalve and a mid stroke position as equilibrium position. Because of the supposed perfect symmetry, this reduction consists of combining the two state equations, related to the chamber pressures, in order to obtain a single state equation associated with the acceleration. Kellal et al.[4] have shown that this reduction is possible for any piston position if the state matrix coefficients a_{11} and a_{22} are replaced by the half sum of these two coefficients,i.e.by the inverse of a mean time constant equal to the half sum of the inverse of the time constant τ_p and τ_n:

$$\frac{1}{\tau_m} = \frac{\gamma r T_s}{2}\left[\frac{C_{pp}}{V_{pe}} + \frac{C_{pn}}{V_{ne}}\right] \quad (7)$$

The frequency responses given in the paper of Kellal et al.[4] show the validity of this approximation. Furthermore Richard [8] has shown that this reduction is associated to an observability loss. The reduced tangent linearized state model obtained by the Kellal's et al. [4] procedure, has the following form:

$$\frac{d}{dt}\begin{bmatrix} y^* \\ \dot{y}^* \\ \ddot{y}^* \end{bmatrix} = \begin{bmatrix} 0 & 1 & 0 \\ 0 & 0 & 1 \\ 0 & -\omega^2 - \frac{b_{vis}}{\tau_m M} & -\frac{1}{\tau_m} - \frac{b_{vis}}{M} \end{bmatrix} \begin{bmatrix} y^* \\ \dot{y}^* \\ \ddot{y}^* \end{bmatrix} + \begin{bmatrix} 0 \\ 0 \\ \frac{\gamma r T_s}{M}\left[\frac{G_{ip}}{V_{pe}} + \frac{G_{in}}{V_{ne}}\right] \end{bmatrix} i^* \quad (8)$$

In the following the coefficient of the control matrix not equal to zero will be designated by b.

The structure of the state matrix shows that this system is composed of a second order system in series with an integrator,whose parameters are:

$$\omega_{ol} = \sqrt{\omega^2 + \frac{b_{vis}}{\tau_m M}} \qquad z_{ol} = \frac{1}{2\omega_{ol}}\left[\frac{1}{\tau_m} + \frac{b_{vis}}{M}\right] \quad (9)$$

If the viscous friction is negligible the damping coefficient z_{ol} has for expression:

$$z = \frac{\gamma r T_s}{4\omega}\left[\frac{C_{pp}}{V_{pe}} + \frac{C_{pn}}{V_{ne}}\right] \quad (10)$$

It can be noticed that each parameter ω, z and b is a function of the equilibrium position y_e because it is function of the equilibrium values of the volumes V_{pe} and V_{ne}.The influence of position on the three previous parameters can be estimated:

-firstly by introducing three adimensional parameters obtained by calculating the ratio between the parameter under consideration and its value for the central position,

-secondly by representing the variations of each of these three ratios as a position function.

The curves shown in figure 3 have been obtained for a cylinder having a diameter of 40 millimeters, a stroke of half a meter, for a null exterior force and negligible friction (b_{vis}=o).On these curves it can be clearly seen that the highest variation is related to the coefficient b of the control matrix.The relative changes of the natural pulsation and of the damping coefficient are the same.As the damping coefficient corresponds to an equilibrium position ,ie.to a closed position of the servovalve, its value is always small but not equal to zero due to the servovalve leakages.Moreover the corresponding flow pressure coefficients C_{pp} and C_{pn} are only significant for a steady state point of view,i.e. for the determination of the steady state position error and steady state stiffness.

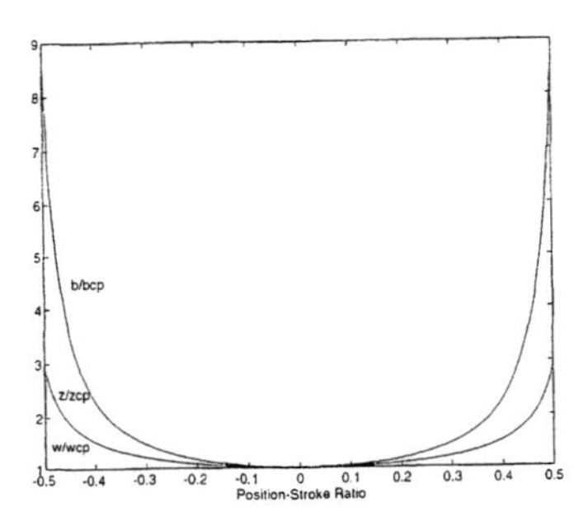

Figure 3: Cylinder parameter changes with the piston position.

5 CONTROL LAWS TAKING ADVANTAGE OF THE LOCAL PROPERTIES

5.1 Linear Control Laws

Usually the control law adopted is a state feedback deduced from the reduced linearised state model corresponding to the central position. The matrix state feedback coefficients are calculated, for example, by a pole placement method :

-firstly by choosing a third order characteristic polynomial in closed loop,for example, the following product of a dominant second order polynomial and a first order polynomial:

$$\left(p^2 + 2z_{cl}\omega_{cl}p + \omega_{cl}^2\right)\left(p + \frac{1}{\tau_{cl}}\right) \quad (11)$$

-secondly, by choosing the coefficients of this polynomial to obtain the desired closed loop behaviour.

The following values are usually adopted [4] [5]: $z_{cl} = 1$, $\frac{1}{\tau_{cl}} = 6\omega_{cl}$, $\omega_{cl} = \alpha\omega_{ol}(0)$, i.e.$\omega_{cl}$ is chosen proportional to the open loop natural frequency of the cylinder for the central equilibrium position, the coefficient α being between 1and 2.

In order to make a best use of the local dynamic properties of the cylinder one can:

- firstly use, for calculating the coefficients of the state feedback matrix, the reduced linearised state model corresponding to the desired position instead of the linearised state model associated to the central position

-secondly choose a value of ω_{cl} prorportional to the value $\omega_{ol}(y_d)$ of the cylinder natural frequency calculated for the desired position.

Then by assuming servovalves symmetry and equality of the volume chambers when the piston is in central position ($V_{pe}(0) = V_{ne}(0) = V_0$) and null exterior effort,we can easely obtain the following expressions for the three feedback coefficients corresponding to the desired equilibrium position y_d:

$$K_p = 6\,\omega_{cl}^3 \frac{1}{b}$$

$$K_v = \left[13\ \omega_{cl}^2 - \omega_{ol}^2(y_d)\right]\frac{1}{b} \qquad \text{with: } \frac{1}{b} = \frac{MV_0}{2G_i A\gamma r T_s}\left[1 - \frac{A^2 y_d^2}{V_0^2}\right] \qquad (12)$$

$$K_a = \left[8\omega_{cl} - 2z_{ol}(y_d)\omega_{ol}(y_d)\right]\frac{1}{b}$$

. It can be noted that the same parabolic function of the desired position y_d appears in the three expressions obtained.This function is due to the inverse of the control matrix coefficient b.The steady state position error and the steady state stiffness are given by:

$$\text{steady state position error:} \quad \varepsilon(\infty) = \frac{F_e^*}{2A\dfrac{G_i}{C_P}K_p} \qquad (13)$$

$$\text{steady state stiffness:} \quad \left|\frac{F_e^*}{y^*}\right| = 2A\frac{G_i}{C_p}K_p \qquad (14)$$

The previous expressions of: the state feedback matrix coefficients, the steady state position error and the steady state stiffness show that ,with the increase of the absolute value of the desired position y_d, this choice leads to:

-an improvement of the local dynamic properties of the closed loop system due to the corresponding increase of the cylinder natural frequency,

-a decrease of the steady state position error and an increase of the steady state stiffness.

The associated problem in discrete time has been studied in the context of state - affine systems, the design of the state affine control law is given in references [5]

5.2 Non-Linear Control

5.2.1 Point To Point Control Law

In order to obtain a non-linear state model linear with the input, Richard.[7] [8] has proposed:

-firstly decomposing the area of each servovalve orifice into a constant leakage area A_l and a current controlled algebric area A_c, then obtaining a non-linear state model under the desired form:

$$\dot{x} = f_a(x) + g(x)A_c \qquad (15)$$

-secondly transforming the system into a triple integrator by using an input - output linearising control law as a function both of the state vector x and the jerk J:

$$u(x) = \Delta^{-1}(x)\left[J - \Delta_0(x)\right] \qquad (16)$$

By using a state feedback, i.e. position ,velocity and acceleration feedbacks, this triple integrator can be transformed into a third order system with a given dynamic behaviour:

$$\dddot{y} + \alpha_2\ddot{y} + \alpha_1\dot{y} + \alpha_0 y = \alpha_0 y_d \qquad (17)$$

A suitable choice of the coefficient α_i enables us to obtain either a constant dynamic behavior (α_i are constants) or a variable dynamic behaviour. If each coefficent α_i is a function of the position, one can profit from the natural dynamic properties of the cylinder] in order to improve the dynamic behaviour and decrease the desired steady-state error when the desired piston position moves near one cylinder end. For a rodless cylinder (stroke 4.23m, diameter 40mm,and an inertial load of 10 kg) Lin et all. [6], have imposed a behaviour described by the following third equation order differential:

$$\dddot{y} + K_a(y)\ddot{y} + K_v(y)\dot{y} + K_p(y)y = K_p(y)y_d \quad (18)$$

The corresponding control scheme is given in figure 4.

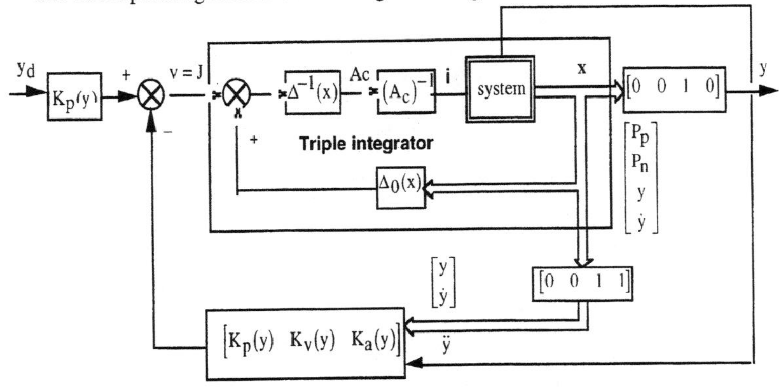

Figure 4: Non-linear point to point control law

One can notice that the input current of the servovalve is obtained by using the inverse function of $A_c(i)$ which is deduced from experimental characteristics.By using the two control inputs ,i.e. the two servovalves currents, two output variables can be controlled, for example: position y and one of the two pressures P_p and P_n. This possibility has not yet been used.The control algorithms previously proposed either by Miyata and Hanafusa [9] or by Scholz [10] can be related to this kind of non-linear control though they are not directly deduced from non-linear control theory.

5.2.2.Tracking.Control Law

More often the user wants to define the time history of one of the following variables: position, velocity, acceleration and jerk, so it is preferable to choose a tracking control law.

This scheme shows that this tracking control law involves the use of [7] [8]:

- a linearizing non linear state feedback which transfoms the real system into a triple integrator

-a triple integrator as reference model,

-a feedback of the state error $-\mathbf{K}(\mathbf{x}-\mathbf{x_r})$ which sets the dynamic behaviour of the state error,

-an input of the real system equal to $$i = \frac{1}{\Delta(P_p, P_n, y)}\left[J_d - J_r - K(x - x_r)\right] \quad (19)$$

. Richard and Scavarda [7], Lin et all.[6] have published experimental results on point to point control and tracking control. It can be noticed that the best use of the cylinder dynamic properties corresponds to relating the dynamic behaviour of the error to the natural frequency of the cylinder. This possibility has not been used.

6 CONCLUSION

This paper shows how different linear and non-linear positioning control laws have been chosen in order to take advantage of the cylinder dynamic properties. The reference papers show experimental results. Our present research is trying to extend this approach to tracking problems.

REFERENCES.

[1] Shearer, J.L., Study of pneumatic processes in the continuous control of motion with compressed air. Parts I and II. Trans. Am. Soc. Mech. Eng. 1956, **78**, 233-249.

[2] Chitty, A.and Lambert, T.H., Modelling a loaded two-way pneumatic actuator. J. Meas. and Control, 1976, **9**, n°1, 19-25.

[3] Burrows, C.R. Non linear pneumatic servomechanisms. Ph.D.:Thesis, University of London, 1968, pp.298.

[4] Kellal, A., Scavarda, S.and Fontaine, J.G., Electropneumatic servodrive for a robot . 16th ISIR, 29-9 au 2-10-1986, Bruxelles, 117-128

[5] Scavarda, S., Thomasset, D., Richard, E.and Bourdat, S., State affine control of electropneumatic cylinder. 113th ASME Winter Annual Meeting, November 8-13, 1992, Anaheim, Californie (USA),n° 92- WA/FPST-12

[6] Lin, X.F., Thomasset, D., Richard, E., Scavarda, S., Non linear position control for long pneumatic actuator, TheThird Scandinavian_International_Conference on Fluid Power, 25-26 may 1993, Linkoping (Sweden), p. 435-444

[7] Richard, E.and Scavarda , S., Non linear control of a pneumatic servodrive. Proc. Second Bath Int. Fluid Power Workshop - 21-22 Septembre 1989 Bath (U.K.).

[8] Richard, E., De la commande linéaire et non linéaire en position des systèmes électropneumatiques Thèse: Sci.: Institut National des Sciences Appliquées de Lyon, 1990. pp. 291

[9]]Miyata, K.and Hanafusa, H., Pneumatic servo control system by using adaptative gain pressure control. In : Prep. First JHPS Int. Symp. on Fluid Power, Tokyo, March13-16, 1989, 161-168.

[10] Scholz, D., Auslegung servopneumatisher antriebssysteme, Dissertation RWTH, Aachen, 1990, 231p

ROBOTIC BUTCHER FOR OVINE CARCASS DRESSING

Malcolm G. Taylor
Industrial Research Limited
PO Box 20-028, Christchurch, New Zealand
email: m.taylor@irl.cri.nz

ABSTRACT

A new robot has been designed and built for use in a wide range of food processing applications. The specific focus for this application has been for use in abattoirs for key operations associated with the pelt removal of sheep and lamb carcasses. The robot is manufactured from stainless steel and its lightweight construction results in high performance characteristics allowing it to operate from an overhead pose for an inverted dressing system to increase productivity. The prototype can readily be reconfigured in terms of tooling and programming for other dressing operations and it introduces innovative technology to the meat processing industry. The prototype robot that has emerged is unique in terms of its configuration and application and has been trialled at a meat export processing plant in New Zealand.

KEYWORDS: robot, abattoir, sheep, dressing, New Zealand, Y-cutting

INTRODUCTION

In this paper, we describe our activities in developing the next generation of technology for wash-down food processing environments, exemplified by the dressing of sheep carcasses, based on pragmatic use of robotics and innovative engineering. Mechanization in the meat processing industry of New Zealand has resulted in a number of technological changes and innovations to manual dressing practices over the past 15 years [1, 2]. From about 1981 a series of developments has led to a successful inverted manual dressing system for sheep and lamb carcass processing that incorporated several existing butchery techniques. In addition it provided a basis for the introduction of various other automated machines including: automatic stunning, neck-breaking, wide-to-narrow transfer, shoulder pulling, final pulling, and trotter removal. The development philosophy was to use a low-medium level of technology able to do the job, an approach that has been very effective [3]. Our research is now opening new opportunities through application of advanced technology.

We have designed and built a new robot for use in a wide range of food processing applications. We have specifically focused this application for use in abattoirs for key operations associated with the pelt removal of sheep and lamb carcasses. The robot is manufactured from stainless steel and its lightweight construction results in high performance characteristics allowing it to be installed overhead in an inverted dressing system to increase productivity. The prototype can be reconfigured in terms of tooling and programming for other dressing operations and it introduces innovative technology to the meat processing industry. The prototype robot that has been designed, developed and tested is a world-first in terms of its configuration and application and has been successfully trialled at a meat export processing plant in the North Island of New Zealand. It is currently being prepared for trials at a larger scale commercial plant in the South Island.

At a meat conference in 1995, Buhot [4] stated that "the concept of a [fully] 'automatic' abattoir has now been displaced by the 'machine assisted' abattoir". He based this statement on the changes that have occurred within the meat industry in the 1990s, namely that the 1980s trend of rapidly increasing labour costs did not continue, while the cost of large scale automation has remained the same. Nidd [5] examined the requirements of a meat processing plant for the 21st century. He comments on automation: "the degree of automation in the plant will be limited due to the variability of raw product and the harshness of the environment". Large scale automation projects such as Fututech [6], which were built on extremely high levels of borrowed capital failed, often for economic rather than technical reasons.

Nevertheless, the variable nature of the raw product (sheep carcasses) and the increasing demands for improved efficiency in terms of productivity and hygiene have opened up an exciting opportunity for small scale and flexible automatic machinery that can withstand the harsh environment of the meat processing industry. A reprogrammable and washable robot with interchangeable tooling capable of performing a variety of functions has emerged as a viable technology for solving the efficiency driver which the industry continually demands.

A conscious decision was made to avoid a capital intensive approach to the robot. The machine described in this paper combines higher levels of technological innovation, a cost effective approach to technology and a pragmatic approach to solving a difficult manual operation for sheep and lamb dressing [7, 8]. A successful introduction of appropriate flexible automation to meat processing will lead to reduced direct labour costs, reduced indirect labour costs, improved quality, improved hygiene and a reduction in the number of menial tasks performed by humans. Hence a robot suitable for the meat processing industry is an appropriate technology solution.

A global search was made for robots suitable for meat-processing applications with limited results. There are few programmable manipulators available that are able to withstand the harsh environments commonly found in abattoirs. Such environments are characterised by high humidity, high-temperature water, high-pressure water, and caustic cleaning chemicals. After researching various options it was decided to proceed with in-house design and local manufacture of a custom-built, but scalable, robot, known as the IRL7L.

Y-CUT APPLICATION

The application of this robot involved carrying out one of the first pre-pelting operations on a sheep carcass prior to other manual cutting operations and eventual pelt removal. Currently the system is programmed to perform a Y-cut. This is an initial cut which slits the pelt without damaging the underlying membrane and is made down one front leg, across the chest and up the other leg of the sheep. See Figure 1.

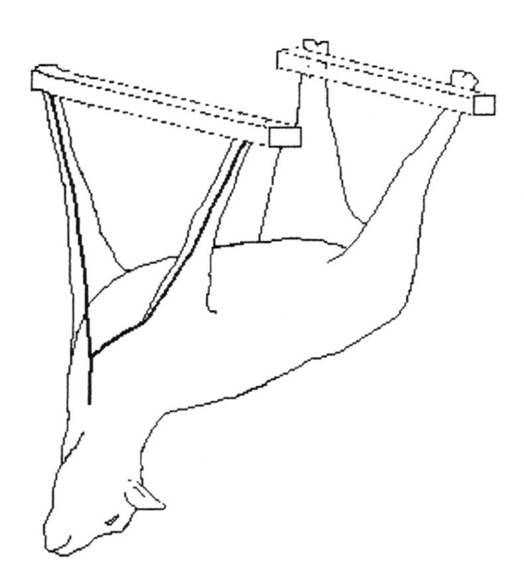

Figure 1. The Y cut path is the bold line which goes down both forelegs and meets at the throat.

DESIGN AND CONSTRUCTION

The robot was developed for meat industry use and in particular pre-pelting operations. Nevertheless, its characteristics are pertinent to a wide range of food-processing applications. The robot was designed to meet the following criteria:

- overhead mounting in order to leave floor space uncluttered;
- smooth surfaces easy to clean;
- lightweight yet rigid construction;
- corrosion resistant exterior;
- free from external umbilical service cabling;
- high-speed performance with maximum velocity greater than 2 m/s;
- continuous operation for two shifts per day in a humid environment;
- minimal noise level.

The majority of pre-pelting operations are performed in a vertical plane. We began with an anthropomorphic design concept, recognising that the human arm has adequate reach and dexterity to perform the required tasks. However, it is neither practical nor necessary to duplicate the human arm exactly. Initial development tests, using an ABB IRB 2000 industrial robot, showed that for the cutting task intended, only four mechanical degrees of freedom were required. This resulted in the linear-revolute-linear-revolute or cylindrical configuration as shown in Figure 2. Although conventionally one thinks of a cylindrical robot having its major axis vertical, in this case it was chosen to be horizontal allowing the arm to sweep with maximum velocity in a vertical plane. The primary linear axis (Figure 2, axis 1) was configured to track the motion of a dressing conveyor chain. The axis was aligned in parallel with the chain but displaced by approximately 900 mm. An auxiliary axis was employed to track the chain motion accurately. As a consequence, robot tasks could be programmed during setup for a stationary carcass and then be performed automatically 'on the fly' with a minimum of programming effort.

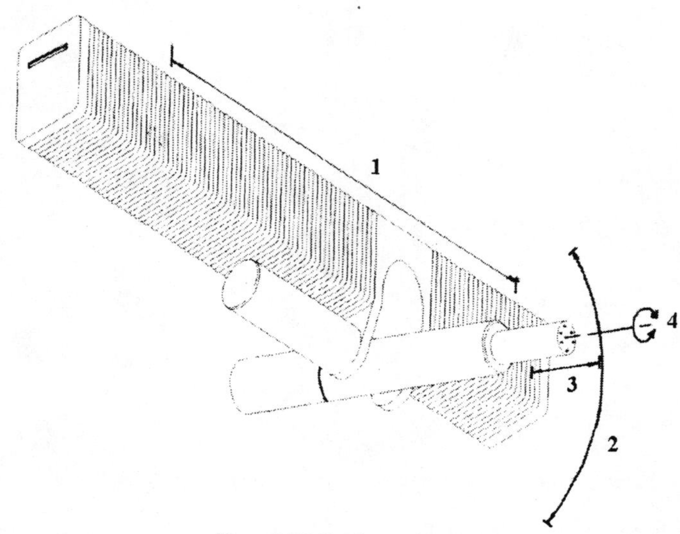

Figure 2. IRL7L robot work envelope.

The main arm (axis 2) is driven through a high ratio harmonic drive providing 90 degrees of angular movement. It carries the third axis, a telescopic arm (axis 3) which can extend up to 300 mm. The fourth revolute wrist axis (axis 4) supports a tool plate which can rotate a full 360 degrees. Different end-effectors can be attached to this plate provided that their mass does not exceed 7 kg and the center of mass is close to the end of the arm.

The robot is manufactured from stainless steel, for corrosion resistance, and aluminium to achieve a lightweight yet rigid construction. The stainless steel body has been formed from sheet metal into tubular sections, which are inherently strong and easy to keep clean. The resulting fabrication is welded and polished. All aluminium components have been machined to a high precision and polished. A key feature of the design was the integration of all electrical and pneumatic services through the internal structure of the arm, enhancing its clean surgical lines. The internal wiring harness of the robot was carefully designed to prevent kinking or premature fatigue.

The robot has been designed to prevent the ingress of water. All joints are sealed by push fits and chemical sealants. Electrical heating elements are mounted internally to prevent condensation occurring within the bellows volume of the robot. The bellows themselves provide a large flexible seal to prevent contamination of the linear bearings or motor drives.

OPERATION AND CONTROL

The operation of the robot is carried out from a personal computer which drives an industrial servo motion control system. The Microsoft Windows™ based controller incorporates an integrated PLC and graphical user interface. The controller interprets G-codes, a widely used NC programming language, and offers on-line programming, graphical tool path display and configurable on-screen control panels. The controller user interface can be either be via a touch screen or a point and click GUI. A typical screen is shown in Figure 3.

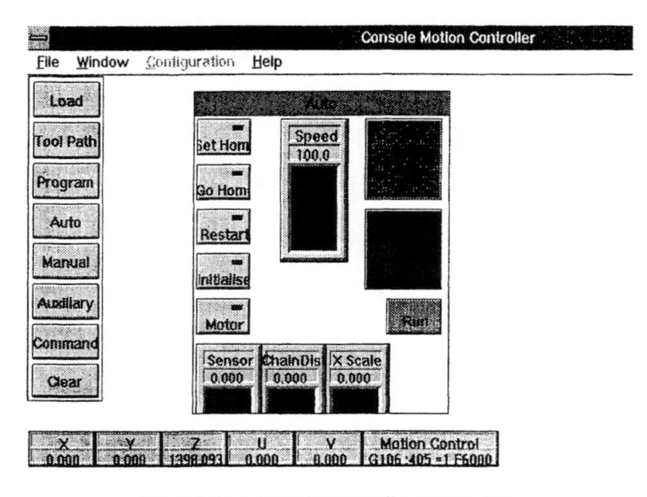

Figure 3. Console Motion Controller user interface.

The control system can be considered as four elements: the motion controller, the operator interface, the programmable logic controller (PLC) and the input/output system. The motion controller uses 4 ms interrupts to update the position of the robot, which is sent to the motion controller by encoders located on the axis motors. The position of the robot is compared with the desired position specified by the operator either through the graphical user interface (GUI) or by a line in the G code program.

Each axis can be jogged separately under manual teach control in order to position the tool. Alternatively, by switching to a setup mode, all axes can be positioned by hand and each axis value read off the screen and keyed directly into a program. Programs for the robot are created using the controller's own text editor as a series of G-code instructions from which a tool path is generated. It is extremely useful to view the path before committing the robot to move in the workspace. This function is provided by a simple tool path viewer. Other controller features include global adjustment of speed, individual scaling and offset adjustments for each axis, continuous display of axes positions, sensor readings and PLC flags.

ROBOT TRIALS IN A MEAT PROCESSING PLANT

End-effector. The robot's end effector is a sophisticated pneumatic cutting tool and its successful operation is critical to the success of this application [9]. A degree of compliance within the tool has been achieved with compressed rubber mountings. The careful design of the highly successful cutting method derives maximum benefit from this compliance, from that provided by the method of sheep support, and from the relatively looseness of the pelt.

Environment. A meat processing plant for all major domestic animal species (pork, venison, beef and sheep/lamb) undergoes a ruthless cleaning cycle every day, using very high pressure water and caustic chemicals. Automation machinery must be able to withstand this onslaught and suffer no degradation in performance. This requirement emphasises the need for high-quality components and seals. The robot has spent several months in the environment and over one hundred hours operating and has proved to be easy to clean and resistant to leaks.

Trials. The robotic butcher (affectionately named Robochop) has successfully completed numerous trials, including a full production day, with a production rate of 2.5 carcasses per minute [9]. During this day over 1000 sheep were processed with a success rate of 99.1%.

FUTURE WORK

The next stage is to trial the robot at a higher capacity meat processing plant in the South Island of New Zealand. This will demand increased operating speed and workload of nine carcasses per minute. A refinement of its design and manufacturing methods is expected to move the robot from its present prototype stage to a full commercial product.

This prototype robot has been used for a specific task but has been designed and built for general-purpose use. The robot's scalable configuration allows it to be potentially used for large scale processing such as beef, or reoriented and used for smaller scale processing such as fish allowing for different end-effectors and tools to accommodate each application.

ACKNOWLEDGMENTS

The author is grateful for the support of the New Zealand Meat Research and Development Council who initially funded the Y-cut and Aseptic projects out of which the robot described herein was conceived. We would also thank Progressive Meats Limited for allowing their facilities to be used for conducting our trials. Appreciation is due to Dr Richard Templer for his invaluable contribution to this work and to Dr Howard Nicholls and Jem Rowland for reviewing drafts of this paper.

REFERENCES

1. Authier, J.F. "Mechanical dressing and beef processing/boning." *MIRINZ 27th Meat Industry Research Conference*, Hamilton, New Zealand (1992), 327-338.
2. Longdell, G.R. "Advanced technologies in the meat industry." *Proc. 38th Int. Cong. of Meat Science and Technology*, Clermont-Ferrand, France, (1992).
3. Frazerhurst, L.F. "Robotics and automation in the meat industry." *Proc. ROBHANZ '86*, Auckland, New Zealand (1986), 59-61.
4. Buhot, J. "Trends in Abattoir Automation: Research and Development Aspects." *Meat '95 The Australian Meat Industry Research Conference*, Gold Coast, Australia (1995).
5. Nidd, M.J. "The Meat Plant in the 21st Century." *Meat '95 The Australian Meat Industry Research Conference*, Gold Coast, Australia (1995).
6. Doonan, R. "Fututech." *Meat '93 The Australian Meat Industry Research Conference*, Gold Coast, Australia (1993).
7. Taylor, M.G. "Automated Y-cutting of sheep carcases." *Meat '93 The Australian Meat Industry Research Conference*, Gold Coast, Australia (1993).
8. Taylor, M.G and Brooking, G.N. "Y-cut dressing of sheep carcases using a robot." *MIRINZ 28th Meat Industry Research Conference*, Auckland, New Zealand (1994), 181-186.
9. Templer, R.G., Taylor, M.G and Hagyard, P.A. "Robotic Y Cutting." *IEEE International Conference on Robotics and Automation: Video Proceedings,* Accepted, (1996).

A COMPARISON OF THREE ROBOT CONTROL ALGORITHMS IN FAULT RECOVERY

YUNG TING
Department of Mechanical Engineering
Chung Yuan University
Chung-Li, Taiwan, R.O.C.

SABRI TOSUNOGLU
Department of Mechanical Engineering
Florida International University
Miami, Florida 33199 USA

ABSTRACT

Fault tolerance in robots may be realized by strategically integrating the sensors, actuators, kinematic structures and robot's degrees of freedom. In the case of the multiple robots, cooperation aspects of robots play a crucial role. For successful implementation of fault tolerance, structural redesign is accompanied by a fault recovery scheme that includes fault identification, system reconfiguration, and a fault-tolerant controller design—the latter being the most critical element for fault rejection. Motivated by the important role of controller design, this work develops fault-tolerant controllers based on computed-torque, sliding control and adaptive parameter estimation techniques. The performance of these controllers are tested numerically under actuator failure which represents the most severe failure for serial robots. The simulation results indicate that sliding control excels as a robust controller immediately after failure. On-line parameter estimation is found to be most useful to fine-tune parameters after the gross system errors are eliminated. After the steady state is reached, switching to the computed-torque method has resulted in satisfactory post-failure system performance.

KEYWORDS: Robotics, manipulators, fault tolerance, fault-tolerant controller, control algorithm, intelligent control.

INTRODUCTION AND BACKGROUND

Fault tolerance attempts to improve a system's reliability and availability not by improving the component reliability, but by rearranging the existing components within the given system. In addition to having better reliability and availability, such systems possess improved maintainability which means reduced operational costs. This concept has been applied to computer hardware and software, airplanes, nuclear reactors for a few decades, and is now being developed for computer-integrated manufacturing (CIM) systems [1], mechanical systems and robotics [2]. This is achieved by developing fault tolerance in system software and hardware.

Fault-Tolerant Software Development

Fault-tolerant software has three main components as shown in Figure 1. The user interface module is a high-level, graphical software which provides easy access of the fault-tolerant system to the user. The second software module, fault detection and identification (FDI), continuously monitors the system to detect any malfunction in the components. If a failure is detected, the same module verifies the error, and then informs the recovery module. Recovery takes place by executing the following four stages:

(i) System Reconfiguration: Consistent with the information provided by the FDI, the corrective action is decided and implemented at this level. This results in the isolation of the failed component.

(ii) Task Replanning: This involves a decision whether the remaining portion of the task should be replanned or not. If necessary, it is implemented in this stage.

Fault-Tolerant Software
- User Interface
- Fault Detection and Identification
- Recovery
  - Reconfiguration
  - Replanning
  - FT Controller
  - Maintenance

Fault-Tolerant Hardware
(Mechanical System and Controller)
  - Sensor
  - Component
  - Sub-system
  - System

Fault-Tolerant System

Figure 1. Fault-tolerant systems consist of fault-tolerant hardware and software

(iii) Design of Fault-Tolerant Controller: The new controller design consistent with the reconfigured system and replanned task is decided in this stage.

(iv) Maintenance: Implementation of the fault-tolerant controller is carried out, and the system performance is monitored for possible intervention or improvements in the system response.

After a failure occurs, a fault-tolerant controller tries to eliminate system errors completely in order to recover the system to its pre-failure state. However, if that is not possible, it aims to ensure graceful system degradation [3]. In the literature, fault-tolerant mechanical architectures were studied in [2], fault detection and identification methods were developed in [4], and fault rejection in serial- and parallel-structured robots were addressed in [2, 5, 6].

The present work studies the performance of three fault-tolerant controllers on serial robots when one of the actuators fails. Under such a failure, the joint driven by the failed actuator is locked and the remaining joints are driven to complete the task. Trajectory replanning is usually needed to maintain the original end-effector path. The system errors are eliminated by applying a fault-tolerant controller for the reconfigured robot. The controllers used in this study are based on three popular methods: computed-torque with PID feedback, robust sliding control and adaptive control techniques.

Fault-Tolerant Hardware Development

At the sensor and component levels, a fault-tolerant architecture employs redundant sensors and actuators to compensate for component failures. At higher levels, it uses parallel-structured mechanisms, makes use of mechanical redundancy and multiple systems.

Figure 2 shows a one-degree-of-freedom (DOF) joint which is normally powered by a single motor. If the same joint is powered by two motors, then the joint will have fault tolerance against actuator failure. Fault tolerance is also introduced by using different kinematic structures as illustrated in Figure 3. While the system shown in Figure 3 (1) is serial, its fault tolerance is increased by using a four-bar mechanism (2), or the level of fault tolerance is further increased by introducing a five-bar mechanism (3). Finally, an alternate design to a serial system is the parallel-structured hybrid robot shown in (4) which is more fault tolerant than the serial robot in (1).

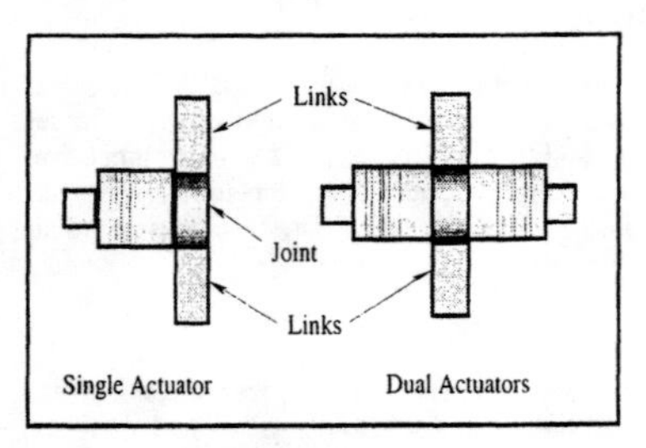

Figure 2. 1-DOF joint driven by single and dual actuators

Actuator failure is considered to be one of the most serious problems a robot encounters particularly in hazardous or remote environments such as the space, nuclear reactors or waste sites. As shown in Figure 4, assume that an actuator of a three-link, n-DOF serial robot fails. Such a failure has three implications in fault tolerance.

If dual actuator are installed at the joint under discussion, then the full or partial failure of one actuator may not result in any capability loss; i.e., the joint may be kept functional after failure.

When a single actuator is installed at the joint, then even a minor failure in any component results in a reduced output torque capacity. In this case, the actuator remains operational. Fault-tolerant controllers based on time dilation and torque distribution were developed [3, 5] to address this type of failures.

When a single actuator at one of the joints completely fails and can not supply any output torque, then we face a full joint failure. This represents the most challenging case especially in serial arms. To recover from failure, one has to apply the brakes to lock this joint as soon as the failure is identified. This means that the robot loses one of its degrees of freedom as shown in Figure 4. Therefore, the post-failure configuration requires a new system model, a new controller as well as a new task description in the joint space even if the end-effector task remains unchanged. In this work, we consider failures of this type.

ROBOT DYNAMIC MODEL

The robot dynamics before and after failure are represented by

$$\tau = M(q)\ddot{q} + C(q,\dot{q}) + G(q) + F(q,\dot{q}) \tag{1}$$

where τ is the n-dimensional input torque vector,
q is the n-dimensional joint displacement vector,
$M(q)$ is the $n \times n$ generalized inertia matrix,
$C(q,\dot{q})$ is the vector of Coriolis and centrifugal loads,
$G(q)$ is the vector of gravity-related forces, and
$F(q,\dot{q})$ is the vector of frictional loads.

If an n-DOF robot loses one of its degrees of freedom due to a complete failure of one of the actuators, it becomes an (n-1)-DOF system as illustrated in Figure 4. In that case, the fault-tolerant controller modifies the system model, adjusts system parameters to the new robot configuration, and then provides the most suitable controller design for the new system. In this study, we consider three controllers based on the computed torque, robust sliding control and adaptive control techniques to recover the system from failure.

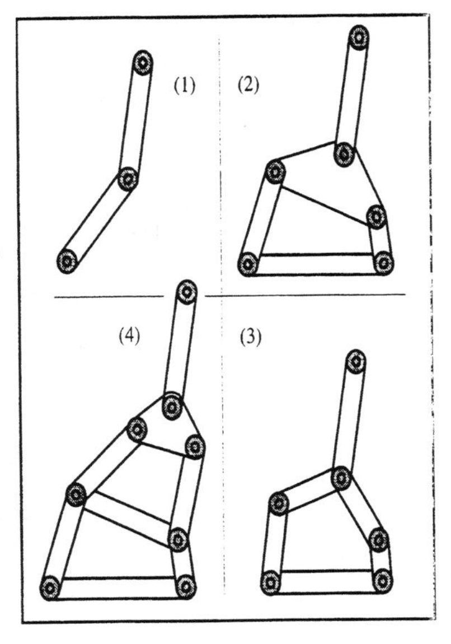

Figure 3. Kinematic structures with an increasing fault tolerance level from (1) to (4)

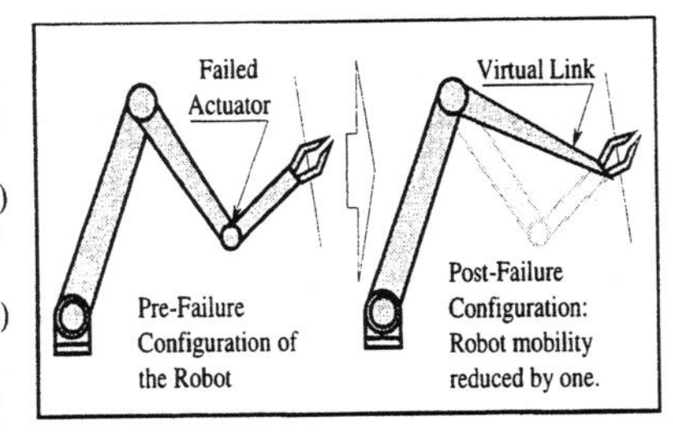

Figure 4. Reconfiguration of a robot after failure

FAULT-TOLERANT CONTROLLERS

Computed-Torque Controller. This controller with the feedforward desired trajectory parameters and feedback control is given as

$$\tau = \hat{M}(q)(\ddot{q}_d + u) + \hat{C}(q,\dot{q}) + \hat{G}(q) + \hat{F}(q,\dot{q}) \tag{2}$$

where u represents the feedback control law. For PID feedback control, u is given by

$$u = K_v\dot{e} + K_p e + K_I \int e\,dt \tag{3}$$

where $\hat{M}(q)$, $\hat{C}(q,\dot{q})$, $\hat{G}(q)$, and $\hat{F}(q,\dot{q})$ represent the on-line calculated values. Here, $e = q_d - q$ is the vector of position errors, $\dot{e} = \dot{q}_d - \dot{q}$ is the vector of velocity errors, and K_v, K_p, K_I are the gain matrices.

Robust Sliding Controller. The variable structure system attempts to drive the errors into the sliding surface so that the tracking errors approach to zero. For this purpose, a sliding surface, $\mathfrak{S}$, is selected with the combination of the tracking errors (including the integral control) as

$$\mathfrak{S} = \{(q,\dot{q}) \mid S = \dot{e} + \lambda_1 e + \lambda_2 \int e\,dt = 0\} \tag{4}$$

whose time derivative is given by

$$\dot{S} = \ddot{e} + \lambda_1\dot{e} + \lambda_2 e \tag{5}$$

where λ_1 and λ_2 specify the response of the error dynamics in terms of the bandwidth or percentage of overshoot. Subtracting (1) from (2) and considering the parameter uncertainties and disturbances acting on the system, the error-driven system equation is described by

$$E_m = \ddot{e} + u = \hat{M}^{-1}(q)\{[M(q) - \hat{M}(q)]\ddot{q} + [C(q,\dot{q}) - \hat{C}(q,\dot{q})] + [G(q) - \hat{G}(q)] + [F(q,\dot{q}) - \hat{F}(q,\dot{q})]\} + D \tag{6}$$

where D is the disturbance acting on the system. Assuming that the sum of model errors and disturbances is bounded by

$$\left| \hat{M}^{-1}(q)\{[M(q) - \hat{M}(q)]\ddot{q} + [C(q,\dot{q}) - \hat{C}(q,\dot{q})] + [G(q) - \hat{G}(q)] + [F(q,\dot{q}) - \hat{F}(q,\dot{q})]\} + D \right| \le F \tag{7}$$

and choosing the control law u as $$u = \lambda_1 \dot{e} + \lambda_2 e + K_s \operatorname{sgn}(S) \tag{8}$$

where the gain matrix K_S is given by $$K_s = F + \eta \tag{9}$$

then, the Lyapunov criterion is satisfied since

$$\frac{1}{2}\frac{\mathrm{d}}{\mathrm{dt}} S^2 = S\dot{S} \le -\eta S \operatorname{sgn}(S) = -\eta |S| \tag{10}$$

where the strictly positive η implies that $S \to 0$ as $t \to \infty$ (that is, $e \to 0$, $\dot{e} \to 0$).
To avoid chattering, the switching function sgn(S) may be replaced by the saturation function sat(S/μ) with a boundary layer thickness μ. For a smoother response, $\tanh(S/\mu)$ may also be used. The sat(S/μ) function is defined as

$$\operatorname{sat}\left(\frac{S}{\mu}\right) = \begin{cases} S/\mu & \text{if} \quad |S/\mu| < 1 \\ 1 & \text{if} \quad S/\mu > 1 \\ -1 & \text{if} \quad S/\mu < -1 \end{cases} \tag{11}$$

Adaptive Controller. On-line identification of parameters is potentially a very useful tool to minimize model errors in a reconfigured, faulty system. Parameter adaptation technique is a popular method in robot control. This method employs the reparametrization of the dynamic model into the product of a known nonlinear matrix based on robot geometry, and an unknown constant vector regarding the estimated parameters such as the payload, center of mass, moment of inertia, coefficient of friction, etc. The reparametrization which relates the robot dynamic equations to the unknown constants P (r × 1 vector) in a linear form is represented as

$$M(q)\ddot{q} + C(q,\dot{q}) + G(q,\dot{q}) + F(q,\dot{q}) = W(q,\dot{q},\ddot{q})P \tag{12}$$

By subtracting (12) from (2), the system error equation is rewritten as

$$E_m = \hat{M}^{-1}(q)W(q,\dot{q},\ddot{q})P \tag{13}$$

The (r × 1) vector P defines the actual parameters, and the estimated parameters are denoted by $\hat{P}$, then the error of estimation is $\tilde{P} = P - \hat{P}$. Note that the control scheme is to combine the parameter adaptation technique (PA) with the sliding control (SC) method such that the PA algorithm measures the parameter uncertainties to reduce model error, and the SC algorithm copes with unmeasurable parameter uncertainties as well as the disturbances. The candidate Lyapunov function composed of the error surface and the parameter estimation error terms is expressed as follows

$$V = \frac{1}{2}\left(S^T S + \tilde{P}^T \Delta^{-1} \tilde{P}\right) \tag{14}$$

where Δ is a diagonal matrix with chosen positive constant elements. By differentiating (14), the stability condition is satisfied with the following selection. The unknown parameters are also estimated by it:

$$\dot{\hat{P}}(t) = -\dot{\tilde{P}} = \Delta W^T(q,\dot{q},\ddot{q})\hat{M}^{-1}S \tag{15}$$

Sliding Controller with Boundary Layer. When the boundary layer is introduced to the sliding control design, the switching function associated with the sliding control remains active outside this boundary. However, as the solution enters the boundary layer, the controller behaves similar to a PID controller. To satisfy the Lyapunov criterion in (10), $\dot{S}$ is designed as $$\dot{S} = -\eta \operatorname{sgn}(S) \tag{16}$$

Then, (5) and (16) yield $$\ddot{e}(t) + \lambda_1 \dot{e}(t) + \lambda_2 e(t) = -\eta \operatorname{sgn}(S) \tag{17}$$

The saturation function sat(·) in (11) is used to replace the sgn(·) in (17). Hence, (17) becomes

$$\ddot{e}(t) + \lambda_1 \dot{e}(t) + \lambda_2 e(t) = -\eta \operatorname{sat}(S/\mu) \tag{18}$$

When the sliding surface enters the boundary layer with small errors, the saturation function in (11) equals to S/μ, which indicates that the Lyapunov criterion is still satisfied with $S\dot{S} \le -\eta/\mu S^2$. Under this condition, however, the sliding control mode switches to perform as a PID controller. Therefore, by substituting S in (4) into (18), and later equating the coefficients of the PID expression in (3) and (6), we have

$$K_v \dot{e}(t) + K_p e(t) + K_I \int e(t)\mathrm{dt} = (\lambda_1 + \eta/\mu)\dot{e}(t) + (\lambda_2 + \lambda_1 \eta/\mu)e(t) + \lambda_2 \eta/\mu \int e(t)\mathrm{dt} \tag{19}$$

$$K_v = \lambda_1 \eta/\mu, \;\; K_p = \lambda_2 + \lambda_1 \eta/\mu, \;\; K_I = \lambda_2 \eta/\mu \tag{20}$$

Hence, the sliding control behaves like a PID controller when the errors enter the boundary layer. This also allows us to compare the performance of controllers which use these methods. For fault-tolerance purposes, sliding control will be preferable to a controller based only on PID feedback.

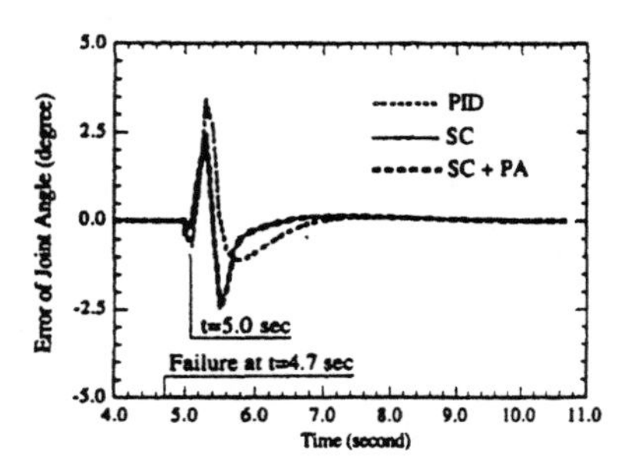

Figure 5. Errors at joint 1 by PID, Sliding Control (SC), and Parameter Adaptation (PA)

The proposed fault-tolerant controller is tested numerically on a four-link, planar robot. Since a planar robot requires three parameters to specify the end-effector position and orientation, this robot is redundant in the 2-D space. This allows full functionality of the robot even after it loses one of its actuators as a result of a full failure. In general, when a partial failure occurs, the torque redistribution and time dilation methods may provide an acceptable solution; however, when a full failure is experienced sliding and adaptive control techniques seem more appropriate as evidenced by the results presented here.

The robot operation is planned for 10 sec in the absence of a failure. The manipulation trajectory and controller implementation for the simulation are selected as follows: The initial joint angles of the serial robot are chosen as $\phi_1(0) = 20°$; $\phi_2(0) = -10°$; $\phi_3(0) = 0°$; $\phi_4(0) = 0°$. The operational range of each joint is given by $\phi_1(t)$: 20° to 40°; $\phi_2(t)$: -10° to -50°; $\phi_3(t)$: 0° to 75°; $\phi_4(t)$: 0° to -75°, each defined by a 4-5-6 polynomial. Taking the computation time for computing kinematic and dynamic parameters into account, and from prior experimentation, the total period for detecting the failure and recovering with a new configuration is conservatively set at 200 ms. Assuming that the joint failure occurs at time t = 4.7 sec, during the time interval t = 4.7 to 4.9 sec the robot drifts, finds the error, and starts to recover. During the time interval t = 4.9 to t = 5.0 sec, the brake is applied to the failed joint which reduces its relative velocity and acceleration to zero. Then, the robot starts to recover back to the failed position from t = 5.0 to 5.5 sec governed by a 4-5-6 polynomial. For the purpose of examining the performance of sliding control and parameter adaptation algorithms, simulations are conducted with large parameter uncertainties (moment-of-inertia: 50% of actual value and center-of-mass: 150% of actual value). This extreme case can be attributed to a case where joint angle at the locked joint can not be measured due to a failed sensor. From the simulation results, the chattering effect is obvious when sliding control is used. System response deteriorates when boundary layer is not used, or when unsuitable boundary layer thickness values are used (for instance, μ=0, 0.025, 0.05 are judged to be too small).

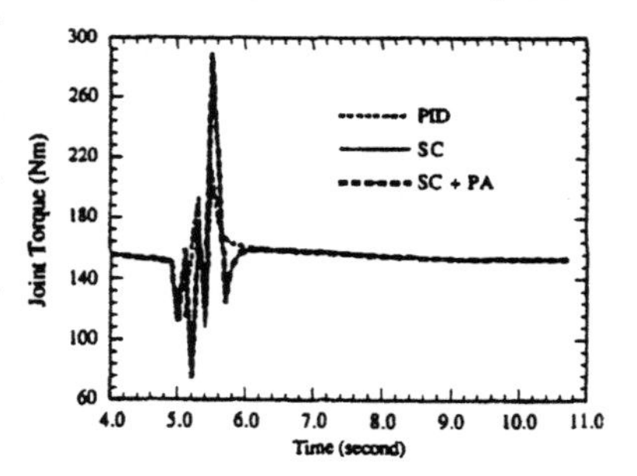

Figure 6. Actuator torque at joint 1

Through a number of trials an optimal thickness μ=0.1 is selected so that the tracking performance became acceptable. The gains (K_v, K_p, K_I) = (9, 26, 24) for PID control assigns the poles in the s-plane at (-2, 0), (-3,0), (-4,0). Thus, the sliding control design parameters are solved as $\lambda_1 = 6, \lambda_2 = 8, \eta/\mu = 3$ according to (20).

The simulation results are shown in Figures 5–8 for the period between t= 4.0 and 11 sec. Figure 5 plots the error in joint 1 by using the control laws based on PID, sliding control (SC), and SC with parameter adaptation (PA). The sliding control with or without parameter adaptation outperforms the PID feedback control; in particular, it reduces the error faster. There is not a significant difference in the tracking performance between the SC and SC with PA. This may be so when the magnitudes of the unknown parameters are not significant, or when the SC or PID component of the controller dominates the total input torques. As shown in Figure 7, the latter is clearly the case in this simulation. A comparison of torque magnitudes in Figures 6 and 7 shows this.

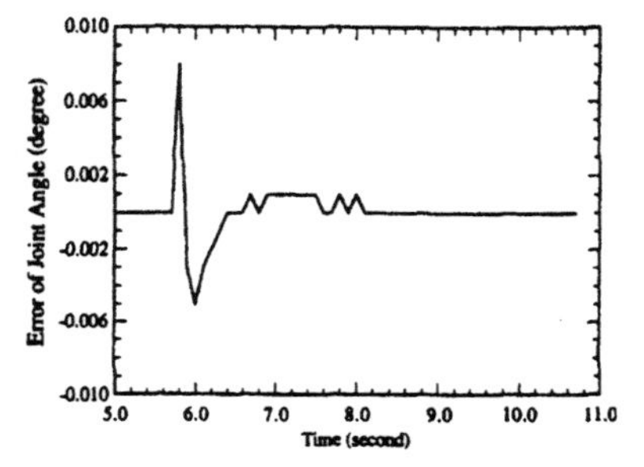

Figure 7. Component of joint 1 actuator torque due to only parameter adaptation

Since the sliding control with a boundary layer μ=0.1 was used, parameter estimation is set to begin when the errors enter the layer. Otherwise, the chattering effect may influence the accuracy of parameter adaptation. The

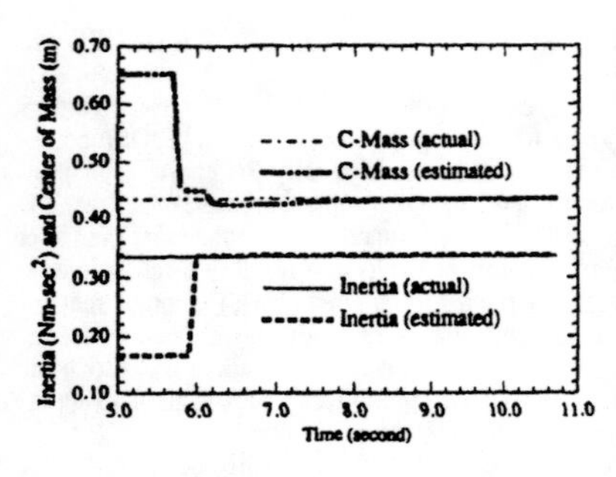

Figure 8. Parameter estimation
(Note: Parameter estimation starts after the solution enters the boundary layer)

gains for parameter estimation, e.g. moment-of-inertia: $\Delta = 0.164$ and the center-of-mass: $\Delta = 0.0425$ for 50% model errors, are required to arrive at the estimated parameters within a 0.5% deviation from their actual values. The estimated parameters of moment of inertia and center of mass shown in Figure 8 successfully reach their actual values with less than 0.5% errors. The joint torque trajectory is shown in Figure 6 for joint 1. The torque requirement for sliding control is much larger than that of PID, especially during failure and at the beginning of recovery when larger tracking errors exist. Although the difference in the required torque between the SC with and without PA is very small, Figure 7 shows that there is a sudden change when the estimation starts. This phenomenon is due to the sudden variations of the system parameters. If the parameter uncertainly is large, e.g., if the payload suddenly increases, this abrupt change will be more serious. Therefore, activation of parameter estimation may not be suitable when the system undergoes drastic changes. A slow estimation scheme may lessen the drastic change, but sacrifice the system performance in this transient period. Nevertheless, the capability to identify parameters in a failed robot is a desirable feature, and it would preserve the ability of the system to recover from other failures by making the controller more robust.

DISCUSSION AND CONCLUSIONS

Fault-tolerant controllers gain importance as more demands are placed on system performance. Fault-tolerant systems represent one such area where a system is expected to function without any interruption even when some of its components fail. This challenging problem is studied in terms of n-link, 3-D robots. When a failure occurs, the problems of model uncertainty and disturbances exist in the system. Selection of a control algorithm which is robust in the presence of these errors and which reduces the tracking errors quickly is a challenging task. With the use of the characteristics of the boundary layer, which is primarily designed to avoid the chattering effect, a feasible solution is obtained by using the sliding control method. That is, the switching function acts on the system to cope with the errors arising from the disturbance and parameter variations outside the boundary layer, and then it behaves like a PID controller when the errors are reduced. Sliding control, a robust control technique, is presented and its merit in reducing errors efficiently is illustrated with simulations. A conclusion can be drawn that sliding control generally gives the system a faster response to a change in parameters and/or disturbance, and the robustness is preserved with bounded model uncertainties and disturbances. Therefore, such a controller is an ideal candidate to explore fault tolerance applications. The parameter adaptation method (PA) identifies the selected system parameters on-line by the use of suitable gains. Although the PA achieves improvements in the simulations presented here, it is believed that following a successful parameter estimation, system performance can improve more significantly, especially when failures cause large model uncertainties.

REFERENCES

1. Adlemo, A., and Andreasson, S., "Failure Semantics in Intelligent Manufacturing Systems," Proceedings of the 1993 IEEE International Conference on Robotics and Automation, Vol. 1, pp. 166–173, Atlanta, Georgia, May 2-7, 1993.
2. Sreevijayan, D., and Tosunoglu, S., "Architectures for Fault Tolerant Mechanical Systems," *Proceedings of the IEEE Mediterranean Electrotechnical Conference; MELECON'94*, Antalya, Turkey, Vol. 3, pp. 1029–1033, April 12–14, 1994.
3. Ting, Y., and Tosunoglu, S., "A Control Structure for Fault-Tolerant Operation of Robotic Manipulators," *Proceedings of the 1993 IEEE International Conference on Robotics and Automation*, Vol. 3, pp. 684–690, Atlanta, Georgia, May 2–7, 1993.
4. Visinsky, M.. L., Walker, I. D., and Cavallaro, J. R., "Layered Dynamic Fault Detection and Tolerance for Robots," *Proceedings of the 1993 IEEE International Conference on Robotics and Automation*, Vol. 1, pp. 180–187, Atlanta, Georgia, May 2–7, 1993.
5. Ting, Y., Tosunoglu, S., and Freeman, R., "Actuator Saturation Avoidance for Fault-Tolerant Robots," *Proceedings of the 32nd IEEE Conference on Decision and Control*, Vol. 3, pp. 2125–2130, San Antonio, Texas, Dec. 15–17, 1993.
6. Maciejewski, A., "Fault Tolerant Properties of Kinematically Redundant Manipulators," *Proceedings of the 1990 IEEE International Conference on Robotics and Automation*, pp. 638–642, Cincinnati, Ohio, May 1990.

KINEMATIC AND STRUCTURAL DESIGN ISSUES IN THE DEVELOPMENT OF FAULT-TOLERANT MANIPULATORS

SABRI TOSUNOGLU
Department of Mechanical Engineering
Florida International University
Miami, Florida 33199 USA

ABSTRACT

Fault tolerance technology promises higher system reliability even under unexpected component failure. Such capability is attained by developing structural system designs that can deliver fault tolerance and by designing controllers that can take advantage of the fault-tolerant structure. This paper reviews fault-tolerant design issues for mechanical and robotic systems from a kinematic and structural design viewpoint. This is accomplished by studying the kinematic design of a fault-tolerant robotic system at four levels: (1) Joint level (Single and dual actuators); (2) Link level (Serial and parallel modules); (3) Sub-system level (Non-redundant and redundant manipulators); (4) System level (Multiple cooperating manipulators). However, each of the above levels introduces challenges to the structural design of a robot. This work addresses the four levels from a structural design standpoint. Relative advantage and disadvantage of various levels are listed and discussed in an effort to identify the most efficient structural modification to enhance fault tolerance of a robot.

KEYWORDS: Robotics, manipulators, fault tolerance, robot architecture, kinematics, design, dual actuators, evaluation of fault tolerance, measure of fault tolerance, levels of fault tolerance.

INTRODUCTION

Fault tolerance was initially introduced for computer hardware and software, and subsequently applied to aerospace systems, nuclear power plants and computer-integrated manufacturing (CIM) where system reliability was particularly important [1–4]. In recent years, this concept is being developed for mechanical systems, and especially for robotics [3, 5–9].

In order to develop a fault-tolerant system, one needs to address two problems. The first is the structural design of the mechanical system, and the second is the design of a fault-tolerant controller which enables the utilization of the fault tolerance capabilities of the hardware. In this paper, we address the development of a fault-tolerant robot architecture which encompasses four distinct levels: (1) Joint Level, (2) Link Level, (3) Sub-system Level, and (4) System Level.

In this architecture, the joint level forms the lowest level fault tolerance and proceeds towards the higher levels through the introduction of tolerances at the link, sub-system and system levels. Although the ultimate aim in fault tolerance is to provide a capability to a system to fully function even under the loss of its components, and each of these levels intends to accomplish that task, we note that the above architecture is arranged such that a lower level is contained in the higher levels. Therefore, the aim becomes to contain a detected fault at the lowest possible level and proceed towards the higher levels if a lower level can not isolate the fault.

In a robotic manipulator, the most important factors influencing fault tolerance are the kinematic structure of the robot and the design of its actuation system. Hence, we present the fault-tolerant robot architecture in terms of these two critical design issues. Later, various manipulator designs are evaluated in terms of the fault tolerance capability they offer.

DESIGN CRITERIA FOR FAULT TOLERANCE

In this section, we review two critical design parameters that fundamentally affect fault tolerance. These parameters are the actuation system of a manipulator, and the robot's kinematic structure as briefly introduced below.

Actuation System Design

Serial Robots. This deals with the way a robot is actuated. The most common design found in today's serial industrial robots is the actuation of each degree of freedom (usually corresponds to a joint) by a single motor. In fault-tolerant robots, however, the concept of dual actuation is introduced. As sketched in Figure 1, this design incorporates two motors to drive a single degree of freedom of the robot. In normal operation, where no fault is involved, these two actuators may apply output torques in the same direction; thus, increasing the payload capacity of the robot. In a different scenario, under the no-fault condition, one of the motors may apply a small amount of counter torque to absorb the joint backlash; thus, drastically reducing or eliminating the backlash. When one of the actuators fails, this motor is disengaged by a clutch mechanism and the remaining motor supplies torque to the joint.

If one of the motors fails in a serial manipulator which does not have this feature, then the joint powered by this motor has to be blocked by applying the brake. This causes the robot to lose one of its degrees of freedom which could substantially reduce the robot's workspace and cause the robot to become practically useless to continue its operations.

However, if the failed joint is driven by a dual actuator system, then the robot can continue its operations—with a reduced load-carrying capacity—even after one half of the actuation system has been disengaged. Hence, incorporation of dual actuator systems in robot design is beneficial to improve fault tolerance capability of a serial robot.

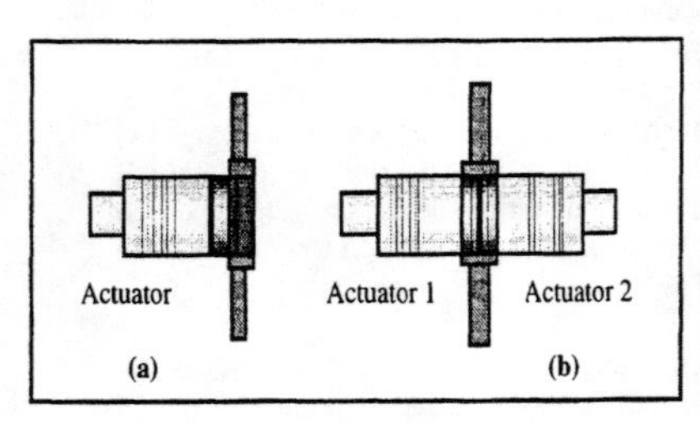

Figure 1. (a) Single and (b) dual actuator systems

Parallel Robots. In order to improve fault tolerance of a parallel-structured robot, a different actuation system is required. Consider, for instance, the four-bar mechanism shown in Figure 2. (Many robots contain either four-bar or slider-crank type substructures even in otherwise serially-designed robots.) As a one-degree-of-freedom mechanism, the four-bar requires only one input; i.e., only a single motor to drive it as depicted in Figure 2 (a). Fault tolerance of this mechanism is improved if another joint is also actuated by a single motor as illustrated in Figure 2 (b). Finally, three single actuators (Figure 2(c)) or a dual actuator with an additional single motor may also be used in this mechanism to further improve the fault tolerance capability as shown in Figure 2 (d).

The use of a dual actuator is an attractive option if the torque transmission characteristics of the four-bar mechanism are significantly inferior at other joints. Otherwise, the use of multiple single motors at various joints is better suited for parallel mechanisms. We note that although the four-bar mechanism is used in this presentation, slider-crank and other mechanisms (including redundant mechanisms such as the five-bar) may equally be considered to improve fault tolerance of mechanical systems.

Kinematic Structure

The kinematic structure of the robot is an extremely important criteria for fault tolerance since the recovery from a faulty condition is often decided based on the robot kinematics. In this category, we identify two major branches in mechanism design. These are the serial and parallel structures of the basic robot kinematics. Serial and parallel structures offer various advantages and disadvantages in the general field of mechanism design. However, each is evaluated under a very different set of requirements when they are used to construct fault-tolerant robots. We review these two classes of mechanisms below.

Serial Robots. In general, serial robots are superior to parallel mechanisms because of the exceptional workspace they enjoy. By comparison, parallel mechanisms have smaller working volumes because the parallel structure of these mechanisms mechanically limits the movement of its links. The serial make-up, however, provides a smaller end-effector payload capacity since the serial structures do not have the superior mechanical advantage of parallel mechanisms. Also, in serial robots, errors in each joint add up towards the end-effector; thus, the error at the robot's hand is maximized. Nevertheless, the simplicity and exceptionally large workspace of serial robots made them quite popular.

Evaluated for fault tolerance, serial robots perform differently. When an actuator of a serial robot fails, control of the arm downstream the failed joint is lost unless that joint is locked following the failure. This causes the robot to lose one of its degrees of freedom, which is not desirable. A solution in that case is to mount a dual actuator system (Figure 1 (b)). Otherwise, the employment of a redundant robot will be beneficial. Even after locking one of the joints, a 7-DOF robot becomes a 6-DOF system which remains fully functional in the 3-D space.

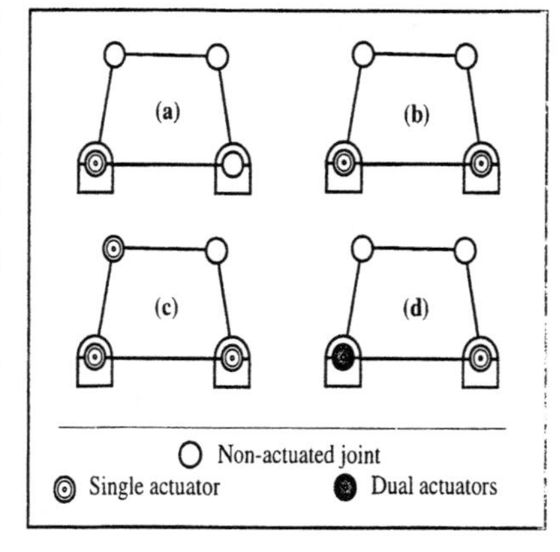

Figure 2. Actuation systems for a four-bar mechanism

Parallel Robots. Parallel systems behave differently than serial robots under the failure of its components. For instance, when one of the actuators fails, locking that joint may result in a 0-DOF structure (in the case of 1-DOF parallel mechanisms). Hence, when a four-bar mechanism is employed, one should release the failed joint and actuate another one as shown in Figure 2 (b). Hence, if no dual actuators are used, then the failed joint should be locked in serial robots, and released in parallel mechanisms. This action reduces the DOF of the serial robot by one, whereas it does not influence that of a parallel mechanism. To provide fault tolerance by designing new actuation systems, the serial robots should have dual actuators, while the parallel robots need to have extra single actuators mounted at its most suitable joints.

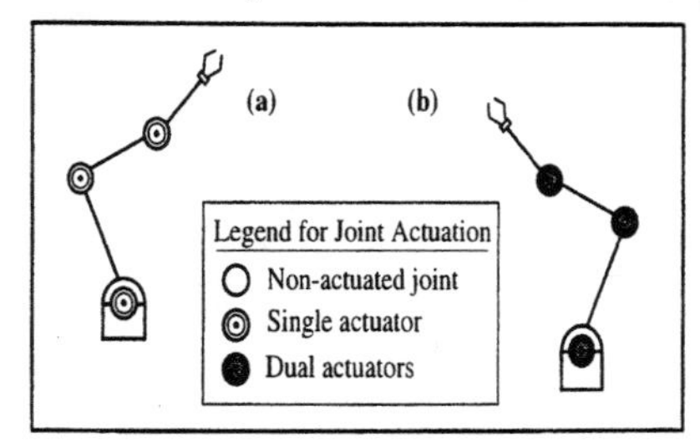

Figure 3. Level 1: Joint level fault tolerance

FAULT-TOLERANT ROBOT ARCHITECTURE

Fault-tolerant robot architecture is presented in four levels. Starting from the lowest, we move towards the higher levels. In this architecture, a higher level encompasses all of the lower levels. Upon detection, a fault should be rejected at the lowest possible level. If that is not physically possible, then fault isolation is carried out at the next higher stage.

Joint Level Fault Tolerance

The joint level fault tolerance incorporates the employment of a dual actuator as shown in Figure 1 (b). While dual actuators can be mounted at every joint (Figure 3 (b)), it is possible to install them at the most critical joints such as the shoulder and elbow joints. This allows simpler wrist designs whose increased weight is especially detrimental to the payload capacity. Since the base joints—the shoulder and elbow—influence the workspace of a robot the most, fault tolerance of these joints becomes more important. Alternatively, an extra degree of freedom at the wrist will provide a fault tolerance capability with a less penalty in the wrist weight. While dual actuators are most suitable for serially-designed robots, they can also be used in parallel mechanisms when the mechanical advantage of other joints is inferior to the one that is actuated. Figure 4 (c) illustrates such a serial robot with inner parallel loops.

Link Level Fault Tolerance

If failure of a motor can not be masked with the use of a dual actuator or if the system is not equipped with one, then the link level fault isolation is sought. This level fault tolerance is available for robots which contain parallel substructures. Shown in Figure 4 (a) is the kinematic structure of such a robot. Each of the three four-bar mechanisms is actuated by a single motor. This robot's fault tolerance is enhanced by installing additional motors (Figure 4 (b)). Fault tolerance is further strengthened if selected single actuators are replaced by dual actuation systems (Figure 4 (c)). As briefly described earlier, mechanical advantage of a parallel mechanism varies as the actuated joint is varied. By denoting the input actuator torque by τ_{in}, and the external load by F_e, the relationship between these two quantities is described by the so-called first-order influence coefficients [10]:

$$\tau_{in} = G\,F_e$$

Since the G-function is a nonlinear function of the joint displacements and the link lengths of the parallel mechanism, different input torques τ_{in} will be required as the position of the input is changed. The average value of the coefficient G throughout the useful workspace of the mechanism determines the mechanical advantage of the mechanism.

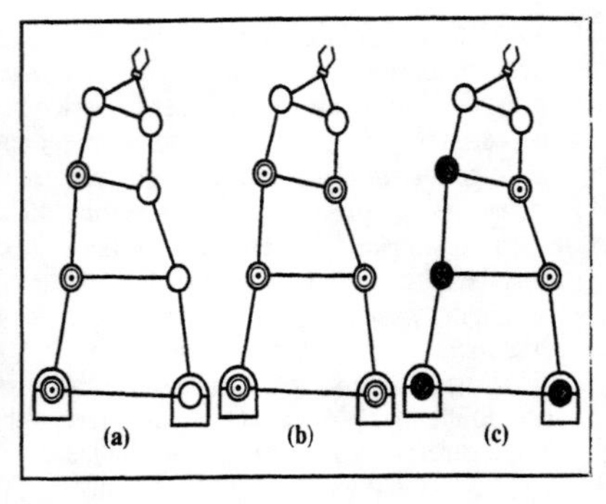

Figure 4. Level 2: Link level fault tolerance

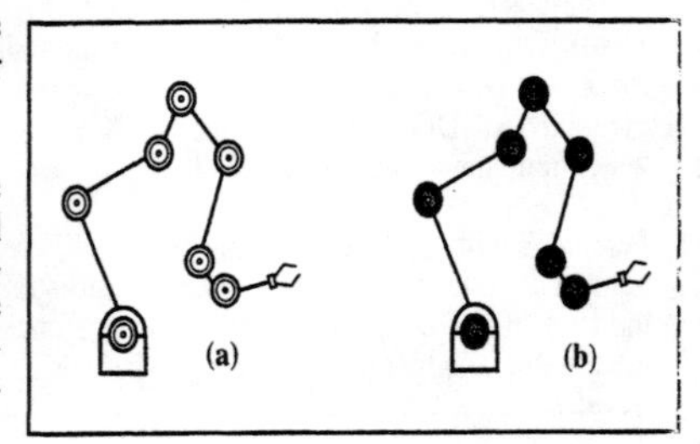

Figure 5. Level 3: Sub-system (manipulator) level fault tolerance

Hence, under the condition in which an acceptable mechanical advantage is not obtained by actuating different joints, installation of a dual actuator at the most suitable joint (with the highest average G value) provides the most advantageous design to enhance fault tolerance. If another joint provides a comparable performance as the original input joint, then the use of additional single actuator(s) will be the most suitable design alternative when the simplicity of single motor installation is considered.

Sub-system Level Fault Tolerance

When a fault is not contained within the first two levels of the fault-tolerant architecture, the third level supplies a sub-system or manipulator level alternative by taking advantage of the kinematic redundancy of the robot. This implies that a robot working in 3-D space will have to have more than six degrees of freedom, and more than three 1-DOF joints if it works in the plane. Figure 5 (a) sketches a 7-DOF robot which can tolerate the loss of one actuator. This robot will have 6 functional joints left even after one has failed. Although theoretically a 6-DOF robot remains fully functional in 3-D space, its workspace may be substantially reduced should one of the shoulder or wrist actuators of the original 7-DOF robot fails. Hence, the tolerance provided by a 7-DOF robot, although redundant, may be of limited value. This is overcome by installing dual actuators as depicted in Figure 5 (b). A hybrid manipulator version of the one in Figure 5 is shown in Figure 6. Also, design of robots with 8 or more DOF enhances fault tolerance capabilities at the sub-system level.

System Level Fault Tolerance

The system level tolerance is required when the sub-system and other lower levels of fault tolerance can not provide an alternative to isolate a fault. At this level of fault tolerance, either a backup robot system or a cooperating robot already installed nearby the faulty one shares the tasks of the failed robot so that the ongoing operation is not entirely shut down as a result of a single robot failure. Two examples for this level of fault tolerance are shown in Figures 7 (a) and (b).

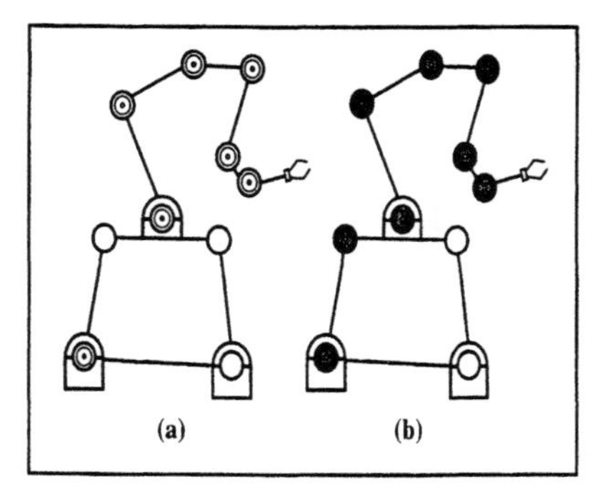

Figure 6. A hybrid manipulator system

EVALUATION OF FAULT TOLERANCE FOR VARIOUS MANIPULATORS

In the previous sections, we have presented four levels of fault tolerance dealing with the kinematic architecture and actuation structure of robotic systems. Although each level enhances fault tolerance of the system, a comparative study becomes necessary to help the designer and the user to decide which level of fault tolerance is most beneficial for a given set of operational requirements. Since the fault-tolerant architecture is presented in four stages, performance of each stage is evaluated considering various criteria such as fault tolerance capability, workspace of the robot, complexity introduced to the design as a result of fault tolerance enhancement, and the cost of the system. Comparative value of fault tolerance is then evaluated for five robots. In Table I, these are indicated as follows:

(1) Level 1: Serial non-redundant robot representing level 1
(2) Level 2: Parallel-structured robot representing levels 1 and 2
(3) Level 3: Redundant serial robot representing levels 1 and 3
(4) Level 4: Dual redundant robots representing levels 1, 3 and 4
(5) Hybrid System: Hybrid redundant robot representing levels 1, 2 and 3

Table I lists evaluation of fault tolerance capability of these robots. Before studying the overall value of design, this table is useful to assess the relative value for each criteria. For instance, when only the fault tolerance capability is considered, we see that robots which contain parallel structures and multiple robots become the most successful designs. This is due to the inherent success of parallel structures in fault tolerance [9], as well as the significant reliability enhancement of a backup system. By contrast, when only the design complexity of the system is considered, serial, non-redundant robots are preferred since they represent the simplest design among the architectures evaluated. Parallel structures and redundant robots do not lend themselves as attractive choices in this category. In applications where workspace is more important, clearly the redundant serial robots become the most attractive choice as indicated in Table I. Parallel mechanisms perform poorly when evaluated by their workspace alone. This is due to the fact that the parallel structure itself imposes mechanical constraints on its workspace. When the cost of the system is factored in, multiple robots become the least favored option, while the non-redundant serial system offers itself as the most attractive option. Hence, each architecture offers advantages and disadvantages when differing preferences are considered. When the cumulative effects of the positive and negative evaluations in each category are isolated for each architecture, a measure of relative merit for the given system is established as shown in Table I.

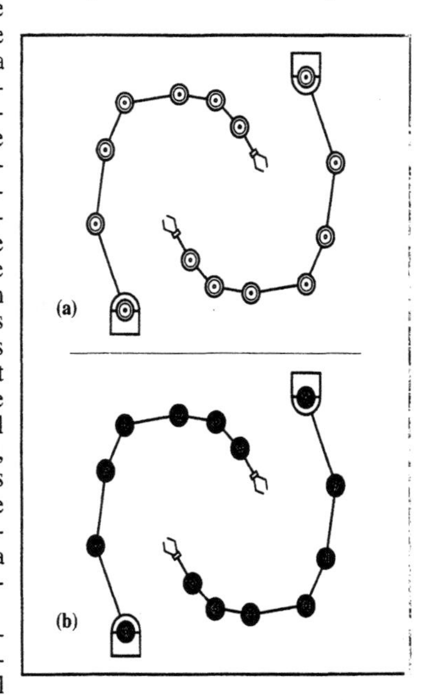

Figure 7. Level 4: System level fault tolerance

When the cumulative effects are normalized, the redundant hybrid robot that contains a parallel substructure becomes the most preferred choice. The serial redundant robot takes the second place while the non-redundant serial robot becomes the third choice. This robot still enjoys a fault tolerant capability of Level I by employing dual actuators. Multiple robots, which present the largest workspace, take the fourth place largely due to the prohibitive price of such systems. (It should be noted, however, that if multiple robots are already installed in a factory assembly line regardless of the fault tolerance issue, then the negative cost element should be removed from the evaluation.) Although potentially useful, extensive use of parallel loops throughout the robot structure increases robot's complexity and reduces its workspace; hence, they become less attractive solutions as fault-tolerant systems.

TABLE I. Evaluation of Fault Tolerance for Various Robot Systems

Figure	3 (b)	4 (b)	5 (b)	7 (b)	6 (b)
Criteria	Level 1	Level 2	Level 3	Level 4	Hybrid System
Fault Tolerance	3	5	4	4	5
Workspace	2	1	4	5	4
Design Complexity	5	1	4	3	4
Cost	5	2	4	1	4
Total	15	9	16	13	17
Normalized %	88	53	94	76	100
Standing (out of 5)	3	5	2	4	1

Note: (1) Each score out of 5. (2) Higher scores are better.

CONCLUSIONS

Fault tolerance of robotic systems is addressed from a structural design viewpoint. This envisions fault tolerance integrated into the system design at the joint, link, sub-system and system levels. While at the joint level dual actuators enhance system reliability, link level improvement is seen when parallel-structured mechanisms are used. Sub-system level reliability corresponds to the use of redundant robots while the system level fault tolerance makes use of back-up robots. When critical issues in design and fault tolerance are evaluated for this mechanical architecture, it is observed that a hybrid system that contains both serial and parallel structures is the most promising design for fault tolerance.

REFERENCES

1. Isermann, R., "Model Based Fault Detection and Diagnosis Methods," *Proceedings of the American Control Conference; ACC'95*, Seattle, Washington, June 21–23, 1995.
2. Von Neumann, J. "Automata Studies," *Annals of Mathematics Studies*, No. 34, edited by C. E. Shannon, and E. F. Moore, Princeton University Press, Princeton, New Jersey, 1966.
3. Sreevijayan, D., and Tosunoglu, S., "Architectures for Fault Tolerant Mechanical Systems," *Proceedings of the IEEE Mediterranean Electrotechnical Conference; MELECON'94*, Antalya, Turkey, Vol. 3, pp. 1029–1033, April 12–14, 1994.
4. Andreasson, S., Adlemo, A., Fabian, M., Gullander, P., and Lennartson, B., "Database Design for Machining Cell Level Product Specification," *IEEE 21st International Conference on Industrial Electronics, Control, and Instrumentation*, Orlando, Florida, Vol. 1, pp. 121–126, November 6–10, 1995.
5. Visinsky, M., Walker, I., and Cavallaro, J., "Layered Dynamic Fault Detection and Tolerance for Robots," *Proceedings of the 1993 IEEE International Conference on Robotics and Automation*, Vol. 1, pp. 180–187, Atlanta, Georgia, May 2–7, 1993.
6. Maciejewski, A., "Fault Tolerance Properties of Kinematically Redundant Manipulators," *Proceedings of the IEEE International Conference on Robotics and Automation*, Vol. 1, pp. 638–642, September 1982.
7. Kelmar, L., and Khosla, P., "Automatic Generation of Kinematics for a Reconfigurable Modular Manipulator System," *Proceedings of the IEEE Conference on Robotics and Automation*, Philadelphia, Pennsylvania, Vol. 2, pp. 663–668, April, 24–29, 1988.
8. Chladek, J., "Fault Tolerance for Space Based Manipulator Mechanisms and Control System," *First International Symposium on Measurement and Control in Robotics*, Houston, Texas, June 1990.
9. Ting, T., Tosunoglu, S., and Fernández, B., "Control Algorithms for Fault-Tolerant Robots," *Proceedings of the 1994 IEEE International Conference on Robotics and Automation*, San Diego, California, Vol. 2, pp. 910–915, May 8–13, 1994.
10. Tosunoglu, S., and Lin, S., "Accessibility and Controllability of Flexible Robotic Manipulators," *ASME Journal of Dynamic Systems, Measurement and Control*, Vol. 114, No. 1, pp. 50–58, March 1992.

A PROGRAM FOR THE ECONOMICAL RADIATION HARDENING OF ROBOTS

JAMES S. TULENKO, R. DALTON, G. YOUK, H. LIU, H. ZHOU
Department of Nuclear Engineering Sciences, University of Florida

ROBERT M. FOX
Department of Electrical and Computer Engineering,
University of Florida

ABSTRACT

The Nuclear Engineering Sciences Department and the Electrical and Computer Engineering Department of the University of Florida has been actively researching and developing radiation hardened robotic systems components for use in nuclear facilities. The program has been supported both by the Department of Energy and by robotic component manufacturers. Under this program the University of Florida has developed a radiation hardened ANDROS robotic system for operation in radiation environments with accumulated exposures to one megarad. Additionally, the research team has developed a monitoring technique to detect the approach of radiation induced failure of integrated circuits. As a by-product of its extensive radiation testing program, the Florida Team with Department of Energy funding, has developed an on-line (Internet-accessible) database of radiation effects testing for typical robotic system components termed UFRED. Using this database the University of Florida is working to include radiation effects in failure analysis studies being carried out for the Department of Energy.

KEYWORDS: radiation, hardened, robotic, nuclear, database, reliability

INTRODUCTION

The University of Florida has a broad program to develop the technology to enable the safe and effective use of robotic systems in nuclear facilities. Radiation levels found in many nuclear facilities present a hazardous environment for humans necessitating the use of remote or highly shielded operations. The radiation levels also can present a severe environment to the electronic systems/sensors in commercial robotic systems. The failure of a robotic system in a nuclear facility would be highly disruptive. To avoid the possibility of a failure and ensure reliable robotic operations, the University of Florida has undertaken a careful evaluation of the radiation tolerance of commercial solid-state electronic components and to investigate methods to extend the useful life of the components.

The University of Florida Team is seeking to specify, predict, design, test, demonstrate, and monitor robotic operations to meet the designed reliability level for radiation environments. The key item is that all of the above must be achieved at an acceptable and competitive price. This effort is being driven by the fact that with the end of the cold war, radiation hardened components are not being restocked. When the current stock of rad-hard components runs out, it will be necessary to use naturally radiation resistant components such as bipolar components. [The database can be used to choose radiation resistant substitutes for soft radiation components.]

RADIATION EFFECTS DATABASE

The University of Florida has developed a radiation effects database for robots called the University of Florida Radiation Effects Database (UFRED). UFRED is currently on line for local (at the University of Florida) uses. The database utilizes the Microsoft ACCESS database with software being written to connect the database to the internet. An example of the database screen is shown in Figure 1. The database has querying capability by type, device number, manufacturer, log number, and a combination of these categories. A program of comprehensive data entry is underway with World Wide Web availability of UFRED anticipated by the end of the calendar year.

The database is used by the University of Florida to determine the reliability of robotic components under radiation environments. Reliability is the probability that parts, components, products, or systems will perform their designed-for function without failure in specified environments for desired periods at a given confidence level. For example, if we are looking for an operational amplifier that can operate with a minimum gain of 1500, the database would identify under operational amplifier, the irradiation response of OP27 manufactured by MPS. When the gain data versus irradiation is fit as in Figure 2, the data shows that the device should function until a radiation level of 150 Krads. When we fit the data (failure dose histogram) with a Gausian probability (Figure 3), a failure probability is developed which can be then supplied to the reliability engineers who add this data to their other failure probability data to predict maintenance and replacement schedules for the robot's operation.

On-line Monitoring of Radiation-Induced Degradation

The radiation testing conducted by the University of Florida falls into two categories, either functional or parametric. The functional test determines how long a system works under radiation. Functional tests represent a minority of tests run by the University of Florida and are usually used for system testing (e.g., amplifiers and cameras). Another set of tests, the parametric, tracks the radiation-induced degradation of a key parameter (metrics). These tests are the basis of the input for the database.

The University of Florida thrust is to find a method to make parametric measurements that will reliably predict the failure of an individual component before it occurs, thus a part can be used until failure is imminent. By this measurement technique, the effective life of components could be greatly increased from the conservative failure probability method discussed earlier in the paper. In the case cited earlier for the operational amplifier, a ninety percent confidence level of performance would mean retiring the component at 100 kilorads whereas the average lifetime is 200 kilorads. Thus, considerable savings could be accrued from a method that would allow operation until the approach of a failure level.

Current Work on Simple Integrated Circuits (ICs)

Integrated Circuits (ICs) fail primarily because of (1) an inability to switch at system clock speed or (2) excessive leakage currents that keep transistors from turning off completely. These effects are caused by shifts in transistor parameters under irradiation, especially threshold voltages (V_T). The missing element has always been knowledge of how individual parameters in an IC are varying (i.e., what is going on "inside" the IC). The University of Florida, under Dr. Fox, has set up a "Parameter-Extraction Circuit System." For the study of NMOS FETs, the system applies inputs that set V_a to V_{ao}, turning the PMOS off and then the operational amplifier measures current while sweeping $V_D = V_D'$. To study the PMOS FETs, the system applies inputs setting V_G to zero which turns the NMOS off, and the PMOS on. Then the PMOS currents are measured for varying V_D. Varying V_{DD} allows complete parameter extractions in NMOS and PMOS FETS. The research work currently underway on single ICs measure the effect of radiation on input switching level. This data is then compared to computer models of the circuit (RADSPICE) to monitor intra-device tracking and verify self-consistency. The system should be applicable to virtually any CMOS digital logic IC. Figure 4 shows results for a CD4076 quad D flip-flop IC at a dose rate of 3.6 Krads/hr. The data shows that this IC is approaching failure due to excessive leakage current as the NMOS V_T drops below zero volts. Research in this area is continuing.

SELECTION OF COMPONENTS FOR RADIATION ENVIRONMENTS

The University of Florida has conducted several programs to harden commercial prototype robot components for operation in moderate to severe environments.[1][2][3] Basically, the components can be categorized into (1) radiation hardened, (2) radiation tolerant, and (3) radiation soft. A radiation hardened component is defined as "an electronic component built and tested to ensure operation within specification after exposure to a specified total dose of radiation." In our design of the ANDROS robotic system, many radiation-hardened components were used, particularly from Harris Corporation because of our design requirement to ensure operation to one megarad. For example, in the EPROM/EXPANDER eight radiation hardened equivalent components (seven from Harris and one from Honeywell) were used to replace the original radiation soft CMOS components.

Radiation tolerant components are defined as "an electronic component is radiation tolerant if it is built and/or tested to ensure satisfactory operation within requirements under exposure to a specified level of operation." The use of commercial-grade radiation tolerant parts is a recent development and is a current trend resulting from the increasing unavailability (not restocking) of radiation hardened parts due to greatly reduced military and space needs with the ending of the cold war. It is well known[4] that bipolar devices can be several orders of magnitude harder than the widely used CMOS devices. This increase in hardness for bipolar devices results from the fact that bipolar devices are dependent on the electronic behavior at its junction, while the gamma radiation effect is near a surface or interface effect. In our ANDROS redesign, the bipolar Digital to Analog (D/A) converters (AD558) were tested to 10 megarads without functional failure, thus these components were kept in the ANDROS design. Similarly, the six UCL5801A CMOS relay drives used in the ANDROS robot were replaced with six bipolar ULM2803 relay drives. Five of these devices were tested for successful operation to 1.5 megarad. Junction Field Effect Transistors (JFET) are even harder than bipolar and when available present an excellent radiation tolerant replacement.

SUMMARY

The University of Florida has completed an extensive research and development program to develop technology to measure, model, mitigate, and understand radiation effects on electronic components for robotic systems. The University of Florida has developed the framework for a database system (UFRED) to transfer the radiation resistant information to the operational field. The developed technology, including the precursor monitoring method have been verified by testing parts and systems in a radiation environment. The research programs have been funded by the Department of Energy, particularly SANDIA contract # AL-3016.

REFERENCES

1. Tulenko, J.S., D. Ekdahl, S. Toshkov, H. Liu, K. Phillips, G. Youk, F. Sias, S. Jones, T. Cable, and H. Harvey. "Development of a Radiation Hardened ANDROS Robot for Operations in Severe Environments." *Proceedings of the ANS, Sixth Topical Meeting on Robotics and Remote Systems, Vol 3, pp. 165-168*, Monterey, CA, Feb. 1995.
2. Tulenko, J.S., S. Toshkov, D. Ekdahl, R.M. Fox, and F. Sias. "Development of a Radiation Hardened Robotic System." *Intelligent Automation and Soft Computing: Trends in Research, Development and Applications, Vol 1*. Edited by M. Jamshidi, et al. ISI Press, Albuquerque, NM (1995) pp. 373-376.
3. Tulenko, J.S., G. Youk, D. Ekdahl, H. Liu, H. Zhou, K. Phillips, F. Sias, S. Jones, T. Cable, and H. Harvey. "Development of a Radiation Hardened ANDROS Robot for Environmental Restoration and Waste Management." *Proceedings of the Fifth International Conference on Radioactive Waste Management and Environmental Remediation, Vol. II, pp. 1725-1728* (ICEM '95) Berlin, Sept. 1995.
4. Adams, A., L. Holmes-Siedle. *Handbook of Radiation Effects*. Oxford University Press (1993).

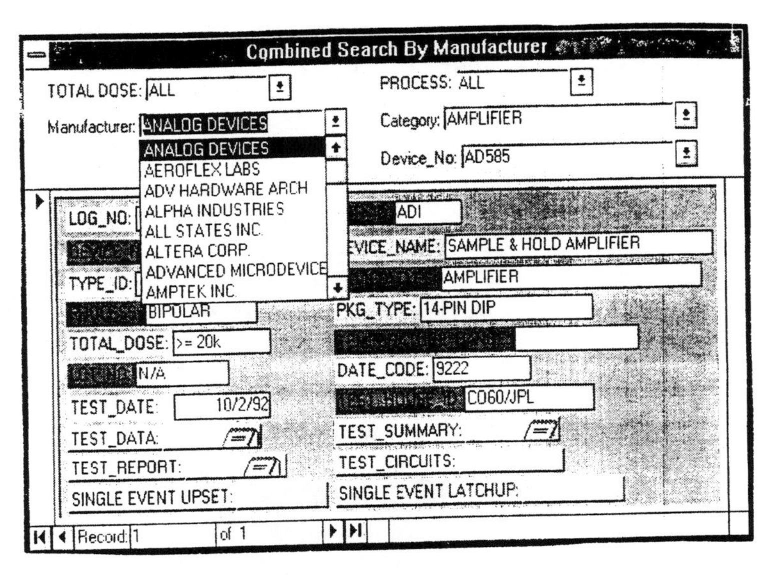

Figure 1. UFRED computer screen for combined search for analog devices amplifiers.

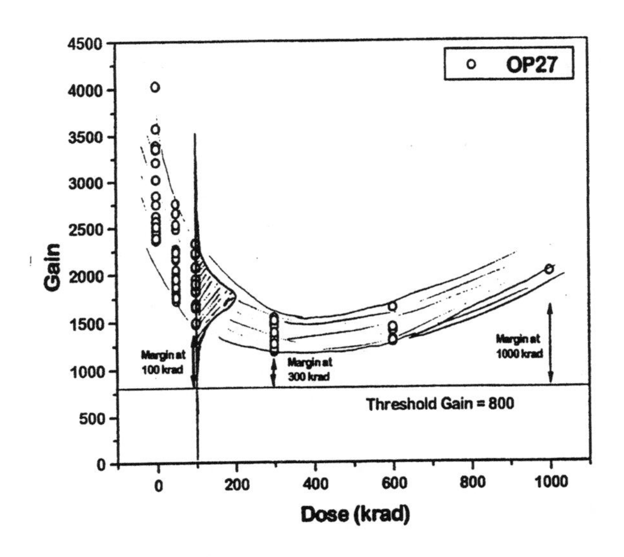

Figure 2. The gain degradation of the operational amplifier OP27 due to radiation exposure.

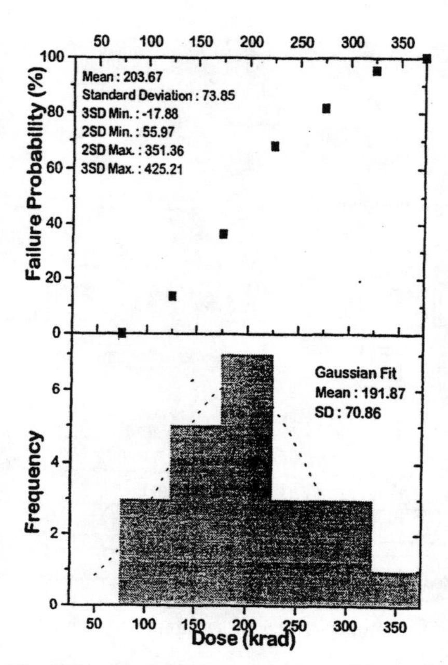

Figure 3. The failure dose histogram and the failure probability of Op27 as a function of dose.

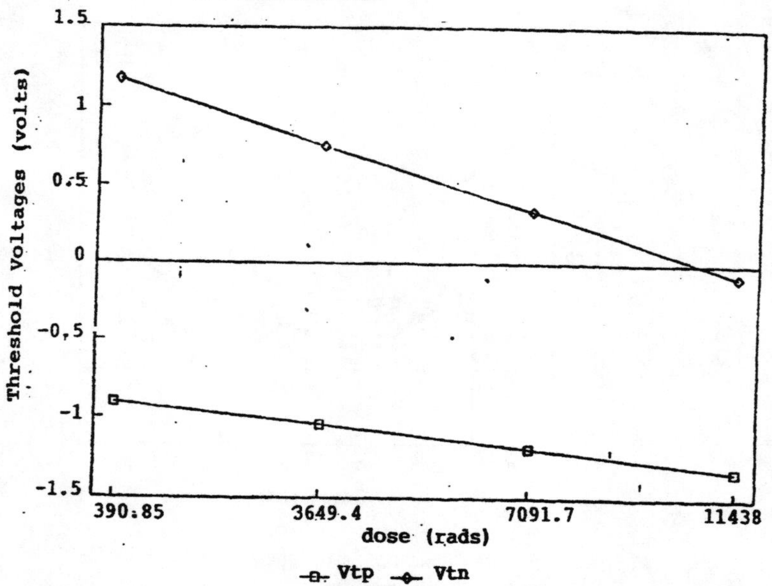

Figure 4. Variation in threshold voltage versus radiation to CD4076 Quad D flip flop IC for PMOS and NMOS.

MOBILE ROBOT AUTONOMY VIA HIERARCHICAL FUZZY BEHAVIOR CONTROL

Edward Tunstel*
NASA Center for Autonomous Control Engineering
Department of Electrical and Computer Engineering
University of New Mexico, Albuquerque, NM 87131 USA
tunstel@robotics.jpl.nasa.gov

ABSTRACT

Realization of autonomous behavior in mobile robots, using fuzzy logic control, requires formulation of rules which are collectively responsible for necessary levels of intelligence. This collection of rules can be conveniently decomposed and efficiently implemented as a hierarchy of fuzzy-behaviors. This paper describes how this can be done using a behavior-based architecture. The approach is motivated by ethological models which suggest hierarchical organizations of behavior. A behavior hierarchy and mechanisms of control decision-making are described. Simulation predicts performance and reveals characteristics of behavior interaction.

KEYWORDS: mobile robot, fuzzy-behavior, behavior hierarchy

INTRODUCTION

Robust behavior in autonomous robots requires that uncertainty be accommodated by the robot control system. Fuzzy logic is particularly well suited for implementing such controllers due to its capabilities of inference and approximate reasoning under uncertainty. Many fuzzy controllers proposed in the literature utilize a monolithic rule-base structure. That is, the precepts that govern desired system behavior are encapsulated as a single collection of *if-then* rules. In most instances, the rule-base is designed to carry out a single control policy or goal. In order to achieve autonomy, mobile robots must be capable of achieving multiple goals whose priorities may change with time. Thus, controllers should be designed to realize a number of task-achieving behaviors that can be integrated to achieve different control objectives. This requires formulation of a large and complex set of fuzzy rules. In this situation a potential limitation to the utility of the monolithic fuzzy controller becomes apparent. Since the size of complete monolithic rule-bases increases exponentially with the number of input variables [1], multi-input systems can potentially suffer degradations in real-time response. This is a critical issue for mobile robots operating in dynamic surroundings.

---

*On leave from NASA Jet Propulsion Laboratory, Pasadena, CA. USA

Hierarchical rule structures can be employed to overcome this limitation by reducing the rate of increase to linear [1, 2].

This paper describes a hierarchical behavior-based control architecture. It is structured as a hierarchy of fuzzy rule-bases which enables distribution of intelligence amongst special-purpose *fuzzy-behaviors*. This structure is motivated by the hierarchical nature of behavior as hypothesized in ethological models. A fuzzy coordination scheme is also described that employs weighted decision-making based on contextual behavior activation. Performance is demonstrated by simulation highlighting interesting aspects of the decision-making process which arise from behavior interaction.

HIERARCHICAL BEHAVIOR CONTROL

The behavior control paradigm has grown out of an amalgamation of ideas from ethology, control theory and artificial intelligence [3, 4]. Motion control is decomposed into a set of special-purpose behaviors that achieve distinct tasks when subject to particular stimuli. Clever coordination of individual behaviors results in emergence of more intelligent behavior suitable for dealing with complex situations. The paradigm was initially proposed by Brooks [4] and realized as the 'subsumption architecture' wherein a behavior system is implemented as distributed finite state automata. Until recently [5, 6, 7], most behavior controllers have been based on crisp (non-fuzzy) data processing and binary logic-based reasoning. In contrast to their crisp counterparts, fuzzy-behaviors are synthesized as fuzzy rule-bases, i.e. collections of a finite set of fuzzy if-then rules. Each behavior is encoded with a distinct control policy governed by fuzzy inference. Thus, each fuzzy-behavior is similar to the conventional fuzzy controller in that it performs an inference mapping from some input space to some output space. If X and Y are input and output universes of discourse of a behavior with a rule-base of size n, the usual fuzzy if-then rule takes the following form

$$IF\ x\ is\ \tilde{A}_i\ THEN\ y\ is\ \tilde{B}_i \tag{1}$$

where x and y represent input and output fuzzy linguistic variables, respectively, and $\tilde{A}_i$ and $\tilde{B}_i$ $(i = 1...n)$ are fuzzy subsets representing linguistic values of x and y. Typically, x refers to sensory data and y to actuator control signals. The antecedent consisting of the proposition "x is $\tilde{A}_i$" could be replaced by a conjunction of similar propositions; the same holds for the consequent "y is $\tilde{B}_i$".

Behavior Hierarchy

The proposed architecture is a conceptual model of an intelligent behavior system and its behavioral relationships. Overall robot behavior is decomposed into a bottom-up hierarchy of increased behavioral complexity in which activity at a given level is dependent upon behaviors at the level(s) below. A collection of *primitive behaviors* resides at the lowest level which we refer to as the primitive level. These are simple, self-contained behaviors that serve a single purpose by operating in a reactive or reflexive fashion. They perform nonlinear mappings from different subsets of the robot's sensor suite to (typically, but not necessarily) common actuators. Each

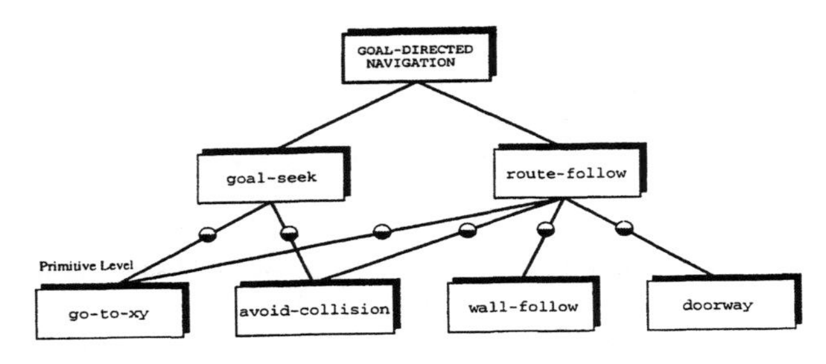

Figure 1. Hierarchical decomposition of mobile robot behavior.

exists in a state of solipsism, and alone, would be insufficient for autonomous navigation tasks. Primitive behaviors are building blocks for more intelligent *composite behaviors.* They can be combined synergistically to produce behavior(s) suitable for accomplishing goal-directed operations.

A behavior hierarchy for indoor navigation might be organized as in Figure 1. It implies that goal-directed navigation can be decomposed as a behavioral function of `goal-seek` and `route-follow`. These behaviors can be further decomposed into the primitive behaviors shown, with dependencies indicated by the adjoining lines. `Avoid-collision` and `wall-follow` are self-explanatory. The `doorway` behavior guides a robot through narrow passageways in walls; `go-to-xy` directs motion along a straight line trajectory to a particular location. The circles represent weights and activation thresholds of associated primitive behaviors. As described below, fluctuations in these weights are at the root of the intelligent coordination of primitive behaviors. The hierarchy facilitates decomposition of complex problems as well as run-time efficiency by avoiding the need to evaluate rules from behaviors that do not apply.

Note that decomposition of behavior for a given mobile robot system is not unique. Consequently, suitable behavior repertoires and associated hierarchical arrangements are arrived at following a subjective analysis of the system and the task environment.

COORDINATING FUZZY-BEHAVIOR INTERACTIONS

Complex interactions in the form of behavioral cooperation or competition occur when more than one primitive behavior is active. These forms of behavior are not perfectly distinct; they are extremes along a continuum [8]. Coordination is achieved by weighted decision-making and behavior modulation embodied in a concept called the *degree of applicability* (DOA). The DOA is a measure of the instantaneous level of activation of a behavior and can be thought of in ethological terms as a motivational tendency of the behavior. Fuzzy rules of composite behaviors are formulated such that the DOA, $\alpha_j \in [0,1]$, of primitive behavior j is specified in the consequent of *applicability rules* of the form

$$IF\ x\ is\ \tilde{A}_i\ THEN\ \alpha_j\ is\ \tilde{D}_i \tag{2}$$

where $\tilde{A}_i$ is defined as in (1). $\tilde{D}_i$ is a fuzzy set specifying the linguistic value (e.g. "*high*") of α_j for the situation prevailing during the current control cycle. This feature allows certain robot behaviors to influence the overall behavior to a greater or lesser degree depending on the current situation. It serves as a form of motivational adaptation since it causes the control policy to dynamically change in response to goals, sensory input, and internal state. Thus, composite behaviors are meta-rule-bases that provide a form of the ethological concepts of inhibition and dominance. Behaviors with maximal applicability ($\alpha_{max} \leq 1$) can be said to dominate, while behaviors with partial applicability ($0 < \alpha < \alpha_{max}$) can be said to be inhibited. These mechanisms allow exhibition of behavioral responses throughout the continuum. This is in contrast to crisp behavior selection which typically employs fixed priorities that allow only one activity to influence the robot's behavior during a given control cycle. The coordination scheme includes behavior selection as a special case when the DOA of a primitive behavior is nonzero and above its activation threshold while others are zero or below threshold. When this occurs, the total number of rules to be consulted on a given control cycle is reduced. The reduction in rule evaluations is not as dramatic or static as in the strict rule hierarchies proposed in [1, 2] since we are dealing with a *behavior* hierarchy that achieves interacting goals. As such, the number of rules consulted during each control cycle varies dynamically as governed by the DOAs and thresholds of the behaviors involved.

Coordination and conflict resolution are achieved within the framework of fuzzy logic theory — via operations on fuzzy sets [9]. Fuzzy rules of each applicable primitive behavior are processed yielding respective output fuzzy sets. These fuzzy sets are equivalent to the result produced by rule-base evaluation in conventional fuzzy controllers *before* applying the defuzzification operator. Following consultation of applicable behaviors, each fuzzy behavior output is weighted (multiplied) by its corresponding DOA, thus effecting its activation to the level prescribed by the composite behavior. The resulting fuzzy sets are then aggregated using an appropriate t-conorm operator (e.g. *max*), and defuzzified to yield a crisp output that is representative of the intended coordination. Since control recommendations from each applicable behavior are considered in the final decision, the resultant control action can be thought of as a consensus of recommendations offered by multiple experts.

NAVIGATION EXAMPLE

In order to demonstrate the operational aspects of the controller in the simplest manner possible we consider only the composite behavior — `goal-seek`. As illustrated in Figure 1, its effect arises from synergistic interaction between primitive behaviors, `go-to-xy` and `avoid-collision`. When more behaviors are involved the approach proceeds in a straightforward manner by appending additional DOAs and any necessary antecedents to applicability rules accordingly.

The simulated mobile robot is modeled after LOBOt, a custom-built base with a 2-wheel differential drive and two stabilizing casters. It is octagonal in shape, 75 cm tall and 60 cm in width. The sensor suite includes optical encoders on each driven wheel and 16 ultrasonic transducers arranged primarily on the front, sides, and forward-facing obliques. The simulated "world" is a hypothetical indoor layout not unlike a warehouse or office

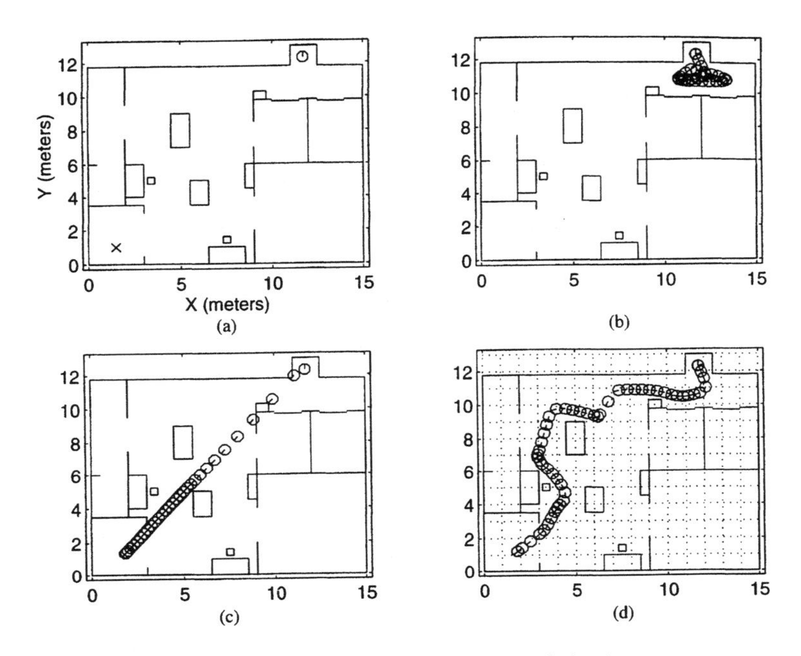

Figure 2. Simulation of goal-seeking behavior.

building. The initial state of the simulation is shown in Figure 2a with LOBOt located at its docking station with pose $\mathbf{p} = (11.7\ 12.3\ \frac{\pi}{2})^T$. Its task is to navigate to the goal located at $(1.5, 1)$, marked by the X. The `avoid-collision` and `go-to-xy` behaviors are each shown acting alone in Figure 2 b and c, respectively. Recall that these behaviors are only capable of exhibiting their particular primitive roles. Thus, `avoid-collision` merely displays cyclic collision-free motion in the immediate vicinity of the robot's initial location, while `go-to-xy` displays a taxic reaction that propels the robot toward the goal irregardless of obstacles in its path. Successful completion of the task, resulting from adaptive coordination of the primitive behaviors, is shown in Figure 2d.

In Figure 3, the behavioral interaction during the run is shown as a time history of the DOAs of each primitive behavior. The interaction dynamics shows evidence of brief bouts of competition (overlapping oscillations) and cooperation with varying levels of dominance. Initially, `avoid-collision` has the dominant influence over the robot due to the close proximity of walls at the docking station. It virtually maintains dominance throughout the task due to the relatively uniform clutter in the environment. The first bout of competition corresponds to the robot's approach toward the obstacle located at $(5, 8)$; a second bout occurs as it enters the goal room. Elsewhere, applicabilities vary continuously reflecting levels of activation recommended by the behavior control system.

CONCLUSION

The hierarchy of fuzzy-behaviors provides an efficient approach to controlling mobile robots. Its practical utility lies in the decomposition of overall behavior into sub-behaviors that are

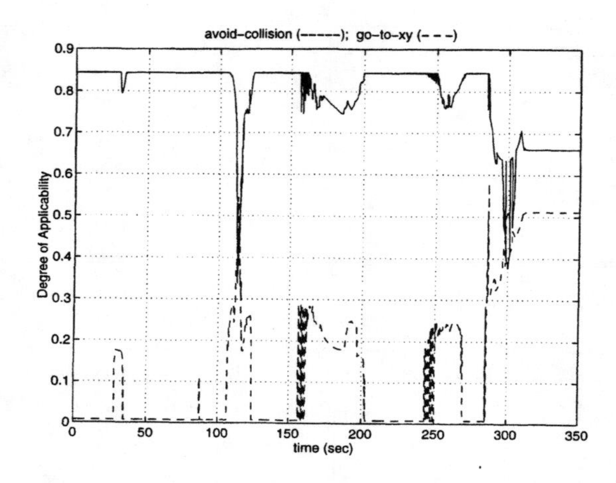

Figure 3. Behavior interaction during goal-seeking.

activated only when applicable. When conditions for activation of a single behavior (or several) are satisfied, there is no need to process rules from behaviors that do not apply. This would result in unnecessary consumption of computational resources and possible introduction of "noise" into the decision-making process. The modularity and flexibility of the approach, coupled with its mechanisms for weighted decision-making, makes it a suitable framework for modeling and controlling situated adaptation in autonomous robots. To date, simulations have been used to predict the performance of a real robot on which real-time experiments are currently being prepared.

REFERENCES

1. Raju, G.V.S., Zhou, J. and Kisner, R.A. "Hierarchical Fuzzy Control." *Intl. J. of Control*, 54 (5) (1991), 1201–1216.
2. Bruinzeel, J., Jamshidi, M. and Titli, A. "A Sensory Fusion-Hierarchical Real-Time Fuzzy Control Approach for Complex Systems." Technical Report No. 95347. LAAS-CNRS, Toulouse, France (August, 1995).
3. McFarland, D.J. *Feedback Mechanisms in Animal Behavior.* Academic Press, New York (1971).
4. Brooks, R.A. "A Robust Layered Control System for a Mobile Robot." *IEEE J. of Robotics & Automation*, RA-2 (1) (1986), 14–23.
5. Pin, F.G. and Watanabe, Y. "Navigation of Mobile Robots Using a Fuzzy Behaviorist Approach and Custom-Designed Fuzzy Inferencing Boards." *Robotica*, 12 (6) (1994), 491–504.
6. Saffiotti, A., Ruspini, E.H. and Konolige, K. "Blending Reactivity and Goal-Directedness in a Fuzzy Controller." *IEEE Intl. Conf. on Fuzzy Systems* (1993), 134–139.
7. Tunstel, E. and Jamshidi, M. "Fuzzy Logic and Behavior Control Strategy for Autonomous Mobile Robot Mapping." *IEEE Intl. Conf. on Fuzzy Systems* (June, 1994), 514–517.
8. Staddon, J.E.R. *Adaptive Behavior and Learning.* Cambridge University Press, New York (1983).
9. Tunstel, E. "Coordination of Distributed Fuzzy Behaviors in Mobile Robot Control." *IEEE Intl. Conf. on Systems, Man and Cybernetics.* (October, 1995), 4009–4014.

AN APPROACH TO MODELING A KINEMATICALLY REDUNDANT DUAL MANIPULATOR CLOSED CHAIN SYSTEM USING PSEUDOVELOCITIES†

M.A. UNSEREN
Oak Ridge National Laboratory
Center for Engineering Systems Advanced Research
Oak Ridge, TN 37831-6364

ABSTRACT

The paper discusses the problem of resolving the kinematic redundancy in the closed chain formed when two redundant manipulators mutually lift a rigid body object. The positional degrees of freedom (DOF) in the closed chain are parameterized by a set of independent variables termed pseudovelocities. Due to the redundancy there are more DOF and thus more pseudovelocities than are required to specify the motion of the held object. The additional "redundant" pseudovelocities are used to minimize the distance between the vector of unknown joint velocities and a vector of "corrective" joint velocities in a Euclidean norm sense. This leads to an optimal solution for the joint velocities as a linear function of the Cartesian object velocities and the corrective velocities. The problem of determining the corrective velocities to avoid collisions of the links with a wall located in the workspace and to avoid joint range limits is illustrated by an example of two redundant planar revolute joint manipulators mutually lifting a rigid object.

KEYWORDS: interacting redundant manipulators, redundant pseudovelocities, collision and joint limit avoidance, corrective action

INTRODUCTION

When two serial link manipulators possessing N_1 and N_2 joints, respectively, mutually lift a rigid body object the values of the joint velocities of the manipulators are restricted by M rigid body kinematic constraints‡ [1]. M configuration degrees of freedom (DOF) are lost and the closed chain system has $(N_{12} - M)$ DOF, where $N_{12} = N_1 + N_2$. In our previous work [1], the configuration DOF were parameterized by $(N_{12} - M)$ independent scalar variables termed pseudovelocities. The joint velocities were expressed as linear functions of the pseudovelocities. When each manipulator is kinematically redundant $(N_i > M)$, there are more DOF than are required to control the translational and rotational motion of the object at its center of mass. In [1], M of the pseudovelocities were viewed as nonredundant, and were selected to be the components of translational and angular velocity of the object at its center of mass. On the other hand, the remaining $(N_{12} - 2M)$ pseudovelocities were viewed as being redundant. A dynamical model comprised of $(N_{12} - M)$ second order differential equations governing the motion of the closed chain was derived, where each of the equations is a linear function of an $((N_{12} - M) \times 1)$ vector containing the time derivatives of the pseudovelocities. Based on the model, a control scheme was proposed where each of the pseudovelocities was explicitly controlled to track a

† Research sponsored by the Engineering Research Program, Office of Basic Energy Sciences, of the U.S. Department of Energy, under contract DE-AC05-96OR22464 with Lockheed Martin Energy Research Corp.

‡ $M = 6$ and $M = 3$ for spatial and planar dual manipulator configurations, respectively.

reference trajectory. Please note that with this approach the M object DOF and the $(N_{12} - 2M)$ redundant DOF are treated in the same way. They are equally important. An open problem not addressed in [1] is how to select reference trajectories for the redundant pseudovelocities.

This paper takes a different approach where the $(N_i - M)$ redundant pseudovelocities associated with manipulator i are used to induce joint self motions (i.e., motions of the joints that do not contribute to the motion of the held object) to make the vector of joint velocities tend towards a known vector of "corrective" joint velocities in some optimal sense. We propose to calculate the redundant pseudovelocities to minimize the distance between these vectors in a Euclidean norm sense. This leads to an optimal solution for the joint velocities containing a component that contributes to the object's motion and a self motion component that is a function of the corrective velocities. Since the object's motion is constrained to follow a reference trajectory whereas inducing joint self motions is optional, the control of the Cartesian pseudovelocities takes precedence over calculating the redundant pseudovelocities to minimize the aforementioned distance.

In our earlier work [2], an algorithm for calculating the corrective joint velocities was proposed to establish a joint limit avoidance capability for an *un*constrained manipulator. Here it is investigated if the corrective velocities can be determined to give each manipulator in the closed chain an additional capability via joint self motions. This will be illustrated by an example where the algorithm in [2] is extended to give each of two redundant planar revolute joint manipulators mutually holding a rigid object the complimentary and simultaneous capabilities of avoiding joint limits and avoiding collisions of the links with a wall located in the workspace. The planar configuration is shown in Figure 1. There are other approaches to collision avoidance when dual manipulators share a common workspace such as applying reflexive action [3], but kinematically redundant manipulators were not considered in the analysis. Other approaches to utilizing kinematic redundancy in the closed chain are discussed in [4, 5].

KINEMATIC REDUNDANCY RESOLUTION

Let the $(M \times 1)$ vector v_o denote the the Cartesian velocity of the object at its center of mass with respect to a stationary world reference coordinate frame. In the closed chain, v_o can be expressed as a linear function of the joint velocities of either manipulator [1]:

$$v_o = A_i \, \dot{q}_i \tag{1}$$

where $\dot{q}_i$ denotes the $(N_i \times 1)$ vector of joint velocities of manipulator i and $A_i(q_i)$ is a $(M \times N_i)$ matrix. It is assumed that A_i has full rank $M \, (< N_i)$.

An underspecified solution to eq. (1) is given by:

$$\dot{q}_i = E_i \, v_o + F_i \, \nu_i \tag{2}$$

where E_i and F_i are $(N_i \times M)$ and $(N_i \times (N_i - M))$ full rank matrices, respectively, which satisfy the matrix identities $A_i \, E_i = I_M$ and $A_i \, F_i = 0_{M \times (N_i - M)}$. Here I_k signifies a $(k \times k)$ identity matrix and $0_{k \times l}$ a $(k \times l)$ matrix of zeros. The components of v_o constitute the Cartesian pseudovelocities in the closed chain. The components of the $((N_i - M) \times 1)$ vector ν_i are the redundant pseudovelocities associated with manipulator i. They parameterize the null space of A_i. Therefore $(F_i \, \nu_i)$ induces joint self motions which do not affect object motion. It is assumed that the quantities $\{E_i \, , \, F_i\}$ $(i$=1,2) are known (see [2] for methods to determine them) and that a reference trajectory for the center of mass of the held object has been specified. Thus v_o is known and the only unknown quantity in eq. (2) is ν_i.

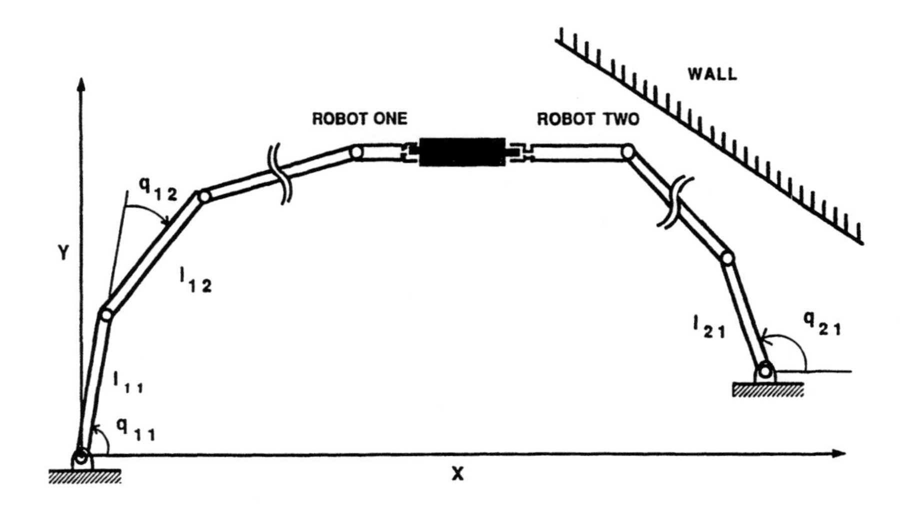

Figure 1. Closed Chain With Two Planar Manipulators

Let $\dot{q}_i^c$ denote an $(N_i \times 1)$ vector of "corrective" joint velocities which represents a desired target or goal that the joint velocities of manipulator i should tend to in some optimal manner. It must be emphasized that $\dot{q}_i^c$ is a *not* a "reference" trajectory to be explicitly tracked by $\dot{q}_i$ using servo control techniques. Besides, the N_i components of $\dot{q}_i$ cannot be independently controlled to track reference trajectories because their values must satisfy M rigid body kinematic constraints $A_1\,\dot{q}_1 - A_2\,\dot{q}_2 = 0_{M\times 1}$ arising from eq. (1) . It is assumed that $\dot{q}_i^c$ has a component lying in the null space of A_i but it may also have a component which is orthogonal to every vector in the null space in the general case. $\dot{q}_i^c$ is a known function of the measured or sensed variables associated with manipulator i and is calculated by an algorithm furnished by the designer. An example of such an algorithm to give two interacting planar manipulators the capabilities of avoiding collisions with a wall located in the workspace and avoiding joint limits is presented in the next section.

In this paper we propose to determine ν_i to minimize the distance between $\dot{q}_i$ and $\dot{q}_i^c$ in a Euclidean norm sense. To accomplish this objective, a performance index is introduced:

$$P_i = (\dot{q}_i - \dot{q}_i^c)^T\,(\dot{q}_i - \dot{q}_i^c) = (E_i\,v_o + F_i\,\nu_i - \dot{q}_i^c)^T\,(E_i\,v_o + F_i\,\nu_i - \dot{q}_i^c) \quad (3)$$

where superscript T denotes a transpose and where eq. (2) has been applied. The necessary optimality condition is obtained from $d\,P_i/d\,\nu_i = 0_{(N_i-M)\times 1}$. Solving this equation for ν_i and substituting the result into eq. (2) yields the symbolic solution for the joint velocities of manipulator i:

$$\dot{q}_i = E_i\,v_o + F_i\left(F_i^T\,F_i\right)^{-1}F_i^T\,(\dot{q}_i^c - E_i\,v_o) = A_i^T\,(A_i\,A_i^T)^{-1}v_o + F_i\left(F_i^T\,F_i\right)^{-1}F_i^T\,\dot{q}_i^c \quad (4)$$

Eq. (4) was obtained by choosing E_i to be:

$$E_i = A_i^T\,\beta + F_i\,\gamma \quad (5)$$

where β and γ are $(M \times M)$ and $((N_i - M) \times M)$ parameter matrices, respectively. Using this definition, $(E_i\,v_o)$ always contains a component $(A_i^T\,\beta\,v_o)$ which contributes to the held object's motion, but it may also contain a component $(F_i \gamma v_o)$ which induces joint self motions in the general case. Interestingly, the self motion component of $(E_i v_o)$ vanishes from the final solution for $\dot{q}_i$ in eq. (4) . It is easy to verify that $\beta = (A_i A_i^T)^{-1}$ by premultiplying eq. (5) by A_i.

Although the vector of corrective velocities does not lie entirely in the null space of A_i, eq. (4) reveals that $\dot{q}_i^c$ has been projected into the null space (of A_i) by its coefficient matrix. Therefore any "corrective action" applied to the system using $\dot{q}_i^c$ does not affect the motion of the held object.

CALCULATION OF CORRECTIVE ACTION

Let $\dot{q}_{i,j}^c$ denote the corrective velocity corresponding to the jth joint of manipulator i. In this section an algorithm is presented for calculating $\dot{q}_{i,j}^c$ for each and every joint that, when used in conjunction with eq. (4) , induces self motions of the joints of each manipulator to avoid collisions of the links with a wall located in the workspace and to avoid joint limits. It is assumed that the dual-manipulator closed chain is a planar system, and that all manipulator joints are of the revolute type.

Wall Collision Avoidance Strategy

Suppose there is a wall that is perpendicular to the the plane of motion located in the workspace of the closed chain system (see Figure 1). This wall is modeled by a straight line:

$$y = a\,x + b \tag{6}$$

Further, let the position of the outer end of the jth link of manipulator i be signified by the coordinates $(x_{i,j},\, y_{i,j})$ with respect to a stationary world reference frame. The Cartesian coordinates are related to the joint coordinates by:

$$(x_{i,j}\,,\, y_{i,j}) = \left\{ x_{i,j-1} + l_{i,j}\cos\left(\sum_{p=1}^{j} q_{i,p}\right)\,;\; y_{i,j-1} + l_{i,j}\sin\left(\sum_{p=1}^{j} q_{i,p}\right)\right\} \tag{7}$$

where $(1 \le j \le N_i)$ and $(x_{i,0},\, y_{i,0})$ is the position of the base of manipulator i. $q_{i,j}$ and $l_{i,j}$ signify the jth joint angle and jth link of manipulator i, respectively.

The line passing through the point $(x_{i,j},\, y_{i,j})$ that is perpendicular to the wall is:

$$y = -\frac{1}{a}\,(x - x_{i,j}) + y_{i,j} \tag{8}$$

and the distance from $(x_{i,j},\, y_{i,j})$ to the wall is:

$$d_{i,j} = (a\,x_{i,j} - y_{i,j} + b)\;/\;\left(\pm\sqrt{a^2+1}\right) \tag{9}$$

where the sign in the denominator is chosen such that $d_{i,j}$ is nonnegative.

Let $\alpha_{i,j}$ signify the angle between the "perpendicular line" defined by eq. (8) and link $l_{i,j}$, which is measured positive in the counterclockwise sense with respect to the perpendicular line (see Figure 2). $\alpha_{i,j}$ can be expressed as a function of the slopes of $l_{i,j}$ and the perpendicular line:

$$\tan(\alpha_{i,j}) = \{a\,(y_i - y_{i-1}) + (x_i - x_{i-1})\}\,/\,\{a\,(x_i - x_{i-1}) - (y_i - y_{i-1})\} \tag{10}$$

Let the positive quantity tol^d denote a constant threshold distance from the wall. If the distances from the tips of one or more of the lower $(N_i - 2)$ links of manipulator i to the wall are less than tol^d, it is regarded that a shutdown or damage to the manipulator are imminent due to those links colliding with the wall. Accordingly, it is desired to compute corrective velocities for the joints corresponding to these links that, when

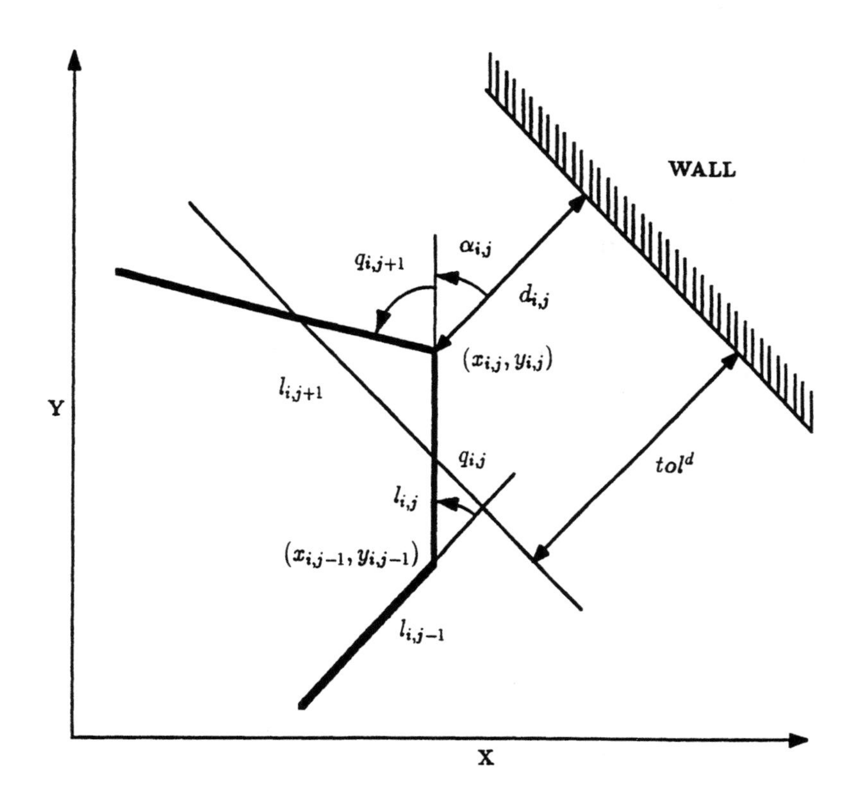

Figure 2. Parameters for Wall Collision Avoidance

used in conjunction with the optimal solution for the joint velocities given by eq. (4) , tends to move them away from the wall using self motion of the joints.[§]

Joint Limit Avoidance Strategy

The corrective velocities for the two outermost joints of manipulator i and for those joints (among its lower $(N_i - 2)$ joints) whose corresponding link tips are located at a distance greater than tol^d from the wall can be calculated to provide a joint limit avoidance (JLA) capability via joint self motions when used in conjunction with eq. (4) .

Let $q_{i,j}^{max}$ and $q_{i,j}^{min}$ signify the absolute hardware limits in the range of joint j of manipulator i. In our strategy, the positive, constant angles tol_i^{hi} and tol_i^{lo} are

[§] It is important to note that the trajectory of the outermost N_ith link of each manipulator is determined because the trajectory of the held rigid object is specified and each manipulator securely holds the object. Therefore, self motions of the joints do not affect the trajectory of the N_ith link, and the corrective velocities for the outermost two joints of each manipulator are not computed in the wall collision avoidance strategy.

introduced to define the ranges $\left(q_{i,j}^{max} - tol_i^{hi}\right) < q_{i,j} < q_{i,j}^{max}$ and $q_{i,j}^{min} < q_{i,j} < \left(q_{i,j}^{min} + tol_i^{lo}\right)$. When $q_{i,j}$ lies within either of these ranges, it is regarded that a shutdown and damage to the manipulator are imminent due to the joint reaching a limit. Accordingly, a corrective velocity $\dot{q}_{i,j}^c$ is desired to drive joint j back into the range $\left(q_{i,j}^{min} + tol_i^{lo}\right) \leq q_{i,j} \leq \left(q_{i,j}^{max} - tol_i^{hi}\right)$.

Based on the aforementioned strategies, the algorithm for wall collision avoidance (WCA) and JLA is presented next.

The Proposed Algorithm

$\dot{q}_{i,j}^c$ is calculated for each and every joint ($i = 1, 2;\ j = 1, 2, \ldots N_i$) by the following conditional algorithm, where distance $d_{i,j}$ is computed as a function of the measured joint angles $\{q_{i,1}, q_{i,2}, \ldots, q_{i,j}\}$ using eqs. (7) and (9) whenever $(j \leq N_i - 2)$:

$0 < d_{i,j} < tol^d$ **and** $j \leq N_i - 2$: Calculate $\alpha_{i,j}$ by eq. (10) and $\dot{q}_{i,j}^c$ for WCA using:

$$0 \leq \alpha_{i,j} \leq 90^o \text{ and } q_{i,j} \leq \left(q_{i,j}^{max} - tol_i^{hi}\right) : \dot{q}_{i,j}^c = \frac{z_i^d\, \dot{q}_{i,j}^{max}}{tol^d}\left(tol^d - d_{i,j}\right) \quad (11)$$

$$0 \leq \alpha_{i,j} \leq 90^o \text{ and } q_{i,j} > \left(q_{i,j}^{max} - tol_i^{hi}\right) : \dot{q}_{i,j}^c = 0$$

$$-90^o \leq \alpha_{i,j} < 0 \text{ and } q_{i,j} \geq \left(q_{i,j}^{min} + tol_i^{lo}\right) : \dot{q}_{i,j}^c = -\frac{z_i^d\, \dot{q}_{i,j}^{max}}{tol^d}\left(tol^d - d_{i,j}\right) \quad (12)$$

$$-90^o \leq \alpha_{i,j} < 0 \text{ and } q_{i,j} < \left(q_{i,j}^{min} + tol_i^{lo}\right) : \dot{q}_{i,j}^c = 0$$

$\left(d_{i,j} \geq tol^d\right.$ **and** $\left.j \leq N_i - 2\right)$ **or** $j > N_i - 2$: Calculate $\dot{q}_{i,j}^c$ for JLA based on the measured value of $q_{i,j}$ using:

$$q_{i,j}^{min} < q_{i,j} < \left(q_{i,j}^{min} + tol_i^{lo}\right) \;:\; \dot{q}_{i,j}^c = \frac{z_i\, \dot{q}_{i,j}^{max}}{tol_i^{lo}}\left(q_{i,j}^{min} + tol_i^{lo} - q_{i,j}\right) \quad (13)$$

$$\left(q_{i,j}^{max} - tol_i^{hi}\right) < q_{i,j} < q_{i,j}^{max} \;:\; \dot{q}_{i,j}^c = \frac{z_i\, \dot{q}_{i,j}^{max}}{tol_i^{hi}}\left(q_{i,j}^{max} - tol_i^{hi} - q_{i,j}\right) \quad (14)$$

$$\left(q_{i,j}^{min} + tol_i^{lo}\right) \leq q_{i,j} \leq \left(q_{i,j}^{max} - tol_i^{hi}\right) \;:\; \dot{q}_{i,j}^c = 0$$

where $\dot{q}_{i,j}^{max} (> 0)$ denotes the maximum or peak time rate of change of joint $q_{i,j}$. In eqs. (11) and (12), z_i^d is a constant scaling factor whose value is restricted to the range:

$$0 < z_i^d \leq 1 \quad (15)$$

z_i^d is introduced to enable to designer to specify a scaled peak velocity $(z_i^d\, \dot{q}_{i,j}^{max})$ in the WCA scheme. Scalar z_i in eqs. (13) and (14) is defined in a similar manner for the JLA scheme.

The WCA portion of the algorithm is conditioned on the value of $\alpha_{i,j}$ *and* the sub-range that the measured value of $q_{i,j}$ lies in. Consider eq. (11), which is the equation of a line segment connecting (but not including) the points $\left(d_{i,j} = 0,\ \dot{q}_{i,j}^c = z_i^d\, \dot{q}_{i,j}^{max}\right)$ and $(tol^d, 0)$. It is easy to see that $\dot{q}_{i,j}^c$ is positive and that its magnitude is based on the distance from the tip of link $l_{i,j}$ to the wall when calculated by eq. (11). Observing Figure 2, the basic idea here is to apply a corrective action to induce $q_{i,j}$ to rotate counterclockwise which moves the outer tip of link $l_{i,j}$ away from the wall. However, it is logical that eq. (11) be applied only when $q_{i,j}$ does not lie in its upper prohibitive subrange. Indeed, to calculate $\dot{q}^c_{i,j}$ using eq. (11) when $q_{i,j} > q^{max}_{i,j} - tol^{hi}_i$ may reduce the possibility of $l_{i,j}$ colliding with the wall at the expense of increasing the already high possibility of $q_{i,j}$ reaching its upper range limit. No corrective action can be applied to joint $q_{i,j}$ to alleviate this situation, and we set $\dot{q}^c_{i,j}$ to zero.

By observing Figure 2 and applying the same reasoning, $\dot{q}^c_{i,j}$ is calculated as a negative quantity using eq. (12) when $(-90^o \leq \alpha_{i,j} < 0)$, provided that $q_{i,j}$ does not lie in its lower prohibitive subrange. But if $q_{i,j} < q^{min}_{i,j} + tol^{lo}_i$, $\dot{q}^c_{i,j}$ is set to zero.

The logic of the JLA portion of the algorithm is now explained. Eq. (13) is the equation of a line segment connecting (but not including) the points $\left(q_{i,j} = q^{min}_{i,j} + tol^{lo}_i,\ \dot{q}^c_{i,j} = 0\right)$ and $\left(q^{min}_{i,j},\ z_i\,\dot{q}^{max}_{i,j}\right)$. Likewise, eq. (14) is the equation of line segment between the points $\left(q^{max}_{i,j} - tol^{hi}_i,\ 0\right)$ and $\left(q^{max}_{i,j},\ -z_i\,\dot{q}^{max}_i\right)$. It is easy to see that $\dot{q}^c_{i,j}$ is positive when calculated by eq. (13). This is logical since it is desired to rotate joint $q_{i,j}$ counterclockwise away from the lower hardware limit. By the same reasoning $\dot{q}^c_{i,j}$ is negative when evaluated using eq. (14). It should be mentioned that the JLA algorithm given here, unlike the WCA algorithm, is directly applicable to the spatial case.

When $d_{i,j} > tol^d$ (if applicable, i.e., if $j \leq N_i - 2$) and $\left(q^{min}_{i,j} + tol^{lo}_i\right) \leq q_{i,j} \leq \left(q^{max}_{i,j} - tol^{hi}_i\right)$, no correction action is needed for joint $q_{i,j}$ and $\dot{q}^c_{i,j}$ is set to zero. It follows that if all of the lower $(N_i - 2)$ links of manipulator i are sufficiently away from the wall and all N_i of its joints lie in their respective center subranges, then $\dot{q}^c_i = 0_{N_i \times 1}$ and there is no corrective action applied to manipulator i. This supports our contention that the corrective action component in the joint velocity solution is nonzero only when it is detected, by sensing, that it is required. This reduces the computational burden and yields a minimum Euclidean norm solution for $\dot{q}_i$.

The approach given here contrasts sharply from the result obtained using gradient projection [6, 7, 8], where the solution for $\dot{q}_i$ would contain one or multiple terms that project the gradients of scalar functions into the null space of A_i. In some approaches, each term represents a distinct secondary criteria, e.g., joint limit- and wall collision-avoidance. The problem is that each gradient projecting term is always computed, regardless of whether or not any links are close to the wall or any joints are close to their hardware range limits. In the author's opinion this computation is wasteful when feedback sensing indicates that manipulator i is in a configuration where JLA and WCA are not needed.

The aforementioned gradient projection technique would assign a scalar weighting factor to each gradient projecting term to establish a priority of importance among the multiple secondary criteria. On the other hand, the algorithm presented here for calculating $\dot{q}^c_{i,j}$ is done on a joint by joint basis depending on sensed conditions.

CONCLUSION

The paper proposed an alternative approach to applying the "redundant" pseudovelocities to the modeling of the closed chain motion of two serial link, kinematically redundant manipulators mutually lifting a rigid body object. In our previous work the redundant pseudovelocities were treated in the same way as the "nonredundant" pseudovelocities (i.e., those assigned to be the Cartesian velocities of the held object) and were explicitly controlled to track reference trajectories. Here the redundant pseudovelocities were applied to resolve the kinematic redundancies of the manipulators, where it was assumed that their joint self motions do not affect the motion of the held object. An optimal solution for the joint velocities was determined by applying the redundant pseudovelocities to minimize the distance between the vector of joint velocities and a vector of "corrective" velocities $\dot{q}_i^c$ in a minimum Euclidean norm sense. Based on an example of a planar dual-manipulator closed chain, a novel algorithm for calculating the components of $\dot{q}_i^c$ was proposed where $\dot{q}_{i,j}^c$ is computed to (i) induce joint $q_{i,j}$ to rotate in a direction that moves the tip of link $l_{i,j}$ away from a wall located in the workspace if it is within a critical distance to the wall (however, if $q_{i,j}$ is to be rotated counterclockwise (clockwise) but the joint lies in close proximity to its upper (lower) range limit, $\dot{q}_{i,j}^c$ is set to zero), or if condition (i) tests false or if $q_{i,j}$ is one of the two outermost joints of manipulator i , to (ii) induce joint $q_{i,j}$ to move away from an upper or lower hardware range limit if it is in close proximity to either limit, or if condition (ii) tests false, to (iii) set $\dot{q}_{i,j}^c$ equal to zero. Thus each corrective velocity is computed only when it is needed. An interesting aspect of our approach is that when none of the joints are in close proximity of their range limits and none of the links are in close proximity to the wall, the corrective velocities are set to zero and there is no corrective action (self motion) component in the joint velocity solution.

The future work involves simulating the proposed redundancy resolution method to test its effectiveness and determining how to calculate the corrective velocities to avoid collisions with an object of complex shape that cannot not be described by an analytic expression.

REFERENCES

1. Unseren, M.A., "A New Technique for Dynamic Load Distribution when Two Manipulators Mutually Lift a Rigid Object," *Intelligent Automation and Soft Computing*, (Proc. WAC '94, August 14-17, 1994, Maui, HI) edited by M. Jamshidi, etc.; TSI Press Series, (1994), pts. 1-2, 359-372.
2. Unseren, M.A. and D.B. Reister, "New Insights Into Input Relegation Control for Inverse Kinematics of a Redundant Manipulator," Oak Ridge National Laboratory Technical Report No. ORNL/TM-12813 pts. 1-3, (July, 1995), Oak Ridge, TN, USA.
3. Cellier, L.; Dauchez, P.; Uchiyama, M. and R. Zapata, "Collision Avoidance for a Two-Arm Robot by Reflex Actions: Simulations and Experiments," *J. of Intelligent and Robotic Systems*, 14(2), (Oct. 1995), 219-238.
4. Ramadorai, A.K.; Tarn, T.J.; and A.K. Bejczy, "Task Definition, Decoupling and Redundancy Resolution in Multi-Robot Object Handling," *IEEE Conf. Robotics & Automation*, v1, (May, 1992), 467-474, Nice, France.
5. Tao, J.M. and J.Y.S. Luh, "Robust Position and Force Control for a System of Multiple Redundant-Robots," *IEEE Conf. Robotics & Automation*, v3, (May, 1992) 2211-2216, Nice, France.
6. A. Liegeois, "Automatic Supervisory Control of the Configuration and Behavior of Multibody Mechanisms," *IEEE Trans. Systems, Man, & Cybernetics*, smc-7(12), (1977), 868-871.
7. Klein, C.A. and C-H. Huang, "Review of Pseudoinverse Control for Use with Kinematically Redundant Manipulators," *IEEE Trans. Systems, Man, & Cybernetics*, smc-13(4), (March/April, 1983) 245-250.
8. Nenchev, D.N., "Redundancy Resolution through Local Optimization: A Review," *Journal of Robotic Systems* 6(6), (Dec. 1989), 769-798.

A ROBUST LEARNING CONTROL SCHEME FOR MANIPULATOR WITH TARGET IMPEDANCE AT END-EFFECTOR

Danwei Wang and Chien Chern Cheah
School of Electrical and Electronic Engineering
Nanyang Technological University
Singapore 639798

ABSTRACT

In this paper, a robust learning control algorithm is proposed for control of robotic manipulator with target impedance at the end-effector in the presence of state disturbances, output measurement noises and errors in initial conditions. Design method for analyzing the learning impedance system is developed and sufficient conditions for guaranteeing the convergence are derived.

KEYWORDS: Learning Systems; Robots; Impedance Control; Robustness.

INTRODUCTION

Learning control [1] is a concept for controlling dynamic system in an iterative manner. Many research efforts have been devoted towards defining and analyzing learning control schemes [1-5]. A recent survey of the works can be found in [6]. In most of the learning schemes proposed so far, the learning control system is designed to track a given desired trajectory as the operation is repeated. Examples of such applications are motion control and simultaneous motion and force control of robots employed to perform repetitive tasks. However, in certain applications like impedance control [7] of robotic manipulators, the control objective is specified explicitly by a desired model (or target impedance) rather than a desired trajectory. Impedance control does not attempt to track motion and force trajectories but rather to regulate the mechanical impedance [7] specified by a target model at the robot end-effector. By controlling the manipulator position and specifying its relation to the external forces, the contact force response can be kept in a desired range. Impedance control is one of the two major approaches used in the controller design for force control [6] problem of robotic manipulator. It provides a unified approach to all aspects of manipulation [7]. Both free motion and contact tasks can be controlled using a single control algorithm. The nature of trajectory learning formulation has limited the research of this important problems because in impedance control problem, a desired model is specified rather than the trajectory [7]. Recently, a learning Control law for impedance control of robotic manipulator is proposed by Cheah and Wang [5]. However, it was implicitly assumed in the paper [5] that the manipulator must take the same initial position and velocity at every operation trial and there are no disturbances and measurement noises. From a piratical point of view, it is important to establish the robustness of the learning control with respect to bounded errors of initialization, disturbances of dynamics and measurement noises during operation.

In this paper, we propose a Learning Control scheme for control of robotic manipulator with target impedance at the end-effector. Sufficient condition for selection of learning gain to guarantee the convergence of the learning system is derived. We study in detail the robustness of the proposed algorithm to state disturbances, output measurement noises and errors in initial conditions. Furthermore, we shall show that the impedance error converges to a bound and this bound tends to zero in the absence of the state disturbances, output measurement noises and errors in initial conditions.

ROBOT DYNAMIC MODEL AND PROBLEM FORMULATION

The dynamic equations of motion for a rigid link manipulator with n degree of freedom can be described by in joint space as,

$$M(q_k(t))\ddot{q}_k(t) + V(q_k(t), \dot{q}_k(t)) + d_k(t) = u_k(t) + f_k(t) \tag{1}$$

where $q_k(t) \in R^n$ represents the joint angle at the kth operation, t is the operation time, $M(\cdot) \in R^{n \times n}$ is the inertia matrix, $V(\cdot,\cdot) \in R^n$ contains the centrifugal Coriolis and gravitational forces, $d_k(t) \in R^n$ is the external disturbance, $f_k(t) \in R^n$ is the external force and $u_k(t) \in R^n$ is the control torque. Now, let $X_k(t) \in R^m$ be the cartesian space vector defined by,

$$X_k(t) = h(q_k(t)), \tag{2}$$

where $h(q_k(t))$ is the manipulator kinematics describing the relation between the joint and cartesian space. Then, the derivative of $X_k(t)$ is given as

$$\dot{X}_k(t) = J(q_k(t))\dot{q}_k(t), \tag{3}$$

where $J(\cdot) = \frac{\partial h(\cdot)}{\partial q_k} \in R^{m \times n}$ is the manipulator Jacobian matrix. The external force $F_k(t) \in R^m$ in cartesian space is related to the force in joint space as

$$f_k(t) = J^T(q_k(t))F_k(t). \tag{4}$$

where we assume that the stiffness relation between $F_k(t)$ and $X_k(t)$ at the contact point be dominated by

$$F_k(t) = K_s(X_s(t) - X_k(t)) \tag{5}$$

and $K_s \in R^{m \times m}$ is a symmetric and positive definite matrix that describes the environmental stiffness. $X_s(t) \in R^m$ can be seen as representing the location to which the contact point $X_k(t)$ would return in the absence of contact force. We also suppose that a feedback law [3] has been designed for stability of the closed-loop system as follows:

$$u_k(t) = K_v(\dot{q}_d(t) - \dot{q}_k(t)) + K_p(q_d(t) - q_k(t)) + m_k(t) \tag{6}$$

where $q_d(t)$ is the reference joint angle, K_v, $K_p \in R^{n \times n}$ are the feedback gains and $m_k(t) \in R^n$ is the learning control input added to learning the desired impedance as the action is repeated. When the control input (6) is applied to the robotic manipulator described by equation (1), we have,

$$M(q_k(t))\ddot{q}_k(t) + N(q_k(t), \dot{q}_k(t), t) + d_k(t) = m_k(t) + f_k(t) \tag{7}$$

where $N(q_k(t), \dot{q}_k(t), t) = V(q_k(t), \dot{q}_k(t)) + K_v(\dot{q}_k(t) - \dot{q}_d(t)) + K_p(q_k(t) - q_d(t))$.
For learning control design, we assume that the learning system has the following properties [2,3], :

(A1) Every operation ends in a finite time interval. i.e. $t \in [0, T]$.

(A2) The desired motion is specified *a priori* over the time duration $t \in [0, T]$ by the following target impedance at end-effector,

$$M_m(\ddot{X}_d(t) - \ddot{X}(t)) + C_m(\dot{X}_d(t) - \dot{X}(t)) + K_m(X_d(t) - X(t)) = -F(t) \quad (8)$$

where M_m, C_m and $K_m \in R^{nxn}$ are positive definite matrices that specify the dynamic response of the system.

(A3) The initial state error of the system is bounded such that $\left\| \begin{bmatrix} q_d(0) - q_k(0) \\ \dot{q}_d(0) - \dot{q}_k(0) \end{bmatrix} \right\| \leq b_{q0}$ for any operation k.

(A4) $M(\cdot)$ is positive definite and bounded. Hence, $M^{-1}(q)$ exists and is positive definite and bounded. Furthermore, $J(\cdot)$ and its derivative $\dot{J}(\cdot)$ are bounded for all $t \in [0, T]$.

(A5) $M^{-1}(\cdot)$, $N(\cdot, \cdot, \cdot)$, $J(\cdot)$ and $\dot{J}(\cdot)$ are *local Lipschitzian* functions of their arguments [3] for $t \in [0, T]$.

(A6) The external disturbances $d_k(t)$ is bounded by constants b_d on $t \in [0, T]$. An example of such disturbances is dynamics fluctuations for which the operation is not repeated under exactly the same condition.

(A7) The measured outputs are contaminated by noise in the following way:

$$\hat{\omega}_k(t) = \omega_k(t) + n_k(t),$$

where $w_k(t) \in R^n$ is the impedance error to be defined and $n_k(t) \in R^m$ denotes the measurement noise that is bounded by a constant b_n for $t \in [0, T]$.

(A8) The system dynamics is invertible such that for a given desired model (8), there exists a unique control input $u_e(t) \in R^n$ corresponding to the solutions $X_e(t) = h(q_e(t))$, $\dot{X}_e(t) = J(q_e(t))\dot{q}_e(t)$ and $F_e(t) = K_s(X_s(t) - X_e(t))$ of the desired model.

The objective of Learning Impedance Control design is to develop an iterative learning law such that, as $k \to \infty$,

$$\|w_k(t)\| \leq e \quad (9)$$

where

$$w_k(t) = M_m(\ddot{X}_d(t) - \ddot{X}_k(t)) + C_m(\dot{X}_d(t) - \dot{X}_k(t)) + K_m(X_d(t) - X_k(t)) + F_k(t) \quad (10)$$

is defined as the impedance error and e is a positive constant depending on the measurement noises, disturbances and repeatability of the robot.

Remark 1. In the conventional learning control formulation, the controller is designed to track a desired trajectory as the action is repeated. In general, as $k \to \infty$,

$$\|x_d(t) - x_k(t)\| \leq e$$

where $\| \cdot \|$ denotes a certain function norm [6]. In our learning approach, the control objective can be specified by a target impedance so that the impedance error described in equation (9) converges to zero as the action is repeated. That is, as $k \to \infty$,

$$\|w_k(t)\| \leq e,$$

ROBUST LEARNING IMPEDANCE CONTROL

The iterative learning control input $m_k(t)$ for learning the desired model (8) is updated by the following equation:

$$m_{k+1}(t) = m_k(t) + L(q_k(t))(w_k(t) + n_k(t)) \tag{11}$$

where $L(\cdot) \in R^{n\times m}$ is the learning gain to be chosen.
Theorem *Consider the Learning Impedance Control systems described by equations (7), (4), (5) and (11) with the desired model specified by equation (8). Let $L(\cdot)$ be any bounded learning gain that satisfies the condition:*

$$\| I_n - L(q_k(t)) \cdot M_m J(q_k(t) M^{-1}(q_k(t)) \| \leq p < 1. \tag{12}$$

Then, the impedance errors generated by the control are bounded such that for all $t \in [0, T]$

$$\lim_{k\to\infty} \sup \| w_k \|_\alpha \leq \hat{c}_1 b_{q0} + \hat{c}_2 b_n + \hat{c}_3 b_d, \tag{13}$$

where $\hat{c}_1 \cdots \hat{c}_3$ are constants to be defined.
Remark 2. The convergence of the impedance error are derived in α-norm. The α norm for a function $b(t)$ is defined as

$$\| b(t) \|_\alpha \triangleq \sup_{t\in[0,T]} e^{-\alpha t} \| b(t) \| \tag{14}$$

where α is some arbitrary positive scalar.
Proof :
For clarity of the convergence proof, the dependence of the system parameters on time is implied unless otherwise specified. From equation (9), let us define the desired state $[X_e^T, \dot{X}_e^T]^T$ and desired force F_e corresponding to the desired impedance (8) as,

$$w_e = M_m(\ddot{X}_d - \ddot{X}_e) + C_m(\dot{X}_d - \dot{X}_e) + K_m(X_d - X_e) + F_e = 0, \tag{15}$$

where

$$F_e = K_s(X_s - X_e). \tag{16}$$

Since $w_e = 0$, from equations (9) and (15), we have

$$w_k = w_k - w_e = M_m(\ddot{X}_e - \ddot{X}_k) + C_m(\dot{X}_e - \dot{X}_k) + K_m(X_e - X_k) - (F_e - F_k), \tag{17}$$

where

$$F_e - F_k = -K_s(X_e - X_k), \tag{18}$$

as seen from equation (16) and (5). The desired state X_e and $\dot{X}_e$ are unknown since F_e is an unknown desired force. Here, the definitions of the desired state and force are for analysis purpose and are not used in the control law. Differentiate equation (3) with respect to time, we have

$$\ddot{X}_k = J(q_k)\ddot{q}_k + \dot{J}(q_k)\dot{q}_k \tag{19}$$

From equations (11), (15), (2), (3), (19) and (7), we have

$$\begin{aligned}
\delta m_{k+1} &= [I - \hat{L}(q_k)M_m J(q_k)M^{-1}(q_k)]\delta m_k - L(q_k)M_m(\delta J(q_k)\ddot{q}_e + \dot{J}(q_k)\delta\dot{q}_k + \delta\dot{J}(q_k)\dot{q}_e \\
&\quad - M^{-1}(q_k)\delta N(q_k, \dot{q}_k, t) - \delta M^{-1}(q_k)(\delta N(q_e, \dot{q}_e, t) + f_e + m_e) + M^{-1}(q_k)\delta f_k \\
&\quad - M^{-1}(q_k)d_k) + L(q_k)(C_m(J(q_k)\delta\dot{q}_k + \delta J(q_k)\dot{q}_e) + K_m\delta h(q_k) + \delta F_k + n_k),
\end{aligned} \tag{20}$$

where $\delta m_{k+1} = m_e - m_{k+1}$, $\delta m_k = m_e - m_k$, $m_e = M(q_e)\ddot{q}_e + N(q_e, \dot{q}_e, t) - f_e$, $f_e = J^T(q_e)F_e$, $\delta J(q_k) = J(q_e) - J(q_k)$, $\delta\dot{J}(q_k) = \dot{J}(q_e) - \dot{J}(q_k)$, $\delta q_k = q_e - q_k$, $\delta\dot{q}_k = \dot{q}_e - \dot{q}_k$, $\delta f_k = f_e - f_k$, $\delta F_k = F_e - F_k$, $\ddot{X}_e = J(q_e)\ddot{q}_e + \dot{J}(q_e)\dot{q}_e$, $\delta N(q_k, \dot{q}_k, t) = N(q_e, \dot{q}_e, t) - N(q_k, \dot{q}_k, t)$, $\delta h(q_k) = h(q_e) - h(q_k)$. From equation (4), we have

$$\delta f_k = J^T(q_k)\delta F_k + \delta J^T(q_k)F_e. \tag{21}$$

Substitute equations (18) and (21) into equation (20) and then taking norm, using the bounds and Lipschitz conditions, we can show that,

$$\| \delta m_{k+1} \| \leq p \| \delta m_k \| + b_L c_1 \| \begin{bmatrix} \delta q_k \\ \delta\dot{q}_k \end{bmatrix} \| + b_L b_M b_d + b_L b_n, \tag{22}$$

where $c_1 = b_{Mm}b_{q2}c_J + b_{J1} + c_{J1}b_{q1} + b_M c_N + c_M b_N + c_M b_1 + b_M c_{JT} b_{Fe} + b_M b_{JT} b_{Ks} c_h + b_{Cm} b_J + c_J b_{q1} + b_{Km} c_h + b_{Ks} c_h$, b_L, b_J, b_{J1}, b_M, b_{JT}, b_{Ks}, b_{Mm}, b_{Cm}, b_{Km}, b_N are the norm bound for $L(\cdot)$, $J(\cdot)$, $\dot{J}(\cdot)$, $M^{-1}(\cdot)$, $J^T(\cdot)$, K_s, M_m, C_m, K_m, $N(\cdot,\cdot,\cdot)$ respectively, $b_{q1} = \sup_{t\in[0,T]} \|\dot{q}_e(t)\|$, $b_{q2} = \sup_{t\in[0,T]} \|\ddot{q}_e(t)\|$, $b_{Fe} = \sup_{t\in[0,T]} \|F_e(t)\|$, $b_1 = \sup_{t\in[0,T]} \|m_e(t) + f_e(t)\|$ and c_J, c_{J1}, c_N, c_M, c_{JT}, c_h are the lipschitz constants for $J(\cdot)$, $\dot{J}(\cdot)$, $N(\cdot,\cdot,\cdot)$, $M^{-1}(\cdot)$, $J^T(\cdot)$, $h(\cdot)$ respectively. From equation (7), we have

$$\begin{bmatrix} \delta\dot{q}_k \\ \delta\ddot{q}_k \end{bmatrix} = \delta f(q_k, \dot{q}_k, t) + \delta g(x_k)m_e + g(q_k)\delta m_k \tag{23}$$

where $f(q_k, \dot{q}_k, t) = \begin{bmatrix} \delta\dot{q}_k \\ -M^{-1}(q_k)N(q_k, \dot{q}_k, t) \end{bmatrix}$, $\delta f(q_k, \dot{q}_k, t) = f(q_e, \dot{q}_e, t) - f(q_k, \dot{q}_k, t)$, $g(q_k) = \begin{bmatrix} 0 \\ -M^{-1}(q_k) \end{bmatrix}$ and $\delta g(q_k) = g(q_e) - g(q_k)$. Integrates equation (23) with respect to time and taking norm, we have,

$$\| \begin{bmatrix} \delta\dot{q}_k \\ \delta\ddot{q}_k \end{bmatrix} \| \leq \| \begin{bmatrix} \delta\dot{q}_k(0) \\ \delta\ddot{q}_k(0) \end{bmatrix} \| + \int_0^t \{\|c_2\| \begin{bmatrix} \delta q_k \\ \delta\dot{q}_k \end{bmatrix} \| + b_g\|\delta m_k\| + b_d\}d\tau, \tag{24}$$

where $c_2 = c_f + c_g b_{me}$, c_f and c_g are the lipschitz constants for $f(\cdot,\cdot,\cdot)$ and $g(\cdot)$ respectively, $b_{me} = \sup_{t\in[0,T]} \|m_e(t)\|$ and b_g is the norm bound for $g(\cdot)$. Using Bellman-Gronwall Lemma [8], we obtained

$$\| \begin{bmatrix} \delta q_k \\ \delta\dot{q}_k \end{bmatrix} \| \leq \| \begin{bmatrix} \delta q_k(0) \\ \delta\dot{q}_k(0) \end{bmatrix} \| e^{c_2 t} + \int_0^t e^{c_2(t-\tau)}(b_g \| \delta m_k \| + b_d)d\tau. \tag{25}$$

Substituting equation (25) into equation (22), multiplying both sides by $e^{-\alpha t}$, define $c \triangleq \max\{c_2, b_L b_g c_1\}$ and let $\alpha > c_2$, we have,

$$\| \delta m_{k+1} \|_\alpha \;\leq\; \hat{p} \| \delta m_k \|_\alpha + \;\varepsilon, \tag{26}$$

where $\hat{p} = [p + \frac{c}{\alpha - c}(1 - e^{(c-\alpha)T})],$, $\varepsilon = b_L c_1 b_{q0} + c_3 b_d + b_L b_n$ and $c_3 = \frac{b_L c_1}{\alpha - c_2}(1 - e^{(c_2-\alpha)T}) + b_L b_M$. If we choose p to be less than 1, $\alpha > c$ and large enough so that $0 \leq \hat{p} < 1$, then equation (26) converges such that,

$$\lim_{k\to\infty} \| \delta m_k \|_\alpha \leq (\frac{1}{1-\hat{p}})\varepsilon. \tag{27}$$

Apply the same argument to equations (25) and (23), we have

$$\lim_{k\to\infty} \| \begin{bmatrix} \delta q_k \\ \delta \dot{q}_k \end{bmatrix} \|_\alpha \leq c_4 b_{q0} + c_5 b_n + c_6, \;\; \lim_{k\to\infty} \| \begin{bmatrix} \delta \dot{q}_k \\ \delta \ddot{q}_k \end{bmatrix} \|_\alpha \leq c_8 b_{q0} + c_9 b_n + c_{10} b_d, \tag{28}$$

where $c_4 = 1 + \frac{b_g b_L c_1 c_7}{1-\hat{p}}$, $c_5 = \frac{b_g b_L c_7}{1-\hat{p}}$, $c_6 = c_7 + \frac{b_g c_3 c_7}{1-\hat{p}}$, $c_7 = \frac{1}{\alpha - c_2}(1 - e^{(c_2-\alpha)T})$, $c_8 = c_2 c_4 + \frac{b_g b_L c_1}{1-\hat{p}}$, $c_9 = c_2 c_5 + \frac{b_g b_L}{1-\hat{p}}$ and $c_{10} = c_2 c_6 + b_g + \frac{b_g c_3}{1-\hat{p}}$. Therefore, from equations (17), (18) and (28), we have,

$$\lim_{k\to\infty} \sup \| w_k \|_\alpha \;\leq\; \hat{c}_1 b_{q0} + \hat{c}_2 b_n + \hat{c}_3 b_d, \tag{29}$$

where $\hat{c}_1 = c_{12} c_4 + c_{11} c_8$, $\hat{c}_2 = c_{12} c_5 + c_{11} c_9$, $\hat{c}_3 = c_{12} c_6 + c_{11} c_{10}$, $c_{11} = b_{Mm} b_J$, $c_{12} = b_{Mm}(c_J b_{q2} + b_{J1} + c_{J1} b_{q1}) + b_{Cm}(B_J + c_J b_{q1}) + b_{Kms} c_h$ and b_{Kms} is the norm bound for $(K_m + K_s)$. $\triangle\triangle\triangle$

Remark 3. Equation (29) shows the dependence of the bounds the impedance error $w_k(t)$ on the bounds of the error in initial conditions, disturbances and measurement noises. Note that if b_{q0}, b_d and b_n tends to zero, the impedance error also tends to zero.

CONCLUSION

A learning control algorithm has been developed for control of robots with target impedance at the end-effector. Given a target impedance, the learning controller is able to learn and eventually drive the closed loop response to reach the target impedance as the actions are repeated. The robustness of the proposed algorithm has been analyzed in the presence of error in initial condition, disturbances and measurement noises.

REFERENCES

1. S. Arimoto, S. Kawamura, and F. Miyazaki. "Bettering operation of robots by learning." *Journal of Robotic Systems*, 1(2):440–447, 1984.
2. P. Bondi, G. Casalino, and L. Gambardella. "On the iterative learning control theory for robotic manipulators." *IEEE Journal of Robotics and Automation*, 4:14–22, Feb 1988.
3. S. Arimoto. "Learning control theory for robot motion." *Int. Journal of Adaptive Control and Signal Processing*, 4:543–564, 1990.
4. G. Heinzinger, D. Fenwick, B. Paden, and F. Miyazaki. "Stability of learning control with disturbances and uncertain initial conditions." *IEEE Trans. A.C.*, 137(1):110–114, 1992.
5. C. C. Cheah and D. Wang. "Learning impedance control for robotic manipulator." *Proc. IEEE Int. Conf. on Robotics and Automation*, pages 2150–2155, Nagoya, Japan, 1995.
6. M. W. Spong, F. L. Lewis, and C. T. Abdallah. *Robot Control : Dynamics, Motion Planning, and Analysis.* IEEE Press, New York, 1993.
7. N. Hogan. "Impedance control: An approach to manipulation: Part i, part ii, part iii." *ASME J. Dynamic Syst., Measurement, Contr.*, 107, 1985.
8. T. M. Flett. *Differential Analysis.* Cambridge University Press, Cambridge, 1980.

CONTROL ARCHITECTURES FOR AUTONOMOUS GUIDED VEHICLES AND MOBILE ROBOTS

Peter J. Wojcik
Alberta Research Council
6815 - 8 St. N.E., Calgary, Alberta T2E 7H7, Canada
phone: (403) 297-7569, e-mail: peter@skyler.arc.ab.ca

ABSTRACT
The paper presents the main factors that must be taken into account during the control structure design process for a particular application. Then, the paper describes several different approaches to mobile robot control architecture, and outlines the advantages of possible designs as well as their drawbacks. The content of the paper, all the findings and examples are supported by a very extensive list of references.

KEYWORDS: hierarchical, behavioural, reactive, blackboard, control architectures, mobile robots, autonomous guided vehicles

INTRODUCTION

Autonomous Guided Vehicles and Mobile Robots are built to perform tasks in remote or hazardous locations where human presence is either impossible or undesirable. To complete required tasks successfully such systems must be either teleoperated or equipped with on-board intelligence enabling supervised or fully autonomous operation.

At the current state of technology development, mobile robotic systems are capable of performing only very simple tasks autonomously. As a result, the overwhelming majority of the existing systems are controlled using a teleoperated mode. Consequently, most of the research and development effort in this area is directed towards increasing the autonomy of remotely controlled vehicles. Increased autonomy and a desired system behaviour can be achieved through the appropriate design of the system control structure.

OVERVIEW OF MOBILE ROBOT CONTROL STRUCTURES

To carry out complex tasks in dynamic environments, autonomous robots must decide when and how to plan, when to act, what course of action to take, how to detect and recover from errors, how to handle conflicting goals, etc. On the other hand, advanced robotic systems must effectively manage their limited physical and computational resources. These two major factors must be taken into account during the process of designing mobile robotic systems.

A versatile mobile robot should be capable of performing specified tasks while interacting intelligently with a dynamic environment. To support such a sensor guided control strategy, the research community has paid a lot of attention to the development of different control architectures suitable for a wide variety of applications.

All approaches to mobile robot control architecture design are driven by the system functional requirements, such as reactivity, multiple sensors, multiple goals, intelligence (planning, replanning for unpredictable circumstances), robustness, reliability. There are also some limiting factors affecting the design process of the system structure. These limitations might come from the system design requirements, that impose additional conditions on the physical system such as modularity, flexibility, expandability or adaptability. Such requirements are strongly dependent on a particular application, which the system is being designed for. The final system structure, that best suits an ultimate application, is usually a trade-off between the functional and design requirements.

Out of the entire spectrum of existing approaches, based on functional requirement breakdown, the following classes of control structures are on the two opposite ends of this spectrum:

- classical hierarchical architectures,
- behavioural architectures.

In classical hierarchical architectures, the control problem is divided along functional lines into progressive levels of data abstraction. The functions that need to be performed are used as a basis for system functional decomposition. Data flows from the sensors to a series of perception processes, to a decision-making processes, through the series of action processes to the actuators (sense -> plan -> act). This architecture performs quite well in a structured environment, but reacts very slowly to unpredicted and sudden changes in the external world.

In behavioural systems the basic idea is to break up the control problem into the behaviours that should be exhibited. Thus, in such systems, sensor data is converted directly into control actions. There is no explicit map or model of the environment, and the robot system does not reason about its actions, but merely responds to external stimuli. The direct connection of sensor input to actuator output means that the system can respond rapidly to changes in the environment and requires (due to the lack of a world model) relatively few on-board computing resources. Behavioural control architectures tend to result in robust, relatively cheap mobile robots which survive well in their environments. However, such systems are only capable of executing relatively straightforward exploration and survival tasks. They cannot complete more complex tasks which require planning.

"Real life" mobile robotic systems must be able to perform planned tasks and to survive environmental changes. Therefore, in terms of a control architecture, a trade off must be reached between a traditional, functional, hierarchical, multilayered structure and a behaviour based approach. The following sections of the paper describe hierarchical, behavioural and "mixed" architectures, which have been implemented in real systems.

HIERARCHICAL ARCHITECTURES

The family of classical hierarchical architectures is best represented by NASREM (NASA/NBS Standard Reference Model for Telerobot Control System Architecture) developed in 1986 as a result of more that 15 years of research [24],[31]. The overall control system is represented as a three legged hierarchy of computing modules: a task decomposition hierarchy, a world modelling hierarchy, and a sensory processing hierarchy. These computing modules are serviced by a communication system and a global memory. Goals at each level of the task decomposition hierarchy are divided both spatially and temporally into simpler commands for the next lower level. This decomposition is repeated until, at the lowest level, the drive signals to the robot actuators are generated. In order to accomplish its goals, task decomposition modules must often use information stored in the world model, which always stores the best estimate of the state of the world. The sensory processing hierarchy must update the world model as rapidly as possible to keep the world model as close as possible to the physical world.

This control strategy can be described as a deliberative or feedforward method. Its major limitation is that strict top down constraints may prevent the system from being responsive to changes in the environment. Even if the system is responsive, the reaction time is usually too high.

Practical implementations of a hierarchical control structure depend strongly on the applications and their requirements. Drunk [10] and Song et al. [30] describe hierarchical control systems for Autonomous Guided Vehicles. The first system has been implemented at Fraunhofer Institute for Manufacturing, Germany on the research platform IPAMAR (IPA Mobile Autonomous Robot) operating in an industrial factory environment. Its structure consists of three levels: planning, task execution and motion control. The second AGV control system that has been implemented in Belgium has four levels: task specification, trajectory generation, path following control and servo control.

Dillman and Frolich [8] describe IRAS - a robot control system designed for space application in a supervisory mode of operation. The system structure is hierarchically layered with three levels: mission execution planning and control, task execution planning and control and action execution planning and control. The Field Material-Handling Robot [25] is a heavy-lift pallet handling robotic system for military applications. A hierarchical control structure of the system consists of four levels closely following NASREM model. Lueth and Rembold [23], and also Damm et al. [6] describe KAMRO (Karlsruhe Autonomous Assembly Robot) that consist of a mobile platform with an omnidirectional drive, two PUMA type manipulators and several sensors (CCD cameras, ultrasonic range sensors, force torque sensors). The robot is controlled by a hierarchical, function oriented planning and control architecture, and is capable of executing assembly tasks.

These successful application examples prove how powerful the hierarchical control structure is. However, all of these implementations were designed to operate in relatively well structured and static environments. When robotic systems are required to operate and survive in dynamic and unknown environments, some elements of behavioural control structure must be implemented.

BEHAVIOURAL ARCHITECTURES

Behavioural or reactive control architectures, initially introduced by Brooks [3], are designed for robotic agents that continually sense their environment and compute appropriate reactions to their sense stimuli within bounded time. There is no explicit map or model of the robot environment. Such architectures offer advantages over other approaches because they can react more quickly to changes in their environment, and because they can operate more robustly in worlds that are difficult to model in advance. Since the first work of Brooks there has been a lot of research activity directed towards implementing several variations of this concept.

Eustace et al. [13] describe the Behaviour Synthesis Architecture, a sensor driven behavioural control architecture, which consist of a number of behavioural patterns. Each of these patterns takes a sensory stimulus and generates a motion response together with an associated utility, which is a measure of the importance of the motion response. Multiple motion responses generated by the individual behaviour patterns are resolved using a process of additive synthesis, that uses utility of the individual motion responses to resolve conflicts. This particular control architecture is especially useful for co-operating multi robot devices. A different approach to co-operating multi agents, called a societal architecture is presented by Duffy et al. [11].

Another variation of the behavioural architecture has been implemented by Gat et al. [18] in Rocky IV, a prototype microrover designed for a low-cost scientific mission to Mars. The control architecture implements a large variety of functions displaying various degrees of autonomy, from completely autonomous sequences of actions to very precisely described actions resembling classical AI operators. Experimental results for this particular application are very encouraging.

A very interesting concept of layered behavioural control and its adaptation to autonomous underwater vehicles (AUVs) is presented by Bellingham et al. [2]. This architecture has been successfully implemented on an underwater vehicle, the Sea Squirt.

The practical implementations of behavioural robotic architectures, described above, were designed to complete necessary tasks while surviving unexpected changes in the environment. Consequently, these architectures have some elements of hierarchical structures which enable the robotic systems to plan and execute longer missions. When the missions to be accomplished and tasks to be performed become very complex, the system architecture must be biased more towards hierarchical approach.

Thus, most of the practical applications of the autonomous or semi-autonomous robotic vehicles have "mixed" control architectures, which combine the advantages of both the hierarchical and behavioural structures. Such systems are briefly described in the next section.

"MIXED" AND OTHER APPROACHES

A large number of control architectures implemented in real systems cannot be explicitly labelled as purely hierarchical or behavioural, because they contain features of both approaches working to the advantage of ultimate performance. Examples of such systems are described by Mitchell [26], Firby [15], Simmons [28], Gat [17], Dillmann et al. [9], Fleury et al. [16]. The first four approaches are closer to behavioural architecture, while the last two papers describe layered control architecture with embedded capability of behavioural reactivity.

There is a very large group of autonomous and semi-autonomous robotic systems that have been especially designed for operating remotely. Such systems might support teleoperation, exhibit different degrees of autonomy and be capable of dealing with significant time delays between a local control station and a remotely located robotic system. Applications range from space, underwater, indoor AGVs to all terrain military vehicle systems. Some of these applications are mentioned in the following paragraphs.

Laird et al. [22] describes an interesting concept in the development of reconfigurable, modular robots (MODBOTS). The proposed Modular Robot Architecture (MRA) supports rapid integration and prototyping of evolving sensor and control technologies into demonstrable systems. Space station telerobotic system architectures are presented in [1],[27],[32],[33]. Mobile robot control for indoor use is addressed in [7], [20]. All terrain and military vehicles are dealt with in [4],[12],[19],[21].

Special attention should be drawn to blackboard architectures, which could be thought of as a variation of the behavioural concept, however with increased on-board intelligence. A typical blackboard system contains three main components: a shared global database, called the blackboard, knowledge sources, that embody specialised knowledge and a trigger mechanism, which involves activity in knowledge sources based on changes in the blackboard database. Knowledge sources can interact with the blackboard database, but not directly with each other. This approach offers a good paradigm for the integration of multiple sensors and actuators. However, data flow via the blackboard does not contribute to low-level control efficiency. The advantages and drawbacks of this type of control structures are described in detail by Cassinis et al. [5], Fayek et al. [14] and Skillman [29].

CONCLUSIONS

It can be concluded that there is no definite answer to the question of which control architecture is the best. The final and best control structure design must be based on a detailed analysis of the system functional and design requirements for a particular application. Finally, due to a vast number of publications in the area of robotics control architectures, and the limited length of this paper some important papers might be missed on the list of references. Please, accept my apology for this.

REFERENCES

[1] Backes, P.G. "Ground-remote control for space station telerobotics with time delay" -1992, Guidance and control 1992; Proceedings of the 15th Annual AAS Rocky Mountain Conference, San Diego, CA, Feb 8 - 12, 1992, pp. 285-303.

[2] Bellingham, J. G., Consi, T. R., Beaton, R. M., Hall, W. "Keeping layered control simple" -1990, Proceedings of the Symposium on Autonomous Underwater Vehicle Technology, AUV '90, Washington, DC, 5 - 6 June 1990, pp. 3-8.

[3] Brooks A. R. "A robust layered control system for a mobile robot" -1986, IEEE Journal on Robotics and Automation, vol RA-2#1, March 1986.

[4] Byrne, R. H. "A practical implementation of a hierarchical control system for telerobotic land vehicles" - October 1992, IEEE Aerospace and Electronics Systems Magazine, Vol. 7, No.10 pp. 22-6.

[5] Cassinis, R., Biorli, E., Meregalli, A., Scalise, F. "Behavioural model architectures: a new way of doing real-time planning in intelligent robots" -1988, Proceeding of the Confgerence SPIE, Mobile Robots II, Cambridge, MA, 5 - 6 Nov. 1987, Vol. 852, pp.275-280.

[6] Damm, M., Kappey, D., Schloen, J., Rembold, U., Dillmann, R. "A multi-sensor and adaptive real-time control architecture for an autonomous robot" -1993, Proceedings of the International Conference on Intelligent Autonomous Systems IAS-3, Pittsburgh, PA, 15-18 Feb. 1993, pp.439-448.

[7] D'Orazio, T., Ianigro, M., Stella, E., Lovergine, F. P., Distante, A. "Mobile robot navigation by multi-sensory integration" -1993, Proceedings of 1993 IEEE International Conference on Robotics and Automation, Atlanta, GA, 2 - 6 May 1993, Vol.2 pp. 373-9.

[8] Dillman, R., Frolich, J. "IRAS-interactive remote A & R servicing" -1993, Proceedings of Telemanipulator Technology Conference, Boston, MA, Nov 15 - 16, 1992, pp. 102-112.

[9] Dillmann, R., Kreuziger, J., Wallner, F. "The control architecture of the PRIAMOS mobile system" -1994, Control Engineering Practice, Vol. 2, No. 2, April pp. 341-6.

[10] Drunk, G. "Free Ranging AGV" -1988, International Journal of Machine Tools & Manufacture Vol.28 No 3, pp. 263-272.

[11] Duffy, N. D., Herd, J. T., Eccles, N. J. "Agents get framed in novel architecture" -1991, Industrial Robot Vol.18, No 2, 1991, pp. 23-26.

[12] Eirich, R., Kramer, A. "A Generic Semi-autonomous Ground Vehicle Control System" - Proceedings of the 3rd Conference on Military Robotic Applications - Military Robotic Vehicles MRV91, September 9-12, 1991, Medicine Hat, Alberta, pp. 68-73.

[13] Eustace, D., Barnes, D. P., Gray, J. O. "A behaviour synthesis architecture for co-operant mobile robot control" -1994, Proceedings of the International Conference on Control, Coventry, UK, 21 - 24 March 1994, Vol. 1 pp. 549-554.

[14] Fayek, R. E.; Liscano, R.; Karam, G. M. "A system architecture for a mobile robot based on activities and a blackboard control unit" -1993, Proceedings of 1993 IEEE International Conference on Robotics and Automation, Atlanta, GA, 2 - 6 May 1993, Vol.2 pp. 267 - 274.

[15] Firby, R.J. "Task directed sensing" -1990, Proceedings of the Conference Sensor Fusion II: Human and Machine Strategies, Philadelphia, PA, 6 - 9 Nov, 1989, vol. 1198 pp. 480-489.

[16] Fleury, S., Herrb, M., Chatila, R. "Design of a modular architecture for autonomous robot" -1994, Proceedings of the 1994 IEEE International Conference on Robotics and Automation, San Diego, CA, 8 - 13 May 1994, pp. 3508-3513.

[17] Gat, E. "On the role of stored inernal state in the control of autonomous mobile robots" - Spring 1993, AI Magazine, Vol. 14, no. 1 pp. 64 - 73.

[18] Gat, E., Behar, A., Desai, R., Ivlev, R., Loch, J., Miller, D. P. "Simple sensors for performing useful tasks autonomously in complex outdoor terrain" -1992, Proceedings of the Conference Sensor Fusion V, Boston, MA, 15 - 17 Nov. 1992, Vol. 1828 pp. 367-72.

[19] Goel, A., Donnellan, M., Vazquez, N., Callantine, T. "An integrated experience-based approach to navigational path planning for autonomous mobile robots" -1993, Proceedings of 1993 IEEE International Conference on Robotics and Automation, Atlanta, GA, 2-6 May 1993, Vol.1 pp.818-25.

[20] Hu, H., Brady, M., Probert, P. "Transputer architecture for sensor-guided control of mobile robots" - 1993, Proceedings of the 1993 World Transputer Congress. Transputer Applications and Systems '93, Aachen, Germany, 20 - 22 Sept. 1993, pp. 118 - 33.

[21] Lacroix, S., Chatila, R., Fleury, S., Herrb, M., Simeon, T. "Autonomous navigation in outdoor environment: adaptive approach and experiment" -1994, Proceedings of the 1994 IEEE International Conference on Robotics and Automation, San Diego, CA, 9 - 13 May 1994, Vol. 1, pp. 426-432.

[22] Laird, R.T., Smurlo, R. P., Timmer, S. R. "Development of a Modular Robotic Architecture" - 1991, Report of Naval Ocean Systems Center -Technical Document 2171, September 1991, 198 pages.

[23] Lueth, T. C., Rembold, U. "Extensive manipulation capabilities and reliable behavior at autonomous robot assembly" -1994, Proceedings of the 1994 IEEE International Conference on Robotics and Automation, San Diego, CA, 8 - 13 May, 1994, vol. 4 pp. 3495 - 3500.

[24] Lumia, R. "Integrating sensors into a standard control architecture for robotic applications" -1990, Proceedings of the IEEE Instrumentation and Measurement Technology Conference, San Jose, CA, 13 - 15 Feb 1990, pp. 120-125.

[25] McCain, H. G., Kilmer, R. D., Szabo, S., Abrishaminan, A. "A hierarchically controlled autonomous robot for heavy payload military field applications" -1987, Proceeding of the Conference on Intelligent Autonomous Systems, Amsterdam, Netherlands, 8 - 11 Dec 1986, p. 372-381.

[26] Mitchell, T. M. "Becoming increasingly reactive" -1990, AAAI-90 Proceedings. Eighth National Conference on Artificial Intelligence, Boston, MA, July 29 - 3 Aug 1990, Vol.2, pp. 1051-1058.

[27] Nechyba, M.C., Yangsheng Xu "SM2 for new space station structure: autonomous locomotion and teleoperation control" -1994, Proceedings of the 1994 IEEE International Conference on Robotics and Automation, San Diego, CA, 8 - 13 May 1994, Vol. 2 pp.1765-70.

[28] Simmons, R. G. "Structured control for autonomous robots" -1994, IEEE Transactions on Robotics and Automation, Vol. 10, No. 1, February, pp. 34-43.

[29] Skillman, T.L. "Distributed Cooperating Processes in a Mobile Robot Control System" -1988, Proceedings of the Second Conference on Artificial Intelligence for Space applications, August 1988, pp. 325 - 336.

[30] Song, K.T., De Schutter, J., Van Brussel, H. "Design and implementation of a path-following controller for an autonomous mobile robot" -1990, Proceedings of an International Conference. Intelligent Autonomous Systems 2, Amsterdam, Netherlands, 11-14 Dec 1989, Vol.1 pp. 253-263.

[31] Wheatley, T., Michaloski, J. "Configuration and performance evaluation of a real-time robot control system: the skeleton approach" -1990, Proceeding of the IEEE International Conference on Systems Engineering, Pittsburgh, PA, 9 - 11 Aug 1990, pp. 268-271.

[32] Wojcik, P., Chrystall, K., Paquin, N. "A semi-autonomous robot manipulator system with increased dexterity" - Proceedings of the Fifth International Symposium on Robotics and Manufacturing ISRAM'94, Aug. 14-18, 1994, Maui, HI, pp. 305-314.

[33] Zimmerman, W., Backes, P. "Supervisory Autonomous Local-Remote Control System Design: Near-Term and Far-Term Applications" - 1993, Proceedings of The Sixth Annual Workshop on Space Operations Applications and Research (SOAR 1992), February 1993, pp. 28-40.

A Replanner toward Assembly Motions in the Presence of Uncertainties: Design and Testing*

Jing Xiao and Lixin Zhang
Computer Science Department
University of North Carolina - Charlotte
Charlotte, NC 28223, USA
xiao@uncc.edu, lizhang@uncc.edu

Abstract

High-precision assembly tasks cannot be successfully done by robots without taking into account the effect of uncertainties. Often a robot motion may fail and result in some unintended contact between the part held by the robot and the environment. To automatically recover a task from such a failure and to ensure its success in spite of uncertainties, Xiao and Volz introduced a systematic *replanning approach* [7, 8, 9] consisting of *patch-planning* based on contact analyses and *motion strategy planning* based on constraints on nominal and uncertainty parameters of sensing and motion. The replanning algorithms have since been implemented and tested under a general geometric simulator SimRep [10]. With the feedback from testing, we further improved and extended the replanner. This paper introduces our newly extended replanner, demonstrates its testing under SimRep, and discusses its performance.

Keywords: assembly, motion, replanning, uncertainty

1 Introduction

Various uncertainties, such as sensing, modeling, and motion uncertainties, can be crucial enough to make a low-tolerance robotic assembly task fail. Thus, an important problem is how to enable a robot to accomplish an assembly task successfully in spite of the inevitable uncertainties.

*This research is supported by a National Science Foundation grant under grant number IRI-9210423.

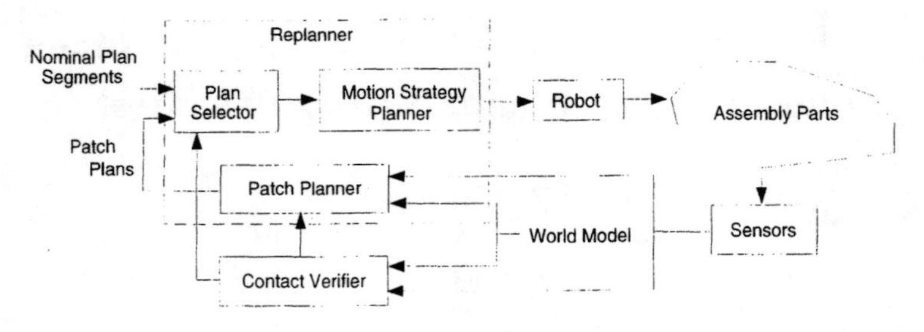

Figure 1: The replanning scheme

There are generally three types of approaches to tackle the problem. One is to model the effect of uncertainties in the off-line planning process, such as the preimage approach [4, 3]. This method requires the search of configuration space regions with uncertainties taken into account. Computability is the crucial issue. The second approach is to rely on on-line sensing to identify errors caused by uncertainties in a motion process and to replan the motion in real-time based on sensed information, such as the replanning approach proposed by Xiao and Volz [7, 8, 9]. The third approach is to use task-dependent knowledge to obtain efficient strategies for specific tasks rather than focusing on generic strategies independent of tasks. Examples are many strategies dealing with different types of insertion tasks (e.g., [1, 6, 5]).

The replanning approach is the compromise between the other two approaches. It decomposes the fine motion planning problem into three levels, i.e., the global and off-line *nominal planning* assuming no uncertainties, the on-line and local *patch-planning* based on sensing feedback, and the low-level motion strategy planning based on uncertainty models. Such decomposition simplifies the problem and also provides a general and flexible framework for assembly motion planning in the presence of uncertainties. The conceptual model of the replanning approach is shown in Fig. 1, where the replanner mainly consists of patch-planner and motion strategy planner, and certain forms of nominal plans are assumed as the initial input. In Section 2, we will review these major components along with discussions on new extensions. we will discuss the testing results on the newly extended replanner in Section 3 and conclude the paper in Section 4.

2 The Extended Replanner

We will first introduce the basic models assumed in the replanner before discussing the planning strategies.

2.1 Uncertainty and task models

The error bound model is used to model various uncertainties, i.e., an uncertainty in a parameter is defined as the magnitude of the maximum possible difference between a sensed or commanded value of the parameter and the actual value. In particular,

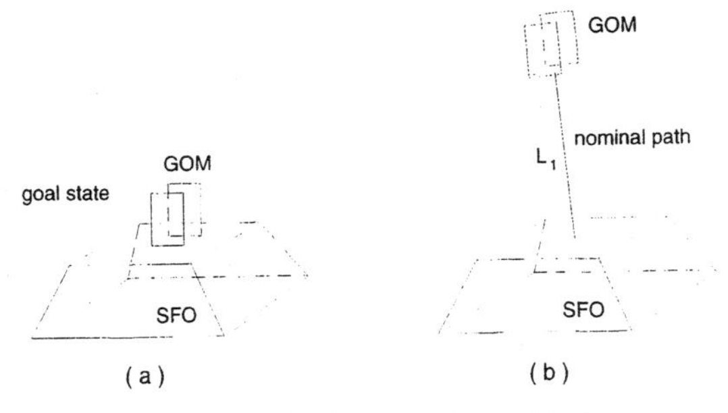

Figure 2: **Task 1:** stack two blocks

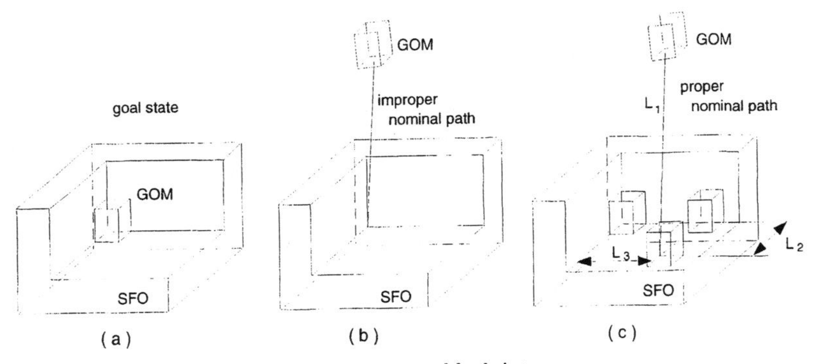

Figure 3: **Task 2:** insert a block into a corner

the following uncertainties are considered in the replanner: *position sensing uncertainty* ϵ_p, *orientation sensing uncertainty* ϵ_o (as an angular bound), *linear velocity uncertainty* ϵ_v, and *angular velocity uncertainty* ϵ_ω.

An assembly task is defined as to bring a gripper-object-model (GOM), which describes a part rigidly attached to (or held by) a manipulator, into assembly with a single-fixed-object (SFO), which describes the task environment (i.e, other parts and fixtures). Both objects are polyhedral with known geometric models.

The goal of an assembly task is defined in terms of the contact states between the GOM and the SFO where they achieve final assembly. We denote such contact states as the *goal contact states.* In addition, in each goal contact state, there may be a neighborhood of relative locations of GOM w.r.t. SFO which result in the same assembly, called a *goal neighborhood.* The assembly task is to lead the GOM into one of its goal neighborhoods.

2.2 Nominal plans

A nominal plan does not consider the effect of uncertainties. It was assumed to be given as an input to the replanner in terms of a *nominal path,* which was simply the straight-line path connecting a sensed start position of the GOM to a sensed target position where the GOM and SFO were expected to be assembled [7, 8, 9], assuming that the GOM was properly pre-oriented[1].

The above version of the nominal path,.however, was found to be too coarse for certain tasks during our testing of the replanner under SimRep. For example, consider the two tasks in Fig. 2 and 3. Suppose in both cases, the GOM is separate from the SFO at the start position (Fig. 2b and Fig. 3b). The single straight-line nominal path was appropriate for **task 1** but not for **task 2**. This is because in **task 2**, such path tolerates no uncertainty, which makes the corrective motions offered by the replanner to combat uncertainties ineffective. Whereas, the replanner was very effective for the nominal path shown in Fig. 3c, which consists of three straight-line segments and two subgoal positions connecting different segments.

In general, we have discovered that although nominal planning is a separate activity from replanning, a nominal path generated must provide some room of tolerance so that the assembly task can be accomplished by the replanner in spite of uncertainties. Specifically, if the goal contact state of the task constrains the GOM to one or zero translational degree of freedom, and if the GOM is collision-free at the start position, then the nominal path should consist of two or three straight-line segments connected by one or two (sensed) subgoal positions, such that each subgoal defines a contact state, and by reaching the contact state, the GOM loses one translational degree of freedom (see Fig. 3c).

2.3 Patch planning

When the GOM is stopped before reaching one of the goal (or subgoal) neighborhoods due to uncertainties, a patch plan is needed to bridge the gap, i.e., to bring the GOM from its current state to some desirable state. A patch plan, which can be either translational (specifying guarded or compliant translations) or rotational (specifying compliant rotations), is a *local* plan generated by the patch planner based on the relationship between the nominal path and the GOM's sensed contact state and configuration [7, 8, 9]. As such, a patch plan is often not designed to lead the GOM to one of the goal (subgoal) configurations directly: subsequent patch planning is often needed[2]. Thus, the target of a patch plan can be one of the following types:

(1) a target position or orientation of the GOM,
(2) breaking of the current contact state,
(3) a target contact state.

A patch plan motion with a type (1) target is terminated when the GOM reaches certain neighborhood Φ of the target in which, due to uncertainties, no further motion is guaranteed to make the GOM any closer to the target. Such a neighborhood Φ is called *the region of convergence*[3]. On the other hand, a patch plan motion with a

[1]Note that both the GOM's orientation and the two positions defining the nominal path were subject to uncertainties.

[2]Even if a patch plan aims at a goal configuration directly, its execution may be stopped prematurely by an unintended collision due to uncertainties. In such cases, new patch planning is again needed.

[3]Proper motion strategy (Section 2.4) can ensure that the GOM reaches the region of convergence unless it is stopped by another collision.

target of type (2) or (3) is terminated when a change of contact state is detected (by a force/torque sensor).

There are several aspects of patch-planning not discussed in [7, 8, 9] but newly added in our implementation of the patch planner under SimRep. One major addition is the strategy to generate patch plans which lead to a type (3) target, i.e. a contact state. The need for such patch plans arises when the execution of a previous plan (nominal or patch) has led the GOM to stop inside Φ (the region of convergence) of a target *contact* position/orientation but outside the desired contact region. For example, in **task 1** (Fig. 2b), the GOM could stop inside Φ of the target (goal) position without touching the SFO if Φ is too large (due to large position uncertainty). In such case, a patch plan with a type (3) target is needed to enable the GOM to make the desired contact. Another major addition is the strategy to pick one rotational patch plan out of several candidates based on certain criteria deciding if one plan is more constrained (so that there is less uncertainty) or more effective. Details of these additions and other implementation issues can be found in [11].

2.4 Motion strategy planning

A motion strategy controls the execution of the nominal or a patch plan of the GOM so that the GOM can enter the region of convergence (Section 2.3) of its target position/orientation. It regulates the movement of GOM with a sequence of motion steps, such that after each step, the motion is recalibrated based on newly sensed data. Such a motion strategy is different from a conventional feedback control strategy in that the sampling rate is determined on-line taking into account both sensing and motion (i.e. velocity) uncertainties. In particular, the motion strategy planning is based on two types of constraints: the *design constraints* on nominal parameters and uncertainties in the task, whose satisfaction is the precondition for the convergence of a motion, and the *motion constraints* on motion steps to realize the convergence. Constraint analysis is the key. The motion strategy planner as described in [7, 8] only focused on constraints for generating motion strategies to regulate the translational motions of the GOM.

The current motion strategy planner is much extended by new constraint analysis and methods for generating motion strategies to regulate compliant rotational motions and the newly added translational patch motions with type (3) targets (Section 2.3). The major results include the following:

- For rotational motions constrained by a contact face between GOM and SFO, the size of the region of convergence is a function of orientation uncertainty ϵ_o only and is independent of motion uncertainty ϵ_ω of the GOM, which confirms the advantage of compliant rotation.

- In order to guarantee the success of a translational patch plan with a type (3) target, i.e., the GOM will always reach the target contact state, the size of the SFO face to be contacted, measured by the radius r of the circle circumscribed by the face, must be greater than a threshold value which is a function of uncertainties ϵ_p, ϵ_o, and ϵ_v. Under certain amount of uncertainties, this threshold

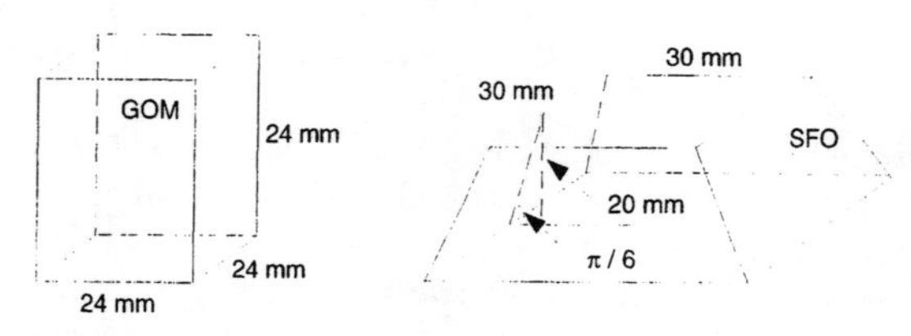

Figure 4: Typical geometric parameters for **task 1**

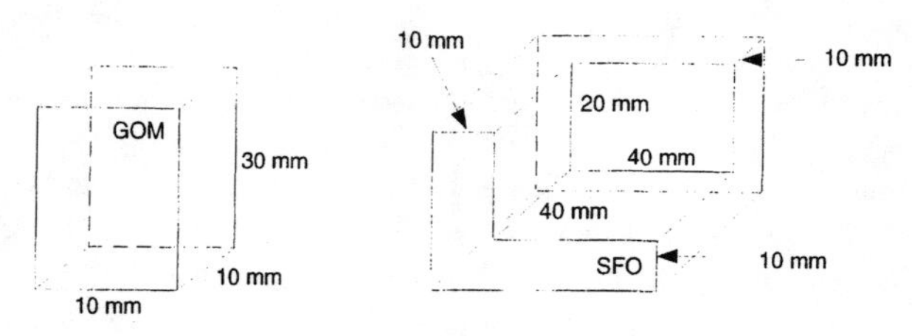

Figure 5: Typical geometric parameters for **task 2**

puts a geometric constraint on the smallest objects which can be assembled reliably by the replanner.

More details can be found in [11].

3 Testing the Replanner

We tested the replanner under SimRep, which allows the user to design and select the assembly task and all related parameters flexibly, runs the replanning algorithms, simulates the motion and displays the assembly process in 3D graphics [10].

The **task 1** and **task 2** shown in Fig. 2 and 3 were used in the testing, whose typical geometric parameters are shown in Fig. 4 and 5. For system parameters, the following typical values based on a commercial robot were chosen:

commanded linear speed: $v^d = 100$ mm/s
commanded angular speed: $\omega^d = 2.0$ rad/s
velocity uncertainties: $\epsilon_v = 10$ mm/s, $\epsilon_\omega = 0.2$ rad/s
sensing uncertainties: $\epsilon_p = 0.025$ mm, $\epsilon_o = 0.01$ rad
segment length of nominal path: $L_1 = 200$ mm, $L_2 = L_3 = 15$ mm

Fig. 6 records a sample execution of the **task 2** under large uncertainties $\epsilon_v = 26.0$ mm/s and $\epsilon_\omega = 0.8$ rad/s, with the other parameters assuming the typical values. Fig. 6a through e show the contact states traversed to reach the first sub-goal. Fig. 6f and g show the contact states traversed to reach the second sub-goal. Fig. 6h shows the final goal contact state.

We tested the performance of the replanner by varying uncertainties and part sizes for both **task 1** and **task 2**. In particular, we varied the parameters in the given

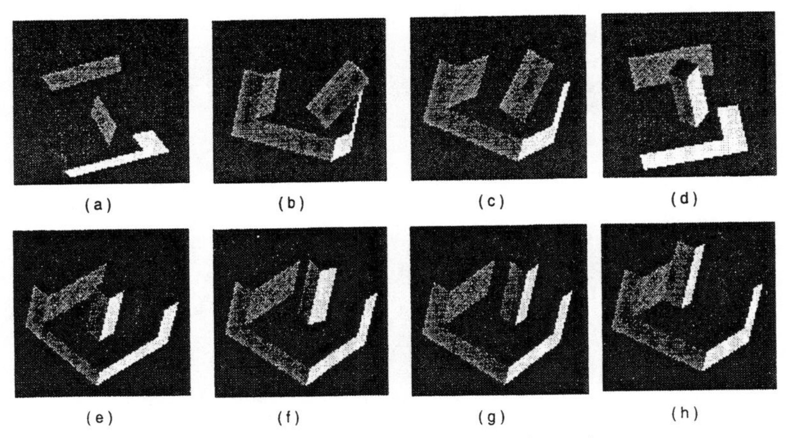

Figure 6: Typical execution of the **task 2**

ranges below one at a time and fixed the values of the rest as the typical ones: $\epsilon_v \in [4.0, 48.0]$(mm/s), $\epsilon_\omega \in [0.1, 0.9]$(rad/s), $\epsilon_p \in [0.025, 10]$(mm), $\epsilon_o \in [0.01, 0.16]$(rad), and $k \in [0.01, 8.0]$, where k is the ratio of the tested geometric parameter values to the typical values shown in Fig. 4 and 5.

For every set of the above values, we ran the replanner 1000 times and obtained the following statistical results:

- The GOM reached the region of convergence of the final target configuration as well as the goal contact state 100% times.
- For larger motion (i.e. velocity) uncertainties, the above result was achieved by the replanner by generating, on average, more patch plans and more motion steps in executing the nominal and patch plans.
- For **task 1**, as sensing uncertainties increase, the difference between the final configuration actually reached by the GOM and the final target configuration increases on average, as the result of larger region of convergence.
- In addition, as k decreases, the replanner generated more patch plans and more motion steps to accomplish the **task 1**.

4 Conclusions

We have extended, implemented, and tested the replanner under SimRep on two typical assembly tasks. The test results confirmed the effectiveness of the replanner in the presence of large sensing and motion uncertainties.

Although the extended replanner is more powerful than ever, it can be further extended by adding the capability of generating optimized plans. Currently, patch-plans are generated opportunistically with very limited search effort. To optimize plans may require not only the knowledge of the present contact state but also that of its certain neighborhood of contact states, i.e., a contact state graph. How to acquire and use such knowledge is one of the major future research topics. In addition, the replanner should be tested with a real robot.

References

[1] H. Asada, "Teaching and Learning of Compliance Using Neural Nets: Representation and Generation of Non-linear Compliance," *IEEE Int. Conf. Robotics & Automation*, pp. 1237–1244, 1990.

[2] B. R. Donald, *Error Detection and Recovery in Robotics.* Springer-Verlag, 1989.

[3] M. Erdmann, "Using Backprojections for Fine Motion Planning with Uncertainty," *Int. J. Robotics Res.*, 5(1):19–45, Spring 1986.

[4] T. Lozano-Pérez, M. T. Mason, and R. H. Taylor, "Automatic Synthesis of Fine-motion Strategies for Robot," *Int. J. Robotics Res.*, 3(1):3–24, Spring 1984.

[5] R. H. Sturges, and S. Laowattana, "Fine Motion Planning through Constraint Network Analysis", *IEEE Int. Symp. Assembly & Task Planning*, pp. 160–170, August 1995.

[6] D. E. Whitney, "Quasi-static Assembly of Compliantly Supported Rigid Parts," *J. Dynamics Sys. Measurement, and Control*, 104:65–77, March 1982.

[7] J. Xiao and R. Volz, "On Replanning for Assembly Tasks Using Robots in the Presence of Uncertainties," *IEEE Int. Conf. Robotics & Automation*, pp. 638–645, 1989.

[8] J. Xiao, *On Uncertainty Handling in Robot Part-Mating Planning*, PhD thesis, the Univ. of Michigan, 1990.

[9] J. Xiao, "Replanning with Compliant Rotations in the Presence of Uncertainties," *IEEE Int. Symp. Intelligent Cont.*, pp. 102–107, August 1992.

[10] J. Xiao and L. Zhang, "A Geometric Simulator SimRep for Testing the Replanning Approach toward Assembly Motions in the Presence of Uncertainties," *IEEE Int. Symp. Assembly & Task Planning*, pp. 171–177, Pittsburgh, August 1995.

[11] L. Zhang, *Extension and Testing of the Replanning Approach Toward Assembly Motions in the Presence of Uncertainties*, M.S. thesis, the Univ. of North Carolina – Charlotte, 1996.

CHARACTERIZATION OF CIRCULAR CYLINDERS ON THREE GIVEN POINTS

P. J. ZSOMBOR-MURRAY

Department of Mechanical Engineering & Centre for Intelligent Machines
McGill University, 817 Sherbrooke Street West, Montreal, Quebec, Canada H3A 2K6
paul@cim.mcgill.edu

ABSTRACT

A method is developed to determine the two parameter set of circular cylinders whose surfaces contain three given points. An efficient algorithm to compute radius and a point where the axes intersect the plane of the given points is implemented. The geometry of the surface of points, whose position vectors represent cylinder magnitude and axial orientation, is revealed and described in terms of symmetry and singularity inherent in the triangle with vertices on the given points. This strongly suggests that general problem is reducible to a sixth order polynomial.

KEYWORDS: cylinder, radius, axis, orientation, symmetry, singularity, pose, recognition

INTRODUCTION

The assembly of cylindrical parts and the avoidance of collision among manipulator components approximated by cylinders of revolution (particularly in the case of parallel architecture) are important issues in robotics. Such tasks, if they are to be carried out in real time, require minimization of data and computation to encode and establish the pose of these parts or components. In this regard, ideas introduced by Schaal [SC86, SC85] and Strobel [ST91, ST89] may be adapted to obtain the points where the axes, of a two parameter set or *congruence* of cylinders, pierce the plane of three given points on the surface of all cylinders in the congruence. There are two angular orientation parameters and one of cylinder radius associated with each piercing point. Therefore the plane can be contour mapped with these three sets of parametric characteristics. Consider some insertion or avoidance scenario between a cylinder of known pose and another which has been surveyed with three surface point measurements. One might imagine defining a patch on a characteristic map, representing limiting pose conditions on the surveyed cylinder. The patch represents the feasible pose set. If the known pose corresponds to a point in the patch then insertion or collision is valid. Before attempting to implement such a procedure it is necessary to explore the general properties of the congruence of cylinders on three points. Unfortunately, this is the limited objective of the work reported below. Nevertheless quite a few interesting geometric properties of this congruence are revealed. Some are obvious but others are not. Furthermore it is shown that up to six cylinders, whose axes are perpendicular to a given direction and whose radius is specified, may contain the three given points.

ANALYSIS

$\{P_i\}_1^3 \in \mathcal{Y}$: Three given points, P_i, are on the surface of every cylinder in a two parameter set $\mathcal{Y}$. Let $\mathbf{p}_i \equiv P_i$, the position vector of point, P_i.

P_i is projected perpendicularly as $P_{i\pi}$ onto planes, Π, on P_1. Furthermore let $P_1 \equiv O$, the origin. Π is a two parameter set of planes with outward normal unit vector, $\mathbf{n}$, which makes angles α, β, γ with respect to axes x, y, z, respectively. These axes are fixed to the rigid body on P_i. Note that $\Pi = \Pi(O, \alpha, \beta, \gamma)$ where $\cos^2\alpha + \cos^2\beta + \cos^2\gamma = 1$. Moreover choose $\mathbf{p}_2 = \{1, 0, 0\}^T$, where P_1P_2 is the longest side of the triangle whose vertices are P_i. Clearly $\mathbf{n} = \{\cos\alpha, \cos\beta, \cos\gamma\}^T$ and let $\mathbf{p}_{i\pi} \equiv P_{i\pi}$ be the position vector of the projection of P_i on Π. We obtain the projected points with equation $\mathbf{p}_{i\pi} + k_i\mathbf{n} = \mathbf{p}_i$, where k_i is the length of the projector joining $P_{i\pi}$ to P_i, and with the perpendicularity condition between $\mathbf{n}$ and all lines in Π, expressed as $\mathbf{p}_{i\pi} \cdot \mathbf{n} = 0$. This can be expanded as the matrix multiplication below.

$$\begin{bmatrix} 1 & 0 & 0 & \cos\alpha \\ 0 & 1 & 0 & \cos\beta \\ 0 & 0 & 1 & \cos\gamma \\ \cos\alpha & \cos\beta & \cos\gamma & 0 \end{bmatrix} \begin{bmatrix} x_{i\pi} \\ y_{i\pi} \\ z_{i\pi} \\ k_i \end{bmatrix} = \begin{bmatrix} x_i \\ y_i \\ z_i \\ 0 \end{bmatrix} \tag{1}$$

Using the first three equations, to eliminate $x_{i\pi}, y_{i\pi}, z_{i\pi}$ from the fourth, yields k_i.

$$k_i = \cos\alpha x_i + \cos\beta y_i + \cos\gamma z_i \tag{2}$$

so

$$\mathbf{p}_{i\pi} = \begin{bmatrix} \sin^2\alpha x_i - \cos\alpha(\cos\beta y_i + \cos\gamma z_i) \\ \sin^2\beta y_i - \cos\beta(\cos\gamma z_i + \cos\alpha x_i) \\ \sin^2\gamma z_i - \cos\gamma(\cos\alpha x_i + \cos\beta y_i) \end{bmatrix} \tag{3}$$

Each plane in Π will cut one cylinder in $\mathcal{Y}$ perpendicular to its axis, which is parallel to $\mathbf{n}$, so that the centres, $M \equiv \mathbf{m}$, of the set of circles with circumference on the point projections, $P_{i\pi}$, will define a surface, $\mathcal{M}$.

$$\mathcal{M}(x, y, z) = \mathcal{M}(P_{i\pi}) \tag{4}$$

It seems appropriate to characterize this surface with three sets of curves on it, i.e., characteristics of constant $\alpha, \beta, (\gamma), r$ where r is a cylinder radius.

The position vector, $\mathbf{m}$, of a point on $\mathcal{M}$, can be conveniently obtained by intersecting two right bisecting planes of the segments $P_{i\pi}P_{i+1,\pi}$ with a plane in Π. An equivalent process is represented by intersecting right bisecting lines of two sides of the projected triangle.

$$[\mathbf{m} - (2\mathbf{p}_{i\pi} + \mathbf{p}_{i,i+1,\pi})/2] \cdot \mathbf{p}_{i,i+1,\pi} = 0 \tag{5}$$

where $i = 1, 2$ and $\mathbf{p}_{i,i+1,\pi} \equiv \mathbf{p}_{i+1,\pi} - \mathbf{p}_{i\pi}$. Recall that $\mathbf{m} \cdot \mathbf{n} = 0$.

Since the triangle $P_1P_2P_3$ is normalized to $\|\mathbf{p}_2 - \mathbf{p}_1\| = 1$, which is the longest side, the other two may be nondimensionalized as $\rho_1 = \|\mathbf{p}_3 - \mathbf{p}_2\|$ and $\rho_2 = \|\mathbf{p}_1 - \mathbf{p}_3\|$, $\rho_2 \leq \rho_1 \leq 1$ which gives

$$\mathbf{p}_3 = \{(\rho_2^2 - \rho_1^2 + 1)/2, \sqrt{\rho_2^2 - x_3^2}, 0\}^T = \{x_3, y_3, z_3\}^T \tag{6}$$

In this way the multiplicative effects of point pattern size, hand and circulation, i.e., up/down normals, $\mathbf{n}$, are eliminated from the mapping of $\mathcal{M}$ which can proceed with the solution below. First, note that $\mathbf{p}_{1\pi} = \{0, 0, 0\}^T$ and that from Eq. 3 expressions for $\mathbf{p}_{2\pi}$ and $\mathbf{p}_{3\pi}$ simplify to

$$\mathbf{p}_{2\pi} = \begin{bmatrix} x_{2\pi} \\ y_{2\pi} \\ z_{2\pi} \end{bmatrix} = \begin{bmatrix} 1-a^2 \\ -ab \\ -ca \end{bmatrix} \tag{7}$$

and

$$\mathbf{p}_{3\pi} = \begin{bmatrix} x_{3\pi} \\ y_{3\pi} \\ z_{3\pi} \end{bmatrix} = \begin{bmatrix} (1-a^2)x - aby \\ (1-b^2)y - abx \\ -c(ax - by) \end{bmatrix} \tag{8}$$

The two equations, Eq. 5, $i = 2, 3$, simplify to

$$(\mathbf{m} - \mathbf{p}_{i\pi}/2) \cdot \mathbf{p}_{i\pi} = 0 \tag{9}$$

Together with the condition $\mathbf{m} \cdot \mathbf{n} = 0$ these provide three simultaneous equations which can be solved for $\mathbf{m} = \{x_m, y_m, z_m\}^T$.

$$x_{2\pi}x_m + y_{2\pi}y_m + z_{2\pi}z_m = (x_{2\pi}^2 + y_{2\pi}^2 + z_{2\pi}^2)/2 \tag{10}$$

$$x_{3\pi}x_m + y_{3\pi}y_m + z_{3\pi}z_m = (x_{3\pi}^2 + y_{3\pi}^2 + z_{3\pi}^2)/2 \tag{11}$$

$$ax_m + by_m + cz_m = 0 \tag{12}$$

where $a = \cos\alpha$, $b = \cos\beta$, $c = \cos\gamma$, $x = x_3$ and $y = y_3$. Clearly, cylinder radius, r, can be obtained immediately as $r^2 = x_m^2 + y_m^2 + z_m^2$. It might be convenient to compute $\mathcal{M}$ with the spherical coordinate pair, (θ, ϕ), where $a = \cos\theta\cos\phi$, $b = \sin\theta\cos\phi$ and $c = \sin\phi$. Note that θ and ϕ correspond to angles of meridian and latitude, respectively, which define the direction of cylinder axes.

COMPUTATION

To show that the solution is computationally efficient, a BASIC program containing twelve scant lines of code, which generates points in the sequence necessary to form a family of characteristics in θ, is tabulated below.

```
100 DTR=3.141592654/180:INPUT R1,R2:
    X=(R2*R2-R1*R1+1)/2:Y=SQR(R2*R2-X*X)
110 FOR T=0 TO 71
120 TH=5*T*DR:CT=COS(TH):ST=SIN(TH)
130 FOR P=1 TO 17
140 PH=5*P*DTR:CP=COS(PH):W=SIN(PH):
    U=CP*CT:V=CP*ST:REM U=a, V=b, W=c
150 A=1-U*U:B=-U*V:C=-W*U:D=A*X-U*V*Y:
    E=(1-V*V)*Y-U*V*X:F=-W*(U*X+V*Y)
160 K=(A*A+B*B+C*C)/2:L=(D*D+E*E+F*F)/2
170 DM=A*E*W+B*F*U+C*D*V-U*E*C-V*F*A-W*D*B:
    IF DM=0 THEN GOTO 200
180 XN=K*E*W+C*L*V-V*F*K-W*L*B:
    YN=A*L*W+K*F*U-U*L*C-W*D*K:XM=XN/DM:YM=YN/DM
190 ZN=B*L*U+K*D*V-U*E*K-V*L*A:ZM=ZN/DM:
    R=SQR(XM*XM+YM*YM+ZM*ZM)
200 XQ=XM-ZM*U/V:YQ=YM-ZM*V/W:NEXT P
210 NEXT T
```

To form a characteristic family in ϕ one merely interchanges the nested `FOR/NEXT` loops. So one parameter families of cylinder radii and axes, characterized by directions on circles of latitude and meridian with respect to an equatorial plane on $\{P_i\}$, may be readily provided. Similarly it is easy to compute the cylinders whose axes are parallel to the spokes of any great circle whose normal direction is (θ_w, ϕ_w). Here the angle parameter of the spokes, ψ, will span an arc subtending π on the spoke diameter parallel to the equatorial plane. The algorithm above is used with a single loop iterated in ψ and with θ and ϕ are computed as follows.

$$\sin\phi = -\sin\psi\cos\phi_w, \quad \cos\phi = \sqrt{1-\sin^2\phi}$$

$$\cos\theta = (\sin\psi\cos\theta_w\sin\phi_w + \cos\psi\sin\theta_w)/\cos\phi$$

$$\sin\theta = (\sin\psi\sin\theta_w\sin\phi_w - \cos\psi\cos\theta_w)/\cos\phi$$

SINGULARITY

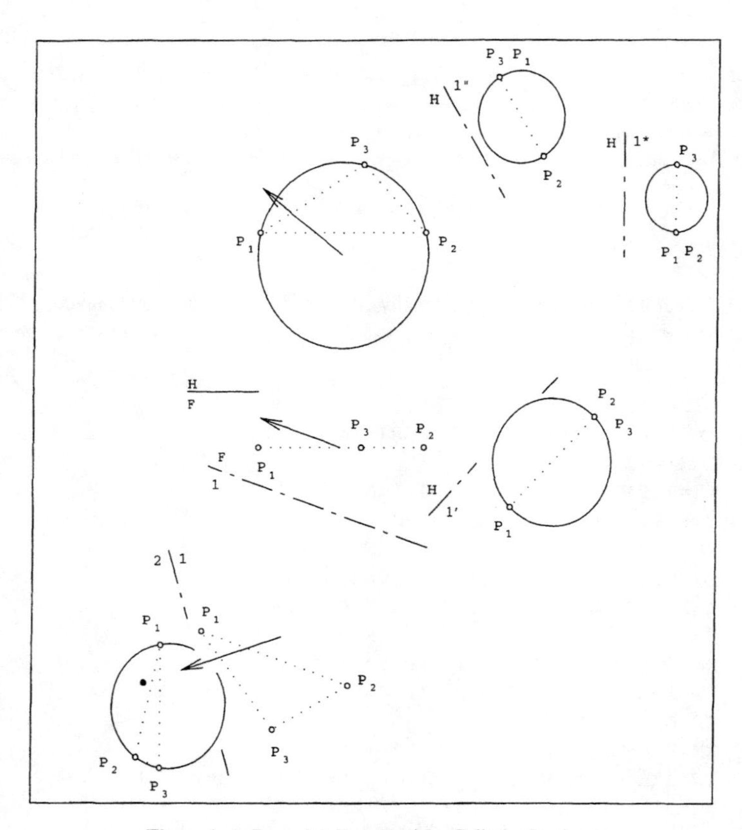

Figure 1. A General & Four Singular Cylinder Sections

Fig. 1 shows various views of P_i, chosen arbitrarily. A projection of the equatorial plane appears as P_i^H where one sees a section of the cylinder circumscribing these point projections. This view represents a singularity because here, at $\phi = \pi/2$, the section and hence the cylinder radius remain constant for all values of ϕ. Below this is a front view where the points P_i^F appear in a line and the circumscribing circle's axis lies in the equatorial plane and the cylinder has infinite radius. This occurs in *all* axial directions parallel to the equatorial plane except those shown as P_i', P_i'' and P_i^* where the circumscribing circles may appear as having diameters of the three triangular heights. Of course a circle of infinite radius is also a valid solution in these three cases as well. The arrow direction represents an arbitrary axial direction which produces the cylindrical section on P_i^2. The axis of this circle can be projected to locate M in the reference frame of P_i, *i.e.*, P_i^H, P_i^F. Returning to the projections P_i', P_i'' and P_i^*, we see the three cylinders of minimum radius and these views must be separated by maxima. Therefore it seems that any trajectory of axis directions on the sphere which spans $-\pi/2 \leq \theta \leq \pi/2$ should be of sixth order.

POLAR PLOTS OF CYLINDER RADIUS

Having dealt with directions at $\phi = 0$, plots of cylinder radius, $r = r(\theta)$, for an equilateral and a $60^o/30^o$ triangle at $\phi = 1^o, 2^o, 5^o, 10^o, 20^o$ and for spoke directions, ψ at $\theta_w = 30^o$, $\phi_w = 60^o$, are shown below.

Figure 2. Cylinder Radii for Equilateral and $60^o/30^o$ Triangles, Axes at $\phi = 1^o, 2^o, 5^o, 10^o, 20^o$

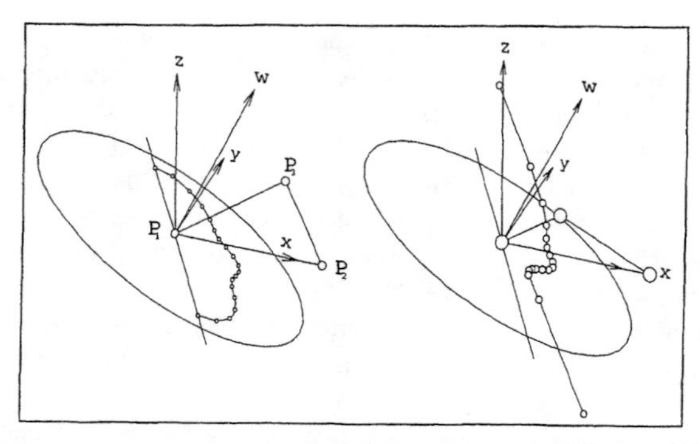

Figure 3. Cylinder Radii for Equilateral and $60^o/30^o$ Triangles, Axes Perpendicular to w, $\theta_w = 30^o$, $\phi_w = 60^o$

CONCLUSION

The sixth order nature of the cylinder radius as a function of axis orientation is shown clearly in Fig. 2 and less clearly in Fig. 3. Furthermore it may be seen that at latitudinal directions $\phi > 1^o$, approximately, a semicircle of constant radius will intersect a unidirectional characteristic less than six times, maybe not at all. So it would be useful to be able to compute axial direction as a function of some given value of r.

Therefore before embarking upon the development of avoidance/insertion procedures for cylindrical objects there remains this crucial characterization task. It is not hard to formulate the problem. Say that, along with P_i, r and a vector w in the direction w perpendicular to a desired axis are specified. Then the following ten equations can be written.

$$\mathbf{u} = \mathbf{s} + u(\mathbf{t} - \mathbf{s}), \;\; \mathbf{s}^2 = r^2, \;\; (\mathbf{p}_2 - \mathbf{t})^2 = r^2, \;\; (\mathbf{p}_3 - \mathbf{u})^2 = r^2$$

$$(\mathbf{t} - \mathbf{s})\cdot\mathbf{s} = 0, \;\; (\mathbf{t} - \mathbf{s})\cdot(\mathbf{p}_2 - \mathbf{t}) = 0, \;\; (\mathbf{t} - \mathbf{s})\cdot(\mathbf{p}_3 - \mathbf{u}) = 0, \;\; (\mathbf{t} - \mathbf{s})\cdot\mathbf{w} = 0$$

Notice that the first equation expands to three, in x, y, z, which specifies the colinearity of three points on the axis, given by position vectors $\mathbf{s}, \mathbf{t}, \mathbf{u}$, in terms of the scalar parameter u. This approach unfortunately yields a univaritate polynomial of 64th degree; not very satisfactory. Current refinements indicate that a homgeneous line coordinate formulation holds promise. No success has been achieved so far in this regard and the struggle continues.

REFERENCES

[SC86] Schaal, H., 1986, "Konstruktion der Drehzylinder durch vier Punkte einer Ebene", *Sber. d. österr. Akad. d. Wiss.*, v.195, pp.405-418.

[SC85] Schaal, H., 1985, "Ein geometrisches Problem der metrischen Getriebesynthese", *Sber. d. österr. Akad. d. Wiss.*, v.194, pp.39-53.

[ST91] Strobel, U., 1991, "Die Drehkegel durch vier Punkte. Teil II", *Sber. d. österr. Akad. d. Wiss.*, v.200, pp.91-109.

[ST89] Strobel, U., 1989, "Über die Drehkegel durch vier Punkte", *Sber. d. österr. Akad. d. Wiss.*, v.198, pp.281-293.

AUTHOR INDEX

AUTHOR INDEX (cont'd)

SUBJECT INDEX

SUBJECT INDEX (cont'd)